PC28a

Validation of Pharmaceutical Processes

Validation of Pharmaceutical Processes

Third Edition

Edited by

James Agalloco

Agalloco & Associates

Belle Mead, New Jersey, USA

Frederick J. Carleton

Carleton Technologies Incorporated

Boynton Beach, Florida, USA

informa

healthcare

New York London

Informa Healthcare USA, Inc.
52 Vanderbilt Avenue
New York, NY 10017

Printed in the United States of America on acid-free paper
10 9 8 7 6 5 4 3 2 1

International Standard Book Number-10: 0-8493-7055-8 (Hardcover)
International Standard Book Number-13: 978-0-8493-7055-7 (Hardcover)

Library of Congress Cataloging-in-Publication Data

Validation of pharmaceutical processes / edited by James P. Agalloco, Frederick J. Carleton.
– 3rd ed.
p. ; cm.
Includes bibliographical references and index.
ISBN-13: 978-0-8493-7055-7 (hardcover : alk. paper)
ISBN-10: 0-8493-7055-8 (hardcover : alk. paper)
1. Sterilization.
2. Pharmaceutical technology–Quality control.
3. Pharmaceutical technology–Standards.
I. Agalloco, James P., 1948- II. Carleton, Frederick J., 1925-
[DNLM: 1. Technology, Pharmaceutical. 2. Quality Control. 3. Sterilization–methods. QV 778 V172 2007]

RS199.S73V345 2007
615′.19–dc22

2007018225

Visit the Informa Web site at
www.informa.com

and the Informa Healthcare Web site at
www.informahealthcare.com

Validation of Pharmaceutical Processes

Third Edition

Edited by
James Agalloco
Agalloco & Associates
Belle Mead, New Jersey, USA
Frederick J. Carleton
Carleton Technologies Incorporated
Boynton Beach, Florida, USA

informa
healthcare

New York London

Informa Healthcare USA, Inc.
52 Vanderbilt Avenue
New York, NY 10017

Printed in the United States of America on acid-free paper
10 9 8 7 6 5 4 3 2 1

International Standard Book Number-10: 0-8493-7055-8 (Hardcover)
International Standard Book Number-13: 978-0-8493-7055-7 (Hardcover)

Library of Congress Cataloging-in-Publication Data

Validation of pharmaceutical processes / edited by James P. Agalloco, Frederick J. Carleton.
– 3rd ed.
p. ; cm.
Includes bibliographical references and index.
ISBN-13: 978-0-8493-7055-7 (hardcover : alk. paper)
ISBN-10: 0-8493-7055-8 (hardcover : alk. paper)
1. Sterilization.
2. Pharmaceutical technology–Quality control.
3. Pharmaceutical technology–Standards.
I. Agalloco, James P., 1948- II. Carleton, Frederick J., 1925-
[DNLM: 1. Technology, Pharmaceutical. 2. Quality Control. 3. Sterilization–methods. QV 778 V172 2007]

RS199.S73V345 2007
615′.19–dc22

2007018225

Visit the Informa Web site at
www.informa.com

and the Informa Healthcare Web site at
www.informahealthcare.com

The complexities of developing this third volume on validation involved a dependence on many sources.
We were helped and encouraged by a great many people who listened to our ideas
and gave us the benefits of their thoughts. Foremost among these were our wives,
Helen Carleton and Linda Agalloco,
as well as our children,
Brant and Penny Carleton, and Stephen, Andrea, and Adam Agalloco,
who, throughout these many years of preparation,
encouraged us, propelled us to think more aggressively,
played devil's advocate, and considered ideas and topics with us.
Most of all, we are grateful for their love and understanding.

We do wish especially to thank Drs. Sol Motola, Clarence Kemper, Mark Litchman,
and Theodore Meltzer whose assistance were vital to this endeavor.

Preface

We are excited to introduce the third edition of *Validation of Pharmaceutical Processes*.

With the goal of ensuring pharmaceutical quality of the final manufactured product, the editors have introduced to the industry a subject so extensive that it impacts on good manufacturing practices, research and development, quality assurance, and facility design. Validation inevitably leads to process optimization, increased productivity, and lower costs in man-time and manufacturing.

The objective of the third edition is to encompass the changes that have taken place in the pharmaceutical industry during the past several years. George Bernard Shaw once wrote that "science is always simple and always profound. It is only the half-truths that are dangerous."

The first and second editions of *Validation of Pharmaceutical Processes*, which filled the void of both aseptic and non-aseptic processes, are classics in the field of validation. The editors continue to build on the success of the previous books with significant updating of this third edition with information that must be made available in current practices. This new edition serves as a guide to validation methodologies for the preparation of all pharmaceutical products.

This edition now includes practices on managing validation in multinational and small companies, and regulations and validation discussions relative to the interpretation in the Code of Federal Regulations 21. Process analytical technology, real-time monitoring, practitioner compounding and European and Japanese approaches to validation are explored. The fundamental concepts of validation are explicitly discussed including calibration and metrology, temperature measurements, validation of water systems, as well as qualification and significant aspects of change control.

Introducing organization to validation, the editors bring to practitioners the fundamentals of those processes which deal with the microbiology of sterilization, validation of air systems, validation of all heat sterilization processes including steam, terminal, and dry heat sterilization. Detailed sterilization processes are described for the validation of products employing ethylene oxide, chlorine and radiation processes, each in separate chapters. Statistical analysis as well as Six Sigma reviews are examined at length. Since the preparation of pharmaceutical products requires distinct process steps, the authors and editors identify critical process control points to reach acceptable results. The fundamentals of validation are brought into full view for the reader.

The methods that are developed are guides and are not intended to establish standards. The impact of validation on the world pharmaceutical health care business has radically influenced the changes of processing in the past 30 years.

The third edition of *Validation of Pharmaceutical Processes* is a detailed book and is considered a must for the scientist in both industry and academia.

Frederick J. Carleton
James Agalloco

Contents

Contributors

J. Robert Adamson Foster Wheeler, Reading, Berkshire, U.K.
James Agalloco Agalloco & Associates, Belle Mead, New Jersey, U.S.A.
James Akers Akers Kennedy & Associates, Kansas City, Missouri, U.S.A.
Suraj B. Baloda Millipore Corporation, Bedford, Massachusetts, U.S.A.
Michael C. Beckloff Beckloff Associates, A Cardinal Health Company, Overland Park, Kansas, U.S.A.
Thomas J. Berger Department of Microbiology and Global Process Sterilization Engineering, Hospira, Inc., Lake Forest, Illinois, U.S.A.
Robert S. Bottome Genentech, Inc., San Francisco, California, U.S.A.
Rebecca Brewer Dober Group, Midlothian, Illinois, U.S.A.
Göran Bringert Independent Consultant, North Andover, Massachusetts, U.S.A.
Carolyn Broughton Quality Control Microbiology, Genentech Inc., South San Francisco, California, U.S.A.
Rory Budihandojo Boehringer-Ingelheim Chemicals, Petersburg, Virginia, U.S.A.
Laurie B. Case Pharmaceutical Project Management, Sea Girt, New Jersey, U.S.A.
Timothy S. Charlebois Wyeth Biotech, Andover, Massachusetts, U.S.A.
Warren Charlton Consultant, WHC Bio Pharma, Inc., Greenville, North Carolina, U.S.A.
Mark A. Czarneski ClorDiSys Solutions, Inc., Lebanon, New Jersey, U.S.A.
Roger Dabbah Tri-Intersect Solutions, Rockville, Maryland, U.S.A.
Phil DeSantis Schering-Plough, Inc., Union, New Jersey, U.S.A.
Pamela D. Deschenes Wyeth Global Pharmaceuticals, Frazer, Pennsylvania, U.S.A.
Franco De Vecchi Vectech Pharmaceutical Consultants International, Inc., West Bloomfield, Michigan, U.S.A.
Denis Drapeau Wyeth Biotech, Andover, Massachusetts, U.S.A.
Norman Elder Genentech, Inc., South San Francisco, California, U.S.A.
Hélène Gazzano-Santoro Quality Control Biological Technologies, Genentech Inc., South San Francisco, California, U.S.A.
John R. Gillis SGM Biotech, Inc., Bozeman, Montana, U.S.A.
Eric Grund Fast Trak Biopharma Services, GE Healthcare, Protein Separations, Uppsala, Sweden
Klaus Haberer Compliance, Advice and Services in Microbiology GmbH, Cologne, Germany
Mark Hallworth Particle Measuring Systems, Inc., Boulder, Colorado, U.S.A.
Gayle D. Heffernan Pharmaceutical Consultant, Middlesex, New Jersey, U.S.A.
Tsutomu Hinomoto Santen Pharmaceutical Co., Ltd., Higashiyodogawa, Osaka, Japan
Tom Ingallinera Consultant, Charleston, South Carolina, U.S.A.
Geoffrey P. Jacobs Dr. Geoffrey P. Jacobs Associates, Pharmaceutical Consultants, Jerusalem, Israel
Kevin Jenkins Pfizer Inc., Terre Haute, Indiana, U.S.A.
Susan Dana Jones BioProcess Technology Consultants, Inc., Acton, Massachusetts, U.S.A.
Maik W. Jornitz Sartorius Biotech, Inc., Edgewood, New York, U.S.A.
Clarence Kemper Independent Consultant, Winchester, Massachusetts, U.S.A.
Charles S. Levine Levine Pharmaceutical Consulting Inc., Linwood, New Jersey, U.S.A.
Howard L. Levine BioProcess Technology Consultants, Inc., Acton, Massachusetts, U.S.A.
William G. Lindboe, Jr. Consultant, East Brunswick, New Jersey, U.S.A.
Paul Lorcheim ClorDiSys Solutions, Inc., Lebanon, New Jersey, U.S.A.
Matthew T. Lord Eli Lilly and Company, Indianapolis, Indiana, U.S.A.
Sheila Magil BioProcess Technology Consultants, Inc., Acton, Massachusetts, U.S.A.
Steven I. Max Wyeth Biotech, Andover, Massachusetts, U.S.A.

David W. Maynard Maynard & Associates, Ewing, New Jersey, U.S.A.
R. D. McDowall McDowall Consulting, Bromley, Kent, U.K.
Theodore H. Meltzer Capitola Consultancy, Bethesda, Maryland, U.S.A.
Deborah E. Mentel Pfizer Global Research and Development, Groton, Connecticut, U.S.A.
Joshua D. Morton Eli Lilly and Company, Indianapolis, Indiana, U.S.A.
Gregg Mosley Biotest Laboratories, Inc., Minneapolis, Minnesota, U.S.A.
Terry E. Munson PAREXEL International, Waltham, Massachusetts, U.S.A.
Toshiaki Nishihata Santen Pharmaceutical Co., Ltd., Higashiyodogawa, Osaka, Japan
Steven Ostrove Ostrove Associates, Inc., Elizabeth, New Jersey, U.S.A.
David A. Porter Vectech Pharmaceutical Consultants, Inc., Farmington Hills, Michigan, U.S.A.
Tsuguo Sasaki Musashimurayama, Tokyo, Japan
Dieter Schmalzing Quality Control Analytical Technologies, Genentech Inc., South San Francisco, California, U.S.A.
Bradley S. Scott McNeil Consumer Healthcare, Guelph, Ontario, Canada
John Shirtz Consultant, Columbia, Maryland, U.S.A.
Danny Shive Metrics, Inc., Greenville, North Carolina, U.S.A.
Ronald J. Simko Manalapan, New Jersey, U.S.A.
Martin S. Sinacore Wyeth Biotech, Andover, Massachusetts, U.S.A.
Jeffrey M. Singer PharmaBasics, Inc., Douglaston, New York, U.S.A.
Christopher Smalley Wyeth Pharmaceuticals, Collegeville, Pennsylvania, U.S.A.
Leslie M. Snyder Eli Lilly and Company, Indianapolis, Indiana, U.S.A.
Gail Sofer Fast Trak Life Sciences, GE Healthcare, Piscataway, New Jersey, U.S.A.
Saeed Tafreshi Intelitec Corporation, Irvine, California, U.S.A.
Stephen C. Tarallo Lyne Laboratories, Inc., Brockton, Massachusetts, U.S.A.
Rodney E. Thompson BioProcess Technology Consultants, Inc., Acton, Massachusetts, U.S.A.
Edward H. Trappler Lyophilization Technology, Ivyland, Pennsylvania, U.S.A.
Kevin D. Trupp Department of Microbiology and Global Process Sterilization Engineering, Hospira, Inc., Lake Forest, Illinois, U.S.A.
Anne Wilcock McNeil Consumer Healthcare, Guelph, Ontario, Canada
Fumi Yamamoto Ministry of Health, Labor and Welfare (MHLW), Chiyoda, Tokyo, Japan
Alex M. Zislin DSCI, Freehold, New Jersey, U.S.A

List of Abbreviations

AA Amino acid
AAMI Association for the Advancement of Medical Instrumentation
AAMI American Association of Medical Instrumentation
ACPS Advisory Committee for Pharmaceutical Science
ADR Adverse drug reaction
AHU Air-handling units
ANDA Abbreviated new drug application
ANN Artificial neural network
ANOVA Analysis of variation
APA Aseptic processing areas
APE Antimicrobial preservative efficacy
API Active pharmaceutical ingredient
APIC Active Pharmaceutical Ingredients Committee
APR Annual product review
AQL Acceptable quality level
ASME American Society of Mechanical Engineers
ASTM American Society for Testing and Materials
ATCC American type culture collection
AZT International Association for Pharmaceutical Technology
ATR Attenuated total reference
BB Bioburden
BET Bacterial endotoxin testing
BFS Blow–fill–seal
BI Biologic indicator
BIER Biological indicator evaluator resistometer
BLA Biological license application
BMS Building management systems
BP Bubble points
BPC Bulk pharmaceutical chemical
BS British Standard
BSA Bovine serum albumin
BSC Biological Safety Cabinet
BSE Bovine spongiform encephalopathy
CA-HCl Cure-all hydrochloride
CAPA Corrective and preventive action
CAS Chemical abstract service
CBE Changes being effected
CBER Center for Biological Evaluation and Research
CCD Charge-coupled device
CCP Critical control points
CDER Center for Drug Evaluation and Research
CDRH Center for Devices and Radiological Health
CDS Chromotography data system
CEN European Committee for Standardization
CFC Chlorofluorocarbon
CFM Cubic foot per minute
CFP Criteria for Forward Processing
CFR Code of Federal Regulations
CFU Colony-forming units
cGMP Current good manufacturing practice
CHO Chinese hamster ovary
CIP Clean in place
CIs Configuration items
CJD Creutzfeldt–Jakob disease
CMC Chemistry, manufacturing, and controls
CMO Contract manufacturing organization
CNC Computer(ized) numerical(ly) control(led)
COA Certificate of Analysis
COC Cyclo-olefin copolymer
COTS Commercial off-the-shelf
CPD Cumulative population doubling
CPG Compliance policy guide
CPK Capability index for process average
CpK Process capability, average
CpM Process capability, target
CPP Critical process parameter
CQA Critical quality attribute
CRT Cathode ray tube
CSQMP Computer System Qualification Master Plan
CSV Computer system validation
CTM Clinical trial material
CTQs Critical to quality attributes
CV Coefficient of variation
CVM Center for Veterinary Medicine
DCS Distributed control systems
DEC Dynamic environmental conditioning
DEHP Di-2-ethylhexyl-phthalate
DGHM German Society for Hygiene and Microbiology
DHFR Dihydrofolate reductase
DHR Development History Report
DI Direct impact
DIN German Institute of Standardization (Deutsche Industry Norm)
DMADV Define–Measure–Analyze–Design–Verify
DMF Drug master file
DO Dissolved oxygen concentration
DOE Design of experiment
DOP Dioctyl phthalate
DQ Design qualification
DS Design specification
DUT Device under test
DVM Digital voltmeter
EC European Commission
ECSs Environmental control systems
EEPROM Erasable electronically programmed random only memory
ELA Establishment license application
ELISA Enzyme-linked immunosorbent assay
ELN Electronic Laboratory Notebook
EM Electron microscope
EMA European Medicinal Agency
EMEA European Medicines Evaluation Agency
EMI Electromagnetic interference
EN European standards
EO, EtO Ethylene oxide
EOP End-of-production
EP European Pharmacopoeia
EPA Environmental Protection Agency

EPR Electron paramagnetic resonance
EPROM Electronically Programmed Random Only Memory
EQ Equipment qualification
ERP Enterprise Resource Planning
EU European Union
EU-GMP European Union good manufacturing practice
F/N Fraction/negative
FAT Factory acceptance test
FD&C Food, Drug, and Cosmetic act
FDA Food and Drug Administration
FDIS Final Draft International Standard
FEQ Facility and equipment qualification
FFS Form, fill and seal
FIT Filter integrity test
FMEA Failure mode and effects analysis
F-N Fraction–negative
FRS Functional requirement specification
FS Functional specification
FTIR Fourier Transform Infra-Red Spectrometry
FTM Fluid thioglycollate media
GAMP Good automated manufacturing practices
GC Gas chromatography
GCP Good clinical practice
GLP Good laboratory practice
GMP Good manufacturing practice
GPLOT Global Post Launch Optimization Team
GQS Global Quality Standards
GRAS Generally recognized as safe
GUI Graphical user interface
GUM Guide to the Expression of Uncertainty in Measurement
GXP Good practices (manufacturing, practice, and laboratory practice)
HACCP Hazard analysis and critical control plan
HAP Hamster antibody production
HCP Host cell proteins
HDPE High-density polyethylene
HEPA High-efficiency particulate air
HEPA High efficiency particulate air filter
HETP Height equivalent to a theoretical plate
HIC Hydrophobic interaction chromatography
HIMA Health Industry Manufacturers Association
HMI Human–machine interface
HPLC High-performance liquid chromatography
HTST High temperature–short time
HVAC Heating, ventilation, and air-conditioning
I/O Input/output
IAEA International Atomic Energy Agency
ICH International Conference on Harmonization
IDEF Integrated definition
IEC Ion exchange liquid chromatography
IEEE Institute of Electrical and Electronic Engineers
IES Institute of Environmental Sciences
IFO Institute for Fermentation, Osaka, Japan
II Indirect impact
IMS Ion mobility spectrometry
IND Investigational new drug
IOQ Installation and operation qualification
IP Inoculated product
IP Internet protocol
IPA Isopropyl alcohol
IPP Independent process parameters
IPT International press tooling
IPTG Isopropyl-beta-D-thiogalactopyranoside
IQ Installation qualification
IQP Installation qualification plan/protocol
IR Infrared
ISO International Organization for Standardization
ISPE International Society for Pharmaceutical Engineering
IT Information technology
JCM Japan Collection of Microorganisms
JP Japanese Pharmacopoeia
LAF Laminar air flow
LAL Limulus amebocyte lysate
LAN Local area network
LC Liquid chromatography
LCL Lower control limit
LDPE Low density polyethylene
LIF Light-induced fluorescence
LIMS Laboratory Information Management System
LL Like-for-like
LOD Loss on drying
LPS Lipopolysaccharide
LSL Lower specification limit
LTSF Low-temperature steam and formaldehyde
LVP Large volume parenteral
MAC/MACO Maximum allowable carryover
MAP Mouse antibody production
MAS Microbiological air sampler
MCA Medicines Control Agency
MCB Master Cell Bank
MEF Murine Embryonic Fibroblasts
MES Manufacturing execution system
MFG Manufacturing
MHLW Ministry of Health Labor and Welfare
MLR Multiple linear regression
MMI Man–machine interface
MMV Murine minute virus
MOS Maintenance of sterility
MPN Most-probable number
MPPS Most-penetrating particle size
MRP Materials resource planning
MS Mass spectrometry
MS&T Manufacturing Science and Technology
MSC Minimal sporocidal concentration
MSDS Material safety data sheet
MVP Master validation plan
MW Molecular weight
NaOH Sodium hydroxide
NCE Non-conforming event
NDA New drug application
NDIR Nondispersive infrared
NF National Formulatory
NF/EP National Formulary/European Pharmacopoeia
NFR Nonfiber releasing
NI No impact
NIOSH National Institute of Occupational Safety and Health
NIR Near infrared
NIST National Institute of Standards and Technology
NLT No less than
NMR Nuclear magnetic resonance
NMT No more than
NMWCO Nominal molecular weight cutoff
NOEL No observed effect level
NP Nonperishable
OECD Organization of Economic Cooperation and Development
OOS Out-of-specification
OPCs Optical particle counters
OQ Operational qualification

OQP Operational qualification plan
ORP Oxidation–reduction potential
OS Operating system
OSHA Occupational Health and Safety Act
OST Oxidizable substances test
OTC Over-the-counter
P&D Penetration and distribution
P&ID Process and instrumentation drawings
PAFSC Pharmaceutical Affairs and Food Sanitation Council
PAG Polyacrylamide gel
PAGE Polyacrylamide gel electrophoresis
PAI Preapproval inspection
PAO Polyalpha olefin
PAR Proven acceptable range
PAT Process analytical technology
PBS Phosphate-buffered saline
PC Personal computer
PCA Principal components analysis
PCR Polymerase chain reaction
PD Particulate detection
PDA Parenteral Drug Association
PDV Positive displacement volumetric
PEG Polyethylene glycol
PEL Permissible exposure limits
PERT Product-enhanced reverse transcriptase
PET Polyethylene terephthalate
PFD Process flow document
PFSB Pharmaceutical and Food Safety Bureau
PhRMA Pharmaceutical Research Manufacturers Association
PIC/S Pharmaceutical Inspection Cooperation Scheme
PID Piping and instrument diagram
PLC Programmable logic controller
PLS Partial least squares
PM Preventative maintenance
PMA Pharmaceutical Manufacturers Association
PMDA Pharmaceuticals and Medical Devices Agency
pMDI Pressurized metered-dose inhaler
PNSU Probability of a non-sterile unit
PNU Protein nitrogen units
PP Polypropylene
PQ Performance qualification
PQA Preferred quality attributes
PQP Performance qualification plan/protocol
PQRI Product Quality Research Institute
PS Pure steam
PSLR Predicted spore logarithmic reduction
PV Process validation
PVC Polyvinyl chloride
PVP Process validation package
QA Quality assurance
QACs Quaternary ammonium compounds
QC Quality control
Q-PCR Quantitative polymerase chain reaction
QU Quality unit
R Electrical resistance of product
R&D Research and development
R/S Requirements and specifications
RABS Restricted access barrier system
RAID Redundant array of inexpensive disks
RCS Reuter centrifugal sampler
RD Rhabdomyosarcoma
RF Radio frequency
RFI Radio frequency interference
RFID Radio frequency identification
RFP Request for proposal
RH Relative humidity
rhBMP Recombinant human bone morphogenetic protein
RID Radial immunodiffusion
RL Risk level
RMS Root mean square
RO Reverse osmosis
RODAC Replicate organism detection and counting
RPM Revolutions per minute
RSD Relative standard deviation
RTD Resistance temperature detector
RTM Requirements traceability matrix
RTP Rapid transport port
RT-PCR Reverse transcription polymerase chain reaction
RVLP Retroviral-like particles
RZE Reject zone efficiency
SAC Static atmospheric conditioning
SAL Sterility assurance level
SAN Storage area network
SAS Surface air system
SAT Site acceptance test
SCADA Supervisory control and data acquisition
SDI Silt density index
SDLC Software development life cycle
SDS System design specification
SEI Software Engineering Institute
SF Safety factor
SILC System implementation life cycle
SIP Sterilization in place
SIPOC Suppliers, inputs, process, outputs, consumers
SISPQ Safety, identity, strength, purity and quality
SLR Spore logarithmic reduction
SMA Sterilizable microbial atrium
SME Subject matter expert
SOP Standard operating procedure
SP Sulfopropyl
SPC Statistical process control
SPVP Sterile process validation package
SQ System qualification or Specification qualification
STA Slit-to-agar
STEL Short-time-exposure-limit
SVP Small volume parenterals
T–P Time–pressure
TAR Test accuracy ratio
TC Thermocouple
TDS Total dissolved solids
TEM Transmission electron microscopy
THM Trihalomethanes
TLD Thermoluminescence-dosimetry
TM Trace matrix
TNTC Too numerous to count
TOC Total oxidizable carbon
tPA Tissue plasminogen activator
TQM Total quality management
TR Technical report
TS Tensile strength
TSB Trypticase soy broth
TSE Transmissible spongiform encephalopathies
TSS Total suspended solids
TUR Test uncertainty ratio
TWA Time-weighted average
UAD Unidirectional airflow devices
UAT User acceptance testing
UCL Upper control limit
UF Ultrafiltration

UF water Ultrafiltered water
UFRS User functional requirements specification
UHF Ultra high frequency
UID Unique identifier
UK MCA United Kingdom's Medicines Control Agency
URS User requirements specification
USLx Upper specification limit
USP United States Pharmacopeia
UV Ultraviolet
VHP Vapor-phase hydrogen peroxide
VMP Validation Master Plan
VOC Volatile organic carbons
VPHP Vapor phase hydrogen peroxide
WCB Working Cell Banks
WD Weight dosing
WFI Water for injection
WHO World Health Organization
WHO-GMP World Health Organization-Good Manufacturing Practice
XE Xenon

1

Why Validation?

James Agalloco
Agalloco & Associates, Belle Mead, New Jersey, U.S.A.

INTRODUCTION

The origins of validation in the global healthcare industry can be traced to terminal sterilization process failures in the early 1970s. Individuals in the United States point to the LVP sterilization problems of Abbott and Baxter, while those in the U.K. cite the Davenport incident (1). Each incident was a result of a non-obvious fault coupled with the inherent limitations of the end-product sterility test. As a consequence of these events, non-sterile materials were released to the market, deaths occurred, and regulatory investigations were launched. The outcome of this was the introduction by the regulators of the concept of "Validation":

> Documented evidence which provides a high degree of assurance that a specific process will consistently produce a product meeting its predetermined specifications and quality attributes (2).

The initial reaction to this regulatory initiative was one of puzzlement; after all, only a limited number of firms had encountered difficulties, and all of the problems were seemingly associated with the sterilization of LVP containers. It took several years for firms across the industry to understand that the concerns related to process effectiveness were not limited to LVP solutions, and even longer to recognize that those concerns were not restricted to sterile products. Perhaps most unfortunate of all was the lack of enthusiasm on the part of industry in adopting this concept. From its earliest days, validation was identified as a new regulatory requirement to be added to the list of things that firms must do, with little consideration of its real implications. The first efforts reflected what can be termed the "scientific method" of observation of an activity, hypothesis/prediction of cause/effect relationship, and experimentation followed by new observations in the form of the experimental report. In the pharmaceutical validation model this has evolved into the validation protocol (hypothesis and prediction), field execution (experimentation), and summary report preparation (documented observations).

By 1980 when it was evident to all that validation was here to stay, pharmaceutical firms began to organize their activities more formally. Ad hoc teams and task forces that had started the efforts were replaced by permanent Validation Departments whose responsibilities and scope varied with the organization but whose purpose was to provide the necessary validation for a firm's products and processes. The individuals in these departments were the first to grapple with validation as their primary responsibility, and their methods, concepts, and practices have served to define validation ever since:

> *Validation*: Establishing documented evidence which provides a high degree of assurance that a specific process will consistently produce a product meeting its pre-determined specifications and quality attributes (3).

The first efforts at validation were rather crude and limited in their understanding of the full implications. For example, the first sterilization validations at most firms were performed without prior qualification of the equipment. Once validation had been established as a discipline and something more than a passing fad, methods for its execution became substantially more formalized and rigorous.

The validation community made significant strides in clarifying the various components of a sound validation program. Perhaps most important of all was the separation of activities into two major categories: Equipment Qualification and Process Qualification. The former (sometimes sub-divided into Installation and Operational Qualification) focused on the equipment in which the product was being processed. It is predominantly a documentation exercise in which details of the physical components of the system are recorded as definition of the equipment. Equipment operational capabilities are also established. Process Qualification (also known as Process Validation or Performance Qualification) confirms the acceptability of the product manufactured via the equipment, and relies heavily on the results of physical, chemical, and microbial tests of samples.

It was soon apparent that validation had to be more closely integrated into the mainstream of cGMP operations in order to maximize its effectiveness in larger organizations. A number of areas can be identified as pre-requisites for process or system validation. The origins of these elements can be identified in the cGMP requirements for drugs and devices (Table 1) (4).

With this understanding of its dependencies, validation is more easily assimilated into the overall cGMP environment rather than something apart from it. While a firm will likely continue to have a validation

---

Abbreviations used in this chapter: cGMP, current good manufacturing practice; FDA, Food and Drug Administration; GAMP, good automated manufacturing practice; LVP, large volume parenteral; PAT, process analytical technology; PMA, Pharmaceutical Manufacturers Association.

Table 1 Pre-Requisites for Validation

Process Development [21 CFR 820.30—Design Control]. The activities performed to define the process, product or system to be evaluated
Process Documentation [21 CFR 211 Subparts F—Production and Process Controls and J—Records and Reports]. The documentation (batch records, procedures, test methods, sampling plans) and processes (software) that define the operation of the equipment to attain the desired result
Equipment Qualification [21 CFR 211 Subparts C—Buildings and Facilities and D-Equipment]. The specifications, drawings, checklists and other data that support the physical equipment (hardware) utilized for the process
Calibration [21 CFR 211 Subparts D—Equipment]. The methods and controls that establish the accuracy of data
Analytical Methods [21 CFR 211 Subpart I—Laboratory Controls]. The means to evaluate the outcome of the process on the materials
Cleaning—[21 CFR 211.67 Equipment Cleaning and Maintenance]. A specialized process, the intent of which is to remove traces of the prior product from the equipment
Change Control—[21 CFR 211.100(b) Equipment Cleaning and Maintenance]. A formalized process control scheme that evaluates changes to documentation, materials, and equipment

department, it must be supported by the activities in other parts of the organization. For example, a poorly developed process performed using uncalibrated equipment to make a product that has no standard test methods could never be considered validated. All of the supportive elements must be properly operated in order to result in a compliant product, and one that can be validated. A later definition that addresses the larger scope of validation within the overall organization is:

> Validation is a defined program which, in combination with routine production methods and quality control techniques, provides documented assurance that a system is performing as intended and/or that a product conforms to its predetermined specifications (5).

APPLICATION OF VALIDATION

Beginning with its first association with LVPs in the early 1970s, the application of validation spread quickly to other sterilization processes. It was also applied for the validation of other pharmaceutical processes, albeit with mixed success. In sterilization and, to a slightly lesser extent, in processes supporting the production of sterile products using aseptic processing, there is little difficulty applying validation concepts. The apparent reasons for this are the common and predominantly quantitative criteria for acceptance of the quality attributes of sterile products. Building consensus on validation of sterile products has been achievedbut not without debate. There are numerous excellent guidance documents outlining validation expectations on the various sterilization processes, as well as numerous publications from individuals and suppliers. The only relatively deficient areas in sterile product validation are elements unrelated to sterility, e.g., endotoxin and particulate matter.

Validation of non-sterile products and their related processes is less certain. Despite the obvious importance of cleaning procedures, cleaning validation was not publicly discussed until the early 1990s. To this day there is still confusion regarding the requirements for validation of this important process. The difficulties with validation are even more complicated for pharmaceutical dosage forms. There are no widely accepted validation requirements for the important quality attributes of drug products. While the key elements are known (dissolution, content uniformity, and potency), there are no objective standards upon which to define a validation program. The compendial standards of the various pharmacopeia are poorly suited to validation. The small sample size and absolute nature of the acceptance criteria are extremely problematic for direct application to large scale commercial production. After more than 30 years, the absence of universal criteria for dosage forms is unfortunate and problematic.

Applying validation requirements to water and other utility systems is somewhat easier than for pharmaceutical products. Equipment qualification of utility systems is relatively easy to perform, and samples of the supplied utility (water, steam, environmentally controlled air, compressed gas, solvent, etc.) taken across the system can directly support the acceptability of the preparation, storage (where present), and delivery system. Classified and other controlled environments have proven relatively easy to validate. Their physical elements readily lend themselves to equipment qualification, and sampling affords confirmation of their operational capabilities directly.

Biotechnology first came of age in the late 1980s into a regulatory environment that expected validation of important processes. Since the first biotech products were injectable drugs, it was quite natural for these firms to validate their processes from the onset. As a consequence, cell culture and purification processes of all types have always been subject to validation expectations. There is a substantial body of validation knowledge on these processes available. In marked contrast, the bulk pharmaceutical chemical segment of the industry has been relatively slow to embrace validation concepts. While the rigorous environmental expectations associated with many dosage forms and virtually all biotechnology processes are not present, the important considerations of impurity levels, byproduct levels, racemic mixtures, crystal morphology and trace solvents all suggest that there are important quality attributes to be controlled (and thus validated) as well.

Computerized systems became subject to validation requirements when they were first applied for cGMP functions in the 1980s. For ease of understanding, the parallels between computerized systems and physical systems are utilized. The computer hardware can be qualified like the process equipment to which it is often connected, while computer software has some similarities to the operating procedures utilized to operate the equipment. This approach may be an over-simplification of the required activities for the software, but it provides some clarity to the uninitiated. Computerized system validation is still a subject of substantial interest, but is no longer the misunderstood behemoth task it appeared to be when first encountered. The early efforts of PMA's Computerized Systems Validation Committee and the later development of GAMP have reduced the uncertainty associated with the use of computerized systems substantially (6).

One useful concept taken from the validation of computerized systems as it evolved was the "life cycle model" (7). Originally utilized for computer software, it was later applied to the entire computerized system. It suggests that considerations of system qualification, maintenance and improvement be incorporated at the onset of the design process. Its utility for computerized systems is substantial; however it may have even greater functionality for pharmaceutical processes. In the early 1990s, the FDA launched an initiative related to the demonstration of consistency of processes and data from clinical lots through to commercial manufacture (8). They mandated the conduct of Pre-Approval Inspections to affirm that commercial materials had their basis in the pivotal clinical trial materials. The utility of the "life cycle model" in this context is clear. Its application to pharmaceutical development, scale-up, and commercial production allows for a coordination of supportive information in the same manner as software and computerized systems validation. A landmark publication in this area was Kenneth Chapman's paper entitled "The PAR Approach to Process Validation" (9). It addressed the developmental influence on the ability to successfully validate commercial operations, a message that has been somewhat forgotten until just recently. Ajaz Hussain, then of the FDA, voiced concerns relative to the lack of process knowledge on the part of many pharmaceutical firms (10). That the FDA believed that such a missive was necessary supports the lack of appreciation for Chapman's earlier effort:

> The goal of development is to identify the process variables necessary to ensure the consistent production of a product or intermediate (11).

Application of the "life cycle model" to pharmaceutical operations addresses the compliance and quality expectations of the industry in an appropriate manner and should be a near universal goal.

Another regulatory development of some importance is that of PAT (12). The concept was well articulated by Dr Hussain while he was with the FDA. To many in the industry, PAT seems like an advance of some magnitude that could seemingly replace validation. To those well versed in automation, PAT is nothing more than the extension of long-standing control practices into pharmaceutical batch production. Engineers familiar with process control will recognize PAT as the installation of feedback control relying on sensors in the process equipment. This is by no means startling, except to those unfamiliar with control loops. PAT has its utility and will improve the quality of products produced by it—of this there can be little doubt. It will not, however, replace validation. In order to use a PAT system, the designer must assure that the installed sensor accurately reflects the process conditions throughout the batch otherwise it will provide no benefit. The need for that assurance means that the PAT system, rather than replacing validation, will actually have to be validated itself!

WHY VALIDATION

First, and certainly foremost, among the reasons for validation is that it is a regulatory requirement for virtually every process in the global health care industry—for pharmaceuticals, biologics, and medical devices. Regulatory agencies across the world expect firms to validate their processes. The continuing trend toward harmonization of requirements will eventually result in a common level of expectation for validations worldwide.

Table 2 Benefits of Validation

Increased throughput
Reduction in rejections and reworks
Reduction in utility costs
Avoidance of capital expenditures
Fewer complaints about process related failures
Reduced testingin process and finished goods
More rapid and accurate investigations into process deviations
More rapid and reliable startup of new equipment
Easier scale-up from development work
Easier maintenance of the equipment
Improved employee awareness of processes
More rapid automation

Utility for validation beyond compliance is certainly available. The emphasis placed on compliance as a rationale has reduced the visibility of the other advantages a firm gleans from having a sound validation program. Some years ago this author identified a number of tangible and intangible benefits of validation realized at his employer at the time (Table 2) (13). In the intervening years, there has been repeated affirmation of those expectations at other firms, large and small. Regrettably, there has been little quantification of these benefits. The predominance of compliance-based validation initiatives generally restricts objective discussion of cost implications for any initiative. But once a process/product is properly validated, it would seem that reduced sample size and intervals could be easily justified, and thus provide a measurable return on the validation effort. Aside from utility systems, this is hardly ever realized and represents one of the major failings relative to the implementation of validation in our industry.

Validation and validation-like activities are found in a number of industries, regulated and unregulated. Banking, aviation, software, microelectronics, nuclear power, among others all incorporate practices closely resembling validation of health care product production. That such verification activities for products, processes, and systems have utility in other areas should not be surprising. The health care industrys fixation on compliance has perhaps blinded us to the real value of validation practices.

CONCLUSION

Validation is here to stay; it has become an integral part of regulatory requirements and everyday life in the global health care environment. There are millions of pages of validation documentation across the world. The presence of such a mountain of information is not justification for its continued existence. Its presence affords a level of confidence in the quality of products for human health. The extent that the risk to the patient is reduced by a validation effort (or any other activity impacting product quality) will ultimately determine its continued utility. If risk-based thinking is adopted across the industry, as it appears it might be, then certain validations will be become more rigorous, others less so, and others unchanged. If the considerations associated with the implementation of validation for a process become financially driven, there may be additional opportunities. Validation for its own sake seems unlikely for the foreseeable future.

REFERENCES

1. Chapman K. A history of validation in the United States—part I. Pharm Technol Validation Comput Relat Syst 1991; 15(10).
2. Fry E. Presentation at PDA Annual Meeting. Nov 1980.
3. FDA, Guideline on General Principles of Process Validation, May, 1987.
4. Agalloco J. Course Notes on Principles of Validation, 1985.
5. Agalloco J. Validation—yesterday, today and tomorrow. In: Proceedings of Parenteral Drug Association International Symposium. Basel, Switzerland: Parenteral Drug Association, 1993:1–11.
6. Chapman K. A history of validation in the United States—part II. Pharm Technol Validation Comput Relat Syst 1991; 15(11).
7. Agalloco J. The validation life cycle. J Parenter Sci Technol 1993; 47(3):142–7.
8. FDA, Process Validation Requirements for Drug Products and Active Pharmaceutical Ingredients Subject to Pre-Market Approval (CPG 7132c.08, Sec 490.100), 2004.
9. Chapman K. The PAR approach to process validation. Pharm Technol 1984; 8(12):24–36.
10. FDA, Pharmaceutical cGMPS for the 21st Century—A Risk-based Approach, 2004.
11. Agalloco J. Course Notes on Principles of Validation, 2005.
12. FDA, Guidance for Industry PAT—A Framework for Innovative Pharmaceutical Development, Manufacturing, and Quality Assurance, 2004.
13. Agalloco J. The other side of process validation. J Parenter Sci Technol 1986; 40(6):251–2.

2

Organizing for Validation

Ronald J. Simko
Manalapan, New Jersey, U.S.A.

INTRODUCTION

Process validation and its role within a pharmaceutical organization have come a long way from its inception in the 1970s. At that time, the effort was primarily focused on sterilization validation and demonstrating that the conditions to achieve sterility were met. As a result, the mission was often managed from within the sterile manufacturing organization using a small team.

In the 1980s validation organizations were created and began interacting with the other traditional groups such as Research, Engineering, Production, Manufacturing, and Quality Assurance. At that time, several papers describing specific organizational strategies were presented at a Pharmaceutical Manufacturers Association Seminar (1–4). Just as it was then, there still is no single way that validation issues are addressed. Validation still depends on the overall organization and how it defines its mission. Where there were large multinational corporations in the 1980s, there are now even bigger multinational organizations resulting from the merger of some of the largest corporate names of the 1980s. While this has been occurring, there has also been the continual creation of new pharmaceutical companies. Some of these have a research focus that applies new technologies, while other firms have been created simultaneously with the purchase of commercial products from the aforementioned large multinational companies. Since annual product sales of $100 million are often no longer large enough for large pharmaceutical companies to retain interest, these products are often sold off to emerging firms at six times or more of annual sales. The net effect of all this is that there are pharmaceutical firms of all sizes, and some of them have commercial products, while others do not. The matter of organizing for validation becomes a different thought process in each of these circumstances.

Overlaying this wide spectrum of companies has been the continual evolution of the validation mission as it now has become a key component in managing the supply chain for those companies that are commercially active. For those other firms with products still in various clinical phases, the need to address process validation is also a key issue. This chapter will discuss these issues and provide an update on how the pharmaceutical industry is organizing for validation.

Abbreviations used in this chapter: FDA, Food and Drug Administration; GMP, good manufacturing practice; IQ, installation qualification; OQ, operational qualification; PQ, performance qualification; R&D, research and development; SOPs, standard operating procedures; TQM, total quality management.

ESTABLISHING THE MISSION

Formulating a mission is essential to ensure proper definition of a department role in the organization. This is necessary so that not only process validation staff members understand the breadth of their jobs but also that other corporate groups with whom they interact understand it as well.

Although there is broad diversity of Validation department missions within the pharmaceutical industry, the mission that is germane to all Validation departments is the satisfying of the regulatory requirement to have processes validated. The industry discovered this in the mid-1970s when the FDA demanded that the sterilization process for injectable drugs be validated. This began an era in which the focus of validation expanded at the FDA's request into other pharmaceutical processes: sterile filtration, solid dosage manufacturing, topical ointments and creams, cleaning validation, and computer system validations. During this period, the pace of new construction also accelerated and such activities as IQ, OQ, and PQ became fundamental elements of a comprehensive validation program. The prevailing mindset in the industry was, however, that validation was an obligation, a burden imposed by the FDA, which only added to the cost of manufacturing. In the earlier editions of this chapter, it was projected that "validation need not be an expense of doing business in the pharmaceutical industry," but that "it had the potential for being a key contributor to a cost efficient business strategy." This thinking has continued to take hold and, in fact, has become the basis for the ultimate successful management of the supply chain.

The successful conclusion of a validation effort simply proves that the process step being evaluated reliably does what was expected of it. A successful process validation effort thus builds quality into the process so that reliance on end-product testing can be minimized. It, in effect, provides greater statistical confidence about the process than does final testing. This, too, is the hallmark of TQM philosophies, which speak of continual improvement. Once such confidence about the reliability of the process is achieved, the company can schedule the production of product with the commensurate confidence that a given delivery date will be met. This means that inventories can be managed more precisely, and costs can be controlled. In essence, validation becomes a tool in the successful management of the supply chain.

Since the mid-1980s, the movement to embrace the quality principles, espoused by Deming et al. (5) and which has formed the foundation upon which these supply chain advances have been achieved, has begun to find a receptive audience in the pharmaceutical industry and influence its mission. The notion of constantly trying to improve one's work processes is really an expansion of the concept begun in the 70s—that of validating a process.

What does this mean to the effort to organize for validation today? For companies in which TQM philosophies are well-ingrained, the validation effort becomes indistinguishable from the corporate-wide mission of constant improvement. The process validation effort is just one of the tools. In the large multinational corporations in which these initiatives play a major role, the validation organization is typically a part of either the Manufacturing or the Quality organization. Where matrix management and cross-functional teams are often employed, there are inevitably close working relationships established with the other functions attempting to support the TQM approach to management. In these large pharma situations, validation efforts are often managed at the site level while oversight is provided by those in corporate who ensure a common approach across all of the company's locations.

Certainly the validation mission is influenced by the size of the company as well as its product lines. Companies, often the large multinationals, who have incorporated validation concepts into the supply chain business model, approach this issue differently than the large number of emerging companies. These smaller firms, sometimes with only a few products, often need to rely on outside firms for validation support. This support is provided to fulfill the fundamental reason for validation, that of regulatory compliance. Because of the critical regulatory impact of validation, it is important that the small pharmaceutical firm understands clearly what is being done by the contractor. In some situations, that contractor may actually be a manufacturer or a packager, who is contract manufacturing or packaging the product. In these circumstances, both the small firm (product owner) and the manufacturer are vulnerable if the validation effort is flawed. As a result, it is imperative that they collaborate to ensure that validation protocols and reports have been reviewed and approved by both parties. In examining the organizational structure for validation in this example, the contract manufacturer probably has an organizational structure not too unlike what has been addressed for larger firms. That is, there is a validation unit reporting either into Manufacturing or Quality that defines strategy, writes protocols, implements the plan, and writes up the final reports. These reports get circulated throughout both the contract manufacturing organization and the client's organization for approval.

For the typical small firm that may be contracting out the manufacturing and validation, these functions may be handled by only one individual or a small team. Because, in these situations, a person wears many hats, process validation being just one of them, there is a need to rely on consultants to provide back up support for projects as needed.

STAFFING ISSUES

When staffing a Validation group, the mission and the organization do exert a degree of influence, primarily in the academic backgrounds of the members. Because of the aforementioned diversity, a considerable variety of academic backgrounds are usually found among validation professionals, with members having degrees in chemistry, microbiology, pharmacy, statistics, computer science, biochemistry as well as engineering disciplines. When the mission is directed toward a sterile products focus, having a microbiology degree would be quite beneficial, as an example.

In a general sense, probably more important than the actual area of academic background are these three skills: problem-solving capability, interpersonal skills, and oral and written communication abilities. The technical talent to recognize and solve problems is fundamental to validation. Because of its pivotal role in the company, considerable interactions are required with others. Strong interpersonal skills are required for maximum effectiveness. Finally, unless validation objectives and concerns are effectively expressed both orally and in written form, the best of efforts in the field may be wasted. What emanates from the field work must be attractive written documentation. By presenting the documentation clearly in written form, a well thought-out and organized effort will be conveyed. If the validation professional can successfully communicate orally, especially during an FDA visit, the strength of the validation package is even greater.

A position that can also be used effectively, especially in a large pharmaceutical company is that of Validation Technician. These individuals are usually experienced production operators who have been promoted to the next job classification. The position provides the technician with an opportunity to contribute to problem solving, which may have been inhibited in a production environment. To the Validation department, there is the benefit of providing a work force of competent people who provide stability while others in the department may be in more dynamic career paths. The existence of technicians also provides validation professionals with the opportunity to develop their supervisory skills.

While the above describes what one might find in the Validation group of a large firm, the staffing at a small firm is distinctly different. As indicated before, one typically wears many hats in a small firm, and while the academic backgrounds that may have prepared the "small company" validation professionals are similar to those in the large firms, there is often a lot of learning "on the job." For these folks, it can be a rewarding and broadening career opportunity. The individuals pressed into validation duty may be from research, production, quality, or anyone possessing an interest in the subject. That is the way it often is in a small firm. For firms who outsource the manufacturing and validation, there is a need for one to not only have the technical competency but also some business acumen. The individuals have to have the willingness to learn and also the courage to tell their management what they do not know, both from a technical and from a business perspective. The consequences of failing to do this could lead the firm down a treacherous path. That is where the use of experienced consultants can play an important role in the staffing support for validation at these smaller firms.

With a large number of consulting firms to choose from, the selection criteria become ever so important. There are large entities that have assembled many associates to provide support for their clients. There are also the one- and two-person organizations. Whichever type of firm is chosen, be certain that the individuals assigned to the project fully understand the task at hand and possess the expertise to accomplish it. There are a diversity of manufacturing processes, facility installations, computer systems, and analytical methods that require validation, and it is rare that one individual has expertise in all areas. Additionally, it is critical that someone within the company has an indepth understanding of all validation activities that are contracted out. This is essential to ensure that the company can assess the quality of the validation service being provided, and furthermore, be able to fully defend the validation when the FDA comes to inspect. After all, the consultant will probably not be present at the inspection to answer for what was done. The validation knowledge must be in-house. This leads to what should be a fundamental truth for the use of a consultant. Do not look for the consultant to possess expertise that the organization does not have. Instead, hire the consultant to meet periods of peak activity. The other basic demands on the consultant are to meet the time and budget constraints imposed on the project. To ensure this, the internal expert should really serve as a project manager. Regular meetings should be held verifying that the project is still on target from a budget, time, and technical competence perspective. This is essential, even for those organizations that are small in size.

DEPARTMENT INTERACTIONS

Once department missions have been formalized and the validation operation organized, the challenge is to implement the plan. That implementation requires the validation organization to interact with many peer groups. Within the company, those other departments include the following:

1. R&D: involved with new product development and new process improvement.
2. Engineering: involved with new or modified equipment or facilities.
3. Production: concerned with processes that require validation.
4. Maintenance: concerned change control, calibration, and preventative maintenance.
5. Quality Control: involved with the testing laboratories.
6. Quality Assurance: concerned with GMP compliance.

Additionally, for those companies that outsource the manufacturing or packaging of their products, these interactions occur with the contracting firm's Validation department. This poses an additional set of dynamics.

R&D

The research organization is involved with new product introductions and often existing process improvements. It should be the Validation department's key objective in interacting with R&D to ensure the acceptability (and thus validatability) of new products or "improved" processes in the manufacturing area. Some firms have "older" processes that perform at a less than optimum level. A successful accomplishment of validation objectives ensures that new products or processes do not fall victim to the same fate.

Communication is critical in accomplishing this. The R&D organization must be made aware of the validation plan and resulting acceptance criteria. The awareness of these expectations should prompt an R&D testing regimen that will enhance the probability of acceptance of the product (or process) to manufacturing. It also affords the analytical R&D laboratories the opportunity to develop and validate analytical methods during the development phase. The results from this plan are products or processes that are expected to be validatable in the production plant. Of course there can always be surprises, especially when a product does not scale-up as expected. In this situation, the Validation, Manufacturing, Quality Assurance, and R&D departments must work out a suitable solution, because the product should not be introduced to production if it cannot be first validated in the pilot plant.

Engineering

The relationship that, along with R&D, possesses the greatest potential for long-term validation benefits is with the Engineering groups involved with new facility or equipment start-ups. In the initial stages of capital projects, there exists the ideal opportunity to ensure the acceptability of the processes later on. The concern of validation must be built in at the design phase and continued through construction. It is one thing for a water-for-injection system to be designed properly, and quite another for it to be constructed properly. Thus, in this example, it is necessary for the validation effort to include such activities as the documentation of weld quality and distribution piping slope verification during the construction.

In applying this to biotech production, Hill and Beatrice discussed the qualification and validation challenges in 1989. They said "the validation of the actual processes (operations) should not be an afterthought in the design of plants producing biotech-derived products. Rather, plans for the qualification/validation programs should be an integral part of the preliminary plant concepts and specifications." They continued by saying that validation "should not be looked upon solely as a burdensome requirement for complying with FDA requirements, but as a means of increasing productivity through consistently producing lots of products that meet all specifications" (6).

Once construction is complete, the qualification phase can begin. Qualification protocols defining design and operating criteria need to be developed and signed off by all parties involved. This ensures that there are no misunderstandings as to what is expected of the facility. Production, Quality Assurance, Engineering, and Validation all need to approve this plan in writing.

It should be noted that one of the ways the evolution of validation has manifested itself has been the increasing role that validation has played at equipment manufacturers. They often stand ready to provide qualification and/or validation support services for their products. This often takes the form of equipment testing being done at their production site. The purchaser of this equipment should take advantage of the opportunity to be on site for this test phase of the equipment before it is shipped. The best time to uncover flaws is then, and not after it arrives following a trans-Atlantic shipment.

Production

Interactions with production personnel should stress the benefits of a validation program. If the benefits are really understood, production personnel will be supportive of these efforts rather than skeptical. This is probably less of an issue in the 21st century than it was in the early days of validation. Production personnel today often see the contributions a well thought-out validation program can bring to the supply chain, and the resulting cost savings. There are fewer rejects, retests, and reworks—meaning dollar savings. Production schedules can be created with a higher level of confidence that they will actually be met. These positive effects of a validation program justify the efforts for economic reasons, rather than just regulatory compliance.

After the completion of a validation study, the results are presented in a written report which is then approved by all protocol signatories. Signed off reports should then be distributed to all affected operations so that procedural changes or acceptable process parameter ranges discerned from the validation report can be incorporated into SOPs. This procedure further ensures that production SOPs reflect validated conditions.

Maintenance

Without the support and cooperation of the maintenance organization, the best designed and implemented validation study will soon be rendered worthless. This will occur the instant that an undocumented change is made to a validated piece of equipment. As a result, an education program is essential to make maintenance personnel understand the effect of their preventative or emergency maintenance activities. Once this is understood, the documentation of any changes made to a system must be communicated so that an assessment can be made. One must realize that changing a belt on a dry-heat sterilizer fan motor could affect airflow within the sterilizer and thus change "cold spot" locations and perhaps sterilizing capabilities within the oven.

Quality Control

Because there is a reliance on Quality Control laboratories for testing support, effective communication is extremely important. This communication may be with the Quality Control department within a large firm, or with a contracted Quality Control Lab. Certainly validation protocols that require laboratory support should require laboratory management sign-off. This ensures that laboratory personnel know not only the number and type of tests required for the study but also how the testing fits into the overall validation program. This affords them the opportunity to understand how the data will be used and to avoid situations in which the laboratory personnel's test invalidates the intent of the validation. Additionally, validation staff members should acquire an understanding of laboratory testing procedures. What results is two-way communication that ensures good understanding of both organizations' intentions. Ensuring this understanding and buy-in by Quality Control makes good business sense. If Quality Control is not prepared to meet the validation plan's timetable, the otherwise best-laid plan will not be accomplished on schedule, and product launches could be delayed.

Quality Assurance

Significant interactions also occur with Quality Assurance. These interactions are designed to ensure a firm's regulatory compliance. Through the technical competency of the validation staff and the GMP compliance expertise existing within the Quality Assurance group, these efforts should be successful. The key point is to communicate so that the regulatory compliance objective of validation is met.

Professional Associations

Interactions of validation professionals with counterparts at other pharmaceutical firms prove extremely valuable. Most frequently, this is accomplished through meetings sponsored by societies, universities, or seminar organizations. These gatherings are extremely beneficial, not only because of the knowledge imparted during the structured presentations, but also because of the opportunities available for informal discussion of problems and concerns.

Food and Drug Administration/European Medicines Evaluation Agency

No discussion of interactions of validation personnel can be complete without the inclusion of the regulator. These interactions are part of any validation professionals work experience. Although heightened anticipation exists when the inspector is in the plant, the potential is certainly there for a valuable learning experience. The interactions during these investigations frequently prove challenging and can be constructive both for the firm as well as the individual defending the validation package.

Contract Manufacturers

With many firms relying on the use of contract manufacturers and packagers for the production of both commercial and clinical products, the challenges of communication are many and varied. These interactions are typically guided by both a supply agreement and a quality agreement. Once defined in these agreements, it will be imperative from a business sense that these roles be dutifully adhered to. Certainly the company that owns the products and has contracted its manufacturing must have enough validation expertise to engage the contractor in important validation discussions. Protocols as well as final reports will need to be reviewed and signed off. It is also recommended that the contracting firm be in attendance when the process validation study is being implemented. The contracting firm must have the validation expertise to fulfill all of these roles. Employing consulting support is highly recommended if resources are stretched to ensure validation issues are properly addressed. In reality, the interactions between the contracting firm and the contractor are similar to that between two "internal" departments. However, in this case both companies have vulnerability if things do not go well, as batches not getting shipped will certainly affect both firms. The best approach is to establish a close working relationship when it comes to validation issues. An ounce of medicine is worth a pound of cure, as they say.

MAINTAINING THE ORGANIZATION

Continuing Education

For the continued realization of validation objectives, the quality of the staff must be maintained. A program of continuing education is critical to achieving this. It is necessary for the organization to provide staff members with opportunities to take courses that can aid them in remaining current technologically. At the same time, it is the employees' responsibility to avail themselves of those opportunities made available. Job-related courses and seminars are frequently sponsored by various trade associations and are often offered in conjunction with meetings designed to keep their membership technically current in their related fields. Universities and professional seminar organizations also add to the complement of technical education courses available.

In general, courses should be chosen to bring staff members to a basic level of understanding of the skills needed to do the job. The most effective learning experience, though, is found on the job. If, for example, a validation professional is assigned to sterile products validation, then a course on microbiology and engineering of sterilization processes would be appropriate. It is probably best scheduled, however, after several months of on-the-job training. This would set the stage for the course to have maximum effectiveness.

Beyond those specific training courses are general courses of study that should be made available on a broader scale to validation staff members. All non-pharmacy majors should be exposed to a short course on pharmaceutics, for example. These are typically offered by pharmacy schools and provide attendees with a general overview of the subject. Computer education has also become critically important, while a course that teaches problem-solving and decision-making skills surely will assist Validation Engineers in their work. With cost consciousness around us at all times, opportunities to teach larger number of employees by purchasing courses that are on DVDs should be considered. In-house or interactive training is still another option.

Because continuing education is a necessary aspect of everyone's career, it must be treated as such and used to supplement on-the-job learning. Because the nature of this education is both varied in content and in presentation format, its inclusion in the career development of a validation professional cannot be ignored. With the proliferation of information available on the Internet, it also behooves those in Validation departments to take it upon themselves to search out information. Many trade journals have websites accessible to subscribers, for example. The key is to search out these opportunities, as it has become an essential way of staying current in today's world.

Organizational Transfers

Another way of building the strength of the entire organization is through interdepartmental personnel transfers. Validation professionals are conscientiously aware of quality manufacturing procedures and can apply these concepts in a Production or Quality Assurance organization. Technical areas within Engineering or R&D may also find validation talent helpful in filling openings. Because the validation operation interacts so closely with all of these other areas, a smooth transition to other assignments can be achieved. Certainly situations such as this are a two-way street, as professionals from other departments within the company can also transfer into the Validation department in an attempt to broaden their own career. This can provide the staff greater depth and strengthens each person's ability to qualify for promotional opportunities.

BROADENED REGULATORY FOCUS

Just as the emergence of TQM philosophies in the pharmaceutical industry can be traced to the gradual building-in of quality consciousness, which started in the days of sterilization validation, one can also look at the expansion of validation concepts in another way.

In the late 1970s, it used to be enough to validate the sterilization processes performed in an autoclave or dry-heat oven. Then it was all other manufacturing processes, including cleaning. The mid-1980s saw the coming of computer system validations. Then there was bulk chemical production, and other bulk processes, such as fermentation and purification. Recognizing this, Sawyer and Stotz, when they outlined the "Validation Requirements for Bulk Pharmaceutical Chemical Facilities" (7) alluded to validation organization strategies stating the importance of "broad pharmaceutical/FDA/cGMP and strong chemical process experience." They recommended an on-site team with a designated on-site project manager, a client/contractor steering committee for the project, and the need for "good interpersonal skills" for all involved.

More recently, the need to validate training and inspection processes has been identified. As with the processes mentioned earlier, all of these are important to the quality of the finished product. They need to be in a state of control and need to be conducted with consistent quality. This is still good advice in the 21st century.

The requirement for validation from a regulatory viewpoint has expanded in still another way. Formerly, it was sufficient to validate only those products marketed commercially. However, for those whose responsibilities include clinical products, there must be recognition that these processes must be characterized in such a way as to assure reproducibility, and can be correlated to the process that is ultimately commercialized. The sterility aspects of clinical supplies must be validated to the same extent as commercial materials.

So how is this validation effort to be organized to achieve this expanding regulatory focus? The answer is once again dependent upon the size of the firm. For the small firm that is producing product solely for clinical studies, the Manufacturing unit (or person) is probably charged with the responsibility of validating both bulk pharmaceutical processes as well as those "finishing" processes. These firms may also be employing contractors requiring the need to manage this outsourced service as stated earlier. This is important when the product is in the clinical phases, and becomes vital as the first products approach commercialization. Staffing the organization to fulfill this role becomes critically important.

As these small firms grow, their facilities will also vary from rather small, dedicated facilities to large multi-product plants with a diversity of processing capabilities. At these large installations, the validation efforts more closely resemble those conducted at the traditional pharmaceutical firms. This makes sense because of the diversity of products and processes. What should be in place, however, to guide this growth is some corporate overseer of validation, who ensures a common approach to validation within the company. By serving as a liaison between groups, information beneficial to one can be transmitted to the other saving the "reinventing the wheel" syndrome.

CONCLUSION

Despite the events that have influenced validation since the early days, the issues of organizing to meet the validation challenges have not changed appreciably. Whether the process is a bulk process or one of the finishing steps; whether the process is a proprietary purification process, a steam sterilization process, or a conventional nonsterile process; whether the focus is a clinical manufacturing lot or commercial production; or whether the effort is accomplished within the firm, contracted out in conjunction with an outsourced manufacturing agreement, or with the assistance of a consultant, the validation staff must possess four things:

1. Technical expertise, allowing a thorough understanding of the process being reviewed.
2. Understanding of the fundamentals of validation and the ability to apply them to the process.
3. Interpersonal skills necessary to deal with all of the organizations within and outside of the firm. Some business acumen would be a great attribute for those involved with contract manufacturers.
4. Support from management, which positions the validation effort as a critical element in the company's success.

These are the basics and are what will ensure that the validation effort is successfully accomplished. This was true in the early days, and will remain true in the future.

REFERENCES

1. Primm TM. Organizing for validation—engineering division. Presented at the PMA Seminar on Validation of Solid Dosage Form Products. Atlanta, GA, May 1980.
2. Papariello GJ. Organizing for validation—process development. Presented at the PMA Seminar on Validation of Solid Dosage Form Products. Atlanta, GA, May 1980.
3. Guillemoteau JY. Validation of solid dosage forms—an organization within the manufacturing division. Presented at the PMA Seminar on Validation of Solid Dosage Form Products. Atlanta, GA, May 1980.
4. Green W. Organizing for validation—quality assurance viewpoint. Presented at the PMA Seminar on Validation of Solid Dosage Form Products. Atlanta, GA, May 1980.
5. Gitlow H, Gitlow S. The Deming Guide to Quality and Competitive Position. Englewood Cliffs, NJ: Prentice Hall, 1987.
6. Hill D, Beatrice M. Facility requirements for biotech plants. Pharm Eng 1989; 9(4):35–41.
7. Sawyer CJ, Stotz RW. Validation requirements for bulk pharmaceutical chemical facilities. Pharm Eng 1992; 12:44–52.

3

Validation and Facility Design

J. Robert Adamson
Foster Wheeler, Reading, Berkshire, U.K.

INTRODUCTION

The design, construction, and commissioning of a new facility for the pharmaceutical industry is a complex process that involves the interaction of a wide variety of engineering, process and QA, and control disciplines and may proceed through a series of different phases from a conceptual, feasibility study, through to the final detailed design, construction, commissioning, and final site validation activities. The FDA's risk-based approach to GMPs for the 21st century has changed the industry's perspective to validation and qualification. This new initiative allows the facility designer, constructor and commissioning group to take a risk-based approach to the qualification of facility and equipment. The basic requirements for validation of facilities and equipment are defined in both the European community's *Guide to Good Manufacturing Practice* Vol. IV (Medical Products) and the United States's GMPs CFR Title 21 (1,2). These documents clearly define the need for a whole system that is based on QA. This is essential for pharmaceutical companies to ensure that products meet their quality and marketing authorization requirements. cGMP is a key element of an overall QA system and GMP extends through people, production, premises, and equipment. Both the U.S. GMP and the EC Guide emphasize that premises and equipment should be designed to be appropriate and fit for the purpose. Interpretation of this statement implies that it is essential that facilities must be built to standards that meet the requirements of the GMPs and be demonstrated to meet these requirements. The process of validation is a key component within the concept of QA and GMP. The consequence of this for the facility designer is that he or she must use design and engineering methods that will comply with and demonstrate that the facility, when complete, does meet the requirements of cGMPs.

Abbreviations used in this chapter: cGMP, current good manufacturing practice; CIP, clean in place; DQ, design qualification; EC, European Commission; EMEA, European Medicines Evaluation Agency; FAT, factory acceptance test; FDA, Food and Drug Administration; FEQ, facility and equipment qualification; GMPs, good manufacturing practices; HVAC, heating, ventilation, and air-conditioning; IOQ, installation and operation qualification; IQ, installation qualification; NDA, new drug application; OQ, operational qualification; PID, piping and instrument diagram; PQ, performance qualification; PV, process validation; QA, quality assurance; SIP, sterilization in place; SOPs, standard operating procedures; UK MCA, United Kingdom's Medicines Control Agency; URS, user requirements specification; USP, *United States Pharmacopeia*; WFI, water for injection.

This chapter describes an approach that can be used by the designer to ensure that the design, engineering, and construction process can meet the GMP requirements. This approach is sometimes referred to as Validation Master Planning. More commonly, industry develops a Validation Master Plan to cover all aspects of the validation of an operating unit such as: Process Validation, Cleaning Qualification, Automation Validation, and Facility and Equipment Qualification. This chapter shall refer to it as the Facility and Equipment Qualification or FEQ plan. The key basis to successfully qualify a facility is to plan the qualification from the earliest stage of the facility design by the development of a clear validation strategy that will develop into a plan for validation throughout the project. The main focus of this chapter is on new facilities; however, a separate section discusses how this approach can be adapted to meet the needs of revamp or refurbishment projects. A complete section is devoted to the development of an FEQ Plan. The FEQ section outlines the whole of the qualification requirements both in scope and for all stages of the project.

THE ENGINEERING DESIGN PROCESS FOR A FACILITY

The engineering or feasibility design process typically follows a series of phases.

- Conceptual design
- Design development, front-end design or basic/preliminary engineering
- Detailed engineering
- Procurement
- Construction
- Precommissioning
- Commissioning

Each of these phases has its own engineering objectives and, consequently, the qualification requirements have both a different scope and extent at each phase. The concepts for qualification will be described for each phase.

Conceptual Design

Introduction

The actual process design commences at a much earlier phase than the engineering design. Pharmaceutical drug discovery, design, and production are key elements of the industry. At a stage during the drug development and clinical trial phases, it will become apparent whether the company has a new product that it wishes to bring to the market. This is usually the point when first considerations for the engineering and manufacturing needs for the production of the drug will be addressed.

Production of clinical trial material will have moved from laboratory facilities to pilot-scale operations. Experience gained at this pilot-scale production will normally give sufficient information to enable a process definition to be prepared. The marketing organization will also have some early projections for demand levels and the type of formulations that will be required. These key elements will give a basis for a conceptual design study. The collection of process data for subsequent full-scale PV will also already have begun. Clearly, the current regulatory bodies emphasis on proof of drug equivalence, i.e., final production batches must be equivalent in biological and chemical activity to those used in the clinical trial and any subsequent submissions (typically for the NDA) will already have some significant effect on the manufacturing route, engineering design, and equipment selection.

The conceptual study must consider all these aspects and incorporate their requirements into this early design. Consequently a plan is required to ensure that GMP, qualification, and process requirements are incorporated.

Purpose

The main purposes of a conceptual study is to provide:

1. An agreed basis for the design philosophy to be able to proceed to the next phase of development (frequently called Front-End Design by the engineering contracting organization or sometimes known as Design Development or Basic Engineering).
2. To provide an initial capital cost estimate, usually for a preliminary budget sanction by senior management. Often a conceptual study is used as a feasibility study (i.e., should we proceed or not?).
3. Deliverables.

The typical deliverables of this phase are as follows:

- Statement of basis of design
- GMP statement
- Process block flow diagrams or schematics
- Major equipment item list
- Conceptual layout and accommodation schedule
- Building and HVAC philosophy
- Outline of utility systems
- Outline of control philosophy
- Safety considerations
- Budget estimate

Approach

Usually, the conceptual study will be run as a mixed disciplinary team bringing together research and development, production, and engineering disciplines led by a study manager. Although QA does not have a major role to play at this stage, it is important that the team has access to appropriate personnel.

In the design of a pharmaceutical facility, one of the most important aspects of the development is the layout. A typical approach that has been useful is to develop an accommodation schedule (Fig. 1), which shows a typical example for an aseptic suite.

This shows the flow of personnel, materials, and products. Figure 2 shows the variety of data that goes into the development of this schedule, which usually brings together specialist disciplines, including an engineer who understands layout development. This early process is an iterative phase during which all disciplines will have their input into the accommodation schedule, although it is best if a single individual, skilled in layout development, coordinates the activities and provides the preliminary drawings for review by the team. Once a first layout is agreed on, it must then be formally reviewed for GMP compliance. Figure 3 shows such a preliminary layout. The process may be repeated as the layout is developed and, consequently, GMP principles are built into the design from an early stage. The whole process

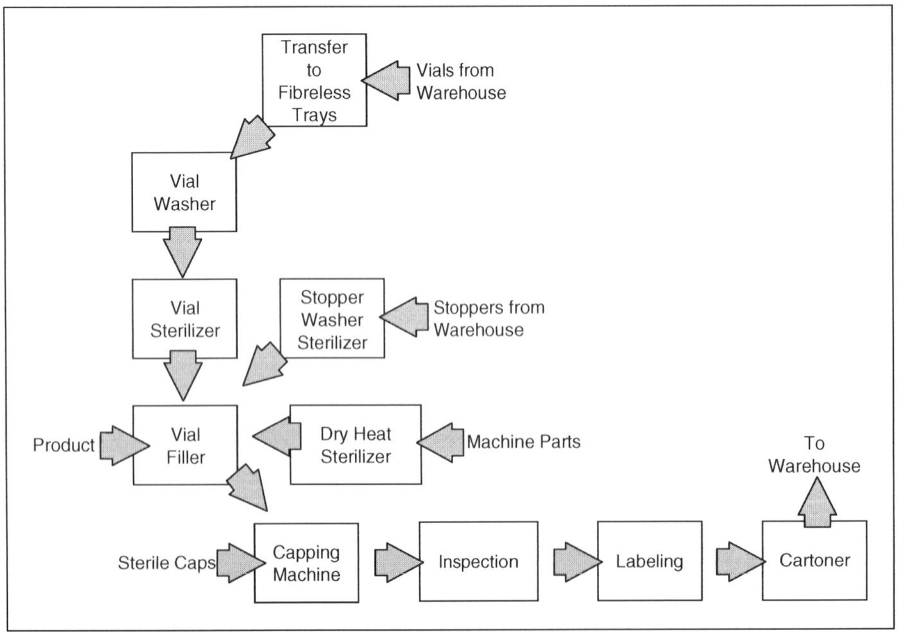

Figure 1 Accommodation schedule.

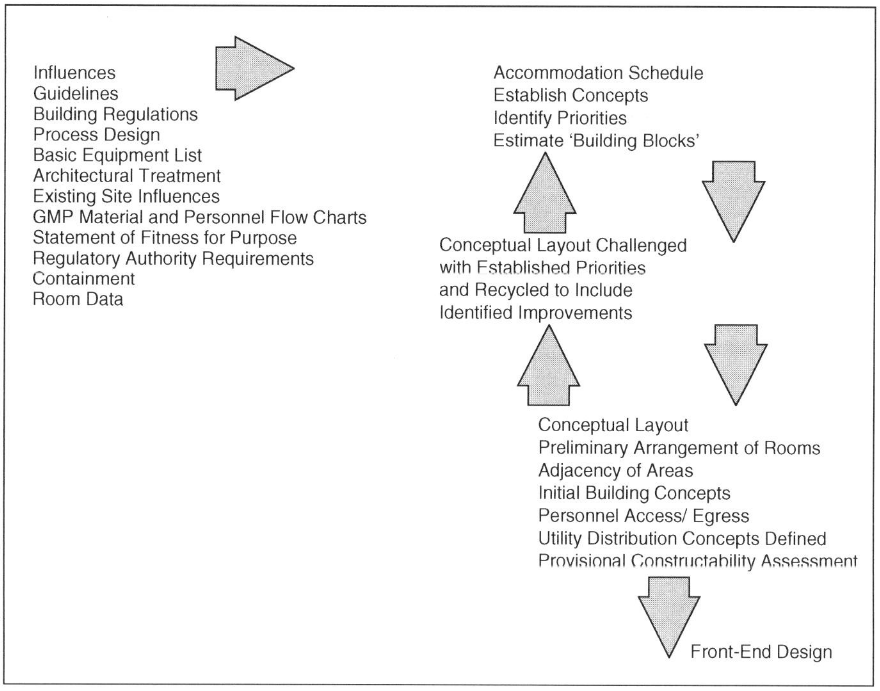

Figure 2 Conceptual layout development.

ensures that the final layout meets GMP and is documented. This is part of the qualification of the design and is key "DQ documents" and must be approved by the appropriate team members.

Qualification Activities

At this stage, the qualification of the facility is in its earliest phase and the emphasis must be on the qualification of the design. This can be completed by reviews of the proposed design against defined user requirements criteria. The preliminary nature of the study limits the depth of review. It should address critical issues against the user specification and the GMP requirements.

Qualification Cost

Clearly if the conceptual phase is to provide a cost estimate for the project, then the qualification must be similarly estimated at this stage. Some form of qualification statement and policy is required to at least determine

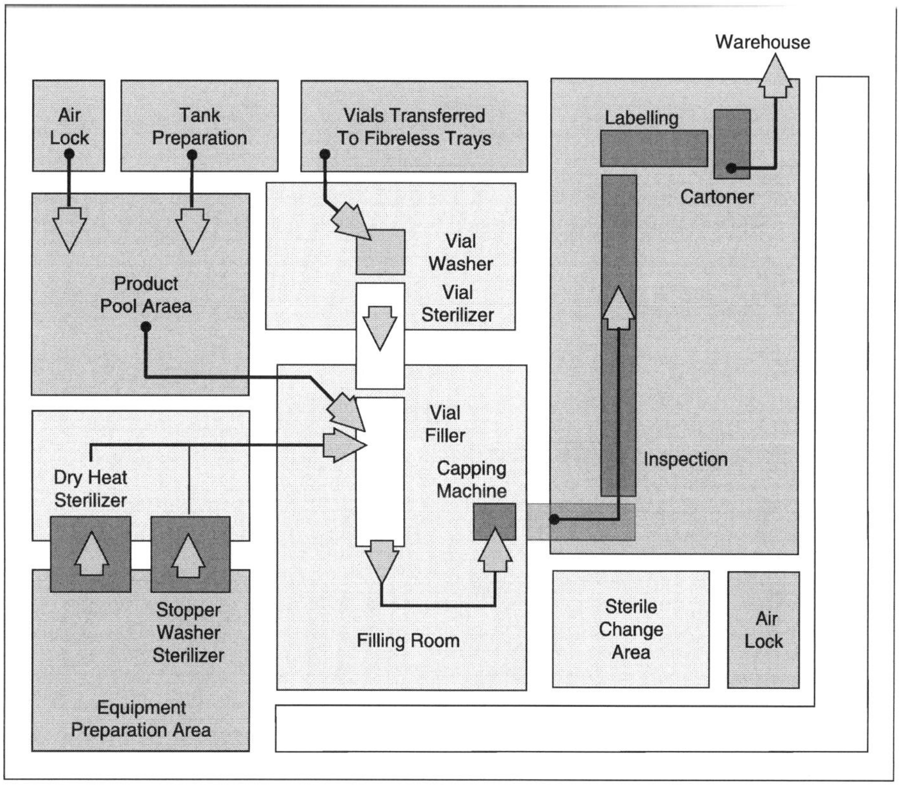

Figure 3 Preliminary conceptual layout.

its future scope. At this stage, this may involve only a flowchart (Fig. 4). Some may prefer to develop a very preliminary facility and equipment plan (see the section entitled Facility Qualification Plans). The decision of which route to take may be determined by the extent of the study and company policy. Without significant details of the facility and its contents, specific costs for the key qualification tasks cannot be easily determined unless access to similar projects' costs is available. At this stage, it is probably more normal to make an allowance based on in-house or the design engineers' experience. It is important to have an estimate that reflects that of the study. If the study is $\pm 25\%$, then it is reasonable for the qualification estimate to fall within similar limits.

Design Development

Introduction

Usually, by this phase of the project, the pharmaceutical company believes that it is highly probable that the project will proceed subject perhaps to certain restrictions, usually based on schedule and total final cost. The first key decision is (*i*) should this phase be done in house? (*ii*) involve an external design construction consultancy? (*iii*) an Engineering Management Contractor? Frequently, the choice is very dependent on organization culture.

Clearly, whatever the choice, some key questions are "Can the designer meet and demonstrate that the design complies with GMP?" "Are you going to use a single engineering organization to manage the project through design, procurement, construction, commissioning, and qualification?" "Are the systems in place to aid qualification?" Choosing your contractor is discussed in more detail elsewhere (3). The answers to these questions have significant bearing on the route adopted.

Purpose of Design Development

The main objectives of this design development phase are as follows:

1. To establish a basis for detailed design
2. To progress the design to establish the technical, capability, and safety aspects of the project
3. To provide the necessary design data to evaluate and, subsequently, comply with the regulatory, environmental, and planning requirements of a project with the relevant authorities
4. To provide an improved cost estimate and so enable sanction of the project.

Deliverables

Typical design development deliverables are as follows:

- Process flow diagrams
- Process and equipment specifications
- Utility specifications
- Control and automation user requirements specification
- Preliminary process PID
- Floor plans and equipment layouts
- Facility and equipment qualification plan
- List of systems
- Building evaluation
- Building finishes
- HVAC schematics and routings
- Safety and GMP reviews
- Environmental considerations
- Project schedule
- Estimate

Approach

Once the choice of management for the project is made, a team must be assembled under a project manager, who

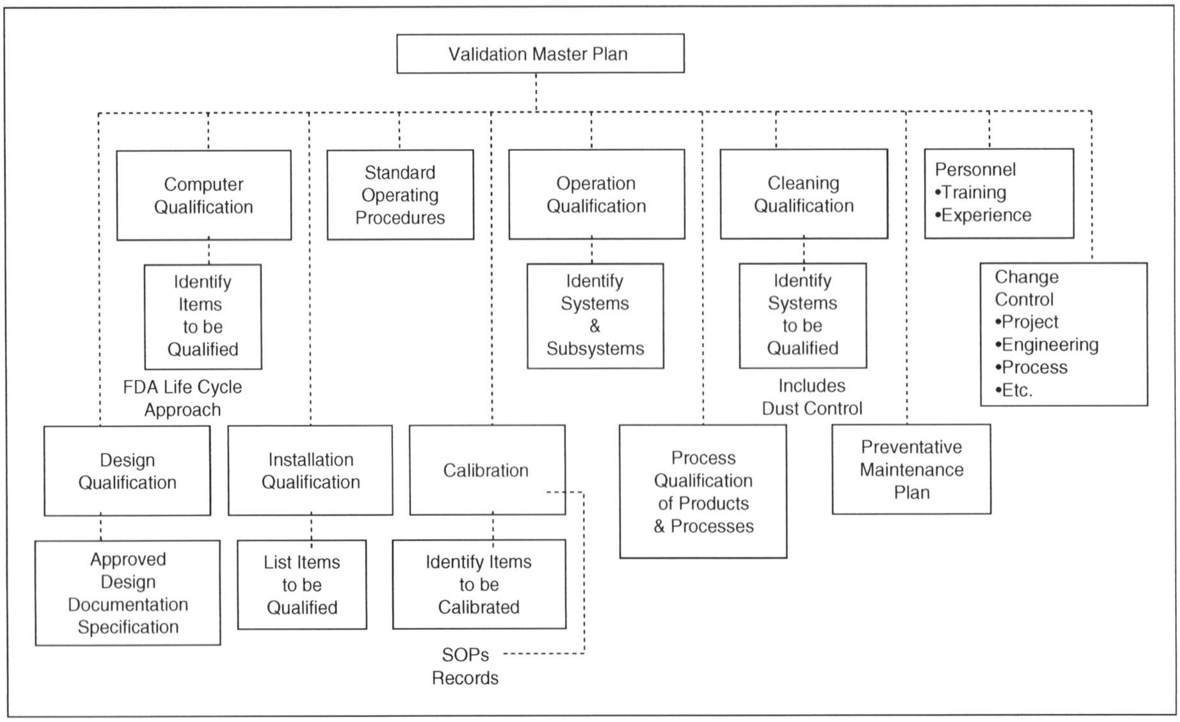

Figure 4 Facility and equipment qualification plan.

will preferably see the project through all the subsequent phases to provide a high degree of continuity. This is an important factor to consider for this key position.

The key areas of development during this phase are the following:

1. The layout, to define and fix the building size
2. Define all major items of equipment
3. Define piping and instrument requirements, as shown in the PID
4. Identify the key process services to equipment (e.g., pure steam, WFI, air, nitrogen, and so on)
5. Establish philosophy for process control and automation, containment, and safety
6. Identify the utility services, HVAC, drainage, electricity, and others
7. Identify preliminary architectural details and building structure and foundations
8. Identify any long-lead items: usually equipment (e.g., major items can be on delivery times as long as 12 months)
9. Ensure the design meets GMP and can be demonstrated (validated) to do so
10. Develop a cost estimate at a defined accuracy (usually 10–15% is required at this stage)

Layout Development

From the conceptual design, the materials, personnel and product flows will have been agreed on, and the philosophy determined. During this phase, each of these needs to be challenged and developed in detail. Once completed, no further changes should be made during detailed design other than minor accommodations to permit interfacing with final equipment installation requirements.

Typically decisions made affect both the DQ and the subsequent validation process. We consider two examples: aseptic changing facilities and the options that might be chosen for sterile stoppers used for vials.

Three schematic options are shown for aseptic changing facilities (Figs. 5–7); all have an appropriate use, depending on specific requirements. The designer must be aware of the implications of their choice. The simplest (Fig. 5) is suitable only for low-traffic-flow areas and may need some form of traffic control to prevent an exiting operator passing an ingoing operator at a critical point. The option in Figure 6 is straightforward and preferred (i.e., separate in and out), but is more expensive to build, whereas Figure 7 is a compromise, but reuse of the garments will require validating. The option in Figure 5 requires validation of the traffic flow procedures and cleanup rates between exit and reentry; perhaps automatic systems may be considered to prevent personnel who are moving in opposite directions from meeting; more normally the firm would rely on procedure. The option in Figure 6 clearly eliminates this potential adverse consequence and so makes the subsequent validation of operations simpler. Each option presents its specific challenge in design and in subsequent validation requirements, and an evaluation of capital cost versus validation costs should be a part of the decision process.

Again, two options are shown schematically in Figures 8 and 9, for handling stoppers for an aseptic vial filling process. The designer's choice has significant effect both on layout and subsequent validation requirements. The option in Figure 8 may initially appear very attractive, the use of prewashed and sterilized stoppers reduces the need for expensive equipment to be purchased and installed. However, QA must audit the supplier, and the designer must devise an aseptic means of transfer to the filling line. The solution is frequently a manual transfer by a pass-through hatch and manual loading into the stopper bowl. Each operation will have to be validated. The route shown schematically in

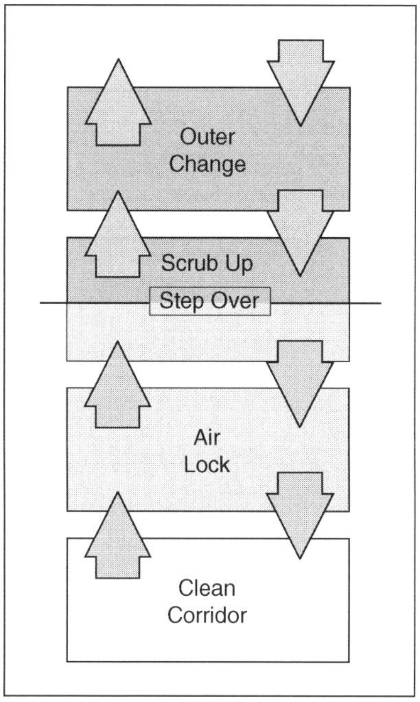

Figure 5 Aseptic changing facilities—low traffic flow.

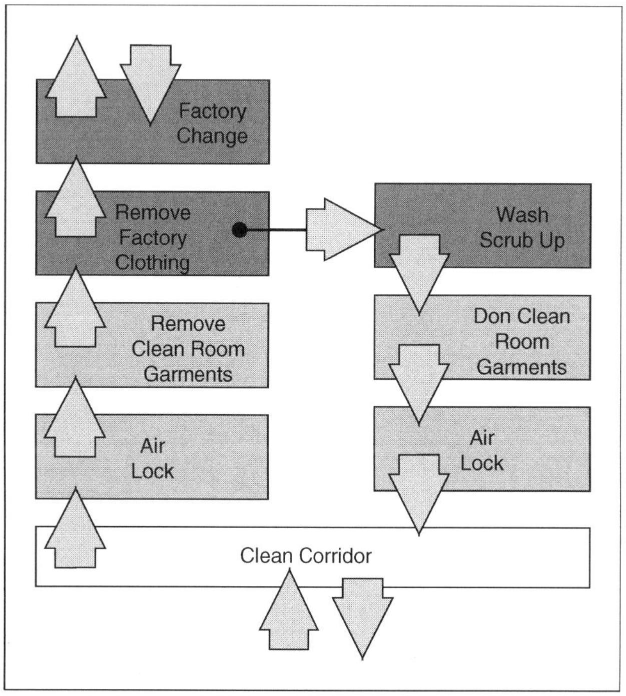

Figure 6 Aseptic changing facilities—separated flow in and out.

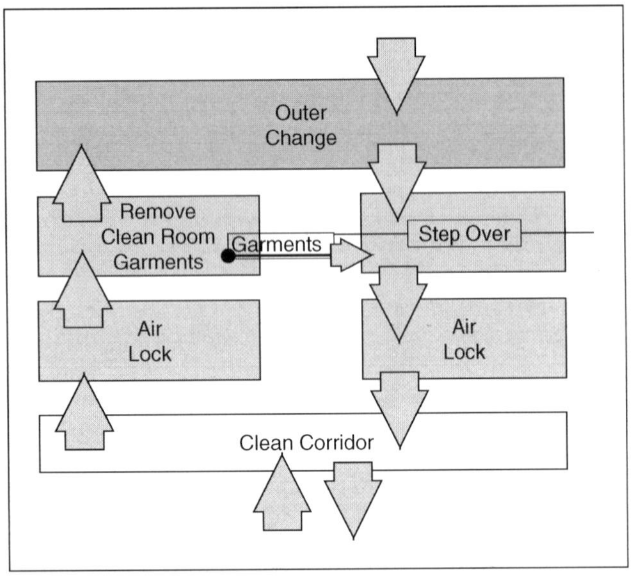

Figure 7 Aseptic changing facilities—separated flow with garment reuse.

Figure 9 shows stoppers being washed and sterilized on site and then being transferred from the clean side of the stopper washer–sterilizer to the vial stopper hopper. This can be achieved in closed containers minimizing manual contact and operations. Options are available from some suppliers that make use of isolation technology. Clearly, this latter route has very significant influence of validation requirements and design.

The two examples demonstrate that the development of the design choices at this stage has implications for the layout, the layout's qualification, and the validation requirements to be later conducted by the operations group. These requirements can be covered in the GMP review that is a key part of the DQ and can be used as part of an evaluation of the options.

Equipment

Usually, within the scope of this phase, the major equipment specifications are developed. These specifications will form part of the DQ and should be related to the user requirements specification. For some major items, with long-lead times, it may be necessary to develop these into requisitions or tender documents to meet the overall project schedule. The requirements for validation must be developed concurrently with these specifications and requisitions. These requirements include identifying all types of documentation that will be necessary to execute the qualification. This documentation will typically include the following generic topics:

- Equipment suppliers' documents and drawings
- Engineer's documents and drawings
- FAT documents
- Delivery and installation documents and drawings
- Protocols IQ, OQ, PQ, and associated documents.

An approach that can be used to assemble the list of detailed documents and drawings is to develop the lists in a matrix form (Fig. 10). It is important to incorporate the document drawing requirements into a requisition, for this can represent a significant proportion of the cost. Negotiating for documents post-delivery of an item can prove costly and, in some circumstances, result in no documentation being received. The implications of this for the completion of qualification are potentially severe. Many of the vendor documents are also essential to commence and complete the IQ and OQ protocols.

Further details on the protocol contents and associated documents are found in the section entitled Facility Qualification Plans. Ensuring availability of relevant documents at the correct time in the program is critical to the validation program. Delays in the supply of, or inadequate documentation, provided by the equipment suppliers can significantly delay the validation and, consequently, adversely affect the final target dates for production of the saleable product.

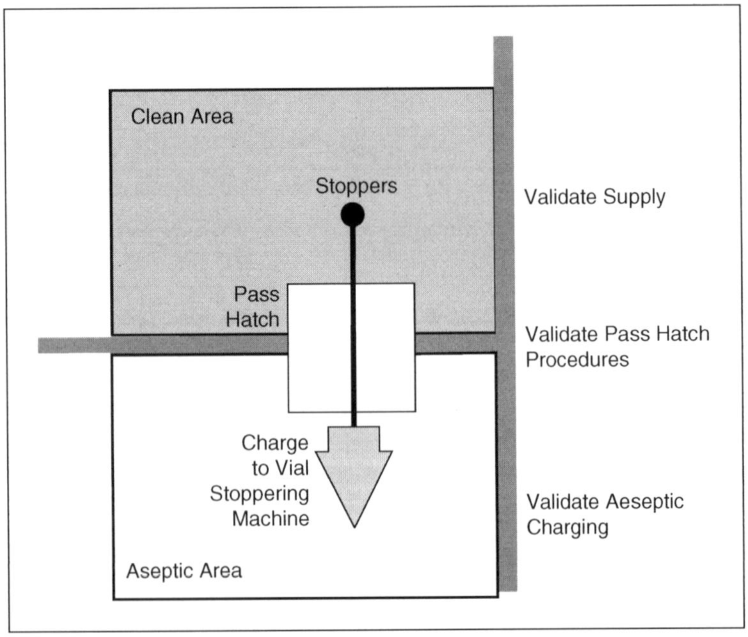

Figure 8 Transfer of prewashed and presterilized stoppers.

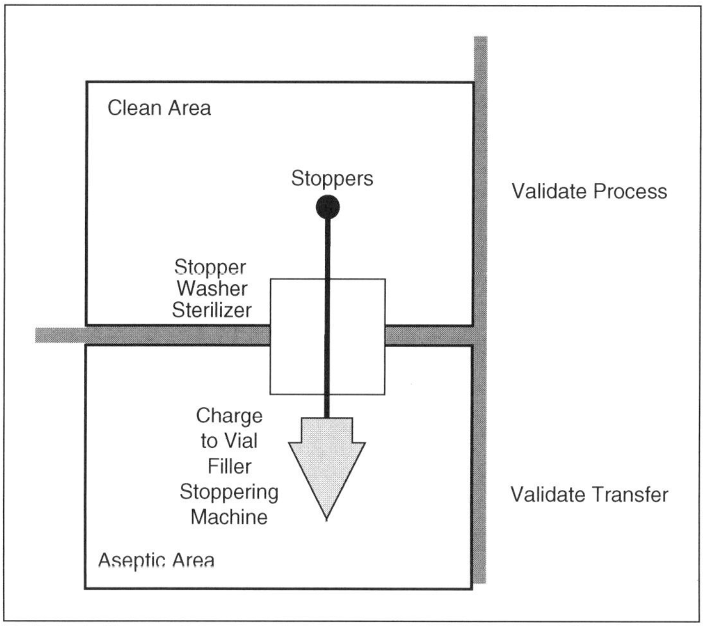

Figure 9 Washing and sterilizing stoppers in situ and transfer to filling.

Validation Planning

If a preliminary plan was not developed at the conceptual stage, then qualification planning must commence during this stage, usually toward the completion of this phase or at the commencement of detailed design.

Clearly, from the aims of this phase, qualification planning must be developed to be consistent with that of the design development to be able to define resource requirements, schedule, and costs. A Facility and Equipment Plan is a good means of focusing on these elements. Developing a preliminary list of systems allows the validation engineer to conduct a preliminary risk and impact assessment (4); (see also the section entitled Facility Qualification Plans for further details on risk and impact assessment). This approach allows the qualification estimate to be more precise by focusing the qualification activity on those systems that will impact on the process and product quality. Clearly, the aim is to enable qualification costs, resources, and planning estimates to be set down for management sanction of cost and to give a basis for the detailed design phase (see the section entitled Facility Qualification Plans).

Documents/ Activity	Qualification Phase								
	DQ			IQ			OQ		
Note 1	P	A	E	P	A	E	P	A	E
Building & Utility Services									
Process Services									
Process Equipment									
Complete System									

Figure 10 Validation document matrix.

Detailed Design and Procurement

Introduction

The main objectives of this phase are as follows:

1. To provide a detailed design for issue to construction
2. To provide a detailed equipment list and specification for each item
3. To place orders for the purchase of major items based on a schedule for delivery to site
4. To develop a construction strategy and program
5. To ensure that all aspects of GMP are adhered to in the design
6. To further develop a strategy for start-up through precommissioning, commissioning and validation
7. To ensure that all aspects of DQ are met and to set the basis for IQ

Approach

The project team will grow rapidly during this phase; this presents problems of both organization and assurance that all members of the team understand the GMP requirements. This must be addressed by either careful selection or training, with a particular focus on the engineering team leaders.

Documentation becomes of key importance, and documentation management is frequently used as means of controlling the project and can be used to control the design and its qualification (see documentation later). A change control procedure should be in place for the design. It must be a key element of the DQ to check the design against the original user requirement specification. This should be part of a GMP audit of the design. This can be done in a series of reviews toward the end of design development or early in the detailed design, as appropriate to the project.

These planned reviews of key documents should be done by an independent auditor(s), all of which is part of the DQ. They should be documented and be a part of the validation record.

Layout

Only minor changes should occur at this stage, and these should be a part of the developing detail. The final construction issue drawings are key, and a final issue of flows of personnel, raw materials components, and products can be completed. These are important documents that confirm good design and are frequently used as part of a presentation to authorities, typically FDA, EMEA, or UK MHRA to obtain their views before construction commences. This is not an essential item, but is a recommended course of action where possible.

Equipment

During this phase most of the equipment is placed on order. A typical activity flow for equipment purchase and procurement is shown in Figure 11. Key to the equipment qualification is the technical specification that must initially begin as a URS. This can be developed into an initial inquiry requisition by the engineers. It is advisable in the early phases to develop a list of potential equipment suppliers and approve them. From an iterative process of discussions with potential suppliers and preliminary evaluations, a preferred supplier can be selected.

Before placing an order, it is important to ensure that any development of the specification and user requirements are fully discussed and that the supplier is aware of all requirements pertaining to the supply of the equipment, necessary documentation, testing procedures, installation, maintenance, and operation procedures. A more exhaustive list is given in the matrix in Figure 10. Procedures must be in place to approve the documents and to conduct the necessary supplier inspections during fabrication. Many of these documents will be key, both to the engineering design and to the final equipment qualification and operation.

The final stage in this process is the predelivery checkout and inspection. This is sometimes referred to as the FAT. This can be an important step in the project and the subsequent validation phases. The FAT is not just a physical examination of the item to check that all components are present and to ensure contractual obligations have been fulfilled, but also a time to ask the question: does the item meet the agreed specification and requisition, and is all the documentation in place?

There are two basic approaches to answering the question: who should supply the documentation? One option is to rely on the vendor's works checkout sheets, or alternatively, to prepare a set of checkout documents within the project scope. The choice is dependent on the quality of documents that are normal from the vendor. It would be a good policy to identify the need for and extent of FAT offered when preparing the requisition and in subsequent discussions with potential vendors before order placement. It may be a factor in the choice of the vendor. Usually, the documentation and scope of testing offered falls below the requirements. A strategy needs to be evolved to determine who will prepare, review, approve, and execute each part of the FAT.

A well-executed FAT can contribute significantly to the IQ at site and some of the OQ, particularly, for packaged items (e.g., proving a liquid vial filling line for speed and accuracy of fill). If done in the vendor's works, under an approved protocol, this would certainly reduce the work at the site.

Planning

Detailed planning for the installation and construction phases commence in this phase. Much of the validation execution will commence toward the middle and end of the construction phase.

Validation planning needs to be developed concurrent with the main project planning to ensure that key goals and milestones are met and to identify the resources

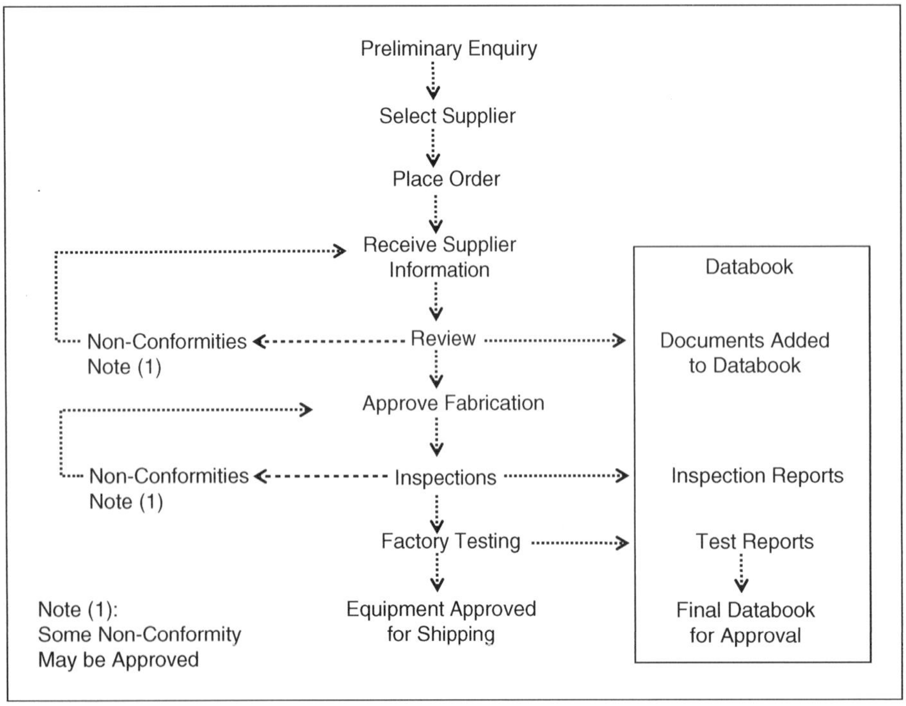

Figure 11 Equipment purchase and procurement flowchart.

that will be required in the execution phases. A typical overview schedule is shown in Figure 12.

As the detailed design progresses, planning for construction commences. Usually this will be done by area, using specialist discipline contractors (e.g., civil and structural, mechanical, and building services; instrument, electrical, and specialist contractors for installation of process equipment packages). Many of these have a role to play in the IQ. The approach should be covered in the master plan and follows that developed for equipment vendors. Again, it is essential to identify the scope of validation services required, particularly documentation, drawings, test procedures, and certificates.

In many cases, these will be required to provide key documents in the IQ (e.g., the ductworker installer for a sterile clean room suite will need to have documented installation and test procedures to qualify the critical parts of the installation.).

Similarly, the specialist contractor for walls, floors, and ceiling will need to provide test procedures and certificates to qualify that the finishes meet the standards laid down in the specifications and requisitions. This level of detail needs to be considered for each area, and its appropriateness and relevance to qualification must be determined.

Facility and Equipment Qualification Plan

The FEQ is developed at the start of detailed design to cover all aspects of the design and installation phase. As many of the detailed requirements are specified throughout the detailed design phase, the FEQ has to either become a document that will undergo change and revision or be at a level that states only intent. The continued requirement for senior management approval probably dictates the latter option. In this case, it is more appropriate to have additional but separate support documents (such as schedule, systems lists, templates for protocols, etc.) that will be easier to modify and approve.

This plan will specify the requirements for the qualification of the facility and equipment and may form part of a series of subplans under a Validation Master Plan. It will define the requirements for DQ, IQ, OQ, and PQ. It will define responsibilities and the risk and impact assessment process that defines which systems will be qualified and which will not.

The assessment process used most commonly is that described in the ISPE Baseline Guide to Commissioning and Qualification. This process considers the facility and equipment as systems and reviews each system for impact on product quality attributes. System are identified as direct, indirect or no impact. Direct impact systems are further evaluated to determine which components are critical. These latter components are those that require qualification. Care needs to be exercised with the indirect impact systems as occasionally there are, for example, cascade control systems that connect to the adjacent direct impact systems. The impact assessment process should form part of an organization's procedures and be approved and documented. All systems should be engineered and commissioned using good engineering practices. Those system that are direct impact are qualified.

Design Qualification

DQ is not a regulatory requirement although the European GMPs refer to DQ in Annex 15 (5) and ICH

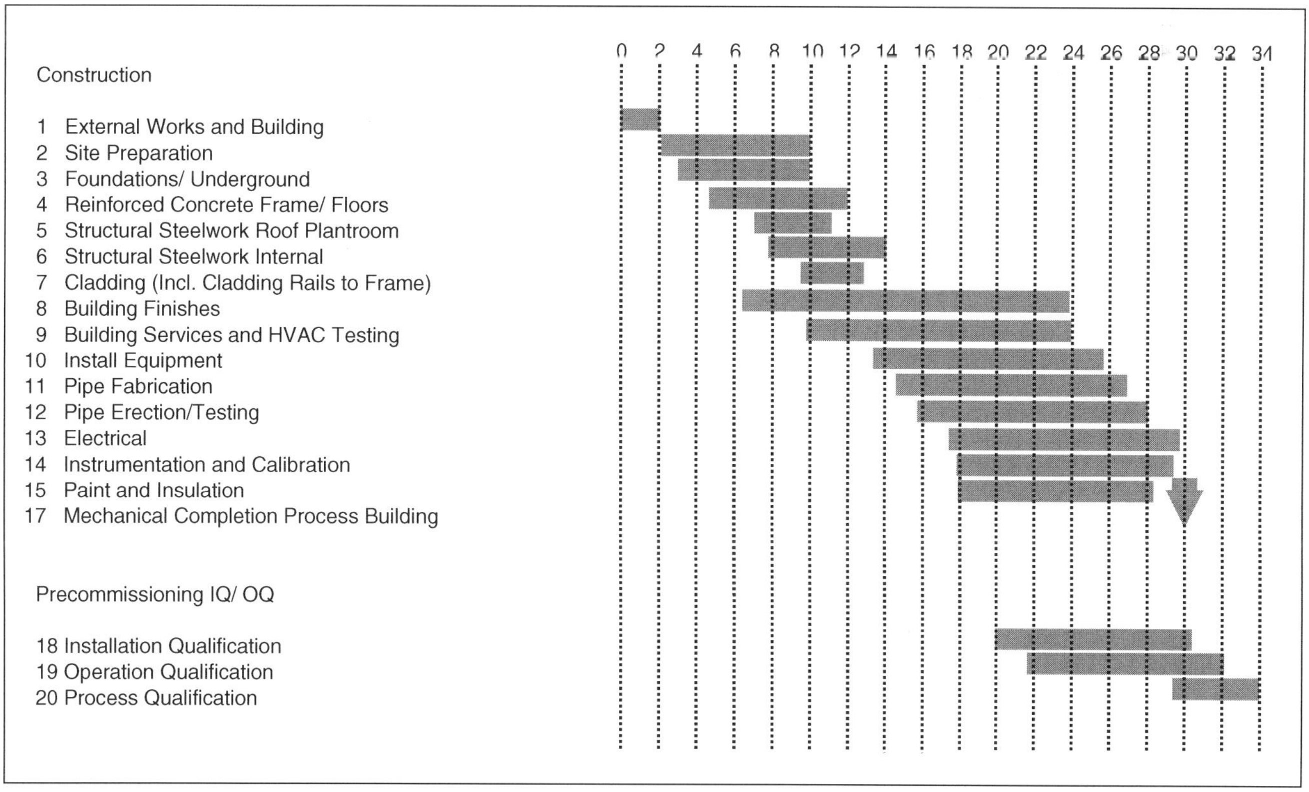

Figure 12 Qualification schedule from concept to validated facility.

Q7A (6) would normally be conducted during this phase. It essentially confirms that the design meets the user requirements with particular reference to those requirements that impact on product quality attributes and process controls that impact on critical parameters.

Construction IQ

As discussed earlier in the section entitled Detailed Design and Procurement construction will normally commence part way through detailed design, with site clearance, foundations, and drainage being laid down in the early phases. The building will be erected and the fit out will commence as soon as the facility is weatherproof. Construction will normally proceed on an area a trade (civil, piping, mechanical electrical and instruments) basis while the commissioning group requires completed systems to be able to commence precommissioning activities such as walkdowns to ascertain completeness prior to commencement of the commissioning activities. At some stage construction will need to complete systems in order that commissioning can commence. Initial commissioning activities involve start-up to confirm basic functionality of the systems for by more extensive testing and set up to adjust and regulate the system such that it performs as intended. The IQ can be completed once the system is confirmed as construction complete and the precommissioning activities are complete. The OQ can be executed once the system has been commissioned. A typical construction program is shown in Figure 13.

The construction group will offer systems as completed to the commissioning engineers for checkout. The engineer will check the system against the design and construction drawings and provide a punch list. The list will identify where there are anomalies that require rectification. The importance of change control at this phase becomes evident. The checking should be completed against specifications approved during detailed design. Changes must be evaluated to check whether there are any implications that impinge on GMPs or final product quality. Those impacting on quality or GMP aspects would require quality approval. Once changes are agreed to, they can then be executed and approved. This process is normally controlled using an approved change control procedure.

Change control would be the normal practice in construction management, but it is particularly important for a facility intended for the manufacture of medicinal products. For example, changes to room finishes that may impact on the cleaning of critical surfaces in a suite of clean rooms for sterile product manufacture would need to be reviewed and approved.

Delivery and installation of equipment at site is again all part of the IQ and should have had all its groundwork of documentation completed in advance. The completion of a FAT before shipment will often simplify the IQ and some of the OQ activities. Frequently, final documentation is not available from vendors until the FAT is complete. This tends to delay the preparation of OQ documents. However, the need to complete as much before this stage is essential. The speed of execution activities and schedules set for completion and commencement of production draw near at an alarming pace. The more of this work completed during the detailed design phase, the easier the task at site will be.

Site OQ

The site OQ phase has two key objectives:

1. To ensure that the system or subsystem works and performs as intended
2. To ensure that Operations personnel receive the relevant training and experience.

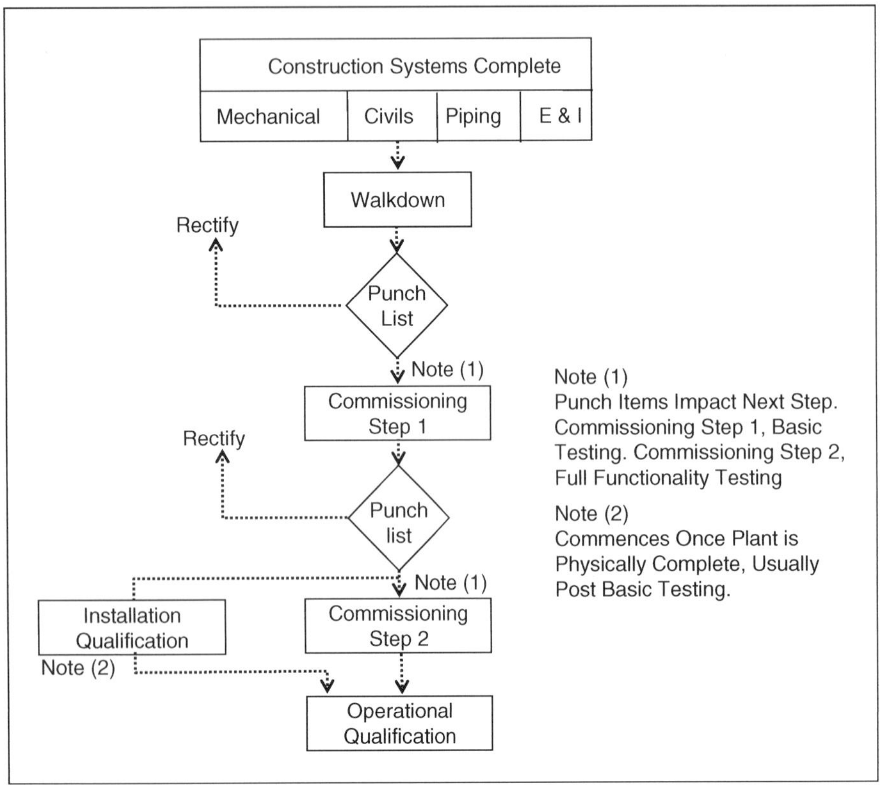

Figure 13 Schedule of qualification activities during site construction phase.

The protocols and procedures prepared are now used in the execution of this phase. This execution will be a joint exercise, conducted by the engineers, operations personnel, and quality control colleagues. The systems will be tested using the approved protocols, and this serves as a means of training. It is essential that a system and procedures for this are in place and that it is fully documented. Commissioning should be complete and the system should be fully functional. Critical instruments will have been calibrated and control loops regulated such that they perform as intended. The OQ will then ensure that those Operations personnel who will conduct the process qualification are themselves qualified for this later exercise. Without this, it would be reasonable to question whether the subsequent process qualification was itself valid if conducted by untrained personnel.

FACILITY QUALIFICATION PLANS

Validation and qualification have been defined many times and typical examples are as follows:

The EC definition: "Action of proving, in accordance with the principles of Good Manufacturing Practice, that any procedure, process, equipment, material, activity or system leads to the expected results (see also Qualification)" (1). Note EC Guide (1) defines *qualification* as "Action of proving that any equipment works correctly and actually leads to the expected results. The word VALIDATION is frequently widened to incorporate the concept of qualification" (1).

The U.S. definition: "Process Validation is establishing documented evidence which provides a high degree of assurance that a specific process will consistently produce a product meeting its predetermined specifications and quality attributes" (7).

Each of these definitions emphasizes the need to demonstrate that a system does what it purports to do. To be able to execute this, a plan is essential. For a single system, this is achieved through a *protocol*, which in simple terms is a plan, followed by its execution. For a whole facility and its operation, we require a plan that encompasses all aspects of validation and qualification and this is usually termed a Validation Master Plan. It would cover facility and equipment, automation, cleaning, process and laboratory and analytical systems. These would often have their own subplan and in the case of our facility we have termed this the facility and equipment qualification or FEQ plan. Qualification normally pertains to systems and not processes as defined on the EC guide (1). The actual choice of name validation master plan or FEQ is very much an individual or corporate preference, what is important is that the scope is clearly defined in any such plan. This section describes the controls of a typical facility and equipment qualification plan or FEQ.

Contents

The typical contents of an FEQ are as follows:

- Introduction
- Methodology
- Qualification
  - DQ
  - IQ
  - OQ
  - PQ
- Personnel and responsibilities
- Schedule
- Preventative maintenance
- Change control
- Procedures
- Documentation
- Appendices

The foregoing list may vary depending on the project phase for which an FEQ is written. At the conceptual stage, it will be very preliminary, whereas at the detailed engineering phase it needs to have substantial detail and address all aspects of the qualification. It may be high-level document with supporting documentation. How this is applied at each of the project phases is discussed in the following.

Introduction

This section should be written primarily as an introduction to the qualification process, the facility and equipment and is intended to set the scene. Awareness of the potential audience is very important, because the plan may be used for various purposes (e.g., as a corporate document or an introduction to the qualification for the inspection and regulatory bodies). The latter use is probably more typical. It should include a description of the facility, its premises and equipment, and its purpose. The intention and scope of the qualification should be set down. It is in this section that other relevant site policies and plans should be referred to and how this particular plan relates to these. These will probably include factory or corporate policy statements on, for example, GMP, QA, and such.

Methodology

The plan needs to be developed and to focus on those standards that must be met including regulatory requirements. This section of the plan should address these requirements by identifying the standards that are to be applied to the facility. These will subsequently be used in the development of the acceptance criteria that are used to judge the validation.

The standards will normally comprise the following three elements:

- Regulatory and guidance documents
- National standards (or equivalent)
- Company standards

Regulatory and guidance documents would encompass, for example,

- FDA CFR Title 21 specific sections
- EC *Guide to Good Manufacturing Practice* vol. IV *Medicinal Products Guide to the Manufacture of Drugs by Aseptic Processing*
- GMP regulations of Japan
- Pharmaceutical Inspection Convention GMPs

International and national standards could include the following examples:

ISO	13408, 14644
British standards	BS5950, BS5750
U.S. military standards	AFM 84-4 to TO 25-203
Engineering standards	Company or contractor standards, IEEE

Some organizations (operating firms, A&Es, equipment manufacturers and contractors) have developed their own internal standards, for example:

- Surface of equipment finish for sterile products
- Valves and piping for use in USP and WFI systems
- Finishes for walls and floors in clean rooms

It is normal during the planning phase of any project to set standards that will apply to the engineering design; these are usually referred to as the *basic engineering design data*. This phase of the FEQ identifies those standards that will be critical to the validation and its implementation. They are applied through the project to ensure compliance with the GMPs.

It is also important to develop clear policies on documentation standards that will be applied. These may be planning and execution documents, such as protocols, records, reports, or construction and commissioning documents that will or may be used in support of qualification. Many organizations have developed procedures and standards for these types of documents. This section should either refer to these procedures or, when necessary, either augment or provide new, as required. These procedures should include:

- Layouts for each type of document
- Authorization procedures for review and approval

These procedures and formats, if not already in existing corporate documents, for completeness can be appended to the FEQ, or be in supporting documentation.

All systems will be commissioned using good engineering practices (GEP). There should be a commissioning plan and schedule of activities that is integrated with the construction schedule. GEP requires commissioning documentation and records. Part of the process of documentation is to develop a Turnover Plan and documentation known as ETOP for equipment. Much of the testing and records if executed using GEP will be used in support of the qualification.

Qualification

The phases of qualification have included, for example, design, installation, operational, and performance. For the purposes of this chapter, the elements of qualification and their scope are defined as follows:

DQ is defined as "Providing documented verification that all key aspects of the design, procurement and installation adhere to the approved design intention and that all the manufacturers' recommendations have been suitably considered."

IQ is defined as "Providing documented verification that all key aspects of the design, procurement and installation adhere to the approved design intention and that all the manufacturers' recommendations have been suitably considered."

OQ is defined as "Providing documented certification that the system and subsystems operate as intended throughout all anticipated operating ranges."

PQ is defined as "Providing documented verification that the system performs and does what it purports to do."

The foregoing definitions are so written that they encompass all aspects of the design, procurement, installation, and commissioning process. Some authorities have defined other phases of qualification (e.g., receipt, this being a checkout of equipment band systems as delivered to site). Others have linked some areas together (e.g., IOQ), and produced unified protocols, or have even called this phase "engineering or equipment qualification." Clearly a flexible, but formalized, approach is required, and it may be appropriate to adapt the approach to the specific project's needs. The important issue is to ensure that definitions in the organization and for a specific project are consistent and cover all aspects of the qualification process, and that the validation structure and organization is clear to any inspecting authority.

Design Qualification

DQ covers all aspects of the design and procurement of facility and equipment. It is intended to encompass all those activities that might take place in the design phase, detailed and development, including activities associated with procurement of equipment and checkout at the supplier's works. DQ is a verification that the design meets user requirements with a particular focus on those requirements that relate to GMP and product quality. The extent of DQ may depend on the contract arrangements. Design may be subcontracted to suppliers or subcontractors and how this is covered should be defined in the plan. DQ is not a regulatory requirement but is a smart activity to include in the qualification process. Where DQ is not identified as a specific step, it is still essential that aspects of design are demonstrated in the qualification process as the existing regulations require that facility and equipment are of suitable design and appropriate to purpose.

Installation Qualification

The IQ element of the FEQ should clearly define those areas and items of equipment systems that are to be qualified. The lists will vary depending on the nature of a facility. A sterile filling unit might include the following.

Premises	Layout
	Flow of personnel, product, raw materials, and such
	Finishes of walls, ceilings and floors
Utility services	Drains
	Water systems (e.g., cooling, hot and cold)
	Services gases (e.g., instrument air)
	Electrics
	HVAC class 100 systems
	Class 10,000 systems
	Class 100,000 systems
Process services	USP and WFI
	Process gases: nitrogen, propane, and others
	Clean steam
Equipment	Steam sterilizer
	Stopper washer—sterilizer
	Tray washer—autoclave
	Dry heat sterilizer
	Vessels
	Hot air tunnel sterilizer
	Ampoule or vial washing machine
	Filling and capping machines
	Lyophilizer
	Inspection line
	Labeler
	Packing (primary)

After identifying the systems that will be qualified, the next stage is then to develop a qualification plan. This is the protocol. A protocol will contain the following:

- A clear definition of purpose
- A plan for execution
- Who will compile the execution
- How it will be conducted
- What procedures are required
- Acceptance criteria
- Defined methods for recording the results
- A final acceptance review
- Means of certification of the qualification

Operational Qualification

The pattern established for IQ is followed in OQ. The key approach in OQ is to identify and define those systems that are to be qualified. The facility should be split into system with clearly defined boundaries. This should cover the whole of the facility, and it is usually appropriate to be consistent with those developed for IQ. The types of systems identified will be dependent on the nature of the facility, but a typical example list a for a secondary sterile facility are given as follows:

- Facility
- HVAC class 100, 10,000, 100,000
- WFI water
- Process gases: air, nitrogen, CO_2
- Propane
- SIP systems
- CIP systems
- Vial washer
- Vial tunnel sterilizer
- Vial filler and stopper machine
- Lyophilizer
- Vial capper
- Vial inspection
- Vial primary packing
- Autoclave
- Dry heat sterilizer
- Stopper washer–autoclave
- Solution preparation system

Once all the systems have been defined, then specific protocols for each can be prepared. These have a form similar to that described for IQ. The information for the IQ and OQ is frequently presented in a matrix form identifying those systems to be qualified. OQ tests the systems throughout their normal operating range.

Performance Qualification

This is generally applicable to those systems that require extended testing over a period of time such as water systems, heating, and ventilation systems such as those applicable to clean rooms and the actual performance of the clean room to meet the defined standards of operation over periods of time. Some organizations may include this type of testing in the OQ.

Process Validation

Process qualification is the phase during which the manufacturing process and procedures are qualified. It would not normally be an area for which the engineering organization—either internal or external—would be involved. It is a primary responsibility of the Production and Quality Control departments. It is not the intention of this chapter to cover this in detail, but to suggest that the approach already proposed for IQ, OQ, and PQ is valid for PV. The processes and process systems are to be identified. Protocols can then be prepared and executed.

It would be normal to draw up a matrix identifying all the systems and whether they require validation and to what extent (Fig. 14).

Personnel

People are the key to success of any validation exercise. The CFR 21 [see Sec. 211.21(2)] states "Each person engaged in and each person responsible for supervising the manufacture, processing, packaging or holding of a drug product shall have the education, training, and experience, or a combination thereof, to enable that person to perform the assigned functions." It is reasonable to imply from this that persons involved in the PV, which is part of the GMP and QA process, must also fulfill this requirement. Hence, if the process is executed with inappropriately qualified and trained personnel, then the validation could be deemed invalid.

The FEQ should lay down the principles for personnel requirements. It must address these aspects for each phase of the validation process. Personnel will change throughout the engineering design, construction, and commissioning program. The task of ensuring each is appropriately qualified and trained and has relevant experience may fall on different organizations (e.g., the engineering contractor of the pharmaceutical manufacturing company). The experience can be demonstrated by written biographies or curriculum vitae. The extent of detail will vary with the phase of the project. When training is required to augment experience and qualification, it can be provided in-house or externally. Courses should be run by specialists or equipment suppliers. A combination of these is probably most desirable. Documenting the training is essential and is a requisite of the GMPs (1,2).

Schedule

A work program is essential and should be prepared at an early stage. It sets out the milestones for the validation process and incorporates them into the overall project schedule. This will normally be in the form of bar charts and critical path networks, and it needs to be planned to the same depth as the overall project.

The importance of a plan becomes evident as the complexity grows. A good plan will contain all the necessary features to identify when various activities are due for execution and demonstrates to the outside that the project is under control. This enables resources to be allocated at an appropriate time to achieve the activity. A typical example would be the completion of process specifications to enable requisition placement, with the subsequent delivery of documentation from the vendor to allow the design and protocol preparations to proceed (Fig. 10). It ensures that all parties with an interest in the project are aware of not just the engineering targets, but of the validation targets, and it has a tendency to assist in gaining commitment from all who are involved, from those conducting the execution, to top management. It

VALIDATION MATRIX

Equipment Description	Validation Required	URS	DQ	FAT	IQ	Calibration	OQ	SOP Operation	SOP Cleaning	SOP Training	SOP Maintenance	Computer Validation
Vial Washer	✓	✓	✓		✓	✓	✓	✓	✓		✓	
Stopper/Capper Washer	✓	✓	✓	✓	✓	✓	✓	✓	✓		✓	
Vessels	✓	✓	✓		✓	✓						
Filters And Integrity Test	✓	✓	✓		✓	✓	✓	✓	✓		✓	
Oinment/Cream Homogeniser	✓	✓	✓		✓	✓	✓	✓	✓		✓	
Tube Filling Machine	✓	✓	✓		✓	✓	✓	✓	✓		✓	
Filling Machine	✓	✓	✓	✓	✓	✓	✓	✓	✓		✓	
Depyrogenation Oven	✓	✓	✓	✓	✓	✓	✓	✓	✓		✓	✓
Label Printers	✓	✓	✓	✓	✓	✓	✓	✓	✓		✓	✓
WFI	✓	✓	✓	✓	✓	✓	✓	✓	✓		✓	✓
Pure Steam Generator	✓	✓	✓	✓	✓	✓	✓	✓	✓		✓	
Monobloc Filler/Capper	✓	✓	✓	✓	✓	✓	✓	✓	✓		✓	
Freeze Drier	✓	✓	✓	✓	✓	✓	✓	✓	✓		✓	✓
Vial Filler	✓	✓	✓	✓	✓	✓	✓	✓	✓		✓	✓
Ampoule/Vial Inspection	✓	✓	✓	✓	✓	✓	✓	✓	✓		✓	
Capping Unit	✓	✓	✓	✓	✓	✓	✓	✓	✓		✓	
Laundry Equipment												
Balances	✓	✓	✓		✓	✓	✓	✓	✓		✓	
PH Meter	✓	✓	✓		✓	✓	✓	✓	✓		✓	

Figure 14 System qualification matrix.

can also show due dates for that all-important final inspection.

Preventive Maintenance

This element is frequently considered to be the responsibility of the Site Maintenance and Operations department and often is given a low priority within an engineering design team. There is a clear requirement to keep a facility in a state of qualification. A preventative maintenance program is an essential component of a schedule of work to achieve this objective. The Validation Master Plan must identify the need for this program and, hence, to flag its importance to the designers. The role of vendors and suppliers is very important in this area. Operation and maintenance manuals should be considered as a key part of the specification program. This activity should be conducted during the design phase, and the documentation required should be included in the requisition. The execution of a preventive maintenance program can take on greater relevance within the precommissioning and commissioning phases demonstrating that, once qualified, a unit has been maintained both in a proper manner and in accordance with the supplier's instructions.

Change Control

The frequently asked question is "When is qualification/validation complete?" The process is never finished; it is an ongoing exercise as the facility, its services, equipment, and processes must always be in a state of validation to comply with the regulatory requirements. Change control applies not only to the ongoing manufacturing processes but also throughout the whole of the project. Change control should address all aspects of the facility and its design—through construction to operation—and should be addressed in the FEQ.

This section of the FEQ should then lay down the requirements for a set of procedures for change control that cover:

1. The project through design, construction, and commissioning
2. The ongoing change that will inevitably occur in both the process, the equipment, and the engineering aspects
3. Identifying how to determine which changes require QA approval and which require only Engineering approval

The link between this section and previous is very strong. Both preventive maintenance plans and change control are intimately linked.

Procedures

Procedures are an essential part of any system of validation. These cover engineering standards used in the project design, through to commissioning phases, and the facility's SOPs. Usually, the FEQ will identify the commitment to written procedures and identify an approval procedure for formats, preparation, and authorization of these procedures.

Documentation

The documentation section of the FEQ is usually used to identify the documentation that will be produced. Depending on the stage in a project when the plan is produced, the detail will vary. A preliminary plan may identify only the broad areas of documents that will be produced; for example:

- Engineering drawings
- Equipment supplier drawings and documents
- Factory acceptance documents (works qualification)
- IQ documents
- OQ documents
- PQ documents

Appendices

Much of the outcome of the execution will be written documents. The appendix section is commonly used in more detailed master plans to hold examples of the types of documents and formats that will be used in the execution stage.

REVAMP OR EXPANSION PROJECTS

Introduction

Projects of this nature usually take the form of upgrades of exiting facilities and can include the following:

1. Environmental upgrades to fine chemical, pharmaceutical, and microbiological facilities
2. Expansion of existing pharmaceutical operations and facilities
3. Infrastructure work in fine chemical and pharmaceutical businesses, such as waste management and utilities
4. Plant and buildings demolition and disinvestment
5. Plant and buildings maintenance and repair
6. Facilities replacement, refitting, and redevelopment
7. Safety and GMP upgrades
8. Decommissioning of plant, equipment, buildings, or facilities

It is important at an early stage to have a clear definition of both the purpose and scope of the upgrade, and an understanding of where the project is coming from and what are the main drivers. Frequently, the major factors driving the project will be a need to provide a modern facility to meet the latest GMP standards, an upgrade in capacity, improvements in working methods and technology, or a reorganization to take a new product line. A knowledge and understanding of these drivers helps and enables a validation plan to be developed.

Approach

Once a scope and purpose are set, the methodology is very similar to that for detailed design and construction (see the section entitled Detailed Design and Procurement). Some additional points need to be considered.

Location

It is important to review the location of the project and evaluate its effects on the surrounding operations.

How will you manage interruptions in services? Or segregation to avoid contamination to adjoining operations? What if decontamination is necessary if changing products?

Equipment

Is new equipment to be used or will it be reuse of existing equipment? What is the current state of validation IQ, OQ, and PQ? Will it need upgrading, and do you need to involve the original vendor? Is the model still manufactured; what is the current spares situation?

Services

Are you proposing new services or reuse of existing services? Do the existing services have adequate validation documentation, IQ, OQ, and PQ? Are the existing services adequate for their new role?

Some of the common areas of concern do revolve around reuse of equipment. Frequently, this equipment does not have adequate validation records, and obtaining documentation to support its qualification is difficult, especially if it is being considered for a new use. Suppliers often no longer support that specific model range. All of these factors increase the validation effort. Similarly, existing services may not meet current standards (e.g., a USP water system running for many years requires extension to the ring main). What standards do we apply to the new section, and what strategy should be taken to the validation?

Validation Aspects

The approach should be similar to other projects, as discussed earlier. A Qualification Plan is essential, and it is important that it encompasses all aspects of the project and its effect on other systems. It should involve all parties, Engineering, Production, and QA. The Validation Steering group should consider all the key factors and ensure that these are addressed in the plan.

Reviews and audits of both the design and execution strategy are important to the services of a revamp or an upgrade project. Changes to either of these must be examined for effect on quality, not only just in the project and its intended scope, but also on surrounding activities. It is not uncommon for changes to affect adjacent processes. Clear strategies for evaluation must be incorporated into the Validation Master Plan and then be executed. Reviews should consider additional requirements.

- Decontamination
- Cross-contamination
- Operation and process protection during decommissioning and construction phase
- Any breach to GMP integrity of the system
- Any effect on existing procedures or protocols

The foregoing consideration should be in addition to those already discussed in previous sections of this chapter (The Engineering Design Process for a Facility and Facility Qualification Plans).

SUMMARY

Validation is an essential part of GMP (8) and as a key element must be incorporated into the design and building of pharmaceutical facilities for the manufacture of pharmaceutical products. It should be considered from the earliest phase (i.e., conceptual study) and be a key feature of the project.

The extent of the validation or qualification requirements will vary with the project phase. The responsibility for its execution should be clearly defined and allocated to the appropriate discipline functions of engineering, production, and QA. Its execution is best achieved by having a fully defined scope that is then incorporated into the project plan and schedule. This is best achieved by an FEQ that may become a subset of plans linked to a validation master plan. This can become a living document and identify the qualification/validation requirements at each of the project phases. This plan can then become a vehicle for demonstrating a structured and organized approach to the regulatory and inspecting bodies.

REFERENCES

1. The governing of medicinal products in the European Community. Good Manufacturing Practice for Medicinal Products. Vol. IV. Luxembourg Office for Official Publications of the European Communities, 1992.
2. Code of Federal 21 (CFR 21) Parts 200–299. Office of the Federal Register National Archives, and Records Administration.
3. Simmons PL. The design construction and commissioning of a new facility in accordance with GMP regulations. Pharm Engin 1992; 12(5).
4. Baseline Guide to Commissioning and Qualification ISPE 2001.
5. European GMP Annex No Validation.
6. ICH Q7A.
7. Center for Drugs and Biologics, and Center for Devices and Radiological Health. Guideline on General Principles of Process Validation. Rockville, MD: Food and Drug Administration, 1987.
8. Adamson JR. An approach to validation. Pharm Technol 1992; 12(5).

4

Validation of Environmental Control Systems Used in Parenteral Facilities

Franco De Vecchi
Vectech Pharmaceutical Consultants International, Inc., West Bloomfield, Michigan, U.S.A.

INTRODUCTION

The ECSs play a central role in preventing microbial contamination into sterile products delivered by the parenteral route. Maintenance of the product's sterility during manufacturing is of particular importance when products are aseptically manufactured as there is no subsequent process step, such as terminal sterilization, that provides a higher degree of sterility assurance. The industry recognizes the potential for microbial cross contamination risks associated with the manipulation of sterile products or sterile process devices and components. The risk increases while the sterile product or components are directly exposed to the environment. The highest risk for cross contamination is perceived at those steps where personnel is close or intervene in handling sterile product or components in operations of the so-called "aseptic process." The design of a comprehensive ECS that covers the facilities used for the formulation and packaging of sterile products presupposes a clear understanding of the principles outlined above. On those bases, a careful selection of methodologies and structural components directed to the control are necessary for controlling and in some cases eliminating the microbial cross contamination risk. The ECS validation program should be designed to provide assurances that all the components of this ECS meet the approved performance environmental specification in a consistent and reproducible way. In one word—it should demonstrate the robustness of the system. There are various components of the comprehensive ECS; the purpose of this chapter is to outline the basic concept for a validation program for key components: Cleanrooms and HVAC and associated systems such as unidirectional air flow devices (formerly laminar flow). The validation approach for other components are outlined in other chapters.

Abbreviations used in this chapter: AHU, air-handling units; ANDA, abbreviated new drug application; ASTM, American Society for Testing and Materials; BMS, building management systems; CCP, critical control points; CFU, colony-forming units; cGMPs, current good manufacturing practices; CIP, clean in place; DOP, dioctyl phthalate; DQ, design qualification; ECSs, environmental control systems; ELA, establishment license application; EMEA, European Medicines Evaluation Agency; EU, European Union; FDA, Food and Drug Administration; FMEA, failure mode and effects analysis; GMP, good manufacturing practice; HACCP, hazard analysis and critical control plan; HEPA, high-efficiency particulate air; HP, power requirements; HVAC, heating, ventilation, and air-conditioning; IES, Institute of Environmental Sciences; IQ, installation qualification; ISO, International Organization for Standardization; ISPE, International Society for Pharmaceutical Engineering; LAF, laminar air flow; LAN, local area network; NDA, new drug application; OQ, operational qualification; PDA, Parenteral Drug Association; PLC, programmable logic controller; PQ, performance qualification; RABS, restricted access barrier system; RL, risk level; SCADA, supervisory control and data acquisition; SIP, sterilization in place; SOP, standard operating procedure; UAD, unidirectional airflow devices; URS, user requirements specification; USP, *United States Pharmacopeia*; VMP, Validation Master Plan.

VALIDATION APPROACH FOR CLEANROOMS AND HVAC SYSTEMS

The validation program for a pharmaceutical facility is part of a sequential list of the events that start at the time the manufacturing process and facility are designed. This logical sequence is part of the "Quality by Design" approach currently employed by the industry and required or recommended by regulatory agencies worldwide. The concept consists in the integration of all the elements that participate in the manufacturing of a safe, effective and pure pharmaceutical product. Validation steps are usually carried at the time the systems are completed and ready to perform their assigned duties. The scope of the validation program is to demonstrate and document that the system meets the original and approved "user" and "functional" specifications developed at the time the system was conceived in a consistent and reproducible manner.

"User and Functional Specifications" developed and documented by system owners constitute the basis for structuring the validation program. These requirements are translated in the systems conceptual design, detail component specifications and methods for construction and testing of all the components of the system. These documents usually outline the critical environmental parameters needed to produce product of the specified quality, and indicate their acceptable ranges of operation. The acceptance criteria to be used in the various installation verifications (IQ), operational testing (OQ) and performance challenges (PQ) qualification protocols should be extracted from these documents. These documents should provide the appropriate and applicable reference standards to be used as part of the validation testing activities.

Validation is a formal process that also has regulatory implications and thus the structuring and adherence to a specific plan of action is important to avoid discrepancies and potential regulatory issues. The use of a VMP is required to implement and streamline the validation program. Such a document should: outline the criteria for the validation program structure, define the responsibility and authority of all the participants, define the structure and approval procedures for documentation, define the basis for the execution of the various phases of the validation program, define the actions to be taken while managing nonconformities or deviations to approved construction operation or performance specifications, and, at the end of the program, define the methodology for project closure and approval etc.

The validation program for critical utilities such as HVAC and Cleanrooms should start when there are reasonable assurances that the system is ready for operation. Costly delays and cumbersome realms of documentation needed to document corrections to nonconformities or deviations can be avoided if "Good Engineering Practices" are implemented during the facility and critical utilities construction. The appropriate application of commissioning principles could assist. Guidance on the implementation of commissioning pharmaceutical systems can be found in the documents issued by the ISPE. Its reading is highly recommended for all of those involved in validation of critical utilities. It should be noted that validation activities are not the place to correct or fine tune a critical system operation. Validation is the confirmation that all the preceding steps have been completed as planned. In synthesis validation offers the confirmation that demonstrates that overall design, construction, commissioning and startup of the system has been executed as apart of a well defined "Quality Plan."

To facilitate the understanding of some of the proposed approaches, this chapter provides some references and extracts from regulatory guidelines and standards that impact the design and testing of the HVAC and Cleanrooms systems. For proper understanding the reader should obtain these documents from the source. Study and careful analysis is important for a successful validation program. It should be noted that their final implementation many times depends on the nature of the facility and the hardware and software selected.

Validation verifies, tests, and challenges the installation, operation, and performance of all of the components of the critical system. This program is to be executed by qualified individuals using approved techniques and adequately qualified equipment (when needed). The program should use documents and protocols that have been prepared, reviewed, and approved by personnel representing the Quality, Manufacturing, and Management areas of the pharmaceutical plant to be validated.

A systematic and documented approach is needed to demonstrate that the functional requirements established for the system are met on a consistent basis by installed components and systems.

IMPLEMENTATION OF RISK ASSESSMENT PRACTICES AND TECHNIQUE IN A VALIDATION PROGRAM

Validation should have an added value and not just be a bureaucratic exercise that produces realms of documents with limited use and in some cases value.

As indicated, the user and functional requirements along with the engineering specifications provide the basis or "input" for validation. In some cases the extent of the verifications, tests, and challenges are defined by existing guidelines or standards, but in some cases the definition on where and to what extend these practices are to be carried out is not so clear. In those cases, individuals involved in the structuring of the validation program need to refer to methodologies that provide a technical and in some cases scientific rationale for the program to be carried out.

The use of "Risk Assessment" techniques is now considered as part of the validation approach for critical utilities. These principles have been used in as part of validation of medical devices, computers systems etc. Methodologies such as the ounces outline by the FMEA and HACCP serve as a good outline.

In principle the risk is assessed based on these basic principles:

- What can go wrong?
- How serious is the event in terms of the system or process performance?
- How probable is it for the event to occur?
- How detectable or predictable is the event?

The complexity of models for risk detection and management vary. For the objective of validation an HVAC systems and Cleanrooms the user can use a reduced model. One has to measure the Risk Priority. For that effect values are given to each of the fields:

- Gravity (G): Grave (4) to not serious (1)
- Probability (P): Very probable (4) not probable (1)
- Predictability (Pp): Not predictable (4), predictable (1)
- etc.

The RL is obtained by multiplying the value:

$$RL = G \times P \times Pp$$

In the case of HVAC systems one has to breakdown all the components of the system to determine the individual impact such component has in achieving the system specified performance. Let us use a hypothetical case for the AIU. Components are blowers, prefilters, heating coils, cooling coils, drives (belts), electrical motors, automated dampers etc.

Component	Blower/fan	Drives/belts	Electric motor
What can go wrong	Cage breakage	Belt rupture	Overload/stop and go
Performance impact	No air flow	No air flow	Disruption of air flow in critical area
Gravity	Serious (4)	Grave (3)	Grave (4)
Probability	Reduced (2)	Moderate (3)	Reduced (2)
Predictability	High (1)	High (2)	Moderate (2)
Risk index	$4\times2\times1=$ 8 (low)	$3\times3\times2=$ 18 (high)	$4\times2\times2=$ 16 (high)

(*Continued*)

(*Continued*)

Component	Blower/fan	Drives/belts	Electric motor
IQ validation impact	Low Manufacturer reports and verifications would be sufficient	High Verify materials of construction and belt pulley installation	High Verify loads/ check blower performance curves

Using this technique, it is easier to determine in each case the activities to be included as part of the protocol (IQ in this case). It will provide a better understanding of the rationale to be followed for the verifications, tests, and challenges.

ENVIRONMENTAL REQUIREMENTS DEFINITION

Although important, microbial contamination is not the only environmental parameter affecting product strength, safety, purity, and effectiveness. Environmental parameters, such as the concentration of airborne nonviable contamination (particulate matter), chemical and gaseous materials, as well as less frequently listed factors, such as vibration, radiation, lighting levels, and the like, may have an effect on the overall quality of the product. Therefore, for every product or group of similar products to be manufactured in a pharmaceutical facility, a definition of such environmental parameters is required, taking into account that these may vary at every phase of the process. These are known as "User and Functional Requirements." As indicated they constitute the basis for the conceptual and detail engineering of the facility and systems installed within.

Once the design phase is completed, these requirements eventually are expressed in the engineering and equipment specifications to be furnished for the construction.

Because of its importance, an effort should be made by designers, users, constructors, and validators to clearly identify the process and end user's requirements. The validation program should be designed to demonstrate that these requirements are met by the design, construction, operation, and performance of the facility and the systems installed within.

These process requirements are often defined on product registration or licensing documents, such as the NDA, ANDA, ELA, and such, and constitute the legal framework for health authorities and regulatory compliance inspections.

ENVIRONMENTAL CLASSIFICATIONS

The environment can impact the quality of a pharmaceutical product differently at each manufacturing step. Certainly, the most critical environment is the one in which sterile products or components (vials, ampoules, stoppers, and such) are directly exposed. This is true for all aseptic-processing operations.

To link the environmental needs for every phase of the process, the critical environments are classified. These classifications follow diverse sets of criteria that are linked either to room airborne cleanliness levels, use of the environment, containment or segregation levels, and the like (Table 1).

Defining parameters used for classification include particles of a specific size per unit volume of air, microbial levels, and type of microorganisms (biohazard). The designation or nomenclature changes within industries and among countries. These classifications are set as part of standards or guidelines prepared by national health authorities such as the U.S. FDA, EMEA and other governmental institutions (e.g., U.S. National Institutes of Health), professional associations (e.g., PDA, IES) and ISO. In addition, many corporations have internal standards and guidelines that apply to all of their associated companies.

FDA Guidance for Industry Sterile Drug Products Produced by Aseptic Processing Guideline, Published September 2004

Critical Environments

A critical area is one in which the sterilized drug product, containers, and closures are exposed to environmental conditions that must be designed to maintain product sterility CFR 21 [section 211.42(c)(10)]. Activities conducted in such areas include manipulations (e.g., aseptic connections, sterile ingredient additions) of sterile materials prior to and during filling and closing operations.

Supporting Areas

Supporting clean areas can have various classifications and functions. Many support areas function as zones in which nonsterile components, formulated products, in-process materials, equipment, and container/closures

Table 1 Standards for Aseptic Processing

Activities and environmental risk		Designation of cleanliness class according to: EMEA	Designation of cleanliness class according to: FDA[a]
Aseptic processing of sterile products or materials	Exposure of sterilized materials and primary packaging in contact with the product, exposure of the sterile product	Grade A[b] (with grade B background) ISO 5	Critical area (aseptic core) ISO 5
Processing in the vicinity of critical aseptic processing environments (A)	Exposure of sterilized materials not in contact with the product	Grade B, ISO 5	ISO 6, supporting environments (see below)
Preparation (weighing, dissolving)	Exposure of non-sterilized materials and products	Grade C[c]	ISO 7, supporting environments (see below)

[a] Limiting values for airborne particles and microorganisms under working conditions.
[b] Should be maintained in the zone immediately surrounding the product whenever the product is exposed to the environment.
[c] Limiting values for airborne particles and microorganisms under unmanned conditions.

are prepared, held or transferred. These environments are soundly designed when they minimize the level of particle contaminants in the final product and control the microbiological content (bioburden) of articles and components that are subsequently sterilized.

The nature of the activities conducted in a supporting clean area determines its classification. FDA recommends that the area immediately adjacent to the aseptic processing line meet, at a minimum, Class 10,000 (ISO 7) standards under dynamic conditions. Manufacturers can also classify this area as Class 1,000 (ISO 6) or maintain the entire aseptic filling room at Class 100 (ISO 5). An area classified at a Class 100,000 (ISO 8) air cleanliness level is appropriate for less critical activities (e.g., equipment cleaning).

Area Classifications and/or Cleanliness: Class Requirements Indicated in the September 2004 FDA Guidelines

Clean area classification (0.5 μm particles per ft³)[a]	ISO designation[b]	> 0.5 μm particles per m³	Microbiological active air action levels[c] (cfu per m³)	Microbiological settling plates action levels[c,d] (diam. 90 mm; cfu per 4 hr)
100	5	3,520	1[e]	1[e]
1,000	6	35,200	7	3
10,000	7	352,000	10	5
100,000	8	3,520,000	100	50

[a] All classifications based on data measured in the vicinity of exposed materials/articles during periods of activity.

[b] ISO 14644-1 designations provide uniform particle concentration values for Cleanrooms in multiple industries. An ISO five particle concentration is equal to Class 100 and approximately equals EU Grade A.

[c] Values represent recommended levels of environmental quality. You may find it appropriate to establish alternate microbiological action levels due to the nature of the operation or method of analysis.

[d] The additional use of settling plates is optional.

[e] Samples from Class 100 (ISO 5) environments should normally yield no microbiological contaminants.

EMEA Environmental Grade Definition and Cross Reference to Particle Concentrations

Annex 1

The airborne participate classification for these grades is given in the following:

Grade	At rest[b]: 0.5 μm[c,d]	At rest[b]: 5 μm	In operation[b]: 0.5 μm[c,d]	In operation[b]: 5 μm
	Maximum permitted number of particles/m³ equal to or above[a]			
A	3,500	1[e,f,g]	3,500	1[e,f,g]
B[h]	3,500	1[e,f,g]	350,000	2,000
C[h]	350,000	2,000	3,500,000	20,000
D[h]	3,500,000	20,000	Not defined[i]	Not defined[i]

Notes included in the EMEA GMP Annex 1 (current revisions are underway as of September 2005, ref: EMEA/INS/GMP?318222/2005/correction):

[a] Particle measurement based on the use of a discrete airborne particle counter to measure the concentration of particles at designated sizes equal to or greater than the threshold stated.

[b] A continuous measurement system should be used for monitoring the concentration of particles in the grades A zone, and is recommended for the surrounding grade B areas.

[c] The particulate conditions given in the table for the "at rest" state should be achieved after a short "clean up" period of 15 to 20 minutes (guidance value) in an unmanned state after completion of operations.

[d] The guidance given for the maximum permitted number of particles in the "at rest" and "in operation" conditions correspond approximately to the cleanliness classes in the EN/ISO 14644-1 at a particle size of 0.5 μm.

[e] The particulate conditions for grade A "in operation" given in the table should be maintained in the zone immediately surrounding the product whenever the product or open container is exposed to the environment. It is accepted that it may not always be possible to demonstrate conformity with particulate standards at the point of fill when filling is in progress, due to the generation of particles or droplets from the product itself.

[f] These areas are expected to be completely free from particles of size greater than or equal to 5 μm. As it is impossible to demonstrate the absence of particles with any statistical significance the limits are set to 1 particle/m. During the cleanroom qualification it should be shown that the areas can be maintained within the defined limits.

[g] Characteristics such as temperature and relative humidity depend on the product and nature of the operations carried out. These parameters should not interfere with the defined cleanliness standard.

[h] For routine testing the total sample volume should not be less than 1 m³ for grade A and B areas and preferably also in grade C areas.

[i] In order to reach the B, C and D air grades, the number of air changes should be related to the size of the room and the equipment and personnel present in the room. The air system should be provided with appropriate terminal filters such as HEPA for grades A, B and C.

For more details, the reader should consult the official version of the EMEA Annex 1 and associated documents as they are currently under revision.

EMEA Recommended Limits for Microbial Contamination

Grade	Air sample cfu/m³	Settle plates (diam. 90 mm), cfu/4 hr[b]	Contact plates (diam. 55 mm), cfu/plate	Glove print five fingers, cfu/glove
	Recommended limits for microbial contamination[a]			
A	<l	<1	<1	<1
B	10	5	5	5
C	100	50	25	
D	200	100	50	

[a] Particle measurement based on the use of a discrete airborne particle counter to measure the concentration of particles at designated sizes equal to or greater than the threshold stated.

[b] A continuous measurement system should be used for monitoring the concentration of particles in the grades A zone, and is recommended for the surrounding grade B areas.

COMPONENTS OF COMPREHENSIVE ENVIRONMENTAL CONTROL SYSTEM

A comprehensive ECS is needed to achieve the targeted environmental conditions. The system typically comprises the following components:

Architectural Components

These components are designed to isolate the process from surrounding environments while permitting personnel access and materials. The use of architectural barriers is necessary to consistently maintain the desired levels of environmental control. Selected materials of construction and workmanship influence the final results thus they should be selected and verified as part of the validation program. Architectural materials are selected in accordance with the criticality of the environment. The architectural components to be verified include but not limited to:

1. Walls and partitions
2. Flooring materials
3. Wall-finishing materials

4. Lighting components
5. Doors and windows
6. Door interlocking devices
7. Product-transfer port

RABS, Glove Box Barrier, or Isolator Systems

These elements are designed to physically isolate the process from surrounding environments by means of a barrier and prevent operating personnel from coming in direct contact with the product during processing. The structural integrity of the barriers need to be verified and challenged as to determine the level of segregation offered by the devices. These tests and verifications are part of the validation program. As in the case of architectural components, the quality and workmanship are fundamental to provide the required assurance of integrity.

Heating, Ventilation, and Air-Conditioning Systems

HVAC include all the air filtration, electrical, and electronic devices required to create and maintain specific levels of cleanliness, temperature, humidity, pressure differentials, and airborne cleanliness in controlled environments.

Personnel Contamination Control System

These encompass all the training and qualification procedures and devices used to prevent product microbial and chemical cross contamination in a controlled environment by operating personnel. Its components include but are not limited to: personnel training, specially designed coveralls and uniforms, transfer devices, etc.

Cleaning and Disinfection System

This system groups all the methods and materials used for controlling, reducing or eliminating microbial contamination on working surfaces, operating equipment, and in rooms by using disinfecting agents.

Monitoring and Control System

This system includes the devices and software used for the operation, control, and monitoring of the components of the environmental control system. In general, it includes the devices and software used for the operation and monitoring of the HVAC system as well as for the monitoring of critical parameters such as pressure differentials, temperature, relative humidity, nonviable particle concentrations in the air and viable monitoring of particles in airborne and surfaces (Fig. 1).

A combination of the ECS components is needed to control and maintain specified environmental conditions in a controlled environment. Cleanliness classifications are typically achieved by using a combination of hardware and software components. Hardware components include: architectural components, HVAC systems, unidirectional air flow devices (previously designated as laminar flow devices), RABS, isolators, etc. (Fig. 2). Software may include all the SOPs used for controlling personnel behavior in controlled environments, procedures for cleaning and disinfection, and all the procedures required for environmental monitoring. Various combinations of hardware and software are used to deliver the degree of protection needed by the product at every stage of the process.

The following section provides a basic system description for those ECS components. These components may have an effect on controlling the environmental conditions control system and the parameters they affect. Further knowledge of the construction and operation of these devices can be found in other texts (1).

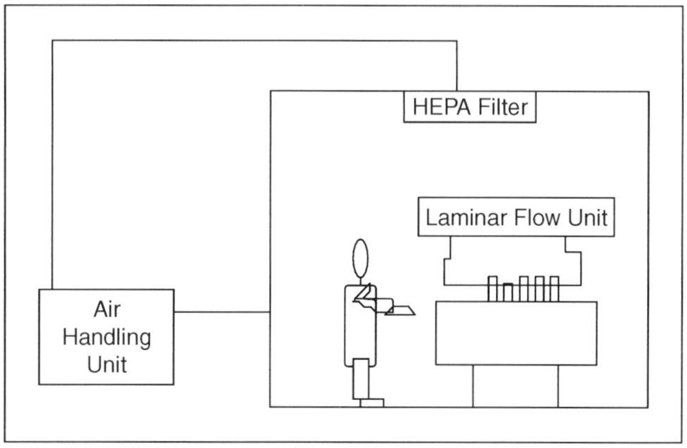

Figure 1 Macroenvironments involve the whole manufacturing room environment, which includes the air-handling system, HEPA filtration, and laminar flow units.

HEATING, VENTILATION, AND AIR-CONDITIONING

The HVAC system function is to condition (heating and cooling), replace (makeup, fresh air, oxygen replacement), pressurize (containment), and clean (filter) the air in the environment to meet the required operational conditions.

To achieve this objective, electrical, mechanical, and electronic components are arranged in several configurations such that they produce the expected results. The following represents some of the primary components of an HVAC system:

Air-Handling Units

In an HVAC system, one or several AHU can operate in parallel, in series, or in tandem, servicing various environments to precondition the air and fine-tune the environmental conditions.

The AHU houses the major components used for moving, filtering, and conditioning the air to be distributed throughout controlled environments. These components include:

1. *Blowers* (fans) deliver the airflows at specified pressures to every point of use throughout the system.
2. *Heating and cooling coils* condition the air to the specified temperature and humidity ranges.
3. *Air filters* control airborne contaminants to prespecified levels.

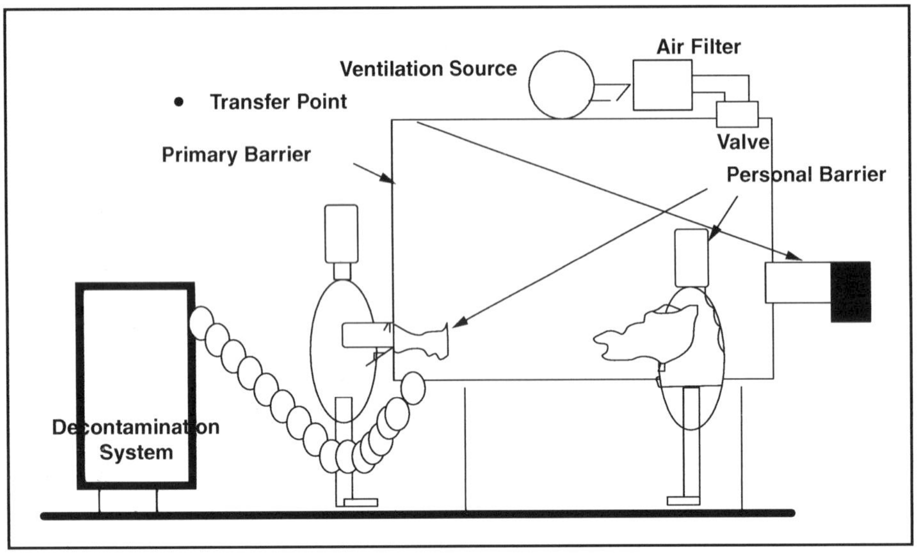

Figure 2 Microenvironments involve the components of the manufacturing environment, such as laminar flow devices, barriers, isolators, and decontamination systems.

4. *Humidifiers* bring air humidity to specified levels.
5. *Control and monitoring devices* control the operation of the mechanical devices installed within the unit and deliver information on the status of these devices.

Chemical Dehumidifiers

These devices are installed as separate units to reduce and control the humidity in controlled environments.

Typical components include the following:

1. *Chemical Desiccants.* Air-drying or dehumidification can be achieved by the use of chemical compounds. These compounds retain water on their surface and can easily be reactivated by applying heat. Depending on the design of the system, the air stream can be exposed to desiccants such as silica gel beads, granulated activated alumina, lithium chloride impregnated in a rotating drum or liquid lithium chloride solution.
2. *Regeneration-Heating Elements.* This group of devices is used to evaporate the water from the compound surface. Its operation is critical for the proper achievement of the controlled humidity range.
3. *Blowers.* Blowers are used as airflow-moving devices.
4. *Section Seals.* The particular importance of this device is the correct operation of the seals that separate the processing and regeneration sections of the dehumidifier. Incorrect operation could affect the operation of the devices and in some cases, the pressurization of the environments served by these units.
5. *Heat Transfer Devices and Mechanisms.* Various devices are used to cool or heat the air. This may include compressors, heat exchangers or coils, water chillers of various types, and steam boilers.

   These devices provide the fluids (refrigerants, chilled water or steam) to the coils normally located on the AHU.
6. *Controls.* All of these devices are equipped with a series of electric, pneumatic, or electronic devices to control and monitor the operation of its components. They are interconnected to the control and monitoring system for the facility.

Ductwork-Air Distribution Network

Components include the following:

1. Duct network, which is a network of ducts used for distributing the air from the air handlers to the points of use in the controlled environments and from these controlled environments back to the air handlers or to extraction points.
2. Insulation materials, used to maintain the air temperature and humidity along the network.
3. Dampers or valves, used for the proper balancing of the system.
4. Reheat coils installed along the system to maintain system temperature.
5. Air diffusers and terminal filter housings, used for the installation of HEPA filters at the point of delivery in clean or controlled environments.

Air Filters

Terminally Installed Air Filters. Such filters are generally high-efficiency filters (US 99.97% DOP test HEPA, MERV 20, MERV 19 (Minimum Efficiency Reporting Value) ASHRAE (3); Eurovent EN 1822 European designation H13 and H14, installed at the point of delivery in clean and controlled environments. They are mounted in a special housing that assures filter integrity.

In some instances, HEPA filters are installed at both supply and return points within a clean room to prevent or control room-generated contamination from entering the ductwork.

Proper operation of these devices is critical for obtaining the desired levels of environmental cleanliness both from the viable and nonviable point of view.

Intermediate or Low-Efficiency Filters. These filters (MERV 14 MERV 15 Minimum Efficiency Reporting Value) ASHRAE (3); EN 1822 F5–F9 are generally installed as part of the main or intermediate air handlers (when these are used in tandem). Their purpose is to reduce the particle burden on more efficient and expensive filters, and to reduce the cost of maintenance of the components of the AHU and the duct distribution network.

Other filters, such as adsorbent cells, may be required for specific applications in the control of chemical or gaseous contamination.

Dust Control and Containment Systems. These systems are designed to prevent dust spreading and potential cross-contamination. They are installed in areas where weighing of powdered materials takes place. These systems are also used where compounding and preparation of solutions takes place.

Critical Parameters. The humidity, temperature, pressure differentials, and air cleanliness classification (both for viable and nonviable levels) are affected by this system.

HVAC CONTROL AND MONITORING SYSTEMS

The control and monitoring group includes those devices used for controlling and monitoring the HVAC hardware and the environmental quality.

Upon measuring a specified environmental parameter such as temperature, relative humidity or pressure differentials, the controlling and sensing instrumentation send, feedback signals that actuate part of the HVAC systems to readjust to the prespecified setting. Monitoring devices provide visual or recorded (paper or electronic) reports on the system or controlled environment status condition. These systems may operate devices to signal malfunctions or alarms. Reporting can be via preset timing or real-time basis. Monitoring devices do not exert any corrective or resetting actions on the hardware.

Certain operations of the monitoring (pressure differential reporting) and control (valves adjustments) system can be done manually employing approved SOPs. In general, control and monitoring systems are operated automatically by microprocessor control devices actuated by built in software or by coded software.

The components of the control and monitoring system vary depending on the general architecture of the system and the facility in which it operates. Systems hardware includes input and output devices such as sensors, transducers, input/output boxes, multiplexers, and such. The operation of these devices should be based on the type of parameters measured and the overall configuration of the system.

In general, HVAC control and monitoring systems are known as building management systems. Their availability and cost make them an attractive alternative to electromechanical or pneumatic management devices used in conjunction with manually operated systems.

The elements typically constituting an automated HVAC control system include:

1. Sensing and transducer devices (temperature, humidity, pressure, and airflow)
2. Transmission network, which is sometimes enhanced with intermediate computers for data storage and transmission
3. Processing device, which can be purely mechanical, such as thermostats, humidistat, and manometers, or automated through a microprocessor such as a PLC or a SCADA supervision systems.
4. Data storage and processing stations
5. Personal computers
6. LANs
7. Hardware and software, which require extensive validation work

CONTINUOUS PARTICLE—OR AIRBORNE PARTICULATE—MONITORING SYSTEMS

The use of devices designated as particle counters is widely used in modern pharmaceutical plants to monitor critical areas on a real-time basis. The basis for selection of this system is based on the fact that these systems can detect particles of sub-micron size level (less than 1 μm), offering the possibility of detecting immediate changes on the particle concentration (of specific size) within the controlled environments at specified locations. The selected particle size in general is of 0.5 U and larger due to its proximity to bacterial size. This method does not constitute the sole or optimum proof to demonstrate the microbial (sterility) suitability of the environment, as there is no scientific correlation between the particle concentration and the presence of microorganisms. Nevertheless, the current conventional agreements on non-viable organism concentrations (particles) measured by these devices, along with microbial environmental methodologies, are sufficient to demonstrate the adequacy of an environment for aseptic processing. Particle counting technologies are constantly being updated. Not all the particle counting systems operate under the same scientific principles or use the same software to process the information collected. The user should be aware of the differences and provide for the proper validation of these devices prior to using them either for testing during validation or for routine monitoring. Monitoring systems designed to continuously monitor the level of airborne particulates in general consist of dedicated or remote sensors and a network either of tubing (where a sample is transported to a remote location) or wiring (where a signal from the sensor is transmitted to a processing unit). Both types of data gathering and transmission present advantages and disadvantages. Transport of air samples could be difficult via tubing if the transport path and the tubing installation is not well designed, executed and protected. Potential losses or particle accumulation can occur creating data inconsistencies. Locating particle counter sensors in the proximity of critical locations could be difficult because of the device's size and production equipment interaction; nevertheless, the transmission and integrity of the data can be of better quality. These devices are normally set up to report predefined particle concentrations as alert and action alarms. They can also continuously or periodically report the system and room "status." By interaction with computers they make possible the tracking of deviations as well as the determination of trends by using statistical models, thus documenting the performance of the overall system. These systems can also be converted into full data-acquisition systems for all the other environmental parameters such as temperature, relative humidity and pressure differential among controlled environments.

The software and hardware operating these devices require extensive validation.

PERSONNEL CONTAMINATION CONTROL SYSTEM

Personnel are considered to be the prime source of contamination during aseptic processing. Separation or isolation of personnel from the process is critical at the aseptic-processing stages. Training is a fundamental factor to reduce the risk represented to the process by personnel. The use of adequate uniforms as barriers helps achieve the desired levels of environmental control. The components of this system are as follows.

Devices, such as air locks and dressing rooms are used to control personnel flow and contamination tracking by personnel displacement. This concept applies to all manufacturing areas. Uniforms for the low-risk areas partially cover the body and are lint-free and clean, but these uniforms are not typically sterilized.

Uniforms for Aseptic Processing Areas

These totally enclose the human body. These uniforms are sterilized, lint-free, and free of chemical or particulate contaminants.

Training Program

The training program is an indirect, but very important, component of the contamination control program. The lack of adherence to proper operating procedures will render the most complex environmental control system useless. Evaluation of the personnel's understanding of the methods and procedures employed for environmental control is fundamental for the achievement of overall results.

Personnel Bioburden Qualification Monitoring

All personnel who work in aseptic environments should be assessed for their ability to perform gowning procedures as well as the proper wearing of the uniform during operations. A program aimed at evaluating the dressing techniques should be employed. It consists of sampling the uniform in several sections of the body after completion of the dressing procedure. This sampling is done using RODAC, or similar devices, which are incubated and the CFU counted. The results obtained are compared against suggested or internal standards. From this comparison, individuals are considered qualified to work in the aseptic areas.

Similar procedures are used to evaluate the individuals after leaving a simulated work shift. The results are also compared against the internal or suggested standard; after acceptance, the individual is qualified to work in the controlled environment.

These procedures are repeated on a routine basis during the normal operation of the environment. Individuals failing to meet the requirements are to be retrained. Professional associations such as PDA have prepared documentation for personnel training and qualifications. The United States Pharmacopoeia has issued informative chapters where suggested limits for personnel microbial contamination are listed.

CLEANING AND DISINFECTION

Cleaning procedures are needed to eliminate residual materials that could contaminate future product lots. Disinfection is needed to reduce or eliminate the bioburden on surfaces to an acceptable level. Depending on the system used for disinfection, it may be considered as a sterilization procedure. However, disinfection techniques are not generally considered a sterilization technique because of the difficulty in assessing the results of the procedure in situ. The primary components for this system include

- Cleaning and disinfection agents are selected based on the spectrum of contaminants and microorganisms to be eliminated or controlled
- Cleaning and disinfectant application and filtration devices
- Cleaning and disinfection procedures

Critical environmental parameters affected by this system include the surface bioburden.

ARCHITECTURAL COMPONENTS

The components of this system are used to provide adequate working conditions for the manufacturing operations. They also contribute to facilitating isolation or separation, as needed, to guarantee the integrity of the process. The components of the architectural system are very closely related to the design and construction of the HVAC systems, as well as with the implementation of the personnel operation and disinfection practices:

- HVAC systems are typically integrated into the architectural structure.
- Building materials should be compatible with the cleaning and disinfection agents.
- Personnel flow and safety, process segregation, cross-contamination control, and such, are to be considered as part of the facility layout.
- Control and monitoring devices should be mounted in the appropriate locations within the controlled environments to properly reflect the operational conditions.

The components of this system typically include the following:

- *Architectural components such as walls, floors, and doors, windows, and ceilings*: These elements should provide the adequate degree of structural integrity needed to achieve and maintain the room environmental conditions. They should be easy to clean, and the construction materials should be compatible with disinfectant agents
- *Lighting systems*: The light intensity level in the facility or isolator should be adequate, such that one is able to visually identify debris and materials within the controlled environment. Lighting fixtures should be of the appropriate type to guarantee cleanability and environment integrity
- *Power and utilities distribution*: Power and utility distribution system should be adequate to satisfy the requirements of the systems installed in the controlled

environments. Penetrations and electrical receptacles should be of the adequate type to guarantee the proper integrity of the environment.

ISOLATORS

This subject is described in greater detail in other chapters. The following are general comments on the use and applicability of these systems.

One of the primary objectives of the environmental control system is to assure the integrity and the quality of the process by preventing sources of contamination that could alter the purity, strength, and safety of the product. The use of isolation technology with built-in or attached sterilization systems provides a high degree of sterility assurance levels, for it eliminates one of the primary sources of microbial contamination: people. The design of these units should achieve practical results for that purpose. Consideration of all of those elements interacting within, requires careful study during design phases. This is especially true when a large volume of product is to be processed aseptically. Integration with machinery, monitoring equipment, transfer devices for product and containers, sterilization devices, cleaning devices, as well as personnel requires careful consideration.

In general, these units are prefabricated, self-contained chambers, or glove boxes, that are capable of being sterilized in situ and are designed to prevent the direct exposure of the sterile products to personnel and the surrounding environments.

To provide proper isolation, these devices are equipped with microbial retentive filters (HEPA or better) to maintain pressurization and integrity of the chamber. Transfer devices are to be designed to avoid direct contact of the chamber with the surrounding environment during the introduction or removal of components that are needed in the aseptic process.

Construction materials should be compatible with the cleaning and sterilization procedures employed. The surrounding environment should not directly impinge on the efficiency of the isolator in maintaining the inner-environmental conditions.

Integrity of the chamber must be verified after construction and monitored continuously during operation. These systems are typically prepared with connections to accept the introduction of a sterilizing gas or fluid needed for the SIP or CIP of the chamber and the components within.

Considerations for the Qualification and Validation of Isolators In Situ Sterilizers

Sterilization systems employed as part of barriers or isolators are subject to specific validation programs designed to demonstrate their proper installation, operation, and performance. Although the installation and operational structures of the validation program are similar to those used for the HVAC system, the PQ requires the use of appropriate microbiological challenges. For this purpose, bioindicators are located in sufficient numbers inside the controlled environment to demonstrate the efficacy (lethality) and penetration of the sterilizing agent at all critical locations within the unit, especially those considered hard to reach.

Qualification of the Isolator Integrity

Various methodologies exist for testing and verifying the integrity of the glove box or isolator components. The validation program should consider challenges to demonstrate that the "direct isolation" protection to the product from the surrounding environments and "reverse isolation" protection from isolator process activities to surrounding environments (in bio hazardous processes) can be achieved. The use of chemical agents as well as pressurization of the chamber are used as std tests for integrity

Qualification of CIP Systems in Isolators

The validation program should consider specific challenges to demonstrate the ability of the cleaning-in-place system to meet the prespecified cleanliness levels after execution of the cleaning cycle. The typical "cleaning validation approach" applies here.

Biohazard Contamination Control

With the advent of biotechnological processes, the need to consider segregation of processing areas has been of importance. Some of the considerations made in the design and operation of these facilities may be imposed for safety as well as for aseptic-processing requirements. In some cases, the relevance of these factors may vary along the process, but generally they are closely interlinked (e.g., processes that involve viral inactivation require total segregation of viral-free materials from the upstream phases of the process before viral inactivation). In this case, biocontainment is relevant to the final product quality, as well as safety. Thus, the design, construction, and operation of this bioprocessing facility is to be validated to comply with cGMPs and safety at the same time.

HVAC AND CLEANROOMS VALIDATION PROGRAM

Validation Modeled as a "Feedback" Loop

As indicated, the validation process can be modeled as a feedback loop with the *input* comprising the process functional requirements, its correlated design engineering, and its operational and performance specifications; and the *output* comprising verifications and test results (Fig. 3). Validation constitutes the feedback portion of the loop and is determined by the comparison of the input and output. If the data obtained demonstrate congruency between the input and output over a time period, then the system can be considered *validated*. If not, the system can be considered out-of-specifications. Thus, it cannot be considered to be validated.

From this analogy, one can determine the importance of properly defining what constitutes the input. *Input*, in the case of the ECS, is established based on the "User and Functional Requirements" and the system engineering construction and operation specifications. Process functional requirements are normally defined in URS during the developmental phases of the manufacturing process or the facility and are part of its conceptual design. Part of these are the required environmental conditions needed to meet the quality attributes of the product. The *output* is

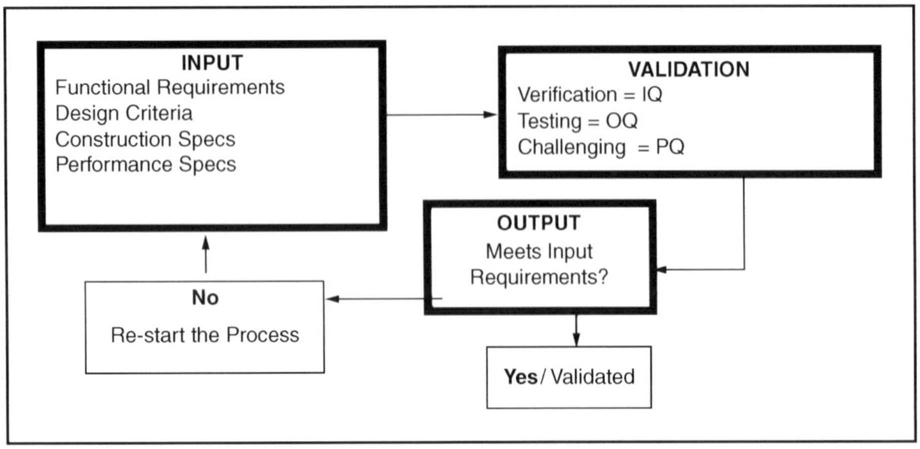

Figure 3 Validation modeled as a "feedback" loop.

documented on the protocols for verifications (IQ), tests (OQ) and challenges (PQ).

The process functional requirements represent the basis for the system's design, as well as the detailed engineering and construction specifications. The latter defines the type of systems and devices and the performance characteristics needed to satisfy the environmental control requirements. These are documented by design and equipment selection specifications, drawings, performance criteria specifications, monitoring and control device specifications, and the sequence of operation, process, and instrumentation diagrams, operating procedures, and commissioning and testing procedures. These documents are considered the input of the validation loop and are prerequisite for the validation of the system.

The expectation is that once the design and engineering phases are properly reviewed and approved by those responsible, the system can be constructed, operated, and validated as originally intended. The validator will then be responsible for determining if the original design philosophy has been carried throughout the whole process.

Validation Program Organization

A systematic approach should be followed to properly organize the validation process and obtain and document the results of the verifications, tests, and challenges executed as part of the validation program.

As indicated, it is customary to define the validation program organization and structure in a VMP. This comprehensive document defines the scope of a specific validation effort, as well as its organization, methods for data collection, document format, acceptance criteria for specific systems, and the preparation and content of validation summaries. A chapter on VMP can be found elsewhere in this book.

Validation of the HVAC systems and cleanrooms should follow the approach defined within the VMP. In the absence of such a plan, the following section provides the basic elements of a validation program organization.

Validation Task Force

The task force is a working group of individuals, from either internal or external resources, representing the various specialties and disciplines that are affected by the environmental control system. This includes, but is not limited to, the following departments:

- Process Development
- Manufacturing
- Engineering
- Construction Management
- Quality Assurance
- Quality Control
- Facilities Maintenance
- Research and Development
- Regulatory Affairs
- Maintenance and Metrology
- Validation

The coordination of the task force group is generally conducted by a task force leader, typically a member of the validation department, with responsibility for the coordination of:

- Task force activities
- Validation protocol approval process
- Protocol execution activities
- Recording of the task force activities
- Communications among the members of the task force
- Retrieval and organization of the validation data and documentation

Members of the task force could be selected from internal or external resources, but their qualifications should be commensurate with the requirements of duties assigned. Duties performed by nonqualified personnel will render the results of their work questionable and could seriously jeopardize the integrity of the validation program.

The task force duties can be assigned as required by the specific organization. These typically include:

- Review and approval of the validation schedule
- Assignments for validation protocol preparation and inherent documentation
- Review and comment on the protocols prepared by the other members
- Review of all proposed changes to approved system specifications and related documentation
- Coordination of the preparation, review, and approval process for all the validation documents
- Preparation of the final protocol summaries

- Preparation and circulation of all supplemental documentation needed to document changes to or deviations from the validation program

Task force members can be assisted by other individuals or organizations as they see fit. Nevertheless, all personnel involved in the validation program should meet the proper qualification requirements.

The earlier the task force is organized the better. The task force should remain active at every phase of the validation program. During this period, the task force will be part of the system change control through its involvement in the review and approval of modifications to original system specifications. Once changes are approved, the task force is responsible for amending and implementing the required changes to the validation program.

Validation Documentation Issuance and Control

Documentation is the core of any validation program. To have an organized program, all validation-related documents should be prepared, reviewed, and approved following a clearly defined procedure. All documents should be issued using an approved format, defined before the start of the validation program. The more uniform the format, the easier it is to review, modify, and approve the documents.

Validation Protocols

Protocols are the main validation documents. They define the extent of the verification, testing, and challenging activities, along with their appropriate acceptance criteria, testing methods, and data recording methods.

The validation protocols are generally prepared to define the approach and methodology to be used to implement and document the validation program in all its phases included but not limited the following phases:

1. *Commissioning* is a well planned, documented, and managed engineering approach to the start-up and turnover of facilities systems and equipment to the end user that results in a safe and functional environment that meets established design requirement and stakeholder expectations. (ISPE Pharmaceutical Guides for new and renovated facilities, Volume 5 "Commissioning and Qualification" March 2001).
2. *Design Qualification Protocol* is a formal part of the validation program in alignment with the ISO series, ISO 9000. It includes the requirement of verifying the proper implementation of the process functional requirements into the design and engineering activities.

   This document should generally provide the means to verify that the proper process and functional requirements have been incorporated into the basis for design for all the engineered systems and are included as part of the performance criteria for the system hardware and software components.
3. *IQ Protocol*. The intent of this document is to provide basis for verifying the proper installation of the designed system, in accordance with the design and engineering specifications.
4. *OQ Protocol*. This document provides the basis for testing the components of a installed system to demonstrate conformance with the approved operational criteria.
5. *PQ Protocol*. The intent of this document is to provide the basis for challenging the proper performance of the whole-total system while operating as an integral part of the process.

Depending on the size and the complexity of the system or piece of equipment to be validated, the decision may be made to combine commissioning protocols into IQ–OQ protocols. This format tends to be a practical alternative for small projects and can expedite the execution of the protocol. It is also particularly useful for some verifications, such as those for computers and controls, that do not fall neatly into IQ or OQ categorizations. In any event, if protocols are combined, care should be taken to execute the protocols according to required sequences. This is especially true in areas affecting personnel safety or equipment protection, or those that directly affect the validity of testing, such as calibration.

Safety Considerations

Designers of a validation program should carefully study the systems to be tested and take the appropriate precautions to insure personnel safety. The material safety data information for all strenuous or chemical agents should be investigated prior to use. Protocol execution should be always conducted under conditions that permit the assistance of personnel executing potentially dangerous tasks. It is suggested that a section on the protocol be dedicated to provide safety considerations and advice to the technicians.

Validation Program Coordination and Execution Procedures

Execution of the validation program should be based on a preapproved program, assuring that all the sequential steps are properly stated (i.e., the execution of IQ should precede the OQ, and it in turn should precede the PQ phase, unless clearly stated and approved otherwise).

Ideally, systems and equipment should be installed, started up, and commissioned before beginning validation. In addition, a formal change control procedure should be developed and implemented before beginning the validation process. At the very least, the IQ should start after the system or part of it has been fully installed and no changes are anticipated. Likewise, the OQ activities should start after the system start-up procedures are completed and no changes in its operation are foreseen. Finally, the PQ activities should not begin until all the systems and procedures are fully implemented and the operating personnel have been adequately trained.

Procedures for handling program deviations and modifications to approved protocols or activities, thus provide proper traceability.

A system for the development and maintenance of validation documentation should be defined as part of the validation activities, usually as a part of the VMP.

The intent is to guarantee the integrity of the documentation and data obtained during the validation activities.

Validation Data Review and Final Summary Preparation

A procedure should be established for the final review, interpretation, and approval of the collected documentation. Likewise, an approved approach for the preparation of the final summaries should be deemed as part of the protocol. Although not mandatory, this approach will facilitate the validation process.

The proper organization and structuring of the validation program is fundamental to assure compliance with the requirements imposed by the cGMPs from which the validation program derives.

ENVIRONMENTAL CONTROL SYSTEM VALIDATION SEQUENCE

All of the components of the environmental control system should be validated. Because of their interaction, a logical sequence for systems' validation should be followed. The following sequence is indicative of some of the validation activities. Note, however, some activities can be executed in a parallel fashion before the final ECS PQ. These activities include

1. Architectural components
2. Cleaning and disinfection agent qualification
3. Personnel operating techniques validation
4. HVAC validation, and control and monitoring systems validation

The following section provides the scope and structure for the protocols needed to validate ECS.

Architectural Systems Validation

The structure of the protocol follows the same lines of any other system or utility. In general, it can be outlined as follows:

DQ Protocol

Scope. The scope of this document is to demonstrate that the approved process functional requirements have been incorporated into the design specifications and the architectural components selection.

Objective. This section provides verification of the conformance of the selected components' design specifications, with their intended operational and performance specifications. This includes verifying that the construction and workmanship specifications meet the requirements originally specified by the user and that all changes to user requirements or functional specifications have been reviewed, justified, and approved, as established by the VMP.

Authority and Responsibility. As in any other part of the validation process, it is important to define the authority and responsibility for those individuals performing the tasks outlined in the protocol. The tasks typically included in this protocol follow:

1. Verification and documentation procedures required to fulfill the protocol requirements
2. Protocol execution and collection of the data
3. Review and interpretation of the data for accuracy, completeness and cGMP compliance
4. Approval of original protocol formats
5. Approval for the final summaries and system qualification statement

Although the organization of the protocol execution and data collection process may vary depending on the corporation, in this phase, more than any other, the execution of the protocol should be entrusted to personnel with a profound knowledge of the architectural components' design, construction, and testing, as well as with the overall process to be executed within the controlled environment.

System Description. This is typically the first section of any protocol and it should provide

1. Classification of the environments served
2. A description or references to specifications for products to be manufactured in the environment, either specifically or by category
3. Room design criteria clearly specifying the conditions required for the process
4. Design and guidelines for architectural components and materials selection
5. References to construction codes (local, state, and federal) as they may apply
6. Test and qualification requirements for the system or components, either in situ or at the manufacturers
7. Personnel qualification requirements for the construction, installation, testing, and validation of critical systems or devices

IQ Protocol

Scope. The scope of this protocol is to provide the elements to verify that all the architectural components installed are

1. The ones specified in the design and engineering documentation
2. Serviced properly by the required utilities (electrical power, compressed air systems, or other, as they may apply)
3. Installed at the specified locations
4. Constructed of the same materials as specified
5. Installed using the workmanship level specified

It also verifies that system components properly reflect the as-built system drawings and that operation manuals, catalogues, guarantee materials, and related documentation are available.

Objective. The objective section provides an indication of the architectural components to be validated. Typically, this includes those described in the design section, or in the supplements added after changes have been authorized.

Authority and Responsibility. In the process of validation, it is important to designate individuals responsibility for the following:

1. The verification and documentation procedures required to fulfill the protocol requirements
2. Protocol execution and data collection
3. Review and interpretation of the data for accuracy, completeness, and cGMP compliance
4. Approval of original protocol formats
5. Approval for the final summaries and system qualification statement

Manufacturer's Component Verification. The intent of this section is to provide an accurate inventory of the components as installed and compared with the approved specifications. These sections should reflect the design features of the equipment or components, as well as the most important operational characteristics.

The following documents are to be provided as part of the documentation:

1. *Purchase Orders*
   a. Component specifications reference list indicating
      i. Manufacturer
      ii. Model number
      iii. Serial number
      iv. Electrical requirements
   b. Other specifications or forms deemed appropriate
   c. Verification that the equipment has been installed as specified in the approved drawings
2. *Construction Workmanship Verification.* The scope of this section is to provide evidence that the architectural components have been installed in accordance with the design and manufacturer's specifications. This section typically provides a checklist to verify that finishes meet the requirements set for the controlled environments.
3. *SOP Verification.* The intent of this section is to verify that all required operational SOPs for the maintenance and cleaning of the architectural components exist in either draft or final form.

   Each SOP must be current and approved in accordance with the documentation control systems established by the system's owner. If drafts are used, they should state a limited validity, an expiration date, and be approved by those properly authorized.
4. *Utility Connections Verification.* All utility connections supporting the architectural components, such as electric light fixtures, automatic doors, electrical interlocking devices, video cameras, or other, are installed properly and in accordance with listed specifications and drawings.
5. *Change Parts or Replacement Parts.* The intent of this section is to provide a list of the critical replacement and change parts needed to keep the system operational in accordance with its specifications.
6. *Maintenance Procedures Verification.* The intent of this section is to provide evidence that a maintenance program for a safe and reliable operation exists.
7. *Lubricants.* The use of nonapproved lubricants may cause potential sources for adulteration of the environment. Thus, a clear definition is to be provided of the type, brand, and place of application of these products.
8. *Critical Systems Change Request Procedure Verification.* Once this phase of the validation program is completed, for a system to remain validated, it is critical that proposed changes be reviewed and approved. The intent of this section is to provide evidence that the validated system has been integrated as part of the routine Critical Systems Change Request Procedure. This procedure governs changes to validated critical systems or components and ensures a thorough review of the proposed change and its effect on the validation status of the system, as well as its conformance with regulatory requirements.
9. *As-Built Drawings Listing and Verification.* An "as-built" drawing is a drawing that is physically verified through inspection of all the system components, then signed and dated by the person(s) performing the inspection.

There is a need to provide a list of all the actual as-built drawings referenced in the IQ section. This is necessary because sometimes there are differences on the number and quality of the original design drawings and those finally providing evidence on how the system was built.

OQ Protocol

The purpose of this section is to verify that the specified architectural components operate as specified and are in agreement with the acceptance criteria for critical systems.

Scope. All of the system components described in the final design and specifications, or authorized changes to that design or specifications, need to be qualified to demonstrate their adequate operation.

In general, the OQ scope is to test the individual components of the system, such as light fixtures, automatic door openers, interlocking devices, and such.

Operational System Description. A brief description of the operations of the architectural components is typically included in this section of the protocol. There is no need to provide great detail if references are provided in the specification documents in which a detailed explanation is given.

SOP Verification. The SOPs should have been verified during the execution of the IQ. If all of the documents were not available before start the OQ tests, all the SOPs for the architectural components system should have at least been drafted. This is to assure that the operation of the system is conducted in an approved and systematic way. The lack of this approved systematic approach will render the OQ efforts invalid, because it will not be possible to reproduce the working conditions of the system. For each operation performed, the SOP must be current and approved. Documentation attesting to the fact that these documents exist should be included as part of the protocol.

OQ Tests. These are a series of tests designed to prove the proper operation of the architectural components system. The extent and limitations of them are determined by the systems served. The following is given as an example of the structure to be followed in the protocol.

Lighting Levels Verification Test:

1. Test objective: Demonstrate that the lighting levels within the controlled environment meet the design criteria.
2. Acceptance criteria: The acceptance criteria is to be set in accordance with the design specifications.
3. Test procedure
   a. The testing procedure is designed as a function of the control system. This section of the protocol should outline the conditions and areas to be tested concurrently to assure that illumination conditions correspond to normal operational conditions.
   b. A photoelectric illumination meter of a type approved for field measurement in accordance with the IESNA (6) lighting handbook is used.
   c. Before starting the test fluorescent lights must be operational for at least 100 hr. Take the readings after the lights have been in operation for at

least two hour. Measure the lighting levels at working levels.

4. Reporting requirements
   a. Locate the measurement positions on a diagram.
   b. Calculate the lighting levels as a function of the measuring device.
   c. Analyze the test data collected and write a conclusion on the acceptability of the test results based on the acceptance criteria provided.

Another OQ test for architectural components is the room or enclosure integrity test. It could include the following sections:

Room or Enclosure Integrity Test:

1. Scope: Determine if there is intrusion of unfiltered air into the clean work areas from the surrounding environments through joints, conduits, utilities penetrations, or other sources.
2. Acceptance criteria: No leaks should be present from less critical to critical environments.
3. Methods and procedures
   a. Equipment:
      i. Aerosol generator
      ii. Particle counter having a sampling rate of 28.3 L/min (1.0 ft^3/min and particle size discrimination capability of 0.5 μm).
4. Method:
   a. Measure the particle concentration outside the environment to be evaluated adjacent to the penetration to be evaluated. This particle concentration should be at least $3.5 \times 10^6/m^3$ (1.0/ft^3).
   b. Scan at a distance of 5–10 cm (2–4 in.) from the joint for vertical or horizontal joints, or seal scan it at the rate of 3 m/min (10 ft/min).
5. Reporting: Report readings greater than 10^{-3} the outside measured concentration.

Performance Qualification

No specific PQ protocol exists for the architectural components because it is executed as part of the HVAC protocol.

Cleaning and Disinfectant Agents Validation

Cleaning validation procedures will be validated as part of the HVAC system PQ. Before that, the cleaning and disinfection agents to be used are to be qualified to demonstrate their effectiveness when used in the strength and concentration specified over the surfaces present in the controlled environment. Their lethality rate is to be demonstrated for this purpose with several challenges of known bacterial spore strains. Typical challenge organisms are listed in the *USP*. Testing methodology is defined in several publications from the PDA and the American Association of Official Chemists. These procedures are explained elsewhere in this book.

Personnel Operating Practices Validation

Personnel qualification is executed in several phases. Evaluation of personnel training should be conducted in accordance with cGMP concepts as well as the specific operational practices to be executed within the controlled environment. This assessment can be made by the use of direct questioning, supervised exercises, or other, to determine the individual level of understanding. If understanding is considered acceptable, then gowning procedure evaluation will take place. The dressing procedures may be evaluated by using the contact plate (RODAC) sampling method to determine the level of dexterity of the individual. If the results obtained are within limits, aseptic practice evaluation will take place. The individual is to perform his or her duties in simulated conditions within the controlled environment. At conclusion of this simulation, a uniform contact plate (RODAC) sample test is to be conducted while exiting the clean room. If the operations were performed as specified and the "body samples" are acceptable, the individual is considered qualified.

Personnel qualification activities should precede the final HVAC PQ test.

HVAC Control and Monitoring System Validation

The control and monitoring systems are considered the core of the environmental control system. The final performance of the environmental control system can be evaluated only after these have been validated.

PQ protocols covering this system, in general, include the performance challenges for the all the components of the environmental control system described in the foregoing. The system is challenged by simulating the process or at the time the aseptic processes qualification or media fills are performed.

The validation protocols follow the same format structure outlined before for the architectural components. The following section provides a generic checklist approach used for the HVAC controls and monitoring systems validation.

As indicated earlier, the HVAC systems could be either manually, electric, pneumatic, or electronically controlled, operated or monitored. Computer- or microprocessor-controlled equipment is subject to the software and hardware validation procedures that apply to computerized systems.

HVAC System DQ Protocol

Scope. The scope of this document is to demonstrate that the approved performance specifications expressed in the functional requirements have been incorporated into the HVAC design specifications and equipment selection.

Objective. This section will provide for the verification of the conformance of the selected components design specifications with their intended operational and performance specifications. It should be verified that the approved operational and performance ranges meet the requirements originally approved by those responsible. Changes to operating and performance specifications that took place during the design phase should be reviewed, justified, and approved by authorized personnel.

Authority and Responsibility. As in any other part of the validation process, it is important to define the authority and responsibility of the involved individuals.

The verification and documentation procedures required to fulfill the protocol requirements include the following:

1. Protocol execution and data collection
2. Review and interpretation of the data for accuracy, completeness, and cGMP compliance
3. Approval of original protocol formats
4. Approval for the final summaries and system qualification statement

Although the organization of the data collection process may vary, depending on the corporation, in this phase more than in others, the execution of the protocol should be entrusted to personnel with a profound knowledge of the HVAC engineering, the process, and the components of the environmental control system.

System Description. This is typically the first section of any protocol, and it should include the following:

1. Classification of the environments served by the HVAC system.
2. A description or references to specifications for products to be manufactured in the environment, either specifically or by category.
3. Room design criteria clearly specifying the environmental conditions (temperature, humidity, process isolation, and such) required by the process.
4. Design and guidelines for equipment and materials selection (such as the FDA Aseptic Processing Guideline; IES recommended practices (see Appendix I), or other such).
5. References to codes (local, state, and federal) as they may apply, such as the Code of Federal Regulations, Section 21 (cGMPs), Federal Standard 209E (published by (IES), *U.S. Pharmacopeia* Forum (microbiological evaluation of clean rooms and other controlled environments).
6. Testing and qualification requirements for the system and its components.
7. Personnel qualification requirements for the construction, testing, and validation of critical systems or devices.

IQ Protocol

Protocol Scope. To provide the elements that will verify the following for all the new components of the HVAC system that have been installed:

1. That those components are included with their approved design and engineering specifications.
2. They are properly served by the required utilities, such as electric power, chilled water, pure steam, plant steam, compressed air, and such.
3. The components are installed at the specified locations.
4. All critical measuring instruments and gauges are calibrated against traceable primary instrumentation.
5. You are able to obtain operation manuals and spare part lists to assure the proper and continuous operation of the system.
6. They are properly reflected in as-built drawings.

Authority and Responsibility. Define in this section the names, positions, and responsibilities for those the executing and approving this protocol.

Manufacturer's Equipment and Workmanship Verification. This section should provide an accurate inventory of the equipment as installed so it can be compared with the approved specifications. It should reflect the design features of the equipment or components, as well as the critical operational characteristics.

It also should provide the means to verify the proper installation of the system components in accordance with the manufacturers' recommendations and the workmanship standards set in the engineering specifications.

The following documents are to be provided as part of the documentation pertaining to this section:

1. Purchase orders for major components
2. Equipment specification reference for all the listed components as well as verification of their proper installation and identification.
3. The list should include, but is not limited to, the following HVAC system components:
   a. Air-handling units:
      i. Manufacturer
      ii. Model number
      iii. Serial number
      iv. Fan HP
      v. Electrical requirements
      vi. Supply fan installed
      vii. Cooling coil
      viii. Condensate collection pan
   b. Air handler heating section components
      i. Control valve type
         - Model number
         - Serial number
      ii. Steam coil
   c. Air handler humidification section components
      i. Humidifier
         - Manufacturer
         - Model
      ii. Pure steam connection
      iii. Control valve type
         - Model number
         - Serial number
   d. Air handler filtration section components
      i. Prefilters (usually 95% ASHRAE efficient cartridge filter)
      ii. Final filters (usually high-efficiency or HEPA filters)
   e. Air handler electrical, pneumatic, or electronic control devices
      i. Type
      ii. Location
      iii. Range and accuracy
      iv. Manufacturer
      v. ID number
   f. Air distribution network components, ductwork, noise attenuators, dampers, and such
      i. Develop the checklist based on the final approved engineering specifications. Pay attention to changes or modifications after issuance of specifications. These may be considered as deviations from the original design.
      ii. Provide cleaning inspection reports.
   g. Verify the actual equipment
      i. Provide ID number for valves, when possible

  ii. Installed versus the reference specifications
  iii. Is properly reflected in the as built drawings
- h. Filters and terminal filter housings
  - i. Provide the specification used as a reference(s), include authorized change orders
  - ii. Verify that the specified terminal filters have been installed versus the reference specifications
  - iii. Provide a list of the terminal filter locations and actual serial numbers
- i. Chilled and hot water distribution systems
  - i. Provide the specification reference(s) including authorized change orders
  - ii. Verify that the actual equipment has been installed versus the reference specifications
- j. Controls
  - i. Provide a detailed description of the operation of the automatic control system
  - ii. Verify installed controls against approved specifications
  - iii. Verify that all control wiring and tubing has been installed in accordance with approved drawings
  - iv. Execute a point-to-point verification to confirm correct installation and identification of field control devices and wiring and tubing

Calibration Verification. The intent of this section is to provide a list of critical instrumentation and control panels and to document that they have been identified and calibrated in accordance with an approved procedure.

Provide a list of calibrated instruments for the control system and file a copy of the calibration record (see the chapter on calibration for further detail).

SOP Verification. The intent of this section is to verify that required operational SOPs exist in either draft or final form. Each SOP must be current and approved for use on the systems involved. They must represent the methods to be used in the operation of the system.

Utilities Connections Verification. The utilities connections supporting the HVAC system need to be verified to assure that they are installed properly and are in accordance with listed specifications and drawings. The following list represents many elements covered in this section:

1. Power supply verification: All the electrical connections and power supply for all of the components of the system should be verified to determine if they are in accordance with the unit requirements and system specifications.
2. Steam supply piping verification: Steam is used as a utility for the humidifiers. Inadequate selection and installation of components used for steam transport can be a potential source of contamination. In general, "pure steam" is considered the appropriate choice. For this reason, it is necessary to verify the correct source of steam and its installation in accordance with specifications of the components of the steam system servicing the air handler, humidifiers, heaters, and other equipment.

Change Parts and Replacement Parts Verification. The intent is to provide a list of the critical replacement and change parts needed to maintain the system in operation in accordance with specifications. This list includes, but is not limited to, blowers, air filters, belts, and monitoring equipment and devices.

Maintenance Procedures Verification. The intent of this section is to provide evidence that a maintenance program for the system exists to assure proper, safe, and reliable operation.

Lubricants Verification. The use of nonapproved lubricants may cause potential sources for adulteration of the environment. Thus, a clear definition is to be provided of the type, brand, and place of application of these products. It must be clear in which cases the lubricant is authorized for air stream contact and when it is not.

Critical Systems Change Request Procedure Verification. For a system to remain validated, it is critical that a review and approval process take place before the execution of any change considered to be critical for its performance. The intent of this section is to assure that once the IQ validation activities have been concluded, the system remains under the Critical Systems Change Request Procedure.

As-Built Drawings Listing and Verification. This is an important part of the IQ activities, as it provides the record for the components that were covered and verified under this protocol. An *as-built* drawing is a drawing that is physically verified through inspection, signed and dated by the person performing the inspection. Many times differences are present from the original design drawings and those finally providing evidence of how the system was built.

OQ Protocol

It is suggested that testing methodologies included in the Operational and PQ protocols be cross referenced with those described in the ISO standard ISO/FDIS 14644-3 Cleanrooms and Associated Controlled Environments Part 3: Test Methods.

Objective. The OQ objective is to verify that the specified components of the HVAC system operate within their specified range and meet the acceptance criteria set by the system designers.

Scope. The HVAC system components described in the final design and specifications, or authorized changes to that design or specifications, need to be qualified to demonstrate their adequate operation. In general, the OQ scope is to test the individual components of the system such as AHU, ductwork, blowers, and others. The final performance of the system in terms of environmental quality, such as temperature, humidity, airborne cleanliness (viable and nonviable), can be evaluated only under dynamic conditions whether real or simulated and when the other components of the environmental control system are in place.

HVAC Operational System Description. A brief description of how the system operates and its sequence of controls is required. Those involved in the execution of the protocol of the intended method of operation and ranges of operation, such as pumps' flow rates, alarms and such, should be specified. More detailed information is given in the engineering specifications. There is no need to provide great detail if references are provided to the specification documents in which a detailed explanation is given.

Instrumentation Calibration Verification. Before beginning the OQ testing, it is necessary to verify that the instrumentation specified has been installed and remains in calibrated status (original calibration should be verified during the IQ). In this way, the readings taken from these instruments will be considered valid and acceptable.

Testing Equipment Calibration Verification. The calibration status of test equipment to be used in the execution of the OQ protocol shall be verified before performing any of the tests included as part of the protocol. Copies of all the calibration certification should be obtained and attached as part of the documentation.

SOP Verification. The SOPs should have been verified before the execution of the IQ. If all of the documents were nor available at the time, before the start of the OQ tests, the SOPs for the HVAC system should at least be drafted. The idea behind this is to assure that the operation of the system is conducted in an approved and systematic way. The lack of this approach will render the OQ efforts invalid, for it will not be possible to reproduce the working conditions of the system. Each SOP must be current and approved for use for the operation performed. Documentation attesting the fact that these documents exist should be included as part of the protocol.

OQ Tests. These are a series of tests designed to prove the proper operation of the HVAC system. The extent and limitations are determined by the area served by the HVAC system, in particular, as well as the design itself.

Air System Balancing and Pressure Differential Test

Air system balancing is one of those activities that could be executed during commissioning. Data obtained can be imported into the OQ when the procedure and recording instruments recommended by recognized balancing bureaus are used.

The distribution of the air among the various areas served by the HVAC system is fundamental to achieve proper environmental conditions because it affects the ration of heat exchange needed for the control of temperature and humidity, the ex-filtration levels needed for pressurization control, and the dilution levels (changes per hour) needed to achieve and maintain cleanliness levels.

To achieve balance, the system dampers (valves) located in the ductwork are closed and opened as needed to achieve the volumes dictated in the design. This is a trial-and-error procedure and may take a long time, depending on the configuration of the HVAC system, the number of AHU interacting within the areas served, and the control system.

The design volumes are usually considered targets, rather than absolutes, because the conditions of the system may need changes difficult to foresee during the design stages. To avoid unnecessary delays in validation, it is important that this test be executed once the final air balance has been achieved. The prebalancing attempts could serve as a good indication of the potentially problematic areas that should be resolved before entering into validation. Validation is not designed to resolve start-up or problems or fine tune the system; the intent of validation is to verify compliance with specifications.

1. Test objective: To verify that the HVAC system meets the design criteria pertaining to air volumes delivered to the controlled space and that it is able to maintain the pressure differential gradients between adjacent environments, as specified. The air balancing is to be conducted by a qualified contractor.
2. Acceptance criteria: The supply and return air volumes should conform with the range specified. Pressure differentials between rooms should be maintained as indicated in the specifications.
3. Test procedures
   a. The data should be obtained by the contractor using adequate and calibrated equipment, such as thermo anemometers, electro manometers, airflow meters, and such, as specified by the organization certifying the contractor. Airflow should be measured under full HVAC operational conditions at
      i. AHU supply and return sides
      ii. Duct network primary and secondary branches
      iii. Terminal air filters or supply air diffusers
      iv. Exhaust grills and diffusers in the room
      v. Target values and final values should be reported
   b. Pressure differentials should be measured under full-operational conditions. The system should be challenged by operating the exhaust systems that may affect the pressurization schedule.
   c. Pressure differentials should be measured and reported for all adjacent and/or interconnected rooms (environments) of different classification levels unless otherwise specified.
   d. Target values and final values should be reported. Once obtained, the actual air-balancing data and pressure differential data should be compared with the design specifications.
4. Reporting requirements: Report results on the approved forms specifying the target engineering values before balancing first, second, and third trial data (as needed) until the final air balance and pressurization control has been achieved. Indicate the basis for the system acceptance or rejection, as the case may be based on the obtained data.

Air Volume for Non-Unidirectional Airflow Terminal HEPA Filters Test

Reporting the volumes of air supplied by individual terminal HEPA filters is important because it allows the verification that the filter operates within design range. Because of their design, HEPA filters should operate at or below the specified output; otherwise, collection efficiency of small particles could be impaired. That is why this data is to be collected concurrently with the filter integrity test.

The environment cleanliness is a function of:

1. The number of times the air is re-circulated through the air filters
2. The efficiency of the filters
3. The number of filters in a series
4. The quality of the incoming makeup air

5. The particle generation within the controlled environment

The volume of re-circulated air through the HEPA filter defines the dilution ratios or so-called air changes per hour which, in turn, determine the environmental cleanliness. Thus, verification of proper air changes is an important part of the validation studies.

The data collected create a record that serves as a baseline for the specific system. This data is reviewed during the semiannual maintenance to assure that the system operates properly, the pressure drop across the HEPA filter has not changed, and the system's original balance and airflow deliveries are maintained.

1. Test objective
   a. To verify that the air volumes are within specifications before performing the HEPA filter integrity test
   b. To verify the individual HEPA filter pressure drop at the operational airflow
   c. To verify the air volume of supply HEPA filters at the time of the HEPA filter integrity test
2. Acceptance criteria: This report is prepared for informational purposes only
3. Test procedure: Refer to test method, "Air Volume for Nonunidirectional Air-flow," attached (see Appendix).
4. Reporting Requirements: Report results on the approved form. Analyze the test data collected and write a conclusion on the acceptability of the test results based on the acceptance criteria specified.

HEPA Filter Integrity Test (ISO 14644-3 Part B.6)

Absolute air filters (those having an efficiency of at least 99.97% or greater for particles 0.3 μm and larger) or HEPA filters are of paramount importance in obtaining viable and nonviable cleanliness levels in controlled macro- or microenvironments. These filters are rated for their efficiency and operational characteristics by the manufacturer before delivery, typically using the method recommended in Military Standard Mil-std-282 "DOP Smoke Penetration and Air Resistance of Filters", IES RP-CC-001-83 and European Standard EN 1822-1. Test results are reported in an attached label or certificate provided for every filter.

These filters are delicate, even though modern design makes them less susceptible to damage. Potential damage can occur during shipping and installation. Because of that, filters must be tested after installation to guarantee integrity. For this purpose, an air-generated or mechanically generated aerosol challenge to verify the integrity of the filters by using a light-scattering photometer has been devised and accepted by the industry as adequate to determine defects in HEPA filters or HEPA filter systems. In general, the condition under which this aerosol is generated provides a consistent concentration of 10 μg/L of air, which is considered as an adequate challenge. This concentration is equivalent to approximately 3×10^{10} droplets per cubic meter (10^9 droplets per cubic foot) when the air is generated by a Laskin nozzle.

The use of a challenge aerosol is justified because, in general, the recirculation levels (number of times the same volume of air is passed through the filter) are so high that it is practically impossible to detect minor leaks using the room air as a challenge. Until recently DOP has been considered the primary aerosol choice. However, owing to its nature and origin, a possibility exists, even though it has not been proved, that DOP may represent some health risk. Substitutes have been developed that have the same particle distribution as DOP when a cold aerosol is generated. A synthetic oil, EMERY 3000, has been selected as an approved replacement.

The quantities of aerosol needed as a challenge for this test are minimal. A single nozzle can provide enough airflow to accommodate approximately 28 m^3/min (1000 ft^3/min). Normally, a compressed air source capable of delivering 75 L/min of air at standard conditions at 138 kPa (2.7 ft^3 min at 20 lb/in^2) is required for each nozzle. Thermally produced aerosols may have greater capacities.

Usually, no more than 100 g/L of air are used to obtain the full-scale response in the scanning photometer, which is the device typically used for this test.

The use of outside air as a challenge is possible, as long as particle concentrations beyond 10 million/ft^3 of air are present. The process of challenging HEPA filters with outside air can be more difficult and lengthy owing to the variations in the challenge concentration and the need for a particle counter instead of a simple light-scattering photometer.

Occasionally, particle counters are used instead of light-scattering photometers. The procedure is implemented by using dissolution chambers and other devices that minimize the exposure of the delicate optical part of the particle counter to the challenge aerosols and to within the response capabilities of the instrument. When the intent is to use a particle counter as the detection device, it is advisable to consult the manufacturer of the unit before the implementation of the test method because improper settings may permanently damage these units. On the other hand, the scan probe needs to be designed such that it assures that the scanning velocity is adequate to obtain the proper response on the unit.

The HEPA filter integrity-testing procedure is clearly described in the IES-RP-CC006.2 and ISO 14644-3 sec B.6 documents. The following is a summary of this document.

1. Test objective: To provide evidence of the integrity of the HEPA filters and seals in situ.
2. Acceptance criteria: An unacceptable leak is defined as a penetration of 0.03% or more of particles, 0.3 μm and larger than the reference calibration curve for 99.97% efficient filters, or as penetration of 0.01% or greater of particles 0.3 μm and larger than reference calibration curve for 99.99% efficient filters.
3. Test procedure
   a. Introduce the challenge aerosol upstream from the filter to be tested at an adequate distance to assure proper air and aerosol mixing, a minimum of 10 to 15 cm (4–7 in.) from the filter face.
   b. Calibrate the aerosol photometer as given in the manufacturer's procedures.
   c. Scan the filter face at an appropriate and approximate rate.
   d. Locate and repair the filter leaks in accordance with the approved procedure.

e. Retest after repairs have been completed.
f. Replace HEPA filters if necessary, and retest.

4. Reporting requirements
   a. Report results on the appropriate forms.
   b. Analyze the test data collected and write a conclusion for the acceptability of the test results based on the acceptance criteria specified.

Pressure Differential Stress Test

In an interactive system or in areas where several doors can be opened simultaneously, it is important to institute challenges that determine how the system performs under stress conditions. For this purpose the designers and users of the system should design experiments that represent possible, although perhaps unlikely conditions, to determine the capabilities and boundaries of the system.

Planning for this type of test requires a profound understanding of the HVAC system, its controls, and the operational methods within the area in question. Improper design of this test can cause permanent and potentially serious damage to the equipment. Thus, before instituting these challenges, consult with the designers and contractors familiar with the system and obtain their approval. See ISO 14644-3 section B.5.

1. Test objective: To establish baseline pressure differentials when several doors accessing the same area are open or in the event of air system failures. This test is to be conducted under static conditions.
2. Acceptance criteria: This test is conducted to determine the point at which the system is not capable of coping with the simulated stress conditions. The point of failure is established at the time when the pressure differentials between rooms of different classifications go out of limits.
3. Test procedure
   a. The testing procedure is designed as a function of the facility design. This section of the protocol should outline the sequence to be followed for the opening and closing of the doors. A matrix-type chart facilitates the simulation of opening and closing events.
   b. All doors should be sequentially opened, and various combinations employed to simulate possible or even unlikely conditions. Pressure differentials are to be measured and recorded between interconnecting environments by using a calibrated manometer. Pressure readings should be compared against the approved specification. If these are within specifications, continue opening the rest of the doors until the system fails to maintain the pressure.
   c. Airflow is to be videotaped by generating visual smoke across the door opening. If the generated smoke (normally theatrical smoke, Titanium tetrachloride fumes, or dry ice are use for this experiment) follows the incorrect pattern (from less critical to critical environments), then the condition should be noted and recorded; at that point the system will be considered out of specification. These procedures require proper safety protection for the operators, because inhaled smoke could be dangerous.
4. Reporting requirements
   a. Record the results and provide additional comments or description of unexpected test results.
   b. Identify the videotapes and prepare a written description outlining the findings.
   c. Analyze the test data collected and write a conclusion on the acceptability of the test results based on the acceptance criteria provided.

Start-Up and Shutdown Test

1. Test objective: To test the start-up and shutdown sequence of the operation of the AHU as controlled by the control system. Prior to implementation consult with the system designers and constructors improper design of this test can be un-safe and cause permanent damage to structures and equipment.
2. Acceptance criteria: The AIU start-up and shutdown sequence operates in accordance with the design specifications.
3. Test procedure: The testing procedure is designed as a function of the control system. This section of the protocol should outline the sequence to be followed and the devices that intervene.
4. Reporting requirements
   a. Record the results of the start-up and shutdown sequence and provide additional comments or description of unexpected test results.
   b. Analyze the test data collected and write a conclusion on the acceptability of the test results based on the acceptance criteria provided.

Control and Monitoring Devices Test

Calibration, testing and operational verification of the control and monitoring systems devices employed as part of the HVAC system is a standard practice during commissioning. both for safety and sound operational reasons. Documenting the quality systems design and construction of both software and hardware during commissioning can serve precious time and resources during validation. Validation testing can be designed based on the type and extent of coverage of the control and monitoring system and the test conducted and documented during commissioning. Functional test that should be implemented as part of validation to verify the proper performance of critical monitoring operations such as pressure differential monitoring and operational alarms.

1. Test objective: Verify the proper performance of the HVAC control, monitoring and alarm systems.
2. Acceptance criteria: The HVAC control and monitoring devices as well as the alarms should operate in accordance with the design specifications.
3. Test procedure
   a. As indicated, this test is dependent on the configuration of the control system for both the software and hardware components. All microprocessor controlled systems software and hardware should be validated as required for computer control systems.
   b. Functional test by reproducing control, alarm, or out-of-range conditions are used to test the

operation of electrical, mechanical, and pneumatic monitoring and control systems.

4. Reporting requirements
   a. Record the proper response and identification of all signals generated at the sensing and receiving points (point-to-point response verification).
   b. Record the set conditions used for a functional response of alarms and control systems.
   c. Record the responses to functionally induced conditions.
   d. Record the compliance or noncompliance with the approved specifications.
   e. Analyze the test data collected and write a conclusion on the acceptability of the test results based on the acceptance criteria provided.

Power-Fail and Recovery Test

Because of the nature of the air-handling system and its controls, potential problems may occur if the system's power supply fails. The intent of this test is to verify that the control system can maintain the components of the AHU within the specified range after a power failure. Prior to implementation consult with the system designers and constructors improper design of this test can be un-safe and cause permanent damage to structures and equipment.

1. Test objective: To test the operation of AHU pneumatic, electric, or electronic control system during a power-fail and recovery cycle.
2. Acceptance criteria: The controlled environment should recover to the original setup after loss of power.
3. Test procedure: This procedure is to be designed as a function of the system tested. Because the power fail and recovery test is a major system test *great care must be exercised to prevent damage to personnel or equipment*. The tests should be designed in accordance with the HVAC and controls design engineers. There-fore, the following should be conducted:
   a. Proceed to simulate the failure.
   b. Bring system to a complete stop.
   c. Wait the required time before restart to prevent mechanical or electrical (overcharge) damage to the system.
   d. After completion of the foregoing three steps restart the system.
4. Reporting requirements
   a. Record the time it takes for the system to reestablish the approved conditions.
   b. Record the monitored environmental parameters (air volume, pressure differentials, temperature, humidity).
   c. Compare the data that had already been acceptable for the environment tested. Particular attention should be given to pressure differentials as the best indicator of the system capability to regain control.

UAD (Formerly Laminar Flow Systems LAF) Systems Validation

The LAF devices are used to protect and separate the critical and control environments by using a HEPA filter unidirectional air. Airflow uniformity throughout the work zone covered by the LAF device is needed to achieve a continuous cleansing effect. This is achieved by maintaining a prespecified flow velocity and by minimizing the impaction of the air mass against surfaces that are not considered aerodynamic. The lack of the appropriate conditions may induce turbulence which, in turn, is a potential cause of cross-contamination.

It is vital that the airflow within the enclosure maintains a unidirectional velocity across the whole working area. Most working areas in the pharmaceutical manufacturing environment are not aerodynamically designed and are turbulence-inducing owing to the displacement of containers. Therefore, the selected air velocities may vary. Adequate airflow distribution at the filter face will obtain this result.

The IES-RP-CC006.2 provides a set of recommended practices for the verification of airflow uniformity and airflow parallelism.

The following represents a summary of the procedures and recommendations made in those documents.

Airflow Velocity and Uniformity Test

For reference see ISO 14644-3 section B.4.2

1. Test objective: To determine that the UAD meet the criteria for airflow velocity and uniformity set in the design specifications.
2. Acceptance criteria: The airflow velocity in a unidirectional airflow device should not exceed the limits set in the design criteria. Typically the highest and lowest reading should not be more than 15% to 20% from the unit average velocity.
3. Test procedure
   a. In general the environment is divided by a grid, depending on the configuration of the room and classification of the environment. A 60×60-cm (2×2-ft) grid is typically used. Velocities are measured by using a thermal anemometer, vane-type anemometer, or its equivalent.
   b. Readings are taken at the center of each square. If a single-point-sampling anemometer is used or a one-reading, multiport matrix-type sampler is used, the probe should be located at a distance equivalent to 15 cm or 6 in., from the filter face or before the air encounters an obstacle.
   c. Readings at work height are taken for informational purposes only. Turbulence induced by non aerodynamically designed objects upstream may impede the proper and accurate measurement.
4. Reporting requirements
   a. Report results in the approved forms.
   b. Specify the basis for the acceptance or rejection as a function of the acceptance criteria.

Airflow Parallelism Test

For reference see ISO 14644-3 section B.7

1. Test objective: Verify the parallelism of the airflow throughout the work zone.
2. Acceptance criteria: Air should flow in a parallel stream to prevent the flow of outside air into the critical environment.

3. Test procedure
   a. Generate visible smoke (theatrical smoke or similar) upstream from the work zone.
   b. Establish a reference point by using a plumb bob or similar device.
   c. Generate smoke over the manufacturing environment.
   d. Videotape the direction of the flow in both cases.
   e. Determine the direction of the airflow. Depending on the source of smoke, protection gear may be needed to protect personnel.
4. Reporting requirements
   a. Report results in the approved forms.
   b. Inventory and label the videotapes.
   c. Specify and report the basis for the acceptance or rejection as a function of the acceptance criteria.

Performance Qualification

The final and real challenge for the environmental control system and HVAC system is represented by the process that is to be executed within the areas it is serving. Thus, the PQ tests are to be executed either by simulating the process or by actually conducting it.

In the final analysis, all data collected under static conditions, although good for determining a baseline, do not provide proof of compliance of the system. Therefore, tests should be conducted under dynamic, "in use," conditions either real or simulated as close as possible to the actual process.

Sometimes changes in the process have occurred at the time of validation. This is generally, because of the time span from the design of the system to the actual construction and commissioning. In these cases, the new input is to be analyzed and a determination should be made as to how the changes may impinge on the operation of the environmental control system. If, for instance, more or fewer people are used, this may have an effect on the size and operation of the HVAC system. With more people, the system may not be sufficient to handle the new sensible and latent heat loads, or under another circumstance, the lack of these loads may cause freezing of the cooling coils.

Once a determination is made of the new approved conditions then, if needed, changes to the system are to be executed and revalidated before proceeding to the PQ phase.

As with the OQ, the PQ procedures involve a series of tests and measurements. For this, as indicated, the system should be fully operational.

Temperature-Humidity Control Test

The temperature-humidity control test provides verification of temperature and humidity under dynamic conditions, as well as indicating that the system is capable of maintaining the design conditions. It also provides a good basis for determination of the general status of the system, for its malfunction can be used as diagnostic of the inadequate operation of the HVAC.

Seasonal conditions can affect the temperature and humidity levels. This may vary with the system design and the amount of external non conditioned air supplied to the AHU serving the controlled environment. Occasionally, seasonal variations can be simulated during validation conditions. Clear notations in the validation protocol should be made if additional testing is required.

Testing for these parameters does not require that the environments be clean, but it must be operational. For reference se ISO 14644-3 section B.8:

1. Test objective: To demonstrate the ability of the HVAC system to control temperature and humidity during operating conditions. This test is to be executed while the process or operations are simulated or executed.
2. Acceptance criteria: The system shall be capable of maintaining
   a. The specified temperature range (in general 18–20°C or 60–75°F).
   b. The specified relative humidity range (normally 45% to 55% in aseptic-processing areas, unless otherwise specified by the process requirements).
3. Test procedure
   a. In general, the environment is divided by a grid depending on the configuration and classification of the room. The size of the squares varies from 60×60 cm (2×2 ft) or larger, depending on the parameters for the performance of the operations within a specified environment. Sampling locations are typically at work height and as close to the operators as possible.
   b. In many cases a multiple-point chart recorder is used as this facilitates the data collection and documentation procedures.
4. Reporting requirements
   a. Report results in the approved forms.
   b. Specify the basis for the acceptance or rejection as a function of the acceptance criteria.

Air Cleanliness Test (Nonviable Particle Counting)

For reference see ISO 14644-1.

The concentration of nonviable particles is used as a quick estimate of the overall cleanliness of the environment because of the versatility and fast response offered by the airborne particle counters. Although there is no proved correlation between viable and nonviable particle concentration, the assumption is frequently made that the level of nonviable particles provides a good indication of the overall environmental airborne bioburden.

Airborne concentrations are measured with white light, laser, or condensation nuclei particle counters. Hardware and provided system software vary, depending on the manufacturer, although all of these systems are standardized following the IES and ASTM standards. As indicated, air cleanliness classification for controlled environments has occasionally been based on the concentration of particles of a specific size per unit of volume. The standard currently in use for this characterization is the ISO 14644-1 Cleanrooms and associated controlled environments-Part 1: classification of air cleanliness published by the ISO. Although a nonmandatory standard for pharmaceutical industry, the sampling methods expressed in this standard, as well as the statistical manipulation of the data obtained can be applicable and approvable for the FDA, EMEA and other regulatory agencies around the world.

The acceptance criteria, sampling location, and sample size should be carefully set. The number of samples suggested by the ISO Standard and sample volumes may not reflect the criteria used for a system that is primarily dedicated to avoid microbial contamination. The expectations are that the conditions prevailing during the media fills constitute the real operational standard for the critical environment. Therefore, more than two sampled points are suggested during qualification, preferably in those areas where the containers, stoppers, or closures remain exposed to the environment for a longer time. The principles and methodology suggested in the HACCP and depicted in ISO standard provide a better guidance for the number and location of sampling points.

The statistical handling of the data suggested by the ISO Standard is acceptable for the FDA, as long as no counts exceeding the class level are averaged. When a single count exceeds the set limit for the environment, the FDA considers it a failure.

The use of grids for sample site selection is often employed as an alternative method that provides a systematic approach. Locations can be easily identified and used for other tests, such as the airborne bioburden and surface bioburden assessments. The typical size of the squares varies from 2×2 ft in the critical environments to 5×5 ft in the controlled environments. This approach provides extensive support for statistical determinations. The induction leak test using smoke generation and particle counters with personnel present permit a better identification of potentially critical locations as a function of personnel practices and equipment operation.

1. Test objective: To determine that the complete as-built, operational facility meets the air cleanliness requirements specified in the process functional requirements.
2. Acceptance criteria
   a. *Critical environments:* The particle concentration under dynamic conditions should not be more than 3.5 particles of 0.5 μm and larger per cubic meter (100 particles of 0.5 μm and larger per cubic foot).
   b. *Other environments:* Typically a tenfold gradient is used from critical to less critical environments (i.e., 1000 for environments adjacent to critical environments, 10,000 for those adjacent to it, and so on).
3. Test procedure
   a. In general, the environment is divided by a grid, depending on the configuration of the room and classification of the environment. A 60×60cm (2×2ft) grid is typically used for critical environments.
   b. Larger squares (twofold projection) are used for less critical environments in ascending order. Sampling locations typically are at work height and as close to the operators as possible.
   c. Frequently, a chart recorder attached to the particle counter is used to facilitate data collection and documentation procedures.
4. Reporting requirements
   a. Report results in the approved forms.
   b. Specify the basis for the acceptance or rejection as a function of the acceptance criteria.

Airborne Bioburden Test

For reference see ISO standard ISO 14698 biocontamination control.

As indicated, the achieving of predetermined levels of microbial contamination in aseptic-processing and supporting environments is a primary goal of the environmental control system. The scope of this test is to determine if all the components of this system are capable of performing in accordance with the process functional requirements for airborne bioburden.

An adequate and validated microbiology laboratory is needed to support the requirements of this test. The proper methods should exist for the preparation and preincubation of the culture media to be used. Proper methods should exist for determining the adequate performance of the media by applying growth promotion and sterility evaluation test. Suitable neutralizers should be added to avoid inadequate interaction with disinfectant agents employed in the environments.

This can be a quantitative or qualitative test. Thus, all the sampling devices employed should meet this requirement. Impaction-type airborne microbial air samplers are advisable. These devices are of various types and are characterized by the fact that the air sample is directly impacted over a culture medium at a pre specified flow rate.

Organisms found in samples obtained from critical environments should be quantified and identified. Organisms found in other less critical environments should be quantified, although it is suggested that, at least in the beginning, they be identified to establish the prevalent flora in the environment. As with any microbial-sampling methods, the variability is great. Thus, particular attention should be given to the handling of the sampling devices as well as the receptacles containing the culture media.

The selection of sampling sites is also critical, and special attention should be given to areas near personnel, as well as areas of transfer and connection to less critical environments. The sampling during validation must be extensive because the collection of microorganisms in a such a clean environment is extremely difficult.

Some insight to the level of sampling and methodology required for this can be found in the ISO 14698 Cleanrooms and Associated Controlled Environments—Biocontamination Control Part 1: General Principles and Methods, from the *Guide for Microbial Evaluation of Controlled Environments* recommended practice RP-0023 published by the IES and the HACCP (7) techniques.

The HACCP program was proposed by the US Department of Agriculture and the FDA in the late 1960s. The countries of the EU have adopted this method in their regulatory bodies, and it also appears that it will be part of the proposed ISO TC-209 standard currently under development. HACCP provides a systematic, organized approach to controlling safety hazards in general. The suggested approach can be applied to both airborne and surface microbial contamination control. This is a two-part technique. First, a hazard analysis is implemented to define the conditions that are most likely to cause failure to comply with product quality standards. Second, CCP along the process are identified at which these events may occur.

The overall criteria is to establish controls to reduce, prevent or eliminate the product risk.

This process of implementation of a HACCP system defines eight basic rules as follows:

1. Define and classify the risk to the process (low, medium, high).
2. Define the phases or steps of the process where the product could be at risk.
3. Identify the CCP.
4. Define the environmental levels or limits required to eliminate or reduce risk at the CCP.
5. Define the monitoring approach for every CCP.
6. Define a corrective action plan when the specified limits are reached or exceeded.
7. Establish an adequate record-keeping system to document results.
8. Establish a data evaluation system to determine the status of the system implementation.

The selection of the CCP during validation can provide a congruent and systematic approach for the future routine monitoring of the environment. Importantly, the validation of a controlled environment will be seriously compromised if there is a lack of data demonstrating the overall condition of the environment. This takes on special significance in light of the fact that some documents, such as the recent *FDA Guideline for Aseptic Processing*. This document indicates that sampling at the filling point during aseptic processing is acceptable. Although this may represent the most critical point in the process, it may also represent the point with the least potential for cross-contamination (i.e., containers on unscramble tables become exposed to the environment for long periods and, many times, there is direct human intervention to correct problems). Thus, these areas represent a point of higher risk.

1. Test objective: To determine that the complete as-built, operational facility can meet the air bioburden requirements specified in the process functional requirements.
2. Acceptance criteria
   a. *Critical environments:* no more than 1 cfu/m^3, or 0.03 cfu/ft^3 (*USP* monograph).
   b. *Typical for other environments:* Adjacent to critical environments: 5/m^3 or 0.15/ft^3.
   c. Controlled environments: 87/m^3, or 2.5/ft^3.
3. Test procedure
   a. In general, the environment is divided by a grid depending on the configuration of the room and classification of the environment. A 60×60-cm (2×2-ft) grid is typically used for critical environments.
   b. The larger squares (twofold projection) are used for less critical environments in ascending order.
   c. Sampling locations typically are at work height and are as close to the operators as possible.
4. Reporting requirements
   a. Report results in the approved forms.
   b. Specify the basis for the acceptance or rejection as a function of the acceptance criteria.

Surface Bioburden Test

The final evaluation of the cleaning and disinfection procedures is done by determining the bioburden content on the surfaces either in the facility, or on particular equipment and devices. The use of adequate cleaning and disinfection procedures is fundamental to achieve a pre specified level of environmental quality. As part of validation, it is important to determine the effectiveness of these procedures at the same time the rest of the environment is being qualified. Assessing the surface bioburden in pre cleaning conditions, as a challenge, and testing after cleaning and disinfection in a repetitive fashion, it is possible to establish the ability of these methods to perform.

The use of grids is also used. Squares are sized as a function of the environment classification. The selection of grids to be sampled can be also based on a risk using the HACCP approach. The methods used for sampling include swabbing and the use of contact plates, such as RODAC.

The same microbiology laboratory-supporting requirements indicated for airborne bioburden testing apply for surface bioburden testing.

1. Test objective: To determine that the complete as-built, operational facility can meet the surface bioburden requirements specified in the process functional requirements.
2. Acceptance criteria
   a. *Critical environments:* no more than 1 cfu/12.9 cm^2 or 2 in^2 (FDA *Aseptic Processing Guidelines).*
   b. *Typical for other environments:* adjacent to critical environments: 5/12.9 cm^2 or 2 in^2.
   c. *Controlled environments:* 20/12.9 cm^2 or 2 in^2
3. Test procedure
   a. In general the environment is divided by a grid, depending on the configuration of the room and classification of the environment. A 60×60-cm (2×2-ft) grid is typically used for critical environments.
   b. The larger squares (twofold projection) are used for less critical environments in ascending order. Sampling locations typically are those in contact with the sterile product or surfaces upstream from the product. Standards for floors may vary with the room classification. Air curtains are considered critical surfaces when located above the critical environments. Samples should be taken in areas difficult to clean and as close to the operators as possible
4. Reporting requirements
   a. Report results in the approved forms.
   b. Specify the basis for the acceptance or rejection as a function of the acceptance criteria.
5. System Qualification Statement and Qualification

SUMMARY

After conclusion of all the verifications, tests, and challenges indicated and after complying with the approved specifications and parameters, the system can be considered validated.

A final report or summary should be prepared for every section and for the overall protocol. Deviations should be noted and explained if approved, the rationale should be included in the final protocol summaries.

The documents should be verified for completeness, accuracy, and compliance with cGMP requirements before its final approval by the members of the Validation Task Force.

Validation of the environmental control system should serve as a basis for not only approval and commissioning of all the components of the system, but also to establish the basis for routine monitoring.

The validity of the validation program is often determined by the proper organization of the documentation, the selection of the individuals performing the critical tasks, the overall analysis of the results obtained during the execution of the protocols, and the proper adherence to cGMP practices. Thus, proper attention should be given to all of these details to assure the acceptability of the program by those in Quality Assurance and the respective regulatory agencies.

REFERENCES

1. International Organization for Standardization. ISO 14644-1 Part 1: classification of air cleanliness; ISO 14644-2: specifications for testing and monitoring to prove continued compliance with ISO 14644-1; ISO/FDIS 14644-3 Part 3: test methods; ISO 14644-4 Part 4: design construction and start-up; ISO 14698-1 Biocontamination control, Part 1: general principles and methods. Institute of Environmental Sciences and Technology (IEST), www.iest.org
2. U.S. Pharmacopeia. Microbiological evaluation of clean rooms and other controlled environments, 1995. Pharmacopeia Forum 21 <1116>. Rockville, MD: United States Pharmacopeial Convention, 1995.
3. American Society of Heating, Refrigerating and Air Conditioning Engineers. Gravimetric and dust spot procedures for testing air cleaning devices used in general ventilation for removing particulate matter, Standard 52.1-1992; Method of testing general ventilation air-cleaning devices for removal efficiency by particle size, Standard 52.2-1999.
4. Pharmaceutical Inspection Convention. Isolators used for aseptic processing and sterility testing. PIC/S Secretariat P.O. Box 5695 CH-1211 Geneva 11, http://www.picscheme.org
5. European Standard EN 1822-1. High efficiency air filters (HEPA and ULPA) Part 1: classification performance testing, marking. Central Secretariat, Rue de Stassart, 36, B-1050 Brussels.
6. IESNA Lighting Handbook.
7. Pierson MD, Corlett DA, Jr. HACCP: Principles and Application. New York: Van Nostrand Rein-hold, 1992.

BIBLIOGRAPHY

Agnew B. Laminar Flow-Clean Room Handbook. Los Angeles: Angew-Higgins, 1965.

American Conference of Governmental Industrial Hygienist. Industrial Ventilation. Ann Arbor, MI: Edwards Brothers, 1980.

American Society of Heating, Refrigerating and Air Conditioning Engineers (ASHRAE). Guide and Data Book. New York: ASHRAE, www.ashrae.org

Associazione Farmaceutici dell'Industria. Buona Fabbricazione. Milan, 1983.

Banker GS, Rhodes CT. Modern Pharmaceuticals. New York: Marcel Dekker, 1990.

Department of Defense. Military specification filter, particulate high efficiency, fire resistance. MIL-F-51068C, June 8, 1970.

DeVecchi F. Training Personnel to Work in Sterile Environments. Los Angeles: Pharmaceutical Technology, 1978.

DeVecchi F. Clean room analysis. In: Pharmacy and Technology Conference Proceedings, 1981.

DeVecchi F. Air systems validation. Center for Professional Advancement, Course Notes, 1982.

Dimmick RL, Wolochow H. Problems of measuring numbers of microbes in occupied spaces. In: Institute of Environmental Sciences Proceedings, 1981.

Federal Standard 209-D. Clean Room and Work-Station Requirements. Washington, DC: GBA, 1966.

Food and Drug Administration. Current good manufacturing practice in manufacture, processing, packing or holding. Fed Reg 1976; February 13.

Fuscaldo AA, Erlick BJ, Hindman B. Laboratory Safety. New York: Academic Press, 1980.

Hansonn J. Managing Indoor Air Quality. Lilburn, GA: Fairmont Press, 1991.

Heinsohn RJ. Industrial Ventilation: Engineering Principles. New York: Wiley, 1991.

McCrone W, Braftz R, Delly JG. The Particle Analyst. Ann Arbor, MI: Ann Arbor Science Publishers, 1968.

Meckler M. Indoor Air Quality Design Guideboo. Lilburn, GA: Fairmont Press, 1991.

NASA. Symposium on Clean Room Technology in Surgery Suites. St Louis: Midwest Research Institute, 1971.

NASA. Handbook for Biological Engineers. Washington, DC: Technical Information Services.

U.S. Air Force. Technical Order 00-25-203: Standard and Guidelines for Design and Operation of Clean Rooms and Clean Work Stations. Washington, DC: Technical Information Services, 1963.

U.S. Department of Health and Human Services. Safety in Microbiological and Biomedical Laboratories. Washington, DC: U.S. Government Printing Office, 1988.

5

Validation of Critical Utilities

David W. Maynard
Maynard & Associates, Ewing, New Jersey, U.S.A.

INTRODUCTION

This chapter will review the utilities used in clinical trial and production facilities, whether these facilities are finished drug, APIs, whether the products are biological, solids, liquids, creams, ointment or sterile products. It is important to understand the methods needed to first determine what types of utilities are critical to the processes and second those on the fringe or outside the domain of qualification. This effort will focus on the critical utilities but provide guidance on the noncritical utilities.

This chapter will assist the pharmaceutical engineer, clinical trial scientist, quality assurance professional, qualification engineer, etc., and the management team at the facility in determining which system needs to be addressed and how these systems will be commissioned, qualified and in some cases validated. It will suggest a team approach that is established to advance the project to its swift and successful conclusion.

Further, the decisions that need to be made in first ascertaining which utilities need to be commissioned or qualified and how this commissioning/qualification of utilities is to be accomplished are best made with a cross-functional team that will be assembled to address the project. This is sometimes termed a risk assessment. This team may be augmented by outside contractors or consultants and will be discussed later in the chapter.

The chapter will first establish some definitions that are important to the process and then use these definitions throughout the document to explain the why and how of the qualification of utilities.

TERMS USED IN THIS CHAPTER

Noncritical (No Impact). Utility that has no impact on process or product quality (i.e., water—source for cleaning/sanitizing of non-product contact surfaces; steam—used for heating of vessels).

Noncritical Point of Use. No direct impact on quality of product/process for which it is being used.

Support (Indirect Impact). Utility that supports a Direct Impact utility but does not have a direct impact on the quality of a product.

Critical (Direct Impact). Utility that is in direct contact with the product or that could have a direct impact on the quality of the product.

Critical Point of Use. Direct impact on quality of product/process for which it is being used (i.e., water—used for cleaning of surfaces with direct contact to product, used in formulation processes, used in supply to pure steam generator; steam—used in sterilization processes).

Critical Process Parameter. A process parameter that is controlled within a predetermined range to ensure product meets its CQA.

Critical Quality Attributes. A set of measured characteristics inherent in the product that describes the products acceptability for use.

Commissioning. A well-planned, documented, and managed engineering approach of inspection and testing of equipment and systems to ensure they are installed according to specifications and are ready for operation in a safe and functional environment that meets established design requirements or qualification when required (1).

Installation Qualification. Documented evidence that the equipment, system or utility meets all critical installation requirements (2,3).

Operational Qualification. Documented evidence that the equipment, system or utility operates as intended throughout all required ranges.

Performance Qualification. Documented evidence that the equipment, system or utility perform as intended and meets all preestablished acceptance criteria.

Sampling Plan. Written procedure describing the physical location of sample points, the frequency of samples taken to ensure system is in control, and the equipment to be used in taking the sample.

PLANNING ACTIVITIES FOR THE CRITICAL UTILITY

The first step is to list all of the utilities at the facility or site and determine the criticality of the system. This can be accomplished by performing impact assessment that presents the risks to product posed by the utility. It can also be determined by following a series of questions that continue to refine the analysis until it is clear which path needs to be followed—no commissioning, commissioning only or commissioning and qualification required.

Abbreviations used in this chapter: APIs, active pharmaceutical ingredients; ASME, American Society of Mechanical Engineers; CPP, critical process parameter; CQA, critical quality attribute; GMP, good manufacturing practice; I/O, input/output; IQ, installation qualification; ISO, International Organization for Standardization; NF/EP, National Formulary/European Pharmacopoeia; NLT, no less than; NMT, no more than; OQ, operational qualification; P&ID, process and instrumentation drawings; PLC, programmable logic controller; PQ, performance qualification; SOPs, standard operating procedures; USP, *United States Pharmacopeia*; WFI, water for injection.

An example of the questions that can be used to determine if a system needs commissioning and/or qualification is as follows:

Is the utility supporting GMP activity? If "No" then there is probably no need to even Commission the system, but if "Yes" then it needs more clarification as in the next question.

Does the utility or direct output come in direct contact or primary packaging contact? If "Yes" then the system must be Commissioned and Qualified. If "No" then it needs more clarification as in the next question.

Is the direct output of the utility used in the environment surrounding an exposed product? If "Yes" then the system must be Commissioned and Qualified. If "No" then it needs more clarification as in the next question.

Is the utility or output used in final cleaning steps (equipment with direct product contact or the primary packaging components)? If "Yes" then the system must be Commissioned and Qualified. If "No" then it needs more clarification as in the next question.

Are the utility and its direct output used within a sterilization or sanitation process? (4) If "Yes" then the system must be Commissioned and Qualified. If "No" then it needs more clarification as in the next question.

Does the operation or control of the utility have a direct impact on the CQA of the product or the CPP of the production systems? If "Yes" then the system must be Commissioned and Qualified. If "No" then the system needs only to be commissioned.

The second step in any qualification or validation process is to develop a plan of what is to be accomplished. This can be a complex qualification plan that addresses many different utilities in multiple areas of the site or could be a specific qualification strategy or plan that addresses only one specific utility. Each of the plans will include how the system will be commissioned or qualified, who will perform the effort, what type of protocol is to be used and what approval signatures are required. If the commissioning documents are to be leveraged into the qualification, the copies of those documents must be integrated into the qualification documents prior to the approval of the qualification protocol.

The plans should be developed by a cross-functional team consisting of Engineering, Operations, Quality Control, Quality Assurance and the Commissioning or Qualification personnel. Each of these participants will have their own roles and responsibilities that are important to the outcome of the project. This group allows all GMP functions to participate early in the project and help assure a satisfactory outcome to the testing.

The protocols will include the system limits, the physical parameters and attributes to be tested, the acceptance criteria to be met and the signatories required to approve or certify the qualification and validation actions. The reports will include the synopsis of the testing and verify the acceptance criteria have been met.

Third step is the actual commissioning, qualification and validation of the individual utility. These activities will verify the design, installation and operation of the equipment or systems. As part of the qualification and validation activities, the CPPs that have been established for the utility will be verified to ensure the ongoing control and certification of the systems in the daily activities of the facility. Prior to performing any validation testing the sampling routines for the utility must be established in order to ensure the validation activities will be the same as those used in routine operation. Note: one of the most frequent comments by regulatory agencies occurs when the validation sampling does not accurately reflect the routine use of that particular point of use, including flushing times and methods of use (i.e., if a hose is used between the point of use, then the sampling should be from the hose and not directly from the point of use). The validation will test the Critical points of use of the system and set in place the routine monitoring of these points.

SYSTEMS TO BE DISCUSSED IN THIS CHAPTER

The utilities in typical facilities include gases (compressed air, nitrogen, oxygen, and carbon dioxide); liquids (process water and solvents); steam (process and clean); house vacuum; electrical and drains (process and waste). There may be other utilities encountered within the facility and the same or similar validation processes can be adapted to the other systems.

Gases

The most common gases used in pharmaceutical industry are *compressed air* used for instruments or product contact, and *nitrogen* used for providing an inert gas in the vial, ampoule or WFI tanks and used for creating an inert pressure pad in processes where solvents are present.

The validation of each of these is similar in that the equipment used to generate, store and distribute the gas must be first Commissioned and then qualified. Once the IQ and OQ Summary Reports for the equipment have been approved, then the distribution system can undergo Qualification to ensure the delivery of the gas to the acceptance criteria established that verifies the specifications for the gas. It is noted that instrument air need only be tested through the OQ, as it does not come into product contact.

Some of the instruments used to gather compressed gas samples include SAS Microbiological Air Sampler, Mattson Garvin Compressed Gas Sampler—Model P-320, and SMA Compressed Air Sampler.

Liquids

The most prevalent liquid utilities used are *process waters* (e.g., soft, deionized, USP purified, WFI) that are used in cleaning operations and product batching, and water used in the heating and cooling processes that are not in product contact and thus does not require anything more than commissioning of the system.

The validation of each of these is similar in that the equipment used to generate and store the water must be first Commissioned and then Qualified (5). Once the IQ and OQ Summary Reports for the equipment have been approved, then the distribution system can undergo qualification to ensure the delivery of the water to the acceptance criteria established that verifies the specifications for the water. USP Purified Water and the WFI need to have an extensive testing of all points of use over a one-month period to ensure the quality of water is delivered routinely. The points of use tested during the PQ will typically be those used for routine sampling. In addition, an extended Qualification is executed wherein the distribution system is monitored over a one-year period (including the 30-day period included in the initial PQ) to ascertain any seasonal differences that impact the quality of the water.

Steam

Typically there are three types of steam used in our industry—*plant steam* that typically has boiler chemicals entrained; *chemical free steam*, which is without boiler additives; and *pure steam* that, when condensed, meets the water requirements of USP purified water. Plant steam and chemical free steam distribution system validations are typically completed at the end of the OQ. Each of the systems must be of appropriate design (including steam trap location), be properly maintained and operated using approved procedures.

Pure steam is typically produced by specially built steam generators or from the first effect of a multiple effect still. The feed water is typically purified water or WFI or from other sources of known chemical quality as specified by the vendor of the generator. The steam contact surfaces, including the generator and distribution system, must be corrosion-resistant material (316 L stainless steel is the most common material used).

A properly designed and constructed steam generation and distribution system that is operated and maintained correctly will negate the concern expressed by some regulatory agencies. The three physical attributes of noncondensable gases (dryness, fraction and superheat) appear to have been concerns in older installations, including hospitals wherein the systems may not have been properly designed, installed or maintained. However, those pharmaceutical facilities that use pure steam for product sterilization purposes should consider the performing of these tests as they may be required by regulatory agencies. These steam quality tests are detailed in U.K. Department of Health and Social Security document, Health and Technical Memorandum 2010 part 3 (6) and in ISO 11134, "Sterilization of Healthcare Products—Requirements for Validation and Routine Control—Industrial Moist Heat Sterilization" (7,8). The steam quality limits are also included in the European standard EN 285 (9).

The qualification of each of these is similar in that the equipment used to generate the steam must be first Commissioned and then Qualified (Note: Plant steam does not require more than an IQ) to ensure the equipment operates properly. Pure steam will undergo extensive PQ testing beyond the OQ to show the quality of steam is maintained over an extended period of time, typically one month.

House Vacuum

House vacuum systems are used for many services, but the distribution systems that come into contact with the product or the primary container require attention. These systems will require that the equipment used to generate the vacuum must be first Commissioned and then Qualified. Once the IQ and OQ Summary Reports for the equipment have been approved, then the reservoir tank and distribution system can undergo Qualification.

Electrical

The electrical systems are often overlooked in the verification/certification of the facility and utilities. These systems require similar commissioning and qualification activities are performed to ensure the continued deliverance of the power to operate the other utilities and the facility processes. Each of the electrical systems in the facility has specific requirements that will include both quality and quantity attributes that are established by the requirements of the facility. These attributes will include frequency, phase and voltage requirements as well as sufficient capacity to enable the full load required to operate the facility. Other systems may include battery backup, standby generator capacity, clean lines for computer operation, and voltage surge protection.

The electrical system to be tested requires complete documentation including monitoring system identification, electrical schematics that include all pertinent information including wire size, circuit identification, switching equipment and backup systems. There should be written instructions on operation and maintenance of the systems as well as emergency procedures that will come into effect in case of natural or man-made disasters.

The qualification will consist of monitoring the systems to ensure voltage, phase and load conditions can be maintained while the plant is being started and maintained during production usage. The portions of the facility that does not have backup or emergency power generation systems will need to be tested after the loss of supplied electrical power to ensure the facility can come back on line safely, both from a temporary loss of power and from a sustained loss of power. Procedures should be put in place to cover these situations, as typically there will be a need to have a sequential restart to a facility to ensure the safe operation of the plant and systems.

In areas that are subject to "brown out" or reduced voltage conditions, the electrical supplies to primary process equipment will need to be verified that the equipment can operate in this reduced energy level and continue to meet all performance attributes established.

All alarm monitoring and display systems as well as where there is emergency power equipment, the production facility should be tested using both the primary and secondary power systems. The switchover capability and operation of the equipment must be tested to ensure the smooth transition between the power sources and to verify the operation of the equipment on the backup power system. Computer equipment must be fully tested to ensure no loss of data during a transition from one power source to another.

All protection equipments (overload, safety switches, voltage stabilizers, line suppressors) should be tested for both normal operation and peak-load or worst-case conditions.

Drains

The drains in facilities are often overlooked and sometimes are the source of unanticipated problems (i.e., contamination, backflow, means of causing flooding from external storm sewer systems) and as such require full understanding of there design and connections. For this discussion consider only the drains to process and sanitary. The sanitary drains remove various wastes from the areas whereas process drains remove process specific fluids. The design and construction of the facility will need to be verified to ensure there is not an interconnection between the two systems. A simple dye test where dye is placed into the process drain and shown that it does not appear in the sanitary waste exit from the facility will verify the systems are not interconnected. There is a need for complete IQ and OQ of drains in process areas to ensure that all drain points are interconnected and drain to process waste.

The specific qualification will be similar to other distribution systems and the drawings, material of construction, pipe size, valves, leak testing, safety features, etc., must be verified. Hard connections between water/steam systems must be avoided and an air break must be verified to ensure there is no back siphoning. The use of check valves between process systems should be tested to ensure the systems remain separated. Where solvents are used the drainage system will be verified to be explosion proof and properly vented.

TESTING

Common Steps in Commissioning, Qualification and Validation of Utilities

All of the utility systems during the various stages will have a common series of evaluations and tests that ensure the installed equipment and systems meet the required specifications and design elements that assist in ensuring the long-term operation of the utility. These will include the verification of the materials of construction to the design specifications and engineering/construction drawings; the verification of the welding and of the welder performing the welding of the generation and storage systems by the use of video boroscope to verify the individual welds and the use of test coupons that certify the welder as being competent of performing the welding operations; the cleaning and, if required the passi vation of the generation, storage and distribution systems to remove the residual fragments of material, welding material, oils and greases used in the construction of the metallic systems and also the removal of other materials in the system; pressure testing or pressure hold tests to ensure the systems are integral and meet the various code requirements including ASME, etc.; various safety tests to ensure the systems are protected from over pressure or over temperature conditions. Each of these individual verifications will depend on specific procedures and test equipment that are described in other chapters.

In order to have a traceable inventory of completion of these tasks, each of the above will require written and approved documents that certify the veracity and accuracy of the tests that were performed. These tests can be performed in-house or by qualified engineering, consulting or testing organizations that specialize in these activities. The required testing activities will be compiled in preapproved protocols that will list all of the tests along with their acceptable results or acceptance criteria. The completed protocol will be executed and checked by the performing organization and then approved or certified by the Quality organization.

The documentation will be organized in files that can be retained in a site library or depository of information along with the Commissioning, IQ, or OQ documents.

General Tests for All Utilities

Installation Qualification

1. Verification of qualification prerequisites including successful completion of all Commissioning activities.
2. Verification of system documentation that could include technical data sheets, functional specification requirements, material of construction, welding documentation, piping insulation documentation, cleaning and passivation reports, and pressure testing reports. If these were included in the Commissioning activities, reference copies may be attached or the section of the commissioning report referenced.
3. When sanitary piping is used, a separate welding documentation will be required to indicate the weld number, weld log, welder certification, piping isometric verification will be verified.
4. Verification of preventive maintenance documentation and spare parts lists including verification of proper entry into a computer-controlled preventive maintenance management system if used at the facility.
5. Verification that the P&IDs, wiring and cabling drawings are accurate and reflect the installed system. These drawings are commonly referred to as the "as-built" drawings.
6. Verify that all generation and distribution components and piping are properly identified.
7. Verification of all components in the system is in compliance with the specifications and design.
8. Verification of all instruments that are critical to the operation of the system is calibrated and is properly entered into the computer-controlled calibration management system if used at the facility.
9. Verification of all supporting utilities that are properly installed and are in compliance with requirements of the system.
10. Verification of all the unit operations that are operating properly and meet all specification and testing requirements.
11. Verification of proper software version electronic copies that are available for backup capability, PLC source logic is complete and clear, there is no dead code in the system and that hard copies of the control logic are available and made part of the protocol.
12. I/O verification is performed to ensure that all I/O points were addressed and properly connected to field devices per specification.

Operational Qualification

1. Verify that all OQ prerequisites are complete in that all IQ sections are completed or that any un-executed sections would not impact the execution activities of the OQ.
2. Verify that all generation and distribution components and piping are properly identified.
3. SOPs are in place and approved and all required individuals training is documented and available.

4. Prepare list of all SOPs, including reference number, title and effective date related to the generation or distribution systems.
5. Obtain copies of all current SOPs.
6. Challenge all Operation SOPs to verify the document is suitable and allows the system to operate properly.
7. Challenge all Preventive Maintenance SOPs to verify the document is suitable and allows the system to be properly maintained.
8. Obtain copies of all current Preventive Maintenance documentation including work orders or evidence of maintenance having been performed in the proper frequency.
9. Verify that all instruments that are critical to the operation of the system are in current calibration and calibration certification available.
10. Verify that all testing instruments are in current calibration and the calibration certificates available.
11. Verify that environmental conditions surrounding the system do not adversely affect the operation of the generation and distribution systems.
12. Verify any environmental conditions surrounding a PLC driven system to not impact the operation of the controller.
13. Radio frequency and electromagnetic interference tests are performed close to the control panels to verify the system is not adversely affected by these disturbances.
14. Power failure and recovery tests are performed to document the effects of these events on the control of the system.
15. Alarm and interlocks are tested to verify the proper operation of the system.
16. Software security access levels are verified to ensure system cannot be modified without specific authorization.
17. Sequence of operations are verified to challenge the operational sequence of the control system to assure that the systems functions are followed as described in specifications.

Performance Qualification

1. General description of the system to be tested including specific information in sufficient detail to create a verbal picture of the system and its component parts as well as its location in the facility and any special requirements of the system.
2. Develop the specific CPPs to be reviewed or certified as part of this phase of qualification.
3. Verify that all PQ prerequisites are complete in that all OQ sections are completed and reviewed, documents are completed and that any un-executed sections would not impact the execution activities of the PQ.
4. Verify that all sampling procedures are current and reflect the testing that is to be accomplished in the PQ.
5. Verify that all personnel that will be executing the PQ, or involved with the testing of samples, have been trained in the expectations of the protocol and the need to accomplish the testing in the proper sequence and in the correct time period.
6. Verify that all generation and distribution sampling points are properly identified and are listed in the current sampling procedure.

Extended Qualification

The high purity water systems and pure steam systems will require testing at the sampling ports for one full year to indicate the ability of systems to maintain proper quality throughout one complete cycle of the seasons.

Specific Tests—Gases

In addition to the general tests, all gas systems CPPs, including pressure, flow rate, and capacity must be verified during the OQ. Specific tests must be established to ensure the systems are delivering the gas at the required conditions.

Compressed Air

The typical compressed air system consists of an oil-free air compressor that compresses the ambient air and an air dryer connected to an air receiver surge tank that supplies air throughout the distribution system. Where a non-oil-free air compressor is used, the compressed air passes through a coalescing filter and oil vapor absorber filter where the hydrocarbons are removed then passes into the surge tank and distribution system.

1. All use points will be sampled for a minimum of seven consecutive days including the compressed air exit point from the air compressor.
2. Testing will be in accordance with documented procedures.
3. Environmental tests (microbiological and nonviable particulate) will be taken for information only for non-sterile use.
4. Environmental tests for sterile use must meet the criteria for the area of use.
5. Suggested acceptance criteria for the compressed air generation system include:

Parameter	Parameter specifications
Compressor oil temperature	<123°F reference only (based on specific system)
Cooling tower differential pressure	<7 psi (based on specific system)
Maximum discharge pressure	Reference only
Maximum dew point	Reference only

6. Suggested acceptance criteria for the Compressed Air distribution system at product point of use (using a 0.22-μm in-line filter in controlled areas) include:

Parameter	Parameter specifications (10,11)
Water and oil	None detected (10)
Oil	0.1 mg/m^3 (11)
Odor	No smell (10,11)
Atmospheric dew point	<50°F (20 mg/m^3)
Microbiology for total count	Reference only (no growth for aerobic organisms for sterile)
Particulate (nonviable) (total count of particles <0.5 μm per cubic foot)	Reference only (meet area quality standards)
Carbon monoxide	NMT 10 ppm; 5 ppm (11)
Carbon dioxide	NMT 50 ppm (10); 500 ppm (11)
Nitric oxide/nitrogen dioxide	NMT 2.5 ppm (10); 2 ppm (11)
Sulfur dioxide	NMT 5 ppm (10); 1 ppm (11)

Nitrogen

The typical nitrogen system consists of a generation plant, liquid nitrogen backup tank and distribution header. The basis of operation is for ambient air to be treated and separated creating NF/EP quality nitrogen that is delivered to the distribution header. In an event of high demand peaks or when the generation plant is shutdown for maintenance activities, the liquid nitrogen backup tank will provide gaseous nitrogen to the distribution header.

1. All use points will be sampled for a minimum of seven consecutive days including the nitrogen exit point from the generation plant.
2. Flush sample lines through the stainless steel sampling collector to ensure all ambient air is removed from the container prior to collecting test sample.
3. Testing will be in accordance with documented procedures.
4. Environmental tests (microbiological and nonviable particulate) will be taken for information only for nonsterile use.
5. Environmental tests for sterile use must meet the criteria for the area of use.
6. Suggested acceptance criteria include:

Parameter	Parameter specifications (10,11)
Identification match test	Extinguish in absence of oxygen
Purity (nitrogen assay)	>99.00% (10); >99.50% (11)
Carbon monoxide	≤0.001% (10); ≤5 ppm (11)
Carbon dioxide (11)	≤300 ppm
Oxygen	<1% (10); ≤50 ppm (11)
Water (11)	≤67 ppm (11)
Oil	Not discussed (10,11)
Microbiology for total count	Reference only (no growth for aerobic and anaerobic for sterile)
Particulate (nonviable)	Reference only (meet area standards)

Specific Tests—Liquids

In addition to the general tests, all water systems CPPs, including temperature, pressure, flow rate, and capacity must be verified during the OQ. Specific tests must be established to ensure the systems are delivering the water at the specified rates to minimize the buildup of contamination in the piping systems.

Water Pretreatment

The water pretreatment system is designed based on the local water conditions and the intended use of the water. Typically the system consists of units dedicated to specific functions including removal of suspended solids, chlorination removal, hardness and metal removal, removal of microbiological impurities and control of microbial growth. Each of these has their specific tests based on the end use of the water.

Suggested tests for deionized water with the acceptance criteria include:

Parameter	Parameter specifications (10)
Microbiology for total count	<500 CFU/mL
Microbiology	Absence of coliforms
Conductivity	≤1.3 µS/cm @ 25°C

Purified Water

1. All use points will be sampled for a minimum of seven consecutive days including the water exit point from the generation plant.
2. Testing will be in accordance with documented procedures.
3. Suggested acceptance criteria include:

Parameter	Parameter specifications (10)
Microbiology for total count	≤100 CFU/mL
Microbiology	Absence of coliforms
Microbiology	Absence of *Pseudomonas aeruginosa*
Microbiology	Absence of *Burkholderia cepacia*
Endotoxin	≤0.25 EU/mL
Conductivity	≤1.3 µS/cm @ 25°C USP <645>
Total organic carbon	0.5 ppm (500 ppb) maximum USP <643>

Water for Injection

1. All use points will be sampled for a minimum of 30 consecutive days including the water exit point from the generation plant to release the system for use. The testing will continue for an additional 11 months to verify the system is capable of maintaining specification conditions over seasonal changes.
2. Testing will be in accordance with documented procedures.
3. Suggested acceptance criteria include:

Parameter	Parameter specifications (10)
Microbiology for total count	≤10 CFU/mL
Microbiology	Absence of gram-negative rods
Endotoxin	≤0.25 EU/Ml
Conductivity	≤1.3 µS/cm @ 25°C, USP <645>
Total organic carbon	0.5 ppm (500 ppb) maximum, USP <643>

Specific Tests—Pure Steam

1. All use points will be sampled for a minimum of seven consecutive days including the water exit point from the generation plant to release the system for use.
2. Testing will be in accordance with documented procedures.
3. Acceptance criteria for pure steam condensate must meet the WFI quality and include:

Parameter	Parameter specifications (10)
Microbiology for total count	≤10 CFU/mL
Microbiology	Absence of gram-negative rods
Endotoxin	≤0.25 EU/mL
Conductivity	≤1.3 µS/cm @ 25°C, USP <645>
Total organic carbon	0.5 ppm (500 ppb) maximum, USP <643>
Noncondensable gasses	NMT 3.5% expressed in terms of mL of gas per 100 mL of condensate, EN 285
Dryness fraction	NLT 0.90, EN 285
Superheat	NMT 25°C, EN 285

Specific Tests—Vacuum

1. Testing will be in accordance with documented procedures.
2. Suggested acceptance criteria include:

Parameter	Parameter specifications
Pressure hold test	≤2 in. Hg in 30-minute period for a tightly sealed system with minimum use points
Maximum vacuum pressure	Reference only

CONCLUSION

This chapter has outlined methods to ensure that critical utilities, equipment and systems meet specifications. The specifications shown in the tables above are combinations of industry standards in conjunction with compendial (USP, ISO) specifications and guidance. The tables give a starting point for discussions within the individual corporate organizations. Each corporation must establish their acceptance criteria for the utilities.

In order to maintain the qualified and validated state requires ongoing awareness of changes to the equipment/systems including accumulative effects of small changes made to enhance performance. All changes to qualified and validated systems must be reviewed to ensure changes do not compromise the qualification and/or validation.

Periodic verification should include a documented review of history of equipment/systems since the qualification or validation exercise. This review should include changes made by preventive maintenance, calibration, equipment/system modifications. The history review should be discussed with and approved by the same functional areas that were involved with the qualification/validation, including the quality organization.

REFERENCES

1. Pharmaceutical Engineering Guides for New and Renovated Facilities. Vol. 5: Commissioning and Qualification. ISPE Baseline Pharmaceutical Engineering Guide. 1st ed. International Society for Pharmaceutical Engineering, March 2001.
2. United States Code of Federal Regulations, Food and Drug Administration. 21 CFR: Parts 210 and 211—Current Good Manufacturing Practice in Manufacturing, Processing, Packing, or Holding of Drugs, 1978.
3. United States Code of Federal Regulations, Food and Drug Administration. 21 CFR: Part 820—Good Manufacturing Practice Regulations for Medical Devices, 1996.
4. Parenteral Drug Association, Technical Report Number 1.
5. Pharmaceutical Engineering Guides for New and Renovated Facilities. Vol. 4: Water and Steam Systems. ISPE Baseline Pharmaceutical Engineering Guide. 1st ed. International Society for Pharmaceutical Engineering, January 2001.
6. Health Technical Memorandum 2010 (HTM 2010). Part 3: Validation and Verification Sterilization.
7. International Standard ISO 11134. Sterilization of health care products—Requirements for validation and routine control—Industrial moist heat sterilization.
8. International Standard ISO 17665. Sterilization of health care products—Moist heat—Development, validation and routine control of a sterilization process for medical devices.
9. European Standard: Sterilization—Steam sterilizers—Large sterilizers (EN 285).
10. United States Pharmacopeia (USP 29)—National Formulary (NF 24), 2006.
11. European Pharmacopoeia, January 2005.

6

The Validation of Pharmaceutical Water Systems

Theodore H. Meltzer
Capitola Consultancy, Bethesda, Maryland, U.S.A.

BACKGROUND

The need for process validation may seem obvious to the technically oriented. Validation is the essential "proof of the pudding"; an attestation that the process in question can be depended upon to consistently produce a drug product of stipulated quality and purity. The scientific exercise focuses upon the surety of drug identity and the reliability of its manufacture. Dependability, reliability, consistency, certainty and other virtues as well are seen as the accompaniments of the validation practice. Given what now seems so obvious a requirement, it may be puzzling to some that it took years to develop the concept as posed by the FDA and as modified and accepted in concert with the pharmaceutical industry. The history of this development has been detailed most helpfully (1).

The FDA's focal point was and is the high level of safety and quality desired for drug products. The requirement for process validation was established on the proposition that "quality cannot be tested into a product." In effect, it was premised that a process of product selection based on culling the acceptable from the unacceptable by analytical testing was unsuited to the need for certainty in production. A sufficient statistical reliability could not practically be attained by sampling. Therefore, what the FDA aspired to was the "built-in quality" of a drug preparation that would be the invariant result of the process designed to produce same.

The inherent assumption was that processing systems could be so designed. The thinking was, and remains, that defining by experimental inquiry the proper operation of each purification unit, followed by the experimentally verified performance of their assembly into a holistic system would inevitably lead to a product that would be invariant and high in its desired qualities. The logic held that, experimentally determined and defined, each individual operation would be documented so appropriately as to make the necessary processing manipulations plainly evident. They could then be flawlessly repeated, even by different operators, in accord with SOPs.

The advocacy of process validation had and has its critics. Presenting an action plan less vulnerable to criticism is not always possible. It should be recognized that regulatory authorities are often called upon to make decisions based on the incomplete knowledge of the time. Most would agree that the stated prerequisite of the FDA that processes used in preparing drug products are to be validated is sufficient reason to comply with the requirement.

Validation, while straightforward in its goals, is, however, a concept of some complexity, given its many aspects. A frequently asked question concerning validation may explain the high interest associated with it, namely: What does the FDA require of the validation practice? The question invites different interpretations, gives rise to various opinions, and to much advice and many recommendations. Indeed, written expositions, lectures and presentations, and the number of consultancies dedicated to its answer have come to constitute a robust cottage industry. It is in the pursuit of surety in meeting the agency's expectations that this question is so frequently asked. The serious manner in which the implementation of validation is sought is a compliment both to the FDA and to the pharmaceutical industry. The FDA's requirement does have the force of law. However, the importance of the FDA's views regarding validation is overemphasized by its practitioners to the point where individual thought and interpretation is too often discouraged. Too often the guidance of others is solicited, as if the practice demanded an exceptional insight or understanding beyond that of the technical community. To differ with the FDA in matters of technical concern is not yet an act of *lèse majesté*.

It may be helpful to consider process validation as if it were a practice that is independent of FDA's interpretations. This freedom from the concerns of officialdom may stimulate original thinking and generate new approaches to the betterment of the practice. Perhaps process validation can be thought of simply in terms of performing a company's job assignment.

Consider the undertaking by a pharmaceutical company to plan, construct, equip, and operate a compendial water manufacturing facility. Having been given the assignment, the individuals and groups responsible would be expected by management periodically to

---

Abbreviations used in this chapter: DI, direct impact; DQ, design qualification; EMA, European Medicinal Agency; EPA, Environmental Protection Agency; EQ, equipment qualfication; FDA, Food and Drug Administration; GMP, good manufacturing practice; IQ, installation qualification; ISPE, International Society for Pharmaceutical Engineering; LAL, limulus amebocyte lysate; NIST, National Institute for Standards and Technology; OQ, operational qualification; ORP, oxidation–reduction potential; OST, oxidizable substances test; P&ID, process and instrumentation drawings; PDA, Parenteral Drug Association; PhRMA, Pharmaceutical Manufacturers Association; PQ, performance qualification; QC, quality control; RO, reverse osmosis; RTDs, resistance temperature detectors; SDI, silt density index; SOP, standard operating procedure; TDS, total dissolved solids; THM, trihalomethanes; TOC, total organic carbon; TSS, total suspended solids; USP, *U.S. Pharmacopeia*; WFI, water for injection.

report on the progress being made and on the status of the undertaking.

Declaratory statements in periodic reports would be illustrated and confirmed by relevant data obtained using reliable, hence calibrated, instruments. The system design necessary to produce consistently water of the required quality in amounts sufficient for peak demands and for total quantities would be decided on. The correct constructions and dependable functioning of installed units would be checked and substantiated. Proof would be offered that the system designs, their constructions, and safe operations were on target. Pertinent documents relating to safety, drawings, and descriptions of equipment, operational protocols, sampling and testing procedures, all would be collected and retained. The reports and documented presentations would be intended to demonstrate to management by way of data that the assignment was being carried out correctly. Ultimately, the successful completion of the task would be attested to in a final report. The several SOPs necessary to operations and maintenance would be detailed. The claims of an operationally functional water purification system, dependable over long durations, would be supported by an adequacy of appropriate data and documentation. Consistency in the stipulated quality of the product water over prolonged periods would be the proof of the attained goal.

It can be useful to consider such a company report as the essential system validation required by the FDA. Its purpose would be the twofold one of ensuring that the subject purification system is indeed capable of producing compendial quality water, in ample quantities, and that it does so consistently in a dependable fashion. Because this twofold goal is the very aim of the FDA's validation requirements, its orientation to address the FDA's regulatory structures then becomes an important, but secondary, exercise. Validation, then, eventuates as an expression of good engineering.

The need to validate the pharmaceutical water system to ensure its dependable performance carries a regulatory burden. Consequently, as stated, the question is: "What does the FDA require of the validation activity?" The answer can also be posed in question form:

Does the system work? For how long a time does it do so consistently? Can its operation be prolonged by maintenance, refurbishing, replacement, renewal, etc? When and how are these to be implemented? Supportive, experimentally secured, documented data in substantiation of positive responses are central to the validation of the pharmaceutical water system.

THE KEY ROLE OF EXPERIMENTATION AND DOCUMENTATION

The fields of scientific endeavor have long depended upon experimental investigation. Lord William Thompson Kelvin's aphorism regarding the primacy of measurement as the key tool of technical exploration still holds true: "When you can measure what you are speaking about and express it in numbers, you know something about it." It is equally important that the technical findings, the fundamental data on which conclusions are based, must be fully documented.

To emphasize the importance of documentation, one is often advised to "Document everything!" This statement indicates the importance of documentation; but its too broad overstatement detracts from its utility. Findings and occurrences pertinent to claims and explanations should be recorded fully. Inconsequential matters should be ignored. The problem, as usual, requires the exercise of individual judgments, of which there may be many. Keer (2) summarizes the documentation needs as follows:

> Documentation of a water system is a continuous exercise that starts at the very beginning of the project and ends when the facility is closed.

He continues,"A systematic approach to the task will yield the proper documentation to give the Owner and Regulatory Authorities the confidence in the system's control. The Owner's objective is to meet all regulatory requirements in a cost-effective manner. The regulatory agencies want to ensure there is no compromise or adulteration of products. Full and organized documentation satisfy the inspector's concern with minimal interruption to a facility's operation and at a small relative cost." In short, the claim of having successfully accomplished a process validation importantly requires substantiation by documented experimental evidence.

INTRODUCTION

The purification of water for pharmaceutical applications is an extensive subject. The validation of its process of manufacture is managed by implementing a series of qualification steps that logically lead to its fulfillment. Its DQ encompasses designing the system in terms of the purifying units that are required to meet its intended goals. Among the system design considerations to be discussed are the decision points for choosing among alternative purification units and their proper operations. This includes the materials of construction, system sanitizations, and clearing of foulants. The engineering requirements of water volume and flow balancing include definitions of pumps and pipe sizes. There are also the several considerations, documentary, microbiological, regulatory, and unit process operational, relating to the validation issue that requires attention and description.

The technical reasons in support of the selected units are to be explained. Choices and expectations are to be justified. The depth of detail of the system design may well be taken to indicate the sophistication of the designers' knowledge. Judged by experts, the conclusions may be justified. However, with regard to system design, even experts have different views of the available choices.

Even before the formal validation begins, a description and drawing of the system must be at hand. It must indicate all of the equipment items involved, from the water intake to the points of use. It should also plainly identify the sample port locations from which specimens of the water may be withdrawn for analyses. This is part of the DQ and part of the validation plan that sets forth the proposed operational approaches of the system.

Before the system can undergo testing of its various operations, it is also necessary to verify that the assembled units, piping, and ancillary equipment have been correctly installed. This is part of the IQ step that follows the DQ. It is a prelude to the OQ of the system. In the OQ phase of the validation exercise, it is ascertained that each purification unit functions in the way it is supposed to do. The investigation of the various facets of the OQ should be detailed, extending even to making certain, for instance, that the effluent of a given pump has a flow rate of *X* gpm at a differential pressure of *Y* psi.

Once the installation and OQ have been successfully completed, the PQ of the system can be undertaken. As stated, its purpose is to verify that the system does consistently produce water of the requisite pharmaceutical specifications under the standard operating conditions developed and established by experimental trial.

VALIDATION

There are many different definitions of validation. A U.S. FDA definition enunciated in 1987 stated: "Validation is the attaining and documenting of sufficient evidence to give reasonable assurance, given the current state of science, that the process under consideration does, and/or will do, what it purports to do" (3).

A less eloquent but very serviceable definition was given in the FDA Guidelines on Sterile Drug Products by Aseptic Processing (1987): "Establishing documented evidence which provides a high degree of assurance that a specific process will consistently produce a product meeting its predetermined specifications and quality attributes" (4).

In essence, validation seeks experimentally obtained answers to: Does the process or device do what it is intended to do? If so, for how long does it do it? For water systems, Artiss (5, 6) defined validation as ensuring that the particular system will consistently produce water of predictable quality when operated in the prescribed manner.

The FDA requires each of the manufacturing processes whereby drugs or drug components are prepared be validated. Each step or piece of equipment utilized in the process must be demonstrated and documented to be performing the function that it is supposed to do.

DOCUMENTATION AND INFORMATION

The importance of proper documentation has already been stated. As part of the validation requirements, a documentation and information master file will be established. It will include a full description of the system, specifying its acceptable ranges and limits. It will contain schematics of the electrical, mechanical, and water flow details. This will enable subsequent verification of the proper installation of the several purification units, of the control devices, of the safety and alarm systems, and of the provisions for instrument calibrations.

The documentation will list the activities necessary for the consistent production of the stipulated grade of pharmaceutical water. Perhaps the most important of these will be the SOPs that set forth in detail the operational protocols necessary for the dependable production of product waters of requisite quality using the installed water purification system. A companion set of instructions and protocols that will also be developed will be the maintenance procedures for the given water purification system. They will detail the replacement, regeneration, renewal, sanitization, and maintenance operations that are necessary to extend the system's reliability. Another subset of the SOPs will set forth the protocols and procedures relative to sampling and testing, and to equipment calibration. Documentation is an important part of the validation exercise. It is the adherence to the pharmaceutical company's policy on validation, as set forth in the SOPs and their allied protocols and as revealed by the documented data and the conclusions drawn, that the FDA inspectors will investigate.

Angellucci (7) offers the appended lists common to the concerns of validation documentation (see Appendixes A–G). The exercises covered range from critical instrument checklists to flow viscosity specifications involving flow rates and pipe diameters. Although not necessarily typical, some examples derived from pages of operating, maintenance cleaning procedures, and system descriptions are presented in Figures 1 and 2.

Tables 1 and 2 present Keer's views of certain of the basic design documentation desired, and of submittal documents to be supplied in conjunction with design operations.

The documentation requirements, as set forth by Artiss (5,6), include the written validation procedures and protocols necessary for the dependable long-term operation of the water system, the written authorized approval of these, the written validation reports, and the relevant physical report forms.

It is possible, however, to be critical of certain of the current practices relevant to validation documentation. One authority states, "Some in the pharmaceutical industry believe that validation is a paper chase, that it is a costly endeavor providing little quality or value to the product or process." Indeed, the gathered documented information sometimes seems to have less bearing on controlling and maintaining quality processes than on recording serial numbers and pipe finishes. Validation involves more than documentation, whether as protocols or checklists. Validation is intended to be a process whereby a company demonstrates and documents a mastery of its operations.

There are architectural and engineering firms and other consultant services that, on the basis of long experience, are competent to deal with the designing of plants and systems for the production of pharmaceutical water. They and the equipment manufacturers can be counted on to be helpful in collecting the information that should constitute the necessary document and information master file. Inevitably, however, that file will bear the imprint of the end user and, most importantly, must reflect the operational knowledge gained from that company's particular proceeding.

Documentation is a GMP requirement. Extensive records are to be kept as checks on procedures. They constitute a paper trail of information demonstrating process control. This permits an audit of the company's practices, and pinpoints a fault in the event of product

VALIDATION DOCUMENT			
Pharma-Renfrew Co. **Parenterals Division** **Arlington, Virginia** *Title*: **Performance Qualification of the Deionized Water System** *Study Number*: 009 Date Prepared: **April 15, 20XX**		Approved May 1, 20XX	**Section 9**
		Effective May 1, 20XX	**Page 8 of 82**
Name-Title-Date	Name-Title-Date Countersigned	Name-Title-Date Quality Assurance	

9. **PROCEDURE**

SAFETY - Plant safety rules shall be strictly observed to prevent the possibility of injury, in particular all procedures contained in any appropriate Pharma Renfrew chemical handling Instructions and Material Safety Data Sheet (MSDS).

9.1 General

The DI water produced and circulated throughout the facility will be tested for conformance to the acceptance criteria listed in this document. The initial, intense validation testing will consist of frequent sampling of water at various stages of the pretreatment and intermediate treatment process and sampling of each point use and sample port in all the loops at least twice a week for a five-week period.

The pretreatment system is defined as the portion of the process from the incoming water through the Reverse Osmosis stage. The intermediate treatment system is defined as the process from the RO treated water feed to each loop through the final filters of the individual loops. Testing in these portions of the process is designed to determine proper performance of the elements of the pretreatment and intermediate treatment systems. The testing program for the loops is designed to insure that the loop can maintain the requisite quality of water throughout the loop in a regular and consistent manner. The testing program is defined in the *Sampling / Testing Schedules* below. All tests shall be conducted using the appropriate EPA or USP methodology, where such methodology exists. Acceptance criteria for each test is described in Section 18 of this document. Laboratory results of the testing will be appended to this document and entered on the Water Test Logs. (Attachment M)

9.2 Sampling Procedure

9.2.1 Prepare the required number of containers necessary for the scheduled water samples, ensuring that each sample container is identified with the date, time on taking the sample.

9.2.2 For microbial samples, spray the port with a 70% isopropanol solution.

9.2.3 For microbial samples, wear latex gloves when handling sample containers.

9.2.4 For microbial samples, attach a length of sterile tubing to the port, if possible. In all cases, flush the port being sampled to drain for at least 3 minutes.

9.2.5 Collect the required quantity in the container, insure the container is properly identified, and remove to testing laboratory.

Figure 1 Validation document. (*Continued*)

failure. Proper documentation prevents the issuance of regulatory warning letters and maintains a quality reputation for the company.

The documentation aspect of the validation process is extremely important. A very large proportion of FDA investigations are involved with the documentation. It enables the inspector to examine data that disclose the activities that have taken place. The documentation enables an after-the-fact review of the process. It permits judgments to be made about what has been done, against the background of what the company said they would do, as expressed in the SOPs, for instance. Such retrospective examinations of data permit the FDA investigators to judge the appropriateness of the conclusions reached concerning the water system and its operation. To enable this to be done, the raw but recorded data and the written conclusions drawn by the system operators must be available, in clear, documented form, for the FDA investigators to examine and inspect.

In devising the water system and in establishing its operational requirements, by way of the SOPs, the pharmaceutical water producer has defined the company's position on this activity. The company's adherence to this policy should clearly become evident to the FDA inspector from an examination of the documented data and the conclusions derived.

Meshnick (8) has useful observations concerning general documentation procedures.

First, there should be a short discussion of how records and documents should be assembled and the data should be certified by the technicians.

One of the goals of the validation documents is to provide evidence of when, how, and by whom the work was performed, and a record of what results were obtained. In other words, what procedures were intended to be done, what actions were actually performed, why there is a difference (if there is), and what raw data support the results? To accomplish this documentation, it is a good practice to keep an accurate chronology of a project, and to follow standard GMP procedures for recording and counter-signing of data. For example, the original recordings of data, even handwritten notes

VALIDATION DOCUMENT (Continued)			
Pharma-Renfrew Co. **Parenterals Division** **Arlington, Virginia** *Title*: **Performance Qualification of the Deionized Water System** *Study Number*: 009 Date Prepared: **April 15, 20XX**		Approved May 1, 20XX Effective May 1, 20XX	**Section 9** **Page 9 of 82**
Name-Title-Date	Name-Title-Date Countersigned	Name-Title-Date Quality assurance	

9.3 Sampling plan

9.3.1 Pretreatment
The pretreatment system will be sampled and tested as follows:

9.3.1.1 Feedwater from sample port, SP-01 will be sampled and tested for conformance to EPS drinking water standards once every two weeks.

9.3.1.2 Water from sample port, SP-02 after the Multimedia Filter, G-1, will be sampled and tested twice each week for particulate content.

9.3.1.3 Water from sample port, SP-03, after the Carbon Filter, G-2 will be sampled and tested twice each week for residual chlorine.

9.3.1.4 Water from sample port, SP-04, after the Softener, G-3 will be sampled and tested twice each week for residual hardness expressed as $CaCo_3$.

9.3.1.5 Water from sample port, SP-05, after the Reverse Osmosis, G-4 will be sampled and tested twice each week for microbial and chemical levels.

9.3.2 Manufacturing Intermediate Treatment
The Manufacturing intermediate treatment system will be sampled and tested as follows:

9.3.2.1 Feedwater to the Manufacturing intermediate treatment, sample port will be sampled and tested twice each week for microbial and chemical levels.

9.3.2.2 Water from the mixed bed bank, sample port, SP-06, will be sampled and tested twice each week for microbial and chemical levels, and resistivity/conductivity.

9.3.2.3 Loop supply to the Manufacturing loop, sample port, SP-07, will be sampled and tested twice each week for microbial and chemical levels.

9.3.3 R&D Facility Intermediate Treatment
The R&D Facility intermediate treatment system will be sampled and tested as follows:

9.3.3.1 Feed water to the R&D facility intermediate treatment, sample port, SP-08, will be sampled and tested twice each week for microbial and chemical levels.

Figure 1 (*Continued*) Validation document.

or calculations, should be kept, initialed and dated by the technician performing the work. Work should be recorded in ink, not pencil. Corrections to this information should be made by drawing a line through the incorrect entry, and the correction dated and initialed. This information must be maintained as part of the final report.

Handwritten notes and drawings are acceptable, but must be legible and signed. An investigator may request to see your original data, as typewritten observations were obviously not taken in the field. If used, the final copy should be checked to guarantee against transcription errors. Do not complete all the checks at the end of the day, or when the report is completed. Initial what you have verified, on the day you verify the work.

It is common practice, and regularly cited in audits, that criteria for tests are too general and not specific enough for the tests being performed. Statements such as "Ensure that the system is installed according to manufacturer's specifications…" may sound sufficient, but do not provide the specifics necessary to judge the effectiveness of the test. Make your statements clear, concise, and easily tested. Avoid broad and sweeping terms such as "performs as intended" or "surface appearances are good." Subjective, general evaluations such as these cannot be proven, and are not supportive of a qualified condition.

VALIDATION STEPS

The logical sequence of qualifying steps involved in the validation of the pharmaceutical water system has already been mentioned. The definition and design of the total system; hence, of its constituent purification units, comes first. The DQs of each separate unit are ascertained; to be succeeded by the IQ wherein the correct linking of the purification units is made. Each of the linked modules is tested and challenged in an OQ that demonstrates its and the overall system's operational capability. The long-term suitability of the system's functioning, the establishing of its dependable reproducibility is attested to in the performance or process qualification step that follows. Traditionally, these steps are identified as DQ, IQ, OQ, and process qualification.

STANDARD CLEANING PROCEDURE		
Agate-Biopharm-Arlington, Inc.	Procedure Number 001-064-10-14	Page of Pages 12 of 18
	Supersedes None Previous	Date Approved May 1, 20XX
	Department Maintenance - Sub 2	Date Effective May 1, 20XX

STANDARD CLEANING PROCEDURE FOR THE CARBON BED STEAMING IN PLACE

Name - Title - Date	Name - Title - Date	Quality Assurance (Name-Title-Date)

Purpose

The purpose of this document is to define the Standard Steaming-in-Place Cleaning Procedure for the carbon bed.

8. Scope
   - Schedule
   - Steam-in-place

9. Responsibility

It is the responsibility of the maintenance - sub 2 group to properly implement this standard cleaning procedure.

10. Procedure

A. Schedule.

1. This standard cleaning procedure is to be performed at seven-day intervals.

B. Steam-in-place procedure.

Caution: Before initiating this procdure, ensure the purified water storage tank is at a level of 0% or greater and the reverse osmosis unit is not operating. Turn off the reverse osmosis unit at control panel, EP-001-1

Caution: Record at necessary information on the carbon column SIP log 6B6.

Caution: Ensure the clean steam generator is operating as per standard operating procedure for the clean steam generator SOP 40-A,B, and C-1.

Caution: Signal the carbon bed to be hot, by hanging red tag.

1. Turn off soft water/reverse osmosis recirculation pump P-74-A at the local BB6 switch. Close discharge valve V-10-10B
2. Open the following valves:

   V-10-12-B (Softener feed to deaerator)

Figure 2 Standard cleaning procedure for carbon bed steaming in place. (*Continued*)

Some groups add a PQ before the process qualification; others call the process qualification the validation. Some combine the first two and call them I/O/Qs or EQs. But the terminology and distribution of activities between these categories are not the important issues.

However performed, it is required that there be a logical progression from one qualification to another. For example, it is inappropriate for there to be a process qualification before completion of an installation or OQ. If the system has not been properly installed and verified to operate over the required range of conditions, then water testing of the system PQ is not warranted. The final validated condition is a sum total of the preceding qualifications.

Reestablishment of control of the system's reliable operation is required following the implementation of alterations and changes that may be instituted from time to time. The question to be answered is whether the changes involved are substantive enough to significantly alter the quality of the product water. If so, revalidation is required.

What is being sought is proof of consistent process control. If the supportive data, however positive, are scattered, the need for additional trials or tests may be indicated.

VALIDATION SEQUENCE[a]

Equipment Design

The design of the equipment constituting the water purification system obviously comes first. It derives

[a] The material contained in this section and subsections is liberally based on the information and text supplied by Dan Meshnick, Validation Project Manager at Foster Wheeler U.S.A., Perryville Corporate Park, Clinton, NJ.

STANDARD CLEANING PROCEDURE FOR CARBON BED STEAMING IN PLACE	Procedure Number 001-064-10-14	Date Apprroved May 1, 20XX	Page of Page 13 of 18

10. Procedure (Cont'd)

B. Steam-In-Place Protocol (Cont.)

3. Close the following valves:
V-10-13 (Carbon Bed to Reverse Osmosis System), V-10-14 (Carbon Bed to Deaerator), (Softener to Carbon Bed).

4. Open vent valve 10-15 and valve V-12-2A and then drain valve V-10-18.
Allow the carbon bed to drain completely

5. Close valves V-10-15, V-10-18 and V-12-2A

6. Open valve V-10-20 to drain Soft Water Inlet Pipe. Close V-10-20 after drained.

7. Open the steam condensate valve V-10-23. Cautiously open the carbon bed clean steam supply valve V-10-26. Allow low pressure steam to enter the carbon bed. Monitor the carbon bed temperature at 35°C or lower and the inlet steam pressure at PI-C-100 and PI-C-160.

NOTE: Steam pressure should not exceed 18 PSIG as indicated at PU-C-100. Manually regulate steam pressure at drain valve V-10-15.

8. When TI-A-01 reaches 65°C, close inlet steam valve V-10-26. Close V-10-15. The carbon bed should continue to heat to a temperature of 95-115°C

MAINTAIN THE CARBON BED AT >99°C FOR AT LEAST TWO HOURS.

CAUTION: **The carbon bed temperature must not exceed 120°C.**

9. Record the pressure at PI-C-100 and PI-C-160 and temperature at TI-A-01 every 30 minutes.

10. Connect a hose from the Clean Steam Line at valve V-10-201 to the R.O Feed Line at valve V-10-167. Verify V-10-43 is closed and open V-10-32. (Both valves located on the R.O. skid) Open valves V-10-33 and V-10-34 located on the 5-micron filter housing. Open V-10-17-A.

11. After the required two-hour hold time, open the following valves: V-10-15, V-10-18. Open V-10-168 for five minutes. This will allow the carbon bed to cool before initiating the required backwash.

12. Monitor the temperature of the carbon bed at TI-A-01. When the temperature is 38°C or lower, the backwash must be instituted. Refer to Standard Cleaning Procedure for the carbon column.

Figure 2 (*Continued*) Standard cleaning procedure for carbon bed steaming in place.

from the requirements of the water purification process. With a water system, this generally means that the quality of the water will minimally meet either USP Purified Water or WFI specifications (or other similar compendial requirements in other jurisdictions), depending on its usage. It is the design documents that set the standards and goals of the hardware.

Next, there must be definition of the process capacities, such as the total volume of water needed per hour or day, what reserve capacity is necessary, and whether elevated temperature storage is necessary. In other words, somewhere early on in the project, someone should write a functional definition of the project or process, so that there is a clear understanding of how the system must perform. This document, after it has been approved by the responsible groups, becomes the basis of the system design.

The succeeding protocols and reports will then document how the water system was designed to address these requirements. This link between the intended purpose and the final design of the system is important. Too often, the qualification process begins with the system after it has been purchased, and there is no clear statement of user requirements. The end user is ultimately required to provide the criteria to judge the system. Do not leave these to the system manufacturer or fabricator; what is important is not that a system meets an equipment manufacturer's specifications, but that what the vendor has provided meets the process requirements of the water user.

Where this functional definition is included is a matter of preference. It is sometimes referred to as a Master Plan document, or a DQ package, or even as an Introduction to an IQ protocol. However, because it is the basis of the qualification package, it should be consistent with normal company practices, and it must be easily retrieved when needed.

This is a valuable document for describing the system during an inspection, upgrading, or repairing the system, and especially for controlling the validated

Table 1 Submittal Documents

Document	Preparation	Input
Process and instrumentation drawing	Design	Design, operations
Instrument loop diagrams	Supplier	Design, operations
Equipment data sheets	Supplier	Design, operations
Instrument data sheets	Supplier	Design, operations
Control panel face layout	Supplier	Design, operations
Control screen presentations	Supplier	Design, operations
Sequence of operation	Supplier	Design, operations
Pipe-routing plans	Supplier	Design, operations
Pipe isometrics	Supplier	Design, operations
Skid pipe arrangement	Supplier	Design, operations
Junction box wiring	Supplier	Design, operations
Terminal layout	Supplier	Design, operations
Electrical conduit routing	Supplier	Design, operations
Construction plan	Supplier	Design, operations
Inst/equip/valve tag numbers	Operations	Design, supplier, validation
Instrument list	Supplier	Design, operations
Equipment tag numbers	Operations	Design, supplier, validation
Valve tag numbers	Operations	Design, supplier, validation
Pressure test procedure	Supplier	Design, operations
Flushing/cleaning procedure	Supplier	Design, operations
Passivation procedure	Supplier	Design, operations
Sanitization/sterilization procedures	Supplier	Design, operations, validation
Software documentation	Supplier	Design, operations

Source: From Ref. 2.

Table 2 Basis of Design Documentation

Document	Preparation	Input
Raw water sample	Operations	Design
Product water quality	Operations	Design, validation
Facility operation	Operations	Design, construction
Quantity	Operations	Design
Diversity	Operations	Design
Pressure/temperature	Operations	Design
Microbial control	Design	Operations, validation
Purification technology	Design	Operations, validation
Monitoring requirements	Validation	Corporate, design
Documentation for validation	Validation	Corporate, design
Design codes and standards	Corporate	Validation, operations, design
Current good manufacturing practices	Validation	Design, operations
Mechanical space limitations	Operations	Design, construction
Budget	Operations	Design, construction
Schedule	Corporate	Design, construction, validation
Environmental	Operations	Design, corporate
Control philosophy	Operations	Design, validation, corporate

Source: From Ref. 2.

condition of the water system as part of a change-control procedure.

This functional design basis is easily included for new construction, but it is not so simple to obtain or develop with existing water systems. Still, some description of the construction and design of the system must be committed to paper for an existing water system. This is important for future reference, especially to individuals who may not have been involved with the validation, but who may have to redesign and revalidate it at a later date. It is especially useful to someone who may have to explain the design and validation during an inspection some years later. It is difficult to defend a report or procedure without a clear statement of the design basis and functional goals.

There are, regretfully, no meaningful design or construction standards presently used for water production in the drug industry. All too often, the information and recommendations forthcoming from equipment suppliers are relied on. These are necessarily limited to their own expertise and are not always objective. General engineering consultants may be experienced in accepted engineering practices, but these may not sufficiently apply to the particular requirements of specific pharmaceutical situations. Long-accepted design concepts suitable for process and potable waters may not suffice when extended, even by the use of sanitary components, to pharmaceutical water applications. The different design requirements dedicated to the critical needs of pharmaceutical water systems may necessitate a new approach. This design will reflect some engineering principles in its functionality, safety, and code requirements. However, it should not be the sole burden or responsibility of the engineering department. The critical considerations of operational suitability, microbial control, and adherence to regulatory needs should first be set forth. The engineering design should then be formulated to meet these needs. Therefore, DQ should, from the first, include the participation of all appropriate groups such as engineering design, production operations, quality assurance, analytical services, and others. The need for a team approach is necessitated by the complexity of the undertaking. Materials selection, equipment suitability, operational controls, construction techniques, cleaning and sanitization procedures, component compatibility, preventive maintenance, sterilization programs, and sampling and regulatory requirements, all are involved. It is essential that an adequate address to all these considerations are "designed in" the designation of the system.

Where an insufficiency of guidance from other disciplines is involved, "add-ons" usually result in an effort to correct an inadequate system design.

The DQ document will list the activities necessary to the consistent production of the stipulated grade of water. It will contain a full description of the system, specifying its acceptable operating ranges and limits. It will supply full schematics of the electrical, mechanical, and water flows for subsequent verification of their proper installation. It will identify the specific purification units, the various control devices, and the safety and alarm systems. It will also provide for the calibration of critical instruments and set the microbial action and alert limits, which will specify sampling plans and ports for chemical and microbial testing, stipulate sanitizing methods, and define procedures for the analysis and plotting of data.

Artiss (5,6) states, "the basic design package should include the following:

1. Flow schematics for the proposed water system showing all of the instrumentation, controls and valves necessary to operate, monitor, and sterilized the system. All major valves and components should be numbered for reference.
2. A complete description of the features and function of the system. This is of critical importance to enable production and quality assurance personnel, who may be unfamiliar with engineering terminology, to fully understand the manner in which the system is to be designed, built, operated, monitored, and sterilized.
3. Detailed specifications for the equipment to be used for water treatment and pretreatment.
4. Detailed specifications for all other system components such as storage tanks, heat exchangers, pumps, valves, and piping components.
5. Detailed specifications for sanitary system controls and a description of their operation.
6. Specifications for construction techniques to be employed where quality is of critical importance. These techniques should be suitable for exacting sanitary applications.
7. Procedures for cleaning the system, both after construction and on a routine basis.
8. Preliminary SOPs for operating, sampling, and sterilization. These procedures will be cross referenced to the valve and component numbers on the system schematics.
9. Preliminary SOPs for filter replacement, integrity testing, and maintenance.
10. Preliminary sampling procedures to monitor both water quality and the operation of the equipment.
11. Preliminary system certification procedures.
12. Preliminary preventive maintenance procedures.

The design package should be as complete as possible to enable all disciplines involved to understand what the final system will entail.

Validation Plan

As stated earlier, the functional definition is often included as part of a Validation (Master) Plan. This document is not a requirement of the FDA (it is an EMEA expectation), but it has become almost an industry standard. It is a good idea to include such a document as part of the validation, as it sets the overall goals and limits that will be followed during the validation, and can be referred to throughout the project, but especially much later, well after the study has been completed. As a reference document, the plan permits a reviewer immediately to understand the scope of the validation, and so avoid misconceptions.

The validation plan should contain all the information relevant to the water system. It will be a repository for the basic design information, drawings, specifications, procedures, and protocols. It will state the reasons for equipment selection, for cleaning and sanitization frequencies, and for component replacements and renewals. It will contain the records for equipment modification and of procedural alterations. It will have the equipment and filter logs and any recertification data. In short, it will constitute the major reference file for the entire water production system. As such, it will serve internal investigatory purposes, and form the basis for outside regulatory reviews.

The validation plan is used to set the limits of the validation, to define the scope of the project, the systems included and not included in the qualifications, and what the project will attempt to prove. For example, if the project includes the use of deionized water to feed a clean-steam generator, the validation plan would define which components would be involved in the preparation of such a water; what general quality attributes each purification unit would be expected to achieve; and the length of time the system will undergo sampling at what frequency. Issues involving choices should be addressed in the validation plan, including the reasons for the choices. It must be made apparent why the selected decisions are appropriate. The validation plan must be consistent with the company QC policies, and should be included in the SOPs.

Such a validation plan will be much appreciated when reviewing the validation at a later date, such as in response to an out-of-tolerance condition, in a quality audit setting, or when performing a revalidation.

Installation Qualification

The IQ protocol will consist of a system description followed by a procedures section. Before the operational characteristics of the system can be investigated, the proper installation and assembly of the various items of equipment require verification. This follows a careful check that each piece of equipment ordered and received is identical with that stipulated in the system design. As sagely advised by Artiss (5,6), "Consideration should be given to conducting an inspection of the equipment before it is shipped from the supplier. Features of operational function and compliance with specifications can be verified and any deviations can be corrected without incurring the cost and time delay of reshipment."

The IQ ascertains that all the unit components are installed as per specifications and according to the design drawings. It is also required that the support systems such as instrument calibration programs, preventive maintenance procedures, and operating SOPs are addressed. It provides verification in that the established specifications have been complied with during construction and installation. Included in this operation is a review of P&ID and isometric representations, verification of materials of construction, examination and documentation of welds, inspections for dead-legs and for pipe slopes, verification of stainless steel passivation, and any other pertinent information. The IQ confirms the "as-built" drawings, and ensures the suitability of the completed system. The absence of leaks, which may provide pathways for invading organisms, can be ensured by vacuum testing, or by the use of pressurized air or water.

As stated in the FDA guidelines (3): "This phase of validation includes examination of equipment design; determination of calibrations, maintenance, and adjustment requirements; and identifying critical equipment features that could affect the process and product. Information obtained from these studies should be used to establish written procedures covering equipment calibration, maintenance, monitoring, and control."

The first step in the IQ document should be a general statement covering the why and how of the water purification system. This will constitute the system description. It will center on the major design criteria and operational parameters. A detailed description of the water purification system has already been provided in the equipment design section. It would be well to repeat at this stage, detailing the succession of purification unit processes (e.g., pretreatments, principal purification devices, storage vessels, control units) that will affect the product water quality. This should be accompanied by an exposition of the operational parameters necessary to the major design criteria. The feedwater sources should be identified along with factors for which variations would affect the operation of the system. If ion-exchange-treated water is to be used as feedwater for stills, its bacterial endotoxin content control must be considered in the system's description.

This is to be followed by a procedure section, the setting for a protocol on how to proceed in performing what is specifically required. The protocol should define the procedures to be performed, the documents to be assembled, and the items to be checked and verified. The plan should be approved before the qualification work begins. Subsequent changes should be quantified, recorded, and approved in the final report.

The IQ execution is usually in the form of checklists that verify components or details critical to the validated condition of the equipment. Confirmation of design items such as materials of construction, surface finishes, weld mapping and inspection documents, major equipment inventories (pumps, filters, UV lights, control valves, and such), process instrument lists, utility connections (including drains), and other such equipment, define the system as installed.

In developing these IQ check sheets, individual components must first be identified, as in the system description or in a comprehensive equipment listing. Vital characteristics, necessary to the proper operation of the components, are included, along with the specific criteria that must be met. Spaces must be provided for each item to be verified, by date and initials of the individual performing the checks.

This documentation should concentrate on the process and the components that are critical for process control. Focusing the checklists on these will help in limiting the scope of work required in the validation. The report should provide information on the items that affect the operation of the system, especially the components that change the performance of the equipment if they were to be replaced or adjusted. Information should be gathered that would be useful in evaluating changes to the operations such as design and capacities of steam traps, surface areas of heat exchangers, or communication setting on input/output boards. Items that are not critical to process operation need not be addressed. When in doubt, the process should be reexamined. If the item does not have a critical effect, it is safe to eliminate it. Record the decisions, because they may be questioned in later audits.

There must be stated criteria that describe the expected acceptable condition and sufficient space for comments and observations. The raw data gathered as part of the verification must be included as part of the final report, along with a description of the procedures used.

These checklists should be prepared with the intent that they will be taken into the field and completed as the work is performed. If raw data are recorded on separate sheets, these must be included with the completed checklists.

P&IDs are ideal documents to provide a clear description of where critical instrumentation and major system components are located. As-built construction drawings showing actual measured piping layouts, filter locations, drafting stations, dead-legs, pipe slope, etc., are necessary as verification of how these critical items are installed. Initialing the drawings along with the date that a component is verified is a common practice. Any notes or comments on the verification should be recorded right on the drawing, or attached and referenced.

Areas that are often overlooked during IQ are items that are generally contracted to service groups. Cleaning and passivation documents, especially procedures, types and concentrations of acids and neutralizers, and the pH results of the various rinses are often neglected. Request these items beforehand and make sure they are signed and dated by the technicians performing the work. This is also important for documents such as weld certifications, for which quality procedures are sometimes lax. Make sure the welders document welds as they are made, not at the end of the day or at end of project. The purpose of the inspections and verification is to ensure careful, precise welds.

The actual construction techniques used for system installation should be carefully monitored to ensure compliance with the written specifications.

The IQ protocol should be well-documented relative to its flow of logic. This can prove critical to change control, serving as a basis on which subsequent changes to the water purification system can be explained and justified.

Instruments and Controls

One category of items that falls between the IQ and the OQ are instrumentation and control systems. These two groups contain issues that could be included in either or both.

The first step in qualifying instruments and controls is to make a list of all system instruments. This must be available for inclusion as part of the IQ document. Such a comprehensive list should be available as part of the P&ID. At this point, classification determinations must be made for each instrument as to whether it is critical or non-critical to the process.

Critical instrumentation (needed for direct process control and monitoring or recording) will require periodic calibration under CFR21, Section 211.68 and 211.160 of the GMPs. These calibrations must be traceable to a recognized standards organization, such as the NIST. Critical instruments include such items as temperature controlling RTDs, tank level sensors, chart recorders that provide documents for batch records, resistivity meters and controls, and flow meters used to control resin bed regeneration. Procedures must be available for the calibration of these instruments, which include the method of calibration, the range and accuracy of the instruments, and an appropriate schedule for performing these calibrations. Records of these calibrations must be kept to comply with the GMPs. Instrument identification numbers and a sticker indicating date of calibration and the date of next calibration must be clearly visible on all critical instruments.

Non-critical instruments such as instrument air regulator gauges, or redundant pressure or temperature instruments, do not need rigorous calibration schedules. However, non-critical instruments must still be identified and logged into a calibration program. There must be a clear identification on such an instrument that it is not used for process control.

Instruments and controls require careful installation and identification. Generally, on new systems, the instruments should be left uninstalled until most of the heavy construction has been completed. Wiring, instrument air lines and supplies, and transmitters can be checked and verified for proper installation, but instrument calibration and tuning of controls should be left for the end of the construction process. Some companies like to postpone the installations and include these functions in the OQ procedures for this reason.

Calibration of instrumentation can be performed either at the end of the IQ process, and recorded as part of the IQ, or at the beginning of OQ. Either way, before operational testing is begun, all system instruments must be verified as calibrated.

When completed, this information is included not only as part of the qualification package, but also as part of the company's metrology program documents. This program is to ensure that the controls and recordings from the system will be accurate and reliable. In it are documented calibration procedures, schedules, calibration results, and response to out-of-specification calibrations.

Operational Qualification

When the installation of the equipment assemblage has been verified as being correct, it becomes possible to undertake the OQ documentation of the system. The system should be carefully cleaned and all construction debris removed to minimize any chance of contamination or corrosion. Once the cleaning has been completed, the equipment should be started up and carefully checked for proper operation. The purpose is to demonstrate that each component of the system functions. The OQ protocol of the water purification system assumes defined acceptable product specifications.

Equipment qualification verifies the capability of the processing units to perform satisfactorily within operational limits. Considerations of feedwater quality, of system capacity, temperature controls, and flow rates are involved in the OQ protocol of the water purification process. It involves an examination of the equipment design to identify features critical to the process and product. The goal of OQ is to evaluate the limits of control within which the validated system is expected to perform. The focus is on defining the critical items and practices. Alarm conditions for utilities such as low steam pressure or instrument air, diverter conditions resulting from low condensate resistivity, and differential pressure limits (high or low) are just several examples of events that should be confirmed as functioning correctly.

In this operation, reliance solely on information from the equipment supplier is inappropriate. However, the manuals relating to each piece of equipment are parts of the documentation. A review is made of the maintenance and adjustment procedures. Post-repair and calibration requirements are stipulated in advance to avoid confusion following emergency repairs. Emphasis is placed on avoidance of inadequate water quality caused by the use of incorrect procedures. Documented procedures are developed for the maintenance, adjustment, calibration, monitoring, and control of the equipment involved.

The protocol should begin with an introduction and a description section, this time describing operation of the system. This will entail repeating what may already have been written previously. Meshnick (8) has the following to say about repeated material or references thereto:

> Some may think that it is much easier to simply add references to reports, allowing the reader to access procedures or design information from other protocols or SOPs. While there is no set format as to how to reference other documents, this should be carefully considered, since this practice guides a reviewer into other documents which may not be the focus of his

interest, and will generally confuse the audit. If it is an item specific to the Qualification, then include it. Better to repeat the information at the point required, than trek over the entire paper trail. Also, it is important that any documents which are subject to updating and revision such as general procedures or protocols be included or referred to as the official revision number in effect when the study was conducted.

Subsequent to writing the OQ protocol's requirements and procedures, the specific acceptance criteria are to be set forth. It is by way of these that the system's results will be judged. The specifying of the acceptance criteria is key to the protocol development effort. It defines the bounds within which the system is to be controlled. What are required are acceptance criteria which, while they challenge the system, are appropriate to the system's operation. Exaggerated standards are unnecessary. However, the criteria that are set must adequately assure the proper water quality. The system's specifications should be precisely defined and adequate for its operation.

The OQ protocol provides for functional testing of the system components. The OQ of each purification unit leads to the PQ of the system as a whole.

Performance Qualification

The purpose of the PQ protocol is to provide a rigorous testing to demonstrate the effectiveness and reproducibility of the total, integrated system. The system's set points, control sequences, and operating parameters are probed. The process is challenged repeatedly to prove its consistent performance. All the acceptance criteria are to be met under "worst-case" process conditions. When failures occur they should be identified and corrected. Tests should be re-run to vouch for the elimination of the causes of failure. Consistency of acceptable product water quality is sought. Documentation of every operation involved is necessitated.

According to the FDA's advice: "The observed variability of the equipment between and within runs can be used as a basis for determining the total number of trials selected for the subsequent PQ studies of the process."

The performance verification ensures the suitability of the system's function. The proper operation of the system's equipment and controls should lead to repeatability in the system's characteristics in the product it produces. This is established by repeated start-ups and shut-downs, simulating manual, automated, and emergency conditions. In the process of so doing, operational SOPs are reconfirmed and procedures for operation are finalized. The goal is to achieve the production of a dependable product in a continuous mode. A procedural verification of the written SOPs is sought. The limits of the product quality are explored by sampling and analytical testing, chiefly of electrical resistivity, TOC content, and bacterial and endotoxin levels. Long-term trends and evaluation are explored. The general strategy of how the system is challenged should be a part of the validation plan, and the PQ protocol should elaborate on the specifics of the program.

As will be seen, the customary plan of the PQ would include an intensive pattern of sampling for a relatively short time, one or two months, while the system is operated under normal conditions. As much information as possible concerning operating conditions should be gathered during this phase. Some firms attempt to challenge the system to the limits of the operating ranges during this testing, whereas others run the system as close to the center of the operating ranges as possible. Both have their merits; however, challenging the limits may have detrimental effects on the study, such as random failures, and certainly presents difficulties in determining which combination of variables would represent the most appropriate challenges. Thus, it is easier to run under conditions that represent the normal system. The water produced by the system may be used, provided this intensive period of investigation so indicates.

After the intensive monitoring phase has been successfully completed, the system should undergo a long-term evaluation, for perhaps a year or more. The system is considered qualified or validated, based on the data from the first phase of monitoring, but because of time-dependent effects such as the variability of feedwater supply, the effects of wear or deterioration of components, such as UV lights, and the ability of organisms to adapt to changing conditions, longer-term evaluation is appropriate before a system is considered fully validated.

The developed program should be described in a validation plan and detailed in the PQ protocol. If the two-phase (or longer) program is implemented, the data review and summary reports must be approved when each phase is completed. Do not wait until the end of an extended monitoring program before determining the validated condition of the system, especially if the water is being used.

Change Control

A smoothly operating water system may undergo departures for reasons other than alterations in its water supply. Time-dependent changes are involved. These are to be elucidated, defined, and documented. Given purification units such as ion-exchange beds may become exhausted, RO membranes will require cleaning, tanks and pipes may need re-sanitization, and so on. In general, the devices and accoutrements constituting the system will periodically require such maintenance-related activities as replenishment, refurbishing, cleaning, sanitization, replacement, and renewal of different kinds. Furthermore, the various items will require attention on different time schedules. The necessary system documentation will, therefore, also include a body of information relating to the proper maintenance of each piece of equipment. Much of this will be initially forthcoming from the equipment suppliers and may, indeed, constitute stipulations connected with their guarantees of equipment performance. The relevant documentation composes the standard maintenance procedures necessary to the system's correct handling.

On a time-line basis, the IQ and OQ of the system are performed parallel to the installation of the purification and ancillary equipment and to the preparation of the product water (Fig. 3). This leads to the commissioning of the system and to the qualification of its performance. As will be seen, the validation exercise can be divided into three phases. On the time-line basis,

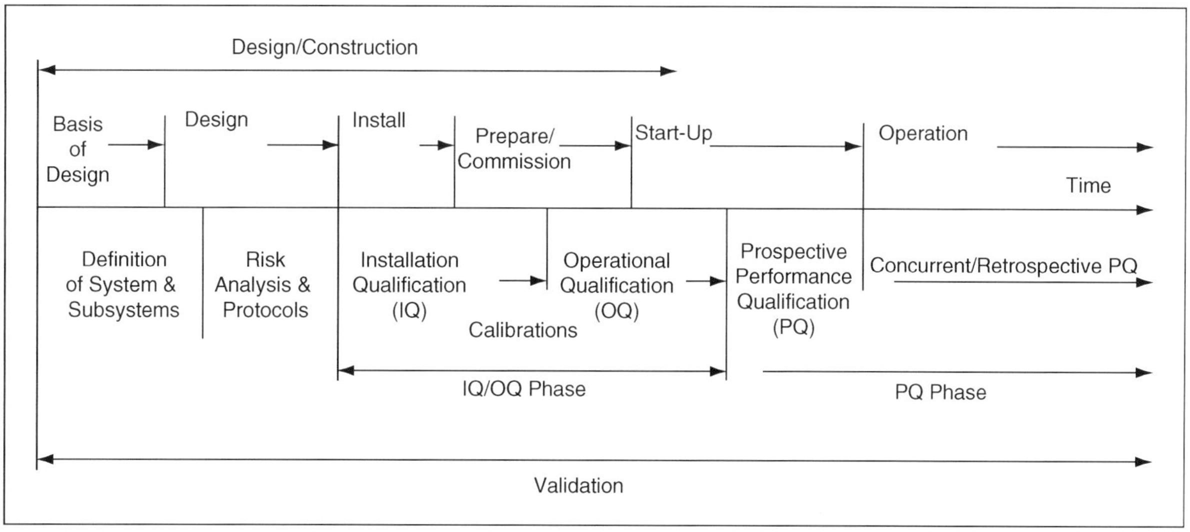

Figure 3 The validation timeline. *Source*: From Ref. 11.

the system's start-up overlaps both the OQ and the prospective phase of the validation. The system's operation parallels the concurrent and retrospective phases. Another view of the overall process is shown in Figure 4. A review of the entire operation should be performed at least annually and preferably more frequently to ensure the ongoing appropriateness of the product water and operational specifications. The system

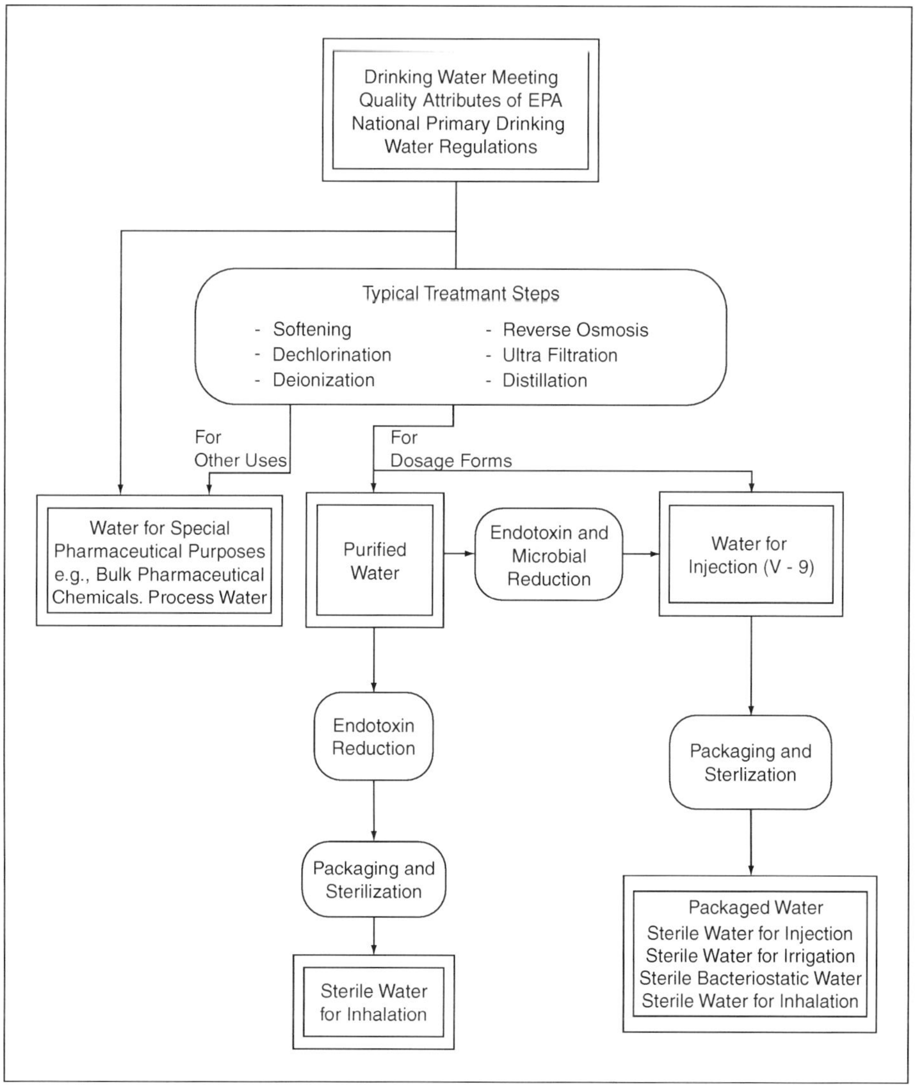

Figure 4 Water for pharmaceutical purposes. *Source*: From Refs. 5,6.

must be revalidated after any significant mechanical or operational change.

Qualification and Validation Final Reports

Once the validation has been planned, the design documented, the protocols written and executed, and the data collected, the last and most important step of the process is the evaluation and reporting of the study. Meshnick (8) advises as follows:

> The final report is your opportunity to focus the study. It is where you describe what the data means. Take this opportunity. Results are not self explanatory. The report should include a section which summarizes the raw data; in tables, figures and drawings, graphs or other means. Review, explain and finally, conclude what the data support, based upon the acceptance criteria in the protocol. The review section should specifically review each point in the protocol acceptance criteria, and state whether the criteria were met. Then, based upon this review, the conclusion section should state whether the system is considered qualified or validated. In any event, do not just present the raw data in the report, or fail to make a conclusion. Do not allow the reviewer to judge the data alone since he may not come to the same conclusions as yourself. In a properly formatted and concluded report, there may still be disagreements, but at least there is a clear stand as to what the data mean. Without this, there is an open invitation to a differing view.

The final report must be written as an adjunct to the individual protocols, relating the observations recorded during the study, to the procedures and acceptance criteria in the protocol. Most often, the procedure employed during an audit review, is to begin with the protocol and perform a step by step comparison with the report. It is critical that the protocol be followed, and that deviations documented. Often, when the procedure is being implemented, it becomes obvious that it cannot be performed as planned, or a better procedure becomes apparent. At this point some groups feel a protocol rewrite is needed, with the resulting effort for circulating the documents for review and approvals. An easier approach is to address such issues as well controlled and documented deviations.

Deviations from an approved protocol should be dealt with carefully. They must be the exception, and not the rule. Consider what effect deviations have on the protocol and the validation plan. If the protocol is substantially changed by the deviations, then a rewrite may be in order. Even major excursions from the protocol can be addressed in a deviations section of the report, provided they are justified and approved by authorized individuals. But consider the consequences. Lost or destroyed information, failures to follow SOPs and other faux pas may indicate to investigators an insufficiency in the required technique.

THE VALIDATION EXERCISE

In the U.S.A., the FDA and the USP are the main factors in defining the activities appropriate to pharmaceutical water system validation. In addition, there are professional organizations engaged in the activity. The PDA and the ISPE are perhaps the leading organizations. Several consultancies are notable contributors, and the personnel of most equipment suppliers are well grounded in their relevant specialties. A profusion of lecture presentations are offered, albeit targeted and provided by different levels of expertise. Almost all engaged in pharmaceutical water system design and/or evaluation see water systems as consisting of a series of sequential purification stages.

The validation exercise has been characterized as consisting of a prospective phase, a concurrent validation phase, and a retrospective phase. According to spokespersons for the FDA, in the prospective phase of the validation, daily samples are to be assayed for their chemical and microbiological quality at each unit of the water purification system and at each point of use. This is to be done for a minimum one month period. The data obtained during this first period should be used to develop the SOP. During the second period of a month or so, the concurrent validation phase, the same frequency of testing is to be observed. The resulting data serve to refine the SOP, to establish it firmly except for the long-term effects yet to be explored. These will be investigated during the remainder of the year in phase 3. During phase 3, the retrospective validation step, microbiological testing of the WFI systems should be performed daily, with each point of use being tested at least weekly. For Purified Water systems, each point of use should also be tested weekly. It is to be emphasized that the analytical numbers are not to serve as pass/fail values. Their usefulness is to establish trend lines. The USP chapter on the subject agrees that this one year period should normally suffice for the validation exercise (10,11).

As stated, even when there are many points of use, each should be tested at least once a week. This is true, regardless of whether water for drug formulation is drawn from that particular point of use or not. The condition of each point of use, part of the overall water system, reflects on the status of the system in general.

As Cooper (12) points out, however, assumptions are inherent in accepting this one year duration. Is it sufficiently long to disclose all the possible alterations in the source water supply, seasonal and otherwise? Will the second year's characteristics mimic that of the first? The systems are dynamic; they do change. Cooper correctly observes, "The application of the term 'validation' to this ever-changing situation surely adds a new and dubious dimension to the mystique of the validation concept." It must be concluded that the parameters of the water purification system may continue to change and that an ongoing, continual monitoring of the system is necessitated.

Underlying the setting of the time period for the initial phase of the validation is the assumption that the constancy of the feedwater composition is such that its management in the purification process can be defined and described in an SOP within a period of one month or so. But source water, particularly when processed by a municipality using mixes from different origins, may alter in character, even on a daily basis. One American biotechnology company depends on a municipal source that in answer to its own urgencies may, on very short notice, mix well waters with river waters containing large quantities of TOC of unknown compositions derived from industrial sources and farm runoffs. This particular source water changes often, with dramatic frequency and suddenness. The system is made manageable only by the use of pretreatment purification units sufficiently exaggerated in size and scope that they can handle peak contamination loads. Several months were required to define an acceptable SOP for this water purification system, and intensive ongoing daily monitoring was involved to ensure that the developed SOP really suffices from day-to-day, the source water being of such inconstant quality.

The three-phase validation scheme is the one approved of by the FDA. Some perform validations consisting of only the concurrent phase, a practice wherein the process is evaluated simultaneously with manufacture of the product water. In this approach, individual batches may be used even before the entire validation has been completed. Such concurrent validations are not favored by the FDA.

Historical data may be solely relied on to establish a retrospective validation in instances where the water has been produced for many years by a process that remains unchanged. In such instances, it is required that documented evidence firmly establishes that there is a significant ongoing experience reflecting a constancy of practice.

Microbiological Validation

It is the FDA whose views must be accorded when performing the microbiological aspects of the validation process. It should be repeated, however, that the pharmaceutical water processor is responsible for ascertaining that the validation exercise is sufficient and proper for their purposes. Complying with FDA regulations, advices, guidelines, and such should lead to reassuring results. However, if despite the FDA's "approval" (or lack of disapproval) the "validated" system does not prove suitable for drug processing, the responsibility is that of the pharmaceutical manufacturer. The FDA imprimatur confers no immunity for the drug processor against untoward consequences.

Munson (13) divides the validation exercise into the three phases already discussed.

In phase 1, daily water samples are taken downstream from each unit in the treatment system and from each point-of-use in the holding/distribution system to assess the chemical and microbiological quality of the water. The data from the daily samples taken over a one month period should be used to develop SOPs, appropriate maintenance and cleaning protocols, and analytical schedules for each unit in the system.

Daily sampling is continued in phase 2 for another month. During this phase, the water system is operated according to the protocols and schedules developed during phase 1. The data from this phase are used to confirm that the operating and maintenance protocols and schedules are adequate and that the system can consistently produce water meeting its specifications. The data can also be used as the base-line data for trend analysis of the system.

At the end of phase 2 the sampling schedule changes to that of routine monitoring. Chemical testing of pharmaceutical water systems should be performed at least weekly. Microbial testing of WFI systems should be daily, with each point-of-use being sampled at least weekly. For Purified Water systems, microbial testing should be performed on each point-of-use at least weekly.

Phase 3 of the validation program consists of reviewing the routine of monitoring data for at least 6 to 10 months. This time period will demonstrate that the operating protocols are adequate to handle variations in the quality, both chemically and microbially, of the incoming feed water. At the end of phase 3, if all data indicate that the water system when operated according to its SOP consistently produces water that meets its specifications, the water system can be considered validated.

The validation program described is only a suggestion. It should not be interpreted as the only acceptable program that FDA will accept. Each water system, including the validation data and program, will be judged on its own merit.

The water generated during the (second and third) validation phases can be used to manufacture drug products. You do not have to wait for a whole year before you can use the water.

Microbiological Levels

The bacterial endotoxin contents of WFI are set at the 0.25 EU/mL level. The microbiological content of Purified Water is not to exceed 100 cfu/mL and that of WFI is not to exceed 10 cfu/100 mL. These, however, are not rejection limits, but rather, are alert levels (10,11,13). The organism action level for WFI permits little room for maneuver.

The Purified Water action level is, however, amenable to modification, depending on the use to which it is put. When the Purified Water is designed for antacid preparation, the level should be reduced. This is necessitated by the ease with which organisms grow at the alkaline pHs of such preparations in which preservatives are generally ineffective. A lower level would reduce the risk caused by the potential for organism growth in the product. Oral medications might be permitted higher counts than otic, nasal, and wound area topicals. Where one water system is depended on in the preparation of several products, the action level should be set in accordance with the needs of the product offering the highest risk for microbial growth. Thus, room is left for individual judgments on the action levels for Purified Water, depending on its ultimate use.

Munson, as spokesperson for the FDA, advises: "Failure to meet these action levels does not mean automatic rejection of products. As the definition indicates, action levels are points which signal a drift from normal operating conditions and which require action on the part of the firm. When these levels are exceeded you should conduct an investigation designed to determine why the action level is being exceeded. Then identify and implement the corrective action needed to restore the system to normal operation. You should also recheck products made prior to the corrective action to determine if the contamination has affected the quality of the product. You should increase your sampling rate for a period after the corrective action is implemented to ensure that the system has returned to a state of control. This also does not mean that if you get a count of 110 cfu/mL for your Purified Water that you must shut down the system during the investigation."

Because microbial test results are already two to five days old you should not wait for two consecutive samples to exceed the action level before you perform an investigation. This is when control charts or trend analysis can be a very useful tool. If the organism(s) isolated do not represent a potential problem and the historical profile of the system indicates that this single result is unusual and not part of an upward trend, then the follow-up action may simply consist of resampling the use point or stepping up the rate of sampling for a short period so that a more accurate determination of whether the system is out of control can be made. The important thing is to document is what follow-up action was taken and that the problem was corrected. No documentation means no investigation, no follow-up action and no correction (13).

Testing for Specific Organisms

The types of organisms present in the Purified Water must be considered. Previously, the FDA had advised that no pseudomonads should be present in the water system. That is not the present FDA position (15); *Pseudomonas* spp. do not have to be specifically monitored for unless such organisms represent potential hazards to the product. The burden for that, as well as the consequences attendant on it, devolves upon the drug manufacturer. It is the responsibility of the drug manufacturers to understand the situation for their products. It is known that the presence of *Pseudomonas aeruginosa* in topical products can produce infections in people with abraded skin or wounds. Therefore, the presence of pseudomonads in waters used to prepare topicals is objectionable. Water systems are required to be free of particular organisms only if they represent pathogens or potential pathogens in the products being produced, when their presence in the drug preparation poses a potential health hazard.

The number of microbes restricted in the compendial waters by the alert and action levels is defined by a prohibition against objectionable organisms, such as *P. aeruginosa*. The presence of opportunistic pathogens also needs to be considered, however. These may be pathogenic when applied to patients with compromised immunities, a situation not known in advance to the drug preparer, who is, nevertheless, liable for untoward consequences. It may be prudent, therefore, to maintain even Purified Water under self-sanitizing storage, as at 80°C or in the presence of ozone. The use of a sterile Purified Water could eliminate the presence of undesirable organisms from the drug preparation.

Microbiological Assay Methods

There is no unanimity on how the microbiological assays are to be performed. There are different views concerning the methods, whether by direct count, or pour plate; the nutrient media to be employed; the incubation period and the incubation temperatures. The elucidation of the various views is beyond the scope of this writing.

The Water Quality Committee of the Pharmaceutical Research and Manufacturing Associations recommendations on this subject are as follows (9,14,15):

Purified water:	Pour plate method
Minimum sample: 1.0 mL	
Plate count agar	
Minimum 48-hr incubation at 30°C to 35°C	
WFI: Membrane filtration method	
Filtration using 0.45-μm-porosity filter	
Minimum sample: 100 mL	
Plate count agar	
Minimum 48-hr incubation at 30°C to 35°C	

It should be remembered, however, that the responsibility for selecting, or even for devising culturing techniques suitable for revealing organism types that may be present in a particular water is that of the pharmaceutical processor.

FDA does not have any significant problems with the PhRMA proposal except that the sample size for water injection should be 250 to 300 mL to obtain a more accurate determination of the microbial count. These methods are not the only methods that can be used and they may not work for all water systems. It is simply a starting point; you may have to develop a method appropriate for your site and water system (13).

Alert and Action Levels

It is intended that pharmaceutical manufacturers set and utilize alert and action levels to guard their WFI and Purified Water from exceeding the specified microbial levels. In this exercise, the alert levels indicate that a process may have drifted from its normal operating condition. It merely provides an early warning of a potential problem. The action level signals such a departure from the normal range that corrective action is required.

It is helpful and prudent for the waters to normally maintain an even greater purity than that stipulated. For example, consider a Purified Water for which the action level is set at 100 cfu/mL, but which is normally purified to better than that amount, say, 50 cfu/mL. If a periodic analysis indicated a level of 70 cfu/mL, the firm would be alerted to check the accuracy of that finding by promptly repeating the analysis. Were the recount to affirm the higher level, action could be undertaken promptly to learn the cause of the excursion and to implement the remedial steps that should be taken to bring the system back into full compliance without any adverse consequence.

The numerical values for alert and action levels are often set arbitrarily. It may be preferable to set them on a statistical basis. Occasionally, the levels are established as multiples of the standard deviation from the normal; namely, two times the standard deviation for the alert level, and three times the standard deviation for the action level. There is also room for ambiguity in how the alert levels are responded to. The alert level may be addressed by the retesting of samples to make sure the higher numbers are real. The alert response may also include corrective actions in advance of those taken when the action level is reached. The water processor is

Table 3 Numerical Interpretation of USP Standards

Component	Purified water	Water-for-injection
pH	5.0–7.0	5.0–7.0
Chloride (mg/L)	0.2	0.2
Sulfate (mg/L)	1.0	1.0
Ammonia (mg/L)	0.1	0.1
Calcium (mg/L)	1.0	1.0
Carbon dioxide (mg/L)	5.0	5.0
Heavy metals (mg/L)	0.1 as Cu	0.1 as Cu
Oxidizable substances	Pass USP permanganate test	
Total solids (mg/L)	10.0	10.0
Pyrogens (EU/mL by limulus amebocyte lysate)	—	0.25

expected to set their own alert and action levels. FDA inspectors will insist, however, that the records show that these levels are respected and adhered to in practice.

Conductivity Measurements and pH

The standards of chemical acceptability for Purified Water and WFI are shown in Table 3, as translated and quantified from the relevant analytical procedures set forth and described in USP (10,11). A single conductivity measurement in conjunction with pH values, at specific, direct, not compensated temperature readings, is substituted for the total of the specific measurements for chloride, sulfate, ammonia, calcium, and carbon dioxide. (The several containerized sterile compendial waters are still to be analyzed in terms of their specific ionic contents, the long-term effects of the containers on conductivity being unknown.)

A conductivity characterization becomes possible because the relevant impurities are all ionic, or ion producing and, therefore, can be detected in electrical conductivity measurements (or in their reciprocal function, electrical resistivity). The electrical conductivity value read for water will be ascribed to the ions having the specific lowest conductivity. This approach will maximize the presumed concentration of these ions. This has the effect of ensuring that the ionic concentrations within the water do not exceed stipulated limits. Different ions have different conductivity values at different pHs and temperatures.

Conductivity curves modeled on the chloride and ammonium ions overlap. The chloride ions have the lower limiting conductivities from pH 5.0 to 6.2, the ammonium ions from 6.3 to 7.0; the pharmaceutical waters having acceptable pHs in the range 5.0 to 7.0. Temperature, as stated, also affects the equilibrium concentrations and the specific conductance of each ionic species. The actual mix of ions not being known, temperature compensation cannot be used. Direct temperature readings are necessitated.

Testing the suitability of water will be possible at each of three stages. Stage 1 will be assayed by online conductivity tests, a situation presumably free of the influences of carbon dioxide and its ionic pH and conductivity consequences. The temperature of the water will also be read directly, not compensated for. Comparisons will then be made of the conductivity/temperature values with those presented in an official table of acceptable levels. The conductivity at the temperature value equal to or less than the measured temperature defines the acceptable limit.

If the conductivity is equal to or less than the tabulated value, the water quality will be acceptable. If it is greater, a stage 2 determination is made to see whether the higher conductivity is occasioned by the presence of carbon dioxide. In the stage 2 assay, the water is stirred vigorously at 25°C ± 1°C to permit its complete equilibration with atmospheric carbon dioxide. The attainment of equilibrations is measured by the leveling off of the periodically determined change in conductivity. When the net alteration becomes less than 0.1 μS/cm over five minutes, the sample conductivity will be recorded. If it is no greater than 2.1 μS/cm, the water will be deemed to be acceptable. If the conductivity value is greater, the possible influence of pH will be ascertained in a stage 3 assay.

Saturated potassium chloride will be added to the sample examined in the stage 2 test, to enable its pH to be measured. Reference will be made to a predetermined pH/conductivity requirements table to find the acceptable conductivity limit at the measured pH level. Unless either the conductivity of the water is greater than the acceptable limit, or the pH is outside the 5.0 to 7.0 range, the water quality is judged to be proper.

By the same token, if at either of the earlier stages of testing the conductivity is acceptable, then the pH must be within its proper range. Indeed, the original role of pH was to limit the concentration of ions not otherwise specifically identified. Therefore, the specific requirement to directly determine the pH of the water would be redundant where the conductivity is suitable.

Total Organic Carbon Measurements

It is possible to utilize TOC measurements as a substitute for the USP potassium permanganate OST. The acceptability standard is set at 500 ppb.

The TOC-measuring devices available in the marketplace differ significantly in their abilities to detect (by oxidation and its consequences) organic molecules of varying complexities. There is a desire not to exclude from consideration any TOC monitor capable of measuring the presence of organic molecules likely to be found. A suitable reference compound, therefore, must be defined. The criteria are water solubility, non-ionicity, non-volatility, and a sufficient molecular complexity to be difficult for some TOC instrument technologies to accommodate. The reference compound selected for the instrument suitability test was 1,4-benzoquinone. It has the useful properties of being a powder at room temperature, readily available in pure form, relatively safe to handle, and well-defined chemically. In the multi-laboratory testing, its average recovery was the lowest among the organic compounds examined. Its use as the standard for the TOC suitability test, therefore, suggests the greatest challenge to oxidation and is a prudent choice for TOC determinations. The 1,4-benzoquinone recovery is to be within the test limits of 80% to 115% for the TOC instruments to be acceptable. Sucrose serves as the TOC test standard.

Sampling Program

The defining of suitable protocols for sampling and testing, and a rigid adherence to the scheduling is part of the validation program. As Artiss states, "The sampling procedure must be concisely written and then adhered to absolutely to ensure that there is no variability caused by different personnel or procedures" (5,6).

The locations of the sampling valves should be evident on the drawings of the system layout. It is best that sampling ports be installed before and after each purification unit, and before and after storage tanks. All sampling valves should be of the same design. Each valve should be numbered or otherwise unambiguously identified. The valves should have small interior dimension to permit easy opening and flushing under high velocity to ensure the removal of organisms presumed to be contaminating the downstream side of the valve.

A description of the actual manner in which the samples are taken, sample size, containers to be used, method of identifying sample with sampling location and time, equipment employed, and points to be sampled, as well as time frames for analysis to be initiated, and disposition/approval of sample results all should be indicated.

Inherent in sampling is the assumption that the sample is representative of the entire bulk of the water being characterized. Care should be taken not to violate that assumption. Artiss wisely points out that point-of-use samples should be drawn using the actual hose or pipe employed in delivering the water for manufacturing or rinsing. In this way, contamination problems possibly inherent in that pipe or hose will become reflected in the microbiological assay results (5,6).

Figures 5–10 illustrate typical sampling sites, the water components being tested for, and the frequency of the testing and offer certain relevant comments (5,6). It is precisely such sampling schemes that are required as part of the water system validation.

Cleaning and Sanitization

Periodic cleaning or sanitization–sterilization and flushing of the water lines, sampling points, unused legs, and hoses off water transmission lines should be performed. If sanitization of the water system is performed using hot water, such sanitization must be validated. A protocol, record of time and temperature, adequate raw data, formal review, and an evaluation of the final report should be prepared.

If sanitization is performed using a chemical agent (e.g., peracetic acid, hydrogen peroxide, formalin, ozone, plant steam or other antimicrobial agent) analysis for chemical residues should follow sanitization. Residues should be eliminated by subsequent flushing, and residual analysis performed. Details of sanitizations are beyond the bounds of this writing. They are, however, amply described in the literature (16). Documentation, consisting of formal logs, should be kept of these activities. A record of sanitization, equipment replacement, and maintenance is required.

WATER CONSTITUENTS

Natural Waters

Water having the chemical and microbiological qualities required for compounding into drugs is seldom, if ever, found in nature. Being an excellent solvent, it dissolves the ionic species it encounters, its high dielectric constant separating their ionic lattices. It also brings into solution the numerous organic molecules susceptible to hydrogen bonding. Additionally, it suspends and mechanically incorporates colloids and solids. Even falling as pure rain water, it scrubs gases, such as carbon dioxide, from the atmosphere and similarly acquires other impurities from the air.

Purification processes are, therefore, required to eliminate undesirable contaminants from the water intended for drug preparations. Such purification exercises have their price. Therefore, the removal of impurities is made selectively, and only to the extent necessitated by the pharmaceutical application. Degrees of purity in excess of such requirements are not sought because of economic reasons.

Regional Differences

Waters found in nature differ in their extents and kinds of impurities. Geographic and climatic effects in the form of

Location Sample Point	Component	Frequency		Comments
		Validation	Operation	
Raw Water (Potable)	Microbial	Daily	Daily	Review Together to Determine Contact Time
	CL_2 Residual	Daily	Daily	
(1)	Chemical TDS	Daily	Weekly	Fast, Low-Cost Test
(2)	Full Chemical	Weekly	6 Months	
	pH	*	*	Depends On Equip. Use
Sand Filter	Microbial	Daily	Daily	
	CL_2 Residual	Daily	Daily	

Figure 5 Sampling program. (1) TDS, total dissolved solids (by conductivity); (2) may vary considerably depending on source and season. *Source*: From Refs. 5,6.

Location Sample Point	Component	Frequency		Comments
		Validation	Operation	
Carbon Filter	Microbial	Daily	Daily	
	Cl_2 Residual	Daily	Weekly	
D.I. Equipment (3)	Conductivity	Continuous	Continuous	
	Total Solids USP	Daily	Daily	Depends on Use of This Water
	pH	Daily	Daily	Depends on Use of This Water
	Microbial	Daily	Daily	
	Pyrogen	Daily	Weekly	Depends on Use of This Water
	Silical - Colloidal & Dissolved	Daily	Weekly	Depends on Use of This Water
	Resin Analysis	Initial	6 Months	

Figure 6 Sampling program. Will vary depending on service cycle. *Source*: From Refs. 5,6.

Location Sample Point	Component	Frequency		Comments
		Validation	Operation	
Reverse Osmosis Equipment	Microbial	Daily	Daily	
	pH	Continuous	Continuous	Critical on Some Equipment
	CL_2 Residual	Continuous	Continuous	
	Pyrogen	Daily	Daily	Depends on Use
	Conductivity	Continuous	Continuous	
	Chemical USP	Daily	Daily	Depends on Use
	Feedwater Hardness	Daily	Daily	Critical on Some Equipment

Figure 7 Sampling program. Check individual modules during validation period and weekly thereafter. *Source*: From Refs. 5,6.

Location Sample Point	Component	Frequency		Comments
		Validation	Operation	
Distillation Equipment (Assume USP - Water For Injection)	Microbial	Multilple Times in Cycle	Daily	
	pH		Daily	
	Pyrogen		Daily	
	Conductivity	Continuous	Continuous	Inlet And Outlet
	Chemical - USP	Multiple Times in Cycle	Daily	
	Blowdown - TDS		Weekly	
	Particulates		Weekly	

Figure 8 Sampling program. For reactive distillation and distillation equipment, establish repeatability and time for system stabilization. *Source*: From Refs. 5,6.

Location Sample Point	Component	Frequency		Comments
		Validation	Operation	
Storage	Microbial	Multiple Times in Cycle	Daily	
	pH		Daily	
	Pyrogen		Daily	If Req'd for WFI
	Chemical USP		Daily	

Figure 9 Sampling program. Sample period to be daily or to coincide with batch of water. *Source*: From Refs. 5,6.

Location Sample Point	Component	Frequency		Comments
		Validation	Operation	
Distribution Use Points	Microbial	Daily	Weekly	On Rotation
	Pyrogen	Daily	(6)	
(4)	Chemical TDS	Daily	Monthly	Fast-Low Cost Test
(5)	Chemical USP	Weekly	(6)	
	Particulates	Daily	Monthly	
	pH	Weekly	(6)	
Clean Steam Generator	Blowdown Chemical TDS	Daily	Weekly	To Prevent Scale Build-Up

Figure 10 Sampling program. (4) TDS, by conductivity; (5) TDS, by evaporation; (6) sample only when indicated by failure to satisfy other tests. *Source*: From Refs. 5,6.

rainfall, geological makeup, and industrial or agricultural activities have their influences. Well waters are generally constant in their compositions, but may have a high hardness, often approaching 20 grains. If the water has encountered limestone deposits it may have TDS contents of 300 to 400 ppm and may consist of 75% hardness and alkalinity. Such ground waters may also contain iron and manganese. Generally, however, well waters are low in organic components.

Source waters in many areas may be largely surface waters. Because the water is a product of land run-off, it may exhibit large TOC, seasonally as high as 10 ppm, and high TSS, except where first processed by a municipal authority. Such surface waters generally are limited in their hardness, some six to seven grains, and their TDS is often below 300 ppm.

Although well waters are relatively constant in composition, surface waters are less so. Lakes and reservoirs are subject to seasonal turnovers during which their upper strata, densified by a lowering of temperature, slip to the bottom. A water roiled with sediments results. According to Cooper, it is not unreasonable to assume that microbial counts, and perhaps endotoxin levels as well, increase during summer months (12).

To an extent greater than wells and aquifers, surface waters exhibit the effects of diluting rainfalls, and their colloidal and suspended solids-engulfing run-off. Crits gives a generalized characterization of various source waters relative to the impurities contained (17).

Site Specificity

Because there is no standard source water, there can be no single or universal purification treatment, nor any single protocol adequate to represent the problems of water purification. The first rule in pharmaceutical water purification is that purification requirements are uniquely site-specific. This reflects not only the individuality of the water impurities, and their constancy or lack thereof, but also the flow rates, usage volumes, and peak loads of the prepared pharmaceutical water.

Source Water Analyses

Obviously, to properly purify water, the identities and extents of its impurities must be known. In broad classifications, its TDS, TSS, and TOC require analytical quantification. Table 4 provides a limited list of ingredients for which analysis should be made, and Table 5 illustrates a more expanded listing. Even more detailed listings are possible. Thus, although calcium, magnesium, and alkalinity generally serve to characterize the hardness of water, it may be prudent, for reasons soon to be considered, to also analyze for barium and strontium.

The impurities that are the subject of the source water analysis are of concern for two reasons. In the first,

Table 4 Analysis of Well Water and Surface Water for Scale-Forming Elements[a]

Well water		Surface water	
Element	**Concentration**	**Element**	**Concentration**
Calcium	49.50 ppm	Calcium	15.75 ppm
Magnesium	13.60 ppm	Magnesium	2.47 ppm
Sodium	156.60 ppm	Sodium	10.90 ppm
Potassium	3.59 ppm	Potassium	2.30 ppm
Bicarbonate alkalinity	35.00 ppm	Bicarbonate alkalinity	10.00 ppm
Sulfate	22.06 ppm	Sulfate	7.49 ppm
Chloride	62.75 ppm	Chloride	12.84 ppm
Phosphate	0.01 ppm	Phosphate	0.68 ppm
Nitrate	3.46 ppm	Nitrate	0.45 ppm
Free CO_2	23.59 ppm	Free CO_2	3.53 ppm
Silica	9.71 ppm	Silica	3.18 ppm
Iron, as Fe	0.03 ppm	Iron, as Fe	0.10 ppm
Manganese, as Mn	0.01 ppm	Manganese, as Mn	0.01 ppm
pH	6.4	pH	6.8
Total dissolved solids, as NaCl	128.26 ppm	TDS, as NaCl	34.70 ppm
Turbidity, nephelometric turbidity unit	0.46	Turbidity, NTU	1.35
Color, Pt-Co units	11.80	Color, Pt-Co units	21.00
Organics, as O_2 consumed	8.10	Organics, as O_2 consumed	8.10

[a] Expressed in ppm $CaCO_2$, unless otherwise stated.
Source: From Ref. 20.

the removal, at least to certain levels, is required to accord with the analytical specifications of Purified Water or WFI as stipulated for these compendial waters in various national or regional pharmacopeias. Table 3 identifies ingredients listed in the USP. Calcium, sulfate, and chloride ions are examples of such ingredients. Secondly, another group of impurities requires elimination because their presence would compromise the efficiency or service lives of the water purification units (e.g., ion-exchange, RO, or distillation) employed to prepare the compendial water. Silica is an example of such an impurity. Its presence, while not proscribed by the USP, can cause fouling of RO membranes and vitrification of still surfaces, much to their detriment.

There may be a third reason for analyzing for given elements or compounds. In the U.S.A., the USP specifies that pharmaceutical waters must be prepared from source waters of potable quality. This stipulation is made in the USP monograph section, has the force of regulations, and is enforceable by the FDA. The specifications for potable water are the province of the EPA. That agency's stipulations on heavy metals in potable waters are more restrictive than are the specifications by USP on Purified Water and WFI. Therefore, analysis of the source water for heavy metals would seem to be necessary. Not all pharmaceutical source waters are of the potable quality. That is usually a consequence of prior processing by a central water authority. Non-potable source waters, therefore, have to be brought to potable water quality in the process of being purified to the suitable compendial standard. The original source water may contain heavy metals, for example, in unacceptable concentrations; the finished pharmaceutical water should not. It is not necessary to segregate the water during its purification into a potable water stage. It must, however, be ascertained that at some point in the water purification process water of potable quality has been achieved.

In the case of *Escherichia coli*, its absence must be ensured, as mandated by EPA potable water standards. This is usually achieved by the early chlorination of the source water. The absence of *E. coli* thus provided and tested for, the subsequent testing for *E. coli* in the prepared compendial waters need not be reaffirmed by analysis. The same logic should apply to the testing for heavy metals.

REASONS FOR PRETREATMENTS

Ion exchange, RO, and distillation are the principal methods of water purification in pharmaceutical system design. More recently, electrically driven purification units such as continuous deionization have been applied. In principle, any of these unit purification processes can furnish water of requisite pharmaceutical quality. (Source waters having high chloride ion contents may require two pass, product-staged RO treatments.) The question is whether such purifications will offer realistic service lives, or whether the processes may be curtailed, overwhelmed by the quantity of impurities normally present in the feedwater. Pretreatment techniques are instituted to prolong the useful service lives of the principal purification units to extents that are economically practical. The appropriate pretreatment selectively removes or diminishes part of the impurity burden to the point at which compromise of the principal purification unit is eliminated or becomes tolerable. For example, the presence of certain types of TOC in the feedwater may prematurely foul a RO purification unit. The use of an activated carbon bed in a pretreatment mode to remove all, or some, of the TOC by absorption can prolong the life of the RO to an extent that is operationally practical. The definition of what is "practical" in water purification contexts has an economic component. Water purification exercises are

Table 5 Analysis of a Source Water

	Analysis ppm as such	Conversion factor (*x*)	Analysis ppm as $CaCO_3$	MEQH
Cations				
Calcium (Ca^{2+})	60.0	2.50	150	3.00
Magnesium (Mg^{2+})	7.3	4.12	30	0.60
Sodium (Na^{+})	50.5	2.18	110	2.19
Potassium (K^{+})	7.8	1.28	10	0.20
Hydrogen=fredericamycin A (H^{+})		—		
Total cations			300	5.99
Anions				
Bicarbonate (HCO_3^-)	183.0	0.82	150	3.00
Carbonate (CO_3^{-2})	—	0.5	—	—
Hydroxide (OH^-)	—	2.94	—	—
Sulfate (SO_4^{-2})	53.8	1.04	56	1.12
Chlorate (Cl^-)	63.8	1.41	90	1.79
Nitrate (NO_3^-)	2.5	0.81	2	0.04
Phosphate(Ortho) (PO_4^{-3})	1.3	1.58	2	0.04
Total anions			300	5.99
Total hardness ($CaCO_3$)	—	—	180	
MO alkalinity ($CaCO_3$)	—	—	150	
pH alkalinity ($CaCO_3$)	—	—	—	
Carbon dioxide (CO_2)	211	1.15	2.4	
Silica (reactive) (SiO_2)	30	0.83	24.9	
Silica (non-reactive)	5			
Iron (Fe)	2			
Manganese (Mn)	0.1			
Chlorine, free (Cl)	0.5			
Total dissolved solids	360			
Chemical oxygen demand by permanganate	5			
Total organic carbon	8			
Turbidity	2			
Color	—			
Specific conductance (μohm/cm at 25°C)	660			
Specific resistance (ohm/cm at 25°C)	1520			
Temperature (°F)	55–68			
pH	8.1			
pH^s	7.62			
Langelier index	0.48			

Source: Courtesy of Continental Penfield Water.

not simply technical practices; they are technicoeconomic undertakings.

The practice of pretreatment is an essential part of the pharmaceutical water purification operation, particularly so when the composition of the source water is so sufficiently variable over short-term intervals that it sporadically threats to overwhelm the principal purification unit. In such cases, outsized pretreatment units may become necessary. The proper matching of pretreatment arrangements to whatever principal purification operations are employed, be they ion-exchange, RO, distillation, or other, is essential in establishing the design and operatinal SOPs for the water purification system. Because the pretreatment unit operations may greatly influence the principal purification results, they require validation along with the principal units.

Specific Impurities

Total Suspended Solids

It is necessary to reduce or remove suspended and colloidal matter from the feedwater entering the principal purification units. Ion-exchange beds, in addition to their demineralizing functions, serve as deep-bed filters. They are composed of a packed volume and depth of resin beads, generally between 16 and 50 mesh size, although the range of 20 to 40 mesh is often preferred. They accumulate particulate matter within their interstices precisely in the manner of deep-bed filters. The accumulated particulate material, the suspended matter present in the feedwater, causes elevated pressure differentials within the ion-exchange beds. This occurrence slows the flow of water, may foul macroreticulated resin beads, and otherwise offers spatial interferences that will detract from the intended ion-exchange function.

In RO operations, suspended matter will foul by blockage of areas of the RO membrane, effectively removing the affected area from useful contributions. Unlike the ion-exchange beds, RO devices will retain colloids, further interfering with the RO ion-removal function.

The accumulation of suspended matter in distillation equipment interferes with the designed heat-transfer effects and increases the likelihood of particle entrainment in the vapor. It is this very occurrence that is the object of the blow downs, largely automatic, that are part of still operations.

To forestall the fouling occasioned by suspended matter, such materials are removed by the use of sand filters or multimedia deep-bed filters, or both, in a pretreatment step. Colloid removal is accomplished by the use of coagulation and flocculation, usually by the addition of alum or a polyelectrolyte. (Such treatment is usually standard in potable water system.)

The colloidal content and nature of a water should be known, particularly if the water will be fed to a RO unit and especially if hollow-fiber RO units, with their more easily blocked small bores are involved. The SDI test, however fallible, provides such information. Feedwater exhibiting SDIs in excess of values of five for spiral-wound ROs and of three or less for hollow-fiber units require coagulation and flocculation pretreatment.

Scale-Forming Elements

In Tables 4 and 5, analyses are cited for pH, calcium, magnesium, alkalinity, sulfate, silica, and other impurities. These are entities that in certain concentrations and combinations produce precipitates and deposits that will interfere with the principal purification unit operations. A chief concern is the hardness of the elements. Calcium ions combine with sulfate ions to form insoluble calcium sulfate deposits. The avoidance of such scale formation should be particularly sought in RO operations. Calcium carbonate can also be deposited as interfering scale, and its formation from soluble calcium bicarbonate is a consequence of a shift in the pH of the water to the alkaline side. The Langelier Solubility Index provides a measure of a water's tendency to form calcium carbonate scale and indicates pH adjustment toward the acidic is necessary to avoid such formation (19).

Because the precipitation of salts of limited solubility reflects the solubility products, the concentrations of the relevant ions should be known through analyses. The TDS content of a feedwater, therefore, should be quantified.

The formation of scale and deposits of other compositions are also possibilities; calcium fluoride, voluminous magnesium hydroxide, and silica deposits are among them. Appropriate pH management, and the use of anti-scalants that are promotive of calcium sulfate supersaturation can be effective in preventing scale. The use, mechanism of action as well as water system design considerations, in general, are beyond the scope of this writing. Interested parties are directed to pertinent sources for such information (17).

The most effective way of avoiding scale formation by the bivalent alkaline earth elements that represent permanent water hardness, and the bicarbonate ion that defines temporary hardness, is to rely on a water-softening operation. Usually, ion-exchange resins in their sodium form are used to remove calcium, magnesium, barium, and strontium. The latter two elements, if not removed in the water-softening pretreatment, can so irrevocably combine with cation-exchange resins that they impair the resins' capacity for regeneration. Alternatively, the deposition of insoluble barium sulfate scale and, to a somewhat lesser extent, strontium sulfate on a RO membrane, will prove recalcitrant in the extreme to removal. Therefore, one should analyze for the presence of barium and strontium, and if present, their removal in the water-softening step should be affirmed.

Water softening will also serve to remove soluble iron and manganese. These elements are commonly removed in a separate pretreatment step involving their oxidation, whether by chlorine or green sand zeolite, followed by the filtrative removal of the insoluble oxidized product, usually by deep sand beds or by multimedia deep-beds. Where smaller quantities of iron or manganese are involved, their removal by way of cation-exchange resins may be more economical.

Temporary water hardness caused by bicarbonate ion is eliminated by the addition of acid to pHs below 4.4. This converts the bicarbonate to water and carbon dioxide; the latter is removed from the system by a degasification or de-carbonation unit. Such acidification becomes necessary when RO units are composed of cellulose acetate, to minimize the hydrolysis of that polymer. Bicarbonate ions are also eliminated by rejection, by polyamide RO units at pHs above 8 or 8.5. However, care must be taken not to expose the polyamide RO membrane to the ruinous effects of chlorine.

The alkalinity of water is described as the sum of its titratable bases. It is chiefly a measure of the bicarbonate, carbonate, and hydroxide ions present: there is a pH-dependent equilibrium among carbon dioxide, bicarbonate, and carbonate. The strength of the bicarbonate concentration can be measured by titrating the water with acid to bring it to a methyl orange endpoint of approximately 4.4.

Soluble silica can be removed from water by the use of a strong base ion exchange or by rejection in the RO operation. Soluble silica should not be permitted to enter stills, although in certain successful water purification operations it is actually permitted to do so. Curiously, soluble silica is not removed in any pretreatment step. None has been designed for that purpose, unless strong-base cation exchange is considered.

Total Organic Carbon

In formal terms, TOC means total organic carbon. As such it is a misnomer. At most, it signifies total oxidizable carbon, as defined by an automated TOC analyzer. There are several such TOC instruments, each with different oxidizing capabilities. The definition of TOC may vary among them since organic compounds each have their own susceptibility to oxidation, and very little is known about the TOC constituents of any water. The readings of the various TOC instruments will be standardized by the use of reference compounds (20). These present developments are an improvement over the USP's traditional "Oxidizable Substances Test," an analysis noteworthy for its insensitivity.

There is no universal means for removing TOC from waters; it depends on the specific nature of the organics involved. Some organic compounds cannot be removed with an assured efficiency. Largely, adsorption of TOC to activated carbon granules is relied on. The flow rates involved in such an exercise are low, about 1 gpm/cubic foot of carbon. By contrast, the removal of chlorine by activated carbon is possible at double or triples these flow rates. Organics, with some notable low molecular weight exceptions, such as phenols, acetic acid, and alcohols, are rejected by RO membranes. Ultrafiltration can also be used to retain organic compounds. Some still designs incorporate the oxidative means to destruct organic materials in the still pots. Ion-exchange resins are also relied on to remove TOC. Organic molecules containing carboxylic acid moieties, usually the consequences of oxidative alterations, can be removed by anion exchange. Advantage can be taken of this to subject the TOC to oxidation by ozone, ultraviolet light, or hydrogen peroxide, or by various combinations of these agencies. The resulting oxidized TOC, bearing carboxylic acid groups that normally characterize one stage of the TOC oxidative degradation chain, is then amenable to removal by anion exchange. The THM, which are themselves the end products of organic matter oxidized by chlorine, are exceedingly difficult to remove from water by any means, especially chloroform. Adsorption to activated carbon has only an indifferent success; TOC can be removed by adsorption to ion-exchange resin beads because of the large surface areas these present. The danger is, however, that the beads may become excessively fouled, thereby compromising their intended function of ion exchange. Therefore, sacrificial resin beds are sometimes used in a pretreatment mode to achieve TOC removal.

Microorganisms

Microorganisms are perhaps the most insidious of the impurities present in source waters. Other impurities, once removed, remain removed, but organisms, even when removed to the state of sterility, can reinvade the water and multiply to significant populations. Organisms and their derivative pyrogenic lipopolysaccharides, the bacterial endotoxins, require attention throughout the system and limitation in their concentrations in water intended for pharmaceutical applications.

Once freed of microorganisms to the specified acceptable degree, the pharmaceutical water, whether Purified Water or WFI, should be maintained during its storage and transfer, to the same degree of microbiological and bacterial endotoxin purity. A common practice is to leave chlorine in the water as long as possible in the pretreatment system to help control the microbial population.

Ionic Constituents

The chemical impurities, the presence and quantification of which are the subjects of source water analyses, are almost all ionic. Calcium, magnesium, barium, strontium, sodium, and potassium are all cations. The pH measurement reveals the concentration of the hydrogen ion (more properly, hydronium ion), also cationic or, at its higher values, of the hydroxyl anion. Sulfate, nitrate, and chloride are anions. Carbon dioxide, the acid anhydride of carbonic acid, relates to the existence of bicarbonate or carbonate ions. These latter two anions, along with the hydroxyl ions, also anionic, constitute the water alkalinity. Ammonia is the anhydride base of ammonium hydroxide. When added to water, it yields hydroxyl ions through the feeble dissociation of ammonium hydroxide into ammonium cations and hydroxyl anions.

Ion-exchange practices or RO processes are the principal purification methods relied on to remove these ionic impurities. Distillation is the other well-established purification method commonly employed for this purpose. (Certain electrically motivated techniques may also be used to effect de-ionization or de-mineralization. Their use, although small, is on the increase. Chief among these is an electrodeionization process called continuous deionization by its manufacturer.) However, distillation does not suffice for the removal of carbon dioxide or of other volatile impurities having significant water solubilities and which, in consequence, may remain to some degree in the condensed distillate.

SYSTEM DESIGN CONSIDERATIONS

The Purification Unit Processes

As the concept of process validation has it, if each unit process of a water purification system is demonstrated and documented to be operating as designed and expected, then the sum of those units processes, the total system, must, of necessity, be dependable in its production of the water quality that is proper for its intended purpose.

The design of water systems being quite site-specific, it is difficult to generalize concerning the unit process components of a "typical" system. Broadly speaking, however, pharmaceutical water purification arrangements could involve the following sequence:

1. A chlorination unit to supply the means of controlling the organism content of the feedwater (this may be already be present if the source water is a municipality).
2. A sand or multimedia bed to remove suspended solids down to 10 to 40 μm in size (also potentially provided by a municipality).
3. A water softener to remove scale-forming ions. These are principally the divalent cations that can yield insoluble sulfate, carbonate, hydroxide, or fluoride salts that preempt ion-exchange sites or block RO or other membrane areas, or that add scale to stills, all of which interfere with later purification activities. When the alkalinity content of the water would so indicate, the presence of bicarbonate must be taken into account, lest its generation of carbon dioxide unfavorably influence the pH of the product water.
4. An activated carbon bed to remove chlorine and TOC: The presence of TOC in the water could result in the choice of anion-exchange resin traps to remove it. Otherwise the water could be flowed directly into activated carbon beds for the same purpose, but more especially to remove the chlorine that had previously been added. The chlorine could also be removed by the addition of chemical reducing agents. The removal of volatile TOCs, and particularly of the THM could well depend on the use of vacuum degasifiers.

The foregoing pretreatment steps having been accomplished, some principal means for deionizing the water would next be selected. This would call for the use of ion-exchange, of RO, or a combination of the two. When RO is used, it could be in the two-pass product-staged mode. In parenteral water manufacturing contexts, distillation is usually the principal purification unit process used.

A decision to use RO would involve a selection of the membrane type. The use of cellulose acetate RO, often selected in pharmaceutical contexts, would entail acidification to about pH 5.5 to 6 and, if bicarbonate is present, carbon dioxide removal. This would involve the employment of degasification equipment.

In addition to these, various accoutrements could be used, such as ozone, ultraviolet units, filters of various sorts and ratings, chemical additions of anti-scalants or of the acids or bases necessary for pH adjustments.

Chlorination

Because organisms, particularly the gram-negative varieties native to aqueous habitats, can proliferate in water, their control is sought early on. Municipally treated waters generally arrive chlorinated, or containing a given concentration of some biostat, generally chloramine. Otherwise, the source water is chlorinated as soon as it is acquired by the pharmaceutical plant.

Usually, the chlorine content of the feedwater is adjusted to a residual of 0.5 to 2 ppm. On occasion, higher concentrations are used: 2 to 6 ppm, when there is concern over *Legionella pneumophilia* (21). Whatever chlorine level is selected, its concentration is maintained at that level to compensate for the quantity expended on chlorine-oxidizable substances present in the water. When sand beds are employed to remove suspended matter, they are often periodically sanitized by being subjected to the "superchlorination" of 5 to 10 ppm chlorine for from 15 to 20 minutes.

The purpose for introducing chlorine into the water may be to eliminate live organisms completely, or to regulate them to a particular count. It is important that the goal intended by the chlorination be defined with some exactness, for that objective has significance in the OQ step of the validation process. The protective chlorine umbrella is maintained over the water for as long as possible as it is being processed, usually through the water-softening pretreatment. Eventually, however, it must be removed lest its oxidizing action cause harm to certain of the water purification units downstream. Polyamide RO membranes are ruinously susceptible to chlorine; stainless steel stills are corroded thereby, and even by the chloride ion that is a product of chlorine in water.

$$Cl_2 + H_2O \cdot HOCl + H^+ + Cl^-$$

Ion-exchange resins are also oxidatively degraded by chlorine. They suffer a loss of capacity and generate TOC (24). Nevertheless, it is not unusual for the chlorine to be removed by permitting its entry into the ion-exchange beds. Whether this action is wise or not depends on how much TOC generation and how much ion-exchange degradation is tolerable. If the ion-exchange unit has a polishing function, chlorine should not be allowed to contact the resin, because the negative consequences of the resin degradations would be excessive.

Chlorine removal is normally accomplished by its adsorption to and reaction with activated carbon. Carbon beds have been used successfully for this purpose for almost 30 years in at least one pharmaceutical company. However, the proper maintenance of carbon beds is often seen to be onerous. Moreover, all too commonly, the carbon bed construction, usually in the interests of economy, does not permit its sanitization by hot water or steam. The result is that the carbon bed can then serve as a focal point of organism contamination. Chlorine is, therefore, in some 30% of all instances, removed by chemical injection, usually of sodium bisulfite or metabisulfite, into the water. The chlorine undergoes destruction, its oxidizing power being neutralized by reaction with the reducing agent. Chloride ions are a product of this chemical reaction. Their subsequent removal is necessitated.

Chloramination

The EPA decrees that the potable waters leaving water purification facilities must contain some biocidal residuals to offer antimicrobial protection during the water's distribution to consumers. Chlorine is eschewed because its oxidative powers can turn TOC into the carcinogenic THMs. Therefore, increasingly the chlorine is converted to chloramines, of lesser oxidation potential incapable of THM formation (or less so), as the biocidal residue. Because the chloramines are less reactive chemically than chlorine, they react at lower rates or find fewer molecular species with which to react. As a result, they last longer. In this regard, they are more stable, there being a reciprocity between stability and reactivity, usually expressed as a time/concentration relationship. However, over time the oxidative strength of the chloramides is enough to deteriorate RO membranes.

The removal of the chloramines is managed by their adsorptive contact with activated carbon. The preferred mixture of the chloramines consists of two-thirds monochloramine and one-third dichloramine. This composition adsorbs evenly to carbon, the individual adsorption rates of the components becoming balanced by their 2:1 ratio. The uptake is quite slow, a flow rate of about 0.5 gpm/ft^3 or 3.785 liters per 0.0283 m^3. Ammonia is a product of the adsorption reaction. The very high solubility of ammonia in water makes its removal problematic. The distillation of ammoniated waters for its removal is unreliable because of its high water solubility. RO does not remove ammonia whether in its NH_3 or NH_4^+ form. At higher pHs ammonia is converted to the ammonium ion. As ammonium ion, NH_4^+, it can be removed by use of cation exchange, but only with difficulty because it is just above sodium in the displacement hierarchy. It is lower, however, in the series than the bivalent hardness elements. In practice, dual sodium-form ion-exchange softener units in series are utilized to effect its removal.

The newest means of removing chlorine and chloramines is by ultraviolet destruction. The same 254 nm wavelength that is used in UV sterilizations is employed. The rule of thumb is that 10 times the dosage required for sterilizations is needed for the removal of 1.0 ppm chlorine. The system is available in modular form suitable for the removal of 0.5 to 2.2 ppm of chlorine. Although successful in many trial and application settings, the technique is considered as being in its development stage.

Ozone

Ozone is a more efficient biocide than chlorine. It is more rapid in its killing action and is more effective against a wider variety of microbes, including viruses. Its action against *E. coli* is 3125 times as rapid as that of chlorine (23). Indeed, no microorganisms seem immune to its cidal effects. Additionally, ozone offers the significant advantage of being removable in seconds by 254-nm-ultraviolet radiation. Unlike the removal of chlorine, the elimination of ozone requires no difficult-to-maintain carbon beds, chemical injections, or prolonged effort. In quantitative terms, 90,000 rads/cm^2 per second of UV light energy are required to eliminate completely 1 ppm of ozone. An ample dosage of UV radiation can be supplied to waters moving at 40% the velocity used for recirculation when the germicidal effects of ultraviolet light is sought. For example, if a flow rate of 60 gpm is set to achieve germicidal effects, then a flow rate of 24 gpm should be used to attain ozone destruction. Ozone concentrations of less than 1 ppm can be removed at even higher flow rates.

Different concentrations of ozone over different contact times are used to eliminate viruses, spores, and bacteria. Residual ozone levels of 0.4 mg/L have been maintained for 8 to 12 minutes for the primary disinfection of potable water supplies; 0.2 to 0.5 mg/L concentrations have been relied on over a 10-minute contact time, when spore populations are involved; and 0.01 to 1.0 mg/L amounts of ozone have been employed over five minute durations for the destruction of a concentration of vegetative organisms of 10^6 to 10^7/mL (24).

Ultraviolet Radiation

Ultraviolet wavelengths have germicidal effects. They produce photochemical reactions involving biomolecules in organisms. The resulting molecular alterations inhibit the growth of microorganisms, and in higher doses will kill them. The germicidal effectiveness of UV radiation depends on its wavelength, and different organisms show different sensitivities to various parts of the UV spectrum. Table 6 shows the UV resistance data to 100-μW/cm^2 intensities for selected organism strains. Figure 11 illustrates the roughly Gaussian germicidal action curve whereby germicidal effectiveness is plotted against wavelength. Disinfection action depends not only on the UV emission spectrum, but also on the radiation intensity, the duration of the organism exposure, the sensitivity of the organism involved, and the UV transmission of the medium that suspends the organism. Suspended matter will reflect and block UV transmission, but even

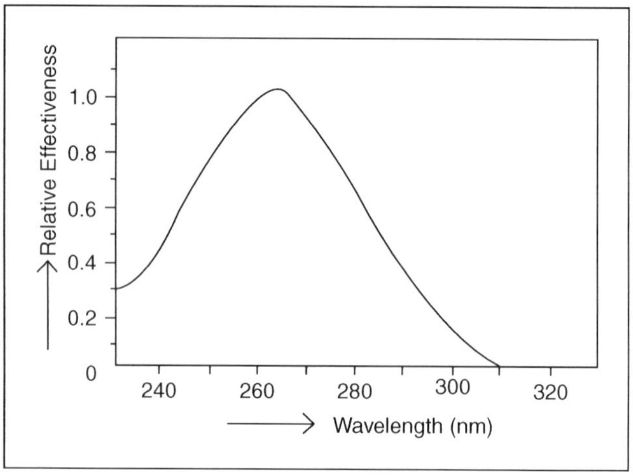

Figure 11 Germicidal action curve in the ultraviolet region of the energy spectrum. *Source*: From Ref. 26.

Table 6 Resistance Data on Selected Organisms Exposed to 100 U.W/cm^2 UV Intensity

Organism	*D*-value[a] Test 1	Test 2	Test 3
Micrococcus radiodurans	33.25	32.99	33.06
ATCC 13939	(0.974)	(0.988)	(0.981)
Staphylococcus aureus	0.90	0.97	0.92
ATCC 6528	(0.988)	(0.986)	(0.986)
Streptococcus faecium	2.00	1.97	2.02
ATCC 10541	(0.988)	(0.986)	(0.988)
Candida albacans	7.31	7.51	7.55
ATCC 10231	(0.993)	(0.995)	(0.989)
Escherichia coli	1.36	1.38	1.33
ATCC 8739	(0.983)	(0.989)	(0.998)
Pseudomonas aeruginosa	0.63	0.57	0.60
ATCC 9027	(0.992)	(0.994)	(0.987)
Pseudomonas diminuta	1.23	1.27	1.20
ATCC 19146	(0.984)	(0.981)	(0.983)
P. diminuta	1.94	2.02	1.97
ATCC 11568	(0.992)	(0.990)	(0.993)
Pseudomonasmaltophilia	1.14	1.17	1.19
ATCC 13637	(0.978)	(0.981)	(0.985)
Pseudomonas cepacia	0.96	0.95	1.00
ATCC 25416	(0.988)	(0.982)	(0.987)
Pseudomonas Putrefaciens	1.48	1.41	1.44
ATCC 8071	(0.990)	(0.987)	(0.988)

[a] UV dose obtained using UV intensity of 400 mW/cm^2 for various exposure times.
Source: From Ref. 25.

optically clear solutions may contain entities, such as certain hydrocarbons, sugars, colored materials, and iron or manganese salts, that can adsorb UV emissions and so prevent their reaching the suspended organism. Moreover, UV lamps do not have standard outputs, change in their radiation qualities over time, and are subject to fouling during their use. For all these reasons, the performance of a UV installation is hardly a constant exercise, the variables requiring continual reassessment and attention. Therefore, UV radiation is not considered an absolute method of killing organisms. In particular, the radiation exposure time is an expression of the water flow velocity and of the geometry of the radiation chamber. These, in addition to the quantity of energy delivered by the UV lamp, are the determinants of the effectiveness of the UV installation. It is the UV dosage reaching the organism that is the real consideration. It is expressed as the product of the radiation intensity and the exposure time, in terms of microwatts-seconds per square centimeter ($\mu Ws/cm^2$). The control of the several factors governing the performance of the UV operation must be assured in its validation exercise.

The oxidation of TOC by UV light, particularly of the 185-nm wavelength, proceeds through the agency of free hydroxyl radicals. These become generated by the action of the UV radiation on the water molecules. The action is heightened by the presence of hydrogen peroxide, and by the synergistic effects of ozone (16). The oxidative alteration of the TOC, as previously stated, culminates with the appearance of carboxylic acid groups at one stage of the oxidation chain reaction. The presence of these permits the removal of the altered TOC by anion-exchange.

Multimedia Deep Beds

The use of deep-bed sand filters and of multimedia deep-bed filters for the removal of suspended matter has already been alluded to. Silica sand is commonly used in sand bed constructions (27). Such beds have nominal porosities of 10 to 40 µm; newer beds have lower porosities. The size of the sand grains determines the packing densities. Suspended matter retention takes place in the top 6 inches or so of the bed. The remaining depth of the bed serves to regulate the flow of water through the bed. At moderate rates of flow (4–16 gpm/ft^2 for rapid sand beds, 2–3 gpm/ft^2 for slow sand beds), the flow volumes are defined by the extent of bed surface. Actual flows will depend on the SDI of the feedwater. Pre-RO sand beds require flows as low as 1 gpm/ft^2 to ensure the sufficient removal of the suspended particles.

In the construction of a sand bed, the sand grains stirred into the water settle, in response to Stoke's law, in such a fashion that they progressively segregate the particles with the smallest sizes in the top layers. The smallest particles are hydraulically floated to the top where they form the densest layer. The result is a deep-bed filter, with the finer pores upstream and the coarser pores downstream. In consequence, the filtration performed by the bed occurs largely in its top layers. If, however, different granular media are mixed, for instance sand, anthracite, and garnet rock, then the formation of the bed will comprise three strata. Anthracite, the least dense at 1.5, would form the top layer; the silica sand, having a density of 2.5, would constitute the intermediary layer; and the garnet, with a density of 3.5 to 4.5, would constitute the bottom layer. Such a multimedia bed would perform filtrations at three levels. If, in addition, the succeeding two denser layers were ground progressively finer, they would pack closer to give the multi-structure an overall funnel shape, conducive to accepting higher particulate loadings (Fig. 12). The factors to control in multimedia bed construction are the densities of the granules and the fineness of their size. Interestingly, a multimedia bed construction of five layers, consisting of five different sizes of silica sand, has been reported.

The accumulation of suspended matter by the deep-bed filters occasions a progressive buildup of differential pressure, leading eventually to an unacceptably low rate of water flow. Usually at the development of about 15 psi, differential pressure, the bed is backwashed to a "quick-sand" consistency to rid it of its accumulated matter. It is then allowed to drain and resettle into its original configuration for reuse.

During the operation of the deep-bed filter, its SOP is established. It describes the manipulations and operational procedures necessary to secure the proper deep-bed performance, including the backwash exercise, and the sanitization by chlorine (previously mentioned). During the validation of the deep-bed operation, the adequacy of the SOP becomes attested to by trial.

Carbon Bed Operation

One example of a successful carbon bed operation may serve to focus what can be a diverse and sometimes contradictory experience. One pharmaceutical company

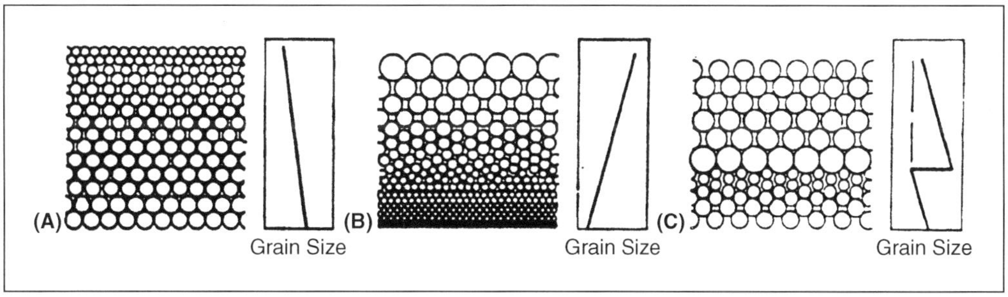

Figure 12 Cross-sections of representative filter particle gradations: (**A**) A single-media bed, such as a rapid sand filter. The bottom half of a filter of this type does little or no work. (**B**) An ideal filter, uniformly graded from coarse to fine from top to bottom. (**C**) A dual-media bed, with coarse coal above fine sand, which approaches the goal of the ideal filter. *Source*: Courtesy of the American Water Works Association.

has, for some three decades, been using a carbon bed downstream from sand beds and separated by filters from downstream ion-exchange beds. The carbon tank or shell is six feet high and 54 inches in diameter, and contains about 50% of freeboard. The normal effluent rate is about 120 gpm. The beds are backwashed twice daily, largely to cleanse them of iron deposits. The backwash is at the rate of 200 to 250 gpm. Microbial assays are performed on alternate days, thrice weekly. Microbial alert limits are set at 600 to 700 cfu/mL. The action limit is 1000 cfu/mL for three consecutive days on the cold water system, ascertained as total heterotrophic plate counts. This action limit invokes hot (65°C) water sanitization. The heated water is flushed into the bed and is then trickled to a total volume of 500 to 1000 gal in an overnight operation during a weekend. Carbon fines are removed from new beds by an upward flush (backwash), barely vigorous enough to overflow the fines to drain. This backflow/fines-removal is done overnight. The successful operation of the carbon bed is ascribed to its continuous recirculating flow from its inception. The flow, through a 1- to 1/2-inch line capable of delivering about 30 gpm, is at a minimum of from 25 to 40 gpm (approximately 10–2 gpm/ft^2) regardless of whether water is being supplied to the downstream ion-exchange beds or not. The return loop to the carbon bed is by way of the preceding sand beds. In summary, the three elements of this carbon bed's maintenance are continuous recirculation, twice-daily automated backwashes, and weekly sanitizing with 65°C hot water.

The Water Softener

The brine makeup tank may itself serve as a haven for organism proliferation. To minimize this possibility, the brine should be maintained in a clean area under closed conditions, at a saturated concentration, agitated (preferably by recirculation), and should periodically be prepared fresh.

It is good practice to use two softeners that are out-of-phase by design, so that one is being regenerated while one is operative. To avoid organism growth, softeners not in use should be kept in recharged condition with 26% brine, ready to be flushed free of the brine and thus made water-operative on signal. The addition of calcium hypochlorite pellets to the salt supply helps keep the latter sanitized. Also, wherever possible, hot water sanitization of the water softener should be performed at 65°C to 90°C. The cation-exchange resin survives heating at 90°C.

Water contaminated with organisms derived from the water-softening operation will inoculate the ion-exchangers that follow in sequence. For this reason, use is made of ultraviolet light units and of organism-retaining filters to minimize such possibilities.

Principal Unit Purification Processes

As stated, generalized pretreatment design would consist of chlorination of the source water; the removal of iron, manganese, and suspended matter by coagulation and flocculation or deep-bed filtration; water softening; and the removal of TOC, followed by elimination of the chlorine from the treated water. Each of these pretreatment steps will require documented experimental verification or validation to make certain it conforms to the operational SOPs devised for it, and to ensure that its purported action is indeed attained.

The pretreated water is then ready for purification by any or all of the principal processing units.

Ion Exchange

Ion-exchange resins are poly-electrolytes immobilized by the cross-linking of their large organic moieties. The cation-exchange resin has as its functional group anchored sulfonic acid substituents in which mobile hydrogen ions (actually hydronium H_3O^+) can exchange with other cations. The law of mass action, as modified by specific ion selectivities, governs these exchanges. The result is an uptake of cations by the spatially fixed resin and a release of hydrogen ions in exchange. The anion-exchange resins, of which like the cation-exchange resins there are several types, use spatially fixed quaternary amine groups associated with mobile hydroxyl ions. As anions are acquired by the anion-exchangers, hydroxyl ions are liberated. The released hydrogen and hydroxyl ions interact to form water, largely undissociated into ions. The overall effect of cations and anions being removed from solution is one of demineralization, or of deionization. This is the very purpose of the ion-exchange purification unit. The chemical reactions whereby ion-exchange resins function, and are regenerated are depicted in Figures 13 and 14. It is part of designing the ion-exchange procedure to

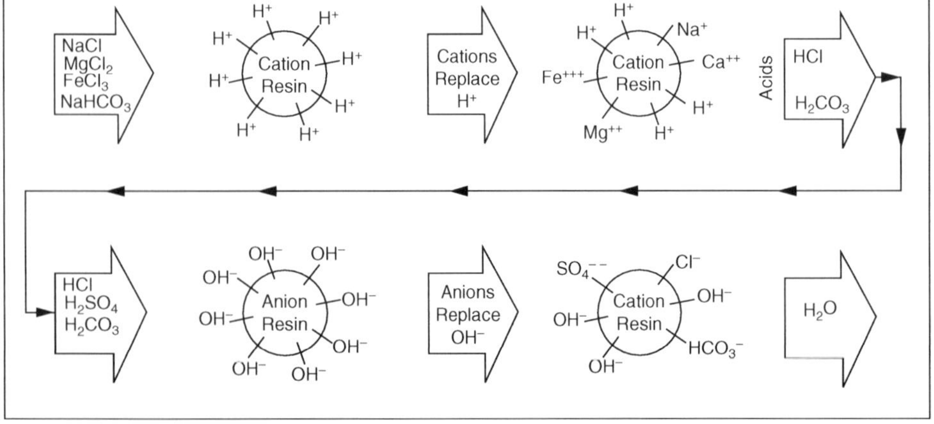

Figure 13 Deionization process. *Source*: Courtesy of G. Zoccolante, U.S. Filter Co.

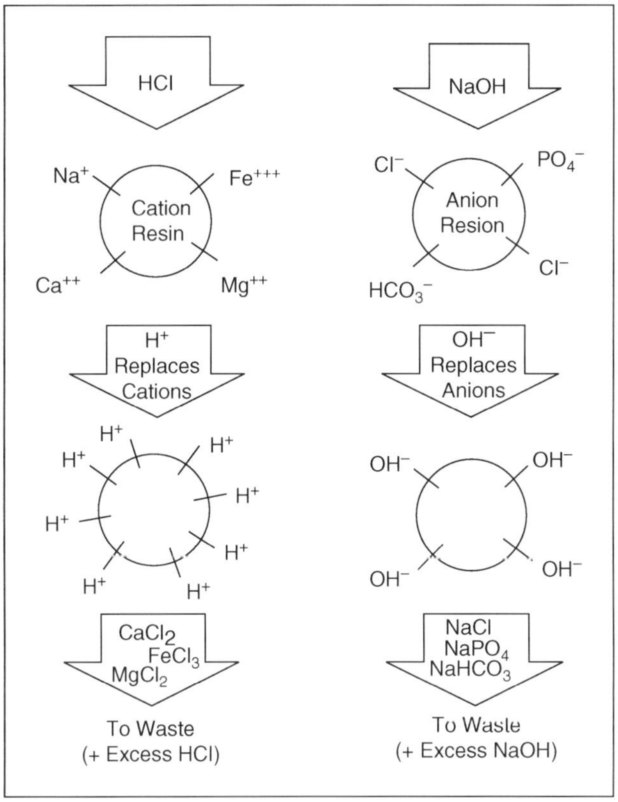

Figure 14 Deionizer regeneration. *Source*: Courtesy of G. Zoccolante, U.S. Filter Co.

elucidate the operational protocol the practice of which, when formalized in an SOP, will ensure the degree and quality of the deionization being sought. In the validation exercise, documented proof of adherence to that SOP and verification of its expected results will be investigated.

Ion-exchange beds may also be employed for the removal of TOC. If that is so, the extent of such intended TOC reduction must be examined as part of the validation of the ion-exchange unit process.

Strong-base, ion-exchange resins are required to remove weakly acidic entities, such as silica. The regeneration of these resins requires the action of heated (50–60°C, usually no higher) sodium hydroxide. However, the carbon–nitrogen bond that is a feature of the quaternary amine group is thermo-labile. On being heated, it ruptures in a reaction called the Hofmann degradation. In fact, these resins are unstable enough to lose some of their capacity even at room temperature. The type I strong-base resin will give rise to a few parts per billion of trimethylamine, the fishy smell of which is objectionable. Type II resins undergo the Hofmann degradation to release ethanol or acetaldehyde; a more pleasant odor. Accordingly, it is the type II strong-base resin that is more widely used in pharmaceutical contexts.

Organisms can thrive in ion-exchange beds, the moist interiors of which contain an abundance of nitrogenous, carbonaceous, and other nutrients. The sanitization of these beds is difficult, the application of heat being limited owing to the thermal lability of the strong-base, anion-exchange resins. The best means of sanitizing the ion-exchange beds is by way of regenerating the resins: organisms are not resistant to the alkali or to the mineral acids (hydrochloric acid or sulfuric acid) used in the regeneration process. There is a history of organisms that have been released from ion-exchange beds contaminating povidone-iodine preparations (28). The FDA has accordingly warned that pharmaceutical water processors should regenerate their ion-exchange beds not on an engineering basis that reflects loss of exchange capacity, but at frequencies dictated by the microbiological counts emanating from these ion-exchange beds (31). This being so, it becomes necessary for pharmaceutical water preparers to conduct periodic microbiological assays of the waters issuing from the ion-exchange columns, to detect departures from the trend lines that would signal the need for action to keep the organism counts under control.

The SOP for operating the ion-exchange columns would define the frequency of such regenerations. The validation exercise for the ion-exchange process unit would include documented evidence of conformity with the SOP protocols and of the results achieved thereby.

Reverse Osmosis

The USP, until recently in its monograph section, permitted the use of ion exchange or, indeed, any suitable unit process for preparing Purified Water when that substance is directed to the compounding of non-injectables. However, ion exchange is interdicted for WFI manufacture because the organism counts released from the resin beds would likely be excessive for application to injectable preparations. The USP until quite recently designated distillation and/or RO as being suitable for WFI preparation in the U.S.A. Given the opposition to membrane processes by the Europeans, different wording is substituted, namely, that distillation or other suitable method may be used. The FDA had permitted RO usage with the modification that two-pass product-staged RO is required. In such a disposition, the product or permeate water that is effluent from the first RO unit is used as feedwater for the second RO unit that follows. More generally, the FDA, in effect, now holds that any method that can be validated may be essayed; subject to its actually being validated. Such may also be used for the preparation of Purified Water.

The *Japanese pharmacopeia* permits the use of ultrafiltration as well. Most European authorities permit WFI to be prepared only by distillation

Even in their solid or glassy state, long-chain polymeric molecules, in their normally coiled configurations, contain intersegmental spaces. These are present in all polymeric materials, and provide separating distances among the different segments of the long polymer chains. At temperatures above absolute zero, these spaces continually undergo dynamic alterations. These intersegmental spaces are peculiar in their dimensions to each given polymer structure. Some polymeric materials, notably the cellulose esters and some polyamides, are characterized by intersegmental spaces that have dimensions that permit permeation by water molecules, connected as these are to one another by hydrogen bonding. However, these intersegmental distances are too small to permit the passage of hydrated ions whose small crystallographic dimensions are made overlarge by their skirts of waters of hydration. Such polymeric entities dispersed in very thin-film form, to minimize resistance to flow, can be used to permit the passage of water under applied pressures; however, they will discriminate against permeation by hydrated ions. It is this activity that characterizes the usefulness of RO membranes in de-mineralizing operations.

There are different types of RO membranes available in the field. The oldest type is composed of cellulose acetate. RO units based on polyamide membranes are newer and have the advantage of exhibiting better rejection qualities; however, the polyamide RO membrane is subject to catastrophic ruination by chlorine

Cellulose acetate is also oxidatively affected by chlorine, but at an acceptably slower rate. What is desirable is that the cellulose acetate RO membrane passes chlorine. In doing so, it provides protection for the RO permeate against microbial contamination. The chlorine is removed subsequently, usually by chemical (bisulfite) addition. The polyamide RO, although superior in its rejection powers, cannot do this. As a result, in pharmaceutical settings, when freedom from organism contamination is a leading consideration, cellulose acetate RO units are often preferentially employed.

Cellulose acetate RO units have their own limitations, however. The polymer undergoes hydrolysis at all pH levels, but minimally at pH 4.5. Its practical use dictates, therefore, that it be employed at acidic pHs. Usually, a pH range of 5.5 to 6 is used; cellulose acetate rejecting best at pH 6, as measured on mixed ions. However, if bicarbonate is present in the feedwater, carbon dioxide will be released. Its passage through the cellulose acetate membrane will impart an unacceptably low pH to the product water. To avoid this, the generated carbon dioxide will require removal by a de-carbonator, usually a forced-air de-gasser. These devices risk exposing the waters to air-entrained organisms. By contrast, polyamide RO membrane rejects bicarbonate ions at pH 8.5. Acidification and its consequences in terms of carbon dioxide generation are avoided.

Extent of Recovery

RO processing of water by means of tangential flow involves the discarding of some percentage as the reject stream. It is now realized that being profligate with water can be a costly error. Therefore, to minimize the water loss, part of the reject stream is mixed into the feedwater. Also, high recoveries of the feedwater in the form of permeate or product water is usually aimed for. A 75% recovery is the general practice. The problem is that the greater the recovery, the more impure the reject stream and the more its use necessitates the disruptive cleaning of the RO membrane. Recoveries of 50%, 75%, 80%, and 90% result respectively in 2, 4, 5, and 10-fold increases in contamination concentrations.

It is advised that a well-run RO operation should initially be operated at 50% recovery. Larger recoveries involve the reuse of highly contaminated reject waters. Such an operation should be trouble-free over a time period sufficient to demonstrate that stabilization of the RO and pretreatment units has occurred. Not needing to clean the RO unit can be taken as the acceptable endpoint. At its attainment, a higher recovery becomes justified. Further increases in recoveries can be explored in the same fashion. Larger RO units of capacities above 15 gpm must operate at higher recoveries, but careful system design and extensive pretreatments are needed to achieve penalty-free recoveries of 60% and 70%, or even higher.

Distillation

Stills vaporize water, which permits the separation of water from its non-volatile impurities. The water vapor is then condensed, allowing its separation from volatile impurities. Unfortunately, the need for stills to furnish large volumes of distillate involves high heat inputs that provide opportunities for vapor entrainment of droplets. A consequence is the carryover of non-volatile impurities. This is addressed by still design and by the proper operation of the still to avoid such occurrences. The validation of the still operation, therefore, must involve documented experimental evidence that the SOP protocols appropriate for the still were adhered to. Still design is also concerned with minimizing the heating costs by optimization of their functions.

Still manufacturers will specify the purity of the feedwater suitable for their stills; therefore, evidence must be provided attesting to conformity with these requirements.

One of the chief concerns in still operations is the possibility of the carryover of endotoxins from the feedwater. Still operations are not absolute in their performance. Rather, they can be depended on to provide log reductions of impurities. Still manufacturers may be reluctant to guarantee more than a 3- or 4-log reduction in endotoxin, although better results can be obtained (30). The FDA requires, therefore, that the endotoxin content of the still feedwater not attain concentrations for which management by distillation cannot be assured. Consequently, it is common for the feedwater to be treated with ultra-filters or with charge-modified microporous filters to eliminate or reduce the endotoxin content. Whatever quality parameters are set in the relevant SOPs for the still feedwater and for the operational manipulations of the still must, as part of the validation exercise, prevail in practice.

GOOD MANUFACTURING PRACTICES RELATED TO WATER SYSTEMS

The FDA has established a number of GMPs that pertain to the preparation of pharmaceutical waters. Since their promulgation several have undergone modification in practice, as already discussed, a consequence of the FDA's ongoing and evolving understanding of what pharmaceutical water systems require. The relevant GMPs are discussed in the literature (5,6,31).

SPECIFIC UNIT PROCESS VALIDATIONS

The operation of each individual purification unit process requires its own focus and assessment. The system having been correctly designed and installed, how is it to be validated? Several questions must be answered in the validation exercise:

- What component or unit of the system is being addressed?
- What is its function?
- What is the measure of its performance?
- What are its normal maintenance requirements?
- What additional maintenance may occasionally need to be performed?
- When are sanitizations to be carried out?
- What sanitation method is to be utilized?

Consider, for example, the purification unit, that is, the carbon bed. Its chief purpose is to de-chlorinate the water. Secondarily, it may be relied on to reduce the TOC load. To assess the performance of its purported function, one would measure the chlorine content of both the incoming and effluent waters. One could do likewise for the TOC, using an online TOC monitor. What would be the measure of the carbon bed's performance? In the case of de-chlorination, complete chlorine removal would be expected. Therefore, the effluent water should contain zero chlorine. The performance standard for the TOC could be set variously. It could, for instance, be judged against a goal of less than 1 ppm, or the target could be a reduction of 50% of the original TOC burden, whatever seems appropriate. Next would be to list and carry out the normal maintenance needs. For instance, consider the backwashing, bumping (to eliminate channeling), and rinsing. Shall it be performed, as a multimedia bed would be, on the basis of pressure drop, or in response to a reduction in the TOC uptake, in other words, in accordance with the achievement of a measured value? Alternatively, maintenance could be instituted on a time basis, say, daily. The time basis can be set from experience against a historically measured value, or in keeping with a prophylactic philosophy.

Are there occasionally additional maintenance protocols that must be invoked? The replacement of the carbon bed is an example.

When is the carbon bed sanitization to be accomplished? Shall it be in response to specific levels of organisms in the effluent water, or periodically on a time basis? How is the sanitization to be performed; by hot water, by steam?

Safety protocols must also be observed. In the design of the water purification system, the installation of the necessary test ports, isolation valves, and protocols for microbiological and chemical assays are to be provided. All performed operations and the data they generate will require being analyzed, documented and countersigned.

The Chlorination Unit

Chlorine is usually added to a water supply to a residual concentration of 0.5 to 2 ppm. Municipally treated waters drawn into a pharmaceutical plant ordinarily reflect such a treatment. However, municipally treated waters may, for a variety of reasons, also fall short in their chlorine content. Therefore, it is best to analyze the incoming waters for their actual chlorine content and to adjust as necessary.

The targeted chlorine concentration is measured by a chlorine monitor. The automated chlorine analyzer is responsive to the concentration present in the water. Chlorine being an oxidant, its concentration can also be determined by an ORP analyzer. This type of instrument senses the oxidizing status of the water. In effect, then, it assays the chlorine concentration that is present. It is used in conjunction with a metering device that permits a proportioned addition of chlorine as necessary to maintain a preset concentration.

The purpose of the chlorine treatment being the addition of chlorine, the validation of its function is obtained from comparisons of the before and after chlorine concentrations. This includes demonstration that the requisite chlorine concentration is maintained.

The biocidal efficacy of the chlorine treatment is usually assumed. However, its validation can be confirmed by microbiological assays of the feedwater, before and after chlorination.

The Deep Bed Filter

For the water exiting the media beds, the free chlorine content should be held at 0.1 to 1 ppm, measured weekly. The corresponding microbiological levels, to be assessed weekly, should not exceed 100 cfu/mL. The bed sanitizations are commonly performed once or twice yearly using 200 ppm sodium hypochlorite solutions. Generally, the beds are functional for a five-year period.

The deep-bed filters, of whatever construction, require flow rates of between 5 and 15 gpm/ft^2, depending on the application. When they accumulate enough particulate matter to boost their pressure differential to about 15 psi (1 bar), they are back-flushed in a maintenance operation.

The function of the deep bed filters is to remove suspended matter. Validation of their performance is obtained by comparing the suspended solids contents of the before and after filtered waters. It usually suffices to measure the proper flows by way of flow meters, and to arrange for timely automated back-flushes in response to pressure buildup in the beds. Backwash operations in this context, as also for any deep bed, can be assessed using a backwash turbidity meter.

Softening Operations

Because organisms may proliferate in the water-softening unit, the chlorine content of the feedwater is usually not removed until after the water is softened. Indeed, a free chlorine residue is advised. On the other hand, the water-softening resin will undergo degradation by the oxidizing action of the chlorine. Some firms choose, therefore, to remove the chlorine before the water enters the softener.

Periodic sanitizations of the water softener should be performed using water at 82°C (180°F) over a two-hour contact period. Bacterial assays should be carried out weekly. The water-softening resin should last for about three years.

Prior analyses should have revealed whether barium, strontium, iron, or manganese are among the more conventional elements present. The water softener's function in reducing or removing hardness can then be readily validated by comparison with the analytical results shown for the water emerging from the softener.

The water-softening ion-exchange resin periodically needs to be regenerated by brine. Organism growth is possible in the salt solution. Periodic microbial assays should be used to establish a sanitizing schedule. Also, the validation exercise should define the frequency of the resin regeneration, as indicated by a drop in the softening efficiency, as measured by water-hardness element concentrations. Having this point defined as a function of time or gallonage permits the regeneration to be accomplished automatically. Ultimately, the resin used in the water softener will need to be replaced. The validation of this activity will be signaled by a decrease in the ability to regenerate the resin.

The Carbon Beds

The validation of the carbon beds has already been considered. Briefly, their ability to remove chlorine should be affirmed by an analysis of their effluent water. The chlorine content should be zero. As already stated, the chlorine can be analyzed for in a number of ways.

The backwashing, bumping, and rinsing procedures normal to carbon bed maintenance should be implemented daily. Recirculation of the water is indicated.

The sanitization of the carbon beds should be performed in accordance with the SOP schedule developed during the pre-validation stage of the process. Eventual replacement of the activated carbon should be made in keeping with the requirements, as defined in the SOP. Sanitizations using water at 82°C (180°F) for at least two hours should be performed at least weekly.

The RO Operation

The proper way to validate the operation of the RO units depends on their intended functions in the purification system. Their purpose in 99% of RO usages is to reduce the ionic content of the treated waters. Single-pass ROs are usually expected to reduce the ionic concentrations by more than 95%. Reductions of more than 99% are usually the goals of two-pass systems, as measured daily.

Depending on the qualities of the feedwater, two-pass, product-staged RO operations may be needed to yield waters that have acceptable chloride ion levels. If the feedwater contains 400 to 500 ppm of chloride, and the single-pass RO gives a 5% leakage, the two-pass, product-staged RO becomes mandatory.

Whether the performance of the RO function is as purported can be gauged by measuring the salt rejection of the single RO unit. Instrumentation determines the conductivity of the influent and effluent streams and displays the results in terms of salt rejections. For the two-pass units, where the ionic reduction is expected to attain some 99%, the performance is judged by the direct measurement of the effluent water's resistance. The RO operation should also reduce the colloidal and endotoxin loads of the water. For the RO function or that of ion exchange, conductivity-resistance measuring instruments are used.

Another function planned for the RO may be the removal of TOC. Depending on the source water, the TOC reduction can be 80% to 90%. Rarely will the TOC reductions be at the optimum of 99%. The validation of the TOC reduction function by the RO can be assessed by the use of TOC monitors. Usually, one is mounted on-line at the RO outlet. However, the analyses can be performed using grab samples in conjunction with off-line TOC analyzers.

RO is also used to remove bacteria and bacterial endotoxins. The validation of those functions to the levels designed can be assayed using suitable microbiological analytical techniques and LAL testing. The target for bacteria-diminishing actions can be the 10 cfu/100 mL required for WFI, the 100 cfu/mL desired for Purified Water, or any other designed standard, as determined periodically.

In some instances, albeit rarely, the second RO of the two-pass arrangement is assigned the sole function of assuring the removal of organisms and bacterial endotoxin. Ionic removal is not involved. That has presumably been accomplished adequately by the first of the two ROs. In these instances, the validation of the operation can be determined by suitable assays of the product water. The water exiting the second RO unit should have resistance in excess of 300,000 ohm/cm.

The normal maintenance of RO operation should include the automatic inspection of the O-rings. Damaged O-rings can cause loss of rejection. Cleaning of the RO should be undertaken following a 25% increase in differential pressure, after a loss of flow by as much as 10%, or when the rejection decreases by 1% (32).

Ion-Exchange Operations

The performance of ion-exchange units of whatever construction, whether twin-beds, mixed-beds, or other, depends on two considerations: bed capacity and the onset of ion leakage. Therefore, in the validation of the ion-exchange operation, these two items require assessment. The ion-exchange capacity must be determined in establishing the operational SOP so that regeneration can be undertaken in a timely fashion. The exhaustion of capacity is defined by the onset of ion leakage. For the cation exchanger, the first ion to breakthrough is the sodium ion. For the anion-exchange bed it is silica that first manifests leakage. Therefore, the measurement of these ions in the influent and effluent waters will provide evidence, to be documented, that the ion-exchangers are doing what they are purported to do. Sodium ions can be measured using sodium ion-specific electrodes. Ion-specific electrodes are expensive, and they require periodic maintenance, but they are very useful. Silica can be analyzed colorimetrically using an online silica analyzer. More generally, information concerning ion leakages can be obtained from conductivity measurements, increasing ionic concentrations, yielding higher electrical conductivities. Conductivities are the easiest of these assays to perform. Although the least specific, they may suffice. The baseline conductivity should be measured as an indication of conductivity. After the bed is regenerated, it is washed until its effluent waters show a flat conductivity area. This continues until ion breakthrough manifests itself. The flat area is the baseline conductivity.

Conductivity measurements can be taken every six hours or so in the search for early manifestations of ion leakage.

Although the essentials of the ion-exchange operation involve the analyses just indicated, other tests can usefully be conducted as well. The chlorine content of the water where such is permitted entry into the beds can be measured, as by amperometric titrations, on the incoming and outgoing waters. It is helpful to have periodic microbiological assays performed on the waters flowing out of the ion-exchange unit. Indeed, the FDA requires that the ion-exchange beds be regenerated as a means of sanitizing them whenever the effluent organism counts depart from the normal trend lines. Also, because ion-exchange beds can serve to reduce TOC, when such action is part of the system design, TOC analyses should be conducted to make sure that the purported performance is realized.

Implementation of the SOP should be keyed to the flow diagram of the system. Which way the water is flowing, to where, and which valves should be opened or closed should clearly be indicated during every step of the ion exchange and ion-exchange regeneration process.

The standard maintenance procedure dealing with ion-exchange bed upkeep should have its own analytical protocols necessary to the regeneration step. Safety procedures will direct the proper handling, use, and disposal of the regeneration chemicals. Measuring instruments should be kept in proper calibration. How to perform such calibrations should be specified. The resins themselves should receive at least yearly examinations to determine their moisture content or uptake (a measure of their cross-link deterioration), and to determine the wholeness of the beads. Although twin ion-exchange units can tolerate about a 20% portion of fragmented or split beads, the mixed-bed resins become more difficult to separate at that level. Additionally, the cation-exchange beads may be soaked in dilute hydrochloric acid to obtain a slow elution of iron. Other bead examinations may be conducted as well.

Sanitizations of the DI resins should be attempted using 70°C (160°F) water. (The specific upper temperature limits are defined by the thermal lability of the particular anion-exchange resins.) This should be performed weekly. Recirculation through the bed should be part of the design. The water exiting the bed should have resistance greater than 10 Mohm/cm. The bacterial counts should preferably be fewer than 50 cfu/mL. The mixed-bed life-to-renewal depends on the feedwater quality, but is usually three months or so.

In particular, the characterization of the product water should be an ongoing activity. Signed daily logs of the various operations are kept, and the data should be tracked, and trend lines should be drawn frequently.

Distillation

The proper action of a still depends on the use of feedwater of a requisite purity, coupled with operation in accordance with the defined SOP to avoid impurity carryover by vapor entrainment. Some still manufacturers argue, therefore, that the validation of the still operation, given the proper feedwater quality, entails no more that the certainty that the established SOP is being adhered to. By the same token, using feedwater of the requisite quality, the SOP necessary to proper still operation can be defined by the still manufacturer at their factory.

Distillation is not an absolute method for effecting bacterial endotoxin removal; instead, a defined log reduction is accomplished, depending on the still, but more especially on the mode of its operation. What is required, therefore, is to ascertain by LAL analysis that the feedwater to the still is sufficiently low in endotoxin content to ensure that the product water emanating from the still will not exceed 0.25 EU/mL, or whatever standard is set for the operation. To validate that this purported goal is reached, LAL testing of the distilled water is required.

Although distillation operation will almost invariably include the use of an online conductivity meter, it should be emphasized that its readings do not reflect the presence or absence of microbes or of bacterial endotoxins in the water. In any event, when the waters are as pure as they should be on both entering and exiting the still, their electrical resistance (the reciprocal of conductivity) should show no difference.

The distilling process is often the last unit process in the water purification train. Its feedwater, therefore, is usually very pure. Indeed, Kuhlman (35) advocates that multi-effect stills use feedwater of 1 Mohm resistivity and that they be free of chlorine and chloride ions; that they contain not more that 1 ppm silica, and no amines, for these have volatilities so like that of water as to be inseparable by distillation. No TOC content is specified. Water of that purity can be used at the still manufacturer to develop the SOPs necessary to define its proper operational conditions. These will preclude misting, flooding of the still, and corrosive and scaling influences. The observance of the defined SOPs in the water pretreatment would, in this view, constitute the validation of the process. This approach to the validation of the still operation is not prevalent in the industry. It is expected that validation of the still operation will be performed in the pharmaceutical water production setting.

There is a competition among the various still manufacturers over the most suitable type of still. It is more costly to prepare purer waters. Thus, an advantage of the vapor compression still may be its operation on softened water, as against the need for 1 Mohm feedwater for the multiple effect stills. This is said to be possible because of the vapor compression still's lower operation temperature. It is fair to add, however, that the marketing of such stills may sometimes be advanced competitively on the basis that less pure and, therefore, less expensive, waters are adequate for their operation. The question raised is whether less pure feedwater does not eventually manifest operational or maintenance problems. The purity of the feedwater may define the frequency at which periodic cleaning and de-scaling will be required. (Blow downs are now continuous in most stills). The validation requirements in such cases can be set forth only in accordance with the particular operation and the purity of its feedwater.

The distilled water should be free of organisms. This condition can be validated by employing microbiological assays. Even in a down condition were the still to acquire organisms and spores in its operation while being brought up to temperature and then being maintained for two hours or so at the elevated temperature before product water is collected, these foreign entities should be killed or removed by flushing. The microbiological testing of the stored distilled water should be performed routinely.

Ozone

The validation of the ozone system involves the regulation of the production of the gas and of its introduction into the water by way of a contactor. The control of these two activities governs the concentration of the ozone in the water, as measured by an ozone analyzer. The almost universal application of ozone is for the destruction of organisms (although it can serve also oxidatively to degrade TOC). It would seem, therefore, that the validation of its function would entail microbiological work. Yet, so certain is the destructive action of ozone on microbes that the expected end is often assumed, provided that the ozone concentration sought is assured by measurement. Confirmatory microbiological analysis should be performed.

Ozone can be prepared by corona discharge using oxygen or air as the source of oxygen. Although the latter is sometimes done for economy's sake, complications are caused by the presence of humidity. Moreover, the nitrogen in the air generates oxides of nitrogen. These give rise to acidic components with unacceptable consequences to the pH of the water. Therefore, the generation of ozone from air is seldom, if ever, used in the pharmaceutical industry.

When the ozone is electrolytically prepared, no gas flow is involved. The monitoring of the ozone concentration and its maintenance at a given level are then straightforward. When ozone is prepared from oxygen, the purity of the oxygen can be ascertained by analysis. When this is held constant, the measurement of three factors serves to ensure a constant ozone concentration: the gas pressure, the gas flow rate, and the current being supplied the ozone cell. The ozone production can be varied by changing the current. Keeping the ozone concentration constant may require adjustments to the current. This can be signaled and controlled by a dissolved-ozone monitor. Changes in the purity of the feedwater may lead to different rates of ozone consumption. This may cause variations in the residual ozone concentration; therefore, the cleaner the water the better.

An online analyzer can be used to measure the ozone concentration. Analog signals can be used to provide continuous recordings of the ozone concentration. When the water is very clean, one can relate the ozone function to the residual ozone concentration. When, however, the water contains entities that can foul the analytical probe, that relation becomes uncertain. One way of addressing the problem is to feed enough ozone to make sure that a residual is present in the off-gases, and to relate that ozone concentration to the function sought.

Online ozone analyzers or monitors can operate in various ways. In clear water systems, double-beam ultraviolet spectroscopy is very satisfactory. Alternatively, amperometric determinations can be made using probes that are shielded against fouling by membranes selective for ozone penetration. A third method for analyzing for the ozone concentration is to sparge the sample with nitrogen gas to strip the ozone into the gas phase, wherein it can be detected spectroscopically. Wet chemistry involving colorimetry based on dye loss can serve as an analytical backup method.

In some biotechnology operations, for which the drug product is extremely costly, the possible de-gradative effects of ozone concentrations, even below detectable limits, may raise concerns and insecurities. In one such operation, in an ultra-conservative approach to the removal of ozone, a manufacturer employed two ozone-destruct units in series. The reading of zero ozone by the ozone monitor is not fully relied on; therefore, the compatibility of the processed water with the product is monitored carefully during stability studies.

One of the concerns often expressed by FDA investigators is the real-time measurement of the ozone concentration. Ozone monitors are degraded by ozone over time, perhaps to the point of becoming inexact. Confirmatory wet-chemistry analysis should be used. The FDA concern seems overstated. Consider the validation of the ion-exchange exercise. The remaining capacity of the resin at any moment is never measured. Its sufficiency is assured over time from the longer-term studies relating the ion-exchange performance as a function of time. That the ion-exchange capacity of the resin has not been overwhelmed is revealed by conductivity measurements of the water effluent of the ion-exchange bed.

For ozone, the concentration sufficiency, even when inexactly measured, can be ensured by maintaining the ozone concentration above its minimum required level. This can be ensured by the long-term studies with the goal of defining the systems SOP. There is a reliable relation between ozone concentration and its organism-killing propensity. Precise ozone concentration readings, although desirable, are not essential to the process, notwithstanding FDA investigator views to the contrary. Confirmatory microbiological work is always an option, and the speed of such assays can be hastened by the epifluorescence technique.

Ultraviolet Installations

The validation of the ultraviolet light units consists of two distinct activities. The UV lamp manufacturers need to make constant and sufficient the wavelength and radiation outputs of the UV lamps. The user's obligation is to ascertain that the SOPs, defined experimentally to ensure proper UV action, are conformed in practice.

It is necessary that the UV radiation be supplied in its proper dosage. The UV lamp outputs, initially high, decrease during the first two days or so of use. Some UV lamp suppliers, therefore, recalibrate their lamps after 100 hours of use, when the output decline has leveled off to a relatively constant value. Continuous intensity meters, based on recordings of the voltage across the lamp, are available. A fixture is available whereby each individual lamp, from multi-lamp arrangement, after removal from the unit, can be inserted into a test device to have its intensity measured.

The ultimate validation of the UV unit rest on the user. The user must ensure, by documented measurements, that the water being treated is sufficiently free of radiation-attenuating particles, as by turbidity measurements, and of UV-adsorbing entities, as by chemical analysis, in accord with the SOP established for the UV operation. By utilizing test ports, water samples can be withdrawn to assess whether the organism reductions expected from the UV instrument were actually attained. Microbiological analyses before and after the water is treated by the UV light are central to the validation effort.

The UV devices are not absolute in their killing of organisms. Their contribution to the water purification system could be the reduction of the microbial content by 1, 2, or 3 logs, or to some particular level, say, 100 cfu/mL. The validation of their designed action is supported by appropriate microbiological testing.

The normal maintenance of the UV lamps will entail periodic replacements, or when indicated by the online intensity meters. The periodic wiping of the sleeves to remove radiation-attenuating film is required.

The validation methodology appropriate to other units, devices, and portions of the water purification system should by now be apparent. The validation consists of answering by documented measurements such questions as why the arrangement is present in the system, what its purported function is, and whether the performance standard set for it are met. The operational concerns relevant to other appurtenances such as chemical additions, iron and manganese removal, and so forth, and their maintenance should be forthcoming from the technical information available on these subjects (17).

Filter Validations

The subject of filter validation is one of complexity caused in no small part by the lack of agreement among filter manufacturers on the proper way to conduct integrity testing. It is certain that filters require integrity testing periodically, particularly after being newly installed and on some fixed schedule of a frequency, one hopes, keyed to the possibilities of potential problem development. Micro-porous filters are traditionally replaced, not cleaned and reused. Filter change-out intervals should be justified by pertinent data. An exception is made for filter refurbishing and reuse in the case of air filters, the use of which, even over prolonged periods, leaves their capacities so little impaired that their frequent replacement, however justifiable on the grounds of avoiding cross-contamination, would be judged economically wasteful. In these instances, however, it is periodically necessary to retest them for their integrity. This is most conveniently done on-line, as by use of the water intrusion test method (35,36).

It is questionable whether RO or ultra-filters can be integrity tested, at least in any way practical to their reuse. They cannot be integrity tested by surrogate measurements for their retention of microbes. Ultra-filters may have their functionality for the reduction of microbial counts, of particle numbers, and of bacterial endotoxin levels evaluated. Their performances and validations can be judged accordingly. Ultra-filters should be cleaned and sanitized on a monthly schedule, the latter using a 1% or so solution of peracetic acid to keep the colony-forming units per milliliter level at or fewer than 10. This should be ascertained by weekly microbiological analyses.

Considerations for the cleaning, refurbishing, and replacing membranes of these types are also to be defined as their periodic sanitizations. Careful inspection of O-rings for possible replacement should be made. This should be done at every installation of a filter.

However, for the validation of specific filter functions, there need be no doubts. Analytical measurements are made, upstream from the filter and down, to gauge whether the filter has performed to the standards set for it. Bacterial endotoxin elimination can be assessed by LAL testing, organism reductions, or sterilizations involving microbiological investigations; the retention of sand, carbon, and resin fines, to the designed degree, can be validated by the use of particle counters.

Flow Rates, Pressures, Temperatures

The various deep-beds constructed of multimedia, activated carbon particles, resin beds, depth filters, and others, will have been sized to yield particular flow rates in conjunction with their purification activities. Whether these intended functions are indeed achieved may depend on adequate flow rates, as defined in the operational SOP. These, in turn, are consequences of specific pressure levels and bed dimensions. Too rapid flows may detract from exchange or adsorptive efficiencies. Too diminutive a flow rate may cause channeling and other improper water distribution problems within the bed, and the overloading of localized areas.

All of the unit operations require certain permissible ranges of flow rates to be effective. This includes water softening, ultraviolet effects, and ozone applications, in addition to those of the deep-bed varieties. Certainly, ion exchange and RO require given flow rates for optimal performances. Temperature is also often of significant influence in a given unit process, particularly for RO. However, these operational factors are defined in the SOPs that govern the purification operations; their proper documented observance forms part of the validation operation.

Therefore, the validation exercise, aided by flow meters, pressure gauges, thermometers, and such, must ascertain that the flows, pressures, and temperatures defined as necessary in the SOPs are indeed adhered to in the actual operation.

AN APPRECIATION

In the foregoing, a technical explanation of process validation was set forth. Justification for its requirement derives from its intention to assure the safety and efficacy of drug preparations. Process validation is a legal requirement. In addition, this worthy endeavor achieves a societal objective as well. Perhaps in all cultures, life and well-being are valued highly. Indeed, there is the tradition that he who saves a single life saves creation entire. Those engaged in drug processing are, therefore, privileged to have so worthy a vocation. It is an obligation we can fulfill with justified professional pride.

UNIGENE LABORATORIES, INC. INSTALLATION QUALIFICATION PROTOCOL

TITLE: CHILLED WATER SYSTEM GENERATION, STORAGE AND DISTRIBUTION	PROTOCOL NO: VP039.IQ REV . NO: DRAFT EFF . DATE: NEW SUPERSEDES: PAGE 23 of 38

14. **FIELD VERIFICATION**

C. INSTRUMENTS/DEVICES

2.] Critical Instruments

Control Parameter: V-910 Level, Continued

Equipment Number: V-910 P/M Number: ________

Date of Last Calibration: ________ Frequency: ________

Reference Drawings: FU-901

Devices in V-910 Level Loop:

Tag Number: LALL-01		
Manufacturer: Panalarm	Model: w/ Ann-7	S/N:
Function: Low Low Level Alarm	Size: NA	Range:

Tag Number: HS-04		
Manufacturer: Allen Bradley	Model: 800T-J2A	S/N:
Function: HOA Switch for P-910	Size: NA	Range: NA

Comments: ________

Verified by: ________ Date: ________

Reviewed by: ________ Date: ________

APPENDIX A

UNIGENE LABORATORIES, INC. INSTALLATION QUALIFICATION PROTOCOL

TITLE: CHILLED WATER SYSTEM GENERATION, STORAGE AND DISTRIBUTION	PROTOCOL NO: VP039.IQ REV . NO: DRAFT EFF . DATE: NEW SUPERSEDES: PAGE 24 of 38

14. **FIELD VERIFICATION**

C. INSTRUMENTS/DEVICES

2.] Critical Instruments

Control Parameter: Chilled Water Supply Pressure

Equipment Number: NA P/M Number: ________

Date of Last Calibration: ________ Frequency: ________

Reference Drawings: FU-901

Devices Controlling Chilled Water Supply Pressure:

Tag Number PCV-09		
Manufacturer: Fisher	Model: 63EG	S/N:
Function: Chilled Water Supply Pressure	Size:	Range: Set @ 60 PSIG

Comments: ________

Verified by: ________ Date: ________

Reviewed by: ________ Date: ________

APPENDIX B

UNIGENE LABORATORIES, INC. INSTALLATION QUALIFICATION PROTOCOL

TITLE: CHILLED WATER SYSTEM GENERATION, STORAGE AND DISTRIBUTION	PROTOCOL NO: VP039.IQ REV . NO: DRAFT EFF . DATE: NEW SUPERSEDES: PAGE 25 of 38

14. **FIELD VERIFICATION**

C. INSTRUMENTS/DEVICES

2.] Critical Instruments

Control Parameter: ______

Equipment Number: ______ P/M Number: ______

Date of Last Calibration: ______ Frequency: ______

Reference Drawings: ______

Tag Number:		
Manufacturer:	Model:	S/N:
Function:	Size:	Range:

Tag Number:		
Manufacturer:	Model:	S/N:
Function:	Size:	Range:

Comments: ______

Verified by: ______ Date: ______

Reviewed by: ______ Date: ______

APPENDIX C

CRITICAL INSTRUMENT CHECKLIST

Attachment #8A
Doc. # 23–2100–1
Edition:
Page 60 of 65

Company: ______ Date: ______

Equipment: ______	Function: ______
Manufacturer: ______	Model No: ______
Serial No.: ______	Calibration Date: ______
Range: ______	Scale Division: ______
Equipment: ______	Function: ______
Manufacturer: ______	Model No: ______
Serial No.: ______	Calibration Date: ______
Range: ______	Scale Division: ______
Equipment: ______	Function: ______
Manufacturer: ______	Model No: ______
Serial No.: ______	Calibration Date: ______
Range: ______	Scale Division: ______
Comments:	

Performed by: ______ Date : ______

Reviewed by: ______ Date : ______

APPENDIX D

REFERENCE INSTRUMENT CHECKLIST

Attachment #9A
Doc. # 23–2100–1
Edition:
Page 61 of 65

Company: ______________________ Date: ______________

Equipment: ______________ Function: ______________
Manufacturer: ______________ Model No: ______________
Serial No.: ______________ Calibration Date: ______________
Range: ______________ Scale Division: ______________

Equipment : ______________ Function: ______________
Manufacturer: ______________ Model No: ______________
Serial No.: ______________ Calibration Date: ______________
Range: ______________ Scale Division: ______________

Equipment: ______________ Function: ______________
Manufacturer: ______________ Model No: ______________
Serial No.: ______________ Calibration Date: ______________
Range: ______________ Scale Division: ______________

Comments:

Performed By: ______________________ Date : ____________
Reviewed By: ______________________ Date : ____________

APPENDIX E

XXX UNIVERSITY
XXX INSTITUTE

INSTALLATION QUALIFICATION
AUTOCLAVE #2
PROTOCOL NO. P10-1Q

13.2.2 VALVES Listed below are the Valves found in the Autoclave.

DESCRIPTION								
Valve I.D.No.	Manufacturer	Model No.	Mat'l	Type	Location	Yes/No	Initialed by	Date

Comments: ______________________

____

Complied by: ______________________ Date : ____________
Reviewed by: ______________________ Date : ____________

APPENDIX F

OPERATIONAL QUALIFICATION
WATER FOR INJECTION SYSTEM

PROTOCOL NUMBER

FLOW VELOCITY – "NO LOAD" CONDITIONS

Study -Distribution Loop	1	2	3
[FR] = Flowrate (GPM)			
[D] = Pipe inner diameter (in)			
[FV] = Flow velocity (fps) FR(409) = D^2			
[SV] = Specified flow velocity	≥ 5 fps	≥ 5 fps	≥ 5 fps
Pass/Fail			
Comments:			

Calculation derived:

NOTE: $\text{Area (ft}^2) = \pi\left(\left[\frac{\text{D in}}{2}\right]\left[\frac{\text{ft}}{12\text{ in}}\right]\right)^2 = .00545D^2$

Flow Velocity (fps) $= \left(\frac{\text{gal}}{\text{min}}\right)\left(\frac{\text{min}}{60\text{ sec}}\right)\left(\frac{\text{ft}^3}{7.48\text{ gal}}\right)\left(\frac{1}{\text{Area(ft)}^2}\right) = \frac{\text{GPM}(.409)}{\text{SV}^2} = \frac{\text{FR}(.409)}{\text{D}^2}$

Comments:

Performed by: ____________________ Date: ____________________

Reviewed by: ____________________ Date: ____________________

FLUOR DANIEL VALIDATION SERVICES

APPENDIX G

REFERENCES

1. Badagliacca R. Regulatory aspects of validation. In: Carleton FJ, Agalloco JP, eds. Validation of Pharmaceutical Processes: Sterile Products. 2nd ed. New York: Marcel Dekker, 1999, (chap. 28).
2. Keer D. Gathering Validation Documentation for Pharmaceutical Water Systems. Bensalem, PA: Rust, 1995.
3. Food and Drug Administration. Guideline on General Principles of Process Validation. Rockville, MD: Food and Drug Administration, 1987.
4. Food and Drug Administration. Guideline on Sterile Drug Products Produced by Aseptic Processing. Washington, DC: Center for Drugs and Biologics/Office of Regulatory Affairs, 1987.
5. Artiss DH. Water system validation. In: Carleton FJ, Agalloco JP, eds. Validation of Aseptic Pharmaceutical Processes. New York: Marcel Dekker, 1986, pp. 207–251.
6. Artiss DH. Pharmaceutical water systems: design concepts. J Parenter Drug Assoc 1978; 32(2):89–95.
7. Angelucci LA, III. Documentation Relative To Pharmaceutical Water System Validations. Berkeley Heights, NJ: John Brown Co, 1995.
8. Meshnick D. The Role of Documentation in the Validation Process. Clinton, NJ: Foster Wheeler U.S.A. Corporation, 1995.
9. Water Quality Committee/Pharmaceutical Manufactures Association. Updating requirements for pharmaceutical grades of water: validation and technology selection. Pharmacopeial Forum 1993; 19(6):6633–45.
10. Food and Drug Administration. Guide to Inspections of High Purity Water Systems. Rockville, MD: Division of Field Investigation, Office of Regulatory Affairs, 1993:4.
11. U.S. Pharmacopeial Convention. Water for pharmaceutical purposes. Official Monograph. In: United States Pharmacopeia, Rockville, MD, 1995:1636–7; 1996; (Suppl. 5):3442–4; 3547–55; 3464–7.
12. Cooper MS. Purified pharmaceutical waters. Microbiol Update 1994; 12(7):1–4.
13. Munson TE. FDA Views on Pharmaceutical Water. Rockville, MD: Food and Drug Administration, 1993.
14. Water Quality Committee/Pharmaceutical Manufacturers Association. Updating requirements for pharmaceutical grades of water: microbial considerations. Pharmacopeial Forum 1992; 18(6):4397–9.
15. Water Quality Committee/Pharmaceutical Manufacturers Association. Updating requirements for pharmaceutical grades of water: proposed revisions. Pharmacopeial Forum 1994; 20(3):7526–44.
16. Meltzer TH. Pharmaceutical Water Systems. Littleton, CO: Tall Oaks Publishing, 1996.
17. Crits GJ. Pretreatment methods for ion-exchange and reverse osmosis. Ultrapure Water 1989; 6(6):48–57.
18. Abshire RL, Schlech BL, Dunton H. The development of a biological indicator for validating ultraviolet radiation sterilization of polyethylene bottles. J Parenter Sci Technol 1983; 37(5):191–7.
19. Kemmer J, ed. The Nalco Water Handbook. New York: McGraw-Hill, 1988.
20. Water Quality Committee/Pharmaceutical Manufacturers Association. Updating requirements for pharmaceutical grades of water: total organic carbon. Pharmacopeial Forum 1993; 19:5858–62.
21. Muraca PW, Yu VL, Goetz A. Disinfection of water distribution systems for Legionella: review of application procedures and methodologies. Infect Control Hosp Epidemiol 1990; 11:79–88.

22. Hoffman BJ, Kasahara M, Gavaghan M. The effect of chlorine on mixed beds in ultrapure water systems. In: Transcripts of Sixth Annual Semiconductor Pure Water Conference, Santa Clara, CA, January 15–16, 1987:59–74.
23. Meltzer TH, Rice RG. Ultraviolet and ozone systems. Mittelman MW, Geesey GG, eds. Biological Fouling of Industrial Water Systems: A Problem Solving Approach 1987, pp. 99–137.
24. Meltzer TH. High Purity Water Preparations for the Semiconductor Pharmaceutical and Power Industries. Littleton, CO: Tall Oaks Publishing Co., 1993.
25. Gagnon S, Cheney J, Drake H. Low TDS, high TOC and their effects on high purity water production in New England. In: Proceedings of the Ultrapure Water Expo 1991, East Conference on High Purity Water, Philadelphia, PA, April 29–May 1, 1991:53–62.
26. Meuleman CCE. The basic principles of UV sterilization of water. In: Proceedings of Ozone and Ultraviolet Water Treatment. Massehelein WJ, Coordination, Paris, France, International Ozone Association European Committee, B.1.1–1.13.
27. Ogedengbe O. Characterization and specification of local sands for filters. Filtration Separation 1984; 21:331–4.
28. Berkelman RL, Anderson RL, Davis BJ, et al. Intrinsic bacterial contamination of a commercial iodophor solution: investigation of the implicated manufacturing plant. Appl Environ Microbiol 1984; 47:752–6.
29. Michels DL. Letter to the Pharmaceutical Industry. Re: Validation and Control of Deionized Water Systems. Rockville, MD: Food and Drug Administration/Bureau of Drugs/Assoc. Dir. of Compliance, 1981.
30. Saari R, Suomela H, Haahti E, Raaska E. Production of pyrogen-free distilled water for pharmaceutical purposes. Bull Parenter Drug Assoc 1977; 31:248–53.
31. Tankha A, Williford H. GMP considerations in the design and validation of a deionized water system. Pharm Eng 1987; 7(4):17–20.
32. Zoccolante GA. Course on pharmaceutical water—design and validation. In: Parenteral Drug Associates, Raleigh, NC, September 26–27, 1994.
33. Kuhlman H. Kuhlman Technologies, Inc., Kirkland, WA (Personal communication to TH Meltzer, 1994).
34. Tarry SW, Henricksen J, Prashad M, Troeger H. Integrity testing of ePTFE membrane filter vents. Ultrapure Water J 1993; 10(8):23–30.
35. Waibel P. Integrity testing hydrophobic membrane filter cartridges with water intrusion test. In: Parenteral Drug Association Meeting, Chicago, March 10, 1994.
36. Glaze WH, Kang JW, Chapin DH. The chemistry of water treatment processes involving ozone, hydrogen peroxide and ultraviolet radiation. Ozone Sci Eng 1987; 9:335–52.
37. Leeds & Northrup. Application Data Sheet, Two Cell Conductivity Analyzer Applied to Softening Monitoring. North Wales, PA: Leeds & Northrup, 1987.

7

Calibration and Metrology

Göran Bringert
Independent Consultant, North Andover, Massachusetts, U.S.A.

INTRODUCTION

The regulatory authorities in the U.S.A. and Europe have expanded the scope of regulations by reference to international standards for Quality and Risk Management. The GMP regulations are still in effect, however the enforcement focuses on the critical risk factors determined by risk assessments.

In September 2004, FDA issued three documents describing the FDA's current thinking:

- Draft Guidance for Industry—Quality Systems Approach to Pharmaceutical Current Good Manufacturing Practice Regulations (1)
- Guidance for Industry—Sterile Drug Products Produced by Aseptic Processing—Current Good Manufacturing Practice (2)
- Pharmaceutical CGMPs for the 21st Century—A Risk-Based Approach Final Report (3)

And Medical Device–specific international standards:

- ANSI/AAMI/ISO 13485:2003 Medical Devices—Quality management systems Requirements for regulatory purposes (4)
- ANSI/AAAMI/ISO 14971:2000/A1: 2003 Medical devices—Application of risk management to medical devices (5)

All of the above documents refer either directly or indirectly to the following documents as standard requirements for calibration:

- ISO 9001:2000 Quality management systems—Requirements (6)
- ISO 10012: 2003 Measurement management systems—Requirements for measurement processes and measuring equipment (7)
- ISO/IEC 17025:2005 General requirements for the competence of testing and calibration laboratories (8)
- GUM Guide to the Expression of Uncertainty in Measurement, (corrected and reprinted 1995) issued by BIPM, IEC, IFCC, ISO, IUPAC, IUPAP, and OIML (9)
- ANSI/NCSL Z540-2-1997 U.S. Guide to the Expression of Uncertainty in Measurement (10)

The GMPs in Europe and the U.S.A. have common objectives but differ to some degree in their approaches and have specific regulatory requirements for human and veterinary drugs and medical devices.

Validation verifies that processes perform to specifications. Specifications serve to define what performance is needed for a consistent quality output from the process. Measurements of critical parameters are needed to judge the performance of the process. The measurements have to be accurate and repeatable. Accurate and repeatable measurements require adequately calibrated good quality measurement equipment. Current regulations and standards do not specify accuracy requirements as processes and accuracy specifications vary widely across the industry. The regulations hold the organizations of the regulated industries responsible to set specifications and tolerances for calibrations and to verify that calibration laboratories/providers have the competency required for compliance. The reader should keep in mind that the standards and regulations cover implementation of Quality Management Systems and that calibration is only one of the several components of the Quality Management System.

Validation of Pharmaceutical Processes is not possible without reliable and repeatable measurements. The "Predicate Rules" (G×Ps) require that critical measurements be performed with adequately calibrated measuring devices.

Heat Penetration Studies are performed to calculate the accumulated lethality, F_0, in the load. The accumulated F_0 is the time integral of the lethality function:

$$L = 10^{(T-T_b)/z}$$

At a base temperature $T_b = 121°C$ and $z = 10°C$, the effect of 1°C error in measured temperature at 121°C results in approximately 26% error in the lethality calculation.

This is why in section IX sub-clause C 2 *Equipment Controls and Instrument Calibration,* of the 2004 Guidance for Industry—Sterile Drug Products Produced by Aseptic Processing—cGMPs Validation of Aseptic Processing and Sterilization (2), FDA states:

"For both validation and routine process control, **the reliability of the data generated by sterilization cycle monitoring devices should be considered to be of the utmost importance.** Devices that measure cycle parameters should be routinely calibrated. Written procedures should be established to ensure that these devices are maintained in a calibrated state.

Temperature and pressure monitoring devices for heat sterilization should be calibrated at suitable

---

Abbreviations used in this chapter: CAPA, corrective action, preventive action; CFR, Code of Federal Regulations; cGMPs, current good manufacturing practices; FDA, Food and Drug Administration; GMP, good manufacturing practice; GUM, Guide to the Expression of Uncertainty in Measurement; NIST, National Institute of Standards and Technology; RTD, resistance temperature detector; SOP, standard operating procedures; TUR, test uncertainty ratio.

intervals. **The sensing devices used for validation studies should be calibrated before and after validation runs."**

ANSI/ISO/DIS 17665:2004 Sterilization of health care products—Moist heat—Development, validation and routine control of a sterilization process for medical devices (11), *requires that calibration of the validation system, "measuring chain,"* shall be verified before and after each stage of validation.

These statements are a clear indication that regulatory authorities consider the integrity of temperature measurements a critical part of the validation of thermal sterilization processes. It is important that the validation SOP reflects the theoretical and practical aspects of how to achieve and maintain high-accuracy temperature measurements in conjunction with thermal validation.

Due to the large number of topics covered by the current standards and regulations, this chapter will focus on the practical issues of calibration based on the assumption that the organization has a quality management system that is compliant with applicable requirements. The reader should refer to the complete regulatory documents, referenced in this chapter, to fully assess what has to be in place within the reader's organization for a competent and compliant calibration function for the organization's specific application area. Depending on the origin, some of these documents are available free from the Internet while all ISO standards and regulations are copyright protected and can be purchased as hard copy or downloadable from any national ISO-affiliated organization.

APPLICATION SPECIFIC REGULATION (PREDICATE RULES)

U.S. FDA GMP

- 21 CFR Part 58 Good Laboratory Practice for Nonclinical Laboratory Studies (12)
- 21 CFR Part 211 Current Good Manufacturing Practice for Finished Pharmaceuticals (13)
- 21 CFR Part 606 Current Good Manufacturing Practice for Blood and Blood Components (14)
- 21 CFR Part 820 FDA, Subchapter H—Medical Devices, Quality System Regulation (15)

For full text, go to (16) and under Reference Room select Code of Federal Regulation.

Europe GMP

- The European GMP is published in EUDRALEX Volume 4—Medicinal Products for Human and Veterinary Use: GMP (17).

Medical Devices

There are three current standards specific to medical devices recognized by the FDA and EU:

- ANSI/AAMI/ISO 13485:2003 Quality System Requirements for Medical Devices (4)
- ANSI/AAMI/ISO 14971:2000/A1:2003 amendment, Medical Devices—Application of risk management to medical devices (5)
- ANSI/ISO/DIS 17665:2004 Sterilization of health care products—Moist Heat—Development, validation and routine control of a sterilization process for medical devices (11); a draft standard expected to revise and replace ISO 11134:1994, ISO 13683:1997, and CEN 554:1994 to keep current with technology and moist heat sterilization practices
- ISO 17025 has two main sections, Management requirements and Technical requirements. Section 4, Management requirements, has 15 sub-clauses covering eight pages and Section 5, Technical requirements, has 10 sub-clauses covering 14 pages. It is beyond the scope of this chapter to review the detailed requirements, which will be left to the reader to review as needed.

Corrective Action/Preventive Action (CAPA)

ISO 9001 (6), ISO 10012 (7), ISO 17025 (8), and GMPs have mandatory requirements for CAPA.

GMPs (PREDICATE RULES)

Regulatory Requirements for Calibration in Europe and the United States

In "Guidance for Industry—Sterile Products Produced by Aseptic Processing—cGMP, September 2004" (2), the FDA has included text boxes with quotes from the CFR and stated:

> The quotes included in the text boxes are not intended to be exhaustive. Readers of this document should reference to complete CFR to ensure that they have complied, in full, with all relevant sections of the regulations.

GMP Calibration Requirements

This section will identify calibration requirements, as of April 2004, defined by CGMP regulations, the Predicate Rules including the 21CFR 820 Quality System Regulations, and EU regulations.

- **21 CFR part 58**—Good Laboratory Practice for Nonclinical Laboratory Studies (12)
- **21 CFR part 211**—Current Good Manufacturing Practices for Finished Pharmaceuticals (13)
- **Sec. 211.68** Automatic, mechanical, and electronic equipment.
- **Sec. 211.160** Laboratory Controls—General requirements.
- **Sec. 211.165** Testing and release for distribution.
- **Sec. 211.194** Laboratory records.
- **21 CFR part 606**—Current Good Manufacturing Practice for Blood and Blood Components (14)
- **Sec. 606.60** Equipment
- **Sec. 606.100** Standard Operating Procedures
- **Sec. 606.160** Records
- **21 CFR part 820** Quality System Regulation (15)
- **Sec. 820.72** Inspection, measuring and test equipment

ISO/DIS 17665:2004

Sterilization of health care products—Moist heat—Development, validation and routine control of a sterilization process for medical devices (11).

Summary, Standards, and Regulations

The standards and regulations, in the preceding summary, require that organizations, involved in the validation of processes, have personnel with a thorough knowledge and understanding of what is needed to achieve and maintain compliance. Calibration is critical, as *"reliability of the data generated by sterilization cycle monitoring devices should be considered to be of the utmost importance."* (FDA 2004 Guidance for Industry—Sterile Drug Products Produced by Aseptic Processing—cGMP Validation of Aseptic Processing and Sterilization) (2).

ISO 17025-2005 (8) provides general requirements for the competence of testing and calibration laboratories and the GUM (9) clause 1.1 states:

"This *Guide* establishes general rules for evaluating and expressing uncertainty in measurement that can be followed at various levels of accuracy and in many fields—from the shop floor to fundamental research. Therefore, the principles of this *Guide* are intended to be applicable to a broad spectrum of measurements, ..."

An organization that intends to comply has to define and document, in SOP, how the organization achieves and maintains compliance.

Estimation and Expression of Uncertainty (An Overview)

The *Guide* (9) is a complex document that covers far reaching requirements and it should be the responsibility of the metrology function within the organization to define the criteria for uncertainties in measurement that are needed for specific compliance requirements. Calibration is used to maintain measurement errors within acceptable limits needed to ensure that product quality consistently meets predefined product quality specifications. Historically this was done by the comparison of how much a measured value differed from a "true" value. The difference was defined as the measurement "error." The international metrology community has agreed that true values are by nature indeterminate and that is why the term "true value" is not used in the ISO GUM (9). In other words, there will always be a degree of uncertainty in any measurement.

GUM 2.3.1 defines standard uncertainty as:

> uncertainty of the result of a measurement expressed as a standard deviation.

This illustrates standard deviation in a normal distribution, based on definitions in the U.S. GUM ANSI/NCSL Z540-2-1997 (10).

ANSI/ISO/DIS 17665:2004 (11) requires, and FDA Guidance for Industry—Sterile Drug Products Produced by Aseptic Processing 2004 (2) recommends that the measurement system used for validation should be calibrated before and after validation runs. Calibration requires the use of a temperature standard and a stable thermal source. Each of these items contributes some degree of uncertainty to the calibration result in addition to the uncertainties that are inherent in each of the components of the measuring chain (the measurement system used for validation). The *Guide* (10) defines how to combine these uncertainties into a combined standard uncertainty for the measurement system after calibration. A coverage factor is then used as a multiplier of the combined standard uncertainty in order to obtain an expanded uncertainty. If Figure 1 represents the combined standard uncertainty of a measurement system with normal distribution, the three confidence intervals shown in the graph represent coverage factors 1, 2 and 2.57 for 68%, 95% and 99% confidence intervals respectively.

European Standard EN 554:1994 required a minimum TUR of 3:1 for the validation of thermal sterilization processes. This means that the validation system should have an expanded uncertainty that is at least three times less than the process specification limits. This is illustrated in Figure 2. The process under validation has specification limits $\pm 0.5°C$ which, with an assumed normal distribution, can be represented by the standard uncertainty $1\sigma = 0.1667°C$. The validation

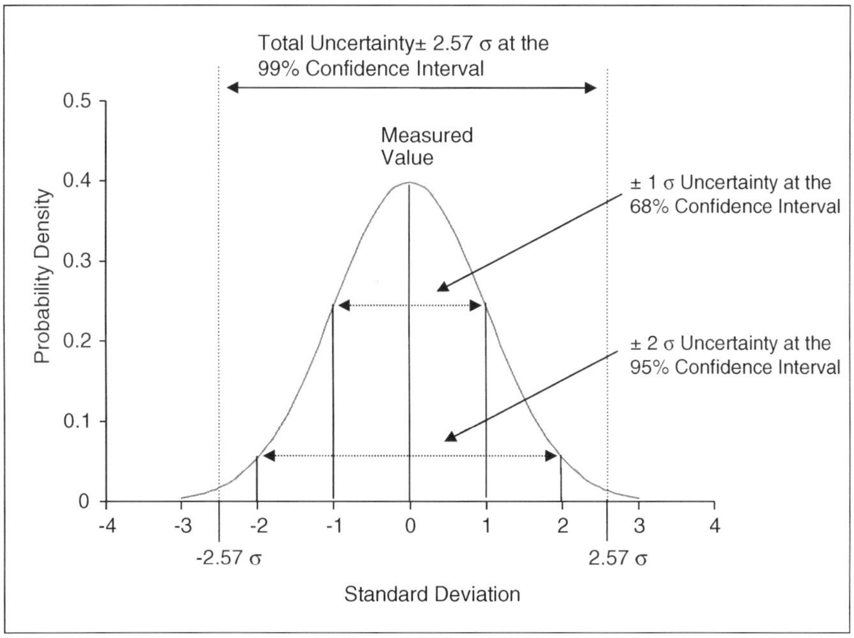

Figure 1 Standard deviation in a normal distribution.

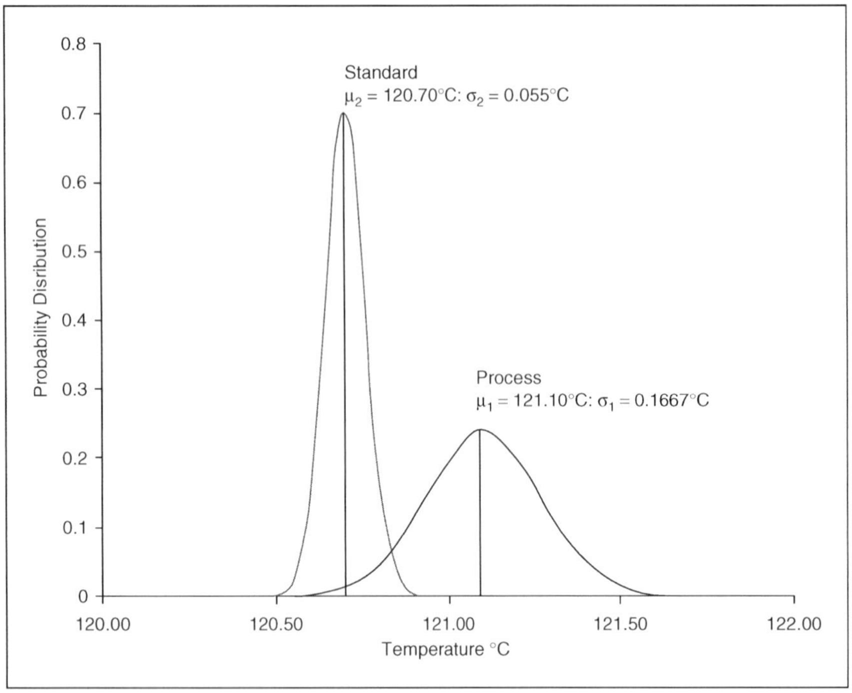

Figure 2 Adequate test uncertainty ratio for validation.

system has to have a standard uncertainty at least three times smaller or $1\sigma=0.055°C$. This is an example of an adequate TUR for validation of a thermal sterilization process.

A larger expanded uncertainty of the validation standard increases the probability for an erroneous validation result that could have severe consequences. This is illustrated by Figure 3.

The U.S. version of GUM, ANSI/NCSL Z540-2-1997, U.S. GUM (10), is an adaptation of the ISO GUM (9) to promote consistent international methods in the expression of measurement uncertainty within the U.S. standardization, calibration, laboratory accreditation, and metrology services. The U.S. Guide is identical to the ISO Guide (corrected and printed, 1995) with the exception of minor editorial changes to facilitate its use in the United States.

ISO 17025:2005 (8) makes reference to GUM "for further information on estimation of uncertainty in measurement."

ISO 10012:2003 (7) Guidance to clause 7.3.1 Measurement uncertainty refers to the "*GUM*" for methods that can be used to combine uncertainties and present the results. It also states that other documented and accepted methods may be used.

To fully understand and master the methods referred to above, it is necessary to study the documentation and integrate the information and requirements into the user's Quality Management and Risk Management Systems, a topic beyond the scope of this chapter.

Practical Discussion of How to Calibrate the Measuring Chain

The integrity of temperature measurements is a critical part of validation of thermal sterilization processes. It is important that the validation SOP reflects the theoretical and practical aspects of how to achieve and maintain high-accuracy temperature measurements in conjunction with thermal validation.

Heat penetration studies are performed to calculate the accumulated lethality, F_0, within the load items. The accumulated F_0 is the time integral of the lethality function:

$$L = 10^{(T-T_b)/z}$$

At a base temperature $T_b=121°C$ and $z=10°C$, the effect of 1°C error in measured temperature at 121°C results in approximately 25% error in the lethality calculation.

The FDA definition of Process Validation is:

> Establishing by objective evidence that a process consistently produces a result or product meeting its predetermined specifications.[a]

The required temperature uniformity in the chamber, according to contemporary industry standards for terminal sterilization, should be better than or equal to 1°C or 0.5°C depending on the application. The combined standard uncertainty for the instrument used for validation measurements, including temperature sensors, should be at least three times less than the specified for the process variable. This means that the overall system combined standard uncertainty should be better than or equal to ±0.33°C or ±0.17°C, respectively. All components involved with the measurement, (from the tip of each sensor, via the connecting wires, cold junction reference, signal interface, analog to digital conversion, conversion from mill volts to temperature, to display and printout of the measured values) are referred to as the measuring chain. Figure 4 shows a measuring chain using thermocouples.

[a] (GMP/Medical Device Quality System Manual, 4. Process Validation, 1997).

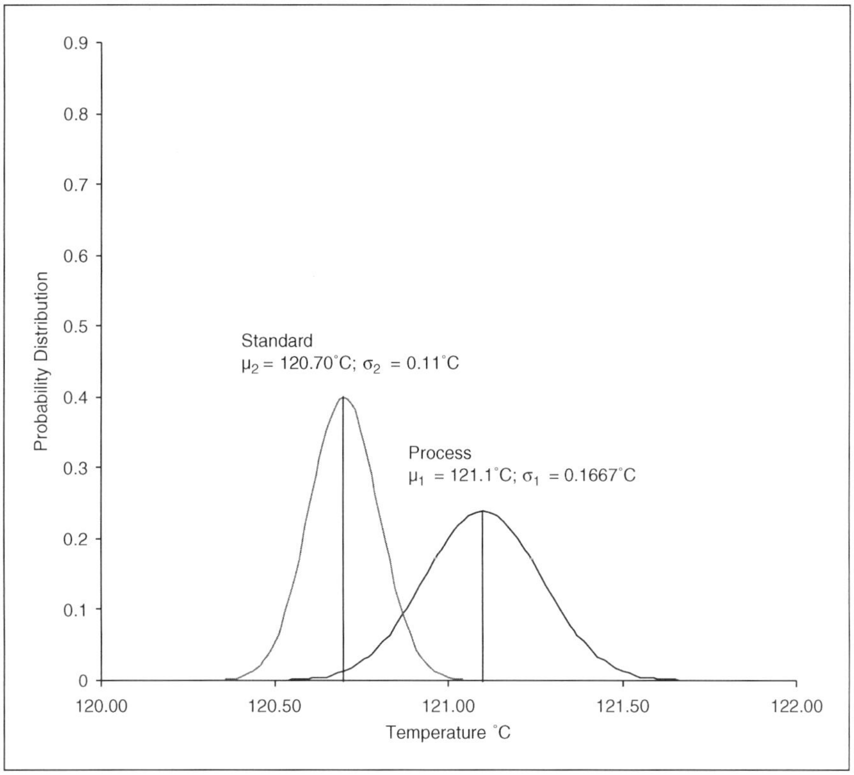

Figure 3 Inadequate test uncertainty ratio for validation.

Error Sources

Several variable error sources can affect the temperature measurement accuracy in validation. Control and management of these error sources should be recognized as the responsibility of the people who perform the validation. Individuals responsible for validation should have the competency to adequately perform the validation studies.

It is important to distinguish between systematic and random errors. Systematic errors are eliminated by calibration, while random errors are not eliminated by calibration, and can only be minimized through the application of knowledge and proper procedures. The operator has to understand how to minimize the influence of random temperature measurement errors to consistently achieve the accuracy required for thermal validation of steam sterilization processes. Procedures documented in the validation SOP and individual training of validation personnel are necessary to maintain competency of the validation team.

Electronic temperature measurements for validation are acquired using temperature sensors connected

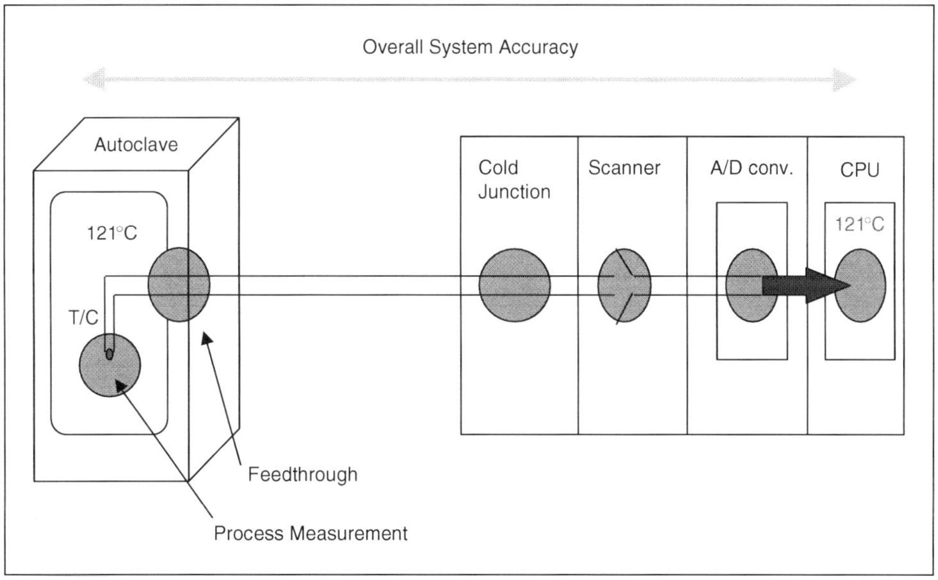

Figure 4 The measuring chain is composed of all components involved in the measurement system.

to an electronic data logger or recorder. The components of the measuring chain contribute errors, systematic or random, that contribute to the Overall System Accuracy. Figure 5 identifies the most significant Random Error Sources (in red) in the measuring chain.

Significant random errors can occur in the following areas:

- Sensor design
- Sensor location
- Sensor wire nonhomogeneity
- Thermal scatter at the cold junction reference

Temperature Sensors

Thermocouple type T (copper/constantan) is the most commonly used thermocouple for temperature measurements in validation applications due to its high accuracy and low cost. Temperature measurement is affected by several ambient conditions, that is why the ASTM Manual on The Use of Thermocouples in Temperature Measurement, Series MNL 12, 1993 (18) has the following statement on its first page:

> Regardless of how many facts are presented herein and regardless of the percentage retained, all will be for naught unless one simple important fact is kept firmly in mind. The thermocouple reports only what it 'senses.' This may or may not be the temperature of interest. Its entire environment influences the thermocouple and it will tend to attain thermal equilibrium with this environment, not merely part of it. Thus, the environment of each thermocouple installation should be considered unique until proven otherwise. Unless this is done, the designer will likely overlook some unusual, unexpected, influence.

Calibration

The measuring chain shall be calibrated prior to and after calibration runs. Adequate calibration equipment shall be used when in calibration of the measuring chain. A temperature transfer standard, traceable to a primary standard, and a stable thermal source are required to perform calibration of the measuring chain. The combined expanded uncertainty of all components of the measuring chain, temperature transfer standard and stable thermal source shall be determined by a competent calibration facility and documented to be adequate for the validation of the process equipment.

Measurement standards are classified based on metrological qualities.

- The primary standard has the highest metrological qualities and is accepted without reference to other standards of the same quantity. Primary standards are normally kept in national measurement laboratories and designated as national standards.
- The secondary standard has its value assigned by comparison with a primary standard of the same quantity.
- The transfer standard is used for the comparison of standards of the same quantity.

This means that the validation study has to have documented evidence that the measuring chain was in calibrated state before and after the validation study was performed. The documented procedures shall describe the criteria chosen for the calibration procedures. This should be backed by historical documentation that gives the rationale for the procedure. The regulations require three consecutive successful validation runs for a successful validation. Based on the company's risk assessment and risk management it seems realistic to calibrate before the validation study begins and to verify that the measuring chain remains in calibration at the end of the third successful run.

Be patient. A frequent mistake in calibrating instrumentation is to take measurements and make adjustments before conditions have stabilized. It may take much longer than expected for a system to become completely stable, because thermal equilibration takes place exponentially and the output may seem to be stable even though it is still changing slowly. Automatic detection of stability to preset stability criteria, offered by modern calibration equipment and software, is the most reliable and repeatable method for stability determination.

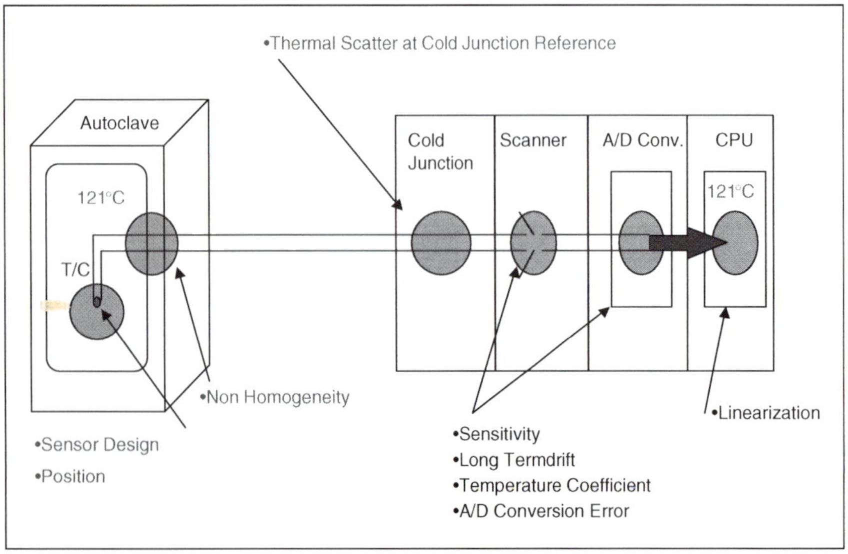

Figure 5 Random and systematic error sources in the measuring chain.

Temperature Transfer Standard

The accuracy of the transfer standard must be better than that of the instrument being calibrated. This would seem obvious, but it is amazing how often a voltage calibrator is used that has a greater error than the system being calibrated. It is important to recognize that the accuracy of the calibration can be no better than the standard used, and it is a mistake to change the adjustment of a measuring system if it is already more accurate than the standard.

The characteristics of the transfer standard must have been determined by a procedure that is traceable to accepted primary standards. In the U.S.A., NIST is the accepted source of primary standards. The transfer standards need to be calibrated by NIST relative to their primary standards or by a qualified Standards laboratory relative to standards calibrated by NIST. In either case, the test results and test numbers should be known so the calibration procedure can be traced to the primary standards.

The transfer standard must be independent of the measuring system. Because the output of a thermocouple depends on the entire circuit, it is not a desirable transfer standard. An RTD is a device that indicates changes of temperature by a change of resistance. Because the resistance of an RTD is only a function of its temperature, and the resistance can be measured independently of the system being calibrated, RTDs are ideal temperature transfer standards.

The transfer standard must be stable in shipment and tolerate other handling. As its name implies, the purpose of the transfer standard is to transfer a measured characteristic from one laboratory to another. The characteristics of the standard must be the same when received from NIST as when it was calibrated relative to their standards. Liquid in glass thermometers may be damaged or develop small voids in the liquid during shipment, and therefore are not reliable temperature transfer standards. RTDs are sensitive devices that maintain their characteristics only with careful handling and shipment.

Stable Thermal Source

Significant random calibration errors that can occur in a dry block temperature reference:

- Transfer calibration
- Stem conduction
- Uneven heat transfer
- Immersion depth
- Well inserts not used
- Stability
- Time needed for stabilization

Reference Error, Using Thermocouples

When calibrating a thermocouple T_1, against an RTD transfer standard T_2, a key contribution to error is the difference in temperature between these devices when placed in the reference (Fig. 6). This difference is called *transfer calibration error* and is potentially the largest contribution to calibration errors in dry block references.

Transfer calibration error contains two components:

1. Stem conduction error, which cools the thermocouple tip (Fig. 7).
2. Uniformity of the reference wells relative to the standard well.

A dry block with one common large diameter well is not suitable for calibration of a measuring chain using multiple thermocouples (Fig. 8). The stem conduction will cause heat losses that are greater than the heat radiated from the walls of the single well. Figure 8 illustrates this inadequacy.

A dry block with smaller diameter wells and inserts provides closer thermal coupling between well walls and sensors under calibration, minimizing the effect of stem conduction and transfer calibration error. A dry block designed for maximum transfer calibration accuracy has small diameter wells with inserts that fit the size of the sensors under calibration (Fig. 9).

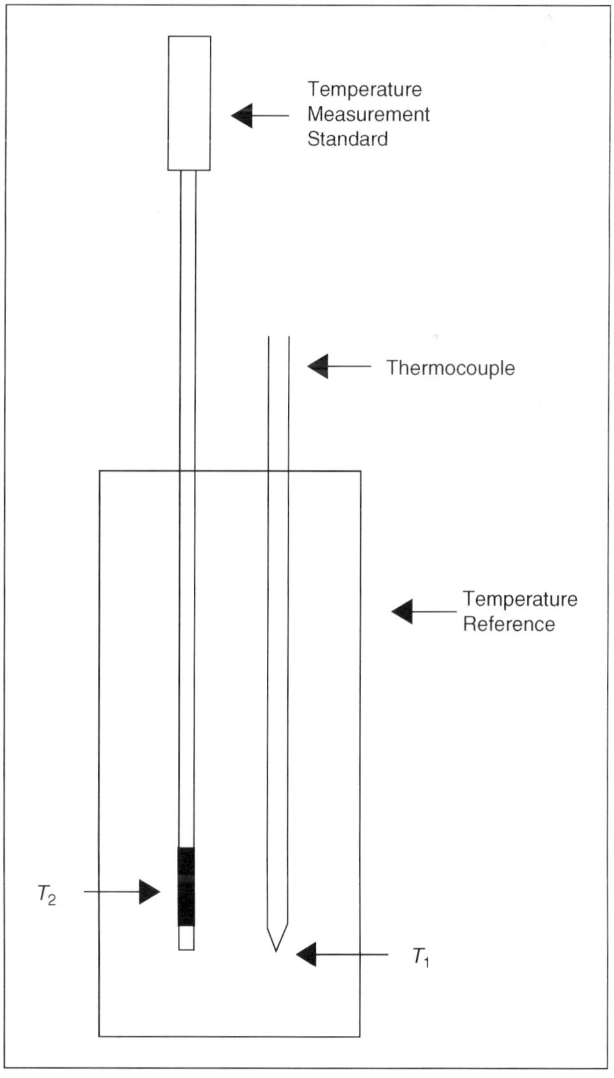

Figure 6 Transfer calibration error is the temperature difference between thermocouple tip (T_1) and the measurement standard (T_2).

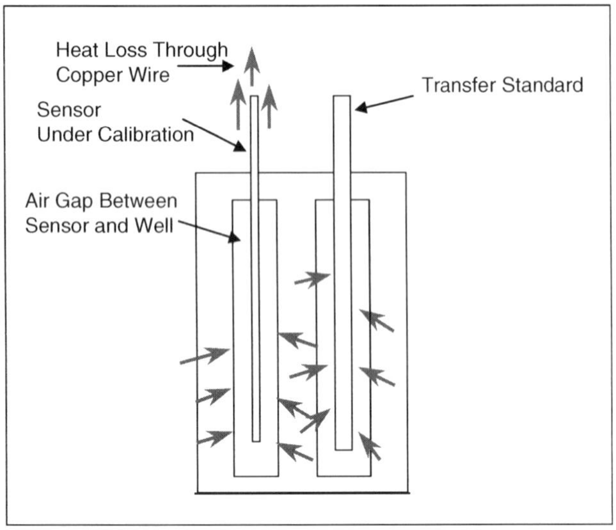

Figure 7 Stem conduction causes heat loss and generates calibration error.

Calibration Procedure

Calibration of the measuring chain (Fig. 3) should be done with the sensors connected to the data logger/recorder and installed in the sterilizer via the feed through and out through the open sterilizer door. The sensors should be inserted into the temperature reference bath or dry block, just outside the open sterilizer. Calibration should be performed prior to a validation study and a calibration check should be performed at the conclusion of the validation.

Prestudy Calibration

A two-point calibration should be used with calibration points bracketing the sterilization temperature for the process under validation, e.g., 100°C and 130°C, and calibration checkpoint should be, e.g., 121°C, between the two calibration points to verify the calibration.

Poststudy Verification

A two-point comparison between the temperature standard and the temperature sensors should be performed to verify that the calibration of the measuring chain is intact.

The calibration should be documented, to provide evidence that the temperature of the reference, the transfer standard, and the sensors were stable before the determination of *calibration correction values*. The calibration documentation should include data on the deviation between the temperature standard and each temperature sensor before and after calibration. To ensure traceability, the documentation must list the calibration parameters and equipment including serial numbers and last calibration dates.

SUMMARY

Validation of thermal sterilization processes requires accurate temperature measurements to provide reliable results. In order to ensure measurement integrity it is necessary that the validation personnel has adequate training and well-defined processes to follow.

Risk assessment and risk management are now mandatory for processes that are used for manufacture or production of products with critical tolerances. The result of the risk assessment should serve as a basis for definition of process tolerances and the corresponding measurement tolerances for the process control system. Appropriate measurement tolerances vary by application, while a moist heat sterilization process needs to be measured to ±0.5°C a depyrogenation process would be adequately measured with a tolerance of ±1.0°C.

Internationally accepted good metrological standards and process control engineering practice call for the application of TUR between the equipment under calibration/validation and the calibration equipment itself. For most applications the minimum TUR should be 3:1 and preferably higher (Figs. 1–2). The higher the TUR the more expensive calibration equipment has to be used. This is an area where risk assessment and risk management is used to determine the level of

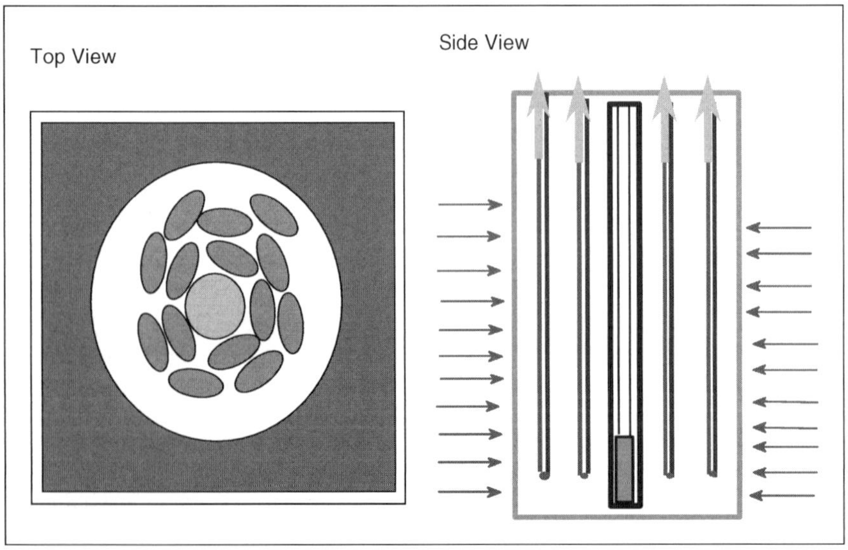

Figure 8 Dry well with large diameter wells.

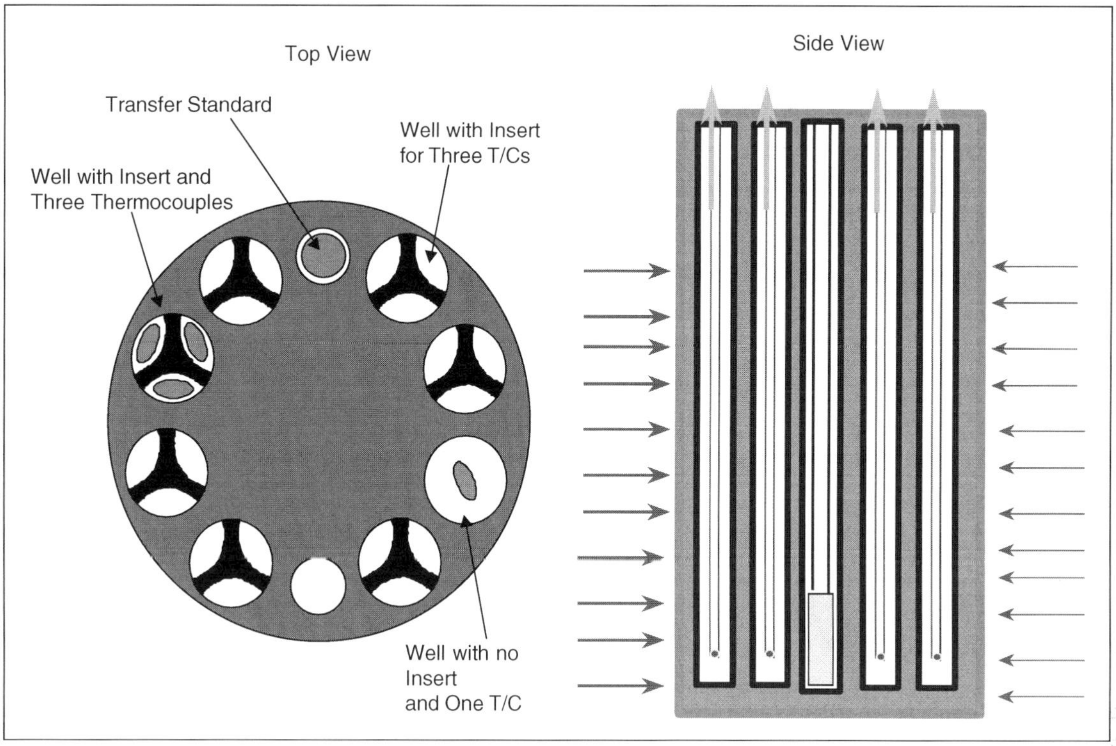

Figure 9 View of dry block with one large well and no insert.

compromise needed to balance the level of acceptable risk versus the cost of more sophisticated calibration equipment. According to ISO 17025 and ISO 10012 the management is responsible for the determination and justification of the balance of risk versus cost.

Calibration/verification of the sensors monitoring and controlling the process is as critical as the calibration of the sensors used for validation of the process. Each individual measuring chain has to be calibrated/verified against a transfer standard, i.e., sensor, wiring, and measurement system, to at least the same expanded uncertainty as required for the calibration of the validationsystem.

The pressure sensor in the autoclave must be calibrated in place under standard operating conditions. A two-point comparison between the installed pressure sensor and a temporarily connected pressure transfer standard, traceable to a national standard, should be performed. Based on the comparison, zero and span adjustments are done on the installed pressure sensor.

REFERENCES

1. FDA Draft Guidance for Industry—Quality Systems Approach to Pharmaceutical Current Good Manufacturing Practice Regulation, September 2004, U.S. Department of Health and Human Services Food and Drug Administration, Center for Drug Evaluation and Research (CDER) Center for Biologics Evaluation and Research (CBER), Center for Veterinary Medicine (CVM), Office of Regulatory Affairs (ORA).
2. FDA Guidance for Industry—Sterile Drug Products Produced by Aseptic Processing—Current Good Manufacturing Practice, September 2004, U.S. Department of Health and Human Services Food and Drug Administration Center for Drug Evaluation and Research (CDER) Center for Biologics Evaluation and Research (CBER) Office of Regulatory Affairs (ORA).
3. FDA Pharmaceutical CGMPs for the 21st Century—A Risk-Based Approach—Final Report, September 2004, Department of Health and Human Services, U.S. Food and Drug Administration.
4. ANSI/AAMI/ISO 13485:2003 Medical Devices—Quality management systems—Requirements for regulatory purposes. Association for the Advancement of Medical Instrumentation, Arlington, VA.
5. ANSI/AAMI/ISO 14971:2002/A1:2003—Application of risk management to medical devices. Association for the Advancement of Medical Instrumentation, Arlington, VA.
6. ISO 9001:2000 Quality management systems—Requirements. International Organization for Standardization, Central Secretariat, Geneva, Switzerland.
7. ISO 10012:2003 Measurement management systems—Requirements for measurement processes and measuring equipment. International Organization for Standardization, Central Secretariat, Geneva, Switzerland.
8. ISO/IEC 17025:2005 General requirements for the competence of testing and calibration laboratories. International Organization for Standardization (ISO) and International Electrotechnical Commission, Geneva, Switzerland.
9. ISO Guide to the Expression of Uncertainty in Measurement (GUM), issued by BIPM, IEC, IFCC, ISO, IUPAC, IUPAP. International Organization for Standardization, Central Secretariat, Geneva, Switzerland, 1995.
10. ANSI/NCSL Z540-2-1997 American National Standard for Expressing Uncertainty—U.S. Guide to the expression of Uncertainty in Measurement. National Conference of Standards Laboratories, Boulder, CO.
11. ANSI/ISO/DIS 17665:2004 Sterilization of health care products—Moist heat——Development, validation and routine control of a sterilization process for medical devices. International Organization for Standardization, Geneva, Switzerland, 2004.

12. FDA, Good Laboratory Practice for Nonclinical Laboratory Studies, Title 21, Part 58.
13. FDA, Current Good Manufacturing Practice for Finished Pharmaceuticals, Title 21, Part 211, Code of Federal Regulations (U.S.A.).
14. FDA, Current Good Manufacturing Practice for Blood and Blood Components, Title 21, Part 606, Code of Federal Regulations (U.S.A.).
15. FDA, Subchapter H—Medical Devices, Title 21, Part 820, Quality System Regulation, Code of Federal Regulations (U.S.A.).
16. http://www.fda.gov, 2006
17. European Good Manufacturing Practice, EDURALEX Volume 4—Medicinal Products for Human and Veterinary Use, August 2004 (update of 1998 edition), European Commission, Enterprise Directorate—General Pharmaceuticals: Regulatory Framework and Market Authorizations. http://ec.europa.eu/enterprise/pharmaceuticals/eudralex/homev4.htm, 2006
18. ASTM Manual on the Use of Thermocouples in Temperature Measurement, Manual Series MNL 12, American Society for Testing and Materials 1916 Race Street, Philadelphia, PA, 1993.

8

Temperature Measurements

Clarence Kemper
Independent Consultant, Winchester, Massachusetts, U.S.A.

Göran Bringert
Independent Consultant, North Andover, Massachusetts, U.S.A.

INTRODUCTION

Temperature is the most common of all industrial process measurements, and in thermal sterilization processes it is the most critical. Verification in which all temperature measurements are accurate and reliable, is one of the most important requirements in the validation of these processes.

> The reliability of the data generated by sterilization cycle monitoring should be considered to be of the utmost importance (FDA 2004 guidance for Industry—Sterile Drug Products Produced by Aseptic Processing) (1).

The output of a temperature-measuring system is a result of its entire thermal environment and the indicated value of temperature may change as the system attains thermal equilibrium with this environment. There are several factors that can affect the accuracy of temperature measurements, and all such factors must be considered in calibrating the system. Each temperature measurement installation should be considered unique until proven otherwise. If this is not done, some unusual or unexpected factor may be overlooked.

Prior to calibrating a measuring chain (See Chap. 7, "Calibration and Metrology"), it is important to verify that the entire measuring chain, sensor, wiring, measuring equipment, and its installation is designed for the application at hand. Calibration and validation SOPs should define application- and installation-specific requirements.

The two most commonly used temperature sensors in pharmaceutical processes are the T/C and the RTD. It is necessary to be aware of their different physical properties, as the error sources which affect the use of either may differ or have similarities. The discussion in this chapter will describe the physical properties of T/Cs and RTDs and address the broader issue of assuring that the indicated value of temperature is an accurate representation of the value being measured.

T/Cs are the most satisfactory sensors for conducting heat penetration and temperature distribution studies in validation, whereas RTDs are the most satisfactory transfer standards for temperature calibration and Pt 100 RTD sensors are commonly installed in processing equipment for control and monitoring.

THERMOCOUPLES

A T/C is a simple, versatile temperature sensor constructed by joining two wires of different composition to form a "T/C junction." When a T/C is connected to a well-designed reference and measuring system, the indicated output is a unique function of the junction temperature. It will be shown that the total output of a T/C circuit is not a sensor characteristic; however, the entire measuring system must be considered in a proper calibration procedure.

The primary reasons for choosing T/Cs for validation of heat penetration and heat distribution are that T/Cs are small, flexible, easily interchangeable, and mechanically resistant and are more convenient to place in difficult-to-reach locations in the load than RTDs.

The inaccuracies in most T/C systems do not occur in the sensors; they occur in the instrumentation used to measure the outputs and in the circuitry connecting the T/C sensors to the measuring system. Additional errors may be caused by sensor designs that are not fit for the specific application and by a location of sensors in an area that is not representative of the temperature of interest. A simplified explanation of thermoelectric theory is included in this chapter as a guide to proper installation of T/C circuits. By understanding the source of thermoelectric output, it is easier to avoid the mistakes most often encountered in the use of T/Cs, thereby assuring better measurement accuracy.

THERMOELECTRIC THEORY

During the 180 years since T. J. Seebeck discovered that a current flows in a circuit of two dissimilar conductors whenever the junctions of the conductors are at different temperatures, many investigators have developed theories to explain thermoelectric phenomena. Some, such as Thomson and Bridgman, have based their explanations on thermodynamic considerations; others, such as Mott and Jones, have employed the electron theory of

Abbreviations used in this chapter: BIER, biological indicator evaluator resistometer; DIN, German Institute of Standadization (Deutsche Industry Norm); DVM, digital voltmeter; FDA, Food and Drug Administration; LVP, large volume parenteral; NIST, National Institute of Standards and Technology; RF, radio frequency; RTD, resistance temperature detector; SOP, standard operating procedure; T/C, thermocouple..

solids (2,3). The following explanation of thermoelectric phenomena might be objectionable to both thermodynamicists and atomic physicists, but is a concept that can be understood easily and employed to avoid many of the errors encountered in T/C circuits.

1. The energy level of an electron in any conductor increases as the temperature of the conductor increases.
2. The amount of energy change for a given temperature change depends on the composition and molecular structure of the conductor.

The material property that expresses the amount of energy increase for a given temperature increase is called the thermoelectric power. The value of thermoelectric power is given in units of microvolts of energy increase for each degree Celsius of temperature increase in the material.

Figure 1 depicts a simple T/C circuit consisting of two external conductors, A and B, which are connected at a junction where the temperature is T_2. For simplicity, it is assumed that the entire circuit of the voltage-measuring device is at a uniform temperature, T_1. It will be shown that under this assumption the net thermoelectric potential difference is generated only in the external circuit. If the temperature of the voltage-measuring device is equal to T_1 throughout, then the two terminals to which the conductors A and B are connected must also be at the same temperature T_1.

The lower portion of Figure 1 gives a graphic representation of the thermoelectric potential of the circuit shown in the upper position. The horizontal ordinate is the temperature at a given location in the circuit and the vertical ordinate is the corresponding electrical potential at that location. Because the thermoelectric power is the amount of energy increase for a given temperature increase, the slope of each line is equal to the thermoelectric power of that conductor. If the thermoelectric powers are different, then the slopes of the two lines will be different.

Because the potential at the junction where the two conductors are joined together must have a singular value, Figure 1 shows that there will be a net potential, difference between the two terminals of the voltage-measuring device.

Consider the energy levels of the electrons in the conductors as the circuit is traversed in a clockwise direction, starting at the terminal where material A is connected to the voltage-measuring device.

Assuming that the temperature T_2 is greater than the temperature T_1, the energy level of the electrons in material A will increase as the junction with material B is approached because the temperature of the material is increasing. Assuming that the thermoelectric power of the conductor is constant, the amount by which the energy increases is equal to the thermoelectric power of material A, P_A, multiplied by the change in temperature $(T_2 - T_1)$. As the circuit is traversed from the junction of the two conductors to the terminal, where material B is connected to the voltage-measuring device, the energy level of the electrons will decrease by an amount equal to the thermoelectric power of material B, P_B, multiplied by the change in temperature $(T_1 - T_2)$.

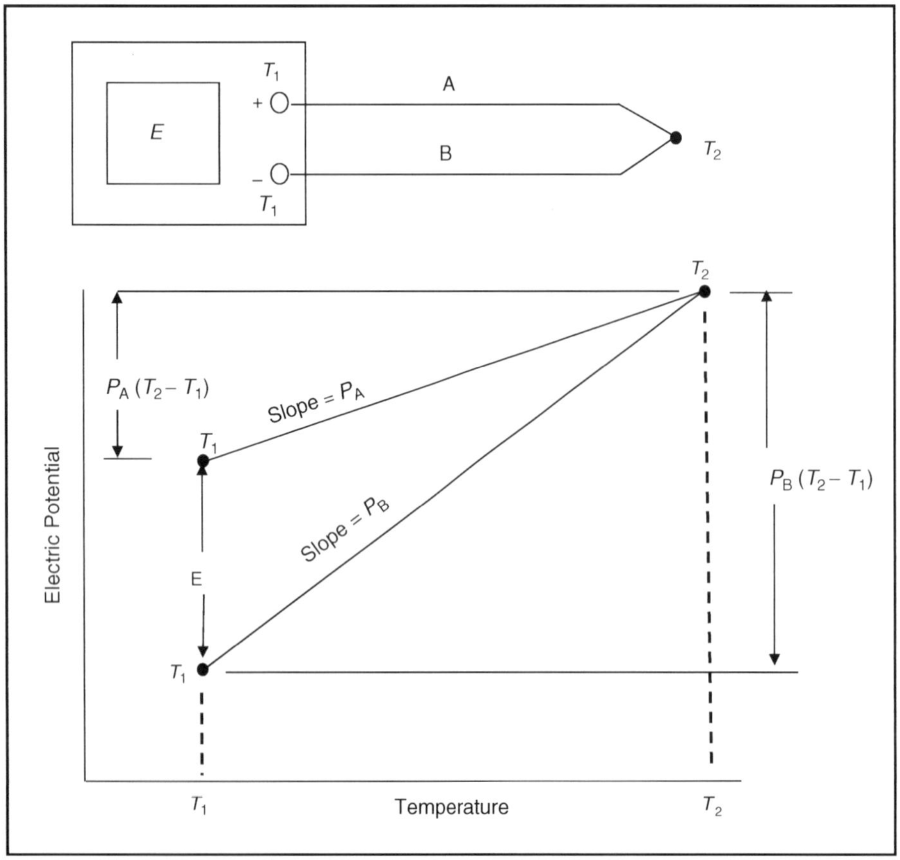

Figure 1 Basic thermocouple.

In this example, the thermoelectric power of material B will be greater than that of material A, so the change in energy level in B will be greater than that in A for the same change in temperature. This simple circuit illustrates a characteristic of T/C materials that causes some confusion. When the temperature of the external junction is greater than that of the junctions at the voltage-measuring device, the material having the greater thermoelectric power will be the negative lead. Because material B is assumed to have the greater thermoelectric power in the following examples, the terminal to which it is connected will be the negative terminal. The circuit summations are expressed starting at the terminal to which material B is connected and traversing the circuit in a counterclockwise direction to yield a positive potential difference.

If P is the thermoelectric power and T is the temperature at any location in the circuit, the *gradient* explanation of thermoelectric output states that the net potential generated by the circuit is equal to the cyclic integral of the product of the thermoelectric power and the differential change in temperature (4,5). That statement is expressed mathematically by equation (1), in which E is the net electric potential difference generated by the circuit:

$$E = \oint P \, dT \tag{1}$$

Although the thermoelectric powers of all conductors change slightly with a change in temperature, for the purpose of this discussion the thermoelectric powers are assumed to be constant within any length of a homogeneous conductor. In reality, the lines in Figure 1 would be curved slightly, rather than being straight, as they are when it is assumed that each homogeneous conductor has a constant thermoelectric power. If the internal circuits of the measuring instrument are uniform in temperature and the thermoelectric power of each conductor is constant, the integral of equation (1) can be evaluated by equations (2) and (3), where P_A is the thermoelectric power of conductor A, P_B is that of conductor B, and P_I is that of the internal circuit.

Equation (2) represents the *conductor* explanation of T/C output, which states that the net electrical potential difference generated by each conductor in the circuit is equal to the thermoelectric power of the conductor multiplied by the temperature difference between the ends of the conductor. The net electrical potential difference generated by the total circuit is equal to the sum of the differences of each conductor. The conductor explanation is a simplified form of the gradient explanation.

$$E = P_B(T_2 - T_1) + P_A(T_1 - T_2) + P_I(T_1 - T_1) \tag{2}$$

It is obvious that the last term in equation (2) is zero and the contribution of the internal circuit is zero if the temperature is uniform. The net output of the entire circuit under these assumptions is given by equation (3):

$$E = P_B(T_2 - T_1) + P_A(T_1 - T_2) \tag{3}$$

An alternative to the gradient or conductor explanation of thermoelectric output is the *junction* explanation (2,3). It states that the electrical output of each junction of two conductors is equal to the product of the temperature of the junction and the difference between the thermoelectric powers of the two conductors. The net electrical potential difference generated by the total circuit is equal to the sum of the outputs of all junctions in the circuit. Equation (4) is the mathematical expression of the junction explanation for the circuit of Figure 1:

$$E = T_1(P_I - P_A) + T_2(P_B - P_A) + T_1(P_A - P_I) \tag{4}$$

It is not quite as obvious in equation (4) that the contribution of the internal circuit is zero, but the two terms containing P_I do cancel and the other terms may be rearranged to yield equation (5). It is also not obvious that the subtraction of the thermoelectric powers at each junction must be performed in a direction consistent with cyclic integration. Thus, the difference of thermoelectric powers in the second term in equation (5) is the negative of that in the first term. This requirement is often confusing to the inexperienced investigator applying the junction explanation.

$$E = T_2(P_B - P_A) + T_1(P_A - P_B) \tag{5}$$

Both equations (3) and (5) may be rewritten to yield equation (6), showing that, for this simple circuit of homogeneous conductors having constant thermoelectric powers, the output predicted by either explanation is the same. It is equal to the difference of thermoelectric powers of the two conductors multiplied by the difference between the temperatures at their junctions.

$$E = (P_B - P_A)(T_2 - T_1) \tag{6}$$

Many persons focus on the junctions in evaluating T/C circuits, so they often fail to recognize phenomena such as regions of stress within a conductor. When wires are flexed repetitively at one location, the resulting cold-working can create regions of nonhomogeneous thermoelectric power, thereby changing the net electric output of the circuit. By using the gradient or conductor explanation to evaluate a circuit, it will be seen that the electrical output is generated where temperature gradients exist in the conductors and that the thermoelectric power of the conductors in those regions must be known (5). This is particularly important when T/Cs are used to measure the temperatures of elements within a chamber in which the temperature is different from the surrounding ambient temperature.

Figure 2 depicts a slightly more complex T/C circuit that adds a third conductor C between the voltage-measuring instrument and each of the other conductors. Because most T/C circuits are constructed of duplex wire with the two conductors in close physical proximity, the error introduced by assuming that both conductors are at the same temperature, at any location in the circuit, will be negligible. Thus, in the circuit of Figure 2, it is assumed that both of the junctions to the C conductors are at a reference temperature T_r. Because the internal circuit of the voltage-measuring device will have no net output if the temperature is equal to T_1 throughout, the output of the circuit of Figure 2 is given by equation (7):

$$\begin{aligned} E = {} & P_C(T_r - T_1) + P_B(T_m - T_r) + P_A(T_r - T_m) \\ & + P_C(T_1 - T_r) \end{aligned} \tag{7}$$

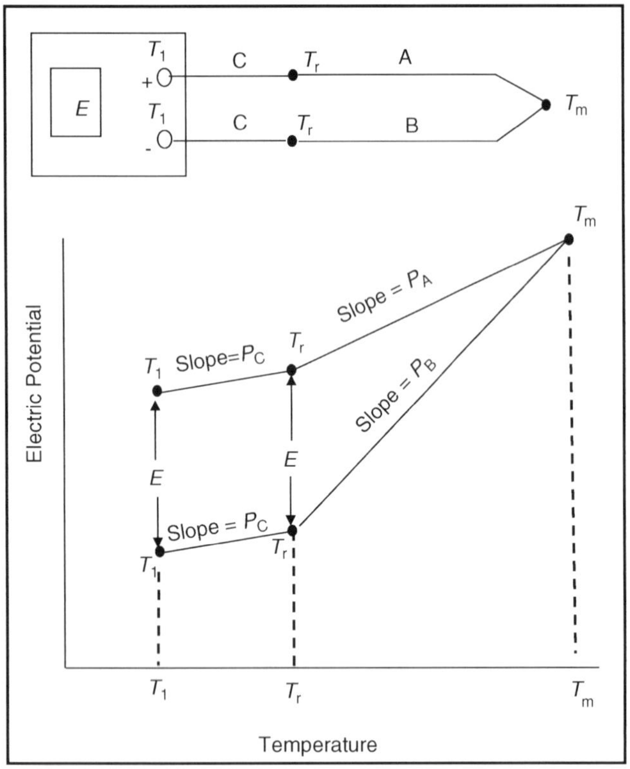

Figure 2 Simple thermocouple circuit.

The terms containing P_C cancel, illustrating an important characteristic of T/C circuits: If at all points along the length of a duplex T/C pair the temperature is the same in both conductors, the output of that portion of the circuit will be zero if the thermoelectric powers of the two conductors are the same. Duplex copper leads do not contribute to the net output of a circuit if they are made of pure, instrument-grade copper and have not been stressed to create cold-worked regions. Equation (7) can be rewritten to yield equation (8):

$$E = (P_B - P_A)(T_m - T_r) \tag{8}$$

That the portion of the circuit makes no net contribution to the output is also shown graphically in the lower portion of Figure 2. Because the slope of each line is equal to the thermoelectric power of the conductor, the output curves of the two C conductors are parallel. Therefore, even though the temperature changes from T_r to T_1 in that portion of the circuit, there is no change in the net potential difference.

From equation (8), it may be observed that if the temperature T_r is maintained at some known reference value, and the thermoelectric powers of conductors A and B are known, the electrical output of the circuit is proportional to the temperature of the measuring junction T_m. It should be remembered that the thermoelectric powers of conductors vary slightly with temperature, thereby giving all T/Cs a nonlinear output versus temperature rather than the linear output of this simplified explanation. For the purpose of understanding how to avoid circuit errors, however, the linear assumption is adequate.

The fact that the duplex conductors, C, make no net contribution to the electrical output is important, because this allows duplex copper leads to be used in connecting the voltage-measuring device to the junctions at the reference temperature, T_r. Then the temperature of the terminals on the voltage-measuring device can be any value without changing the net electrical output of the circuit. It is relatively easy to maintain the junction of two conductors at a constant, known temperature, but it would be extremely difficult to do so at the terminals of a voltage-measuring device.

The value that has been chosen universally as the standard reference temperature for T/C circuits is the equilibrium temperature between ice and air-saturated water, or 0.00°C. A few instruments are sold with oven-controlled reference temperatures of higher values, but all standard tables give the output of T/Cs as a function of the measured temperature, T_m, on the assumption that the reference temperature, T_r, is at 0.00°C (6).

Figure 3 depicts a T/C circuit composed of several lengths of duplex T/C wire. This type of circuit that might be used in a typical validation study. Section 1 could be the length of T/C wire that goes from the measuring system to a connector outside the autoclave; section 2 could be the connector; section 3 could be the length of T/C wire that goes from the external connector, through the wall of the autoclave, to the junction inside the autoclave. Because the circuit from the reference junctions to the voltage-measuring device makes no net contribution, the electrical output of the circuit in Figure 3 is given by equation (9):

$$E = P_{B1}(T_1 - T_r) + P_{B2}(T_2 - T_1) + P_{B3}(T_m - T_1) + P_{A3}(T_2 - T_m) + P_{A2}(T_2 - T_2) + P_{A1}(T_r - T_1) \tag{9}$$

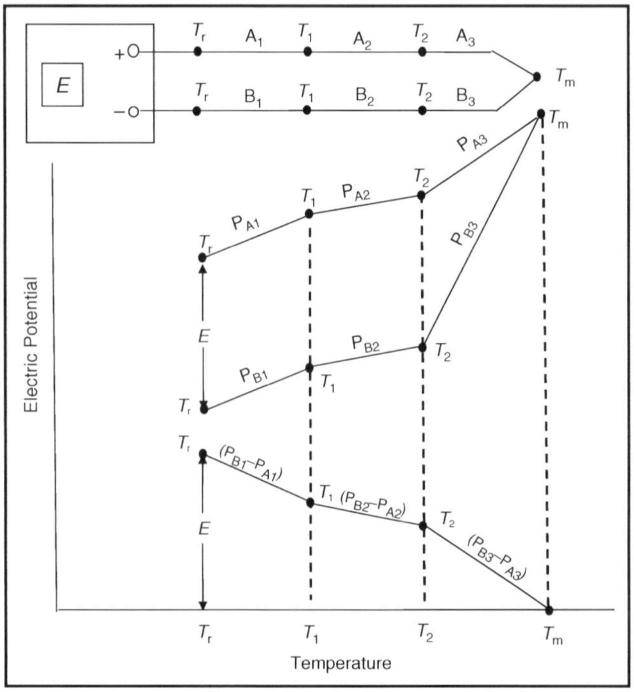

Figure 3 Typical thermocouple circuit.

Table 1 Approximate Seebeck Coefficients of Common Thermocouple Materials Relative to Platinum-67 at 0.0°C

Material name	ASTM E-20 letter code	Approximate composition	Seebeck coefficient (μV/°C)
Chromel	EP and KP	90% Ni, 10% Cr	25.8
Iron	JP	99.5% Fe	17.9
Copper	TP	100% Cu	5.9
^{90}Pt^{10}Rh	SP	90% Pt, 10% Rh	5.4
^{87}Pt^{13}Rh	RP	87% Pt, 13% Rh	5.3
Alumel	KN	95% Ni, 2% Al, 2% Mn, 1% Si	−13.6
Constantan	JN[a]	55% Cu, 45% Ni	−32.5
Constantan	EN and TN	55% Cu, 45% Ni	−32.9

[a] JN is similar to EN and TN, but will generally have a slightly different output.

Equation (9) may be rewritten to yield equation (10):

$$E = (P_{B1} - P_{A1})(T_1 - T_r) + (P_{B2} - P_{A2})(T_2 - T_1) + (P_{B3} - P_{A3})(T_m - T_2) \quad (10)$$

It may be observed from equation (10) and the graphic representation of the output in Figure 3 that, when at all points along the length of a T/C the temperature is the same in both conductors, the output of that portion of the circuit depends only on the difference between the thermoelectric powers of the two conductors and the temperature change along their length. The difference between the thermoelectric powers of two conductors is known as the Seebeck coefficient of the pair (5,6).

The Seebeck coefficient for a single material is always given relative to some reference material. The early evaluations by Peltier, Seebeck, and others were done relative to lead, and recent evaluations are relative to platinum-67. Table 1 gives the approximate Seebeck coefficients of the most common T/C materials relative to ^{67}Pt, and Table 2 gives the corresponding values of the most frequently used T/C pairs at temperatures near the ice point. The Seebeck coefficient for any pair of conductors is equal to the difference of the Seebeck coefficients of each conductor relative to a standard material. If the Seebeck coefficient of copper relative to ^{67}Pt is +5.9 μV/°C and that of constantan is −32.9 μV/°C, the Seebeck coefficient of a copper–constantan duplex pair (type T T/C) is 38.8 μV/°C. The material having the most positive Seebeck coefficient relative to ^{67}Pt will be the positive lead, which is the copper lead in the previous example.

If the Seebeck coefficient S is substituted for the difference in thermoelectric power, $P_A - P_B$, throughout the circuit in Figure 3, equation (10) can be rewritten to yield equation (11):

$$E = S_1(T_1 - T_r) + S_2(T_2 - T_1) + S_3(T_m - T_2) \quad (11)$$

Table 2 Approximate Seebeck Coefficients of Common Thermocouple Pairs at 0.0°C

Thermocouple name	ASTM E-20 letter code	Seebeck coefficient (μV/°C)
Chromel-constantan	E	58.7
Iron-constantan	J	50.4
Chromel-alumel	K	39.4
Copper-constantan	T	38.8
Platinum-^{90}Pt^{10}Rh	S	5.4
Platinum-^{87}Pt^{13}Rh	R	5.3

Figure 4 depicts a typical T/C circuit of duplex leads for which the output is expressed in terms of the Seebeck coefficients of the conductor pairs. If the T/C wire in every section of the circuit is obtained from a single homogeneous length of wire, the Seebeck coefficient in every section will be the same. That condition is expressed by equation (12):

$$E = S_1 = S_2 = S_3 = S \quad (12)$$

Substituting the condition of equation (12) into equation (11) yields equation (13):

$$E = S(T_m - T_r) \quad (13)$$

If the temperature is expressed in degrees Celsius and the temperature of the reference junctions is maintained at 0.00°C, and equation (13) is equal to zero, then the net output of the circuit is given by equation (14):

$$E = ST_m \quad (14)$$

Equation (14) is valid only when the Seebeck coefficient is constant. The Seebeck coefficient of any pair of T/C wires changes slightly with temperature, but in a homogeneous length of wire it is a unique function of temperature. Therefore, when the temperature is expressed in degrees Celsius and the reference junctions are at 0.00°C, the true output of a homogeneous

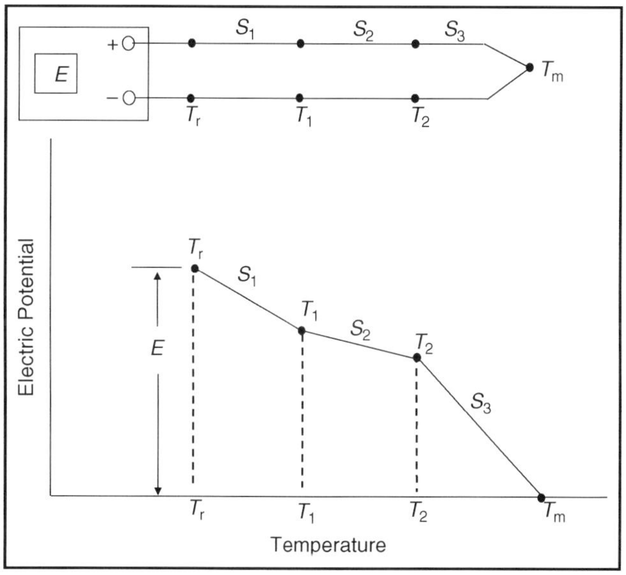

Figure 4 Typical duplex-lead thermocouple circuit.

T/C is a unique function of the measured temperature as given in equation (15):

$$E = \int_{0}^{T_m} S(T)dT \tag{15}$$

Standard values of voltage output as a function of measuring junction temperature, with reference junctions at 0.00°C, have been developed for all commonly used T/C pairs. Those values are given by NIST Monograph 175 in the United States and DIN standards in Europe.

T/C REFERENCE TEMPERATURE

Equation (13) shows the importance of establishing an accurate reference junction temperature. Any difference between the actual reference temperature and the standard value produces an error that is equal to the temperature difference multiplied by the Seebeck coefficient at the reference temperature. The ice-point has been chosen as the "standard" T/C reference temperature because it is a known value of temperature that can be established quite accurately with relatively little effort.

Ice Bath References

Figure 5 depicts an ideal T/C circuit in which a pair of continuous, homogeneous conductors extend from the measuring junction to their junction with copper. The junctions to copper are immersed in an ice bath and are called the reference junctions of the circuit.

The reference junctions must be inserted to a sufficient depth in the bath to avoid conduction errors (7).

The temperature of the ice bath will be 0.00+0.01°C if the following procedures are followed:

1. Use a Dewar flask that is at least 10 in. (25.4 cm) deep and 4 in. (10.1 cm) diameter.
2. Make ice using distilled water and crush it finely.
3. Fill the Dewar flask completely with the crushed ice and fill the voids between the ice particles with distilled water.
4. Insert the T/C leads into the central portion of the bath to a depth of at least 4–8 in. (10.1–20.3 cm) depending on the size of wire.
5. Allow approximately 30 minutes for the ice and water to reach thermal equilibrium.
6. Pack the ice down into the Dewar flask, removing the excess water and adding additional crushed ice to maintain a solidly packed bed of ice with the voids filled by water.
7. Repeat step 6 as required.

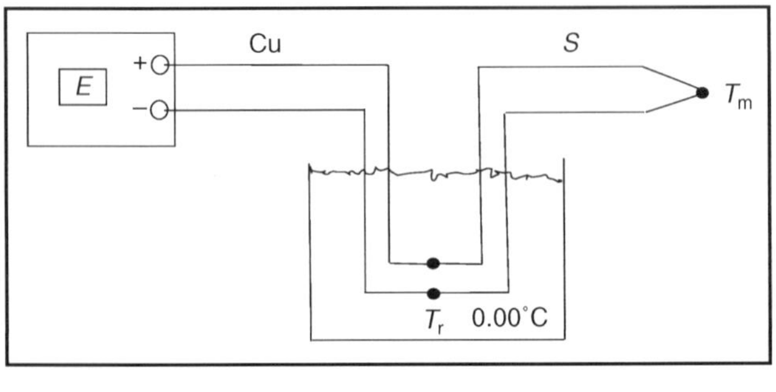

Figure 5 Thermocouple circuit with ideal reference.

Automatic References

Although the mixture of ice and water in a Dewar flask is an ideal reference, it is not very practical outside the Standards laboratory. An excellent alternative to the Dewar flask is an automatic ice bath that maintains a mixture of ice and water in a sealed chamber by means of thermoelectric cooling. Immersion wells extend into the chamber and the reference junctions of the T/C circuit are inserted to the bottom of the wells. When used in accordance with the manufacturer's operating instructions, the reference temperature provided by automatic ice bath is typically 0.00±0.30°C (8). The output of the circuit depicted in Figure 5 is given by equation (14).

In many applications it is not convenient to construct reference junctions on each lead of a T/C. Automatic ice baths are available with built-in reference T/Cs attached to terminals to which the external T/C leads are connected (8). Figure 6 depicts a circuit using this type of reference.

Each pair of "input" terminals is for a specific T/C type. Internal wires of the same type form reference junctions to copper that are maintained at 0.0°C in the ice bath. The copper leads from the reference junctions are connected to the "output" terminals of the reference.

The output of the circuit depicted in Figure 6 is given by equation (16). T_t is the temperature of the terminals, S is the Seebeck coefficient of the external T/C, and S_i is the Seebeck coefficient of the internal T/C. Even though the internal T/C is of the same type of material as that in the external portion of the circuit, its Seebeck coefficient may be slightly different because it may be from a different production lot of wire.

$$E = S(T_m - T_1) + S_i(T_t - T_r) \tag{16}$$

The external T/C produces a voltage equal to $S(T_m - T_t) + S_i(T_t - T_r)$ and the internal T/C produces a voltage equal to $S_i(T_t - T_r)$. The total voltage is the sum of the two. Because T_r is equal to 0.0°C, equation (16) can be rewritten to yield equation (17):

$$E = ST_m + T_t(S_i - S) \tag{17}$$

The second term of equation (17) may be considered an error term. If the Seebeck coefficient of the internal T/C is exactly equal to that of the external T/C, the second term of equation (17) is zero and the output is the same as that of the ideal circuit of Figure 5 and equation (14).

T/C Compensators

In many industrial temperature-measuring applications, even an automatic ice bath with built-in reference T/Cs may not be practical. Automatic ice baths are expensive and do not operate reliably in ambient temperatures below 0.0°C or above 40.0°C. All instruments and systems being sold today for T/C temperature measurement provide an electronic circuit for determining the temperature of the terminals to which the T/Cs are attached. An appropriate reference voltage is added by the system to that produced by the external T/C. Early versions of such circuits were called compensators, because they compensated for the fact that the terminals to which the T/Cs were connected were not at the ice-point temperature. Figure 7 depicts such a circuit.

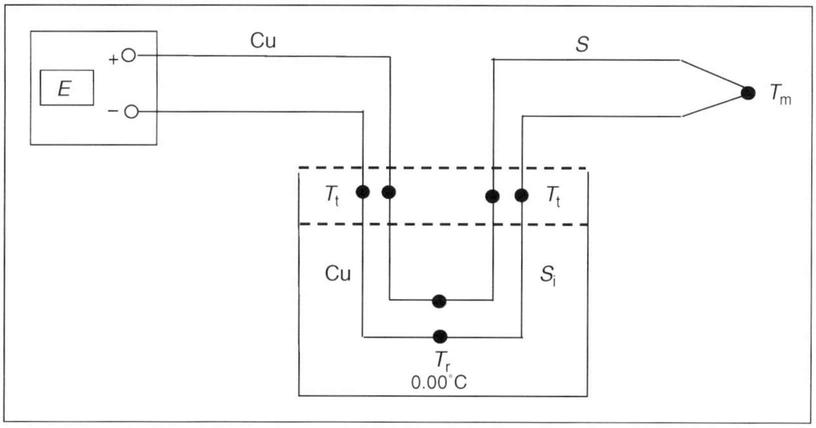

Figure 6 Thermocouple circuit with reference having internal thermo-couples.

The compensator produces a voltage that is a function of the terminal temperature. A typical compensator is a resistance bridge with the temperature-sensitive resistor installed near the T/C terminals. The bridge is adjusted to have a zero output when the temperature of the resistor is 0.0°C and to produce the proper voltage for the specified T/C at a normal ambient temperature (9). The compensation voltage is added to the voltage produced by the T/C and the total voltage is measured by the voltage-measuring device. Equation (18) gives the output of the circuit depicted in Figure 7:

$$E = S(T_m - T_1) + E_r \tag{18}$$

The perfect compensator would have an output equal to that which would be produced by the external T/C when its reference junctions were at ice point and its measuring junction was at the terminal temperature. That characteristic is expressed by equation (19):

$$E_r = S(T_t - T_r) \tag{19}$$

Substituting this ideal compensator output into equation (18) and setting T_r equal to zero yields the same total output as that given by equation (14) for the ideal circuit.

Multichannel T/C Systems

In older multichannel systems, the compensation voltage is added electrically, as depicted in Figure 8. When compensation voltage is added electrically to the output of a multipoint scanner, all of the T/Cs in the group must be of the same type. The compensator must be designed to produce the output required for a given T/C type, and its output should be adjustable in calibration to match the Seebeck coefficient of the external T/Cs.

Modern multichannel T/C-measuring systems use microprocessor capability to add the proper compensation voltage to the measured T/C output (Fig. 9). Rather than adding a compensation voltage electrically, the temperature of the terminals is measured by the system and the value stored in memory. When a channel is programmed to be a T/C input, the system automatically computes the appropriate compensation voltage for that type of T/C, adds it to the measured voltage, and converts the total voltage to the corresponding temperature.

Whether the system reference voltage is added electrically as in Figure 8, or mathematically as in Figure 9, it is based on a single measurement of the terminal temperature, which may be different from the temperature of each individual pair of terminals in the group. The total voltage output of the first T/C in Figure 9 is given by equation (20):

$$E_1 = S(T_{m1} - T_{t1}) + S(T_t) \tag{20}$$

The first term of equation (20) is the measured voltage produced by the external T/C and the second term is computed by the system based on the measured

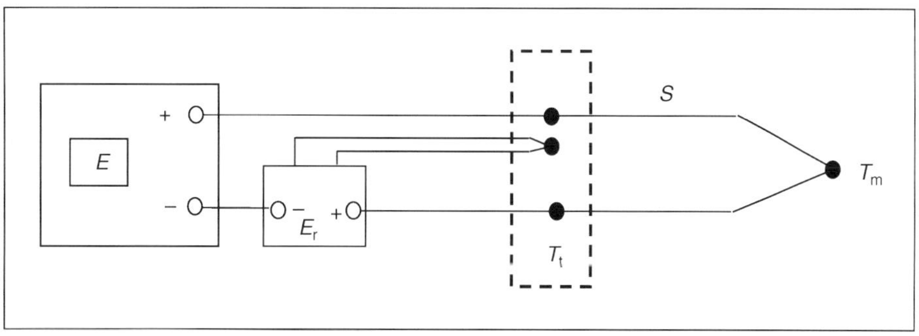

Figure 7 Thermocouple circuit with compensator.

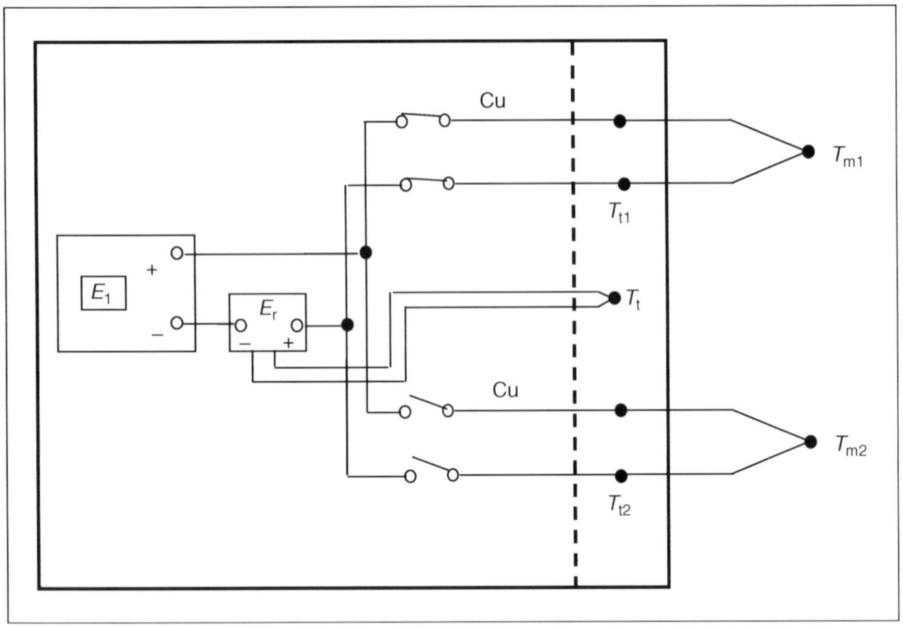

Figure 8 Multichannel thermocouple system with internal compensator.

value of terminal temperature. Equation (20) may be rewritten to yield equation (21):

$$E_1 = S(T_{m1}) + S(T_t - T_{t1}) \tag{21}$$

The second term of equation (21) is an error term. If the actual temperature of the terminals to which a T/C is attached is different from the terminal temperature measured by the system, an error is introduced that is equal to the temperature difference multiplied by the Seebeck coefficient at the terminal temperature.

In some installations it is not possible, or desirable, to run T/C wire from the measuring junction to the recording system. Figure 10 depicts one solution that is similar to the multichannel computer system of Figure 9. In this case, however, the conversion to copper is at the terminals of a remote uniform temperature reference, the temperature of which is measured by some independent means (8). If the logic of the multichannel data system is designed to operate with a remote reference, the output of each T/C is computed in the same fashion as when an internal reference is provided. Equations (20) and (21) apply to this type of installation as well.

SOURCES AND TYPES OF ERROR

The dictionary defines accuracy as the absence of error, but accuracy is a term that has many different meanings. Any discussion of temperature measurement accuracy

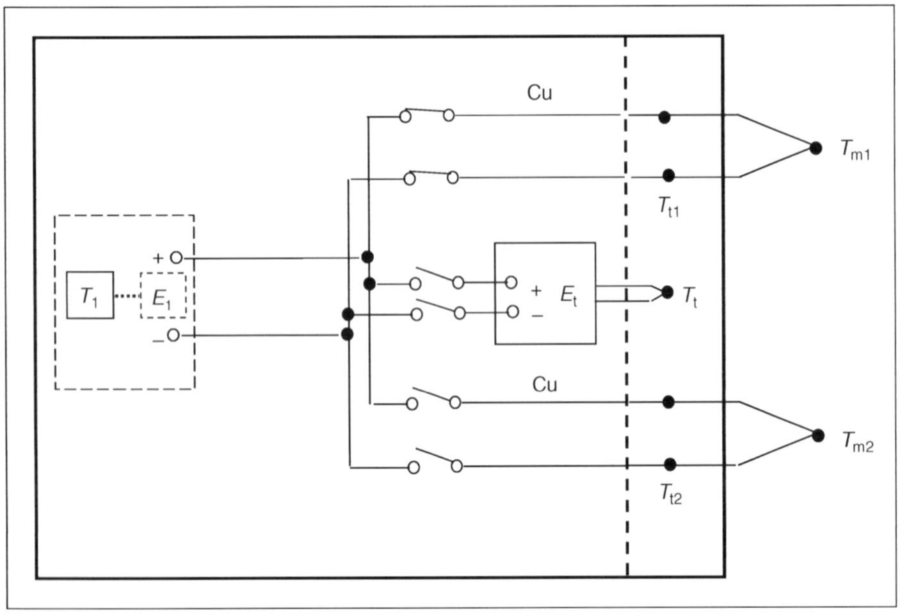

Figure 9 Multichannel thermocouple system with computer and internal reference.

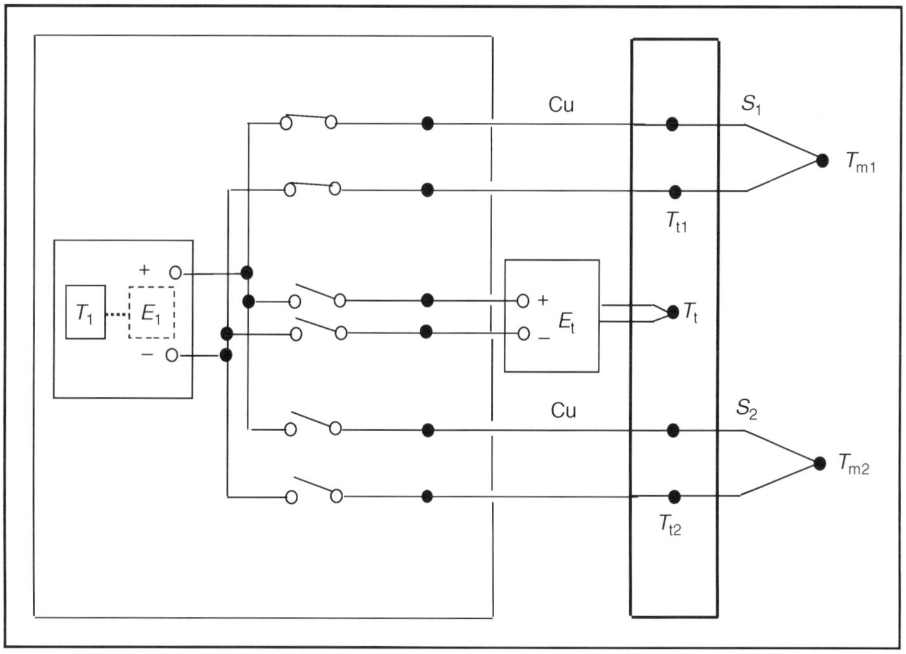

Figure 10 Multichannel thermocouple system with computer and external UTR reference.

must focus on the various sources and types of errors. In a typical T/C installation, the three primary sources of error are the T/C sensors, the circuit that connects the T/C sensors to the measuring system, and the measuring system (10,11).

In discussing errors and accuracy it is important to distinguish between relative accuracy and absolute accuracy. *Relative accuracy* is the degree to which temperature measurements at different locations can be compared or the degree to which the measurement of a single temperature is repeated. *Absolute accuracy* is the degree to which a measurement gives the absolute thermodynamic value of temperature. In many processes relative accuracy is sufficient, but in thermal sterilization processes absolute accuracy is essential.

The rate at which microorganisms are destroyed is a strong function of the temperature, so the time required to produce a sterile product depends on the temperature of the product. If the true value of temperature is less than the indicated value, improper sterilization may result.

Another important distinction to make is that of systematic errors and random errors. Systematic errors can be eliminated from the final results by calibration, but random errors can be minimized only by proper selection and installation of the measuring instrumentation. The lack of interchangeability, conformity, and uniformity produces systematic error; nonhomogeneous regions in the circuit and the lack of repeatability produce random errors.

Sensor and Circuit Errors

In T/C systems, it is difficult to draw a clear distinction between sensor errors and circuit errors, because a T/C is a total integrator of the temperature change from the measuring junction to the reference junction. Conformity and interchangeability are the characteristics generally attributed to the sensors; nonhomogeneous effects are attributed to the circuit.

Conformity to Standard

Conformity error is the difference between the actual voltage produced by a T/C and the standard output voltage for that T/C type at the same measured temperature. The reference junctions of the T/C circuit are assumed to be at 0.00°C. One specification that is often quoted for T/Cs is the maximum conformity error that T/Cs can have and still meet accepted industrial standards. For standard grade type T (copper constantan) T/Cs, that error is the greater of $\pm$1.0°C or $\pm$0.75%. For special grade type T T/Cs it is the greater of $\pm$0.5°C or $\pm$ 0.4% (12). Selected grade T/Cs supplied by GE Sensing, formerly Kaye Instruments, have a maximum conformity error of $\pm$0.25°C or $\pm$0.2% at 120°C (13).

It must be emphasized that the conformity error is not indicative of the total measurement error in any particular installation. Conformity errors can be eliminated by calibration at a number of temperatures over the operating range, and there are many other system errors that may be larger than the conformity error.

Interchangeability

The degree to which a number of T/Cs all have the same output at the same measured temperature is known as the interchangeability of the T/Cs. Interchangeability is important when comparing two temperatures in an uncalibrated system. When a number of T/Cs are made from the same production lot of wire, the maximum interchangeability error is typically the greater of $\pm$ 0.1°C or $\pm$0.1%. As with conformity errors, interchangeability errors can be eliminated by calibration. In both cases, it is often sufficient to calibrate the sensors at the two extreme temperatures of the operating range and apply a linear correction to the measurements. If the measuring system does not provide the capability of applying individual calibration corrections to each input, interchangeability error becomes an important consideration, and all T/Cs used at one time should be made from the same production lot of wire.

Nonhomogeneous Regions

The thermoelectric power of a conductor is a function of the composition and structure of the material. Most T/C conductors are alloys of several elements. Among the commonly used T/C materials, only copper and platinum are essentially pure elements; even copper wire must be checked to be sure it has the proper characteristics. The Seebeck coefficients of T/Cs will vary slightly between production lots of wire because of variations in composition and annealing. Annealing affects the thermoelectric power because it alters the grain structure of the conductor. Similarly, the thermoelectric power of a conductor can be changed slightly if it is stressed to the point of permanent distortion. The phenomenon known as cold-working changes the thermoelectric power as well as the physical characteristics of a metal (14). When a T/C circuit is constructed of continuous, homogeneous wire from the measuring junction to the terminals of the measuring system, calibration can eliminate most errors associated with the sensor and the circuit. Tests have shown conclusively that the output of a homogeneous length of T/C wire depends only on the total change in temperature from one end to the other; the location of the change within the wire does not matter. This characteristic is extremely important in calibrated systems, because the location of the gradient in the wire during operation will generally be different from the location of the gradient during calibration.

Connectors introduce a section of nonhomogeneous conductors in a T/C circuit. When they must be used, connectors should be made of the same materials as the wire and located away from regions of large temperature gradients. Although the materials of T/C connectors are essentially the same as the wire, the annealing process used to make a rigid connector pin must be different from that used to make flexible wire. The resulting Seebeck coefficient is usually slightly different.

Repetitive flexing of T/C wire at one location can also cause a nonhomogeneous region due to cold-working. In the validation process, T/Cs are normally installed through fittings in the walls of sterilizers where they are clamped rigidly. In placing the T/Cs at different locations within the sterilizer, some amount of flexing at the fitting is unavoidable. Since solid wire is much more susceptible to cold-working than stranded wire of the same size, only stranded wire should be used in this application, and great care should be exercised to avoid flexing the wire more than necessary. The sterilizer wall is the region of maximum temperature gradient during operation, so even a small change in Seebeck coefficient in that region can cause a significant error.

The effect of nonhomogeneous regions in a circuit is illustrated in Figure 11 and the following example. All of the wire in the circuit has a Seebeck coefficient S, but the connector in the circuit has a Seebeck coefficient S_c. The temperatures at the ends of the connector are T_1 and T_2. The output of the circuit shown in Figure 11 is given by equation (22):

$$E = S(T_m - T_2) + S_c(T_2 - T_1) + S(T_1 - T_r) \tag{22}$$

Equation (22) can be rewritten to yield equation (23):

$$E = S(T_m - T_r) + (S_c - S)(T_2 - T_1) \tag{23}$$

The second term of equation (23) is the error caused by having a connector in the circuit. The error will be zero if the Seebeck coefficient of the connector is equal to that of the wire or if there is no temperature difference across the connector. It is unlikely that the Seebeck coefficient of a connector will match that of the wire exactly, so it is important to avoid using connectors where they will have large temperature differences imposed on them.

To illustrate the error that would be caused by installing a connector in the wall of a sterilizer, assume the following values for the circuit of Figure 11 and equation (23):

$T_m = 120.0°C$ $\quad T_2 = 100.0°C$ $\quad T_1 = 50.0°C$

$T_r = 0.00°C$ $\quad S = 40.0\ \mu V/°C$ $\quad S_c = 42.0\ \mu V/°C$

The output according to equation (23) is

$$E = 40.0(120.0 - 0.0) + (42.0 - 40.0)(100.0 - 50.0)$$

$$= 4800\ \mu V + 100\ \mu V = 4900\ \mu V$$

$$\text{Error} = 100\ \mu V \text{ or } 2.5°C$$

The values employed in this example are typical of those that would be experienced if a connector in a type T (copper–constantan) T/C circuit was installed in the wall of a steam autoclave. The error is 100 μV, or about 2.5°C. A similar error could be caused by a cold-worked region at the wall, but the magnitude of the error would be less. For a connector to have a Seebeck coefficient 5% greater than the wire it is designed to match is typical, but the change due to cold-working will be much less.

A second type of nonhomogeneous circuit is illustrated in Figure 12. Many T/C probes are constructed using lengths of T/C wire swaged into stainless steel tubes. This type of material may be purchased in long

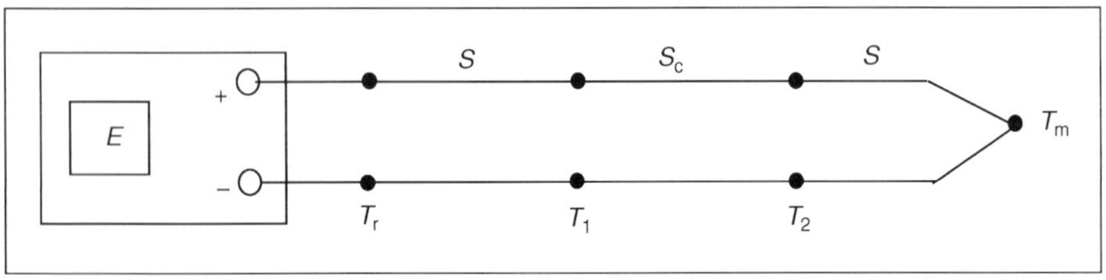

Figure 11 Thermocouple circuit with connector.

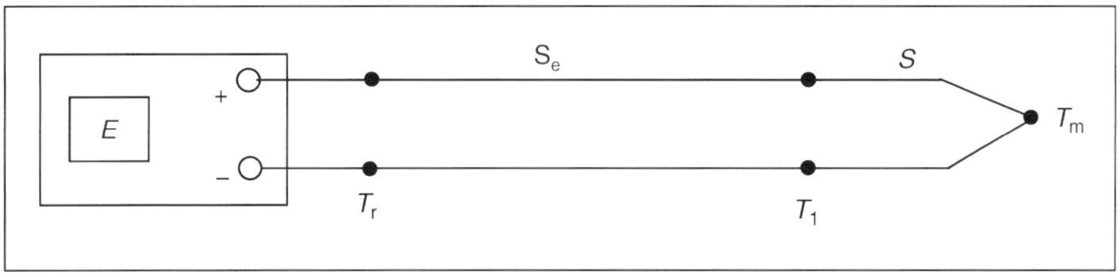

Figure 12 Thermocouple circuit with two sections.

sections and cut to form stainless steel T/C probes of the desired length. One end is welded to form the measuring junction and the two wires at the other end are attached to extension leads of matching T/C wire. The wire in the stainless tip has a Seebeck coefficient S and the extension wire has a Seebeck coefficient S_e.

The output of this circuit is given by equation (24):

$$E = S(T_m - T_1) + S_e(T_1 - T_r) \tag{24}$$

Adding and subtracting the term ST_r and rewriting equation (24) yields equation (25):

$$E = S(T_m - T_r) + (S_e - S)(T_1 - T_r) \tag{25}$$

The second term of equation (25) is an error term. The error will be zero if the Seebeck coefficient of the extension wire is equal to that of the wire in the tip or if the junction between the two is maintained at the reference temperature.

This type of probe should be avoided, unless it can be calibrated under the same conditions encountered in normal operation. Specifically, the value of T_1 must be the same during both calibration and operation, or an unrecognized error will be introduced. Assume the following values for the circuit of Figure 12 and equation (25), when the probe is calibrated in a laboratory:

$T_m = 120.0°C \quad T_1 = 30.0°C \quad T_r = 0.0°C$

$S = 40.0\ \mu V/°C \quad S_e = 40.2\ \mu V/°C$

The output of the circuit according to equation (25) is

$$E = 40.0(120.0 - 0.0) + (40.2 - 40.0)(30.0 - 0.0) = 4800 + 6$$

$$= 4806\ \mu V$$

Assuming that the standard Seebeck coefficient is 40.0 μV/°C, the calibration correction is 6 μV or 0.15°C when the probe is measuring 120.0°C.

When this probe is used inside a steam autoclave, the temperature of the junction between the tip and the extension wire will be at autoclave temperature. Assume that all other values are the same as in the calibration example, but the value of T_1 is 120.0°C. The output will be

$$E = 40.0(120.0 - 0.0) + (40.2 - 40.0)(120.0 - 0.0) = 4800 + 24$$

$$= 4824\ \mu V$$

Applying the calibration correction of 6 μV still leaves an error of 18 μV, or almost 0.5°C. This example, even more than the previous one, shows the importance of using T/Cs that have continuous length of homogeneous wire from the measuring junction to the region outside the autoclave.

Diffusion of Steam

All insulating materials are permeable to steam after extended exposure.

When stranded wire is installed through the wall of an autoclave, steam will eventually diffuse through the insulation, flow to the lower pressure outside the autoclave through the small passages formed between the strands of wire, and condense to form drops of moisture where the insulation ends. This diffusion of moisture along the wire will not cause an error in the output of a T/C, but it should be prevented from collecting on terminals or connectors where corrosion could cause problems. Diffusion of moisture along the wire to the outside the autoclave will not occur if solid wire is used instead of stranded wire, but solid wire is more susceptible to cold-working. The flexing of solid wire at the wall of an autoclave may introduce a serious error, whereas moisture dripping from stranded wire is only an inconvenience.

Some T/C assemblies are constructed using flexible hose to protect the T/C wire inside the autoclave. One end of the flexible hose fits over a length of stainless steel tubing that forms the T/C probe and the other end of the flexible hose connects to a stainless steel tube that provides a pressure seal at the wall of the autoclave. This design guarantees that there will be no cold-working of the homogeneous wire that runs continuously from the measuring junction to a connector outside the autoclave.

Unfortunately, the steam that diffuses through the flexible hose will condense inside when the assembly is cooled down. If some of the moisture collects in the stainless steel probe near the measuring junction, it will cause an error if the probe is used subsequently to measure temperatures above 100°C. Since the passage from the inside probe through the hose is open to the atmosphere, any moisture in the probe will boil at 100°C, absorbing energy from the surrounding material and reducing the temperature of the probe tip. Depending on the amount of moisture, the distance of the moisture from the measuring junction, and the rate of heating at the outer surface of the probe tip, the magnitude of the error caused by moisture in this type of probe can vary from a few tenths of a degree to several degrees.

The presence of moisture in a probe tip is detected readily in calibration, so it should never cause an error in a validation run if the probes are calibrated before each run. If a large amount of moisture is present it will prevent the tip from ever reaching the calibration temperature, and the steam condensing inside will make the portion of the probe that extends above the calibration bath extremely hot. If a small amount of moisture is present it will boil away, permitting the tip to achieve the proper temperature, but it will retard the rate of heating during the time it is boiling. A probe with moisture will take a noticeably longer time to reach the calibration temperature than a dry one. If a maximum acceptable time to reach calibration temperature is specified, the presence of moisture in a probe will be detected.

Attempts to fill the probe tip with a solid material to prevent moisture from collecting near the measuring junction can cause cold-working of the wire due to differential thermal expansion of the wire and the filling material. The resulting errors are more serious than the presence of moisture. Recent tests with a new filling technique indicate that moisture errors may be eliminated in future probes without causing other problems (13).

Circuit Resistance

The resistance of a T/C circuit has no effect on the voltage generated. The indicated outputs of early industrial T/C meters were inversely proportional to the resistance of the external circuits because galvanometers were used to measure the current flowing in the circuits rather than the voltage potential. Null balance potentiometers generate a balancing potential so no current flows in the circuit, and modern T/C meters have measuring circuits with extremely large input impedance compared to that of the circuit. When using either of the latter types of instrumentation, normal levels of T/C circuit resistance will not affect the indicated temperature.

Cracked wire or poor electrical contact at connectors can introduce extremely high resistance in a circuit, affecting the accuracy of the voltage-measuring device and giving erratic values of the indicated temperatures. The wire in a circuit can be broken but held together by the insulation. When the wire is stretched, the ends come apart and cause an open circuit; when it is relaxed the ends of the wire may touch, again completing the circuit but with a high resistance at the point of contact. The surface of the copper contacts in a copper–constantan connector can become oxidized, thereby creating a high resistance. If an ohmmeter is used to measure the contact resistance, it may indicate a fraction of an ohm because the excitation voltage of the ohmmeter can break through the oxide film. With only the small potential generated by a T/C imposed, however, the resistance may be thousands of ohms. If erratic readings are experienced in a T/C circuit having a connector, cleaning the connector contacts may solve the problem. Oxidation of copper contacts can be prevented by plating them with gold (13).

Measuring System Errors

Different manufacturers state the accuracies of T/C-measuring systems in different fashions. Some give detailed breakdowns of the error sources; some simply state total error when operating within a limited range of ambient temperatures. Changes in ambient temperatures are the most significant sources of error in T/C-measuring systems, particularly in multichannel systems with internal references.

Resolution

The resolution of a measuring system is the ability to read the output. In analog chart recorders, resolution is determined by the width of the chart paper and the temperature range corresponding to the total width. Since the width of the chart is fixed, a smaller temperature range must be set if better resolution is required. This type of recorder can be purchased with plug-in cards to set the temperature range.

In digital systems, resolution is the value of the least significant digit. The resolution of temperature measurements may be 0.01°, 0.1°, or 1.0°, Fahrenheit or Celsius. Some meters even give a resolution of 0.001°. Measurement accuracy can be no better than the resolution, but it should never be assumed that the accuracy of a measuring instrument is as good as its resolution.

Conformity to Standard

All modern T/C-measuring systems use microcomputers to add compensation voltage to the measured voltage generated by the external T/C and to convert the resulting total voltage to the corresponding temperature for that type of T/C. The conversion from voltage to temperature typically utilizes a series of straight lines or polynomial functions that approximates the standard tables. The difference between the calculated temperature and the standard temperature at a given voltage is the conformity error at that temperature. At any given measured temperature, the conformity error will always be the same. Maximum conformity error ranges from $\pm$ 0.02°C in high accuracy systems to as large as $\pm$1.0°C in some systems.

Uniformity

Uniformity is the degree to which the measuring system indicates the same value when exactly the same input is applied to different channels of a multichannel system. The largest error in most multichannel T/C systems is the uniformity error caused by differences in the temperatures of the terminals to which the T/Cs are attached. It is not unusual to have terminal temperatures differ by 1.0°C. A difference in terminal temperature causes an error equal to the temperature difference multiplied by the ratio of the Seebeck coefficient at the terminal temperature to that at the measured temperature. Even when the terminals are insulated to protect them from external heating and cooling effects, they will be heated nonuniformly by the internal electronics of the measuring system. Once a system has warmed up completely in a steady ambient temperature, the terminal temperatures will be stable. If each T/C is calibrated at the ice point (0.00°C), the uniformity error due to the terminal temperature difference will be included in the calibration correction. If the ambient temperature subsequently changes, the terminal temperature difference may also change. Although the systematic uniformity error was eliminated by calibration, an additional random uniformity error may be introduced by a subsequent ambient temperature change.

Repeatability

Repeatability is the degree to which the measuring system will indicate the same output over a period of time when exactly the same input is being measured. Repeatability errors can be classified as short term (seconds), medium term (minutes), and long term (weeks). Short-term errors in the indicated output are caused by electrical phenomena. Continuous fluctuations in the output are usually caused by instabilities in the measuring circuit of the system. Sudden jumps of brief duration in the output are usually caused by common-mode voltage differences. The common-mode voltage difference is the potential difference between the sensor and the ground of the measuring system. In steam autoclaves, large static potential differences can be created between ungrounded probes and the ground of the measuring system, particularly when the probes are installed in plastic containers. Proper grounding of the probes can minimize the error caused by this phenomenon. The ratio of the maximum measurement error to the common-mode voltage difference is called the common-mode rejection of the system and is expressed in decibels. A decibel is a measure of voltage ratio or current ratio equal to 20 times the common logarithm of the ratio. The common-mode rejection varies from better than 140 dB (10 million to 1) in high-accuracy systems to less than 100 dB (100,000 to 1) in some systems. Medium-term errors in the indicated output are caused by thermal phenomena. Temperature changes in the measuring circuit, in the T/C reference, and in the input terminals all cause errors in the indicated output. The magnitude of the measurement error caused by a change in ambient temperature is given by the temperature coefficient of the system. All manufacturers specify the temperature coefficient based on the system being stable before and after the change in ambient temperature; transient errors that occur during the temperature change may be much larger. Temperature coefficients vary from 0.01°C/°C for high-accuracy systems to 0.1°C/°C in some systems.

Long-term errors in the indicated output result from component aging. Invalidation studies this type of error is not important, because the system is calibrated with sufficient frequency to account for any long-term variations.

CALIBRATION PROCEDURE

T/C systems used to measure temperatures in the validation process should be calibrated before and after each use. Typically, neither the measuring system nor the T/Cs will change their characteristics between calibrations, but the calibration process assures proper operation of the entire system. Because corrections applied to each T/C also include the uniformity error of the measuring system, each T/C should be connected to the same channel in calibration as in operation. To the extent possible, the entire system should be calibrated under the same ambient temperature and other conditions as it will experience during operation.

Calibration Basics

There are a few basic rules that should be followed in any calibration procedure.

1. *Challenge all results.* No single measurement should be accepted as being correct unless it is verified by other results. The transfer standard used to determine the temperature of the calibration bath could have an error. If two standards agree, the probability that they both have the same error is extremely low.
2. *Be patient.* A frequent mistake in calibrating instrumentation is to take measurements and make adjustments before conditions have stabilized. It may take much longer than expected for a system to become completely stable, because thermal errors decay exponentially and the output may seem to be stable even though it is still changing slowly. Computer-based validation systems are available that provide automatic two-point calibration including automatic stability determination. This eliminates calibration errors caused by operator inconsistencies.
3. The accuracy of the transfer standard must be better than that of the instrument being calibrated. This would seem obvious, but it is amazing how often a voltage calibrator is used that has a greater error than the system being calibrated. Rules such as being 10 times as accurate or even twice as accurate are not absolute: It is only important to recognize that the accuracy of the calibration can be no better than the standard used, and that it is a mistake to change the adjustment of a measuring system if it is already more accurate than the standard.
4. The characteristics of the transfer standard must have been determined by a procedure that is traceable to accepted primary standards (15). In the United States, the National Bureau of Standards (NIST) is the accepted source of primary standards. The transfer standards used should have been calibrated by the NIST relative to their primary standards or by a qualified Standards laboratory relative to standards that they have had calibrated by the NIST. In either case, the test results and test numbers should be known so the calibration procedure can be traced back to the primary standards.
5. The transfer standard must be independent of the measuring system. Because the output of a T/C depends on the entire circuit, it is not a desirable temperature transfer standard. An RTD is a device that indicates changes of temperature by a change of resistance. Because the resistance of an RTD is only a function of its temperature, and the resistance can be measured independently of the system being calibrated, RTDs are ideal temperature transfer standards.
6. The characteristics of the transfer standard must be stable in shipment and other handling. As its name implies, the purpose of the transfer standard is to transfer a measured characteristic from one laboratory to another. The characteristics of the standard must be the same when received from the NIST as when it was calibrated relative to their standards. Liquid in glass thermometers may be damaged or develop small voids in the liquid during shipment, and therefore are not reliable temperature transfer standards. RTDs are fairly rugged devices that maintain their characteristics in normal handling and shipment.

Measuring System Calibration

The first step in calibrating a T/C system is to check the operation of the measuring system in the voltage mode and adjust it if necessary. Each manufacturer has a recommended procedure and calibration interval, which should be followed. A precision low-level voltage source having accuracy better than $\pm 1.0\ \mu V \pm 0.01\%$ in the range of 0.0 to 20,000 μV should be employed in the voltage calibration. The measuring system should be turned on several hours before starting the calibration process to be sure that it has become completely stable. If the system is to be used for important voltage measurements, a second voltage source should be used to check the results of the adjustments. If the only important measurements are T/C temperature measurements, the calibration of the sensors will correct for any small voltage errors.

Once the voltage-measuring circuits have been adjusted, the T/C reference of the system should be checked by connecting T/Cs to the proper input terminals and placing several of their measuring junctions in an ice bath. If a crushed ice bath is used, it should be made and maintained as described in the Section entitled Ice Bath References. If an automatic ice bath is used, the measuring junctions should be inserted to the bottom of the wells. In either case, allow 10 or 15 minutes for the temperature to stabilize before making any adjustments.

The operation of the T/C reference in a multichannel, computer-based system is discussed in the Section entitled Multichannel T/C Systems. When the input terminal temperature of the system is above 0.0°C, a T/C with its measuring junction in an ice bath will generate a negative voltage. If the internal reference is adjusted until the indicated temperature is 0.0°C (32.00°F), the output of the internal reference is adjusted to equal the output that is generated by the external T/C when its reference junction is at 0.0°C and its measuring junction is at the temperature of the input terminals. The external T/C is generating a negative voltage of the same magnitude. As discussed in the Section entitled T/C Compensators and shown by equations (18) and (19), this procedure provides the perfect internal reference or compensation voltage for that external T/C.

Since the input terminal temperatures and the Seebeck coefficients of each T/C in a multichannel system may be slightly different, other T/Cs connected to the measuring system may not indicate exactly 0.0°C when the internal reference is adjusted as described in the previous paragraph. For best overall accuracy, the internal reference should be adjusted until the average of the indicated temperatures of all T/Cs in the ice bath is 0.0°C. If the measuring system can be programmed to compute the average of the outputs of a group of T/Cs, that value can be used directly in the calibration procedure. It should be emphasized that calibration of the internal reference is a measuring system calibration and not a calibration of the external T/Cs.

T/C Calibration

In order to assure absolute accuracy of every temperature measurement, each T/C must be calibrated by determining its output when its measuring junction is at two or more known temperatures. Electronic T/C calibrators are quite useful in checking systems for proper operation, but they do not provide temperature calibration of the T/Cs being used with the systems.

All temperature sensors should be calibrated at the ice point if 0.0°C is within their normal range of operation. As was discussed in the Section entitled Thermoelectric Theory, the ice point is a known temperature that can be established quite accurately with relatively little effort. Measuring the ice-point temperature is an ideal check for any temperature indicator. It is also important to calibrate a temperature sensor at, or near, the maximum and minimum temperatures to be measured. Some T/C-measuring systems provide a feature that permits the automatic application of a two-point correction on each T/C. These software-controlled validation systems provide fully pre-programmed calibrations, including selective setpoint control of dry block temperature reference and automatic stability determination.

In steam autoclave measurements, the recommended minimum calibration temperature is 90°C and the recommended maximum calibration temperature is 130°C, with a post-calibration verification at 121°C. When selected grade T/C wire (13) is calibrated at 90.0°C and 130.0°C, and a linear correction is applied between those temperatures, the maximum conformity error relative to the NIST standard output (5) will be less than +0.1°C. This result has been verified by thousands of calibrations of the selected grade wire (13).

Typical operating temperatures in hot air ovens are in the vicinity of 200.0°C, and depyrogenation tunnels may be operated at temperature above 300.0°C. In validating those processes, the T/Cs should be calibrated at a temperature near the maximum expected operating temperature of the process. If the ice point is used as the second temperature of a two point calibration of selected grade T/C wire and a linear correction is applied, the maximum conformity error relative to the NIST standard output may be as large as ±0.30°C between 0.0°C and 200.0°C and as large as +0.50°C between 0.0°C and 300.0°C. This level of error is normally acceptable in these higher temperature processes, and the error becomes much smaller near the maximum calibration temperature, which is also the normal operating temperature.

If better accuracy is required at higher temperatures, the T/Cs must be calibrated at intermediate points. The maximum expected error in any temperature measurement increases at higher temperatures. When a T/C is calibrated at two temperatures and a linear correction is applied between the two temperatures, the maximum expected error due to the T/C's characteristics is less than ±0.05°C between 100.0°C and 150.0°C, approximately ±0.10°C between 150.0°C and 200.0°C, and approximately ±0.20°C between 250.0°C and 300.0°C.

The type of equipment and instrumentation that must be used in a temperature calibration facility, and the amount of personnel training required to operate it, depend on the level of accuracy desired. To achieve calibration accuracies of ±0.01°C requires very expensive, elaborate instrumentation and highly trained personnel. Calibration accuracies of better than ±0.1°C can be achieved with relatively inexpensive instrumentation and simple procedures (16). The less elaborate calibration facility is actually preferred in most validation processes because the level of accuracy is better than required and it is less likely that an error will be introduced by faulty procedure.

The following equipment and instrumentation is required in a basic temperature calibration facility to achieve total calibration accuracy of better than 0.1°C at temperatures up to 150.0°C and ±0.20°C at temperatures between 150°C and 300.0°C:

1. An automatic ice bath (7) or a Dewar flask filled with crushed ice and distilled water as described in the section entitled Thermoelectric Theory.
2. A high-temperature reference block (15) or a stirred oil bath with temperature uniformity better than ±0.03°C in the working region.
3. At least three RTDs that have been calibrated traceable to NIST standards to an accuracy of +0.03°C at the minimum and maximum temperatures in the calibration range, and at intervals no larger than 500°C if ±0.1°C accuracy is required or 100.0°C if ±0.2°C accuracy is required.
4. An independent instrument to measure the resistance of the RTDs to an accuracy corresponding to ±0.03°C.
5. A precision resistor with calibration traceable to NIST standards to calibrate the resistance-measuring instrument.

The RTDs should be of a four-wire design, which provides independent leads for the excitation current and for measuring the voltage difference across the resistor. The same excitation current must be used in transfer calibrations as was used in the original calibration of the RTD, because the self-heating error of an RTD is a function of the current. The most common excitation current for a Pt 100 RTD is 1 mA. At least three RTD transfer standards should be available, because two standards must agree at each calibration temperature, and the third is required to determine which of the first two is correct if they do not agree.

A 25 Ω platinum RTD is the primary standard temperature sensor used by all primary calibration laboratories. It is quite expensive and delicate. An industrial grade, 100 Ω, platinum RTD is quite acceptable as a transfer standard, and its resistance can be measured to an accuracy of ±0.01 Ω with relatively inexpensive instrumentation (16). A resistance change of 0.01 Ω corresponds to a temperature change of approximately 0.025°C. The resistance-measuring instrument must be calibrated at two values in the range to be measured. One of the values can be zero resistance, or a shorted input, and the second value should be approximately equal to the maximum RTD resistance to be measured. When 100 Ω RTDs are used to measure temperatures between 0.0°C and 300.0°C, a 150 Ω precision resistor is recommended as the second point. The resistor calibrations should be independently traceable to NIST standards and accurate to ±0.005 Ω.

The resistance-measuring instrument should be capable of measuring the resistance of up to three RTDs and the precision resistor at the same time. The current leads of the precision resistor and the RTDs should be connected in series, so that the same excitation current passes through the precision resistor and the RTD whose resistance is being measured. Adjusting the current to make the instrument indicate the proper value of the precision resistor automatically calibrates it for the RTD reading. In effect, the instrument compares the resistance of the RTD to that of the precision resistor.

The following detailed procedure is recommended for calibrating T/Cs to be used with multichannel-measuring system in a validation procedure:

1. Connect all T/Cs to the channels of the measuring system to which they will be connected in the validation run. Each T/C must be labeled clearly and a record made of the channel to which each is connected.
2. Turn on the measuring system and the resistance-measuring instrument at least two hours before taking any measurements. If an automatic ice bath and a high temperature reference block are to be used, they should be turned on at the same time. If a crushed ice and distilled water bath is to be used, it should be prepared at least half-an-hour before being used. Most stirred oil baths require about 15 to 20 minutes to stabilize.
3. Once the measuring system has stabilized, it should be calibrated according to the procedures of the Section entitled Measuring System Errors.
4. Place two RTD transfer standards in the wells of an automatic ice bath or in a crushed ice bath. If an automatic ice bath is used, the RTDs should be inserted to the bottom of the wells and the wells filled with water. At least one manufacturer of automatic ice baths recommends filling the wells with silicone oil having a specific gravity greater than unity (7). Oil is recommended to prevent the possibility of ice forming in the wells, but water is much more convenient, and the formation of ice in such units is an extremely rare occurrence. The temperature accuracy is the same in either case. If a crushed ice bath is used, the RTDs should be inserted to a depth of approximately 30 probe diameters. A 3/16 in. diameter RTD should be inserted to a depth of 6 in. and a 1/4 in. diameter RTD to a depth of 8 in. After the probes have been inserted for a few minutes, all excess water should be removed and additional crushed ice added to create a solidly packed bed of ice with the voids filled by water.
5. After the RTDs have reached equilibrium, check the calibration of the resistance-measuring instrument by measuring the value of the precision resistor and make an adjustment if necessary. Then measure the resistance of each RTD and compare the measured value to the calibrated value of resistance at 0.00°C. The measured resistance of a 100 Ω RTD should agree with the calibrated value to within ±0.01 Ω at 0.00°C. If the RTDs indicate the same temperature, but both indicate that the ice bath is not 0.00±0.03°C, check the ice bath. If one of the RTDs has a resistance more than 0.01 Ω different from the calibrated value, it should be removed from service or recalibrated by a Standards laboratory.
6. Place both RTD transfer standards in the high temperature reference block, or oil bath, and adjust the temperature to the desired value. Allow at least 10 to 15 minutes to stabilize if a reference block is used and about 5 minutes when using an oil bath. Measure the resistance of each RTD and determine the

corresponding temperature of each from the appropriate calibration tables or equations. The RTDs should indicate the same temperature to within ±0.05°C if the temperature is below 150°C and to within ±0.1°C if the temperature is between 150°C and 300°C. If they do not, a third RTD should be used to determine which of the other RTDs is in error, and the faulty RTD should be removed from service or recalibrated by a Standards laboratory. When proper operation of both transfer standards has been verified, continue to monitor the high-temperature reference with one of the standards.

7. Place the T/Cs in the ice bath and allow at least 10 minutes for them to stabilize. This part of the procedure can be done at the same time as step 6. Once step 6 is complete and the T/Cs have become stable at the ice point temperature, their values at 0.00°C should be recorded for future correction. If the measuring system provides the capability to incorporate calibration corrections in the indicated output, the correction at the first point should be entered. In some systems this can be done automatically by pressing the appropriate keys on the operator's panel.
8. Place the T/Cs in the high-temperature reference and allow sufficient time for them to stabilize at the new temperature. The stabilization time will be approximately 10 minutes if a reference block is used and about 5 minutes in an oil bath. Once the indicated temperatures have become stable, the difference of each from the temperature indicated by the standard should be recorded for future corrections. If the measuring system provides the capability of incorporating calibration corrections in the indicated output, the correction at the second point should be entered.
9. If more than a two-point calibration is to be employed, steps 6 and 8 should be repeated for each calibration temperature.

There are several complete validation systems available that have fully automatic programmable multipoint calibration, improving the repeatability and reliability of the calibrations and providing significant time savings.

Documentation is an important aspect of any calibration procedure. A record must be made of the probe number attached to each channel and the location of each probe in the autoclave or oven during the validation test. The calibration corrections for each T/C must be recorded even when they are applied automatically by the measuring system. The calibration certificates of each RTD transfer standard and the precision resistor must include the actual data values obtained. If the calibrations were performed by the NIST, the certificates will contain a test number. If the calibrations were performed by another Standards laboratory, the certificates must contain the NIST test numbers of the instrumentation used by that laboratory and provide traceable calibrations of the transfer standards. Every transfer calibration must be documented in order to provide traceability to the primary standard and proof of the accuracy of the final measurement.

T/C Summary

One of the most important steps in obtaining accurate temperature measurements with T/Cs is the proper design and installation of the T/C circuit. If possible, a continuous length of stranded homogeneous wire should be used from the measuring junction to the terminals of the measuring system. When two or more sections of wire are required by operational considerations, the connections between the sections must be in locations where the temperature in the circuit does not change significantly along its length. Ideally, each section of wire should be from the same production lot. If that is not practical, the wire should be selected to have the best interchangeability possible.

The measuring system must be designed specifically for high-accuracy T/C measurements. The input terminal section should provide a uniform temperature of all terminals and a means of measuring that temperature accurately. The system's voltage-measuring accuracy must be ±1.0 μV or better, and the computation of temperature from the measured voltage should deviate from the standard value by no more than ±0.06°C over the entire measurement range. Most importantly, the T/C reference must track changes in ambient temperature accurately and the voltage measurement must not be affected by such changes, so that the calibration factors determined in the laboratory will still be valid on the production floor.

Finally, the entire system must be calibrated before each use. While it is not necessary to do a full calibration after each use, it is good practice to verify proper operation by calibrating the system at the process temperature after the validation run. When a properly designed and installed T/C system is calibrated by the procedures described in this chapter, the total measurement accuracy should be better than ±0.1°C at 120°C, ±0.2°C at 200°C, and±0.4°C at 300.0°C.

RESISTANCE TEMPERATURE DETECTORS

Most temperature sensors, permanently installed by the manufacturer of the sterilization equipment, are RTD sensors used for process control and monitoring of production runs. These RTD sensors are components in measuring chains and therefore the sensors shall be part of the calibration and verification of the measuring chain with calibration intervals defined in the calibration and validation SOPs.

RTDs are also used in wireless loggers (no real-time display) and in RF measurement transducers for remote real-time sensing. These are battery-powered, self-contained measuring systems.

The wireless logger simplifies access to hostile, remote, and hard-to-reach environments by eliminating the need to hard-wire sensors, greatly reducing study set-up time and associated cost. Loggers are available for measuring temperature, humidity and pressure, and come in a wide range of standard configurations to simplify data acquisition. On the other hand, the wireless logger has some minor disadvantages: (*i*) It is battery operated, the battery life is a function of sampling rate, study duration and operating temperature. (*ii*) There is no real-time indication, only historical data obtained after completion of the study.

Wireless logger systems are made up of three components: Wireless Loggers which measure and record process conditions; a Reader station for communicating with the Loggers; and Proprietary Software through which process studies, calculations and reports are generated. Many manufacturers' software is 21 CFR Part 11 compliant.

The Loggers are precision measurement and recording devices, designed for validating and monitoring the most severe temperature, humidity, and pressure applications:

- Steam sterilizers
- Depyrogenation tunnels
- EtO sterilizers
- Retorts
- Freeze dryers
- Dry heat ovens
- Washers
- Incubators
- Stability chambers
- Warehouses

RF measurement transducers are currently used in storage and stability applications. The transmission reliability from the inside of an autoclave is not yet fully acceptable. In most cases the calibration interval recommended by the manufacturer is 6 to 12 months. The relatively long recommended calibration interval puts the responsibility for risk analysis and preventive action on the user. It is imperative that the annual/semi-annual calibration procedure includes an as-found report to verify that the unit has met its calibration tolerances during the period preceding the calibration. In many cases, interim verification of the unit calibration status is made an integral part of the SOP.

The use of RTDs requires measurement circuits that are different from the T/C circuits. The T/C electrical output is a direct function of temperature as discussed earlier in this chapter. RTDs, on the other hand, produce a resistance change as a function of temperature variations. The RTD requires a current source to generate a voltage

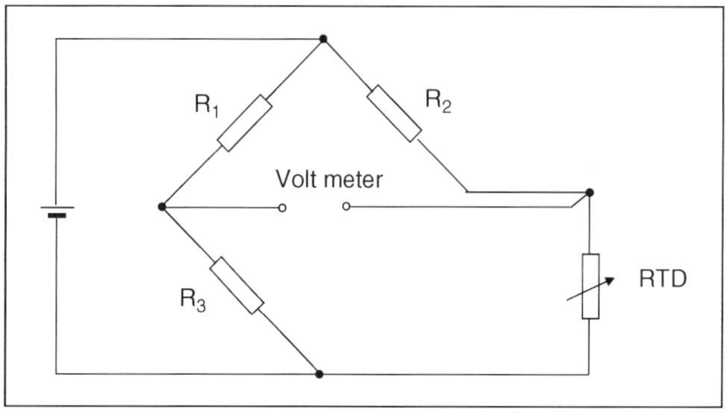

Figure 13 Three-wire Wheatstone bridge.

drop across the RTD. The most frequently used resistance value for platinum RTD is 100 Ω at 0°C and its temperature coefficient α is +0.385 Ω/°C at 0°C according to DIN standard. As both the slope and the nominal resistance is small, lead wire resistance can contribute significant measurement errors. As an example, 2 Ω lead resistance will cause a $2/0.385 \cong 5.2$°C measurement error and the temperature coefficient of the lead wires can contribute measurable error. An early method to compensate for these errors was the use of a Wheatstone Bridge in a three-wire configuration that minimized the errors generated by the lead wires.

If wires A and B are perfectly matched in length, the effect of the lead resistance will cancel as the two leads are in opposite legs of the bridge. Wire C carries no current and acts as a sense lead only for the bridge's output voltage-measuring device. The bridge shown in Figure 13 has a nonlinear characteristic between resistance change and bridge output voltage change. This means that a second equation is needed to convert the bridge output voltage to equivalent RTD resistance that is then converted to temperature.

To meet the uncertainty requirements for critical temperature measurements it is necessary to use the RTD in a four-wire configuration (Fig. 14).

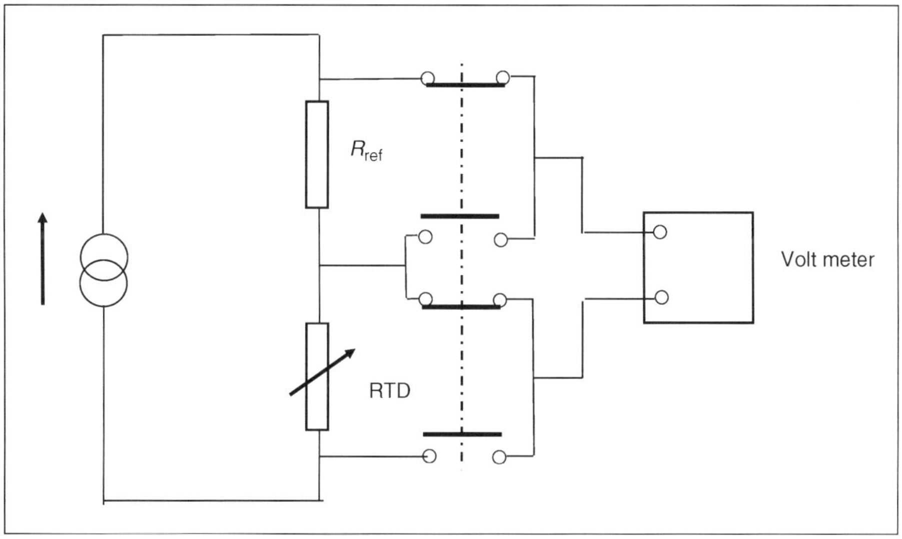

Figure 14 Four-wire resistance temperature detector measurement circuit.

A current source drives a current through the RTD via two wires, and a high impedance DVM senses the voltage dropped over the RTD, via a second pair of wires. The voltage registered by the DVM is directly proportional to the RTD resistance, therefore only one conversion equation is needed to generate the temperature data. A precision reference resistor is connected in series with the RTD to provide the actual current value needed to calculate the momentary resistance of the RTD. The DVM is insensitive to the resistance of the lead wires as no current flows through them.

This solution requires a fourth extension wire but that is a small inconvenience compared with the improved accuracy of the measurement.

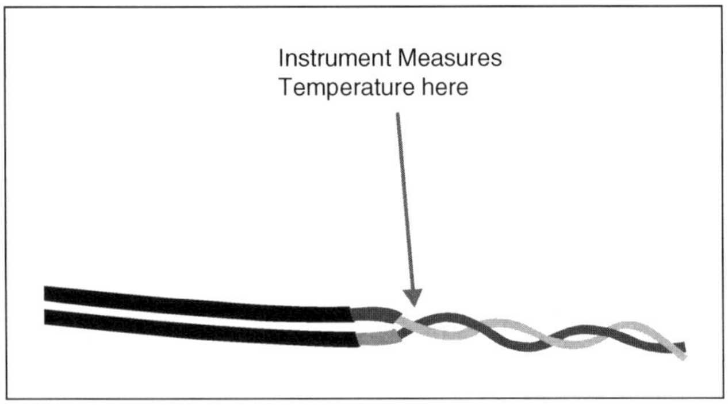

Figure 15 Twisted thermocouple.

SENSOR DESIGN

The temperature sensor should be designed for the application. A sensor designed for measuring the temperature in a LVP bag cannot be used for measuring the temperature in a 1 mL ampoule. Several factors have to be considered when specifying the design of a temperature sensor for a particular application.

> Regardless of how many facts are presented herein and regardless of the percentage retained, all will be for naught unless one simple important fact is kept firmly in mind. The thermocouple reports only what it "feels." This may or may not be the temperature of interest. Its entire environment influences the thermocouple and it will tend to attain thermal equilibrium with this environment, not merely part of it. Thus, the environment of each thermocouple installation should be considered unique until proven otherwise. Unless this is done, the designer will likely overlook some unusual, unexpected, influence (ASTM International) (17).

The statement above is valid for all types of temperature measurements. The examples below are in some cases relevant for both T/Cs and RTDs; a few are specific to the type of sensor used.

Examples

T/C Specifics

Twisting bare wires together increases the contact between the leads over the length of the twisted portion. The instrument measures the temperature at the first point of contact, i.e., the furthest point from the tip (Fig. 15).

Using a twisted T/C to measure air temperature in a steam sterilizer would not significantly affect accuracy because the difference in air temperature between the tip and the last point of contact is negligible. Twisted conductors could produce incorrect data, however, when the T/C is used to measure the temperature of liquid in a vial. Inserting this T/C (Fig. 16) causes the instrument to indicate a temperature somewhere between the air and liquid temperature.

Avoid this problem by reducing the junction to the smallest practical size. Use an argon welder to create a T/C junction, resulting in a small bead that joins the wires at the tip (Fig. 17). Strip the wires no more than necessary to create a weld. The insulation that is left on each wire separates the unwelded bare lengths of wire.

Heat conduction. The copper wires in type T T/Cs can conduct heat into or out of the temperature sensor depending on the cross-sectional area of the copper wire and the temperature difference between the tip and the environment (18).

RTD Specifics

Self-heating. RTDs and Thermistors are subject to self-heating from the current used for excitation. A current of 1 mA through a 100 Ω resistor generates 0.1 mW which does not create a significant self-heating error.

Common Issues

Size. A long or large sensing element will report an average temperature over the length of the element. In penetration studies, a small sensor will give a more true reading of the cold spot.

Shape. A sensor for measuring surface temperature needs to be flat and adhere to the surface.

Thermal shunting. The size of the temperature sensor should be small relative to the object being measured in

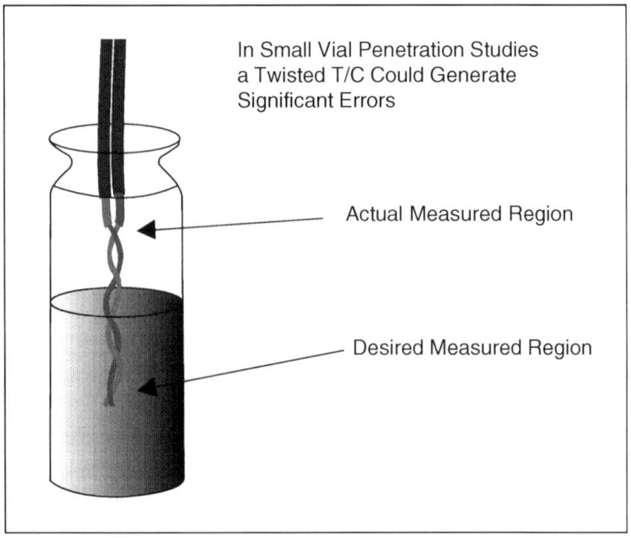

Figure 16 In a small volume for penetration studies, a twisted T/C could generate significant errors. Indicated temperature somewhere between the two regions.

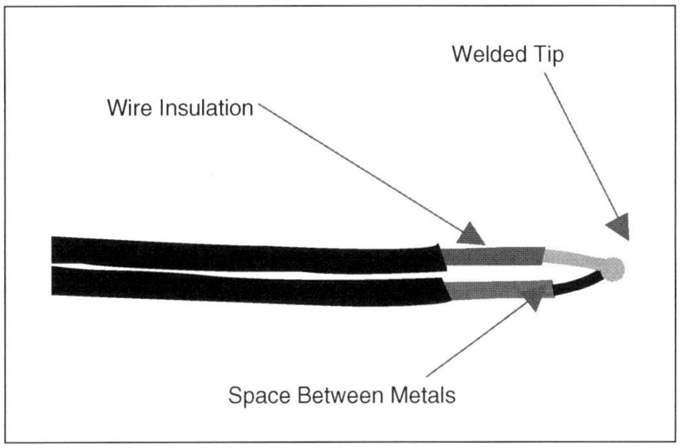

Figure 17 A thermocouple with welded tip provides secure contact at a single point, allowing it to be used in many different applications.

order to minimize the influence on the thermodynamic properties of the object of measurement.

Response time. The response time of the sensor is size and mass dependent. The response time should be at least five times shorter than the fastest rate of change in the process to be recorded in order to give a true representation of the process dynamics (19). This is especially important for determination of *D* and *z* values using ampoules in BIER vessels.

Sensor position. The temperature sensor reports the temperature it "feels." Therefore, the sensor must be positioned in an unambiguous thermal environment.

- A sensor-measuring temperature distribution in a sterilizer must be freely suspended in the chamber. If the sensor touches the chamber wall, it will report some temperature that lies between the actual chamber temperature and the temperature of the chamber wall.
- A sensor-measuring heat penetration must be fixed in position relative to the walls and content of the container.

REFERENCES

1. FDA Guidance for Industry—Sterile Drug Products Produced by Aseptic Processing—Current Good Manufacturing Practice, September 2004, U.S. Department of Health and Human Services Food and Drug Administration Center for Drug Evaluation and Research (CDER), Center for Biologics Evaluation and Research (CBER), Office of Regulatory Affairs (ORA).
2. Finch DI. General Principles of Thermoelectric Thermometry Publication DI.1000. North Wales, PA: Leeds & Northrup Company, 1969.
3. Roeser WF. Thermoelectric thermometry. J Appl Phys 1940; 2(6).
4. Moffat RJ. The gradient approach to thermocouple circuitry. Temperature, Its Measurement and Control in Science and Industry. Vol. 3, Part 2. New York: Reinhold, 1962.
5. Bentley E. The distributed nature of EMF in thermocouples and its consequences. Aust J Inst Control 1982.
6. Powell RL, et al. Thermocouple Reference Tables Based on the IPTS-68, NBS Monograph 125. Washington, DC: US Department of Commerce, 1974.
7. Caldwell R. Temperatures of thermocouple reference junctions in an ice bath. Res Natl Bur Stdt 1965; 69C(2).
8. Thermocouple Reference Systems. Bedford, MA: Kaye Instruments, Inc.
9. Muth, Jr. Reference junctions. Instruments and Control Systems. Vol. 40, No. 5. Philadelphia, PA: Reinbach Publications Division of Chilton Company, 1967.
10. Gray WT, Finch DI. Accuracy of temperature measurement. Temperature, Its Measurement and Control in Science and Industry. Vol. 4, Part 2. Pittsburgh, PA: Instrument Society of America, 1972.
11. Howard JL. Error accumulation in thermocouple thermometry. Temperature, Its Measurement and Control in Science and Industry. Vol. 4, Part 3. Pittsburgh, PA: Instrument Society of America, 1972.
12. American National Standard for Temperature Measurement Thermocouples, ANSI-MC96.1. Pittsburgh, PA: Instrument Society of America, 1975.
13. Copper–Constantan Thermocouple Wire, Probes and Accessories. Bedford, MA: Kaye Instruments, Inc., 1983.
14. Fenton AW. The traveling gradient approach to thermocouple research. Temperature, Its Measurement and Control in Science and Industry. Vol. 4, Part 3. Pittsburgh, PA: Instrument Society of America, 1972.
15. Cooper MH, Jr., Johnston WW, Jr. Traceability, what and how, relating temperature measurements at ORNL. Temperature, Its Measurement and Control in Science and Industry. Vol. 4, Part 2. Pittsburgh, PA: Instrument Society of America, 1972.
16. Practical Temperature Calibration Standards. Bedford, MA: Kaye Instruments, Inc., 1983.
17. Manual on the Use of Thermocouples in Temperature Measurements. ASTM Manual Series: MNL 12. 4th ed. Philadelphia, PA: ASTM, 1993.
18. Kemper CA, Harper GF. Probe-induced errors in temperature measurement. Pharm Technol 1978.
19. Murrill PW. Fundamentals of Process Control Theory. Instrument Society of America, 1981.

9

Qualification and Change Control

Steven Ostrove
Ostrove Associates, Inc., Elizabeth, New Jersey, U.S.A.

INTRODUCTION

The terms Qualification and Validation are often used almost interchangeably when dealing with meeting the GMPs. According to the definitions found on the ISPE web page[a]:

- Qualification: *Action of proving and documenting that equipment or ancillary systems are properly installed, work correctly, and actually lead to the expected results. Qualification is part of validation, but the individual qualification steps alone do not constitute process validation*
- Validation: *Establishing documented evidence which provides a high degree of assurance that a specific process will consistently produce a product meeting its predetermined specifications and quality attributes.*

Obviously, they are similar but not the same. The intention of both is the same; the process (validation) and the equipment (qualification) must be reproducible, meet predetermined attributes and be well documented as to their state and conditions. In short, equipment or equipment systems are Qualified and processes are Validated. There is a further subdivision to the distinction between Qualification and Validation. Current trends in the industry further divide the Qualification into the IQ and OQ protocols while the PQ usually belongs to Validation. These documents, in particular the IQ and OQ, will be discussed in more detail later in this chapter.

According to the definition, Qualification is that process whereby both the physical and operating characteristics of the equipment are documented. This means that the important or critical components are tested to document and "prove" that they are:

- What was ordered
- Received as ordered (specified)
- What was required for the process and
- Operate as specified (for the specific process or use)

All of the equipment used in the production of a pharmaceutical product must be qualified before it can be used in production. This includes not only the process equipment but also the process test instruments and supporting utilities. There are three main qualification documents or protocols[b] used in the industry. They are the IQ, OQ, and PQ. This chapter will concentrate on the IQ and OQ documents and some comments will be made about the PQ. These three documents may be combined or kept as single entities depending on the corporate philosophy and the complexity of the equipment or system being qualified. The first two (IQ and OQ) are considered qualification and the PQ is grouped into validation. If they are combined, then the IQ and OQ can be called or considered an EQ. Recently, a fourth document, DQ, has come into use. This last document began in Europe and is now being used in the United States for both medical devices and pharmaceuticals.

The purpose of any qualification protocol is to document the original installation conditions and establish that the equipment is suitable for the task it is assigned. The protocols need to be organized, easy to follow and must "test" each major component or operation.

Testing during the IQ and OQ stage is performed either on systems or on individual pieces of equipment. In general, it is best to use the system approach to the qualification program. Systems are established by combining individual units that must work together as one. An example of this is a refrigerator (e.g., condenser, cooling coil, fan, chamber) or a reactor system (e.g., reactor, feed tank, receiving tank, pumps). Testing during the PQ usually takes a working group or the entire process line (e.g., a packaging line) and demonstrates that it functions as expected as a unit.

Why do we need to qualify all of the manufacturing and associated equipment? Usually it is said that it is required by the FDA and/or the other regulatory agencies throughout the world. However, this should not be the reason to do qualification. The real reason to perform the qualification is that it makes good business sense, that is, it will save the company money and time. Yes, there is an upfront cost to the qualification program; but if executed correctly it will save more than that cost during the life of the equipment, will help assure that the equipment is functioning as required, and will be easier to track and thus fix problems (if they were to occur) during operation. While this chapter will concentrate primarily on the FDA

[a] International Society for Pharmaceutical Engineering.

Abbreviations used in this chapter: cGMP, current good manufacturing practice; CFR, Code of Federal Regulations; DQ, design qualification; DS, design specification; EQ, equipment qualification; FAT, factory acceptance test; FDA, Food and Drug Administration; FRS, functional requirement specification; GMP, good manufacturing practice; HVAC, heating, ventilation, and air-conditioning; IQ, installation qualification; ISPE, International Society for Pharmaceutical Engineering; LL, like-for-like; OQ, operational qualification; PLC, programmable logic controllers; PQ, performance qualification; PV, process validation; QU, quality unit; SAT, site acceptance test; SOP, standard operating procedure; TM, trace matrix; URS, user requirements specification; V, vendor activity.

[b] Protocols: the formal design or plan of an experiment or research activity (from ISPE web page).

requirements, the European Union, Japan, etc. all have similar requirements.

The term "Qualification" appears twice in Title 21 of the CFR[c] :

- 21 CFR 211.25—Personnel Qualifications
- 21 CFR 211.34—Consultants

These two sections deal with the qualification of the personnel implementing the study. This does not mean that qualification does not apply to the facility or equipment used or to other sections in the CFR, but it is expected and understood that the tasks necessary to assure the company, and the regulatory agencies, that the equipment will function as expected is suitable for its intended use and will not alter or adulterate the product in any way are properly executed.

The second part of any qualification program is the implementation of Change Control. This program is instituted so that a qualified system is maintained in a qualified state during its useful lifetime.

This chapter will discuss each of the qualification documents and describes how a change control program can be effectively implemented and maintained. Cost, while not a concern of the regulatory agencies, is certainly a concern for each manufacturing site. The ideas expressed in this chapter should help minimize the cost and time allocated to complete these necessary functions. The examples presented here are intended only as examples and should not be used as "absolute" answers to your qualification programs. It is not meant to be all-inclusive, but to serve as a guide to the development of a cost-effective, compliant cGMP qualification program that will satisfy regulatory requirements. This chapter will not discuss PQ (except to touch on its relationship with the IQ and OQ) or the PV.

WHAT GETS QUALIFIED

Anything that is used in, or for, the production of a pharmaceutical product or medical device (i.e., anything that comes into contact with the product) should be qualified. However, there really are limits on this. This means that equipment, or systems, actually used as part of the production process for the production or manufacturing of a pharmaceutical or medical device product must be qualified prior to its use. This includes supporting utilities as well as all process equipment. Examples of some items that usually do not require qualification are the plant electrical system (excluding the emergency power), water used in the wash rooms, and the scoops and related items used for manually transferring material to another vessel (e.g., in the weigh room).

Utilities

All process utilities that have direct product contact or a direct or indirect impact on the product quality or efficacy need to be qualified. If there is no direct impact on the product, the utility may be commissioned only (refer below and Chapter 2). Some examples of utilities that often have no direct impact on the product include the HVAC system (except in sterile area operations), electrical power, boiler steam (if used for general heat in the facility and not to heat a reactor), or non-process compressed air (i.e., for actuating non-process control valves). Again, the thought must always be focused on their impact on the product.

Certainly these systems impact the overall production. Without them there would be no production. However, the key here is they do not directly affect the product quality. These systems may affect the operator's comfort (e.g., HVAC) or the availability of "power tools" (e.g., electricity), but if they are absent, product safety, efficacy, or stability is not impacted.

Equipment

As stated before, all process equipment or systems need to be qualified prior to their use in production, but what about equipment that has been transferred from another facility or process. The answer again is "YES"; it must be either qualified or re-qualified as the case may be. If the equipment was not used in a similar operation, or if the product was different from the current intended use, the equipment must be qualified for its new use. It is best to qualify a piece of equipment over its full operating range and functionality. This will allow its use in a variety of potential future operations, not just the one for which it is currently being qualified.

Support equipment such as pumps, agitators, and heat exchangers also need to be fully qualified. These units are typically qualified along with and as part of the major unit with which they are associated.

Laboratory Equipment

Laboratory equipment, with the exception of basic research laboratory equipment, must be qualified prior to its use. This equipment is used to determine the status or release of the product (or intermediates) either to the next process step or for release for commercial distribution. Although test equipment often stands alone, there are some instruments that are "on-line." The product passes through these on-line instruments and the results are available immediately. In other cases, the product samples are taken to a lab for analysis and the results are available at a later time.

THE ORDER OF A QUALIFICATION PROGRAM

The following is just one approach to a qualification program. There are several things that must be completed prior to beginning a qualification program. These items are usually considered as prequalification. During the prequalification phase of the project, the qualification team is established. The person leading the qualification team must be able to set priorities, interface effectively with the engineering and construction teams, and organize the qualification program to meet its desired goals. The Team Leader must assign qualified personnel to collect, review, and organize all of the documents that will be needed for the program.

To have an effective, well-managed and cost-effective qualification project, the leader should follow the general order shown in Figure 1 and Table 1.

After collecting and organizing the documents by system, the next important step is to develop the qualification schedule. Like all project schedules, this

[c] Code of Federal Regulations—Title 21 Parts 210 and 211.

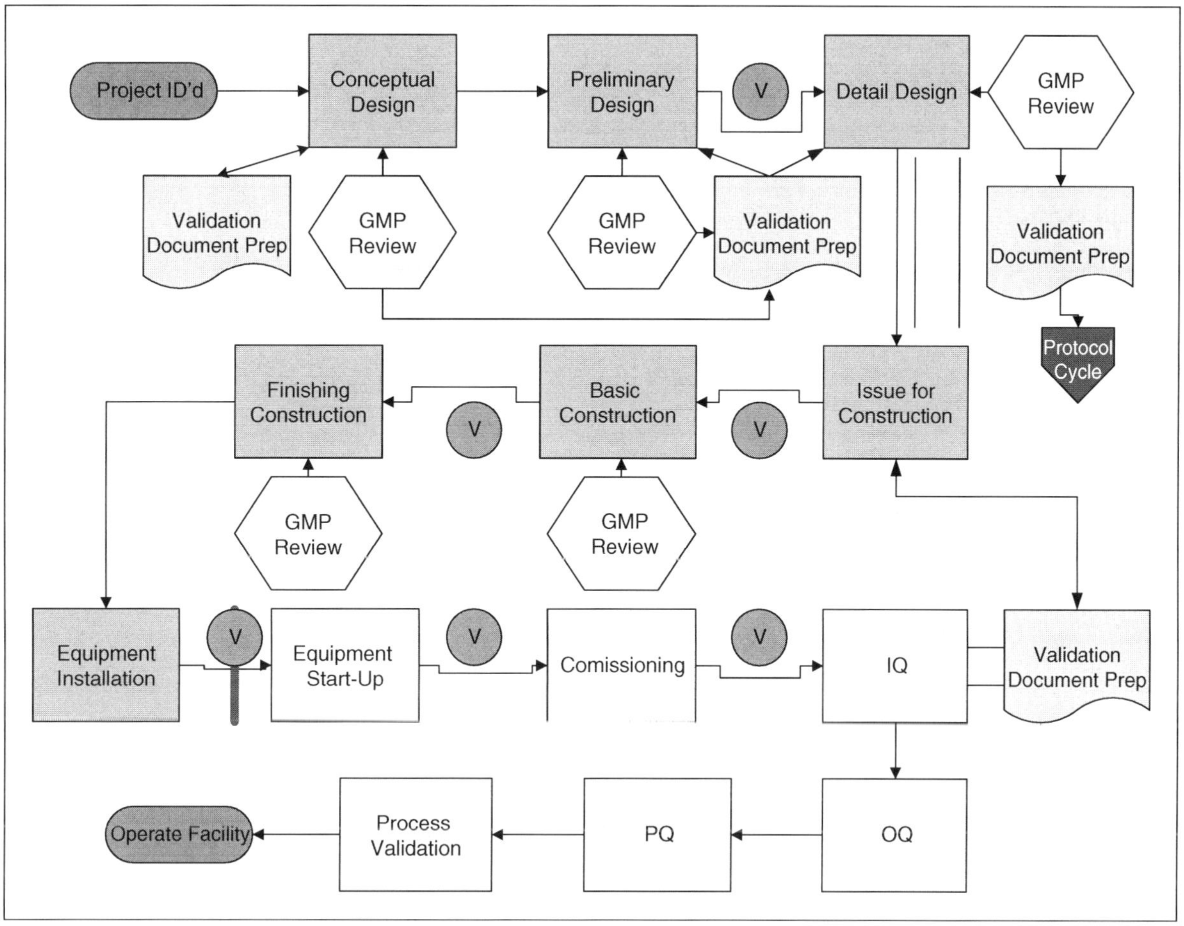

Figure 1 General outline of a qualification project. Dark grey denotes the responsibility of the qualification team and light grey denotes engineering or plant functions.

schedule is likely to change and thus should be kept up-to-date by the manager or a dedicated scheduler.

By the time you get to organize the qualification project, the vendor should have submitted the equipment manuals, the certified equipment drawings, and their own EQ protocols. Today, as compared to a few years ago, most vendors have a validation/qualification package available which can certainly be used as a starting point in developing your own qualification protocols. The vendor protocols should not be used as presented, or as the only qualification documents since your use is unique, and the vendor documents are designed to "pass" the equipment in a general application. You should also use the same protocol format for each of your qualification documents, which will not be similar to those provided by the vendor.

Table 1 Steps in Qualification Program

Define the project
Determine what equipment or systems need to be qualified
Determine when the equipment must be qualified in relation to other equipment and to the overall project
Select the team
Train or brief the team on the project/process requirements
Obtain the relevant documents from the client/users
Review for good manufacturing practice compliance
Questions need to be asked and addressed
Establish the project schedule
Set or obtain protocol formats, using existing protocols where available
Begin document (protocol) preparation
Review prepared documents (protocols)
Submit for client review and comment
Edit and update as necessary
Execute the protocol
Draft the final report
Submit report to client for comment
Finalize reports and close the project

Now that you have the qualification team, the reference documents organized by system and the project schedule established, you are ready to begin writing the qualification protocols.

Table 2 lists some key documents that should be collected, in steps 1 through 3 in Table 1, prior to preparing any qualification document. This is a "Wish List" since not all of the documents will be either available or needed. If all are available, then little difficulty will be encountered in the following steps. Certainly, in any given project, there may be other documents that pertain to the specific equipment or system that need to be added to this list.

After the protocol is prepared and approved, execution plans need to be made. This involves notifying the affected areas that qualification will be performed over a range of dates. Obviously, if a new area is being

Table 2 Document Wish List

- Drawings
  - Process flow diagram
  - Piping and instrumentation diagram
  - Vendor equipment drawings
  - Electrical drawings
  - Piping isometric drawings
  - Equipment layouts
  - Airflow diagrams
  - Personnel flow diagrams
  - Material flow diagrams for
    - Raw materials
    - Wastes
    - In process
    - Finished products
- Manuals
  - Operation
  - Cleaning
  - Preventive maintenance
- Standard operating procedures (Note: Controlled drafts are acceptable for the purposes of qualification efforts)
  - Operating
    - Process equipment
    - System controls
  - Preventive maintenance
  - Cleaning
  - Emergency shutdown
  - Area emergency procedures (e.g., fire, flood)
  - Backup for computer system data
  - Change control
  - Training
  - Calibration
- Reports
  - Weld
  - Cleaning (prior to first use)
  - Balance (air/heating, ventilation, and air-conditioning)
  - Vendor tests—factory acceptance test (site acceptance test)
  - Passivation
  - Instrument loop checks
- Specifications
  - Preliminary design specification
  - Operating ranges
  - User requirements specifications
  - Functional requirement specifications
  - Detailed design specification
- Other
  - Purchase orders
  - Standards used (traceability)
  - Ladder logic (where appropriate)
  - Source code (where appropriate)
  - Batch records (where appropriate)
  - List of critical/non-critical
    - Instruments
    - Components
    - Spare parts
  - Catalog cut sheets
  - Traceability matrix (cross-reference all documents so that all items are accounted for in the qualification)
  - Instruments
    - Contractor certifications
      - Personnel qualifications
      - Equipment certifications
    - Calibration certifications
- Logs
  - Equipment use (History)
  - Cleaning
  - Preventative maintenance

renovated or built, this notification is not necessary since none of the equipment may be used for production purposes until qualification and validation are complete. Arrangements must be made with needed specialists, test equipment suppliers, and the operators to assure that during the qualification testing there will be no delays due to missing or unavailable materials, equipment, or skills.

The people performing the tests need to review the protocols before execution starts. They are to check that all required materials, personnel and test instruments are ready and that they know how to perform or direct the testing to be done. They must understand the principle of the test as well as its execution.

Now that the entire program has been organized and the documents needed for the qualification are being prepared or collected, the next phase of the prequalification can take place. These activities are the factory acceptance and site acceptance testing followed by the commissioning program.

FACTORY ACCEPTANCE TEST AND SITE ACCEPTANCE TEST

FATs are critical tests that should be performed prior to the vendor shipping the unit to the production site. The FAT should be completed on all major equipment before they leave the vendors' facility. Thus, systems that are fabricated on-site, e.g., HVAC ducting, or process piping do not receive a FAT despite the fact that they are part of a large and important system. FATs are designed to demonstrate that the vendor has produced a unit that will operate as expected when delivered to the production area. The vendor, the receiving company or a third party designated by the receiving company, may prepare the test documentation for the FAT. As with all qualification activities, the test script should be prepared in advance and be accepted by the receiving company. The tests performed should cover all of the major functions throughout their full operating ranges. In all testings, the software or PLC code aspects of the functions will need to be verified. This "code review" should be part of the vendor audit program and will not be discussed further in this chapter.

If documented correctly, and signed by trained observers, some of the data collected in the FAT may be used to support the IQ or OQ protocols. Caution is needed here in that QA and others must agree which test results may be used, and to what extent, since the FAT may not be considered a GMP-controlled document (if not approved or reviewed by QA), although the qualification protocols are controlled.

Examples of testing that may be performed as part of the FAT are the following:

- Temperature control
  - Ramp up/down
  - Hold or dwell times
- Speed control
- Stability or lack of vibration during max run conditions
- Pressure hold or leak tests
- Function of interlocks and other safety items.

These are critical operations for the equipment (e.g., blenders, granulators, reactors) and should be tested before the unit is shipped so that any errors may be corrected prior to installing and starting up the unit at the production area. Successful completion of the FAT will save time in the end and gives the client the opportunity to make corrections to the specification or correct any miscommunication, or misinterpretation, of the user requirements.

After the equipment is installed and the contractor is ready to "turn" the unit over to the plant, the SAT is usually performed. This test may include the same test functions as the FAT, except that it is conducted at the client's site after the equipment or system is installed. It should be completed before the vendor leaves the site and the final payment is made. The SAT may include additional testing not performed at the vendor's factory. These tests should demonstrate that the unit is installed and operating as expected with the equipment owner's utilities. This should not be confused with the IQ or OQ testing, which is to be performed later. The focus of the prequalification testing is to make the vendor demonstrate that all components work properly so that commissioning and qualification can be started. One last point to keep in mind regarding the SAT is that with proper and complete documentation (e.g., signing and dating observations) and agreement with QU, some of the data may be used to support the IQ or OQ testing that will be performed later in the qualification sequence.

COMMISSIONING (START-UP, SHAKEDOWN RUN)

Commissioning is the step used to "optimize" the equipment and set (or reset) for its function in the plant. For example, the control points, valves, and other functional units and are adjusted and "fixed" according to specifications. Commissioning is a written set of procedures used to prepare the equipment for qualification. Commissioning documents are not usually considered a cGMP document (i.e., approved by QU). Commissioning usually follows the SAT; however, it may be combined with the SAT. Some companies perform the commissioning after the IQ, while others do it before the IQ. The reason for the differences is the company's philosophy on the commissioning document. The IQ is a GMP document and thus its control, data entry and all other aspects must be carefully monitored. The changes performed during commissioning usually do not lend themselves to such control. Thus, it is often performed prior to the execution of the IQ.

All critical components of the system are to be tested to ensure that the equipment will meet its expected operating criteria. Electrical or mechanical adjustments may be necessary in order to reach design operating condition (e.g., air pressures may be adjusted, valves replaced, or wiring fixed), and these adjustments or changes must be recorded to provide an accurate record of the starting set points or operating conditions. The adjustments provide assurance that the system will meet the IQ or OQ requirements. Following the completion of the SAT and commissioning steps, the unit can be considered ready for qualification. This is why Commissioning is considered a start-up operation and not a cGMP-regulated function.

PROTOCOL CYCLE

As will be discussed in the rest of this chapter, the qualification protocols have a distinct pattern for their generation and execution. The chart in Figure 2 gives one such cycle and the relative time usually required for each step. Note that the colors are coordinated to indicate the responsible person or group for the activity. The times indicated are approximate and will vary from project to project.

APPROVALS

All qualification documents need to have two sets of signatures. The first preapproval set of signatures verifies that the information in the protocol is correct to the best knowledge of those signing. The parties signing represent the various plant functions that have responsibility for the units use or function. It is usual for all qualification documents to be signed by at least the following:

- Protocol author
- Engineering
- Operations/manufacturing
- QU

Others groups may be needed (e.g., safety) to approve the information in the protocol and their signatures may be added, according to the company's policy, on qualification.

Each person reviewing and signing the preapproval should be reviewing the document from the perspective of his or her own specialty, that is, engineering reviews for compliance to the engineering aspects, operations/manufacturing reviews it for its impact on the operators and the process (are all the buttons, controls, etc. there and accounted for), and QA reviews it for adherence to company and regulatory requirements.

The second set of signatures, by the same disciplines that signed the preapproval (above), is at the completion of the execution work. All testings are complete, all reports written and all deviations closed. Only when all of the protocol components are complete will the approvers again review the document and attest that all data are correct to the best of their understanding (i.e., it appears correct, all completed and any corrections are appropriate) and that the protocol is considered complete and ready for the final report.

REPORTS

Summary reports must be prepared upon successful execution of the protocols. The reports may be combined for the IQ and OQ, or they may be independent. They may be attached to the protocols, or prepared as standalone documents, to be presented to the regulatory agencies without the encumbrance of the supporting data. Of course, either way, all supporting data and tests must be available upon request. The report must also be approved by the same people who approved the protocol.

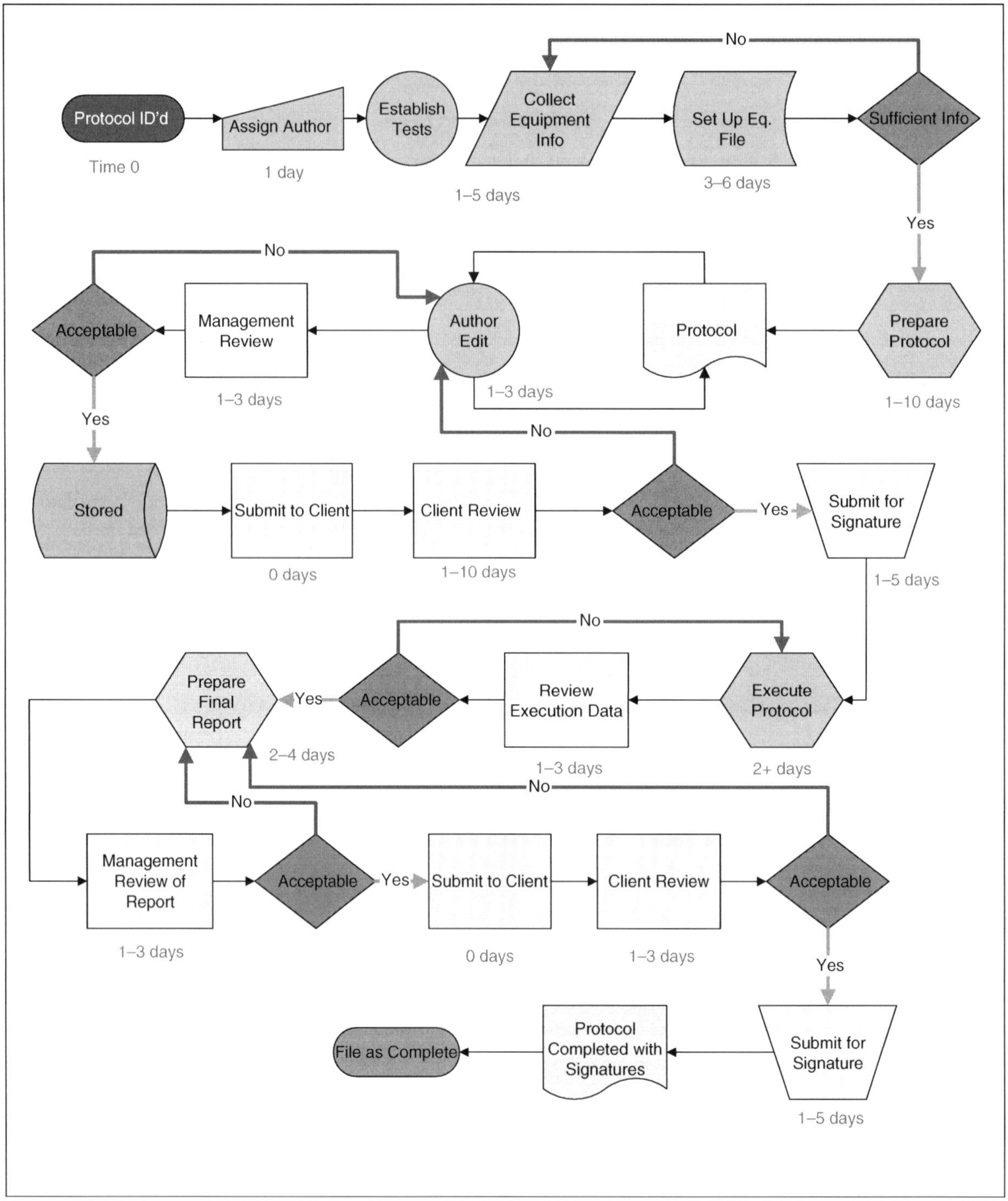

Figure 2 Protocol life cycle.

DETERMINING CRITICAL FACTORS/PARAMETERS

Qualification protocols need to test those parameters considered critical to their operation or function as well as those functions necessary for their assigned function. Some of these may not be critical functions or controls but "secondary" gauges or controls that are used either to make a cGMP decision in the process or to adjust the process flow. Thus, if a unit has 10 functions but only five will ever be used, then only those five need to be qualified. However, all of the other unused functions must show that they do not interfere or compromise the functions that are to be used. An important point here is that unless or until the unused functions are qualified, the functions cannot be used in production. Critical parameters, as well as the non-critical parameters, are usually determined during the protocol development stage but need to be reviewed and updated (others may be added) when scaling up to production size.

A critical parameter or factor is usually defined as a characteristic

- that is necessary to produce the product
- whose change will affect the product efficacy, safety, or stability

DOCUMENTS

Design Qualification

The DQ is an activity that has been in place for many years, but the documentation is relatively new. In its original form, it was known as the GMP review (Fig. 1). The DQ document is really a checklist for all parts of the engineering design. It breaks the entire design/build/validate project into sections and lists the expected documents that are to be produced in each phase. The following is an example of a typical facility design and build project. The associated validation work is also listed.

Project phases are as follows:

- Conceptual design: GMP philosophy established
- Preliminary design: Validation Master Plan started
- Detailed design: qualification protocols prepared
- Construction: on-site GMP inspection
- Installation: execution of protocols
- Qualification: completion of execution and final reports prepared.

The DQ reviews each section of the project and its required documentation to provide an organized "checklist," verifying that all critical aspects of the project are met. For example, it will verify that all environmental and building permits are in place and have been correctly applied, that the vendors have submitted their drawings and specifications and the contractors are licensed, trained and sufficiently experienced in the area in which they are to work.

The DQ is equivalent to the design or GMP review that has been ongoing for years. The DQ establishes an order and methodology for reviewing the engineering drawings, assuring that all drawings are accounted for and have been appropriately reviewed.

In the past, the DQ has been performed as the "Equipment History File" or other similar names. The purpose again was to collect all relevant documents and materials for each system so that the protocol preparation and its execution will move smoothly. All components of a system are listed and support material collected.

Equipment Qualification

The EQ is another term for the IQ and OQ respectively combined into one protocol. Each of these protocols is discussed in the following sections; however, keep in mind that they may be combined and presented as an EQ.

Installation Qualification

Preparation

The IQ document specifies all of the physical attributes of the system or equipment requiring qualification. This means that all physical parameters that may affect the machine operation, product safety, density, strength, or purity need to be listed and verified as being appropriate for its intended use. The IQ provides a record of the equipment ordered and verifies that it was correctly received. It also provides a baseline record of the starting point and serves as a reference if another unit must be ordered to replace the current one.

The protocol should be designed so that it is easily followed and that the information is easily compared to the specification. An example of an IQ format is found at the end of this chapter. The basic format, or information, should be along the following lines and is generally in a fill-in-the-blank format. Keep in mind that not every column must be included. For example, the "Expected Results" may not be needed or required especially if a more risk-based approach is used for the qualification.

Item #	Expected result	Actual result	How determined	Initial/ date
Doors				
Material of construction	Solid vinyl		Vendor documentation	
Size	2′10″ × 8′8″		Measured	
Hinged	On left		Visual	

All items should be easily identified. This information is to be included in the protocol. Tag numbers should match the drawings. Each major component should list the following information:

1. Tag number
2. Manufacturer
   a. Model
   b. Serial Number (usually not included in the "as expected" column)
3. Asset number (or equivalent)
4. Location
5. Capacity (as applicable)
6. Instruments
   a. Critical
   b. Non-critical
7. Materials of construction (for all product contact surfaces)
8. Utilities
   a. Electrical requirements (volts/amps/phase)
   b. Weld reports

Additional items need to be included for specific specialty items such as filters, chromatographic columns, etc.

The following represents a generalized Table of Contents for an IQ:

1. Approvals
2. Purpose
3. Responsibilities
4. References
   a. Purchase orders
   b. SOPs (expected list only)
   c. Manuals (reference location)
5. Equipment description
   a. Equipment data sheet with sketch, as required
   b. Instrument list
   c. Spare parts list
6. Acceptance criteria
7. Deviations
8. Appendix/attachments
   a. Drawings
   b. Lubricants

The purpose of the system is described, personnel responsibilities assigned (the task of each department or group), and reference documents recorded. The predetermined acceptance criteria are listed for each component. Lastly, if there are any deviations from the expected results, they must be listed and explained. The explanation should include the results of an investigation or determination of the actual reason for the exception or deviation.

There are two major types of deviations or exceptions: (*i*) deficiencies, i.e., not meeting the expected result or condition, but not impacting or adversely affecting the unit's use as defined and (*ii*) deviation, i.e., a failure of the unit to meet its acceptance criteria that may affect its use as defined. These are usually just grouped into the term "deviation" for sake of clarity in the protocol. However, a deficiency is more serious since it may affect the process due to a "fault" in the design or construction of the equipment. For example, if the acceptance criteria is "unit is red" and the color is blue, this is a deficiency while a deviation would be having two switches or tanks when the design calls for three.

Execution

Depending upon the order of the qualification work (commissioning and IQ), the IQ execution may be started at the time the equipment is being installed. However, this is usually not practical because the vendor or installers do not have the time, ability, knowledge or even the desire to do more paperwork at the time when they are trying to put a piece of equipment into place and to transfer ownership to the operating company in order to get their final payment. This last statement is not meant as a criticism of vendors, but as a point where improvement can be made in the qualification process.

During the execution phase, the actual installation (as built/installed) is recorded. It is necessary to document that the unit was installed according to the manufacturer's specifications (level, airflow requirements verified, utilities correctly attached and more). In addition to the manufacturer's requirements, the engineering department will often have its own specifications. These should be found in the DSs and/or the functional specifications developed during earlier phases of the project.

Execution of the IQ will involve the vendors, engineering and validation departments and third party contractors (depending on the project size). In addition, some specialists may be needed, such as electricians or plumbers, so that the measurements may be safely conducted.

Operation Qualification

Preparation

The OQ is the document that tests all of the operating characteristics of the equipment. It verifies that all parameters demonstrating that the equipment operates over its entire range, as specified by the manufacturer, are correctly installed. Testing must be established in a logical sequence so that the data, test equipment and staff can be utilized to their optimum efficiency. Thus, testing may not necessarily be conducted in a "production" order, but in a sequence that allows the unit to be tested and test results documented in the most efficient manner.

The unit should be tested over its entire operating range, not just the range to be used during production. The reason for this is simple: facilities never know what product or conditions will be required in the future. It is very costly to stop production to test additional speeds, etc. The primary use of the OQ is similar to that of the IQ and serves as the basis for the future. It sets a boundary for the operation of the equipment and serves as a benchmark for the status of a unit's operating parameters.

The tests that are to be conducted should include those operations and functions listed in the user requirements. These should be referenced against the functional specifications as well as vendor requirements. In no case should a test be required or performed that may damage the equipment (e.g., run a pump dry).

Parameters include:

- Speed
  - Range
  - Speed control
  - Ramp up/down
- Temperatures
  - Ramp up/down
  - Operating range
- Flows
  - Maximum and minimum

Water, or placebo materials, not product, are usually used in the OQ testing. However, in some cases where a critical parameter must be met, such as mixing at high viscosities, it may be necessary to use the product or a placebo with similar viscosity to mimic the specific conditions.

The protocols should be written so that each test can "stand alone" and is easy to execute. An example of an OQ format is shown at the end of this chapter. It should spell out the procedure for testing and list the necessary test instruments, for example:

Test instruments:

- Tachometer

Test procedure:

1. Start the motor according to SOP #XYZ1
2. Allow the motor to run for 30 seconds to 1 minute
3. Measure revolutions per minute and record on table X of this section.

The following lists a typical Table of Contents for an OQ protocol. As usual, additions will need to be made to assure that the tests fit the equipment to be qualified.

- Approvals
- Purpose
- Responsibilities
- Reference documents
- Equipment description
- Individual tests with test procedures
- Test equipment needed
- Acceptance criteria
- Results
- Deviations

As in the IQ, the system is described, responsibilities are assigned and the specific tests are delineated. Each test needs a stated purpose, the method to be used in the testing, a list of test instruments, and a place to record the results. The results page, or section, will then be used to prepare the final report as discussed above. In addition, each test should have its own predetermined set of acceptance criteria that must be met in order to be successful. Lastly, if there are any deviations from the expected results, then they must be investigated and explained. The explanation should include the results of the investigation or determination as the actual reason or root cause for the deviation.

Execution

Execution of the OQ should be started only after the completion of the IQ execution. While this statement is the generally accepted approach, there are always other possibilities; for example, if the qualification is performed as an EQ, this would not be necessary since both would be included in the same protocol. In addition, there are times when it may be necessary to start the OQ prior to final IQ signoff. QA must approve these exceptions which cannot, in any case, be started without their involvement. For instance, it may be possible to start the execution of the OQ if the missing item does not affect the test or operation of the equipment (e.g., the test is for temperature control and the missing part is a mixing blade). Another example of this may be that the "as-built" drawings have not been completed but have been marked up (red-lined) and signed as correct. In both cases, these do not impact the operation of the system or the specific test to be conducted.

In the execution of the OQ, the relevant SOPs need to be in place at least in draft form and the operators who will operate the equipment during the testing should be fully trained and competent on the equipment they are to operate, with training verification in the operator's file. Engineering may be part of this testing, but it often falls to the validation group or contractor. Another point to be aware of is that the OQ test sequences may not follow actual operation but, again, will be designed to allow the best use of staff and equipment.

Performance Qualification

Note: This is often considered postqualification and becomes a part of the validation teams' efforts.

Preparation

The PQ is performed on those critical units or systems that usually function as a group and not individually. However, some single units may also be PQ tested depending on their impact on the product. Examples of these are water systems (utilities), packaging lines (production), or autoclaves/steam sterilizers (production). Now that each major component has been qualified in the IQ and OQ stage, the complete system must be run as expected. The format for a PQ protocol is similar to an OQ protocol and is as follows:

- Approvals
- Purpose
- Responsibilities
- References
- Process description
- Test for process parameters
- Acceptance criteria
- Results
- Deviations

In the PQ, the system is run using typical operating parameters. In addition, "worst-case" testing is also performed. Worst-case testing is not destructive to the equipment; it is only meant to be those conditions that are the worst the process or materials can encounter during normal operation. These tests are considered limit tests or boundary testing. A major difference between OQ and PQ testing is that often the PQ tests are repeated and/or run for extended periods to demonstrate reproducibility or ability to operate over an extended period of time. PQ tests can be performed with production materials rather than placeboes and the effect is directly measured on the materials processed in the equipment.

Execution

The execution of the PQ protocols is again similar to that of the OQ. Trained operators are needed to run the equipment, and sufficient test materials are needed for the testing. It is best to run the tests starting with the lowest concentration or speed and build up to maximum operation. This allows for conservation of materials. Also, it may minimize cleaning or set-up times between runs. All data must be carefully reviewed by the appropriate individuals and a final report prepared. This report can be incorporated into the IQ and OQ report or may stand alone to represent the final acceptance of the unit(s).

CHANGE CONTROL

Change control is used to maintain the qualified, GMP or "validated" state of the equipment. Validation or qualification is an ongoing process that does not stop upon completion of the final qualification or validation report. Any change that is made to the equipment or its documentation after it is qualified must be done under a formal change control program.

A change control program should be in place before the qualification starts. It becomes active upon completion of the commissioning of the equipment since any change in the unit from that time will impact its GMP readiness. Some companies start their change control program upon the completion of the EQ activities. This, however, leaves open the possibility that a change, however small, may take place and not be properly recorded.

The purpose of change control is to have a written record of any and all changes that occur to the equipment, process or any of the supporting documentation, including computer software or PLC code. An outcome of a working change control program is that all documents related to the changed items are updated and reapproved if necessary. This includes the URS, FRS, DS, TM or other life cycle validation documents.

There are several categories of change control. Each one has its own method for implementation. These can generally be grouped as:

- Major—FDA should be notified immediately
- Secondary major—reported to the FDA in the annual report
- Required—by a regulatory agency
- Emergency
- Local
- Deferred

Let us take a brief look at each type.

Major and secondary major changes are those changes that may occur due to a change in the equipment or use of a piece of equipment. This may be the result of an emergency change or may be due to replacing aged equipment. An example of a major change would be the replacement of a "ribbon blender" with a "V blender" even if the capacities are the same. Another example would be a change in the production step. An example of a secondary major change, i.e., one that can be reported in the annual report, would be a change in the motor size. In any case, it is up to the QU and the regulatory affairs groups to determine when and how the FDA should be notified of the change.

Required and emergency changes occur periodically. Those changes mandated by any of the regulatory agencies should be made as quickly as possible and documented as to what and how the change was made. Emergency changes occur at times when least expected. These changes can be minimized by following a complete and comprehensive preventative maintenance program. In an emergency, it is not usually necessary to prepare and file the document for change control at the time of the emergency. The emergency must be rectified first and then the paperwork MUST be submitted. This allows the QA department to make a determination about the status of the batch if one was involved. Most companies allow 24 hours or the next business day from the emergency to filing the papers.

The local type is by far the most common of all changes, that is, any change that takes place in an SOP, or other GMP document related to the operation or use of manufacturing equipment, must be controlled and documented under the change control program. This includes even minor corrections such as a spelling error in an SOP. Other examples of local changes may be renaming or renumbering the SOPs, replacing a label on a piece of equipment or improvements related to safety.

Deferred changes are minor changes to a system that might be identified and then placed on hold until a more convenient time for execution, or required change is made and then implemented. This allows for their inclusion in the change evaluation assessment.

A typical approach to change control is outlined as follows:

1. A change is thought to be necessary (e.g., a line operator thinks that changing a manual valve to any automatic valve will not only make his/her job easier but also makes the product more consistent as it will not be dependent on an operator's decision).
2. The operator writes up his idea and submits it to his/her supervisor.
3. The supervisor decides whether it is a worthwhile idea and sends it to his/her supervisor or to the line manager.
4. The line manager reviews it and sends it to engineering.
5. Engineering reviews it and sends it to QA/regulatory/validation.
6. QA (regulatory and/or validation) approves and sends it to upper management.
7. Management reviews and finally approves.

If the change request is approved, then the affected departments will be notified. Usually this means engineering, validation (QA) and the requesting group. However, at any point in the review process, the idea can be stopped for any reason: cost, time, value, or how it will affect the GMP status of the equipment (the line will have to be re-qualified or a change may be needed in the papers filed for the product with the FDA). This general outline is shown in Figure 3. The flow chart represented here is broken into the major areas as discussed above, but the basic review cycle is the same. Thus, the following departments/groups are usually involved in the change control:

- Originating (any)
- Engineering
- QA and/or regulatory
- Management/finance.

It was stated earlier that any change must be done under the change control program. This is true even when there is a LL exchange. In this case, an item that is *identical* with the item requiring replacement is changed. Companies should have a list of approved LL replacements. It is possible that the LL replacement may not be from the same vendor, but the characteristics must be identical to the part to be replaced.

A change control form has the following information:

Part 1: Origination

1. Name of originating person
2. Date requested
3. Date needed
4. Reason for request
5. Justification (as necessary)
6. Supervisors approval

Part 2: Engineering

1. Date received
2. Reviewer
3. Disposition
4. Reason for disposition
5. Date sent back or forwarded

Part 3: Validation

1. Date received
2. Reviewer
3. Disposition
4. Reason for disposition
5. Date forwarded or rejected

Part 4: Management

1. Date received
2. Reviewer
3. Disposition
4. Reason for disposition
5. Date forwarded or rejected

Part 5: To File

1. Date received

When supporting a change control request, all necessary engineering, validation, and production information must either be attached, or its location referenced, for easy retrieval and review.

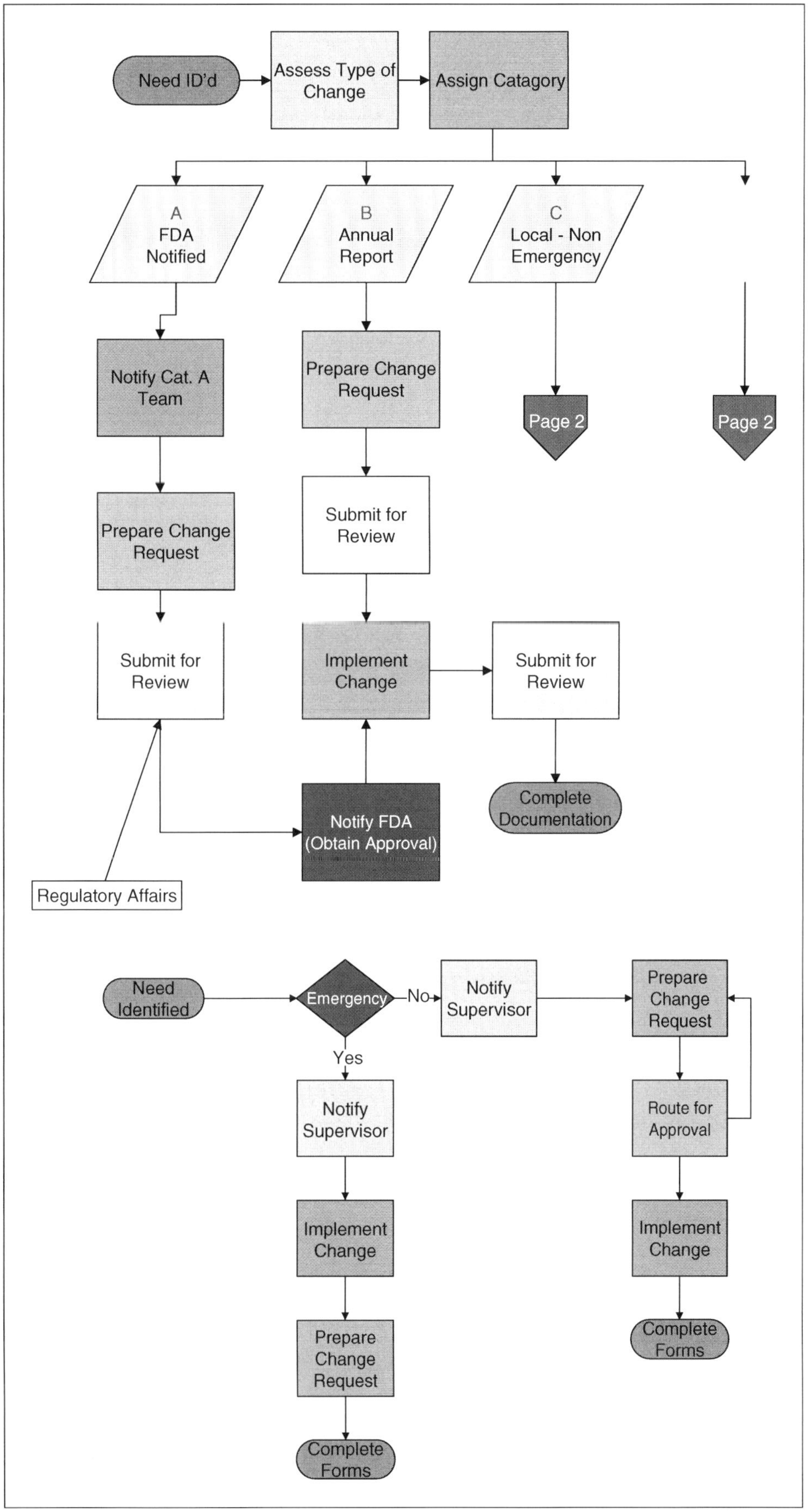

Figure 3 General change control flow plan.

FORMAT EXAMPLES

CHANGE CONTROL REQUEST FORM

SECTION 1: Initiating Department ____________________

Originator: (Supervisor/Designee)

Description of Change: (Include the current state and the proposed change)

Product(s) Affected and/or Used: **Material Code Number(s):**

Validation Required? (IQ/OQ/PQ, Process, Calibration, Cleaning, Methods, Computer)?See Section 2

☐ **Yes**?Enter steps in attached action plan

☐ **No**?Provide justification and describe below

Printed: ____________________ Signature: ____________________ Date: __________

System Owner (if different from Originator): (Enter steps in attached action plan as needed.)

Department:

Printed : ____________________ Signature: ____________________ Date: __________

Department Head Approval: (If different from Originator, the System Owner Department Head should sign.)

Department:

Printed: ____________________ Signature: ____________________ Date: __________

SECTION 2: Validation/Stability Required?

☐ **Yes**?Enter steps in attached action plan and attach stability protocol

☐ **No**?Provide justification and describe below

Printed: ____________________ Signature: ____________________ Date: __________

SECTION 3: Others (Engineering, etc.) (Enter steps in attached action plan as needed.)

Printed: ____________________ Signature: ____________________ Date: __________

SECTION 4: Compliance (Enter steps in attached action plan as needed.)

Regulatory Submission Strategy

☐ Prior Approval Required (PAR) Supplement

☐ Changes Being Effected (CBE) Supplement (30 Day Wait Period)

☐ Changes Being Effected (CBE) Supplement (No Wait)

☐ Annual Report (AR)

☐ No Registration Required

Comments:

Printed: ____________________ Signature: ____________________ Date: __________

CHANGE CONTROL DECISION

SECTION 5: Final Approval of Change (Resulting in Implementation)

Initiating Department:

Engineering	Printed:	____________	Signature:	____________	Date:	______
Compliance:	Printed:	____________	Signature:	____________	Date:	______
Validation	Printed:	____________	Signature:	____________	Date:	______
QA:	Printed:	____________	Signature:	____________	Date:	______

SECTION 6: Change Control Implementation Plan

Page ________ of ________ Add additional page(s) if needed

ACTION STEPS All Steps and Include Documentation Updates	**Requestor**	**Who's Responsible**	**Target Completion Date**	**Actual**		**Notes and/or Attachments**
				Begun	**Completed**	

Installation Qualification

Instructions

1. For each data sheet, record the requested information in ink. A "Y" or "Yes" answer is required for acceptance, and all "N" or "No"replies, and/or "N/A" replies, must be explained in the comments section.
2. When more than one unit of the same type exists, replicate the corresponding data sheets to match and uniquely identify each.
3. When a list of acceptable options is presented, circle, underline or otherwise indicate the option that is actually present.
4. Initial and date each verification.
5. Sign and date at the end of each section. Record any unusual/additional information in the comments. Record any deviation and its explanation in the Comments area of each section.
6. Each Description and/or Verification box must be completed, signed and dated in the assigned space using an ink pen.

1 ACTIVE ISOLATOR (Isolator Number x)

PARAMETER	DESCRIPTION	VERIFIED BY/DATE
Identification		
Location		
Manufacturer		
Model Number		
Serial Number		
Illumination	ft candles	
Electrical	V type/1 phase /60 Hz	
Motor Manufacturer		
Motor Model Number		
Motor Serial Number		
Motor RPM		
Motor Horse Power	hp	
Electrical	V type/3 ph/60 Hz	
Spray Wand Manufacturer		
Spray Wand Model #		

DESIGN	REQUIREMENT	HOW VERIFIED	ACCEPTANCE			
			Y	N	N/A	VERIFIED (INIT. / DATE)
Design	Isolator designed for containment leve IIII	Check with product literature				
	Rigid 316 L stainless steel base with upper and lower frame of 304 stainless steel	Check with product literature and visual inspection				
	Totally enclosed gas-tight construction with three (3) pairs of glove ports	Visual inspection				
	Work surface constructed of 316 L stainless steel, reinforced to support loads of 300 lbs	Product literature, materials certification report and visual inspection				
	Lights externally mounted at the work surface	Check with visual inspection				

	REQUIREMENT	HOW VERIFIED	ACCEPTANCE			
			Y	N	N/A	VERIFIED (INIT. /DATE)
Design	Easy replacement of parts such as light bulbs, HEPA filter housings, blower motors, etc., with out breaking containment	Check with product literature and visual inspection				
	Glove ports and sleeves made of	Check with product literature and visual inspection				
	Enclosures designed to maintain 100 fpm through a glove port in the event of failure	Check with product literature and visual inspection				
	Spray wand provided	Check with visual inspection				
	Drainage connection with a sanitary flush bottom valve and sanitary connection	Chock with product literature and visual inspection				
	Isolators designed for multiple operators	Check with visual inspection				
	Framework will be round tubular and on locking castors with height adjustable legs	Check with visual inspection				
	A minimum of four (4) sanitary ports in which electrical cords can pass	Check with visual inspection				

MECHANICAL	REQUIREMENT	HOW VERIFIED	ACCEPTANCE			
			Y	N	N/A	VERIFIED (INIT. /DATE)
Mechanical	All foreign objects have been removed	Check with visual inspection				
	Properly braced and supported	Check with visual inspection				
	Controls are properly identified	Check with visual inspection				
	Bottom surface of the isolator is sloped in the direction of the drain	Check with visual inspection				
	Gauges, control valves and/or pneumatic operators installed	Check with visual inspection				
	Verify that the unit can be locked out/tagged out	Check with visual inspection				

ELECTRICAL	REQUIREMENT	VERIFICATION	ACCEPTANCE			
			Y	N	N/A	VERIFIED (INIT. /DATE)
	460 V type/3 ph/60 Hz	Physical verification with certified equipment				
	Grounded					

DOCUMENTATION	REQUIREMENT	VERIFICATION	ACCEPTANCE			
			Y	N	N/A	VERIFIED (INIT. /DATE)
General Documentation	Product literature is on file in the System Description located in Engineering	Check with visual inspection				
	Product literature contains spare parts list	Check with visual inspection				
	Product literature contains electrical schematics and diagrams	Check with visual inspection				
SOPs	Operation Maintenance Cleaning Emergencies					

Operational Qualification

Comments: ____________________

Inspected By: ____________________

Validation Specialist Date

8.5 HEPA FILTER INTEGRITY TESTING (Continue)

HEPA FILTER LEAK TEST REPORT

Isolator Number: ____________________ Manufacturer: ____________________

Filter Model #: ____________________ Serial #: ____________________

Circle One: Supply Air Return Air

Filter Size: __________ x __________ x ______ Repair Sq. Inches: ____________________

Filter Sq. Inches: ____________________ Filter Restricted: ____________________ %

Mark location of test points

O Indicates Leakage

⊗ Indicates Successful Repair

Leakage: Yes: __________

No: __________

Aerosol Photometer

Manufacturer: ____________________ Model Number: ____________________

Serial Number: ____________________ Date of Calibration: ____________________

BIBLIOGRAPHY

Biotechnology Inspection Guide Reference Materials and Training Aids, 1991.

Code of Federal Regulations (21 CFR 210), 2006.

Code of Federal Regulations (21 CFR 211), 2006.

Commissioning and Qualification Baseline Guide ISPE, Volume 5, 2005.

FDA Guide to the Inspection of Quality Systems, 1999.

FDA Guideline on General Principles of Process Validation, May 1987.

FDA Guideline documents (www.fda.gov/opacom/morechoices/industry/guidedc.htm).

Gibson W, Powell-Evens K. Validation Fundamentals—How to, What to, When to Validate. Interpharm/CRC–1998.

Good Drug Regulatory Practices, A Regulatory Affairs Quality Manual Helene I. Dumitriu–Interpharm Press, Inc. 1998.

Nally, JD (ed). Good Manufacturing Practices for Pharmaceutical 6th Ed. Informa Healthcare, 2007.

Quality Systems Approach to Pharmaceutical Current Good Manufacturing Practice Regulations, 2006.

www.emea.europa.eu/.

www.picscheme.org/index.php.

www.who.int/en/.

10

Microbiology of Sterilization Processes

Roger Dabbah
Tri-Intersect Solutions, Rockville, Maryland, U.S.A.

David A. Porter
Vectech Pharmaceutical Consultants, Inc., Farmington Hills, Michigan, U.S.A.

GENERAL INTRODUCTION

Sterilization processes for pharmaceutical products, medical devices and allied medical products are at the end of a long process that essentially minimizes the microbial contamination at every step of the manufacturing process prior to sterilization. If one looks at the manufacturing of medical products as a continuum from the selection of raw materials, excipients and other additives to the development of products. Then each step of the continuum is important, microbiologically speaking.

This chapter will not discuss the engineering aspects of sterilization processes, but will concentrate on the microbiologic aspects. The purpose of sterilization of medical products is to ensure that a product that is to be administered to patients is sterile.

Sterility is defined by the absence of microorganisms. This is a simple concept, but is difficult to establish in absolute terms. It must be expressed in probabilistic terms such as the probability of one contaminated unit out of 1,000,000 units following sterilization. The concept of sterility is also complicated by processes of sterilization that do not involve sterilizing agents that would destroy microorganisms, but by processes that remove the microorganisms from the products, such as in filtration sterilization. The combination of containers being sterilized and product being sterilized separately then combined under aseptic processing is another complication of the sterilization process, and from a microbiologic point of view presents different problems and solutions.

The goal of this chapter is to give a general understanding of microbiology, its role in the pharmaceutical and biotechnologic industry, and its control of the quality of the finished product.

After reviewing the various methods of sterilization from a microbiologic point of view we will discuss issues related to the use of BIs and physicochemical integrators, the special cases of biologicals and biotechnologic products sterilization, and the use of parametric release for some of these processes. Finally we will review the regulatory arena, national and international, for guidance documents as well as harmonization of microbiologic procedures among the *USP*, the *Japanese Pharmacopoeia*, and the *European Pharmacopoeia*, including sterility testing and microbial limit testing.

INTRODUCTION TO MICROBIOLOGY

Microorganisms are ubiquitous in nature and as such they will be present in air, floors, ceilings, personnel, raw materials, excipients, water, instruments and equipments. Microorganisms can be useful when used in fermentation for the manufacture of antibiotics or when they are genetically modified through DNA recombinant processes to yield proteins of therapeutic value. Microorganisms are objectionable when their presence or their by-products, such as toxins or pyrogens, might result in deleterious effects for patients. Pharmaceutical products that are sterile are injections, IVs, most ophthalmic products, and even some oral products that are to be used by immuno-compromised patients.

Microorganisms are characterized by their taxonomy, structures, functions, metabolism, and conditions under which they can be detected and identified. Microbiology is the science of microscopic forms of life, unicellular as well as in some cases, multicellular, with sizes ranging from 0.5 to 50 μm. They are composed of protoplasm bounded by a cell membrane and are composed of water, proteins, lipids, and nucleic acids organized structurally in organelles. The DNA within microorganisms controls all the biochemical processes of the microorganism, and if removed or destroyed, the microorganism will die. That characteristic of DNA will be used in some sterilization processes such as radiation sterilization that destroys DNA. DNA is found in the nucleus of eukaryotic cells (e.g., fungi) or free in the protoplasm of prokaryotic cells (e.g., bacteria).

Microorganisms in order to grow require a source of energy to drive the biochemical processes, and a source of carbon for the production of biomass. Depriving microorganisms from either a source of energy or a source of carbon can be used as control mechanisms to inhibit the growth and proliferation of microorganisms. Microorganisms can either be aerobic (utilize O_2) or anaerobic (cannot utilize O_2). The growth of anaerobes can be controlled by replacing air with a nitrogen blanket. However, life is not simple and some microorganisms, facultative anaerobes will grow anaerobically if O_2 is

Abbreviations used in this chapter: BI, biological indicator; BSE, bovine spongiform encephalopathy; cGMP, current good manufacturing practice; EPA, Environmental Protection Agency; EtO, ethylene oxide; FDA, Food and Drug Administration; IV, intravenous; SAL, sterility assurance level; USP, *U.S. Pharmacopeia*; WFI, water for injection.

not available. There are more than 3000 species of bacteria and they reproduce primarily asexually, with genetic variations within species done via exchange of small strands of DNA between two bacterial cells. Depending on the processes used to obtain energy, most microorganisms are chemoheterotrophs (oxidizing reduced organic/inorganic molecules to obtain energy, and utilizing complex organic compounds pre-synthesized by other organisms for biomass production); few are photoautotrophs (use light energy and CO_2 and simple salts for biomass); few are photoheterotrophs (energy obtained from light, and biomass from preformed organic compounds by other organisms).

The shape of bacteria—coccus, bacillus, and spiral—can be used to microscopically determine the type of microbiologic contaminant detected. The presence or absence of flagella, cell wall, or the Gram reaction (negative or positive) are also used to rapidly determine, microscopically, the type of contaminant encountered. Taxonomy is also determined by the presence or absence of spores. In general spores are more resistant than vegetative cells to sterilizing agents and one would like to know prior to sterilization if the spore content of the product is within the spore load that the validated sterilization cycle is capable of handling.

In order to detect, differentiate, and quantitate microorganisms one has to able to grow microorganisms. A culture medium must provide the elements necessary for growth. These elements can be well defined or undefined. Example of undefined medium is Soybean–Casein Digest Medium, where undefined components such as pancreatic digest of casein or papaic digest of soybean meal are part of the composition. Microorganisms have various requirements for optimal growth and it is not possible to develop a medium with all the various optimized components. On the other hand if one wants to differentiate microorganisms one can selectively remove or add components that will inhibit the growth of the undesirable microorganisms and enhance the growth of the desired microorganisms.

Microorganisms need water in order to grow. If you restrict the amount of water you can control the proliferation of microorganisms to a minimum or totally. Water activity of a product if under a certain value will restrict the growth of microorganisms. A corollary is that if in an aseptic processing suite you restrict the presence of water, you can control the growth of microorganisms. Too much dryness in a sterility suite or an aseptic processing suite will however be very uncomfortable for personnel involved in processing.

The detection and quantitation of microorganisms will also depend on the pH of the medium and the temperature of incubation. For bacteria, USP indicates a temperature of incubation of 30°C to 35°C, while it indicates a temperature of 20°C to 25°C for yeasts and molds. Microbiologic procedures in USP will not detect psychrophiles such as some *Pseudomonas* species nor thermophiles such as *Bacillus thermophilus*. Since *B. thermophilus* spores are resistant to some sterilizing agents, one would like to know the spore contents prior to sterilization in order to make sure that the validated sterilization cycle is adequate. A sample incubated at 60°C will give an indication of the spore content.

The issue of "U.S. Indicator microorganisms" has been controversial for years. USP chapter <61> Microbial Limit Tests (USP 28, 2005) lists the following indicator microorganisms: *Staphylococcus aureus*, *Salmonella* species, *Escherichia coli*, *Pseudomonas aeruginosa*. These will be increased by the addition of the following microorganisms: *Clostridium* species, *Candida albicans*, and bile-tolerant gram(negative) microorganisms when the harmonization process with the European Pharmacopoeia and the Japanese Pharmacopoeia is completed. The controversy is that these "indicator microorganisms" that will be called "Specified Microorganisms" in the harmonized draft give only a general view of the microbiologic quality of the materials that will be sterilized subsequently. These microorganisms can indeed be pathogenic, especially if they encounter patients with immune-deficient systems. These microorganisms are not the only ones that a manufacturer should assure are absent. The absence of these microorganisms does not relieve the responsibility of the manufacturers to tests for other relevant indicator organisms if the nature of the product, its origin, and the type of patients that will receive the finished products warrants it. Finally, not all products need to be tested for the four indicator microorganisms. In general, a monograph in USP for the compendial product will give direction on the requirement for one or more of the absence of these microorganisms.

This microbiology primer should prepare the readers and practitioners for prepare the remainder of the chapter where the role of microbiology in the manufacture of products is detailed.

Microorganisms in Pharmaceutical Processes

The diversity of microorganisms and their relative resistance to sterilizing agents or processes is wide. Attempt to list all types of microorganisms in this chapter is counterproductive and not relevant to the objective of the chapter. Suffice to say that for the purpose of this chapter the microbiologic world is divided into viruses, bacteria, yeasts and molds (fungi), and prions. The recent introduction of the issue of prions into raw materials of bovine origin might necessitate the reevaluation of sterilization processes, using sterilizing agents as well as filtration sterilization.

Viruses, in general, are not heat resistant, EtO resistant, or radiation resistant. However, the classical filtration sterilization using filters of 0.2 μm is not adequate if one has to rely solely on filtration for sterilization. There are however specialized filters that could be used to retain viruses which are smaller than 0.2 μm.

Bacteria are ubiquitous, can live under extreme conditions at both ends of the temperature spectrum and can grow under a variety of conditions such as in the presence of simple inorganic compounds as well as complex organic compounds, at a variety of pHs, and also grow in logarithmic fashion. Most bacteria, given the right host conditions, can be undesirable, but in general most bacterial species are harmless and do not cause disease. Diseases are caused by pathogenic bacteria that when injected into the body can produce septicemia, and if not treated can result in death of the patients. The presence of pathogenic bacteria in raw materials and excipients is an indication of the unsanitary manufacture of these products or human contamination during manufacture of these raw materials. From a sterilization process point of view, the characteristic of some bacteria is the formation of spores that occur when conditions of growth are not favorable. The resistance of spores to sterilizing agents can cause problems in sterilization, thus spores have to be taken into consideration in the development of a sterilization cycle. For filtration sterilization, since bacteria can range from fractions of a micrometer to several micrometers, the use of specialized filters smaller than 0.2 μm will depend on the prevalent microflora in the product to be filtered sterilized.

Fungi that include yeasts and molds are larger in size than bacteria. Some yeasts form spores similar to bacteria but their resistance to sterilizing agents is low and is not a problem in sterilization processes. Molds also produce spores but they are not resistant to sterilizing agents.

Prions are modified forms of a normal cellular protein found primarily on the surface of neurons and on other tissues in kidneys, liver, spleen, and lung among others.

Prions are resistant to acid hydrolysis and denaturation by heat and chaotropic agents. The spread of BSE in cattle in a number of countries and the capability of prions to infect species other than cattle, such as humans, has created a need for processes designed to inactivate or destroy these prions. Humans are affected and the disease is characterized by spongy degeneration of the brain with severe neurologic symptoms. Risk reduction strategies have been reviewed by Ian DeVeau, Roger Dabbah, and Scott Sutton in an article in *Pharmacopeial Forum* (30:1911–1921). They include precautions during manufacturing and collections of bovine materials, process validation to ensure that materials of bovine origin, if it contains prions will inactivate or destroy the prions. Cleaning of equipment used in handling materials of bovine origin has to be validated as well. The issue of prions in sterilization processes has not been addressed directly because of a lack of procedures for identification of prions in suspected materials.

Microorganisms in pharmaceutical processes originate from a variety of sources. They do originate in raw materials, excipients, and ancillary materials. One would expect that they originate in water used in formulation, in manufacturing, and in the cleaning of equipment. They certainly will originate from the environment, unless the environment is controlled for aseptic processing. The predominance of gram-negative microorganisms can give rise to bacterial endotoxins that as pyrogens will cause an increase in temperature in patients when injected IV, and these microorganisms generally originate in water systems. The final verification of sterility and non-pyrogenicity is done using the compendial sterility test and bacterial endotoxins test. However, sterility assurance is not established by a sterility test, but by the validation of sterilization cycles. The often misunderstood compendial sterility test will also be discussed in some detail in this chapter as well as the microbiologic continuum that results in the sterilization of products to be injected in patients. The role of the USP will also be discussed, since at every level of the continuum USP has monographs, general chapters, or general information chapters that address microbiologic procedures.

Microbiology of Raw Materials, Excipients, Drug Substances, and Biologic Substances

These materials when used in manufacture of pharmaceutical/biotechnologic products are generally not sterile when received by a manufacturer. The materials to be used should have been produced under cGMPs. When this is not possible, especially for products of plant origin or other natural products, USP chapter <1111> Microbiological Attributes of Non-Sterile Pharmaceutical Products indicates that special treatment can be given to reduce the microbial bioburden to an acceptable level. Regardless of the method used to reduce the bioburden (EtO, radiation, steam, or dry heat), the potential for generation of undesirable and/or toxic microorganisms must be investigated and determined.

If the finished product is to be a sterile product, the magnitude of the bioburden and the species distribution will impact on the severity of the sterilization cycle. The endotoxins content of these materials will have to be controlled since the finished product must be sterile and non-pyrogenic, unless the manufacturing process prior to sterilization contains treatments validated for the removal of endotoxins.

The methods of bioburden determination are the pharmacopeial methods in current USP under chapter <61> Microbial Limit Tests. This chapter, which will be divided into two chapters under the harmonization process of the *European Pharmacopoeia*, the *Japanese Pharmacopoeia*, and the USP, describes in detail procedures, methods and interpretations of microbial limit tests and the determination of specified microorganisms. We will discuss global harmonization of microbiologic methods later in this chapter. According to the USP, alternative methods to the compendial methods can be used provided that data showing equivalence or better are available for inspection by FDA inspectors. Chapter 51 in this book on "Validation of Alternative Microbiological Methods" gives additional detail.

Since, in general the development a sterilization cycle is based on the bioburden of the product prior to sterilization, the characteristics of the microbial flora of the various formulation components need to be controlled, and determined. If some products prior to sterilization have a large proportion of bacterial spores that are resistant to the sterilizing agent, that information is critical. Special tests need to be devised to assess the spore bioburden of the product and to control their numbers if the sterilization cycle developed is to be successful.

Microbiology of Water

Water is used in manufacturing of sterile products and is the most widely used excipient in the pharmaceutical/biotechnologic industry. Furthermore, water is also used in cleaning and sanitizing equipment used in the manufacture of sterile products. Water to be used as an excipient in pharmaceutical formulations that are to be sterilized is USP WFI. The source water for the preparation of WFI is drinking water that fulfills the requirements of the EPA for drinking water. The process to obtain WFI is through distillation or other process that have been shown to be equivalent to distillation. The quality of WFI is governed by a USP monograph on WFI and includes specifications for conductivity (see chap. <645> Conductivity in the current USP) and for total organic carbon (see chap. <643> Total Organic Carbon in the current USP).

The issue of microbiologic guidelines for water is discussed in USP chapter <1231> Water for Pharmaceutical Purposes. In this chapter guidelines on the microbial levels are indicated, and those have become the de facto specifications that an FDA inspector will want to see. From the sterilization point of view waterborne contaminants are generally gram-negative microorganisms that are not resistant to the sterilizing agents. However the elaboration of endotoxins by gram-negative microorganisms can present problems in the sterile finished product that needs to be non-pyrogenic. Methods for the evaluation of the microbial burden of water are recommended in chapter <1231> but from a compliance point of view, the USP chapter references the methods in the current edition of Standard Methods for the Examination of Water and Waste Water that is prepared and published jointly by the American Public Health Association, the American Water Works Association, and the Water Environment Federation.

Microbiology of the Environment

The environment can contribute to the microbial burden of a product prior to sterilization. Sterilization processes are not necessarily designed to correct microbiologically unsanitary conditions resulting in large bioburden counts or infractions to cGMPs.

The microbial contamination of products from the environment should be minimized to ensure that the validated sterilization cycles are not overburdened. In essence, if one controls all the sources of contamination of a product prior to sterilization, the finished sterile product will be of high microbiologic quality. If the various sources of contamination are not controlled, then the validated sterilization cycles might not be appropriate.

The issue of microbiologic environmental control for aseptically processed product will be reviewed in more detail later in this chapter. Even if a product will be terminally sterilized, it is important to determine the bioburden from that environment as well as the distribution of microbial species, especially their resistance to the sterilizing method used.

Bacterial Endotoxins

The presence of bacterial endotoxins in a sterile product is the result of the endotoxin burden of the various formulation components, which when added will exceed the specification for endotoxin levels in the sterile product. The acceptable level of endotoxins in a parenteral sterile product is calculated on the basis of the dose administered and the threshold of human pyrogenic dose of endotoxin per kg of body weight. The endotoxin levels for each product are indicated in the USP monographs on these products. The various sterilization agents are not very effective against endotoxins and they should not be relied upon to inactivate or destroy the endotoxins. Of all the modes of sterilization, dry heat is effective for depyrogenation and it is used for removing pyrogens from equipment and glassware (minimum of 30 min at 250°C).

Sterility Testing and Sterility Assurance

The classical demonstration of the sterility of a product is through the USP Sterility Test (see chap. <71> Sterility Testing in the current USP). This statement unfortunately is not correct, from a microbiologic as well as a statistical point of view. The testing of 20 units from a large batch of product cannot predict the sterility of each and every unit of the batch. The compendial sterility test is a procedure used by the regulatory agencies to determine compliance with the sterility requirement. Since the sample size is not a statistical sampling, the projection of sterility from 20 units to a large batch is not warranted. Sterility is assured through validation of sterilization cycles and control of the microbial bioburden of the formulation prior to sterilization. Sterility assurance is expressed in probabilistic terms as indicated in USP chapter <1211> Sterility and Sterility Assurance of Compendial Articles as follows: "It is generally accepted that terminally sterilized injectables articles purporting to be sterile, when processed in an autoclave, attain a 10^{-6} microbial survivor probability, i.e., assurance of less than one chance in a million that viable microorganisms are present in the sterilized dosage form." This probability is expressed in a SAL of 10^{-6}. This SAL should not be confused with the aseptic processing maximum contamination rate that one unit in 5000 could be nonsterile while using aseptic processing.

The compendial sterility test by itself, and because of the types of media used, the temperature of incubation and the amount of sample tested, is a compromise. The USP Sterility Test (see USP chap. <71> in the current USP) describes the types of media to be used (Fluid Thioglycollate Medium and Soybean–Casein Digest Medium), the number of containers to be tested, the amount of product to be tested, the temperature of incubation, and the time of incubation (14 days). It is conceivable that there are some survivor microorganisms that are injured and cannot recover under the conditions of the test, but this is unlikely in most cases.

Sterilization Processes

Regardless of the sterilizing agent or process used there is a certain common perspective that should be understood by all, especially by those who will select the mode to be used. The principal reason for selection of a mode of sterilization depends on the nature of the product. If the product is heat labile it could be sterilized by filtration or even a heat sterilization cycle could be devised based on the utmost control of the microbial bioburden that will provide the appropriate sterility assurance without affecting the stability, integrity and effectiveness of the product. The economics of sterilization can also play an important role in the decision of one mode of sterilization versus another. The availability of a certain mode of sterilization equipment can also play an important role in the selection process. The nature of the containers used could also dictate the selection process, as well as the interaction of the container with the product as they are affected by the sterilizing mode.

There are a number of sterilization processes that can be used. These are: steam or moist heat, gases, ionizing radiations, dry heat, and sterilization by filtration. Combinations of sterilization methods are also used as in aseptic processing, flow molding approaches, as well as manufacturing under isolators conditions.

Steam Sterilization

Thermal sterilization uses saturated steam under pressure in an autoclave. This is the most common method of sterilization used in the pharmaceutical industry, because it has a very predictable and reproducible effect on the destruction of bacteria, and the parameters of sterilization are time and temperature that can be easily controlled and monitored once the cycle has been validated. Generally, most heat sterilization is performed at 121°C under 15 psig. At this temperature one can invoke the lethality concept of F_0 that is used if the temperature of sterilization is different from 121°C. The F_0 of a process that is not run at 121°C is the time in minutes required to provide a lethality equivalent to that provided at 121°C for a stated time.

Dry Heat Sterilization

In general dry heat is used for sterilization of materials through the process of incineration of microorganisms. The process is either continuous or a batch process in an oven, with temperatures around 170°C for sterilization or 250°C for depyrogenation. Dry heat is used for sterilization/depyrogenation of glassware that will be used for products aseptically processed. The temperatures used are too high for plastic containers. As for steam sterilization the process is predictable and reproducible thus it can be well controlled. Often dry heat sterilization conditions are used for depyrogenation of glassware and other components capable of resisting the temperatures used. And in general the validation of a depyrogenation cycle for a dry heat process inherently includes sterilization as well.

Gases Sterilization

The most commonly used gas is EtO in its pure form or in combination with inert gases. The gas is very volatile and highly flammable. Since it is an alkylating agent it provides destruction of microorganisms including spores and vegetative cells. The sterilization is done in a pressurized chamber.

Sterilization produces toxic materials such as ethylene chlorohydrin that are generated in the presence of chlorides ions in the materials. It is used for sterilization of medical devices and clothing materials as well as disposable pipettes and Petri dishes used in microbiology laboratories. EtO residues are toxic materials that need to be removed from the sterilized material through post-sterilization degassing that can be enhanced by the application of temperatures higher than room temperature. This also contributes to the need for protection of personnel from harmful effects of the gas.

The factors involved in gas sterilization include humidity, concentration of gas, temperature and appropriate distribution of the gas within the sterilizer chamber. This complicates the monitoring and control of the process, but the destruction of bacteria is as predictable and reproducible as for steam sterilization. Since the kill of bacteria depends on the presence of moisture, gas and temperature within the innermost parts of the articles, the penetration of these through packaging, either primary or secondary package, must be allowed, requiring special design of these packaging materials.

Ionizing Radiation Sterilization

The destruction of bacteria to render an article sterile can be done using ionizing radiations, since they affect the nucleic acids of the microorganisms in a nonreversible way. The formation of free radicals and peroxides that are highly reactive compounds also contribute to the lethality of the sterilization process. There are two type of ionizing radiation process that can be used: gamma irradiation and electron beam irradiation. Radiation sterilization is used for medical devices when they are heat sensitive or when residues of EtO are not appropriate. The precise measurement of radiation dose, which is not affected by temperatures, is the controlling factor in radiation sterilization along with the time of irradiation. Monitoring and control of the process is simple, but precautions have to be taken for the safety of the operators.

Ionizing radiations have also been used in the sterilization of drug substances and formulations. The compatibility of the articles to be radiation sterilized is a factor that needs attention since materials and substances are affected by radiations, perhaps not immediately following the treatment but later when the product stability can be affected. For medical devices and plastics, changing from EtO sterilization to radiation sterilization requires the determination of short-term and long-term effect of radiations and sometimes requires the modification of the manufacture of plastics and rubber materials to make them compatible with radiation sterilization.

Filtration Sterilization

This mode of sterilization differs from the others cited above because it involves the removal of microorganisms through filtration and does not rely on the destruction of the microorganisms. The physical removal of microorganisms relies on filters that are constituted by a porous matrix with pore sizes that will not allow microorganisms to go through. There is more than a simple sieving effect that controls the effectiveness of a filters, it also includes the absorption of microorganisms into the substrate of the filter. This mode of sterilization is mainly used for liquid products that can be filtered or that are heat labile and unable to be sterilized by another mode of sterilization. Membrane filtration technology advances have allowed the increased use of filtration sterilization, especially when coupled with an aseptic processing system.

The effectiveness of filtration sterilization may be a function of the magnitude of the microbial bioburden, since clogging of the filters can occur at high concentration of microorganisms. Pressure, flow rate, and characteristics of filters are parameters that need to be controlled to achieve sterilization of product in a predictable and reproducible fashion. This also requires that the bioburden prior to filtration be determined and controlled. Nominal pore size for sterilizing filter is 0.2 μm or less and filters are made of a variety of materials such as cellulose acetate, cellulose nitrate, fluorocarbonate, acrylic polymers, polycarbonate, polyester, polyvinyl chloride, vinyl, nylon, polytef, and many more types of materials including metal membranes.

Aseptic Processing

This process is a combination of filtration sterilization of the product, sterilization of containers, and aseptic assembly of the product in the container in an environment that is microbiologically controlled.

The product could be sterilized by filtration or by heat, if it is not heat labile. The container is sterilized by heat, EtO, dry heat, or radiation depending on the nature of the container and the economics of each method of sterilization. Closures can be sterilized by a number of methods. The aseptic processing success depends on the microbiologic quality of the environment and the operating practices employed by personnel where that processing occurs. The environment where the container is filled with the product must be free of microorganisms and trained operators must conduct the aseptic processing. The facility where the components are assembled in the final product is one of the important considerations that help ensure the sterility of the final product. The facilities and the equipment have to be easily cleaned and sanitized before use to minimize microbial contamination. Furthermore, since personnel are generally the source of microorganisms in the environment, strict gowning procedures will have to be followed. The environment around the aseptic processing area should not contribute to the microbiologic load of the processing area and this can be controlled through pressure differential between the processing area and the areas just outside the aseptic core. The use of airlocks and air showers in combination with gowning ensure that the contamination of the environment in the processing aseptic core is minimized.

The use of isolators is a special case of aseptic processing. Since personnel are the primary source of potential microbiologic contaminants in an aseptic processing core, the removal of personnel from that area will reduce the likelihood of the microbial contamination of the finished product. This can be accomplished by the use of isolators. Isolators have been used mainly for sterility testing, but they are now used by a number of manufacturers in aseptic processing.

Another form of aseptic processing includes the blow/fill/seal approach. In this case, the plastic container is molded and then filled with sterilized product, then sealed in a continuous uninterrupted way within an unbroken sterile environment. Microbial contamination of the finished product is minimized and even eliminated in a properly functioning system.

Validation of Sterilization Processes

Validation of a sterilization process is based on the following principles:

1. The process equipment has the capability of operating under a controllable set of conditions. These conditions depend on the sterilizing agent used.
2. The control equipment can operate within the limits needed to ensure reproducibility and accuracy of the parameters of the sterilization equipment.
3. Replicate sterilization cycles are used to test the operational ranges of the equipment and the impact on the probability of survival of micro-organisms.
4. Validated process will have to be monitored during routine operation and also needs to be requalified at periodical intervals.
5. Sterilization cycles are developed then validated with the help of BIs specific for the sterilizing agent that is used.
6. Document and archive data from all the steps above in a retrievable fashion.

Biologic Indicators

A BI is defined as a preparation of a specific microorganism that provides a defined and stable resistance to a specific sterilization process. The main characteristics of a BI are:

1. First and foremost, that the microorganism used has an intrinsic and measurable resistance to the sterilization process for which it is designed. It does not have to be the most resistant microorganism to the particular sterilization process.
2. The resistance of the microorganism to the sterilization process must be stable or must not vary significantly.
3. The resistance of BIs is determined under very specific conditions using specialized equipment not normally available to a manufacturer of pharmaceutical/biotechnologic derived products. In general, manufacturers of BIs determine their resistance and certify to the users of the level of resistance under these sets of conditions.
4. Microorganisms used for BIs must not present safety problems to those who will handle them. Pathogenic microorganisms are not to be used as BIs.
5. In general, spores of microorganisms are used for BIs since they provide a higher resistance to the sterilization process than the vegetative cells.
6. A method for the recovery of microorganisms must allow for growth of the treated BIs under the conditions specified.
7. There are a number of types of BIs. For example the most commonly used are carriers that have been inoculated with spore suspensions. The carriers could be strips or discs made of filter paper, glass, plastic, and metals. The carrier should not interfere with the resistance of the microorganisms and should be protected from microbial contamination by being packaged properly in envelopes. The carrier or the packaging should not interfere with the performance of the BI or with the penetration of sterilizing agent and other parameters of a specific sterilization process. Another type of BI is a spore suspension that can be inoculated on or into representative samples of the product to be sterilized. If this is not practical or possible a simulated product mimicking the real product can be used. Another form of BI is a self-contained BI. In this case, the primary package of the BI will contain a growth medium for recovery of microorganisms that have been subjected to a specific sterilization process.
8. A number of BIs are described in detail in monographs in the current USP. These include BI for Dry

Heat Sterilization, Paper Carrier; Biological Indicator for Ethylene Oxide Sterilization, Paper Carrier; Biological Indicator for Steam Sterilization, Paper Carrier; and Biological Indicator for Steam Sterilization, Self-Contained. Monographs for BIs with metal carriers are in preparation.

9. The preparation of BIs and their resistance to sterilizing agents are affected by the growth and sporulation conditions under which the spores are grown.
10. From the standpoint of regulations, BIs are considered to be Class II devices, and as such require that manufacturers of BIs obtain a device 510 K approval prior to commercial use.

Filtration sterilization, since it involves not the destruction of microorganisms but the removal of microorganisms from solutions uses a BI suspension of *Pseudomonas diminuta* a small bacterium that challenges the sterilizing filters.

Physicochemical Indicators and Integrators

The Federal Code of Regulation, Part 211 on Good Manufacturing Practices for Finished Pharmaceuticals in section 211.165 reads as follows:

> There shall be appropriate laboratory testing, as necessary, of each batch of drug product required to be free of objectionable microorganisms

This statement opens the door for the use of alternative tests for the sterility test performed on each batch of sterilized product. These indicator/integrators can thus be used as part of the testing of batches of sterilized products that are to be released under a parametric release program.

1. The performance of these physicochemical indicators/integrators must be consistent from lot to lot.
2. The composition of these indicators/integrators must not interact physically or chemically with the products being sterilized.
3. The safety of the operators and laboratory personnel in contact with these items should be ensured at all times.
4. These indicators/integrators are considered as Class II devices and it is necessary to obtain a 510 K approval prior to commercial use.
5. Physicochemical indicators are devices that respond in a measured fashion to one or more critical sterilization parameters. They are used to monitor a physical parameter that indicates that the load has been exposed to that factor, for example temperature.
6. Physicochemical integrators are devices that respond to one or more sterilization critical parameters in a measurable quantifiable manner that can be correlated to microbial lethality. It integrates, for example, temperature of the sterilization process with the time of exposure and concentration of the sterilizing agent. It is not a substitute for a BI but is useful in indicating if the sterilization cycle is too long and if over- or under-exposure by the sterilizing agent has occurred.
7. Physicochemical integrators for steam sterilization react in a predictable fashion to a specified combination of physical parameters of sterilization such as temperature, steam pressure, and time of exposure. Deviations of the preset parameters will be captured by the physicochemical integrator.
8. Physicochemical integrators for EtO sterilization react to a preset combination of parameters such as temperature, humidity, gas concentration, and time of exposure.

Parametric Release of Sterilized Products

The use of parametric release for sterilized pharmaceutical products requires prior approval by the FDA. The principle of parametric release of sterilized products is simple, but the practical application is more difficult. Few organizations in the pharmaceutical industry have used parametric release because it involves the following:

1. The mode of sterilization is very well understood and predictable.
2. The lethality of the cycle to be used has been microbiologically determined and the cycle validated using BIs for steam sterilization and EtO sterilization; for radiation sterilization, validation is done using precise dosimeters. Some organizations have used in addition to dosimeters, BIs with microorganisms resistant to radiation.
3. When a validated sterilization cycle operates consistently, a combination of critical parameters can provide accurate and repeatable data that ensure that the preset lethality has been achieved.
4. A parametric sterilized product will be released without the need to perform a sterility test. However, for compliance purposes the expectation of the FDA is that the product, if tested by the sterility test will pass. This is not a critical requirement, since the nature of the sterility test, its sample size, and the limitations of that test that are too well known to be repeated here gives less assurance of sterility than product that has been parametrically released.
5. Since parametric release is based on preset parameters for a validated cycle, the manufacturers should ensure that the autoclaves used function as intended and within the preset parameters. This requires the qualification of all production autoclaves to ensure that the critical parameters are always controllable and controlled. Changes in production autoclaves must be assessed in relation to their performance of a validated sterilization cycle. It is necessary to establish a well-planned program of change control. No changes can be made to autoclave physical parameters unless there is an assessment of the impact of the change on the validated cycle. Not all changes are significant in their impacts on the validated cycle, but it is a good policy to ask that all changes be communicated to a change control system for assessment.
6. Development and validation of cycles are based on bioburden, or based on an overkill approach, or on a combination of bioburden and use of BIs. When the cycles are based on bioburden, it is imperative to have a comprehensive microbiologic program to assess the various stages of manufacture and of components prior to the sterilization. Bioburden might change with the season for a given supplier, and with different suppliers. This needs to be taken

into consideration when parametric release is contemplated and the cycles are based on bioburden. This is one of the reasons why bioburden-based cycles are not used often, except for radiation sterilization of products that have a very low bioburden.

When sterilization cycles are based on the overkill principle, the importance of the bioburden and of microbiologic control is much less. Overkill is used when the quality and function of the product to be sterilized will not be affected by the sterilization conditions. For these cycles one will use BIs with 10^6 microorganisms resistant to the sterilization process and achieve an endpoint of the process. Then the process is doubled to achieve an SAL of 10^{12}.

The bioburden/BI approach to develop sterilization cycles is used when an overkill cycle would damage the sterilized product and the desire is to have a cycle that will ensure the sterility of the product through the destruction of a large number of microorganisms resistant to the particular sterilization process. This approach will require the knowledge of the bioburden and their resistance to the sterilization agent. Using BIs with 10^6 microorganisms and performing fractional exposure cycles will determine the relative resistance of the bioburden to the resistance of the BI.

Container/Closure Integrity Validation

Regardless of the mode of sterilization utilized, including filtration sterilization and aseptic processing, the integrity of the container/closure system is an important factor during the sterilization process as well as during shelf life of the product. Microbiologic contamination of the products through the closure system has been known to occur during sterilization since containers and closures are subjected to various physical forces that can distort their interface. If the container/closure system cannot maintain its integrity during storage, microbiologic contamination could occur. In the current USP an information chapter <1207> Sterile Product Packaging—Integrity Evaluation discusses the maintenance of sterility of sterilized products until the product is to be used for patients. Physical and microbiologic approaches to test the integrity of container/closure system are available and should be correlated. This is generally done at the development level and verified during the actual manufacturing and the shelf life of the product. The selection of an evaluation method will depend on the nature of the system, its design, the manufacturing method, and the intended purpose of the product.

Physical tests for integrity of the container/closure system include, among others, pressure and vacuum decays rates, dye emersion tests, liquid chemical tracer tests, gas ionization of evacuated containers, high-voltage leak detection, visual examination, or package headspace analysis.

Microbiologic tests for integrity of the container/closure system include immersion of the container/closure system in a suspension of actively growing microorganisms, inoculated shipment testing, and spraying of microorganism suspensions of the closure/container interface.

The validation of the integrity of the container/closure system must be redone if major changes occur in the design of the system or when conditions of sterilization are changed significantly.

The Role of BIs in the Validation of Sterilization Processes—A Theoretical Approach

Validation of a sterilization process using BIs is based on the assumption that the sterilization process destroys microorganisms in an orderly and predictable fashion that is also reproducible. The general principles involved in validation of sterilization processes are as follows:

1. The equipment used in a sterilization process and the instrumentation for the control of the process are properly designed and calibrated, and that documentation exists and is retrievable for each of the equipment and instrumentation used.
2. The quality of the sterilizing agents and water if used in the process is of the appropriate quality for the purpose intended.
3. Since each and every product in a batch needs to attain specified conditions of the critical parameters that will ensure sterilization, the distribution of these parameters within a given sterilizer must be uniform and consistent. For example a temperature profile for a given process must be uniform within the sterilizing vessel. This can be mapped using thermocouples placed in strategic locations within a sterilizer. The temperature profile is done in an empty chamber as well as in a fully loaded chamber.
4. The effectiveness of an EtO sterilization process is based on the effectiveness of heat, humidity, and concentration of gas and their penetration in a uniform manner in all products in the batch regardless of their geographic location in a sterilizer chamber.
5. The physical measurements above are to be accompanied by either BIs or inoculated products with standard suspensions of microorganisms to test directly the effectiveness of the sterilization process, which after all is defined in terms of destruction of microorganisms.
6. The concept of *D* value, which is the time, in minutes, to reduce the microbial load by 90% or 1-log cycle at a specific temperature. If the *D* value of a microorganism is 1.5 min at 121°C and the product is treated for 12 min, the lethality of the process is 8*D*.
7. The *D* value of typical microorganisms used in BIs for various sterilization processes are shown in the table below:

BI	Sterilization mode	*D* value
Geobacillus stearothermophilus	Saturated steam, 121°C	1.5 min
Bacillus subtilis var. niger	Dry heat, 170°C	1 min
Bacillus subtilis var. Globigii	EtO (600 mg/L) 50% relative humidity at 54°C	3 min
Bacillus pumilus	*Gamma radiation*	
	Wet	0.2 mrad
	Dry	0.15 mrad

8. The *D* value of a BI is determined using the survivor curve, which is the linear curve obtained by plotting survivor numbers on a logarithmic scale versus exposure times with the sterilizing agent.

The number of survivor microorganisms is plotted on semilog graph paper (y-axis) and results in a linear curve. The *D* value is the slope of the linear survivor curve. If one starts with a BI with 10^6 microorganisms and subjects it to fractions of the sterilization process and measures the survivors, one can develop the survivor curve and calculate the *D* value. Specific instructions for the determination of *D* values for each of the sterilization process are detailed in the current USP under chapter <55> Biological Indicators—Resistance Performance Tests.

9. The concept of F_0 is used in saturated steam sterilization validation. The F_0 value is the lethality, expressed in terms of the equivalent time in minutes at a temperature of 121°C delivered by the saturated steam sterilization process in its final container with reference to microorganisms possessing a Z value of 10. The total F_0 of a process takes account of the heating up and cooling down phases of the cycle and can be calculated by integration of lethal rates with respect to time at discrete intervals (1). If the spore population is 10^6, and its *D* value at 121°C is two minutes, then the F_0 is 12.
10. The concept of *Z* value is also important since it is defined for a given microorganism as the rate of change of the *D* value in function of the temperature. This is useful since not all steam sterilization processes will be run at 121°C. *Z* value can be determined experimentally by determination of *D* values at different temperatures and plotting of the *D* values versus the temperatures. *Z* value is expressed in degree centigrade. The curve obtained is the thermal resistance curve and the slope is the *Z* value. The use of *D* value, *Z* value, and F_0 in the validation of steam sterilization processes that are all determined based on BI's resistance to the sterilization agent is part of the validation scheme for products. Other chapters in this book will cover in detail the use of these concepts.

Each sterilization mode will have a particular approach to validation of cycles for achieving sterilization of products. A good example is filtration sterilization, where the concept is to remove a BI from the product, not to destroy it as for the other methods of sterilization. Since filtration is dependent on the effectiveness of filters in retaining the microorganisms, the physical characteristics of the filter to accomplish that purpose must be tested. In addition, challenging the filter with BIs will validate the capability of the filters used to produce sterile product. This cannot be done in situ and must be accomplished outside of the manufacturing area, preferably in a laboratory, since the test is destructive. Surrogate to the microbiologic testing are physical measurements that can be done in situ such as a bubble point test or diffusive air flow test.

Validation of Aseptic Processing for the Manufacture of Sterile Products

Aseptic processing is a hybrid mode of sterilization with the purpose of ending with sterile products with an expectation of sterility assurance that is sufficient in most cases. In subsequent chapters in this book the details of validation of aseptic processing will be given. In this section, we will concentrate on the microbiologic aspects of the validation. The range of aseptic processing system is wide and we will not review all of them in this section. A combination of sterilization modes can be part of an aseptic processing. For example, the containers and closures can be sterilized using EtO, while the product can either be steam sterilized in bulk or can be sterilized by filtration. Each one of the modes of sterilization used must be validated.

Common Microbiologic Characteristics of Aseptic Processing

1. The filling of sterile containers with sterilized product is performed in a microbiologically controlled environment. The environment ranges from clean rooms, to partial barriers, to isolators.
2. The closures are applied to the containers in a microbiologically controlled environment to protect from microbiologic contamination.
3. The environment for aseptic processing must be monitored and controlled on a continuous basis or at regular intervals to ensure that potential microbial contamination is minimized.
4. Personnel constitute the major source of microbiologic contamination in a controlled environment. Removing personnel from direct contact with the aseptic environment is desirable, and this occurs with the use of isolators. In other aseptic environments, the personnel must be gowned completely and must be trained in aseptic handling of materials.
5. The cleanliness of the aseptic environment requires that materials and equipment used be easy to clean and sanitized. In essence, it also requires validation of cleaning and sanitizing of these areas.
6. The air quality in the aseptic environment must conform to the requirements of ISO class M3.5 (Class 100).
7. The integration of the various modes of sterilization used in aseptic processing combined with aseptic filling of sterile product into sterile containers that are then aseptically closed is validated using media-fills.

Microbiologic Evaluation of Clean Rooms and Other Controlled Environments

The current USP chapter <1116> Microbiological Evaluation of Clean Rooms and Other Controlled Environments reviews the various microbiological issues involved in aseptic processing, especially the issues relating to clean rooms and other controlled environments.

1. Classification of clean rooms is based on particulate count limits: it is understood by most that the relationship between particulate count and microbiologic counts in clean rooms are not correlated. However, the classification based on particulates gives to the manufacturers of clean rooms specifications of the performance of these rooms that will allow them to design, build, test, and maintain clean rooms in a state of control. U.S. Federal Standard 209E has been supplanted by the ISO 14644 requirements for clean rooms.
2. The design, development and implementation of a microbiologic evaluation program are an integral part of the assurance that the aseptic processing system is under microbiologic control.
3. The microbiologic evaluation program should include the determination of the number and the type

of microorganisms in the environment. It should also include a review of trends over an extended period to evaluate the effectiveness of the microbiologic program.

4. Alert and action levels should be part of the microbiologic program as well as standard operating procedures when these levels are exceeded.
5. Microbiologic evaluation of clean rooms should include a sampling plan, the frequency of sampling, and the critical sites for sampling. Sampling sites include air, surfaces of equipments, floors, walls and other critical sites. Sites that are in direct contact with product or containers should be tested more frequently than those away from the critical sites.
6. The importance of personnel training should not be underestimated. The need for strict procedures and supervision of personnel involved in aseptic processing is directly related to the sterility assurance of the final product.
7. Media-fills: Uses microbiologic growth medium instead of a product and follows a simulated aseptic processing to assess the overall aseptic process in terms of microbiologic contamination. Following aseptic processing of the medium it is incubated at 20°C to 35°C with a range of ± 2.5°C for 14 days. Visual examination for growth is done and if growth is detected it is identified to the genus to pinpoint the probable source of contamination. Some of the critical issues to be considered include the number of units per run, the number of runs, and the interpretation of results. The number of units per run is used in media-fills in order to assess that not more than a contamination rate of one positive unit is encountered in 5000 units. However, the number of units per run seems to go up in order to increase the confidence that the results are extrapolated to the operational runs with product.

Validation of Isolator System

Validation of isolators for aseptic process will have in common with isolators for sterility testing the following characteristics:

1. Isolators not only protect the product from contamination, but also the personnel from contamination by the product.
2. Isolator internal environment is sterilized by decontaminating agents such as peracetic acid and/or hydrogen peroxide. Sterilization of the internal environment is a misnomer, since the internal section is decontaminated, that is the contaminants inside the isolators are rendered nonviable.
3. The sterile products and containers are introduced into the internal section of the isolator via transport ports that have been validated to ensure that contamination will not occur during transfer.
4. Operators are not in contact with the product and manipulate the filling and closures of the containers from the outside using gloves or half suits.
5. The air in the isolator is either sealed or supplied through a microbially retentive filter.
6. Isolators need not be placed in a classified environment, thus no microbiologic monitoring of the outside environment is needed. However, the environment outside the isolator should be restricted to only essential personnel.
7. The decontamination process used must be able to produce a 6-log reduction of an appropriate BI. However, the maintenance of the sterility of the internal portion of the isolator cannot be guaranteed over time. Fraction negative approach for validation or survivor counts can be used as for other methods of sterilization.
8. The maintenance of sterility within the isolator enclosure depends on the operational parameters that control the various barriers to microbial contamination from the outside of the enclosure. In general, if contamination is detected within the enclosure through a microbiologic monitoring system, the most likely source of ingress is during the introduction of the sterilized containers, closures and product into the isolator. Validation of the process used to ensure decontamination before ingress of the material is critical. Also potential sources of contamination are the gloves or half suits that the personnel use to manipulate the products and containers/closures. Small leaks in gloves have been shown to be a source of contamination, but they are difficult to detect.

Summary

This chapter was designed to review briefly the microbiology of sterilization process and the role of microbiology in validation of sterilization processes, including aseptic processing. The common characteristic of all sterilization processes is their endpoint, the assurance that the final product is sterile with a high degree of confidence.

Sterilization is only one part of a continuum that starts with raw materials and excipients, and ends prior to sterilization. Sterilization is not designed to take care of microbiologic contamination that has accumulated in the processing continuum, but the processing of pharmaceuticals must be under microbiologic control from beginning to end. It also involves the development of microbiologic monitoring of raw materials, water, and environments to present to the sterilization process a product that is microbiologically safe and of high quality.

REFERENCE

1. http://www2.pharmweb.net/plugins/processtech/sht.pdf

BIBLIOGRAPHY

The United States Pharmacopeial Convention, Inc., <55> *Biological Indicators-Resistance Performance Tests*, USP 28-NF 23, 2005:2244.

The United States Pharmacopeial Convention, Inc., <61> *Microbial Limit Tests*, USP 28-NF 23, 2005:2246.

The United States Pharmacopeial Convention, Inc., <71> *Sterility Tests*, USP 28-NF 23, 2005:2251.

The United States Pharmacopeial Convention, Inc., <85> *Bacterial Endotoxins Test*, USP 28-NF 23, 2005:2264.

The United States Pharmacopeial Convention, Inc., <151> *Pyrogen Test*, USP 28-NF 23, 2005:2289.

The United States Pharmacopeial Convention, Inc., <1035> *Biological Indicators for Sterilization*, USP 28-NF 23, 2005:2537.

The United States Pharmacopeial Convention, Inc., <1050> *Viral Safety Evaluation of Biotechnology Products Derived from Cell Lines of Human or Animal Origin*, USP 28-NF 23, 2005:2607.

The United States Pharmacopeial Convention, Inc., <1111> *Microbiological Attributes of Non-Sterile Pharmaceutical Products*, USP 28-NF 23, 2005:2681.

The United States Pharmacopeial Convention, Inc., <1116> *Microbiological Evaluation of Clean Rooms and Other Controlled Environments*, USP 28-NF 23, 2005:2682.

The United States Pharmacopeial Convention, Inc., <1196> *Pharmacopeial Harmonization*, USP 28-NF 23, 2005:2730.

The United States Pharmacopeial Convention, Inc., <1207> *Sterile Product Packaging—Integrity Evaluation*, USP 28-NF 23, 2005:2734.

The United States Pharmacopeial Convention, Inc., <1208> *Sterility Testing—Validation of Isolator Systems*, USP 28-NF 23, 2005:2735.

The United States Pharmacopeial Convention, Inc., <1209> *Sterilization—Chemical and Physicochemical Indicators and Integrators*, USP 28-NF 23, 2005:2738.

The United States Pharmacopeial Convention, Inc., <1211> *Sterilization and Sterility Assurance of Compendial Articles*, US 28-NF 23, 2005:2740.

The United States Pharmacopeial Convention, Inc., <1222> *Terminally Sterilized Pharmaceutical Products—Parametric Release*, USP 28-NF 23, 2005:2746.

The United States Pharmacopeial Convention, Inc., <1227> *Validation of Microbial Recovery from Pharmacopeial Articles*, USP 28-NF 23, 2005:2752.

The United States Pharmacopeial Convention, Inc., <1231> *Water for Pharmaceutical Purposes*, USP 28-NF 23, 2005:2755.

The United States Pharmacopeial Convention, Inc., *Biological Indicator for Dry Heat Sterilization, Paper Carrier*, USP 28-NF 23, 2005:257.

The United States Pharmacopeial Convention, Inc., *Biological Indicator for Ethylene Oxide Sterilization, Paper Carrier*, USP 28-NF 23, 2005:258.

The United States Pharmacopeial Convention, Inc., *Biological Indicator for Steam Sterilization, Paper Carrier*, USP 28-NF 23, 2005:258.

The United States Pharmacopeial Convention, Inc., *Biological Indicator for Steam Sterilization, Self-Contained*, USP 28-NF 23, 2005:259.

11

F, *D*, and *z* Values

John Shirtz
Consultant, Columbia, Maryland, U.S.A.

INTRODUCTION

The science of microbiology and the sterilization process often involve several interrelated variables, each of which operates under a unique set of circumstances. Over the years, patterns of response have been noted in many of the more routine processes, and scientists, together with mathematicians, have developed mathematical relations for some of these activities, primarily as a mechanism for the analytical determination of sterilization values. The object of this chapter is to help clarify the often complicated circumstances involved with the sterilization process, and to condense these variables into terminology that will allow us to use them for comparison against each other or against predetermined limits of acceptance.

The three factors that play the most important roles in the sterilization process are temperature, time, and resistance:

- *Temperature.* The temperature can be the measurement of the sterilizing chamber, the sterilizing medium, or the item being sterilized. In any event, consideration needs to be given to the temperature throughout the entire process.
- *Time.* The process time at the process temperature is the key element of the sterilization process. The mathematical calculations include those portions of the cycle that are not at the optimal temperature, but owing to their magnitude, impart some measurable lethality to the microbial population.
- *Resistance.* As is common with all living things, microorganisms have an inherent resistance to death. Some are relatively weak, but others, such as spores, have a relatively strong resistance; this is why spores are usually selected as the microbiological challenge for sterilization processes.

The use of these terms and concepts is not difficult, once understood. To expect to memorize all of the formulas and tables is unreasonable; what is reasonable is to develop an understanding of the concepts, to know where to access the necessary information, and to know how to use the proper variables and formulas for the specific process application under consideration.

Abbreviations used in this chapter: BIs, biologic indicators; BIER, biological indicator evaluator resistometer; FDA, Food and Drug Administration; cGMP, current good manufacturing practice; HTST, high temperature–short time; LTSF, low-temperature steam and formaldehyde; MPN, most-probable number; MSC, minimal sporocidal concentration; PDA, Parenteral Drug Association; RH, relative humidity; RTD, resistance temperature detector; SAL, sterility assurance level; SOP, standard operating procedure; USP, *U.S. Pharmacopeia*; VHP, vapor-phase hydrogen peroxide; WFI, water for injection.

TERMINOLOGY

To consolidate the essential variables into a system that permits evaluation of the thermal destruction capability of a specific sterilization process cycle, three unique terms have been developed:

- *F value.* Equivalent time at a specific temperature delivered to a container or unit of product.
- *D value.* Time in minutes required to inactivate 1 log of a challenge microorganism.
- *z value.* Number of degrees of temperature change necessary to change the *D* value by a factor of 10.

Heat sterilization is a function of *probability* that is dependent on (1):

1. The number of challenge microorganisms
2. The heat resistance of these microorganisms
3. The amount of heat exposure

Use of the *F*, *D*, and *z* values allows sterilization scientists and engineers to compare the effectiveness of various sterilization cycles using a mathematical model. This terminology is used extensively in both the pharmaceutical and food industries and is well accepted by regulatory agencies as demonstration of sterilization effectiveness. In many cases, *F*, *D*, and *z* terminologies may also be used to construct a mathematical rationale for the adequacy of sterilization, in lieu of extensive challenge testing.

F VALUE

Introduction

The *F* value is used as a measurement of sterilization effectiveness. $F(T,z)$ is defined as the equivalent time at temperature *T* delivered to a container or unit of product for the purpose of sterilization, calculated using a specific value of *z*. The term F_0 is defined as the number of equivalent minutes of steam sterilization at temperature 121.1°C[a] delivered to a container or unit of product calculated using a *z* value of 10°C (2).[b] Therefore, wherever a value is stated in terms of F_0, it is referring to the equivalent time at precisely 121.1°C. If, for example, there is a stated F_0 value of 9, it is saying that the process being described is equivalent to exactly nine minutes at precisely 121.1°C regardless of the process temperature

[a] Published literature on steam sterilization cites both 121.1°C and 250°F as reference process temperatures. In actuality, 121.1°C is equivalent to 249.98°F and 250°F is equivalent to 121.11°C. In this discussion, the value of 121.1°C will be used for all F_0 calculations.

[b] For the purpose of process calculations involving the moist heat resistance of indigenous microorganisms, it is appropriate and acceptable to assume a *z* value of 10°C (18°F) unless an alternative value has been determined from resistance studies (1). An expanded discussion of *z* values appears later in this chapter.

and time used in the cycle. If, on the other hand, there is a stated $F_{115.0}$ value of 9, it is saying that the process being described is equivalent to exactly nine minutes at precisely 115.0°C regardless of the process temperature and time used in the cycle.

The term F_H is similar to F_0 and is used to describe the number of equivalent minutes of dry-heat sterilization at temperature 170°C delivered to a container or unit of product calculated using a z value of 20°C. Although dry-heat z values ranging from 13°C to 28°C have been reported in the literature (3–9), most of these have been in the range of 17°C to 23°C, so 20°C is usually considered an acceptable assumption.

Dry heat is a process usually designed for the destruction of endotoxin, but it is also used throughout the world for sterilizing hospital supplies, such as powders, oils, petroleum jellies, glassware, and stainless steel equipment that cannot be sterilized with saturated steam (10,11). Dry-heat sterilization processes are generally less complicated than steam processes, although higher temperatures or longer exposure times are required because microbial lethality associated with dry heat is much lower than that for saturated steam at the same temperature.

Although moist heat (steam) sterilization is a relatively easily defined condition in that the RH is 100%, dry-heat sterilization involves an RH of any value between 0% and 100%. Because the destruction rate of dry microbial cells is a function of their water content (as determined by the atmosphere surrounding the cells), the destruction rate will vary with the RH of the system. For example, a 1968 study (12) showed that an increase from 0% to 20% in the humidity of air passing over microbial spores caused an increase in the D value by a factor of 100.

For dry-heat sterilization to be effective, the USP (13) specifies that containers for pharmaceutical products should be held at 170°C for two hours; for depyrogenation USP recommends 250°C for not less than 30 minutes. Foreign compendia state other time–temperature combinations. Most pharmaceutical applications of dry heat are primarily designed for depyrogenation and, although USP suggests that validation of dry-heat methods include challenges with both highly resistant spores and measured quantities of endotoxin, dry-heat-resistant spores, such as *Bacillus atrophaeus*, have D values of only a few seconds at temperatures used for depyrogenation (14). Therefore, a process designed to provide for adequate depyrogenation should be adequate for sterilization, because microbial reductions well in excess of 10^{100} can be expected. The actual processing temperature for dry-heat sterilization is arbitrary. Generally 170°C to 250°C or higher (so-called HTST means just what the name implies: sterilization at a very high temperature for a relatively short time period). Similar to steam sterilization, any reference temperature besides 170°C can be used in the F-value determinations.

Other popular methods for sterilization involve the use of either gas or ionizing radiation. Monitoring of these processes is usually conducted using resistant *Bacillus* spores as BIs: *B. pumilus* for radiation, *B. atrophaeus* for ethylene oxide as well as for dry heat, *B. subtilis* ATCC 5230 for steam cycles that operate at less than 121.1°C, and *G. stearothermophilus* for LTSF (12). VHP has also been shown to be an effective sterilant, and its use in pharmaceutical applications has been explored. Microorganisms that are commonly used as a challenge to the VHP process are spores of *B. subtilis*, *G. stearothermophilus*, and *Clostridium sporogenes*.

Gaseous sterilization is considerably more complex than other sterilization procedures and depends on the interaction of several factors including temperature, pressure, RH, time, gas concentration and distribution within the sterilizer chamber, and moisture and gas penetration throughout the sterilizer load. Consequently, any statement of the SAL achieved in a gaseous cycle must be qualified by a detailed definition of the sterilization conditions. Some recommended practices also caution that resistance characteristics of the process monitors used for evaluation of the adequacy of the sterilization cycle should be examined using inoculated product or inoculated simulated product (15) and that the user should not rely simply on BI paper strips unless they are correlated to the inoculated product or inoculated simulated product.

The two types of ionizing radiation in use are radioisotope decay (gamma radiation) and electron beam radiation. In either case, the radiation dose to yield the required degree of sterility assurance should be established such that within the range of minimum and maximum dose sets, the properties of the article being sterilized are acceptable. For gamma radiation, an absorbed dose of 25 kGy (2.5 Mrad) has historically been used as a target, but lower or higher levels may be acceptable depending on the material being sterilized (13). Consideration must be given to the packaging materials, radiation resistance of the inherent bioburden, product-loading pattern, and dose mapping in the sterilization container before deciding on an appropriate radiation dose. A material being irradiated will not require a "come-up" time, as in thermal heating, but will be instantaneously penetrated by the ionizing track. Nevertheless, the materials being sterilized will act as a shield, which will deter the penetration of the gamma or beta rays. Several different methods of using fraction-negative experimental cycles are recognized for the examination and establishment of an acceptable dose range for the intended sterilization cycle (16).

Because the U.S. FDA had issued a proposal in 1991 for terminal sterilization as the preferred method of product sterility (*Use of Aseptic Processing and Terminal Sterilization in the Preparation of Sterile Pharmaceuticals for Human and Veterinary Use.* Federal Register Oct 11, 1991; 56(198):51354), the pharmaceutical industry has moved toward reducing the overall size of manufacturing facilities in many cases as a compromise position. Placing the product-filling or testing equipment within an isolator is becoming a very popular method of minimizing the aseptic environment. Along with this rapid development in the area of isolation technology, other methods for sterilization have become the subject of sterilant resistance. The two primary agents, peracetic acid and hydrogen peroxide, have been studied for their relative sporocidal properties as agents for the sterilization of heat-sensitive equipment. Some of these studies concluded that peracetic acid has the highest sporocidal activity when compared with hydrogen peroxide (both alone and in combination with peracetic acid), chlorine, and formaldehyde (17–20), *Bacillus* spores, known for their ability to hydrolyze cellulose and for their great resistance to adverse conditions, such as heat, chemicals, drying, poor medium, and others, are usually chosen as the challenge microorganism in peracetic acid-resistance studies. These studies have characterized the resistance of the spore to peracetic acid in terms of MSC, with the contact time in terms of hours as opposed to the usual minutes normally associated with other means of sterilization.

Other studies have touted a patented process of using 31% aqueous hydrogen peroxide that is vaporized into a sealed enclosure (21,22). Although aqueous hydrogen peroxide has been recognized for years as a bacterial and sporicidal agent, the use of hydrogen peroxide vapor as a sterilant has also been proven to be effective. Studies have reported *D* values in the range of 18.2 minutes at 0.9 mg/L to 0.1 minute at 2.9 mg/L using spores of *G. stearothermophilus*, which has been determined to be the most resistant microorganism to hydrogen peroxide gas (22,23). Temperature, volume, and initial RH of the enclosure are critical parameters for successful sterilization, but unlike many other sterilants, hydrogen peroxide gas can be catalytically broken down to nontoxic water vapor and oxygen. It has also been demonstrated that the resistance characteristics are relatively unchanged when the challenge microorganism is inoculated onto stainless steel coupons in the presence of organic soil (5% bovine serum).

Microorganisms such as bacteria and viruses have a very high radiation resistance compared with multicellular organisms, with bacterial spores being the second highest resistant and yeasts and fungi of intermediate resistance (24). Several exceptions exist, however, so generalizations are difficult to state. The experimental conditions for testing the radiation resistance of a specific bacterial species must be clearly defined, for the results may be influenced by conditions in the environment (particularly the presence of oxygen and water), before, during, or after the irradiation. Even different strains of a species may have widely different resistance; in one study involving 48 strains of *Streptococcus faecium*, the dose required for an inactivation factor of 10^8 varied from 30 to 50 kGy (3.0–5 Mrad) (24).

For all thermal methods of sterilization, the *F* value is used along with the number of challenge microorganisms, the type of suspending medium, the *z* value, and the *D* value to determine the microbiological effectiveness of the sterilization cycle. Although steam sterilizers are designed to operate at a specific designated temperature (usually 121.1°C), the actual temperature controllers used allow the temperature to oscillate within a tolerance range near the target temperature. Depending on the precision and responsiveness of the recording device, the oscillation may not be clearly evident on the cycle record. Although this oscillation is usually minimal, it can have a significant affect on the *F*-value determination, especially if the process temperature is predominantly on the high or low side of the target temperature. The *F*-value calculation takes into consideration all of the fluctuations around the target temperature by reducing the individual observations to a common equivalent value.

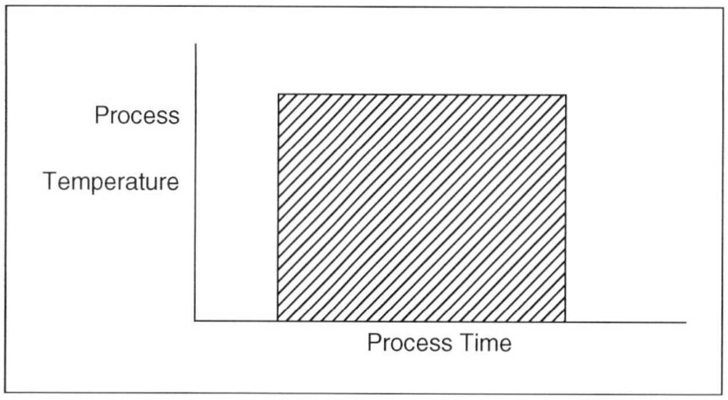

Figure 1 Graphic image of the cycle produced by a typical BIER vessel showing the square-wave cycle.

In a perfect world, the *F* calculation would be fairly simple. In that perfect world, the sterilizer chamber would come up to temperature immediately, the cycle would run precisely at that temperature for the desired time period, following which the temperature would immediately drop to ambient. The *F* value could then simply be the number of minutes the cycle ran at the process temperature times, a value called the lethal rate (see later) for that temperature. Unfortunately that perfect world does not exist. There are, however, some small sterilizers designed to minimize the come-up and come-down times associated with steam sterilizers. These units, known as mini-retorts, or BIER vessels, are miniaturized versions of production-sized sterilizers, so although the thermodynamics of microbial destruction are comparable, their load capacities are very limited. Consequently, these units have a minimal come-up and come-down time and are generally referred to as a square-wave unit from the shape of the graphic image of the cycle (Fig. 1).

Laboratory- and production-sized sterilizers are generally of a size for which the come-up time for the chamber temperature is relatively slow, primarily due to the necessity for temperature uniformity within the chamber and the mass of the load. If the load consists of a large volume of liquid, the thermodynamics of transferring the heat energy from the steam into the load by means of conduction can be a long procedure. A primary feature of the *F* value is the "compensation" it provides for these lower temperatures experienced during the come-up and come-down periods. The *come-up* time is the phase in the process when the sterilizer has been turned on, but the load has not yet reached the designated temperature. The *come-down* time is that period of time when the cycle has been terminated and the temperature is descending from the designated operating temperature (Fig. 2).

The *F* calculation takes into consideration the additional thermal microbial destruction derived from these periods; temperatures at these times are not at the optimum processing temperature, but nevertheless their effects contribute a detrimental effect to the microbial bioburden. These detrimental effects are not really significant below the temperature of 100°C (as will be seen later in the lethal rate calculation) so no "credit" is achieved unless the chamber or product temperature is at least 100°C (212°F). As the process temperature of the sterilizer continues to increase with time during

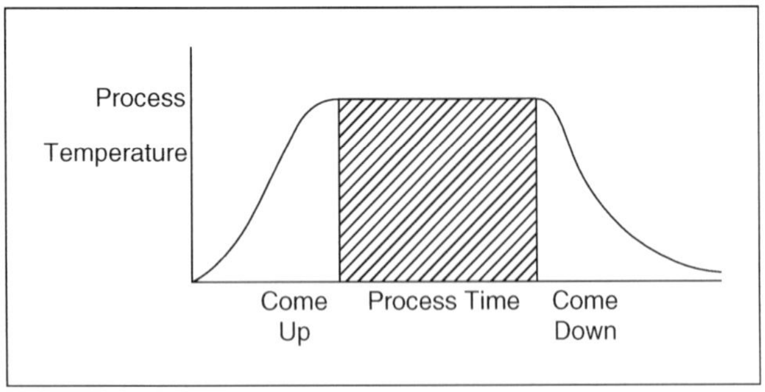

Figure 2 Graphic image of the cycle produced by a typical sterilizer. The time required for the load or chamber to reach maximum operating temperature and to cool off to ambient is usually a function of the chamber load.

the come-up phase, the F_0 value of each incremental time period also increases and is added to a running total for the process up to that point in time.

Lethal Rate

To understand the development of the F_0 calculation, it is necessary to first understand the primary element used in the determination of lethality. This term is known as the *lethal rate*.

The lethal rate is defined as the equivalent time for any specific temperature relative to another temperature. *Example*: the lethal rate for 117.0°C relative to 121.1°C (assuming a z value of 10) is 0.398. This means that for every full minute (60 seconds) of process time at a temperature of 117°C, the process is "credited with" the equivalent of only 0.4 minute at 121.1°C.

This sounds reasonable but how does one determine this lethal rate for every temperature likely to be encountered in the course of process development and validation? Actually, it is very simple. The lethal rate value can be calculated quite easily by the following formula, or could simply be referenced from one of the numerous publications in which the lethal rates have been calculated (1,25,26):

$$L = \log^{-1}\frac{T_0 - T_b}{z} = 10^{(T_0 - T_b)/z}$$

where T_0 is the temperature within the item being heated; T_b, the reference temperature; z, 10°C (assumed).

As an example, assume a process that shows a consistent minute-by-minute temperature of 118°C in a product vial being processed in a terminal sterilization cycle in which the sterilizer is maintained at 121.1°C. To examine the microbial lethality within that particular unit, it is necessary to know the lethal rate for 118°C:

$$L = \log^{-1}\frac{T_0 - T_b}{z} = 10^{(T_0 - T_b)/z}$$

$$L = 10^{(118-121.1)/10}$$

$$L = 10^{-3.1/10}$$

$$L = 10^{-0.31}$$

$$L = 0.489$$

Therefore, for every minute of the process in which the contents of the subject vial remain at 118°C, 0.489 minute of equivalent time at 121.1°C is added to the total accumulated process.

What is the lowest temperature that can be used for lethal rate? Temperatures below 100°C generally add insignificant credit to the overall sterilization assessment. For example, the lethal rate calculation for 100°C:

$$L = \log^{-1}\frac{T_0 - T_b}{z} = 10^{(T_0 - T_b)/z}$$

$$L = 10^{(100-121.1)/10}$$

$$L = 10^{-21.1/10}$$

$$L = 10^{-2.11}$$

$$L = 0.008$$

The formula can also be used to calculate the lethal rate in terms of degrees Fahrenheit. Substituting the values in the previous example for their equivalent in °F results in the same calculated lethal rate value of 0.008:

$$L = \log^{-1}\frac{T_0 - T_b}{z} = 10^{(T_0 - T_b)/z}$$

$$L = 10^{(212-250)/18}$$

$$L = 10^{-38/18}$$

$$L = 10^{-2.11}$$

$$L = 0.008$$

There is no upper limit for the lethal rate calculation. An F value can be calculated for any temperature above 100°C; however, in all practicality, this calculation may be limited only by the capability of the scientific calculator being used. For example, a number such as 150.0°C, which is nearly 30° hotter than the reference temperature of 121.1°C, shows an extremely high lethal rate:

$$L = \log^{-1}\frac{T_0 - T_b}{z} = 10^{(T_0 - T_b)/z}$$

$$L = 10^{(150.0-121.1)/10}$$

$$L = 10^{28.9/10}$$

$$L = 10^{2.89}$$

$$L = 776.2$$

What is that figure of 776.2 saying? It says that every minute of process time at 150.0°C is equivalent to 776.2 minutes (or 12.9 hours) at 121.1°C. Seems kind of far-fetched, but sterilizers are not usually designed to operate at this temperature.

The same formula can be used for calculating the lethal rate of a dry-heat sterilization process, the only difference being the reference temperature and assumed z value which, for dry heat, are 170°C and 20°C, respectively. For example, a dry-heat sterilization process that operates at 155°C:

$$L = \log^{-1}\frac{T_0 - T_b}{z} = 10^{(T_0 - T_b)/z}$$

$$L = 10^{(155-170)/20}$$

$$L = 10^{-15/20}$$

$$L = 10^{-0.75}$$

$$L = 0.18$$

Therefore, every minute at the process temperature of 155°C is equivalent to 0.18 minute at 170°C.

Mathematical F_0

In mathematical terms, F_0 is expressed as follows:

$$F_0 = \Delta t \times 10^{(T-121.1)/z}$$

where Δt is time interval between measurements of T; T, temperature of the sterilized product at time t; z, z value.

What this says is that the F_0 value is equal to the number of minutes of exposure time at the observed temperature (Δt) times the log of the value calculated by subtracting the reference temperature from the observed (process) temperature and dividing this quotient by the z value.

The best way to understand this calculation is to run through a few examples. For the sake of simplicity, assume these cycles were conducted in a BIER vessel, so come-up and come-down times were negligible.

For a process that ran for 12 minutes at exactly 121.1°C:

$$F_0 = \Delta t \times 10^{(T-121.1)/z}$$

$$F_0 = 12 \times 10^{(121.1-121.1)/10}$$

$$F_0 = 12 \times 10^0$$

$$F_0 = 12 \times 1$$

$$F_0 = 12 \text{ minutes}$$

For a process that ran for 12 minutes at exactly 120.1°C

$$F_0 = \Delta t \times 10^{(T-121.1)/z}$$

$$F_0 = 12 \times 10^{(120.1-121.1)/10}$$

$$F_0 = 12 \times 10^{-1.0/10}$$

$$F_0 = 12 \times 10^{-0.10}$$

$$F_0 = 12 \times 0.79$$

$$F_0 = 9.48 \text{ minutes}$$

For a process that ran for 12 minutes at exactly 119.1°C:

$$F_0 = \Delta t \times 10^{(T-121.1)/z}$$

$$F_0 = 12 \times 10^{(119.1-121.1)/z}$$

$$F_0 = 12 \times 10^{-2.0/10}$$

$$F_0 = 12 \times 10^{-0.20}$$

$$F_0 = 12 \times 0.63$$

$$F_0 = 7.56 \text{ minutes}$$

It should be fairly evident that a pattern is emerging: the lower the process temperature, the lower the F_0 value of the overall process. The same calculation applies if the process temperature is above the reference temperature.

For a process that ran for 12 minutes at exactly 122.1°C:

$$F_0 = \Delta t \times 10^{(T-121.1)/z}$$

$$F_0 = 12 \times 10^{(122.1-121.1)/z}$$

$$F_0 = 12 \times 10^{1.0/10}$$

$$F_0 = 12 \times 10^{0.10}$$

$$F_0 = 12 \times 1.26$$

$$F_0 = 15.12 \text{ minutes}$$

This same F-value calculation can be used to express equivalent time relative to any temperature. If, for example, the requirement is to determine the $F_{115.0°C}$ value for a cycle in which the temperature ran a slightly cooler 114.8°C. In this case, the formula for lethal rate would be the same as before, the only exception being the use of 115.0 instead of 121.1 as the reference temperature:

$$L = \log^{-1}\frac{T_0 - T_b}{z} = 10^{(T_0 - T_b)/z}$$

$$L = 10^{(114.8-115)/10}$$

$$L = 10^{-0.2/10}$$

$$L = 10^{-0.02}$$

$$L = 0.954$$

In this case, it is easy to see that a process that has run slightly less than the reference temperature would "acquire" nearly the full 1.0 F value. If the temperature runs slightly hotter at 115.6°C, what then is the $F_{115.0}$ value?

$$L = \log^{-1}\frac{T_0 - T_b}{z} = 10^{(T_0 - T_b)/z}$$

$$L = 10^{(115.6-115.0)/10}$$

$$L = 10^{0.6/10}$$

$$L = 10^{0.06}$$

$$L = 1.148$$

Again, the 1.148 minutes equivalent time for a process 0.6°C hotter than the reference temperature is reasonable. The total $F_{115.0}$ for this process would simply be the Δt times the lethal rate. If using the "near-perfect world" BIER unit and the process ran for 30 minutes, the $F_{115.0}$ in the latter example would be:

$$30\ (\text{minutes}) \times 1.148\ (\text{lethal rate}) = 34.44$$

or somewhat more than that which is necessary to "claim" a minimum $F_{115.0}$ value of 30. (In the example with the temperature of 114.8°C, the calculation would be $30 \times 0.954 = 28.62$, or slightly less than the $F_{115.0}$ of 30.)

Use of Calculated F_0

Why is F_0 used to describe the sterilization process? The answer lies in the fact that the world of sterilization is not perfect so there are truly variations in even the most precisely controlled cycles. For sterilization processes, there is a need to more precisely characterize the capability of the process to enable the process engineer to compare the resultant effects of one cycle with other process cycles with the same intended outcome, or to compare the results with a predetermined level that has been defined as "adequate sterilization."

Most modern sterilizers are equipped with microprocessors that not only serve to control the time, temperature, vacuum, and pressure but also integrate the lethal rate calculations into real time F_0 determinations, and either report this value along with the other parameters or use this information to control the termination of the cycle. By insertion of the controllers' RTD or thermocouple into a representative product (or simulated product) unit that has been shown to be equivalent to the remainder of the load, the sterilizer operator (or the microprocessor itself) can visualize the on going status of F_0 accumulation in a typical unit and can terminate the process once the desired level of F_0 is achieved. By doing so, the variabilities of temperature, load pattern, container penetration, etc. are all negated because the operator needs only to monitor the primary objective of this process—delivery of sufficient microbial lethality to the representative product unit.

In the past few years, many companies have been evaluating and using terminal sterilization for products that may be considered somewhat "heat sensitive." Much of this activity is in response to the FDA issuance of their proposed rule in which they served notice of their preference for terminal sterilization over aseptic processing. To do this, manufacturers must now consider terminal sterilization for all products, not just those products that have traditionally been insensitive to the rigorous thermal stresses of an autoclave.

These terminal sterilization cycles for heat-sensitive products must be precisely engineered to avoid overprocessing (as well as underprocessing) the product. Some sterilizers, such as those that operate in the manner of a superheated water shower, may have more ability to control product unit temperatures during come-up and come-down to avoid the addition of heat lethality during these periods. Traditional steam autoclaves can be used for these products, but they generally do not have the capability to control product temperature as well, especially if the target range is relatively narrow. In either case, a process must be designed that will allow for the imprecision of the sterilization equipment such that it delivers, with some degree of assurance, a uniform measure of heat lethality to every unit within the load.

In these situations of marginal terminal sterilization suitability, the process designer must first determine the level of terminal sterilization to which the product can be safely exposed. This is usually done through stability studies wherein groups of representative units of the product would be exposed to a range of F_0 levels, then tested for quality attributes normally impacted by excessive heat, such as potency, development of degradants, particulates, color, etc.

Once this level has been determined, a decision on the routine amount of exposure must be made. The process should be designed such that it remains within the validated limits, yet allows for unforeseen variations in the cycle to avoid overprocessed product. Some pharmaceutical manufacturers have elected to employ a routine terminal sterilization cycle at a level of roughly one-half the validated safety level, on the premise that if the initial cycle is interrupted for any reason (power failure, inadvertent termination, etc.), a second cycle may be performed and the product would not be exposed to lethality greater than that which has been validated. In these processes, variations in load size could likely have a significant influence on the lethality imparted during come-up and come-down times, so provisions need to be made to accommodate these variations. This approach potentially subjects the product to increased heat and is therefore somewhat limited in its application by product stability considerations.

Determination of Minimum Required F_0

Quite often the process designer will need to know precisely how much F_0 to provide for in a new sterilization cycle to meet a desired sterility assurance objective. In this case, that specific F_0 amount can be determined by evaluating the desired level of sterility assurance required, together with the bioburden of the product being sterilized, and the resistance of indigenous microorganisms in the bioburden using the formula:

$$F_0 = D_{121.1}(\log_{10}A - \log_{10}B)$$

where $D_{121.1}$ is D value (at 121.1°C) of the bioburden; A, bioburden per container; B, maximum acceptable SAL.

Example. The product being sterilized has a bioburden of 100 spores per container, the D value of the spore is 3.3 minutes, and the desired SAL is 10^{-6}, i.e., assurance that no more than 1 unit in 1 million units will be non-sterile:

$$F_0 = D_{121.1}(\log_{10}A - \log_{10}B)$$

$$F_0 = (3.3)[2-(-6)] = 26.4 \text{ minutes}$$

Therefore, for this sterilization process to achieve the desired destruction of all 100 spores to the extent of a 6-log SAL, all units in the batch must receive at least 26.4 F_0, i.e., the equivalent of 26.4 minutes at 121.1°C.

Calculation of Delivered Process Lethality

Now that the lethal rate can be calculated, this value can be used to determine the amount of lethality delivered in the process using the following formula based on the assumption that the lethal effect obtained at different temperatures is additive (2):

$$F_0 = \sum_{T=1}^{x} L\,dt$$

where $T=1$ is the first time increment with an F_0 value, and x is the last time increment with an F_0 value.

Integration of the lethal rates can best be calculated using the Trapezoidal Rule Computer Program that measures the area under the curve by dividing it into equally spaced parallel cords. The lengths of the cords are $y_1, y_2, y_3, \ldots, y_n$ and the distance between the cords d is the time between successive temperature measurements (27).

Is it really necessary to do all of these calculations manually? The answer is a resounding no. Most modern data loggers come with mathematical packages that can be programmed to perform the calculations automatically. Most of them can also be formatted to calculate and report the incremental F value for the current reporting interval, and the cumulative F value for the process up to that specific time point. From a mathematical perspective the tabulation is simply an expression of the cumulative F value—in this example, in Table 1, the sterilizer started with a cool chamber and was programmed to run a five-minute cycle at 121.1°C.

The accumulated lethal rate will add each incremental F_0 value to the running total, thus allowing the process operator (or the properly programmed microprocessor) to visualize the accumulating total. Most modern data loggers and many sterilizer control systems have this capability.

Table 1 Example of the Minute-by-Minute Temperature Observations of a Sterilizer Cycle with a 5-Min Hold Time Along with the Incremental and Cumulative F (Lethal Rate) Values

Time (min)	Temperature (°C)	Lethal rate	Total lethal rate
1	25.0	0.000	0.000
2	55.0	0.000	0.000
3	85.0	0.000	0.000
4	100.0	0.008	0.008
5	110.0	0.079	0.087
6	115.5	0.282	0.369
7	119.0	0.631	1.000
8	121.1	1.000	2.000
9	121.1	1.000	3.000
10	121.1	1.000	4.000
11	121.1	1.000	5.000
12	121.1	1.000	6.000
13	116.0	0.316	6.316
14	108.0	0.200	6.516
15	96.0	0.000	6.516
16	80.0	0.000	6.516
17	65.0	0.000	6.516
18	42.0	0.000	6.516
19	30.0	0.000	6.516
20	25.0	0.000	6.516

Rahn Semilogarithmic Survivor Curve

Simple *linear* rates of change are usually represented by a constant rate of change, e.g., $2X$, where whatever the value of X is, its value is doubled. *Nonlinear* or geometric rates of change are represented multiplicatively or exponentially, e.g., 2^X, where whatever value X is, is the number of "times 2" multiplied together. When microorganisms replicate, they increase in an exponential manner, i.e., each parent has two siblings that go on to become parents that further subdivide and so on. Between any given parent and sibling, the relation is linear ($2X$), but the overall population grows at an exponential rate of 2^X where X is defined as a generation. The Rahn Semilog Survivor Curve represents the exponential relationship that exists between spore survivorship and time at lethal temperature (28).

Before delving into the mathematics of thermal destruction of microorganisms, it may help to first review the principles of logarithms. Because this microbial growth and destruction occur logarithmically, it is important to have at least a general understanding of the fundamentals.

A *logarithm* is defined as an exponent of a stated number called the base and is used to represent powers of that base (29). Logarithms were first invented in the 17th century to help simplify the arithmetical processes of multiplication, division, expansion to a power, and extraction of a root, but they are now used extensively for many other purposes in applied mathematics.

The method of logarithms can be illustrated by considering a series of powers of the number 2: 2^1, 2^2, 2^3, 2^4, 2^5, and 2^6, corresponding to the series of numbers 2, 4, 8, 16, 32, and 64 (2 to the "1" power is 2, 2 squared is 4, 2 cubed is 8, ...). The exponents 1, 2, 3, 4, 5, and 6 are *logarithms* of these numbers to the base 2. Each logarithm consists of a whole number and a decimal fraction, referred to as the *characteristic* and *mantissa*, respectively. By mathematical processes, the logarithm of any number to any base can be calculated, and exhaustive tables of these logarithm calculations have been prepared.

The most commonly used and easiest to understand system of logarithms has the base 10. In this system, the logarithm is an exponent of the number 10. As an example, the logarithm of the number 5 has the characteristic 0 and the mantissa 0.69897 and is thus written 0.69897. The logarithm of the number 50 is 1.69897, the logarithm of the number 500 is 2.69897, and so on. The logarithm of the number 1 is 0 and the logarithm of any number less than 1 is a negative value, e.g., the logarithm of 0.5 is -0.3010. Because any practical system for treating microbial destruction must be relatively simple in concept and use, logarithms to the base 10 are used because they are easier for the nonmathematician to understand and use.

Thermal degradation of microorganisms by means of steam sterilization has been experimentally shown to obey the laws of mass action (26) and chemical kinetics (30). The primary interest is in killing microorganisms and in the number that die, but what is more interesting is the number of microorganisms that actually survive. Using N to indicate the number of viable (surviving) microorganisms present in the system at any given time, the change in the number of viable spores with time is a function of the number of viable spores present and can be represented in mathematical terms as follows:

$$\frac{dN}{dt} = -KN$$

where K is a constant that is typical of the species and conditions of the chosen microorganism. The degradation (sterilization) reaction develops similar to a first-order chemical reaction in which the reaction rate is proportional, at each moment in time, regardless of the number of microorganisms remaining to be degraded or decomposed. Therefore, there is a constant percentage reduction of viable microorganisms for each arbitrary multiple of time t. The time required, then, to reduce the microorganism challenge population to any preset value is a function of the initial concentration of that microorganism.

Having expressed the number of survivors in equation form, the equation can then be rearranged into differential equation form and integration of the differential equation is as follows:

$$\frac{dN}{dt} = -K - dt$$

and by converting to base 10 logarithm:

$$\log N = -Kt + C$$

At time zero, $t=0$ and $N=N_0$; therefore $\log N_0=C$. The final equation can then be derived:

$$\frac{N_t}{N_0} = 10^{-Kt}$$

where N_t is number of microorganisms at time t; N_0, initial number of microorganisms; K, reaction time constant (which depends on the species and condition of the microorganism); t, reaction time of steam sterilization, which simply says that the number of survivors of a steam sterilization process will decrease in an exponential (geometric) manner. As shall be seen later, this exponential decrease is not quite the same as the exponential growth seen during the development of the spore crop. If something is known about the resistance of a microorganism, more specifically a BI, and D can be defined as the time it takes to kill 1 log of this BI, then using the foregoing model, $1/D$ (or $-1/D$) can be substituted for K. A more resistant BI ($D \gg 1$) will have more survivors; therefore, the relation between K and D is inverse. Furthermore, if the formula is standardized by making $N_0=1$, then $N_t=10^{-1/D}$.

Table 2 Probability of Survival of Process Times from 0.01 to 10 Min with *D* Values Ranging from 100 to 0.1 Min

	N_t			
Time (min)	$D=100$	$D=10$	$D=1$	$D=0.1$
0.01	0.999	0.997	0.977	0.794
0.1	0.997	0.977	0.794	0.100
0.2	0.995	0.954	0.630	0.010
0.4	0.990	0.912	0.398	0.0001
0.6	0.986	0.870	0.251	0.000001
0.8	0.981	0.831	0.158	0.00000001
1	0.977	0.794	0.100	1.00×10^{-10}
2	0.954	0.630	0.010	1.00×10^{-20}
3	0.933	0.501	0.001	1.00×10^{-30}
4	0.912	0.398	0.0001	1.00×10^{-40}
5	0.891	0.316	0.00001	1.00×10^{-60}
6	0.870	0.251	0.000001	1.00×10^{-60}
7	0.851	0.199	0.0000001	1.00×10^{-70}
8	0.831	0.158	0.00000001	1.00×10^{-90}
9	0.812	0.125	1×10^{-9}	1.00×10^{-90}
10	0.794	0.100	1×10^{-10}	1.00×10^{-100}

Example. By using this formula of $N_t=10^{-1/D}$ for a range of D values from 10 to 0.1, a process time ranging from 0.01 to 10 minutes, and $N_0=1$, a probability of survival may be created as shown in Table 2.

Graphically, this probability of survival (N_t) can be displayed as shown in Figure 3.

D VALUE

Introduction

Before discussion of the various aspects of the D value, it may help to take a moment to examine the fundamental principles involved with the death of a microorganism. Extensive research on the subjects of disinfection and sterilization has provided a great deal of information on the death of microorganisms. Microbiologists differ in their views on the most essential properties of living organisms, but the criterion almost universally used to

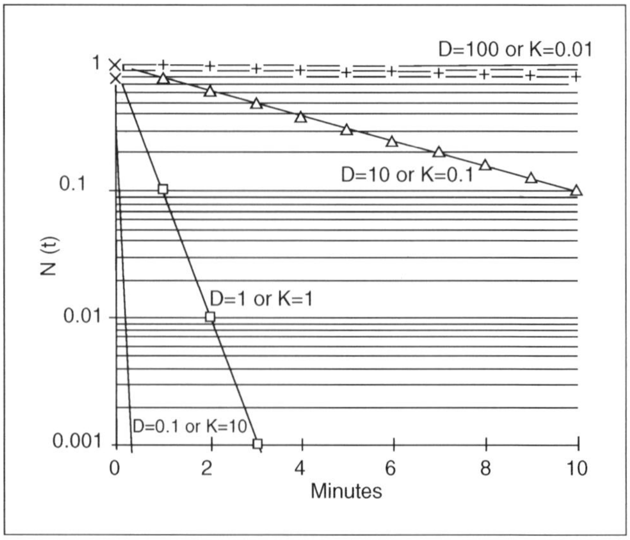

Figure 3 Graphic representation of the probability of survival of process times from 0.01 to 10 minutes with *D* values ranging from 100 to 0.1 minutes.

define the death of a microorganism is *the failure to reproduce when suitable conditions for reproduction are provided* (26,31).

For the microbiologist, this would mean the failure of a transferred isolate to produce colonies on a solid culture medium, or the inability of the isolate to render a liquid culture medium turbid. This is obviously not true with all living organisms, for there are many living and breathing individuals whose inability to reproduce has absolutely no bearing on their definition of "living." The primary reason for this difference lies in the fact that this is a comparison of *uni*cellular survivors versus *multi*cellular survivors.

The challenge microorganism chosen for this task is usually a bacterial spore, because they are among the most resistant microorganisms to wet- and dry-heat destruction. Nonsporing bacteria are more sensitive to inactivation, along with the vegetative forms of yeasts and molds. The larger viruses show resistance similar to vegetative bacteria. Spores formed by molds and yeasts are generally more sensitive to inactivation than bacterial spores (12). Bacterial spores are also much more stable than vegetative cells, an important consideration for commercial development in that the spore crop may be cultivated, counted, evaluated for resistance, packaged, and distributed without concern for variation within the spore lot. Bacterial spores are also representative of the environmental bioburden, for they are widely distributed in soil and are found at times in air samples.

Most of the resistant microbiological spores known are found in two genera: *Bacillus*, an aerobe that may be facultatively anaerobic, and *Clostridium*, which is usually a strict anaerobe. The most commonly used heat-resistant species are *B. subtilis*, *G. stearothermophilus*, *Bacillus coagulans*, *C. sporogenes*, and *Clostridium thermosaccharolyticum*. *B. pumilus* is commonly used to monitor radiation sterilization and *B. subtilis* var. *niger* is commonly used to monitor ethylene oxide and dry-heat sterilization owing to their respective resistance to those methods of sterilization.

The *D* value is a term used to describe the relative resistance of a particular microorganism to a sterilization process. D_T (or $D(T)$) is defined as the time required at temperature *T* to reduce a specific microbial population by 90%, or, as the time required for the number of survivors to be reduced by a factor of 10. The letter "*D*" stands for the fact that *D* value is also referred to as the decimal reduction time, the word *decimal* being defined by most dictionaries as "pertaining to or founded on the number 10." There are several ways to visualize this graphically. One would be to use simply straight arithmetic graph paper; this would require a rather tall piece of graph paper, however, because one axis would be disproportionately more extensive than the other axis owing to the extensive range of challenge microorganism populations necessary to develop a thermal death pattern. A second way would be to use the same arithmetic graph paper but simply plot the log value of the challenge population.

The third, and preferred, method is to use semi-logarithmic paper on which the *y*-axis is arranged in logarithmic format and the *x*-axis is arranged in arithmetic format. This type of graph paper enables the user to envision a reaction that causes a substantial change (at logarithmic proportions) over a constant rate of time.

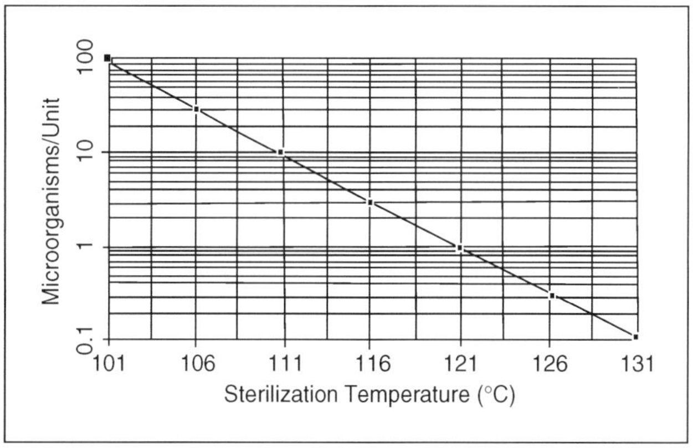

Figure 4 Typical thermal death survivor curve on semilog graph paper with log level of survivors on the *y*-axis and temperature on the *x*-axis.

As shown by the graph in Figure 4, this semi-logarithmic arrangement allows for the use of two data factors in relative proportion, despite the fact that one factor would take up considerably more space if both were plotted on simple arithmetic graph paper.

The *D* value can also be determined from a straight line on semilog paper as the negative reciprocal of the slope of the line fitted to the graph of the logarithm of the number of survivors versus time (26). One cycle on the logarithmic scale represents a 10-fold change in the number of survivors; therefore, the *D* value is the time for the straight line to traverse 1 logarithmic cycle (Fig. 5).

This *D*-value time element is a critical parameter used in both the validation of a process as well as in the routine monitoring of validated processes. Selection of a microorganism with the appropriate *D* value for the intended application should be performed only by a qualified microbiologist.

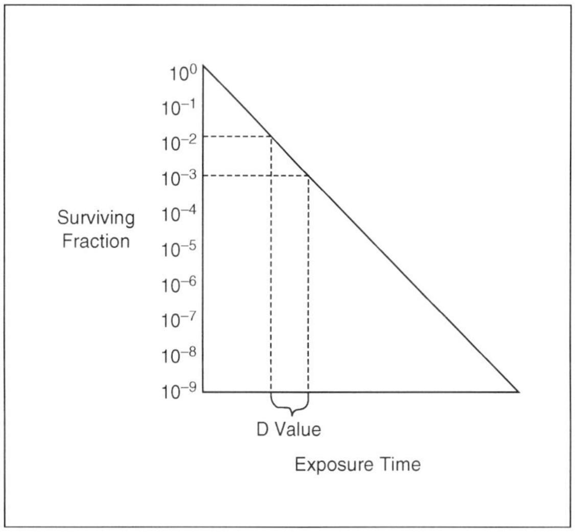

Figure 5 Simplified method for graphic estimation of *D* value using log level of the survivors on the *y*-axis and exposure time on the *x*-axis. The *D* value is the time required to reduce the number of survivors by one log.

Cultivation of Bacterial Spores

Cultivation of bacterial spores differs from that of vegetative cells. The bacterial spore originates from a vegetative cell, wherein a unique process takes place to produce a profound biochemical change that gives rise to the structure known as a spore or endospore. This process is not a part of the reproductive cycle, and it has been characterized as the ultimate example of adaptation of bacteria to starvation. The resultant bacterial endospore is highly resistant to adverse environments, such as lack of moisture or essential nutrients, toxic chemicals, radiation, and high temperatures (32).

Although sporulation has been studied for decades, it has only been in the last few years that significant progress has been made in isolating the developmental genes of spores and in unraveling their functions and interactions (33). The process of sporulation is generally induced by starvation: in a good growth medium the vegetative rod-shaped cells proliferate and multiply by doubling in length followed by a central division to produce two identical daughter cells; sporulation begins with an asymmetric division producing these two daughter cells that differ markedly in size. The smaller of the daughter cells is called the prespore and is engulfed by the mother cell. When this engulfment has just been completed, the membranes surrounding the cytoplasm of the prespore (now called the forespore) take on a very amorphous appearance and gradually start to develop the oval shape of the mature spore. A modified form of a cell wall known as the cortex begins to develop at the same time a proteinaceous spore coat begins to be deposited on the outside surface of the spore. In the final development step, the characteristics of resistance, dormancy, and germinability begin to appear in sequence, and the spore is released by lysis of the mother cell.

Biologic Indicators

Resistant bacterial spores are commercially available as BIs primarily in the form of:

1. *Spore strips*. A narrow strip of fibrous paper impregnated with a bacterial spore suspension and contained within a glassine envelope. In some processes other substrates, e.g., aluminum, stainless steel, fiberglass, may be used.
2. *Spore dots*. Circular pieces of fibrous paper impregnated with the spore suspension.
3. *Spore suspensions*. A pure spore suspension of the desired challenge microorganism which can then be inoculated onto the surface of a material.
4. Self-contained units containing spores strips and the media in which they are to be incubated, allowing for greater simplicity of use.

These devices are prepared by many vendors and are used extensively in the pharmaceutical industry as challenge microorganisms to monitor sterilization processes and to verify the actual destruction of viable microorganisms. Guidance on their use can be found in the USP (13). The primary advantage of using these vendor-supplied materials is that the resistance profile of the microorganism is well characterized and, if used as intended, will provide a realistic representation of the survivability of a highly resistant microorganism under conditions designed to stress and incapacitate it.

Users of BIs purchased from a vendor should establish a program designed to verify the manufacturers' label claims of population and resistance on each batch of BIs purchased. It is important to understand that the population verification count *must* be performed in a manner similar to that of the vendor to achieve similar results—some vendors use a blender to break up the fibrous paper of a paper strip, whereas others use sterile glass beads in a sterile test tube and there are other methods as well. It is incumbent on the user to verify the level of challenge population by using the same method as the vendor. Internal limits of acceptance should be established by the user in concert with the vendor; these limits should provide for a fairly wide margin of variability inherent to this type of testing.

The name of the challenge microorganism is usually not as important as its resistance. Traditionally, *G. stearothermophilus* is used in processes that rely on saturated steam, and *B. subtilis* var. *niger* is used to monitor processes that use ethylene oxide and dry heat. In both applications, the spore is usually selected based on its resistance to the process sterilant and the assumption that all other microorganisms are less resistant. Other spores or other microorganisms are acceptable for use as process monitors, provided their resistance characteristics to the process sterilant have been thoroughly examined.

Another item to consider is the application of the challenge spore. *D* values for spore strips are determined by exposure of the paper strip (still contained within the glassine envelope) to the sterilization process in perhaps a "best-case" situation, i.e., the spore strip is directly exposed to the sterilant. As a routine process monitor then, spore strips should not be used in a manner in which the sterilant can more easily kill the spore, e.g., without the glassine envelope, to avoid false-negative results. Most users will place the spore strip in a predetermined worst-case location, on the premise that destruction of the spore under these extreme circumstances can assume destruction of spores in all other less-challenging circumstances.

D values for most commercial spore suspensions are determined by applying the spore to a paper strip that is then tested for resistance just like all the other spore strip lots. How many people are going to use spore suspensions in this manner? The answer is probably none. Most applications using spore suspensions would usually involve the inoculation of something, e.g., a liquid product, an empty vial, a stopper, etc. Unfortunately, spore suspension manufacturers cannot predict or control this use so they provide the resistance information in unbiased means by inoculating the same type of fibrous paper as used for spore strips and performing the resistance studies as though the object was the spore strip itself. To verify the vendor's reported *D* value, the user would need to inoculate samples of the same paper used by the vendor, and perform *D*-value studies of the inoculated strips.

The user must then also develop resistance data for the spore suspension in the configuration in which it is intended to ultimately be used. For example, if the spore will be used to inoculate liquid product containers, resistance studies should be conducted to examine the thermal resistance of the challenge spores in representative units of this product. There should also be studies conducted to measure the microbiocidal effects of the product itself on the challenge spore to verify minimal spore reduction from simply the product–spore interaction. In the event the product exhibits a sporicidal effect on the spore, an alternative nonsporicidal medium with characteristics similar to the product should be chosen.

If the process is one in which only a narrow range of resistance can be tolerated, adjustments to the resistance to precisely target the desired range may be achieved through modification of the suspending medium. For example, a study reported in 1977 (34), resistance studies using *B. stearothermophilus* at approximately 120.9°C in dextrose 5% in saline, dextrose 5% in lactated Ringer's solution, dextrose 5% in water, WFI, Sorensen's buffer, and Butterfield's buffer showed *D* values of 1.30, 2.12, 2.42, 2.98, 3.36, and 4.70 minutes, respectively; in other words, more than a threefold difference in resistance by simply changing the suspension solution. In the same study, *C. sporogenes*, at approximately 105°C in the same solutions as the foregoing showed *D* values of 2.68, 1.14, 1.34, 13.7, 42.6, and 21.2 minutes, respectively, nearly an eightfold difference! These two datasets demonstrate the wide range of resistance characteristics available by using different suspending solutions as well as the lack of correlation among challenge microorganisms, thereby reinforcing the necessity to verify the challenge microorganism under identical conditions to those in which it is to be used.

Kinetics of Microbial Inactivation

When homogeneous populations of microorganisms are exposed to a lethal process, they lose their viability in a consistent manner. The rate of this inactivation is directly proportional to the number of microorganisms present at any given time and thus a constant portion of the surviving population is inactivated for each increment of exposure to the lethal agent. Mathematically, the inactivation process can be described in the same way as a first-order chemical reaction (12):

$$N_t = N_0 e^{-kt}$$

where N_t is number of surviving microorganisms after time t; N_0, number of microorganisms at time zero; k, microbial inactivation rate constant; t, exposure time.

If the logarithm of the fraction of survivors (N/N_0) is plotted against exposure time, the resulting survivor curve will be linear, with a negative slope (Fig. 6). The slope of the line is $k/2.303$ from which the microbial inactivation rate can be calculated.

D-Value Determination

In sterilization microbiology, the *D* value is frequently used instead of *k* as a measure of the rate of microbial death. Again, the *D* value is the exposure time required for the number of survivors to change by a factor of 10, or the time for the microbial population to be reduced by 90%. The *D* value will vary with

1. The challenge microorganism
2. Type of suspending medium
3. Process temperature

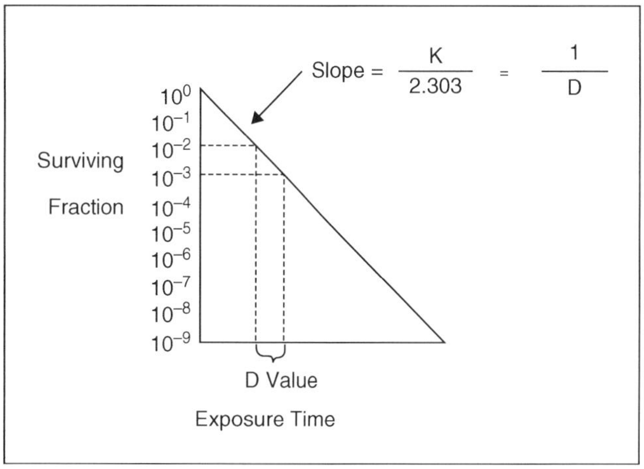

Figure 6 Simplified method for graphic estimation of the slope of the survivor curve and *D* value using log level of the survivors on the *y*-axis and exposure time on the *x*-axis.

4. Holding time of the BI after sterilization
5. Preconditioning of the BI
6. Outgasing of BIs used in gaseous sterilization

There are several means with which to determine the *D* value. It may be estimated from the previous graph using the equation:

$$D = \frac{t}{\log N_0 - \log N_t}$$

or

$$D = \frac{2.303}{k}.$$

There are two primary mathematical models used for the experimental determination of *D* value: the Spearman–Karber method (sometimes referred to as the Holcomb–Spearman–Karber method) in which the analysis is based on the mean time to survival, and the Stumbo, Murphy, and Cochran method which uses the MPN approach. Both of these methods require data in the quantal range with a mix of positive and negative results, while the Spearman–Karber method requires the results to be in the form of a dichotomous response: at short heating times all replicates show growth, at longer heating times both positive and negative units result from the same heating time, and at still longer heating times all results are negative.

In the *D*-value studies it is essential to provide conditions in which consistency from cycle to cycle can be assured. The ideal mechanism for providing this consistency is one that will produce a square-wave format, with minimum come-up and come-down times. This unit may also be referred to as a mini-retort or a BIER unit (as previously described); various models and styles are available from several manufacturers. If one of these units is not available, a small-chambered laboratory sterilizer could be used; in this case, it is more of a challenge to maintain consistency between subsequent cycles, but it can be accomplished.

The first step in the process is to ensure an adequate supply of the inoculated spore challenge. Prior knowledge of the expected lethality of the spore is helpful in targeting for the quantal range, because total kill or total survival of all units in the individual groups should be kept to a minimum. Groups of the inoculated challenges (typically 5 to 10 units per group is considered adequate) are then exposed to the same stress environment within this sterilizer in a pattern of increasing exposure times. Following these exposures, the subject units can be cultured in one of two ways:

1. The individual units can be suspended in a diluting medium, diluted to the appropriate level, and cultured. The estimated number of survivors can be determined by multiplying the number of colony-forming units on the plates by the dilution factor and the D value then determined by the survivor curve method.
2. Alternatively, the individual units can simply be cultured in individual containers of an appropriate growth medium and evaluated as either positive (growth) or negative (no growth) and the D value determined by the fraction-negative method.

Survivor Curve Method

If the number of challenge units is small, enumeration of the survivors on a survivor curve may be the best way to proceed. A survivor curve is a plot of the logarithm of the number of surviving microorganisms on the y-axis versus the heating time on the x-axis (35). During a survivor curve test, the number of viable microorganisms is reduced by the sterilizing agent from a large initial population, e.g., 10^6, to a final small population, e.g., 10^1 as determined by plate count. A change in this number of microorganisms cannot be meaningfully presented on an arithmetic scale, so an accepted method of presenting data is to plot the logarithm of the number of microorganisms surviving versus heating time on semilogarithmic graph paper. Researchers have observed that when the logarithm of the estimated number of surviving microorganisms is plotted versus heating time (or the numbers of surviving microorganisms are plotted directly on semilogarithmic paper), the survivor curve is often represented by a single line.

To develop statistical confidence in the results, the studies should be repeated over the course of several days. Linear regression may then be used to determine the slope of the line using the following formula (26):

$$\text{Slope } (k) = \frac{\sum U_i(\log N_i) - [(\sum U_i)(\sum \log N_i)/n]}{\sum U_i^2 - (\sum U_i)^2/n}$$

where U is the equivalent heating time at heating medium temperature; N, the number of survivors; n, the number of (U, log N) values.

Example

No.	Heating time (min) (U)	No. of survivors (N)	log N	(log N)^2	U^2	(U) (log N)
1	10	2.2×10^7	7.342	53.905	100	73.42
2	10	2.3×10^7	7.362	54.199	100	73.62
3	10	2.4×10^7	7.380	54.464	100	73.80
4	15	1.3×10^6	6.114	37.381	225	91.71
5	15	1.1×10^6	6.041	36.494	225	90.62
6	15	1.4×10^6	6.146	37.773	225	92.19
7	20	2.2×10^5	5.342	28.537	400	106.84
8	20	2.1×10^5	5.322	28.324	400	106.44
9	20	2.7×10^5	5.431	29.496	400	108.62
Total	135		56.480		2175	817.26

$$\text{Slope } (k) = \frac{\sum U_i(\log N_i) - [(\sum U_i)(\sum \log N_i)/n]}{\sum U_i^2 - (\sum U_i)^2/n}$$

$$= \frac{817.26 - \frac{135(56.480)}{9}}{2175 - \frac{135(135)}{9}}$$

$$= \frac{817.26 - 847.2}{2175 - 2025}$$

$$= \frac{-29.94}{150} = -0.1996$$

The D value, then, is the negative inverse of the slope:

$$D = -\frac{1}{k} = \frac{-1}{-0.1996} = 5.01 \text{ minutes}$$

Fraction-Negative Method

D-value determinations may also be calculated using the fraction-negative (or survival/kill) method. To do this, an orderly series of heating times is created using separate groups of challenge units subjected to the same environmental stress except the exposure time duration of each group. Ideally, the time intervals should form an ordered set of five to seven durations, e.g., 5, 7, 9, 11, 13, ... minutes, but there are also ways for accommodating for random exposure times.

Spearman–Karber Method

The Spearman–Karber analysis procedure provides an estimate of the mean time until sterility (U_{sk}), or the expected time until a sample containing microorganisms becomes sterile. This method also provides an estimate of the variance of U_{sk}, and the confidence interval around the number. This method requires:

1. The heating times and number of replicates at each heating time be consistent
2. One set with all positive units, another set with all negative units, and several (minimum of two to three is recommended) sets of units in the quantal range.

The general equation for computing the Spearman–Karber value is

$$U_{sk} = U_k - \frac{d}{2} - \frac{d}{n}\sum_{i=1}^{k-1} r_i$$

where U_{sk} is Spearman–Karber estimate; U_k, the first heating time for which all units are negative (at all longer heating times all units are negative); U_1, the longest heating time for which none of the units are negative (at all shorter heating times all units are positive); d, time interval between heating times; n, number of replicates at each heating time; r_1, number of replicates negative at each heating time.

Once the Spearman–Karber estimate is known, the D-value estimate can be calculated using the equation:

$$D = \frac{U_{sk}}{\log N_0 + 0.2507}$$

Example

Cycle no.	Heating time (U)	No. of replicates (n)	No. of replicates negative (r_i)
1	0	10	0
2	2	10	0
3	4	10	1
4	6	10	3
5	8	10	7
6	10	10	9
7	12	10	10
8	14	10	10

$$U_{sk} = U_k - \frac{d}{2} - \frac{d}{n}\sum_{i=1}^{k-1} r_i$$

U_1 is the longest heating time where none of the units are negative=2; U_k, first heating time where all units are negative=12; U_{k-1}, heating time before U_k=10; d, time interval between heating times=2; n, number of replicates at each heating time=10; $\sum_{i=1}^{k-1} r_i$, sum of replicates negative from U_1 to U_{k-1}=20.

$$\frac{d}{2} = \frac{2}{2} = 1 \qquad \frac{d}{n} = \frac{2}{10} = 0.2$$

$$N_0 = 10^8 \qquad \log N_0 = 8$$

$$U_{sk} = U_k - \frac{d}{2} - \left(\frac{d}{n}\sum_{i=1}^{k-1} r_i\right)$$

$$U_{sk} = 12 - 1 - (0.2 \times 20) = 7$$

$$D = \frac{U_{sk}}{\log N_0 + 0.2507} = \frac{7}{8 + 0.2507} = 0.85 \text{ minutes}$$

Stumbo–Murphy–Cochran Method

The Stumbo–Murphy–Cochran method is considered a MPN method of statistical analysis that requires both knowledge of the initial number of microorganisms per replicate, and one or more data results in the quantal range. This method uses the semilogarithmic model

$$D = \frac{U}{\log A - \log B}$$

where U is the heating time at specified temperature; A, initial number of microorganisms on each replicate.

$$B = 2.303 \log\left(\frac{n}{q}\right)$$

where n, total number of replicates; q, number of negative replicates.

The number of surviving microorganisms at each heating time is calculated from the quantal data using a method in which the MPN of surviving microorganisms is calculated from the fraction of replicate units that are negative. When the fraction of replicate units negative is small, the relative numbers of survivors per positive unit are comparatively large. The Stumbo–Murphy–Cochran method uses this information to develop an estimate of the D value at each heating time with dichotomous results.

The Stumbo–Murphy–Cochran method can be used with limited data, but the precision of the estimate is increased by the use of additional heating times, which also have dichotomous results. Pflug (26) has developed a format that can be used to estimate a D value from fraction-negative data by the Stumbo–Murphy–Cochran method:

N_0=___; $\log N_0$=___

Calculating D_1:

$$N_{U_1} = \frac{n}{r} = \ln\_\_\_ = \_\_\_ \quad \log N_{U_1}\_\_\_$$

$$D_1 = \frac{U_1}{\log N_0 - \log N_{U_1}} = \_\_\_ = \_\_\_$$

Calculating D_2:

$$N_{U_2} = \frac{n}{r} = \ln\_\_\_ = \_\_\_ \quad \log N_{U_2}\_\_\_$$

$$D_2 = \frac{U_2}{\log N_0 - \log N_{U_2}} = \_\_\_ = \_\_\_$$

Calculating D_3:

$$N_{U_3} = \frac{n}{r} = \ln\_\_\_ = \_\_\_ \quad \log N_{U_3}\_\_\_$$

$$D_3 = \frac{U_3}{\log N_0 - \log N_{U_3}} = \_\_\_ = \_\_\_$$

$$D_T = \text{average of } (D_1 + D_2 + D_3 \cdots) = \_\_\_$$

Example. A series of cycles was performed with BIs population 10^6 in a BIER vessel at 121.1°C. After each cycle, the BIs were cultured in media and incubated for 48 hours.

Heating time (U)	No. of replicates (n)	No. of replicates negative (r)
2	10	0
4	10	1
6	10	2
8	10	7
10	10	10
12	10	10

$N_0 = 10^6$; $\log N_0 = 6.0$

Calculation D_1:

$$N_{U_1} = \frac{n}{r} = \ln\frac{10}{1} = 2.302 \qquad \log N_{U_1} = 0.362$$

$$D_1 \approx \frac{U_1}{\log N_0 - \log N_{U_1}} = \frac{4}{6.0 - 0.362} = 0.71$$

Calculation D_2:

$$N_{U_2} = \frac{n}{r} = \ln\frac{10}{2} = 1.609 \qquad \log N_{U_2} = 0.207$$

$$D_2 \approx \frac{U_2}{\log N_0 - \log N_{U_2}} = \frac{6}{6.0 - 0.207} = 1.04$$

Calculation D_3:

$$N_{U_3} = \frac{n}{r} = \ln\frac{10}{7} = 0.357 \qquad \log N_{U_3} = -0.448$$

$$D_3 \approx \frac{U_3}{\log N_0 - \log N_{U_3}} = \frac{8}{6.0-(-0.448)} = 1.24$$

$$D_T \approx \text{average of } (D_1 + D_2 + D_3 \cdots) = \frac{0.71 + 1.04 + 1.24}{3}$$
$$= 1.0 \text{ minutes}$$

All of the preceding methods work equally well for all methods of sterilization such as saturated steam, ethylene oxide, dry heat, and irradiation. This is because the method of sterilization usually causes a breakdown in a critical microbiological component or function (such as destruction of the cell membrane) that leads to the inability of the microorganism to reproduce, and ultimately, death. Microorganisms that die as a result of these sterilization methods do so in an orderly, predictable way (31). All of the terms and methods used in the microbiological engineering of wet-heat sterilization processes also apply to dry-heat and ethylene oxide sterilization. For wet heat, the factors of time and temperature are used; for dry heat: time, temperature, and RH; and for ethylene oxide: time, temperature, RH, and gas concentration.

z VALUE

The *D*-value determinations are generally carried out under isothermal conditions, but it has been proven experimentally that the resistance of a microorganism may change with alterations in temperature. This change in the rate of microbial inactivation with a change in the temperature is known as the *z* value (1,12) and is defined as the number of degrees of temperature change necessary to change the *D* value by a factor of 10. The *z* value is only relevant to the thermal sterilization processes; the temperature dependence of radiation and gas sterilization procedures is not widely defined in this manner. It is the *z* value that allows integration of the lethal effects of heat as the temperature changes during the heating and cooling phases of a sterilization cycle; the *z* value is a necessary component of the calculations that allow comparison of the spore lethality at different temperatures.

Although the *z* value is a fundamental characteristic of a microorganism, it is not truly independent of temperature, and is constant only for small temperature differences of the order of 20°C to 25°C. Heat sterilization processes, however, are usually carried out within a small temperature range, e.g., 110°C to 135°C; therefore, the *z* value is usually considered constant (12). In the absence of alternative data, the generally accepted *z*-value assumptions are:

Steam sterilization $z = 10°\text{C}$ (18°F)
Dry-heat sterilization $z = 20°\text{C}$ (36°F)

Theoretically, the *z* value can be determined by one of two methods:

1. Experiments performed at different temperatures to determine the thermal death times
2. Experiments performed to determine *D* values at different temperatures

The first method requires a large number of homogeneous replicate units to be processed at several different *F* values, each at several different temperatures. Survival (the largest *F* value in which at least one replicate is positive) and destruction (the smallest *F* value in which all replicates are negative) times are plotted on semilogarithmic graph paper and a line of best-fit is drawn. Use of this method is rarely mentioned in the literature.

The second (and more popular) method involves comparison of two thermal death resistance curves plotted on semilogarithmic graph paper. *D* values for the microorganism are determined at several (minimum of two) different temperatures, and then plotted on the log scale on the ordinate and the temperature is plotted on the abscissa. A straight line is drawn through the data points; the *z* value is the number of degrees of temperature for the *D* value to change by a factor of 10, e.g., from 3.0 to 0.3, or from 0.7 to 7.0. Because it could change the calculations significantly if used incorrectly, it is always important to consider the temperature scale when using *z* values. If the calculation involves sterilization expressed in Celsius (Centigrade), the *z* value must be in terms of Celsius. If the calculation is in Fahrenheit, the *z* value must be in terms of Fahrenheit.

Example

$$D_{105°\text{C}} = 20.0 \qquad D_{115°\text{C}} = 4.5 \qquad D_{125°\text{C}} = 1.0$$

Once determined, these *D* values are plotted on semilogarithmic paper and a straight line is drawn through the points (Fig. 7).

To determine the *z* value, identify the temperature at two *D*-value coordinates one log apart. For example, according to the graph, the *D* value of 10 falls on the graph at about 109°C and the *D* value of 1.0 falls on the line at 125°C. The difference between these two numbers (125°C − 109°C = 16.0°C) is the *z* value (Fig. 8).

Because this is a straight line, any set of numbers selected on the line will give us the same *z* value.

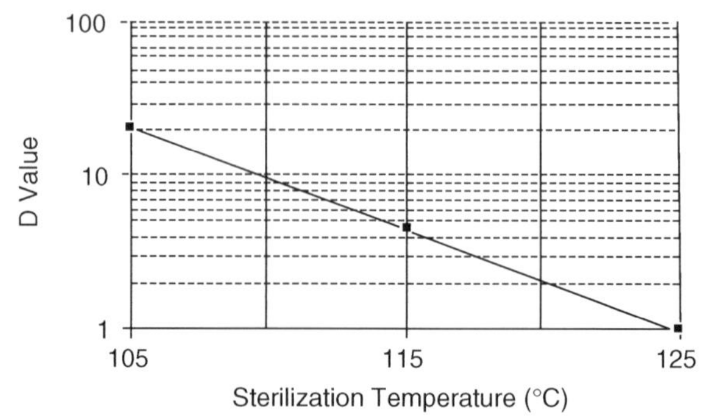

Figure 7 Plot of the *D* value determined at several different temperatures. This response curve is then used to determine the *z* value.

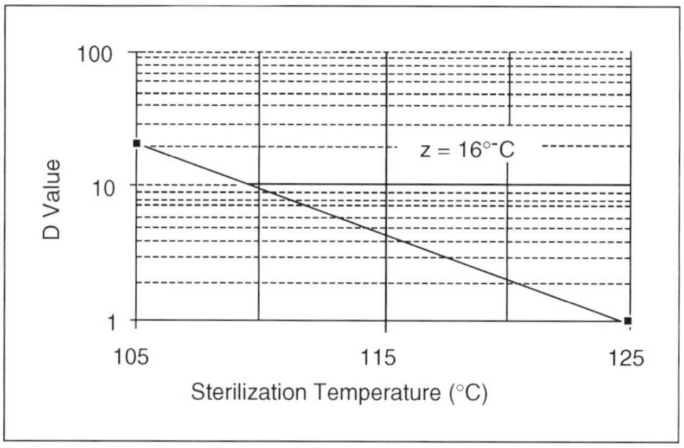

Figure 8 *z*-Value estimation created by plotting the *D* value determined at several different temperatures. As shown by the bold Line, the *z* value is the difference in temperature required to change the *D* value by a factor.

CONCLUSION

The concepts and calculations used for *F*-, *D*-, and *z*-value determinations are best understood if they are used on a regular basis. It is also helpful (and, in many cases, a cGMP requirement) to have the essential details of these calculations clearly described within the format of an approved SOP so that the calculations are used consistently in the same manner each and every time.

This chapter has described the primary calculations that are routinely used for the determination of *F*, *D*, and *z* values. For a more comprehensive experience, one should consider the instructional course entitled "Microbiology and Engineering of Sterilization Processes" which is conducted periodically by Dr. Irving J. Pflug of the University of Minnesota. This course includes a step-by-step examination of all of the elements involved with the use of *F*, *D*, and *z* values and provides the participants with the opportunity to work through numerous example calculations. It also provides the participants with several reference texts that can be valuable resources for the process of sterilization cycle development and quality control.

REFERENCES

1. Parenteral Drug Association. Validation of steam sterilization cycles. Technical Monograph No. 1. Philadelphia, PA: PDA, 1978.
2. Pflug IJ. Heat sterilization. In: Phillips GB, Miller WS, eds. Industrial Sterilization. Durham, NC: Duke University Press, 1973.
3. Angelotti R, Maryanski JH, Butler IT, Peeler JT, Campbell JE. Influence of spore moisture content on the dry-heat resistance of *Baccilus subtilis* var. *niger*. Appl Microbiol 1968; 16.
4. Fox K, Pflug IL. Effect of temperature and gas velocity on the dry-heat distribution rate of bacterial spores. Appl Microbiol 1968; 16.
5. Health Industry Manufacturers Association. Sterilization cycle development. HIMA Rep 78, Washington, DC, 1978.
6. Molin G, Ostlund K. Dry heat inactivation of *Bacillus subtilis* var. *niger* spores with special reference to spore density. Can J Microbiol 1976; 22.
7. Pflug IJ. Sterilization of space hardware. Environ Biol Med 1971; 1.
8. Pheil CG, Pflug IJ, Nicholas RC, Augustin JAL. Effect of various gas atmospheres on destruction of microorganisms in dry heat. Appl Microbiol 1967; 15.
9. Winans L, Pflug IJ, Foster TL. Dry heat resistance of selected psychrophiles. Appl Environ Microbiol 1977; 34.
10. Block SS. Disinfection, Sterilization, and Preservation. Philadelphia, PA: Lea & Febiger, 1983.
11. Parenteral Drug Association. Validation of dry heat processes used for sterilization and depyrogenation. Technical Report No. 3. Philadelphia, PA: PDA, 1981.
12. Silverman GJ. The resistivity of microorganisms to inactivation by dry heat. Contract Ns G-691, Cambridge, MA: Massachusetts Institute of Technology, 1968.
13. United States Pharmacopeia. Easton, PA: Mack Publishing.
14. Tsugi K, Harrison S. Dry heat destruction of lipopolysaccharide: dry-heat destruction kinetics. Appl Environ Microbiol 1978; 35.
15. Association for the Advancement of Medical Instrumentation. Guideline for Industrial Ethylene Oxide Sterilization of Medical Devices. Arlington, VA: AAMI, 1980.
16. Ball CO. Thermal process time for canned foods. Natl Res Counc Bull 1923; 7(Part 1):37.
17. Alasri A, Valverde M, Roques C, Michel G, Cabassud C, Aptel P. Sporicidal properties of peracetic acid and hydrogen peroxide, alone and in combination, in comparison with chlorine and formaldehyde for ultrafiltration membrane disinfection. Can J Microbiol 1993; 39.
18. Baldry MGC. The bactericidal, fungicidal and sporicidal properties of hydrogen peroxide and peracetic acid. J Appl Bacteriol 1983; 54.
19. Doll JP, Trexler PC, Reynolds LI, Bernard GR. The use of peracetic acid to obtain germfree invertebrate eggs for gnotobiotic studies. Am Midl Nat 1963; 69.
20. Hussaini SN, Ruby KR. Sporicidal activity of peracetic acid against *B. anthmtis* spores. Vet Rec 1976; 98.
21. Lewis JS, Rickloff JR. Inactivation of *Bacillus stearothermophilus* spores using vaporized hydrogen peroxide. In: 91st Annual Meeting of the ASM, May 9, 1991.
22. Rickloff JR, Orelski, PA. Resistance of various microorganisms to vaporized hydrogen peroxide in a prototype tabletop sterilizer. In: 89th Annual Meeting of the ASM, May 16, 1989.
23. Klapes NA, Vesley D. Vapor-phase hydrogen peroxide as a surface decontaminant and sterilant. Appl Environ Microbiol 1990; 56.
24. Christensen EA. Radiation resistance of bacteria and the microbiological control of irradiated medical products. In: Sterilization and Preservation of Biological Tissues by Ionizing Radiation. Report of a panel, Budapest, IEAE, Vienna, 1970 STOPUB/247, June 1969.
25. Pflug IJ. Industrial sterilization. In: Phillips GB, Miller WS, eds. The Proceedings of the International Symposium, Amsterdam, 1972. Durham, NC: Duke University Press, 1973.
26. Pflug IJ. Syllabus for an Introductory Course in the Microbiology and Engineering of Sterilization Processes. 4th ed. St. Paul, MN: Environmental Sterilization Services, 1980.
27. Patashnik M. A simplified procedure for thermal process evaluation. Food Technol 1953; 7.
28. Okinczyc T. Quality Assurance Statistician, Burroughs Wellcome Co., Greenville, NC, Personal Communication.
29. Phillips RS, Dickey NH, eds. Funk and Wagnalls New Encyclopedia, Vol. 15. New York: Funk and Wagnalls, 1979.
30. Pistolesi D, Mascherpa VF. What It Means, How to Calculate It, How to Use It for Adjustment, Control and

Validation of Moist-Heat Sterilization Processes. Fedegari Autoclavi SPA 1989.

31. Rahn O. Injury and death of bacteria by chemical agents. In: Luyet BJ, ed. Biodynamica Monograph No. 3. Normandy, MO: Biodynamica, 1945:9–41.
32. Hugo WB, Russell AD. Pharmaceutical Microbiology. Oxford, U.K.: Blackwell Scientific, 1977.
33. Errington J. *Bacillus subtilis* sporulation: regulation of gene expression and control of morphogenesis. Microbiol Rev 1993; 57:1.
34. Pflug U, Smith GM. Survivor curves of bacterial spores heated in parenteral solutions. In: Barker AN, Wolf J, Ellar DJ, Dring GJ, Gould GW, eds. Spore Research 1976, Vol. 2.
35. Holcomb RG, Pflug IJ. The Spearman–Karber method of analyzing quantal assay microbial destruction data. In: Microbiology and Engineering of Sterilization Processes. 3rd ed. St. Paul, MN: Environmental Sterilization Services, 1979.
36. Association for the Advancement of Medical Instrumentation. Process Control Guidelines for Radiation Sterilization of Medical Devices. Arlington, VA: AAMI, 1980.
37. Ball CO. Mathematical solution of problems on thermal processing of canned foods. Univ Calif Publ Public Health 1928; 1:2.
38. Bigelow WD, Bohart GS, Richardson AC, Ball CO. Heat Penetration in Processing Canned Foods. Washington, DC: National Canners' Association, Bulletin 16L, 1920.
39. Bigelow WD. Logarithmic nature of thermal death time curves. J Infect Dis 1921; 29.
40. Carleton FJ, Agalloco JP. Validation of Aseptic Pharmaceutical Processes. New York: Marcel Dekker, 1986.
41. Haas PJ, Melgaard HL, Lysfjord JP, Pflug II. Validation concerns for parenteral filling lines incorporating barrier isolation techniques and CIP/SIP systems. Available through Despatch Industries or TL Systems, Minneapolis, MN.
42. Pflug IJ, Bearman JE. Treatment of sterilization-process microbial survivor data, Appendix E. In: Environmental Microbiology as Related to Planetary Quarantine. Semi-Annual Progress Report 9, NASA Grant NGL 24-005-160, Division of Environmental Health, School of Public Health, University of Minnesota, 1972.
43. Soper CJ, Davies DJG. Principles of sterilization. In: Denyer SP, Baird RM, eds. Guide to Microbiological Control in Pharmaceuticals. London: Ellis Horwood, 1990.

BIBLIOGRAPHY

Alters MJ, Attia IA, Davis KE. Understanding and utilizing F_0 values. Pharm Technol, May 1978.

Parenteral Drug Association. Parametric release of parenteral solutions sterilized by moist heat sterilization. Technical Report No. 8. Philadelphia, PA: PDA, 1987.

Pflug D, Holcomfa RG. Principles of thermal destruction of microorganisms. In: Block SS, ed. Disinfection, Sterilization and Preservation. Philadelphia, PA: Lea & Febiger, 1983:751–810.

12

Steam Sterilization in Autoclaves

Phil DeSantis
Schering-Plough, Inc., Union, New Jersey, U.S.A.

INTRODUCTION

The validation of steam sterilization in autoclaves constitutes perhaps the most-studied validation problem faced by the pharmaceutical industry. Indeed, it was failure to sterilize certain LVP solutions that resulted in several patient deaths, prompting the U.S. FDA to call for the "validation" of sterilization processes. Because of this, autoclave sterilization was the first validation program undertaken by the industry. This requirement soon spread to other pharmaceutical processes. Sterilization in autoclaves remains a universal issue in nearly all facilities where sterile operations occur and continues to be of paramount concern to both the industry and the various international drug regulatory agencies.

The initial and time-honored FDA definition of *validation* is roughly abbreviated to "proof that a process does what it purports to do." This definition tends to foster an emphasis on testing of the process as required to provide that "proof." For many years, the validation of autoclave sterilization was focused on testing the sterilization cycle, with the goal of achieving repetitive successful results. Little attention was paid to the sterilization mechanism or process, the equipment, or the controls applied.

The modern validation discipline recognizes the need for an integrated program of development, design, testing, operation, and maintenance. This program must always be based on the established relationship among function, structure, appropriate tests and acceptance criteria, as well as ongoing operation. For sterilization, among all processes common to pharmaceutical manufacturing, it is easy to bypass the integrated or "life cycle" approach and concentrate on the testing aspects only. This may be because the tests generally applied to this process are rigorous and, in themselves, provide substantial assurance of reliability. Still, the ultimate achievement of validation is dependent upon control of the process. This control, in turn, is dependent upon an understanding of the process and the equipment that facilitates it. Note the characterization of validation as a state to be achieved, not a task to be carried out. Failure to follow this approach reduces our assurance and presents the risk of unexpected failure.

The objective of this chapter is to provide a basis for understanding sterilization using moist heat and the sterilizers employed. The approach to validation presented is geared to a practical application of a large body of experience. It is meant to give the user the means to understand the principles of microbial death and the meaning of sterilization. This understanding will be extended to

- The design and operation of reliable autoclaves
- The characteristics of the loads to be sterilized
- Design of effective sterilization processes or cycles
- The testing of these cycles to provide a high degree of assurance that they will be reliable
- Maintenance of the state of control necessary to ensure the quality of all sterilized materials and products

Mechanism of Sterilization

The microbiology and mathematical modeling of sterilization is described in detail in many references. It is presented in brief here for completeness.

Steam sterilization under pressure is the most effective sterilant (1). It is the method of choice when heat and moisture damage is not a problem. The temperature range for the growth of most living organisms is −50°C to 80°C (1). Exposure to temperatures outside this range usually results in the death of the organism, except for some heat-resistant spore formers.

The mechanism responsible for the death of microorganisms is not clearly understood. To date, the most commonly employed criterion for describing microbial death remains the loss of the cell's ability to reproduce. Iff a sample incubated at suitable conditions in a suitable medium did not exhibit growth within a specified time, it was assumed to be sterile. Unfortunately, this test destroys the sample. Thus, this simplified view is not effective in predicting the sterility of lots or batches of medical products. Traditionally, the sterility of a batch or lot of products was certified by tests such as those described in the *USP* 29. These tests use a small sampling of the sterilizer load to determine the presence of viable (reproducing) organisms in the entire lot. Frequently, this may have been done without consideration for the

Abbreviations used in this chapter: ASME, American Society of Mechanical Engineers; BB, bioburden; BI, biological indicator; DIN, German Institute of Standardization; EQ, equipment qualification; ETO, ethylene oxide; FDA, Food and Drug Administration; GAMP, good automated manufacturing practice; GMP, good manufacturing practice; HMI, human–machine interface; I/O, input/output; IQ, installation qualification; LVP, large volume parenteral; MES, manufacturing execution system; NIST, National Institute of Standards and Technology; OQ, operational qualification; PC, personal computer; PID, piping and instrument diagram; PLC, programmable logic controller; PNSU, probability of a nonsterile unit; PQ, performance qualification; PV, process validation; RTD, resistance thermal device; SAL, sterility assurance level; SCADA, supervisory control and data acquisition; TC, thermocouple; USP, *U.S. Pharmacopeia*.

mechanism of microbial death or the conditions required to facilitate that mechanism. The finished product test in *USP* 29, which requires 20 samples, can detect a contamination level of only 15% with 95% confidence. This corresponds to an approximate probability of survivors of 10^{-1} (10%) (2). Sample size may be increased to improve confidence of detection. This can be costly and wasteful. Obviously, for this method to ensure absolute sterility all samples would have to be tested.

An alternative approach to predicting sterilization is the definition of sterility as a probability of survival. This probability is related to knowledge of the mechanism of microbial death and the conditions causing it. The most prevalent description of sterility used today is the reduction of anticipated levels of contamination in a load to the point at which the PNSU is less than 10^{-6} (one in one million) (2).

This transition to thinking of sterilization as a probability function is one that is now firmly accepted throughout the industry and among its regulators. It is closely linked to the concept of validation. Once the levels of microbial contamination and resistance to the sterilizing process are known, probability of survival can be calculated. Validation involves the measurement of sterilizing conditions and challenge of the sterilizing cycle to ensure that these conditions have been met.

It is generally believed that microbial death can be linked to the denaturizing of critical proteins and nucleic acids within the cell, although clear proof of the theory has not been attained. This denaturizing is a result of the disruption of the intramolecular hydrogen bonds that are partially responsible for spatial orientation of the molecule. Proteins are specifically ordered chains of amino acids, linked by polypeptide bonds. Nucleic acids are polycondensations of ribose sugars joined by phosphate linkages. Each is dependent on a specific spatial orientation to perform its function. As the hydrogen bonds are broken, the structure, and thus the function, is lost. However, the denaturizing may be reversible or irreversible. The functional structure of the molecule is lost in stages. If halted before a critical number of hydrogen bonds are cleaved, it is possible for the molecule to return to its original state. For example, DNA gradually changes from a helix to a random coil.

Significant research data support the theory that microbial death may be described as a first-order chemical reaction. This leads to the conclusion that death is essentially a single-molecule reaction. We are probably dealing with the denaturizing of a critical molecule within each cell (1).

Bacterial spores are the forms most resistant to thermal death. The spore is the normal resting state in the life of certain groups of organisms, namely, bacilli and clostridia. During this stage, the processes of the cell are carried out at a minimal, albeit not stagnant, level. Spores are the most resistant of all organisms in their ability to withstand hostile environments. Their thermal resistance has been linked to the relative absence of water in their dense central core. There is considerable disagreement on the subject. Some investigators attribute this heat resistance to the existence of the spore core as an insoluble gel or the presence of lipid material. The dry-heat resistance of spores is greatly influenced by the history of the spore relative to water as well as the water content of the spore during the heat treatment. All of the foregoing highlights the importance of moisture in thermal death. Bacterial spores are much more rapidly destroyed in the presence of saturated steam than by dry heat. It is possible that the water causes the hydration of a stabilizing polymer (calcium dipicolinate) within the spore. Furthermore, water is linked directly to the denaturizing of proteins and nucleic acids by hydration (1).

Mathematical Modeling

The consideration of microbial death, and more specifically spore inactivation, as a monomolecular reaction with water is consistent with first-order reaction kinetics. That is, the rate of reaction is governed by the concentration of the reactant (spores). Mathematically this is expressed as

$$\frac{dN_a}{dt} = kC_a$$

where t is the time, N_a is the number of spores, C_a is the spore concentration, and k is a reaction rate constant at constant temperature. In integral form,

$$\log\left(\frac{C_a0}{C_a}\right) = k(t-t_0)$$

where the subscript 0 indicates initial conditions. Also, $(t-t_0)$ is usually simplified arbitrarily by setting $(t_0=0)$. Thus, a semilogarithmic plot of concentration versus time will yield a straight line of slope k, as shown in Figure 1. The rate constant k is expressed in min^{-1}. In turn, the negative reciprocal of the rate constant is equivalent to the number of minutes required at a given temperature to destroy 90% of the organisms present (i.e., a 1-log reduction). The reciprocal of the rate constant is referred to as the D-value and is expressed in minutes. D is the measure of the relative heat resistance of an organism at a constant temperature.

As shown in Figure 1, the simple logarithmic model yields a straight-line survivor curve. Although it does not fit all experimental data, its use is recommended because of its wide applicability and simplicity.

In general, sterilization takes place over a range of temperatures. Therefore, the sterilizing effect must be integrated over a range of temperatures and requires a temperature-dependent model.

A common measure of the temperature dependence of an ordinary chemical reaction is the Q-value. Q is defined as the change in the reaction rate constant k for a change of 10°C. This can be written as

$$Q = \frac{k_{(T+10^\circ\text{C})}}{k_T}$$

The Q-value for many chemical reactions is close to 2. For spore destruction with saturated steam it is much higher, from 10 to 18.

Another common temperature coefficient model for chemical reactions is the Arrhenius equation. This is written as

$$k = A\exp\frac{E_A}{RT}$$

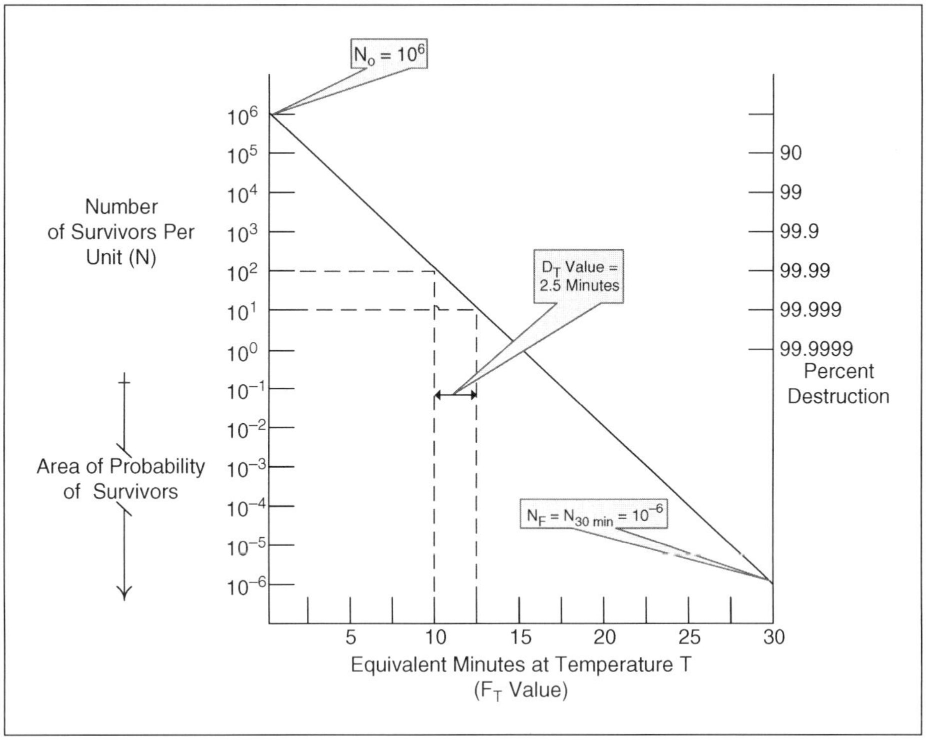

Figure 1 Microbial death (survivor) curve.

where k is the rate constant, A is a constant, E_A is the activation energy, and R is the universal gas constant.

A plot of ln k (determined experimentally) versus $1/T$ will give a straight-line slope.

This model is consistent with empirical data gathered on the temperature dependence of spore D-values. A typical plot of the effect of temperature on D, the thermal resistance, is shown in Figure 2. The negative slope of this thermal resistance curve is called the z-value. z is defined as the temperature change required to cause a 1-log decrease in the D-value and is expressed in °C. Remembering that D is the reciprocal of the rate constant k, z can be related to Q as follows:

$$z = \log \frac{Q}{10}$$

Similarly, it can be related to E_A in the Arrhenius model

$$z = \frac{2.33RT^2}{E_A}$$

The use of the D- and z-values to predict microbial death over time and temperature should be approached with caution. The straight-line relations predicted by these models will hold over a limited temperature span and then only for a homogeneous culture of a single species of microorganism. Mixed populations of several levels of heat resistance will produce a curve determined by the relative populations and D-values of the organisms. The usefulness of D and z, though, is that in nature one subpopulation, by virtue of its high thermal resistance (D) and initial concentration, is usually controlling to reach sterility. This subpopulation will follow the model (3).

Thermal Death Time Curve

The usefulness of the temperature-dependent model in the steam autoclave is to calculate the lethality of the cycle over a range of temperatures (including heat-up and cooldown). To do this a new variable, closely related to D, is introduced. This is called F, the thermal death time. F_{121} is defined as the time in minutes to produce a sterilization effect equivalent to that of the reference temperature of 121°C (250°F). This reference temperature is chosen as a base because it is an economical and effective one for saturated steam sterilization. It may be considered a benchmark, similar to a 0.300 batting average in baseball. It should not be assumed that 121°C is required to achieve effective sterilization.

Both the thermal resistance curve in Figure 2 (log D vs. T) and the thermal death time curve (log F vs. T) are dependent on z as follows:

$$\frac{D_T}{D_{121°C}} = \frac{F_T}{F_{121°C}} = 10^{(T-121)/z}$$

where T is the measured temperature. The curves are parallel, both with a slope of z.

The most commonly used value of z for the destruction of microbial spores is 10°C (18°F). This is based on experimental observations for *Geobacillus stearothermophilus* and *Clostridium botulinum*, both highly heat-resistant organisms. These organisms are chosen for divergent reasons. *C. botulinum* was the subject of the pioneering experiments by food scientists attempting to destroy this deadly cause of botulism in canned foods. *G. stearothermophilus* is a readily available and safe indicator organism for use in sterilization studies and has similar resistance.

When the assumption of $z = 10°C$ is made, F may be written as F_0. This is the most commonly used measure of

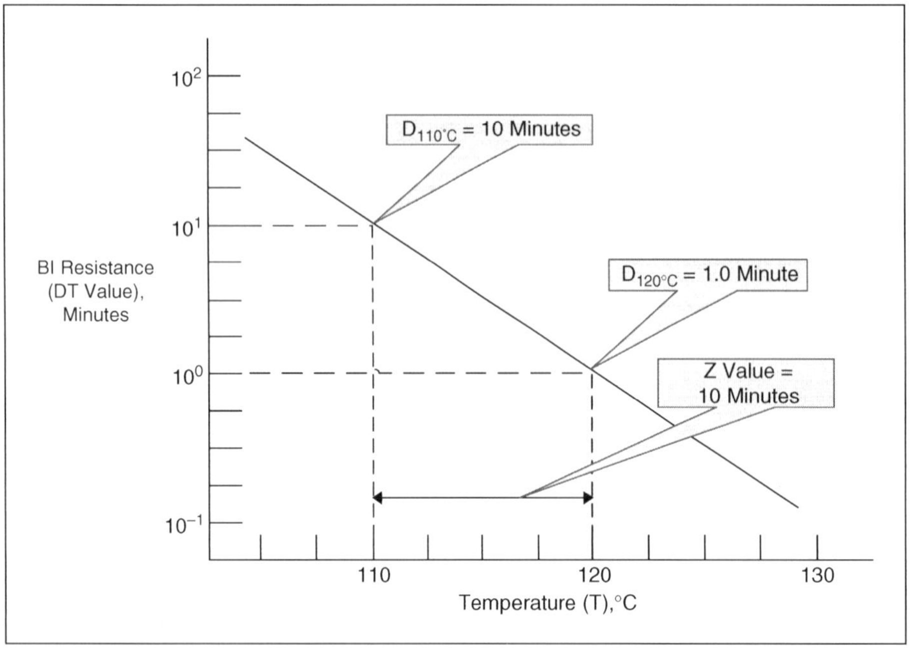

Figure 2 Thermal resistance curve.

the lethality of a sterilization process spanning a range of temperatures. (Note that, out of respect for the scientists who developed this concept, this is read "*F* subzero" not "*F* sub-Oh.")

F_0 is a summation over time of the instantaneous lethal rates at a series of temperatures. In integral form this is

$$F_0 = \int 10^{(T-121)/10} dt$$

which approximates to

$$F_0 = \sum 10^{(T-121)/10} \Delta t$$

where Δt is the chosen time interval and T is the average temperature over that interval. The smaller the interval chosen, the more accurate the calculation will be.

Use of the important value F_0 enables one to simply measure the relative effectiveness of any steam sterilization process. This value, along with an understanding of the number and thermal resistance of the microbial population to be sterilized (bioburden), allows us to determine sterility. Although sterility is in fact an absolute condition (it is not suitable to refer to an item as being "almost" sterile), it is beyond our means to ascertain absolutely, short of total destructive testing. Therefore, it is a common practice to use the F_0 equation to determine the probability of sterility or SAL. Of more common use, the PNSU may be found. As has been previously discussed, the most widely accepted target by the industry and international regulatory agencies for sterilization in autoclaves is a PNSU of 10^{-6} (one in one million), corresponding to a SAL of 10^6.

STERILIZER DESIGN FOR PARTS AND HARD GOODS

The validation of a steam sterilization cycle is dependent on the equipment chosen. The sterilizer and its support systems must be designed and constructed to deliver the effective cycles repeatedly and consistently. Qualification of the sterilizer consists of proper design, installation according to design, operational testing to ensure that design criteria and operational requirements are met and, finally, PQ to confirm that the product or materials and equipment are sterilized per specification.

The usefulness of saturated steam for parts/hard goods sterilization has been well documented (1,4). The sterilizing effect is accomplished by the heat transfer from the steam to the load and by the hydrating effect of the resultant condensate. Condensate is formed because of the return of the steam to the lower energy liquid state. This phase change requires the transfer of the latent heat of the steam (that which was required to change it from liquid to vapor: 970 Btu/lb or 1 kcal/kg) to the surroundings, thus heating the sterilizer and its load. The heat transferred by the condensation of saturated steam is many times greater than that which would be transferred from steam above its boiling point, called superheated steam. This heat amounts to only 1 Btu/lb °F (1 kcal/kg °C). Also, superheated steam is sometimes known as "dry steam," as it does not form condensate as it cools. Thus, the important hydrating effect is not present.

Sterilization with superheated steam is a dry-heat phenomenon, less efficient than a saturated steam process. Superheat may be avoided by maintaining steam in equilibrium with water at the boiler or steam generator. Also, supplementary heat sources, such as jacket heat, must be controlled so as not to drive the system above the vapor–liquid equilibrium line. Condensation to water causes a volume decrease in excess of 99%. This would result in a pressure decrease if the condensed steam were not immediately replenished, as it is in the sterilizer. It is the condensation–replenishment cycle that allows the steam to penetrate to all the surfaces to be heated until they reach an effective sterilization temperature. Sterilizers and sterilization cycles are designed to ensure that saturated steam reaches all of these surfaces.

There are several characteristics common to all modern steam sterilizers in use in the pharmaceutical industry. These include:

- A pressure vessel constructed according to a recognized national or international code (e.g., ASME). This must withstand at least 50% in excess of the required internal steam pressures. It may be rectangular or cylindrical in cross section.
- A steam jacket and insulation: these are energy-conserving features designed primarily to heat the metal mass of the vessel and to limit heat loss from within the vessel. Some laboratory and small special-use sterilizers are unjacketed. Where jackets are employed, they should be operated at lower pressure than the chamber to avoid superheat.
- A safety door mechanism to prevent opening while the unit is under pressure (the term "autoclave" means self-closing). The locking device may be actuated directly by internal pressure or indirectly through an automatic switch. The door itself may be of the swing-out or sliding type.
- A thermostatic steam trap to efficiently remove air or condensate from the chamber: this is open when cool and closed when in contact with steam. As air or condensate collects, the trap opens owing to the slight temperature reduction and the condensate is discharged. There is also a trap to remove condensate from the steam jacket.
- Process control system (typically a PLC for controlling and monitoring the process). These are discussed in a later section.
- A Process Data Recorder or Data Collection system.
- A microbial retentive vent filter.
- A chamber pressure indicator.
- Pressure relief valves for both the chamber and jacket.
- A vacuum pump or eductor to remove air from the chamber and load.

STERILIZER CONTROL SYSTEMS

A key to effective sterilizer operation lies in the automated process control system. By eliminating the dependence on operator intervention and data recording, automatic temperature and sequence control provide assurance that the "validated" sterilization cycle is consistently and repeatedly delivered.

A typical control system for a new sterilizer includes the following hardware components:

- PLC
- Operator Interface Panel(s)
- Data Recorder/Data Collection System
- Process Variable Sensors
- I/O Devices

The PLC is the most commonly used primary component of the automated process control system as it typically does all the sequential control of the process, provides PID control of all proportional valves, controls all devices, receives operator input via the Operator Interface Panels and provides process information (such as process variable information and alarms) to the operator via displays and/or Operator Interface Panel. The PLC also typically contains specific recipe information for the various cycles to be utilized. In some cases the PLC can be used for data collection, but it is much more common to use a separate Data Recorder/Data Collection system.

The operator interface panel can be as simple as switches and displays to as complex as a stand-alone PC running a SCADA HMI software package. These devices are typically used to select the recipe, start the cycle and display process information during the cycle. The higher level PC-based SCADA-type operator interface panels can provide detailed cycle reports and trending information.

The data recorder/data collection system can be as simple as a strip chart recorder to a full-blown MES-type data collection system. In many cases, the PLC can also provide batch data logging functionality. The minimum variables to record for steam sterilization processes are typically time, temperature and pressure with pressure only being a critical parameter for saturated steam process.

Typical sensors include temperature measurement devices (RTDs or TCs), pressure measurement devices, and where applicable level and flow measurement devices.

The temperature sensor used to control the process temperature shall not be used to provide the batch record process data. A secondary temperature sensor for batch reporting provides a high degree of insurance that the cycle actually ran within its defined limits. Heavy wall thermowells should not be used, as this will affect the time response of the measurement. Thin walled thermowells or temperature elements with stainless steel sheaths should be used for temperature measurement.

The pressure sensor should be equipped with a sanitary-type diaphragm and connect to the sterilizer using a sanitary-type connection. A sanitary diaphragm introduces additional errors to the pressure measurement due to the stiffness of the diaphragm. This stiffness is related to the size of the diaphragm. This effect is negligible for diaphragms above a 3-inch diameter. This should be considered when sizing the connection to the sterilizer.

For I/O devices, there are analog and discrete types. The analog inputs are typically from process sensors and the analog outputs are typically for control of proportional valves. The discrete inputs are typically from switch-type (operator and process) devices and the discrete outputs are typically for activating hardware such as valves, pumps, and lights.

I/O devices using buss connections (Profibus, ControlNet, Hart, Fieldbus, LonWorks, Ethernet, etc.) are not really considered a true I/O point. These devices constitute more of a Network and have a whole host of communications capabilities, diagnostics, and maintenance functions available. Typically, these devices provide much more than just the measured variable. Many sensors having these buss-type connections are available.

The design and development of the software should follow the principles of GAMP 4 for automated process control systems. GAMP details a software life cycle from conception through all phases, from development to maintenance of the software in a validated state.

STERILIZATION CYCLES FOR PARTS AND HARD GOODS

Removal of air is a common problem for parts/hard goods sterilization. Air entrained within the load depresses temperature and prevents the penetration of steam to all the required surfaces. The efficiency of heat transfer (heat transfer coefficient) from the steam to the load is thereby reduced. The most well-known sterilization processes, or cycles, have been designed to remove air. Originally, air removal cycles were designed for loads of porous materials, like hospital surgical packs. It has been shown that vacuum as low as 15 to 20 mmHg (less than 0.4 psia), applied for 8 to 10 minutes, is required to remove air from some porous loads (1). This level of vacuum is very difficult to achieve and even if applied as described may be inefficient in removing entrapped air from more complex parts and hard good loads. Therefore, by far the most common cycle used for these loads uses pulsed vacuum.

Because of the difficulty in obtaining the high-vacuum conditions needed for efficient air removal, manufacturers developed pulsing systems. These employ a series of alternating steam pulses followed by vacuum excursions. The maximum and minimum pressures are variable. In general, the pulsing system removes air effectively without achieving the level of vacuum required in simple prevacuum cycles. The steam provided serves to rapidly fill the voids attained by the vacuum pulses, forcing out residual air.

The pulsing cycle is among the most prevalent in use throughout the pharmaceutical industry. Many of the loads requiring sterilization contain items for which air removal is difficult. These include coiled hoses, filter housings, and densely packed containers of stoppers. It is important to recognize this difficulty and to specify cycle parameters effective in overcoming it. Figure 3 is an example of a multiple, or pulsed, prevacuum cycle.

MICROBIOLOGICAL STERILITY ASSURANCE DEVELOPMENT

The designation of a sterilization cycle and development of that cycle is dependent upon product characteristics, specifically heat stability and bioburden. Cycles that have been classified above based upon their "mechanical" modes may be further classified according to stability and microbiological characteristics as follows:

- Overkill cycle approach
- BI/Bioburden cycle approach
- Bioburden approach

Bioburden and BI/Bioburden cycles are predominantly utilized in the sterilization of liquid-filled containers whether for final products (terminal sterilization), or for the sterilization of laboratory/production media and certain in-process liquids. The methods and practices associated with their sterilization are defined in the following chapter. The remainder of this chapter will only discuss the validation of items using the overkill method. Overkill sterilization can be utilized for certain stable liquids in which case the methods utilized are a hybrid of those presented in this chapter and the succeeding one.

Overkill Cycles

When sterilizing heat-stable materials, an overkill approach may be adopted. Loads sterilizable by overkill include filters, container closures, hoses, filling parts and other hard goods, and soft goods, such as gowns. Remember that the accepted criterion for sterility is the probability of survival of no greater than 10^{-6}. The objective of the overkill cycle is to assure that level of assurance, regardless of the number and the heat resistance of the organisms in the load. Extremely high F_0 values are generally used. Because the load is heat stable, thermal degradation is of no concern and only the minimum F_0 in the load is considered. This may be chosen to provide at least a 12-log reduction of microorganisms with a *D*-value of one minute at 121°C (5). A quick calculation will arrive at $F_0=12$ as a minimum overkill cycle.

Most microorganisms found in production environments have *D*-values ranging from 0.5 minutes or less. Thus, a population of up to 10^6 (one million) spores per unit of even the most heat-resistant strains of these environmental/materials microbial isolates will be reduced to a PNSU of 10^{-6} by the minimum overkill cycle. Less resistant organisms and the much smaller populations usually found in the clean room environments employed in preparing loads will be reduced to a much greater extent.

Despite the high degree of sterility assurance provided by the minimum overkill cycle, most validation teams will often choose to increase the cycle to ensure the deactivation of the bioindicators used to confirm sterilization during validation studies. These bioindicators are most often strips or suspensions containing from 10^4 to 10^7 spores of a highly heat-resistant organism (usually *G. stearothermophilus*). *D*-values for these organisms have been observed to range from 1.5 to as high as 4 minutes. Therefore, overkill cycles as high as $F_0=52$ (4 minutes × 13-log reduction) will be required to provide a PNSU of 10^{-6} for the bioindicator. Note that it is not required that the extent of overkill attain these levels. Nevertheless, because the failure to inactivate even one bioindicator may cast doubt over the validation study, it is common to adjust the cycle to effect total kill of the BI.

Bioburden studies are not required, nor are they usually carried out when an overkill cycle is planned. An assumption of *D*-value (either $D=1$ minute or the bioindicator *D*-value stated by the manufacturer) and a minimum target of a 12-log reduction are adopted. These are used to calculate the F_0 to be delivered to the cold spot in the load. Cycle parameters of time and temperature, as well as the location of the cold spot, are determined during the validation studies.

STERILIZATION PROCESS CONFIRMATION

When viewed as a life cycle activity, the most important key to validation is to understand the sterilization process

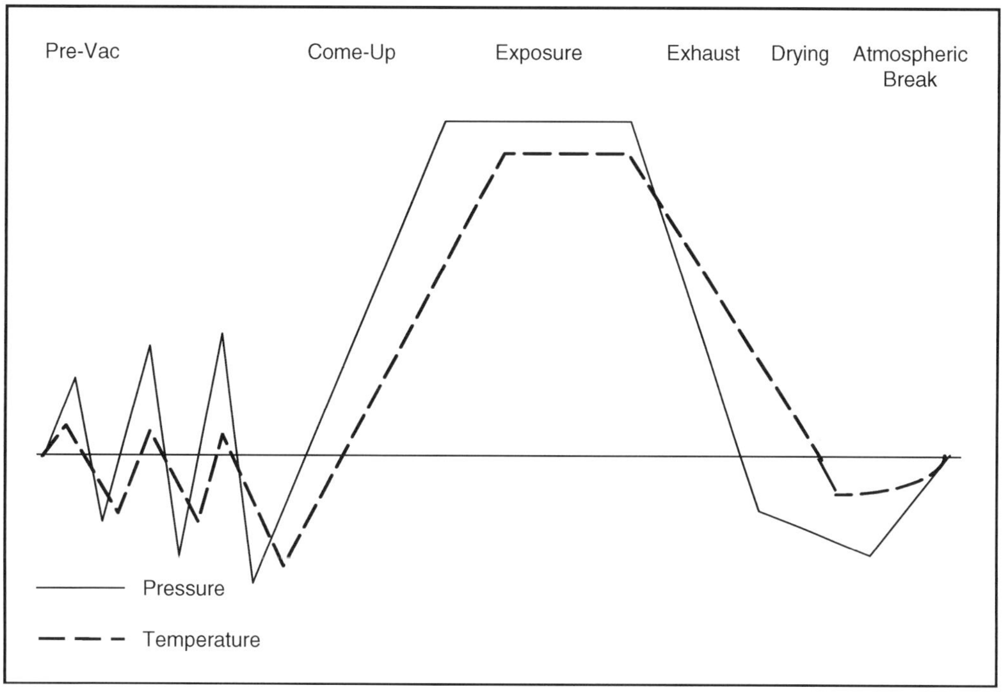

Figure 3 Typical pulsed vacuum cycle.

as related to both physical (thermal, mechanical) and microbiological parameters. This is most effectively done during the developmental program on both the R&D and Production scales. These studies form an important foundation for the final testing and documentation that has been commonly referred to as "validation." Sometimes this is called *process qualification* or PQ or PV. The author adheres strongly to the life cycle approach; thus, references to *validation* in this section shall not be limited to those activities that are commonly included in this final confirmatory phase. For this phase, it is preferable to use the term *process confirmation*. This may become the more broadly accepted jargon in the future.

With this in mind, it is futile (or at least very risky) to initiate a process confirmation study for a sterilization process that has not been shown to work. Whether this evidence is determined on an R&D scale, on a Production scale, or both, is dependent upon the load and the sterilizer used. For example, hard goods and parts cycles in a pulsing-vacuum-type sterilizer can be predicted fairly successfully from any scale of preliminary study. In fact, because sophisticated temperature measurement and monitoring systems can actually calculate lethality instantaneously, some of these cycles may forego developmental studies. (The author does not recommend this). On the other hand, terminal sterilization of product in containers utilizing a bioburden-based cycle should always be demonstrated on a full Production scale prior to formal validation (see chap. 13).

In addition to an understanding of the process or cycle, the sterilizer and associated equipment must be qualified; that is, determined to be suitable for its intended purpose and able to meet all critical requirements, as defined by the sterilization process.

Measuring Temperature

In all qualification and process confirmation studies, the ability to accurately measure temperature is critical. Several items will be required to measure and record temperature effectively. The most versatile temperature-sensing devices for validation are TCs. These are constructed from wires of two dissimilar metals. They can be encased in flexible sheaths, PTFE (polytetrafluroethylene) being widely used. Type T (copper–constantan) TCs are most applicable in steam sterilizer confirmation work. Their working temperature range is wide, and they are resistant to corrosion in moist environments. A high-grade TC wire should be chosen. Standard grades have an inherent error as high as 1°C. This is very significant when calculating the experimental lethality. Premium grades of wire, accurate to as close as 0.1°C at 121°C, are recommended. These must then be calibrated against a temperature standard traceable to the NIST (formerly National Bureau of Standards), DIN, British Standard, or an acceptable national standard.

The temperature standard may be a mercury-in-glass thermometer or platinum RTD. The RTD standard is recommended because of its greater durability and accuracy. In fact, this is the same device most often specified to measure the chamber temperature used to control the sterilization cycle. The TCs to be calibrated are placed in a highly stable temperature source (controlled ice-point device, hot reference device or controlled temperature bath) along with the reference standard. The differences in the readings between the TC and the reference device are recorded. The acceptable error should be no greater than the sum of the TC wire accuracy (e.g., +0.1°C to −0.3°C) and the degree of traceability of the reference instrument (e.g., +0.1°C to −0.3°C). TCs that do not meet this criterion should be replaced.

Calibration of TCs should be carried out at two temperatures. One of these is an ice point reference at 0.0°C. The other should be a hot point slightly higher than the expected sterilization temperature (130°C is commonly chosen). If the TC meets the accuracy criterion specified above, it is then permissible to apply a correction factor to bring the TC reading to the same as that of the reference RTD. Data loggers and recording devices are often programmed to perform this correction automatically. Once correction factors are applied at both calibration temperatures, the response of the TC over the temperature range can be linearized. The corrected temperature measurements are used in lethality calculations. Calibration should be repeated after a series of validation runs. Experience has shown that weekly recalibration is perfectly adequate.

TC access into the sterilizer should be considered in the design of the sterilizer. Most sterilizer manufacturers routinely include one or more unused ports in their pressure vessels. These can be tailored to the specific needs of the validation team. All penetrations must be made before the vessel's code compliance is stamped. To make modifications at a later date is troublesome. These must be made by a board-certified welder and are subject to reinspection and test by a code inspector. Special gland adapters are joined to the access ports that allow the TCs to pass into the chamber without developing pressure leaks. The adapter can be a pressure gland made of two mated flanges separated by two flexible gaskets. The TCs pass between the gaskets and the flanges and are bolted together tightly to prevent leaks. Another method is to use special-purpose fittings made for the express purpose of TC access (e.g., manufactured by Conax industries, Buffalo, New York). This is a specially drilled rubber gland within a compression-type housing.

Validation runs usually involve numerous temperature measurements. These can be recorded in a number of ways. Because of the frequency and number of recordings, a data logger is often employed. This is a digital output, multichannel device capable of frequent printouts or data transmission of many temperature measurements. These can be very sophisticated and can be preprogrammed to make TC calibration corrections, store data, and even calculate F_0 and print or electronically record cumulative values. The best data loggers have large capacity (32–48 channels), precision to 0.1°C or better, fast scan rate of all TCs (one cycle per second or better), and ability to interface with a computer, either by way of stored data or in real time through an output connector.

Commissioning and Equipment Qualification

Prior to the initiation of process confirmation studies, it is important that the sterilizer be suitably qualified to perform its function. This qualification ensures that the system meets critical requirements of the sterilization process as defined by the sterilization scientists (i.e., the "users"). These requirements should be clearly defined in terms of the sterilization process to be executed. The authors recommend a specific User Requirements document for this purpose. These requirements always include the ability to achieve and maintain sterilizing conditions throughout the chamber. Typical critical requirements that are considered to affect the sterilization process (e.g., "quality-critical" requirements) are:

- Accurate temperature and pressure measurement
- Air removal to some predefined level of vacuum (not always required for terminal sterilizers)
- Temperature distribution and uniformity throughout the chamber. (Uniformity requirement may vary depending upon the cycle defined. For bioburden-based terminal sterilization, this is often ±0.5°C. For overkill cycles, this limit may be more flexible.)
- Precision of temperature control (usually the ability to maintain a control point temperature of ±0.5°C around the set point)
- Precise control of the sequence of operations and timing
- Alarms to indicate out-of-specification conditions

User Requirements should not be engineering specifications. The technical approach to achieving requirements is best left to engineers and sterilizer manufacturers. Therefore, design issues such as choice of instrumentation, line sizing, and chamber configuration are not subject to formal EQ, but to commissioning by the vendors and engineers who have developed the specifications. A commissioning study is a rigorous series of inspections and tests to ensure that specifications are met. There is, however, an inherent flexibility within commissioning that allows for adjustments, corrections and even modifications to bring the system to an acceptable state. These, of course, are thoroughly documented and reported. However, approval beyond the engineering and user groups is not required.

Once a sterilizer has been commissioned it will be ready to be formally qualified. There is a school of thought emerging in the industry that qualification may be reduced to a paper exercise that merely confirms that the commissioning study has clearly proved that all quality-critical requirements have been satisfied. This probably achieves the level of assurance necessary to proceed with process confirmation studies. However, in the current regulatory environment, this approach is not universally accepted and the sterilizer qualification protocol may need to address some level of confirmatory inspection and testing.

As a minimum, a sterilizer qualification protocol should confirm both the critical installation (IQ) and the operational (OQ) requirements, as defined by the user. It is not, however, necessary to perform a detailed analysis against engineering details, as this has been completed in commissioning. Using the above list of typical critical requirements for reference, the qualification of a sterilizer should include the following (note that installation and operational requirements may be covered in a common protocol or in separate protocols, as the investigators see fit):

- Calibration of temperature and pressure sensors (traceable to an accepted national or international standard)
- Air removal (where required; usually measured by vacuum level achieved vs. defined requirement)
- Demonstration of the sequence of operations, including cycle timing
- Confirmation of alarms and interlocks

- Precision of temperature control
- Temperature distribution and uniformity

Empty Chamber Temperature Distribution

This study has traditionally been considered a critical aspect of sterilizer qualification. The intent of this study is to demonstrate the temperature uniformity and stability of the sterilizing medium throughout the sterilizer. Temperature distribution studies should initially be conducted on the empty camber. Temperature uniformity may be influenced by the type, size, design, and installation of the sterilizer. A satisfactory empty chamber temperature uniformity should be established by the User Requirements. A narrow range is required and is generally acceptable if the variation is less than $\pm 1°C$ ($\pm 2°F$) the mean chamber temperature. Sterilizers to be used for terminal sterilization of products utilizing a bioburden or BB/BI cycle approach may require enhanced temperature uniformity. In these cases, the sterilizer may be specified to maintain a uniformity of $\pm 0.5°C$ ($\pm 1°F$) of the mean or even better.

With modern sterilizers, temperature deviations greater than $+2.5°C$ ($+4.5°F$) may indicate equipment malfunction (5). Stratified or entrapped air may also cause significant temperature variations within the sterilizer chamber. The investigator is cautioned to determine that the sterilizer has been specified to maintain the temperature uniformity profile required by the most-demanding sterilization cycle. It is of no value to assign arbitrary acceptance criteria to a sterilizer that has not been specified or built to meet them. It is important to note that the drain of a sterilizer is expected to be its coldest point and is outside the sterilization zone. For this reason, it is recommended that the drain not be included in calculations of the mean chamber temperature and not be subject to chamber distribution requirements. It should also be noted that uniformity across the chamber may be expected only at steady state. Uniformity measurements are meaningful only after the control point temperature has stabilized at the desired set point.

Initially, a temperature distribution profile should be established from studies conducted on the empty chamber. Confidence may be gained through repetition; therefore, empty chamber studies are often conducted in triplicate to obtain satisfactory assurance of consistent results. After more than 30 years of sterilizer advancement, uniformity has become a virtual certainty. When qualifying sterilizers the design of which a firm has previously qualified, it may be permissible to conduct a single empty chamber distribution study. Subsequent to the empty chamber studies, maximum load temperature distribution studies should be conducted to determine whether the load configuration influences the temperature distribution profile obtained from the empty chamber studies. This is normally done during the process confirmation trials.

The TCs used in the heat distribution studies are distributed geometrically in representative horizontal and vertical planes throughout the sterilizer. The geometric center and corners of the sterilizer should be represented. An additional TC should be placed in the exhaust drain adjacent to the sensor that controls vessel temperature, if possible. The number of TCs used in the heat distribution study will be dependent on sterilizer size. In a production-size sterilizer, 15 to 20 TCs should be adequate. The TCs used for loaded chamber heat distribution studies should be positioned in the same locations used for empty chamber heat distribution studies. It is the uniformity and stability of the sterilizing medium that is monitored in the distribution studies. Consequently, the temperature probes should be suspended to avoid contacting solid surfaces and should not be placed within any containers. Temperatures must be obtained at regular intervals (e.g., each minute) throughout the time duration specified for a normal production cycle.

Container and Object Mapping

For overkill cycles, certain non-product load items may pose concerns relative to consistent and effective heating throughout the item. Filter housings, hoses, containers filled with stoppers, and small filling assemblies present both air removal and steam penetration problems. Developmental studies should be performed on these and similar objects with several TCs to determine the slowest to heat zone within an object. Once this has been determined, sensor placement within objects may be specified to probe the slowest to heat zone (i.e., the "cold spot"). Note that this type of study need not be repeated for every cycle, or even for every specific object, as long as classes of objects (e.g., hoses, filters) have been characterized adequately to determine appropriate probe location.

Heat Penetration Studies

Heat penetration studies comprise the core of sterilization process confirmation. The intent of these studies is to confirm that the slowest to heat objects within a specified load has achieved the requisite lethality. Cold spots originate because of the varied rate of heat transfer throughout the load. Therefore, it is imperative that developmental heat penetration studies be conducted to determine slow-to-heat items within a loading pattern and assure that those items are probed during confirmation studies to ensure that they are consistently exposed to sufficient heat lethality (5). Penetration TCs are positioned at points within the process equipment suspected to be the most difficult for steam heat penetration. For homogeneous loads (e.g., a load of stoppers or packs of gowns), TC placement in the load should cover the entire profile of the autoclave, including geometric center, corners, and near the top and bottom of the chamber. Temperature data are obtained from representative maximum loads to establish temperature profiles depicting load cold spots. Equipment load configurations may be designed to allow reasonable flexibility for the operating department by permitting the use of partial loads. For this, partial loads would be defined as a portion of the established maximum validated load. Thus, minimum load studies are not required. They are often run, however, to provide additional assurance.

Another question on heat-stable loads is the geometric configuration of heterogeneous or mixed loads. In these cases, many studies have shown that the cold spot in the load is related to a specific object (the most difficult for steam to penetrate), rather than to a point within the autoclave. Therefore, when performing heat penetration studies in mixed loads it is important to probe each type of component in the load. Triplicate studies will determine the hardest component to penetrate. For added assurance, it is a good practice to reconfigure the load between runs. This will add evidence to the "hardest to penetrate" theory. Having determined this, future loads will not be subject to rigid configuration, as long as the cycle chosen provides adequate lethality to sterilize the most difficult parts.

The cold parts established for a specified load or configuration will eventually be used to control the exposure time in subsequent routine production runs. The temperature sensor(s) that control sterilization cycle exposure time at process temperature may be positioned within the load adjacent to or within the previously detected cold item. This procedure will assure that the coldest item and, consequently, the entire load is exposed to sufficient heat lethality and achieves the desired temperature. It may not always be possible or convenient to position a temperature sensor at the coldest point. Here, it is necessary to know, through the validation study, the relation between the coldest point and the control point. Then the cycle can be adjusted to provide adequate time for the coldest item to reach the desired value.

Lethal rates can be determined from the temperature data obtained from the heat penetration studies. The temperature data are converted by the following formula:

$$L = \log^{-1} \frac{T_0 - T_b}{z} = 10^{\frac{T_0 - T_b}{z}}$$

where L is the instantaneous lethality, T_0 is the temperature within the object or container, T_b is the process temperature (121°C) and z is the temperature required to change the D-value by a factor of 10. (For this calculation z is set to 10, based on the experimental values for highly resistant organisms.)

The total lethality of the cycle is then determined by integrating over time as follows:

$$F_0 = \int 10^{(T-121)/10} dt$$

which approximates to

$$F_0 = \sum 10^{(T-121)/10} \Delta t$$

where Δt is the chosen time interval and T is the average temperature over that interval. The smaller the interval chosen, the more accurate is the calculation, so typically intervals of less than one minute are chosen.

Using this formula it is possible to measure the theoretical lethality of the entire cycle despite the fact that the temperature may deviate from the traditional reference sterilization level.

Microbiological Challenge Studies

Because heat penetration studies can only confirm temperature and not the other conditions required for effective moist heat sterilization, microbiological challenges are employed to provide additional necessary assurance that adequate lethality has been delivered to all parts of the load. These are most often conducted in parallel with heat penetration studies. Calibrated BIs used for this purpose function as bioburden models providing data that can be employed to calculate F_0 or to substantiate and supplement physical temperature measurements obtained from TCs (5). The microorganisms most frequently used to challenge moist heat sterilization cycles are *G. stearothermophilus* and *Clostridium sporogenes*. These spore-forming bacteria are selected because of their relatively high heat resistance. For the bioburden cycles, in addition to the selection of an appropriate organism for use as a BI, the concentration and resistance of the indigenous microbial population is established.

Modest reductions in sterilizing conditions used for the microbial challenge studies are a common practice in parts sterilization to afford and additional safety measure. Reduction of sterilizing dwell time by one to three minutes and/or reduction in sterilizer set point temperature of 1°C can have been successfully utilized for years. The half-cycle approach originally developed for ETO sterilization validation is an extreme example of this practice, and while effective it extends processing times unnecessarily.

When inoculating solid materials, the spores can be introduced onto the surface of the item. Subsequent to inoculation, the spore suspension is allowed to dry on the surface. Recovery counts should be conducted on selected inoculated components to verify the delivered concentration of spores. Commercially available spore strips may also be used when the confirmation loads are composed of devices and solid materials.

Microbiological challenge studies are typically conducted concurrently with the heat penetration studies. Similarly, when spore strips are used they should be placed adjacent to a TC probe. To expedite recovery and eliminate possible confusion, any directly inoculated items should be identified by markings or other suitable means.

After the sterilization cycle is complete, the inoculated items or spore strips are recovered and subjected to microbiological test procedures. Strips are immersed in a suitable growth medium (soybean casein digest medium is typical) and incubated for up to seven days. Incubation temperature for *G. stearothermophilus* is 50°C to 55°C. For overkill cycles it is expected that all spore strips will be negative (not exhibit growth). To provide further assurance, both positive (unsterilized strips) and negative (growth medium with no spores) controls should be incubated along with the challenge samples.

Sterilizer Filter Evaluation

Microbially retentive filters are employed on most parts sterilizers to ensure that loads are not contaminated by air used to vent the chamber as it cools or dries. Product loads are protected from such contamination by their primary containers (vials, bags) and many nonproduct loads are protected by wraps to provide a microbial barrier. Nevertheless, because of the possibility of pressure differentials between the chamber and the sterilized article during cool down or vacuum-drying, filters are valuable.

For filters, two issues are of concern: sterility and integrity. Filters are typically sterilized during the load sterilization cycle. Filters should be probed with TCs upstream and downstream of the membrane. A suitable microbial challenge should be applied to the filter itself.

Because filters are resistant to heat, this should be similar to the challenge applied in overkill cycles. If the load will undergo a bioburden cycle, it may be necessary to sterilize the filter in a separate phase of the cycle. Bioindicators may be treated as in overkill cycle challenges.

The integrity of the filter must also be evaluated according to recommendations by the filter manufacturer. To ensure that filters will remain functional under all expected conditions, the integrity tests should be done following the maximum cycle time and temperature allowable under standard operating procedures. Triplicate studies are recommended.

The Validation Report

Record keeping is a prime requirement of current GMPs. The records required for a validated steam sterilization cycle follow. They are usually stored in a secure common central file, but they must be readily accessible. It is wise to assign the task of organization and retrieval of records to a single group. These records are as follows:

- Qualification reference documents (specifications, drawings, and calibration records)
- OQ protocol and record
- Approved process confirmation (validation) protocol
- Raw calibration and validation data
- Approved validation report

The validation report is the guideline to maintenance of a validated sterilization process. It describes the cycle and the operating conditions that have been proved to give adequate assurance of sterility. It explains in detail how the manufacturing group can obtain results consistent with the validation study.

Several formats and degrees of complexity are used in report writing. However, all reports should contain some common elements as follows:

- Identification of the task report by number.
- Reference to the protocol under which it was carried out.
- A brief summary of the range of operational conditions experienced and how they were controlled.
- A procedure for maintaining control within the approved range. This may be in the form of a standard operating procedure.
- A summary and analysis of the experimental results. This will include the range of lethality and degree of sterility assurance.
- A brief description of any deviation from the expected results.

The range of lethality is calculated directly from the temperature data. It is important that a range be reported in the case of heat-labile products. The upper range of temperature exposure is critical to product stability. A sterilization cycle is also a product-processing step. Its effect on the product, as well as on the microbial population, must be considered. The description of such effect need not be included in the validation report itself. It should be the subject of adjunct analytical or stability studies. A discussion of the importance of this consideration is included later.

Cycle development reports are not usually a part of the validation report, although these are important to the life cycle validation concept preferred by the authors. Some reference to how the cycle was chosen may be included in the validation report. This can be the title of the cycle development report or a brief summary of the results of that report. This should include the type of mechanical cycle recommended (high prevacuum, air–steam mixture, or others), heat resistance and bioburden data or assumptions thereof, and level of lethality required.

Bioindicator data are the ultimate proof that the sterilization cycle has been successful. As such, it should be highlighted in the validation report. The microbiology section of the report should include the methods used, a summary of results, and conclusions. On completion, the final report is circulated for approval. This is generally done by the same people who approved the protocol.

Maintenance of Validation

The last, and often overlooked, step in validating any process is the program to ensure that conditions established in cycle development and confirmed in process confirmation or validation studies are controlled and maintained. If this is done, and no major changes in equipment or process are made, periodic repetition of selected validation studies is not really required.

What is required is a periodic review of the system for adherence to the validation criteria. This may be very simple if a good program of *validation maintenance* is established. This term is used rather than "revalidation" to emphasize the continuity of the program. A validation maintenance review report may be issued to commit to record the attention being paid to this critical aspect of validation. Some key points of a good validation maintenance program are as follows:

- A routine calibration program for all instruments critical to the operation of the sterilizer and its support systems.
- A preventative maintenance program for other system components. This should include periodic operational rechecks and comparison to the OQ Record.
- Routine monitoring of bioburden and (optionally) periodic bioindicator challenges.
- Well-maintained and accessible operating records and equipment logs.
- Process and equipment change control procedures.

These subject proposed changes to prior review to establish whether additional validation experiments are required.

Because of the critical nature of sterilization, it is recommended that studies be performed on an annual basis to supplement the validation maintenance program. These should entail representative loaded chamber heat penetration and microbiological challenges. Many firms choose to perform only one such study on a selected load.

The basis of continued validation maintenance is communication among the various operating groups (Manufacturing, Quality Assurance, Validation, or others). Sterilization processes were the first for which validation was emphasized. They continue to be the most heavily reviewed. It is important that the state of control of these processes be strongly maintained and the subject of concern to all these groups.

CONCLUSION

The proposed approach to validation of steam sterilization in autoclaves follows the basic life cycle concepts applicable to all validation programs. Understand the function sterilization process, develop and understand the cycles to carry out the process, and define a suitable test or series of tests to confirm that the function of the process is suitably ensured by the structure provided.

Sterilization of product and components and parts that come in direct contact with sterilized product is the most critical of pharmaceutical processes. Consequently, this process requires a most rigorous and detailed approach to validation. An understanding of the process requires a basic understanding of microbial death, the parameters that facilitate that death, the accepted definition of sterility, and the relation between that definition and sterilization parameters. Autoclaves and support systems need to be designed, installed, and qualified in a manner that ensures their continued reliability. Lastly, the test program must be complete and definitive.

Fortunately, steam sterilization in autoclaves is very effective. Failure of autoclave cycles in modern pharmaceutical manufacturing operations is rare. Nevertheless, the consequence of failure is so great that it is easy to justify the effort required to validate this critical operation.

REFERENCES

1. Perkins JJ. Principles and Methods of Sterilization in Health Sciences. Springfield, IL: Charles C Thomas, 1973.
2. Phillips GB, Miller SW, eds. Industrial Sterilization, International Symposium, Amsterdam, 1972. Durham, NC: Duke University Press, 1973.
3. Pflug IJ, ed. Microbiology and Engineering of Sterilization Processes. 3rd ed. St Paul, MN: Environmental Sterilization Services, 1979.
4. Ball CO, Olson FC W. Sterilization in Food Technology. New York: McGraw-Hill, 1957.
5. PDA. Validation of Steam Sterilization Cycles. Technical Monograph No. 1. Philadelphia, PA: Parenteral Drug Association, 1978.

13

Validation of Terminal Sterilization

Thomas J. Berger and Kevin D. Trupp
Department of Microbiology and Global Process Sterilization Engineering, Hospira, Inc., Lake Forest, Illinois, U.S.A.

INTRODUCTION TO PARENTERAL PRODUCT STERILIZATION

The previous chapter discussed the steam sterilization approach for the processing of hard goods or porous loads. This chapter will discuss the sterilization validation approach that can be used in the processing of parenteral products by terminal sterilization using moist heat. The underlying principles of steam sterilization are applicable to both hard goods and terminal sterilization of parenteral products, but both have their unique characteristics.

An organized sequential flow of activities must occur as new parenteral formulations are developed, and subsequently processed in the manufacturing facility. The moist heat sterilization of pharmaceutical solutions is established and verified through a series of activities that confirm the product has received a defined thermal exposure that renders the product sterile. R&D activities can include sterilization developmental engineering studies consisting of sterilization cycle development; container thermal mapping; microbial closure validation, *D*- and *z*-value analysis; container–closure integrity validations as well as final formulation stability studies.

The subsequent production phase activities must include initial sterilization vessel qualification which demonstrates that the vessel will deliver the defined sterilization process in a consistent and reproducible manner. Also, solution and container–closure microbial validation studies must be conducted at subprocess production sterilization conditions employing heat-resistant microorganisms. Equipment validation, filtration studies and assessment of the bioburden on component parts, as well as the environment, must also be ascertained.

The developmental and production phases of sterilization technology activities are then drafted into documents that are submitted as part of a new drug application for the particular parenteral formulation. These reports must follow applicable regulatory requirements for products that are terminally sterilized. Such studies allow one to establish, with a high level of sterilization assurance, the correct sterilization cycle (F_0, temperature, product time above 100°C, etc.) to be used for the steam sterilization of a specific parenteral formulation in a particular container–closure system.

STERILIZER DESIGN

The validation of a steam sterilization cycle is dependent on the equipment chosen. The sterilizer and its support systems must be designed and constructed to deliver the effective cycles repeatedly and consistently. Qualification of the sterilizer consists of proper design, installation according to design, operational testing to ensure that design criteria and operational requirements are met and performance qualification to confirm that the product is sterilized per specification.

Sterilizer design is geared to the type of product or materials/equipment to be sterilized. All steam sterilization cycles are based on contact with saturated steam, steam–air mixtures or superheated water. Saturated steam is water vapor in equilibrium with liquid water. The values of temperature and pressure at which pure saturated steam can exist are shown by the phase diagram in Chapter 12, Figure 3. Saturated steam can exist only along the phase boundary for liquid and gaseous water; that is, the relation between its temperature and pressure is fixed. An increase or reduction in the temperature of saturated steam must result in a corresponding increase or decrease in its pressure and vice versa. Steam–air mixtures can be used when overpressure is required to maintain product shape or container integrity. Superheated water cycles require air overpressure and the water is either heated by direct injection of steam or indirectly via a heat exchanger.

Parts and hard goods are typically steam sterilized using a saturated steam process whereas the trend for product sterilization is towards the use of superheated water or steam–air mixture processes. These processes are needed as a majority of the new products require air overpressure during the sterilization process to maintain desired container characteristics and integrity.

Abbreviations used in this chapter: AAMI, Association for the Advancement of Medical Instrumentation; APE, antimicrobial preservative efficacy; API, active pharmaceutical ingredient; ASME, American Society of Mechanical Engineers; BET, bacterial endotoxin testing; BI, biological indicator; BIER, biological indicator evaluator resistometer; DP, direct plate; EMA, European Medicinal Agency; F/N, fraction/negative; GAMP, good automated manufacturing practice; HMI, human–machine interface; I/O, input/output; ICH, International Conference on Harmonization; ISPE, International Society for Pharmaceutical Engineering; LAL, limulus amebocyte lysate; LVP, large volume parenterals; MES, manufacturing execution system; MOS, maintenance of sterility; P&D, penetration and distribution; PLC, programmable logic controller; PSLR, predicted spore logarithmic reduction; R&D, research and development; RTD, resistance temperature detector; SCADA, supervisory control and data acquisition; SLR, spore logarithmic reduction; SVP, small volume parenterals; TC, thermocouple.

Products in glass containers can utilize the saturated steam processes as described in Chapter 12, but many products and containers require the use of air overpressure during the sterilization process. This section will discuss some of the key design considerations for terminal steam sterilizers and provide some specifics for the various type steam sterilization processes utilizing air overpressure.

Typical Design Considerations for Steam Sterilizers

1. A pressure vessel constructed according to the ASME or equivalent international code. This must withstand at least 50% in excess of the required internal pressures.
2. A safety door mechanism to prevent opening while the unit is under pressure: the locking device may be actuated directly by internal pressure or indirectly through an automatic switch. The door itself may be of the swing-out or sliding type. Separate entry and exit doors are preferable.
3. Process control system (typically a PLC for controlling and monitoring the process).
4. Process data recorder or data collection system.
5. Product racks designed to hold/support the sealed product containers and to provide adequate heating/cooling media flow throughout the product zone.
6. Pressure safety relief valves for both the chamber and jacket (if equipped with jacket).

Note. A microbial retentive vent/air filter would not typically be required for processes used for terminal sterilization as there is no direct contact between the heating/cooling media and the contents of the containers.

Steam–Air Mixture Sterilization

The primary benefit to the steam–air mixture process over a superheated water process is the product is not subjected to direct contact with water (except as condensate), which in some cases can cause cosmetic issues with the container. Steam–air mixture processes typically utilize large recirculating fans to prevent the formation of cold/hot spots in the sterilizer. The steam–air mixture process typically uses an indirect cooling method such as cooling of the jacket or with cooling coils within the sterilizer. Because of this indirect cooling method, the cooling rate of the product is typically much slower and less efficient than direct exposure of the product containers to cooling water.

Some of the specific sterilizer design considerations for a steam–air mixture process include the following:

1. A jacket and insulation: the jacket would utilize steam during heating and exposure phases of the cycles and cooling water can be introduced to the jacket during the cooling phase of the process.
2. A thermostatic steam trap to efficiently remove the condensate from the chamber: this is open when cool (in contact with air or condensate) and closed when in contact with steam. As condensate collects, the trap opens owing to the slight temperature reduction and the condensate is discharged. There is also a similar steam trap to remove steam condensate from the jacket.
3. Fan(s) to continuously recirculate the steam–air mixture during heat-up and exposure and to recirculate the air during cooling.
4. Cooling provisions (e.g., cooling coils) to cool the air/product.

Recirculated Superheated Water Sterilization

Sterilization with recirculating superheated water (sometimes referred to as a water cascade or raining process) is more efficient than a steam–air mixture and is therefore more common. There are many types of recirculating superheated water processes, the most common is a process where the bottom portion of the sterilizer (below the product zone) is filled with water and a recirculation pump is used to continuously recirculate water from the bottom of the sterilizer to spray nozzles above the product zone. A slight modification to that process is the use of a water distribution pan in lieu of spray nozzles. Another version of the recirculating superheated water process is to completely submerge the product in water but this process is inefficient from a utilities consumption standpoint. All of these recirculating superheated water processes utilize air overpressure and the overpressure can be controlled during the sterilization process to minimize most types of container deformation. There is no limit to the maximum overpressure used but it would typically be limited by the chamber pressure rating. The minimum overpressure will be driven by the temperature being used, the pressure needed to maintain the desired product characteristics and the required overpressure needed to prevent the recirculation pump from loosing prime. These recirculating processes are typically heated and cooled indirectly with external heat exchangers located in the recirculating water loop but direct injection of steam and cooling water can also be used.

In addition to the typical sterilizer design considerations mentioned earlier, a superheated water sterilizer would also include a large recirculating water system (e.g., pump, pipes, heat exchangers, headers, spray nozzles) including specific water level control valves and monitoring devices.

Rotary and Shaker Sterilization

In some cases, certain products (i.e., suspensions and emulsions) require agitation during the sterilization process. For those types of products, it is typical to use a rotating rack within the sterilizer but other agitation methods such as an internal shaking device are available. Refer to Figure 1 for the typical design of a rotary sterilizer and Figure 2 for the typical design of a sterilizer using a shaking mechanism. It is possible to use any of the sterilization processes listed above with product agitation.

Continuous Sterilization

For this version of the superheated water process, containers are terminally heat sterilized in a continuous sterilizer by a process where the containers move through a constantly controlled environment in carriers with individual compartments. The time, temperature, and pressure requirements are set to predetermined values and are automatically and continuously controlled, monitored, and recorded. Refer to Figure 3 which depicts the pattern

Figure 1 Photo of sterilizer with a rotary mechanism. *Source*: Photo Provided by Fedegari Autoclavi SpA, Albuzzano, Italy.

that containers (e.g., parenteral flexible product containers) follow as they move automatically through the continuous sterilizer. The water lock of the pressure vessel is used to provide product entrance into and out of the overpressure environment. The overpressure environment is constantly maintained within predetermined limits.

The sterilizing phase begins as the product enters the hot water environment within the pressure vessel. The hot water environment may be a superheated water spray, which is circulated over the top of the continuously moving carriers. The residence time of the product within this sterilizing environment and the water temperature are controlled within predetermined limits to assure the required heat input.

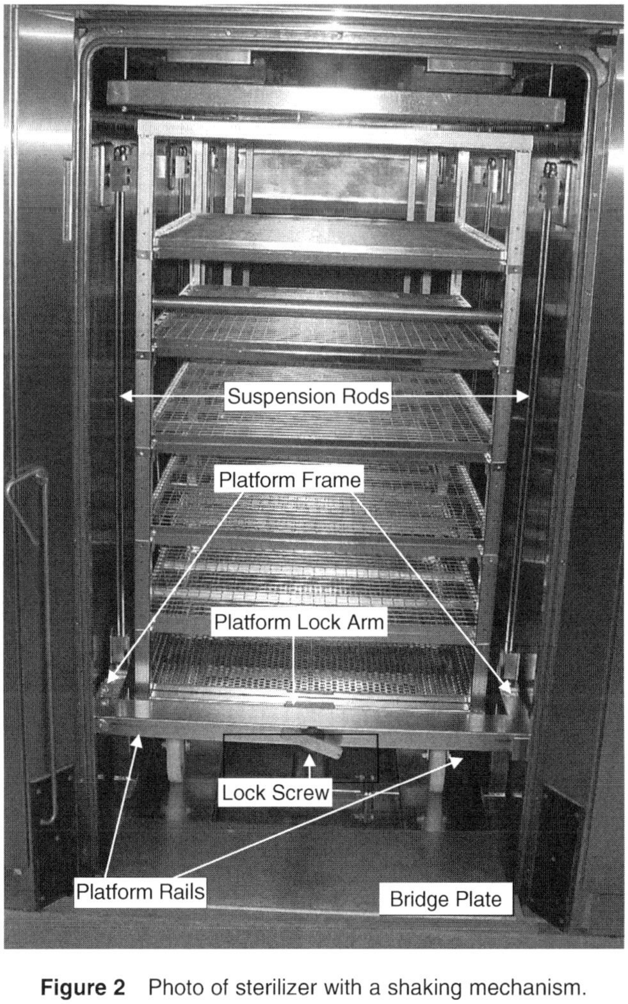

Figure 2 Photo of sterilizer with a shaking mechanism.

Cooling begins as the product transfers from the sterilizing environment and enters the cooling environment, which is also within the pressure vessel. The cooling water environment is a cool water spray that is circulated over the top of the continuously moving carriers. The temperature of this water is controlled within predetermined limits to assure that the required degree of cooling is achieved before the product leaves the cooling environment.

A system of fixed temperature sensors located in the entering and exiting recirculating water for both heating and cooling continuously monitors, records, and controls the temperature of the process water.

Air overpressure is required to protect the container from stress while exposed to the high sterilizing temperatures.

STERILIZER CONTROL SYSTEMS

A key to effective sterilizer operation lies in the automated process control system. By eliminating the dependence on operator intervention and data recording, automatic temperature and sequential control provides assurance that the "validated" sterilization cycle is consistently and repeatedly delivered. A typical control system for a new sterilizer includes the following hardware components:

- PLC
- Operator interface panel(s)
- Data recorder/data collection system
- Process Variable Sensors
- I/O devices

The PLC is most commonly used as the primary component of the automated process control system as it provides sequential control of the process, provides control of all proportional valves, controls all devices, receives operator input via the operator interface panels and provides process information (such as process variable information and alarms) to the operator via displays and/or operator interface panels. The PLC typically contains specific recipe information for the various cycles to be utilized. In some cases the PLC can be used for data collection, but it is much more common to use a separate data recorder/data collection system.

The operator interface panel can be as simple as switches and displays or as complex as a stand-alone PC running a SCADA with a HMI software package. These devices are typically used to select the recipe, start the cycle and display process information during the cycle. The higher level PC based SCADA type operator interface panels can provide detailed cycle reports and trending information.

The data recorder/data collection system can range from a simple strip chart recorder to a full-blown MES type data collection system. In many cases the PLC can also provides batch data logging functionality. The minimum variables to record for steam sterilization processes are typically time, temperature, and pressure.

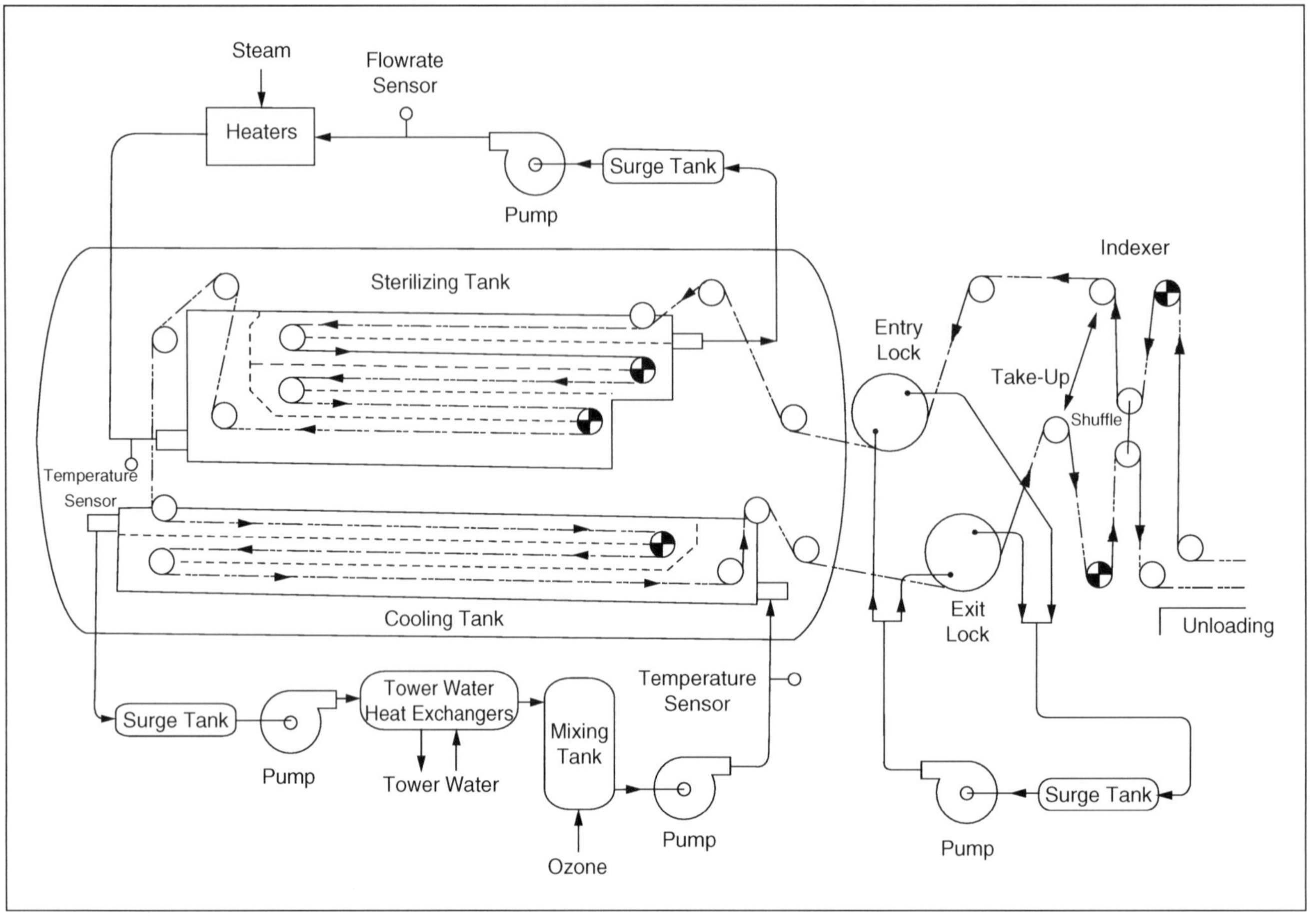

Figure 3 Schematic of a sterilizer for "continuous" processing of flexible containers.

Typical sensors include temperature measurement devices (RTDs or TCs), pressure measurement devices, and where applicable level measurement devices and flow measurement devices. It is customary that the temperature sensor used to control the process temperature not be used to provide the batch record process data. An independent/secondary temperature sensor for batch reporting provides a high degree of insurance that the cycle actually ran within its defined limits. Heavy wall thermowells should not be used, as this will affect the time response of the measurement. Thin-walled thermowells or temperature elements with stainless steel sheaths should be used for temperature measurement.

The pressure sensor should be equipped with a sanitary-type diaphragm and connected to the sterilizer using a sanitary fitting. A sanitary diaphragm can introduce errors to the pressure measurement due to the stiffness of the diaphragm. This stiffness is related to the size of the diaphragm. The impact is negligible for diaphragms above 3 inches diameter. This should be considered when sizing the connection to the sterilizer.

Sterilizers that maintain a specific water level (i.e., recirculated water process) should be equipped with liquid level sensors. These sensors may be in the form of single-point-level-type probe or a continuous level sensor. Regardless of what type sensor is used a separate high-level sensor must also be provided. The separate high-level sensor provides greater assurance that the collected water at the bottom of the vessel remains below the product level.

Sterilizers that rely on recirculated water as part of the sterilization process can include a flow sensor. The flow sensor may be a direct measurement such as a flow meter (i.e., coriolis, ultrasonic, magnetic, etc.) or an indirect measurement such as differential pressure sensor across the recirculation pump. Direct measurements are always preferred.

For I/O devices, there are analog types and discrete types. The analog inputs are typically from process sensors and the analog outputs are typically for control of proportional valves. The discrete inputs are typically from switch type (operator and process) devices and the discrete outputs are typically for activating hardware such as valves, pumps, lights, etc.

The design and development of the sterilizer control system software should follow the principles of ISPE GAMP 4 Guide for Validation of Automated Systems (1). This guideline details a software life cycle from conception thru decommissioning.

STERILIZATION CYCLES

The type of steam sterilization cycle to be utilized is dependent on product needs and equipment availability. As discussed in Chapter 12, the sterilization of hard goods or porous loads typically require the use of a pulsed pre-vacuum cycle as it is preferable to remove the air from the porous materials being sterilized whereas in the terminal sterilization of aqueous solutions in sealed containers, the major concern is to provide rapid heat transfer to the wall of the filled product containers and air removal is not required (nor even desirable as the hydrating moisture is contained within each container). Parenteral products may be filled into rigid or flexible containers. In either there is typically air or nitrogen present in the headspace above the liquid. As the solution is heated, this gas expands and adds to the internal pressure increase resulting from the evolution of water vapor from the aqueous vehicle within the heated container. Thus, the pressure within the container will exceed the chamber pressure during steam process for sealed containers.

Glass vials can be sealed with special closures to withstand this pressure. As long as the pressure differential between the chamber and the containers does not become too great during the steam exhaust portion of the cycle, the vials will not burst. If rapid cooling of the load is desired, the pressure differential might become significant enough to cause closure integrity to be lost.

Plastic bags, semi-rigid containers and syringes present a greater problem because they do not have the inherent strength of glass and may burst or deform as the pressure differential increases. To prevent this, air must be injected into the chamber to raise the pressure above the saturation pressure of the steam. This is particularly important during the cooling cycle, when the chamber pressure is reduced at a much faster rate than that within the container.

The following section provides a description of the various steam sterilization cycles used for parenteral products in sealed containers.

Saturated Steam–Pre-Vacuum Cycle

For a saturated steam process, the most common (and perhaps most effective) method to remove the entrapped air from the sterilizer is to remove it mechanically before the actual sterilization begins. This is done by means of a mechanical vacuum pump or steam eductor. This cycle can be used for products in glass containers. A sketch of a typical pre-vacuum cycle is shown in Chapter 12, Figure 7.

Saturated Steam–Gravity Displacement or Steam Purge Cycle

Other means for eliminating air without a vacuum source include the use of a gravity displacement cycle or a steam purge cycle. For the gravity displacement cycle, steam is introduced on the side or top of the vessel and the cold air is forced out via the drain. The steam purge cycle uses large quantities of steam distributed via headers under the entire product zone with numerous large vents located at the top of the vessel. For these types of cycles, the appropriate vents or drains should be fully open and the large steam supply valve fully open for an extended time and temperature to ensure that the air is adequately removed from the sterilizer. Once the vents and drains close, the process runs like a traditional saturated steam process. It is important to determine that the measured temperature and pressure are consistent with the steam saturation curve in Chapter 12, Figure 3. This process can be used with glass containers.

Steam–Air Mixture Cycle

It is important to understand the physical principle involved in a mixture of steam and air. The fixed relationship between temperature and pressure seen in Chapter 12, Figure 3 no longer applies. Dalton's law states that the pressure of an ideal mixture of gases is equal to the sum of the partial pressure of the gases, or

$$P = P_A + P_B + P_C.$$

Raoult's law further states that, for ideal mixtures, the partial pressure of the gas is equal to its vapor pressure multiplied by the mole fraction in the liquid. For steam in equilibrium with pure condensate, this reduces to

$$P_A = p_A^*$$

where P_A is the partial pressure of steam and p_A^* is the vapor pressure of the condensate. The difference between the observed chamber pressure P and P_A is the partial pressure of air.

The presence of air, although necessary for the maintenance of container integrity, can reduce the heat transfer efficiency. The objective of the design in the "air overpressure" cycle is to maintain a well-mixed chamber. This assures that the heat transfer to the load will be uniform regardless of the presence of air. Mixing may be accomplished in several ways. The air may be injected directly into the incoming steam. Usually, though, some mechanical means is selected.

Most steam–air sterilizers use a fan built into the top or end of the chamber, which circulates and mixes the air and the steam (Chapter 12, Fig. 9). Some steam–air sterilizers are capable of using water during the cool-down process to cool the containers more rapidly. This rapid cooling may also be necessary for product stability. Various methods (i.e., direct injection, recirculation through a heat exchanger, etc.) for introducing the cooling media can be utilized.

Recirculating Superheated Water Cycle

The typical recirculating superheated water process (sometimes referred to as a water cascade or raining process) begins by the addition of water to the sterilizer to a predefined level (below the product zone). Then a water recirculation pump is started to continuously recirculate water from the bottom of the sterilizer to spray nozzles or a water distribution pan above the product zone. The recirculation pump is on throughout the heat-up exposure and cool-down phases. During heat-up, the water is heated at a pre-defined rate via a heat exchanger in the recirculation loop or with the direct injection of steam. Also during heat-up, compressed air is added to the chamber to attain the desired overpressure levels. Once the temperature set point is achieved, the controller steps into the hold portion of the cycle and the temperature and pressures are maintained at the desired levels. For cooling, the steam supply is shut off and the recirculating water is cooled at a controlled rate by introducing cooling media to a heat exchanger installed in the water recirculation loop or by the direct injection of cooling water into the recirculating loop. This type of process does not require the use of a jacket but does require specific water level controls.

The recirculating superheated water process is very efficient and the temperatures and pressures can be tightly controlled during the entire process, thus minimizing container stresses.

STERILIZATION CYCLE DEVELOPMENT

This section will address sterilization and associated microbiological activities that occur in R&D areas as well as the production environment when using the BI/bioburden approach in support of a parenteral product. The list below depicts some of the sterilization engineering and microbiological activities associated with a parenteral product as it moves through development. These studies or similar ones are ordinarily conducted in developmental sterilizers or may occur as investigative engineering studies in a production sterilizer as appropriate. The overkill method can be used for some of the more stable parenteral formulations, and its validation is accomplished as described in Chapter 12.

Sterilization development activity	Activity statement
Cycle development	Develop preliminary container sterilization specifications with engineering parameters such as temperature, time and F_0
Container thermal mapping	Determine cold spot and assess heat penetration within finished container
Formulation development	Perform analytical feasibility studies prior to product finalization with method's validations
Parenteral solution microbiological evaluations:	
Moist heat *D*- and *z*-value analysis	Perform triplicate *D*-value analysis on each parenteral formulation at three temperatures, e.g., 112°C, 118°C and 121°C and then calculate the *z*-value
APE	Perform on final product if it contains a preservative or if there is a multidose claim for the container
In-process bioburden analysis	Perform studies with a panel of microorganisms to validate 70% recovery for the filtration process
Spike hold time studies	Inoculate parenteral product with bioburden and growth promotion compendia organisms to evaluate the product's ability to support microbial growth
Container closure evaluations:	
Microbial closure inactivation	Perform kill curve kinetics using bioburden and BI (spores) inoculated onto the worse case closure site
Container integrity	Perform dye ingress, microbial challenge or physical integrity tests following exposure to maximum sterilization conditions that stress the container
Stability runs	Perform analytical chemistry and microbiological evaluations at various temperatures and times per ICH and or compendia requirements
BET	Perform test method validation of API, excipients and final product per compendia requirements

Sterilization engineering personnel primarily focus their efforts in determining whether a parenteral formulation packaged in a particular container configuration can be sterilized in a current cycle or whether a new cycle must be developed. The referenced EMEA (2) decision tree is followed when evaluating a new parenteral product in an LVP or SVP container. Sterilization feasibility studies are conducted in a sterilizer to ascertain the physical effects of the cycle on the product in question. Product attributes that can be affected by a cycle are closure integrity, product potency, pH, color, shelf-life stability, visible, and subvisible particulates as well as final product sterility. Once the basic engineering parameters (e.g., temperature, time and F_0) are established, then engineering thermal container mapping studies can be performed (3,4).

Container Thermal Mapping Validation Studies

An R&D sterilizer is smaller than a production facility sterilizer, but can simulate the sterilization cycles conducted in the larger production vessels. Container thermal mapping studies (when applicable) are typically performed in a laboratory sterilizer:

1. To locate the coldest zone or area inside a container.
2. To determine the cold zone in the container and its relationship to the location monitored during validation studies.
3. To generate data that may be used during the setting of production sterilization control parameters.

When conducting thermal mapping studies, there are various factors to be considered, and these are dependent upon the:

1. Type of container (flexible or rigid)
2. Container orientation, size and fill volume
3. Cycle type and temperature
4. Viscosity
5. Autoclave trays/design/surface contact
6. Autoclave spray patterns/water flow.

Typical container mapping data obtained for lipid emulsions contained within a 1000 mL glass container are shown as an example in Tables 1 and 2. The following summarizes the process for obtaining heat map data from the glass intravenous container filled with approximately 1000 mL of lipid emulsion.

Table 1 Heat Input (F_0 Units)

	1000 mL glass I.V. containers-heat mapping study (lipid emulsion)				
	Run CLHK00.049		Run CLHK01.050		
TC number	**btl 1**	**btl 2**	**btl 1**	**btl 2**	**Average (SD)**
1,12	7.91	C7.28	8.13	C7.36	7.67 (0.415)
2,13 (PC)	7.79	7.49	8.02	7.64	7.74 (0.226)
3,14	C7.46	7.40	C7.71	7.47	C7.51 (0.137)
4,15	7.64	7.80	7.87	7.96	7.82 (0.135)
5,16	12.66	12.90	12.95	12.91	12.86 (0.132)
6,17	12.73	12.46	12.77	12.68	12.66 (0.138)
7,18	12.78	12.69	12.95	12.91	12.83 (0.120)
8,19	13.32	13.33	13.42	13.78	13.46 (0.223)
9,20	14.21	14.33	14.03	14.56	14.28 (0.222)
10,21	H15.87	H17.24	H15.18	H16.09	H16.10 (0.856)
11,22	15.47	16.56	14.77	16.07	15.72 (0.773)
H–C	8.41	9.96	7.47	8.73	8.64 (1.028)
PC–C	0.33	0.21	0.31	0.28	0.28 (0.053)

Note: H denotes hottest TC location; C denotes coldest TC location; PC denotes approximate location of the production profile TC; Data from TC#9 used with a postcalibration variance of +0.25°C at 100°C; All heat input values are calibration corrected.

TC probes (Copper Constantan, type T, 0.005 in. diameter) were used to monitor 11 locations within the 1000 mL container. The TC probes were positioned at various distances (in inches) as depicted (Fig. 4). Each container was filled with approximately 1000 mL of the lipid emulsion, evacuated to 20 in. of mercury and sealed with an aluminum overseal.

A flat perforated rack on a reciprocating shaker cart was used in the autoclave. The cycle's target temperature was 123°C, recirculating water spray cycle with 70 rpm of axial agitation and 30 psig (pounds per square inch) of air overpressure.

When the sterilization cycle was controlled to give a heat input of approximately 7.5 F_0 minutes in the coldest emulsion area, the average coldest emulsion area was found to be measured by TC number (TC#) 3,14. The average hottest emulsion area was measured by TC# 10, 21. The difference between the hottest and coldest emulsion areas ranged from 7.5 to 10.0 F_0 minutes with an average of 8.6 F_0 minutes. Therefore, when the coldest emulsion area registered 7.5 F_0 minutes, the hottest emulsion area would average 16.1 F_0 minutes.

The emulsion area approximating the validation TC location was measured by TC # 2,13 and averaged 7.7 F_0 minutes when the coldest emulsion was approximately 7.5 F_0 minutes (Fig. 5).

Solution/Product Moist Heat Resistance *D*- and *z*-Value Analysis

A BIER vessel meets specific performance requirements for the assessment of BIs per American National Standards developed and published by AAMI (5). One important requirement for a BIER steam vessel is the capability of monitoring a square wave heating profile.

Refer to Figure 6 for a schematic of the steam BIER vessel used to generate the D and z value data. D-value is the time in minutes required for a one log or 90% reduction in microbial population (Refer to Chapter 12, Fig. 1). The z-value is the number of degrees of

Table 2 Solution Heat Rates (Minutes)

	1000 mL glass I.V. containers-heat mapping study (lipid emulsion)				
	Run CLHK00.49		Run CLHK01.050		
	btl 1	**btl 2**	**btl 1**	**btl 2**	**Average (SD)**
Coldest location					
Thermocouple number	3	12	3	12	–
Time to 100°C	19.0	19.0	19.0	19.0	19.00 (0.000)
Time ≥100°C	21.0	21.0	21.0	21.0	21.00 (0.577)
Time ≥120°C	4.0	3.0	4.0	3.0	3.50 (0.577)
Time ≥120–100°C	4.0	5.0	4.0	5.0	4.50 (0.577)
Maximum temperature (°C)	120.82	120.77	120.92	120.77	120.82 (0.071)
Heat input (F_0)	7.46	7.28	7.71	7.36	7.45 (0.187)
Production profile TC location					
Thermocouple number	2	13	2	13	–
Time to 100°C	19.0	19.0	19.0	19.0	19.00 (0.000)
Time ≥100°C	22.0	21.0	22.0	21.0	21.50 (0.577)
Time ≥120°C	4.0	3.0	4.0	3.0	3.50 (0.577)
Time ≥120–100°C	5.0	5.0	5.0	5.0	5.00 (0.000)
Maximum temperature (°C)	120.91	120.82	120.91	120.92	120.89 (0.047)
Heat input (F_0)	7.79	7.49	8.02	7.64	7.74 (0.226)

Note: H denotes hottest TC location; C denotes coldest TC location; PC denotes approximate location of the production profile TC; Data from TC#9 used with a postcalibration variance of +0.25°C at 100°C; All heat input values are calibration corrected.

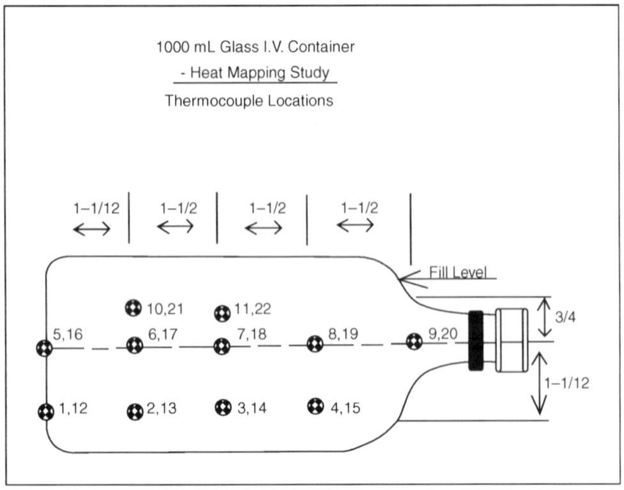

Figure 4 Heat mapping study using a 1000 mL glass container.

temperature required for a 10-fold change in the *D*-value. (Refer to Chapter 11 for additional details on *F*-, *D*- and *z*-values.)

Master Solution/Product Concept

The family category of lipid emulsions and their respective $D_{121°C}$ and *z* values as well as classification in terms of microbial resistance is shown in Table 3. A categorization of parenteral formulations with associated $D_{121°C}$ and *z*-values and their potential impact on microbial resistance using the BI, *Clostridium sporogenes* were previously reported (6). In addition, the methodologies used for *D*- and *z*-value analysis were likewise cited. The data in Table 3 indicate that the # 1 emulsion is at the top of the list, since it affords the most microbial moist heat resistance. It is therefore the emulsion that should be microbiologically challenged (inoculated with spores) as part of the emulsion validation scheme. *D*- and *z*-value data have been reported for other BIs such as *Geobacillus stearothermophilus* (6–8) and *Bacillus subtilis* 5230 (9). There are many factors that can affect moist heat resistance including a BI's age, sporulation media used, as well as the particular spore strain employed (10).

PSLR Values

Lipid emulsion moist heat resistance values ($D_{121°C}$ and *z*-values) were generated in the steam BIER vessel using the BI *C. sporogenes* as shown in Table 3. The columns in the Table list the representative code or list number of the product, the emulsion or product name, its average $D_{121°C}$ value and *z*-value and finally the PSLR value. Those parenteral formulations with the lowest PSLR value(s) are those that should be used for the microbial validation at subprocess conditions, since these provide the most microbial resistance (6).

Accumulated F_{bio} for Lipid Emulsions

Accumulated F_{bio} and *z*-values (Table 4) were used to construct the PSLR ranking for lipid emulsions as previously discussed for Table 3. The F_{bio} is the heat input for the biological solution based on the emulsion's moist heat *D*- and *z*-values. By inputting the sterilizer temperatures from the coldest TC of an engineering run for a particular container/sterilization cycle, the emulsion can be ranked according to PSLR values. The combined $D_{121°C}$ and *z*-value allows comparison of moist heat rankings between emulsions.

The data in Table 4 demonstrate that the #1 Emulsion has the lowest PSLR (7.105), thereby affording the highest moist heat resistance upon inoculation. Generation of this table allows prediction of which emulsion to microbiologically challenge as part of validation in the production sterilizer.

Microbial Closure Inactivation Validation in a Developmental Sterilizer

In lieu of using the large type steam sterilizers in the production environment, microbial inactivation at the closure/bottle interface of an emulsion container can be assessed in a developmental sterilizer. The closure microbial inactivation (kinetic) studies can determine how the size of the container, type of closure compound used as well as closure preparatory processes (e.g., leaching, washing, siliconizing, autoclaving) influence microbial inactivation. Microbial closure kinetic studies are conducted at various time intervals in a given sterilization cycle. DP count or F/N methodologies are used to

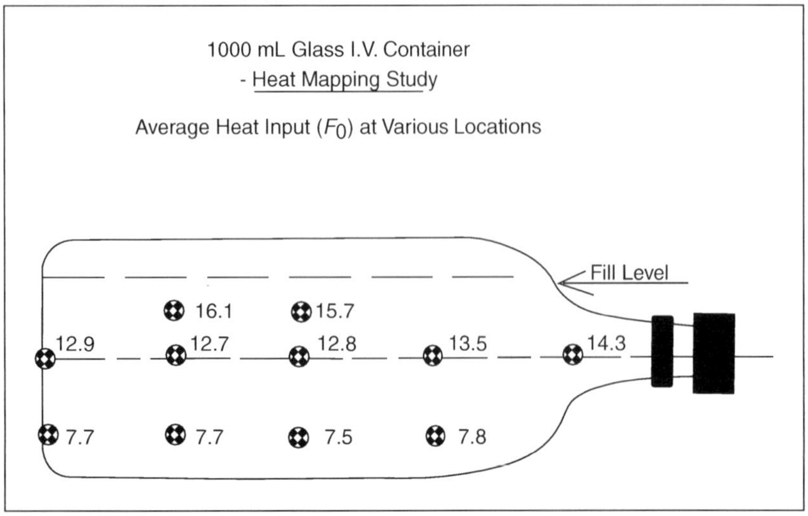

Figure 5 Heat mapping study with average heat input (F_0) at various locations in a 1000 mL glass container.

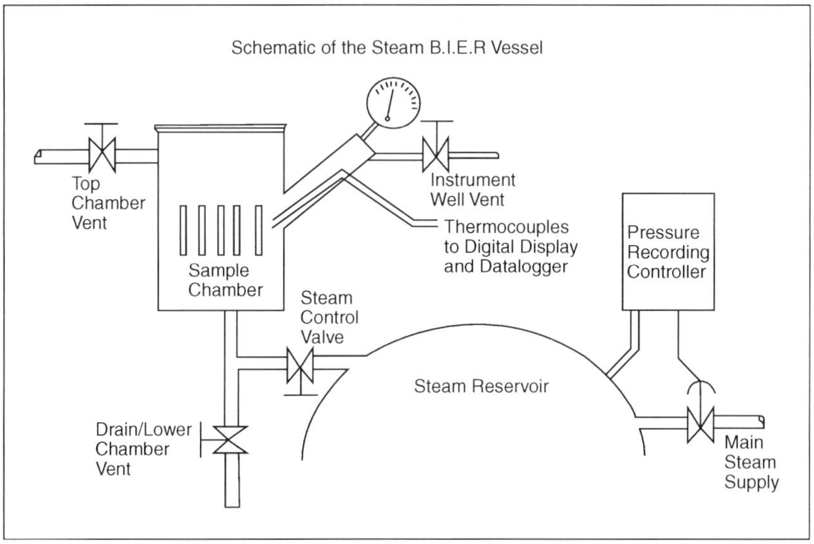

Figure 6 Schematic of a steam biological indicator evaluator resistometer vessel used for generating moist heat resistance *D*- and *z*-values for inoculated parenteral solutions or for biological indicators.

Table 3 IV Lipid Emulsions Ranking

List #	Solution	D121	*z*-value	Predicted spore log reduction
1	20% Emulsion	0.7	10.6	7.1
2	10% Emulsion w/increased linolenate	0.7	11.4	7.5
3	10% Emulsion w/100% soybean oil	0.6	10.1	8.0
4	20% Emulsion w/100% soybean oil	0.7	12.8	8.2
5	20% Emulsion w/increased linolenate	0.6	10.6	8.3
6	10% Emulsion w/50% safflower & 50% soybean oil	0.6	10.7	8.4
7	20% Emulsion w/50% safflower & 50% soybean oil	0.6	12.7	9.5
8	10% Emulsion	0.4	11.1	12.9

Table 4 Accumulated F_{bio} by List Number and *z* Value

Solution										
Temperature (°C)	Time (min)	*F* (PHY) $z=10.0$	1 $z=10.6$	2 $z=11.4$	3 $z=10.1$	4 $z=12.8$	5 $z=10.6$	6 $z=10.7$	7 $z=12.7$	8 $z=11.1$
105.4	1	0.0269	0.0330	0.0419	0.0278	0.0592	0.0330	0.0340	0.0579	0.0384
110.1	1	0.0793	0.0915	0.1082	0.0813	0.1380	0.0915	0.0935	0.1359	0.1019
114.1	1	0.1991	0.2181	0.2427	0.2023	0.2834	0.2181	0.2212	0.2806	0.2336
116.2	1	0.3228	0.3442	0.3709	0.3265	0.4134	0.3442	0.3476	0.4106	0.3611
118.1	1	0.5000	0.5200	0.5445	0.5035	0.5819	0.5200	0.5232	0.5794	0.5356
119.1	1	0.6295	0.6462	0.6663	0.6324	0.6966	0.6462	0.6489	0.6946	0.6591
119.4	1	0.6745	0.6897	0.7079	0.6772	0.7352	0.6897	0.6921	0.7334	0.7014
119.2	1	0.6442	0.6604	0.6799	0.6470	0.7092	0.6604	0.6630	0.7073	0.6729
118.5	1	0.5483	0.5672	0.5903	0.5515	0.6253	0.5672	0.5703	0.6230	0.5819
117.8	1	0.4667	0.4872	0.5124	0.4702	0.5513	0.4872	0.4905	0.5487	0.5033
116.2	1	0.3228	0.3442	0.3709	0.3265	0.4134	0.3442	0.3476	0.4106	0.3611
114.1	1	0.1991	0.2181	0.2427	0.2023	0.2834	0.2181	0.2212	0.2806	0.2336
110.6	1	0.0889	0.1020	0.1197	0.0911	0.1510	0.1020	0.1042	0.1487	0.1130
105.9	1	0.0301	0.0367	0.0463	0.0312	0.0648	0.0367	0.0379	0.0634	0.0426
101.7	1	0.0115	0.0148	0.0198	0.0120	0.0305	0.0148	0.0153	0.0296	0.0178
Total *F*		4.7436	4.9734	5.2646	4.7826	5.7366	4.9734	5.0107	5.7044	5.1573
D value			0.70	0.70	0.60	0.70	0.60	0.60	0.60	0.40
PSLR			7.105	7.521	7.971	8.195	8.289	8.351	9.507	12.893

evaluate the surviving organisms. Test data has been generated demonstrating the value of using both a moist heat organism (*C. sporogenes*) and a dry heat organism (*B. subtilis* now known as *Bacillus atrophaeus*) as BIs for the sterilization validation of closure systems (11,12). A typical graphic representation of the inactivation kinetics is illustrated in Figure 7. The above studies may also be performed in a production sterilizer as engineering or feasibility studies.

Container–Closure Integrity Validation

Container–closure integrity or MOS validations are run on all moist heat terminally sterilized products with closure systems of a parenteral container. This validation is performed to demonstrate that the closure system of a container is capable of maintaining the emulsion and fluid path in a sterile condition throughout the shelf life of the product.

In a typical MOS study, the product container is sterilized at a temperature which is higher than the upper temperature limit of the chosen sterilization cycle and for a time that is greater than the maximum time limit for the cycle or producing an F_0 subzero level greater than the maximum F_0 level for the cycle. The rationale for the selection of the maximum temperature and heat input level for the prechallenge sterilization is

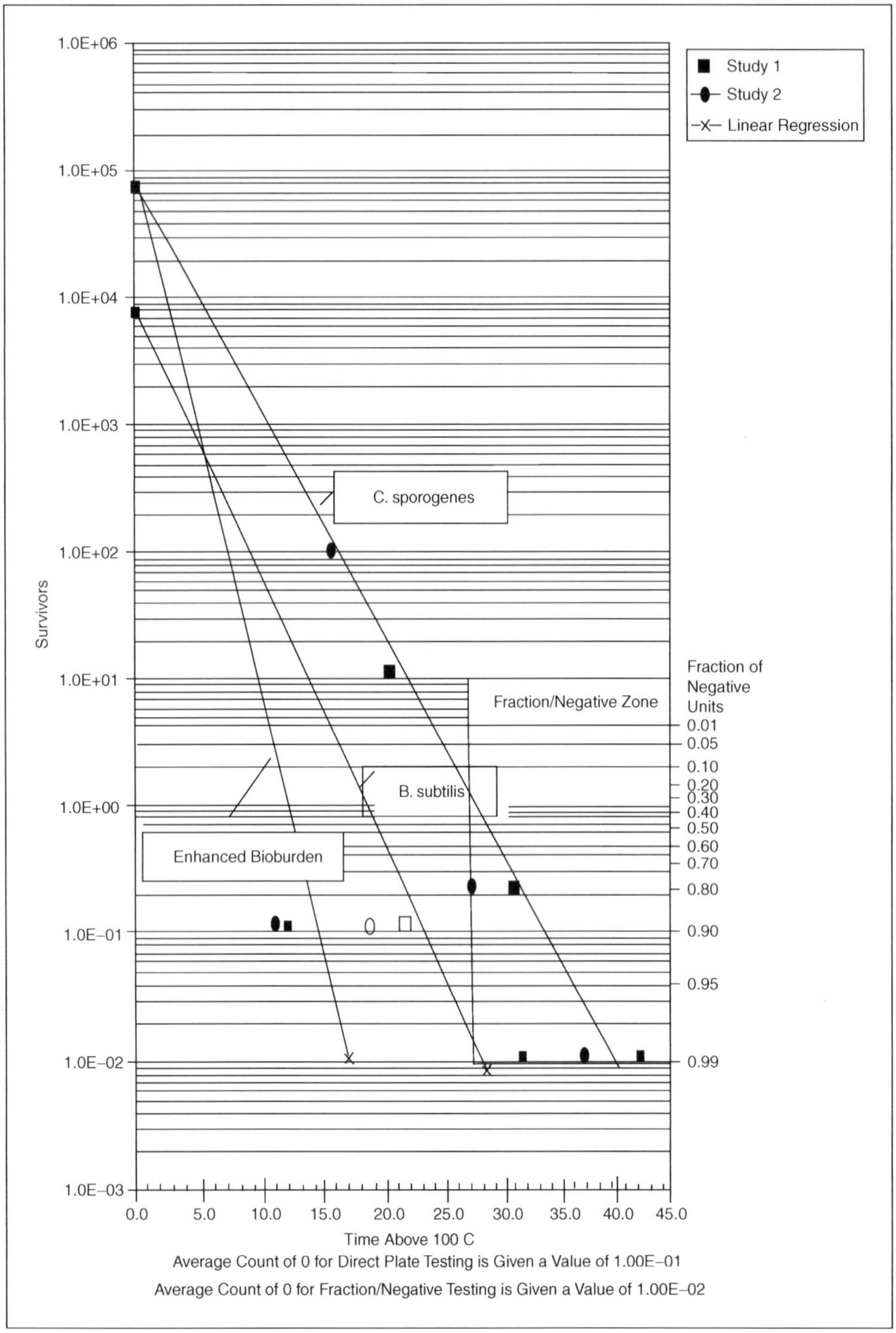

Figure 7 Microbial kinetic inactivation of bioburden as compared to biological indicators, *Bacillus atrophaeus* (formerly named *Bacillus subtilis*) and *Clostridium sporogenes*.

that rubber and plastic closures are subjected to thermal stresses during sterilization and those stresses are maximized at the highest temperature and the longest time allowed.

In some cases, the closures, e.g., administration or additive port are claimed to be sterile by a radiation process. In such cases, the closures are sterilized in bulk exceeding the maximum end of the radiation process e.g., 40 kGy, then fabricated to the flexible container and exposed to steam sterilization cycle conditions exceeding the maximum temperature end of the cycle. Thus, the closures are stressed by a joint process of radiation as well as steam prior to performing the closure integrity test.

Product Validation for Endotoxin

Endotoxins are lipopolysaccharides from the outer cell membrane of gram-negative bacteria. Endotoxins can be detected by the manual gel-clot method known as the LAL test. There are also various quantitative methods (turbidimetric and chromogenic) which use more rapid automated methodologies. All final product formulations have regulatory requirements to be tested for endotoxins and the method must be validated using three different lots of final product. LAL testing should be performed on final product formulations per FDA Guidelines and other regulatory compendia. Emulsion formulations, if colored or opaque cannot be tested by the turbidimetric method and therefore may use a comparable test e.g., LAL, chromogenic or kinetic.

The LAL test is for products other than oral and topical products (e.g., parenteral solutions, some devices, etc). Endotoxin testing is usually required at three different times in the cycle of the product. First, endotoxin testing should be performed on the lot of drug being used in clinical studies to ensure that the product is safe for the patients with respect to endotoxin. Second, in the developmental stages, endotoxin testing is usually required at the beginning and end of the stability studies. Finally, once the product is ready to be marketed, each lot of the product requires endotoxin testing prior to release.

To improve in-process control, a process should also be in place to decide if endotoxin testing should be performed on the APIs and/or excipients used in the product. In order to determine this, the ICH guidelines for quality should be used; i.e., Q7A "Good Manufacturing Practice Guidance for APIs."

PRODUCTION FACILITY STERILIZATION DEVELOPMENT

The production list below depicts some of the sterilization engineering and microbiological activities associated with a parenteral product as it moves into the production environment. These studies occur in a production environment as appropriate.

Production facility activity	Activity statement
Heat P&Ds	Perform triplicate studies for minimum and maximum loading conditions using temperature probes within the product containers and outside the containers to measure the sterilizer heating medium temperature
Solution (master) microchallenge validation	Perform microbial validation of a parenteral solution or master solution at sub-process conditions in the production sterilizer
Container–closure microchallenge validation as applicable	Perform microbial validation of the container–closure system at sub-process conditions in the production sterilizer
Hold time studies	Microbial, chemical and endotoxin studies are performed to establish the longest time that a product can be held following manufacture but prior to filling and sterilization

Engineering P&D Validation

Perform triplicate studies with minimum and maximum loading configurations with temperature probes penetrating the product containers as well as temperature probes distributed outside the product containers in a production sterilizer at nominal operating process parameters.

Microbial Solution Validation in a Production Sterilizer

Table 5 shows the microbial solution validation conducted at subprocess conditions in a fully loaded

Table 5 Lipid Emulsion Microbial Solution Validation

Fraction negative method						
Organism	Code	Average no. spores/bottle	No. positive[a]/ no. positive controls	No. positive[a]/ no. negative controls	No. positive[a]/ no. test samples	Spore[b] logarithmic reduction
C. sporogenes	5C6	4.8×10^5	2/2	0/4	0/20	>7.0
C. sporogenes	15C6	6.4×10^5	2/2	0/4	0/20	>7.1
G. stearothermophilus	5B2	7.6×10^1	2/2	0/4	0/20	>3.2
G. stearothermophilus	15B2	7.7×10^1	2/2	0/4	0/20	>3.1

F_0 Range[c]: 5.8–7.6; Temperature Range: 120–125°C; Agitation: 67–73cpm

[a] Positive for the indicator microorganism.

[b] Spore logarithmic reduction = log *a* − log *b*; where *a*, initial population of spores; *b*, 2.303 log (N/q) = ln (N/q); where *N*, total number of units tested; *q*, number of sterile units.

[c] *F*, integrated lethality or equivalent minutes at 121.1°C for hottest and coldest thermocoupled containers.

Table 6 Lipid Emulsion Microbial Closure validation

Microorganism	Initial population/stopper	No. positive[a]/ no. positive controls	No. positive[a]/ no. negative controls	No. positive[a]/ test samples	Spore[b] logarithmic reduction
C. sporogenes	8.4×10^3	2/2	0/4	0/20	>5.2
B. subtillis	3.0×10^4	2/2	0/4	0/20	>5.8

F_0 Range[c]: 5.8–7.6; Temperature Range: 120–125°C; Agitation: 67–73 cpm

Sterilization validation of 200 mL bottle inoculated closure surface coated with I.V. fat emulsion in cycle with agitation.

[a] Positive for the indicator microorganism.

[b] Spore logarithmic reduction = log *a* − log *b*; where *a*, initial population of spores; *b*, 2.303 log (*N*/*q*) = ln (*N*/*q*); where *N*, total number of units tested; *q*, number of sterile units.

[c] *F*, integrated lethality of equivalent minutes at 121.1°C for hottest and coldest thermocoupled containers.

production sterilizer. The acceptance criteria of 6 SLR was setup for the BI *C. sporogenes* and a 3 SLR for the higher moist heat-resistant BI, *G. stearothermophilus*. Each emulsion (20 containers) is inoculated with the appropriate BI at a target level of 1.0×10^6 and 1.0×10^2 for *C. sporogenes* and *G. stearothermophilus*, respectively. The 20 inoculated containers are distributed throughout the production sterilizer for sterilization at subprocess conditions. The test containers are then returned to the lab for testing by the F/N test method.

Microbial Closure Validation in a Production Sterilizer

Table 6 shows the microbial closure validation at subprocess conditions in a fully loaded production sterilizer. The BIs used were *C. sporogenes* and *B. subtilis*. Acceptance criteria of three SLR must be achieved for moist heat (*C. sporogenes*) and dry heat (*B. subtilis* indicators). The surface of the stopper that comes into direct contact with the sidewall of the bottle was inoculated with the appropriate BI, dried and then a few drops of emulsion were placed over the inoculum to simulate manufacturing conditions. The inoculated closure was assembled to the finished container, exposed to subprocess steam conditions in the production sterilizer and subsequently tested in the lab by the F/N test method. The data demonstrate that a >3 SLR was achieved at subminimal process conditions. Replicate test samples (e.g., 3 to 5) should be considered for use to verify microbial kill in cold zone locations.

ANCILLARY SUPPORT PROCESS TESTING

Bioburden Analysis for Closures and Commodities

Determine microbial load on closures and commodities as well as their moist heat resistance analysis.

As part of the microbiological quality control program, products and commodities are routinely sampled during the production process in order to assess the microbial load. This assessment is performed via the bioburden test for terminally sterilized product. The bioburden test method is developed during the product development stage prior to transfer to the production plant. This test assesses the microbial load of a solution prior to terminal sterilization (In-Process Bioburden Test). Micro R&D is responsible for the validation of the bioburden method prior to transfer to the production plant. The validation will demonstrate that

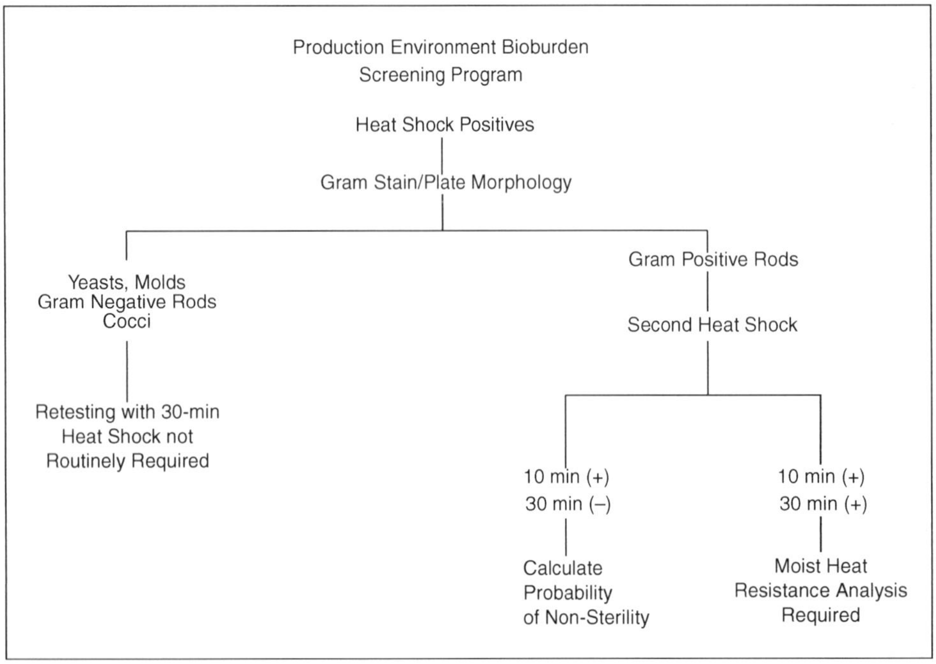

Figure 8 Representative production environment bioburden screening program.

recovery of microbial load at a relatively low level can be achieved.

The Microbial Limits Test is essentially a bioburden test of raw materials used to make the final product. The test method and validation are conducted in much the same manner as the bioburden test. The limit for the microbial limits test is calculated as follows: final Product Action Level/maximum concentration of API in the final product. This limit is then "normalized: by dividing by the total amount of APIs in the final product."

In addition, the production bulk solution is monitored for total bioburden load including spore formers. The screening allows the plant quality lab to ascertain if there are any moist heat-resistant microflora present in the bulk solution prior to the terminal sterilization of the parenteral solution in its finished container (Fig. 8).

Antimicrobial Preservative Efficacy

Perform on those formulations containing a preservative and those container configurations that have a multidose claim. This validation is performed per compendial requirements.

Sterility Testing (if Required)

Once parametric release is approved by regulatory authorities, then sterility testing is no longer required nor can it be used as an alternative in case parametric release parameters are not used.

Biological Testing Support of R&D and Marketed Product Stability Programs

There are a number of analytical and microbiological tests performed over the shelf life of a product. A number of microbiological tests include BET, Container–Closure Integrity and APE if applicable.

CONCLUSION

As one reviews the final configuration that a terminally sterilized parenteral product is packaged in, it is not surprising that a similar evaluation occurred when one was contemplating how to present the new product as being sterile and non-pyrogenic. The product development team focused on the various designs of sterilizers and the various manufacturing site locations for the support of currently marketed products. Once the team decided the appropriate facility for manufacture, then the various sterilization cycles discussed in this chapter were evaluated in order to select the appropriate one best suited for that parenteral product in its final container configuration. If a product is destined for the international market, then R&D personnel will follow the EMEA decision tree to determine if the product can be sterilized at 121°C for 15 minutes minimum. If it cannot, then a justification is documented explaining the reason for selection of an alternate sterilization cycle. Personnel perform the applicable studies in a developmental sterilizer as detailed in this chapter, as well as feasibility studies in development or production sterilizers to monitor and test the physical attributes of the final designed container. Once the parenteral product's designs as well as sterilization processes have been finalized, then the plant/site can perform their standard penetration, distribution and microbiological studies in the production sterilizer.

REFERENCES

1. ISPE, GAMP 4 Guide for Validation of Automated Systems. December 2001.
2. Committee for Proprietary Medicinal Products (CPMP) Decision Trees for the Selection of Sterilisation Methods (CPMP/QWP/054/98), Annex to Note for Guidance on Development Pharmaceutics, EMEA 2000, 1–3.
3. Young JH. Sterilization with steam under pressure. In: Morrissey RF, Phillips GB, eds. Sterilization Technology. New York: Van Nostrand Reinhold, 1993:120–51.
4. Owens JE. Sterilization of LVP's and SVP's. In: Morrissey RF, Phillips GB, eds. Sterilization Technology. New York: Van Nostrand Reinhold, 1993:254–85.
5. AAMI. BIER/Steam Vessels (ST45). Arlington, VA: American National Standard for the Advancement of Medical Instrumentation, 1992.
6. Berger TJ, Nelson PA. The effect of formulation of parenteral solutions on microbial growth-measurement of *D*- and *z*-values. PDA J Pharm Sci Technol 1995; 49(1):32–41.
7. Feldsine PT, Schechtman AJ, Korczynski MS. Survivor kinetics of bacterial spores in various steam heated parent eral solutions. Dev Ind Microbial 1977; 18:401–7.
8. Moldenhauer J. How does moist heat inactivate microorganisms?. In: Moldenhauer J, ed. Steam Sterilization, A Practitioner's Guide. Bethesda, MD/Godalming, UK: PDA, 2003:1–15.
9. Caputo RA, Odlaug TE, Wilkinson RL, et al. Biological validation of a sterilization process for a parenteral product-fractional exposure method. J Parent Drug Assoc 1979; 33:214–21.
10. Pflug IJ, Holcomb RG. Principles of thermal destruction of microorganisms. In: Block SS, ed. Disinfection, Sterilization and Preservation. 3rd ed. Philadelphia, PA: Lea and Febiger, 1983:759–66.
11. Berger TJ, May TB, Nelson PA, et al. The effect of closure processing on the microbial inactivation of biological indicators at the closure–container interface. PDA J Pharm Sci Technol 1998; 52(2):70–4.
12. Berger TJ, Chavez C, Tew RD, et al. Biological comparative analysis in various product formulations and closure sites. PDA J Pharm Sci Technol 2000; 54(2):101–9.

14

Steam Sterilization-in-Place Technology and Validation

James Agalloco
Agalloco & Associates, Belle Mead, New Jersey, U.S.A.

INTRODUCTION

Steam SIP is in daily use in the parenteral industry as a result of the desire for enhanced sterility assurance for aseptically produced materials. Along with this increased usage, there has been concomitant interest in the validation of SIP procedures. It is essential that the readers have a fuller understanding of SIP concepts in order to properly apply the technology and subsequently validate SIP. The available literature on the subject of SIP has been extremely limited when compared with other sterilization processes. Articles which have been published on this subject have focused on issues such as process fundamentals, heat penetration, and filter sterilization (1–10). Each of these articles has proven valuable in helping SIP become better understood and in advancing the industry's awareness of the subject. Only limited assistance has been provided to the individual who is responsible for the detailed design of the SIP system. The criticality of system design to achieving sterility with SIP systems is such that it must be considered more closely than any other aspect of the SIP validation effort.

The majority of the papers on steam sterilization have focused on products, materials and equipment positioned inside an autoclave. The autoclave provides the means for control of the sterilization process parameters. Correlation of the autoclave documentation to the process lethality delivered to the materials inside the chamber is achieved through the validation effort. There are large pieces of process equipment utilized in the production of parenterals whose size and configuration will not allow them to be placed inside an autoclave for sterilization. To assure a higher degree of sterility assurance for these items, they should be sterilized in situ rather than sanitized. Steam-in-place sterilization enables the entire processing system to be sterilized as a single entity, thereby eliminating or reducing the need for aseptic connections. Manufacturing tanks, lyophilization chambers, processing equipment, filling lines and other large systems are normally sterilized in this manner.

The subject of steam sterilization has been so ingrained in the minds of validation specialists that certain key aspects are often overlooked. SIP, which employs the same moist heat mechanism as steam sterilization in autoclaves, forces an attention to detail in system design that exceeds that of steam sterilizers. The reason for this increased emphasis is straightforward. When a firm applies SIP to its systems and equipment it becomes the designer of the sterilizer itself, a role usually adopted by the autoclave manufacturer. Autoclave manufacturers have had many years of experience in designing their equipment and design differences in sterilizers are relatively minor. Most of the important features of sterilizers that assure their effectiveness are required in the design of SIP systems. What may not be evident to practitioners of SIP is the application of sterilizer design concepts to the more flexible circumstances that arise in SIP systems. In order to better understand the design and qualification effort requires a review of the nuances of SIP system engineering as it relates to the physical elements. A brief summary of SIP fundamentals is provided by way of introduction to the subject.

SATURATED STEAM AND SIP

Saturated steam is a steam–water mixture in which the vapor phase (steam) is in equilibrium with the liquid phase (water or condensate). Saturated steam can exist at only one temperature and pressure along the saturation curve (Fig. 1). The addition of heat to saturated steam can result in its de-saturation (or superheating). The loss of heat from saturated steam will result in its condensation. Steam sterilization occurs most effectively when *saturated steam* contacts a surface or organism. The presence of liquid water is required for the effective sterilization through denaturation of proteins in the cell wall at temperatures in the range of 121°C. Saturated steam is far more effective as a sterilizing medium than superheated steam where the liquid water is absent. In order to raise the temperature of an object with saturated steam, the steam must undergo a phase change to the liquid state at which time the heat of condensation is released. Of necessity, this produces a large amount of condensate especially at the beginning of the sterilization process when the process starts with the equipment at ambient temperature (Fig. 2). *Superheated steam* is steam that has been heated above its saturation temperature. The presence of this additional thermal energy converts any liquid condensate in the steam to the vapor phase. The absence of a liquid phase in superheated steam markedly reduces its lethality to microorganisms. In effect, superheated steam behaves similarly to dry heat as a sterilizing vehicle. Dry heat is significantly less effective at the conventional temperatures (115–125°C) at which moist heat sterilization is employed. Thus caution must be

Abbreviations used in this chapter: APA, aseptic processing areas; CIP, clean in place; IQ, installation qualification; OQ, operational qualification; PAR, proven acceptable range; PQ, performance qualification; SCDM, Soybean–Casein Digest Medium; SIP, sterilization in place; SOP, standard operating procedures..

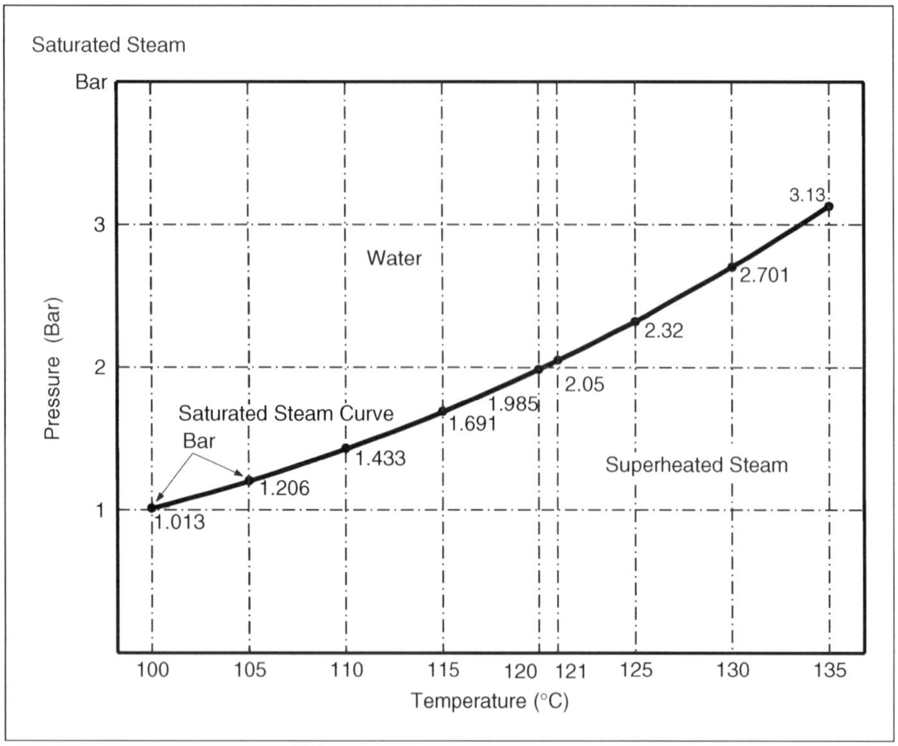

Figure 1 Saturated steam temperature versus pressure diagram. *Source*: Courtesy of Fedegari Spa.

exercised to assure that the steam utilized in an SIP process is saturated and not superheated.

Consider the difficulties in effecting the steam sterilization of a system or piece of equipment in situ. In order to be effective at the conventional temperatures for steam sterilization of approximately 121°C, the process must use saturated steam. The need to heat large masses of stainless steel from ambient temperature to 121°C and the loss of radiant heat to the surrounding room will result in the creation of large quantities of condensate especially during the start of the process. While the condensate will initially be in equilibrium (exist at the same temperature–pressure as the steam), it will continue to transfer heat to the surrounding cooler surfaces, and continue to drop in temperature (and become less effective as a sterilizing agent at these lower temperatures). Supplying additional saturated steam to the system in an attempt to raise the temperature of this condensate will only result in the formation of additional condensate! Clearly the only solution to maintaining systems at the proper temperature for effective moist heat sterilization is through the elimination of condensate from all parts of the system. This can only be accomplished by the positioning of condensate drains at every low point in the system. With this discussion as background, the emphasis placed on condensate removal in SIP processes found later in this chapter will be better understood.

SIP FUNDAMENTALS

SIP differs only slightly from steam sterilization in autoclaves. The major difference is that for effective SIP, the sterilization scientist must ensure that the elements necessary for process effectiveness inherent in the autoclave design and operation are provided in the SIP system. The following measures are of particular importance in SIP and they must be properly addressed in the design if the sterilization process (and its ultimate validation) is to be successful:

1. Complete displacement and elimination of entrapped air.
2. Constant bleeds of steam at all low points to eliminate condensate build-up.
3. Strict adherence to the sterilization procedures.
4. Proper maintenance of the sterility after the process.

Each of these plays a major role in the design of the system and an expanded discussion of each is necessary to understand their importance. In order to better understand these concerns relative to the implementation of an SIP process, a parallel review of relevant steam autoclave technology is beneficial. Several essential features of steam autoclave design will be contrasted with the parallel aspects of SIP technology (Table 1).

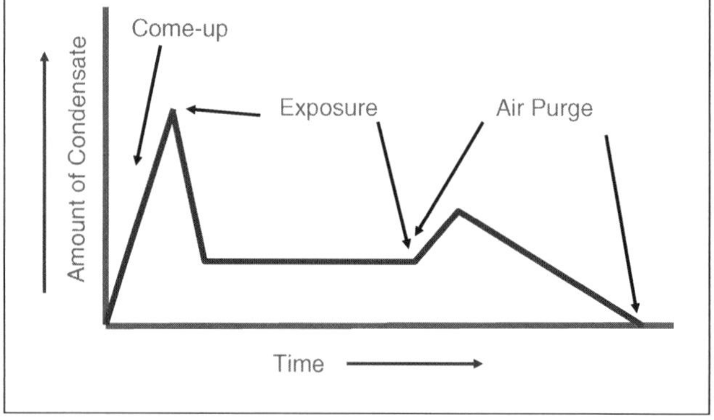

Figure 2 SIP condensate formation.

Table 1 A Comparison of Steam Autoclave and SIP Design Features

Steam autoclave design feature	SIP system equivalent
Vacuum pump	Usually not utilized
Low drain	Multiple drains required
Vent filter	Required
Sequencing controller	None (majority are manually operated)
Temperature controller/recorder	Temperature recorder
Pressure recorder	Pressure controller/recorder
Insulation	Required
Steam heated jacket	Not recommended
Atmospheric break to drain	Atmospheric breaks on all bleed lines

The contemporary steam sterilizer is much more than a large pressure vessel fitted with a clean steam supply. It is a sophisticated system designed to rigid specifications intended to achieve a set of restrictive conditions reproducibly. Modern steam sterilizers include a number of design features that make them far more reliable and effective when compared to their predecessors (see the earlier chapter in this text for additional details on steam sterilizer design).

Air Removal

The first step in many steam sterilization cycles is the removal of air from the load (a notable exception are many of the autoclaves designed for the terminal sterilization of filled containers where the presence of air is sometimes necessary to maintain container integrity). If excessive air remains within the chamber, steam penetration is slowed and the development of cold spots within the load is more likely. Sterilizer manufacturers commonly utilize mechanical removal of air via multiple vacuum cycles to improve air removal. Air removal serves to shorten come-up times (time to sterilizing temperature), improve temperature uniformity, increase steam penetration and consequently increase sterility assurance.

In contrast, relatively few SIP systems employ prevacuums to assist in air removal. It is the responsibility of the SIP system designer to provide other means for the elimination of air. The most common method employed is the addition of bleeds (either drain valves or steam traps) to the system. Consider for a moment, a large fixed tank with multiple inlet lines for vents, filter housings, rupture disks, process fluid, pressure gauges, etc. in the headspace of the vessel (Fig. 3). Each of these offers a potential location for air entrapment and should be evaluated to determine if an air bleed is required. Bleeds should be added at the end of each leg and at each low point in the system to facilitate air removal. Positioning outlets at the points in the system farthest from the steam supply facilitates steam penetration to those locations. The selection of bleed locations is facilitated by the need to remove condensate from the system during the process (see the following discussion).

Caution must be exercised in displacing the air inside the equipment too rapidly, as it may result in entrapment of air in locations that might be purged of it under a slower pressurization of the system. Steam introduction is usually through filters and then lines in the upper portions of the system, using downward displacement of the colder (and denser) air. This step has many similarities to the gravity displacement of air in

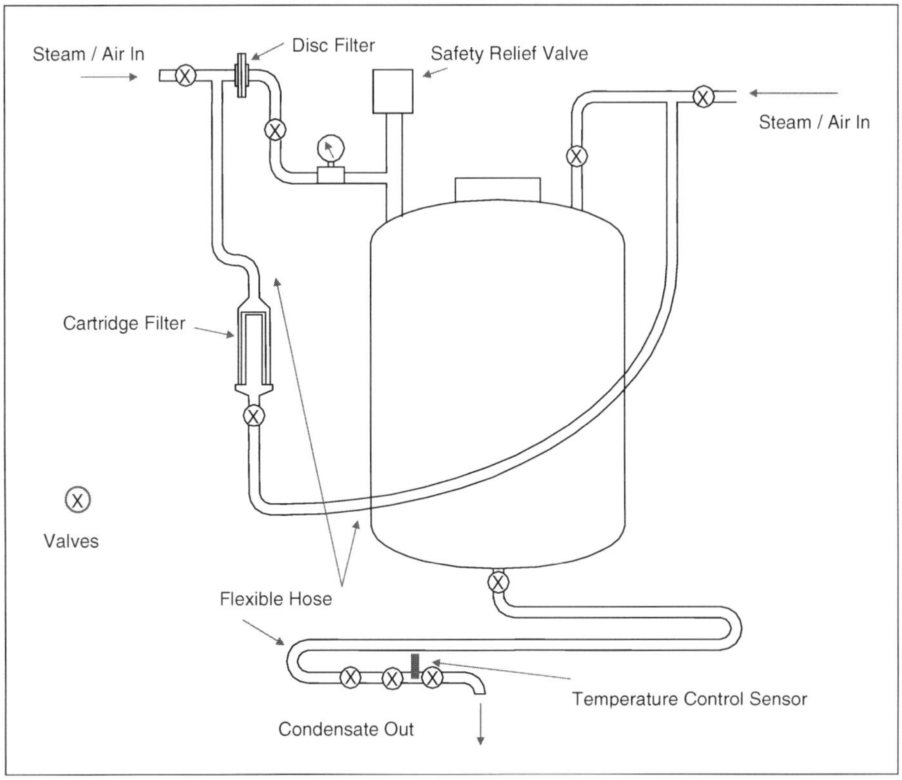

Figure 3 Portable tank with ancillary equipment.

autoclaves that was universally used for many years and is still available in many new sterilizers.

It was noted that pre-vacuums are not the norm in SIP systems. Lyophilizers are a frequent exception as their physical design and operational characteristics make the use of pre-vacuums an attractive option. Pre-vacuums have been successfully utilized in other SIP systems where the additional complexity of the vacuum system is deemed to be offset by the more rapid steam penetration achieved. Where vacuum is utilized in any SIP system it would be drawn using a water-filled liquid ring pump, and not the more capable, but incompatible oil-filled vacuum pump used for deep vacuums.

Condensate Removal

Another aspect of steam autoclave design that is of importance in SIP systems is the reduction and removal of condensate. In modern steam sterilizer design this is accomplished through several design elements: the use of a surrounding jacket, the presence of a thermostatic steam trap and the application of insulation to the external surfaces of the sterilizer. These features serve to reduce the steam requirements for the chamber, and to facilitate the removal of condensate formed in the chamber.

Steam autoclave design conventionally includes a steam jacket, operating at a temperature and pressure slightly lower than that of the sterilizing chamber. Steam sterilizers are also generally well insulated to avoid excessive heat loss. One of the purposes of the external jacket and insulation is to reduce the quantity of chamber steam needed and create a corresponding reduction in the amount of condensate formed (with the additional benefit of a slight reduction in overall cycle time). SIP systems are either un-jacketed (to reduce their complexity) or in cases where a jacket is already present, the jacket is generally disregarded in the SIP process due to concerns with superheating of the internal steam. SIP systems are also less likely to have insulation on all of the exposed piping (a result of size, weight, location, and set-up time considerations). The absence of a jacket and insulation on large parts of the system means that the typical SIP system will produce considerably more condensate than an autoclave of similar internal volume.

Steam sterilizers have carefully sized and positioned thermostatic traps to maintain internal pressure while effecting rapid removal of air and condensate. Extra care is taken during the installation of sterilizers to level the unit, thus preventing the accumulation of condensate at other than drain locations. The condensate in SIP systems is removed via drains at all low points in the system. These drains may be thermostatic steam traps, adjustable (manual or automatic) valves or even fixed orifices. Condensate, unless removed from the system promptly, is always detrimental to sterilization process effectiveness. The condensate will be at a temperature lower than that of the steam due to the loss of radiant heat to the un-insulated piping, and can become cold enough to prevent adequate sterilization (recall that the rate of death for microorganisms is an exponential function of temperature and small differences in temperature can mean large differences in sterilization process lethality). There is documented evidence that the resistance of spores (especially those on paper strips) is increased when in the presence of water relative to a steam–water mixture at the same temperature (11). A well-designed SIP system will have drains installed in each horizontal leg and at every low point in the system.

In the design of a tank and piping system, which is to be sterilized-in-place, the proper sloping of lines will assist in conveying condensate to the appropriate drain location. The size of each bleed in an SIP system, whether for air or condensate, is an important consideration in system design. Bleeds in SIP systems should vary in size in relation to the amount of condensate expected to collect at a particular location. An overly large bleed will result in the use of additional steam to maintain system pressure, while a bleed that is too small risks non-sterility due to condensate build-up or air retention. Air retention and condensate accumulation are so detrimental to the execution of an SIP process that system design should err on the side of caution, using a greater number and larger bleeds (and consequently more steam), rather than a lesser number and smaller ones.

Procedural Conformance

An important part of contemporary steam autoclaves is the control system. The control system regulates the temperature within the sterilizing chamber and performs the sequencing of steps that brings the unit through its process cycle. The control system of the autoclave assures that even the most complex cycle can be carried out reliably and consistently. The proper positioning of valves and regulation of temperature are assured by the presence of a well-designed process control system.

In contrast, many SIP systems have no control system and must rely on an operator's conformance to a detailed SOP and careful monitoring of process variables to achieve success. An operator is responsible for the execution of the process steps in the correct sequence at the appropriate time. This task is made more difficult by the large number of air and condensate bleeds whose manipulation at the appropriate time is essential for the proper completion of the sterilization process.

Where an SIP system is automated, consideration of procedural conformance must be factored into software development. The correct sequence of operation must be established in the software, and changes to the software may be necessary after the completion of the validation effort. Manufacturers of large pieces of process equipment such as lyophilizers, fermentors, etc. will often provide a microprocessor-based control system that can increase process reliability markedly, a major concern given the greater complexity of those systems. In these larger and more sophisticated systems, automation of the SIP procedure is more common. A control system makes the successful execution of SIP processes roughly comparable to the operation of an autoclave. However, it is still safe to say that the majority of SIP systems have no automation and are wholly dependent on the operator. For these manual systems, a comprehensive SOP is essential to a successful sterilization procedure, as a mistake in timing or sequence could result in a compromise to sterility.

Post-Sterilization Integrity

A key element of steam sterilizer design that has relevance for SIP systems is post-sterilization integrity.

In an autoclave, the exposure period is generally followed by a vacuum drying cycle and an eventual return of the chamber to atmospheric pressure just prior to unloading. Safeguards inherent in all new autoclave designs are leak-tested chambers and vent filters to maintain the sterility of the load between the end of the steam exposure and the unloading of the sterilizer.

SIP systems must have similar capabilities. It is essential to maintain sterile conditions in the vessel from the start of cool-down until the system is ready for use. Maintenance of sterility is often accomplished by the introduction of a pressurized gas into the system through an appropriate filter at the end of the steaming step. In SIP processes, a high-pressure gas (air or nitrogen) is introduced to the system through a sterilizing filter, while the system is still under positive steam pressure. The system must then be purged of the residual steam and condensate and maintained under positive gas pressure until ready for use. Additional fittings, valves, and piping may have to be added to the system to protect connection points and other components from microbial contamination prior to use. The introduction of a gas purge can also serve to dry the system, an issue of importance if the product or material to be manufactured in the equipment is non-aqueous in nature. The system should be maintained under positive pressure until ready for use. If the compressed gas supply to the vessel is not maintained until the tank has completely cooled, there is a potential for the development of an internal vacuum in the system that could result in post-sterilization contamination of the internal surfaces.

SIP systems and steam sterilizers share similarities in their critical functions: air removal, condensate drainage, procedural conformance and post-sterilization integrity. The difference between a steam sterilizer and an SIP system are straightforward. The autoclave is designed to perform sterilization processes almost exclusively, whereas SIP systems are designed for other processes under aseptic conditions that necessitate their sterilization. The autoclave is not in direct product contact, it merely serves as a means to sterilize the wrapped parts that will be in product contact later in the process. Major parts of the SIP system will be in direct product contact and their sterility must be assured. A comparison of the key physical elements of steam sterilizers and typical SIP systems are provided in Table 1. Despite these physical differences, the functions necessary for sterilization effectiveness can be achieved in both, however the accomplishment is effected in distinctly different ways. Success with SIP is largely based upon proper attention to design details. Focusing on air elimination, condensate removal, proper sequencing and post-sterilization integrity should lead to greater success with SIP implementation.

SIP SYSTEM DESIGN

The application of the SIP system design concepts can be achieved through a review of the equipment elements that make up that system. As most SIP systems are combinations of many smaller components, the reader is encouraged to review the relevant parts of the text for items present in their system. SIP system designs are largely based upon the field experience rather than rigorous engineering designs. There have been efforts to provide a more rigorous scientific basis for the empirical nature of the design, unfortunately the gulf between theory and practice is still quite large (12–17). The recommendations that follow are derived from experience with a range of SIP system designs, with consideration given to the underlying heat and mass transfer limitations found in the referenced materials.

Pressure Vessels

The term "pressure vessel" includes equipment such as fixed and portable tanks, fermentors, blenders, centrifuges, freeze dryers, crystallizers and other equipment that can be sterilized-in-place. For ease of discussion, the term "tank" is utilized in this section to represent these types of equipment. The majority of SIP systems consist of pressure and full vacuum rated tanks, with associated piping. A typical tank will have numerous nozzles that are used for manholes, sight glasses, lights, rupture disks, pop-off valves, pressure gauges, temperature wells, dip legs, etc. (Fig. 3). These items are supplementary to any process piping, vent and process filters and associated valves required on the tank. Design concerns in this area are numerous.

1. All inlet lines to the tank should be kept as short as possible to minimize air hold-up and reduce condensate formation. Valves should be placed as close to the tank as possible. Lines entering the headspace of the tank should be vertical if at all possible. If "horizontal" lines are utilized, they should be pitched (typically 1/100), to assist condensate flow out of the system.
2. Adherence to the "6D" rule is recommended for all piping connected to the tank (18). The 6D rule relates the length of the branch pipe to its diameter. The 6D rule evolves from the principles of fluid flow, more specifically the Reynolds' number. A common error in applying the 6D rule is to utilize the diameter of the larger pipe rather than the smaller pipe (Fig. 4). For SIP systems, adherence to the 6D rule is not always sufficient to adequate sterilize the "dead leg" segment. Any portions of the system that protrude from the body of the tank or out of the main process

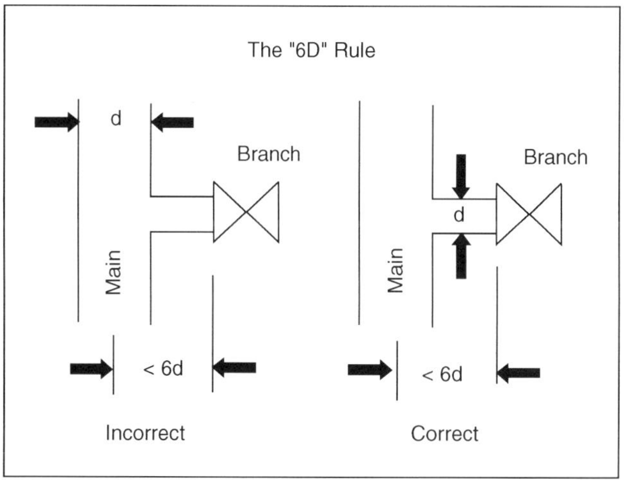

Figure 4 6D rule in pictures correct and incorrect usage.

flow stream in a pipe must be minimized. As a general rule, these legs should be either kept to a minimum length or provided with a bleed or trap at the far end of the "dead leg." Comprehensive discussion of the sterilization difficulties associated with "dead legs" has been discussed in the literature (14–16).

3. Where lines must extend some distance from the tank, bleeds are usually required at the end of each line to assure adequate sterilization.
4. Rupture discs are generally preferable to relief valves because of the reduced dead volume and cleaner interiors. In addition, rupture discs can be positioned closer to the tank proper, thus minimizing the distance that the steam has to travel from the tank proper. If a relief valve must be employed, it may be necessary to open it slightly during the sterilization to assure steam penetration.
5. The use of pre-vacuums to assist in the removal of air prior to the introduction of steam can be employed in an effort to expedite the air removal process. While this is an effective technique, whether it is beneficial is debatable. The use of pre-vacuums is most prevalent on lyophilizers where vacuum rating of the equipment is built in, and the ability of the system to maintain a tight vacuum can be used to advantage. The use of vacuums on tank SIP sterilization is less common, though those firms that employ pre-vacuums feel strongly about their utility in aiding steam penetration.
6. Condensate bleeds should be placed at every low point in the system to facilitate condensate removal. Preference as to the use of traps, valves or fixed orifices as bleeds in an SIP system varies from firm to firm. Each can be successfully employed in an SIP system. The size of the bleed required at any location will vary with the amount of condensate expected at that location. As a general rule, the bleed should be somewhat oversized, as retained condensate is a far more serious problem than excessive steam consumption (a detailed discussion of the choice between valves, traps and orifices is provided later in this chapter).
7. The amount of steam required for the sterilization of a system is directly related to the size and mass of the system. Bleed locations and size should be selected to facilitate drainage accordingly. A 6 in. difference in the location of a bleed can make the difference between an acceptable and unacceptable system.
8. Caution should be exercised in the use of jacket steam in the sterilization of equipment. While providing heat to the jacket may reduce the internal steam requirements to bring the system to sterilizing temperature, it may create problems with superheating of the steam in the vessel proper. Jacket steam is generally superheated and at pressures considerably higher than that required for sterilization in order to facilitate the transfer of heat into a liquid-filled tank through a limited surface area. If jacket steam is utilized, the internal saturated steam may be heated to a temperature above saturation thus rendering the process ineffective as a means of sterilization.
9. A frequent question with regard to SIP systems concerns the use of superheated water in lieu of saturated steam as a sterilizing medium. This was addressed by J. Carlson in which the advantages of superheated water over saturated steam were outlined with regard to the energy savings possible through the use of superheated water (19). The substitution of superheated water for steam is possible, provided one could assure adequate flow throughout the system. If the objective is to sterilize a pipeline then superheated water may be preferable. If the system includes complex piping such as found above a typical piece of parenteral manufacturing equipment, then steam systems are generally simpler and easier to control. In discussions with Mr. Carlson there was general agreement with this distinction as to sterilizing medium preference (20).
10. An atmospheric break should be provided between the discharge of the bleeds and any collection point. Direct discharge of the condensate lines into a sealed header risks potential competition among the lines for clear discharge, and the build-up of excessive backpressure that could inhibit condensate removal. Immersion of the bleed(s) in a collection vessel or drain sump is not recommended.

Piping Systems

Piping is utilized to connect multiple pieces of equipment into a process train (common in larger systems such as sterile bulk and biotechnology facilities), or for such systems as compressed air where the piping constitutes virtually the entire system. Attention to detail in the design of the piping system is essential if SIP is planned.

1. Pipe runs should not be perfectly horizontal; the lines should be pitched (approximately 1/100) to provide for condensate drainage. Steam inlets to piping systems should be located at the highest points in the system. Bleeds must be placed at all low points in the system. Newer elbows and tees have become available in which the angle between the legs is 89° or 91°, so that horizontal lines are properly sloped.
2. Valves should be placed in vertical runs of pipe to the extent possible, to minimize the potential for condensate retention during SIP. Where a valve must be closed during SIP, provision for air and condensate removal as close as possible to the sealing surface should be made.
3. Sanitary diaphragm valves should be used wherever possible. Where tight shutoff is required, ball valves or other types may be required, notwithstanding the limitations in sterilizing the hidden surfaces of the ball. Functionality of the intended valve should prevail over the desire to have the system as sanitary as possible. A similar situation results with certain types of needle or other valves that cannot conform to sanitary design concepts. Sterility is essential, but if the equipment cannot be operated properly with a sanitary valve, then a non-sanitary valve that performs properly should be installed with whatever changes in materials of construction, design, etc. are necessary to minimize its impact on the sterility of the system. The use of a diaphragm valve as a shutoff valve on a lyophilizer will result in an unacceptable leak rate; thus a ball valve would be utilized despite its non-sanitary design.

4. Piping configurations shaped like the letters "W," "M," "V" or "N" should be avoided as each of these will have a low point at which condensate can collect. Condensate removal (using valves, traps or orifices) must be provided for at any low points in the system.
5. Where pumps are utilized in the system, they should be capable of being sterilized in situ. Most centrifugal pumps are acceptable, while piston, vane and similar pumps with multiple chambers cannot be easily sterilized and should be avoided in SIP systems. There should be provision for condensate drainage from the low point of the pump housing.
6. Where flexibility in piping arrangement is required, permanent piping manifolds with appropriate valves or spool pieces to make temporary connections should be employed. The use of flexible hoses to make temporary connections may result in the inadvertent creation of low points that can retain condensate and reduce the effectiveness of the SIP process. Permanent welded systems are generally preferable to systems assembled from individual fittings, as control over system configuration is essential for proper SIP performance.
7. The use of insulation on the exterior of piping is recommended to reduce condensate formation due to radiant heat loss. Where insulation is not provided the resultant loss of heat will increase condensate formation.
8. Minimize dead legs in accordance with the 6D rule to avoid inadequate sterilization of the divergent leg.
9. Where a low point in the system is not bled, condensate build-up can occur despite the presence of acceptable temperatures farther along in the system. Sufficient steam can flow along the top of the pipe to sterilize portions of the system farther from the steam source (Fig. 5). Alternatively, "bubbles" or "bursts" of steam can pass through condensate flooded areas to provide steam for downstream piping. The presence of sufficient temperature downstream of a potential collection point for condensate should not be interpreted as an indication that the intermediate point is adequately sterilized.
10. Gauges and instrumentation installed on the system should be of a sanitary design, with wetted parts having minimal surface area, easily cleaned surfaces, etc. The evolution of instrumentation design is such that virtually all major instrumentation types are available in sanitary designs.
11. In liquid processing or fluid distribution systems, valves must be utilized to avoid fluid loss during use, even if traps or orifices are utilized during SIP to remove air and condensate. The choice between valves, traps and orifices is largely one of personal preference. Each has advantages and disadvantages in their application in an SIP system (Table 2). The size, location, amount of condensate, ease of operation, system usage, etc. dictate selection of the appropriate item. The case can also be made for a system that utilizes both a trap and a valve at each location where air or condensate must be removed. The use of both items in the same location affords the user the advantages of both, with the disadvantages of increased cost and complexity. The practitioner should view the choice between traps, valves or orifices as a choice among near equals, with the preference in each situation dictated by the specific circumstances involved.

Filters

In order to maintain the post-sterilization integrity of an SIP system the presence of a filter is usually mandatory. In considering how filters are to be integrated into an SIP system design, consideration of the design details

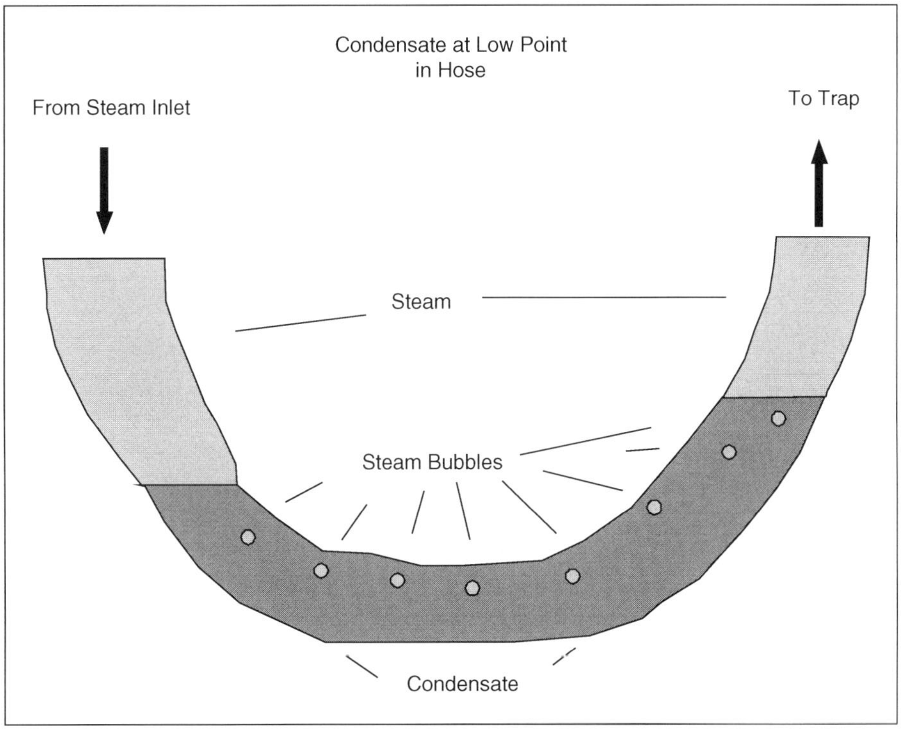

Figure 5 Low point in hose with steam bubbles passing through condensate.

Table 2 Advantages and Disadvantages of Valves, Orifices, and Traps

Advantages	Disadvantages
Valves	
No additional piping for liquid systems	Positioning can have an effect on condensate removal
Frequently needed for fluid handling anyway	Wastes steam compared to same size traps
Can be easily automated	More expensive than comparable sized trap
Some valves can be regulated between full open and full closed	Must be manually or automatically operated
True sanitary design possible	
Can provide feedback to control system	
Orifices	
No manipulation necessary for operation	Wastes more steam than valves or traps
No moving parts	Proper sizing is essential
Little maintenance required	Requires valve for tight shut off
Least expensive	Less sanitary than some valves
More sanitary than steam trap	
Traps	
No manipulation necessary for operation	Must be utilized with a valve in liquid systems
Easier application in automated systems	Periodic maintenance required for proper operation
Operate to remove air and condensate only when necessary	Can fail without obvious fault
Conserve steam	Do not provide tight shutoff
Usually less expensive than a valve	Either full open or fully closed
Smaller size than valve	No true sanitary design available

provided above for the other major system components is essential. There are some additional aspects that bear further explanation with regard to filter and filter housing sterilization.

1. Filter housings may require modification to provide for upstream and downstream bleeds to facilitate air and condensate removal. The bleeds that are positioned on the filter housing by the filter manufacturer are ordinarily intended solely for use during the filtration of the product and may not reflect the appropriate locations for an SIP process. The proper number and position of bleeds is shown in Figure 6.
2. Cartridge filters must be positioned so that the open end is down to facilitate condensate removal (Fig. 7). It is useful to think of the two major parts of the filter housing as the base and the dome rather than the head and the bowl when positioning the housing (this terminology was changed by the filter industry in the late 1980s without any discussion or defined rationale).
3. The filter cartridges utilized in the housing should be examined to see if they are suitable for use in an SIP process. Subtle differences in cartridge design that allow condensate accumulation against the membrane surface can inhibit effective sterilization. The dead volume where condensate can be retained in the cartridge must be kept to a minimum if the filter is to be successfully sterilized in situ.
4. Disc filters are often utilized as vent filters on smaller tanks. When employed for this purpose, the filter housing should be oriented in a vertical plane (Fig. 8).
5. The use of a "loop" to facilitate sterilization of the membrane filter in its housing may be necessary to equalize pressure across the filter and allow the use of lower steam pressures (a detailed discussion of the

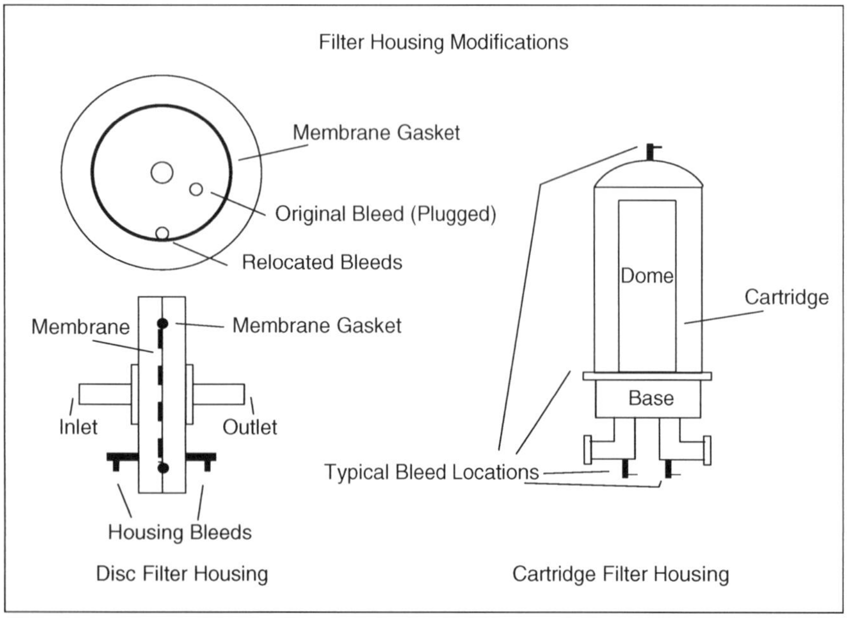

Figure 6 Disc and cartridge filter housing with modifications.

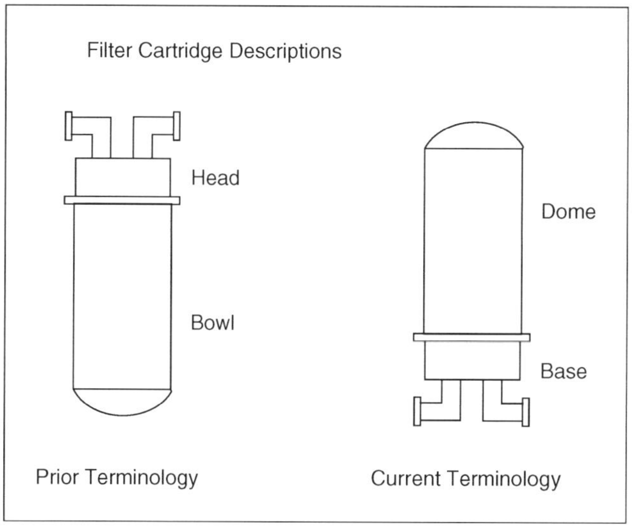

Figure 7 Cartridge housing (bowl/head) base/dome.

"loop" method is provided later in this chapter) (Fig. 9).

6. Depending upon the size of the system, it may be advantageous to utilize more than one steam inlet. Attempting to introduce all of the required steam for an SIP system through the process and/or vent filter(s) may not be possible without damaging the filter(s) due to the high pressures (and correspondingly high temperatures) needed to deliver sufficient steam to a large system. For large tanks, the use of a direct steam inlet to the tank is becoming commonplace, with secondary steam inlets for each filter (this is another application of the "loop" method).
7. When using multiple steam inlets to a tank, caution must be taken to avoid introducing steam at a higher pressure to the downstream side of the cartridge filter than the upstream side. Most cartridge filters cannot withstand a differential pressure of more than 2 to 3 psi in the reverse direction and steaming a filter in the reverse direction will generally result in the loss of filter integrity. If disc filter membranes are employed, the use of back pressure support screens as a routine measure is recommended. Problems with excessive back pressure can be overcome through proper procedures in starting and completing the sterilization cycle. Similar problems can occur during drying or purge cycles if appropriate precautions are not taken.

STERILIZATION OF SYSTEMS

The SIP of larger systems follows the same principles established earlier, with the added complications associated with a more complex arrangement of tanks, lines, filters, valves, etc. Any arrangement of piping which results in a low point for condensate collection must be treated as described earlier. In large systems, particularly those where the vessels are located on the same floor of a facility, there are numerous opportunities for this type of arrangement to occur. The system must be designed in a way that condensate can be readily removed. In order to achieve this objective, it may be necessary to sterilize the system in multiple patterns, in which each pattern sterilizes a portion of the larger system. When using this type of an approach, some portions of the system must be sterilized more than once to assure that all portions of the system are fully covered.

An example of this is portable equipment that must be sterilized in one location and connected aseptically in another. Consider a portable holding tank with three valves installed in series so that the interior surfaces of the first two can be sterilized by regulating pressure within the system by adjustment of the last valve (Fig. 10). After completion of the SIP process the second valve is closed and the piping after it is removed. While the system remains sealed prior to use, the closed second valve is utilized to maintain the sterility in the upstream lines. When the system is ready for use, the second valve is also

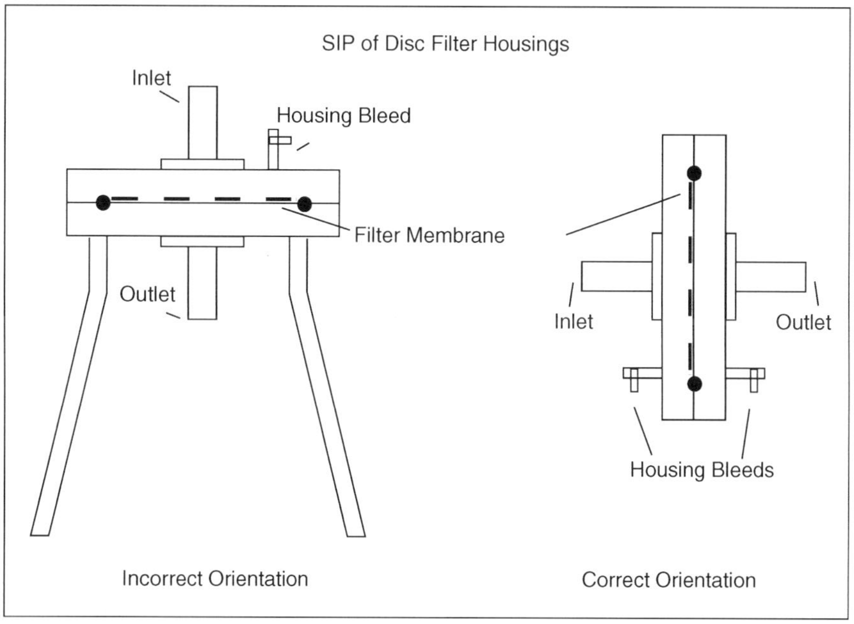

Figure 8 Disc membrane in vertical plane.

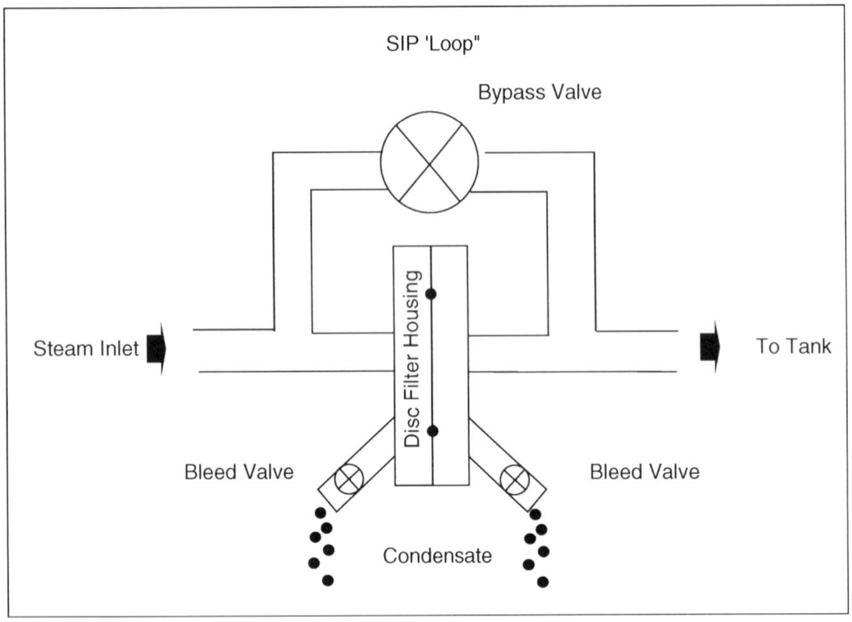

Figure 9 Loop method for disc filter sterilization.

removed and an aseptic connection is made to the first valve (now the only one remaining on the line) and that valve is utilized to regulate flow through the line. In this instance, a single aseptic connection is required to connect the line. This arrangement might be utilized on a portable tank that is transported to the filling machine and aseptically connected to it. It is also possible using specialized valve designs to re-sterilize the connection after making the aseptic connection.

A simple system that depicts how SIP can be utilized is depicted in Figure 11. In this system, any condensate formed in the holding tank (where the majority of the system condensate will be created) has no easy means of egress from the piping system. The condensate must be forced upwards from point B past point E before exiting the system at point D. A simple modification of the piping system incorporating an additional steam inlet and a condensate drain below the tank can be utilized to eliminate the condensate retention problem. In the new system (Fig. 12), there are two separate sterilization patterns. The first sterilization pattern passes steam from the tank via points A, B, and C, while the second pattern sterilizes the two lengths of piping via E, B, C, and E, D. The use of a second pattern allows for the steam to enter the piping at the top of the system, with condensate being removed at the low points. A small portion of the piping near B will likely be sterilized twice in this installation. The addition of a second pattern to be sterilized causes no loss in cycle time, as the second pattern can be sterilized while the tank is receiving material through the liquid filters. The system can be further improved by modifying the piping delivering the solution to the filling machine. Note how, in Figure 12, a series of aseptic connections must still be made to add the polishing filter to the system and bring the piping to the filling machine. A modest refinement results in Figure 13, in which the polishing filter and its associated piping are sterilized-in-place and a single aseptic connection is made under the laminar flow hood. In an ideal installation (Fig. 14), the filling machine itself could be sterilized in situ and there would be no aseptic connection required (21,22)!

A common arrangement of equipment is shown in Figure 15, in which several vessels are shown connected in parallel to a common line leading ultimately to a single bleed. This type of a piping arrangement is common on lyophilizers where the chamber and condenser are piped to a common drain. The use of the single drain is intended to simplify the control of the sterilization process by allowing the process to be regulated by temperature control based upon conditions at a single point in the system. What has been created is an interactive system, in which the two vessels will compete for the use of the drain, and what in effect happens is that each vessel uses the drain intermittently and there may be long periods of time when one vessel or the other is operating without an effective condensate drain! While this might seem to be of little consequence if the temperatures in the overall system are acceptable, in fact this type of problem can have serious adverse consequences due to condensate retention and the resultant temperature reduction. This type of vessel and piping arrangement is quite common, and is often seen when tanks or filters are installed in parallel. A proper SIP design would avoid this type of

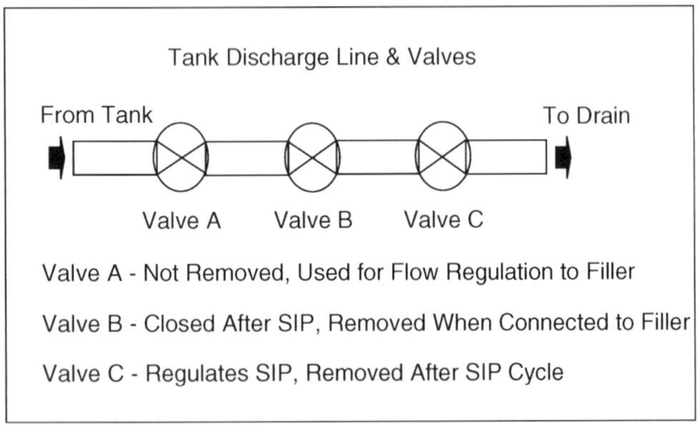

Figure 10 Hose with three valves for aseptic connections.

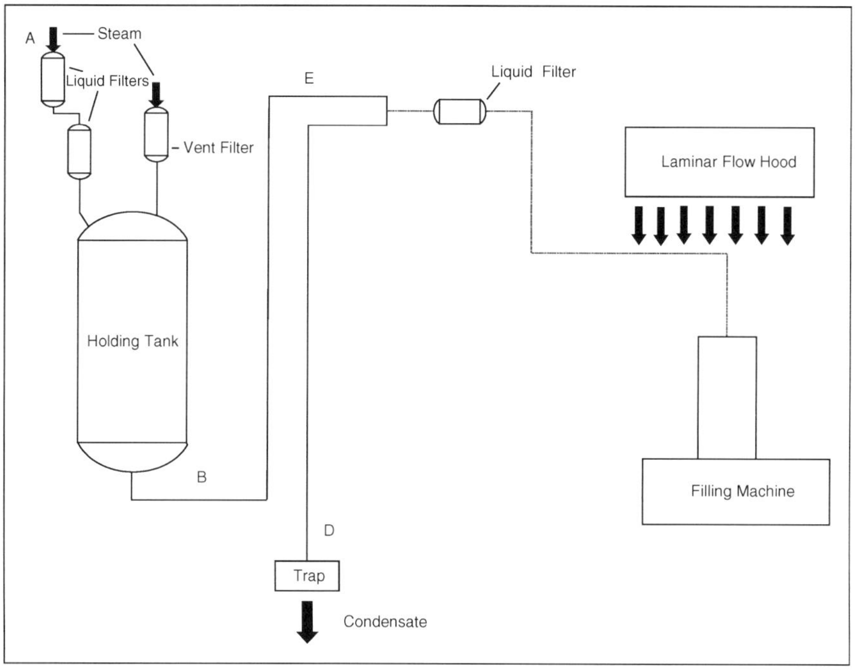

Figure 11 Fixed tank and fill line SIP.

piping arrangement. A modification is shown in shown in Figure 16 that eliminates the interaction between the vessels. A similar appearing arrangement is shown in Figure 17, in which two steam supplies are provided to the upstream of filters installed on the same tank (one filter might be for the process fluid, while the other might be the vent filter). This type of a system generally does not operate as an interactive system because the tank acts as a buffer between the steam supplies and both of them can successfully share a common drain.

An underestimated activity closely related to the SIP sterilization of systems, especially complex ones is the development of a detailed drawing of the system. In order to properly determine how the SIP procedure is to be accomplished an accurate drawing of the system must be available. Only when the full extent of the piping arrangement is known and understood can a SIP process be developed. It is helpful during this activity to ensure that each valve and other major component in the system are uniquely identified on both the drawing and the component itself. The time spent in these activities will facilitate the development of an effective SIP procedure.

A common question with SIP systems is whether they should be located in an APA at all. It might seem obvious that an SIP system should be in an APA, but if the basic concepts of SIP are adhered to and aseptic connections are completely eliminated then the enclosure

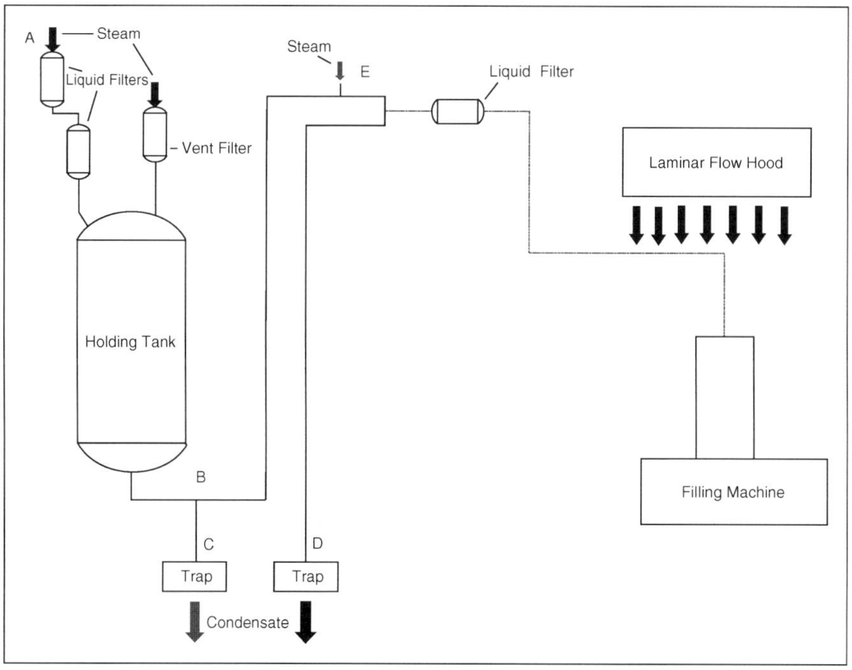

Figure 12 Fixed tank and fill line SIP improved.

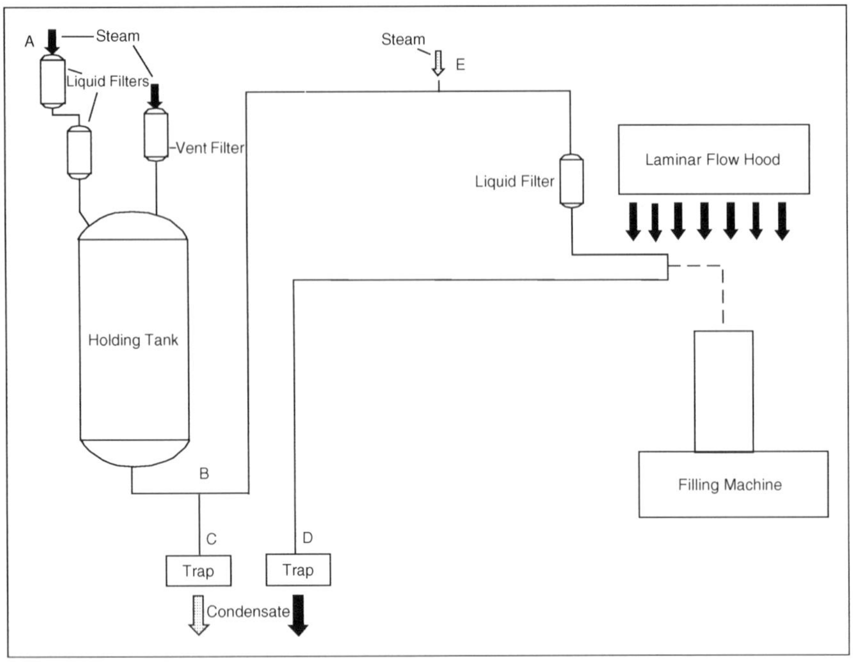

Figure 13 Fixed tank and fill line SIP further improvement.

of the SIP system in a controlled environment is an unnecessary expense. By strict adherence to design principles and operating procedures, it is certainly possible to effect aseptic processing within sealed vessels outside an APA without fear of contamination. Consider lyophilizers, fermentors and even steam sterilizers, where a large percentage of the surface is not contained within an APA. The application of SIP to large process trains where the majority of the train is outside of the APA is common in sterile bulk production. These systems have successfully produced sterile products without difficulty with only minimal portions of the system in an APA. Proper attention to the nuances of SIP system design is of greater importance than any additional sterility assurance provided by a controlled environment in the surrounding room to which the product is never exposed (23).

SIP STERILIZATION OPERATING PROCEDURES

The details of any individual SIP procedure will vary according to the specific configuration. Despite the uniqueness of each SIP procedure, following the design concepts outlined above with regard to air elimination, condensate removal, procedural compliance and system integrity will usually result in a fair degree of similarity among different SIP sterilization procedures.

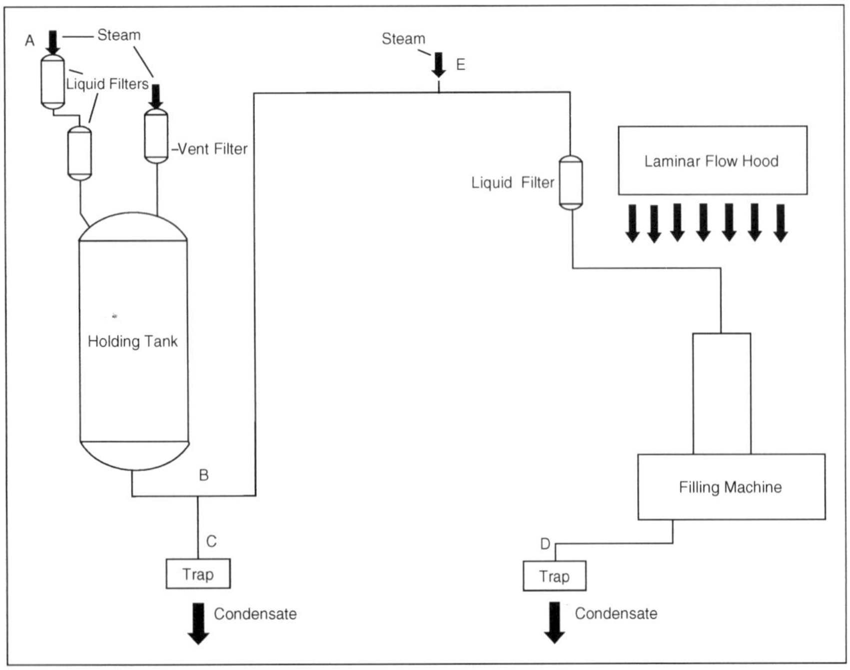

Figure 14 Fixed tank and fill line SIP preferred design.

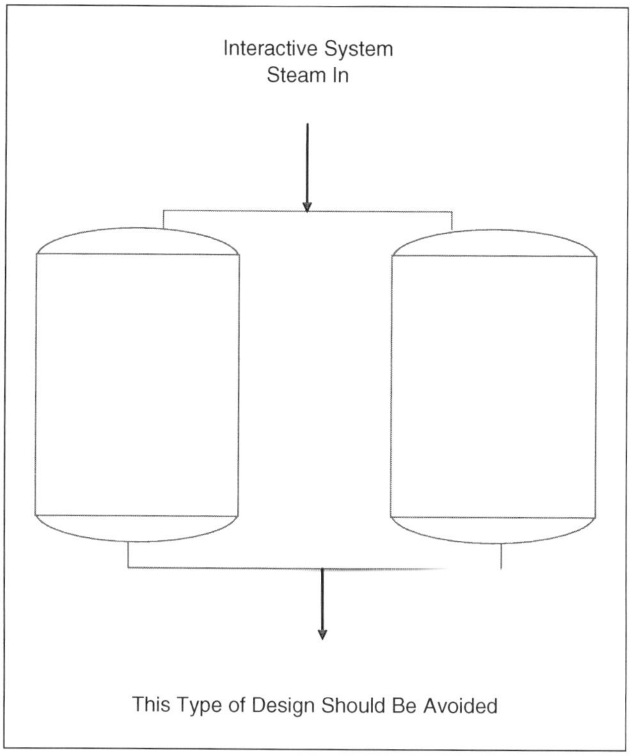

Figure 15 SIP of tanks in parallel (interactive system).

The following section will describe some of the more common aspects of SIP procedures. The user must of course develop an SIP procedure appropriate for their specific system. This discussion can serve as a general guide to the preparation of such a procedure.

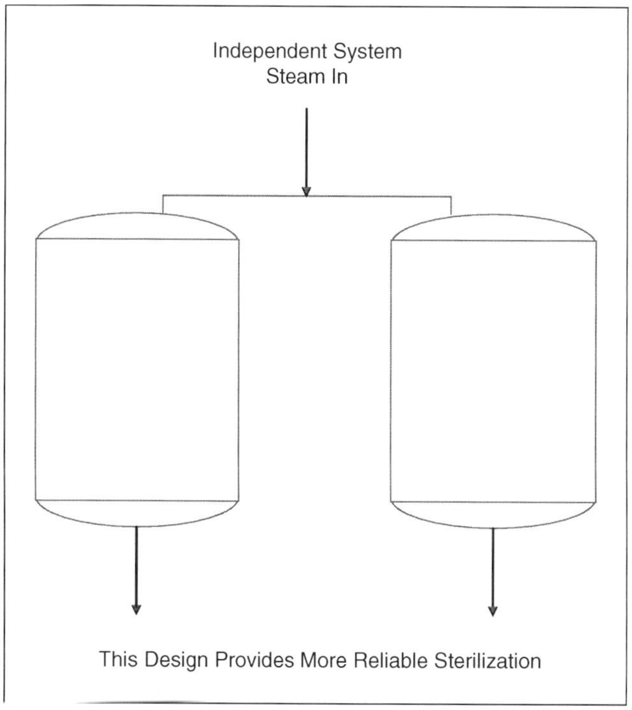

Figure 16 SIP of tanks in parallel (independent system).

Prior to the start of any SIP process, the starting position (open or closed) of each valve in the system should be specified. For convenience, all valves should be in the same initial position, and the closed position is generally preferred. The procedure should then indicate exactly when each individual valve is manipulated.

Just prior to the introduction of steam, the major drain valves, typically valves or traps below the major vessels and at other low points should be opened. Steam is then introduced into the system usually at points high in the vessel or piping system. The steam can be fed directly into the system or it can enter through a membrane filter(s). After steam has been introduced, minor bleeds are opened one at a time, until steam or condensate is observed at the outlet. Eventually wisps of steam and condensate will be observed at all of the outlets of the system. The pressure (and therefore the temperature) in the system is raised by throttling back on the main drain valves. Throttling back on these valves is also necessary to build pressure in the system, as the amount of condensate formed will drop as the internal surface approaches the temperature of the steam. Adjustment of either the steam supply to the system or the main drains will be necessary to raise the internal temperature high enough to where effective sterilization can be accomplished. Adjustment of bleed valves may be necessary once steady state has been attained, as the amount of condensate formed will be reduced once the entire system is at the sterilizing temperature. During the dwell portion of the sterilization process, the amount of condensate formed will be constant and the system will come to a steady state condition where the need to adjust valves on the system will be minimal. Under this condition, a small amount of steam or condensate should be observed being discharged from each low point in the system. In the positioning of valves, the loss of steam from the system due to a valve that is open slightly more than necessary should be tolerated. The alternative situation where the valve is open slightly less than is required will result in the retention of condensate and the potential compromise of the entire SIP process. It is far preferable to waste some steam than risk process failure.

A note of caution must be exercised in bringing the system to the desired sterilizing temperature/pressure. During the initial stages of the process, the bleed locations are required to permit both air and the large amount of condensate formed during the heating of the system. This may require a substantial period of time especially in larger systems or those with small openings. The temptation may be to open the steam input completely in an attempt to shorten the come-up time. This can result in inadequate air removal from the system and reduced cycle effectiveness, despite the apparent successful temperatures attained. A simple means to confirm the effectiveness of air removal is to compare the temperature/pressure relationship to the desired saturation conditions immediately after attaining the setpoint temperature. A slower come-up is generally preferable as it allows sufficient time for air removal.

Once the desired time at temperature has been achieved, the system must be shut down in an orderly manner. The simplest approach to shutting the system down involves the simultaneous closure of the steam supply valve(s) and introduction of a high pressure

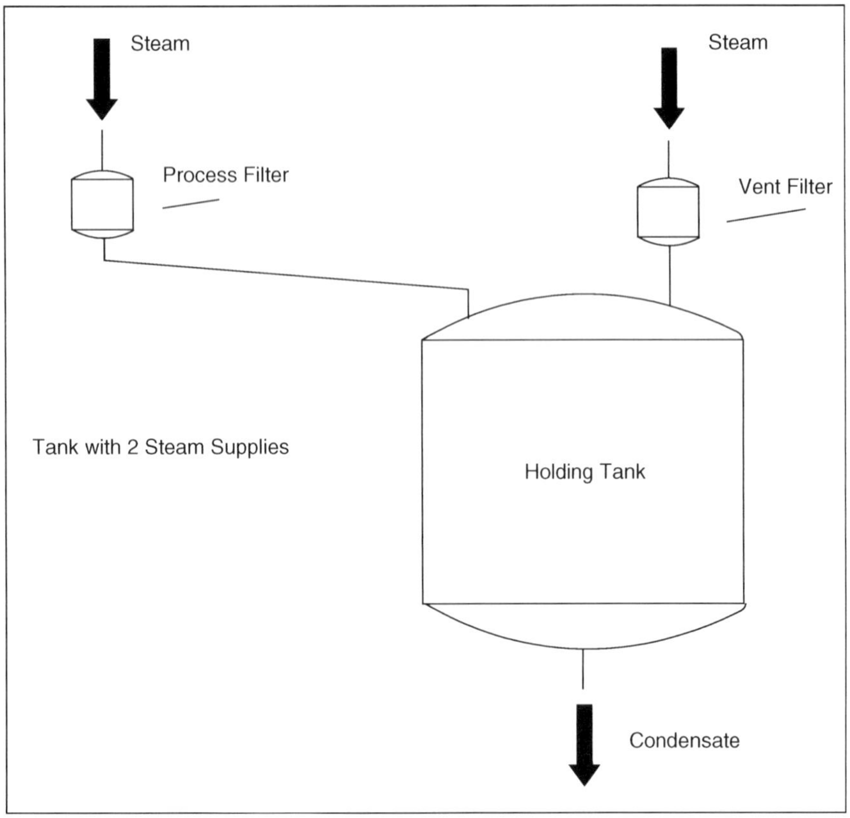

Figure 17 Fixed tank with two steam supplies.

supply of a sterile purge gas. The use of a gas supply pressure higher than that of the steam is necessary to ensure the continued outflow of condensate and eventually the purge gas itself from the system as the system cools. All portions of the system must be supplied with a sufficient volume of purge gas to maintain a positive pressure at all points. Then as each point in the system drops in temperature, the purge gas will prevent the recontamination of the system from the surrounding environment. As the steam is replaced with the purge gas, the temperature in the system will drop and the amount of condensate formed may increase slightly (Fig. 2). The bleeds are typically closed first in the upper portion of the system, allowing the condensate to drain from bleeds in the lower portion while that is accomplished. The last bleeds to be closed are those that served as the major condensate discharge locations, as these may continue to discharge condensate for some time. These bleeds can be left open for some period of time while the gas purge continues. The usually low dew point of the purge gas will assist in drying out the system prior to use. The heat retained by the metallic components in the system will also assist in drying out the last quantities of condensate, as the hotter metal components will transfer heat back to the remaining condensate and facilitate its evaporation.

Once the system has been adequately dried, the last open bleeds can be closed while the purge gas remains on and a slight positive pressure is allowed to build-up in the system. This pressure can be maintained in the system until ready for use thereby preventing the ingress of contaminants through any opening. The addition of the purge gas through a 0.2 μm filter allows for the maintenance of this positive pressure over an extended period of time. In some applications, a portable vessel can be successfully transferred from one APA to another while the pressure is maintained without contamination of the contents.

Process Control

To this point, the control methodology utilized for the SIP process has not been described in detail. Earlier in this chapter the point was made that SIP systems are frequently controlled by pressure rather than temperature. This is a considerable departure from the approach utilized in most steam sterilizers. Most steam sterilizers have a single outlet through which the air and condensate must pass in order to exit the chamber. Thus, a single temperature probe located in the drain line can conveniently serve as a measurement of the coldest temperature in the entire sterilizer and be utilized to control the sterilization process. In contrast, the SIP system may have many locations where air and condensate are eliminated from the system. Recognizing that the pressure in the system will equilibrate throughout enables the use of pressure measurement in the system to control the steam supply for the SIP system. The singular relationship between temperature and pressure for saturated steam makes this possible. While the ability to control pressure will not assure effective sterilization across the entire system, it affords a workable parameter for steam regulation. Choosing a single point within the system in order to control the SIP process is possible, but places a larger burden on the system designer to assure that the location chosen is correct. In a simple tank system, a single point may be both identifiable and usable in SIP process control. In more complex systems such as a bulk antibiotic process train control of the system by the use of internal pressure may be both simpler and more effective. In order to confirm sterilization effectiveness over a complex system multi-point temperature measurement is frequently utilized as an adjunct to pressure control on the steam supply to the system.

Multi-Point Temperature Measurement

In performing SIP procedures on large systems with multiple condensate discharge locations, the use of multi-point temperature recorders is frequently necessary to confirm that the appropriate temperatures are realized throughout the system. This is especially important where the SIP process is manually operated and the manipulation of valves is left to process operators. What more effective way to establish that a large number of valves were operated properly during the SIP procedure than through the use of a multi-point temperature recorder? The appropriate locations to monitor would be determined during the PQ portion of the validation program. It should be noted that the temperature probe inside a tank is almost never the coldest location in the system. The temperature in the discharge line of the tank only a few feet away is almost always lower and therefore of greater importance in routine sterilization process confirmation. In the monitoring of temperature on SIP systems, the introduction of the temperature probe into the fluid stream may not be necessary. Temperature sensors can be attached to the outside of piping and insulated to minimize heat loss to the surrounding environment (24). Where temperature probes positioned outside the piping in this fashion achieve the required temperature, there is reasonable assurance that the interior of the piping is at least that hot.

In measuring temperature in the system, the goal should be to measure the surface temperature and thus thermocouples should be in contact with the surface. Unlike an empty chamber study in an autoclave where uniformity of steam temperature is desired, the objective in SIP is effective sterilization of the product contact surfaces, and thus surface temperatures should be measured. Note also that there is no requirement to have a narrow temperature distribution in an SIP system; heat loss to the surrounding environment will generally preclude a tight range. The primary objective is attainment of lethal conditions at the equipment surface and provided that is accomplished, a wider temperature range is of little consequence.

MEMBRANE FILTERS AND SIP

All systems that are sterilized-in-place include membrane filters. It is difficult to conceive of an SIP system that does not include at least one microbially retentive filter or how a system could be operated without one. The necessity to maintain a system sterile subsequent to SIP will ordinarily result in the introduction of a least one 0.2 μm filter that is utilized to introduce an air or nitrogen purge into the system. The filtered gas is initially utilized to purge the system of steam and then to establish and maintain a positive pressure on the system prior to use. This ensures that any leakage on the system will be in a direction away from the sterile surfaces. If a filter is required for this purpose, it seems logical that the filter should be sterilized with the system. After all, why bother to use SIP at all, if an aseptic connection is needed just to render the system secure from microbial contamination prior to use?

Filter Housing Configuration and Orientation

With the presence of a microbially retentive filter in the system almost certain, what special concerns must be addressed? The basic concerns cited earlier for tanks and other vessels are directly applicable to filter housings. Thus air and condensate removal must be considered, as well as the sequence of events utilized in the sterilization procedure and sterility must be maintained post-sterilization. The removal of air and condensate at first appears to be quite a simple task, since filter housings are supplied with a number of bleed valves. It is frequently necessary to modify filter housings, whether for disc membranes or for cartridge filters to facilitate the removal of condensate (3).

An example based upon the simplest of all filter housings, the disc membrane filter housing will serve to illustrate this point. Most disc housings available are designed with legs to support the housing that maintains the membrane in a horizontal plane for use on a bench top. If the filter is to be sterilized-in-place, how will the condensate formed on the upstream of the membrane surface be removed? There is no effective way to remove this condensate other than by passage through the membrane, which may not be easily accomplished. A workable solution is to position the membrane in the vertical plane. Once oriented vertically, relocation and/or addition of bleeds on both the upstream and downstream sides of the filter at locations just inside the O-ring at the lowest portion of the filter housing has been shown to be necessary to permit effective discharge of the condensate (3).

Having solved the condensate retention difficulties for flat stock membranes and their housings, applying the same principles to pleated cartridge filters is straightforward. The most basic recommendation that can be given for cartridge filters is to always position the cartridge with the open end down. This will allow any condensate formed on the interior of the cartridge to exit the housing without requiring it to pass through the membrane. Additional bleeds on the upstream side of the filter housing are usually required at the lowest points in the housing. This is accomplished in housings where the inlet and outlet are at the same end. The filter housing is most easily adapted to SIP when the housing has the appearance of an inverted "T". When the inlet and outlet are at opposite ends of the housing, sometimes identified as an in-line housing, the need for modification is usually greater.

A further consideration with filter sterilization is the direction of steam flow. Disc membranes should be provided with both upstream and downstream support screens. The presence of the support screen reduces the stress on the membrane during sterilization and should result in lessened problems with filter integrity post-sterilization. For pleated cartridge filters, the general rule is to sterilize them from the outside (upstream side) in. In this manner the differential pressure developed by the introduction of steam to the filter serves to push the filter together rather than to blow it apart. The application of steam in this manner keeps the filter properly seated in the housing, where flow in the reverse direction might tend to lift it. Several filter manufacturers incorporate a locking arrangement that serves to hold the cartridge in the housing. While this feature is helpful, it shouldn't be relied upon to hold the filter in place when steam is traveling upwards through the membrane.

The "Loop" Method

No discussion of filter sterilization would be complete without some mention of the so-called "loop" method (Fig. 9). When the sterilization of large systems employing filters was first attempted, damage to the filter often occurred. The majority of filters in use at that time were made of cellulose acetate, cellulose nitrate or a mixture of the two esters, which become brittle when exposed to excessive heat (around 125°C). With larger systems, the steam pressure needed to bring the entire system to sterilizing temperature would sometimes require the filter to be exposed to temperatures that would impair its integrity. To overcome this limitation, a "loop" or "by-pass" would be added to allow the vessel to receive steam directly, without having to pass all of the required steam through the filter (1–3). This "loop" would of course be closed prior to the end of the steaming process to preserve the integrity of the system during use. The use of a "loop" became common where these earlier types of membranes filter materials are sterilized-in-place. This practice is still necessary when using filter membranes with limited thermal stability. Most of the newer filter media on the market today have significantly greater thermal stability. For these types of filter media, the "loop" is not required and all of the steam necessary to sterilize even large systems may be passed through the filter membrane without difficulty. The advent of ever larger systems has seen a return to the "loop method" as the increasing demands for steam and shortened process time have made the even today's more heat resistant filters rate limiting.

Filter Sterilization

In reviewing many of the observations made above regarding sterilization of filters, the obvious question to be raised is: How were these problems with condensate retention in filters found to be a problem? The basis for much of the information regarding difficulties with filter sterilization was a series of validation studies which this author personally directed during the early 1980s. When faced with the necessity to confirm the sterilization of filters that were to be steam sterilized, the decision was made to utilize a spore suspension and apply the spores directly to the membrane surface. Considering that only the downstream side of the filter was required to be sterile, and recalling that 0.2 μm filters are microbially retentive, the biological challenges were placed on the downstream side of the membrane. It was soon discovered that when a resistant bioindicator such as *Geobacillus stearothermophilus* was utilized that sterilization was not as easily achieved as would be assumed from the time-temperature or F_0 values observed (1–4,8). In resolving these sterilization problems, it became clear that accumulated condensate on or near the filter was the primary cause for the inability to sterilize the membrane surface. In subsequent efforts, the lessons learned from these original studies were confirmed repeatedly in SIP systems of varying size and complexity.

Spore inoculated membrane filters were found to be more reliable indicators of SIP system effectiveness than ordinary spore strips (2,3). The validation of SIP systems utilizing filters was approached in the same manner as other sterilization validation studies. Once filter inoculation was found to be a superior indicator compared with spore strips in the filter housing, they were utilized together over a long series of validation studies. This practice was continued over a period of nearly four years with the following results. There were nearly 90 individual validation trials in which the filters were positive after sterilization while the spore strips present during the cycle were negative for growth. The number of spore strips present in the system during these trials varied from 2 to 10 depending upon the size of the system. The spore strips were placed throughout the system, including in the housing with the inoculated filters. Over this same time period there were over 100 individual validation trials in which neither the filters nor the spore strips were found to be viable after sterilization. The validation trials comprised a range of system sizes ranging from individual tanks with a single vent filter to larger systems containing one or more tanks and multiple filter housings. Filters and filter housings used in these studies included all of the major manufacturers, different membrane materials, several housing designs and both hydrophilic and hydrophobic filters in a range of sizes from 47 mm disc membranes to 10 in. cartridges. In the majority of these studies, the coldest location or minimum F_0 location was not the filter housing, but some other location in the system. In fact, the filter housing was often the hottest location because of its proximity to the steam inlet. In every case, the organisms present on the spore strip were destroyed while those present on the filter surface demonstrated growth or no-growth depending upon the local conditions present at the filter surface. The inadequacy of the spore strip as a means for establishing the effective sterilization of a membrane filter surface became evident. Note that while there were many trials in which the organisms present on the spore strips were destroyed while organisms survived on the filter surface, there was never an incident where the reverse was found.

This is likely due to the spore strips (and thermocouples) being placed in the housing remote from the filter surface (the item being sterilized), whereas the inoculation of the filter surface places the challenge directly upon the surface of critical interest. Clearly, *sterilization of the filter surface is the critical concern* for these types of systems and the use of a microbial challenge on the filter surface is a "worst-case" challenge of the most difficult to sterilize location in the system. The results summarized above led to the conclusion that the filter was the "worst-case" location in any SIP system, and the use of spore strips in SIP systems was subsequently discontinued.

Two examples of the enhanced sensitivity of inoculated filters relative to spore strips are described below:

1. In sterilizing a multi-cartridge vent filter on a steam sterilizer, positive filters were repeatedly found despite F_0 values in the filter housing that exceeded

300 minutes! When condensate retention problems were resolved, the filters were readily sterilized.

2. In the sterilization validation of a large system containing numerous filters, one filter cartridge was found to be non-sterile after repeated trials despite F_0s greater than 60 minutes. When the supplier of the cartridge was changed, no difficulty was encountered in sterilizing the new cartridge, which was able to drain more easily than the original cartridge.

Several publications have included references to the difficulties associated with the sterilization of filters (25,26). While these mentions of similar experience have not been explicit with regard to the details of the studies performed, independent verification of the difficulties encountered in filter sterilization further substantiates the author's already strong beliefs in this area.

There is an indirect benefit to the use of filter inoculation rather than spore strip testing in SIP systems. Where spore strips are utilized in SIP validation, the system must be cooled, tank entry permits obtained, one set of strips recovered and a new set put in place. The accomplishment of these tasks can take several hours especially if the system is quite large. Where inoculated filters are utilized, tank entry is not required and quite large systems can be readied for the next run in a relatively short period of time. In time critical validation activities, the use of inoculated filters can save considerable time over the use of spore strips.

Filter Inoculation and Testing

In order to utilize inoculated filters in SIP, suitable methods must be available for the inoculation and testing of the challenged filters. Inoculation is straightforward, the filter to be challenged is wetted; water for hydrophilic membranes or water–alcohol mixture for hydrophobic membranes, using a pressure can (Fig 18). After complete wetting of the filter surface, a spore suspension of *G. stearothermophilus* in water–alcohol is added to the wetting fluid, to provide the equivalent of approximately 100 spores/cm^2 of filter area. The suspension is filtered through the membrane in the reverse direction leaving the spores on the downstream membrane surface. An additional quantity of wetting fluid is added to rinse the pressure can and complete the transfer of the spores to the filter. The filter is then purged with a gas stream to remove the residual fluid from the housing and lines. Care is taken throughout this procedure not to exceed 2 to 3 psi on the filter to avoid causing physical damage to the membrane when filtering in the reverse direction (an important concern with pleated cartridges). The ends of the filter are covered with a sterile wrap and the filter is ready for use as a biological indicator. The sterile wrap is removed from the filter just prior to installing the filter on the equipment.

Testing of the filter is equally simple. After completion of the sterilization process, the filter is removed from the equipment and the end caps are covered with a sterile wrap. The filter is transported to the laboratory where it is removed from the filter housing. In the case of disc filters, a sterile knife is utilized to cut the membrane just inside of the O-ring that seals the filter membrane in the housing. The entire center portion of the membrane is placed into SCDM and incubated at 55°C to 60°C for seven days. Testing of cartridge filters requires a slightly different procedure. The cartridge housing is opened and the entire cartridge is separated from the housing using sterile gloves. The cartridge is placed into a container with the open end up, and SCDM is aseptically added to the container. The container is covered with a sterile lid and incubated as described above. It may be necessary to fabricate special containers to accommodate 20 and 30 in. filters.

An often-asked question in regard to filter sterilization when the filters are inoculated on the downstream side is: "Can the absence of the spores after sterilization be attributed to physical action rather than microbial death?" The answer to this is simple: the number of times that inoculated filters were found to be non-sterile after completion of the process indicates that viable spores remained on the surface of the filter. Further confirmation of this was established in a study in which the condensate from the downstream side of the filter was collected. Less than 10 viable colony-forming units were

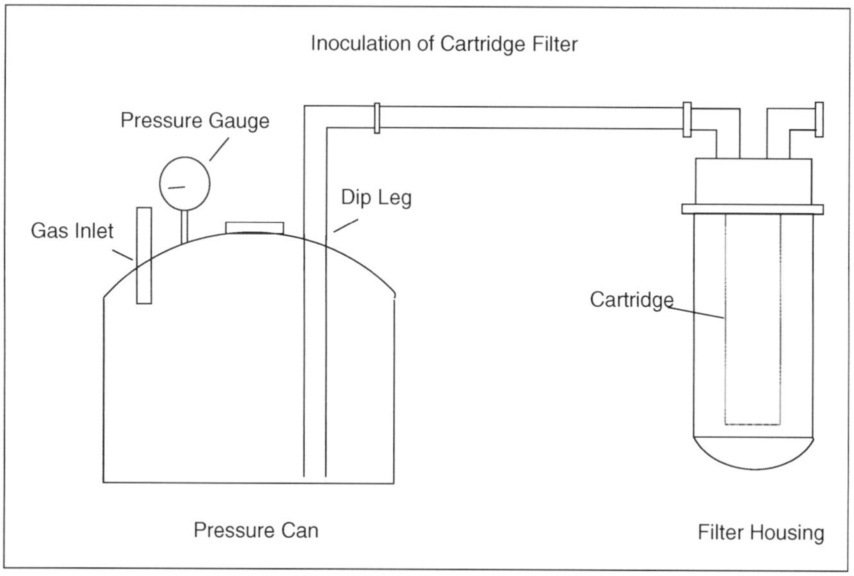

Figure 18 Inoculation of filters.

found in the condensate after sterilization, and the challenged filter was confirmed to be sterile.

Extraordinary precautions are not generally required when conducting challenges with *G. stearothermophilus* in the manner described above. Any incidental contamination that might find its way into the media prior to the incubation is unlikely to grow at the 55°C to 60°C incubation temperatures utilized. The ordinary mesophilic organisms present in the environment and on personnel are unable to grow at the temperatures utilized.

Filter Sterilization in Autoclaves

One last point remains with regard to the sterilization of membrane filters. The description of this effort is somewhat anecdotal, but bears careful consideration given the broad implications. Several years ago, after preparing a filter cartridge as described above for an SIP validation study, the run was canceled and rather than discard the filter it was placed into a steam sterilizer and tested as part of the validation of a multi-vacuum sterilization cycle. The cycle was completed normally and achieved a minimum F_0 in excess of 30 minutes. Much to the investigator's surprise the filter was found to be non-sterile, while the spore strips in the load were sterile. Puzzled by this unusual turn of events, the investigator made a series of additional trials using additional time, additional vacuums, alternative configurations, removal from the housing and higher temperatures all to no avail. Apparently sterilization of filters in a sterilizer where a resistant indicator is located on the surface of the membrane is not straightforward.

The most plausible explanation for these results is condensate retention in the cartridge. In every SIP process where filter sterilization was found to be a problem initially, the problem was resolved by eliminating condensate retention in the system. In a steam sterilizer with steam able to contact both sides of the filter membrane, the condensate formed will have nowhere to exit the system. As steam penetrates the membrane surface to reach the spores, condensate forms and is retained in the filter matrix. As the steam is saturated, the addition of more steam will not return the condensate already formed to the vapor phase. In an SIP mode, the steam pressure on the upstream side will force condensate through the filter and out of contact with the filter. It appears that condensate retention in the filter matrix as a result of the absence of a driving force would explain why SIP works and sterilization in an autoclave (when tested in this more rigorous manner with the spores on the downstream surface of the membrane) would not. Testing of the filter in its housing with spore strips is common and given that condensate is unlikely to accumulate where the strip is located, the absence of growth on the strips is not surprising. There are references to similar results having been observed by others (25,26).

The sterilization of filter cartridges in autoclaves in steam sterilizers is such a widespread practice it is surprising to find that it is not nearly as easy to accomplish as the industry would believe. That the failure to sterilize these filters in an autoclave is not the cause of significant sterility failures in the industry is probably due to the much lower resistance of ordinary organisms that might be on the filter surface to steam sterilization relative to *G. stearothermophilus*.

VALIDATION

Installation and Operational Qualification

The validation of SIP follows the conventional approach utilized for nearly all validation programs including IQ, OQ, and PQ. The emphasis placed on system design and procedural conformance earlier in this chapter should reinforce the importance of the IQ and OQ activities to the reader. During the IQ, the SIP system should be carefully scrutinized for the physical features described earlier. As a general rule, if there is a concern that air or condensate may be retained at a given location then a bleed should be added to the system at that point. Another useful part of the IQ is to ensure that all valves in the system are clearly identified, which makes procedural conformance more certain. It is conventional during the IQ of systems and equipment to define the locations of all instruments on the system. This is sound advice, but must be approached with some degree of flexibility with SIP systems. The results of the PQ study may result in the need to relocate some of the permanent temperature probes to more important locations.

The preparation of an extremely detailed procedure for the sterilization of the system is essential to success in validating the system and is often a part of the OQ. The procedure should include step-by-step directions for the manipulation of each valve in the system from the start of the procedure through to completion. The installation of any additional lines, hoses, fittings, temperature probes, etc. should all be detailed in the SOP. The use of diagrams to facilitate adherence to the procedure is certainly beneficial. Any procedure developed prior to the completion of the PQ should be considered a draft until its effectiveness has been confirmed. The cycle development that follows the OQ utilizes thermocouples to confirm temperature prior to the introduction of biological indicators.

Cycle Development

In conducting PQ studies for SIP, the basic approach is to thermally map the entire system with thermocouples and to parallel the temperature distribution studies with biological indicators. Thermocouple placement in the system should include all low points (where condensate may accumulate), at the end of each length of piping (where steam penetration must be established), at permanent temperature measurement locations (to establish correlation with routine system documentation), on the upstream and downstream side of filters and elsewhere in the system where temperature is of interest. The objective in thermocouple placement is to attempt to locate portions of the system where condensate will remain and the temperature will be reduced. In conducting studies on tanks and similar pieces of equipment, thermocouple placement should focus on the piping entering and leaving the vessel, and any permanent temperature location inside the vessel. The placement of multiple probes inside the vessel has little utility; those points are easily accessible to steam and probes in the discharge of the vessel are nearly always the coldest (a result of the large amount of condensate which must pass out of the vessel). Where temperature is being measured in piping systems, consideration can be given to placing the thermocouples outside the pipe rather than inside. This not only eases the task of thermocouple placement, but is also a "worst-case" measurement of the temperature inside the pipe. The placement of thermocouples must be accomplished with care. The effectiveness of the sterilization must be unaffected by the addition of these additional temperature probes in the system. The use of appropriate pressure fittings for thermocouple ingress is suggested to assure that air and/or condensate removal is not enhanced. Equal care must be taken in placing the probes to ensure that condensate removal is not restricted due to the presence of the thermocouple in the system. In small diameter piping, the use of external or miniature thermocouples should be considered if there is a possibility that the use of conventional probes may have an adverse effect.

Performance Qualification

Once a defined design/procedure is identified that demonstrates sufficient temperature throughout the system, the PQ can commence. In the author's experience, temperature measurement is not always a good predictor of microbial death, especially with filters, but the thermocouples can help to identify problem locations which if microbially challenged would surely result in sterilization difficulties. It is recommended to utilize biological challenges on filters, and not to use spore strips. If employed, spore strips should be placed adjacent to each thermocouple in the system. The placement of spore strips in SIP systems is not always an easy task. One of the simplest ways to ensure that strips are not lost is to place them in a permeable container that can be firmly affixed in the equipment. Good success has been achieved with stainless steel tea infusers (perforated containers with a light chain attached) in which a spore strip is placed and the chain utilized to secure the infuser to the equipment. This technique is only adaptable to locations large enough to accept a relatively large object without obstructing the fluid flow. Using tape to secure spore strips to the side of the vessel is also possible, but care must be taken to ensure that steam can access the entire surface of the strips. The use of spore strips in closed piping is of such difficulty that many practitioners employ strips only in tanks, filters, and other easily accessible locations. The use of loose or poorly secured strips is not recommended, as they can be swept away by steam or condensate and lost. Wherever strips are placed, care must be taken to avoid obstructing steam entry or condensate removal due to their presence.

Possible alternatives to spore strips are inoculated stainless steel coupons that can be placed in the system just as spore strips would. The coupons have the advantage of having sufficient mass that they would remain where they are placed despite the passage of steam or condensate over their surface. Direct inoculation of other stainless steel components in the system, i.e., cups placed over filling needles for CIP/SIP, coupons held by wire in the vessel is also possible. Where these alternates to spore strips are utilized, the end user must confirm the count and resistance of the spores on the substrate.

In performing the PQ studies, consideration should be given to an experimental design that maximizes data utility while minimizing the number of required studies. Good success has been obtained with a bracketing approach that utilizes two different "worst-case" assumptions (Fig. 19). The first series of runs (typically three) are performed using time–temperature conditions that exceed that intended for use in operation, such as 127°C for 60 minutes. After completion of the full

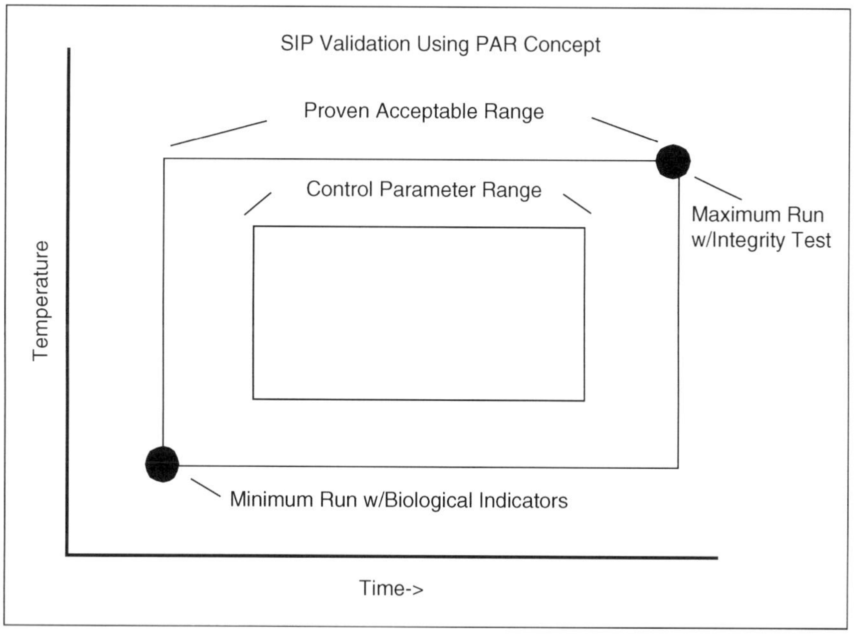

Figure 19 SIP validation using PAR.

series of runs, all of the filters installed in the system, whether for liquid filtration or gas filtration (tank vent or compressed gases) are tested for integrity in their housing. In critical installations, the testing may be performed in situ, if that is the normal means for integrity testing. With the types of membranes in vogue today, there is little necessity to test the filters after each individual SIP run. If the filters are integral after three consecutive studies, there should be little concern that they would be damaged in a single run. Minimum studies with microbial challenges are performed next, with a possible test condition of 122°C for 30 minutes. Three complete sets of filters will be required for the inoculation program. If the conditions cited above were representative of the "worst-case" efforts, and proved successful in the PQ studies, then the routine sterilization SOP might provide for conditions of 124°C to 126°C for 40 to 45 minutes. Thus both the time and temperature conditions utilized in the PQ studies would provide a considerable safety margin over that employed in routine sterilization.

Limits for the range of temperature across the system being validated should be extremely flexible. In large systems temperatures may vary in excess of 10°C from the hottest to coldest location. More appropriate is a requirement that all monitored locations exceed a minimum F_0 value determined from the cycle development effort.

AUTOMATION

The application of computer control systems in the pharmaceutical industry has been on the increase for the past decade. The application of a computer control system to a system subject to SIP can facilitate procedural adherence and ultimately sterility assurance. To achieve this, the application engineer must delay the completion of the software (and possibly the hardware) necessary to automate the SIP process, until the sterilization validation has been completed. Only then it can be assured that the control system will utilize the proper sequence of valve actuation to effect sterilization. As the timely and correct manipulation of the valves in the system is of critical importance to success, the use of computer control provides considerable benefits over a manual operation. The use of a control system also eases the review of multi-point temperature measurements allowing for more precise timing of the SIP cycle. Control systems for SIP are commonly found on equipment that already has a control system present for other operations, i.e., freeze dryers. The application of automated SIP on an otherwise non-automated piece of equipment is unusual due to the extra expense entailed.

There are obstacles to the expanded use of automation for SIP systems. In a large piping system, the sheer number of automated valves required will result in a considerable expenditure in software and hardware to properly execute the SIP procedure. The cost associated with this additional complexity can be a deterrent to automation. A second difficulty is the small bleeds necessary at some locations; automated valves of appropriate design are sometimes not available in the necessary sizes.

BULK STERILIZATION

A subject often discussed in conjunction with SIP is bulk sterilization of liquids. In bulk sterilization, an aqueous-based fluid that cannot be filter sterilized due to the presence of solids (either active materials or other excipients in the formulation) is sterilized in a closed vessel. The methods utilized to perform this sterilization are derived from those utilized for conventional SIP procedures.

1. Prior to the start of the bulk sterilization, the empty vessel and its associated piping should be sterilized-in-place in accordance with the methods described earlier.
2. The fluid material should be introduced into the vessel with a minimum degree of splashing against the upper portions of the tank. All valves in the system other than the vent and on the line utilized to introduce the fluid should remain closed.
3. Agitation of the vessel contents should be started and heat applied to the jacket (in bulk sterilization elevated pressures of steam in the jacket are necessary in order to achieve rapid heat-up of the vessel contents, contrary to the advice provided earlier for SIP).
4. As the temperature of the fluid rises (as measured by a fluid product probe) small amounts of the aqueous phase will be lost. The agitation and increase in temperature will also assist in the expulsion of air from the fluid and the headspace above it.
5. When the temperature of the fluid approaches 98°C to 100°C, the vent and fluid entry line should be closed. The temperature of the fluid will continue to rise, and some adjustment of the steam to the jacket may be necessary to control the temperature in the desired range for the required time period.
6. Upon completion of the sterilization portion of the cycle, the jacket of the vessel is emptied of steam and cooling water is applied to reduce the temperature prior to further processing. Filtered air or nitrogen is introduced to maintain a positive pressure in the vessel as the contents cool.

If the amount of liquid to be sterilized in the vessel does not reach to the upper portion of the vessel's jacket, thermocouples in the headspace may show unusually high temperatures compared to those immersed in the liquid. These temperatures appear anomalous, but may be correct and result from superheating of the steam in the headspace by the exposed jacket. This will be most often observed with the minimum batch size in the tank.

Validation of bulk sterilization is a hybrid of terminal sterilization of fluids in sealed containers and SIP. Considerations prevalent in terminal sterilization validation such as bioburden determination, *D*-value determination in product and stability of the formulation at elevated temperatures must all be considered. Similarities with SIP procedures are found in the placement of multiple temperature probes, and strict adherence to sterilization procedures.

It is tempting to proceed directly to the bulk sterilization of the liquid in the vessel without the initial SIP of the empty tank. This approach is possible if all of the vessel headspace connections are bled to allow for air and condensate removal. Using this approach there will be some loss of liquid volume in the vessel that could affect the formulation. This approach does not allow for the sterilization of the bottom discharge of the tank and for this reason is not recommended.

CONCLUSION

The development and validation of SIP procedures are among the most challenging of all validation activities. The increased attention that SIP has evidenced will ultimately result in further refinement of the concepts and opinions provided in this chapter. The reader is encouraged to remain current as new developments in SIP technology are published, for despite the interest in the subject, the existing base of published information is minimal and significant improvements in our knowledge basis are certain to occur.

REFERENCES

1. Myers T, Chrai S. Design considerations for development of SIP sterilization processes. J Parenter Sci Technol 1981; 35(1):8–12.
2. Myers T, Chrai S. Steam-in-place sterilization of cartridge filters in-line with a receiving tank. J Parenter Sci Technol 1982; 36(3):108–12.
3. Kovary S, Agalloco J, Gordon B. Validation of steam-in-place sterilization of disc filter housings and membranes. J Parenter Sci Technol 1983; 37(2):55–64.
4. Berman D, Myers T, Chrai S. Factors involved in the cycle development of an SIP system. J Parenter Sci Technol 1986; 40(4):119–21.
5. Agalloco J, Shaw R. Steam-in-place sterilization. Parenter Drug Assoc Course Notes 1987.
6. Perkowski C. Controlling contamination in bioprocessing. Biopharmaceutical 1987.
7. McClure H. How to sterilize liquid filling equipment at point of product contact. Pharm Eng 1988.
8. Chrai S. Validation of filtration systems: considerations for selecting filter housings. Pharm Technol 1989.
9. Agalloco J. Steam sterilization-in-place technology. J Parenter Sci Technol 1990; 44(5):254–8.
10. Cappia JM. Principles of steam-in-place. Pharm Technol 2004; (Filtration Suppl.):S40–6.
11. Pflug IJ. Microbiology and Engineering of Sterilization Processes. Minneapolis, MN: Environmental Sterilization Laboratory, 1999.
12. Noble PT. Modeling transport processes in sterilization-in-place. Biotechnol Prog 1992; 8(4):275–84.
13. Noble PT. Transport considerations for microbiological control in piping. J Pharm Sci Technol 1994; 48(2):76–85.
14. Young JH, Ferko BL. Temperature profiles and sterilization within a dead end tube. J Parenter Sci Technol 1992; 46(4):117–23.
15. Young JH. Sterilization of various diameter dead-ended tubes. Bioeng Biotechnol 1993; 42:125–32.
16. Young JH, Ferko BL, Gaber RP. Parameters governing steam sterilization of deadlegs. J Pharm Sci Technol 1994; 48(3):140–7.
17. Bacaoanu A. Steam sterilization of stationary large-scale freeze-dryers (lyophilizers). Am Pharm Rev 2004:40–7.
18. Food and Drug Administration. 21 CFR 212 Proposed Current Good Manufacturing Practice in the Manufacture, Processing, Packing or Holding of Large Volume Parenterals for Human Use, June 1, 1976 (rescinded December 31, 1993).
19. Carlson VR. Response to question & answer. Biopharmaceutical 1990.
20. Agalloco J, Carlson VR. Private communication, 1990.
21. Yoshida S. Development of a new concept for time/pressure filling. PDA J Pharm Sci Technol 2003; 57(2):126–30 (see also 57(4): 316–7).
22. Agalloco J, et al. Process simulation testing for sterile bulk pharmaceutical chemicals. PDA Technical Report #28. PDA J Pharm Sci Technol 1998; 52(Suppl. 4).
23. Odum JN. Closed systems in bioprocessing: impact on facility design. Bioprocess J 2005:41–5.
24. Thorp G, Zwak J. Measuring process temperature in small diameter line. Pharm Eng 2004; 24(5):8–18.
25. Meltzer T, Steere W. Operational considerations in the steam sterilization of cartridge filters. Pharm Technol 1993.
26. Christopher M, Keating P, Messing J, Walker S. Inoculation and sterilization of hydrophilic filters. In: Proceedings of Parenteral Drug Association International Symposium. Basel, Switzerland, Parenteral Drug Association, 1994.

15

Dry Heat Sterilization and Depyrogenation Validation and Monitoring

Laurie B. Case
Pharmaceutical Project Management, Sea Girt, New Jersey, U.S.A.

Gayle D. Heffernan
Pharmaceutical Consultant, Middlesex, New Jersey, U.S.A.

INTRODUCTION

Dry heat is one of the most commonly used methods to sterilize and/or depyrogenate pharmaceutical components and products. Dry heat sterilization is often used for heat-stable oils, ointments and powders. Most often, depyrogenation of parenteral containers is performed utilizing a dry heat oven. The depyrogenation process is also utilized on certain heat-stabile components, glass containers, metal equipment, etc. to render the item and final parenteral product free of pyrogens. The equipment utilized to provide the dry heat medium must be validated to ensure that the system is able to provide sterile and/or depyrogenated components, on a reproducible basis.

The validation of a dry heat sterilization and depyrogenation process involves approaches and procedures which parallel those utilized for steam sterilization. The efficiency of any heat treatment is determined by the design and source of the heat. Hot air is substantially less efficient in a thermal transfer medium as compared to steam.

The validation effort must include heat distribution, heat penetration, bioburden and pyroburden determination, filter integrity, and microbial/endotoxin challenges. This chapter will detail the steps of a program that may be employed to properly validate a dry heat process. The topics discussed will include:

1. Types of Dry Heat Sterilizers
2. Principles of Heat Transfer and Circulation
3. Validation Test Equipment
4. Installation Qualification
5. Operational Qualification
6. Pre-Calibration of Validation Test Equipment
7. Process Qualification Cycle Development
8. Qualification Protocol
9. Qualification Testing
10. Post-Calibration of Validation Test Equipment
11. Qualification Report
12. Routine Monitoring After Validation
13. Documentation
14. Conclusion

TYPES OF DRY HEAT STERILIZERS

The types of dry heat sterilizers commonly employed in the pharmaceutical industry are forced-convection batch sterilizers, infrared tunnel sterilizers, forced-convection tunnel sterilizers, continuous flame sterilizers, microwave, and laser/plasma sterilizers.

Batch Sterilizer Ovens

Batch sterilizers are the most commonly used type of dry heat sterilizers in the industry due to their flexibility in unit and load size (Figs. 1 and 2). The unit operates on the principles of convective or radiation heat transfer to the components. The ovens can employ a range of cycles (by varying time and temperature settings) for utensils, glassware, stainless steel equipment, or products. The items are prepared in a classified non-aseptic area, with controls in place such as limited access, reduced particulate levels, known air quality, gown covering, and hair covering. The preparation of glass containers consists of washing with high-quality water, steam and/or filtered air, prior to loading the containers on racks or carts and into the sterilizer chamber.

The USP recommends that validation of sterilization cycles for heat stable components include a microbial survival probability of 10^{-12} of *Bacillus subtilis* spores (1). It also recommends that to validate depyrogenation cycles, appropriate items should be charged with a minimum of 1000 EU of purified endotoxin, where the LAL test is used to demonstrate the endotoxin has been inactivated to not more than 1/1000 of the original amount (3-log reduction). The cycles are no longer defined by a minimum time and temperature requirement. Historically, the dry heat sterilization cycles were defined as 170°C for not less than two hours, while depyrogenation cycles were defined at a minimum of 250°C for not less than 30 minutes. A typical cycle might employ temperatures in the range of 180°C to 300°C. The temperatures at the lower end of this range will sterilize,

---

Abbreviations used in this chapter: AAMI, Association for the Advancement of Medical Instrumentation; CEN, European Committee for Standardization; DOP, dioctyl phthalate; EU, endotoxin units; HEPA, high-efficiency particulate air; HIMA, Health Industry Manufacturers Association; HVAC, heating, ventilating and air-conditioning; IQ, installation qualification; ISO, International Organization for Standardization; LAL, limulus amebocyte lysate; LPS, lipopolysaccharide; NIST, National Institute of Standards and Technology; OQ, operational qualification; PAO, polyalpha olefin; PDA, Parenteral Drug Association; PM, preventative maintenance; PMA, Pharmaceutical Manufacturers Association; RTD, resistance temperature detector; SOP, standard operating procedure; USP, *United States Pharmacopeia*.

Figure 1 A commercial dry heat batch sterilizer. (Lytzen Class 100 Dry Heat Sterilizer Model H1F/1650). *Source*: Courtesy of Bosch Process Technology, Minneapolis, MN.

while the higher temperatures in the range are suitable for depyrogenation. The cycle effectiveness will also be dependent on cycle time. The total time for batch cycle completion is often greater than three hours including cooling of the load.

Conventionally, there is a cooling phase at the completion of the heating cycle, which serves to minimize component thermal shock and increase handling safety. In most installations, a double-door oven is employed and the load is removed from the oven into the aseptic processing area. The glass containers may be processed in the oven in an inverted orientation to facilitate moisture drainage from washing and to decrease particulate accumulation. In loads where the glass container is upright, lids or protective plates can be situated over the container opening to decrease particulate accumulation.

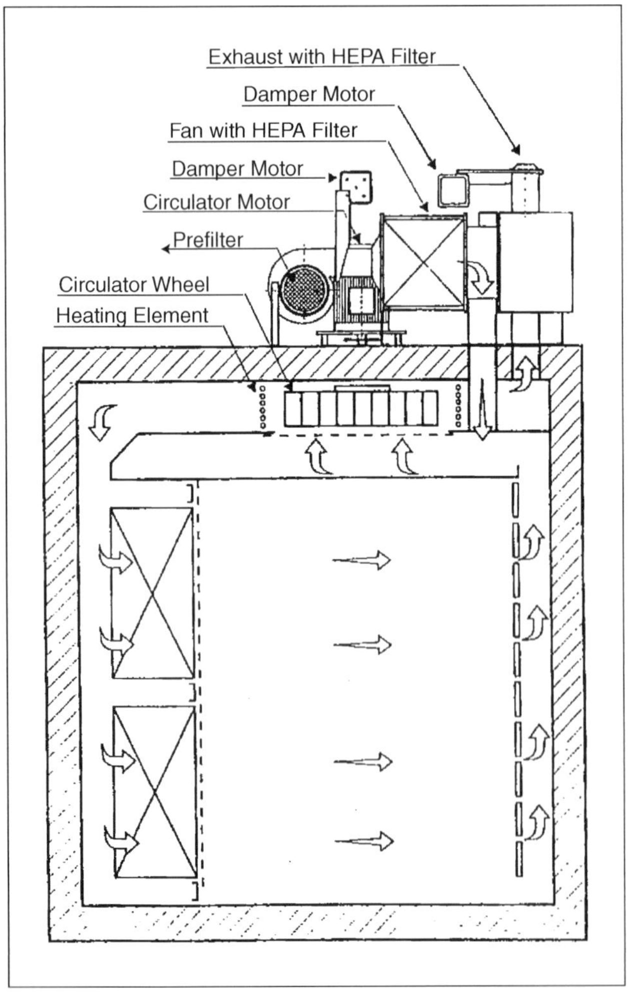

Figure 2 Schematic of a typical batch sterilizer.

Tunnel Sterilizers

The different types of tunnel sterilizers include forced-convection, infrared radiation, and flame sterilizers for ampules. Tunnel sterilizers operate continuously and are typically kept hot for long periods of time with reduced temperatures at nights or weekends to conserve energy. Dry heat tunnels have the capability to process a larger quantity of glass vials and ampules than batch sterilizers (Figs. 3 and 4). The tunnels operate to simultaneously sterilize and depyrogenate glass containers. The continuous tunnel sterilizer has the advantage over the batch sterilizer in being capable of processing only the glass containers which are necessary for the production lot, since the processing time is relatively short. The requirements for additional containers can be reevaluated and updated while filling is performed, in contrast to the batch sterilizer which must have an excess of glass containers processed to account for line breakage and other filling problems.

The forced-convection tunnel is usually heated by electric coils and employs the same principles as forced-convection ovens. Bottles, vials, or ampules are washed and loaded on the non-aseptic end of the tunnel. Most tunnels are designed to work in conjunction with a washer for continuous processing.

The containers are conveyed along the length of the tunnel (10–25 ft). The glassware is heated in the initial heating and center "hold" portions of the tunnel (3–5 ft) at 250°C to 450°C, and gradually cooled down by HEPA filtered air prior to leaving the tunnel at the aseptic end. The belt speeds typically run 2 to 4 in./min with a container temperature above 250°C. The time for the forced-convection type tunnel to process containers is approximately 40 to 60 minutes (2). As shown in Figure 5, the tunnel may be directly connected with a filling line. The bottles are inverted and washed on the inside and outside, then inverted upright and depyrogenated in the tunnel. The vials are then immediately filled and capped.

The infrared tunnel sterilizer is equipped with a source of infrared radiation provided by either a resistance wire or quartz tube which can be covered with polished reflector plates. Temperature sensors to control heating are located within the sterilizing zone. Heat-up and exposure time can be affected by the geometry, color, surface, and composition of the item being treated, as well as the air temperature and air velocity (3). The damper controlled heating zone is

Figure 3 A commercial depyrogenation tunnel with vial washer. (HQL 5000 Series Depyrogenation Tunnel with RRU Vial Washer). *Source*: Courtesy of Bosch Packaging Technology Inc., Minneapolis, MN.

supplied with HEPA filtered air from the hold and cooling zones. The heating zone is set at a higher temperature than the hold zone to bring the glassware to the desired hold zone temperature. The cooling zone drops the container temperature to avoid glass cracking from thermal shock upon exiting the tunnel. Tunnels, such as those manufactured by Despatch and Bosch, are now available with the capability to automatically sterilize the cool zone when the tunnel is not in use and are becoming more common.

The flame sterilizer utilizes conduction and convection heat transfer in the continuous processing of ampules (Fig. 6). It can process up to 10,000 ampules per hour. Ampules are placed on a conveyor belt, washed with water-for-injection, and channeled onto spokes of a rotating wheel. As the wheel rotates, the ampules are heated to 425°C by natural gas heat for approximately one minute. The flame sterilizer has a series of baffles in the sterilizing chamber to increase the uniformity of the heating. The ampules then pass from the heating chamber into a cooling chamber, where they are gradually cooled by HEPA filtered air. The cooled ampules are then filled and flame sealed.

Microwave Sterilizers

Microwave sterilizers use electrical or electromagnetic energy to materials by conductive, near-field coupling, or radiative techniques (4). With microwave technology, the microorganisms are heated without intense heating of the container. Studies have been performed on *B. subtilis* and *B. stearothermophilus* applied to glass vials and heated for various periods of time with microwaves. A 10^{12} reduction of *B. subtilis* was determined to occur within three minutes of treatment with microwaves (5).

Studies with dry *B. subtilis* spores treated in a dry heat oven and a microwave at the same temperature demonstrated that the mechanism of sporicidal action of the microwaves was caused by thermal effects. A dwell time of 45 minutes at 137°C was required for both the dry heat oven and microwave oven, however the microwave come-up time to sterilization temperature was shown to be four times faster than the dry heat oven (6). A spore temperature above the vial temperature is achieved in a microwave oven and results have been reported of a spore reduction of 10^{12} in four minutes with the vial temperature of 160°C as compared with a dry heat oven cycle of 90 minutes at the same temperature (5).

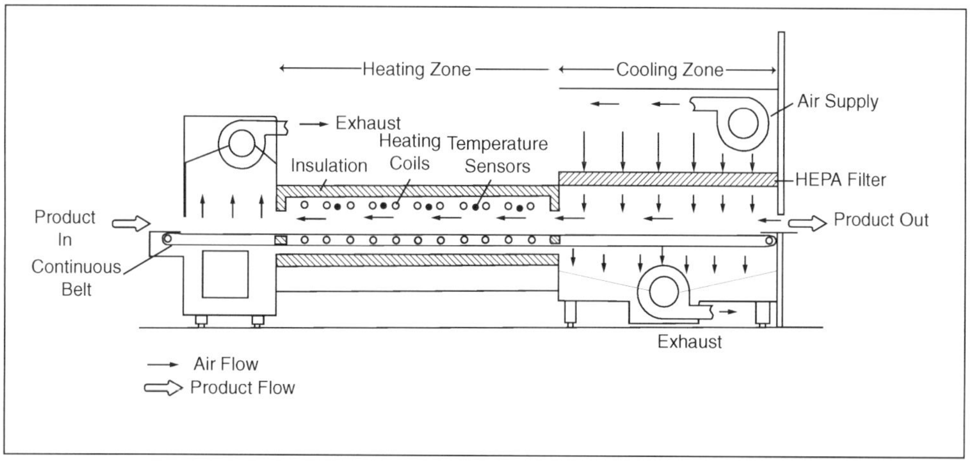

Figure 4 Schematic of a typical tunnel sterilizer.

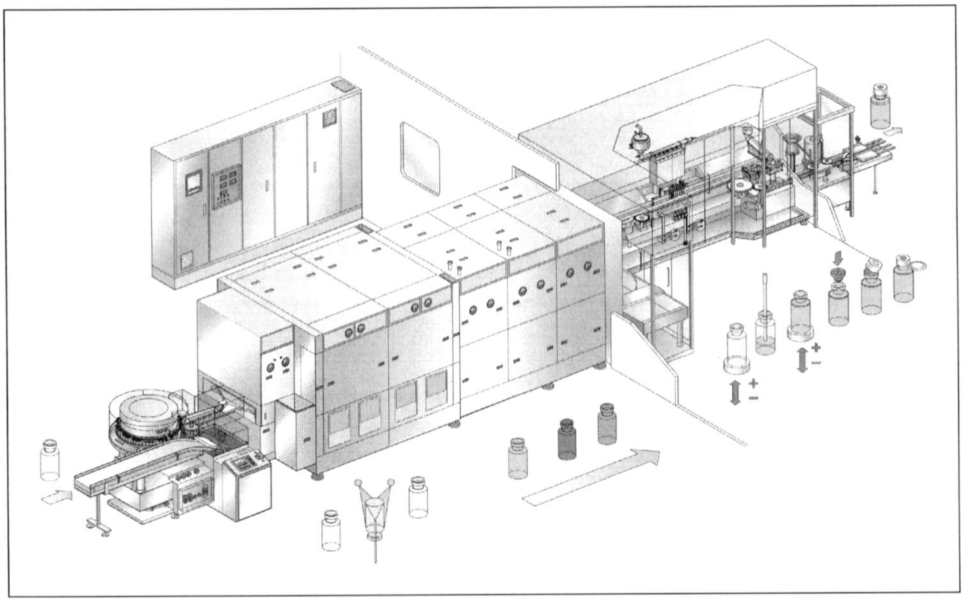

Figure 5 Schematic of a depyrogenation tunnel with washer and filler. (HQL 5000 Series Depyrogenation Tunnel with RRU Vial Washer, FLM Filler, and VRK Capper). *Source*: Courtesy of Bosch Packaging Technology Inc., Minneapolis, MN.

The fluctuations of temperatures and power output, cost, and lack of production designed equipment have prevented microwave technology from becoming a viable alternative to dry heat oven processing at this time.

PRINCIPLES OF HEAT TRANSFER AND CIRCULATION

The dry heat process must effectively heat the article, and the air surrounding the article, to achieve sterilization or depyrogenation. While there are similarities between dry heat and moist heat validation approaches, there are several differences in the processes that direct the emphasis in validation. In moist heat, the condensation of the steam sterilizer releases large amounts of heat energy that serves to heat the items in the sterilizer. In dry heat processes the hot air carries significantly less heat energy than an equivalent volume of saturated steam. Because hot air has both a low specific heat and poor thermal conductivity properties, there is a necessity for long sterilization periods at higher temperatures than those required in steam sterilization. In dry heat processing, the penetration of heated air is not facilitated with pre-vacuums as are often utilized in steam sterilization. The sterilization load is generally slow in heating and cooling, and has a tendency toward temperature stratification (causing temperature variations often greater than 10°C during the cycle). Despite these limitations, dry heat is chosen as the preferred method to induce sterilization and/or depyrogenation over moist heat or other methods in certain instances. Some items are ideally sterilized by dry heat methods. These include glass, stainless steel equipment having surfaces difficult to penetrate with steam, and items that may corrode with moisture. Some products are damaged or contaminated by the

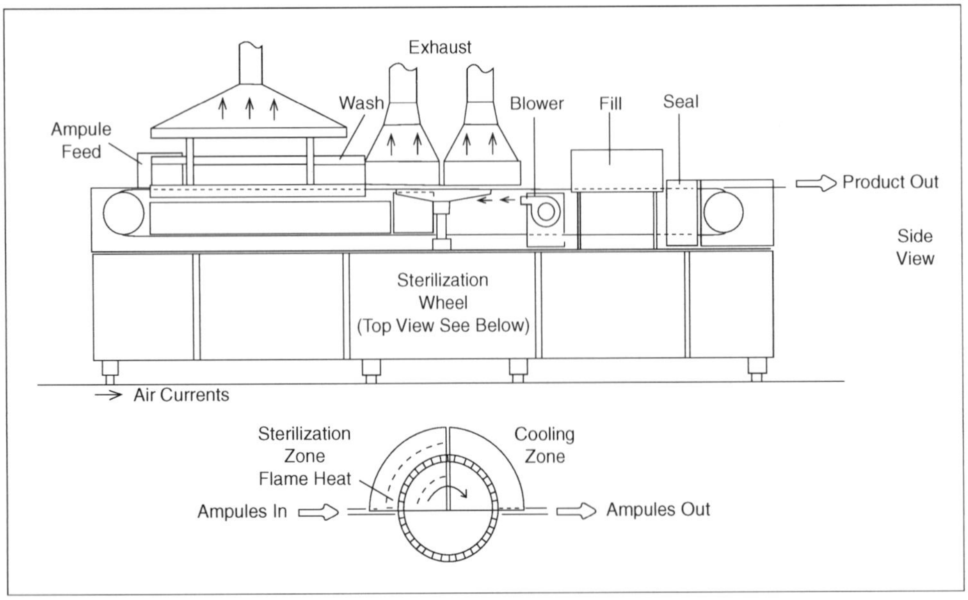

Figure 6 Schematic of a typical flame sterilizer.

presence of water (e.g., petrolatum, oils, nonaqueous vehicles, fats, and powders) (7).

Convection

Dry heat sterilizers use convective heat methods to increase the temperature of the product. Convection is a form of heat transfer whereby heat flows from one body to another due to the temperature difference between them. In the sterilizer, air is heated by convective methods by passing it across heating elements. Energy is transferred to the air from the heating coils. The heated air transfers energy to the items being treated, because those items are at a lower temperature than the air. The rate of heat transfer (how fast the items will heat up) is related to the specific heat of the various materials. Air has the disadvantage of having a relatively low specific heat; therefore, it transfers energy at a slow rate. Saturated steam is an excellent material to use for heat transfer, as it has a relatively high specific heat ($c_v = 1.0$ Btu/lbm °F) as compared with air ($c_v =$ 0.1715 Btu/lbm °F) (8).

Conduction

Conduction of heat consists of the flow of thermal energy through a material from a higher- to a lower-temperature region. Conduction is another means of transferring energy, achieved by molecular interaction where atoms at high temperatures vibrate at high energy levels and impart energy (as heat) to adjacent atoms at lower energy levels. Adjacent materials (such as air surrounding the product) will transfer energy from the higher-temperature material to the colder material because electrons in an excited state bombard and collide with electrons of the lower-energy (colder-temperature) material. The excitation of the molecules of the object increases the level of molecular energy, which increases its temperature (9). Again, the rate at which heat conduction occurs is dependent on the materials involved. A substance with a high thermal conductivity is a good heat conductor, while one with a lower thermal conductivity is a poor heat conductor. While air is not considered a good heat conductor, the contact of hot air with good heat conductors, such as metal objects, will provide a fairly rapid heat transfer rate. Metal items such as stainless steel equipment will heat up faster than other materials (such as glass), because of their greater thermal conductivity.

Radiation

Radiation is the third method commonly used for dry heat sterilization processes. Photons, which are concentrated bundles of energy, may occupy different energy levels. Photons travel as an electromagnetic wave from the emitting material to the object material. This photon propagation will transfer thermal energy to the object and increase the surface temperature. Radiation may be used as the sole source of heating an item, or it may be employed in combination with the convective and/or conductive methods. The radiant heat can be uniformly distributed with properly placed "mirror" polished reflectors over the quartz tubes.

Circulation

To aid the heating process a system to increase the circulation of the heated air is employed. During the heat-up cycle, air circulation helps by removing cool air from the chamber or heating zone and preventing temperature stratification. Higher temperature air replaces the cool air, and the load is heated more rapidly. The circulation of air is also useful at the end of the heating cycle for cooling the load, and protecting the sterile load by maintaining a positive pressure (>0.127 mbar) between the sterile load inside the chamber and the non-aseptic load preparation area.

Booster fans or blowers are used in the oven or tunnel in order to increase the circulation of the heat throughout the load (Figs. 2, 4, and 6). Determining the flow rates of the air within the sterilizer is essential, since it is an important factor in the rate of heat transfer. An air velometer (or hot-wire anemometer) may be used to determine the flow rate of the intake, exhaust, and circulating air. If the sterilizer is installed in an aseptic environment, balancing of the air systems in the room is necessary. A slight positive pressure should exist from the aseptic area to the open sterilizer to prevent contamination of the aseptic area. The sterilizer should also be slightly positive in pressure with respect to the non-aseptic area to prevent the flow of dirty air into the sterilizer.

The air is usually supplied by the central HVAC system or directly from the aseptic area. A filtered HVAC air supply is preferred because it has a lower particulate load and is usually both temperature and relative humidity controlled. Room air may be used, but it has a higher particulate count and variable temperature and humidity levels. Relative humidity is an important factor in trying to heat the air and maintain a consistent temperature, and is an important factor in the qualification of open-ended tunnel sterilizers. Validation information should include monitoring relative humidity of the make-up and surrounding air.

HEPA filters may be used for both cleansing the air supply and circulating air, but the filters in the air circulation path must be designed to withstand the high operating temperatures and should be monitored periodically for integrity. The performance of the HEPA filters may be tested using PAO (a synthetic hydrocarbon manufactured under the brand name of Emery 3004® by Henkel Corporation) or by using DOP (10). The HEPA filter can have a ceramic or heat-resistant housing with high-temperature sealing gaskets. HEPA filters that are subject to extremely high temperatures should not be integrity tested with the aerosols, as they outgas and smoke when heated, and may contaminate the contents in the oven or tunnel. In place of integrity testing, air cleanliness testing should be performed.

The air introduced into the sterilizer and circulating within the unit should be tested for particulates at several locations while the fan is operating. It is preferable that the tests be run at the normal operating temperatures if the air particulate monitoring equipment can handle the temperature. High particulate counts may be caused by an improperly balanced air system, missing or dirty filters, vibrating and/or shedding materials, or inadequate sanitization practices. Ideally Class 100 air should be used for the circulating air supply (11). Air cleanliness classes and cleanroom/clean zone requirements are defined by the International Standard ISO 14644-1 with limits set for the maximum number of particles per cubic meter of air based on micrometer size (12). The ISO standard has replaced Federal Standard 209E. ISO 5 (Class 100) particulate counts may not be achieved during come-up and cool-down zones due to moisture evaporation and glass temperature gradients.

VALIDATION TEST EQUIPMENT

Equipment required to conduct the IQ, OQ and PQ includes that listed below. All temperature equipment employed to perform the validation studies must be traceable to an NIST, CEN or other instrument ultimately calibrated to the International Temperature Scale. This is achieved by sending the "primary" calibration equipment (RTD units and probes) directly to a certified laboratory, and calibrating the other "portable" test equipment (thermocouples and data loggers) to the RTD units. Correct usage and calibration of the validation test equipment must be fully documented in an SOP. The equipment used for validation testing of dry heat processes are discussed below.

Resistance Temperature Detectors

The "primary" temperature detector is the RTD monitor and probe. A platinum/copper detector provides the most linear range between 0°C and 400°C (13). RTD is conventionally utilized for calibrating the "portable," less sensitive, temperature measurement equipment used during validation testing. The RTD may be used with assurance to detect variations in temperature up to 0.01°C.

All calibrated instruments must be numbered, logged, and referenced. A calibration decal on the unit should include the calibration date, by whom calibrated, and the date of the next calibration. A calibration file must be maintained for each instrument, including the information cited for the calibration decal, a history file updated with any repairs made on the unit, and a list of the instruments used to calibrate it (such as resistance standards, voltage standards, etc.), identified with serial numbers. The range, accuracy, and calibration due dates of the equipment used to calibrate the calibrated instrument should be identified. SOPs should be written and approved for the calibration of all instruments and included in the calibration file. A master file of all calibration dates and histories should be maintained for easy reference.

Thermocouples

Thermocouples are the most widely used devices for "portable" temperature measurements. The choice of thermocouple type and insulation surrounding the wires is dependent on the operating temperature and required temperature accuracy. For dry heat sterilization or depyrogenation processes, both type T (copper and constantan) and type J (iron and constantan) thermocouples can be used. The insulation most commonly chosen for high-temperature work is Kapton-H by Dupont. This insulation is rated to 350°C, sufficient for depyrogenation use. Thermocouples generally have a level of sensitivity of 0.1°C. The thermocouple bundle should be tightly sealed to minimize any air leakage out of the dry heat sterilizers. Each thermocouple should be numbered approximately 12 to 18 inches from the tip with a metal tag, and also at the data logger end, for easy identification. SOPs should be written and approved for the calibration of the thermocouples with the data logger and included in the calibration file (refer to sections Pre-Calibration and Post-Calibration of Validation Test Equipment).

Data Loggers

Multipoint recorders (such as those manufactured by GE Kaye or Fluke Corporation) are commonly used during validation studies to record the temperatures sensed by the thermocouples or portable temperature devices. The data logger takes the thermocouple voltage output and converts it to a numerical value. The thermocouples must be checked against the more sensitive and NIST-traceable RTD to make certain that it reads correct temperatures. This is done by manually correcting the zero and span adjustments on the data logger, or by the use of an automatic calibration feature available on some models. Calibrations must be performed on the data logger/thermocouple system before and after the validation runs, as detailed in sections Pre-Calibration and Post-Calibration of Validation Test Equipment.

As with the RTD's, the data loggers must be numbered, logged, and referenced. A calibration file must be maintained for each instrument, including the model and serial numbers, and a history file updated with all repairs made on the unit. SOPs should be written and approved for the calibration of the data logger with the thermocouples and included in the calibration file.

Wireless Temperature Logger

Units are now available that combine the temperature sensor and the data logger in one wireless compact unit. The ValProbe™ temperature loggers, manufactured by GE Kaye, can be loaded into insulating canisters, and placed in the dry heat oven or tunnel. At the end of the cycle the data loggers are placed into a multipoint reader station, and the temperature data are downloaded to a computer. As the temperature loggers in the canisters can withstand a temperature of 170°C for 165 minutes, up to maximum temperature of 360°C for 45 minutes, the wireless temperature loggers are best suited for shorter sterilization and depyrogenation cycles, such as those in a tunnel. Similar to thermocouples, the temperature loggers must be pre-calibrated and post-calibrated.

Infrared Thermometer

Infrared thermometers are ideal for determining cold zones within a tunnel. The infrared thermometers, such as those manufactured by Omega and Fluke, can measure temperatures up to 900°C, and have an accuracy of 1°C to 2°C. The thermometers are rated with a "Distance to Spot Size" ratio (e.g., 50:1), which means that if the target is 50 inches away, then a 1 inch area can be measured.

Constant Temperature Baths

Constant temperature baths are needed to calibrate the thermocouples and data logger to the RTD system. The baths must maintain a constant temperature $\pm 0.05°C$ of their set point while being monitored by an RTD monitor and probe. A low-temperature bath, set below the minimum temperature used in monitoring the cycle (typically 0°C) is used to check the low end of the temperature range, while a high-temperature bath set above the maximum temperature used in the cycle (typically 180°C for sterilization studies, and 300°C for depyrogenation studies) is used to check the high end of the temperature range. The wells in the temperature baths are filled with mineral oil. These wells should be cleaned out occasionally, and refilled when needed.

Stopwatch

A stopwatch is used to time the belt speed of tunnel sterilizers or to confirm the timing cycle on batch sterilizers.

Voltmeter or Ammeter

The voltage, amps, and ohms are tested on the oven and ancillary components during the OQ. It is important that spikes or drops in the line voltage be watched, as facilities age and additional production equipment is added.

Optical Tachometer

The velocity of blowers or fans is checked with an optical tachometer.

INSTALLATION QUALIFICATION

The IQ is designed to compare the system against the manufacturer's specifications for proper installation. All ancillary equipment, utilities, and connections must be checked against the manufacturer's recommendations. Records of modifications made on the unit should also be checked against the equipment and the manufacturer's recommendations. A schematic of all utilities supplying the sterilizer should be available to confirm that all connections are as specified and meet design limits, local and state codes, and Current Good Manufacturing Practices. The IQ documents should be reviewed and approved by designated responsible individuals.

The IQ should include copies of or references to the following information: manufacturer's quotations and specifications, purchase orders, unit model number, serial number, corporate and/or department identification numbers, SOPs, PM programs, sanitization procedures, calibration procedures, modifications made, and the identification and location of all drawings pertinent to the unit.

Structural

Check dimensions, presence of identification plates (check for accuracy against records), correct leveling, proper insulation, presence of seals, and inspect for structural damage.

Filters

All filters used within the system must be recorded, such as those used with air (supply, re-circulating, and exhaust), or in other utilities (e.g., steam, water, or nitrogen). Records on the filters should include proper identification, type, size, change frequency, air capacity, flow rate, temperature limits, and integrity testing required (with SOPs referenced). Some HEPA filters may need to be checked periodically by performing an integrity test using PAO (Emery 3004) or DOP. Membrane filters may be checked periodically for integrity by performing a bubble-point or similar test. Filters used on ducts exhausting outside of the building may require environmental air quality permits. The air or liquid downstream of the filter should be tested for total and viable particulates to make certain that the filters are within specifications, are installed correctly, and do not shed or leak particles.

Electrical

Ensure conformance to National Electrical Code Standards, proper identification, safety cutoff, specifics on the service including voltage, amperage, phase, wire size, and type.

HVAC

Ensure the system provides the RH, temperature, and pressure differential required.

Air Supply

Identify source (direct from the HVAC system or room air), duct size, duct material of construction, and air classification (as per ISO Standard 14644-1) (14). The direction of flow using air pressure differential equipment is tested for operations in all operating modes (i.e., heating, cooling, at rest, and standby) where cleanliness of the room is critical.

Ventilation

Check that the ventilation exhaust duct exhausts to an appropriate area (not to an aseptic environment), and identify the method used to prevent back-flow.

Cooling Medium

Identify type, source, pipe size, pipe material of construction, type and size of cooling coil tubes, cooling water temperature, and flow rate.

Air and/or Natural Gas

Check that the source and type of supply are consistent with the manufacturer's recommendations.

Door Gaskets

Check integrity of gaskets and materials of construction.

Instruments

Identify all devices, controllers, and recorders both critical and noncritical to the operation of the unit. These may include temperature, timers, pressure, belt speed, air flow, and indicator lights. Check the presence of any alarms for the unit, both audible and visual. For each instrument note the serial number, corporate identification number, instrument output range, presence of a calibration sticker, and calibration schedule.

Baffles

The integrity of all baffles or louvers must be checked. Ensure that the baffles are not damaged, misaligned, or missing entirely.

Heaters

Record the manufacturer's model number, the number of heating elements, and the voltage, amperage, and wattage of the elements for the heaters.

Lubricants

Make certain that any lubricants used cannot contaminate the material being sterilized or depyrogenated.

Blowers

The blower must be mechanically sound, the volute in place and correctly balanced, and that the blades rotate in the correct direction. Check for use of the correct fan belt, and that it is in good condition.

OPERATIONAL QUALIFICATION

For clarity, the IQ and the OQ have been described separately. It is common practice in the industry to combine the IQ and the OQ together in one study.

After the equipment has been checked for proper installation, as detailed in the IQ, it is necessary to determine that the sterilizer performs as designed. The components of the system must satisfy the operating ranges as determined by the purchase order specifications. The sterilizer must be operated to confirm that it functions correctly on a repeated basis. The OQ document should be reviewed and signed by the required department representatives. Each of the following process components must be identified, and the operating performance and ranges determined.

Temperature Monitors

The temperature controllers, recorders, and sensors on the process equipment must be calibrated before the unit can be operated reliably. The units are generally calibrated at the time of installation by the manufacturer or user, and should be calibrated at set periodic intervals. The calibrations should be performed by measuring actual temperatures in addition to electronic methods, such as checking voltage or resistance readings at various set points. It is also essential that actual temperatures of the unit be checked at the set points, as described under the component mapping studies in section Qualification Testing. The recorder must accurately document the temperature sensor readings on a chart having a readability level consistent with the operating ranges. The controller must prove reliable in maintaining the temperature within the specified set points.

Cycle Timer

The accuracy of the timer must be determined, so that assurance is provided for cycle length. The recorder must accurately display the cycle time.

Door Interlocks

If a unit is equipped with double doors, the interlocks must operate such that the door leading to the aseptic area cannot be opened if the door to the non-aseptic area is open, if the cycle has not been successfully completed, or if the temperature is too high for safe handling.

Heaters

All of the heating elements must be functional. It is preferable to have them monitored continuously with ammeters in order that burned-out elements can be immediately detected. A failed element could cause a substantial change in the operating performance of the oven.

Blowers

Properly adjusted blowers are very important to the effectiveness of the circulation in the sterilizer. The blower should deliver an air velocity consistent with manufacturer's specifications, which may be accomplished by adjusting the speed of the fan. The air velocity and motor speed should be noted in the OQ records. It is essential that the blades are rotating in the proper direction. Air velocity is measured across the air flow direction and should maintain $\pm 25\%$ of the mean unit velocity for even temperature distribution. An airflow switch or other sensor present to detect and alarm upon a failure of a blower, should be tested.

Cooling Coils

To enable a faster cool-down cycle, the air is often circulated across coolant coils. If coils are present, the type and size of the coils and the temperature of the cooling medium at the inlet and outlet of the coils should be recorded. The effectiveness of the cooling coils can be checked by determining the coolant temperature change between the inlet and outlet of the coils.

Belts

The belt speed is a critical operating parameter in both continuous hot-air tunnels and flame sterilizers. Recorders for charting the belt speed are recommended for units with adjustable speed settings. The belt speed and operating temperature are interrelated in these units, so a slower belt speed at a lower temperature will produce the same effect as a faster belt speed at a higher temperature. The motor speed setting and range should be noted in the OQ records.

Chamber Leaks

The perimeter of the doors for batch sterilizers should be checked for air leakage while operating.

Particulate Counts

Particulate counts should be checked within the containers before and after sterilization to quantitate the particle load contributed to the product by the sterilization process.

PRE-CALIBRATION OF VALIDATION TEST EQUIPMENT

All equipment used for validating production equipment must be calibrated (refer to section Validation Test Equipment, for calibration).

Calibration of the thermocouples and data logger against a traceable RTD monitor and probe is a critical step to be performed before and after the validation study. Pre-calibration makes certain that all thermocouples are in working order, and compares each temperature reading against a known standard. The thermocouples are checked against the RTD after the runs in a post-calibration to ensure that the recorded temperature data are valid (see section Post-Calibration of Validation Test Equipment).

The thermocouples, an RTD probe and monitor, a data logger, and two or three constant temperature baths are required to perform the calibration.

Pre-Calibration Methods

The following procedure should be used to calibrate the thermocouples:

1. The RTD and thermocouples are simultaneously placed in the low-temperature bath. The data logger readings are compared against the RTD monitor, and the "zero correction" tuned on the multipoint recorder until all thermocouple readings are ±0.5°C of the RTD temperature readings. The thermocouples must stay within the range for at least three minutes to demonstrate stability of the thermocouples. The data logger is used to record these temperature readings, and the RTD temperature should be noted on the printout.
2. The thermocouples are taken out of the first bath, allowed to warm up to room temperature, then transferred to the high temperature bath. The data logger is adjusted for "span correction" so that the temperature readings for the thermocouples are again within ±0.5°C of the designated temperature range. If the temperature readings are not within the range, adjustments must be made with the zero correction and the first bath sequence repeated.
3. The printouts are designated as the pre-calibration check and maintained with the executed process qualification documentation.

PROCESS QUALIFICATION CYCLE DEVELOPMENT

The dry heat cycle is utilized to inactivate or remove any microorganisms that can cause deleterious effects on the parenteral patient recipient. The sterilization cycle is designed to inactivate the heat-resistant spores (e.g., *B. subtilis*) as well as any vegetative cells which could potentially be present during processing. The depyrogenation cycle is based on removal or inactivation of bacterial endotoxins (e.g., *Escherichia coli* LPS), which are fever-producing substances in the gram-negative cell outer membrane. The endotoxins, or pyrogens, retain their potency even when the cell is damaged or lysed. The mechanism of inactivation of desiccation-resistant spores by dry heat has been suggested to be due to drying or moisture loss, resulting in thermal oxidation and at higher temperatures, and organism incineration. Typically, dry heat processes will concentrate on the more stringent cycle treatment of removing pyrogenic substances.

An appropriate sterilization or depyrogenation cycle must be developed before validation testing commences. Complete records and documentation of the cycle development must be referenced or included within the validation documentation. The cycle development is discussed in detail below.

Operating Parameters

All operating parameters must be defined during cycle development; including temperature settings, cycle time, penetration temperature profiles, and belt speed (for tunnel or flame sterilizers). The laboratory studies should imitate actual manufacturing conditions. The laboratory studies can identify manufacturing equipment design specifications and operating parameters required to deliver an effective sterilization or depyrogenation cycle.

Microbial Lethality Requirements

In addition to the operating parameters specified in the cycle development study, the required F_H value and the bioburden/pyroburden levels of the components being treated must be determined. Any over-sterilization concerns (for heat-labile products) should be also addressed.

Proper cycle development according to the USP can be achieved by verifying a microbial survival probability of 10^{-12} for sterilization. If the cycle is required to render the container free of pyrogens as well as viable microbes, the cycle must demonstrate a 3-log cycle reduction of bacterial endotoxin (1/1000 of the original amount is inactivated). The pyrogen challenge can consist of inoculating an article with a minimum of 1×10^3 USP EU of bacterial endotoxin. A number of guidelines addressing the cycle development and validation of sterilization cycles are referenced in the USP from the PDA, PMA, HIMA and the AAMI (1).

If overkill cycles are utilized (based on bio-challenges), it is generally not necessary to evaluate the bioburden or pyroburden present on the incoming glass containers or equipment. The usual assumption considers that the relatively low numbers of organisms, or minute concentration of contaminants present on the surface of glassware for parenteral use, are well within the challenge destruction of 10^6 spores of *B. subtilis* or a 3-log reduction in endotoxin. The enumeration and identification of bioburden and pyroburden are considered of little value.

For sterilization processes only, the presence of high numbers of gram-negative organisms on a component before sterilization can raise concerns about the presence of endotoxin on the components prior to processing, since the overkill sterilization cycle will only inactivate the vegetative cells and spores without inactivating the pyrogenic activity of the endotoxin. As glass containers are molded at high temperatures (1500°C) and shrink wrapped by the supplier for shipping, they are unlikely to have a problem with gram-negative organisms. Extensive studies performed on the glass containers as received in the shrink wrap and before washing, were tested as free of endotoxin or having very low levels per container volume (<0.003 EU/mL). Washing of the articles will decrease the endotoxin levels, if any are present (15).

The biological indicators of choice for validating and monitoring dry heat sterilization are commercial spore strips of *B. subtilis* spores. When the strips are utilized, the manufacturer's *D* value can be used. Spore strips can be purchased prepared with 10^4 to 10^9 spores. There are certain grades of strips which can be utilized for temperatures between 250°C and 450°C. Any browning of the paper without crumbling does not interfere with usage or testing.

Alternately, a spore suspension can be inoculated on a part of the load (on a container or piece of equipment) to closely represent the surface bio-challenge. If the spores are dried on the surface of the article to be sterilized, the user must establish the *D* value for each type of component to be tested. Heat-resistant organisms have *D* values of only a few seconds at temperatures generally used for endotoxin inactivation and for this reason a spore challenge is not required in any process where depyrogenation is demonstrated (16,17).

D and Z Values

The microorganism used as a biological indicator must have resistance characteristics (*D* and *Z* values) that are documented and appropriate for the sterilization or depyrogenation cycle. The *D* value is defined as the time required to reduce the microbial population by 90% (one logarithm). The relationship of lethality to temperature is expressed in the *Z* value. The *Z* value studies will define the number of degrees that are required for a change in the *D* value by a factor of 10. Laboratory studies are used to define the *D* value of the typical bioburden and/or pyroburden. The microorganism used as a bioindicator must have resistant characteristics (*D* value) that are documented and appropriate for the cycle (18). The bioburden data, *D* and *Z* values, are used to calculate the minimum F_H value required.

Sterilization—Bioburden Calculations F_H

The F_H value for *sterilization* is the integration of lethality at a reference temperature of 170°C. The F_H value for *depyrogenation* is the integration of lethality at a reference temperature of 250°C. A conservative approach to determining a minimum sterilization F_H would utilize the heat-resistance spores of *B. subtilis* (*globigii*) and assume a D_{170} value of three minutes (at a reference temperature of 170°C) and a *Z* value of 20°C. The lethal rate determines the increment of lethal heat effect obtained over various temperatures using the *Z* value (as compared with a reference temperature). The F_H value is derived by integration of the lethal rate with respect to time. The F_H value (equivalent time at the reference temperature) accumulates the total lethality. When sterilization temperatures other than 170°C are used, the F_H value is reported as process equivalent time at the reference temperature of 170°C (19).

The equations used for equivalent time are as follows:

$$F_H = D_{170°C}(\log a - \log b)$$

where *a* is the bioburden per item, *b* the probability of survival, *D* the time at 170°C to reduce the population of most microorganisms in the product by 90%, and F_H the equivalent time in minutes at 170°C and a *Z* value of 20°C and

$$F_t^Z = \frac{F_{170}{}^Z}{L}$$

where F_t^Z is the equivalent time at temperature *t* delivered to a container for the purpose of sterilization with a specific *Z* value, $F_{170}{}^Z$ is the equivalent time at 170°C delivered to a container for the purpose of sterilization with a specific *Z* value (when $Z=20°C$, then $F_{170}{}^Z = F_H$) and *L* is the lethal rate.

$$L = \log^{-1} \frac{T_0 - T_b}{Z}$$

or

$$L = 10^{(T_0 - T_b)/Z}$$

where T_0 is the temperature within the container or item, and T_b is the base temperature of 170°C.

As a supplement to temperature data, spores of *B. subtilis* are used to monitor the lethality of dry heat sterilization during the validation runs. Heat-labile products require strictly controlled sterilization cycles, since under-processing will result in a non-sterile product while over-processing may cause degradation of the product. The cycle development will determine the minimum amount of dry heat required to ensure that the probability of survival of the bioburden is less than 10^{-6}. The equivalent sterilization time and temperature can be described by the *F* value with a reference temperature of 170°C and assuming a *Z* value of 20°C (20).

Depyrogenation—Pyroburden Calculations

Heat-stable materials, such as glassware and stainless steel equipment, can withstand temperatures well in excess of 250°C. Operating temperatures can be very high and loading configuration may be less restrictive than with heat-labile products. The overkill method relieves the requirement for bioburden and bioburden resistance studies during cycle development and validation. Component preparation is still very important, since cleaning and handling procedures can serve to minimize the level of contamination of both viable and nonviable particulates, including endotoxins.

The F_H requirement will ensure a probability of survival of the bioburden of less than 10^{-6}. In this case, the cycle lethality should be defined on the basis of endotoxin inactivation based on the pyroburden and not on the bioburden (6). Calibrated *E. coli* endotoxin challenges are placed in the load. The coldest location in the loading pattern must be challenged. The endotoxin challenge should be based on the pyroburden of the components, taking into consideration the desired safety factor. The presence of residual endotoxin can be detected by the LAL test. The cycle should be established to ensure a probability of survival of the bioburden of less than 1 organism in 10^{12}.

The depyrogenation process is not as well understood as the sterilization process, and the mechanism of endotoxin inactivation by dry heat is still being researched. The mathematical representation of thermal destruction of bacterial spores is well established for sterilization processes. Mathematical representations for LPS destruction or inactivation are not completely established and continue to be explored. The initial studies published by Tsuji et al. used a second-order model based on the sterilization process mathematical approach for spore inactivation to linearly describe the dry heat inactivation of LPS (21). The studies demonstrated the inactivation curves for purified endotoxin could be made linear and that inactivation for a dry heat process can be predicted given the product heating curve. The isothermal kinetic changes were expressed by the following equation:

$$\log\frac{Y}{t} = \frac{1}{n}\log A + \left(1 - \frac{1}{n}\right)\log Y$$

where Y is any parameter that changes with time t and temperature, and A, n are the constants at a specific temperature.

The linear equation was represented as follows:

$$\log Y = A + B(10^{Cx})$$

where A, B, C are is constants at a given temperature, x is minutes of heating time, Y is the percentage of LPS remaining after heating and the same value as Y in the kinetic equation.

Anderson and Kildsig evaluated Tsuji's linear equation within a defined range of temperature values, since depyrogenation does not follow the semi-log standard time/temperature model developed for dry heat sterilization, especially at lower temperatures. The model is a tool to determine the minimum inactivation temperature (22).

Akers et al. continued use of the mathematical equation by developing F value requirements for endotoxin inactivation (23).

Ludwig and Avis have performed various studies to evaluate the minimum temperatures of depyrogenation, the mathematical application, and the type of LPS pyro-challenge (24). The inactivation of endotoxin and purified derivative was viewed to be two linear biphasic slopes at temperatures of 225°C and 250°C and possibly a second-order inactivation which is temperature dependent. The calculated D values by Ludwig and Avis at 225°C ranged from 0.15 to 1.20 minutes. The maximum calculated Z value was 46.73°C (25). The study explored the difficulty in using the standard F_H value concept where an accurate Z value needs to be derived from the D value and in calculating the D value the inactivation rate must be first-order for the entire process (26). Therefore, it is the USP requirement for demonstration of a 3-log reduction of endotoxin which must be demonstrated to validate the dry heat cycle. The inactivation rate of the endotoxin is dependent on the formulation, purification, and concentration of the challenge. A 3-log reduction in 10,000 EU of endotoxin was achieved in 100 minutes at 170°C and in 0.5 minutes at 250°C. A 6-log reduction in 100,000 EU of standard endotoxin was achieved in 105 minutes at 250°C and 1.0 minutes at 325°C (25).

Nakata performed studies with 10,000 EU and determined D and Z values by linear regression analysis. Nakata views the inactivation curves as monophasic and reported D values of 43.8 and 1.7 minutes at 200°C and 250°C, respectively. The calculated Z value was 30.9°C. Nakata discussed the heat-up primary-phase inactivation as insignificant and the secondary-phase inactivation at dwell temperatures as the stage where almost all of the endotoxin inactivation occurs. The data demonstrated a 3-log reduction in 77 minutes at 200°C and an overkill cycle achieved in 30 minutes at 250°C (27).

Tsuji expressed the rate of endotoxin (LPS) destruction at 250°C using a Z value of 46.4°C and a D_{250} value of 4.99 minutes (28). The cycle should be designed utilizing a worst-case assumption, where the required minimum time and temperature parameters are defined. The F_H can be calculated for a depyrogenation cycle using the general F value equation with a reference temperature of 250°C.

Examples of F_H values

The following table shows various examples of F_H values for sterilization and depyrogenation studies; F_H (170°C) with a Z value of 20°C, and F_H (250°C) with a Z value of 46.4°C, at various temperatures (for one minute):

Temperature (°C)	$F_H^{20}{}_{170}$	$F_H^{46.4}{}_{170}$	$F_H^{46.4}{}_{250}$
F_H values at various temperatures for one minute:			
170	1.0	1.0	0.02
210	100.0	7.3	0.14
250	10,000.0	53.0	1.0
270	100,000.0	142.0	2.7
Assuming a cycle of 250°C for 30 minutes, the minimum F_H values for the total cycle would be:			
250	300,000.0	1590.0	30.0

The F_H (250°C) that is determined can be used to calculate the amount of endotoxin which will be reduced. This can be calculated by integrating the heat penetration–lethality curves. The F_H for depyrogenation is used as a method to predict and quantify the endotoxin inactivation.

The endotoxin challenge will require inoculation of articles in the sterilizer load. It is much more difficult to verify the initial expected recovery of the endotoxin once placed directly on the component, since the endotoxin may tend to bind or adhere to the surface, decreasing initial recovery by 30% to 80% before the component is depyrogenated by the dry heat. Recovery studies should be performed on the actual components to be depyrogenated, as the properties of the surface material may affect the endotoxin recovery results and thus the validity of the log reductions established. Recovery studies are not required for sterilization, as destruction of the microbial challenge is a pass–fail test.

The presence of residual endotoxin is usually confirmed by the LAL test. The LAL test is very sensitive for reaction with endotoxin or LPS, which is a component of the gram-negative microorganism outer membrane.

LAL Test

The LAL test is an important monitoring procedure to test for the presence of endotoxin. The LAL test is based on the initiation by endotoxin of a blood-clotting cascade in the horseshoe crab. Clotting is measured and related to endotoxin concentration by one of three common in vitro methods: gel clot, tubidimetric and chromogenic (27).

An acceptable endotoxin challenge should be based on the history of the pyroburden of the container and its contribution to the end product filled in the container on a per milliliter or milligram basis (29). A meaningful challenge would evaluate the fill volume of the final container in view of the limit for the final product Bacterial Endotoxins Testing limit or sterile Water-for-Injection, USP, which has an endotoxin limit of 0.25 EU/mL or approximately 0.05 ng/mL. It has been reported that the threshold pyrogenic dose for man and rabbit is 0.1 ng/mL (approximately 0.5 EU/mL) (30).

QUALIFICATION PROTOCOL

A validation protocol must be written and approved prior to the start of the actual validation work. The protocol is designed to outline the program to be employed, the specific tests that will be made, and the acceptance criteria for those tests. The protocol may be written to overview the process, or the specific piece of equipment. Once the protocol has been written it must be approved by the designated responsible individuals.

Protocol Outline

The following format may be utilized in a validation protocol:

1. *Objective statement*: A concise statement that defines the objective of the validation protocol.
2. *Responsibility*: Identification of specific departments, and their responsibilities in the validation project. This will assure that each group understands the specific information or materials that it is required to provide.
3. *Test Program*: The test program should include a description of the tests that will be performed during the empty-chamber and loaded-chamber studies, as detailed in section Qualification Testing. The equipment to be used to perform the studies must be described. All SOPs for each piece of equipment or testing process must be referenced. The type and form of biological challenges to be used must be stated. The general location of where the individual tests are to be taken, and where the bioindicators (spore or endotoxin challenge) shall be placed, should be described.
4. *Acceptance criteria*: Acceptance criteria must be listed for each test in the test program section, with limits or ranges specifically identified. The limits or ranges chosen should be those commonly used by the firm, determined during cycle development, or referenced from cited literature.

Changes in Scope of Work

Changes in the scope of the work after the protocol has been finalized may be addressed in Protocol Supplements or Addenda. These supplements must be approved by all parties. The validation report must refer to the issued protocol and all supplements.

QUALIFICATION TESTING

Upon completion of IQ and OQ efforts and approval of the protocol, testing may begin. The testing will include empty-chamber testing for heat distribution studies, and loaded-chamber testing consisting of heat distribution and heat penetration studies. Loaded-chamber testing requires the determination of bioburden or pyroburden on the various loads employing appropriate biological indicators or endotoxins during the validation studies.

Component Mapping Studies

Before conducting the loaded-chamber heat penetration studies, component mapping should be conducted. The studies help to determine the coolest point within a specific load and item. In subsequent loaded-chamber studies, penetration thermocouples should be positioned within the component at that location. For example, container mapping will determine if some areas of an item are heating at a slower rate than other areas. In subsequent testing the item should be monitored in the cold area. Mapping studies can be initially conducted in a laboratory scale oven and confirmed in the manufacturing equipment.

Empty-Chamber Testing

The initial testing is performed on an empty oven or tunnel to establish the uniformity of temperature distribution. The thermodynamic characteristics of the empty unit are depicted in a temperature distribution profile. The temperature profile will serve to locate hot or cold areas in the sterilizer by mapping temperatures at various locations.

The temperature profile is obtained by placing a minimum of 16 thermocouples distributed in the batch sterilizer or empty tunnel in such a way as to determine heat profiles. More thermocouples than the protocol requires should always be included, in case there are problems with individual thermocouples. As shown in Figure 7 the thermocouples may be attached to empty racks to qualify a batch sterilizer, with the thermocouples located in the eight (8) corners, and adjacent to any chamber probe. The thermocouple tips should be

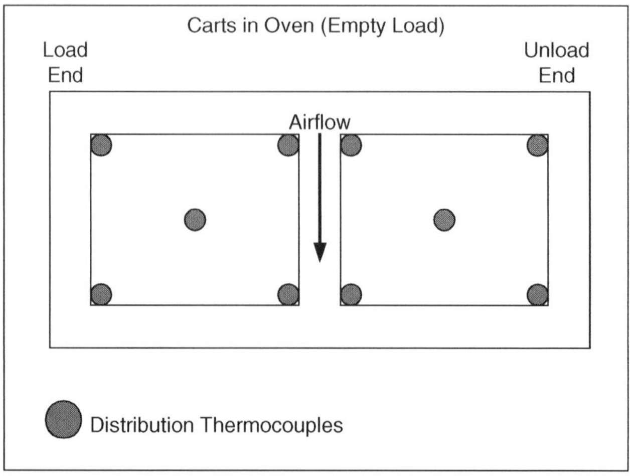

Figure 7 Drawing of empty oven with distribution thermocouples.

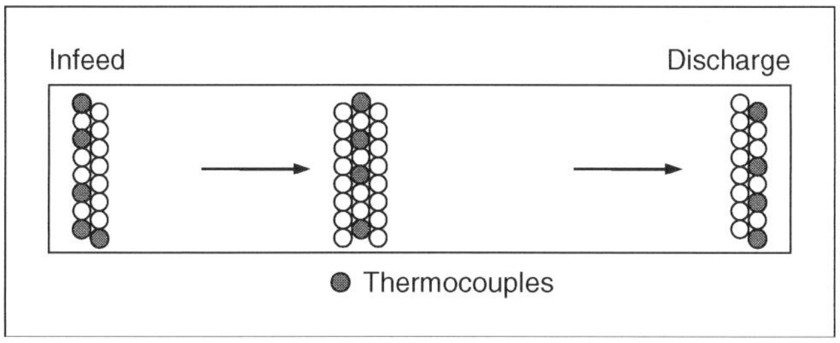

Figure 8 Drawing of empty tunnel with distribution thermocouples.

suspended to avoid contacting any solid surfaces (wall, ceiling, support rods, etc.). For ovens, an acceptable profile should demonstrate uniform temperatures during the last three to five minutes of the cycle (typically ±15°C for an empty chamber). Larger ovens may require a higher temperature range (± 25°C).

For empty tunnel runs, the temperature across the tunnel should be monitored. A single empty tunnel study is sufficient for each unique temperature set point and belt speed combination. As shown in Figure 8, the thermocouples should be spaced evenly across the belt in the first, middle, and last row of glass. Attach the thermocouples to a stainless steel bar to keep them in place. Run a single empty tunnel study for each unique set point temperature and belt speed combination. A temperature range of ±20°C to 25°C is expected. In the flame sterilizer the thermocouples should be placed at the level of the ampules.

All environmental factors should closely represent actual manufacturing conditions (e.g., relative humidity, room temperatures, static air pressure, and balance). All control settings are to be recorded, including any variables that will affect the cycle. Key process variables such as temperature set points, heating elements settings, cycle-timer set point, belt speed, etc. should be recorded. The cycle timer (for batch sterilizers), belt speed (for tunnel or flame sterilizers), controller operating temperature span, and production charts can be verified by a multipoint temperature recorder with an internal timer. Air velocity profiles across the unit can be of significant benefit in correcting distribution temperature difficulties.

In accordance with USP, a sterilization cycle should avoid an accumulation of particulate matter in containers intended for intravenous products (1). The empty chamber runs should be confirmed for ISO 5 (Class 100) conditions at sterilizing temperatures. Sampling probes are positioned within the unit to measure total particulates present. These studies can be conducted at ambient and operating temperatures. The use of sample probe coolers or high-temperature sampling cells of glass or metal is necessary to avoid contamination or damage to the particle counter. Often Class 100 conditions are met at ambient and elevated temperatures, but are not achievable during heat-up and cool-down due to particles leaving the surface of the equipment as the temperature gradient changes.

A thermocouple should be placed adjacent to any heat-controlling temperature sensors to confirm that the operational controls are maintaining the desired heating specifications. It is important to document the come-up time (the time to reach the temperature set point) and the cool-down time (the time from the end of the cycle or dwell period to the time the components are cool enough to remove from the sterilizer), since data variances may indicate electrical or mechanical malfunctions in the batch sterilizer. It is important that the tunnel and flame sterilizer be closely monitored within the sterilizing zone, since temperature variation is most critical at this location. The empty-chamber cycle can be one of maximum time with production operating temperatures or a shorter time period at a predetermined temperature, such as 250°C.

If the empty-chamber temperature distribution profile is not acceptable, then adjustments, modifications, or repairs should be performed and the profile studies repeated.

A detailed diagram of the location of the thermocouples should be included in the empty-chamber data file. This file will be extremely valuable when revalidation of the sterilizer is necessary after any modifications. The empty-chamber data file must include originals (or copies) of all charts, temperature printouts, data calculations, and observations pertaining to the runs.

Loaded-Chamber Studies

For validation purposes, the loads tested must be representative of standard items and quantities. Ideally, each size and type of material should be tested by penetration studies. For ovens, the time and temperature set points should be reduced. For tunnels, the temperature set point should be reduced and the belt speed increased if possible. To reduce the number of test combinations, a selection of representative items is made with consideration to size, number and geometry of the loading pattern. The representative loads should include the smallest and largest items in addition to the items that are most difficult for heat to penetrate (i.e., due to dense mass or tight packing). Detailed loading patterns should be developed; an exact detailed diagram of thermocouple locations must accompany all temperature data. Photographs of the load patterns speed up the loading process and provide clarity in the final report as to what was actually performed. The diagram is necessary to identify where the hot and cold areas are within each specific load. Hot areas in the load are more important for heat-labile items. Cold areas are important to monitor for sterility or depyrogenation assurance.

Load factors will be significant because air has poor conductive and convective properties. As a result, the hot and cold areas may vary for each type of load. This is most likely to occur if each material heats at a different rate (due to size, mass, and packing configuration). In the batch sterilizer it is recommended that the penetration thermocouple locations be moved around after each run, to obtain a broader view of heat penetration.

As in the empty-chamber testing, validation studies for a partially or fully loaded chamber must include heat distribution testing with thermocouples placed near the heat-controlling temperature sensors. Thermocouples used for loaded-chamber heat distribution studies should be positioned in some of the locations used for empty-chamber heat distribution testing. The distribution thermocouple tips should be suspended to avoid

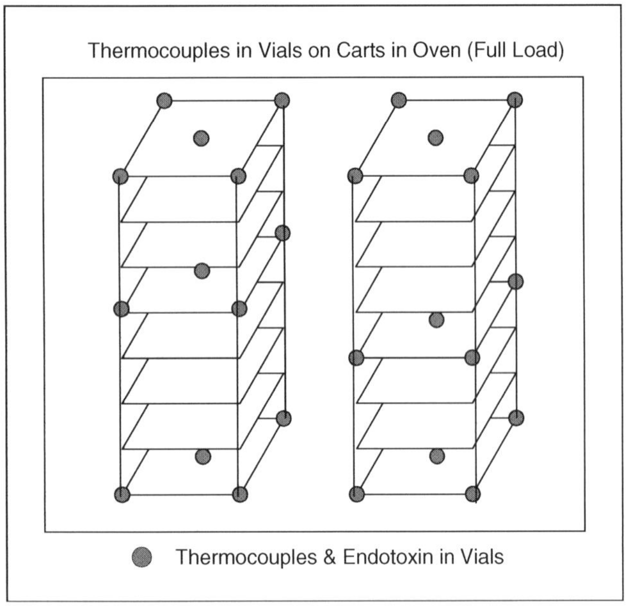

Figure 9 Drawing of loaded oven.

contacting any solid surfaces. Heat distribution studies must be performed to determine the effect of the load on the chamber temperature distribution. It is not uncommon for the presence of a load to improve the uniformity of heat distribution, or for the smaller load to be the "worst-case" load. The system should be tested with a "cold start" to reflect "worst-case" operating conditions and greater assurance of process effectiveness.

Heat penetration studies should be monitored simultaneously with the heat distribution studies. A typical loaded-chamber diagram in a batch oven is shown in Figure 9. Penetration information is critical in a partially or fully loaded chamber, since materials will heat at a rate different from that of the surrounding air. The rate of heat penetration will depend on the type of material in the load, how it is packed (loading configuration) and the distribution temperature uniformity. Heat penetration data are obtained by placing thermocouples inside the container, component, or item in such a way as to ensure contact with the surface (the thermocouple should read the surface temperature not the air temperature).

It is important in the loaded chamber to document both come-up and cool-down rates of the air and product. The come-up time of the distribution thermocouples will describe the time required for the air to reach the temperature-controller set point from ambient temperature. The come-up time of the penetration thermocouples will describe the time required for the load to reach the desired temperature. There is a heating lag as the components in the load reach the minimum required temperature after the air reaches that temperature (as measured by the distribution thermocouples). The heating lag is defined as the difference between the time required for the product to reach the minimum required temperature and the time required for the sterilizer air to reach the minimum required temperature. A heating lag will be magnified during the maximum load of product (i.e., maximum density or mass). The product temperature, as detected by penetration thermocouples, will heat at a slower rate, since convection and conduction are slower in a solid mass (product) than in a gas (air). It is not uncommon for temperature controller settings to be set well above the desired minimum sterilizing/depyrogenating temperature to make certain that the product will attain that temperature for the required length of time. The total time the product is at or above the required temperature is documented in the loaded-chamber study, as well as the final F_H value.

The tunnel or flame sterilizer temperature data may have large variations between runs because the product is heated to high temperatures for a short period of time, as compared with batch sterilizers with long sterilization periods. As a consequence, correlation between different types of sterilizers is difficult to achieve. In tunnels, the worst-case locations are generally the first and last row of glass, and the middle of a dense pack of glass, as shown in Figure 10. Loading and running thermocouples in a tunnel takes a great deal of time and coordination. The thermocouples are placed inside the glassware, and spaced evenly across the width of the belt. A stainless steel bar may be used to hold the thermocouples in place. The thermocouples are fed through the tunnel as the belt is run at the maximum speed and lowest temperature, until they exit the tunnel. If GE Kaye ValProbes™ are used, the canisters should not be placed adjacent to the glassware, as the glass will act as a heat sink and the temperatures will appear lower than they actually are.

Particulate counts of the air must be checked in the dry heat sterilizer and within the items being processed. The particulate counts are highest at the beginning of the cycle "start-up." In some cases, air particulate counters may be measuring the water vapor evaporating from the load articles (if they are loaded wet). Analysis of particulate matter within a container can be performed by adding filtered particle-free water and shaking. The container contents are tested by electronic particle counter or microscopic analysis. The air particulate counts are to be within Class 100 limits and the containers demonstrate results of 5 particles/mL ≥ 5 μm in size (31). These studies are typically performed independently of the thermal studies.

Care must be maintained to observe temperature ranges and fluctuations with awareness of any maximum temperature restrictions. Distribution temperatures (empty and loaded sterilizer studies), penetration temperatures, come-up time, and F_H data can be evaluated for reproducibility between replicate runs using statistical methods.

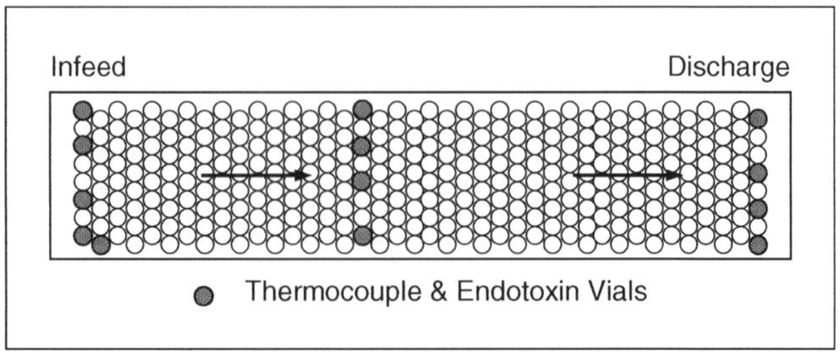

Figure 10 Drawing of loaded tunnel.

In the batch oven, it is often the load items closest to the bottom of the unit and nearest to the doors which are most likely to be the "cold spots." Tunnel and flame sterilizers are particularly sensitive to changes in load configuration. Continuous runs (where bottles, vials, or ampules are flush side to side) are generally the worst-case loads due to the rapid come-up time, short length of sterilization period, and variations in component packing and movement. The "cold spots" tend to be found at the edges of the belt, and the first and last components to go through the unit. The presence of other components at the same temperature generally ensures that the units in the middle of the load are heated more uniformly since they are better insulated from heat loss.

If the temperature profile is acceptable, three consecutive replicate runs in combination with bio-challenge/pyro-challenge studies are utilized to demonstrate loaded sterilizer and cycle reproducibility. The replicate runs must verify that the minimum required F_H value is being achieved within the coldest portion of the load.

Bio-Challenge/Pyro-Challenge Studies

The challenge should demonstrate the lethality delivered by the cycle with either microorganisms or endotoxin. The challenge can be accomplished using commercial strips or suspensions of *B. subtilis* spores for sterilization or *E. coli* endotoxin for depyrogenation. The concentration of the challenge for overkill processes must demonstrate adequate sterility assurance. For dry heat sterilization/depyrogenation to occur, there is little consensus on the minimum times and temperatures to achieve the desired result. A variety of time and temperature combinations can be utilized for dry heat sterilization/depyrogenation. Historically, the dry heat sterilization cycles were defined as 170°C for not less than two hours, while depyrogenation cycles were defined at a minimum of 250°C for not less than 30 minutes. Per USP, a typical empty-chamber temperature range is 250°C ± 15°C (1). A loaded chamber can exhibit an increased range above ± 15°C. The bio-challenge will demonstrate the lethality delivered by challenging the cycle with either microorganisms or endotoxin. A suitable challenge must represent the pyroburden or bioburden for heat-labile materials or exceed it for overkill processes. The pyroburden and/or bioburden calculations were previously determined during Cycle Development (refer to section Process Qualification Cycle Development). The concentration of the challenge for overkill processes must demonstrate adequate sterility assurance.

Bio-challenge studies can be performed concurrently or separately from temperature penetration studies. If studies are performed concurrently, place the challenge items adjacent to items containing thermocouples as shown in Figures 9 and 10.

Studies can be performed by placing the bioindicators in the coldest areas (minimum F_H values) of each load. An alternative to adding the bio-challenge to each load would involve determining the load with the absolute coldest area and minimum F_H value. This load is then considered to be the worst-case load. Successful bio-challenge of the worst-case load would eliminate the need to challenge all other previously tested loads (with higher F_H values).

The bio-challenge work is usually achieved by inoculating components with a known concentration of the challenge microorganism or endotoxin (i.e., *B. subtilis* suspension or *E. coli* endotoxin). In sterilization cycles, a challenge of 10^6 concentration of *B. subtilis* is common. In depyrogenation cycles, there appears to be no general consensus on the challenge level used; concentrations must be recoverable and detectable to demonstrate a greater than 3-log reduction in endotoxin, therefore 1×10^3 EU is typically used as a minimum. The required number of challenged units should be predetermined during cycle development, and cited in the validation protocol. After the sterilization or depyrogenation cycle, the inoculated products are recovered along with unchallenged items (for negative controls), and tested for spore viability or endotoxin inactivation along with positive controls. If the challenge has spore survivors or residual endotoxin, the amount must be quantified and analyzed with respect to the achieved F_H value. The results of this study confirm that the sterilization or depyrogenation process is effective.

POST-CALIBRATION VERIFICATION OF VALIDATION TEST EQUIPMENT

After the validation studies are completed, the thermocouples and data logger must be post-calibrated, to verify that the thermocouples were accurately measuring the temperature throughout the entire period of use. This must be performed without any adjustment of the data logger. This post-calibration verification may be made after any number of validation runs, but there is a risk that is taken that all of the validation tests may have to be repeated if the thermocouples fail to post-calibrate.

Post-Calibration Methods

The following procedure should be used to post-calibrate the thermocouples, to be assured that the temperatures monitored by the thermocouples are accurate:

1. The RTD and thermocouples are simultaneously placed in the low-temperature bath. The data logger readings are compared against the RTD monitor. All thermocouple readings must be within ±0.5°C of the RTD temperature readings. The thermocouples should stay within the range for at least three minutes to demonstrate stability of the thermocouples. Any thermocouple that is out of range on this post-calibration check may not be used as a source of valid data. The data logger is used to record these temperature readings, and the RTD temperature should be noted on the printout.
2. The thermocouples are taken out of the first bath, allowed to warm up to room temperature, then transferred to the high-temperature bath. The temperature readings for the thermocouples must again be within ±0.5°C of the designated temperature for the high-temperature bath. Any thermocouple that is out of range on this post-calibration check may not be used as a source of valid data. Any runs where the coldest thermocouple fails to post-calibrate must be repeated.

3. The printouts are designated as the post-calibration and must be maintained with the validation report.

QUALIFICATION REPORT

After the empty and loaded-chamber studies and bio-challenge studies have been completed, the data must be analyzed to ascertain that all testing requirements have been achieved. The results of the bio-challenge studies and *F* value computation must demonstrate the required degree of lethality (sterilization or depyrogenation) according to the protocol.

The following information should be provided in the process qualification validation report:

1. *Protocol achievement*: A statement reflecting that the acceptance criteria of the identified validation protocol were met.
2. *Summary of data*: A summary of the data collected during the validation runs, including come-up times, minimum and maximum *F* values, and the location of the slowest heating zones. Raw data are generally not included in the report, but should remain in a central file (as described later).
3. *Deviations*: Exceptions to the validation report should be explained, including justifications if certain tests were not performed, or are to be performed in the future under an addendum. Any deviations from expected results should be analyzed and discussed.
4. *Diagrams*: Diagrams and photos showing the load and the placement of thermocouples, and bioindicators and/or inoculated parts should be included. Detailed diagrams of unusual items should be shown. Other data may be included in the report as desired, due to differences in protocol and equipment specifics.

ROUTINE MONITORING AFTER VALIDATION

Once the equipment has been validated for the sterilization or depyrogenation process, the unit must be monitored so that it remains in a state of control. This is achieved by the use of various programs, including sanitization, PM, engineering change control, and revalidation.

Sanitization

The sanitization program should detail the cleaning methods used for the equipment, the SOPs covering each method, and the cleaning materials utilized. Proper sanitization should demonstrate that the level of organisms is controlled, so that the basis for the validation program does not change. Cleaning materials should be nontoxic and leave no residues. Many firms have a policy to change or rotate the cleaning material periodically, to make certain that the organisms do not adapt to the cleaning solution. Changes to the cleaning material will require approval by designated responsible individuals. Revalidation of the worst-case load may be required.

Preventative Maintenance

The PM program provides a schedule by which the equipment is maintained. This includes physical checking of the system, changing of filters, testing of heater elements, calibration of controllers and recorders, etc. The schedule may be that suggested by the sterilizer manufacturer or developed by the user, based on the operating history of the unit. A proper PM program will help to prevent breakdowns during production. Record of specific adjustments that are made to the unit (scheduled or unscheduled) must be maintained.

Change Control

Changes to the equipment that might compromise the validation must be brought to the attention of the group or individual in charge of the change control program. A change control form may be completed by the person requesting the change, outlining the modification or repair required, the reason for the change, and the expected results. The request form should be reviewed by a committee consisting of delegated representatives from the Validation Department, Quality Assurance, Engineering, and Manufacturing. The committee would evaluate the modification to be made (or already completed if done on an emergency basis), and determine if it would alter the validated status of the equipment. The representatives would then recommend specific revalidation checks to be made (if required). A list should be compiled of common repairs made to the sterilizer that do not disturb the validation, to expedite the review. Subsequently, these types of repairs may be made to the unit without prior approval of the committee. The modification should still be noted on a change control form to ensure that good records are maintained.

Revalidation

Revalidation studies may be required after changes or repairs are made on the unit, or at a predetermined periodic interval. Revalidation usually does not include all the original validation studies, but should include the worst-case load (load with the minimum F_H). An addendum to the validated report may be written to include additional loads that differ from the previously validated loads.

DOCUMENTATION

All validation information should be easily identified and kept in a permanent central file, where it can be readily retrieved. The validation file should include the following information.

Qualifications

All information recorded for IQ and OQ for the equipment and/or process. This includes all steps performed in the certification of the equipment. All original data, results, and conclusions must be contained in this file. Information may include blueprints, airborne particulate counts, velocity readings, HEPA testing, etc. All reports should be dated, signed, and approved by the responsible individual(s).

Cycle Development Study

The location of the records of the cycle development study should be referenced, or included in the documentation file if it was an integral part of the testing.

Process Qualification Protocol

The protocol is located in this file.

Chamber Studies

All original data, results, calculations, and conclusions must be retained for empty- and loaded-chamber and bio-challenge studies.

One of the most important records from these studies is the run sheet, a form that is filled out with the appropriate information at the time of the run. Load diagrams (depicting the actual placement of components, thermocouples, and bioindicators) and photos are kept with the run sheets to which the diagrams refer. The run sheets and diagrams assist in making certain that all required information has been recorded, and that the placement of any testing equipment or materials was correct. The diagrams should include the empty-chamber load, and all the different loads used in the loaded-chamber and bio-challenge studies. Other data to be identified with chamber studies would include calibrations, original temperature printouts, equipment temperature charts, bioindicator calibrations and test results, and calculation sheets (such as F_H values and temperature ranges).

Process Qualification Report

The process qualification report is the formal document available for regulatory review. The process qualification report contains the summary data from the various studies (empty and loaded chambers and bio-challenge).

Routine monitoring

All change control information and post-validation mechanical changes are recorded along with any revalidation work. This will prevent the voiding of all previous validation studies.

It is important to consider the validation effort as being protected by proper documentation and permanent files. The initial validation data are necessary for comparison with subsequent validations, and the overall validation program is only as reliable as the traceability of its documentation.

CONCLUSION

Dry heat is a commonly used method to sterilize and depyrogenate. This chapter has detailed the steps of a validation program that may be employed to properly validate a dry heat process. The necessity for this process to be validated has been demonstrated.

Following the outlined methodology will result in a complete program documenting a reproducible dry heat process.

REFERENCES

1. United States Pharmacopeia/National Formulary, USP28-NF23 <1211>. Sterilization and Sterility Assurance of Compendial Articles, Dry-Heat Sterilization. Easton, PA: Mack Pub. Co.
2. Akers MJ, Avis KE, Thompson B. Validation studies of the fostoria infrared tunnel sterilizer. J Parenteral Drug Assoc 1980; 34(5):330–1.
3. Validation of Dry Heat Processes Used for Sterilization and Depyrogenation, Parenteral Drug Association Technical Report No. 3, 1981:19.
4. Kirk-Othmer Encyclopedia of Chemical Technology. Microwave technology. Vol. 15. New York: Wiley, 1981:494.
5. Lohmann S, Manique F. Microwave sterilization of vials. J Parenteral Sci Technol 1986; 40(1):25–30.
6. Jeng DK, Kaczmarek KA, Woodworth AG, Balasky G. Mechanism of microwave sterilization in the dry state. Appl Environ Microbiol 1987; 53(9):2133–7.
7. Perkins J. Principles and Methods of Sterilization in Health Sciences. Springfield, IL: Charles C. Thomas, 1973:286.
8. Halliday D, Resnick R. Physics Parts I and II. New York: Wiley, 1967:549.
9. Parker S. McGraw Hill Concise Encyclopedia of Science and Technology. 5th ed. New York: McGraw-Hill Pub. Co., 1987:455.
10. Moore DM, Marshall JG, Kennedy MA. Comparative testing of challenge aerosols in HEPA filters with controlled defects. Pharm Eng 1994; 54–60.
11. Simmons PL. Hot air and continuous sterilization. In: Proceedings, Validation of Sterile Manufacturing Process. Reston, VA: Parenteral Manufacturers Association, 1978:44.
12. International Standard ISO 14644, Cleanrooms and controlled environments—Part 1: Classification of air cleanliness, 1999.
13. Green D. Perry's Chemical Engineers' Handbook, 6th ed., Vol. 22. New York: McGraw-Hill Co., 1984:32–3.
14. Wood RT. Parenteral Drug Association Short Course on Dry Heat Sterilization Validation and Monitoring on 6/17/82, 1982:12–9.
15. Feldsine PT, Ferry EW, Gauthier RJ, Pisik JJ. A new concept in glassware depyrogenation process validation. I. System characteristics. J Parenteral Drug Assoc 1979; 33:125–31.
16. Guidance for Industry Sterile Drug Products Produced by Aseptic Processing, U.S. Department of Health and Human Services Food and Drug Administration, 2004.
17. Validation of Dry Heat Processes Used for Sterilization and Depyrogenation, Parenteral Drug Association Technical Report No. 3, 1981:43.
18. Validation of Dry Heat Processes Used for Sterilization and Dpyrogenation, Parenteral Drug Association Technical Report No. 3, 1981:28.
19. Validation of Dry Heat Processes Used for Sterilization and Depyrogenation, Parenteral Drug Association Technical Report No. 3, 1981:24.
20. Validation of Dry Heat Processes Used for Sterilization and Depyrogenation, Parenteral Drug Association Technical Report No. 3, 1981:34–9.
21. Tsuji K, Lewis A. Dry heat destruction of lipopolysaccharide: mathematical approach to process evaluation. Appl Environ Microbiol 1978; 36(5):715–9.
22. Anderson NR, Kildsig DO. Alternate analysis of depyrogenation processes. J Parenteral Sci Technol 1983; 37(3):75–8.
23. Akers MJ, Ketron KM, Thompson BR. *F* value requirements for the destruction of endotoxin in the validation of dry-heat sterilization/depyrogenation cycles. J Parenteral Sci Technol 1982; 36:23.
24. Ludwig JD, Avis KE. Dry heat inactivation of rough strain lipopolysaccharide and diphosphoryl lipid A on the surface of glass. J Parenteral Sci Technol 1991; 45(1):35–9.
25. Ludwig JD, Avis KE. Dry heat inactivation of endotoxin on the surface of glass. J Parenteral Sci Technol 1990; 44(1):4–11.
26. Akers MJ. Parenteral fundamentals. Dynamics of microbial growth and death in parenteral products. J Parenteral Drug Assoc 1979; 33:372.
27. Nakata T. Destruction of challenged endotoxin in a dry heat oven. J Pharm Sci Technol 1994; 48(2):59–62.

28. Gould, M. Microorganisms—evaluation of microbial/endotoxin contamination using the LAL test. J Ultrapure Water 1993; 44–6.
29. McCullough KZ. Process control: in-process and raw material testing using LAL. Pharm Technol J 1988; 46.
30. Groves F, Groves M. Encyclopedia of Pharmaceutical Technology, Chapter on Dry Heat Sterilization and Depyrogenation. New York: Marcel Dekker, 1991:468.
31. Akers MJ, Avis KE, Thompson B. Validation studies of the fostoria infrared tunnel sterilizer. J Parenteral Drug Assoc 1980; 34(5):331–6.

BIBLIOGRAPHY

Akers MJ, Avis KE, Thompson B. Validation studies of the fostoria infrared tunnel sterilization. J Parenteral Drug Assoc 1980; 34(5):330–47.

Akers MJ, Ketron KM, Thompson BR. *F*-value requirements for destruction of endotoxin in the validation of dry heat sterilization. J Parenteral Sci Technol 1982; 36(1):23–7.

Avis KE, Jewell R, Ludwig J. Studies on thermal destruction of *E. Coli* endotoxin. J Parenteral Sci Technol 1987; 41(2):49–55.

Baird R. Validation of dry heat tunnels and ovens. Pharm Eng 1988; 8(2):31–3.

Ernst RR. Sterilization by Heat in Disinfection, Sterilization, and Preservation. Philadelphia, PA: Lea & Febiger, 1977:481–521.

Gould MJ. Microorganisms. J Ultrapure Water 1993; 43–7.

Halliday D, Resnick R. Physics. New York: Wiley, 1967.

Holman JP. Thermodynamics. New York: McGraw-Hill Book Company, 1969.

Kirk-Othmer Encyclopedia of Chemical Technology. Microwave Technology. Vol. 15. New York: Wiley, 1981:494–517.

Kirk-Othmer Encyclopedia of Chemical Technology. Sterilization Techniques. Vol. 21. New York: Wiley, 1983:626–43.

Kirk-Othmer Encyclopedia of Chemical Technology. Dry Heat Sterilization and Depyrogenation. Vol. 4. New York: Marcel Dekker Inc., 1991:447–84.

LAL Users Group. Preparation and use of endotoxin indicators for depyrogenation studies. J Parenteral Sci Technol 1989; 43(3):109–12.

Ludwig J, Avis KE. Dry heat inactivation of endotoxin on the surface of glass. J Parenteral Sci Technol 1990; 44(1):4–12.

McBride T. Hot air sterilizer GMP compliance. Pharm Eng 1984; 41–5.

Parenteral Drug Association. Validation of Dry Heat Processes Used for Sterilization and Depyrogenation, Technical Report No. 3, 1981:5–26.

Parenteral Drug Association. Depyrogenation, Technical Report No. 7, 1985:1–13 (see also 101–7).

Perkins J. Principles and Methods of Sterilization in Health Sciences. Springfield, IL: Charles C. Thomas, 1973:286–310.

Pflug IJ. Sterilization of Space Hardware, Environmental Biology and Medicine., Vol. 1. Minneapolis, MN: Gordon & Breach, 1971:63–81.

Remingtons Pharmaceutical Sciences. Easton, PA: Mack Publishing Co., 1979:1393–1467.

Reynolds WC, Perkins H. Engineering Thermodynamics. New York: McGraw-Hill Book Company, 1977:536–81.

Simmons PL. The secret of successful sterilizer validation. Pharm Eng 1981; 38–46.

Swarbrick J, Boylan JC. Encyclopedia of pharmaceutical technology. In: Groves J, ed. Chapter on Dry Heat Sterilization and Depyrogenation. New York: Marcel Dekker, 1991:447–83.

Tsuji K, Harrison S. Dry heat destruction of lipopolysaccharide: dry heat destruction kinetics. Appl Environ Microbiol 1978; 36:710–4.

Tsuji K, Lewis A. Dry heat destruction of lipopolysaccharide: mathematical approach to process evaluation. Appl Environ Microbiol 1978; 36(5):715–9.

United States Pharmacopeia, USP30-AF25. Easton, PA: Mack Publishing Co.

Weary M, Pearson F. A manufacturer's guide to depyrogenation. J Biopharm 1988; 22–9.

Whitiker G. Sterilization, validation, theory and principles. In: Proceedings, Validation of Sterile Manufacturing Processes. Reston, VA: Parenteral Manufacturers Association, 1978.

Wood RT. Parenteral Drug Association Short Course on Dry Heat Sterilization Validation and Monitoring, 1982:12–9.

16

Validation of Ethylene Oxide Sterilization Processes

John R. Gillis
SGM Biotech, Inc., Bozeman, Montana, U.S.A.

Gregg Mosley
Biotest Laboratories, Inc., Minneapolis, Minnesota, U.S.A.

INTRODUCTION

The recent history of EO sterilization has been dominated by advances in engineering technology. These advances include not only the computerized controls for the operation of sterilizers, but also the physical environmental controls that permit safe use of 100% EO gas. Most of these improvements have been driven by economics. Faster, cheaper processes are indeed worthy objectives. However, sometimes it seems we have lost the focus on what we are really attempting to accomplish in the sterilization processes. Any sterilization process must deliver a lethality that kills the naturally occurring bioburden microbes that contaminate the products and materials. If the process does not render the products or materials free from living microorganisms, then sterilization has not been achieved.

The microbiological dimension of the EO process has been overwhelmed by the recent strides in engineering and the physical process controls. The increasing complexity of medical products would be much more difficult to sterilize without these corresponding engineering process improvements. However, failure to properly address microbial lethality renders all of these engineering advancements meaningless if the resulting product is not sterile.

Process validation means establishing by objective evidence that a process consistently produces a result or product meeting its predetermined specifications (1).

The EO sterilization process is expected to deliver sterile products that possess all other specified quality attributes. Validation must document all critical process controls. The products to be sterilized must be challenged with an appropriate microbial system located in the "worst case" or "least lethal" product location. In addition, this microbial challenge product must be positioned in the worst case, least lethal location(s) in the production load. If the microbial challenges are not located in these least lethal locations, then the resulting documented evidence may be biased and result in false conclusions about the adequacy of the sterilization validation program.

The validation of the EO gas sterilization process is one of the more complex programs facing process engineers and microbiologists because some critical process parameters are interactive. EO gaseous sterilization has been shown to be an extremely effective process that can be performed with an infinite number of combinations of parameters. Key parameters that affect sterilization efficacy are (*i*) concentration of EO gas, (*ii*) RH, (*iii*) temperature of the process, (*iv*) accessibility of the product and packaging for these parameters, and (*v*) time.

A validation program must demonstrate that the selected combination of these interactive process parameters result in an effective physical and biological process. The effectiveness of this process is measured by calibrated physical instruments and a calibrated microbial challenge. These process parameters must then be correlated to a calculated SAL for the product. SAL is the probability of a single viable microorganism occurring on a product. The required assurance level may vary depending on the product itself or the end use of the product, but is typically less than one chance in a million of a non-sterile unit or SAL of 10^{-6}.

Another challenge is the task of assuring that the EO gas used does not create a health hazard for the employees in the working area or leave unacceptable residuals in the product delivered to the consumer. Adsorbed EO gas is removed fairly rapidly from processed materials, while absorbed EO gas is released much more slowly. This absorption rate is highly dependent on the specific process conditions, material being processed, as well as the geometry of the product, which affects material surface-to-volume ratios. Appropriate measures must also be taken to assure that EO gas used in the sterilizing environment is controlled and contained so that environmental insult in affected work areas is within acceptable regulated limits.

During the EO gas sterilization process, the gas interacts with the materials processed by reaction, absorption or adsorption. The EO gas is also trapped in the air spaces within the product or material being

---

Abbreviations used in this chapter: AAMI, Association for the Advancement of Medical Instrumentation; BI, biological indicator; BIER, biological indicator evaluator resistometer; CO_2, carbon dioxide; DEC, dynamic environmental conditioning; DUT, device under test; EO, EtO, ethylene oxide; FDA, Food and Drug Administration; GC, gas chromatography; IP, inoculated product; IR, infrared; MW, molecular weight; NIOSH, National Institute of Occupational Safety and Health; NIST, National Institute of Science and Technology; PEL, permissible exposure limits; RH, relative humidity; RTD, resistance temperature detector; SAC, static atmospheric conditioning; SAL, sterility assurance level; SLR, spore log reduction; TAR, test accuracy ratio; TC, thermocouple; TUR, test uncertainty ratio; TWA, time-weighted average.

sterilized. Unreacted residual gas is rapidly removed through evacuation, heated nitrogen or air exchanges. Product that is removed from a sterilizer must be controlled to prevent environmental insult to the workers. The best procedure is to place the sterilized materials in an environment that aids the desorption of the gas and is environmentally controlled to minimize workplace contamination.

CHARACTERISTICS OF EO

Chemical Properties

EO is also referred to as EtO, 1, 2-epoxyethane, and dimethylene oxide (2). It has a formula of C_2H_4O. The following structure is illustrated:

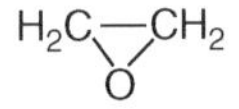

It is a colorless gas, with a molecular weight of 44.05. It has a characteristic ether-like odor at toxic levels. EO has a boiling point of 10.7°C (51.3°F) at 760 mmHg pressure, a melting point of −112.6°C (−170.7°F), a specific gravity of 0.8711 apparent at 20°C (60°F), or a specific gravity of 0.897 at 4°C. EO has a vapor density of 1.5, with dry air being equal to 1.0, and a vapor pressure at 20°C of 1095 mmHg. It is completely miscible in water, alcohol, acetone, benzene, ether, carbon tetrachloride, HCFCs, and most organic solvents, and is a powerful solvent for fats, oils, greases, waxes, some rubber formulations, and paints. It is highly exothermic and potentially explosive when heated or mixed with (*i*) alkali metal hydroxides, (*ii*) highly active catalytic surfaces such as anhydrochlorides of iron, tin, or aluminum, and (*iii*) the oxides of iron and aluminum. The explosive limits are 3% to 97% by volume in air. It has a flash point of −6°C (20°F). It is relatively noncorrosive for materials. EO is relatively stable in neutral aqueous solutions and when diluted with liquid or gaseous carbon dioxide or halocarbons such as HCFCs. EO is relatively unstable in either acidic or alkaline aqueous solutions and may rapidly form ethylene glycol.

Biological Activity

EO reacts irreversibly with numerous chemical moieties on cellular molecules by an alkylation reaction where the [CH_2OH–CH_2–] alkyl group is covalently bonded with the available moiety via an addition reaction. Reactions with –NH_2, –SH, –COOH, and CH_2OH groups are common and illustrated in Figure 1 (3).

Reaction rates vary and depend on the specific pK_a for each moiety and the existent pH. For a more comprehensive review of possible reactions we refer the reader to Russell (4). First-order lethality kinetics require that only one molecule per cell is the critical target (5–7). Reactions other than the critical reaction leading to microbial inactivation must be considered collateral damage reactions. Not all microbial inactivation obeys first-order kinetics. However, even where multiple sites or molecules may be required for inactivation, the concept regarding critical reactions and collateral reactions is the same. Where inactivation is the result of cumulative damage, which is not first-order kinetics, then some damaging reactions must be considered more important to the events leading to microbial inactivation (critical) than other reactions (collateral). Winaro and Stumbo (8) identified EO reactions with DNA as the critical reactions resulting in microbial inactivation. Lawley and Brookes (9) defined specific reactions of EO with the nucleic acid tertiary heterocyclic nitrogen sites (=N-) in numerous experiments resulting in more than 10 publications between 1957 (10) and 1963 where they state (9):

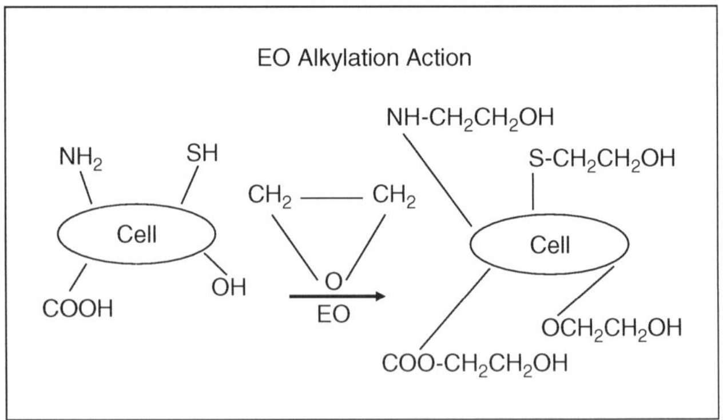

Figure 1 Illustration of the alkylation reaction of ethylene oxide with chemically active moieties in the bacterial cell.

1. Sites in the nucleic acids reactive towards alkylating agents are shown to be, in order of decreasing reactivity: for RNA, N-7 of guanine, N-1 of adenine, N-1 of cytosine and N-3 of adenine for DNA, N-7 of guanine, N-3 of adenine and N-1 of cytosine. Denatured DNA behaves in this respect like RNA.
2. The observed differences between DNA and RNA are ascribed to the involvement of N-1 of adenine and of cytosine in hydrogen bond formation in DNA.
3. In all cases alkylation results in destabilization of the nucleosides or the corresponding moieties in the nucleic acids. At neutral pH, with DNA, 7-alkylguanines and 3-alkyladenines are slowly liberated by hydrolysis, the latter at the greater rate, whereas with RNA slow rearrangements occur, 1-alkyladenine moieties yielding 6-methlaminopurine moieties and 1-alkylcytosines giving the corresponding 1-alkyluracils.

More recent studies suggest that disruption of the DNA molecule may occur differently depending on various repair mechanisms (11). In the case of certain repair mechanisms, the reactions with cytosine may be the injury which ultimately leads to the inactivation of the microbe.

VALIDATION OF EO STERILIZATION PROCESSES

Validation of the EO process is divided into two phases: Engineering Qualification and Process Qualification. When these activities are completed successfully and all aspects of the process are documented, the process can be certified for routine use for manufacturing goods.

Engineering Qualification

Engineering Qualification deals with the sterilizer and associated equipment used in the process. This phase is divided into three segments: Installation Qualification, Calibration, and Operational Qualification.

Installation Qualification

Installation Qualification requires an audit of the equipment as it has been installed in the facility. This audit includes checking all utilities and supplies to the equipment to make sure that they meet the manufacturer's recommended specifications. Engineering drawings must be evaluated to assure that (*i*) the equipment is assembled according to the manufacturer's prints, (*ii*) the equipment is installed according to the installation schematics, and (*iii*) all aspects of the equipment are documented with appropriate engineering drawings or sketches. These drawings are essential for future reference to compare the hardware validated to any future configurations. This segment of the validation program is probably the most abused with sterilizers using nonexplosive containers of EO. Systems using 100% are extremely well documented, which is driven by the safety issue. Once the equipment is hooked up and it "runs," little more is ever documented. With the pressure to get things working, little attention is paid to the documentation for future reference. Inadequately treated items typically include documentation of utilities, spare parts lists and preventive maintenance procedures. Many validations have been performed with all the necessary tests on the hardware relating to product loads, but with no record as to the exact configuration of the equipment when the validation was executed. Since any mechanical device will routinely malfunction, or wear out and require replacement, it is absolutely essential that a well-prepared Installation Qualification document be assembled for each piece of equipment to be validated. If this is not done, subsequent validation data may prove meaningless.

Calibration

The second segment of the Engineering Qualification is the calibration of all process sensing, controlling, indicating, and recording devices on the sterilizer or independent systems associated with it. Recording instruments that appear on the control panel are typically calibrated, but many of the control instruments are often located out of sight and should not be ignored since they may have a tremendous impact on the cycle function. For example with the DEC phase of an EO sterilizing process, it is extremely important to calibrate the stall point of the vacuum pump before the actual pressure or temperature set points are calibrated. This measurement is critical to balance the steam input into the chamber in relation to the capacity of the vacuum pump to remove the steam from the chamber. All critical process control instruments that are recorded and displayed by the control system must be calibrated. This is even more complicated when micro processor control units are employed, because not only are there specific operating set points for those systems, there are also high- and low-limit alarm and other default systems that must be documented and calibrated. The calibration program will also vary depending on the type of computerized system.

The calibration program should be performed with instruments referenced as secondary standards. The secondary or transfer standard is a standard that can be transported to and from the actual sterilization equipment because most instruments associated with the sterilizer must be calibrated at the sterilizer's location. Secondary standards must be traceable to a recognized standard such as those maintained by the NIST.

A measurement or calibration compares a DUT to a standard or reference. This standard should outperform the DUT by a specific ratio, called the "TUR" also known as the TAR. As a rule of thumb, the TUR should be greater or equal to 4:1 (12).

Primary standards should have an even greater sensitivity. It is recommended that these primary standards be submitted to the NIST for calibration and recertification on a periodic basis. Primary standards are usually recertified annually. It is extremely important that detailed procedures be established including limits and acceptable correction variances allowed and calibration frequency for all the instruments on the sterilizer. Adequate records must be maintained. A tracking system is essential to assist metrology, assuring that required calibrations occur at their designated frequencies. A history file should be maintained for each instrument, and the records reviewed to assure established calibration frequencies are appropriate.

Operational Qualification

The third segment of Engineering Qualification is Operational Qualification that deals with the operating parameters of the sterilizer: their function, adjustment, and control. These tests are performed with an empty chamber. The various parameters for the cycle are evaluated to determine if they perform as specified by the manufacturer. Temperature controllers are set and evaluated to determine performance. The temperature distribution within the sterilizer is documented. The unit is sequenced through its operating steps to assure that the sequencing is appropriate. Every operating parameter must be documented to determine its compliance with the manufacturer's operating specification. The Operational Qualification protocol will serve as the basis for developing the Standard Operating Procedure for routine operation of the sterilizer. The Operational Qualification testing specifies in detail how the equipment operates.

Process (Performance) Qualification

The final phase of validation deals with Process Qualification. Even though the unit functions appropriately with an empty chamber, it must now be demonstrated that it sterilizes product. This phase may require repetition with different products and loads.

Load Configuration

There are several key aspects of Process Qualification. First, the specific product and all its packaging must be defined. The next step is to define the way master cartons are arranged into pallets. Pallet arrangement within the sterilizer is also part of the load configuration definition. Many manufacturers have numerous products that must be mixed together in order to achieve effective sterilizer throughput.

Categorizing product for the sterilizing load is an extremely important element. It is important that the particular product mix is configured with a rationale that packaging is similar and products should be of consistent mass and materials and actual product configuration. It is possible that a manufacturer may have in its catalog hundreds of different products. If these all have the same characteristics and packaging, it is possible that they could be sterilized within one or two different sterilization cycles. It is also possible that a manufacturer may produce only a few products, each being so different from the other products manufactured that each product sterilization process will have to be validated in different cycles.

Once product categories have been identified, it is also possible to vary the load configurations. Loads must be extremely specific in the way they are defined. Small tolerances are permissible without changing the overall impact on the biological effectiveness of the sterilization. However, changes in the qualified load must be evaluated and properly documented to determine the potential impact on the biological effectiveness of the process.

Conditions which influence the lethality delivered by the process are mass, density, packaging, product design and materials. External preconditioning of product loads is common and allows easy process measurements. Preconditioning time may vary with different loads. Determining moisture, EO gas and temperature penetration into the palletized load is much more difficult in the sterilizer. If time to achieve acceptable levels of these parameters is similar to the originally defined load, then loads can be considered equivalent. Measured lethality should be similar with similar loads. It should be noted that configuration changes may influence the location of "worst case–least lethal" position. Confirmation will have to be performed and appropriate adjustments may have to be made to assure that a proper monitoring location is documented.

Once the product and load have been defined, then the worst case–least lethal locations in the product, within each pallet and within the vessel must be determined. These locations will have to be monitored physically and biologically to provide data on all critical process parameters.

Pallet Configurations

Pallet construction may depend in part on how much shipping will take place between the time of construction until sterilization. When processing was performed in-house, it was easy to construct pallets with "chimneys" configured between columns of master cartons. These chimneys assured that more surface area of the master cartons was directly accessible for thermal transfer and gas exchange. This type of configuration provides the greatest homogeneity of sterilization conditions across the product load.

Contract EO sterilization is extremely popular today and provides users with "state-of-the-art" systems at reasonable expense. The problem comes not from the sterilizer, but from the logistics involved in transporting the product off-site to the contractor. Pallets are constructed at the product manufacturing site with transportation in mind, not sterilization. Pallets are densely packed because they survive the rigors of overland shipping much better than pallets configured with void spaces (chimneys) for gas permeation. Stretch wrap is commonly used to hold the palletized boxes together. Stretch wrap is exceptional for maintaining pallet integrity during shipping, but it may create a tremendous barrier to sterilizing vapor penetration. Stretch wrap manufacturers are now offering a "net" type of wrapping material which significantly increases the surface of the master cartons directly exposed to the sterilizing vapors. The best technique from a sterilization perspective is to use strapping to band the pallets together. This requires the use of corner protectors so as not to crush the outside corners of the master cartons (Fig. 2). An example of uniformly constructed pallets loaded into a sterilizer vessel appears in Figure 3. More pallet configurations can be sterilized successfully of course, but process times may be longer

Figure 2 An example of a banded pallet of product providing maximum surface exposure to sterilization vapors. *Note*: Corner protectors on pallet protecting the master cartons.

Figure 3 An example of a uniform load configuration. Two identical pallets side by side. All pallets are exactly the same in construction and product.

and variations of microbial lethality and EO residuals across the load may be greater.

CRITICAL STERILIZATION PROCESS PARAMETERS

There are several major considerations to be aware of in order to structure a validation program that will assure that the sterilization process does what it is intended to do.

These considerations include: (*i*) controlled process parameters and their interaction; (*ii*) an integration of the physical process conditions; (*iii*) the selection of appropriate process conditions; (*iv*) the product design; (*v*) how the product is pretreated prior to exposure; (*vi*) how the product is handled following sterilization; (*vii*) how the process is monitored, including physical, chemical, and biological methods; and (*viii*) the effect of residual EO and its reaction products on the material being sterilized.

There are four critical interactive parameters that must be controlled for EO sterilization process: (*i*) EO gas concentration; (*ii*) moisture; (*iii*) temperature; and (*iv*) time. All these parameters interact to affect the lethality delivered by the process.

EO Gas Concentration

General Use Range of EO Gas Concentration

EO gas concentrations below 300 mg/L and above 1200 mg/L are not commonly used in the industry. EO gas concentrations less than 300 mg/L are not effective in practical process times. Concentrations above 900 to 1200 mg/L do not shorten the process times sufficiently to warrant the additional cost of gas. Sterilization effectiveness is dependent on the molecular collision of the EO molecule and the biological entity that is being sterilized. Therefore, more EO molecules lead to more rapid microbial lethality. A sterilizing process using 600 mg/L of EO delivers approximately twice the lethality as a process using 300 mg/L in the same time. However, considering the cost of EO, processes are generally designed toward the lower concentrations of EO. Concentrations of 400 to 600 mg/L appear to be the more popular conditions today for operations to balance the cost of EO, equipment and throughput time.

EO Gas Concentration Controllers

The EO gas concentration is controlled in one of two ways. The most common method of control is the indirect method through the use of a pressure control system. The EO gas concentration desired is calculated as to the corresponding increase in pressure. The desired pressure settings are then maintained by conventional pressure controllers. The direct control method uses analytical instruments that actually detect the EO gas concentration in the environment inside the sterilizer.

The analytical systems are either gas chromatographic, IR or microwave detectors. These instruments are installed directly to the sterilizer. Periodic gas samples are withdrawn from the sterilizer or gas circulation lines and passed through the detector. Some IR or microwave detectors may be mounted on the exterior chamber wall using an access port or in the gas circulation system. Electronic signals are sent to control valves in the gas supply lines allowing makeup charges to maintain the target gas concentration.

Indirect Methods

There are two approaches for the indirect method of measuring EO gas concentration in the sterilizer—they are weight and pressure. The indirect methods are dependent on using gas cylinders containing certified mixtures of EO. When the chamber is pressurized, it is assumed that the mixture contains the given percentage of EO relative to the change in pressure. Therefore, this change in pressure can be equated to an assumed gas concentration. This system is very easy to monitor using pressure transducers and recorders.

The second indirect method measures the weight of the gas cylinder contents dispensed into the vessel. This method assumes that a uniform mixture of the EO and diluent gas was dispersed into the vessel, yielding an assumed concentration of gas in the sterilizing chamber. This system is easy to monitor using acceptably sensitive scales.

These indirect methods are reasonably good estimates for most gas mixtures. Neither method compensates for absorption of EO by the packaging materials or the product. Different materials absorb EO at different rates than they do diluent gases (13). Furthermore, indirect methods do not consider physical leaks in the sterilization system. Thus the indirect method, at best, provides an approximation of the EO gas concentration in the vessel.

Direct Gas Measurement

Direct analysis of the EO in a sterilizing chamber can be performed by specific analytical instruments. Two of the most common analytical methods are the GC and the IR spectrophotometer.

Gas Chromatography. GC has been the most widely used method for determining the level of EO in the sterilizing environment. Some EO processes operate at atmospheric or positive pressure, making withdrawal of a gas sample easy. Sterilization processes that use 100% EO with a nitrogen blanket may operate at slightly subatmospheric pressures and sampling is slightly more difficult. GC is not used in 100% EO processes with no nitrogen overlay because they operate under a deep vacuum. When dealing with explosive mixtures of EO or pure EO, only intrinsically safe instrumentation must be used.

Sample removal is extremely important in order to assure meaningful data. Sample lines must be heated and insulated upon exiting the sterilizer. If cold spots occur in the sampling lines, the EO and water vapor may condense, yielding false data. These samples may be collected using gas collection bottles or with lines attached directly to the GC if an automatic injection system such as a gas sampling loop is used.

Multiple sample sites also present a problem. Representative sites are generally selected throughout the sterilizing chamber. Small capillary tubes serving as sample delivery lines are fitted to the gas sample ports. Care must be taken to permit these sample tubes to be flushed to assure that the sample being extracted is, indeed, from the chamber environment and not a residual in the sample delivery line. For this reason, this method is not acceptable for sampling within the product or product packages. The flushing of the sample lines accelerates the gas penetration into these restricted locations and yields data that are not representative of actual load conditions.

Gas samples can be extracted from the gas recirculation system. This provides a good estimate of the gas concentration in the chamber.

The GC unit must be calibrated prior to sample analysis with a certified standard gas. This certified standard may be either a diluted gas mixture or 100% EO. Most laboratories that are established to perform GC analysis are qualified to use 100% EO as the standard for calibration. However, certified mixtures are available from gas suppliers. The GC is calibrated at one point with this standard gas and expressed as mole percent. These calibration results are independent of temperature and pressure. The mole percent concentration of the sterilization chamber is compared to the standard gas and is then converted into mg/L.

$$\frac{\text{mol\%}}{100\%}\times\frac{44.0\ \text{g}}{\text{mol}}\times\frac{1000\ \text{mg}}{\text{g}}\times\frac{1\ \text{mol}}{22.41\ \text{L}}\times\frac{(14.7\ \text{psia}+Y\ \text{psig})}{14.7\ \text{psia}}\times\frac{273^{\circ}\text{C}}{(273^{\circ}\text{C}+{}^{\circ}\text{C})}=\text{XF}$$

where Y psig, pressure of the sterilizing chamber; °C, temperature in the sterilizing chamber; XF, scaling factor.

Therefore, the EO concentration from the GC data in mole percent multiplied by the scaling factor (XF) yields mg/L.

IR Analysis. Most gases have a characteristic IR spectrum that can be used to identify them. These spectra are usually rather complex; however, each usually contains a small number of strong analytical bands that are used in this analysis.

These IR analyzers incorporate a fixed wave-length filter that corresponds to one of these strong bands. An optical path is also chosen that provides the sensitivity range required for the particular analysis.

The analytical wavelength for 100% EO is 11.8 μm. When HCFC mixtures are used, it has been found that a wavelength of 3.3 μm is more satisfactory and minimizes the interference with the HCFC spectrum. Some systems are theoretically sensitive to 0.4 ppm of EO.

Calibration of these analyzers must also be performed using certified standard gas. Calibration with the standard gas must consider the pressure differential between the calibration gas and the sterilizing chamber. Once the wavelength and path length are set, using the calibration standard, the instrument's response to absorbing the gas is directly correlated to concentration.

EO Gas Monitoring in the Worst Case Location in Product. Gas concentration is generally not monitored inside the product. The reason for this is that the capillary tube necessary to withdraw the gas sample from this location in the sterilizer has sufficient volume to cause erroneous readings in the vicinity of the product. Withdrawing the sample may actually force EO to migrate into the product sample site. If the sample is taken from the environment close to the product, then fewer technical problems are incurred. Samples drawn continuously from the product create a small delta pressure, causing a positive flow of gas from the environment into the sampling locale around the product. If the environment within the sampling area is large and unencumbered, then meaningful gas samples can be withdrawn from the chamber. Samples are generally withdrawn from a spectrum of locations within the chamber, typically warmer as well as cooler than other locations. Samples should be withdrawn from the front, back, top, and bottom of the vessel, so that all geometric areas within the sterilizer are assayed.

EO Gas and Diluents

100% EO No Diluents

The most commonly used form of EO gas for sterilization in the industry is pure EO (100%) with a nitrogen overlay pressure sufficient to reach near-atmospheric pressure within the sterilizing chamber. This process is the most economical and there are no diluent concerns. This process has potentially explosive phases, but the nitrogen blanket minimizes the risk of an explosive mixture with air inside the vessel.

EO Gas Mixtures

Some EO gas mixtures have been created because they are not explosive. Such mixtures do not require expensive safety facilities in which to operate.

Mixtures Diluted with HCFC. The next most commonly used EO is that which has been diluted with halocarbon products, primarily HCFC 124 and HCFC 22. This mixture is normally composed of 10% EO and 63% HCFC 124 and 27% HCFC 22 or 8.6% EO and 91.4% HCFC 124.

Cylinders charged with EO/halocarbon mixtures contain a liquid that is a homogenous mixture of both the EO and the halocarbon. The pressure in these cylinders is low due to the vapor pressure of the liquid at the temperature at which the cylinders are stored. When the sterilizer is charged, a homogeneous blended liquid is drawn off the bottom of the cylinder. The pressure in the cylinder remains virtually constant until the liquid level falls below the level of the cylinder eductor tube. Multiple sterilizer charges can be performed with this mixture yielding consistent EO concentrations.

Mixtures Diluted with Carbon Dioxide. EO may also be mixed with carbon dioxide in concentrations of 10% EO and 90% CO_2, or a 20% EO and 80% CO_2 and 30% EO and 70% CO_2. Since 20% EO and 80% CO_2, 30% EO and 70% CO_2 and 100% EO are explosive, these EO sources must be used only in specially designed sterilizers and buildings that are designed to be intrinsically safe electrically and to withstand potential explosions.

Cylinders charged with EO/carbon dioxide mixtures contain a liquid phase of EO and a gaseous phase consisting mainly of carbon dioxide gas molecules with minor EO. The pressure of these cylinders is much higher than that of the EO/HCFC mixtures. Because of the biphasic condition of a liquid phase/gas phase mixture, it is virtually impossible to achieve multiple charges that are a consistent molecular blend and a consistent EO concentration. These cylinders are designed with an eductor tube with a fixed orifice that draws the liquid EO from the bottom of the cylinder and much smaller openings at the top of the eductor tube to withdraw the diluent CO_2 gas in the upper portion of the cylinder. These two chemicals are mixed in the eductor tube as they are released from the cylinder. In theory, it should work, provided the orifice openings are free-flowing and cylinder pressure remains constant. In practice the system often does not work as intended. To compound this problem, most sterilizers that use this gas mixture also rely only on a pressure reading as an indirect indication of EO concentration. Gas charges from these cylinders will frequently yield mostly CO_2 with little or no EO present. The approach most users of these mixtures take is to select a cylinder size that is equal to a single (unit dose) charge in the sterilizer. Therefore, inconsistencies that occur during the emptying process of the entire cylinder have little effect on the final concentration of EO in the sterilizer. In large industrial sterilizers it may even take multiple cylinders to charge the sterilizer.

This mixture is relatively inexpensive and is becoming increasingly more popular with small sterilizer users. These users are attempting to deliver multiple charges from the same gas cylinder since smaller "single charge" cylinders may not be readily available. The concentration of EO delivered in multiple charges with EO/CO_2 mixtures tend to have slightly higher than expected EO concentration in the first withdrawals. As the tank approaches empty the concentration of EO then tends to decrease very rapidly until the last few pounds are nearly pure CO_2. This problem is further compounded by the suppliers of these cylinders. It appears that these suppliers are companies who fill cylinders for welding gases, which have similar hazardous properties as EO. The design of these cylinders may vary from supplier to supplier. The types of Quality Control requirements may not be the same as manufacturers who focus on sterilizing gases. Little documentation is available on cylinder and eductor tube design for these "custom fillers" of the EO/CO_2 mixtures.

Calculation of EO Concentration

The calculation of the EO gas concentration in the sterilizer is based on the Ideal Gas Law $PV=nRT$. Several assumptions must be made when applying this law. They include that the pressure rise in the sterilizer is due totally to the EO and its diluent, if used, and that temperature is at equilibrium. It also assumes that the mixture inside the sterilizer, which includes residual air and water vapor along with the EO (mixture), behaves as an ideal gas. It assumes that this mixture of components remains unchanged by molecular activity such as adsorption, absorption, condensation or reaction. It assumes that the label on the gas cylinder is accurate and that this molecular ratio remains constant when delivered into the sterilizer. That said the temperature is not at equilibrium. A temperature variance of 10°C is suggested by ISO 11135 (14), but wider ranges may be encountered in practice. The best analytical approach is to use an average chamber temperature value. EO is an active, highly soluble molecule and this allows it to be absorbed by most product and packaging materials (13).

Water vapor will be absorbed by product and packaging materials. As the vessel pressure is raised there will be a corresponding temperature rise in the water vapor in the vessel. This allows any liquid condensate from the humidification phase of the cycle to vaporize and make an additional contribution to the measured change in pressure that occurs during gas charge.

The EO sterilization process is extremely dynamic at the molecule level. The application of the Ideal Gas Law provides a good estimate of the concentration of the EO gas. Conversion factors useful in performing these calculations appear in Table 1.

The total pressure inside the sterilizer minus the change in pressure due to the addition of EO and its diluent (if used) can be expressed as:

$$P=\frac{nRT}{V} \tag{1}$$

Derivation of the EO Gas Concentration Equation

Most sterilizer operations record the pressure change during EO gas injection; therefore, the following equation was derived to allow the calculation of EO concentration from the pressure rise due to this gas injection, with or

Table 1 Useful Conversion Factors

Pressure	Volume	Weight	Temperature
1 atm = 4.7 psia	1 L = 1,000 cc	1 lb = 454,000 mg	°C = (°F − 32) × (5/9)
1 atm = 760 mmHg	1 L = 0.03532 ft^2	1 lb = 454 gm	K = °C + 273.2
1 atm = 29.92 in Hg	1 ft^3 = 28.32 L		
1 atm = 1.013 bar	1 ft^3 = 28,316.9 cc		
1 atm = 101.3 kPa	1 m^3 = 1,000 L		
1 atm = 1,013 hPa			
1 psi = 6,894.7 Pa			
1 psi = 6.8947 kPa			
1 kPa = 0.145 psi			
1 kPa = 7.5 mmHg			
1 in Hg = 25.4 mmHg			
1 in Hg = 3.387 kPa			
1 hPa = 1 millibar			

Pascal (Pa) is an international standard unit of pressure. The Pascal is a unit of pressure equal to one Newton per square meter, or one kilogram per meter per second. Pressure is most commonly measured in kilopascals (kPa).

without diluent gases such as HCFC or carbon dioxide. The purpose of this equation is to provide a simple and rapid method for calculating EO gas concentration in sterilizers.

The pressure rise can be expressed as in equation (2):

$$P = P_{EO} + P_{DG} = \left(\frac{n}{v}\right)_{EO} RT + \left(\frac{n}{v}\right)_{DG} RT \quad (2)$$

$$P = \left[\left(\frac{n}{v}\right)_{EO} + \left(\frac{n}{v}\right)_{DG}\right] RT \quad (3)$$

The above formula can be expressed in mg/L:

$$\left(\frac{n}{v}\right)_{EO} = \frac{g}{MW_{EO}} L = \frac{10^{-3}}{44}\left(\frac{mg}{L}\right)_{EO} \quad (4)$$

$$\left(\frac{n}{v}\right)_{DG} = \frac{g}{MW_{DG}} L = \frac{10^{-3}}{M}\left(\frac{mg}{L}\right)_{DG} \quad (5)$$

where MW_{EO} = molecular weight of EO = 44.0, MW_{DG} = molecular weight of diluent gas = M.

Then the pressure rise can be rewritten as

$$P = \left[\frac{10^{-3}}{44}\left(\frac{mg}{L}\right)_{EO} + \frac{10^{-3}}{M}\left(\frac{mg}{L}\right)_{DG}\right] RT \quad (6)$$

Since the weight percent EO (wt% EO) is usually known and the sterilizer volume remains constant, the expression derived above can be written as

$$\text{wt\% EO} = \frac{\left(\frac{mg}{L}\right)_{EO}}{\left(\frac{mg}{L}\right)_{EO} + \left(\frac{mg}{L}\right)_{DG}} \times 100 \quad (7)$$

Solving for $(mg/L)_{DG}$:

$$\left(\frac{mg}{L}\right)_{DG} = \frac{\left(\frac{mg}{L}\right)_{EO} 100 - \left(\frac{mg}{L}\right)_{EO} \text{wt\% EO}}{\left(\frac{mg}{L}\right)_{EO}\left(\frac{100 - \text{wt\% EO}}{\text{wt\% EO}}\right)} \quad (8)$$

Substituting the above for $(mg/L)_{DG}$, equation (8) becomes

$$P = \left[\frac{10^{-3}}{44}\left(\frac{mg}{L}\right)_{EO} + \frac{10^{-3}}{M}\left(\frac{mg}{L}\right)_{EO}\left(\frac{100 - \text{wt\% EO}}{\text{wt\% EO}}\right)\right] RT \quad (9)$$

Solving for $(mg/L)_{EO}$ in equation (9):

$$P = RT\left[\frac{10^{-3}}{44} + \frac{10^{-3}}{M}\left(\frac{100 - \text{wt\% EO}}{\text{wt\% EO}}\right)\right]\left(\frac{mg}{L}\right)_{EO} \quad (10)$$

Let wt% EO = E and rewrite:

$$P = RT\left[\frac{10^{-3}}{44} + \frac{10^{-3}}{M}\left(\frac{100 - E}{E}\right)\right]\left(\frac{mg}{L}\right)_{EO} \quad (11)$$

The rewrite:

$$P = 10^{-3} RT\left[\frac{1}{44} + \frac{100 - E}{(M \times E)}\right]\left(\frac{mg}{L}\right)_{EO} \quad (12)$$

$$= 10^{-3} RT\left[\frac{(M \times E) + 44(100 - E)}{44(ME \times E)}\right]\left(\frac{mg}{L}\right)_{EO} \quad (13)$$

Then:

$$\left(\frac{mg}{L}\right)_{EO} = \frac{10^{3} P}{RT}\left[\frac{44(M \times E)}{(M \times E) + 44(100 - E)}\right] \quad (14)$$

Table 2 Gas Constants (R)

Pressure	Volume	Temperatures	R
Atm	cm^3	K	82.057
Atm	L	K	0.08205
Atm	ft^3	K	1.3140
Bar	L	K	0.08314
kg/m^2	L	K	847.80
kg/cm^2	L	K	0.08478
mmHg	L	K	62.631
mmHg	ft^3	K	998.90
in Hg	L	K	2.4549
kPa	L	K	8.312

It is important to maintain the proper units when using the ethylene oxide concentration equation and the gas constants.

This equation can be rearranged to:

$$C = \frac{KP}{RT} \quad (15)$$

C is the EO concentration in mg/L, R is the gas constant (Table 2), P is the difference in total pressure due to EO and its diluent (if used), T is the absolute temperature (K) of the EO diluent gas mixture resulting in Pressure (P), K is the constant for a given diluent (Table 3).

K is calculated using the following formula:

$$K = \frac{4.4 \times 10^{4} Mw}{Mw + 44(100 - w)} \quad (16)$$

M is the molecular weight of the diluent or the average molecular weight of the diluent mixture, w is the mass fraction of EO in the diluent.

Moisture

General

Moisture is the most important parameter in the EO sterilization process. Without adequate moisture, the sterilization process is greatly inhibited. When adequate moisture is present, the process will be dependent on the molecular activity of the EO and its interaction with the microbial populations being exposed.

The authors would like to quote Phillips (15) from a 1968 article:

Table 3 Molecular Weights and Gas Constants

Substance	Molecular weight	Gas constants K (mg/g mol)[a]
EO	44.0	4.4×10^4
HCFC 22	86.47	
HCFC 124	136.5	
70% HCFC 124+ 30% HCFC 22	121.49	
CO_2	44.0	
Substance mixtures		
10% EO/27% HCFC 22 and 63% HCFC 124	Diluent MW 121.49	9.989×10^3
8.6% EO/91.4% HCFC 124		9.942×10^3
8.5% EO/91.5% CO_2		3.74×10^3
20% EO/80% CO_2		8.8×10^3
30% EO/70% CO_2		1.32×10^4

[a] Use when calculating mg/L.

Care must be exercised, however, when the objects to be sterilized or, more correctly, the microorganisms contained on them, are equilibrated to lower relative humidities or have been previously exposed to extremely desiccating conditions. Not only is sterilization more difficult at relative humidities below 30%, but once microorganisms have been highly desiccated either chemically or by vacuum, they acquire a resistance that is not completely overcome when the RH is again raised to 30%. Not all of the organisms are resistant, but a few maintain the resistance until they are essentially re-wetted. The phenomenon is not well understood, but it is real.

Moisture is extremely important in making the reactive sites in the microbial cells available to the alkylation action of EO. When cells or spores dry, their proteinaceous and nuclear materials and the active sites are physically withdrawn, making reactions with EO molecules difficult. However, as these materials are hydrated, they swell and expand. This exposes the active sites and makes them available to the alkylation by EO. Without proper humidification, these active sites are protected and impede the lethality of the EO sterilization process.

The moisture take-up of the microorganisms plays an extremely critical role in the sterilization of freeze-dryers used in the pharmaceutical industry. Freeze-drying processes can stabilize organisms in a desiccated state, making them extremely resistant to the EO sterilization process (16). Freeze-dryers are not normally designed as EO sterilizers, and adequate mechanical means for moisturization are not generally supplied. Engineering modifications must be made to these machines in order to sterilize them effectively with EO.

EO sterilization processes must be performed between the adequate levels of moisture. The lower boundary values of 35% to 50% have been referenced (16,17). The upper boundary value appears to be 85%. Once the RH is within the acceptable window for a product and process, increases or decreases in RH within this window do not produce measurable changes in microbial resistance. The boundaries are affected by temperature, load materials, and specific cycle dynamics. Both low RH and dew point conditions during EO exposure phases can produce changes in microbial lethality that are difficult to predict quantitatively. Lower and higher RH levels can produce dramatic and quantum increases in measured microbial resistance. At very low RH conditions, sterilization by EO may not be able to be accomplished in any practical time frame.

The upper limit should be below the conditions where the dew point is reached. Condensed water not only slows down the migration of the EO molecules to the spore, but the EO molecule can react with water or dissolved solutes. Such reactions reduce the available EO for reactions with microbial molecules.

Humidity, Instruments and Controllers

The most problematic parameter to monitor and control in the EO sterilization process is humidity. Humidity is typically measured as RH. The measurement compares the amount of moisture that is present in the air compared to the maximum amount of moisture the air can theoretically hold at that temperature.

$$\%\text{RH} = \frac{\text{Moisture content of the air at the specified temperature}}{\text{Vapor pressure of water at the specified temperature}} \times 100\% \qquad (17)$$

Moisture can be measured indirectly using pressure measurements or directly by analytical instruments that measure either absolute water content or dew point.

Indirect Method

The indirect method uses differential pressure measurements. This differential pressure is a valid measurement only when the pressure change is due entirely to the vapor pressure of water. If liquid water is emitted at the same time as steam pressure, an additional pressure rise will occur when the liquid water is vaporized. If there is an air leak into the vessel an erroneously high indication as to the amount of moisture can result. It is also very difficult to get an accurate measurement of this parameter when a product is in the vessel, because of the moisture-absorbing qualities of various products and packages. Packaging materials many times are of greater mass than the product. They typically have a great affinity for moisture and compete with the product for this moisture. This is a process that works extremely well on an empty chamber, but it is very difficult to assess in a practical manner within the loaded chamber because of moisture exchange between the product and chamber gas environment.

If the indirect method is to be applied, measurements can be calculated using the properties of saturated steam found in engineering handbooks (Appendix I) (18). For example, the vapor pressure of saturated steam at 55°C is 117.85 mmHg. If the sterilization cycle is to be run at 55°C and a RH of 50% is desired, a change in pressure due to the addition of steam will be 58.93 mmHg or 50% of 117.85 mmHg (Table 4).

$$\text{Pressure change required (mmHg)} = (\text{vapor pressure saturated steam in mmHg})T \times \text{desired\% RH}$$

where 58.93 mmHg = 117.85 × 50% RH; *T*, temperature of sterilization process.

Table 4 Comparison of Saturation Moisture Levels and 50% RH

Temperature of air °C	Vapor pressure of saturated moisture (steam) 100% RH (mmHg)	Vapor pressure of moisture at 50% RH (mmHg)
25	23.76	11.88
44	68.26	34.13
50	92.51	46.25
54	112.51	56.25
58	136.08	68.04
64	179.31	89.65

Source: From Ref. 18.

Direct Method

The direct method of assessing moisture uses analytical instrumentation such as an electronic hygrometer, IR analyzer, or GC. Electronic hygrometers can be used inside the vessel. The IR and GC systems require that samples be withdrawn from the sterilizer. When removing samples from the chamber, care must be exercised to assure that sample lines are properly insulated and heated so that the moisture does not condense in the sample lines. Reducing the pressure in the lines is important when using an analytical system that performs only at ambient pressure. If rapid changes in pressure occur, moisture will also be lost from the sample through condensation. These sample lines are usually closed loop-circulating systems. Calibration of these systems is performed with either a saturated water vapor standard or saturated salt solutions that yield very specific headspace water vapor concentration (Table 5).

Moisturization of Load

Product moisturization or humidification generally occur using two distinct operations. Pallets of product may be placed in preconditioning chambers that are typically at ambient pressure. Temperature and moisture are maintained. The pallets are placed into these chambers for a specified time to adequately provide the required moisturization permitting effective sterilization of the load (16,19,20).

The second phase of moisturization occurs within the sterilizer. This phase is generally performed under vacuum. Steam is added to the chamber to create a humid environment for the load.

External Preconditioning of Product Load

A typical controlled preconditioning room will operate at 40°C and a RH of 60%. We find that at 40°C air can hold a pressure of 55.82 mmHg of water vapor (Appendix I). Sixty percent of that value is 60% RH or 33.49 mmHg. If the products are completely equilibrated to that environment, we will have 33.49 mmHg vapor pressure of moisture. Most of the moisture added to the products is done in an external preconditioning room. When product is placed into a sterilizer, operating at 54°C, the air in the sterilizer can hold 112.51 mmHg of moisture (i.e., 100% RH). There is only 33.49 mmHg available, so the RH will drop. The 33.49 mmHg is 29% of the moisture that the air can hold at 54°C. This RH level is dangerously low and as a result the lethality rate produced by EO on the spore may decrease. Re-moisturization of the load in the chamber is thus indicated to ensure that adequate humidity levels are attained.

Table 5 Constant Humidity

Solid phase	T°C	% Humidity	Aqueous tension (mmHg)
$LiCl \cdot H_2O$	20	15	2.60
$CaCl_2 \cdot 6H_2O$	20	32.3	5.61
KNO_2	20	45	7.81
$Na_2Cr_2O_7 \cdot 2H_2O$	20	52	9.03
$NaNO_2$	20	66	11.5
NH_4Cl and KNO_3	20	72.6	12.6
NH_4Cl	20	79.5	13.8
$KHSO_4$	20	86	14.9
K_2HPO_4	20	92	16.0
$CuSO_4 \cdot 5H_2O$	20	98	17.0

The % humidity and the aqueous tension at the given temperature within a closed space when an excess of the substance indicated is in contact with a saturated aqueous solution of the given solid phase.
Source: From Ref. 18.

Moisturization Inside the Sterilizer

SAC Cycle. The SAC cycle is most commonly used in sterilizer moisturization. The SAC cycle employs a pre-vacuum phase at the beginning of the cycle and is held static for a specific period of time. During this vacuum hold, moisture is admitted into the sterilizer in the form of steam. This static hold is commonly referred to as a humidity dwell time. The steam vapor moisturizes or humidifies the product to be sterilized during this dwell period. This dwell process is not efficient and takes many hours to moisturize the product adequately enough for sterilization. This process works most efficiently with goods that have been adequately preconditioned at high relative humidities, prior to entering the sterilizer. The SAC cycle effectively replaces the moisture removed during the evacuation process.

Electronic hygrometers placed inside the vessel actually sense the moisture level inside the sterilizer environment and control the desired level as well.

When low humidity levels are sensed, a steam valve is opened and more steam is emitted into the sterilizer. The hygrometer has a minimum and maximum set point. When the high-level reading is indicated, the steam valve is turned off. The humidity can then be controlled when a drop in environmental moisture occurs which reflects moisture absorption by the load. If this type of control is selected, it is always used in the cycle phases prior to the introduction of EO gas. Since some hygrometers are sensitive to EO gas, the control can only be used prior to the introduction of gas. Systems that are compatible with EO should be used throughout all cycle phases.

DEC Cycle. The DEC cycle does not measure or control actual humidity levels. It relies on pressure controls and temperature controls that indirectly control moisture levels. Large amounts of steam are used and the thermal shock to the hygrometer may render it incompatible with this process. If a hygrometer is compatible it would undoubtedly read saturated or 100% RH after the initial steam pulse because of the condensation of steam on the cool hygrometer. As the hygrometer heats up during the cycle, it will begin to give readings less than saturation. The accuracy of hygrometer readings following this saturated condition should be checked by subsequent calibration.

Humidity control in the DEC cycle is built around the laws of physics regarding temperature and pressure of saturated steam of the cycle design. The steam pulses purge the air from the chamber and goods to be sterilized. A temperature control system measures the temperature of the steam condensate. This condensate is indicative of the temperature within the sterilizing chamber. Therefore, when the product in the chamber has been heated to the steam vapor temperature, it has been moisturized by the steam condensate. At this phase of the cycle the goods are at temperature. The next phase of the cycle removes any excess moisture. Following this steam pulsing phase, EO gas is charged into the sterilizer and a resultant rise in temperature occurs as a result of the rise in pressure in the chamber. This rise in temperature is due to the compression of the steam vapor and causes the condensed moisture to vaporize, thus slightly drying the product.

The DEC cycle is an extremely effective cycle in moisturizing product to be sterilized. The DEC cycle also employs controlled pressure levels based on subatmospheric saturated steam conditions. Humidification and heating, therefore, occur simultaneously. The DEC cycle requires an extremely large vacuum pump for the chamber size and is more popular in smaller sterilizers although it is also used on larger industrial sterilizers.

RH Monitoring in the Worst Case Location in the Product. RH sensors, used to monitor the chamber environment, can also be used to monitor the environment within the package or within the master carton and, in some cases, even within the product. Again, because this is an electronic reading, it does not have the same impact on the parameter that the removal of a gas sample has. However, the humidity element may indeed have an impact on the temperature within that particular environment in the load. Electronic systems may have a mass that has a sufficient heat capacity to cause the water vapor to condense on the sensor, therefore impacting the true environmental conditions within the product. Again, the humidity penetration should be measured at numerous locations within the sterilizer. Specifications should be prepared that detail these specific parameter tolerances. Unfortunately, the degree of biological effectiveness can only be assumed since conditions within the product are not generally measured or monitored.

Temperature

General

Temperature is one of those parameters whose measurement seems quite straight forward. We understand heat and the physics of heat transfer. We use TCs and RTDs to measure temperature in many different pharmaceutical processes. In the case of EO sterilization, we have a very complex situation. The conventional limits on temperature are generally between 20°C (68°F) and 65°C (149°F). Most processes are run between 30°C (86°F) and 54°C (129.2°F). EO gas reactions with cellular molecules correspond to first-order kinetic reactions. First-order reactions are those that proceed at a rate exactly proportional to the concentration of one reactant. Temperature affects the rate of this reaction. An increase in temperature by 10°C (18°F) will approximately double the reaction rate with EO thus affecting sterilization times. This value is expressed as Q_{10}. In a recent publication we empirically derived a Q_{10} for this process as 2.05 (21). A basic doubling of the lethality of the process will occur with this 10°C rise in temperature. The lowest temperature limit is the temperature at which EO is converted from a gas to a liquid, which is 10.4°C (50.5°F). The upper limit is that temperature at which EO gas polymerizes rapidly rendering it biologically inactive.

The laws of physics have created an additional complication in the EO process. Temperature and RH are dependent on each other. If the amount of moisture is fixed, then as temperature rises the percentage RH decreases. The converse is also true.

The international standard for industrial EO sterilization processes (ISO 11135) (14) permits a temperature variance of 10°C in a sterilizer load. The authors believe the temperature spread, if it did exist, would not allow a proper validation program to be performed. This process limit is too large to provide acceptable process controls. The 10°C process temperature variance is very risky with respect to RH conditions. There seems to be a lack of understanding regarding the relationship between temperature and humidity. As stated earlier, RH is the amount of moisture in the air relative to the amount of moisture the air is capable of holding at a specified temperature. Table 4 compares saturated conditions (100% RH) to 50% RH.

Process operating conditions at 54°C and 60% RH are common. At 54°C/60% RH, the vapor pressure of moisture is 67.50 mmHg (Appendix I). If the temperature were to drop to 44°C lower limit which would be allowed by ISO 11135, the moisture level is dangerously close to the dew point (100% RH). The dew point at 44°C (100% RH) is 68.26 mmHg.

Comparing the acceptable high limit allowed by the standard of +10°C or 64°C the vapor pressure (100% RH) is 179.31 mmHg. The moisture level of 67.50 mmHg now drops to 31% which is dangerously low. The rate of lethality decreases significantly below 50%. This change in lethal rate can yield a positive BI spore challenge when it would be expected to be killed.

Temperature increases that lower the RH below 50% are where the real problems occur. The majority of EO sterilization failures are due to insufficient humidification. Thirty years ago it was believed that successful EO sterilization was impossible in the winter months, low ambient humidity, in the northern climates of the U.S.A. and Canada (16,19,22,23). Product warehouses were cold and humidity was very low. Temperature was easily corrected, but humidity was more difficult. The advent of the use of "preconditioning" in rooms external to the sterilizer helped to significantly reduce this problem. Nevertheless, even today "validated" EO cycles yield more positive BIs in the winter or low humidity months than at other times of the year.

Temperature Instruments and Controllers

Process controllers are either the TC-type controllers or RTD controllers. They are compatible with temperature ranges within the sterilization process and give accurate and reliable information. Temperature is controlled primarily by using a jacket around the sterilizer. This jacket may be heated with a hot water/ethylene glycol mixture, or it may be steam heated. The water/glycol mixture operates within a narrower temperature range than steam-heated jackets. Steam heating may be either an atmospheric condition or a subatmospheric condition. The atmospheric steam jackets give the widest spread of temperature, while subatmospheric jackets give the narrowest spread.

The location of the temperature control sensor is much less critical in a glycol-jacketed system. A few degrees of temperature range are generally noted in the glycol system. The temperature controller may even be located outside the sterilization chamber in a glycol recirculating line that heats the jacket.

Steam-heated jackets, however, are usually monitored within the sterilizing chamber. Placement of the control probe is extremely critical to overall temperature control within the chamber because of the hysteresis effect of the controller. There can also be temperature excursions because the controller calls for heat and puts in excess steam into the jacket due to the thermal lag of the chamber mass.

Temperature Monitoring within the Product

TCs may be very small wires and can actually be mounted into the product. Since the TC is an electronic reading coming from the product, it can be placed well into the product to indicate temperature heat-up of a particular surface or environment within the product. Again, the number of TCs will depend on the complexity of the product and the complexity of the loading configuration. ISO 11135 (14) states that no less than 10 TCs should be used in the chamber. As a general rule, no less than 1% of products should be monitored with TCs when mapping the temperature distribution within the load. There are graphic programs available to use the TC data to provide a lethality map of the temperature distribution within the load (24).

Time

General

The sterilization process time is related to: (*i*) moisturization level; (*ii*) EO gas concentration; and (*iii*) temperature. Process time must account for penetration of these critical elements into the worst case or least lethal locations in the product, load, and packaging barriers around the product. Time must be expressed as "Equivalent Process Time" not clock time. This equivalent process time must integrate the lag factors including the come up to exposure as well as the exhaust time effects on total process lethality (25–27). The selection of the best process parameters will result in adequate EO sterilization process times of less than two hours (28). Process times for manufacturing components that are relatively easy to sterilize may even be less than one hour. With palletized loads these times may exceed 8 to 12 hours.

Establishing Process Equivalent Time

Simple clock time is insufficient as a critical process parameter. Time should be expressed as equivalent process time. This equivalent process time must integrate the lethality delivered during gas charge (come up) exposure (hold time) and exhaust (come down) (26).

Equivalent process time directly relates to process lethality. The dynamic conditions that exist in the dynamic phases of the sterilization process (come up and come-down times) deliver lethality that must be accounted for. Mathematical equations have been described which allow the integration of equivalent lethality similar to the *F*-value concept used in steam sterilization (21,26).

The following classically accepted formulas applied to microbial resistance studies are applied here to calculate the lethality equivalent process time (26):

$$D\text{-value} = \frac{U_f}{\text{Log } N_0 - \text{Log } N_f} \tag{18}$$

where U_f is equivalent exposure time, N_0 is the initial spore population, and N_f is the final spore population.

$$\text{SLR}_{\text{FULL PROCESS}} = \frac{U_{\text{FULL PROCESS}}}{D} \tag{19}$$

where SLR is the spore log reduction and $U_{\text{FULL PROCESS}}$ is the equivalent time for the full sterilization process.

The SAL can then be calculated from these values.

$$\text{SAL} = 10^{\text{Log } N_o - \text{SLR}} \tag{20}$$

Calculating Equivalent Time

Annexes to ISO standard 11135 (14) identify methods for calculating process *D*-values, which represent the dose or time at steady state required to reduce a microbial population by 90% or 1 $\log_{10^*}$ (16,29,30). Unfortunately, the document provides little guidance to assist users in actually estimating the equivalent time (*U*) required for such calculations. In the extreme, use of the actual exposure time (which begins after steady-state pressure has been achieved) rather than equivalent time may lead to a gross underestimation of a process's *D*-value and concomitant overestimation of the SLR and an underestimation of the SAL. It is even questionable whether a true steady-state condition ever exists in densely palletized loads. Whenever equivalent time is underestimated for *D*-value calculations, the result will be the same. Figure 4 demonstrates the relationship between *D*-value, SLR, and SAL at steady state when microbial inactivation follows a straight-line log-linear relationship (26).

EO process *D*-value calculations have been used primarily in BIER vessel studies, where the time to steady state approaches zero and the equivalent exposure time approaches the actual exposure time (31,32). However, applying any *D*-value calculation method to EO systems used for actual production sterilization is inappropriate because standard process chambers do not produce square wave cycles and substantial lethality is generated during both their charge (gas injection) phase and gas evacuation phase (which do not fall within the exposure time). This situation accounts for the popularity of the AAMI overkill validation technique and the equivalent ISO half-cycle method, neither of which require calculations of *D*-value, SLR, or SAL (14).

If the actual exposure time is used in equation (18) rather than equivalent exposure time, then as the exposure time approaches zero when Log $N_0 -$ Log N_f is some positive number, then the *D*-value also approaches zero; subsequently, SLR approaches infinity and SAL approaches $10^{-\infty}$. While no one would suggest that a *D*-value would equal zero, this extreme example demonstrates the dangers of underestimating the equivalent time. Adding some arbitrary number to increase *U* does not provide the necessary information to determine *U* and, therefore, the *D*-value, will be underestimated. Overestimating the equivalent time for the full process will similarly result in an overestimation of SLR and an underestimation of SAL, although the percentage error will generally be less than when *U* is underestimated for *D*-value calculations, because the latter error is multiplicative.

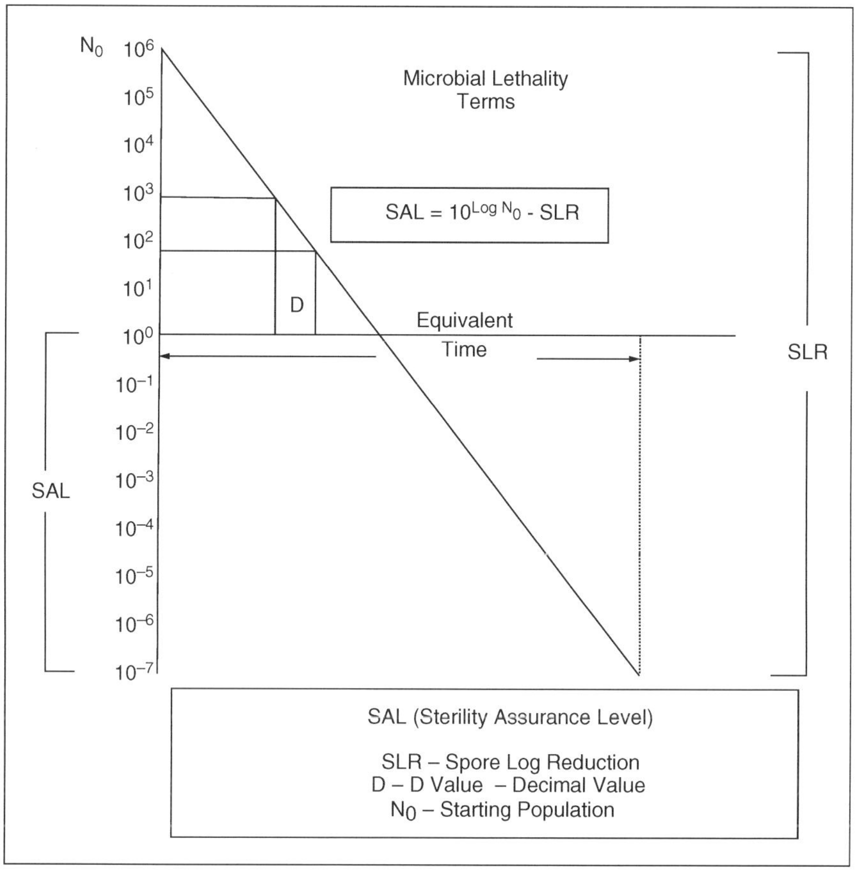

Figure 4 Microbial lethality terms. *Source*: From Ref. 26.

Process Lethality Variations Based on Product Differences. Table 6 represents results from sterilization validations conducted for a variety of medical products (26). These validations used an exposure time of zero minute, yet resulted in few or no positive BIs, which is not surprising if one understands the concept of accumulated lethality. The table also includes estimated equivalent times for these zero-minute exposures, related *D*-values, and full process-cycle SALs. The EO sterilization cycle being validated in most of this testing is depicted in Figure 5.

For this process, the lethality attributed to EO begins with the injection of the gas into the process chamber. Whether pure EO is used, as in the process shown, or a gas mixture (such as or EO/HCFC or EO/CO_2 diluent), lethality increases as the concentration increases, and the concentration increase is proportional to the pressure rise in the chamber (33). For processes with well-controlled pressure ramp-up rates, EO concentration changes also are proportional to time during gas injection and evacuation (exhaust). A cycle's exposure-time phase starts when the control pressure has been achieved, which occurs after gas injection is completed. In practice, absorption, microenvironments, diffusion, and chemical reactions that consume the gas slow the development of steady-state EO concentrations.

The cycle illustrated in Figure 5 indicates that the EO gas injection time is 11 minutes and the exhaust time is 16 minutes, which are common times in EO processing. An 11-minute nitrogen (N_2) overlay immediately follows the EO injection phase; hence EO concentration is at its maximum during that period. Data such as these can be converted to equivalent time for *D*-value, SLR, and SAL calculations using the mathematical model described below (21). The technique is based on lethality rate (L_R), which can be expressed either as a rate function with units of Δ log *N* per minute at specified conditions or as the reciprocal of the *D*-value.

Numerous investigators have shown that microbial *D*-values decline as EO concentration increases (3,22,23,34,35).

Comparative *D*-values are listed in Table 7. Test results are also graphed on a $\log_{10}$/linear plot in Figure 6, which indicates there were reasonable straight-line fits with R^2 values of 0.9695 and 0.9909 for spore strips and the self-contained test, respectively. However, Figure 7, which is a linear/linear plot of both *D*-values and lethality versus EO concentration for spore strips, depicts a more useful relationship.

The Microsoft Excel program for the best fit of data predicts that, when plotted against EO concentration (*C*), the *D*-value predicts a parabolic curve. As *C* approaches zero, then *D* will approach infinity. Logically it follows that EO-associated lethality ($1/D$) must approach zero as *C* approaches zero, creating an intersection on the lethality rate plot at $x=0$, $y=0$; where $D=1/L_R$ approaches zero, then *D* approaches infinity, which is also predicted by the plot of *D*, which is asymptotic in both directions, or hyperbolic. Thus a linear/linear plot of the lethality rate allows a simple approach to calculating equivalent process time if temperature is considered to be constant:

$$L_R \sim C, \quad \text{or} \quad L_R = kC \tag{21}$$

Table 6 Comparison of Different Equivalent Process Times and *D*-Values for Various Products

Product type	Positive BIs/ total BIs	Calculated *U* (min)	Calculated *D*-Value	Calculated full-cycle SAL	Full-cycle process exposure time (hr)
Introducer, delivery, forceps, catheter	1/20 Th[a]	24.65	3.34 Th	1×10^{-66} Th	4
	1/20 SC[b]		3.38 SC	1×10^{-66} SC	
Occluder delivery system	6/20 Th	24.15	3.75 Th	$<1\times10^{-87}$ Th	5
	10/20 SC		3.92 SC	$<1\times10^{-72}$ SC	
Tubing sets and scopes	5/20 Th	24.15	3.69	$<1\times10^{-72}$	4
Cannula	2/20 strips[c]	36.9	5.29	1×10^{-74}	4
Catheters, introducers	22/44 Th	47.75	7.56	1×10^{-32}	2.5
Rotor blade	17/20 Mps[d]	24.95	4.16	$<1\times10^{-51}$	4
Suture anchor	17/20 Mps	25.25	4.00	1×10^{-54}	4
Compass tips and magnets	15/20 Th	25.7	4.07	1×10^{-52}	4
Clamp covers, loops, brush, boots	3/20	24.65	3.63	1×10^{-60}	4
Optical fiber	0/20 SC	24.15	<3.31	$<1\times10^{-74}$	4
Sensor, probe, wire, etc.	0/20 SC	24.65	<3.38	$<1\times10^{-65}$	4
Orthopedic implant product line including bone-harvesting device	0/80 strips	44.9	<6.18 IP	$<1\times10^{-33}$ IP	4
	0/80 SC				
	5/40 IP[e]				
Unassembled bone-harvesting device	0/20 strips	45.15	<6.19 strips	$<1\times10^{-32}$	4
	0/20 SC				
Injectable polymer system	1/19 strips	24.85	4.09	1×10^{-52}	4

[a] Th = 1.5 in. single-strand cotton thread inoculated with $>1\times10^6$ *Bacillus subtilis* (SGM Biotech).
[b] SC = Self-contained test, $>1\times10^6$ *B. subtilis* (SGM Biotech).
[c] Strips = Paper strips, $>1\times10^6$ *B. subtilis* (SGM Biotech).
[d] Mps = Mini paper strip, 2×10 mm, $>1\times10^6$ *B. subtilis* (SGM Biotech).
[e] IP = Inoculated product from a spore suspension.

where k is the rate constant. The equation can also be expressed

$$\frac{L_{R_1}}{C_1} = k \quad \text{or} \quad \frac{L_{R_1}}{C_1} = \frac{L_{R_2}}{C_2} \tag{22}$$

and so on. Solving for L_R

$$L_{R_2} = L_{R_1}\left(\frac{C_2}{C_1}\right) \tag{23}$$

Since the D-value is a reciprocal of the lethality rate, the equation can also be used to solve for D:

$$\frac{1}{D_2} = \frac{C_2}{C_1 D_1} \tag{24}$$

which simplifies to:

$$D_2 = \frac{C_1 D_1}{C_2} \tag{25}$$

L_R may also be used to derive Δ Log N as a function of time (t) where:

$$L_R = \frac{(\text{Log } N_0 - \text{Log } N_f)}{\Delta t} = kC \tag{26}$$

$$\text{Log } N_0 - \text{Log } N_f = kC\Delta t \tag{27}$$

Calculation of accumulated lethality at a constant temperature (T_1) requires each increment to be multiplied

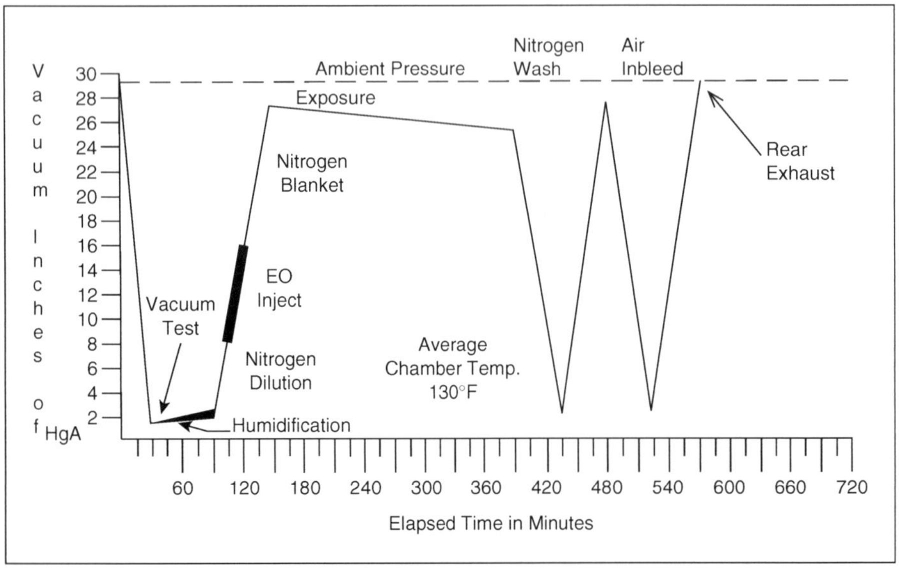

Figure 5 Typical 100% ethylene oxide sterilization cycle with nitrogen overlay. *Source*: From Ref. 26.

Table 7 Comparative *D*-Values at Four EO Concentrations Calculated Using the Holcomb–Spearmen–Karber Method

			EO concentration (mg/L)			
Lot no.	Log spore population	Biological indicator type	300	450	600	750
			D-value			
G-92P	6.531	Self-contained test	5.8	4.2	3.6	2.8
G-103P	6.322		5.6	4.2	3.2	2.8
G-105	6.255		5.2	4.0	3.2	2.6
Average	NA		5.5	4.1	3.3	2.7
BSUB-235	6.398	Paper strips	6.7	4.3	3.5	2.9
BSUB-244P	7.0		6.2	4.4	3.4	2.8
BSUB-249P	6.398		6.1	4.1	3.4	2.8
Average	NA		6.3	4.3	3.4	2.8

These test results are also shown graphically in Figure 6.
Source: From Ref. 26.

by the time at that increment, which is expressed in the summation formula:

$$\text{SLR} = \sum_{i=1}^{n} L_{R_1} = \sum_{i=1}^{n} t_{T_1}\left(\frac{C_i}{C_{\text{ref}}}\right) L_{\text{ref}} \quad (28)$$

The effect of temperature variations on *D*-values is known as the *Z*-value. This effect has been described for steam and dry-heat applications (25). A number of references in the past have indicated a similar correlation in EO sterilization as in steam and dry heat of a *Z*-value. Ernst (19) reported a theoretical lower limit of $Q_{10}=1.8$ for EO sterilization, but a consensus seems to have evolved for a nominal Q_{10} value of 2. (This means that a 10°C change would affect lethality by a factor of 2.) Thus a Q_{10} value of 2 was used for a set of temperature-related tests along with a *Z*-value of 33.2°C, which was calculated using the relationship $Z=10°\text{C}/\log_{10}Q$. This value was intermediate between a recently suggested *Z*-value of 36°C and an older recommendation of 29.4°C (23,33). Test results of Mosley, Gillis, and Krushefski (21) indicate that the best choice of *Z* to fit the experimental data is 32°C, which is essentially the result for a Q_{10} value of 2.05 and very close to the calculated values of ~29°C suggested in earlier studies (26,27,33).

Because they are independent variables, a reference EO concentration (C_{ref}) and temperature (T_{ref}) can be used to calculate the equivalent process time for various temperatures as follows:

$$U_{C_{\text{ref}},T_{\text{ref}}} = [\text{antilog}(\text{Log}\, t_{T_{\text{ref}}})]\frac{C}{C_{\text{ref}}} \quad (29)$$

where

$$\text{Log}\, t_{T_{\text{ref}}} = \text{Log}\, t_T + \frac{(T - T_{\text{ref}})}{Z} \quad (30)$$

For example, using $Z=29°\text{C}$, if the exposure time (t) is 40 minutes, the temperature (T) is 40°C, and the concentration (C) is 300 mg/L, the equivalent process time at $C_{\text{ref}}=600$ mg/L and $T_{\text{ref}}=50°\text{C}$ is 9 minutes:

$$U_{600\ \text{mg/L},50°\text{C}} = \left\{\text{antilog}\left[\text{Log}\, 40 + \frac{1}{29}(40-50)\right]\right\}\frac{300}{600} = 9 \quad (31)$$

In addition, because $D \sim U$, the above equation also can be used to address *D*-value:

$$D_{C_{\text{ref}},T_{\text{ref}}} = \left\{[\text{antilog}(\text{Log}\, D)] + \frac{(T - T_{\text{ref}})}{Z}\right\}\left(\frac{C}{C_{\text{ref}}}\right) \quad (32)$$

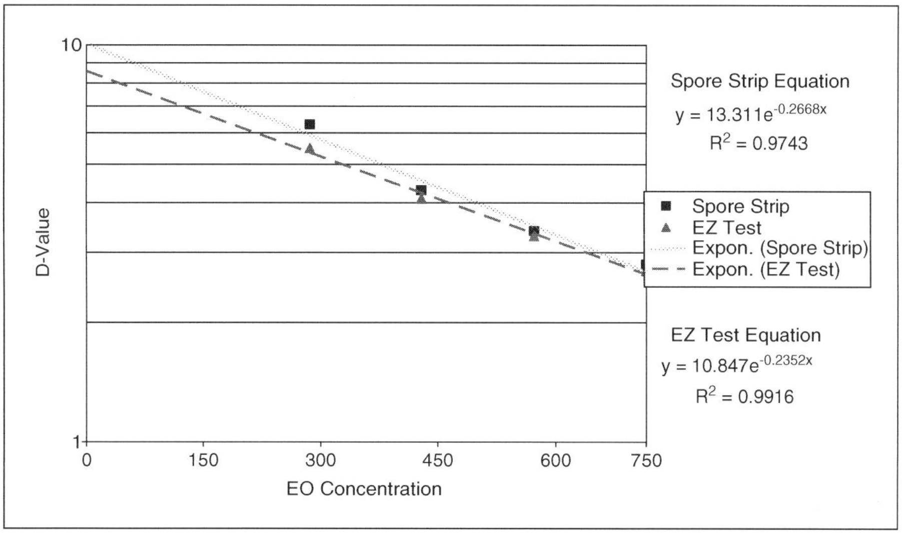

Figure 6 Log_{10}/linear plot of *D*-values versus ethylene oxide. *Source*: From Ref. 26.

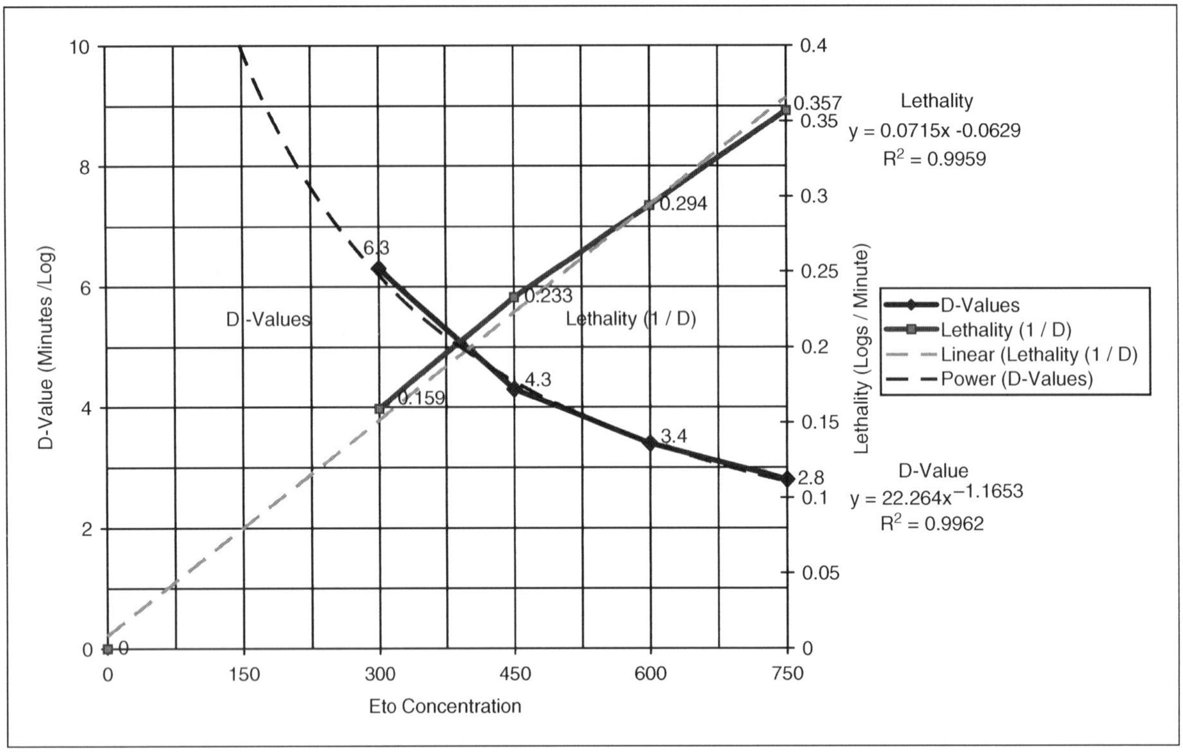

Figure 7 *D*-value and lethality versus ethylene oxide concentration. *Source*: From Ref. 26.

To determine accumulated equivalent process time where conditions are changing for EO concentration and/or temperature, a summation equation can be applied:

$$U = \sum_{i=1}^{n} U_i = \sum_{i=1}^{n} \left\{ \text{antilog} \left[\text{Log}\, t_T + \frac{1}{Z}(T_i - T_{\text{ref}}) \right] \right\} \frac{C_i}{C_{\text{ref}}} \quad (33)$$

The empirical *D*-value results along with the *D*-values that were calculated from BIER conditions of 54°C, 600 mg/L EO, and 60% RH closely agree.

The *Z*-value is the number of degrees of temperature change required to change the *D*-value by 90% or one $\log_{10}$ cycle. The *Z*-value is not an indicator of the rate of microbial lethality, but rather it is a measurement of the rate of change of microbial lethality with respect to temperature. The *Z*-value is therefore a necessary element in the ability to mathematically express equivalent process time *U*.

$$U = \sum_{i=1}^{n} U_i = \sum_{i=1}^{n} \left\{ \text{antilog} \left[\text{Log}\, t_T + \frac{1}{z}(T_i - T_{\text{ref}}) \right] \right\} \frac{C_i}{C_{\text{ref}}} \quad (34)$$

The *Z*-value is not linear over a wide range for process temperatures. Casolari (36) has concluded that the linearity of the *Z*-value is theoretically impossible since, in accordance with the Arrhenius relationship "...*Z*-value can not be regarded as being constant, but varies with temperature..." The consistency of *Z*-values obtained by plotting $\text{Log}_{10}\, D_T$ against temperature is difficult to ascertain in practice, as the evaluation of *D* is not significantly accurate. Several publications reinforce this last assertion of Casolari, particularly at high steam temperatures >132°C (32) where lag factors ensure that *D*-values cannot be accurately determined. It is interesting to note that inaccuracy in *D* due to lag factors at high temperatures was first reported in 1921 by Bigelow (37).

The *Z* is linear over limited temperature ranges and can be appropriately applied to the integration of process lethality. According to the data in Tables 8 and 9 and plotted in Figure 8, the *Z* is quite linear between 40°C and 60°C and EO gas concentration between 300 and 750 mg/L. These limits cover the majority of the commercial EO processes in use today.

These data support the approach that integrated process lethality can be applied to EO sterilization with as much confidence as can be applied to steam processes as long as critical parameters are appropriately controlled. The selection of a universal *Z*-value for EO sterilization appears to be 32°C comparable to the widely accepted value of 10°C for steam. Arguments could be made for *Z*-values ranging from 29°C (21,27,33) to 36°C (23). This range represents a relatively small change in reaction rates when compared to the accepted *Z*-values for steam processes.

BIOLOGICAL MONITORING

Biological Release of Product

Biological monitoring of the sterilization process uses calibrated bacterial spores. The bacterial spores most commonly used are *Bacillus atrophaeus* (28). The *B. atrophaeus* spores are very resistant to the EO sterilization process. These spores are usually placed on a carrier substrate that allows them to be conveniently placed inside product samples (20). The location of choice is the position in the product that is worst-case or least-lethal location. The inoculated product samples

Table 8 *D*-Values of Six New Test Organisms to Various EO Exposure Conditions

Species	EO concentrations (mg/L)	40°C	54°C	60°C
Bacillus atrophaeus ATCC #9372	300	18.11	6.37	4.44
	450		4.30	
	600		3.39	
	750	8.33	2.84	1.94
Bacillus Subtilis "5230" ATCC #35021	300	15.76	6.30	4.44
	450		4.96	
	600		3.98	
	750		3.51	
Bacillus pumilus ATCC #27142	300	13.36	5.40	3.95
	450		4.09	
	600		3.33	
	750	8.29	2.47	1.70
B. subtilis DSM #4181	300	9.26	4.18	3.24
	450		3.11	
	600		2.45	
	750	5.05	2.16	1.50
Bacillus smithii (formerly *coagulans*) ATCC #51232	300	7.69	3.35	2.21
	450		2.55	
	600		2.09	
	750	4.38	1.80	1.19
Geobacillus stearothermophilus ATCC #7953	300	4.09	1.55	1.25
	450		1.11	
	600		0.82	
	750	1.99	0.67	0.56

Source: From Ref. 21.

are then packaged in a similar manner as the product. The samples are placed in positions in the load that also have been identified as worst-case or least-lethal location.

BI systems have been developed using paper strips containing spores. These convenient carriers are placed into least-lethal locations in the product (Figs. 9 and 10). The paper strips may also be packaged in bio-barrier envelopes. Some BIs are packaged in self-contained culture systems. These systems are used in the same manner (Fig. 11). Placement will depend largely on the configuration of the product and package. Sometimes they will not physically fit into the device and must be placed inside the package with the product.

Following the sterilization process, these monitoring systems are removed from the sterilizer and cultured in the laboratory. The *U.S. Pharmacopoeia* recommends culturing in soybean casein digest medium at a temperature of 30°C to 35°C for seven days. Specific culture recommendations may be supplied by the manufacturer of the monitoring system. Some monitoring systems have been challenged using the FDA Reduced Incubation Time protocol with the resulting incubation times of 48 to 72 hours (38).

Manufacturers who intend to run multiple products in the sterilizer load will attempt to define a "master" BI/product combination. Studies must be conducted to demonstrate that the BI/product combination is more resistant when compared to other combinations. For instance, if one has determined that each BI/product combination is a reasonable simulation and that it can be scientifically defended, then it does not matter that they are not directly comparable. The biological challenge becomes the BI/product combination. The type of BI cannot be changed without producing somewhat unpredictable changes in relative resistance. If it has been determined that the BI/product A is the most difficult challenge and a BI strip in glassine is used, then it would be expected that if a second BI lot with a higher *D*-value, of the same type from the same manufacturer was used, then the BI/product combination would yield a higher process *D*-value. However, if one had decided to use direct product inoculation from a liquid suspension as a replacement for the BI strip, the relative results could not

Table 9 *Z*-Values for Six New Test Organisms at Two EO Concentrations

Species	EO concentrations (mg/L)	°C	Average	Mean ± 2 S.D. (±8%)	Mean ± 3 S.D. (±12%)
Bacillus subtilis DSN #4181	300	37.93	40.57	37.32–43.82	35.70–45.44
	750	43.20			
Bacillus subtilis "5230" ATCC #35021	300	37.44	36.79	33.85–39.73	32.38–41.20
	750	36.14			
Bacillus smithii (formerly *coagulans*) ATCC #51232	300	35.47	36.37	33.46–39.28	32.01–40.73
	750	37.26			
Geobacillus stearothermophilus ATCC #7953	300	34.94	36.34	33.43–39.25	31.98–40.70
	750	37.73			
Bacillus atrophaeus ATCC #9372	300	32.40	31.89		
	750	31.38			
Bacillus pumilus ATCC #27142	300	28.60	30.18	27.77–32.59	26.56–33.80
	750	31.76			

Source: From Ref. 21.

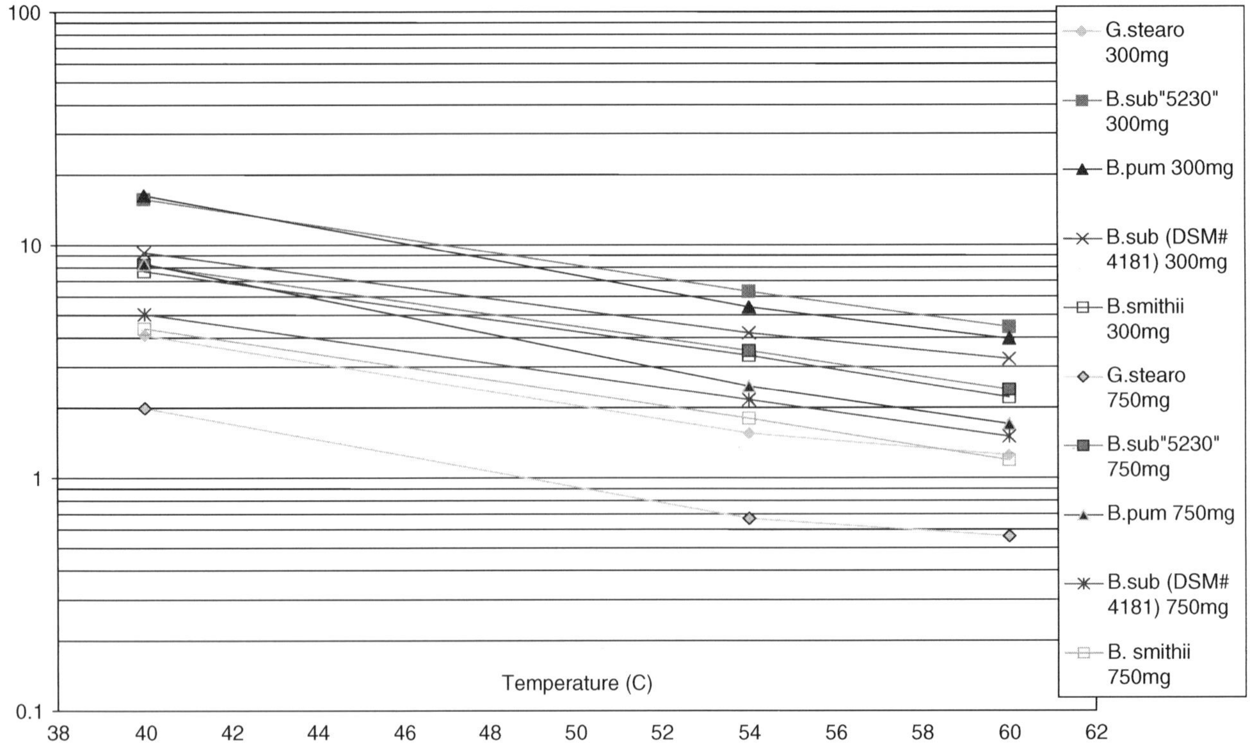

Figure 8 Multispecies composite of *Z*-values—300 and 750 mg/L ethylene oxide and temperatures from 40°C to 60°C. *Source*: From Ref. 21.

be predicted. The new combination could be more or less resistant than the original. The new BI/product A combination might not prove to be the most resistant compared to the other BI/product combinations used in the original study. It is important that the user understand what has been proven and what has not been proven in order that the information can be properly applied. Once a biological master product has been selected, the type of BI used in the BI/product combination cannot be changed without affecting the expected relative resistance. The lot or supplier of the original BI type could be changed, and the overall BI resistance in the supplied BI should create a similar shift in the resistance of the BI/product combination.

BIs are much more convenient than inoculated product or inoculated simulated products (20). A Process Validation program should include product sterility data as well as BI data. Routine process monitoring normally includes the use of BIs only. Normally, a minimum of 10 BIs are used for each sterilization cycle. For extremely large loads up to 1000 ft^3, as many as 30 BIs or more may be tested per cycle. This is dependent on the product application physical size and difficulty to sterilize. The BI data must be integrated into all aspects of the process control program to assure an adequate SAL.

The bacterial spore is the only monitor that can be embedded into the worst case–least lethal location in the product. It is also the only monitor that can integrate all critical process parameters to assess the effectiveness of the sterilization process.

Parametric Release of Product

Details of the current practices for parametric release will not be discussed in this chapter. Parametric release involves accepting or rejecting a load of product from a sterilization cycle based solely on a review of physical and chemical process parameter measurements for the cycle. Once the validation has been completed routine biological testing is not required. This approach has become popular due to potential faster turn around and lower routine sterilization costs. The incubation time for standard BIs has historically been seven days. Although

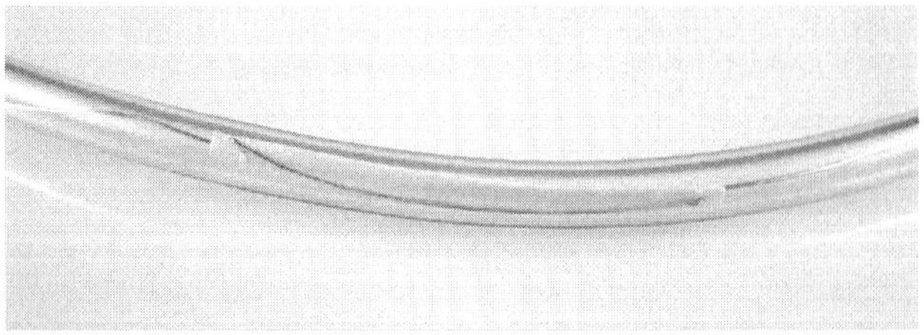

Figure 9 Spore strip biological indicator placed inside process tubing.

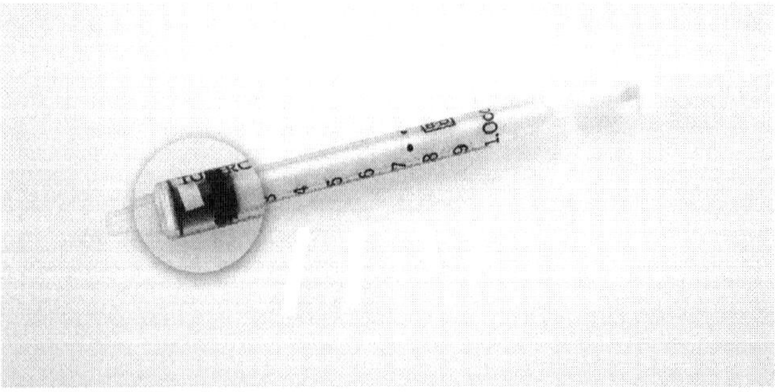

Figure 10 Spore strip biological indicator placed inside syringe.

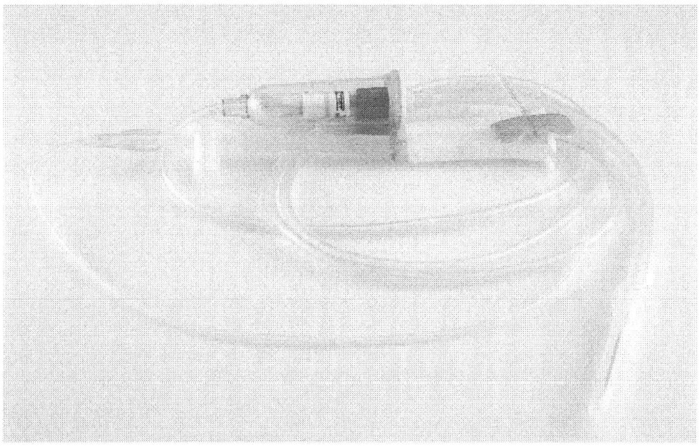

Figure 11 Self-contained ethylene oxide biological indicator placed inside the IV drip chamber of a drug administration set.

the use of BIs with reduced incubation time may reduce the seven-day quarantine time to 5, 3, 2 or less still, there is the cost of BIs and subsequent testing that can be eliminated by a parametric release approach. The cost of a proper validation for parametric release is often significantly more than that of a standard validation because it must be more robust. In addition, the greater amount of routine parametric data may increase review time and associated costs.

However, there are three flaws to the pragmatic implementation of parametric release. First, it is often implemented by companies because they have occasional problems with positive BI results from routine sterilization cycles. Parametric release has been implemented to avoid investigation costs and delays in product flow. This is bad practice and suggests inadequate "root cause" analysis. Since BIs can only detect catastrophic sterilization process failure, a true positive indicates a serious problem. Secondly, most BI positives from routine cycles occur during winter and early spring months, based on our experience. These are the cooler and drier months and suggest problems with material humidification not always detectable using current physical measurements. The complexity of the EO sterilization process should not be underestimated. The Oxborrow et al. (39) report on the AAMI round robin testing of BIER vessels demonstrated significant system bias from one BIER vessel and test lab to another. BIER vessels are designed to operate at control ranges far tighter than routine sterilization systems. However, in the study the unit producing the lowest lethality was 50% less effective than the one with the highest lethality. This suggests that the total variance from calibration, maintenance and routine control for physical measuring systems is greater than often claimed or believed. Total reliance on such controls in the light of empirical evidence seems not to be objectively sound.

Critical process parameters have been discussed extensively in this chapter. The authors know of no instruments that can be placed into the least lethal locations of products to provide meaningful parametric data. Until such instruments are developed, it seems prudent that biological challenge systems should be used to evaluate process delivered lethality.

EO TOXICITY

Residuals

Sterilant residuals and sterilant reaction products must also be considered in the Process Validation program. EO, being a toxic substance, will render a sterile product unusable if excessive amounts remain in the product after sterilization. The EO gas becomes trapped inside product voids. It is also absorbed and adsorbed by the product. Depending on the product material, it is generally easily removed (13). A common approach is to place the post-sterilized product in a heated aeration chamber with very frequent air changes. Ambient storage will also allow the EO gas to dissipate. There are two common EO reaction products that are also considered toxic. The EO gas reacts with chlorine to form ethylene chlorohydrin and with water to form ethylene glycol. The latter compound is much less toxic than the other two chemicals. These reaction products are not easy to remove from materials because their boiling points exceed 100°C. Therefore, it is important to minimize the formation of these reaction products. In the case of ethylene chlorohydrin, product and package materials with chlorinated compounds, such as sodium hypochlorite–bleached paper, are preferably avoided if EO gas is the sterilizing medium. Ethylene glycol formation is dependent on the amount of moisture that is actually present as water. The pH of this water will influence the rate at which the ethylene glycol is formed. The reaction is usually quite slow at neutral pH. The approach is to minimize the EO exposure time and to remove the humidity and EO gas after exposure by evacuation of the chamber and subsequent aeration.

Environmental Exposure

EO is a toxic and hazardous chemical. It is this characteristic that renders it an effective sterilizing agent. Controlling this chemical to minimize and prevent human exposure is an important consideration in the application of EO gas when used to sterilize materials in the pharmaceutical industry. The Occupational Safety and Health Act of 1970 emphasized the need for standards to protect the health and safety of workers (40). The NIOSH has disseminated information about the adverse effects of widely used chemical and physical agents, in an attempt to assist employers in providing protection to employees from exposure to these substances. NIOSH has taken the lead in disseminating information about EO toxicity.

The acute toxic effects of EO in humans and animals include: acute respiratory and eye irritation, skin sensitization, vomiting, and diarrhea.

Known chronic effects consist of respiratory irritation, secondary respiratory infection, and anemia. No definitive epidemiologic studies and no standard long-term study assays are available on which to assess the carcinogenic potential. Limited tests by skin application or subcutaneous injections in mice did not reveal carcinogenicity. However, the alkylating and mutagenic properties of EO are sufficient basis for concern about its potential as a carcinogenic agent. It has since been classified as a carcinogenic agent.

NIOSH is recommending that EO be considered as a carcinogenic agent for humans and that occupational exposure to it be minimized by eliminating all unnecessary and improper uses of EO. The Federal Register on April 21, 1984, proposed that the worker exposure limit be reduced from 50 to 1 ppm in the worker's environment, based on a TWA. This proposal was finalized on September 9, 1985 (Federal Register 50FR9800—March 12, 1985).

At the time of the proposal to reduce the level from 50 to 1 ppm, little scientific evidence existed to support the contention that 1 ppm was necessary to protect the environmental health of the workers. EO was later classified as a carcinogen and is regulated by OSHA Safety and Health Management Guidelines (Federal Register 54:3904–3916, January 26, 1989). When proper control measures are instituted, the escape of EO into the environment is virtually eliminated. These may include catalytic abator systems or acidified aqueous purge tanks that convert EO to ethylene glycol. Under such control, EO can be used as a gaseous sterilant in pharmaceutical facilities with little risk to the health of exposed workers.

Employee exposure is limited to one part EO per million parts of air (1 ppm) measured as an eight- hour TWA. Employee exposure may not exceed the short-term excursion limit of 5 ppm EO averaged over any 15-minute sampling period. These limits are called PELs.

Systems are typically designed to ensure that employees are protected when handling of products containing EO to ensure that the release of airborne concentrations of EO are at or below the standard action level of 0.5 ppm.

Workplaces are exempt from this standard when objective data shows that processing, use or handling of products containing EO cannot release airborne concentrations of EO at or above the action level or in excess of the excursion limit during normal conditions.

APPENDIX I

Example Calculation to Determine the EO Gas Concentration when Using the 10% EO, 27% HCFC 22, and 63% HCFC 124 Blend of Diluent and a Pressure Measurement in kPa

The EO mixture is 10% EO and 27% HCFC 22 and 63% HCFC 124. The pressure change in the sterilizer as a result of the gas charge is 176.98 kPa. The temperature at the end of the gas charge is 54°C.

$$C_{EO} = \frac{KP}{RT}$$

$K^{a} = 9.989$ mg/g mol, $P = 176.98$ kPa, $R^{b} = 8.312$, T K$=$ 54°C$+$273.2$=$327.2 K

$$C_{EO} = \frac{\left(9.989 \times 10^{3} \frac{\text{mg}}{\text{g mol}}\right) 176.98\ \text{kPa}}{\left(8.312 \frac{\text{kPa L}}{\text{g mol K}}\right) 327.2\ \text{K}} \tag{35}$$

[a] Refer to Table 3.

[b] Refer to Table 2.

$$C_{EO} = \frac{9989 \times 176.98}{8.312 \times 327.2} = \frac{1767853}{2719.7} = 650\ \text{mg/L} \tag{36}$$

APPENDIX II

Example Calculation to Determine the EO Gas Concentration when Using 100% EO and a Pressure Measurement In kPa

The pressure charge in the sterilizer is 36.64 kPa. The temperature at the end of the gas charge is 50°C.

$$C_{EO} = \frac{KP}{RT}$$

$K = 4.4 \times 10^{4}$, $P = 36.64$ kPa, $R = 8.312$, T K$=$50°C$+$273.2$=$ 323.2 K

$$C_{EO} = \frac{\left(4.4 \times 10^{3} \frac{\text{mg}}{\text{g mol}}\right) 36.64\ \text{kPa}}{\left(8.312 \frac{\text{kPa L}}{\text{g mol K}}\right) 323.2\ \text{K}} \tag{37}$$

$$C_{EO} = \frac{1611720}{2686.4} = 600\ \text{mg/L} \tag{38}$$

APPENDIX III

Example Calculation to Determine the EO Gas Concentration Using 100% EO and a Pressure Measurement In psia

The pressure change in the sterilizer is 5.13 psia. The temperature at the end of the gas charge is 125°F.

$$C_{EO} = \frac{KP}{RT}$$

$K^{a} = 4.4 \times 10^{4}$, $P = 5.13$ psia (must convert to atm), $R^{b} =$ 0.08205 (atm L)/(g mol K), T F$=$125°F (must convert to K)

$$P = \frac{5.13\ \text{psia}}{14.7\text{psia}} = 0.349\ \text{atm}$$

$$T\ \text{C} = \frac{(125 - 32)9}{5} = 51.7°\text{C}$$

$$T\ \text{K} = 51.7°\text{C} + 273.2 = 324.9\ \text{K}$$

$$C_{EO} = \frac{44000 \times 0.349}{0.08025 \times 324.9} = \frac{15356}{26.66} = 576\ \text{mg/L} \tag{39}$$

APPENDIX IV

EO Gas Concentration Determined by Weight of Gas Dispensed

The gas mixture is 10% EO, 27% HCFC 22, and 63% HCFC 124 percentage by weight. The sterilizing chamber is 100 ft^{3}. The sterilization process requires EO concentration of 475 mg/L. How many pounds of gas mixture must be dispensed?

- Sterilizer volume$=$100 ft^{3}$=$2832 L.
- The percentage of EO in each pound of mixture is 10%.

- The required EO is 475 mg/L.
  - Multiply the sterilizer chamber volume by the mg/L required to determine to total amount of EO required in mg.

$$2832\text{L} \times 475\ \text{mg/L} = 1{,}345{,}200\ \text{mg EO}$$

  - Divide the EO mg required by 454,000 mg/lb to determine the lbs of EO required.

$$\frac{1{,}345{,}200\ \text{mg}}{454{,}000\ \text{mg/lb}} = 2.963\ \text{lb of EO}$$

  - Divide the pounds of EO required by the percentage of EO per pound of mixture to determine the total weight of mixture to be added to the chamber.

$$\frac{2.963\ \text{pounds of EO}}{0.10\ \text{pounds of EO/pound of mixture}} = 29.63\ \text{lb of mixture} \quad (40)$$

REFERENCES

1. FDA, Department of Health and Human Services. Quality System Regulation. 21 CFR part 820 Medical Devices; Current Good Manufacturing Practice (CGMP); Final Rule. Federal Register, Monday, October 7, 1996.
2. Honeywell, 2002, Oxyfume® 2002 Sterilant Gas–Technical Information Sheet.
3. Gillis JR. Ethylene oxide sterilization. In: Carleton FJ, Agallaco JP, eds. Validation of Aseptic Pharmaceutical Processes. New York: Marcel Dekker, 1986.
4. Russell AD. The Destruction of Bacterial Spores. London: Academic Press Inc., 1982.
5. Chick H. An investigation of the laws of disinfection. J Hyg 1908; 8:92–158.
6. Rahn O. Biodynamica monograph No. 3. In: ELuyet BJ, ed. Injury and Death Of Bacteria By Chemical Agents. Normandy MO: Biodynamica, 1945. Republished by Pflug IJ, ed. Selected Papers on the Microbiology and Engineering of Sterilization Processes. 5th ed. Minneapolis, MN: Environmental Sterilization Laboratory, 1988:1–16.
7. Mosley GA. Microbial lethality: when it is log-linear and when it is not! Biomed Instrum Technol 2003; 36(6):451–4.
8. Winaro FG, Stumbo CR. Mode of action of ethylene oxide on spores of clostridium botulinum 62A. J Food Sci 1971; 35:892–5.
9. Lawley PD, Brookes P. Further studies on the alkylation of nucleic acids and their constituent nucleotides. Biochem J 1963; 89(1):127–38.
10. Lawley PD. Submolecular structure of the nucleic acids. Proc Chem Soc, London. 1957; 290.
11. Bhanot OS, Singh US, Solomon JJ. The role of 3-hydroxyethyldeoxyuridine in mutagenesis by ethylene oxide. J Biol Chem 1994; 269(47):30056–64.
12. ANSI/NCSL Z540-1-1994, July 2001, Calibration Laboratories and Measuring and Test Equipment—General Requirements.
13. Manning CR. Controlling EtO residues from the manufacture of medical products. Med Device Diagn Industry 1989; p. 136–144.
14. Medical Devices—Validation and Routine Control of Ethylene Oxide Sterilization, ISO 11135:1994. Geneva: International Organization for Standardization, 1994:11–6.
15. Philips RR. Gaseous sterilization. In: Lawrence C, Block SS, eds. Disinfection Sterilization and Preservation, 1st ed. Philadelphia, PA: Lea and Febiger, 1968.
16. Ernst RR. Ethylene oxide gaseous sterilization. In: Lawrence C, Block SS, eds. Disinfection Sterilization and Preservation, 2nd ed. Philadelphia, PA: Lea and Febiger, 1968.
17. Kaiser U, Heider D, Gömann J, Junghann U. Kill kinetics study of *Bacillus Subtilis* Spores in ethylene oxide sterilisation processes. Zentr Steril 2002; 10:163–7.
18. Weast RC, Selby SM, eds. Handbook of Chemistry and Physics. Cleveland, OH: The Chemical Rubber Co., 1968:E14.
19. Ernst RR. Ethylene oxide gaseous sterilization for industrial applications. In: Phillips GB, Miller WS, Durham NC, eds. Industrial Sterilization, International Symposium, Amsterdam 1972. Durham, U.K.: Duke University Press, 1973:181–208.
20. Gillis JR. Cycle development—microbial challenge systems. In: Industrial Ethylene Oxide Sterilization of Medical Devices—Process Design, Validation, Routine Sterilization, AAMI Technological Assessment Report No. 1–81, Arlington, VA, 1981:21–2.
21. Mosley GA, Gillis JR, Krushefski G. Evaluating the formulae for integrated lethality in ethylene oxide sterilization using six different endospore forming stains of bacteria, and comparisons of integrated lethality for ethylene oxide and steam systems. PDA J Pharm Sci Technol 2005; 51(1):64–86.
22. Bruch CW. Gaseous sterilization. Annu Rev Microbiol 1961; 61:245–62.
23. Joslyn LJ. Gaseous chemical sterilization. In: Block SS, ed. Disinfection, Sterilization and Preservation. Philadelphia, PA: Lippincott/Williams & Wilkins, 2001:337–59.
24. Mosley GA. Overcoming the complexities of instrument sterilization. Mater Manag in Health Care 2005; 14(11):26–8.
25. Pflug IJ. Microbiology and Engineering of Sterilization Processes. 11th ed. Minneapolis, MN: Environmental Sterilization Services, 2003.
26. Mosley GA, Gillis J, Whitbourne J. Calculating equivalent time for use in determining the lethality of EtO sterilization processes. Med Device Diagn Industry 2002; 24(2):54–63.
27. Rodriguez AC, Young B, Caulk K, Zelewski J, Dwasnico S. Calculating accumulated lethality and survivorship in EtO sterilization processes. Med Device Diagn Industry 2001; 23(9):100–7.
28. Kereluk K, Gammon RA, Lloyd RS. Microbiological aspects of ethylene oxide sterilization. Appl Microbiol 1970; 19:157–62.
29. Pflug IJ, Holcomb RG, Gomez MM. Thermal destruction of microorganisms. In: Blockin S, ed. Disinfection, Sterilization, and Preservation. Philadelphia, PA: Lippincott/Williams & Wilkins, 2001:79–129.
30. Holcomb RG, Pflug IJ. The Spearman–Karber method of analyzing quantal assay microbial destruction data. In: Pflug IJ, ed. Selected Papers on the Microbiology and Engineering of Sterilization Processes. 5th ed, Minneapolis: MN environmental Steilization Laboratory, 1988:83–100.
31. Association for the Advancement of Medical Instrumentation, Standard for BIER/EO Gas Vessels, 1982.
32. Mosley GA, Gillis JR. Operating precision of steam BIER vessels and the interactive effects of varying Z-values on the reproducibility of listed *D*-values. PDA J Pharm Sci Tech 2002; 56(6):318–31.
33. Lui TS, Howard GA, Stumbo CR. Dichlorodifluoromethane—ethylene oxide sterilization as a sterilant at elevated temperatures. Food Technol 1968; 22:86–9.
34. Burgess DJ, Reich RR. Industrial ethylene oxide sterilization. In: Morrissey RF, Phillips GB, eds. Sterilization Technology: A Practical Guide for Manufacturers and Users of Health Care Products. New York: Van Nostrand Reinhold, 1993:152–95.
35. Bruch CW. Ethylene oxide sterilization—technology and regulation. In: Industrial Ethylene Oxide Sterilization of

Medical Devices—Process Design, Validation, Routine Sterilization, AAMI Technological Assessment Report No. 1–81, Arlington, VA: AAMI, 1981:3–5.
36. Casolari A. Microbial death. In: Bazin MJ, Prosser JI, eds. Physiological Models in Microbiology. Vol. 2. Boca Raton, FL: CRC Press, 1988:1–44.
37. Bigelow WD. The logarithmic nature of thermal death time curves. J Infect Dis 1921; 29:528–36.
38. The Center for Devices and Radiological Health, FDA Guide for Validation of Biological Indicator Incubation Time, document control number 98984.
39. Oxborrow GS, Twohy CW, Demitrius CA. Determining the variability of BIER vessels for EO and steam. Med Device Diagn Industry 1990; 12(5):78–83.
40. OSHA Fact Sheet, Ethylene Oxide, Title 29 of the Code of Federal Regulations, (CFR) part 1910.1047.

17

Validation of Chlorine Dioxide Sterilization

Mark A. Czarneski and Paul Lorcheim
ClorDiSys Solutions, Inc., Lebanon, New Jersey, U.S.A.

INTRODUCTION

CD is a highly effective sterilizing agent that has many applications in the medical device and pharmaceutical industry including the sterilization of components as well as the medical devices themselves. It has key benefits and characteristics that make it extremely effective and well suited for the use in component and device sterilization. It is a true gas at normal use temperatures and therefore can penetrate into hard-to-reach areas such as that found in lumens or devices with complex geometries. It is efficacious at ambient temperatures so it is excellent for temperature-sensitive materials and devices. It has an yellowish-green color, which allows its concentration to be precisely monitored and controlled by a UV–Vis spectrophotometer. This ensures a very repeatable and robust cycle by providing tight process control from beginning to end.

CD is a single-electron-transfer oxidizing agent with a chlorine-like odor. This odor is the only similarity between CD and chlorine. CD can be generated in a variety of methods in liquid or gas. The following reaction can be used in the solid phase generation process:

$$Cl_{2(g)} + 2NaClO_{2(S)} \rightarrow 2ClO_{2(g)} + 2NaCl_{(S)}$$

CD chemical features can be found in Table 1.

Chlorine dioxide's method of inactivation is different than chlorine (oxidation vs. chlorination); it is far gentler on materials, and provides a highly controllable and reproducible process (Fig. 1). Additionally CD is well suited for sterilizing components and medical devices since it is compatible with many materials found in those components such as stainless steel, aluminum, glass, and most plastics. The rapid sterilizing activity of CD is present at relatively low gas concentrations of 1 to 30 mg/L compared to EtO and present at ambient temperatures compared to steam, EtO, and hydrogen peroxide systems.

Sterilization with CD follows the same general processing steps as with other gaseous sterilants such as EtO. Moisture is required for optimal lethal rate and effective sterilization of spores. Moisture preconditioning is performed at 60% to 75% RH. Gas is introduced to the desired concentration and held for a sufficient period of time to yield the required antimicrobial effect. The process is complete once the CD is removed.

Abbreviations used in this chapter: BI, biologic indicator; CD, chlorine dioxide; EtO, ethylene oxide; EPA, Environmental Protection Agency; HMI, human–machine interface; RH, relative humidity; SMC, Stumbo–Murphy–Cochran; STEL, short-time-exposure-limit; TWA, time-weighted average.

USE

CD has been utilized for almost 20 years as a gaseous sterilization agent. It has, however, a longer and very effective history of use in other industries.

CD is widely used as an antimicrobial agent in many industries. It is used to treat drinking water. In the food and beverage industry, it is used to treat poultry process water and to sanitize fruit and vegetables as well as equipment for food and beverage processing. It is used to decontaminate animal facilities. It is also employed in the healthcare industries to decontaminate rooms, pass-throughs, and isolators and also in various other aspects of the manufacturing process.

About 5% of large water treatment facilities (serving more than 100,000 people) in the U.S.A. use CD to treat drinking water. It is estimated that about 12 million people may be exposed in this way to CD. In communities that use CD to treat water for drinking uses, CD is permitted to be present at low levels in the tap water (2).

It is also estimated that there are 743,015 pounds (337,026 kg) of CD released to the atmosphere from over 100 manufacturing, processing, and waste disposal facilities in 2000 (3).

HISTORY

CD has been recognized for its disinfecting properties since the early 1900s. It is highly soluble in water but does not dissociate. This has led to the widespread use of CD in the treatment of drinking water. By the mid-1990s, CD was used in over 400 drinking water treatment plants in the U.S.A. (4). The Food and Drug Administration allows the use of aqueous CD in washing fruits and vegetables (5). Beyond these and numerous other aqueous applications, the sporicidal properties of *gaseous* CD were demonstrated in 1986.

CD was patented as a sterilant in the mid 1980s. The sporicidal activity of gaseous CD was demonstrated and, in 1988, it was accepted by the United States Environmental Protection Agency for use as a sterilant. Sterilization studies with gaseous CD have demonstrated its effectiveness for medical product sterilization.

EFFECTIVENESS

CD acts as an oxidizing agent and reacts with several cellular constituents, including the cell membrane of microbes. The breakup of the cell results in the death of the organism by breaking their molecular bonds with the removal of an electron (oxidation). The enzymatic function is broken by the CD altering the proteins involved in

Table 1 Chlorine Dioxide Properties

Chemical formula	ClO_2
Molecular weight	67.45 g/mol
Melting point (°C)	−59
Boiling point (°C)	+11
Density	2.4 times that of air

the structure of the microorganisms, causing very rapid bacterial kills. Mutation of the cells to a resistant form is not a concern since CD executes a simultaneous, oxidative attack on many proteins. CDs antimicrobial action is retained longer in the presence of organic matter.

A series of square wave studies was performed in a two-glove 23 ft^3 flexible wall isolator to determine the effect of CD gas concentration on the inactivation rate of *Bacillus subtilis* spores. The *D*-value (the time at a specified CD gas concentration required to reduce the microbial population by 1 log or 90%) of *B. subtilis* spores on unwrapped paper carriers, when exposed *to* CD gas concentrations of 3 and 5 mg/L, was determined using the SMC Method.

Each BI was stored at 75±2% RH prior to entering the isolator and preconditioned in the isolator at 75±2% RH for 30±1 minute prior to CD gas exposure. The decline in %RH during the gas injection and exposure phases of the cycle was recorded for each of the *D*-value runs.

Data calculations for the 3 and 5 mg/L exposure concentrations utilizing the SMC are in Tables 2 and 3. The results can be seen in Table 4.

CYCLE DESCRIPTION

The CD cycle is similar to the EtO cycle, in that humidity is required for sporicidal process efficacy along with gas concentration. The CD cycle can be carried out at pressures from negative pressures (2 KPa) to slightly above atmosphere. Figure 2 shows an example cycle of CD concentration.

The steps in the cycle are as follows:

- Precondition
- Conditioning
- Charge
- Exposure
- Aeration

PreCondition

Precondition is the first step of the CD cycle. At this point the chamber should be leak tested. When using any sterilant it is good practice to perform a chamber leak test prior to each decontamination cycle to ensure chamber integrity. For a vacuum chamber, vacuum is pulled down to a desired level, and then the chamber is held static for a period of time. The pressure difference from the beginning of the dwell time to the end is noted. If the pressure rise is not within acceptable parameters, the chamber must be properly sealed and retested before any sterilant in injected into the chamber.

$$O - Cl^{\bullet} = O$$

Figure 1 Structure of CD. *Source*: From Ref. 1.

Table 2 *D*-Value Determinations Using the SMC Method (3 mg/L)

Gas concentration (mg/L)	*U*	*n*	*r*	*Nu*	Log *Nu*	*D*-value
3	21	10	0	N/D	N/D	N/D
3	24	10	1	2.30	0.362	3.92
3	27	10	1	2.30	0.362	4.41
3	30	10	2	1.61	0.207	4.78
3	33	10	6	0.51	−0.292	4.87
3	36	10	8	0.22	−0.651	5.05
3	39	10	9	0.11	−0.977	5.23
3	42	10	8	0.22	−0.651	5.89
3	45	10	10	N/D	N/D	N/D
					Average:	4.88 min

Once the chamber has been leak tested, the chamber can be brought to the proper RH set point (60–75%). Humidity can be generated in a variety of methods such as steam, fine particle-size atomizers, etc. Steam offers the quickest, cleanest and most efficient way to raise humidity.

Conditioning

Once the humidity is at the proper level (60–75%), the cycle can advance to the next step, conditioning, where the load picks up moisture. During the entire conditioning time, typically 30 minutes, the RH is monitored. If the RH drops by any significant amount, (5%), more steam is added to raise the moisture level. Once the conditioning time is completed, gas can then enter the chamber.

Charge

During charge, CD gas is generated and introduced into the chamber to achieve a set concentration of gas. The target concentration is dependent on different

Table 3 *D*-Value Determinations Using the SMC Method (5 mg/L)

Gas concentration (mg/L)	*U*	*n*	*r*	*Nu*	Log *Nu*	*D*-value
5	18	10	1	2.30	0.362	2.94
5	21	10	2	1.61	0.207	3.35
5	24	10	7	0.36	−0.448	3.46
5	27	10	9	0.11	−0.977	3.62
5	30	10	10	N/D	N/D	N/D
5	33	10	9	0.11	0.977	4.43
5	36	10	10	N/D	N/D	N/D
5	39	10	10	N/D	N/D	N/D
5	42	10	10	N/D	N/D	N/D
					Average	3.56 min

Note: When the number of sterile replicates (*r*) is 0 or 10, the *D*-value is not determined.
SMC formula: *D*-value = U/log No − log Nu.
Key: *U*, time in minutes; *n*, number of replicates tested; *r*, number of sterile replicates out of the number tested; Nu, natural log of n/r [ln(n/r)]; No, population of unexposed 81(3.00×106 CFU/strip).

Table 4 Results and Conclusion: *D*-Value vs. CD Concentration

CD concentration (mg/L)	*D*-value (min)
3	4.88
5	3.56
10	0.75
20	0.27
30	0.12

factors: cycle time, cost, type of load, etc. If cycle time is extremely important, a higher concentration is sometimes selected to achieve a faster kill. At higher concentrations the *D*-Values are much quicker thereby shortening the overall cycle. If cost is the driving factor, then a lower concentration can be selected to preserve the consumables, but the exposure time must be extended accordingly. Under vacuum conditions the penetration of CD gas is quite remarkable. Usually a higher concentration is selected when using vacuum to ensure penetration into complex loads. CD is a surface sterilant and does not have the penetrating abilities of EtO. CD does not penetrate into plastic polymers or through cardboard but it does reach tight areas (inside of syringes, bottles, tips and caps, lumens, stents, etc.). An additional benefit of rapid aeration occurs since CD does not penetrate into the polymers.

Since the CD concentration is easily measurable in real time, the target concentration can repeatably be achieved, thus giving the assurance of a reproducible sterilization cycle. When gas concentration reaches the target concentration, the cycle proceeds to the exposure step.

Exposure

During exposure, the concentration of CD gas is monitored and maintained to keep the concentration at the target concentration for the entire exposure time (typically 30–45 minutes). If gas concentration drops during the cycle, additional gas is injected to ensure gas concentration remains set during the entire decontamination exposure step.

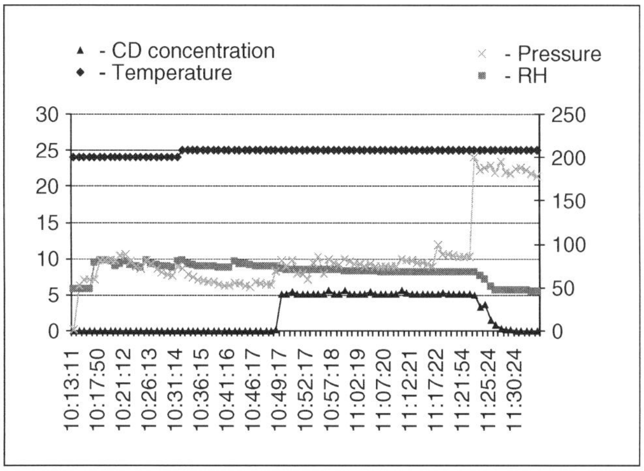

Figure 2 Cycle chart.

Aeration

The aeration step starts once the exposure step is completed. In this step, the CD gas is removed from the chamber. For vacuum chambers, this is accomplished by a series of vacuum pulls and filtered air or nitrogen backfills. Table 5 calculates the amount of CD used for a typical cycle. Table 6 details the same cycle aeration curve. Aeration time is primarily dependent on the rate that the vacuum pump can evacuate the sterilization chamber. Rates of 15 minutes are attainable. This aeration brings the chamber environment to safe levels of 0.1 ppm or less.

CYCLE DEVELOPMENT

Moisture Conditioning

As mentioned above, the presence of moisture in the load is critical to obtaining optimal lethal rates and effective sterilization with gaseous CD. Important points to consider when developing and optimizing the CD sterilization process is as follows:

- What moisture condition has the load been exposed to/stored at prior to sterilization?
- Can the moisture level be affected by seasonal RH variation?
- Are there components or packaging materials that may become desiccated during storage in a dry environment prior to sterilization?
- Could the density of the load or its physical geometry affect the penetration of moisture into least accessible areas?

The choice of a moisture conditioning time in a traditional sterilizer-based application is a function of the issues raised above as well as the approach used to perform the moisture conditioning. Moisture conditioning can be accomplished either in an external chamber or within the sterilization chamber itself. Typical loads can obtain the required moisture with 30 minutes of conditioning as part of the sterilization cycle. Appropriate validation studies are important to assure moisture penetration into the least accessible areas.

Exposure Time/Gas Concentration

Linear inactivation kinetics has been demonstrated with gaseous CD using the BI of choice for this sterilant, spores of *B. subtilis*. As is observed with other gaseous sterilants, the lethal rate increases with increasing gas concentration. Early studies with a traditional sterilizer-based application used a gas concentration of 30 mg/L. This concentration was chosen in this particular situation due to the density and composition of the sterilization load. Rapid inactivation of the BIs was observed with sterilization of 10^6 *B. subtilis* spores occurring in less than 15 minutes in almost all cases. In one application, testing was performed at a CD gas concentration of

Table 5 Typical CD Usage

Chamber volume (ft^3)	100≅28.32 m^3
Target concentration (mg/L)	5
Exhaust rate (CFM)	100≅169.9 m^3/hr
Amount of CD in chamber	14.16 g

Table 6 Typical Aeration (Removal of Gas) Curve

Air exchanges	mg/L	ppm	Time (min)
1	2.5000	892.50	1
2	1.2500	446.25	2
3	0.6250	223.13	3
4	0.3125	111.56	4
5	0.1563	55.78	5
6	0.0781	27.89	6
7	0.0391	13.95	7
8	0.0195	6.97	8
9	0.0098	3.49	9
10	0.0049	1.74	10
11	0.0024	0.87	11
12	0.0012	0.44	12
13	0.0006	0.22	13
14	0.0003	0.11	14
15	0.0002	0.05	15

Note: In each air exchange 1/2 of CD is removed.

3 mg/L with reproducible sterilization of 10^6 BIs. As would be expected, the required total gas exposure time is longer than that used at higher gas concentrations.

Based upon studies using a number of test systems, the following guidance can be given with respect to the choice of gas concentration and exposure time in process development studies.

In a CD sterilizer, the recommended gas concentrations for process development studies are 15 to 30 mg/L. A very large number of sterilization exposures have been performed using a CD gas concentration of 30 mg/L. In almost all cases, complete kill of 10^6 BIs were observed with 15 minutes of gas exposure. At a CD concentration of 5 mg/L, complete kill of 10^6 BIs should be observed with 30 minutes of exposure. It should be noted that these results were obtained in sterilizers that were not densely loaded but did consist of a desiccated load. In the event of a dense, desiccated load where moisture penetration may be impeded, a longer conditioning or gas exposure time may be required for similar sterilization efficacy. Also, the testing described here was performed in sterilizers with well-defined and effective circulation and/or recirculation systems; this ensured uniform moisture and gas distribution as measured by visual inspection, photometric monitoring and inactivation of BIs placed throughout the chamber and/or load. In process development studies, adequate circulation and uniformity of moisture and gas distribution must be carefully considered.

Examples of CD Process Development

Table 7 presents examples of gaseous CD process development studies in sterilizers using 10^6 *Bacillus atrophaeus* spores as the BI. This work evaluated different substrates.

Biologic Indicators

Historical data have shown *B. atrophaeus* (ATCC 9372) spores as the appropriate BI for chemical sterilants such as CD. To confirm the applicability for gaseous CD, tests were done with *B. atrophaeus*, as well as other commonly used BIs.

Four spore-forming organisms were initially selected: *Geobacillus stearothermophilus* (ATCC 7953), traditionally used in steam sterilization activities; *Bacillus pumilus* (ATCC 7953), most often used in irradiation studies; *G. stearothermophilus*, used for hydrogen peroxide systems; as well as *B. subtilis*. A study was developed to expose each type of BI to a standard CD cycle. In each of the three runs, 15 BIs of each type were exposed to the CD standard cycle, removed from the chamber, and aseptically transferred to nutrient media. Microbial growth, as indicated by media turbidity, was recorded as a positive result. This testing was performed in triplicate.

The results are shown in Table 8. As Table 8 shows, *B. atrophaeus* spores were consistently more resistant (highest number of BIs remaining nonsterile) than either or the *G. stearothermophilus* strains or *B. pumilus*. Based on these data, the use of spores of *B. atrophaeus* as the BI for gaseous CD was affirmed.

Table 7 Examples of Gaseous CD Process Development Studies

Preconditioning		CD exposure		No. of non-sterile/tested	Comments
Min	% RH	mg/L	Min		
30	75	10	5	10/10	Spores on unwrapped paper spore strips stored at 23% RH prior to use. Duplicate series of runs on different days
30	75	10	10	5/10	
30	75	10	15	0/10	
30	75	10	5	8/10	
30	75	10	10	1/10	
30	75	10	15	0/10	
30	70	5	30	0/20	Spores on paper spore strips in Tyvek™ envelopes. CD concentration of 5 mg/L
30	75	10	15	0/10	Spores on paper strips, unwrapped
30	75	10	30	0/10	
30	75	10	15	10/10	Spores on paper strips in blue glassine envelopes
30	75	10	30	0/10	
30	75	10	15	0/10	Spores on paper strips in Tyvek envelopes
30	75	10	30	0/10	
30	75	10	15	0/10	Spores on glass fiber discs in Tyvek envelopes
30	75	10	30	0/10	

Table 8 BI Resistance Study Results

BI	Run 1 no. of non-sterile/ total tested	Run 2 no. of non-sterile/ total tested	Run 3 no. of non-sterile/ total tested	Total no. of non-sterile/ total tested
B. subtilis (*globigii*) ATCC 9372	10/15	13/15	15/15	38/45
B. pumilus ATCC 27142	0/15	2/15	1/15	3/45
B. stearothermophilus ATCC 12980	1/15	2/15	2/15	5/45
B. stearothermophilus VHP	9/15	9/15	8/15	26/45

Cycle parameters: 30 mg/L gas concentration, 90% RH prehumidification, 6 min exposure time.

STABILITY OF GAS

CD is produced by the system at the time of use. Throughout the cycle, it is stable and, unlike other decontaminating vapors, it does not need to be continuously fed into and circulated through the chamber. Throughout the Exposure step, the CD just sits in the chamber. Although CD is stable throughout even extended cycles it is not stable enough to be stored in pressurized gas cylinders.

Since CD is a true gas and does not condense out, the stability of CD as the sterilizing agent is greatly enhanced over other methods using vapors and liquids. Because the CD concentration can be monitored and controlled, the concentration is precisely maintained throughout the cycle.

MEASUREMENT/QUANTIFICATION

A UV–Vis spectrophotometer is integrated into the system. It precisely monitors and controls the CD concentration during charge and exposure, and throughout most of the aeration cycle until it gets to approximately 0.1 mg/L. Because CD is a true gas and does not condense, aeration is very repeatable. This repeatability allows for aeration to be validated ensuring when safe conditions are attained for repeat applications. There are also devices such as Draegar™ tubes, which can verify that safe levels are attained prior to opening the chamber.

SAFETY/TOXICITY

The OSHA 8-hour TWA for CD is 0.1 ppm. The 15-minute STEL is 0.3 ppm. CD is a respiratory/mucous membrane irritant. One of the great safety features of CD is that it has a 0.1 ppm odor threshold which makes it self-alerting. Most other sterilants need to be well over their STEL before they can be sensed or smelled. Because CD has such widespread usage in the water treatment and paper and pulp industries, there is a wide selection of environmental monitors and personnel badges available. Also because of this widespread usage, there have been numerous safety studies conducted both for environmental effects, inhalation, as well as ingestion. Since it is also widely used in the food industry for sanitization and disinfection, there are allowable limits from the U.S. government for ingestion. The EPA has set the maximum concentration of CD in drinking water at 0.8 mg/L (6).

Chlorine dioxide's special properties make it an ideal choice to meet the challenges of today's environmentally concerned world. Actually, CD is an environmentally preferred alternative to EtO. The major concerns with EtO centers on its flammability and high reactivity. Acute exposures to EtO gas may result in respiratory irritation and lung injury, headache, nausea, vomiting, diarrhea, shortness of breath, and cyanosis (OSHA Safety and Health Topics). Chronic exposure has been associated with the occurrence of cancer, reproductive effects, mutagenic changes, neurotoxicity, and sensitization (OSHA Safety and Health Topics).

KNOWN INCOMPATIBILITIES

CD reacts with carbohydrates, such as glucose, to oxidize the primary hydroxyl groups first to aldehydes and then to carboxyl acids (7). Ketones are also oxidized to carboxyl acids (8). Although CD has "chlorine" in its name, its chemistry is radically different from that of chlorine. When reacting with other substances, it is weaker and more selective. CD, as with other oxidizers as well as water, causes oxidation to uncoated ferrous materials as well as other materials subject to oxidation. Control of moisture during the decontamination process mitigates the oxidation potential.

IN-PROCESS CONTROLS

Process control is one of the greatest strengths of the CD technology and one that puts it worlds above other methods of sterilization. CD can be precisely monitored with the built-in UV–Vis spectrophotometer because of its yellowish-green color. This technological advantage allows the CD concentration to be precisely monitored and controlled through charge and exposure, and throughout most of the aeration cycle until it gets down to below 0.1 mg/L. CD is a true gas, above 11°C, that distributes rapidly and evenly throughout the chamber. Because it is a true gas, issues with temperature gradients, cold spots, heat sinks due to materials of construction, and other issues that can affect the condensation of vapor decontaminating agents such as hydrogen peroxide and peracetic acid do not affect the decontamination effectiveness of CD. Also due to its properties as a true gas, it can easily penetrate down long lumens and effectively sterilize complex components even while sealed in Tyvek™ bags.

An RH/temperature probe monitors the RH and temperature conditions inside the chamber. A pressure transmitter monitors the chamber pressure.

Figure 3 Cloridox-GMP.

The tight process control and accurate concentration monitoring, along with a detailed run record, can lead to parametric release when used for product sterilization as well as expedite validation efforts for all applications.

DELIVERY SYSTEMS

The ClorDiSys Solutions, Inc.™ Steridox-VP™ sterilizer and Cloridox-GMP™ Sterilization System (Fig. 3) can be used for component or device sterilization. The Steridox-VP is a stand-alone sterilizer. The Cloridox-GMP is a portable CD gas generator system designed for interfacing with an existing steam or EtO sterilization chamber. Additionally, the Cloridox-GMP can be used to decontaminate isolators, clean rooms, processing vessels, biosafety cabinets or any sealed chamber. CD sterilizers can be used in any pharmaceutical, manufacturing, laboratory, or research setting. They provide a rapid and highly effective method for component or device sterilization. Both systems feature a sophisticated sterilant concentration monitoring system to assure a tightly controlled decontamination process. All instrumentation, including the photometer for concentration monitoring, is easily calibrated to traceable standards. The HMI system features a password-protected, recipe management system with historical and real-time trending. The process is easy to validate due to the repeatable cycle, tight process control, and highly accurate sterilant monitoring system. A run record is produced that contains the date, cycle time, cycle steps, as well as the critical operating parameters of RH, temperature, pressure, and CD concentration.

REFERENCES

1. Battisti DL. Development of a chlorine dioxide sterilization system. PhD dissertation, University of Pittsburgh, Pittsburgh, PA, 2000.
2. Agency for Toxic Substances and Disease Registry (ATSDR). Toxicological Profile for Chlorine Dioxide and Chlorite—Draft for Public Comment. Atlanta, GA: U.S. Department of Health and Human Services, Public Health Service, 2002.
3. Draft Toxicological Profile for Chlorine Dioxide and Chlorine, U.S. Department of Health and Human Service, Agency for Toxic Substances and Disease Registry, September 2002:103.
4. Richardsen SD, Thruston AD, Collette TW, Patterson KS, Lykins BW. Future Use of Chlorine, Chlorine Dioxide, and Other Chlorine Alternatives. IFT annual meeting: Book of abstracts, 1996:140.
5. Food and Drug Administration, U.S. Department of Health and Human Services. Secondary Direct Food Additives Permitted in Food for Human Consumption. 21 CFR. Part 173.300 Chlorine Dioxide, 1998.
6. Draft Toxicological profile for chlorine dioxide and chlorite—for public comment, U.S. Department of Health and Human Services, Public Health Service Agency for Toxic Substances and Disease Registry (ATSDR), 2002.
7. Masschelein WJ, Rice RG, eds. Chlorine Dioxide Chemistry and Environmental Impact of Oxychlorine Compounds. Ann Arbor, MI: Ann Arbor Science Publishers, Inc., 1979:111–45, P98.
8. Kaczur JJ, Cawifield DW. Chlorine oxygen acids and salts. In: Kirk-Othmer Encyclopedia of Chemical Technology. 4th ed., Vol. 5. New York: Wiley.

18

Validation of the Radiation Sterilization of Pharmaceuticals

Geoffrey P. Jacobs
Dr. Geoffrey P. Jacobs Associates, Pharmaceutical Consultants, Jerusalem, Israel

The use of gamma irradiation for the sterilization of pharmaceuticals has been a recognized method of sterilization for some 40 years (1,2). However, radiation sterilization may also be carried out using electron beam irradiation or the somewhat innovative application of X-rays.

While high-energy gamma irradiation is used mainly in the healthcare industries for the sterilization of disposable medical devices, there has been over the years a gradual increase in the number of pharmaceuticals being radiation sterilized. Today drugs manufactured by leading pharmaceutical companies are radiation sterilized. These include ophthalmic preparations, topical ointments, parenterals, and veterinary products. Unlike medical devices that are clearly labeled that they are radiation-sterilized pharmaceuticals are not required to be labeled with the mode of sterilization and therefore information on whether a particular drug is radiation sterilized is often unavailable.

Although radiation sterilization may be undertaken using either gamma rays from a radioisotope source (usually cobalt-60) or electron beam or X-ray irradiation, the former is by far the more common.

As with all methods of sterilization, irradiation involves a compromise between inactivation of the contaminating microorganisms and damage to the substrate or product being sterilized. The imparted energy in the form of gamma photons or electrons does not always distinguish between the two.

The usual mechanism for interaction between the high-energy gamma radiation and matter is the formation of ion pairs by the ejection of an electron, leading to free radical formation, and excitation. The free radicals are extremely reactive as a result of the unpaired electron on one of the outer orbitals. Their reactions may involve gas liberation, formation, and scission of double bonds, exchange reactions, migration of electrons and cross-linking. In fact, any chemical bond may be broken and any potential chemical reaction may take place. In crystalline materials, this may result in vacancies, interstitial atoms, collisions, and thermal spurs as well as ionizing effects. Polymerization is particularly common in unsaturated compounds. In microorganisms radiation-induced damage may express itself in various biological changes which may lead to cell death. Although DNA is generally considered the major target for cellular damage, membrane damage may also make a significant contribution to reproductive cell death. In solutions, a molecule may receive energy directly from the incident radiation (the "direct effect") or, for example in aqueous solutions, by transfer of energy from the radiolysis products of water (for example, hydrogen, and hydroxyl radicals and the hydrated electron) to the solute molecule (the "indirect effect").

The process of radiation-induced damage by electrons is similar to that for gamma photons. In electron irradiation, the high-energy electrons produced externally to the target molecule cause ionization of the molecular species as they pass through the medium and release their energy. The ionization process leads to the production of secondary electrons (known as delta rays) with a range of energies capable of bond breakage in the medium in the vicinity of the ionization event. The high-energy electrons are usually produced either by a direct current machine, by accelerating them across a large drop in potential, or by linear or circular electron accelerator.

X-rays are electromagnetic photons emitted when high-energy electrons strike any material and can therefore be produced by an electron accelerator.

For reviews of radiation sterilization the reader is referred inter alia to the Chapters on Gamma Radiation Sterilization (3) and Electron Beam Sterilization (4) in the Encyclopedia of Pharmaceutical Technology.

The advantages of irradiation for sterilization are:

- Its high penetratability, thus allowing the product to be sterilized in its final container—even in its shipping container;
- The very low temperature rise (normally less than 5°C) therefore being compatible with heat-sensitive products;
- Fewer process variables than other methods of sterilization—this improves process control with sterility rejections for radiation-sterilized products being the lowest reported;
- No remaining sterilant residuals.

Contact information: P.O. Box 16352, Jerusalem 91162, Israel.

Abbreviations used in this chapter: AAMI, Association for the Advancement of Medical Instrumentation; ASTM, American Society for Testing and Materials; cGMP, current good manufacturing practices; EPR, electron paramagnetic resonance; FDA, Food and Drug Administration; FDIS, Final Draft International Standard; GMP, good manufacturing practices; HIMA, Health Industry Manufacturers Association; IAEA, International Atomic Energy Agency; IQ, installation qualification; NDA, new drug applications; NIST, National Institute of Standards and Technology; OQ, operational qualification; PDA, Parenteral Drug Association; PQ, performance qualification; SAL, sterility assurance level; TLD, thermoluminescence-dosimetry; USP, *U.S. Pharmacopeia*.

Electron beam irradiation has the added advantages that the sterilization dose can be delivered in just a few seconds, compared to several hours or even days with conventional gamma irradiation. This has an added advantage of easier control of the environmental conditions of the irradiation process, which may be important in radiation-sensitive products (see the section entitled Materials Compatibility). There is also the advantage of flexibility of allowing individual product treatment when required. X-ray sterilization is not as fast as electron beam irradiation. Since electron beam and X-ray machines are electric powered, there are no disadvantages of handling, shipping and disposal of radioisotopes. A disadvantage of electron beam irradiation has been their low penetrating power, although the more modern machines have overcome this problem. X-ray machines may be even more penetrating than gamma rays.

It is usual for irradiation to be carried out by contract sterilizers [for a list of contract irradiation facilities, see (3,5)]. While, many aspects of the validation of the process are usually undertaken by the contract sterilizer, nevertheless, the drug manufacturer bears overall responsibility for the sterility of the product. Essentially, the contract sterilizer is responsible for guaranteeing the delivered radiation dose.

Validation of the radiation sterilization process, as an integral aspect of GMP, comprises the following components which relate either to the irradiation facility itself or the product being irradiated:

- IQ
- OQ
- PQ
- Materials Compatibility
- Selection of Sterilization Dose
- Routine Process Control

It is common practice, because of economic or feasibility considerations for a manufacturer of a radiation-sterilized product to use an outside contractor to provide the irradiation service. The criteria used in choosing such a contractor must be the same as those used for choosing other outside contractors for pharmaceutical processing. It must be shown that the irradiation facility operates in a manner consistent with cGMP, and that it is registered with the appropriate regulatory authority such as the FDA or local health authority, and that it meets all national (or federal) and local regulations.

IQ

IQ, or irradiator commissioning, is to ensure that the irradiator has been supplied and installed in accordance with its specifications. IQ includes plant commissioning, and defined and documented operating procedures for the irradiator and associated conveyor systems, radiation source configuration, for gamma irradiators—the activity of the source, for electron beam and X-ray irradiators—the characteristics of the beam, correct functioning with design specifications of electromechanical systems and associated software, documentation for any modifications, instrument calibration and recalibration, cycle timer setting, choice of dosimeters (see the section entitled Dosimetry), dosimeter placement (including frequency and rationale), and product handling before, during, and after irradiation (as well as process release) in accordance with process specifications.

Some aspects of IQ may be considered as part of the OQ or PQ.

Dosimetry

The essential parameter that has to be controlled in radiation sterilization, particularly when using gamma irradiation, is the measurement of radiation dose. This is achieved using dosimeters—chemical or physical systems that respond quantitatively to absorbed radiation dose. In irradiation practice, although not necessarily at the operational level, four types of dosimeters are used. Three types are used as standards, namely, primary, reference, and transfer dosimeters, and a fourth group, routine dosimeters, are used for routine measurement.

Primary dosimeters are the highest quality dosimeters and are maintained by national standards laboratories. The two most commonly used primary standard dosimeters are ionization chambers and calorimeters (6).

Reference and transfer dosimeters (or secondary dosimeters) are used for calibration of radiation sources and routine dosimetry. The most commonly used reference standard dosimeters are the ferrous sulfate (Fricke) and dichromate dosimeters for gamma and X-ray use, and calorimetry for electron beam applications. In chemical dosimeters (ferrous sulfate and dichromate) the chemical change in a suitable substrate is measured. For example, the concentration of ferric ions formed from the radiation-induced oxidation of an aerated ferrous sulfate solution is determined spectrophotometrically. Calorimetry, probably the most direct method of determining the amount of energy carried by a beam of radiation, is based on the increase in temperature of a block of material placed in the path of the beam. The material must be such that all the absorbed energy is converted to heat. Graphite or metals are used for this purpose. Other chemical reference standard dosimeters are the alanine, ceric–cerous, ethanol–chlorobenzene dosimeters. Most of these reference standard dosimeters may also be used as transfer standard dosimeters. Transfer reference standard dosimeters are usually sealed, packaged dosimeters that are sent to the irradiation facility for irradiation to nominal agreed-upon absorbed dose levels in a prescribed geometrical arrangement. The unopened packaged dosimeters are then returned to the national standardization institute (for example, NIST) to be read and evaluated thus providing calibration of the client's irradiator. For electron beam irradiation, the commonly used reference standard dosimeters are calorimeters, alanine, ceric–cerous, ethanol–chlorobenzene, ferrous sulfate and dichromate systems. However they may be limited by the energy range being used.

Routine dosimeters are used at the irradiation plant level for monitoring and quality assurance in routine irradiation processing. Examples of routine dosimeters for gamma and X-ray use are dyed or clear polymethylmethacrylate, cellulose triacetate, ceric–cerous sulfate, radiochromic dye and ferrous–cupric systems. Most of these systems may also be used for electron beam irradiation.

In selecting a dosimetry system consideration has to be given to inter alia; suitability of the dosimeter for the absorbed dose range of interest and for use with a specific product stability; and reproducibility; ease of calibration; ability to correct responses for temperature, humidity, and dose-rate deviations; ease and simplicity of use; resistance to damage during routine handling; and inter- and intra-batch responses. It is a requirement that dose measurements are traceable to an appropriate national or international standard, and that their level of uncertainty is known.

Practical information on radiation dosimetry can be found in the following ISO/ASTM and ASTM standards (7):

- ISO/ASTM 51608 Practice for Dosimetry in an X-ray (Bremsstrahlung) Facility for Radiation Processing;
- ISO/ASTM 51261: Guide for Selection and Calibration of Dosimetry Systems for Radiation Processing;
- ISO/ASTM 51400: Practice for Characterization and Performance of a High-Dose Radiation Dosimetry Calibration Laboratory;
- ISO/ASTM 51631: Practice for Use of Calorimetric Dosimetry Systems for Electron Beam Dose Measurements and Dosimeter Calibrations;
- ISO/ASTM 51649 Practice for Dosimetry in an Electron-Beam Facility for Radiation Processing at Energies between 300 keV and 25 MeV;
- ISO/ASTM 51702 Practice for Dosimetry in a Gamma Irradiation Facility for Radiation Processing;
- ISO/ASTM 51707: Guide for Estimating Uncertainties in Dosimetry for Radiation Processing;
- ISO/ASTM 51818 Practice for Dosimetry in an Electron Beam Facility for Radiation Processing at Energies between 80 and 300 keV;
- ASTM E 170 Terminology Relating to Radiation Measurements and Dosimetry;
- ASTM E 2303 Guide for Absorbed-Dose Mapping in Radiation Processing Facilities.

More information on specific dosimetry systems including guidance on dosimetry characteristics can be found in the following standards (7):

- ISO/ASTM 51205: Practice for Use of a Ceric–Cerous Sulfate Dosimetry System;
- ISO/ASTM 51275: Practice for the Use of a Radiochromic Film Dosimetry System;
- ISO/ASTM 51276: Practice for the Use of a Polymethylmethacrylate Dosimetry System;
- ISO/ASTM 51310: Practice for the Use of a Radiochromic Optical Waveguide Dosimetry System;
- ISO/ASTM 51401: Practice for Use of a Dichromate Dosimetry System;
- ISO/ASTM 51538: Practice for Use of the Ethanol–Chlorobenzene Dosimetry System;
- ISO/ASTM 51539: Guide for Use of Radiation-Sensitive Indicators;
- ISO/ASTM 51540: Practice for Use of a Radiochromic Liquid Dosimetry System;
- ISO/ASTM 51607: Practice for Use of the Alanine-EPR Dosimetry System;
- ISO/ASTM 51650: Practice for Use of Cellulose Acetate Dosimetry Systems;
- ISO/ASTM 51956: Practice for TLD Systems for Radiation Processing;
- ASTM E 1026 Practice for Using the Fricke Reference Standard Dosimetry System;
- ASTM E 2304 Practice for Use of a LiF Photo-Fluorescent Film Dosimetry System.

OQ AND PQ

These have been included in one section, as opinions may often vary as to whether a particular operation is classified as OQ or PQ. The essential point is that all aspects of the validation are undertaken. OQ is to demonstrate that the installed irradiator can operate and deliver appropriate radiation doses within defined acceptance criteria. PQ is essentially dose mapping.

OQ and PQ at a practical level include information on the dimensions and density of the packaged product as well as orientation of the product within the package, product loading patterns, the effect of process interruption, and dose distribution mapping for assessment of radiation dose ranges within the product package, and reproducibility within products. During dose mapping the location and magnitude of the minimum and maximum delivered doses have to be identified. More specific details of dose mapping can be found in the appropriate ISO guidelines (for example, in section 9 of ISO 11137-1).

Information generated by IQ, OQ, and PQ have to be reviewed and documented. A process specification for each product should be prepared and documented. Details of such a process specification for gamma, electron beam and X-ray irradiation can be found in ISO 11137 1 (section 9.4).

MATERIALS COMPATIBILITY

Any processing, such as sterilization, in the manufacture of a pharmaceutical product must cause no degradation. This also holds for radiation processing. In the first instance, data on the feasibility of irradiating a pharmaceutical can be obtained from the scientific literature. Reviews on the effects of gamma (and electron beam) irradiation are readily available (8–22). Although many of the cited investigations report only superficial examination of the irradiated drug, the reported data give useful insights into overall radiation stability of these products, and indicate whether more extensive testing of the product is worth undertaking.

It is necessary to examine each new compound for assessing its radiation stability, even though data may be available for closely related compounds. A thorough knowledge of radiation chemistry would be necessary to infer the behavior of one compound from another. Furthermore, with a formulated medication, the stability of an individual component may change when irradiated as part of product.

Although sterilization doses of radiation are usually in the order of 25 kGy (see the section entitled Selection of Sterilization Dose), the use of a higher dose such as 50 kGy is useful for feasibility studies as a means of indicating the type of radiolytic decomposition that may be expected at sterilization dose levels.

A number of different analytical tools should be used to detect radiation-induced degradation. Each technique usually reveals a change in a specific moiety of the irradiated molecule, and it is therefore essential to examine all generated data to obtain an indication of the extent of degradation. Wherever possible stability-indicating assays should be used.

As with all stability studies, assays should be carried out over an extended time period to indicate long-term stability of the product. Accelerated aging, under conditions recommended by the appropriate regulatory authority such as the FDA may be undertaken.

Even when radiolysis products are within acceptable compendial limits, it has to be conclusively established that any products formed are without any adverse effect at the concentration found. However, other studies, for example (23), show that such radiolysis products are generally not unique to irradiation. It would often suffice to show that radiolysis products are the same and at no greater concentration than those found when the drug is subjected to other sterilization procedures. In this connection guidance from the FDA/International Conference on Harmonisation Guideline on Impurities in New Drug Products [Q3B(R), issued 11/2003] is useful. It is noteworthy that the FAO-IAEA-WHO Expert Committee (24) has recommended that food items irradiated at doses of up to 10 kGy pose no danger to the consumer and can be unconditionally cleared. Appropriate inferences can be made to pharmaceuticals.

In cases where radiolysis products are formed, these can sometimes be reduced by appropriate action. For example, irradiation may be undertaken in anoxia or at low temperatures, or by incorporation of suitable additives, providing that degradation pathways are known. Of course, such additives must not be toxic or interfere with the efficacy of the drug. They may include energy transfer systems, –SH containing molecules, scavengers of radiolysis products of water, or reagents that convert radiolysis products to the parent compound. One example of such a radiation tailored formulation is that of urea broth, used for identification of *Proteus* spp., and its differentiation from other gram-negative intestinal bacteria (25).

In some cases radiolysis may be reduced by use of electron beam irradiation rather than gamma irradiation. Here *dose rate* may be an important factor. Although there is no general rule, many drugs show less breakdown at the higher dose rate, that is, with electron beam irradiation. This may be due to consumption of all the oxygen (which generally increases radiation damage) with sterilization being completed before oxygen can be replenished, and possibly due to too short a time for production of long-lived free radicals which may increase radiation-induced damage. On the other hand, the high dose rate, could in some cases cause increased damage due to the "high concentration" of gamma photons close to the substrate.

The packaging of a pharmaceutical is an integral part of the product, and therefore the radiation stability of packaging and container materials must never be overlooked when considering radiation compatibility. Lists of radiation-compatible packaging materials are readily available [for example, (3,10,26–29)]. It should be emphasized that to ensure their stability, these materials are often formulated specifically for radiation processing by inclusion of, for example, aliphatic antioxidants rather than aromatic ones that are often responsible for yellowing following irradiation.

SELECTION OF STERILIZATION DOSE

Selection of a radiation dose for sterilization is an integral part of validation of the sterilization process. Any deviation from the selected dose could result in either compromising the sterility of the product (in other words, the predetermined SAL may not be realized), alternatively, an excess radiation dose could result in chemical damage to the product.

A radiation dose of 25 kGy (2.5 Mrad) has generally been accepted as suitable for sterilization purposes (see the section entitled The USP Procedure for Dose Selection) [for example, (30,31)]. The choice of this dose was based on the radiation resistance of the bacterial spores of *Bacillus pumilus*. However, today the choice of radiation dose is based on initial (pre-sterilization) microbial contamination, or bioburden, and the desired SAL of the product.[a] Such considerations are based in part on extensive studies of the effects of sub-sterilization doses on different microbial populations (32,33).

The following demonstrates the various approaches to the choice of dose by the various regulatory and official authorities. Close examination, however, shows the similarity of the different approaches.

The USP Procedure for Dose Selection

The USP 28 (34) states as follows:

> Although 2.5 Mrad of absorbed radiation was historically selected, it is desirable and acceptable in some cases to employ lower doses for devices, drug substances, and finished dosage forms. In other cases, however, higher doses are essential. In order to validate the efficacy particularly of the lower exposure levels, it is necessary to determine the magnitude (number, degree, or both) of the natural radiation resistance of the microbial population of the product.

The USP suggests estimation of the appropriate sterilization dose by one of the methods contained in the guidelines published by the AAMI in the document Process Control Guidelines for Radiation Sterilization of Medical Devices (35). This document formed the basis for the International Organization for Standardization standard, ISO 11137, first published in 1984, which in turn became the new AAMI/ANSI standard. The current International Standard (ISO 11137:2006), also an AAMI/ANSI Standard, has just recently been revised. It has been published in three sections ISO 11137-1, *Sterilization of health care products—Requirements for the development, validation, and routine control of a sterilization process for medical devices—Part 1: Radiation sterilization*, ISO 11137-2, *Part 2: Establishing the sterilization dose for radiation sterilization*, and ISO 11137-3, *Part 3: Guidance on dosimetric aspects for radiation sterilization.*

Selection of Dose by the AAMI/ANSI/ISO Standard

The basis of the dose-setting methods described in the AAMI/ISO standards owes much to the ideas first presented by Tallentire and his colleagues (32,36,37). Subsequently standardized protocols were developed (38,39).

[a] SAL is defined as the probability of a single viable microorganism occurring on a product following sterilization. SAL is normally expressed as 10^{-n}. While the majority of authorities give n a value of 6, the FDA does allow values of less than 6 for non-invasive products.

The first ISO method, designated Method 1, is certainly the most common method used for dose selection for sterilization of medical devices and those pharmaceuticals that are radiation sterilized. The method essentially requires determination of the average microbial contamination of representative samples of the product. Note that the radiation resistance of the microbial population is not determined, and dose setting is based on the resistance of microbial populations originally derived from data obtained from manufacturers. The assumption is made that the distribution of the resistance chosen represents a more severe challenge than that presented by the natural bioburden on the article to be sterilized. This assumption is verified experimentally by irradiating 100 samples at a given verification dose, and accepted if there are no more than two contaminated samples. The sterilizing dose, appropriate for the average bioburden per sample and the desired SAL for the product, is then read from a table.

The second method (Method 2) does not entail enumeration of the bioburden but relies on a protocol for a series of incremental dose experiments to establish a dose at which approximately one in a hundred samples will be non-sterile. A sterilization dose is then established by extrapolation from this 10^{-2} sterility level, using a dose-resistance factor calculated from observations of the incremental dose experiments that characterize the remaining microbial resistance. This resistance is estimated from the lowest incremental dose at which at least one sample is sterile, and from the dose at which the surviving population is estimated to be "0.01 microorganisms" per sample.

In the original AAMI Guidelines other more elaborate procedures (originally known as AAMI Methods B3 and B4) were described for dose setting. These methods were not commonly used because of the extensive experimentation involved.

In the current AAMI/ISO guidelines a relatively new method (Method VDmax) specifically for substantiation of a 25-kGy dose is included. This method was first officially introduced as an AAMI Technical Information Report (40), and is now part of (ISO 11137–2: 2006) (41, 47). This method, put forward by for substantiation of a 25-kGy dose is similar to dose-setting Method 1. Like Method 1 it requires a determination of bioburden and the performance of a verification dose experiment.

In substantiating a 25-kGy dose, this method verifies that the bioburden on the product is less radiation resistant than a microbial population of maximal resistance consistent with the attainment of an SAL of 10^{-6} at 25 kGy. Verification is undertaken at an SAL of 10^{-1} with 10 items irradiated in the performance of the verification dose experiment. The dose corresponding to this SAL (verification dose, VDmax) reflects both the magnitude of bioburden and the associated maximal resistance. If there is no more than one positive test in the 10 tests of sterility, a 25-kGy sterilization dose is substantiated. This method is applied with some modification to both single and multiple batches.

ISO also allows substantiating a 25-kGy dose using Methods 1 and 2. The new ISO guidelines (ISO 11137-2:2006) do allow dose setting by any other method that provides equivalent assurance to the above methods in achieving the specified requirements for sterility.

In accordance to the ISO guideline, all ISO methods require the performance of a periodic audit to confirm the appropriateness of the sterilization dose.

British Pharmacopoeia/European Pharmacopoeia Procedures

According to the British and European pharmacopoeias (42,43):

"A minimum absorbed dose of 25 kGy is generally used for the purpose of sterilization, although other doses may be employed, provided that they have been validated. If doses of less than 25 kGy are used, additional microbiological monitoring of the product before irradiation will be necessary."

These pharmacopoeias give no guidance on how to estimate doses of less than 25 kGy.

PDA Procedures

The PDA has made its own recommendations for dose-setting procedures specifically for parenteral products (44). These procedures, however, are similar to those already in use for other sterilization technologies.

One method is essentially a biological indicator (overkill) method in which the sterilization dose is at least double a radiation dose needed to achieve a six logarithmic inactivation of *B. pumilus* spores on or in the product. In practice, the sterilization dose does not differ much from the classical "25 kGy."

Another method involves determination of the maximum bioburden. The logarithm of this bioburden (with three standard deviations), plus a six logarithm sterility assurance factor is multiplied by the decimal reduction factor (D_{10}) for *B. pumilus* spores to estimate the sterilization dose. The decimal reduction factor is the radiation dose to reduce the number of surviving microorganisms by 90%.

IAEA Procedure for Dose Setting

The IAEA, following an Advisory Group Meeting on the Code of Practice for Radiation Sterilization of Medical Supplies (Colombo, November 1986), adopted a pragmatic approach to the selection of a sterilization dose. The Guidelines developed at this meeting state:

> It is a basic assumption that the product to be sterilized is manufactured under conditions that comply fully with the requirements of GMP. In the present context, it is particularly important that practices be implemented, and actions taken, which ensure that the number of microorganisms on product items destined for radiation sterilization processing is consequently low.
>
> A dose of 25 kGy (2.5 Mrads) has been found to be an effective sterilizing dose. It is generally believed that this dose provides maximally a SAL of 10^{-6}. Where it is not feasible to generate data on the radiation resistance of the natural microbial population present on product items, a minimum sterilizing dose of 25 kGy (2.5 Mrads) can be used.
>
> It is more rational to base selection of a sterilizing dose on a knowledge of the resistance of the natural microbial population present ion product items to be sterilized and

on a reasoned selection of a maximal SAL. Methods of dose selection using this approach are Methods 1 and 2 in Appendix B of the AAMI Process and Control guidelines for gamma Radiation Sterilization of Medical Devices (corresponding to the current Methods 1 and 2 of the ISO 11137).

While it is this author's belief that the methods of dose selection presented in ISO 11137 are the methods of choice, the IAEA approach some 20 years later is still rational particularly for less developed countries.

OTHER DOSE-SETTING PROCEDURES

Other dose-setting procedures have been proposed in the scientific literature, including those of (39,45,46).

ROUTINE PROCESS CONTROL

This includes process specification, pre-irradiation product handling, product irradiation, product loading and unloading, monitoring during irradiation, processing records and documentation, process interruption, and routine and preventive maintenance.

LEGISLATIVE CONSIDERATIONS

Although radiation sterilization has appeared in the USP since 1965, the FDA regards a radiation-sterilized drug as a "new product" (that is, submission of an NDA, albeit abbreviated) with the manufacturer responsible for proving its safety. The current USP 23, in the section entitled Sterilization and Sterility Assurance of Compendial Articles makes the following observations regarding radiation sterilization of drugs:

The rapid proliferation of medical devices unable to withstand heat sterilization and the concerns about the safety of ethylene oxide have resulted in increasing applications of radiation sterilization. It is however applicable also to drug substances and final dosage forms.

... radiation sterilization is unique in that the basis of control is essentially that of absorbed radiation dose, which can be precisely measured.

In the U.K., sterilization by exposure to ionizing radiation has been a recognized method since 1980, when the Ministry of Health agreed to accept materials exposed to a radiation dose of 25 kGy. Medicines controlled under the Medicines Act 1968 are subjected to individual assessment by the Committee on Safety of Medicines of the Medicines and Healthcare products Regulatory Agency. This committee requires in addition to proof of sterility, proof that the potency of the drug is unaffected by the process, and that any degradation products would not be harmful.

Similarly, although the *British Pharmacopoeia* recognizes gamma irradiation as a suitable sterilization process, it is the responsibility of the manufacturer to prove that no degradation of the product has taken place.

Most European countries allow pharmaceuticals to be radiation sterilized, provided that authorization has been obtained from the appropriate health authorities.

REFERENCES

1. British Pharmacopoeia 1963. London: Her Majesty's Stationary Office, 1962 (A209).
2. United States Pharmacopeia XVII. Rockville, MD: The United States Pharmacopeial Convention Inc., 1965.
3. Jacobs GP. Gamma radiation sterilization. In: Swarbrick J, Boylan JC, eds. Encyclopedia of Pharmaceutical Technology. Vol. 6. New York: Marcel Dekker, 1992:302–3.
4. Cleland MR, Beck JA. Electron beam sterilization. In: Swarbrick J, Boylan JC, eds. Encyclopedia of Pharmaceutical Technology. Vol. 5. New York: Marcel Dekker, Inc., 1992:105–36.
5. Nordion, World List of Suppliers of Contract Gamma Services, MDS. Ottawa, Canada: Nordion MDS, 2003. (www.mds.nordion.com).
6. Spinks JWT, Woods RJ. An Introduction to Radiation Chemistry. 2nd ed. New York: Wiley, 1976:67–121.
7. ASTM International. Annual Book of ASTM Standards. Vol. 12.02. West Conshohocken, PA: ASTM, 2006.
8. Association of the British Pharmaceutical Industry. Use of Gamma Radiation Sources for the Sterilization of Pharmaceutical Products. London: ABPI, 1960.
9. Geue PJ. Radiosterilization of Pharmaceuticals, A Bibliography, 1962–1972. Lucas Heights, Australia: Australian Atomic Energy Commission, 1973.
10. Gopal NGS. Radiation sterilization of pharmaceuticals and polymers. Radiat Phys Chem 1978; 12:35–50.
11. Jacobs GP. A review: radiation sterilization of pharmaceuticals. Radiat Phys Chem 1985; 26:133–42.
12. Jacobs GP. A review of the effects of gamma radiation on pharmaceutical materials. J Biomater 1995; 10:59–96.
13. Phillips GO. Medicines and pharmaceutical base materials. Manual on Radiation Sterilization of Medical and Biological Materials. Vienna, Austria: International Atomic Energy Agency, 1973:207–8.
14. Schnell R, Bogl W. Der Einfluss der Strahlenbehandlung auf Artzneimittel und Hilfstoffe. Eine Literaturstudie, Teil VI, Bericht des Instituts fur Strahlenhygiene des Bundesgesundheitsamtes. Berlin: Dietrich Reimer Verlag, 1982.
15. Schnell R, Bogl W. Der Einfluss der Strahlenbehandlung auf Artzneimittel und Hilfstoffe. Eine Literaturstudie, Teil VII, Bericht des Instituts fur Strahlenhygiene des Bundesgesundheitsamtes. Berlin: Dietrich Reimer Verlag, 1984.
16. Schuttler C, Bogl W, Stockhausen K. Der Einfluss der Strahlenbehandlung auf Artzneimittel und Hilfstoffe. Eine Literaturstudie, Teil II, Bericht des Instituts fur Strahlenhygiene des Bundesgesundheitsamtes. Berlin: Dietrich Reimer Verlag, 1979.
17. Schuttler C, Bogl W, Stockhausen K. Der Einfluss der Strahlenbehandlung auf Artzneimittel und Hilfstoffe. Eine Literaturstudie, Teil III, Bericht des Instituts fur Strahlenhygiene des Bundesgesundheitsamtes. Berlin: Dietrich Reimer Verlag, 1979.
18. Schuttler C, Bogl W. Der Einfluss der Strahlenbehandlung auf Artzneimittel und Hilfstoffe. Eine Literaturstudie, Teil IV, Bericht des Instituts fur Strahlenhygiene des Bundesgesundheitsamtes. Berlin: Dietrich Reimer Verlag, 1982.
19. Schuttler C, Bogl W. Der Einfluss der Strahlenbehandlung auf Artzneimittel und Hilfstoffe. Eine Literaturstudie, Teil V, Bericht des Instituts fur Strahlenhygiene des Bundesgesundheitsamtes. Berlin: Dietrich Reimer Verlag, 1982.
20. Trutnau H, Bogl W, Stockhausen K. Der Einfluss der Strahlenbehandlung auf Artzneimittel und Hilfstoffe. Eine Literaturstudie, Teil I, Bericht des Instituts fur Strahlenhygiene des Bundesgesundheitsamtes. Berlin: Dietrich Reimer Verlag, 1978.
21. Wills PA. Effects of ionizing radiation on pharmaceuticals. Aust J Pharm 1963; 44:550–7.

22. Zalewski CH, Schuttler C, Bogl W. Der Einfluss der Strahlenbehandlung auf Artzneitmittel und Hilfstoffe. Eine Literaturstudie, Teil VIII, Bericht des Instituts fur Strahlenhygiene des Bundesgesundheitsamtes. Berlin: Dietrich Reimer Verlag, 1988.
23. Kane MP, Tsuji K. Radiolytic degradation schemes for Co-60 irradiated corticosteroids. J Pharm Sci 1983; 72:30–5.
24. World Health Organization. Wholesomeness of Irradiated Foods, Report of a Joint FAO/IAEA/WHO Expert Committee, Technical Report Series 659, Geneva, 1981.
25. Eisenberg E, Jacobs GP. The development of a formulation for radiation sterilizable urea broth. J Appl Bacteriol 1985; 58:21–5.
26. Association for the Advancement of Medical Instrumentation. AAMI Technical Information Report, TIR 17: Radiation Sterilization-Material Qualification. Arlington, VA: AAMI, 1998.
27. Health Industry Manufacturers Association. Radiation Compatible Materials (Report No. 78-4.9). Washington, DC: HIMA, 1978.
28. Massey L. The Effect of Sterilization Methods on Plastics and Elastomers. 2nd ed. Norwich, NY: William Andrew, 2005.
29. Shang S, Ling MTK, Westphal SP, Woo L. Radiation sterilization compatibility of medical packaging materials. J Vinyl Addit Technol 1998; 4:60–4.
30. British Pharmacopoeia 1988. London: Her Majesty's Stationary Office, 1987.
31. United States Pharmacopeia XXII. Rockville, MD: The United States Pharmacopeial Copnvention Inc., 1989.
32. Tallentire A, Dwyer J, Ley FJ. Microbiological control of sterilized products. Evaluation of model relating frequency of contaminated items with increasing radiation treatment. J Appl Bacteriol 1971; 34:521–34.
33. Tallentire A, Kahn AA. Tests for the validity of a model relating frequency of contaminated items and increasing radiation dose. In: Radiosterilization of Medical Products. Vienna: International Atomic Energy Agency, 1975:3–14.
34. United States Pharmacopeia XXVIII. Rockville, MD: The United States Pharmacopeial Copnvention Inc., 2005.
35. Association for the Advancement of Medical Instrumentation. Process Control Guidelines for Radiation Sterilization of Medical Devices (N. RS-P 10/82). Arlington, VA: AAMI, 1982.
36. Tallentire A. Aspects of microbiological control of radiation sterilization. J Radiat Steroid 1973; 1:85–103.
37. Tallentire A, Kahn AA. The sub-process dose in defining the degree of sterility assurance. In: Gaughran ERL, Goudie AJ, eds. Sterilization by Ionizing Radiation., Vol. 2. Montreal: Multiscience Publications Ltd., 1978:65–80.
38. Davis KW, Strawderman WE, Masefield J, Whitby JL. Gamma radiation dose setting and auditing strategies for sterilizing medical devices. In: Gaughran ERL, Morrissey RF, eds. Sterilization of Medical Products. Vol. 2. Montreal: Multiscience Publications Ltd., 1981:34–102.
39. Davis KW, Strawderman WE, Whitby JL. The rationale and computer evaluation of a gamma sterilization dose determination method for medical devices using a substerilization incremental dose sterility test protocol. J Appl Bacteriol 1984; 57:31–50.
40. Association for the Advancement of Medical Instrumentation. AAMI Technical Information Report, TIR 27: Radiation sterilization—Substantiation of 25 kGy as a sterilization dose—Method Vdmax. Arlington, VA: AAMI, 2001.
41. Kowalski J, Aoshuang Y, Tallentire A. Radiation sterilization—evaluation of a new method for substantiation of 25 kGy. Radiat Phys Chem 1999; 58:77–86.
42. British Pharmacopoeia 2005. London: British Pharmacopoeia Commission, 2005.
43. European Directorate for the Quality of Medicines. European Pharmacopoeia. 5th Ed. Strasbourg, France: Council of Europe, 2005.
44. Parenteral Drug Association. Sterilization of Parenterals by Gamma Irradiation, Technical Report No. 11. Washington, DC: PDA Inc., 1988 (published as a Supplement to J Parent Sci Technol 1988; 42).
45. Darbord JC, Laizier J. A theoretical basis for choosing the dose in radiation sterilization of medical supplies. Int J Pharm 1987; 37:1–10.
46. van Asten J, Dorpema JW. A new approach to sterilization conditions. The IMO concept. Pharm Weekblad Sci 1982; 4:49–56.
47. Kowalski J, Tallentire A. Substantiation of 25 kGy as a sterilization dose: a rational approach to establishing verification dose. Radiat Phys Chem 1999; 54:55–64

19

Isolator Decontamination

James Agalloco
Agalloco & Associates, Belle Mead, New Jersey, U.S.A.

INTRODUCTION

The microbial decontamination of the isolator environment has undoubtedly been the most controversial subject related to their application. It has been the cause of more confusion than perhaps any other aspect of isolator implementation. The reason for this confusion perhaps all related to the expectations regarding what the treatment should accomplish. Should the treatment be considered a sterilization process, or is a less lethal process acceptable? The answer lies in the objective of the process: what is required in order to utilize the isolator for aseptic processing?

Isolators are aseptic processing environments in which sterile materials are handled and in some cases assembled into finished sterile dosage forms. Their application for aseptic processing mirrors that of a manned clean room in which similar procedures have been conducted for many years. Aseptic processing is a term more or less specific to health care manufacturing. It is a processing technology that requires the exclusion of microorganisms from the environment in which sterile materials are assembled. Isolators represent the current pinnacle of aseptic technology: no methodology of comparable capability is presently available. Isolators can afford a degree of reliability with aseptic processing unattainable by other means. However, isolators cannot be shown to provide the same level of sterility assurance as a terminal sterilization of finished product containers. No exclusionary process can ever reliably attain the degree of certainty with regard to microbial elimination as one that is lethal to microorganisms. Thus while an isolator is an extremely capable aseptic processing environment, it cannot equal the performance of a terminal sterilization procedure.

With that limitation in isolator performance recognized, should the isolator itself be sterilized prior to the aseptic process, after all would not that afford greater confidence in the robustness of the aseptic operations to follow? First, one must consider whether that expectation is a reasonable one. To explore this, one must consider the distinction between aseptic and sterile. Aseptic means free of pathogenic organisms, sterile means devoid of all life. We have tended to blur the distinction between aseptically produced and terminally sterilized products. We have long recognized that terminally sterilized products are inherently safer than those which are aseptically produced. Many sterile products will always be produced aseptically because the product cannot withstand a terminal treatment in their final container. The use of an isolator for aseptic processing will not be meaningfully enhanced by treatment of the environment to eliminate (sterilize the isolator) all microorganisms. Manned clean rooms have been successfully used for aseptic processing for over 40 years, and while these clean rooms are largely devoid of microorganisms (especially in recent years), they have never been nor will they ever be "sterile."

In contrast to a clean room, an isolator can be subjected to a sterilization process for its internal surfaces. Just because an isolator can be sterilized, should it be? This question lies at the very heart of the issue. The sterility of the isolator internal surface is an objective not worth attaining. Isolators are utilized to provide an environment for aseptic processing, and just as with manned clean rooms, need not be sterile in order to serve that purpose. In order to produce sterile products an environment largely devoid of microbial contamination is required. Clean rooms attain this state routinely with personnel present, and while isolators are capable of even better performance, perfection in the form of internal sterility is not required. Just because we can sterilize an isolator, does not mean we should. The soft parts of the isolator (gloves, half-suits and other plastic materials) can be damaged by the extended exposure time of these components required to attain sterility. In an effort to attain internal sterility we thus increase the potential for loss of isolator integrity.

The goal is to establish an environment essentially free of microorganisms. This state is accomplished in the clean room using manual sanitization of the internal surfaces by gowned personnel. In an isolator, we realize that same state by treating the internal volume with a gas or vapor that is lethal to microorganisms. The treatment provided to reach this condition need not meet the standards for sterilization. The most prevalent definition of overkill sterilization states:

> A cycle which provides a minimum 12 log reduction of a resistant biological indicator with a known *D*-value of not less than one minute. It requires minimal information on the number and heat resistance of the bioburden (1).

Delivering an overkill process is the objective of sterilization for most items, especially where the materials being processed are intended for human use. The high margin of safety is presumed to provide an

Abbreviations used in this chapter: BI, biologic indicator; EtO, ethylene oxide; F-N, fraction–negative; OSHA, Occupational Health and Safety Act; PDA, Parenteral Drug Association; PNSU, probability of a non-sterile unit; PVC, polyvinyl chloride; RTP, rapid transport port; SIP, sterilization in place; TWA, time-weighted average; VPHP, vapor phase hydrogen peroxide.

acceptable level of risk to the patient. It provides for one chance in a million that an organism might survive the process. That far exceeds the requirements for an aseptic processing environment; the clean room or isolator is after all not going to be injected into a human. All that is required for aseptic processing is an environment safe for the production of sterile products. In clean rooms this is routinely accomplished by manual disinfection of surfaces with a chemical agent. In the majority of these applications, the materials used are not sporicidal (agents capable of killing spores may be used either at less frequent intervals or in response to significant microbial excursions). The use of a sporicidal agent in the decontamination of an isolator providing multiple log reduction (less than 12 logs) is still substantially more effective at the elimination of microorganisms than what is possible in a clean room. The PDA has suggested that the complete kill of a resistant spore with an initial population of 10^3 (effectively a 6- to 7-log reduction when multiple BIs are used) is ordinarily sufficient to render the isolator free of detectable microorganisms (2).

Where a 12-log reduction (sterilization treatment) has been provided to an isolator there are several potentially negative consequences. These include the following:

- It requires longer dwell periods with extended exposure of the materials inside the isolator.
- Longer exposure periods result in increased degradation of exposed materials, risking component failure at inopportune times.
- Longer exposure periods result in potential for increased penetration into porous materials, causing substantially longer aeration periods.
- Longer overall cycles reduce the operational availability of the isolator.

Sterilization processes are expected to deliver a 12-log reduction (3). Sterilization should never have been considered a reasonable objective for isolator microbial control, especially as the isolator is not the item that must be sterile. The isolator only provides the environment in which the aseptic process occurs. A common term for the lesser treatment needed for isolator is decontamination. It differs from sterilization only by the degree and level of documentation. In the pharmaceutical industry we define sterilization as not more than one contaminated unit in a million, but this is an arbitrary risk analysis criterion. An isolator that is decontaminated to an appropriate level is still certainly safer than a clean room, and while it might not be "sterile" to the extent that an injectable product can be considered sterile from a public health perspective it is just as safe as if it were.

A decontamination treatment of the isolator that provides a 6 to 7 log reduction (complete kill of a 10^3 challenge) of a resistant spore is certainly more effective at reducing the microbial population when compared to the manual disinfection of a clean room by gowned personnel. The manual disinfection of a clean room is subject to substantial variations in efficacy, after all the process is only effective if the operators perform the procedure with the reliability of the equipment used for the decontamination treatment. While sporicidal agents are used in clean room disinfection they are a part of a regimen with other non-sporicidal materials, limiting the effectiveness of the treatment. The presence of personnel in the clean room to perform the disinfection is another significant shortcoming; the very personnel charged with applying the disinfecting materials are contributing to the microbial levels in the clean room by their very presence during the disinfection procedure!

In understanding why an aseptic isolator need not be "sterile" in order to be appropriate as a filling environment, an understanding of the other factors contributing to the finished product's sterility assurance is beneficial. Once the isolator has been treated (with either sterilization or decontamination as the goal) in preparation for use, it must operate for an extended period of time while the aseptic process is performed. During the operation of the system, the internals of the isolator are subject to the ingress of contamination from a variety of sources almost none of which have anything to do with the pre-use treatment. Possible sources include component sterilization, glove failure, seal failure, RTP failure, etc. Regardless of the pre-process treatment, contamination from these sources cannot be prevented, and given the adverse material effects of lengthy decontamination processes the potential for contamination ingress may actually be enhanced.

As the goal of the sporicidal treatment is elimination of microorganisms from an operating environment, rather than prepare a surface for human injection, sterility as defined by at least a 12-log reduction in a resistant BI may represent an unrealistic goal. If attaining a sterile state within the isolator were possible without risking material damage, increased aeration time, and decreased capacity, then perhaps it would be a reasonable requirement. As sterility in the classical sense cannot be accomplished without compromising other considerations, then less lethal decontamination appears more appropriate. As decontamination treatments are certainly more effective in elimination of microorganisms than what may be accomplished by personnel in a clean room, they assure the superiority of isolators as aseptic processing environments relative to manned clean rooms even without a "sterilization" treatment.

DETERMINING APPROPRIATE PROCESS OBJECTIVES FOR ISOLATOR COMPONENTS

In the use of isolators for aseptic processing some parts of the system must be sterilized, e.g., product contact parts, while others need only be decontaminated, e.g., isolator internal surfaces (2). This is wholly consistent with the practices employed in manned clean rooms and should not be altered when the aseptic process is conducted in an isolator.

Sterilization of Product Contact Materials

The sterilization requirements for product contact parts or container/closure systems within isolators are identical to those established for conventional aseptic processes: a PNSU of 10^{-6} or better. Direct or indirect connection of autoclaves and dry heat ovens for these materials to aseptic processing isolators is relatively straightforward using either transfer isolators or RTP containers. Aseptic connections of any type for liquid materials product or inerting gases should always be made within the isolator. "SIP"-capable hard piping to a connection point within the isolator is the most desirable form of liquid supply line. Stopper bowls and component delivery systems must be sterilized; however they represent only a small portion of the entire isolator internal surface.

Gas Decontamination of the Isolator and Equipment

Decontamination is a more reasonable objective than sterilization for the non-product contact surfaces of the isolator. A six to seven log reduction of resistant BIs is certainly sufficient (the complete kill of indicators with a population of 10^3 spores) (2). Isolators are an unfriendly environment for microbial survival and colonization; the complete kill of a 10^3 population is far more effective than manual decontamination of a clean room by gowned personnel. Isolator-based aseptic processing is not terminal sterilization.

Methods for Introduction of Wrapped Sterile Materials

Many aseptic processes require that items that have been previously sterilized by either radiation or other processes be introduced for further processing after exposure to a less capable environment. With clean rooms, this is often accomplished by manual decontamination using a sporicidal agent often in conjunction with a double door air lock or pass-through. When using an isolator the wrapped parts or components can be introduced into the isolator through a tunnel or chamber in which surface sanitization of the outer package is accomplished. The goal of this treatment is to protect the isolator environment from potential contamination on the exterior of the packaging. The tunnel or chamber replaces the pass-through or air lock of the conventional clean room. These methods are intended to surface sanitize the exterior of the package only since the interior of the pack and the components are already sterile. Bioburden control is required if operators must handle the wrapped goods during the loading of the tunnel or chamber.

CRITICAL ASPECTS IN ISOLATOR DECONTAMINATION

There are a number of essential requirements for isolator decontamination that must be addressed. These are applicable regardless of whether the objective is sterilization (as some firms have sought) or decontamination (as recommended in this text). The process needs parallel those that have been defined for gas sterilization with EtO (see chap. 16). Sterilization by EtO is a well-established technology, and has formed the basis for nearly all of the other gas sterilization processes in use today including the decontamination processes used for isolators. Sterilization with gases is most affected by the concentration of the agent, the relative humidity of the materials, and the temperature of the process. The influence of these factors impacts the required process dwell time (see Appendix I for a brief treatise on the microbiology of sterilization using various agents).

Concentration of the Decontamination Agent

In order to assure consistent lethality in a sterilization process, it is essential that a constant amount of the sterilizing medium be present. With gas sterilization this is usually accomplished by recirculation of the gas. EtO sterilizers will sometimes use large external blowers to ensure that the gas concentration is uniform across the chamber. When isolators were first decontaminated with hydrogen peroxide it was recognized that internal mixing of the vapor provided clear advantages in decontamination effectiveness, this resulted in the placement of internal fans to assist in achieving uniform lethality. With the introduction of large-scale filling isolators equipped for unidirectional airflow internally, it was sometimes felt that this circulation alone would be sufficient to assure uniformity. As there are few, if any, adverse effects of placing internal fans within isolators, their inclusion in the design is recommended. They certainly do no harm when in operation during the exposure to the agent or during the aeration of the isolator after the decontaminating treatment.

Multipoint concentration measurement across the chamber of the isolator would certainly be desirable. At the present time, this can only be accomplished at reasonable cost for chlorine dioxide (ClO_2). The instrumentation for measurement of H_2O_2 is prohibitively expensive for more than a single-point determination, while in-process concentration measurement of other agents (O_3 and peracetic acid) is not presently available.

The presence of hydrogen peroxide can be confirmed by chemical indicators, and if these can be positioned where they can be observed during the exposure phase they can provide some indication of concentration uniformity. If the distribution of the H_2O_2 is consistent, the chemical indicators should change color at approximately the same time. There are no chemical indicators available for use with the other decontamination agents.

Relative Humidity

Just as achieving a constant gas concentration is necessary, so is constant humidity across the environment. While each of the agents is slightly different, they all require the presence of humidity to affect proper kill, albeit at somewhat different levels. In O_3 and ClO_2 processes, the humidity is introduced prior to the addition of the gas, while H_2O_2 and peracetic acid processes add water in either the vapor or liquid state as part of the decontamination agent introduction.

The same measures used for assuring constant concentration of the decontaminating agent will provide for even distribution of the humidity across the isolator. Its measurement at multiple locations is substantially easier than gas concentration; however, there have been relatively few reports of multipoint measurement having actually been necessary.

Temperature

The temperature of the isolator is also an important factor in the lethality of the decontamination process. Perhaps the greatest impact of temperature is on the relative humidity of the isolator. With H_2O_2 treatments the temperature is often highest close to the inlet point and this can result in a localized lower relative humidity and lower lethality near the inlet. Here again, the presence of circulation fans can help provide constant conditions across the isolator.

Other than its impact on relative humidity, modest temperature differences across the isolator have little impact on the lethality. All of these agents are strong oxidizers and the rate of chemical reaction (and thus microbial inactivation) is relatively constant over a fairly wide temperature range. Increasing the temperature of the isolator was originally thought to be an advantage for H_2O_2 processes, but experience has shown this to be of limited value. As systems get larger it becomes increasingly difficult to assure constant temperatures (and thus constant relative humidity and consistent lethality) across the entire system.

Uniformity of Conditions

To achieve consistent results in a decontamination (or sterilization if that is the objective) process, uniformity of conditions throughout the chamber is highly desirable. The greater the turbulence inside the enclosure the more uniform the process parameters are likely to be and the more consistent the results obtained. Some vendors have attempted to rely on unidirectional airflow (as provided in many filling isolators) to provide uniform conditions. This may be acceptable; however, the reasons unidirectional flow is desired for aseptic processing are exactly why it is undesirable for the decontamination process. It is our strong recommendation that the internals of the isolator be as well mixed as possible, and the use of additional internal fans to promote greater mixing is almost always necessary.

Biologic Indicators

Confirmation of process lethality for gaseous decontamination is established through the inactivation of resistant microorganisms. The most common choices for the agents used for isolators are *Geobacillus stearothermophilus* (for H_2O_2 and peracetic acid) and *Bacillus subtilis* (for ClO_2 and O_3), though the relative resistance of these organisms to any of these agents is not substantially different (4,5). What is apparent is that these organisms are more resistant than the majority of expected bioburden organisms that might be encountered. The relative resistance for the indicator organisms relative to bioburden varies but is at least 1 to 2 logs less.

A variety of BI presentations have been utilized with these agents, ranging from paper strips to fiberglass disks to stainless steel coupons. Regardless of the indicator type selected it is essential that the resistance of the indicator to the process be understood. The supplier of the indicator must play a major role in this process, and close coordination between the supplier and the end user is essential.

Much has been made of so-called "substrate effects" in which firms have exposed resistant indicators on a variety of materials to confirm the effectiveness of the process. This appears to be little more than a "make-work" exercise, because, while the apparent resistance changes from material to material, once a "worst-case" substrate has been identified by one researcher, there is no evidence to support a different result obtained by anyone else. The substrate for the BI is really of little relevance; it is the relative resistance of the spores on the indicator that plays a greater role in cycle effectiveness. One should choose a substrate that provides for consistent lethality over time, and once that has been identified use it for all studies. That is the practice that has been employed successfully for sterilization validation for decades. That substrate effects exist should not come as a surprise to anyone; making too much of the minor differences between one substrate and another is unfortunate.

All of these agents are surface sterilants that penetrate rather poorly, especially relative to steam or EtO. Because their penetration is relatively weak compared to other sterilizing methods, the BI packaging can play a vital role in the resistance. It has been shown that H_2O_2 BIs packaged in Tyvek® are harder to kill than unexposed BIs (6). BIs on different matrixes have been shown to give very different results. Filter paper as used for BIs in steam sterilization resulted in extremely variable results with H_2O_2, and was soon abandoned for other materials. Vendors found that inoculation of glass fiber disks was also a difficult process to control and as a consequence many of the H_2O_2 BIs commercially available today are stainless steel coupons.

There are some cautions with the use of BIs that the reader should be aware of. Proper preparation of the BI is all important, the preparation must be extremely clean; the presence of cellular debris can cause substantial variations in indicator performance. The *D*-value for the challenge organism should not be concentration dependent; that is often an indication of a dirty spore preparation (7). The use of commercially prepared BIs is highly recommended as the vendors have substantially more experience with their preparation than any end user. Preference should be given to the use of BIs on a solid substrate as these have proven to have fewer variations in resistance.

Test each lot of BIs used to verify its performance; counting the population of spores is not enough. The resistance of the spores can be tested in a transfer isolator using an F-N approach at the intended cycle parameters. It should be recognized that the labeled *D*-value of the BI supplier will likely have been determined at conditions that differ from those employed by the purchaser of the BI, and thus resistance testing is essential.

Decontamination Cycle Development

The providers of the decontamination equipment all provide assistance with cycle development for their system. Regardless of the equipment or agent chosen, it is essential that the approach provide assurance that the routine conditions used for decontamination (or sterilization if that is the intent) are defended by appropriate "worst-case" challenges. A well-defined rationale for this is an essential part of the validation report for the treatment.

There have been some concerns raised regarding the use of F-N studies in the validation of isolators (3). That perspective is unfortunate, because F-N studies are essential to properly define the length of the treatment process. If total kill analysis is used, make certain to allow enough lethality to actually attain total kill. The addition of some "cushion" in terms of resistance is recommended. The use of a half-cycle approach as is common with EtO sterilization may be excessive for isolation technology, especially where decontamination is the desired outcome. It can result in very long process times in large isolators, and the security afforded by the half-cycle approach is unnecessary, as the isolator is not required to be sterile, as are the medical devices ordinarily sterilized by EtO.

One of the major concerns in the cycle development is identifying locations for the placement of the BIs. Typical locations include those portions of the isolator that are most important for production use; that, is those proximate to where the aseptic process will take place. If the intent is to sterilize anything in the isolator, those items should have BIs with a population of 10^6 exposed in or on them. Other locations typically used are the corners of the isolator and locations where penetration of the gas/vapor might be difficult. In these locations the use of a lower BI population is wholly appropriate.

There are no defined rules for the number of BIs to be used. In very large isolators, 30 to 50 or more may be required. In transfer isolators 15 BIs are generally adequate, while in double half-suit workstations 25 to 30 BIs is typically enough.

One note of caution must be stated in this exercise, it is certainly possible to place BIs with high populations in locations where they will not be fully inactivated by the process. That these locations should exist may trouble some purists; however the goal of the treatment is to ready the isolator for use in aseptic processing, not to render it sterile. The essential concern is that the critical areas are treated appropriately to fully eliminate microorganisms. This is not only a realistic goal but also routinely attainable.

Decontamination Interval

The interval between isolator decontamination cycles is often a part of the isolator validation effort. This is usually accomplished by performing media fills at the beginning and end of the chosen period, with environmental monitoring over the entire course of the time period. Sterility test isolators are qualified in a similar manner with mock sterility testing or bacteriostasis/fungistasis testing conducted over the course of the decontamination interval. In filling isolators, the current trend for the interval is limited to a number of consecutive batches of the same product. Sterility testing intervals are generally longer, with periods of up to three months reported for the primary sterility test isolator (the transfer isolators used with these units are decontaminated much more frequently).

Revalidation Frequency

Decontamination is considered to be such an important component of the sterility assurance for an aseptic isolator system that annual revalidation is appropriate. In this effort a single confirming study using BIs is used to reconfirm the continued effectiveness of the decontamination treatment. If a more resistant BI than originally employed is utilized for the revalidation, some unexpected positives may result.

The decontamination process has not changed; however, a change in the resistance may suggest that it is not as effective. Before beginning the revalidation, the resistance of the BIs to be used should be determined, and the expectations for its inactivation adjusted appropriately. Caution must be exercised to avoid defining the revalidation effort into failure unintentionally.

Residuals

The user will need to validate a safe level of sanitizing agent residual after the treatment.

Outgassing measurements can be difficult, and are often technique dependent. Residues on non-product contact surfaces are generally non-critical to the process.

Where residues on product contact materials are a concern they must be measured directly. Items that are more susceptible to retaining residual sterilant are often those that are wrapped in Tyvek or sealed in plastic. In considering whether a material will allow the passage of H_2O_2 and thus be a potential problem with residuals post-aeration, the moisture transmission can be a reasonable predictor. Those materials that are permeable to moisture (an easily located piece of information) are often those that are permeable to hydrogen peroxide. Similarly, those materials that readily adsorb water will also tend to adsorb hydrogen peroxide.

Materials Issues

The prevalent decontamination agents are also highly reactive chemicals. It is important to recognize the trade-off between microbial kill and adverse material effects. In attaining a greater kill (as is often the objective when sterilization is the target), the dwell period where materials are in contact with the agent is extended. Longer exposure times are sometimes associated with negative effects on the materials in contact with the agent. The author has experienced embrittlement of gloves, breakdown of some polymers, corrosion of metal surfaces and discoloration of materials after repeated exposure to lengthy cycles. To ensure long life of the equipment, the processing time should be minimized to what is sufficient to provide the required log reduction. The use of half-cycle approaches for isolators that arbitrarily double the exposure time beyond what is needed to attain the needed log reduction can chemically stress the materials exposed to the agent unnecessarily.

Aeration

A major element of each process is the establishment of the aeration period at the conclusion of the treatment. This step is important to process reliability as it assures that materials exposed to the agent are not adversely affected by the treatment. It needs to be established as a part of the decontamination cycle development to define the required aeration period before activities can commence within the isolator. The common (and sometimes necessary) answer to what residual level to aerate to is often the eight-hour TWA defined by OHSA for the agent used. This may appropriate in some instances (e.g., sterility testing) where residual agent can impede microbial recovery; however in many instances operations can commence while the level is still somewhat higher. It must be recognized that the limit is an average acceptable value over an eight-hour period, and that the residual level will continue to decrease over time. Thus a value of 5 ppm of H_2O_2 at the end of the aeration may be fully acceptable as the activities to follow may not be impacted at this level, and by the time process materials are present the value will be substantially lower. The OI ISA TWA limit for H_2O_2 is 1 ppm, while ClO_2 and O_3 both have a TWA of 0.1 ppm. No TWA has been defined for peracetic acid.

The materials present in the isolator chamber can be major contributors to H_2O_2 aeration difficulties. Items that are permeable to moisture are subject to penetration by H_2O_2 and if these materials are exposed in long decontamination periods, desorption of the H_2O_2 in or on these items back into the air is often rate limiting. One of the ways to overcome this is to limit exposure times (thus decontamination as a goal is preferable due to its shorter exposure requirements and consequently lower time for adsorption of H_2O_2).

Aeration times can be improved by increasing the number of air changes in the isolator; however if adsorption is a significant factor with the materials being decontaminated, increasing the number of air changes may have limited impact.

Other Points to Consider

Once the decontamination process is over, the environmental conditions inside the isolator are maintained by the air system of the isolator and "other" factors. It is not reasonable to think that every piece of equipment/item contained in an isolator can or should be "sterilized." There will always be some items within the isolator where the agent is prevented from contacting every part because it is obstructed by another item. Decontamination is really all that should be demonstrated. After all, it is not as if the isolator will be injected into a patient, and thus attaining sterility for it is not an appropriate process objective.

CONCLUSION

The decontamination (or sterilization if that is the intent of the treatment) is readily attainable using any number of different materials that are lethal to microorganisms. The effectiveness of the treatments can be established using resistant BIs. Regardless of what the goal of the process is, it is certain that at the end of it the isolator will contain fewer microorganisms than any manned clean room.

APPENDIX I

Decontamination Methods for Isolators

The decontamination of isolators can use any number of different sterilants—H_2O_2, ClO_2, O_3, or peracetic acid (CH_3COOOH). Hydrogen peroxide has been the "most used" decontamination method and was first introduced in the early 1990s. Various suppliers offer generators to deliver H_2O_2 for reproducible decontamination with substantial documentation. Chlorine dioxide is rapidly lethal to microorganisms and may see greater usage in the future. Ozone has been evaluated for use in isolators; however, at the present time there is no commercially available delivery system for it. Peracetic acid was the first available agent for isolator decontamination and is still used in some sterility testing units (PVC) and a few production applications as well.

Hydrogen Peroxide Decontamination

Hydrogen peroxide is the most widely used method for the decontamination of isolators and as such will be the only decontaminating agent covered in detail in this chapter.

Hydrogen peroxide (H_2O_2) falls into a broad category of sporicidal antimicrobial agents that are classified as oxidants. Other widely used sporicides that fall into the same general category as hydrogen peroxide include ozone, peracetic acid (and other super oxides), and halogens (including chlorine, chlorine dioxide, iodine, and sodium hypochlorite). The sporicidal properties of oxidants in general and H_2O_2 in particular have been recognized for decades. H_2O_2 has been widely used as a surface sterilant at liquid concentrations of $>4\%$ and as a skin disinfectant at concentrations of 3% or less for many years.

Given its common use in medicine as a sterilant and disinfectant the antimicrobial and the sporicidal properties of H_2O_2 have been widely studied. Abundant information is available regarding the sporicidal effectiveness of H_2O_2 at a variety of concentrations (8). Because H_2O_2 alone does not leave toxic residues and decomposes to water and oxygen it is a useful agent for the decontamination/sterilization of materials/surfaces that may come into contact with sterile products.

H_2O_2 decontamination was first introduced by AMSCO (now Steris) with a free-standing generator having a closed, single loop configuration in the late 1980s. This system is still widely used today. The process consists of an H_2O_2 reservoir which contains either 30% or 35% H_2O_2 with the remainder of the water. The target is dehumidified by the dry air stream to a user selectable target value. The liquid H_2O_2/H_2O is pumped from this reservoir at a rate that is controlled gravimetrically—the flow rate can be chosen by the user. The solution is vaporized at temperatures of $\sim 100°C$ into a dehumidified air stream and blown into the target isolator at a user selected airflow rate. The injection of vaporized H_2O_2/H_2O continues for a user-defined time period and the injection rate is chosen to deliver a calculated concentration based upon the volume of the target system and assumptions regarding the decomposition rate of the H_2O_2. In this closed-loop configuration the H_2O_2 in the exhaust "leg" of this loop flows through a mixed bed catalyst decomposing the H_2O_2 into water and oxygen. This air stream is returned through the dehumidifier removing residual water and picks up freshly vaporized H_2O_2 on its way to the inlet point within the target system.

This closed single loop system of VPHP operation was the only technology available through the mid-1990s. In the mid-1990s adaptations of the closed-loop VPHP generator began to appear that operated in an open-loop configuration. This obviated the limitations imposed by the restricted capacity of the desiccant drier systems used in the only commercial generator available. Shortly thereafter, modular generators designed to operate in an open-loop configuration were introduced, as was a generator that employed a continuous duty refrigerant drier that did not require regeneration. Other commercial generators have been introduced that use dual-loop configurations in which the humidity control loop is separated from the H_2O_2 injection loop. There are also generators that are built into the isolator and vaporize H_2O_2 directly onto a hot plate located in the isolator air handler plenum.

At the present time H_2O_2 generators are available from several vendors and some of these systems have unique operational philosophies and process control systems. Although different in many ways, most of these units dehumidify the target to some set moisture level prior to initiating injection, vaporize hydrogen peroxide, and have the ability to control the rate of injection and the total exposure time.

Outline of the H_2O_2 Process

The H_2O_2 process can be considered to consist of up to four "phases" or steps. The various equipment vendors may use different terms to describe these phases. For the purposes of this chapter the four phases are identified as: dehumidification, conditioning, decontamination, and aeration.

Dehumidification

The first phase in all H_2O_2 processes is dehumidification. The purpose of dehumidification is to reduce the ambient moisture level within the isolator enclosure so that the concentration of H_2O_2 can be maximized. H_2O_2 is commonly vaporized from mixtures of 35% H_2O_2 and 65% water, although both higher and lower concentrations of H_2O_2 have been used. Some equipment vendors describe the H_2O_2 process as a "dry" one. These vendors generally recommend that condensation during VPHP process be avoided. Reduction of ambient moisture within the enclosure allows a higher injection rate of the H_2O_2/H_2O mixture while avoiding condensate formation.

It is important to note that reduction of humidity during a gas or vapor anti-microbial process is somewhat atypical. Generally, gas anti-microbial processes require an increase in humidity (sometimes called preconditioning) for optimal kill effectiveness. Not surprisingly, this has become a point of controversy in H_2O_2 decontamination. Other scientists and vendors claim that the H_2O_2 process is a wet process and that effective spore kill occurs only when micro-condensation of the H_2O_2 occurs on the surface. Proponents of the micro-condensation approach often set humidity targets that are higher than those advocated by those whose objective is a dry process.

From a practical perspective initial dehumidification targets of 10% to 40% have been successfully used, where success is defined as a reproducible spore log reduction. Users can use the humidity recommendations of their chosen equipment vendor as a starting point, although they may wish to experiment with a range of values between 20% and 40%. Reduction of humidity concentrations to the point where drying of the spores is possible may actually increase spore resistance.

In some cases the dehumidification process serves a secondary purpose. In closed-loop H_2O_2 processes dehumidification may be accomplished by the circulation of hot, dry air. The dehumidification phase serves to increase the ambient temperature within the target and therefore increase the dew point within the enclosure. Of course, temperature may not increase if refrigerant drying or another form of dehumidification is employed.

When large enclosures are to be decontaminated, some form of external dehumidification is often employed. External dehumidification can substantially shorten the time required to reach the humidity set point, and of course where modular generators are used that lack a integral dehumidification process an external drier is essential.

Process monitoring during dehumidification consists of one or more humidity sensors that provide data to the control system. Dehumidification may be either a timed process based upon empirical data, or the phase can be ended when the set point humidity is reached. In closed-loop systems that rely on desiccant driers the rate of dehumidification will slow as the drying capacity is consumed. This must be considered in setting the dehumidification phase time. Another parameter that can be controlled during the dehumidification phase is the airflow rate. In general, the higher the airflow rate the more rapid the dehumidification phase, although this depends upon the moisture adsorption capacity of the drier.

Conditioning

The purpose of the conditioning phase is to inject H_2O_2 at a rate that enables the calculated concentration to be reached as rapidly as possible. The operating parameters for the conditioning phase are generally not established until the target concentration is determined. During the conditioning phase an injection rate higher than sustainable throughout the process is utilized to increase the internal concentration of H_2O_2 to the desired steady-state condition more rapidly.

Decontamination

The decontamination concentration is determined with consideration of the temperature within the target enclosure. One approach is to analyze the observed or expected temperatures within the enclosure over the expected duration of the decontamination phase. Temperature is the critical parameter in decontamination cycle development because it establishes the H_2O_2 and H_2O dew point. Temperature can be analyzed by placing calibrated thermocouples throughout the enclosure during cycle development. Once the temperature profile during the decontamination phase is known it is possible to determine the appropriate injection rate to achieve the appropriate concentration. Some equipment vendors provide humidity sensors to assist in the process, although it is possible to develop decontamination parameters without the benefit of sensors. Vendors provide cycle development information to assist the user in determining an appropriate injection rate to achieve the appropriate concentration.

Although debate continues as to whether H_2O_2 decontamination is a dry or micro-condensing process, it is accepted that spore killing effectiveness is reduced if the concentration is well below the dew point. The narrower the temperature spread within the enclosure the better, as this assures constant relative humidity. Unfortunately, temperature spreads of 5°C to 10°C are not uncommon in larger enclosures, and effective H_2O_2 processes must be developed with this range of temperature variation. Design features, such as auxiliary circulation fans to maximize turbulence during the process, can narrow the temperature spread and also enhance H_2O_2 distribution.

It is well accepted that the killing effectiveness of sporicides increases as the temperature increases. However, in the case of H_2O_2 since killing effectiveness diminishes at concentrations well below the dew point, the coolest location within an enclosure may be best case in terms of spore log reduction.

Another controversial matter in H_2O_2 cycle development is the decomposition rate of H_2O_2. The continuous injection strategy used by some vendors is predicated on the idea that H_2O_2 has a half-life of <20 minutes and therefore replenishment of the H_2O_2 concentration is necessary. Others hold that the half-life of H_2O_2 is too long to be of practical consequence in the development of H_2O_2 processes. Most importantly, both strategies have resulted in the development of H_2O_2 cycles that are effective, robust and reproducible.

The confirmation of a newly developed cycle can be confirmed directly through the use of chemical indicators and BIs. The purpose of the H_2O_2 process is to kill residual bioburden in the isolator enclosure. Therefore, the direct demonstration of spore-killing effectiveness is the best way to confirm the effectiveness of the H_2O_2 cycle. Chemical indicators are now available that can demonstrate that H_2O_2 vapor is well distributed within the enclosure. Although the chemical indicator is a qualitative method for H_2O_2 analysis, it is nevertheless a valuable tool in cycle development and perhaps the only form of analysis necessary to demonstrate H_2O_2 exposure in routine operation.

After the effectiveness of the H_2O_2 injection strategy and dwell period concentration have been established, it is possible to finalize the conditioning phase. Vendors generally provide guidance in their respective cycle development guidance documents for selecting appropriate conditioning injection rates so that the exposure phase decontamination concentration can be reached relatively rapidly. It is best not to abbreviate the conditioning phase too much, since some generators will demonstrate a rather high variability during the first two or three minutes of injection.

Aeration

There are two critical factors that must be considered in achieving aeration of H_2O_2 vapor from an enclosure. The first and most important factor is the air exchange rate. Studies have shown that the higher the fresh air exchange rate, the more rapid the aeration phase can be completed. For the most rapid aeration fresh air exchange rates of 100 air changes/hr or more should be considered. No other factor will have as great an effect on reducing aeration time. The second key factor is outgassing (desorption) of materials exposed to H_2O_2. Outgassing time can be reduced by selecting materials that are less likely to absorb and retain H_2O_2.

In general, metals and glass are less absorbent of H_2O_2 than plastics and elastomers. Some elastomers, such as Hypalon®, which can be used for glove and sleeve materials are less absorbent than others and therefore outgas more rapidly. There is empirical evidence that items that adsorb moisture will also tend to adsorb H_2O_2. Items wrapped in Tyvek with large internal volumes (such as pre-sterilized filter cylinders for sterility testing) may have substantial penetration of H_2O_2 into the interior volume and thus be difficult to aerate. Where items of this type are present in large numbers, a shorter decontamination phase is desirable as it also results in a correspondingly shorter aeration phase because of reduced adsorption into the materials. If extremely rapid outgassing is desired consideration should be given to constructing the enclosure of non-absorbent materials. In general, it is possible to aerate most large production enclosures to safe operating concentrations of H_2O_2 within three to six hours. More rapid aeration has been achieved with careful equipment design.

Although the starting point for aeration time can be calculated using dilution rate equations, it is also possible to measure concentrations using sensors or test tubes that are relatively accurate to H_2O_2 at levels of <5 ppm. The only certain way to develop aeration processes is through empirical measurement.

Target values for residual vapor levels are highly process dependent. It has been reported that some products that are very sensitive to oxidation require aeration to levels well below 1 ppm. In other cases with less sensitive concentrations of 3 to 5 ppm may be acceptable. Enclosure residual levels do not necessarily correlate with actual levels observed within product containers; therefore studies using actual product containers are highly recommended.

The eight-hour TWA exposure level for personnel to H_2O_2 is 1 ppm. This value has appeared to many users to be a convenient acceptance criterion for residual at the end of aeration. While this acceptance criterion may work for many users, it should not be considered a hard and fast rule. Establishing a set point of less than 1 ppm may result in overall decontamination cycle times that are impractically long, since aeration is in many cases the most time consuming of the four phases of the process.

Decontamination Using the ClO_2 and O_3

The use of halogens in the decontamination of manufacturing equipment and environments must be done with careful attention to concentration and exposure since these agents too can be highly corrosive even to stainless steels. Halogen-releasing compounds such as chlorine dioxide may prove to be a valuable alternative in the decontamination of isolators and in other pharmaceutical applications. However, the use of halogens will generally require careful consideration in the selection of materials and components for the construction of isolators. At a minimum initial passivation of stainless steel and regular repassivation will be essential if chlorine dioxide or other halogenated sporicides are to be used.

Ozone may also prove to be a useful sporicide for the decontamination of isolators. Ozone, like other sporicidal gases including EtO and chlorine dioxide, requires the humidity of the target to be raised to >60% for optimum sporicidal effect. Ozone breaks down into an innocuous substance, oxygen. Thus, there is no concern with toxic residues.

If used in accordance with well-understood principles any of the oxidants can be used safely in the work environment. Oxidants are skin and mucous membrane irritants that require personnel exposure levels on the order of 0.1 ppm over an eight-hour TWA. Some oxidants may have tighter exposure restrictions. In general, industrial sensors are available that can detect any of the commonly used oxidants at the low parts/million levels necessary to ensure that short- and long-term exposure levels can be maintained.

The validation of these gaseous agents often follows the approaches originally developed for EtO with the exception that half-cycle approach is not recommended for the reasons described earlier. The parameters of concern are gas concentration, humidity, temperature and time. As these agents are true gases (as opposed to a condensable vapor like H_2O_2) the adjustment of the parameters from validation and routine operation is straightforward.

Peracetic Acid Decontamination

Unlike the other decontamination processes, peracetic acid decontamination is a wet process in which a liquid mist of peracetic acid is sprayed into the isolator. Peracetic acid is provided as a mixture of peracetic acid, hydrogen peroxide and water. A dehumidification step is not required, given the wet process. Only a three-step process is required—conditioning, sterilization, and aeration.

REFERENCES

1. PDA TR #1. Moist Heat Sterilization in Autoclaves. Draft #13, 2003.
2. PDA TR #34. Design and Validation of Isolator Systems for the Manufacturing and Testing of Health Care Products, 2001.
3. FDA. Guideline on Sterile Drug Products Produced by Aseptic Processing, 2004.
4. Lewis JS, Rickloff JR. Inactivation of *Bacillus stearothermophilus* spores using hydrogen peroxide vapor. In: Proceedings of the 91st General Meeting of the American Society for Microbiology, 1991.
5. Adams D, Hervey L, Woodworth A. Evaluation of biological indicators for vaporized hydrogen peroxide. In: Proceedings of the PDA Spring Meeting, 2000.
6. Kokubo M, Inoue T, Akers J. Resistance of common environmental spores of the genus *Bacillus* to vapor hydrogen peroxide. J Pharm Sci Technol 1998; 52:228–31.
7. Akers J. Private communication.
8. Rickloff JR, Oreliski PA. Resistance of various micro-organisms to vaporized hydrogen peroxide in prototype tabletop sterilizer. In: 89th Annual Meeting of the ASM, New Orleans, May 1989.

20

Validation of Sterilizing-Grade Filters

Suraj B. Baloda
Millipore Corporation, Bedford, Massachusetts, U.S.A.

INTRODUCTION

Sterile filtration is one of the most critical final steps in pharmaceutical manufacturing process, and the challenges of filtration processes have also evolved over time due to constantly changing demands for removal of specific bioburden from the final product. Besides the filter manufacturers, biotechnology and the pharmaceutical industry's own pursuit for perfection in producing safe drugs, the regulatory authorities worldwide also mandate the need for adherence to aseptic processing guidelines. Since the publication of the second edition of this book, a series of guidelines and regulations have been revised or updated. The final draft of the FDA's 1987 *Aseptic Processing Guideline* has also been published in September 2004 (*Sterile Drug Products Produced by Aseptic Processing-Current Good Manufacturing Practice*) (1). Earlier, the PDA published the authoritative summary of best practices in sterile filtration and validation of sterile filtration in its 1998 TR (*Sterilizing Filtration of Liquids-PDA Technical Report No. 26*) (2). The PDA TR 26 is a comprehensive monograph that highlights the history of sterile filtration, provides valuable details on the criteria of selection and functioning of filters and also explains validation considerations and integrity testing methods.

Filter validation is an important aspect of the sterilization process that needs greater understanding to improve sterility assurance and efficiency. Unlike validation of terminal sterilization, which end users generally master with little difficulty, validation of aseptic filtration remains somewhat enigmatic. Thus, before designing a validation protocol, it is essential to understand the objectives of the desired process. This can be accomplished by generating a *filter requirements specification*. From this specification, validation becomes a matter of proving claims made for the filter, as outlined in this chapter.

The investment of initial cost as well as the effort in understanding the scientific rationale of the validation of aseptic filtration process by the end users goes a long way in assuring the quality of the product and reduced operating costs by decrease in scrap or rework. Moreover, the validation of aseptic filtration, apart from being a regulatory requirement, also makes good business sense.

Why Validate?

There are two main reasons to validate processes. The first is obvious, whereas the second is often known, but rarely quantified.

Good manufacturing practice (*regulatory requirements*): Various worldwide regulatory bodies require validation of manufacturing processes for LVPs and SVPs, ophthalmics, veterinary medicines, bulk chemicals, and in vitro diagnostics.

Good business practice: It is essential to understand that an out-of-control process increases the amount of rework or scrap incurred, thereby increasing cost. A controlled process gives reproducibility and product consistency within known limits. A controlled process also aids regulatory compliance and, therefore, provides a license to do business.

Why Validate Filters?

As mentioned above, filtration is an important final step in the manufacturing process and before we discuss the topic further, it is assumed that:

- The reader is familiar with filters, aseptic processing and has a basic understanding of microbiology.
- The reader is familiar with the general principles of validation.
- The filters have been chosen and are correctly sized for the required operation.
- The studies are designed to incorporate all prerequisite filter processing steps. For example, when performing a study to determine potential filter extractables, all normal prerequisite processing steps are first performed on the study filter, such as preflush and sterilization.

The very nature of aseptic processing presumes that the filter is one of the most critical components in the process and that it provides sterility assurance of the final product. It is, for example, as important as an autoclave in a terminal sterilization process. Therefore, a filter requires stringent controls and attention to assure consistent and

---

Abbreviations used in this chapter: ASTM, American Society for Testing Materials; ATCC, American type culture collection; BI, biological indicator; CFR, Code of Federal Regulations; EPA, Environmental Protection Agency; EPROM, Electronically Programmed Random Only Memory; FDA, Food and Drug Administration; FT-IR, Fourier Transform Infrared; GMP, good manufacturing practice; HPLC, high-performance liquid chromatography; ICH, International Conference on Harmonization; ISO, International Organization for Standardization; LAL, limulus amebocyte lysate; LVP, large volume parenteral; NDIR, nondispersive infrared; NFR, nonfiber releasing; PDA, Parenteral Drug Association; PLC, programmable logic controller; SVP, small volume parenteral; TOC, total oxidizable carbon; TR, Technical Report; USP, *U.S. Pharmacopeia*; WFI, water for injection.

reproducible results. Again, both reasons for process validation come into play when talking about filters:

- *Good manufacturing practice*: Validation of filters is a regulatory requirement throughout the world (see References). As a quick reference, a snapshot of the relevant sections of the FDA's Aseptic Processing Guidelines (2004) are also incorporated and depicted as "FDA Guidelines" in the body of the chapter.
- *Good business practice*: Aseptic filters are used to assure sterility of a final product. Just before filtration, a drug product undergoes several value-added steps. After passage through the filter, the drug product is usually placed in its final container. The cost of rework at this stage can be extremely high. Indeed, many manufacturers will scrap product, rather than attempt to rework it at this stage.

What Needs to Be Validated? How Is It Done?

The purpose of filter validation is to ensure that the filter will *reproducibly* remove undesirable components (e.g., microbial bioburden), while allowing passage of desirable components.

Therefore, it is necessary to understand, and include in a validation study, the parameters such as how the filter is manufactured, expected filter operations (how the filter will be sterilized, how many times the filter will be used), the potential for toxic byproducts, and the obvious aspects of retention and inertness. Before any work is initiated to validate a filter, two prerequisites must be considered and satisfied:

1. The properties of the filter itself must be consistent and reproducible from lot-to-lot. This will be dealt with in some detail later in this chapter.
2. The drug product (properties) must be consistent and reproducible from lot-to-lot. (The subject of drug product reproducibility will not be dealt with in this chapter. Reproducibility is a basic underlying assumption on which the entire filter validation will be built.) Without drug product consistency and reproducibility, it is not possible to obtain consistent, reproducible results for flow rate, throughput, retention, filter inertness, and so on. If the validation process uses a consistent, reproducible drug product, then known processing times, flow rates, and throughputs can be set for routine manufacturing. Any changes in drug product (which may be a result of changes in raw materials, processing conditions, concentrations, or the like), may change the filter's performance. This final point cannot be stressed too highly because, not only can the filter performance be affected, but also changes to the drug product or processing conditions can *invalidate* the original study. Furthermore, it is also important to bear in mind and evaluate the impact of the drug product or the pharmaceutical formulation on the viability of the test organism (or bioburden) which would impact the retention test results.

   FDA guidelines:

> Direct inoculation into the drug formulation is the preferred method because it provides an assessment of the effect of drug product on the filter matrix and on the challenge organism. However, directly inoculating *Brevundimonas diminuta* into products with inherent bactericidal activity against this microbe, or into oil-based formulations, can lead to erroneous conclusions. When sufficiently justified, the effects of the product formulation on the membrane's integrity can be assessed using an appropriate alternate method. For example, a drug product could be filtered in a manner in which the worst-case combination of process specifications and conditions are simulated. This step could be followed by filtration of the challenge organism for a significant period of time, under the same conditions, using an appropriately modified product (e.g., lacking an antimicrobial preservative or other antimicrobial component) as the vehicle. Any divergence from a simulation using the actual product and conditions of processing should be justified.

Figure 1 outlines the procedure that can be used to test the viability of the test organism prior to retention testing.

To begin a validation study, it is prudent to first document the requirements of the filter. This should be done in sufficient detail to ensure the proper selection of a filter system that can be readily validated. The next step is to conduct those studies that can be done off-line or

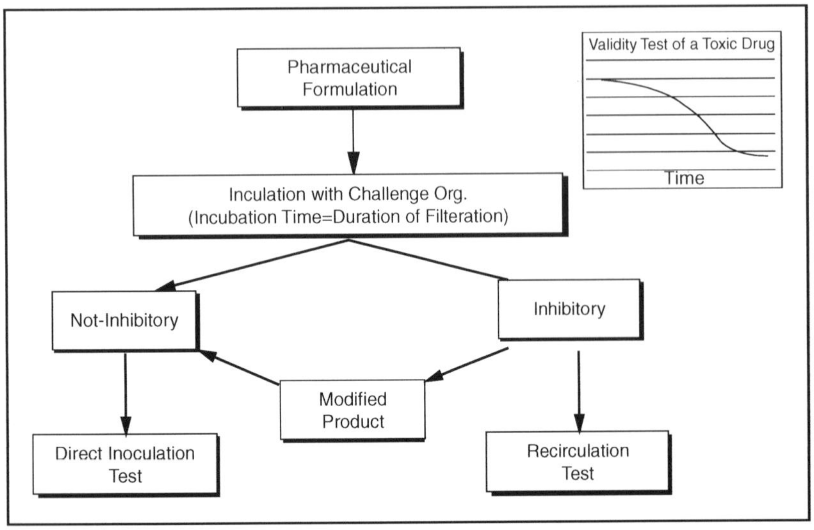

Figure 1 Flow chart for testing the inhibitory effect of a pharmaceutical formulation on bacterial viability.

scaled-down (examples include bacterial retention and filter extractables). Then, full-scale tests are performed to verify correct, consecutive operation and reproducibility. Finally, validation must include ongoing evaluation and provisions that permit change of controls to be implemented over time.

There is considerable debate on whether to use scaled-down or full-scale filters for validation studies. The choice depends on the scientific rationale and the goal of the specific element of the validation study being considered. For example, if bacterial retention is under study, then it should be sufficient to prove that the membrane alone is capable of the required log reduction. This assumption is clearly dependent on an accurate integrity test that can detect a manufacturing defect (such as incomplete filter seaming) in the full-scale filter.

If, however, the ultimate aim of the validation study is the verification of the scaled-down simulation or the validation of the filtration process itself, the validation should be run at full-scale, with a minimum of three consecutive, full-scale runs with three filter lots and three drug product lots. This, however, is neither the beginning nor the end of a filter validation.

Table 1 gives an overview of the elements of a validation study. The elements of interest for the validation depend on the requirements and specification set for the filter. It is not necessary to study each element to validate a filter but the decision to study (or not study) a particular element must be based on the scientific and process rationale. In any case, the final validation data generated must be able to stand on its own merits (and scientific rationale) and be able to pass further regulatory scrutiny.

Table 2 lists some of the tools available to demonstrate that the elements listed in Table 1 are true and fully understood. Some tools can be used to describe more than one goal. Other elements, such as integrity, can themselves be used as tools once they have been qualified for the application.

Who Is Responsible for Validation?

FDA guidelines state:

> When the more complex filter validation tests go beyond the capabilities of the filter user, tests are often conducted by outside laboratories or by filter manufacturers. However, it is the responsibility of the filter user to review the validation data on the efficacy of the filter in producing a sterile effluent. The data should be applicable to the user's products and conditions of use because filter performance may differ significantly for various conditions and products.

Table 1 Validation Study Elements

Physical	Chemical	Biological
Reproducibility	Inertness (i.e., compatibility, extractables, adsorption)	Endotoxins
Sterilization	Activity/stability	Toxicity
Integrity	Consistency and reliability	
Operation		
Shedding		
Particulates		
Fibers		
Particle retention		Microbial retention

Ultimately, the drug manufacturer is responsible for filter validation. Therefore, the drug manufacturer should select a filter manufacturer that not only provides sufficient information but also the services required to facilitate proper validation. Because certain validation requirements are in demand from all filter users, the filter manufacturer should be in a position to provide some of the required information. Thus, some of the burdens of responsibility can be alleviated by choosing a filter manufacturer that can provide this commonly required information.

Table 3 outlines the validation elements and also the subparts of these elements that could be reasonably expected to be either the responsibility of the filter manufacturer or the filter user.

ELEMENTS OF A VALIDATION STUDY

After a filtration process is properly validated for a given product, process, and filter, it is important to ensure that identical filters (e.g., of identical polymer construction and pore size rating) are used in production runs.

The following sections detail the purposes and the background of the various elements and tools listed in Tables 1–3. As mentioned earlier, not all the elements need to be studied for any given filter. Elements and tools should be chosen as a direct result of the filter requirements, such as the purpose, function, or operating limitations of the filter. If a documented filter requirement specification is unavailable, progressing with the filter validation study may result in unnecessary testing, while the functional requirements of the filter remain untested.

Performance Reproducibility

Reproducibility applies to both the drug product and the filter. Success of the validation and the drug manufacturing process will depend on these two key elements.

FDA guidelines state:

> Factors that can affect filter performance generally include (*i*) viscosity and surface tension of the material to be filtered, (*ii*) pH, (*iii*) compatibility of the material or formulation components with the filter itself, (*iv*) pressures, (*v*) flow rates, (*vi*) maximum use time, (*vii*) temperature, (*viii*) osmolality, and (*ix*) the effects of hydraulic shock. When designing the validation protocol, it is important to address the effect of the extremes of processing factors on the filter capability to produce sterile effluent.

Drug Product

The chemical attributes of a drug product (such as formulation concentration, chemical composition and constituents, pH, viscosity, density, ionic strength, and osmolarity) should be known and controlled within defined limits. These limits, upper and lower, will determine important process characteristics, such as flow rate, processing time, throughput for a given filter surface area, and so on. It cannot be stressed too highly

Table 2 Elements and Tools of Filter Validation in the Manufacturing Process

Elements to be demonstrated	Tools in the manufacturing process
Filter performance reproducibility	Filter documentation available in filter validation guides and elsewhere: adherence to good manufacturing practice, change control in the filter manufacturing process, change notification, certificates of quality (C of A), lot release criteria, validation guides
Product reproducibility	Drug product characteristics: consistency, viscosity, particulates, bioburden, concentration, impurities
Sterilization	Heat distribution profile evaluation (thermocouples), heat penetration evaluation (spores)
Integrity test	Correlation to microbial retention, qualification of methods
Operation	Flow rate, throughput, temperature, pressure, integrity, visual
Particulates	Monitoring of effluent, on-line monitoring
Fibers	Filter manufacturer adherence to 21 Code of Federal Regulations, preuse flush
Inertness (compatibility)	Manufacturer's documentation (charts), flow rate, throughput, integrity, weight, visual, pH, conductivity
Inertness (extractables)	Weight change, gravimetric extractables, oxidizable substances, ultraviolet, high performance liquid chromatography, total oxidizable carbon, Fourier Transform Infrared, NVR
Adsorption	Concentration analysis before and after use
Activity/stability	Concentration analysis, activity analysis, stability trials, conformational confirmation
Microbial retention	Challenge with microorganism in drug product
Bacterial endotoxins	Filter lot release, limulus amebocyte lysate testing of in-process samples
Toxicity	Materials of construction, class VI plastics test, cytotoxicity studies, clinical trial data

that, once set and validated, changing any of these chemical attribute limits may affect not only process characteristics but could also negate previous validation work.

Sometimes solution reproducibility is not possible. In this event, it is necessary to define limits and measure incoming raw materials for compliance. Operating with reproducible solutions and filters enables specifications to be set for flow rate, processing times, and throughput. Such specifications provide a very effective tool for in-process monitoring. A significant change in flow rate, throughput, or processing time could indicate a change in a solution's or a filter's consistency.

Defined limits for time, flow rate, and throughput are also necessary to conduct validation studies covering all anticipated operating limits. For example, the longest time is the worst-case for a retention study, because of the potential for bacterial movement through the filter (referred to as grow-through). Although there is some concern over low-flow conditions, it is generally accepted that high-flow rates are worst-case, because they lead to less residence time of a fluid within the filter. High-flow

Table 3 Responsibilities of Filter Manufacturers and Filter Users

Element	Filter manufacturer	Filter user
Filter reproducibility	Validate filter claims and the filter manufacturing process	Review all data and audit manufacturer
Product reproducibility		Ensure consistency and reproducibility
Sterilization	Provide recommended procedures with limits for time, temperature, and number of cycles	Operate within manufacturer's limits Validate the procedure in use
Integrity test	Provide procedures and test specifications Provide test correlation with bacterial retention	Follow manufacturer's procedures Ensure correlation to bacterial retention exists Validate test method Perform integrity ratio work if wetting with product
Operation	Provide limits for operating, temperature, pressure	Design system to meet filter requirements and limits
Particulates	Provide data for removal	Verify required limits are achieved
Fibers	Meet non-fiber-releasing claim [21 Code of Federal Regulations 210.3b(6)]	Ensure NFR, preflush filters according to recommendations
Microbial retention	Provide retention claims, test methods and service	Have the microbial retention test performed with the drug product and microorganism (bacteria, *Mycoplasma*, virus of interest)
Inertness	Provide charts for all materials of construction	Document compatibility
Compatibility		Perform studies
Extractables	Develop methods, identify components	Perform studies
Adsorption	Indicate known problems	Perform studies
Drug activity and stability	Indicate if known problems exist	Verify no conformational changes or activity losses Perform stability studies
Endotoxins	Perform analysis on a per lot basis	Verify low endotoxin levels from filters Ensure process operation does not contribute endotoxin (e.g., prolonged use with intermittent sterilization)
Toxicity	Perform testing and provide results (class VI plastics, cytotoxicity)	Obtain results and reports. Include filter with clinical trials or perform toxicological review

rates have the potential to cause greater differential pressures across the membrane, thereby creating a greater possibility of retention failures from membrane damage. Higher rates of flow frequently lead to higher differential pressures, which in nonsterilizing-grade filters has led to decreased microbial retention and longer times.

Drug product reproducibility must be specified, monitored, and controlled by the drug manufacturer.

Filter Performance Reproducibility

FDA guidelines state:

> After a filtration process is properly validated for a given product, process, and filter, it is important to ensure that identical filters (e.g., of identical polymer construction and pore size rating) are used in production runs. Sterilizing filters should be routinely discarded after processing of a single lot. However, in those instances when repeated use can be justified, the sterile filter validation should incorporate the maximum number of lots to be processed.

There are several questions that must be answered relative to filters used for validation and those that will be supplied long-term by the filter manufacturer.

1. If a scale-down filter is used, is it the same membrane that is used in the full-scale filter?

   Filter manufacturers may manufacture filters of the same polymer by different processes. This could result in a scale-down filter and full-scale filter having significantly different characteristics. The membrane in both devices should be manufactured, tested for quality and performance, and released under the same conditions to ensure that the study is valid.
2. Is the filter being validated representative of the current catalogue item from the filter manufacturer and of all future filters?

   The filter manufacturer should be consulted to ensure that
   a. all filter claims have been qualified and there is control over filter raw materials;
   b. the filter manufacturing process has been validated within specified operating windows;
   c. in-process and final release testing is performed on a per lot basis (particularly critical parameters such as retention and endotoxin);
   d. certificates of quality assurance are available and that validation guides are available;
   e. the filter manufacturing plant may be audited;
   f. there are policies for, and strict adherence to, change control; and
   g. the filter is representative of the current catalogue item—several lots should be obtained from salable inventory.

Although these items are not regulatory requirements, possession of the information is useful during the validation process and in the preparation of marketing applications.

Assessment of a filter manufacturer's policy on implementation of change control during the manufacture of filters or change in filter product characteristics will help access the answer to question 2. It is highly likely, and desirable as technology advances, that changes will be made in a filter manufacturing process. It is necessary, therefore, to ensure that all changes are evaluated before implementation and that requalification takes place as required. Furthermore, because the filter manufacturer cannot judge the subtle effects of filter changes on the drug-manufacturing process, it is necessary that there be a change notification policy. That is, for significant changes, all end users should be notified in sufficient time to evaluate the change. Additionally, end users should be notified of all other changes on a periodic basis.

Sterilization

To validate use of a sterilizing-grade filter, it is not only necessary to prove that the filter is (and will continue to be) adequately sterilized, but also that the sterilization method does not damage the filter.

There are many sterilization methods available, but the preferred method of filter sterilization is with moist heat (steam) (4) because it is relatively easy to use and it minimizes potential sources of residual chemicals.

Important considerations (variables) for steam sterilization of filters are time, temperature, pressure, air and condensate removal, heat-up, cool-down, and the total number of sterilization cycles. Any one of these variables, if uncontrolled, could lead to filter failure.

Current practice for autoclave validation and steam-in-place cycles are that both thermocouples and BI (suspensions or spore strips) be used. The thermocouples verify that adequate temperatures are achieved, and the BI verify kill by moist heat. This chapter does not address steam sterilization validation, as the subject has been addressed in a guideline elsewhere (5). In practice, the validation of the steam sterilization of filters may be summarized as follows:

1. Obtain all relevant, performance specifications of the filter and the filter housing from the manufacturer, such as the maximum recommended operating temperature, thermal resistance, the maximum number of sterilization cycles, the maximum allowable hydraulic pressure resistance, and so on.
2. Install the filters and filter housings to ensure that they self-drain of air and condensate (prevention of cold spots; Fig. 2).
3. Perform cold spot mapping of the filter system (heat distribution studies), followed by thermocouple and BI analysis (heat penetration studies). At a minimum, thermocouples and BIs should be placed both upstream and downstream from the filter (Fig. 3). These locations should ensure that both the upstream and downstream high points and low points are

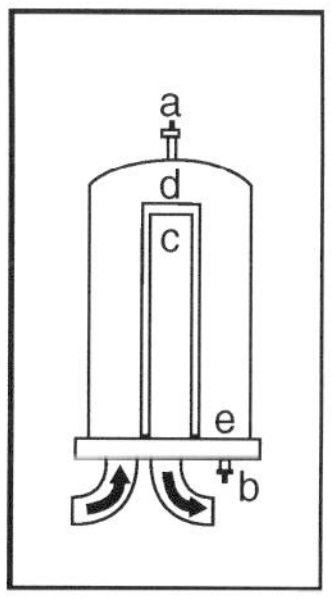

Figure 2 Correct filter installation. a, vent; b, condensate bleed; c, inner core of filter; d, high point air; e, low point condensate.

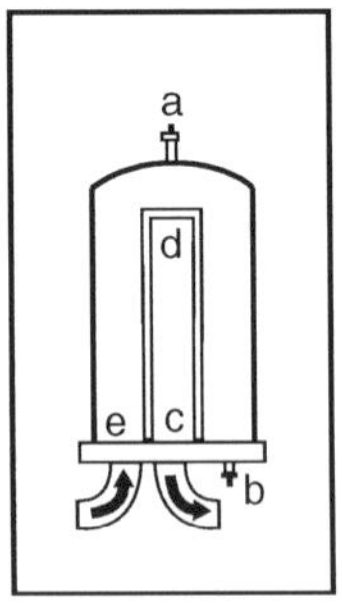

Figure 3 Suggested thermocouple/biological indicator placement. a, vent; b, condensate bleed; c, inner core of filter; d, high point air; e, low-point condensate.

monitored (i.e., verification of low-point condensate removal and high-point air removal).

4. Perform ongoing monitoring of the sterilization for temperature and pressure (to ensure presence of saturated steam conditions) and differential pressure.
5. Ensure that ongoing operating conditions (normal and sterilization) are within the filter manufacturer's defined limits.

Integrity

An integrity test for routine manufacturing use should be nondestructive (bacterial retention tests are by nature destructive), provide an indication of "fitness for use" and, above all, be correlated with bacterial retention. This chapter assumes that the reader is knowledgeable about integrity test methods (such as bubble-point and diffusion). This information will not be presented. The theory of these tests has been the subject of many publications (6,7).

Provision of the test methodology and correlation is the responsibility of the filter manufacturer, and qualification of how the test is used is the responsibility of the end user; that is, it is not sufficient to merely put the procedure into use without adequate operator training and qualification of test equipment. It is the responsibility of the filter manufacturer to demonstrate the correlation of the integrity test value with microbial retention. These correlations should take the form of an integrity test parameter versus retention. The data on the correlation between nondestructive filter integrity test (e.g., diffusion test or bubble point test) and the retention testing (destructive test) by the filter manufacturers provides valuable information to the end user. As a result, the filter users should be able to assess the rigor with which a filter device has been qualified by reviewing the validation guide. In the filter validation guide, the filter manufacturer should provide an integrity test method that clearly cites wetting fluids, test gases, test pressures, test temperatures, test times, and pass/fail criteria.

All filter manufacturers should have established a relationship between a recommended physical integrity test and microbial retention. These data are used by the filter vendors to establish minimum integrity test specifications for their sterilizing grade filters. The pharmaceutical manufacturer is responsible for establishing minimum integrity test values for the filter when it is wetted with the pharmaceutical product to be filtered. These specifications will be used for release of the drug product, since microbially challenging each filter used in production is not practical.

Qualification of Integrity Test USE

Many filter users struggle with the choice of test method. This is often confounded by conflicting expert opinions and by filter manufacturer claims. Regulatory authorities may allow the choice of any of the major tests or may suggest a method but in reality, it is the user's responsibility to document the rationale for their choice (1).

As mentioned above, the user must ensure that the integrity test has been correlated with microbial retention. The manufacturer's recommended test must be strictly adhered to (time, temperature, wetting fluid, pressure, method). Test use must be qualified—in other words, it is necessary to qualify the test accuracy and reproducibility, regardless of whether the test is performed manually or with an automatic tester.

FDA guidelines state:

> Integrity testing of the filter(s) can be performed prior to processing, and should be routinely performed postuse. It is important that integrity testing be conducted after filtration to detect any filter leaks or perforations that might have occurred during the filtration. Forward flow and bubble point tests, when appropriately employed, are two integrity tests that can be used. A production filter's integrity test specification should be consistent with data generated during bacterial retention validation studies.

Qualification of integrity test use, whether manual or automated, is not unlike qualification of any other method or instrument. For manual and automated tests, it is important that the accuracy and precision of the measurement instrument (usually pressure gauges or flow meters) is capable of discerning accurate values within the test time frame. For a manual test, it is important that operators are properly trained and qualified in use of the test. This may be performed through a documented training program, followed by examination of technique with filters of predetermined integrity test values (but unknown to the examinee). In addition, an incorrect filter (either a nonintegral filter or a larger-pore-sized filter) should be included to verify that operators are capable of detecting a filter failure. Qualification of an automated test instrument should also address instrument calibration, verification of test accuracy and reproducibility (usually vs. a manual test), and verification of alarm and security features. In addition, the software driving the unit (whether in disk, EPROM, or PLC format) should be qualified. The latter requirement, obviously, cannot be performed by the end user. Therefore, verification should be obtained from the instrument manufacturer that the qualification has been completed (through data or by an audit). Furthermore, besides the reliance on automatic integrity test instrumentation, it is important to be aware of the fact that the operator training is critical with regards to operation of automatic testers as it is possible to get incorrect results from an improperly operated instrument (Fig. 4).

Drug Product-Based Integrity Tests

Use of the filter manufacturer's recommended wetting fluid as a control is critical to obtain a correct pass/fail value. However, in certain cases (particularly when performing a postfiltration integrity test), ensuring

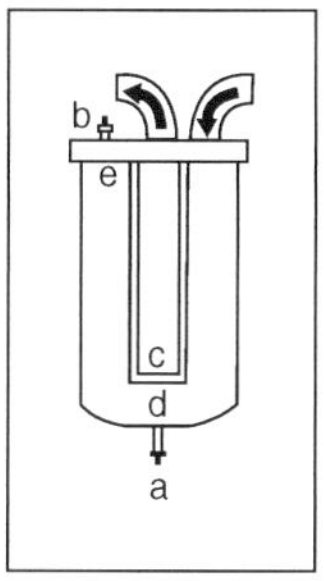

Figure 4 Incorrect filter installation—the setup is not self-draining. a, vent; b, condensate bleed; c, inner core of filter; d, high point air; e, low-point condensate.

purity of the specified wetting fluid can be problematic owing to difficulty in flushing drug product from the filter.

Integrity tests are derived from physical laws and relations and, therefore, are dependent on specific variables. For instance, the bubble-point test is dependent on vehicle surface tension and wetting contact angle with the particular membrane. Diffusion is dependent on the solubility and diffusivity of the gas in the liquid vehicle. Therefore, drug products (or any other nonspecified liquid) can cause enhancement or suppression of integrity test values compared with those that would be obtained with the manufacturer specified liquid (e.g., water). Therefore, if the nonspecified liquid is not completely flushed before integrity testing, the resulting value may be greater than, or less than, the value that would be obtained when using the pure, specified liquid. The end user, therefore, is required to qualify the postuse test in the following manner.

Determine whether the drug product, compared with the specified liquid, suppresses or enhances the integrity test value. This may be performed on a scaled-down version of the filter, for example, a 47-mm disk (2,3). The study should take account of drug product variability, usually by using at least three lots of drug product (assuming that raw materials are carefully controlled). Integrity tests are conducted on the filter membrane wetted with the filter manufacturer's specified wetting liquid. The filter membrane is then allowed to completely dry and is wetted with the drug product. To achieve accurate, reproducible results, a direct scale-down of drug volume to filter surface area should be used (i.e., if production batches are 10,000 L through 5000 cm^2 of filter area, then the scale-down to 10 cm^2 of filter area would use a wetting volume of 20 L). This will determine not only whether the product is an enhancer or suppresses, but also generate a preliminary integrity test ratio:

$$\text{Ratio} = \left(\frac{\text{Value obtained with drug product}}{\text{Value obtained with filter manufacturer specified liquid}} \right)$$

If the ratio obtained is 1 (e.g., the value obtained with the specified liquid is 50 psi and that obtained with the drug product is 50 psi), then the value obtained with product will be the same as the specified liquid and there is no issue with performing the integrity test with product. If the ratio indicates that the product suppresses the integrity test value (that is less than 1.0; i.e., causes the bubble-point to be lowered), then use of the drug product to wet the filter while still using the pass/fail for the specified liquid will be overly conservative. For example, the value obtained with the specified liquid is 50 psi, whereas that obtained with drug product is 45.5 psi. By using adequate validation data, it is possible to accept a drug product-based integrity test value of 45.5 psi. If the criteria are not met, the drug manufacturer would then flush the filter with water and retest (with 50 psi as the minimum acceptable value).

If the ratio indicates that the product is a bubble-point enhancer (i.e., greater than 1.0; causes the bubble-point to be increased), then use of the drug product to wet the filter while still using the pass/fail for the specified liquid will be invalid. For example, the value with specified liquid is 50 psi, whereas that with drug product is 54.5. In other words, if the minimum bubble point for the specified liquid is 50 psi, the product ratio is 1.1 (indicating that a value of 54.5 or higher, with product would be an integral filter), but if the drug manufacturer decides to determine 50 psi as pass/fail for the product, then there is the potential for accepting an out-of-specification filter (e.g., one that when wet with drug product gives a value of 52 psi, which is below the product-compensated specification of 54.5 psi). Therefore, the pass/fail criteria should be 54.5 psi or higher. If the criteria are not met, that is, the value is 54.5 or higher, the drug manufacturer could then flush the filter with water and retest. However, while wetting the filter is an ideal, it is often impractical. Thus, a different approach can also be used. A scaled down test could be run with a set volume followed by confirmation of the values generated on small scale by monitoring the actual test results in the full-scale.

However, the use of 50 psi as the minimum acceptable value for the test after a specified liquid flush may not be valid. The drug manufacturer must validate the flush volume necessary to ensure that only the specified liquid is present in the filter: all products have been removed from the filter. Validation of flush volume can be achieved by measuring product in the flush (concentration assay, spectrophotometric analysis, and so on) or by integrity testing after incremental flush volumes (e.g., flush 50 L then test, flush another 50 L and test—the procedure is repeated until a consistent, stable bubble point value is obtained).

If product integrity ratios are to be used, the scaled-down study is only the initial part of the qualification. The second part would be to obtain additional full-scale data and to monitor the ratio on an ongoing basis. This would be part of the ongoing manufacturing runs. The ongoing ratio of specified liquid preuse to product wet postuse testing should be trended to ensure that the ratio is not changing owing to raw material changes, lot-to-lot inconsistency, and so on.

Qualification of flush volume to ensure product removal should be performed at full scale, because it will be dependent on the filter configuration (size, support materials) and the installation (pipe size, valves, housing). For additional guidance, consult PDA TR No. 26 (1).

Operating Conditions

The validation study must ensure that within the anticipated worst-case operating conditions the filter is not compromised. This is conducted by first obtaining data from the filter manufacturer on maximum recommended operating limits. It is then the responsibility of the end user to operate within these limits. The actual validation must then encompass the anticipated worst-case conditions that may be encountered during the process (i.e., those expected in the end-user's process, rather than the filter manufacturer's maximum limits). In other words, it is only necessary to validate the filter for its proposed conditions, rather than validating for the manufacturer's stated maximum limits. Likewise, the manufacturer's maximum limits should not be exceeded.

Lack of damage to the filter can be verified by performing several tests such as integrity, flow rate, throughput, retention, extractables, or others.

Time

Operating time is a concern for the following reasons:

1. Long processing times could allow bacteria, which have been trapped by the filter, to die, thereby resulting in increased endotoxin levels. This time criteria must be accounted for in endotoxin studies.
2. Long processing times may increase the probability for bacteria to penetrate the filter (8).

Time considerations during validation must ensure that the worst-case is covered. For example, if the drug-manufacturing process results in intermittent filter blockage, then the effect of changing out a clogged prefilter during processing should be incorporated in the maximum time of all experiments. If the sterilizing grade filter is changed intermittently during the filtration process, this must also be considered. Time for the sterilizing grade filter would be the longest period of time that any sterilizing grade filter would be in place. It would not be the full amount of time to process the batch.

The impact of long processing time potentially leading to the increased probability of bacteria penetrating the filter originates from the fact that microporous membrane filters have been demonstrated to allow, over extended periods of time, the penetration of bacteria through the filter and into the effluent (6,8–10). However, filter penetration is simply a term used to describe the depth to which a target particle will travel through the filter before it is removed from the fluid stream. In fact, sterilizing filters, as a rule, allow some penetration into the depths of the filter. This is clearly consistent with both size exclusion and adsorptive removal mechanisms. The term "grow-through" describes a theoretical phenomenon in which organisms initially trapped by filter will multiply on and in the filter. In the process of multiplying, the organisms pass deeper and deeper into the filter matrix until they eventually emerge downstream of the filter. This "phenomenon," similar to the "blow-through," is a hypothesis that has not been rigorously studied and demonstrated to be a real event in the pharmaceutical process. It needs further detailed investigation since several factors, such as lack of proper system sterilization and improper aseptic manipulations (resulting in false positive) may contribute to such observations.

The reliability of sterile filtrations can nevertheless be increased by limiting processing time. Filter manufacturers can provide the data on the retention tests that have been conducted for a specific membrane or a device for extended time which generally suggests that filters should retain bacteria in excess of 48 hours. Accordingly, filter manufacturer's recommendations that aseptic processing is completed within a pre-established period of time is based on their experience. Thus, limiting the filtration time to a time less than that which is recommended by a filter manufacturer provides increased assurance against the phenomenon of grow through (or blow-through) if it is a real phenomenon.

Temperature

The operating temperatures a filter will experience must be carefully reviewed to ensure that the manufacturer's recommended limits are not exceeded. These limits will probably be stated in terms of time at a specified temperature. For example, if a filter is required to process hot oil at 60°C for 12 hours, then the filter must be rated for that use. It must also be rated for the maximum anticipated sterilization temperature and time. Just because a filter is rated for 100°C for 100 hours does not mean it will withstand 121°C for any length of time.

Similarly, it is common practice for vent air filters to be left on WFI tanks for extended periods. These filters should be rated for extended life at high temperatures (typically WFI is held at 70–90°C) for the required times. Filter components may oxidize to varying degrees at elevated temperatures.

Operating temperatures will also have a significant effect on the filter's ability to withstand differential pressure.

For all these reasons, the validation must take into account normal operating temperatures as well as sterilization or sanitization temperatures.

Pressure

There are several aspects of pressure that must be considered. The inlet pressure to the filter must be monitored to ensure that there is no potential for structural damage. The differential pressure across the membrane must comply with the filter manufacturer's recommended limits. Differential pressure is normally stated as a function of temperature. For example, a filter may have a maximum recommended differential pressure of 80 psi at 25°C. This same filter may have a differential pressure recommendation of a maximum of 5 psi at 121°C.

Another aspect that must be accounted for is the direction of applied pressure. Filters are manufactured such that the allowable maximum pressure differential in the forward direction may be different from that in the reverse direction. For example, the aforementioned filter may also have a maximum limit of 50 psi differential in the reverse direction at 25°C and 2 psi reverse at 121°C.

All of these pressure factors must be considered when determining a filter's fitness for use. The anticipated maximum pressures should be incorporated, when possible, into the bacterial retention study and the sterilization study. They should also be considered for normal operating conditions. Fitness for use can be verified by retention, integrity, flow rate, and throughput.

When the limits for pressure are being decided, two areas tend to be overlooked. The first, sterilization pressures, has been discussed. The second, hydraulic stress must also be considered. Hydraulic stress can occur for example, when a valve is opened suddenly, causing a filter to receive full line pressure immediately, or by the action of in-line filling when surge tanks are not in use (causing rapid pressurization and depressurization of the filter). These forward and reverse pressures must be quantified, monitored, controlled, and validated.

Flow Rate and Throughput

This chapter assumes that initial flow decay studies for the given filter and drug product combination have been carried out to ensure that there is adequate surface area to obtain the required flow rate and throughput. These studies are ordinarily performed on a scaled-down filter, for economic reasons. Usually, there is sufficient safety margin added to defer verification until the full-scale validation batches are run.

Full-scale processing times, throughput, and flow rates should be established during the validation batches as a reference for all future batches. These flow rates and throughputs provide an effective tool for in-process monitoring of a validated process. In other words, if throughput or flow rate show intermittent problems, this is an indication of either variation in raw materials, inadequate filter sizing or changes to the filter.

Particulates

When discussing particulates in the context of filter validation, there are two questions that must be answered:

1. Is the filter contributing to the particulate load of the solution?
2. Is the filter specified as reducing the particulate load of the solution? That is, is the reported purpose of the filter to remove particulates of a specified size range?

The USP defines *particulate matter* as follows: "Particulate matter consists of mobile, randomly-sourced, extraneous substances, other than gas bubbles, that cannot be quantitated by chemical analysis due to the small amount of material that it represents and to its heterogeneous composition (7)."

Sources of particulates may be as varied in use as raw materials, process equipment, process environment, or the filters themselves. Indeed, the filter itself may contribute to the particulate load or may provide a particulate load reduction. There are specified regulatory guidelines for particulate limits in pharmaceutical solutions. For example, the *USP* limits (USP29-NF24 <788>) for particulate matter in injections when tested by using the Light Obscuration Test Particle Count method (7), are as follows:

LVPs; i.e., single dose of more than 100 mL:

1. Not more than 25 particulates per mL$\geqq$10 µm
2. Not more than 3 particulates per mL$\geqq$25 µm

SVPs; i.e., single or multiple doses of 100 mL or less:

1. Not more than 6000 particulates per container$\geqq$10 µm
2. Not more than 600 particulates per container$\geqq$ 25 µm

Proof of compliance with these limits is normally performed by sample testing of the final solution by methods such as optical microscopy and light obscuration (7). Because of potential limitations with these techniques, other methods have also been investigated, such as light microscopic image analysis and scanning electron microscopy (11,12).

All methods involve filtering a solution through a suitable analysis membrane and counting or sizing the resulting particulates. In analyzing particulates from a specific filter, it would be necessary to filter the pharmaceutical solution, when practical, through the test filter and the analysis filter. The counts could then be obtained from the analysis filter. It is critical that the background particulate level (i.e., the counts on the test system without a test filter in line) be accounted for in the analysis. It is for this reason that it may be more practical to use a clean liquid, rather than the pharmaceutical solution.

The validation must simulate all anticipated manufacturing operations. For example, if the filter is flushed before use, the initial particulate counts may be reduced. If the filter is then sterilized, the particulate counts may increase. Likewise, particulate counts on a time/volume throughput should be analyzed. This ensures lack of bias in count estimation during the course of a filtration. Examples are counts that start high and quickly drop to below detection limits (indicating that a preflush would be beneficial) or vice versa (indicating a maximum useful life or an incorrect filter type for the function).

If the stated or implied function of a filter is particulate removal, then the removal of a given size and amount of particulates must be validated. This procedure should be performed, when possible, with the pharmaceutical solution. A standard test method (13) is used to challenge the filter with a known quantity and size distribution of particulates. The amount of particulate retention can then be verified. As with all particulate measurement methods, performance of adequate assay controls is critical to test result accuracy.

If the stated function of a filter is not removal of particulates (other than microorganisms), then the need to validate it for this function is obviated (see the section entitled Microbial Retention).

Fibers

Fibers, just as particulates, are of concern for two reasons:

1. Is the filter shedding fibers into the solution?
2. Is it the filter's function to remove fibers?

Fiber is defined by 21 CFR, part 210.3 (5) (i.e., U.S. GMPs; 1) as follows: "… Fiber means any particulate contaminant with a length at least three times greater than its width." Section 211.72 of the same document further states:

> Filters for liquid filtration used in the manufacture, processing, or packing of injectable drug products intended for human use shall not release fibers into such products. Fiber-releasing filters may not be used in the manufacture, processing, or packing of these injectable drug products unless it is not possible to manufacture such drug products without the use of such filters. If use of a fiber-releasing filter is necessary, an additional NFR filter of 0.22 µm maximum mean porosity (0.45 µm if the manufacturing conditions so dictate) shall subsequently be used to reduce the

content of particles in the injectable drug product. Use of an asbestos-containing filter, with or without subsequent use of a specific NFR filter, is permissible only upon submission of proof to the appropriate bureau of the FDA that use of a NFR filter will, or is likely to, compromise the safety or effectiveness of the injectable drug product.

The U.S. GMPs part 210.3, subpart 6 defines a NFR filter as follows:

> "Nonfiber-releasing filter means any filter, which after any appropriate pretreatment such as washing or flushing, will not release fibers into the component or drug product that is being filtered. All filters composed of asbestos are deemed to be fiber-releasing filters."

The first requirement when more than one filter is used in a filtration train is to ensure that, at a minimum, the final filter in line is a NFR filter of 0.22 μm standard (sterilizing grade). It is preferable, where possible, to specify all filters in the filtration train as NFR. Filter manufacturers will certify that their products meet 21 CFR 210.3, subpart 6, based on lack of fibers in their manufacturing components. In addition to this, if fibers are a concern, the end user can test for fibers in the effluent using the particulate measurement systems discussed previously. This is not common practice unless the filter is specified for fiber-particulate reduction.

The German Federal Health Office has published standards for asbestos particulate limits in parenterals (14). The document states that the length of asbestos fibers is the decisive risk factor and has established three size categories:

1. Fibers exceeding 2.5 μm must be eliminated.
2. Fibers between 1 and 2.5 μm "may only be found in low concentrations."
3. Fibers not longer than 1 μm are of no concern.

Furthermore, the USP guideline (<1211> Sterilization and Sterility assurance of Compendial Articles) states the following:

> Fiber-shedding filters, particularly those containing asbestos, are to be avoided unless no alternative filtration procedures are possible. Where a fiber-shedding filter is required, it is obligatory that the process include a nonfiber-shedding filter introduced downstream or subsequent to the initial filtration step.

The test method referenced is the U.S. EPA test, Analytical method for determination of asbestos fibers in water (15).

Considering these limitations for the final dosage form, it would be prudent to either obtain by certification from the filter manufacturer assurance that the filters (regardless of materials of construction) do not exceed these levels of contaminating asbestos fibers, or to perform the measurements. It is more logical to obtain information from the filter manufacturer who can verify filter raw materials at source. In addition, data from the filter manufacturer showing the ability of filters to prevent passage of these fibers would be beneficial for drug manufacturers.

Microbial Retention

FDA guidelines state:

> Filtration is a common method of sterilizing drug product solutions. A sterilizing grade filter should be validated to reproducibly remove viable microorganisms from the process stream, producing a sterile effluent. Currently, such filters usually have a rated pore size of 0.2 μm or smaller (16). Use of redundant sterilizing filters should be considered in many cases. Whatever filter or combination of filters is used, validation should include microbiological challenges to simulate worst-case production conditions for the material to be filtered and integrity test results of the filters used for the study. Product bioburden should be evaluated when selecting a suitable challenge microorganism to assess which microorganism represents the worst-case challenge to the filter. The microorganism *B. diminuta* (ATCC 19146) when properly grown, harvested and used, is a common challenge microorganism for 0.2 μm rated filters because of its small size (0.3 μm mean diameter).

It is necessary to demonstrate microbial retention with a given filter and drug product for the following main reasons:

1. To ensure that the filter is not undergoing degradation, deformation or some other change under the conditions of use.
2. To ensure that the drug product is not causing the organism to shrink, thereby resulting in nonsterilizing conditions.

Initially, a filter is qualified as sterilizing grade by the filter manufacturer, based on its ability to completely retain high levels of microorganisms, as demonstrated by the results of a microbial retention test. By definition, a *sterilizing-grade filter* is one that, when challenged with 10^7 *B. diminuta* ATCC 19146 per square centimeter of filter area will produce a sterile effluent. In addition, each filter manufacturer has manufacturing lot-release criteria, which should include some representative microbial retention testing of each lot of filters.

Filter manufacturers generally base their testing method on the methods published by the (ASTM Committee F-21, 1988) and on guidance provided in the FDA's *Guideline on sterile drug products produced by aseptic processing* (2,17). Furthermore, the ISO 13408-2:2003 also specifies requirements for sterilizing filtration as part of aseptic processing of health care products (18). It also offers guidance to filter users concerning general requirements for setup, validation and routine operation of a sterilizing filtration process, to be used for aseptic processing of health care products. ISO 13408-2:2003 is not applicable to removal of viruses. Sterilizing filtration is not applicable to fluids containing particles as effective ingredient larger than the pore size of a filter (e.g., bacterial whole-cell vaccines). ISO 13408-2:2003 tends to be more conservative and prescriptive than the PDA TR26.

From a regulatory perspective, the pharmaceutical manufacturer is responsible for providing retention data that support the claim of filter validation in their manufacturing process. However, drug manufacturers have neither the experience nor the facilities to perform the test, nor the desire to introduce even small volumes of microbial suspensions into their production facility. Consequently, the pharmaceutical manufacturer has looked to filter vendors for advice on how to best challenge the filters. Indeed, many pharmaceutical manufacturers contract with the filter vendor to perform the test.

The drug product and the filter rating-specific validation of microbial retention, in contrast with those retention tests performed by the filter manufacturer, evaluates the influence of the drug product's physical and chemical attributes on the performance of the filter and efficiency of the filtration under simulated processing conditions. The relevant physical and chemical conditions are listed in Table 4.

It is possible (and often desirable) to perform this product-specific testing on a scaled-down version of the process filtration system. Generally, flow rate (mL/min) per unit surface area is the scaling factor, with the manufacturing process scaled down to 13.8 cm^2 of effective filtration area (a 47-mm-disk filter). The processing temperatures, times, and pressure differentials modeled are those used for full-scale processing. A typical test schematic is shown in Figure 5.

The test's basic components are (*i*) a test filter, (*ii*) a system to deliver the microbial (bacterial) challenge suspension to the test filter, and (*iii*) a system for assaying the test filter effluent. The test filter can range from flat stock disks in filter holders to relatively large fabricated devices, such as stacked disk devices or pleated cartridges. When modeling fabricated device performance in retention testing, it is essential that the membrane used in the devices be the same as the flat stock membrane used in microbial retention studies. Specifically, if flat stock testing is to be used to validate a device, then the two membranes should be identical. Thus, the scale-up retention test data should be compared and be comparable.

Each retention test system should have the following basic features:

1. Valves should be included in appropriate positions throughout the system, and they should not be installed if they are not necessary.
2. Pressure gauges should be placed upstream and downstream from the test filter to accurately determine a pressure differential (ΔP) across the filter. This is especially important if the assay filter is used concurrently with the test filter. A ΔP across the assay filter must be accounted for when ΔP across the test filter is being measured.
3. Hydrophobic vent filters should be part of every reservoir that will require air exchange during emptying or filling of the reservoir. Aseptic connections should always be minimized.

Table 4 Physical and Chemical Processing and Drug Product Attributes

Physical attributes	Chemical attributes
Pressure differential	pH
Flow rate	Viscosity
Duration (contact time)	Osmolarity
Temperature	Ionic strength
Batch size	Surface tension
Surface area	
Filter type and series	

Only those filters for which no test organism (*B. diminuta* or other relevant microorganism) passage is detected downstream are deemed sterilizing-grade. While the colonies may "appear" on the assay filter as a result of false positive or extraneous contamination, the "passage" of even one bacterial colony on an assay filter renders the filter nonretentive. Thus, results of "zero" and "one" are considered significantly different, and size controls (more open filters are usually used) are required to ensure that one does not obtain either of these values illegitimately (16).

Areas of Concern for Product-Specific Bacterial Retention Testing

The following areas are of concern for product-specific bacterial retention testing:

1. The organism must be cultivated to ensure a small and consistent size.
2. The challenge concentration should not be less than $10^7/cm^2$ of filter area unless justified.
3. Drug product must not be toxic to the organism (i.e., it must not reduce the challenge to below $10^7/cm^2$ *B. diminuta* per square centimeter of filter area over the course of the test). This can be ensured by performing toxicity studies.
4. Viability controls should be run before testing.
5. When the drug product is shown to be toxic to the organism the following course can be followed:
   a. Expose the filter to the drug product (to determine if there are any filter-drug product interactions)
   b. Rinse the filter (to remove toxic residue of the drug)
   c. Challenge with the organism suspended in a vehicle that has properties as close to the product as possible (e.g., the product minus the toxic components)
   d. Or, alternatively, expose the filter to the toxic product for the actual period of processing. Then, suspend the challenge microorganism in the toxic product, and challenge the test filter for a shorter period of time (choosing a duration that ensures maximum microorganism survival)
6. The size of the collection or assay filter must be considered. Different testing laboratories will use either a 0.45-μm or a 0.22-μm filter as the assay filter. There are arguments for and against the use of either filter for organism recovery. Basically, it can be argued that the 0.45-μm filter will allow passage of organism. The arguments for use of this filter are that it is the referenced method for sterility testing in the USP (19) and also that it is easier to cultivate organisms on a 0.45-μm filter than on a 0.22-μm filter (20). The arguments for and against a 0.22-μm filter are the opposite of those for the 0.45-μm.

The definition of *worst-case-processing* conditions is continuously under debate and it is recommended that the choice of such conditions be discussed with the regulatory agency before testing.

Filter Inertness

The purpose of an aseptic filter is to remove bacteria and unwanted contaminants from a liquid or gas

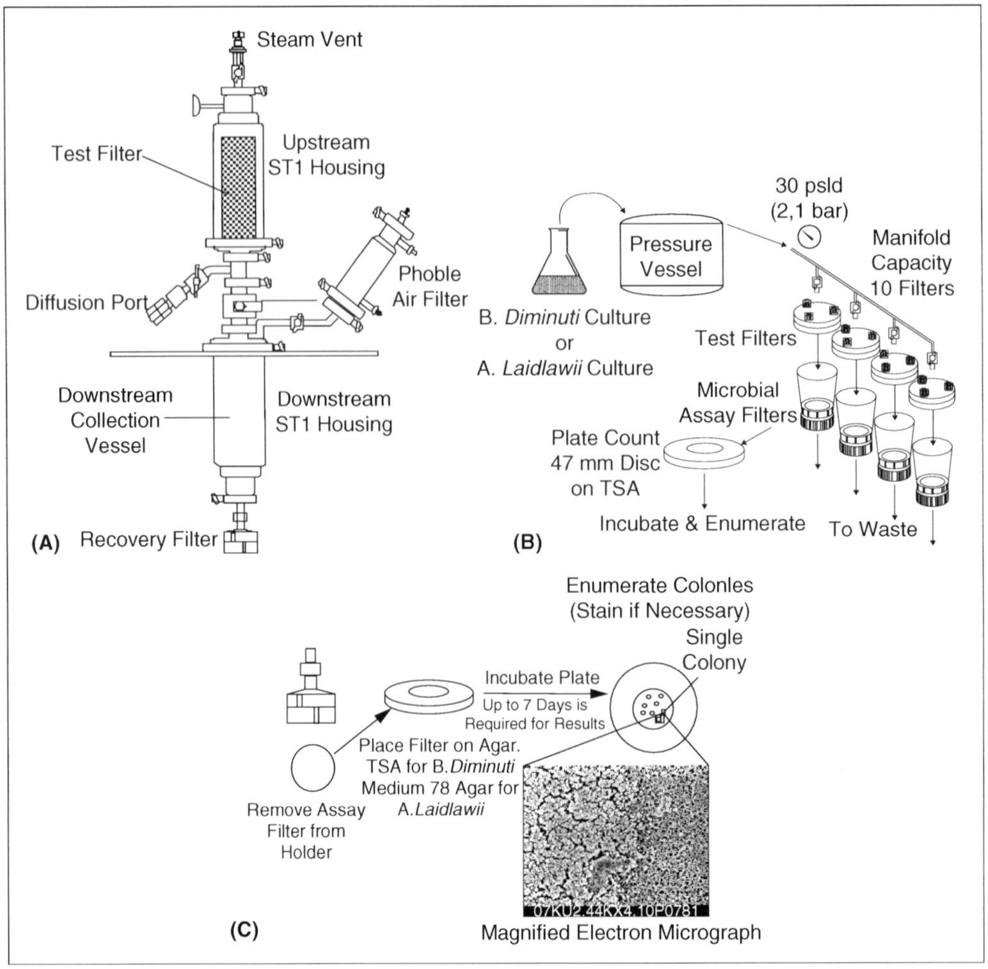

Figure 5 (**A**) Test stand showing filter cartridge setup for retention testing (full-scale version). (**B**) Retention test setup for flat-stock discs (scaled-down version). (**C**) Processing of assay (recovery) filters after retention testing.

stream. Beyond this, the filter should be inert (i.e., neutral). It should neither add anything to the fluid or gas (extract) nor remove anything from it. Obviously, this is an ideal, but may be unrealistic expectation. A small amount of extraction or adsorption is a possibility with any and all commercially available filters, depending on the nature of the drug product and the materials of construction of the filter. Therefore, it is unnecessary to categorically prove or disprove inertness, but rather to quantify its effects. This should be accomplished through empirical studies.

As an example take a drug product filter *extraction* and *adsorption* study. The purpose of an extraction study is to detect trace amounts of filter extractables that may leach into a process on drug solvent stream. Several critical issues must be addressed:

Characterization or identification of the extracted species—Quantification of extracted species (extractables)
Sensitivity of analytical methods for detection and quantification—Potential effect of identified extractables on drug performance or stability
Potential effect of identified extractables on filter performance—Potential toxicity of extractable species.

The purpose of an adsorption (binding) validation study is to determine whether a given filter adsorbs component(s) from a drug product. Clearly, the extent of any adsorption should be determined. The issues are:

1. The effect on the final drug product formulation. If the product is filtered into a bulk holding tank, then the effect on the final product may not be significant. However, if the product is directly filtered into its final package (such as ampoules), then some of the packages may not contain adequate amounts of critical ingredients (such as proteins or preservatives).
2. The effect on drug product stability.
3. The cost (in dollars) of adsorption.

The first step in testing filter inertness is to compare the drug product to be filtered with the membranes proposed for use, to determine if there are known incompatibilities. Historically, this comparison, coupled with integrity testing and product-specific bacterial retention testing, would have been considered sufficient for the purposes of a study on inertness. Today, however, both the drug industry and regulatory authorities are pursuing a course of action to qualify and quantify inertness caused by interaction between any drug product and filter

Table 5 Techniques for Determining Filter Inertness

Extractables	Adsorption
Compatibility	Compatibility
Oxidizable substances	Formulation analysis
pH	
Conductivity	
Gravimetric extractables	
Weight changes	
Advanced analytical techniques	

components. This should not be considered surprising; because filter inertness has always been a concern. Historically, however, only gross measurement techniques, such as weight changes, integrity tests, visual inspections, and so on, could be relied on to warn of any potential problems. Today, more sophisticated and sensitive analytical measurement techniques are routinely used to measure inertness.

Table 5 lists various techniques that have been used to measure filter inertness. A brief description of each follows. Note carefully that these tests must incorporate all materials of construction.

Chemical Compatibility Charts

Chemical compatibility charts are provided by filter manufacturers (21) and are typically included in product literature or validation guides. They should clearly show compatibility and suitability information for all materials of construction of a filter unit across a broad spectrum of solvents. Although these charts are not all-encompassing, they do give an excellent general indication of the scope of potential chemical environments to which a particular filter unit may be exposed. Their real use, however, is not in what is listed as compatible, but rather what is listed as known to be incompatible. (With a properly defined filter requirements specification, this study would have been completed and documented before filter selection). If a solvent stream of interest is not indicated, then the filter manufacturer should be consulted to determine (at least on paper) the compatibility of the primary solvent streams with the filter materials of construction. This study should be documented by a qualified individual with knowledge of the filter materials of construction (such as a filter manufacturer's developmental or analytical chemist). *The compatibility review should be considered a first pass only.* It should be supplemented with a combination of integrity tests, gravimetric extractables, oxidizable substance tests, and advanced analytical techniques.

pH and Conductivity

Either or both of these tests may be performed by taking measurements immediately before and immediately after filtration. This provides indication of gross changes to the filtered fluid. If changes occur, then further investigation is warranted. However, if no changes occur, then this indicates only that no gross change has occurred; it does not by itself indicate filter inertness.

Oxidizable Substances

The USP method (22) measures oxidizable substances levels in various flush volumes of water. Manufacturers will generally provide information on their product in the form of a statement such as: "Meets USP oxidizable limits after a flush of *X* liters of water." It is important for filter manufacturers to provide such flush data. It is equally important for drug manufacturers to perform flush operations before using filters. These flush operations may be carried out in conjunction with the preuse integrity test.

Gravimetric Extractables

A gravimetric extractables test is a USP test method (23) that determines the weight of a dry sample of the filter material pre- and post-static soak in water or an appropriate solvent. Attention to detail while performing this apparently simple procedure is paramount in isolating measurement error. The test will not specify the source of extractables and, therefore, must be tightly controlled; nor will the test give an indication of what the extractables are.

Weight Change

A weight change test is a gross test to determine nonvolatile filter extractables. After determination of dry weight, the filter is soaked in the drug product and in water (as a control). The soak time is chosen to mimic manufacturing conditions. After soaking, the filter is again dried to a constant weight and checked against the original weight and the water control. Any discrepancies warrant further investigation, such as more detailed extractable analysis or adsorption studies. As with the previous tests, an indication of no change pre- and post-test does not indicate lack of filter inertness.

Advanced Analytical Techniques

Several more-advanced analytical techniques may be used to detect and quantify filter extractables in solvent streams (24). These methods require special expertise and are complicated by the fact that the solvent stream being analyzed must be clean initially. If, for example, a drug product with several constituents (and additional trace constituents) is analyzed, potential extractables may be masked by excessive background noise. For this reason, extractable studies are normally performed on the primary solvent stream(s) alone. However, it may be necessary for regulatory purposes to provide additional experimental data that clearly demonstrate that attempts to determine extractable levels in the presence of actual drug product yield meaningless results.

Some of the more commonly used analytical techniques are the following:

High Performance Liquid Chromatography. Samples are analyzed before and after filtration. Additional peaks in the postfiltration chromatogram would indicate the presence of extractables. Extraction is normally performed in a pure solvent stream alone, thereby reducing background noise. It is usually performed under static soak conditions (this tends to concentrate extractables). If performed under static soak conditions, the level of extractables found must be considered in the light of the total process volume to obtain a "per volume of drug product extractable level." To date, the use of dynamic extraction techniques has infrequently been observed.

Fourier Transform Infrared Spectroscopy. This is a qualitative technique that is used to obtain chemical or chemical group information on solutes that are found by preparative techniques such as HPLC. Any extractables found are compared against a spectral library to determine the chemical makeup of the extractable and ensure that the source is a construction material. Positive identification of the extracted species then allows direct reference to previously documented toxicity studies performed by the filter manufacturer on the filtration device.

Total Oxidizable Carbon. This method is designed to detect the accumulated total of most organic carbon present in aqueous solutions and provides greater value. It is a quantitative technique that relies on the oxidation of carbon. There are two components of a TOC analysis: (*i*) the chemical or physical oxidation of the organic material, and (*ii*) the measurement of the products of that oxidation. All current TOC analyzers measure the resultant carbon dioxide. Carbon dioxide is measured using NDIR spectroscopy or by measuring the conductivity of the bicarbonate ion (HCO_3^-). USP 23 chapter 643 provides detail on establishing test requirements for use in a pharmaceutical water TOC monitoring program. These test technologies can be applied to filters.

Adsorption

Adsorption can cause loss of drug product, conformational changes, as well as reduced activity or stability (25). The validation must address all of these issues. Adsorption tests are best carried out by the end user. The issue of adsorption to filters has been covered in the literature (26) and, as such, the theory and mechanisms will not be dealt with in this text.

The method involves analyzing the drug product immediately before and after filtration for the compound(s) of interest. The intent is to detect differences in concentration of the active ingredient or other component of the formulation (such as stabilizers and preservatives). In addition, the drug product activity should be verified after filtration and over the course of a stability study. For proteins, it would also be desirable to ensure that conformational changes are not taking place (27).

If an adsorptive interaction or conformational change is discovered, then the drug manufacturer must determine if the interaction affects drug safety and efficacy. (If safety or efficacy is affected then use of the filter for processing the drug must be rejected). If the adsorptive effect does not affect safety or efficacy, it should be determined if it is possible to compensate for the effect. For example, when directly filtering into vials, it may be discovered that the initial vials lack preservative, owing to adsorption. In such an event, it may be possible to pretreat the filter with preservative and fill all potential binding sites, ensuring that subsequent drug product filtration would not result in the first vials lacking preservative. As with any method, this would require stringent validation controls. If a costly ingredient, such as the drug active or proteins, is adsorbed, such a pretreatment may not be possible. For this, it would be necessary to obtain a more inert filter.

Drug Product Stability

It is important to ensure that drug product stability should not be affected by filtration. The section on adsorption discussed the potential loss of activity, lowering of activity or conformational changes as a result of filtration. These effects may not be readily apparent; therefore, the stability study should investigate any potential adsorptive effects.

Endotoxins

A validation study must address that a filter does not add endotoxin to a drug product. Endotoxin can come from the filter itself, when new, or as a result of the way the filter is used.

The endotoxin content of a new filter will depend on the quality control processes of the filter manufacturer, the filter-manufacturing process itself, and the water used in filter manufacturing. These are all areas that are covered by choice of a filter vendor and through verification of endotoxin levels stated on a filter certificate of conformance or quality. Verification can be achieved by a drug manufacturer through testing incoming filters (28) or by auditing the filter manufacturer's methods and data.

The endotoxin levels of a filter can also be affected by filter use. The drug manufacturer must design and validate a manufacturing process to ensure low or no detectable levels of endotoxin. This is achieved by operating with low initial bioburden (following good aseptic techniques to minimize the source of endotoxins) and ensuring filter change-out frequency based on time, bioburden, and endotoxin analysis. Special caution should be exercised if filters are being used for extended periods or are to be re-sterilized for reuse. In these cases, bacteria that have been concentrated on the filter surface may begin to disintegrate, resulting in higher endotoxin levels downstream from the filter. Therefore, endotoxin determinations on a per-volume and per-time basis should be used to decide on appropriate filter change outs.

Toxicity

A validation study should determine that passage of the drug product through a filter does not cause any toxicological effects. All filter materials of construction should be addressed—not just the membrane. Other components to be covered include any support layers, cage and core (the inner and outer hard plastic supports), end caps (the solid plastic pieces that hold the top and bottom of the filter together), and O-rings. The construction materials themselves should be listed as nontoxic. The filter manufacturer should provide relevant test data, such as a compendial plastics test similar to the USP class VI tests for plastics (29) and the USP mouse safety test (30), for all construction materials.

It is then the responsibility of the drug manufacturer to ensure that contact of the filter and drug product does not result in any toxic by-products. This is achieved by ensuring that any extractables are materials of construction of the filter. These extractables should be quantified for levels in the drug product dose and further qualified as to potential for toxicological effect. This review should be carried out by a qualified toxicologist.

CONCLUSION

Filter validation performed by the drug manufacturer is not just a regulatory requirement; it also makes good business sense. The validation process should start with a filter requirements specification. From this specification, it is simply a matter of choosing the methods that allow verification of the requirements.

Although the regulatory requirements may appear to be more specific in one geographic area than another with the continuing efforts at ICH and future attention of specific pharmacopeias to additional elements of processes validation globally aligned regulatory requirements for sterilizing-grade filters may not be far away.

REFERENCES

1. FDA Guidance for Industry: Sterile Drug Products Produced by Aseptic Processing—Current Good Manufacturing Practice. Pharmaceutical CGMPs. September, 2004.
2. PDA Sterile Filtration Committee. PDA Technical Report No. 26: Sterilizing Filtration of Liquids. Bethesda, MD: Parenteral Drug Association, 1998 (52 pages).
3. Desaulniers C, Fey T. The product bubble-point ratio and its use in filter integrity testing. Pharm Technol 1990; 14:42–52.
4. Steere W, Meltzer T. Operational considerations in the steam sterilization of cartridges. Pharm Technol 1993.
5. DeSantis P. Validation of steam sterilization. In: Carleton FJ, Agalloco JP, eds. Validation of Pharmaceutical Processes, Sterile Products. 2nd ed. New York: Marcel Dekker, Inc., 1998:413–49.
6. Peterson AJ, Hankner DO. Assessment of long-term bacterial retention. Med Device Diagn Ind 1985; 150–5.
7. United States Pharmacopeia 29-NF24, <788>. Particulate matter in injections, 2006.
8. Simonetti JA, Schroeder HG. Evaluation of bacterial grow-through. J Environ Sci 1984; 27:27–32.
9. Wallhäusser KH. Durhwarchs- und durchblaseffekte bei langzeit-sterilifiltration-prozessen (English: Grow-through and blow-through effects in long-term sterile filtration processes). Pharm Ind 1983; 45:527–31.
10. Leahy TJ, Gabler R. Sterile filtration of gases by membrane filters. J Biotechnol Bioeng 1984; 26:836–43.
11. Barber TA, Lannis MD, Williams JG. Method evaluation: automated microscopy as a compendial test for particulates in parenteral solutions. J Parenter Sci Technol 1989; 43(1).
12. Bhattacharjee HR, Paley SJ, Pavlik TJ, Atterbury O. The use of scanning electron microscopy to quantify the burden of particles released from clean-room wiping materials. Scanning 1993; 15:301–8.
13. ASTM. Standard practice for determining the performance of a filter medium employing a single-pass, constant-rate, liquid test. F95-88.
14. German Federal Health Office (BGA). Promulgation regarding measures to reduce the asbestos contamination of parenteral products. November 1993.
15. Chatfield EJ, Dillon MJ. Analytical Method for Determination of Asbestos Fibers in Water. Mississauga, Canada: Ontario Research Foundation, U.S. Environmental Protection Agency, 1983.
16. Carter J. Evaluation of recovery filters for use in bacterial retention testing of sterilizing-grade filters. PDA J Pharm Sci Technol 1996; 50:147–63.
17. American Society for Testing Materials (ASTM). Standard Test Method for Determining Bacterial Retention of Membrane Filters Used for Liquid Filtration. ASTM Standard F838-83. Philadelphia, PA: ASTM, 1983 (revised 1993).
18. ISO 13408-2:2003: Aseptic processing of health care products—Part 2: Filtration. International Organization for Standardization 2003.
19. United States Pharmacopeia 29, <71>. Sterility tests, 2006.
20. Levy RV. Letter to the editor. Pharm Technol 1991; 15(1):18.
21. Filter Manufacturers Validation Guides. Millipore Corporation. Optiseal Durapore Cartridges, LAGL 0.22 (m TP).
22. United States Pharmacopoeia 29. Oxidizable substances, 2006.
23. United States Pharmacopoeia 29. Gravimetric extractables, 2006.
24. Stone TE, Goel V, Leszczak J. Methodology for analysis of filter extractables: a model stream approach. Pharm Technol 1994; 18:116–30.
25. Sarry C, Sucker H. Adsorption of proteins on microporous membrane filters: part 1 and 2. Pharm Technol Int 1992.
26. Meltzer T. Filtration in the Pharmaceutical Industry. New York: Marcel Dekker, 1987:174–5.
27. Truskey GA, Gabler R, Dileo A, Manter T. The effect of membrane filtration upon protein conformation. J Parenter Sci Technol 1987.
28. U.S. Department of Health and Human Services, Public Health Service, Food and Drug Administration. Guideline on validation of the limulus amebocyte lysate test as an end-product endotoxin test for human and animal parenteral drugs, biological products, and medical devices. December 1987.
29. United States Pharmacopeia 29, <88>. Biological reactivity tests—in vivo, class VI plastics test, 2006.
30. United States Pharmacopeia 29, <88>. Biological reactivity tests—in vivo, mouse safety test, 2006.

21

Cleaning and Disinfection in the Control of Pharmaceutical Cleanrooms

Klaus Haberer
Compliance, Advice and Services in Microbiology GmbH, Cologne, Germany

INTRODUCTION

Cleanrooms are environments which are protected by means of suitable segregation measures from contamination by nonviable particles and microorganisms. Such measures include, for example:

- venting segregation, e.g., air filters, pressure cascades, or directed air stream;
- physical segregation by air locks, barriers, or isolators;
- control of materials entered into the environment; and
- access restriction, and gowning rules for personnel.

These measures are intended to prevent particles that might carry microorganisms from entering the protected environment.

Cleanrooms are expensive to build and to maintain and hence, should be appropriately designed for their intended use. A room which is used as a clean environment to manufacture non-sterile products of superior microbial purity, will need less segregation than a room used to manufacture terminally sterilized pharmaceuticals, and an aseptic manufacturing environment will require much higher segregation measures than needed for terminally sterilized products. The better the intended (required) cleanroom classification, the tighter the segregation and control measures that must be applied.

Obviously, some segregation measures are more effective than others. A closed isolator is expected to achieve far better segregation than an area protected by unidirectional air flow in a manned conventional cleanroom background. However, complete physical separation is difficult to achieve where materials have to be transferred into and out of the cleanroom and where operations have to be performed from without the protected environment via flexible gloves. Some risk of contamination will usually remain even though it may be very small.

In conventional cleanrooms, segregation can never be complete, because the personnel entering the protected area are a main source of contamination that cannot be fully controlled. Any manned cleanroom will contain a low level of particulate and microbial contamination that is specific and characteristic for each room and its environmental conditions. The layout of the room, its venting system, number and frequency of people entering, entry procedures, the type of activity performed in the room, and the number and type of equipment operated in the room are among the important factors which determine the cleanliness status of the room. Barriers, directed air streams and other segregation measures within the manned cleanroom are used to specially protect critical working zones within the cleanroom, and these should be designed to give enough segregation to cope with the level of contamination present in the manned environment.

As segregation measures are expensive and involve time-consuming procedures (e.g., for gowning or entry of materials), it does not make sense to use disproportionately high cleanroom protection. A defined low level of microbial contamination is expected and appropriate for pharmaceutical manufacturing areas with the exception of critical aseptic processing zones. Even there, as long as segregation of the zone from the environment is not complete, a low level of contamination must be expected to occur, especially if occasional corrective interventions by personnel into the critical zone cannot be completely avoided. As long as contamination sources like personnel cannot be fully separated from a critical zone, the requirement to maintain a sterile environment therein remains an elusive if not an unattainable goal.

In order to remove any contamination that may have crossed segregation barriers and to maintain the required high level of cleanliness an appropriate routine cleaning and disinfection program must be applied to the environment. Disinfection programs for the pharmaceutical manufacturing environment have been discussed (1,2). However, cleaning and disinfection programs have their own risks of contamination and can even be destructive of the equipment. Also, some disinfection measures are potentially detrimental to the health of personnel and hence, cleaning and disinfection measures should not be out of proportion for the purpose to be achieved.

Environmental monitoring is performed to verify that the environment continues to be under control. Where environmental monitoring results indicate that the environment is not longer at the expected cleanliness level, additional or intensified cleaning and disinfection will be needed to restore the necessary level of control. Cleaning and disinfection should not be used however, to hide a breach in segregation. An investigation should be performed to detect sources of contamination where possible, and to verify that the necessary barriers are still in place and effective.

Abbreviations used in this chapter: DGHM, German Society for Hygiene and Microbiology; FDA, Food and Drug Administration; HEPA, high-efficiency particulate air; IPA, isopropyl alcohol; QACs, quaternary ammonium compounds; QC, quality control.

An appropriate cleaning and disinfection program can only be established and justified if the environmental monitoring program yields sufficient information to reliably judge the cleanliness status of the monitored area. Trends toward higher environmental contamination in the cleanroom must be clearly recognized, as well as an eventual sudden significant change in the level of microbial contamination which may indicate loss of segregation effectiveness or an increased challenge from the outside environment beyond the segregation capacity. Environmental monitoring is a critical tool to verify the state of environmental control and, hence, the basis to judge the effectiveness of cleaning and disinfection.

CLEANING OF CLEANROOMS AND EQUIPMENT SURFACES

Cleaning Methods

Significance of Cleaning

Cleaning in pharmaceutical production is primarily seen as the process to remove residues from a previously filled product from the surfaces of manufacturing equipment in order to render equipment fit for filling the next lot, or a lot of a different product. The principal goal to be achieved is to minimize the maximum amount of the active ingredient of the previous product that might appear in the next product to be processed. However, cleaning is also performed in the hygienic sense as a process to remove inadvertent viable and nonviable contamination of any kind from all surfaces in critical and controlled areas. The purpose of hygienic cleaning is to keep the areas free from all kind of dirt, dust or residues that may be transferred into the product stream or interfere with disinfection or sterilization procedures. It is this latter type of cleaning that this chapter is mainly concerned with.

The two types of cleaning cannot be considered to be completely separate, however. Cleaning of equipment from product residues should also have the effect of rendering the equipment surfaces fit for subsequent sterilization or disinfection, especially where aseptic processing is concerned. If the product manufactured is biological, the organic load of product residues may be considerable, and interference with the effectiveness of disinfectants or sterilization procedures could be seriously impaired by excessive residuals. Also, residues could become a growth substrate for contaminating microorganisms which may be a cause of serious recontamination problems after cleaning. Hence, equipment cleaning has also a hygienic aspect meaning that the critical objective is not only to remove active ingredients and excipients but also any organic load or any other substrate that could interfere with disinfection or be used as an energy source by microorganisms.

The purpose of surface cleaning in controlled or critical areas (as in any other production area) is to remove contamination that has been inadvertently introduced during processes. Such contamination could come from the process, e.g., spilled product or residues from broken vials could be introduced by secondary contamination pathways, e.g., the personnel involved in processing, venting system, or transport vehicles used during entry procedures. As the air entering is HEPA filtered, all materials entering via air locks are tightly controlled, and the personnel is wearing clean gowns and shoes, the level of contamination introduced is typically very low. Unlike a hospital setting, where a high amount of organic material is expected to be present on the surfaces to be cleaned, surfaces in pharmaceutical manufacturing premises are usually very clean at all times. Still it is important to remove surface contamination because residues, e.g., from spilled product, finger (glove) prints, or from media used for environmental monitoring, may interfere with the action of disinfectants or create a microenvironment for the survival of microorganisms.

Cleaning Procedures

Cleaning procedures should be well defined and documented with schedules, responsibilities, cleaning agents and specific techniques applied. While cleaning and disinfection are frequently performed in one step with detergent solutions containing disinfectant agents, cleaning is a step to be considered in its own right.

Cleaning of Surfaces

The solutions and/or solvents used for cleaning should be appropriate for the room or area or specific surface to be cleaned. All cleaning solutions entered into critical areas of aseptic processing should be sterilized whether or not they come into contact with product contact surfaces. Critical surfaces for aseptic filling should only come into contact with sterile cleaning solutions that leave no residues. Sterile 70% ethanol or isopropanol are frequently used for that purpose because these solvents do not leave residues and are cleaning as well as disinfecting.

For cleaning of floors and other non-critical surfaces in controlled aseptic premises, the necessary precautions are less critical but still exacting. As water is the most important prerequisite for microbial growth it must be assured that no humidity remains after cleaning (e.g., a film of liquid under equipment), and no waterborne contamination is entered with the cleaning solutions. Cleaning solutions must be freshly prepared from contamination-free detergents with sterile water if they are not sterilized by filtration. There must be a strict limitation on the time of use for these solutions, as numbers of microorganisms may rapidly increase given sufficient time for exponential growth.

For cleanrooms with lesser cleanliness requirements, tight control should also be in place for all aqueous cleaning solutions, even though they need not be completely free from microorganisms. It should not be assumed that detergents are antimicrobial and per se free from microorganisms. Contamination control is an important problem for manufacturers of washing lotions and hence, the bioburden of detergents must be controlled.

Cleaning and disinfection utensils should also be under an appropriate state of control. Wipes used to clean and disinfect critical surfaces must be sterile. Where confectioned moist alcohol wipes are used, it must be verified that they are free from bacterial endospores. Wipes for cleaning and disinfection of controlled areas should be sterile and disposable or sterilized between each use. Wipes for other clean areas should also be of specified quality. Information should be available concerning the bioburden of microorganisms carried on non-sterile wipes, as wipes have been found to be contaminated by considerable amounts of bacterial endospores.

Mops and buckets should be sterilized before each use. It is not acceptable in any cleanroom to leave wipes or cleaning tools exposed within the environment for drying after a cleaning process.

The time interval acceptable for using a portion of cleaning solution dispensed in a bucket should be tightly limited. It is not acceptable to have a cleaning solution ready to use it during the day, unless it is kept well protected in a closed container. The area that can be cleaned with a single wipe or a portion of cleaning solution dispensed in a bucket should be limited in order to prevent uncontrolled spreading of an eventual contamination over a large area. Cleaning of any room should be performed in a clearly defined sequence which prevents contamination of cleaner areas from less clean areas by the cleaning process.

Cleaning of walls and ceilings should be performed in a defined way so that the critical areas are not affected. Cleaning and disinfection of ceilings will usually involve scheduling where the cleanroom is taken out of commission and brought back to operation with a defined sequence of cleaning and disinfection steps.

Narrow recesses or otherwise hard-to-reach areas should not be present in cleanrooms. Where they cannot be avoided, e.g., underneath equipment, they should be sealed off and/or addressed with special care in cleaning instructions. It poses a serious risk to cleanroom control to access such areas in an uncontrolled way, because contamination could be accumulated in recesses during incomplete cleaning measures. Contamination might then be distributed in the room during a following cleaning cycle which reaches deeper into the problem zone. The cleaning procedures for difficult-to-reach areas should be very specific, to ascertain their complete and thorough cleaning each time. It is preferable to eliminate the potential for this problem by proper design and installation of equipment in the cleanroom.

Cleaning of Equipment

Cleaning of equipment should follow validated procedures to remove active ingredients and any organic or inorganic contamination interfering with disinfection or sterilization. Detergents used (if any) should leave no residues on product contact surfaces, or such residues need to be removed by the rinsing.

It is important to understand that washing fluids remaining in the equipment may serve as a microenvironment in which microorganisms can proliferate. If moisture is not carefully removed, any precautions taken during the cleaning process itself may become useless or even reversed. Cleaned equipment should be carefully dried, and stored for a strictly limited period only, protected from microbial contamination.

Cleaning Validation

Cleaning validation is a mandatory requirement for product contact surfaces or equipment where formal documentation is required, showing that the chosen methods effectively remove the product and any other interfering residues, and cleaning agents do not leave their own residues in the cleaned equipment. More information on the subject of equipment cleaning and cleaning validation can be found in chapter 40.

For cleaning of rooms formal validation studies are not usually required, unless highly active ingredients are processed in rooms where cross-contamination may be a problem. Validity of the cleaning procedures is judged from the results of environmental monitoring during operation of the rooms. It is recommended to install a thorough environmental monitoring program during a one-year validation phase after a room has been taken into operation, in order to recognize any seasonal influence on the microbiological contamination level of the room. After that phase, the data should be evaluated and a routine monitoring program decided upon, which may be more relaxed if the validation data indicate good cleanroom control.

Risk Consideration for Cleaning

It is important to realize that cleaning presents a high intrinsic risk to the cleanroom. An obvious question for risk consideration is residues of cleaning agents like detergents. Such residues must be avoided on product contact surfaces, where they may contaminate product. For this reason, many manufacturers resort to cleaning and disinfection of critical surfaces with 70% alcohol in one step. Even though this is not a sporicidal procedure, there would be no need to apply a more rigorous procedure as long as there is no concern for spore contamination from properly considered environmental monitoring data. On other surfaces like floors the question of detergent residues is of little consequence.

In many cleanrooms, cleaning fluids are the only source of water available to support microbial growth. Thus cleaning fluids must be considered a potential vehicle for contamination. Water from a highly controlled loop is specified to carry microbial contamination of less than 10 CFU/100 mL. This means that one 10-L bucket of water may contain up to 10^3 CFU without considering microorganisms that were introduced with the detergent. These microorganisms would be distributed on the floor of a cleanroom, which is not a problem if the disinfection program is effective. However, the potential for contamination should be kept in mind especially in view of any residual moisture which remains in gaps of joints, underneath equipment or in incomplete seals at junctures. At such positions, a favorable mini-environment for growth of microorganisms can easily be created, and cleaning fluids are a perfect means to supply the necessary water for microbial proliferation. Hence, great care should be taken to seal all such gaps which may come into contact with water during cleaning.

Distribution of contaminants constitutes another risk of cleaning. Spores from a localized source of microbiological contamination or growth have been known to be transported over a wide area by cleaning measures. This can even happen in critical zones, where bacterial endospores from a contaminated juncture have been found to be distributed in the process of wiping the surfaces with 70% alcohol. For this reason, it is strongly recommended to restrict the area that may be cleaned with a single wipe. Critical surfaces in aseptic manufacturing should never be touched with anything else than a fresh sterile wipe, should any cleaning become necessary due to contact with tools or other potential contamination sources in course of an intervention or aseptic assembly.

Access to the cleanrooms by cleaning personnel and entering of cleaning materials must be considered the same way as any other entry into cleanrooms. Personnel must be properly gowned and the same care must be taken when entering cleaning materials as for any other materials used in production.

It is preferable if cleaning and disinfection is considered part of the overall manufacturing process, and performed by the regular cleanroom personnel. However, cleaning is frequently considered an inferior type of work which is performed by less skilled personnel. Cleaning work is often given out to contractors who send in their own personnel after the working hours of the filling personnel. While it is generally understood that such people must be trained in access procedures and general cleanroom behavior, the cleaning procedures themselves are an important issue which must be well understood in order to allow proper cleanroom control. Cleaning procedures must be specific to the requirements of the room and equipment, and the manufacturing processes performed and should be written by the pharmaceutical manufacturer. Cleaning should be supervised by QC to ensure compliance with such procedures irrespective of whether it is performed by the normal manufacturing personnel or by contractors.

DISINFECTION OF ROOMS AND EQUIPMENT SURFACES

Significance of Disinfection

Disinfection is a term which was coined long before the microbial nature of infectious agents was known. It was already used in the 16th century in France for a treatment of an environment that reduced the occurrence of infections. Today, disinfection is frequently defined as a chemical treatment which destroys infectious microorganisms on surfaces. This is partially misleading, as there is no method available that selectively inactivates infectious microorganisms.

The purpose of disinfection is to treat surfaces in a way that destroys most microbial contaminants. However, depending on the disinfecting agent used no assurance of sterility can be given. Effective disinfection methods destroy vegetative forms of bacteria and fungi by at least five logarithmic orders of magnitude. Bacterial endospores are not destroyed by most disinfectants; only a few agents are suitable for sporicidal disinfection in practical use.

Where reasonable assurance can be given that microbial contamination (if any) on the disinfected surface was minimal and did not contain bacterial endospores, the treatment may be considered sufficient to restore aseptic conditions. Such a situation can easily occur during aseptic assembly of equipment, or corrective interventions in aseptic processes where a risk of contamination of a critical or controlled surface cannot be completely excluded. The rare microorganisms deposited would in all experiences be vegetative forms which are sensitive to disinfection measures.

In all other cases, disinfection cannot be accepted as a substitute for sterilization. Where in exceptional cases product contact surfaces are not amenable to other sterilization procedures, chemical treatment of product contact surfaces must be sporicidal and the effectiveness must be shown to be comparable to sterilization processes. The correct term for the procedure would then be surface sterilization and not disinfection.

Resistance to Disinfectants

There are a number of reasons why disinfection may not be successful. One reason is resistance of microorganisms to antimicrobial agents. Such resistance can be intrinsic, physiological or acquired.

Disinfectants are not equally effective against all species of microorganisms. Gram-positive organisms are in many cases intrinsically less susceptible to a variety of disinfectants than gram-negative organisms (3). This is only true, however, for vegetative cells. Bacterial endospores of gram-positive microorganisms are highly resistant to all forms of antimicrobial activity, whether chemical or physical.

Microbial cells attached to surfaces are found to be significantly more resistant to antimicrobial action than planktonic forms (4). Microorganisms colonizing a surface frequently create a layer of organic materials which serves as a matrix in which microbial cells are embedded. Such biofilms form a protective barrier which is very difficult to penetrate by antimicrobial agents and hence, microbial resistance is dramatically increased. When biofilm organisms are detached and exist in a planktonic state, they become more sensitive again. Hence, the resistance can be considered an expression of the physiological state of the microorganisms. Other factors, e.g., pH, temperature, and presence of particular ions are known to influence the physiological resistance of microorganisms toward antimicrobial agents. Multigenic systems have been described in microorganisms which respond to different environmental conditions, e.g., nutrient limitation (5). Observations on population dynamics of disinfectant resistant pseudomonads after isolation (6,7) suggest that there may be similar regulatory systems which result in altered disinfection sensitivity in response to changing environmental conditions.

Acquired resistance properties involve, for example, reduction of cell barrier permeability by the formation of slime layers (capsules) around the microbial cells (8). The difference of rough and smooth forms of bacteria and its genetic origin is well known and was the basis of the classic transformation experiment of Averey which showed that DNA is the molecule that carries genetic information (9). Encapsulated strains have been shown to be more resistant to antiseptics and disinfectants than nonencapsulated strains (10). Differences in cell membrane composition have also been shown to be responsible for altered antimicrobial susceptibility (11). Transmission by plasmids is largely discussed in the context of antibiotic resistance, but there is evidence that resistance toward some disinfectants may also be carried on plasmids (8,12) even though the genetic mechanisms are far less clear. Studies to elucidate the genetics of antimicrobial resistance have frequently not yielded clear results which has been attributed to experimental difficulties and possible instability of properties of the microbial isolates in the absence of the disinfectant (8).

There are several reported cases of isolation of highly resistant mostly gram-negative organisms from antiseptics and disinfectants (13). In a well-documented case, *Serratia marcescens* was isolated from a canister of cotton balls soaked in benzalkonium chloride, which were used for antiseptics and disinfection of the surfaces of multidose vials (14,15).

In summary, resistance of microorganisms toward disinfectants is a complex subject, where multifactorial influences lead to results which are difficult to interpret. It is well known to microbiologists that the resistance characteristics of isolates from an environment treated with disinfectant or a preserved product are exceedingly difficult to characterize. Intrinsic, physiological and genetic factors interact and the exact mechanisms of antimicrobial susceptibility and resistance are mostly not well understood. For reviews on the antimicrobial susceptibility and resistance of microorganisms to disinfectants and preservatives, see Wallhäusser (3) and Russell and Gould (8).

The effectiveness of disinfectants may also be reduced by organic residues on the disinfected surface. Where agglutinated serum or other protein-containing residues are surrounding contaminating microorganisms, successful disinfection may become quite impossible.

Disinfectants can only be effective when they reach the site of contamination. Cracks, recesses, and narrow gaps must be penetrated. The area underneath a piece of equipment or a pallet of material which is cluttering a room cannot be expected to be reached in a disinfection procedure.

Disinfecting Agents

Most Frequently Used Disinfecting Agents

In the following paragraphs, the groups of antimicrobial agents that are most frequently used in disinfectant preparations are briefly discussed. It would be beyond the scope of this text to be all-inclusive or to go into detail with regard to the properties of the agents.

Alcohols

Alcohols are bactericidal within minutes (most vegetative bacteria are killed within 30 seconds) mainly by agglutination of proteins and are also rapidly evaporating (16). Hence, they are frequently used for disinfection of gloved hands where it is impractical to wait for a long period of exposure and drying. They leave no residues on the disinfected surface if used without denaturant and therefore are frequently used to wipe surfaces in critical areas of aseptic production. As un-denatured ethanol is heavily taxed in many countries IPA is frequently used as a cheaper but equally active substitute. It should be noted that 60% to 70% aqueous solutions are more antimicrobially active than concentrated ethanol or IPA.

Alcohols are not sporicidal; in fact, bacterial endospores are preserved very effectively in alcoholic solutions, but spores cannot germinate and proliferate in alcohol. Some lots of alcohol have been found to be contaminated with spores. Hence, it is mandatory to remove (typically by membrane filtration) any spores from alcoholic solutions before they are used for disinfection of surfaces in critical zones for aseptic processing. Solutions can be stored thereafter without problem if protected from spore contamination.

Aldehydes

Aldehydes are excellent antimicrobial agents which act by cross-linking of macromolecules. (Glutaraldehyde is used in electron microscopy to fix biological specimen.) Formaldehyde is sporicidal in aqueous solutions at in concentrations of more than 3%, and evaporated formalin (37% aqueous solution of formaldehyde) has been used for sterilization of rooms by gassing (17). Even though the chemical action of aldehydes on cells is clearly nonspecific, plasmid-mediated acquired resistance to formaldehyde has been described (18). In spite of the excellent antimicrobial activity, the former widespread use of formaldehyde for gassing of cleanrooms has been generally discontinued because of its carcinogenic and immunosensitizing properties (19). Strict safety procedures are mandatory for the application of formaldehyde gassing in those exceptional cases where itis still used. In a well-controlled situation (e.g., to condition a cleanroom after building measures, or to decontaminate a room after a contamination problem has been recognized and corrected), formaldehyde gassing is still used as one of the most highly effective disinfection procedures.

Lower concentrations of other aldehydes continue to be used as a component in the majority of surface disinfectants with a combination of agents marketed. It should be remembered that toxicity for a wide spectrum of microorganisms is the desired activity for antimicrobial agents, and most disinfecting agents are also noxious for human beings. With regard to the safety of the personnel it is important, to use disinfectants prudently and only in the concentrations needed.

In the concentrations normally used for disinfection, aldehydes are not significantly sporicidal.

Quaternary Ammonium Compounds

QACs comprise a large number of ammonium derivatives with the general formula as shown below:

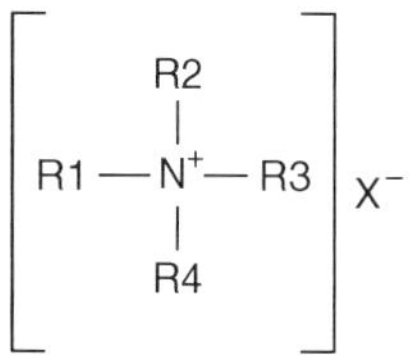

The radicals R1 to R4 represent a wide range of different possible chemical groups or arrangements. It is important to note that for a quaternary compound to be antimicrobially active at least one of these groups has to be an aliphatic hydrocarbon chain with a chain length in the range of 8 to 18 carbon atoms. This clearly indicates a hydrophobic affinity to the bacterial membrane. It has been argued that cationic disinfectants like QACs may act by destabilization of the outer membrane structure of bacteria (20). The activity of different quaternary compounds is of course also dependent on the other three radical groups.

Whereas the spectrum of antimicrobial activity is broad, quaternary compounds are in general more active against gram-positive than against gram-negative organisms. It is well recognized that highly resistant strains of gram-negative organisms do occur, which may even proliferate in solutions of quaternary compounds (21). This is a reason why in many disinfectant preparations quaternary compounds are combined with aldehydes and/or alcohols. Quaternary compounds are not sporicidal.

Guanidine

Guanidine derivatives of the general formula as given below are strong organic bases which are used for disinfection:

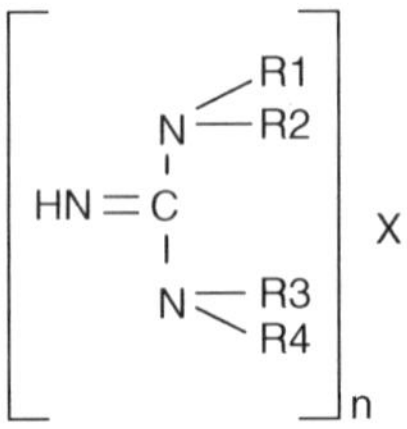

where R1 to R4 are side groups which can be a hydrogen atom or a more complex group.

As for the quaternary compounds, one of the side groups has to be an aliphatic hydrocarbon chain with a chain length in the range of 8 to 18 for a molecule to be antimicrobially active, indicating a mechanism of activity which involves the cell membrane (11). The spectrum of antimicrobial activity is broad, but resistant microorganisms have been reported to occur (22). Guanidines are not sporicidal.

Alkylamine

Alkylamines are amines of aliphatic fatty acids. Again, the aliphatic chain length is critical for the antimicrobial activity. The spectrum of antimicrobial activity is broad. Alkylamines are not sporicidal (23).

Phenols

Phenol and its derivatives are widely used as disinfectants in the United States, but much less so in Europe.

Phenols are strongly cytotoxic, and the antimicrobial action of these molecules is on proteins, which are denatured and precipitated. Lipophilic side chains enhance the activity of the derivatives. Phenols are not sporicidal (24).

Peroxides

Peroxides are the most frequently used disinfecting agents in sporicidal disinfectants, sometimes in combination with organic acids. The antimicrobial action is by unspecific oxidation (25). These disinfectants are aggressive and often corrosive to the equipment and environmental surfaces where they are applied. It is advisable to use them prudently where indicated.

Hydrogen Peroxide

Hydrogen peroxide is used in stabilized solutions frequently in combination with an acid. In low concentrations (about 3%), the agent is used as a relatively mild disinfectant, e.g., for wound dressing. H_2O_2 is also used as a sanitizing agent for water systems, where the instability of the peroxide is an advantage to avoid residue formation in the water system (26).

In much higher concentrations and frequently in combination with peracetic acid, H_2O_2 is the most frequently used type of sporicidal disinfectant today. The action is by the strong oxidative potential of nascent oxygen on the proteins of the microbial cells.

Vaporized or aerosolized H_2O_2 is also frequently used as a surface sterilant in the decontamination of isolators.

Disinfecting Agents in Commercial Disinfectant Preparations

The choice of the disinfecting agent(s) used will depend on the intended use of the disinfectants. A list of marketed disinfectant preparations which have been tested following a published standard procedure (27) and are approved by the DGHM as being effective and recommended for use in hospitals gives a good overview on the use of disinfecting agents and the composition of marketed disinfectants. A summary evaluation of the products listed in 1997 is given in Table 1 for hand disinfectants and in Table 2 for surface disinfectants (28). While the number of products listed was quite large, the number of disinfectant agent classes used is rather limited. For the purpose of this crude evaluation, chemical groups of agents were not further differentiated, e.g., alcohols were grouped together as were other groups of agents or their derivatives.

Skin Disinfectants

For skin disinfectants alcohols are primarily used. Out of 69 products listed, 67 contained alcohol either as a monopreparation in combination with other agents. Almost two-thirds of the products contained only alcohol (mixtures of several alcohols were still considered mono-preparations). The most frequent combination products contained alcohol with quaternary ammonium bases (QACs) (9 out of 24).

Table 1 Hand Disinfectants (69 Products; 1997)

Agent class	Mono	Combinations: 2 Agents	Combinations: 3 Agents	Total
Alcohol	38	24	5	67
Quarternary base	0	9	4	13
Phenol	0	7	2	9
Guanidine	0	4	1	5
Organic acid	0	2	2	4
Iodine	2	1	0	3
Peroxide	0	0	1	1
Pyridine	0	1	0	1
Total	40	24	5	

Table 2 Surface Disinfectants (384 Products; 1997)

		Combinations			
Agent class	**Mono**	**2 Agents**	**3 Agents**	**4 Agents**	**Total**
Quarternary base	95	110	70	1	276
Aldehyde	14	88	51	0	153
Alcohol	13	43	59	1	115
Guanidine	0	25	22	1	48
Alkylamine	24	5	9	0	38
Organic acid	2	6	5	0	13
Peroxide	6	1	3	0	10
Phenol	4	1	3	0	8
Amphotensides	1	6	0	1	8
Pyridine	0	5	0	0	5
Glycol	0	1	3	0	4
Chloramide	3	0	0	0	3
Anorganic acid	0	1	0	0	1
Total	162	146	75	1	

Surface Disinfectants

For surface disinfectants, QACs are on top of the list. Out of 384 products listed, 276 contained QACs either as a mono-preparation or in combination with other agents. Ninety-five products contained QACs as mono-preparation. The most frequent combination of two agents was QACs with aldehydes (66 out of 146), and the most frequent combination of three agents was QACs with alcohol and aldehydes (44 out of 75).

Disinfection Protocols

Disinfection procedures should be well defined and documented with schedules, responsibilities, type or preparation of disinfecting agents, and techniques applied. The measures for routine and additional disinfection measures on occasion (e.g., corrective interventions) should be specified. Intensified emergency disinfection measures in case of monitoring level excursions or other incidents should also be defined.

Disinfection of floors, walls and ceilings should be at scheduled intervals and performed in a defined way so that the critical areas are not affected. Disinfection of walls and floors in cleanrooms is typically done at the end of each shift in critical and controlled areas for aseptic processing. In cleanrooms where less critical work is done, the frequency should be based on the results of environmental monitoring.

Choice of Disinfectants

The disinfectants used should be compatible with the specific surface to be disinfected. Critical surfaces and other surfaces in the critical area for aseptic filling should only come into contact with sterile disinfectant solutions. Sterile alcoholic disinfectants are frequently used for this purpose.

In the 2004 FDA Aseptic Processing Guideline (29), it is stressed that *"a sporicidal agent (should be) used according to a written schedule and when environmental data suggest the presence of sporeforming organisms."* As sporicidal agents are usually corrosive and not without hazard to the personnel, their use should not be exaggerated. On critical surfaces sporicidal agents should not normally be used, because such surfaces should be sterilized and not subject to spore contamination during product manufacturing in a well-segregated critical processing zone.

For floor disinfectants in controlled areas adjacent to critical areas, a sterilization step would not be normally required, but the solutions should not be contaminated with any detectable number of microorganisms. Vegetative forms of microorganisms should normally not survive in disinfectant solutions, and there should be no detectable endospores. Disinfectant solutions must be freshly prepared from contamination-free concentrates with sterile water if they are not sterilized by filtration. There must be a strict limitation on the time of use for these solutions.

For cleanrooms with lesser cleanliness requirements, effective microbiological control should be in place for aqueous disinfectant solutions, used therein, especially if resistant microorganisms to the type of disinfectant used are known to occur (e.g., QACs, phenolics).

Rotation of Disinfectants

Disinfectants are frequently rotated, in order to avoid eventual selection of resistant microorganisms. It is not quite clear whether this is a scientifically solid approach. Even though acquired resistance toward disinfectants and antiseptics has been described, proliferation of resistant strains in pharmaceutical cleanroom environments would not be expected. Conditions in cleanrooms are generally such that proliferation of microorganisms on the surfaces is avoided (e.g., absence of humidity). Without microbial proliferation there can be no selection. Selection of antimicrobial resistant strains can only occur, where in a large population of competing microorganisms antimicrobial resistance presents an advantage for microorganisms to reproduce more effectively in the presence of disinfectants. While such circumstances will occur in clinics where large numbers of microorganisms are proliferating in infected patients which are treated with antibiotics or antiseptics, the environment in controlled pharmaceutical cleanrooms will not allow any microbial growth let alone competitive growth of microorganisms. As long as disinfection serves to remove microorganisms that have been brought into the cleanroom from the outside environment, and not to control growth of a proliferating resident flora, selection of resistant forms would not be expected to occur.

Disinfection Utensils

Information should be available concerning the load of microorganisms carried by wipes and other disinfectant utensils, as it is not unknown that wipes may be contaminated by microbial spores which may remain unaffected by the disinfectant and are then distributed on the surface to be disinfected. Wipes and any utensils used in disinfection of surfaces in critical areas should be sterile. Disinfectant bottles (e.g., for hand disinfection) should be sterilized before they are refilled. The time interval acceptable for using a portion of disinfection solution dispensed in bottles should be defined. Bottles should be labeled with filling and expiration date.

Period of Action

It is important to allow sufficient time for the disinfectants to be fully active as specified by disinfectant suppliers. Even for rapidly acting alcoholic hand disinfectants, at least one minute should be allowed after application of the disinfectant, before the operator proceeds, e.g., to an intervention in a critical zone. For the slower acting surface disinfectants action time can be up to four hours. It is not practicable nor required to keep the surfaces wetted for the full period. It should be understood that a disinfectant agent after entering the microbial cells during the wet period, may continue to act intracellularly even after drying of the surface.

Disinfection Techniques

Disinfectants should not be sprayed over the surfaces but should be applied by wiping in combination with a scrubbing action. Spraying creates an aerosol which is deposited on the surfaces in tiny droplets. Unless a surface is thoroughly wetted to achieve a confluent layer of liquid, there can be no assurance that no dry areas remain between droplets which did not receive the agent. Also, most surfaces are not completely flat on the microscopic level but have some surface roughness, scratches and traces of wear where particles and dirt can remain attached during cleaning. Wiping and scrubbing serves to mechanically remove contamination and to work the agents into the microstructures of the surface in order to achieve better penetration to reach hidden microorganisms. As a health consideration, disinfectants are meant to be noxious to a broad range of microorganisms which means that they generally also have toxic properties toward human beings. Spraying creates potentially harmful aerosols that are breathed by the personnel, an unnecessary health hazard which can be mitigated by direct application of the disinfectant.

Portable equipment and materials should be removed from the rooms wherever possible to allow proper cleaning and disinfection. Where areas are used for staging of materials and cannot be cleared for each disinfection process, it must be assured by detailed procedures that all surfaces are cleaned and disinfected in predefined intervals.

Gassing and Fogging

Routine gassing with formaldehyde is considered an obsolete practice today. While gassing is an effective disinfection method which can effectively reach otherwise inaccessible positions, the associated health hazard of formaldehyde to the personnel is considered too high. In Europe, a specifically trained responsible person for gassing must be registered with the supervising authority before permission is granted to perform formaldehyde gassing. It is mostly used as an exceptional method today, for example, after building measures where the rooms receive an intensive cleaning and disinfection program as a measure to achieve the basic state of cleanliness which is then maintained by segregation and routine cleaning and disinfection measures. Gassing may also be used as a corrective measure after a major contamination event. There are facilities that employ gassing systems using either hydrogen peroxide or ozone as routine measures for environmental control.

Gassing of isolators with vaporized hydrogen peroxide is a more modern technology which is widely used in isolator decontamination.

Validation of Disinfection Effectiveness

Disinfectant Qualification

A number of national and international standard procedures for testing of disinfectant effectiveness have been published (27,30–34). These standards typically require inactivation of a defined number of log scales of a well-defined highly concentrated inoculum of well-characterized reference microorganisms in suspension, as well as on surfaces under carefully controlled conditions. While these procedures may seem deceivingly simple, their execution requires meticulous control of all experimental details, and the amount of work involved is very high. The standards are meant to be performed by disinfectant manufacturers in order to qualify their products for use in public domains, where disinfectant effectiveness needs to be verified. The procedures are also used as reference methods by independent test laboratories as a basis for recognition of a product by certifying organizations.

The standards are not meant as a test method to be applied by every disinfectant user. It is acceptable to rely on certificates of the disinfectant manufacturer especially where they rely on independent qualification testing. Data summaries for their product and certificates of independent test laboratories should be available from reliable suppliers of disinfectant preparations.

Where customized disinfectant preparations are formulated in-house by pharmaceutical manufacturers, a test comparable to the standard methods would have to be performed, however, to verify that the disinfectant preparation is effective.

Verification of Continued Disinfectant Effectiveness

Disinfectant users typically perform verification tests of disinfectant effectiveness by inoculating surfaces of the same material as used in production rooms with a simulated microbiological contamination. Contact plate testing should show that the contamination has been inactivated by the disinfectant after the specified action time. Such tests do not need to employ the same high density of inoculum nor same degree of standardization as the official test methods. The evaluation is typically qualitative, not quantitative. Verification testing should be performed initially, when cleaning and disinfection methods are established and validated. A repeat of disinfectant verification in the course of routine validation of cleanroom maintenance is not normally necessary.

The continued effectiveness of cleaning and disinfection measures is also demonstrated by trending the results of environmental monitoring. As long as the monitoring results continue to remain at the expected low level as established during cleanroom validation, it is obvious that the cleaning and disinfection measures are appropriate. Should there be repeated excursions in environmental monitoring results, the evaluation in case of investigations may well lead to the conclusion that re-verification of disinfectant effectiveness is indicated.

Risk Consideration for Disinfection

Choice of Disinfectant

Disinfectants must be compatible with the surfaces to be cleaned, and procedures must be specific to the requirements of the room and equipment, and the manufacturing processes performed.

Distribution of bacterial endospores or resistant microorganisms across the disinfected surfaces from a contaminated disinfectant solution or a localized source of microbiological contamination or growth is known to be an occasional problem. Critical surfaces in aseptic manufacturing should never be touched with anything else than a fresh sterile wipe should disinfection become necessary due to contact in course of a intervention or aseptic assembly. For controlled areas, the use of disinfectants with a combination of disinfecting agents may minimize the risk of contaminated disinfectant mono-preparations.

Sporicidal disinfectants should be used with care and at a low specified frequency (e.g., monthly) as long as environmental monitoring data indicate that the area is under control.

Residues of Disinfectant

As for cleaning agents, disinfectant residues must be avoided on critical surfaces, where they may contaminate product coming in contact with them. Use of 70% alcohol for disinfection of critical surfaces has been mentioned as an example for a disinfectant which leaves no residues. Even though this is not a sporicidal procedure, there would be no need to apply a more rigorous disinfection procedure as long as there is no hint of spore contamination from the evaluation of environmental monitoring data.

On other surfaces like floors, residues of detergents may even be considered desirable, as proliferation of sensitive microorganisms in the presence of water would be prevented.

Disinfection by Unskilled Personnel

Disinfection is a basic technique intended to keep cleanrooms under control. Usually disinfection of critical areas is performed by the cleanroom personnel. Where disinfection of the floors of controlled areas is given out to cleaning contractors as part of cleaning and disinfection, the considerations about compliance of the contractor's personnel with established procedures and their diligence in performance of the disinfection work must be clearly established as discussed in the section on cleaning. Since lack of effective disinfection may directly influence the microbiological quality of the manufacturing environment, training of the personnel with regard to access procedures and general cleanroom behavior, as well as the detailed disinfection procedures, have to be well documented. Compliance with the written procedures for disinfection irrespective whether performed by the normal manufacturing personnel or by contractors should be verified by QC by frequent observation.

Risks of Gassing with Disinfectants

Gassing with formaldehyde, ozone or hydrogen peroxide poses specific risks. The health hazard to the personnel is considerable, and in Europe specifically trained personnel must be nominated to the authorities as the responsible supervisor for formaldehyde gassing procedures. The rooms must be sealed to prevent uncontrolled leakage of the vapor. After gassing, the rooms must be sufficiently aerated to make sure the residues do not pose a health hazard to entering personnel.

There is also a risk to exposed electronic parts which could be damaged by reaction with the gaseous disinfectant. Condensation of the disinfecting agent on surfaces is dependent on the concentration in the vapor, but also on the temperature of the surfaces. For gassed rooms this can be difficult to control.

ENVIRONMENTAL MONITORING

Sources and Vectors of Microbial Contamination

In order to correctly target cleaning and disinfection measures based on environmental monitoring data, it is important to understand the dynamics of the microflora in cleanrooms, and the limits of the methods used for environmental monitoring. Microorganisms entering a cleanroom can originate from various sources and are carried across segregation barriers by a number of different vectors (e.g., Table 3). The type of microorganism(s) isolated may be frequently correlated to the source of contamination and the vector involved. The number of microorganisms isolated reflects the degree of segregation achieved to retain a microbial challenge by all combined measures. The cleanroom flora reflects the type of challenge, the effectiveness of segregation and the pathways of contamination across the segregation barriers.

By far the most important vector is cleanroom personnel. Cleanroom personnel cannot be effectively decontaminated, and the segregation achieved even by cleanroom gowns used for ISO class 7 (Grade B) cleanrooms is far from perfect. This is reflected in the composition of the flora detected in well-controlled manned class 7 cleanrooms. Typically about 80% to 90% of the organisms discovered with conventional

Table 3 Examples of Types of Microorganisms Found in Cleanrooms, Their Natural Habitat and Possible Pathways and Vectors for Entry into Cleanrooms

Type of organism	Natural occurrence	Pathway	Vector
Gram positive cocci	Body of personnel	Shed with skin scales	Personnel
Gram positive (non-spore-forming)	Ubiquitous in soil	Dust penetrating the cleanroom	Air influx/venting system
		Dust carried on body/scalp	Personnel
		Contact contamination	Transport equipment, tools, electric cables, etc.
Spore-forming bacilli	Soil	Dust penetrating the cleanroom	Air influx/venting system
		Dust carried on body/scalp	Personnel
		Contact contamination	Transport equipment, tools, electric cables etc.
		Solvents/disinfectants	Cleaning and disinfection solutions
Gram negative rods, non-fermenting	Ubiquitous in water, soil	Water system	Water for process or cleaning
		Aqueous liquids	Cleaning and disinfection solutions
Gram negative rods, fermenting	Intestinal system, waste water	Unhygienic behavior	Personnel
		Contaminated sink	Cleaning materials, (e.g. wipes)
Yeasts	Decaying organic materials	Carried on the body	Personnel
Molds	Ubiquitous	Carried on the body	Personnel
		Surfaces of buildings	Venting system, moisture

The list gives only routes which are commonly observed, and is not intended to be exhaustive.

environmental monitoring methods are gram-positive cocci (Fig. 1). About 15% of the isolates are non-spore-forming gram-positive bacilli (35). The rest of the isolates are occasional spore-forming bacilli, gram-negative rods, or yeast and mold. This spectrum of microorganisms is very similar to the spectrum identified from gloves and gowns of the operators, who are the predominant source of contamination in a cleanroom where every other source of contamination is carefully eliminated or minimized. It should be noted, however, that the composition of the flora of microorganisms carried on the body of each human being, as described in textbooks of medicinal microbiology, is far more complex than what is shown in Figure 1. This means that the contaminants detected in environmental monitoring reflect the limitations of the monitoring method applied.

Another source of microorganisms found in cleanrooms is water. The microflora of water is predominantly composed of gram-negative rods, while gram-positive cocci are very rare. The species of gram-negative organisms encountered depends on the amount of nutrients available in the water. Highly purified water hosts different species than water rich in organic matter. While the microflora of wastewater is very complex, a highly purified water system will usually be colonized by only a limited number of species which are adapted to the conditions of the system.

Other potential sources of microorganisms include, e.g., disinfectants, where resistance of the contaminating organisms against the antimicrobial activity is a prerequisite for their survival. Spore-forming microorganisms but also resistant vegetative forms especially of gram-negative organisms have been introduced into cleanrooms by this route on numerous occasions. As resistant microorganisms do not occur frequently in pharmaceutical cleanroom environments, the microflora in a contaminated disinfectant would usually be composed of a single species.

Hence, knowledge of the numbers as well as the composition of the "normal" microflora in a room is critically important to decide which actions are to be taken in case of monitoring data indicating a deterioration of the environmental situation or the occurrence of an adverse event. Has a breach of segregation occurred which needs immediate corrective action, e.g., in the form of altered entry procedures, or has a period of transient increased activity led to increased input of microorganisms which can be reduced back to the normal level by additional or intensified cleaning and disinfection?

Interpretation of Monitoring Results

Evaluation of environmental monitoring samples should always consider the complete body of data collected at a given time and in the recent monitoring history. Isolated high counts cannot be simply dismissed as insignificant but they must always be considered in proper context. It is highly recommended that any consideration of colony numbers should also take into consideration the type of microorganisms isolated.

Isolated High Values

Isolated excursions in environmental monitoring data are very difficult to evaluate. A high value in air monitoring indicates significant input of microorganisms into the air stream at or upstream of the measuring site during the period of monitoring. This may or may not be reflected in other measurements. If the contamination was quantitatively swept away by a directed air stream, no contamination would be found at any other monitoring site or during a later measurement at the same position. Nor would there be any increase in surface counts. The significance of an isolated excursion will always have to take into consideration the exact position of the monitoring site. Is it safe to assume that the contamination was carried away by the directed air stream? If the answer is yes, the risk to the product would be minimal.

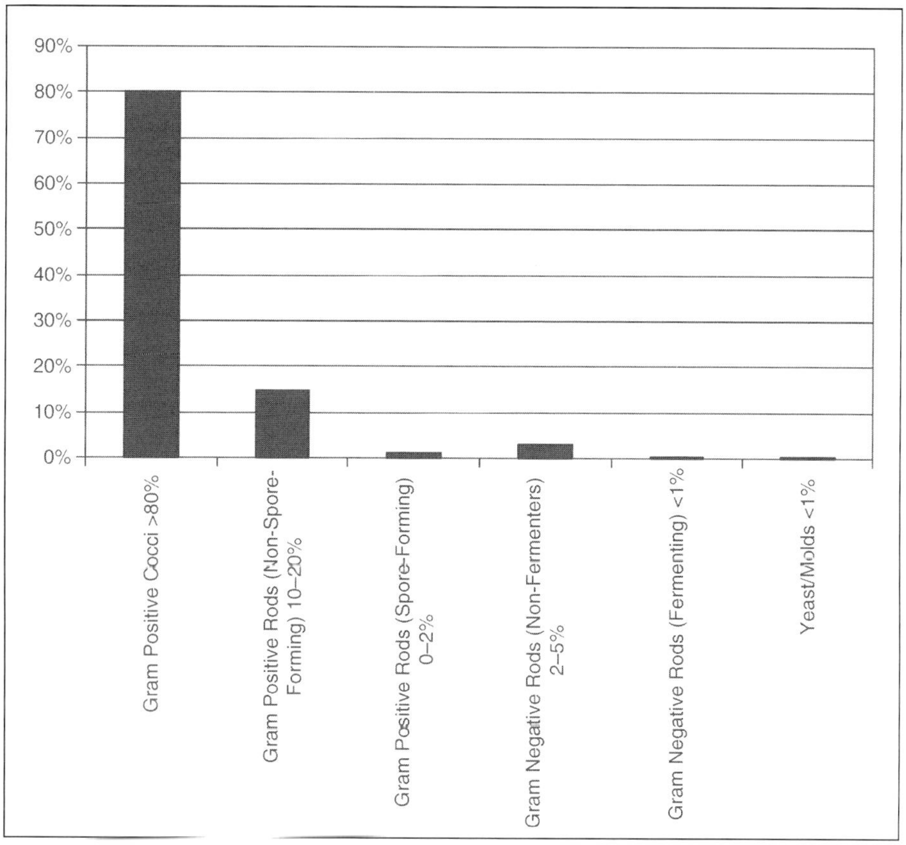

Figure 1 Typical microflora found in manned cleanrooms: relative frequency found occurrence on casein-soy digest agar. A rough estimation of the expected relative frequencies of the major groups of microorganisms collected during environmental monitoring on casein-soy digest media from air and surfaces in manned cleanrooms is given. Figures vary depending on the activities, cleanroom segregation, medium and methods used for monitoring, and many other factors. The figures are derived from monitoring data collected over several years from several pharmaceutical cleanrooms.

An isolated high count in surface contamination indicates a local surface contamination at the site of measurement which is not spread to other surfaces. This may have been caused by contact with a contaminated tool or the body of an operator. Again, the significance would depend on the proximity of the site to product contact and/or the existence or absence of any transport mechanisms to the critical site. Disinfection of the sampled site (if not already performed in the course of routine cleaning and disinfection measures) would be a necessary follow-up measure.

If such high counts remain singular events where product contamination can be deemed improbable, they are typically seen as local events with little bearing on product quality.

Trends

In environmental monitoring, the term trend is understood in a broader sense than a mathematical trend, which usually means a numerical increase or decrease. Increased frequency isolation events of microorganism at the same measuring site or at different measuring sites throughout a processing area should be considered a trend, even if there is no increase in numbers. Multiple occurrences of microorganisms even below the action level should be considered a trend which requires corrective action.

Identification of Isolates

For evaluation of the significance of environmental monitoring data, and for determination of contamination routes, characterization of microbial isolates is of equal or higher importance than collection of numerical data of colony-forming units. There is presently a strong tendency toward in-depth identification of microorganisms. DNA sequencing is advocated in order to achieve identification even below species level. But what is the depth of characterization needed; what is gained by identification of species level?

Even a basic, microscopic characterization in combination with a few orienting metabolic tests will yield very important information. A numerical excursion of the expected level of microorganisms in a manned aseptic environment carries very different significance if the isolated microorganisms are composed of gram-positive cocci, gram-negative rods or spore-forming gram-positive rods. In the first case, there was probably increased personnel intervention; in the second case, a water-related contamination source (e.g., problem of cleaning and disinfection) would be suspected; however in the third case, highly resistant microorganisms have been carried into the cleanroom via an as-yet unidentified vector.

By far the most frequent contaminants detected in a manned aseptic processing environment are *Staphylococcus* species. These microorganisms are typical human commensals and each human being is carrying a number of different staphylococci on their skin. A staphylococcus carried on one skin scale may be different from a staphylococcus on the next skin scale even if deposited by the same person. Hence, identification to species level will not necessarily contribute the information sought. The assumption that the origin of the contaminants was the skin of the cleanroom personnel can be based upon a simple microscopic characterization. On the other side, repeated identification even of low numbers of the same species of spore-forming bacilli would point toward a common source of contamination, hidden somewhere in the cleanroom. This is highly valuable information which may indeed be very helpful in an investigation. The depth of investigation should depend on the type of organism detected and on a sound judgment of the expected gain of information. There can be no one-size-fits-all solution.

Particle Monitoring

Monitoring trends in nonviable particle monitoring should be seen in connection with microbiological monitoring data. A trend toward increased particle counts or bursts of particle counts should be investigated, but unless parallel trends are found in microbiological monitoring, nonviable particle data will not usually by itself prompt corrective cleaning and disinfection measures.

Active Air Monitoring

When reviewing air monitoring data, it should be kept in mind that high volumes of air are circulated through cleanrooms in order to sweep the area free from contaminants. A cleanroom with a volume of 100 m^3 which is operated at 20 air changes per hour will have 16,000 m^3 of air passed through the room within an eight-hour shift. Moreover, if the air flow pattern is well designed, any contaminating microorganisms will not remain in the area but be swept away within minutes and have very little time to settle. With undisturbed unidirectional airflow, microorganisms will not spread in the room, but be carried toward the exhaust of the air stream in the room. With turbulent air flow patterns, microorganisms can accumulate in areas of the room with less than ideal venting characteristics.

An isolated elevated count in an active air measurement means that there was a local input of microorganisms into the air during the sampling time, which was swept away by the air stream as expected in a well-vented cleanroom. The relevance of the finding for product safety must be evaluated in consideration of the criticality of the operations carried out downstream from the sampling site. If elevated counts are encountered repeatedly, frequent input of microorganisms or lack of removal of contaminants by the air stream are probable, and the cause must be investigated. Cleaning and disinfection measures would not be expected to be an effective corrective measure, unless the excursions are also reflected in surface counts.

Passive Air Monitoring

It can be argued that whatever number of microorganisms is carried in the air stream, settle plates reflect the fallout to be expected on a critical surface. In this case, settle plates would be more appropriately considered a form of surface monitoring even though restricted to detection of organisms from the air stream and not from contact contamination. How closely critical surfaces are represented will of course depend on the positioning of the settle plates. If settle plates indicate a fallout of contamination in critical zones, disinfection measures would be indicated in case of level excursions.

Surface Monitoring

Surface monitoring, in contrast to settle plates, will not only detect microorganisms which have settled from a contaminated air stream but also microbial contamination which has been transferred into the cleanroom by other vectors. Contaminants carried on the gloves of operators, on contaminated tools, or on cleaning equipment may never appear in air monitoring and may still be found to contaminate cleanroom surfaces.

If surface monitoring data indicate a trend toward higher surface contamination or a sudden increase indicates occurrence of an adverse event, additional or intensified cleaning and disinfection measures are advocated in order to reduce contamination back to the normal level. Cleaning and disinfection should not be considered, however, to be the only corrective measures, because each contamination must have been introduced by a vector or some other pathway that should be addressed and rectified.

Monitoring in Aseptic Critical Processing Zones and on Critical Surfaces

Monitoring of critical processing zones where open product or components are handled requires in many cases intrusion of the personnel into the most carefully segregated process environment. Unless performed with great care and under application of stringent environmental protection measures, this intrusion can easily be one of the most contamination-prone steps in the whole process. With regard to air monitoring, it must be remembered that the directed airflow in these areas is even more effective in sweeping away any contamination than in turbulently vented cleanrooms. Any microbial contamination detected will reflect the situation during the time of monitoring and contamination introduced in course of the monitoring process will be the most likely one to be detected. For this reason as well as for protection of the product manufactured, it is highly advisable to minimize intrusion during the environmental sampling process.

Monitoring of product contact (critical) surfaces is strongly advocated in the 2004 FDA guide to aseptic processing (29). While it is obviously correct that equipment surfaces that come into contact with aseptically processed product should be sterile, the question must be asked whether it is indeed useful to introduce another less controlled form of sterility test. The problems of sterility tests on product are known all too well. As a form of sterility testing critical surface monitoring can give no positive answer (confirm that the product, or surface is in fact sterile), as these tests are inherently unreliable because they do not detect every instance of contamination, and error prone by introduction of a significant number of false positive results. Unfortunately monitoring of product contact surfaces is favored by the FDA, and protective measures will be needed to avoid that the sampling will not lead to insecurity and unnecessary loss of product while being of very little use to improve the safety of the product.

If microorganisms are detected in critical processing zones or on critical surfaces, the decision how to proceed and whether the product can be released will be very difficult. Monitoring data are generally not sufficient to allow a well-founded decision. Unless trending data indicate that the quality of the environment has deteriorated in general, additional cleaning and disinfection would normally not be the correct solution. Isolated contamination events usually represent errors during an intervention (e.g., for sampling purposes) and corrective actions should primarily address the aseptic technique.

Investigations (Disinfection and Investigation)

Typically a level excursion result will be recognized only after the sample has been processed in the laboratory and incubated for the appropriate time period. With the commonly applied regimes of daily cleaning and disinfection, investigation of the occurrence may be difficult. If the cleaning and disinfection measures are effective, high numbers of microorganisms should have been removed before their growth is recognized in contaminated environmental monitoring samples. It is therefore important to have a sufficient database available to compare the excursion data to the normal situation in the room. Repeatedly occurring elevated counts may point toward elevated input from a known contamination source or input from an additional unknown contamination source. Another possible explanation would be decreased effectiveness of cleaning and disinfection due to some unknown cause.

Frequently attempts to re-isolate a contaminant will not be successful. A sufficient database of identification data to characterize the normal flora would be very helpful in the investigation of an excursion.

CONCLUSION

Cleaning and disinfection are an important part of any manufacturing concept for sterile pharmaceutical products. While cleanroom segregation measures are intended to prevent the ingress of microorganisms into cleanrooms, cleaning and disinfection serves to remove and/or destroy microorganisms that have crossed the segregation barriers and to maintain the required high level of cleanliness in the manufacturing areas. The better the segregation barriers, the less cleaning and disinfection is needed. A highly frequented air lock used for gowning will obviously need more cleaning and disinfection than an isolator, and this should be reflected in the cleaning and disinfection plans. On the other hand, completely impermeable barriers cannot be used in most manufacturing processes and so some defined level of cleaning and disinfection remains necessary even in well-segregated facilities.

Cleaning and disinfection measures also present contamination risks in themselves, and the measures taken may do more harm than good if they are not performed in a carefully controlled manner. Also, their effectiveness can only be assured if their function in the process and their mechanism of action are well understood, and the procedures used are optimized for the given process. The execution of cleaning and disinfection measures is critical and compliance with the written procedures needs to be carefully supervised.

Environmental monitoring data are important to decide upon the necessary frequency and intensity of the cleaning and disinfection measures but also to recognize pathways of contamination. Much more effective and better targeted measures can be taken if the possible pathways of contamination are understood and recognized. Sufficient data are needed for such targeting. It is important to know the basic composition of the cleanroom microflora. This does not mean that each microorganism needs to be identified to subspecies level; more data with a basic differentiation of all isolated microorganisms might be more helpful than very detailed identification of only a few isolates.

Correct dimensioning and execution of a routine cleaning and disinfection program together with careful evaluation of environmental monitoring data and targeting of additional cleaning and disinfection measures are a key to environmental control in sterile product manufacture.

REFERENCES

1. Denny V, Kopis E, Marsik F. Elements for a successful disinfection program in the pharmaceutical environment. PDA J Parent Sci Technol 1999; 53:115–24.
2. Denny V, Marsik F. Disinfection practices in parenteral manufacturing. In: Williams KL, ed. Microbial Contamination Control. New York: Marcel Dekker, 2004:233–82.
3. Wallhäusser KH. Antimicrobial preservatives used by the cosmetic industry. In: Kabara JJ, ed. Cosmetic and Drug Preservation: Principles and Practice. New York: Marcel Dekker, 1984:605–745.
4. Mittelman MC. Bacterial growth and biofilm controlin purified water systems. In: Flemming snf H-C, Geesey GG, eds. Biofouling and Biocorrosionin Industrial Water Systems. Heidelberg: Springer, 1991:113–34.
5. Neidhardt FC. Multigene systems and regulons. In: Neidhardt FC, Ingraham JL, Low KB, Magasanik B, Schaechter M, Umbarger HE, eds. *Escherichia coli* and *Salmonella typhimurium* Cellular and Molecular Biology, Vol. 2. Washington, DC: American Society for Microbiology, 1987:1313–7.
6. Pyle BH, McFeters GA. Iodine sensitivity of bacteria isolated from iodinated water systems. Can J Microbiol 1989; 35:520–3.
7. Pyle BH, McFeters GA. Population dynamics of pseudomonads after iodination. Can J Microbiol 1990; 36:801–3.
8. Russell AD, Gould GW. Resistance of Enterobacteriaceae to preservatives and disinfectants. J Appl Bacteriol Symp Suppl 1988; 17:167S–95.
9. Averey OT, Macleod CM, McCarthy M. Studies on the chemical nature of the substance inducing transformation of pneumococcal types I. Induction of transformation by a desoxyribonucleic acid fraction isolated from pneumococcus type III. J Exp Med 1944; 79:137.
10. Costerton JW, Irwin RT, Cheng K-J. The bacterial glycocalix in nature and disease. Ann Rev Microbiol 1981; 35:299–324.
11. Gilbert P, Pemberton D, Wilkinson DE. Barrier properties of the gram-negative cell envelope towards high molecular weight polyhexamethylene biguanides. J Appl Bacteriol 1990; 69:585–92.
12. Howe TGB. Genetic modifications of the organism and antibiotic resistance. In: Stuart-Harris CH, Harris DM, eds. Control of Antibiotic-Resistant Bacteria. Beecham Colloquium No. 4. London: Academic Press, 1982:97–112.
13. Russell AD, Hammond SA, Morgan JR. Bacterial resistance to antiseptics and disinfectants. J Hosp Infect 1986; 7:213–25.
14. Nakashima AK, Highsmith AK, Martone WJ. Survival of *Serratia marcescens* in benzalkonium chloride and in

multiple-dose medication vials: relationship to epidemic septic arthritis. J Clin Microbiol 1987; 25:1019–21.
15. Nakashima AK, McCarthy MA, Martone WJ, Anderson RL. Epidemic septic arthritis caused by *Serratia marcescens* and associated with benzalkonium chloride antiseptic. J Clin Microbiol 1987; 25:1014–8.
16. Wallhäusser KH. Alkohole. Praxis der Sterilisation, Desinfektion–Konservierung 5. Ausgabe. NewYork: G. Thieme Verlag Stuttgart, 1995:468–89.
17. Wallhäusser KH. Aldehyde. Praxis der Sterilisation, Desinfektion–Konservierung 5. Ausgabe. New York: G. Thieme Verlag Stuttgart, 1995:468–89.
18. Heinzel M. The phenomena of resistance to disinfectants and preservatives. In: Payne KR, ed. Industrial Biocides: Critical Reports on Applied Chemistry, Vol. 22. Chichester: Wiley, 1988:56–67.
19. NIOSH Current Intelligence Bulletin 34. Formaldehyde: evidence of carcinogenicity. DHEW (NIOSH) Publication No. 81–111, 1981.
20. Hancock REW. Alterations in outer membrane permeability. Ann Rev Microbiol 1984; 38:237–64.
21. Dixon RE, Kaslow RA, Mackel DC, Fulkerson CC, Mallison GF. Aqueous quaternari ammonium antiseptics and disinfectants: use and misuse. J Am Med Assoc 1976; 236:2415–7.
22. Wallhäusser KH. Guanidine. Praxis der Sterilisation, Desinfektion–Konservierung 5. Ausgabe. NewYork: G. Thieme Verlag Stuttgart, 1995:584.
23. Wallhäusser KH. Höhermolekulare Aliphatische Amine. Praxis der Sterilisation, Desinfektion–Konservierung 5. Ausgabe. NewYork: G. Thieme Verlag Stuttgart, 1995:598–604.
24. Wallhäusser KH. Phenol und Phenolderivate. Praxis der Sterilisation, Desinfektion-Konservierung 5. Ausgabe. New York: G. Thieme Verlag Stuttgart, 1995:538–50.
25. Block SS. Peroxygen compounds. In: Block SS, ed. Disinfection, Sterilization and Preservation. Philadelphia, PA: Lea & Febiger, 1991:167–81.
26. Wallhäusser KH. Peroxide und sonstige oxidationsmittel. Praxis der Sterilisation, Desinfektion–Konservierung 5. Ausgabe. New York: G. Thieme Verlag Stuttgart, 1995:643–9.
27. Desinfektionsmittelkommission der DGHM. Prüfung und Bewertung chemischer Desinfektionsverfahren—Stand 12.07.1991. Wiesbaden: mhp-Verlag, 1991.
28. Desinfektionsmittelkommission der DGHM. Desinfektionsmittel-Liste der DGHM 17.09.1997. Wiesbaden: mhp-Verlag, 1997.
29. U.S. Department of Health and Human Services, Food and Drug Administration. Guidance for Industry. Sterile Drug Products Produced by Aseptic Processing—Current Good Manufacturing Practice.
30. EN 1499 Chemical disinfectants and antiseptics—Hygienic handwash—Test method and requirements (phase 2/step 2), 1997.
31. prEN 13713 Chemical disinfectants and antiseptics—surface disinfectants used in human medicine, bactericidal activity—test method and requirements (phase 2/step 1).
32. EN Work Item 00216019 Chemical disinfectants and antiseptics—surface disinfectants in the medical area—test method and requirements (phase 2/step 2).
33. prEN 12054 Chemical disinfectants and antiseptics—quantitative suspension test for the evaluation of bactericidal activity of products for hygienic and surgical handrub and handwash used in human medicine—test method and requirements (phase 2/step 1).
34. EN Work Item 00216032 Chemical disinfectants and antiseptics—quantitative suspension test for the evaluation of sporicidal activity of medical and instrument disinfectants—test method and requirements (phase 2/step 1).
35. Haberer K. Identification of environmental isolates. In: Moldenhauer J, ed. Environmental Monitoring: A Comprehensive Handbook, Vol. 2. Bethesda, MD: PDA and River Grove IL: DHI Publishing, LLC, 2005:111–36.

22

Aseptic Processing for Dosage Form Manufacture: Organization and Validation

James Agalloco
Agalloco & Associates, Belle Mead, New Jersey, U.S.A.

James Akers
Akers Kennedy & Associates, Kansas City, Missouri, U.S.A.

INTRODUCTION

The preparation of sterile products often requires that firms utilize aseptic processing, as there are many formulations where terminal sterilization using either steam or radiation is incompatible with the formulation. The inherent and largely unavoidable risks associated with the use of aseptic processing are widely recognized; regulators and practitioners across the globe acknowledge that aseptic processing should only be used where lethal treatments are not possible. A lethal process directed at final containers is always preferable to one that relies on the uncertain exclusion of microorganisms as a final product is assembled from a number of previously sterilized materials. As aseptic processing ordinarily requires activities be performed by personnel, there is the additional concern for introduction of human borne microbial contamination. The regulatory preference for terminal sterilization has been formalized by both FDA and EMEA (1–3). Regrettably, while the expectation and intent is that terminal sterilization will be utilized wherever possible, there are a variety of materials where aseptic processing is currently the only possible means for preparation, i.e., sterile bulk antibiotics, freeze dried formulations, most but not all biological products, and materials without sufficient moisture content (a minimum of approximately 10%). For the production of these types of products, aseptic processing is the only current methodology available.

Users and regulators must recognize that aseptic processing cannot be evaluated parametrically. No amount of physical measurements or microbial or particle monitoring can definitively establish whether an aseptic process is suitable. This fundamental limitation has greatest significance in the"validation" of aseptic processing. As there is no means to directly establish the successful outcome of any aseptic process, it cannot be "validated" in the sense that an autoclave can. The literature may be ripe with references to "validation of aseptic processings"; however, all that has ever been demonstrated by a satisfactory media fill is the capability for successful operation. A media fill with zero contaminated units cannot support the sterility of a production lot made the day before (or the day after). Aseptic processing is an uncertain method as subtle variations in operator practice can profoundly affect the result, and yet there is no ready means to detect, or eliminate those variations. Nevertheless, firms endeavour to "validate" aseptic processing on a regular basis. Awareness is growing that process simulations demonstrate little more than the capability of the firm's facility, practices, operating procedures and personnel on that day. That validation of the aseptic process is expected represents a misinterpretation of the information that a media fill (also called process simulation) or any of the data obtained from the process can provide.

SCOPE

This chapter will review the supportive elements of sterile production using aseptic processing for finished pharmaceuticals. Other chapters in this volume address the similar (and interrelated) subjects of aseptic processing for sterile bulk pharmaceutical chemicals and aseptic processing executed by personnel. The focus will be on the unique considerations associated with aseptic processing for finished pharmaceuticals including process design and process simulation. That this chapter is located within a comprehensive text on validation is fortuitous as the details of many of the supportive activities are described in close proximity. The reader should consult the more extensive treatment provided on these subjects, rather than the brief synopsis of each that appears below, the intent of which is to place these subjects in proper context.

BUILDINGS AND FACILITIES

The aseptic production of sterile products is performed in classified environments supplied with HEPA filtered air. The classification of the environments used for aseptic processing varies with the criticality of the specific activity being performed. Any assembly of

---

Abbreviations used in this chapter: cGMP, current good manufacturing practice; EMEA, European Medicines Evaluation Agency; FDA, Food and Drug Administration; FTM, fluid thioglycollate media; HEPA, high-efficiency particulate air; NMT, not more than; PDA, Parenteral Drug Association; PNSU, probability of a non-sterile unit; SAL, sterility assurance level; SCDM, soybean–casein digest medium; SIP, sterilization in place; TSB, trypticase soy broth; USP, *United States Pharmacopoeia*; WFI, water for injection.

sterile materials should be conducted in ISO 5 environments that have been treated to essentially eliminate the background microbial count. Less critical activities such as washing and component pre-assembly are carried out in ISO 7 environments where microbial control expectations are more relaxed. The microbial requirements for the various classifications will be addressed later in this chapter.

Critical Area

The most important activities are carried out under ISO 5 environments in which environmental control measures have been taken to virtually eliminate microorganisms. This expectation encompasses every location where sterilized and/or depyrogenated components are exposed to the environment. In manned cleanrooms this is accomplished by disinfection of room and equipment surfaces with antimicrobial agents. Aseptic processing isolators are fogged or gassed with similar agents to accomplish the same objective. Depending upon the specific design of the equipment/facility unidirectional air may be provided within the critical area (or zone). Expectations for unidirectional airflow were established years ago when manned cleanrooms were predominant. The transition to isolator based filling has resulted in some questioning of the need for unidirectional air. The very first aseptic isolators successfully employed turbulent air systems, and while many newer designs incorporate unidirectional air there is no definitive proof that it is necessary at all.

Supportive Clean Area

In manned cleanrooms the area immediately surrounding the critical zone is nearly as important. The personnel required for operation of the equipment are present in this environment at all times, and as they must occasionally access the critical zone, preventing contamination is essential. This extends to corridors and gown rooms used for access and are ordinarily classified as ISO 5 as well. Comparable activities in an isolator system are conducted with the same expectation as for cleanrooms. The background environment surrounding an isolator is of substantially less importance due to the more certain separation between the critical zone and the environment where the personnel are located. Current regulatory expectations are for ISO 8 externally, but there is little evidence that external classification of any type is beneficial for aseptic processing.

Environments for Component Preparation

Activities prior to the sterilization/depyrogenation of materials are conducted in a variety of classifications. Once they are washed, items are protected by ISO 5 air until they are either wrapped or enter a sterilization/depyrogenation process. The preparations are commonly defined as ISO 8. The background environment for compounding is almost universally ISO 7, with localized ISO 5 if necessary.

PERSONNEL TRAINING, QUALIFICATION, AND MONITORING

Personnel performance must be the focus of attention in aseptic processing. The operator is often required to perform precision activities (e.g., setup of filling equipment from individual component parts) without introducing microorganisms to any of the product contact parts. That this can be accomplished on a consistent basis is a tribute to the skill of these employees as the gowned human operator is universally acknowledged to be *the primary source* of microbial contamination. The routine accomplishment of such actions without shedding bacteria is accomplished through careful attention to the precepts of aseptic technique. Aseptic technique is a loose assembly of practices originally conceived for laboratory manipulations however, the principles are fully adaptable to aseptic processing of pharmaceuticals. Among the typical aseptic techniques intended to protect sterile materials from contamination by personnel are:

- Every surface of the gowned person is considered non-sterile
- Never touch a sterile object with a non-sterile object
- Never place a gloved hand over an open sterile container
- Use a tool for every activity wherever possible

Additional principles can be found in various microbiology texts, and must be recognized as suggestions rather than hard rules, and of course must be adapted to the specific circumstances of the equipment and materials being handled.

The operators who work in aseptic processing should be cognizant of the basic principles of microbiology, sterilization, disinfection, aseptic technique, gowning practices as well as details of their assigned tasks. Training of personnel should include both classroom session and practicums in which their ability to perform the required procedures or other similar tasks can be evaluated. Training should be near continuous and of course the media fill (process simulation) is perhaps the ultimate evaluation of the operator's proficiency. As with almost any activity in our industry, the aseptic training program should be well documented.

Personnel should be monitored upon exit from the aseptic core, as sampling during the process itself risks residual media on gown or glove surfaces that can prove disastrous to the aseptic process as they could aid the survival of microorganisms shed by the sampled individual. Sampling at the end of the process addresses the potentially weakened integrity of the gowning system after a lengthy and perhaps rigorous period in the aseptic core. Routine monitoring on exit typically focuses on the operator's gloves and forearms, as these are often closest to the sterile materials. Gowning certification is a usual prerequisite for entry into the aseptic core and ordinarily entails sampling of many more surfaces (4). Monitoring requirements for personnel in isolators is generally restricted to gloves impression taken at the conclusion of the process, as gown surfaces are not present.

MATERIALS, COMPONENTS, AND CONTAINERS/CLOSURES

The sterilization and/or depyrogenation of components and materials used in the aseptic process can be performed using any of the many methods available: steam, dry heat, radiation, gas or filtration. With respect to the aseptic process, it is essential that whatever sterilization method is selected, it be validated to attain a minimum PNSU of 1×10^{-6}. An important consideration is the selection of a package configuration that allows for adequate sterilization, yet affords adequate protection of the sterilized materials until ready for use in the process. The maximum time interval between sterilization and production use of the materials (other than the process solution itself) is established in conjunction with the execution of the process simulation by using materials that have been held for the maximum specified time period.

TIME LIMITS

Time limits are imposed to minimize the duration of the aseptic process in order to reduce the risk of microbial contamination. Time is an important consideration for a variety of reasons:

- Operator fatigue can result in poorer adherence to required technique
- Extended use of gowning material can lead to their failure/compromise
- Microbial counts in pre-filtration solution can increase during hold periods resulting in either filtration issues and/or pyrogen concern

Process duration related to operator fatigue is easily factored into the aseptic process simulation (see the following section), while that associated with microbial growth in the formulation is product specific and must be addressed in the validation of the formulation. There is no evidence to support an increase in microbial population within an aseptic environment and given the absence of nutrients and moisture, the potential for such a phenomena in these extremely clean locations should be considered minimal.

ASEPTIC PROCESS SIMULATION

There is a widespread misunderstanding that aseptic processing can be validated such that a defined sterility assurance level can be claimed. This belief is unsupportable, as aseptic processing ordinarily entails the participation of personnel to perform some portion of the process variability in personnel performance limits the certainty with which the process can be considered. A successful media fill on a particular day affords no added confidence in the same procedure performed the next day during a production lot. Process simulations establish that the methods and practices are capable of success; they cannot support that materials produced using identical methods are sterile to the same extent. The oft cited SAL of 1×10^{-3} for aseptic processing is nothing of the sort; it is merely the established maximum projected contamination rate associated with the successful filling of 3000 media units (5). In the intervening years, firms have produced media fills in excess of 100,000 filed units. If such a large fill were devoid of contamination, the firm might claim an SAL of greater than 1×10^{-4} for their operations. Any SAL claim for aseptic processing based upon media filling is without basis; all that is known is the contamination rate of the units filled that day.

The sole use of a media fill is to demonstrate that under the specific circumstances of an individual simulation that the facility is suitable for use in aseptic processing. The inability to consistently achieve the expected result—zero growth in any of the filled containers—is an indication that the process is perhaps overly risky (2,6). Nevertheless it must be recognized that a process simulation ordinarily represents "worst case" challenges of the process in that the increase in interventional frequency associated with media fills and absence of process inherent characteristics (i.e., bacterial inhibition by the formulation) should increase the potential for microbial contamination in the media filled containers relative to a production batch. Industry surveys have established that nearly 10% of all media fills evidence some contamination (7). Provided that the level of contamination remains at low levels within the expected acceptance criteria it should not be considered problematic. The identification of a single positive in a media fill should only rarely trigger a full blown investigation into the source of the contamination as the incidence rate is within expectations and absolute resolution of the contamination source is well nigh impossible.

Study Rationale and Design

Before embarking on an initial media fill program (and periodically thereafter) the firm should prepare a study rationale outlining how its program supports the capabilities of its aseptic operations. For a single product facility, this can be quite easy to prepare as the permutations of lot size, fill volume, fill speed, container-closure, and other process details are likely to be rather limited. In those facilities where each filling line produces a variety of products, the possibilities can increase substantially. The study rationale should provide justification for each filling line indicating how the chosen process simulation studies performed support the product permutations filled on that line. The rationale should be reviewed periodically to ensure its appropriateness consistent with any changes in products, components, practices or equipment that could alter the circumstances. Provided below are some of the more common considerations and choices to be made in developing this rationale.

Media Sterilization

While it may at first appear unusual the sterilization of the media is not a meaningful concern. The media due to its differing formulation from the product(s) being simulated may be more confidently sterilized using a different means than the product(s). Provided that it is introduced into the sterilized system using identical methods and equipment it can be presterilized by steam, alternative filters and even radiation. The intent of the process simulation is to confirm the acceptability of the processing procedures with sterilized equipment, not to validate the sterilization of the product by the filtration system. Validation of the sterilizing filtration must be carried out for each formulation and the ability of the filtration system to sterilize the media is irrelevant to that product by product validation. At the same time, the media utilized in the simulation must be sterile for a valid challenge of the aseptic process, but proof of that sterilization is relevant only to the media process and nothing more.

Frequency and Number of Runs

The initial media fills for a facility are defined in the study rationale and is normally at least three trials per filling line. In larger facilities making a variety of presentations the number of initial studies required may increase substantially. Once the baseline capability has been established, a minimum of two fills per line per year is considered cGMP (2,8). The conduct of additional media fills may be useful for a variety of reasons, i.e., environmental contamination due to an unique event (power loss, water leakage, major breach of asepsis; substantial change in the equipment, processes, components, etc.; adverse environmental trend; or sterility test failure) (2,6).

While it is certainly preferable to await definitive results from a 14-day incubation period (and satisfactory growth promotion), there is no obligation to do so (4). There are firms that conduct their media fills just prior to shutdowns as confirmation of capability at the end of a long operating period and take advantage of overlapping the incubation and shutdown periods. Other firms conduct their media fills immediately after their shutdown period as demonstration of renewed capability including any minor changes to the facility or equipment. For those firms that conduct process simulation post-shutdown practices also vary in relation to whether definitive results are awaited prior to the start of production operations.

Duration of Runs

The seemingly best answer to the required minimum duration of a media fill is that it should exceed the duration of the longest routine filling process, and this is often cited by regulators (2). While that approach may seem the soundest, it presents some substantial problems for those firms making very large lot sizes. The most comprehensive advice on this subject is provided in PDA's TR #22 where recommendations for the complete range of process batch sizes are provided (6). The PDA document is adapted to based 3000 unit media fills; adjustment of that value upwards to suit more contemporary expectations is perhaps the only change required to adapt the approach. For the purposes of this effort, this has been modified to 5000 units, but the advice given could be easily adapted to a different regulatory expectation (10,000 units or more).

Very Small Batches—1000> N

For batches of this size, which are common in certain clinical and radiopharmaceutical operations, a process simulation test at the maximum batch size is recommended. Forcing the production of 5000 or even 1000 units may produce situations so different from the normal operation that the results may be meaningless. For simulation of these batch sizes, the process simulation test must evidence no growth in any of the filled containers to be acceptable.

Small Batches—5000> N> 1000

For this batch size, which might be common for a clinical batch or other developmental situation, the minimum process simulation batch size should be equal to the standard maximum batch size. While this does not afford the level of statistical confidence frequently associated with full process simulation tests, it is a reasonable compromise, given the limitations of the small batch size.

Conventional Batch Sizes—100,000> N> 5000

For these very common production scale process, the number of units to be filled with medium can approach the size of the full production batch, especially with the trend toward larger and larger process simulation tests. Current practice is to produce larger and larger media fills to accommodate the required interventions into the simulation (4).

Large Batch Sizes—N> 100,000

A number of possible approaches have been utilized for very large production batches:

- Fill 5000 units, switch to sterile WFI for an extended period of time, fill an additional 5000 units.
- Fill 5000 units, simulate filling for an extended period of time, fill an additional 5000 units.
- At the completion of a regular production batch, disassemble/reassemble with sterilized equipment; fill 5000 units.
- After an extended WFI fill, disassemble/reassemble with sterilized equipment, fill 5000 units.
- Simulate filling for an extended period of time, disassemble/reassemble with sterilized equipment; fill 5000 units.

These practices can be easily adapted to accommodate batches so large that the filling extends for more than a single shift, or even a single 24-hour day.

What is almost universal in simulation design is that the fill is truly representative of the production process. With larger production fills this forces the media fill to reflect a duration that is a realistic representation of the production process. A lengthy process can hardly be supported by a simulation that is over in less than an hour, nor is there any merit to a 4-hour minimum simulation duration for what might be ordinarily be a 2-hour fill session.

In-Process Media Fills

The conduct of a simulation supporting the production of very large batches can be accomplished in part by the performance of a media fill immediately after the completion of the production fill. The filling line is cleared of the last containers of the production batch, the liquid line is flushed to remove any traces of the product, a vessel of sterile media is connected to the line, and filling is restarted with media into the same components used for the production fill. Alternatively, the product contact parts used for the product can be replaced with a freshly sterilized set of parts. The other aspects of the simulation are essentially unchanged from the other practices described in this chapter. The results of the media fill must be considered in the lot release decision for the production lot.

In-process media fills are particularly useful in the support of very large batches as their successful execution at the end of a long production batch can support that even under the adverse environmental conditions expected after the production that successful aseptic filling can occur. The use of in-process media fills as the sole means of supporting aseptic processing is uncertain, as the impact of the ordinarily highly manipulative system assembly is de-emphasized by the time period between the initial set up and the in-process fill execution. The potential for flush out or inhibition of any setup related microorganisms by the product being filled must be considered as well.

Line Speed

Supporting the full capabilities of a filling line utilized for different containers is easily accomplished. Filling lines will often operate at a variety of speeds with smaller fill volumes associated with higher filling rates, a consequence of the smaller volume being dispensed into each container. One set of media fills for a line should utilize the smallest container operating at the highest speed as this may present the greatest handling difficulty the line may encounter. As handling difficulty is associated with an expected greater need for human intervention in either routine or corrective activity then this is an obvious worst case selection for the process simulation program. The largest container filled at the slowest speed presents the greatest opportunity for airborne contamination to enter from above and is often selected as the other worst case extreme for filling systems.

Container/Component Selection

The largest and smallest containers are often chosen as they represent the extremes of either exposure duration or handling difficulty, but other selections may be appropriate. Consider small vials of similar diameter such as 1 mL, 2 mL and 3 mL units. The 3 mL due to its higher center of gravity may present more of a handling difficulty than its smaller companions, and thus might be a more suitable choice for use in the process simulation. Similarly, the elastomeric closure chosen for use should be the one that presents similar handling concerns. Recognition that excessive handling represents the greatest contamination potential may result in a simulation regimen that includes more than the obvious choices of largest and smallest containers.

There have been regulatory recommendations to replace opaque containers with clear one to aid in the detection of contaminants post-incubation (2). That is an accepted practice provided the removal of the coloring agent or wrapping does not alter the material handling characteristics of the container, in which case the opaque container is preferable despite the added inspectional difficulties (see later section in this chapter).

Media Fill Volume

In the execution of a media fill, the amount of media filled in the container can be modified from that ordinarily filled. The media amount is ordinarily reduced to extend the duration of the fill with a limited media quantity (media quantity is sometimes limited by sterilization constraints on the media that can restrict the maximum amount that can be available for use). There is no minimum media volume that need be utilized provided: there is adequate media to contact the entire sealing surface; there is sufficient media to allow for detection of growth; and there is sufficient media to pass growth promotion. In a few instances the volume filled in the container has been adjusted upwards by firms from that typically used in the container to address one or more of the concerns cited.

Media Selection

Selecting the test media to be utilized is at the core of the simulation process, and in the vast majority of cases is accomplished by the use of SCDM (also called TSB). This general purpose media is the usual choice because of its ability to support the growth of a variety of aerobic environmental and human-derived organisms. In only very limited instances is another media appropriate.

> A firm with persistent low-level microbial contamination in its inventoried products never detected microbial contamination in media fills that utilized SCDM. When media fills were performed with media that resembled its product substrate were conducted the contamination source was identified (9).

Anaerobes/Inert Gassing

Expectations that media fills address anaerobic contamination are only appropriate in limited situations. True anaerobic conditions are not attainable in manned cleanrooms even where inert gassing is utilized. Oxygen levels as low as 0.5% are toxic to true anaerobes and thus anaerobic media fills using FTM are largely unnecessary (10). Anaerobic media fills have been used in nitrogen-filled closed isolators where oxygen levels are much reduced for true anaerobic growth, but that is a relatively rare circumstance.

In ordinary media fills, to facilitate microbial recovery air is often substituted for the inert gas on the filling line. This practice hopes to eliminate the potential microbial inhibition from the inert gas the might impact aerobic organisms that might find their way into the gas distribution system during post-sterilization assembly (4).

Manual Filling

In the preparation of small-scale lots there is often a heavy reliance on personnel to perform many of the functions provided by a filling line, i.e., container movement, closure placement, seal administration, etc. The operator essentially replaces some or all of the filling equipment required for the process. In this instance, each operator assigned to this process should perform triplicate initial and semiannual repeat media fills to demonstrate their aseptic processing proficiency (6).

Aseptic Assembly

The execution of an aseptic process will often necessitate some preparation steps to configure the equipment and materials. The most apparent task of this type is the set up/assembly of the fill line from individually sterilized components into a complete line ready for the fill. Adjustment of conveyor, limit switches, vibratory feeders, and perhaps the fluid material pathway may all be a part of this activity. The aseptic process begins with these steps and they must be performed and evaluated with the same care devoted to the process itself. These activities are an inherent part of the process simulation, as the equipment must undergo the same preparatory steps, nevertheless observation of these activities, and environmental monitoring must be incorporated.

Environmental Monitoring

The aseptic processing environment utilized for the aseptic process should be monitored in accordance with the routine program used for the operation of the facility. The temptation to increase the monitoring during the media fill should be resisted as it may have an adverse affect on the results of the simulation. Environmental monitoring, especially microbial sampling must be recognized as an intervention in the aseptic core, and increasing above normal levels may result in the introduction of microbial contamination that might not otherwise be introduced. Expanded sampling of the surfaces post-filling may be useful in identifying/confirming that the appropriate locations are being utilized during monitoring of the production. The conduct of environmental sampling must be recognized as an intervention, and there must be a balance between the desire to gather information about the conditions proximate to the sterile materials, and the potential introduction of microbial contamination as a consequence of the human presence required to obtain that information. There is "no free lunch"; the gathering of environmental data must not risk product sterility.

Product Contact

As media filled units are removed from the fill line, they should be manipulated to ensure there is contact between the media and the container-closure seal surfaces. Physical contact between the media and the seal surfaces ensure that those surfaces of the container that are more vulnerable to contamination during the process are properly assessed. For syringes and ampules, incubation of the media-filled containers can be performed in a random orientation to maximize the contact between media and the sealing surfaces. Vials are generally inverted briefly prior to incubation (and midway through the incubation period if there is a 7 day inspection of the units). Some firms have chosen to invert vials during incubation but that is not a universal practice (4).

Incubation Time and Temperature

At one time this was among the more controversial and variable practices in effect at many firms (4,11). The selected approaches included incubation at multiple temperatures with transfer from one to another after 7 days of incubation. There was even confusion as to whether it was preferable to begin at a higher temperature (30–35°C) followed by a lower temperature (20–25°C) or begin the incubation at the lower temperature and then move to the higher one. Recognition that growth promotion is required regardless of the actual conditions selected, has led to a more broadly defined practice where the incubation temperature can range from 20°C to 35°C including a single temperature for the entire 14-day period (2). Provided the selected temperature is uniform (the usual range is ± 2.5°C) the use of a single incubation temperature eases the execution of the media fill and coupled with satisfactory growth promotion results is appropriate. This practice allows for flexibility of approach thereby accommodating the greatest potential for microbial recovery.

New Facilities and Lines

The start of operations in an aseptic facility must be supported by initial simulation studies that establish the capability of the facility, equipment, procedures and personnel to manufacture sterile products (2,6). Depending upon the specific circumstances of the products being manufactured, the number of required media fills may be as few as three; as line complexity/capability increases this may entail additional studies. These studies can ordinarily be matrixed to reduce the overall number; however even if the facility has lines comprised of identical components, each line must be evaluated independently of the others.

Suspensions and Aseptic Manufacturing

The process simulation should embrace all portions of the aseptic process from the point of sterilization through closure of the container. All of the interventions (sampling, filter integrity testing, etc.) that are a portion of the formulation process must be included in the simulation. The vessel utilized for the media fill should be identical to that used for commercial operations—the use of a carboy for the media fill where the commercial product uses a stainless steel vessel is inappropriate. This can prove to be more challenging where the formulation includes aseptic steps such as required in the preparation of suspensions, ointments, and other more complex products (2). These processes may require extensive SIP and complex equipment, and thus present some unique issues in the design of the simulation. The practices originally designed for sterile bulk can be adapted for use in these instances (12). In some instances the simulation process may require the use of a sterile solid (generally a placebo material) in portions of the simulation (see following section).

In the preparation of suspensions and in many of those processes described below addressing the less common aseptically produced sterile products and containers, overlapping simulations addressing the overall process may be appropriate with some portions of the process being largely conducted using a sterile powder and the remainder with a sterile liquid media. This is an acceptable practice provided the entire process is covered by overlapping where one part of the simulation ends, and another begins.

Sterile Powders

Sterile powder processing and filling presents a unique difficulty in the conduct of the simulation as the equipment utilized for powder processing cannot easily accommodate the liquid media ordinarily utilized for the simulation. In the majority of instances, execution of the process simulation will require the addition of both sterile liquid media and a sterile powder placebo to the container. The order of the additions, and the extent to which the powder filling process is adapted to accommodate the liquid fill can make this one of the more difficult simulations to execute. The PDA's TR #22 provides a description of the processing options ranked in order of preference (6). TR #28 also from the PDA provides considerations in the selection and preparation of the sterile placebo powder (12). As liquid filling on a powder line is an infrequent event, some firms chose to fill a number of liquid-only containers in conjunction with the powder fill to establish that this activity is not the cause of any detected contamination. As the sterilizing filtration for sterile powders may be conducted in a separate facility (or by a separate firm), simulation concerns at the filling site are generally restricted to the activities performed there, including milling, and blending as appropriate. The sterile bulk powder supplier is responsible for simulation activities of their aseptic process and this subject is addressed elsewhere in this text (see Chapter 23).

Other Aseptic Filled Dosage Forms and Formulations

Process simulation studies are required wherever aseptic processing is utilized for the manufacture of sterile drugs. The base case for all simulations is the solution fill and adaptations to that situation are added to accommodate the equipment and processes used for other products. Some of the modifications are quite simple (the incorporation of a freeze dryer, or filling into a plastic tube), while others may introduce substantial complications (a multichambered syringe with a lyophilized powder with liquid diluent, or a liposome formulation requiring extensive pre-filling processing). The added complexity of these more intricate processes may entail modifications to permit simulation and thus increase the potential for failure. Nevertheless, their association with simple solution filling precludes the use of looser acceptance criteria reflecting the difficulties associated with the simulation. There are instances where process simulation of the types described here for pharmaceuticals have been adapted for the aseptic preparation of medical devices, albeit sometimes with even greater modification to accommodate their rather different processing requirements.

Campaign Production

The campaign filling of a series of batches without intervening cleaning/sanitization (and in rare instances sterilization of all product contact equipment such as stoppers bowls) is a common practice for some large volume sterile products. A media fill program can be developed to support campaign production in this fashion by applying the methods described earlier for large batches and/or in-process media fills.

Interventions

The production of sterile products in either manned cleanrooms or isolators relies on the execution of any number of manual tasks by the operator. These interventions are either routine or nonroutine in nature. A routine intervention is one that is either an inherent part of the process (i.e., setup of the equipment, initial supply/replenishment of components, etc.) or required procedurally (i.e., product sampling, environmental monitoring, fill weight adjustment, etc.). Nonroutine interventions are largely corrective in nature in response to container-closure jams or misfeeds, or other mechanical problem with the equipment. The inclusion of routine interventions in process simulations is relatively simple; they need only be included at the same frequency as they would occur during a production lot. Nonroutine interventions must be integrated into the media fill in the event they do not occur as a natural consequence of the process, and the frequency should match the incidence in the production process. As was noted earlier, the extent of the interventions required whether routine or nonroutine is an important consideration in the selection of the appropriate components/process to be simulated. Practices for all interventions should be carefully defined to ensure consistency between routine production and simulated operations (13).

The most important aspect of interventions is their proper design and execution. First and foremost is the awareness that the best intervention is the one that is not performed at all. The aseptic process should be designed to eliminate interventions (routine or nonroutine) of all kinds, or at the very least minimize the need for their execution. The premium paid for more uniform components, higher quality equipment, and preventive maintenance is well spent if it results in more reliable filling. Pre-assembly of components prior to sterilization, leave behind samples and careful attention of equipment design can eliminate interventions that can impact asepsis completely.

Execution of the Fill

The process simulation should be performed following a defined procedure outlining the various requirements beginning with the sterilization of the media. The use of a batch record at least as detailed as that used for production filling is recommended; however it may be necessary to supplement this in order to adequately document the interventions included during the process. The time of execution for the interventions should be recorded and if possible correlated to a specific portion of the filled units for use in problem resolution. An observer positioned outside the critical area (and preferably outside the aseptic area) can provide a level of documentation well beyond that of the aseptic operator(s) without risking contamination. Firms have found the use of video tape beneficial in media fill execution as it can capture substantially more detail than an observer (the simultaneous use of video tape and an observer/supervisor has also been used); however, in some jurisdictions labor laws may preclude the recording of operators at their jobs.

Initial Inspection of Filled Containers

It is customary to inspect the media-filled units immediately after sealing and prior to incubation to remove nonintegral containers from the test units to be incubated. Nonintegral containers should not be incubated and their removal prior to incubation avoids the unanswerable and inevitable later question when a nonintegral container is found contaminated post-incubation (14). The temptation to discount nonintegral contaminated units must be resisted, as there is no means to establish whether the container was originally nonintegral prior to incubation. Once a container has passed this initial inspection, any contamination detected must be counted against the simulation as these units are intended to represent materials that would be released for distribution. Integral containers that would otherwise be rejected for cosmetic defects (i.e., particle in solution, fibers, marks on container, etc.) are not culled in this preliminary inspection, as their removal in a post-fill inspection is not certain and thus they represent potentially marketed units. The number of units placed into the incubator should be accurately determined.

Post-Incubation Examination of Media-Filled Units

After conclusion of the incubation (currently a 14-day incubation is almost universal) the containers are carefully inspected to detect microbial contamination. This inspection can ordinarily be performed by trained personnel with a qualified microbiologist present to support the selection process. Microbiologists may be preferable for the entire inspection where the media must be removed from opaque containers, as might be the case with plastic tubes or other difficult to inspect items. Units that suspected to be contaminated are counted and set aside for further evaluation. The total number of units inspected should be recorded.

A preliminary inspection is sometimes employed part way through the incubation period to allow for an early assessment, and is used by some firms as support for the commencement of aseptic filling at risk pending the final results.

Growth Promotion

Upon conclusion of the inspection, sterile units are selected randomly from the filled units and individual units are inoculated with less than 100 CFU/container of selected microorganisms. The usual choices for these microorganisms are those identified in USP/EP for the verification of media efficacy, plus some additional microorganism(s) of the firm's choice. Where microorganisms are added to the panel they are usually selected from common environmental isolators not already represented in the compendial panel, or isolated in sterility test failures. The inoculated units are incubated at the same conditions as the test units and must demonstrate satisfactory growth in a limited timeframe, typically 2 to 3 days (6).

Firms would prefer to select units at random immediately after filling and use those in a concurrent growth promotion test in an effort to shorten the timeframe to obtain definitive results; however, regulators frown on this practice as potentially obscuring contamination that might be in the units randomly selected for the growth promotion. Given the low incidence of contaminated units observed in contemporary media fills, such caution hardly seems justified, nevertheless, growth promotion is generally performed post-incubation (4,15).

Microbial Identification

Where positive units are detected in the post-incubation inspection they should be identified to the extent necessary to determine their origin. While the majority of any microbial contamination can be expected to be human derived, it may be possible using genotyping to determine which individual (and perhaps even which activity) was responsible. Any correlation of the speciation information to the environmental monitoring during the simulation can also prove useful. Regardless of what information is gathered about the microorganism, the objective in the identification is to determine the source(s) and take appropriate corrective action to eliminate it.

Accountability

The media fill endeavors to establish that the contamination rate for filled units is less than the firm's acceptance value. Counting the number of positives found post-incubation is generally easy given the generally successful results observed. In a recent survey nearly 90% of all media fills were reported as being devoid of contamination (15). Some aggressive FDA inspectors have raised concerns that unless the accountability of filled units is 100%, missing units must be considered as positives. That perspective seems overly conservative, and accountability for the media fill that is comparable to that of a similar sized production fill should be considered acceptable.

Acceptance Criteria

Selection of an acceptance criterion for process simulation is the province of the regulatory agency. For many years the standard of acceptance for media fill contamination rate was a criterion of not more than 0.1% (8,16). When the first written guidance was published no statistical treatment was provided (17). Over the years, aspects of statistical confidence following a Poisson distribution was added (18,19). Use of a Poisson distribution was considered appropriate as it was believed that microbial contamination in media fills was a random occurrence associated with a variety of possible causes. This expectation reached its zenith in publications that appeared at the end of last century in which the statistical treatment included alert and action levels for the evaluation of aseptic processes (20,21). This approach seemed inappropriate given the growing realization that environmental contamination recovered from aseptic cleanrooms (and media fills as well) is predominantly derived from the human operator, and thus is likely to be associated with operator activities rather than any random source. This perspective was first voiced in PDA's TR #22, where the limitations of statistical treatment were addressed (6). The use of statistics allows a number of contaminated units (9 in 15,710) that is less than 0.1% of the number of units filled. Approached in this manner an aseptic process capable of slightly less than 0.1% contamination would be considered acceptable (under investigation certainly, but acceptable nevertheless). This realization led to changed expectations in newer regulatory guidance in which an expectation of zero contamination as the goal of every aseptic process as first defined by PDA are included (22,23). The latest regulatory word on acceptance criteria for aseptic processing is that provided by the FDA in its Guideline on Sterile Drug Products Produced by Aseptic Processing that extends the most recent thinking and takes it a bit further (2). This guidance has a goal of zero contamination, but accepts no more than one contaminated unit in either 5000 or 10,000 units (the document can be interpreted to require either). Perhaps most troubling of all is the absence of an acceptance criterion that can be applied for larger media fills. The guidance suggests a maximum of one contaminated unit regardless of the number of filled units, an approach that may have the unintended consequence of smaller media fills as filling more units increases the risk of failure as the second positive in any fill is considered a failure. The authors believe an acceptance criterion with a fixed low percentage of contaminated units (0.02% or NMT 1 in 5000 units, 2 in 10,000, etc.) might be more useful as it does not penalize firms that produce larger lots and consequently require larger media fills.

IMPLICATIONS OF RESULTS FOR ASEPTIC FILLING

When the results of the media fill are available, there is certainly no issue when all of the filled containers are free of microbial contamination. Given that microbial contamination is almost always human derived, contaminated units may provide insight into potential sources if they are associated with an intervention. In the absence of linkage to an intervention, any investigative effort is likely to be inconclusive. Nevertheless, an investigation is mandated when any contaminated units are detected, as it is a regulatory expectation (2,8).

In conjunction with the detection of contamination in the process simulation the firm must make a determination on whether action should be taken relative to lots produced in proximity to the media fill. Where the contamination rate is below the acceptance criterion (a unusual condition as limits are becoming ever tighter), the contamination is generally considered as little more than a caution to the firm. In the event of an actual simulation test failure, the first action commonly taken is to place on hold the release of lots produced before and after the media fill. How far to extend this review is a matter of some discussion, and can be based upon either a defined time period or number of lots before and after the simulation (4). In extreme cases, firms have considered the release implications for all lots produced back to the last successful media fill. In an effort to minimize the disruption to operations by potentially failed simulations some firms have chosen to perform periodic media fills in conjunction with planned shutdown periods, scheduling them just before start or immediately after the shutdown period. There are cogent arguments for the use of either approach.

Any lot held pending the investigation is reviewed to determine whether the contamination detected in the simulation could have also contaminated the production materials. A thorough investigation of the media fill contamination and detailed batch records are essential to the release decision that must be made for each lot under quarantine. Fortunately the reduced level of contamination evidenced in contemporary media fills has made instances of quarantine less frequent than in prior years and likely to decrease further as aseptic processing performance continues to improve.

Media Sterilization

Sterilization of the media to be utilized can be accomplished in a number of different ways: i.e., bulk sterilization by steam, radiation or dry heat or by sterilizing filtration. The validation of these methods should be considered as a failure if the media sterilization compromises the entire simulation and will often result in a study that must be repeated. Contrary to what many believe, there is no requirement for the media used in the simulation to be sterilized using the same method as the production process being simulated. What is required is that the aseptic steps in the process be mimicked, and that may not include the sterilization process. Provided the media is introduced into process system before the point of sterilization (which is ordinarily a sterilizing filtration), the simulation is a valid one. One means of accomplishing this is adding pre-filters to the filtration system to contain the large percentage of insoluble fines often found in dry media.

Another common misconception is that a successful process simulation (or media filtration) can be considered to support the filtration efficiency of the product being simulated. As the media is ordinarily quite different from the products being processed on the line, no useful information can be gained from the media filtration.

Placebo Materials

Where a placebo material is required in the process simulation as is necessary in suspension or bulk powder processes, a placebo is commonly used. The selection of the placebo is a compromise between a number of factors: i.e., ease of sterilization, handling properties similar to the production materials, ease of cleanup, lack of microbial inhibition, etc. PDA's TR #28 provides useful information on the selection of the placebo material (12). Regardless of the placebo material chosen for the simulation, the material should be packaged in an identical fashion as the materials they are substituting for. The sterilization of the placebo must be validated to ensure that it does not become a source of contamination in the simulation.

ENVIRONMENTAL MONITORING

A properly conducted process simulation incorporates environmental monitoring as performed during the routine process. While it may seem appropriate to increase the level of monitoring during the simulation, the practice can result in increased contamination potential. Environmental monitoring entails the manipulation of materials and equipment in proximity (or within) to the critical zone where sterile products (and media) are present. These activities can introduce microorganisms to the materials being processed and should not be increased above routine levels during the simulation. Post-process monitoring in the form of increased surface or air samples may prove useful, and as the process is complete do not increase risk. Suggestions that routine monitoring be increased after detection of microorganisms in a media fill should be resisted for the same reason.

Microbiological Monitoring

The conduct of microbial monitoring during the simulation is performed according to the normal regimen for the process being simulated. The hope is that the monitoring results will identify microbial contamination that matches any detected in the filled containers. This is of course the ideal result, as effective corrective actions can then be fairly simply identified and implemented. Some of the newer microbial methods can link a specific organism to a particular operator thus providing conclusive evidence of the problem source.

The monitoring should embrace all of the conventional methods employed in aseptic environments (see the later chapters in this text for more detailed coverage of this subject) and include dynamic air sampling, settle plates (or bottles), surface samples and personnel monitoring. The limits for the monitoring are the same as those used in routine production. If expanded sampling is performed at the conclusion of the fill, those same limits are ordinarily employed. As noted earlier, monitoring during the setup of the equipment (something that can occur several hours or even the prior day) should be conducted as well.

Nonviable Particle Monitoring

Sampling for nonviable particles must be performed during the media fill for no other reason than to ensure that all relevant interventions are a part of the simulation. The relevance of the nonviable monitoring data is often rather minimal, however excursions may provide insight into routes of contamination introduction.

ISOLATOR TECHNOLOGY

The use of isolators (or other advanced aseptic processing technology) alters very little in the conduct of process simulations. All of the aspects and rationale considerations cited above are applicable essentially unchanged. The FDA has indicated that isolators (and perhaps other advanced technologies) by virtue of their superior performance potential may be demonstrated capable using media fills of shorter duration than in conventional cleanrooms (2). No further details on this were provided by the FDA, and firms will have to define their practices in the study rationale.

The ability to process a lengthy campaign in an isolator is highly desirable. The validation of campaign length uses methods similar to that employed for very large batches provided earlier in this chapter. In-process media fills are another means for the establishment of campaign length in isolators. PDA TR #28 provides some useful guidance in means for the establishment of isolator integrity over the campaign duration (12).

CONCLUSION

Demonstration of aseptic processing proficiency as provided by media fills (process simulations) is an integrated exercise incorporating every aspect of the process. Success is only possible when each of the individual elements has been properly defined and controlled. Industry performance in aseptic process simulation has improved substantially over the past 20 years indicating continual improvement in the safety of sterile products produced aseptically (4,24,25). This improvement is the result of the emphasis placed on aseptic processing by everyone concerned. Even when properly designed, executed and controlled aseptic processing is an activity requiring continual vigilance. The process simulation is just one element of the necessary controls.

REFERENCES

1. FDA, Guidance to Industry for the Submission Documentation for Sterlization Process Validation in Applications for Human and Veterinary Drug Products, 1994.
2. FDA. Guideline on Sterile Drug Products Produced by Aseptic Processing, 2004.
3. PIC decision tree.
4. Agalloco J, Akers J, Madsen R. Current practices in the validation of aseptic processing—2001. PDA Technical Report No. 36. PDA J Pharm Sci Technol 2002; 56(3).
5. Agalloco J, Akers J. Letter to the editor. Re: apples, oranges and additive assurance. J Parenter Sci Technol 1992; 46(1).
6. Process simulation testing for aseptically filled products. PDA Technical Report No. 22. PDA J Pharm Sci Technol 1996; 50(Suppl. S1).
7. PQRI 2003 survey results.
8. EU Annex 1.
9. Agalloco J. Personal communication, 1994.
10. Loesche WJ. Oxygen sensitivity of various anaerobic bacteria. Applied Microbiology, 1969; 19:723.
11. Agalloco J, Gordon B. Current practices in the use of media fills in the validation of aseptic processing. J Parenter Sci Technol 1987; 41(4):128–41.
12. PDA. Process simulation testing for sterile bulk pharmaceutical chemicals. PDA Technical Report No. 28. PDA J Pharm Sci Technol 1998; 52(Suppl. 4).
13. PDA TB #2003–02.
14. PDA TB #2003–01.
15. PQRI Survey 2003.
16. FDA. Guideline on Sterile Drug Products Produced by Aseptic Processing, 1987.
17. WHO Guidance.
18. PDA Technical Monograph #2. Validation of Aseptic Filling for Solution Drug Products, 1980.
19. PDA TR #6.
20. Technical Monograph #4. The Use of Process Simulation Tests in the Evaluation of Processes for the Manufacture of Sterile Products. The Parenteral Society, 1993.
21. ISO 13408-1. Sterilization of health care products—aseptic processing. Part 1: General requirements. International Standards Organization, 1998.
22. CEN/TC 204WG 8 N 38. Recommendations on Validation of Aseptic Processes, 1998.
23. PIC/S PE002-1. Recommendations on Validation of Aseptic Processes, 1999.
24. PQRI Survey.
25. Agalloco J, Akers J, Madsen R. Aseptic processing–a review of current industry practice. Pharm Technol 2004; 28(10):126–50.

23

Validation of Aseptic Processing for Bulk Pharmaceutical Chemicals

James Agalloco
Agalloco & Associates, Belle Mead, New Jersey, U.S.A.

INTRODUCTION

The majority of sterile dosage forms are solution products; however, there are a number of sterile products in which the API is not in solution, including both suspensions and sterile powders. In many of these products, the API is aseptically filtered while in solution in an appropriate solvent system, and then crystallized under aseptic conditions. The sterile solid is then separated from the liquid phase, dried, size modified as needed, blended (if necessary) and then bulk packaged for shipment to a dosage form site for final formulation and/or filling. A similar process may be appropriate for sterile additives that are used in these formulations such as arginine hydrochloride and sodium carbonate. In this chapter, sterile BPCs can be either APIs or sterile excipients. Support for the aseptic processing activities utilized in the production of all sterile BPCs is a regulatory requirement (1).

BACKGROUND

Establishing the sterility of aseptically processed sterile drugs is one of the more difficult, if not the most difficult, tasks in the industry. As there are no direct means by which the sterility of aseptically produced materials can be determined, firms utilize a number of measures to support their practices. Solution drug products are supported by practices such as (*i*) sterilization validation for components, (*ii*) filter validation for the production solution, (*iii*) environmental monitoring of air, surfaces and personnel, (*iv*) sterility testing, and (*v*) process simulations. The situation is really not different for sterile BPC manufacturing. On one level, the only difference appears to be the size of the container being filled. In practice, because the sterile materials undergo a phase change after sterile filtration from liquid to solid during the process it is by no means simple. In solution manufacturing, a liquid medium is utilized in the process simulation as a direct replacement for the product formulations. Direct substitution of the medium into the BPC simulation is typically complicated by the presence of both liquid and solid handling equipment in the process train for the sterile BPC. The use of a single material (liquid or solid) from the point of sterilization (typically 0.2 μm filtration) to the end of the process is uncommon. Solids cannot be easily handled in the beginning (liquid portion) of the process stages while liquids are difficult to process in the later (powder handling) stages of the process.

The regulatory expectations for sterile BPC production have been the subject of a rare direct interchange of views between the FDA and industry (1,2). These documents expressed some of the basic differences in perspective between industry and regulators. Among the subjects in their document, the FDA defined its desire that process simulations should be performed in support of every sterile bulk process. In marked contrast, industry felt that there were instances where simulations were either unnecessary or inappropriate. This exchange was followed by a brief hiatus during which PDA and PhRMA developed an industry guidance document defining process simulation practices and methods (3). This document enjoyed only limited success, while it was warmly received by industry, the FDA found several elements objectionable. Concerns were raised regarding the definition of "closed systems," requirements for simulation of closed systems, sampling of materials, and other matters. When these comments were reviewed by the task force that developed the industry guidance, the TR was revised and a meeting with the FDA requested. Unfortunately, the meeting was never held, and the revision process to TR #28 was suspended for a period of almost five years. In 2004, due to continued issues of contention between industry and the FDA, a further revision of the TR was developed (4). This chapter draws upon relevant aspects of that document and interprets them for implementation, while incorporating other relevant facility and operating aspects that are not addressed in the guidance document.

BUILDINGS AND FACILITIES

The facility for sterile BPC production must accommodate the different scales of the process equipment utilized for these processes. The crystallizers, dryers, filter dryers, mills and other equipment that are commonly used can be substantially larger than the typical aseptic manufacturing and packaging equipment used in the preparation and filling of sterile dosage form containers. The ceiling height is higher and the processing rooms are larger to

---

Abbreviations used in this chapter: API, active pharmaceutical ingredient; BPC, bulk pharmaceutical chemical; CFU, colony-forming units; EU, European Union; FDA, Food and Drug Administration; ISO, International Organization for Standardization; LDPE, low density polyethylene; PDA, Parenteral Drug Association; SIP, sterilization in place; TR, Technical Report; WFI, water for injection.

accommodate the larger equipment required. In some instances, the use of solvents in the process will mandate explosion-proof equipment. Those portions of the facility housing process train elements outside the aseptic core may have features more comparable to ordinary BPC processing areas, albeit with somewhat greater attention to microbial and particle control.

Critical Areas

The aseptic processing core of the sterile BPC utilizes materials and finishes like those found in aseptic filling facilities (see other chapters in this book for information on the details of the design). The majority of the aseptic environment is ordinarily designed to maintain ISO Class 7 conditions during operations that will realize EU Annex1 expectations for Grade B under static conditions (5,6). Within this aseptic background environment, localized ISO Class 5 environments (EU Annex 1, Grade A) are provided where aseptic operations such as seed introduction, material sampling, and subdivision into bulk containers are performed. In older facilities, aseptically garbed personnel enter these environments to perform the necessary aseptic interventions to complete the process. Newer designs employ a variety of barrier and/or isolator designs to provide greater separation between the personnel and the sterile materials and components. Facilities using isolation technology throughout could conceivably have these systems located in an ISO Class 8 environment; however, there are no current systems that are operated in that manner.

Closed and Open Systems

The production of sterile BPCs relies on "closed" systems to protect the materials throughout the process within the aseptic core. TR #28 originally defined closed systems in its initial release in 1998, and this was altered slightly in the 2005 revision to clarify the requirements. A closed system can be described as follows (4):

- Is constructed, installed and qualified in a manner which demonstrates that integrity is maintained throughout the full range of operating conditions, and over a time period inclusive of the longest expected usage (i.e., manufacturing campaign). The qualification is done according to a formal protocol, following generally accepted engineering principles, and is documented.
- Is SIP or sterilized while closed prior to use using a validated procedure.
- Can be utilized for its intended purpose without compromising the integrity of the system.
- Can be adapted for fluid transfers in and/or out while maintaining asepsis.
- Is connectable to other closed systems while maintaining integrity of all closed systems (e.g., rapid transfer port, steamed connection, etc.).
- Is safeguarded from any loss of integrity by scheduled preventive maintenance.
- Utilizes sterilizing filters that are integrity tested and traceable to each product lot for sterilization of process streams.

In the most advanced designs, gowned personnel are never located in the same environment as sterile materials and packaging materials because the process utilizes closed systems throughout. The use of closed systems for the liquid handling portion of the train is fairly easy to accomplish. The tanks, sterilizing filters and other liquid equipment are easy to "close," as they ordinarily operate in that manner anyway. As the process transitions to powder handling, and final subdivision, isolators can be used to enclose the powder handling equipment as pressure rated, SIP capable equipment is not generally available for the powder handling steps.

Regulatory comfort with closed systems is not assured. In their review of the 1998 version of TR #28, the FDA explicitly noted, "...that all of the aseptic processing difficulties with sterile bulk pharmaceuticals has been with systems that were 'closed'." Whether that experience preceded the TR #28 definition issuance is uncertain, but some degree of caution is nevertheless warranted. Some of the regulatory caution is associated with an earlier perspective (stated in the 1994 PhRMA position paper) that if a system was fully "closed" then a supportive process simulation is not required (2). This view was restated less firmly in the 1998 industry guidance and removed in the 2005 version.

Barrier and other designs, in which all of the attributes associated with closed systems are not present, are considered "open" and are acknowledged as less capable. Requirements for process simulation in "open" systems have never been a point of contention due to the greater potential for contamination associated with the less certain separation between personnel and sterile materials. The methods utilized for open systems resemble dose filling simulations, albeit on a much larger scale.

Supportive Clean Areas

Dissolution Area

The aseptic operations are ordinarily supported by two distinct environments. The production materials to be made sterile are dissolved in large vessels in which solvents and other items (amorphous carbon for impurity removal/decolorization) may be added. Use of solvents mandates an explosion-proof environment; this area is often distinct from that utilized for the other preparation processes. The processing environment where the dissolution is carried out can be either ISO 7 or 8 depending on the firm's risk assessment.

Preparations Area

The rest of the items utilized in the aseptic core are processed in an ISO 8 environment (with a localized ISO 5 to control particle counts over selected activities). In this area, items are readied for sterilization/depyrogenation through double-door units that can be later unloaded from inside the aseptic core. Air locks and/or pass-throughs are frequently utilized to facilitate the transfer of sealed sterile bulk containers from the core. Final packaging and labeling may be performed in this part of the facility. Wash areas for utensils, wrapping stations, spare parts, and other support systems are located within the preparations area.

Co-location of Bulk Manufacturing and Dose Filling

A few sterile BPC production facilities have been constructed as a single aseptic suite with both bulk manufacturing and final dose filling capabilities. This type of co-location within a single suite can reduce the capital expenditure as many of the infrastructure and support systems can be shared. The single suite permits a degree of integration and cooperation between the two portions of the process that facilitates operations.

PERSONNEL TRAINING AND QUALIFICATION

The personnel who work in sterile BPC facilities must have comparable skills to other aseptic operators. They must be proficient in aseptic gowning, aseptic technique and all of the required process-related skills associated with chemical synthesis operations. The methods and equipment utilized in sterile BPC production are substantially different from those in small container filling, and operating job training must be adapted accordingly. Perhaps the best measure of their performance is their participation in a successful process simulation incorporating all of the relevant interventions required.

CONTAINER–CLOSURE SYSTEMS

Sterile bulk materials are perhaps unique in that nearly every finished container will exit the aseptic core, be warehoused, shipped (perhaps globally), and later be introduced into the critical zone at another aseptic facility. This can place extraordinary demands upon the packaging system, and the best package designs provide maximum protection to the product throughout its shelf life. Further important considerations are the practices and methods utilized for initial filling and sealing of the containers at the manufacturing site, as well as the dispensing of the sterile material at each of the filling sites. The container–closure system is inexorably tied to the equipment and procedures utilized to sterilize, transfer and fill the containers/closures and thus has substantial impact on the facility design as well as the integrity of the aseptic process. This may include means for taking representative samples for sterility and other testing in a noninvasive fashion during the sterile bulk filling/sealing process. The preferred system from a shipping/storage perspective may not be easily filled or discharged. The choice of a packaging system must consider all aspects of its use from preparation through filling, sampling and ultimate discharge. Brief descriptions of the primary packaging systems utilized for sterile bulk materials along with the key advantages/disadvantages of each are presented in the following sections.

Glass

The first container employed for many sterile bulk materials may be a glass vial with conventional stopper, as the scale of production may preclude the use of an alternative package. Scaled up these become wide-mouth screw-capped jars, with tape providing an additional layer of security. These have the advantages of being easily sterilized and/or depyrogenated, impervious to moisture, relatively inert and inspected while still closed. Disadvantages include uncertainties of seal integrity, difficult protection post-sterilization, and susceptibility to physical damage. Heat- or twist-sealed sterilized LDPE over-wrapping can be added to simplify introduction into the aseptic filling suite. Custom-designed shipping containers are often added to enhance the container integrity during shipping. This packaging system is more likely to be used in early development and replaced as the scale of operation increases and as a consequence is less common than the others in commercial operations.

Aluminum/Stainless Steel

The model for these containers is the milk can once so common in the dairy industry. These rely on an elastomeric closure to seal the lid on the body of the container. The lid can be secured with wire or a lever system to ensure its integrity. The canisters can be either single- (aluminum) or multiple (stainless steel)-use design. Advantages are ease of sterilization/ depyrogenation, greater moisture and light protection, and physical strength. Disadvantages include the inability to inspect the contents, awkward protection post-sterilization and large opening size. Stainless steel cans that were once almost universally used have been largely supplanted by single-use aluminum cans. Additional layers of sterilized LDPE are added to facilitate entry into the aseptic core at the filling site. Fiber drums or other protective containers are sometimes provided as an outermost layer during storage and shipping.

Plastic Bags

The use of ultraclean sterilized LDPE (or other plastic) is another common alternative. The bag in contact with the product can be heat or twist sealed (with a surrounding heat-sealed bag). An internal vacuum or pressure can be utilized to provide a direct indication of container integrity to the end user. Additional bags with laminate layers to eliminate moisture or light penetration are commonplace. The complete multilayered package is shipped in corrugated boxes or fiber drums. Advantages include low cost, low weight, and ease of inspection. Among the disadvantages are the absence of structure, potential static and risk of product charring in the heat seal. Package integrity during air shipment can be assured through qualification of the heat sealer.

Sterility/Product Samples

The preparation of a sterile BPC commonly includes the collection of sample materials for sterility and other testing. These samples must be obtained from the batch in a representative fashion that assures the validity of the sample and yet maintains the sterility of the materials being processed. Some adaptation of the product package is often used to collect samples in a manner that provides this confidence. It should be recognized that opening/accessing the sterile bulk containers to collect samples represents an unnecessary risk and is to be avoided. The use of a single composite container inserted into the subdivision process for sample collection is only slightly less risky. The preferred approach is to collect samples at intervals during the process and composite/subdivide them for the various tests in a well-controlled environment.

TIME LIMITATIONS

Aseptic processes often utilize time limits on process steps prior to sterilization to preclude excessive bioburden in the filtration process, as well as prevent endotoxin from becoming a problem. Sterile BPCs are no exception, and appropriate time limits should be individually placed on the dissolution and filtration steps to provide the required controls. Time limitations are also imposed on the length of production campaigns between sterilization of the process train.

PROCESS SIMULATIONS

Study Design and Rationale

A supportive rationale for the process simulation study to be conducted should be developed that outlines the assumptions/choices made in the design of the simulation. The primary considerations in the development of the rationale are presented in the following sections.

Materials/Media Sterilization/Introduction

The materials used in the process simulation should be sterilized using a validated method. There should be no reason to doubt the effectiveness of the sterilization process. The process utilized for sterilization need not be the same as that utilized for the commercial process. The container utilized for the simulation materials should be the same as that employed for the commercial materials so that the handling methods can be identical.

Frequency and Number of Runs

Three initial process simulation studies are conducted for new facilities or after major changes to the process or equipment. As a minimum annual process, simulation studies are used to support the ongoing acceptability of the practices/system for continued production usage.

Duration/Size of Runs

The simulations should include the same number of manipulations as the commercial process being simulated. The amount of material used in the simulation must be sufficient to contact the internal surfaces of the process equipment. The acceptance criteria for these studies correct for differences in the size of the simulation batch relative to the commercial batch size.

Test Methods

The conduct of a process simulation for sterile BPCs must encompass the full breadth of the process from sterilization through filling into the bulk container. In that scope it mimics the conduct of simulations for dosage forms. The most challenging part of the simulation is that sterile BPCs ordinarily start as solutions that are converted to solids during the aseptic process. The transition in material characteristics is accomplished within the process train, which must contain liquids at the beginning of the process and then be capable of handling powders in the later steps. The development of a simulation approach embracing the entire process using a single material throughout introduces concerns unique to the sterile BPC process.

One possible approach is to conduct separate studies for the different steps in the process, allowing the use of a liquid material for the early steps and a powder material near the end. The two simulations must overlap and the results of both are considered in the evaluation of the overall aseptic process.

Test Material Selection

The selection of the test material to be used in the simulation is a major concern. The choices for test material include growth promotion (or growth supportive) media, an inert placebo material (which can be a liquid or powder depending upon the portion of the process to be evaluated), a production material provided it does not inhibit microbial growth (arginine hydrochloride and sodium carbonate are possibilities), and lastly a phantom material (evaluation is performed using post-process environmental and surface monitoring). PDA's TR #28 identified the major considerations relative to the material selection:

> Inherent in the selection of a test material, and the decision to use a test material at all are considerations of potential adverse affects implicit in the use of a material. As a general rule, nothing should be introduced into the system, whether media or placebo, which may present a problem in subsequent processing. The material (if used) must be able to be easily removed from the equipment in order to prevent an increased potential for contamination of production materials that would later enter the system (3).

The choice of test material has profound impact on the simulation design and one of the major considerations is the testing performed on the material after completion of the simulation study. Evaluation of the simulation requires testing of the entire amount of material produced, and this too may influence the material selection process.

The test materials need not be sterilized in the same manner as the production materials provided they are introduced into the system prior to the sterilizing step. When introduced mid-process, as might be the case in a simulation requiring different materials in different portions of the system, the introduction should be made in a manner that minimizes exposure to the environment.

Interventions

One of the essential requirements in the execution of an aseptic process simulation is the inclusion of manual interventions at a comparable level of frequency similar to that encountered in routine process operation. The conduct of a process simulation for a bulk process should include the appropriate number of routine and nonroutine interventions to support their execution during production operations. As bulk processes typically use lower levels of process automation compared to aseptic filling of finished containers, the number of different interventions required may be fewer, but due to the more manual nature of many bulk processes and that they may be more invasive (risky), it may be appropriate to include an identical number of interventions in the simulation as required in the production process.

Testing

Once a process simulation has been completed, testing of the materials produced in the simulation is required. As the packaging for the simulation must be identical to that utilized for the product, there may be instances where direct observation of a medium cannot be performed on the bulk containers. Where the placebo material used for the simulation is not a growth medium, sterility testing of the entire quantity produced is required. This performance of post-simulation testing introduces environments, sterilization processes and interventional activities not required in the preparation of the production materials. It is these activities that introduce additional risk to the process and are an inherent part of the decision to use a nonzero acceptance criterion. While it is essential that this testing be performed in a well-controlled environment with appropriate controls, the testing must be recognized as an aseptic process in its own right with a potential for adventitious contamination.

The testing process may require the dissolution of the entire placebo in a large volume of a suitable diluent followed by membrane filtration of the entire diluent in an adaptation of the membrane sterility test (7). The execution of such a test clearly introduces complexities not present in either the bulk production process or its routine sterility test.

Incubation/Inspection

Aside from the complications associated with obtaining the material to be incubated (either a large number of clean containers that can be inspected or a single membrane filter in a growth medium), the incubation of these materials is easily managed. The items are placed in an incubator at 20°C to 35°C (at a single temperature with all locations within ±2.5°C) and held for 14 days (8). Alternative incubation approaches can be utilized provided growth promotion requirements are satisfied. Post-incubation inspection of the containers is performed using methods similar to those employed for final dosage form containers, with allowances for the larger size of the container. Upon completion of the inspection, samples of the media utilized in the test should have its growth promotion properties confirmed (9).

Interpretation of Test Results

The acceptance criteria for sterile bulk process simulation were first defined by the PDA (3). It was chosen to provide a level of confidence in sterile bulk production comparable to that afforded to finished dosage forms. The basis of these criteria is that the sterile bulk will ultimately be filled into final containers and was set at a maximum contamination level in the bulk simulation that projected to <1 CFU per 10,000 filled final product container. In this instance, the smallest batch size with the largest fill volume represents the worst case. The additional complexities associated with the sterility testing of large quantities, typically multiple kilo amounts, preclude the adoption of no allowable contamination as the acceptance criterion.

Campaigns

The complexities of sterile bulk systems dictate lengthy cleaning and sterilization cycles, and for this reason many sterile bulk manufacturers have adopted a campaign production model. Each campaign is preceded by sterilization of the process train, followed by multiple production batches and an eventual cleaning of the system. Preventive maintenance and other activities are performed during the interval between campaigns. Campaign lengths are selected by the firm based upon their risk tolerance, as the loss of one batch in a campaign to either sterility or other key quality attribute failure may necessitate the rejection of the entire campaign. Where the failure can be absolutely associated with a mechanical failure, action would be only against those lots produced subsequent to the failed lot.

The use of a campaign operational mode must be supported by appropriate process simulation studies. Process simulation studies for open systems (most often the final subdivision of the material) utilized in campaign manufacturing must be supported by full-duration challenge studies in which the length of the campaign and the number of interventions are identical to the production campaign. End of campaign simulations for open systems may be possible, provided trace residuals can be satisfactorily inhibited before the simulation. If this cannot be accomplished, then the process simulation should match the full campaign duration with an equal number of interventions.

For closed systems, a different approach can be utilized. The process simulation can be performed matching a single production batch duration and activity levels. Campaign length is supported by maintenance of the closed system and confirmed by reaffirmation of the system leak rate, matching that at the end of a full campaign. End of campaign simulations for either open or closed systems is not practical with many of these systems, as the production materials are inhibitory of microbial growth to an extent that this cannot be adequately compensated for in the process simulation. As the materials must be fully inactivated prior to the simulation (without compromise to the systems asepsis), this approach is impractical.

STERILIZATION

The process train for the production of the sterile BPC should be subjected to a validated sterilization process. The most widely used method is steam sterilization-in-place (SIP), which has been acknowledged by regulators. SIP is best suited for the closed portions of the system that often coincides with the liquid portions of the train, as the equipment for these steps is ordinarily both vacuum and pressure tight. The caveats stated earlier for closed system are just the beginning of the requirements for a steam sterilizable system. Detailed coverage of SIP is provided elsewhere in this volume (chap. 14). Other sterilization methods for closed systems have been utilized including formaldehyde and peracetic acid; however, these are more difficult to validate than steam and as a result have been viewed with some skepticism by regulators (1,8). Dry heat has been successfully used for sterile BPCs where the conversion from liquid to solid is accomplished by spray drying. Hydrogen peroxide has been successfully utilized for both process equipment and isolators that are increasingly popular in the later more open stages of the process train where the materials are in powder form.

Open systems represent more difficult circumstances. The equipment can be sterilized by a number of methods, usually combinations covering all of the equipment items including incidentals required for the process. In open processes, the surrounding environment may be a conventional clean room or an isolator.

In order to produce a sterile BPC, sterilization of many other items is required. Utensils, tools, sample containers, environmental sampling materials and other items must all be available for use in the process. Sterilization can be accomplished by any of the common methods including steam, dry heat, ethylene oxide or radiation. Most importantly, the container/closure system utilized for the sterile bulk must be sterilized. Provision should be made in all of these processes to ensure that the items can be introduced into the environment in a manner that minimizes the risk of microbial ingress. This may entail multiple layers of protective materials that protect the items from the sterilizer to the point of use.

The last element of the process is the sterilizing filtration used for the liquid product. The validation of filtration and all of the other sterilizations required for the process are essential. This book includes chapters describing the validation of many of these sterilization processes.

Depyrogenation of equipment is accomplished by a variety of means: environmental control, caustic washes, WFI rinses and other methods appropriate for the items and equipment being treated.

LABORATORY CONTROLS

Environmental Monitoring

The conduct of an aseptic process is ordinarily supported by an environmental monitoring that assesses the conditions during the process. In the liquid portions of the process train, this monitoring is restricted to the surrounding environment and thus the results are not indicative of material impact. In open systems, normally associated with the powder subdivision steps, the environmental conditions must be carefully controlled to protect the product quality attributes. The practices required are essentially identical to those defined for final dosage forms. The single biggest caveat is that the presence of fine powders of antibiotics can inhibit the growth of microorganisms in viable samples and interfere with nonviable particle counting as well. The monitoring systems must be adapted to ensure that the results of the monitoring are valid. Details on both viable and nonviable monitoring are provided elsewhere in this book (see Chapters 25 and 26).

Sterility Testing

The performance of sterility testing for sterile BPCs presents no unusual hurdles. Aside from the need to inhibit the antibiotic's antimicrobial activity in order to conduct a valid test, the test is performed in accordance with standard practices. Collection of samples for the testing was discussed in conjunction with the selection of the bulk container and should be performed in a manner that does not compromise the integrity of the materials, in either the bulk container or the sample itself.

Chemical Testing

Samples for chemical and other testing should be obtained in the same manner as those for sterility.

CONCLUSION

The production of sterile BPCs represents perhaps the most difficult of all aseptic processes. The complexity and size of the equipment train, the large openings of the bulk containers and the predominantly manual tasks associated with the more critical operations during the final subdivision of the material all serve to make this process a significant challenge to any firm. The methods described above provide some rudimentary guidance to the practitioner who must be knowledgeable in the unique aspects of bulk chemical processes, sterilization and aseptic processing.

REFERENCES

1. FDA. Guide to Inspection of Sterile Drug Substance Manufacturers, 1994.
2. PhRMA. "Sterile Bulk Pharmaceutical Chemicals: A PhRMA Position Paper," PhRMA QC Section Bulk Pharmaceuticals Committee and Sterile Bulk Pharmaceutical Chemicals Subcommittee, Pharmaceutical Technology, August 1995.
3. Technical Report No. 28. Process simulation testing for sterile bulk pharmaceutical chemicals. PDA J Pharm Sci Technol 1998; 53(S3).
4. Technical Report No. 28 (Revised). Process simulation testing for sterile bulk pharmaceutical chemicals. PDA J Pharm Sci Technol 2005.
5. ISO 14644-1.
6. EU Annex 1, Sterile Medicinal Products, May 2004.
7. USP <71> Sterility Testing.
8. FDA. Guideline on Sterile Drug Products Produced by Aseptic Processing, 2004.
9. PDA TR #22. Process simulation testing for aseptically filled products, PDA Technical Report #22. PDA J Pharm Sci Technol 1996; 50(Suppl. 6).

Validation of Manual Aseptic Processes

James Agalloco
Agalloco & Associates, Belle Mead, New Jersey, U.S.A.

James Akers
Akers Kennedy & Associates, Kansas City, Missouri, U.S.A.

INTRODUCTION

This chapter outlines means for process control and validation of aseptic processing for sterile products manufactured using predominantly manual procedures. Typical processes/products where this type of guidance might prove beneficial include vaccine preparation, cell culture, gene therapy, IND manufacturing, clinical manufacturing, etc., where a substantial portion of the manufacturing process is aseptically performed by operators, and thus susceptible to adventitious contamination. As the aseptic operations in these processes are unique to the specific process and are also substantially different from those employed for either final formulations involving conventional aseptic filling equipment or sterile bulk drugs, none of the existing regulatory or industry guidance is fully appropriate. This chapter addresses aseptic procedures where manual activities constitute the majority of the process; those other manual steps associated with more conventional aseptic processes such as sampling or aseptic connection are readily integrated into the simulation of those processes and will not be addressed further here.

The evaluation of manual aseptic processes is adapted from the better defined methods utilized for aseptic processing on a larger scale with automated or semiautomated equipment. It is expected that the reader is familiar enough with those practices to follow how they have been adapted for execution in a more manipulative setting.

BACKGROUND

Aseptic processes are vulnerable to contamination from a variety of sources, however, the greatest source of microbial contamination is the personnel who participate in it (1,2). An aseptic process wherein personnel perform virtually all of the important processing steps must be carefully designed and executed to minimize the potential for microbial ingress. Like any other aseptic processing activity, a largely manual aseptic process requires an appropriate environment, and effective sterilization of the materials, equipment, and components.

Abbreviations used in this chapter: BSC, Biological Safety Cabinet; cGMP, current good manufacturing practice; EU, European Union; FS, functional specification; IND, investigational new drug; ISO, International Organization for Standardization; SIP, sterilization in place.

BUILDINGS AND FACILITIES

The only meaningful facility concern for manual aseptic processing is the selection of an appropriate environment in which to perform the required activities. In general, the choice is an ISO 5 (EU Grade A, FS 209 Class 100) environment in which unidirectional air flow protection is provided to the materials during the procedure. In practice, this means many different locations are possible including: a portion of a larger environment of the same class, a localized laminar flow hood protecting a specific portion/area within a lower classified environment; a table mounted laminar flow hood (with either horizontal or vertical air flow), and an isolator (open or closed). Each of these systems is in current use for the conduct of manual aseptic processes. In some instances the activity might be limited to a hose connection, sample removal or other brief task, while in other cases it could include the entire aseptic process. Clearly, the more complex the process, the more sophisticated the operating environment that will be required. The manual processes considered in this chapter are ordinarily carried out in either a laminar flow hood, or an isolator. These environs provide appropriate conditions for the execution of predominantly manual operations and are considered the critical zone in the parlance of regulators and industry. Biosafety cabinets are not considered appropriate for manual aseptic processing as they are designed for the containment of microorganisms and there is a net flow of air from the background environment into the BSC to provide that containment during their operation.

The supporting clean area outside the critical zone is typically either ISO 7 or the hybrid EU Grade B. Lesser environments could be considered, but are not recommended. The use of an isolator relaxes the requirements for the surrounding area considerably. An unclassified environment could be used with a closed isolator, but most new installations employ an ISO 8 background. Open isolators are typically installed in ISO 8, though a few have been placed in ISO 7 rooms. The surrounding clean area is where the personnel performing the manual process are located, and appropriate gowning facilities are required which will vary substantially depending upon the background environmental requirements. The execution of the aseptic process is ordinarily supported by various sterilization processes for the materials and equipment required, and these may also be located in the support areas.

The overall design of the aseptic facility mimics that of large-scale environments utilized for equipment-based aseptic filling whether in a cleanroom or an isolator albeit on a smaller scale. In some instances, the manual aseptic process is carried out in a portion of a larger aseptic processing facility.

PERSONNEL TRAINING AND QUALIFICATION

People are the single most important aspect of manned aseptic processing, and when the aseptic process relies on them almost exclusively for success, their proficiency at their assigned tasks must be beyond reproach. A manual aseptic process places such importance on the skill of the operators; their skills must be honed to near perfection.

Training of Personnel

The training requirements for these personnel typically include the usual elements of cGMP, microbiological principles, sterility assurance, sterilization, gowning practices and the other knowledge requirements of ordinary aseptic operating personnel. Theoretical knowledge of these areas is insufficient; the operators must be able to adapt the classroom discussions to the real world environment. Where these individuals must excel is in the execution of those tasks that directly impact sterility assurance: aseptic gowning, aseptic assembly and aseptic technique. They must be able to consistently perform precision tasks without introducing contamination onto the materials they are working with.

Gowning Qualification

Assessment of the operators' proficiency in their assigned tasks can be established through practical exercises in which their skills are challenged and evaluated. The most basic of these, and ordinarily the first which the operator must succeed at is aseptic gowning. This entails repetitive gowning in full aseptic garb under the observation of a fully qualified individual followed by monitoring of gown surfaces. The number of surfaces varies with the firm, but typically includes gloves, forearms, and chest area (the locations closest to any manual activity the operator must perform) (3). The operator must successfully demonstrate their ability to meet the defined monitoring levels after each gowning exercise. Gowning certifications are conducted on a periodic basis to confirm that the operators maintain consistent gowning practices. Once the operator has passed initial gowning certification they are granted access to the aseptic core for the continued instruction in aseptic processing.

Aseptic Handling Challenges

The conventional means for establishing personnel proficiency in aseptic processing is through participation in a media fill (also known as a process simulation) (1,2). These require the operator to perform aseptic interventions during the normal course of the simulation, and for those charged with aseptic assembly, to assemble the sterilized equipment prior to the fill. For individuals who perform manual aseptic processing, these activities may have little or no relevance for a variety of reasons.

- Automated filling practices are less susceptible to human contamination
- Interventions on automated filling systems are infrequent and facilitated by proper equipment design
- Automated filling does not require continuous human intervention
- Automated filling systems are often designed for SIP, requiring minimal set up

Thus an individual who has demonstrated proficiency at aseptic processing must still demonstrate their capabilities in the more rigorous requirements of manual aseptic processing. This is often accomplished by various forms of challenge tests in which the operator must directly handle sterile equipment and materials (usually with media) to affirm their aseptic technique (some firms use this approach for operators conducting automated filling). These tests may bear little resemblance to any specific manual aseptic process, but merely serve to evaluate personnel proficiency. The usual requirement is that the operators achieve perfect results in these evaluations.

EQUIPMENT, COMPONENTS, AND CONTAINERS/CLOSURES

The equipment, raw material components, containers, closures, and other items required for manual aseptic processing vary with the requirements of the process. Perhaps the single most common with these items is their reduced size and number, which may allow them to be supplied to the processing environment in a sealed package after depyrogenation/sterilization. The preparation methods prior to depyrogenation/sterilization mimic those associated with automated aseptic processing though on a smaller scale of operation. The depyrogenation/sterilization methods for all of these items must be validated.

TIME LIMITATIONS

Time limitations with manual aseptic processing are perhaps more important than with more automated aseptic processes. After all the operator's skills may deteriorate with the passage of time. In very short or nonrepetitive processes such as aseptic connection, aseptic manufacture, time may be of little relevance (except as it may relate to material stability), and its impact lessened if not actually ignored. Where the operators must perform repetitive tasks such as container filling/stoppering, egg transfers or similar tasks, the effect of fatigue must be considered in both routine operation and process simulation. A "worst-case" evaluation of fatigue would entail a process simulation equal or greater in time duration to the longest period an operator might perform the task without interruption, exit and reentry to the aseptic environment.

DESIGN OF MANUAL ASEPTIC PROCESSES

The heavy reliance on personnel practices in manual aseptic processes makes the design of the process to minimize the impact of personnel critical to success. While it is impossible to establish detailed design criteria, general process design principles can be followed that will increase the probability of success.

Manual Aseptic Process Design Principles

- Significant aseptic assembly should be avoided through the use of sterilized preassembled items. This will serve to minimize the extent of manual assembly required.
- Tools and utensils should be employed wherever possible rather than the direct contact with the operator's hands. Provide supports for the tool inside the Grade A environment to minimize contact between the tool and horizontal surfaces of the workspace.
- Perform as much of the process inside the Grade A environment as possible to minimize removal and subsequent reentry of sterile items in/out of the Grade A environment. This may require the placement of small equipment within the hood.
- Liquid transfers should be made using peristaltic pumps rather than through the use of automatic pipettes. Containers should be premarked to indicate the amount of material to transfer.
- Materials being introduced into the process should be premeasured prior to sterilization and addition.
- Utilize a second (and third, if required) person to supply/remove items to/from the Grade A environment. The second person should wear sterile gloves and never contact the item. The first person should never contact the wrapping materials.
- Electrical equipment and controls should be located outside the hood if possible. If that is not possible a second operator and not the primary operator should adjust them.
- When items must be removed from the Grade A environment, they should be sterile wrapped with minimal exposure of critical surfaces to the surrounding less clean environment. Upon return to the Grade A environment, the overwrap is removed and discarded.
- Sanitize the operating environment empty, and sanitize each item as it is first introduced. Do not introduce a large item into the environment mid-process.
- Plan the process so that samples can be taken with minimal risk. Take all samples from a container in a single step, and then subdivide that sample as required. If appropriate, leave material for samples in containers from which the remaining materials have been transferred for further processing.
- The entire process should be documented in sufficient detail to ensure continued conformance to the desired practices. The second or third operator should complete the batch record.
- Environmental monitoring practices should be nonintrusive to avoid potential for dissemination of contamination in the Grade A environment. The use of settling plates and postprocess RODACs/swabs to monitor environmental conditions is preferable to active air sampling.
- The process should be rehearsed several times using all of the required items and placebo materials to refine the steps, location of items, etc.
- Steps not required to be aseptic should be performed outside the Grade A environment by the second (or third) operator.
- The operators should work as a team. The primary operator should perform all tasks inside the Grade A environment. The second operator assists in the introduction/removal of items from the Grade A environment, and may assist the primary operator with some tasks inside that environment. A third operator may be necessary in some cases to support activities exclusively in the surrounding environment.
- The hands of the primary operator should remain in the Grade A environment at all times.
- The second operator should don sterile gloves/sleeves prior to any activity by them inside the Grade A environment, or in transfers of items to/from the primary operator.
- The operators should decontaminate their gloves on a frequent basis.
- Extra subassemblies and utensils should be sterilized and available for immediate use in the event a replacement is needed.

VALIDATION OF MANUAL ASEPTIC PROCESSING—PROCESS SIMULATION

Note: This section addresses only those elements of manual aseptic processing simulation that are unique to the extensive participation of personnel in the process. Details of study design provided in the chapters on validation of aseptic processing for either filling or sterile bulk production should be consulted for those activities that are essentially unchanged when manual procedures are employed.

Study Design

The development of a supportive rationale for the manual aseptic process simulation is essential. The rationale must define the adaptations to the production process necessary for the execution of the simulation. The smaller scale of the manual process lends itself rather easily to these adaptations as in many instances only minimal changes to the process are required. Due to the limited number of units filled in these processes, samples of the media during the process can be beneficial in determining at what point contamination was introduced (in the event of a failure investigation) and definition of these sample points should be included in the rationale. The study rationale should be kept current with changes to process, products, components, or equipment that could impact the acceptability of the process.

- Manual aseptic processing can encompass a wide variety of activities, but can be divided into four major categories: filling/subdivision activities, compositing/assembly activities, formulation/compounding activities and manipulative steps performed in conjunction with other processes. The validation of each category is addressed in a different manner.
- Filling/subdivision processes involve repetitive actions in which sterile materials are transferred from a bulk container into smaller containers, closed, and sealed. This practice is common in IND and early clinical stage manufacturing of sterile products, and in the manufacture of extremely small lot sizes. Validation of these processes mimics the practices defined for automated filling (1,2).

- Compositing/assembly processes involve repetitive actions in which sterile materials in smaller amounts are pooled. Such practices are common in vaccine manufacturing in which the contents of incubated eggs are composited early in the formulation process. Adaptation of the validation methods for sterile filling and bulk materials may be appropriate.
- Formulation/compounding procedures in a manual setting might use laboratory glassware and utensils in which a sterile bulk formulation is produced. The smaller scale of the operation mandates changes in transfer methods. The methods utilized for sterile bulk materials are most appropriate in these processes (4).
- Manual activities such as sampling, aseptic connection, etc., are often an integral part of other aseptic processes. As they are an inherent part of those processes, there is no need to address them independently.

Additional details on each of the various validation methods will be provided in conjunction with each of the elements addressed within the overall rationale.

- *Media sterilization*. Preparation of media for use in manual aseptic processing is rarely difficult. The bulk container size is typically small enough where it can be sterilized in a autoclave prior to introduction into the process. For compositing/assembly processes, the availability of suitable sterile materials to use in the process simulation may be difficult and there may be little choice other than the production materials themselves suitably sterilized/adapted (if possible and necessary) for use in the simulation. For the simulation, each of the liquid containers containing sterile materials required for the process should have their contents replaced with media.
- *Frequency and number of runs*. The only existing recommendation for frequency/number of simulation studies with manual processing is that each operator who performs the manual steps in the process should be qualified semiannually.
- *Duration of runs*. Simulation studies should slightly exceed the expected maximum duration of a single working session by a single operator. In compounding simulations, the length of the simulation should mimic that of the commercial manual process with the exception that process hold times without activity can be shortened dramatically.
- *Size of runs*. The size of the process simulation is largely dictated by the time period that a single operator would remain performing the same activity. The actual numbers of units produced in that time period should meet or exceed the production quantity that the operator would normally handle in that time period.
- *Media fill volume*. During the aseptic filling simulations the amount of media transferred should be sufficient to wet the product contact surfaces of the container and be sufficient to detect growth. In compositing simulations, the amount transferred should be identical to that normally handled to mimic the process duration more accurately. In manufacturing simulations, the volumes of media and other fluids (which should all be replaced with media) should be identical to that in the process to be simulated.
- *Anaerobes/inert gassing*. The methods utilized for machine-based aseptic processing are adopted without change. Air should be substituted for inert gases in all systems, except in those rare instances where an isolator providing true anaerobic conditions is utilized for the production process. In those situations, the usual inerting gas would be utilized and the chosen media would most likely be fluid thioglycollate media.
- *Environmental monitoring*. The conduct of environmental monitoring for manual aseptic processing uses methods identical to those for other highly controlled (Grade A/B) environments. The same cautions exercised with monitoring in other applications apply as well. Performing the sampling must not introduce contamination into the environment or into the sample. The smaller size of the environmental systems utilized for manual aseptic processing means that the monitoring methods must be chosen for their lack of impact on the environment.
- *Execution of the simulation*. The process simulation should be performed in a manner that properly documents the activities. A batch record designed specifically for the simulation is a common approach. As with other simulations the presence of an observer who documents the simulation can be beneficial. The methods and principles defined for automated filling or sterile bulk chemical production can be utilized with relatively minor modifications.
- *Additional samples*. Testing of materials (e.g., side--parts, simulated chemistry samples, etc.) other than those representative of the sterile materials being simulated may be useful in the event of failure to determine when/where in the process the contamination may have been introduced. Should these samples be contaminated in the same simulation where the simulated production materials are found sterile, the contamination should not be considered significant.
- *Preincubation inspection*. The methods utilized for machine-based aseptic processing are adopted without change. As the containers in manual aseptic processing are more likely to be hand stoppered/sealed, this inspection must be performed with utmost care, as the potential for a deficient seal might be higher than a machine applied closure. (The same caution should be applied to inspection of hand-sealed production units for the same reasons.) Units with defective or suspected defective seals should be removed from the materials sent for incubation.
- *Incubation time/temperature*. The same considerations as machine-based aseptic processing apply.
- *Postincubation inspection*. The methods utilized for machine-based aseptic processing are adopted without change.
- *Growth promotion*. The methods and considerations relevant to machine-based aseptic processing are adopted without change.
- *Interpretation of test results*. The smaller size of production lots produced by manual procedures is typically less than the current minimum simulation size of 5000 units (5). Thus simulations conducted in support of container filling must be devoid of contamination in any of the filled units. In compositing or

formulation simulations, the simulated bulk material container(s) should be sterile.

CONCLUSION

This chapter may be the first attempt to address this subject in a comprehensive manner. As such, it includes recommendations that establish precedents in the absence of guidance from regulators or industry associations. Firms that perform these types of processes must respect the uncertainties associated with any process that is so heavily reliant on personnel to excel at all times. In new installations, we strongly encourage the use of isolation technology to minimize the microbial contamination potential from personnel however well trained they might be (6).

REFERENCES

1. Agalloco J, et al. "Process simulation testing for aseptically filled products," PDA technical report #22. PDA J Pharm Sci Technol 1996; 50(6) (Suppl.).
2. FDA. Guideline on Sterile Drug Products Produced by Aseptic Processing, 2004.
3. Akers J, et al. "Design and validation of isolator systems for the manufacturing and testing of health care products," PDA technical report #34. PDA J Pharm Sci Technol 2001; 55(5) (Suppl.).
4. Agalloco J, et al. "Process simulation testing for sterile bulk pharmaceutical chemicals," PDA technical report #28. PDA J Pharm Sci Technol 1998; 52(4) (Suppl.).
5. PDA. Guideline on Sterile Drug Products Produced by Aseptic Processing, 2004.
6. Agalloco J, Akers J. The Future of Aseptic Processing. Pharm Technol Aseptic Processing 2005:S16–23.

25 Monitoring of Nonviable Particles

Mark Hallworth
Particle Measuring Systems, Inc., Boulder, Colorado, U.S.A.

INTRODUCTION

Manufacturing pharmaceutical products is a highly controlled process, whether the end product is aseptic, terminally sterilized, lyophilized, or even an originating bulk ingredient. Therefore, the environments in which the activities of manufacture are performed must be controlled and, through monitoring, proven to be in control.

The mechanisms for controlling particles in the manufacturing environments are filters and the cleanrooms or local workstations they supply. HEPA filters are used to clean the air provided to a cleanroom. There are three basic types of cleanrooms used in pharmaceutical manufacturing:

- Conventional, using turbulent flow air circulated at an optimum rate to dilute the concentration of particles in an area to an acceptable limit.
- Unidirectional flow, using the velocity of air as a shower to wash particles from the critical areas.
- A combination of these two technologies.

This chapter will review how monitoring particles within controlled areas (cleanrooms) is performed, the presence and interaction of particles in cleanrooms, fundamental particle counter technologies and design principles, the legislative requirements for monitoring, how monitoring can be performed, and the considerations behind sampling techniques applied to monitoring.

RATIONALE FOR ENVIRONMENTAL MONITORING

There are several reasons why one would approach proving that environmental control has been established. Primarily the contributing factor for contamination is viable airborne particulates; how-ever nonviable particles are also of importance. Some considerations affecting a monitoring rationale are:

- Distinguishing the differences between viable and nonviable particulate material is difficult and any correlation between the two is a useful insight into cleanroom activities.
- Some products and medical devices cannot be terminally sterilized by radiation, thermal, or chemical means and as such environmental conditions become more critical.
- Inert (nonviable) particles are an important indicator to ensure product quality.

Fundamental Requirements

Particle monitoring is required to prove contamination control of an environment. This includes the sampling of air in the cleanroom associated with personnel and process equipment activities, with regard to risk of finished product quality.

USP General Chapters refer to testing a finished product to prove that the product is free from viable contaminants (1). It is the presence of viable organisms in the production environment that are of the greatest concern for contamination of the product.

The average time taken to prove that an environment has been maintained at a suitable level of sterility is between three and five days. This time is needed for sampling, incubation, analysis, and reporting. A real-time indicator of sterility would be the ideal system.

The USP <797> "Pharmaceutical Compounding—Sterile Preparations" refers to the intent of its fundamental requirements as being to "prevent harm and fatality to patients that could arise from microbial contamination, [or] excessive bacterial endotoxins" (2). Although compounding and manufacturing differ, the difference is primarily in the scale of manufacturing as opposed to intent.

Though viable organisms cause the greatest concern in the cleanroom, control over the nonviable particle contaminants is required for several reasons:

1. Proof of control over particulate contaminants
2. Proof of control over viable contaminants
3. Proof of control over cleanroom activities, both personnel and process oriented

Proof of Control Over Particulate Contaminants

Particles present in a cleanroom are largely due to personnel, process, or arise from the atmospheric abundance of particles brought into a cleanroom through the air handling/filtration system. The data in Figure 1 show that a predominance of airborne particle in the

---

Abbreviations used in this chapter: CFM, cubic foot per minute; cGMP, current good manufacturing practice; EC, European Commission; EMEA, European Medicines Evaluation Agency; EPA, Environmental Protection Agency; EU, European Union; GMP, good manufacturing practice; FDA, Food and Drug Administration; HEPA, high-efficiency particulate air; MPPS, most-penetrating particle size; OPCs, optical particle counters; PDA, Parenteral Drug Association; PIC/S, Pharmaceutical Inspection Cooperation Scheme; SOP, standard operating procedure; SPC, statistical process control; USP, *United States Pharmacopeia*; WHO, World Health Organization.

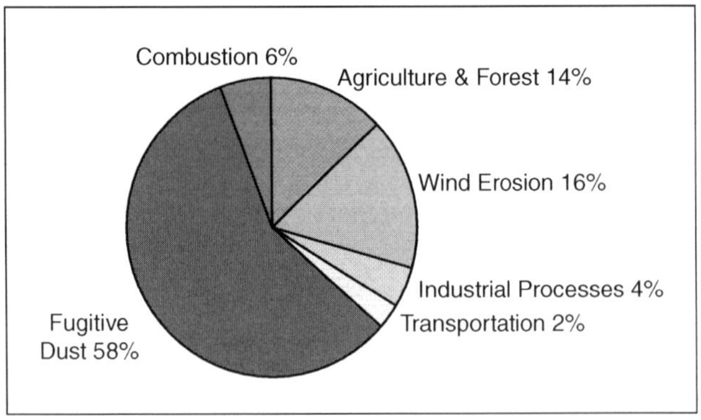

Figure 1 Major sources of particulate matter (PM10). *Source*: From Ref. 3.

atmosphere arises from fugitive dust and wind erosion sources. The particles from these activities are primarily composed of aluminum, silica, and other oxides, are relatively large in size and can be assumed to be inert from a viable activity perspective.

The EPA states that smaller particles (<2.5 μm) are formed by two primary mechanisms:

- Heterogeneous nucleation of vapor phase material
- Homogeneous nucleation of vapor phase material.

These formation mechanisms combine to generate a set of particles that are different from larger sized particulate matter.

The U.S. EPA has classified particles into four size categories shown in Figure 2: Ultrafine, Fine, Coarse, and Supercoarse.

The MPPS of most HEPA filters is between 0.1 and 0.2 μm with increasing efficiency of capture for smaller and larger particles. If it is assumed that all externally sourced particles in the cleanroom are essentially inert (i.e., they are oxides that have passed through the filter media), then their impact on product is either (*i*) chemical reaction/dissolution or (*ii*) of such a magnitude that injury to the patient is caused by

- Sufficient concentration so that the body believes it is under attack and tries to combat the problem using its own defenses, likely resulting in shock, or
- Sufficiently sized particles that lodge within the inner organs, causing a deleterious effect on either the flow of blood through the system or directly on the organ itself.

The control over total particle burden is irrespective of source (operational conditions lead to process-related particles, at rest conditions lead to atmospheric borne particles) therefore important, and international standards have been defined and applied to cleanrooms since the early 1970s with the earliest applications of the Federal Standard FS209 (4). The Federal Standard FS209E (its last iteration) was superseded by the International Standard ISO14644-1 in 2002 (cleanrooms and associated controlled environments—classification of air cleanliness).

Proof of Control Over Viable Contaminants

The above assumption that all particles are chemically inert does not hold true, as most particles within a cleanroom can be attributed to the personnel, or the activities of those personnel, within the cleanroom. Most particles brought into a cleanroom by personnel are skin flakes and therefore are associated with the risk of organic/viable contaminants (5).

It is generally accepted by the industry that a roughly proportional relationship exists between the viable and nonviable particles within a cleanroom that varies with the class of the cleanroom. Therefore, having control over the total particle concentration within a cleanroom also offers a degree of control over the viable proportion of the total burden. It does not, however, provide useful quantitative information on the microbiological content of the environment.

To support the monitoring of all particles to prove environmental control, studies were undertaken to verify that the control of nonviables had a direct influence on the viable contaminant levels. However, the data generated through such studies have been inconclusive and unable to prove that any correlation exists. Recent studies have also maintained that even through improvements in cleanroom design, the base correlation (i.e., that there is no link between viable and nonviable levels), still holds true (6).

This would be disconcerting as a formulated link is an essential element of room validation; however, William Whyte, during his studies, found that a viable contaminant existed between 10 and 15 μm and no smaller, due to the desiccating nature of the dry cleanroom environment (7). Ljundqvist and Reinmueller found a strong correlation between larger particles and

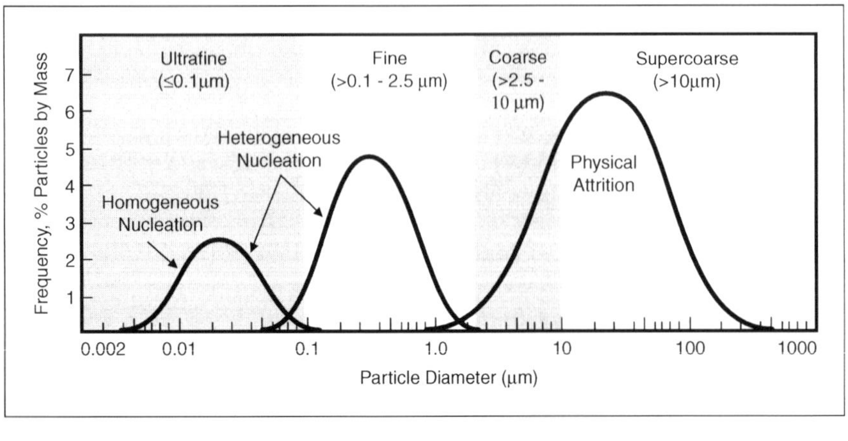

Figure 2 Tri-modal distribution of particle matter in the atmosphere.

the viable risk to process operations (8), and the USP < 1116> Microbial Evaluation of Clean Rooms states that "while airborne micro-organisms are not free-floating or single cells, they frequently associate with particles of 10–20 μm" (9). Therefore proof of control over the macro-particles (ISO 14644-1 definition of particles >5 μm) would also offer a degree of control over the risk of exposure to viable particles.

The cGMP guides support this expected relationship, as particle cleanliness is required to be demonstrated along with a requirement to prove beyond simple cleanliness. By controlling the large particles (>5 μm), it is believed that cleanroom sterility, or risk of losing sterility, can be monitored. Though controlling the general particulate burden offers control over the viable contamination risk, it does not alleviate the requirement for monitoring for the viable fraction using other technologies (dynamic monitoring, settle plates, etc.).

This application of the associated risk between viable particles and macro-particles is discussed later in this chapter when the limits imposed by various regulatory agencies are applied.

Proof of Control Over Cleanroom Activities

A properly designed cleanroom in the "at rest" state, that is no personnel present and no operational machinery, should essentially be completely free from particles (viable and nonviable). It holds, therefore, that should this balance be changed, then so too will the level of particles generated. The correlation of particle activity changes and associated changes to operational conditions is widely known and published, and therefore, mapping the changes in particle concentration to known activities within a room can improve knowledge over process control.

The better informed we are, the better we can control particulate contamination. For example, studies have shown that a poorly gowned operator sheds more particles than one that is properly gowned and that new cleanroom gowns shed fewer particles than those laundered multiple times (8). Using this information can improve the control of the particles created by personnel within a clean space by emphasizing gowning techniques and the frequency at which gowns are replaced.

Another example is using data from monitoring to better manage situations that can occur. A broken vial or a filling line blockage due to machinery or system failure generates particles, and the ensuing intervention will almost certainly generate particles that would not have occurred during "normal" operations. The impact and frequency of the interventions can be monitored using the particle data generated during the period of failure and following SPC standards. An acceptable limit on anticipated interventions over unexpected interventions could assist in the risk assessment of the process.

Requirements to Satisfy Regulatory Standards

To meet the current requirements of nonviable particle counting, a pharmaceutical manufacturer must undertake two components of particle count information to prove regulated use of pharmaceutical grade cleanrooms. The first step is the classification of the room: this is performed to international standards, primarily ISO 14644. Second, the firm needs to prove that the room can maintain its particulate limit through routine monitoring. The guidelines for monitoring are defined in the cGMP guides relevant to the region for which the product is released. Each of these topics is discussed later in this chapter.

CLEANROOM CERTIFICATION

There are two aspects of standards and specifications for nonviable particle counting that are important to pharmaceutical manufacturers: (*i*) the procedures within the standards used to characterize the particles must be accurately defined and (*ii*) the test methods must be carefully determined. When the above are present in an international standard, it allows a "calibration" of the cleanroom to be performed in absolute terms. Once this baseline has been established, then determining the physical nature of the various mechanisms affecting particle generation, transport, and deposition becomes all important in understanding why, when, how, and where to monitor particles in a pharmaceutical manufacturing environment. Knowledge of these mechanisms also assists in understanding the application limits of the monitoring instrumentation.

The original standard universally adopted for cleanroom certification was the Federal Standard FS209. The final version of this was revision FS209E and was replaced in November 2002 by a new international standard, ISO 14644-1.

The certification state of the cleanroom must be determined in advance of testing; three states exist within the context of ISO14644-1:

- *As Built*: a completed room with all services connected and functional, but without production equipment or personnel within the facility.
- *At Rest*: a room where all the services are connected, all the equipment is installed and operating to an agreed manner, but no personnel are present.
- *Operational*: all equipments are installed and are functioning to an agreed format, and a specified number of personnel are present working to an agreed procedure.

The limits for the cleanroom concentration of particles greater than a prescribed size are defined in Table 1.

These limits have been defined in accordance with the calculation from the following standard:

$$C_n = \left(\frac{0.1}{D}\right)^{2.08} \times 10^N$$

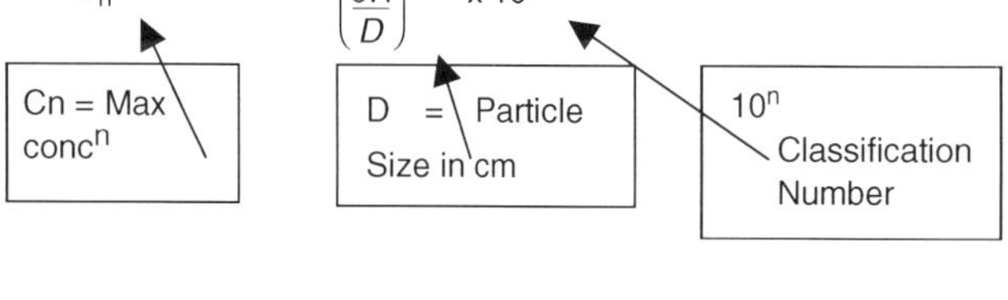

The relationship of particle size to its abundance within a population is therefore a function of $1/D^{2.08}$. If the particle size is plotted against its concentration on a log/log scale, the slope of the curve for each class is 2.08 as shown in Figure 3.

Table 1 Airborne Particulate Cleanliness Classes for Cleanroom and Clean Zones

			Maximum allowable cumulative particles/m³ to meet ISO Certification particle size (μm)					
ISO class	Approx. FS209 class	Minimum size (μm)	0.1	0.2	0.3	0.5	1.0	5.0
1		0.1	10	2				
2		0.1	100	24	10	4		
3	1	0.1	1,000	237	102	35	8	
4	10	0.1	10,000	2,370	1,020	352	83	
5	100	0.1	100,000	23,700	10,200	3,520	832	29
6	1,000	0.1	1,000,000	237,000	102,000	35,200	8,320	293
7	10,000	0.5				352,000	83,200	2,930
8	100,000	0.5				3,520,000	832,000	29,300
9		0.5				35,200,000	8,320,000	293,000

The designation for cleanroom certification should include the following elements:

- The room classification number expressed as "ISO Class N."
- The occupancy state.
- The considered particle size. It is also possible to certify a cleanroom at multiple sizes; if this is the case, then the sample volume requirement for the largest particle size is used.

Example: Room 14 is an ISO Class 5 Cleanroom at 0.5 μm (3520 n/m^3), operational.

The room now needs to be tested to prove the statement. ISO14644-1 standard identifies each of the component steps required to prove compliance. See the example at the end of this chapter.

Room classification will need to be repeated on a frequency defined by ISO14644-2. For a Class 5 or cleaner environment, this is defined as every six months; for a Class 6 or greater cleanroom, the interval is every 12 months. This interval can be extended to the maximum permitted interval providing that the pharmaceutical company shows that "no significant change" has occurred in the control of their cleanroom by evidence of continued compliance. Therefore Class 5 cleanrooms do not need to be recertified on a six-month basis, but may extend the period between reclassifications (Table 2).

CLEANROOM MONITORING

Once a room has been certified as meeting a specific room classification in accordance with ISO14644-1, the room can be used for its intended purpose. However, the interval between recertification tests is insufficient to meet the requirements for cGMPs. Both the United States and the European regulatory agencies require a room to be regularly tested for compliance based upon the risk to finished product quality. The second phase of proving compliance with regulatory agencies requires the monitoring of the cleanroom environment.

Cleanroom monitoring is defined as the observance of the condition within a production area during normal

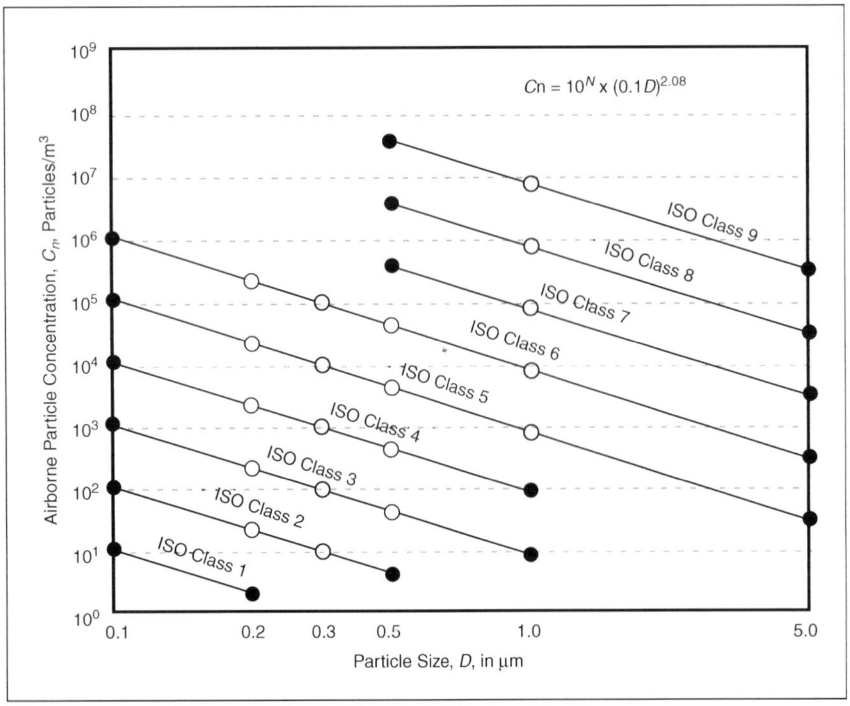

Figure 3 Graphical representation of ISO class concentrations limits for selected classes.

Table 2 Cleanroom Class Certification Requirements

Class	Maximum time interval	Test method
≤ISO 5	6 months	ISO 14644-1 Annex B
>ISO 5	12 months	ISO 14644-1 Annex B

conditions. This means that monitoring is performed in a pharmaceutical production area for at least one of the following reasons:

1. To obtain continuous particle measurements following the operation or performance of a specific machine, or of the conditions at a particular critical location.
2. To obtain particle count results over a long period to establish control limits over a process or SOP.
3. To gather a record of monitoring information that may be required to verify operations in accordance with in-house and regulatory specifications.

The two primary regulatory bodies are the FDA in the United States and the EMEA. These agencies enforce their own requirements for cGMP. In addition, there are regulatory bodies responsible for individual nations. If export of product is intended to reach a locally regulated market, then an audit from one of these inspectors is also required. It is therefore important to be able to satisfy the more stringent requirements. Of these practices, those required for aseptic manufacture impose the greatest level of requirements, especially for proof over both nonviable and viable contaminants.

FDA Requirements

The FDA cGMP for drug products specifies the practices for manufacturing area operations, controls, validation, and documentation (10). Adherence to the cGMP is verified by inspection, and the FDA has issued guidelines in the Code of Federal Regulations on required practices and recommendations for compliance to these requirements (11). The guidelines describe building and facility requirements, especially the critical and controlled areas.

21 CFR 211.42 (c) states, "Operations shall be performed within specifically defined areas of adequate size. There shall be separate or defined areas or such other control systems for the firm's operations as are necessary to prevent contamination or mix ups during the course of the following procedures: Aseptic processing, which includes as appropriate: (*i*) Floors, walls, and ceilings of smooth, hard surfaces that are easily cleanable; (*ii*) Temperature and humidity controls; (*iii*) An air supply filtered through high-efficiency particulate air filters under positive pressure, regardless of whether flow is laminar or nonlaminar; (*iv*) A system for monitoring environmental conditions…" (11). Therefore separate or defined areas in an aseptic processing facility should be controlled to achieve the required air quality that is dependant on the nature and risk of the process.

The FDA describes the area that poses the greatest risk to finished product quality as "critical." This is where sterilized product, glassware, and other associated components exposed to the general supply air are maintained in a sterile environment. All personnel activities conducted in these areas are monitored so that they do not compromise the efficacy of the environment. Table 3 shows the limits defined by the FDA 2004 Guideline on Sterile Drug Products Produced by Aseptic Processing.

This area is deemed critical because product is exposed to potential contamination and is not subsequently sterilized. It is therefore essential that the environment in which aseptic operations are conducted be maintained.

> Air in the immediate proximity of exposed sterilized containers/closures and filling/closing operations would be of appropriate particle quality when it has a per-cubic-meter particle count of no more than 3520 in a size range of 0.5 μm and larger when counted at representative locations normally not more than 1 foot away from the work site, within the airflow, and during filling/closing operations. This level of air cleanliness is also known as Class 100 (ISO 5). We recommend that measurements to confirm air cleanliness in critical areas be taken at sites where there is most potential risk to the exposed sterilized product, containers, and closures. The particle counting probe should be placed in an orientation demonstrated to obtain a meaningful sample. (12)

This ties the guide to the ISO14644-1 cleanroom standard document and to the level of cleanliness required to perform aseptic manufacturing, especially when personnel are present. This also gives insight to the requirements for sample point locations. Sample points should be placed where they are likely to witness any anomalous conditions, closest to where glassware and product are exposed. This adoption of a risk-based approach is echoed in the definition for the support areas and also to the required frequency of monitoring. The guide goes on to state that "regular monitoring should be performed during each production shift. It is recommended to conduct non-viable particle monitoring with a remote counting system. These systems are capable of collecting more comprehensive data and are generally less invasive than portable particle counters" (11).

Therefore, the greater the risk to the finished product quality, then the greater the degree of control required. This can be expressed though the number of

Table 3 FDA Guidance on Room Classifications

Clean area classification (0.5 μm particles/ft^3)	ISO designation	>0.5 μm particles/m^3	Microbiological active air action levels (cfu/m^3)	Microbiological settling plates action levels (diam. 90 mm; cfu/4 hr)
100	5	3,520	1	1
1,000	6	35,200	7	3
10,000	7	352,000	10	5
100,000	8	3,520,000	100	50

allowable particles, the close proximity of the sample to the product, or the frequency with which samples should be taken.

In the case of powder filling, many of the particles are elements of the product and therefore do not pose a risk of being a contaminant. Erroneous results during the process or filling stage lead to false observances of high counts. It is recommended then that the process be monitored in one of two ways:

1. Prove compliance up to the time of batch start, and again after a short clean up period once the batch is finished. Supporting evidence of control will need to be supplied by way of other environmental parameters, viable counts, differential air pressure, and air flow velocity.
2. Select a sample point that reflects the quality of the air in the room or process without being impacted by it. This may lead to a sample point being placed close to the filter face of the area in question.

Also any air monitoring samples from the critical areas should yield no microbiological contaminants. The 5.0-μm particle observations could be used as a means to identify if such exposure risks occur based upon the arguments above.

The critical zone is the central core of the sterile manufacturing environment. This is, however, surrounded by a zone of varying classification possibilities (Fig. 4).

The FDA describes the supporting clean areas as having various classifications and functions in which non-sterile components, products, equipment, and containers are prepared, held, or transferred. These environments should be designed to minimize the level of particle contaminants in the final product. The activity conducted in a supporting clean area determines its classification based on its risk to final product quality. The FDA guidance recommends "the area immediately adjacent to the aseptic processing line meets, at a minimum, Class 10,000 (ISO 7) standards (Table 4) under dynamic conditions. Manufacturers can also classify this area as Class 1000 (ISO 6) or maintain the entire aseptic filling room at Class 100 (ISO 5). An area classified at a Class 100,000 (ISO 8) air cleanliness level is appropriate for less critical activities (e.g., equipment cleaning)" (11).

No information relating to how these areas are to be monitored is given in the guidance. However, as the whole program issued is one of risk assessment, the potential risk to product quality can be reviewed based upon the activities within these areas. The PDA proposed the recommendations given in Table 4 (12).

EMEA Requirements

The EMEA EC enforces its requirements on the manufacture of drug products using the GMPs—Medicinal Products for Human and Veterinary Use, Volume 4, 1998. The applicable annex for sterile manufacture is Annex 1, revised in May 2003. The rationale for the revision was that the guidance had been reviewed following the release of the standard ISO 14644-1 and amended in the interests of harmonization and taking into account any specific concerns unique to the production of sterile medicinal products (13).

The EC recognized that the manufacture of sterile products requires special consideration in order to minimize the risks of both microbiological and particle contamination. The manufacture of sterile products should be performed in clean areas where controls over access and gowning can be enforced. These areas should be maintained at an appropriate level of cleanliness based upon classic cleanroom design principles. The risk associated with the activities in each of the production areas should be classified and based upon the findings the room certified as a particular grade. Each of the assigned

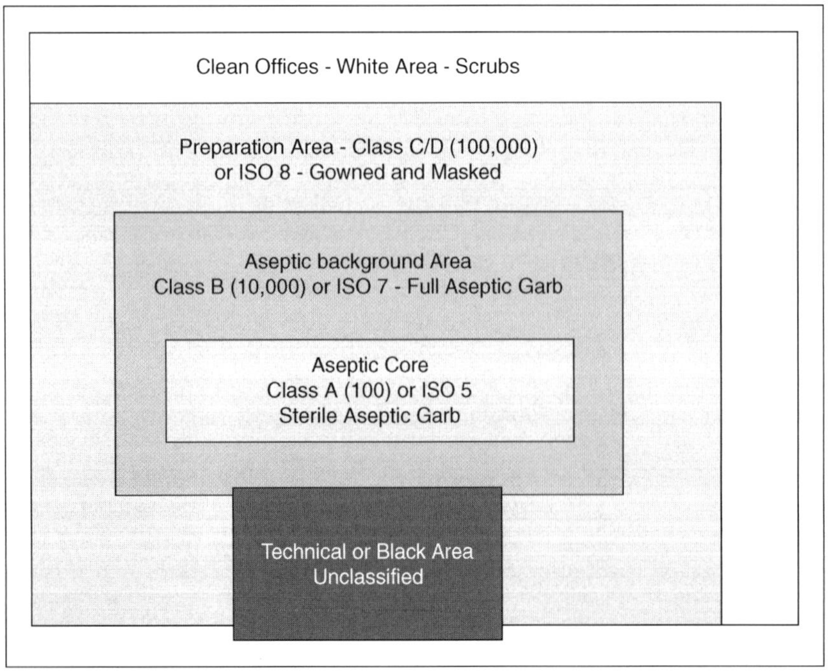

Figure 4 General cleanroom design.

Table 4 Particle Monitoring Requirements for Sterile Support Areas

Air cleanliness classification	Type of operation	Frequency of sampling in operation
Grade A ISO Class 5	Critical aseptic preparation and filling areas	At least once per shift
Grade B ISO Class 7	Areas immediately surrounding the Grade A areas (filling suite). Includes sterile corridors and rooms, areas for sterile product and components storage, gown room exits, etc.	At least once per shift (immediately surrounding Grade A) At least once per day (all other Grade B areas)
Grade C ISO Class 8	Non-sterile filling of terminally sterilized products, areas for equipment and component preparation	At least once per week
Grade D ISO unclassified	Equipment and component washing and handling, gowning rooms, general corridors	At least once per month

Source: From Ref. 12.

grades needs to be monitored in accordance with the assigned grade for particles and microbiological contaminants.

The two operation states are "Operational" and "At Rest." There are also four distinct grades:

- *Grade A*. The local zone for high risk operations, e.g., filling zone, stopper bowls, open ampoules, and vials, making aseptic connections. Such conditions are provided by laminar airflow. The maintenance of laminar conditions should be demonstrated and validated.
- *Grade B*. For aseptic preparation and filling, the background environment for the grade A zone.
- *Grades C and D*. Clean areas for carrying out less critical stages for the manufacture of sterile products.

Table 5 defines the limits set by the EC for maximum particle concentrations.

Again the basic principle of a risk-based approach to environmental control is shown. Areas where the finished product is exposed to the environment must be maintained with a minimum particle exposure rate. Areas where activities offer a reduced risk of contamination by particle contact, or are subsequently cleaned and/or sterilized, are operated in an environment where the maximum permitted particle concentration is significantly higher.

The exposure to the particle risk needs to be monitored in accordance with the recommendations proposed by the regulation. The most critical area, Class A, needs to be monitored with the method that offers the greatest level of control (continuous system of sampling). The EMEA recognizes this in the notes that follow the table of maximum permitted particle concentrations in Annex 1. It states that a continuous measurement system should be used for monitoring the concentration of particles in the Grade A zone and is recommended for the immediately surrounding Grade B areas.

Table 5 EC Annex 1 Particle Monitoring Classifications

	Maximum permitted number of particles/m^3 equal to or above			
	At rest		In operation	
Grade	0.5 μm	5 μm	0.5 μm	5 μm
A	3,500	1	3,500	1
B	3,500	1	350,000	2,000
C	350,000	2,000	3,500,000	20,000
D	3,500,000	20,000	Not defined	Not defined

The limit of 1 particle/m^3 at 5.0 μm is not in line with limits applied by ISO14644-1 guidelines for room cleanliness and also the model of expected size distribution (1/particle diameter ^2.08, ISO146544-1) for a given population of particles within a cleanroom. The EC cGMP requires that these areas be expected to be essentially free from particles of size greater than or equal to 5 μm. As it is impossible to demonstrate the absence of particles with any statistical significance, the limits are set to 1 particle/m^3. During the clean room qualification, it should be shown that the areas can be maintained within the defined limits.

This ties the limits on microbial monitoring (<1 cfu/m^3) to the number of total particles that are greater than 5.0 μm and, as such, pose a risk of being a viable contaminant. The EMEA is therefore looking to prove control over room cleanliness by using ISO guidelines and 0.5 μm particle concentrations, and over room sterility by using the data for the larger particles.

If continuous monitoring is not performed in the critical areas, and routine certification or monitoring is executed using a portable particle counter (see the section entitled Particle monitoring procedures), then the ties to ISO14644-1 (the underlying reason for release of the revision) become apparent and a minimum volume is stipulated for these tests. The equations below offer reasoning as to why a minimum volume is required.

EC cGMP Sample volumes using ISO14644-1 Calculations. In a Class A cleanroom, operational, the limits are 0.5 μm = 3500/m^3 and 5.0 μm = 1/m^3. If we use a particle counter with a flow rate of 28.3 L/minute = 1 cubic foot per minute (1 cfm), the following times are established for testing each location.

$$0.5\ \mu\text{m} = \frac{20}{C\ \mu\text{m}} \times 1000 = \frac{20}{3500} \times 1000 = 5.7\ \text{L}$$

however, a 1-minute sample is required for ISO.

$$5.0\ \mu\text{m} = \frac{20}{C\ \mu\text{m}} \times 1000 = 20 \times 1000$$

$$= 706 \text{ minutes or } 11.78 \text{ hours}$$

ISO dictates that for room certification using multiple sizes, the maximum calculated sample period must be used and this typically reflects the largest particle size. Therefore, to follow ISO would prove impractical in a manufacturing environment, and a requirement of a

1-m^3 sample should be taken during routine testing. It also follows that this routine testing is periodic. Therefore, confidence in control over an area comes not only from the number of samples taken, but also from the volume of that sample.

No definition is set out for monitoring the support areas, whether it is by portable or continuous means. Neither is a limit set at a minimum volume requirement, as the ISO calculations for minimum volume yield an acceptable period of sampling.

Other International Requirements

There are two other international regulatory guidelines which document a need to prove compliance to particle counting limits. The WHO and the PIC/S both state limits for maximum permitted concentrations of 0.5 and 5.0 μm particles.

World Health Organization

The WHO cGMP limits for particles follow the original limits imposed by the EC Annex 1, stating that no macroparticles should be allowed in the critical areas where product is exposed directly to the environment in which it is processed (Table 6). They do not offer guidance on the frequency of monitoring, only that monitoring to prove compliance must be performed.

They also show how the limits enforced are related to other limits imposed by the FDA and EMEA regulations.

The direct relationship between the WHO classifications and the EMEA classifications is evident as is the combination of Class A areas with ISO 5 that the FDA require to be maintained in proving control over a critical area (Table 7). Harmonization between the various standards points toward a common requirement for a risk-based monitoring program. The WHO is also an observer for the PIC/S.

Pharmaceutical Inspection Cooperation Scheme

The Pharmaceutical Inspection Convention, active within Europe and elsewhere, requires a common standard for inspections between member states. This is in an effort to remove trade barriers within a common market. They agreed to harmonize on the rules of a common cGMP; (14) the EU Guide to GMP for Medicinal Products and its Annexes was adopted. The particle counting limits for the PIC/S are therefore the same as those identified in the EMEA cGMP Annex 1 and are shown in Table 5. The scheme is being adopted as a global standard by all major countries for pharmaceutical cGMP.

Table 6 WHO Airborne Particulate Limits for Sterile Manufacture

	Maximum number of particles permitted/m^3			
	At rest		In operation	
Grade	0.5–5.0 μm	>5.0 μm	0.5–5.0 μm	>5.0 μm
A	3,500	0	3,500	0
B	3,500	0	350,000	2,000
C	350,000	2,000	3,500,000	20,000
D	3,500,000	20,000	Not defined	Not defined

Regardless of governing body, particle counting certification and monitoring must be proven, and a system of monitoring needs to be employed. There are two philosophies for sampling that can be adopted:

1. Routine portable testing of an area.
2. Automated monitoring of an area.

The following portion of the chapter describes the implementation of a monitoring system and the selection of sample points that best reflect the activities within a production zone.

PARTICLE MONITORING PROCEDURES

The effective monitoring of a cleanroom involves the measurement of multiple environmental variables. Particle counting, in addition to the measurement of airflow patterns and air velocity, temperature and relative humidity, and of differential air pressure between adjacent rooms, can be vital in ensuring operations are at optimum performance. The monitoring plan should include measurement of all these parameters where necessary.

Routine Portable Testing of an Environment

There are three primary reasons for the portable monitoring of a cleanroom:

- routine verification of performance
- diagnosing particle contamination from a specific machine or an operation used to clean the cleanroom (filter testing) or from a new contamination source (new machine or filter installation)
- providing data showing compliance to regulatory standards.

The testing of a cleanroom will require an operational procedure, which will depend on operational requirements and form testing. All monitoring operational procedures should define:

- the parameters monitored (if other parameters are to be monitored, they should also be listed)
- sampling locations
- sampling frequency
- sampling duration
- target cleanliness levels
- alert and alarm threshold levels (Alert or warning levels mean that operators are prepared to take remedial measures. Alarm levels will require that the cleanroom operation be halted and steps taken to protect product in the area while remedial measures are implemented to control the contamination source.)
- any actions taken against threshold limits, hardware errors or cleanroom comments.

It is necessary to record contamination levels verifying that cleanroom cleanliness levels have been maintained. It is also necessary to maintain detailed records of monitoring device (particle counter, temperature probe, pressure sensor, etc.) validation as well as the monitoring activity results, proving the room is clean and that the instrumentation used is traceable.

Concerns in the use of a portable particle counter are:

- It introduces an additional person into the cleanroom, unless the manufacturing staff is responsible for testing.

Table 7 WHO Comparative Table of Different Particle Standards

WHO (cGMP)	United States (209e)	United States (customary)	ISO/TC (209)	EEC (cGMP)
Grade A	M 3.5	Class 100	ISO 5	Grade A
Grade B	M 3.5	Class 100	ISO 5	Grade B
Grade C	M 3.5	Class 10,000	ISO 7	Grade C
Grade D	M 3.5	Class 100,000	ISO 8	Grade D

- It does not always reflect the actual conditions of the process, or the normal practices of personnel present in the cleanroom due to its intrusive nature.

Portable sampling is less than 5% of the process period compared with continuous data (15). The automation of the monitoring addresses these issues described above.

Automated Monitoring of An Environment

There are two primary means of automated sampling within a cleanroom: continuous or sequential (manifold) monitoring. The decision as to which method to use is based upon the nature of activities within the clean areas.

Continuous Monitoring

Particle sensors are located at each sample collection point and are operated continuously. Data are fed to a central data-processing system. The particle sensor consists of a small enclosure housing an optical system, a light source (laser diode), and signal generation electronics. The sensors often require an external vacuum source and signal communication cable to transmit data to the central monitoring computer. The advantages of such a system are

- the sensors continuously monitor and report data to system, therefore detecting short-lived particle burst situations
- they are simple and have low-cost installation
- the ease of relocation to alternative positions
- provides the highest level of confidence

Manifold Monitoring

Manifold monitoring transfers sample air from each point through tubing to a sample manifold or sequencing valve, which transfers the sample to a single particle-measuring device in a programmed sequence. Manifold systems are very common and consist of a centrally located manifold and single particle counter with up to 32-sample tubes radiating from this central location. Each tube is capable of drawing a sample from a distance up to 38 m (125 ft) from the manifold. The advantages of such a system are

- low cost per sample point monitored
- low maintenance and calibration costs; only a single instrument per manifold to calibrate and service.

The factors that determine the selection of either of the systems include statistical validity of the data that are collected, large-particle sample line losses, differences in response time from individual sensors, requirements for frequency of data collection, system installation and operating/maintenance costs.

Both system types have disadvantages to their selection.

- The multi sensor system (Fig. 5) requires that all sensors are calibrated for correct particle sizing and sample inlet flow rate. The problems of inter-sensor correlation require extremely careful calibration and frequent maintenance to reduce these errors to an acceptable level.
- The manifold system (Fig. 6) uses multiple sample lines throughout the facility. Each of these lines draws a sample through from sample point to manifold. They require the transport of air through this tubing over distances up to 40 m. Particle losses caused by turbulent deposition and electrostatic effects may exceed 50% for particles of 5 μm in diameter. Losses of particles in the size range 0.1 to 1 μm are negligible and within experimental error. Any sample measurement must allow 10 to 15 seconds of purge time to clear a previous sample from a long line.

Combination System

The ideal solution is usually found in a combination of techniques to fully satisfy the monitoring requirements of a manufacturing facility. Various combinations are presented in Table 8.

FUNDAMENTALS OF OPCs

OPCs have been used for counting and sizing particles in air since the mid-1950s. To understand the application of the use of a particle counter, it is important to understand the underlying principle behind its operation.

The original test method for determining the cleanliness of an environment meant employing some classic, empirical techniques. A known volume of air was drawn across a filter and submitted for subsequent analysis. The filter would be either paper or cellulose fiber. To control the volume of air, either a volume flow meter was used for a specified period of time or a total volume meter was used. The sample was taken in the cleanroom and prepared. The deposited particles were then counted by

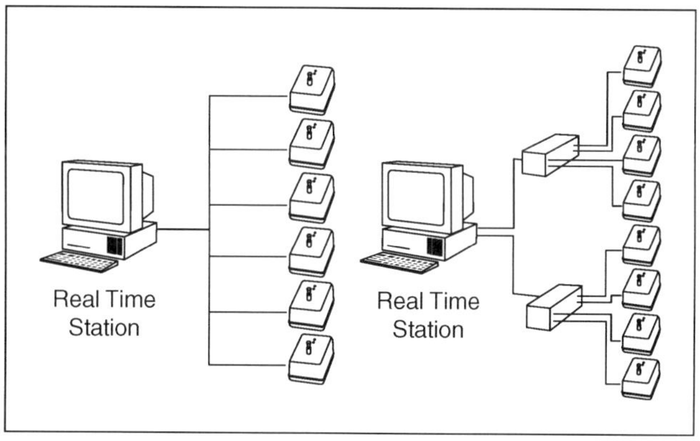

Figure 5 Multi-sensor monitoring system types.

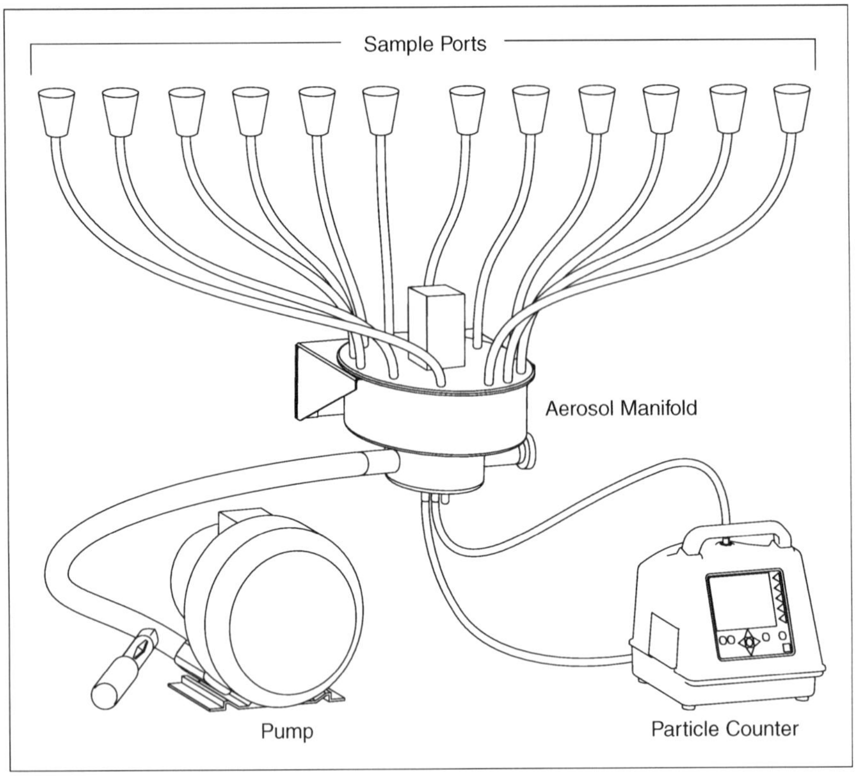

Figure 6 Multiplexed manifold monitoring system.

eye through a microscope. This technique required an extensive period of time and so was only adequate for annual measurements required for a specific area. However, it was unsuitable for routine monitoring.

The onset of instrumentation and OPCs was inevitable. Particle counters provide reliable, accurate, real-time, and repeatable measurements. This is ideal for evaluating the cleanliness of a cleanroom and provides the level of monitoring required to establish control limits over various processes.

Basic Operation

Airborne-particle counters work on a light scattering principle. They utilize a very bright light source to illuminate the particles. The current standard source of the illumination is a laser diode. Previous sources used were gas lasers (Helium–Neon, "HeNe") and "white light" bulbs. This bright light source is focused and shines through an optical block. Within the optical block are collection mirrors and one or more photodetectors. Sampled air is drawn through the laser beam, and entrained particles in the sample air pass through the beam. The laser light interacts with the particles and is scattered (Fig. 7).

The resulting scattered light is collected using parabolic mirrors and focused on the photodetectors. These photodetectors convert the scattered light from each particle into a pulse of electricity. By measuring the signal height of the electrical pulse and referencing it to the calibration curve of the instrument, the size of the particle can be determined. By counting the number of electrical pulses in a sample volume or over a period of time, we can determine quantity. Particle counters are able to size and count the number of particles for a

Table 8 Matrix of Sampling Techniques

Sampling technique	Comments
Continuous	Capable of collecting most comprehensive data. May miss events due to location selection. Cannot be used to certify cleanroom. Expensive to give sufficient coverage for all cleanrooms
Manifold	May miss events due to either location of sample point or frequency or sequential nature of sampling. Particle losses in tubing concern especially in critical areas. Cannot be used to certify cleanroom
Portable	Very labor intensive for both sampling and reporting. Can be used to certify cleanroom
Continuous and manifold	Continuous at critical zones and manifold in surrounding clean zones. Cannot be used to certify cleanrooms
Continuous and portable	Continuous at critical zones and portable in all other areas. May miss events in surrounding rooms. Labor intensive. Can be used to certify cleanrooms. Best solution for cGMP compliance
Manifold and portable	May miss events due to location of sample point or frequency or sequential nature of sampling. Particle losses in tubing a concern especially in critical areas. Frequency of monitoring using portables in critical activities. Labor intensive
Continuous, manifold and portable	Full automation of all cleanroom activities. Continuous at critical sites, manifold in support areas and complimented using portables for fault diagnosis and certification exercises. Total cGMP solution

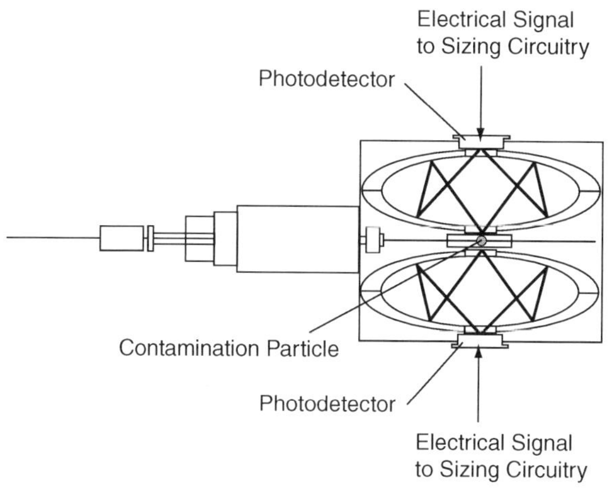

Figure 7 Laser particle counter fundamental principles.

population of particles within a cleanroom giving full quantitative information.

"Light scattering" is a general term and is composed of various different physical phenomena. Scattering is made-up of:

- Reflected light—when a light hits a particle and is angularly deflected.
- Refracted light—when a light goes through the particle and its direction of travel is changed.
- Diffracted light—where light comes close to the particle and is bent around it.
- Phosphorescence—light is absorbed as one frequency and emitted as another.

The interaction of light and particles depends upon the particle composition, its refractive index, and the difference between the particle and the background medium. (For cleanroom particle counters this medium is air.)

During operation, the instrument compares the response from the particle signal to its calibration curve which is previously generated using latex spheres of a known size, shape, and refractive index. The instrument is therefore not counting and sizing particles, it is counting and sizing pulses of light and mapping them to a similar electrical response from the latex spheres. A result of this activity that users should be aware of is that particles with different refractive indices and shapes create scattering responses either smaller or larger relative to the latex standard. For example, an alumina oxide particle, because of its high reflectivity, will scatter a great deal of light, and so will appear larger than a respectively sized latex sphere. A carbon particle that absorbs light will size small relative to the latex standard. These sizing differences (deviations from the latex standard) will assign the particles into larger or smaller size channels, so the only absolute, and therefore the standard for calibration, is latex spheres in clean dry air.

The original technique of passing a volume of air through a membrane filter has not been cast aside. It still remains a valuable method of extracting a sample for subsequent analysis as to particle identification and is an approved method by the EPA for environmental monitoring.

Particle Counter Calibration

A particle counter is calibrated against size thresholds, not count values. This is because the instrument is considered volumetric: only at the extreme range of concentrations is a saturation limit reached or does coincidence, two particles resident in the laser beam simultaneously, occur. The particle counter size calibration procedure is carried out with mono-dispersed (very narrow distribution), spherical latex particles. The ASTM F 50, section 3.1.10 references particle size, and is defined clearly as "the equivalent diameter of a particle detected by an instrument using light scattering. The equivalent diameter is the diameter of a reference sphere having known refractive index and generating the same electrical response in the photo detector of the particle counter as the particle being measured" (16).

The particle counter measures the amount of light scattered by a particle and places the relative size of the particle in a "size bin." A size bin is determined from the boundary thresholds. If a particle counter has several channels, such as a 0.1-μm sensitive instrument, the first channel on the display is shown as 0.1 μm. The second channel is shown as 0.2 μm. The third channel is shown as 0.3 μm. If a particle falls in the first channel, it is sized as being between 0.1 and 0.2 μm. It may be as small as 0.105 μm or as large as 0.195 μm but still falls in the first channel.

Another function of a particle counter is to normalize the counts per unit of flow (counts per cubic foot, counts per cubic meter). Therefore, the flow rate through the counter must be correct.

ASTM F 328 and 649 were written for aerosol particle counters. The F-328 procedure's basic principle is the use of mono-dispersed particles for primary calibration of the instrument and the comparison of the test instrument to a "referee" to determine the counting accuracy. The F-649 procedure's basic principle is the comparison of the test instrument to a reference instrument. This is called secondary calibration.

Error Control and Minimization

Once calibration has been performed, particle counters will still produce data which have inherent errors. To minimize errors, the particle counter performance specifications, beyond calibration, must also be acceptable. The following elements of performance have significant importance for particle counting and are discussed below.

Signal-to-Noise Ratio

The data produced by the particle counter should be produced solely by the detection of particles passing through the optical chamber. If the signal from these particles is not significantly greater than the electrical noise produced by the particle counter's electronic circuits, then some of the data will result from that background noise level, rather than from actual particles. An increased signal-to-noise ratio will reduce electrical noise frequency and allow the detection of increasingly smaller particles; however, the added cost for an improved particle counter may not be required for the cleanroom when only two channels (ex: 0.5 and 5.0 μm) are required. Acceptable particle counter signal-to-noise ratio specification can be determined when more than two particles are recorded in the smallest size channel when sampling clean dry air; this is called the "zero count" capability of the instrument.

Particle-Sizing Accuracy

Particle-sizing accuracy is required because most count information is defined in terms of concentration (number per unit volume) equal to or greater than specified particle size. The particle size distribution usually encountered in cleanrooms will cause an error in particle size measurement, which in turn creates a concentration error that varies with the associated particle size distribution function. Figure 8 shows how a sizing error of 2% can produce a concentration error of 6.25% for a third-power particle size distribution. Sizing accuracy can be converted to counting error based on the particle size distribution in the clean manufacturing area of interest. It is generally assumed to be $1/d^{2.08}$ for cleanrooms based on the ISO 14644-1 standard.

Counting Efficiency

Counting efficiency is an expression of the probability that an OPC will sense, and therefore count, a particle passing through the particle counter's sample volume. This probability is a function of size up to a certain critical size above which all particles are normally sensed and counted. Figure 9 displays the plots of counting efficiency versus particle size. Note that while the signal produced by the particles is distributed symmetrically about the nominal most sensitive threshold, the exponential relationship between particle size and signal returned causes the counting efficiency curve to be asymmetrical.

Sensor Resolution

A particle counter's resolution is its ability to determine small differences in particle size. A number of factors combine to cause the resolution of a particle counter to be other than perfect. These include the uniformity of illumination of the sampling volume, the quality of the optical system, the quality of the electronics in the instrument, and the electrical background noise. If it were possible to introduce particles all exactly the same size to a real-world particle counter, the factors above would cause the reported distribution to be the familiar "bell curve" (normal or Gaussian) shape. Figure 10 shows the reported distributions, which would result from introducing a group of particles, all exactly the same size to a particle counters with "excellent," "average," and "poor" resolution. Note that with excellent resolution, the OPC would always put each of the particles in the same size class regardless of the width of the size class. Note that the minimum possible width of the OPC size classes or "channels" tends to be determined by the fundamental resolution. Thus an OPC with average resolution can have more size classes across the range of the instrument than an OPC with poor resolution. The minimum possible width of the size classes is limited by the fundamental resolution.

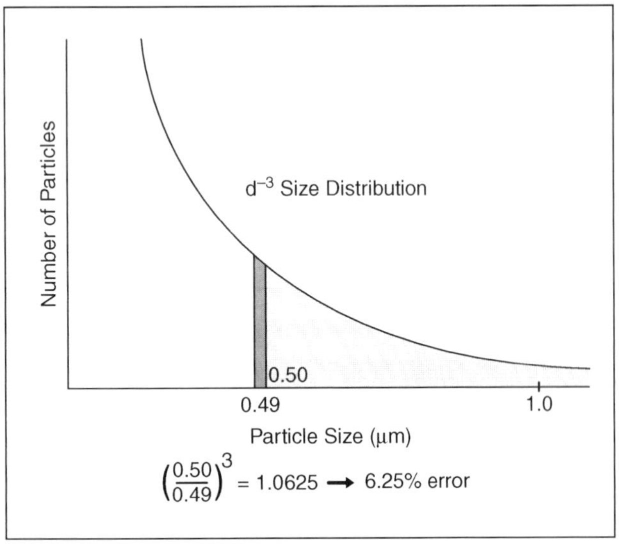

Figure 8 Laser particle counter sizing error.

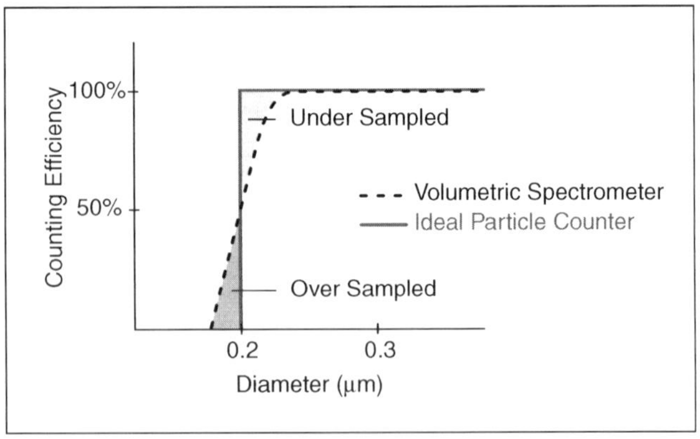

Figure 9 Counting efficiency versus particle size.

Sample Flow Rate

For most critical pharmaceutical activities, very low particle levels are present, especially in the ≥5.0 μm size range, resulting in very little data produced by the particle counter. At these low concentrations, it can be expected that the random occurrence of particles could result in widely varying data if a series of small samples are taken. A particle counter with the largest available sample volume flow rate, or of sampling at the largest possible volumetric sample flow rate, will give statistical confidence over a dataset in the shortest period of time.

Particle Concentration Capability

For the majority of pharmaceutical cleanroom operations, a particle counter's ability to measure high particle concentrations is not necessary. However, when monitoring some cleanroom operations, such as powder filling, it is possible that a burst of particles may appear. If this occurs, the particle counter used to sample the area or the potential local contaminant source should be capable of detecting particle concentrations several orders of magnitude above the normal, with minimum error caused by coincidence. The selection the particle

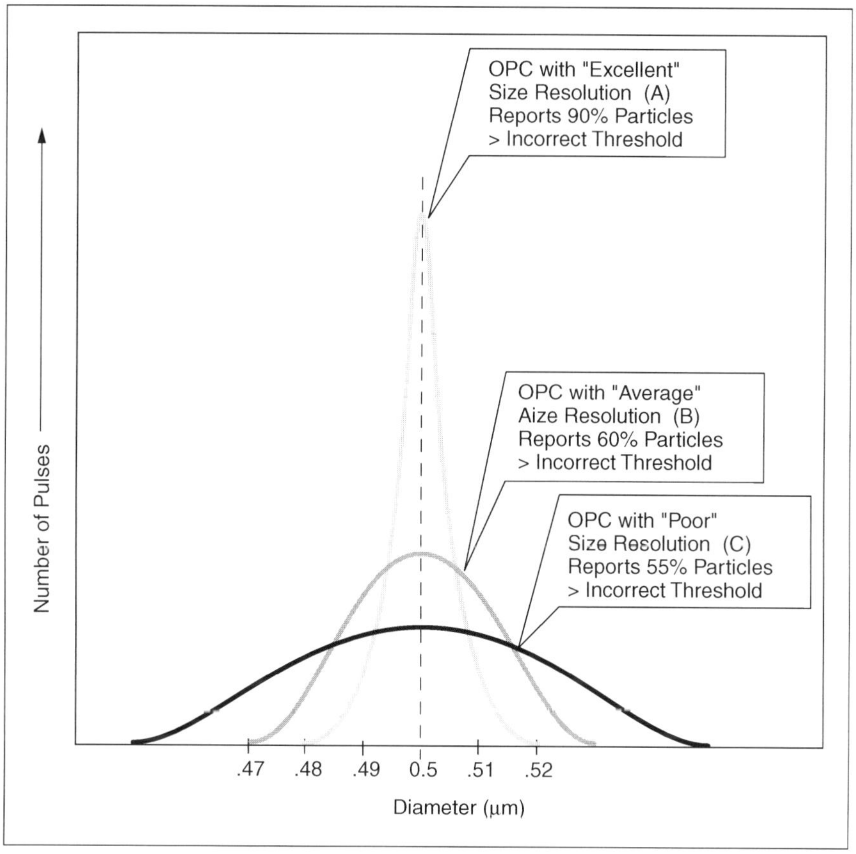

Figure 10 Variances in particle counter resolution.

counter used in such applications may conflict with the previous one of maximum sample size capability.

AEROSOL MONITORING CONSIDERATIONS

The selection of sampling procedures and various ratio nales for monitoring are largely determined by what the system is designed to observe. This has been addressed in previous sections. There is still an underlying principle that governs the selection of sample point location. This involves the physics of how particles move within a body of media and the accuracy to which they can be sampled, irrespective of the particle counting method chosen.

When samples containing particles are taken, it should be considered that the distribution of those particles (even if size is maintained as a constant) within the media (air) is not uniform and may change over time. The concentration of particles within the sample will be low, especially the larger particles, and the direction of particles is not always matched by the direction of the air flow. Also, additional forces are acting on the particles that do not act on the air flow due to the differences in mass of the particles. These additional forces affect particle motion and are discussed below.

External Forces Affecting Particle Migration

Cleanroom monitoring activities can be defined as being performed to qualify and quantify the dynamics of a fluid; that fluid is the body of air within a confined space. This space may be either the air in the general cleanroom, a transport duct, or a laminar flow zone. The following terms describe the mechanisms of how particles behave in air and should assist in the understanding of sampling and management of samples, thus improving the efficiency of sampling.

- The *Stokes Number* is the ratio of a particle's radius to the dimension of an obstacle to fluid flow. This is an important factor in determining when a particle in motion will be collected by an obstacle or will pass around it. An obstacle could be a filter fiber, the sample inlet, or a component that should be kept clean, such as the opening to a vial.
- The *Drag Coefficient* is the ratio of the force of gravity to the inertial force on a particle. It indicates how a particle will resist any force that could cause a change in the particle velocity. Smaller particles have smaller drag coefficients due to their lesser mass.
- The *Relaxation Time* is the time for a particle initially in equilibrium with a moving fluid to match a change in fluid velocity. Large particles have a long relaxation time. When an aerosol stream moves through tubing that contains small-radius bends or elbows, the large particles will deposit on a tube wall because they cannot adapt easily to sudden velocity changes,

Table 9 Settling Velocities of Particles in Stagnant Air

Particle size (μm)	Settling velocity (cms^{-1})
0.00037	
0.01	6.95×10^{-6}
0.1	8.65×10^{-5}
1.0	3.50×10^{-3}
10	3.06×10^{-1}
100	2.62×10^{1}

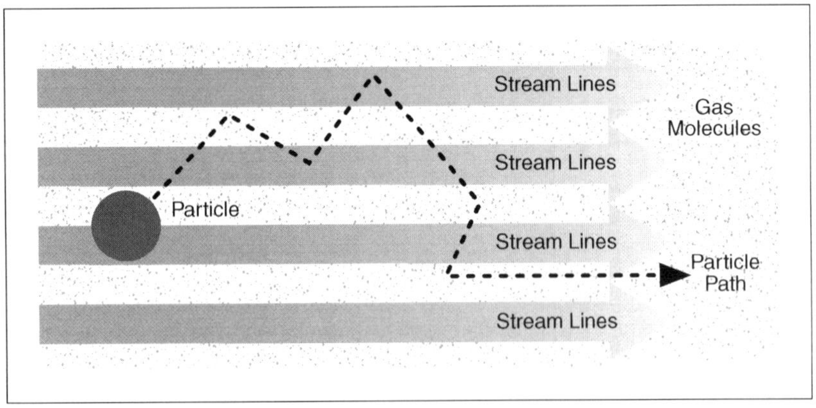

Figure 11 Migration of a particle along a mean free path, due to brownian motion.

though they continue in their original direction until they make contact with the tube wall. A related term to relaxation time is *stopping distance*, which is defined as the distance for a particle initially moving within a gas stream to come to a stop when the gas flow is halted, as by an obstacle.

- The *Deposition Velocity* or *Sedimentation Velocity* is the ratio of particle flux, distance per unit time for sedimentation to occur relative to the ambient particle concentration (Table 9).

There are also additional forces in effect on particles. These forces and their subsequent response to those forces control particles migration through air:

- *Viscous Forces.* The fluid dynamic force from a moving fluid stream. The viscous nature of an air stream will "pull" particles along that flow path.
  - If the flow is *laminar,* then additional forces act upon the larger particles causing settling and deposition. Smaller particles remain buoyant.
  - In *turbulent* flow, when large particles settle, they are re-entrained back into the airflow. Smaller particles are more prone to additional forces acting upon them, preventing them from being transmitted through a tube.
- *Brownian Motion.* As particles migrate through a body of air, random impacts from individual molecules will cause them to veer from course (Fig. 11).
- *Gravitational Force.* Varies with particle mass and the difference between particle and air density; the larger the particle the greater the effect.
- *Electrostatic Forces.* Varies with the particle's electrical charge (surface area controlled) and the strength of the electrical field in which the particle is located. Electrostatic charge can develop as a particle "slips" through the air stream. It is important, therefore, to minimize these interactions to ensure all particles reach the final destination.
- *Diffusion Force.* Varies inversely with a particle's radius. Smaller particles are more prone to interactions due to diffusion.
- *Thermophoretic Forces* (mainly for small particles). Varies with the particle's surface area and temperature gradient.

The particle's response to these forces is controlled by the particle's size, mass, shape, and electrical charge. For essentially all these forces, the major particle parameter is size. The magnitude of the forces varies with particle size squared (or cubed).

Practical Considerations

To overcome the known forces on a particle, it is possible to design a system that minimizes the impact of the forces and the errors that may occur as a consequence of them.

Isokinetic Sampling

In laminar flow environments or in ducts leading to a filter, air flow is considered to be unidirectional. The air must be neither over nor under sampled. This requirement is satisfied with isokinetic sampling which ensures that the velocity in the supply air is the same as the velocity in the particle counter's sample-tubing inlet (Fig. 12).

If the velocities differ, either a positive or negative sample collection error occurs. An isokinetic sample error increases with particle size. This is not of great concern for particles smaller than 1 to 2 μm. FS209E shows that isokinetic sampling errors greater than 5% are not expected for small particles when using a sample probe with an inlet diameter of 2 mm or larger, even when sampling and sampled air velocities differ by an order of magnitude. However, when particles greater than 5 μm are to be measured, isokinetic sampling is required.

Particle Loss in Transport Tubing

When a sample is taken, it is common that the sample probe head is in a separate location from the particle counter. Sample tubing is used to connect the probe head to the particle counter. If the sample is to be transported any significant distance in the tube, some particle loss will

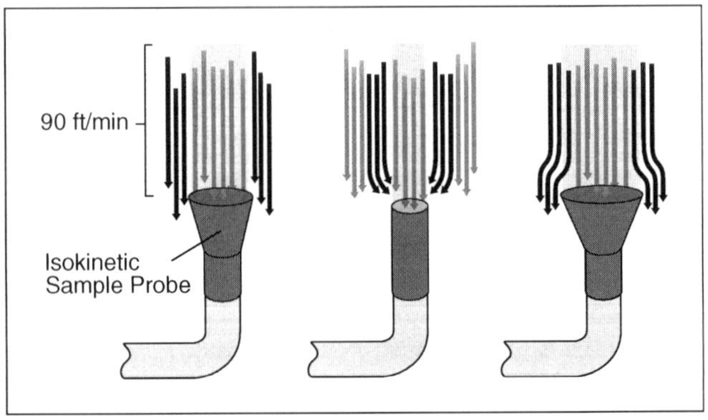

Figure 12 Isokinetic sampling.

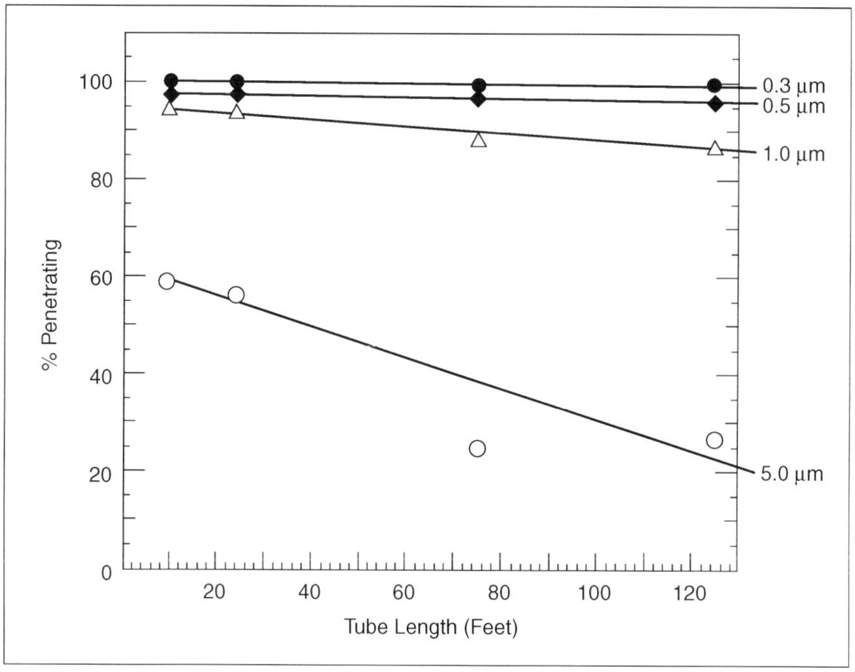

Figure 13 Particle loss in tubing (1/2 inch tubing at 100 L/min).

occur. These losses are dependant on tubing velocity, diameter, and distance. Large particles are lost by a combination of gravitational settling and inertial deposition on the walls of the tubing. Small particles are lost to the duct walls by Brownian motion and diffusion effects.

Figure 13 shows the penetration of different sized particles through a manifold system over distances up to 125 feet. Particles less than 1.0 µm in diameter show no significant losses, and the differences are essentially experimental error. Larger particles show a significant level of loss even over very short distances.

When portable particle counters are used, the flow rate in the tubing is significantly reduced, and so the maximum permissible distance is also reduced. Figure 14 shows a similar pattern to that of manifold sampling but over much shorter distances.

Electrostatic forces also account for a proportion of the losses in a sample. To reduce the effect of these additional forces, various types of material were tested to establish a suitable standard (Table 10). The order, most preferred first, is based on a combination of particle loss rate, electrical conductivity, and potential for oxide or sulfide formation when the tubing is exposed to urban air, with one being the most suitable.

The diameter of the tubing should be selected to ensure the Reynolds number (defined in FS209E) is between 5000 and 25,000. The Reynolds number range is one for which no significant turbulent deposition occurs for particles smaller than 5 to 10 µm. Time in the tubing should be no more than 10 to 20 seconds to ensure the transmission of particles larger than 0.1 µm before any significant losses occur.

Sample Point Selection

The selection of the sample point location is based upon a risk assessment performed for each process. There are

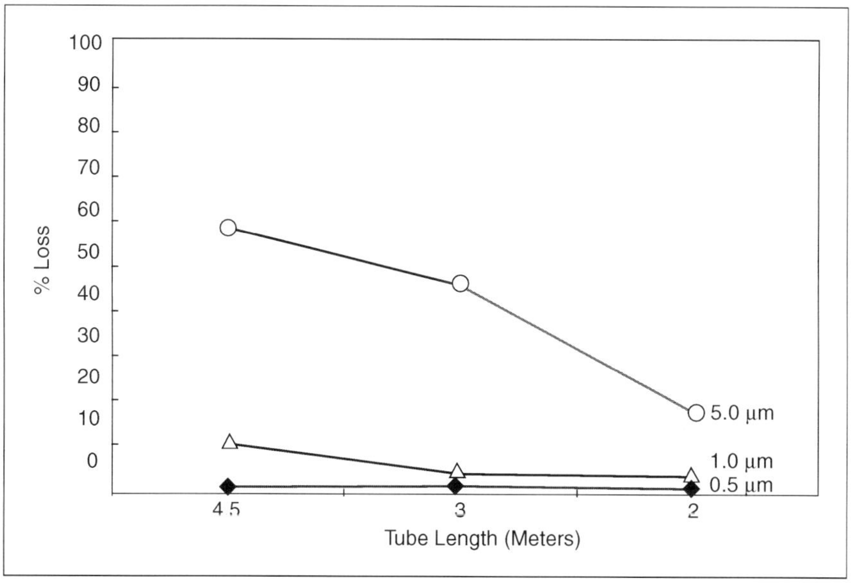

Figure 14 Particle loss in a portable particle counter (3/8 inch tubing at 28.3 L/min).

Table 10 Particle Transport Line Material Preference

Stainless Steel
Bev-a-line (Hytrel lines PVC tubing)
Polyester (as polyurethane)
Polyester lined vinyl
Copper
High density polyethylene
Glass
Teflon

various measures that will influence the final location including

- Pointing the probe into the sample flow where the flow is unidirectional (laminar or iso-axial). In unidirectional flow, it is important to perform isokinetic sampling, especially for macro-particles.
- Analyzing relative risks to product based upon activities within the room environment. Analyze workflow patterns to establish worst-case scenarios for background environments.
- Conducting airflow tests ("smoke tests")
  - Verify flows lead away from product and out of enclosure
  - Verify minimal recirculation
  - Identify particle traps and recirculation zones

It is also possible to perform three-dimensional airflow studies of existing facilities combined with intensive particle monitoring to determine the operational characteristics of the cleanroom and identify any worst-case locations that may exist. The evaluation is best performed in Class B environments due to the number of available particles in air and the variations in air flow patterns. Establish critical contamination risk locations in environmental monitoring by means of three-dimensional airflow analysis and particulate evaluation. (17).

Example. Assume there is a cleanroom to be used as an aseptic filling area. This room needs to meet ISO Class 5 at 0.5 μm, operational. The following steps show how to determine whether the room meets its classification requirements.

The room is 6.5 m by 4 m and has an enclosed filling machine in the center (Fig. 15).

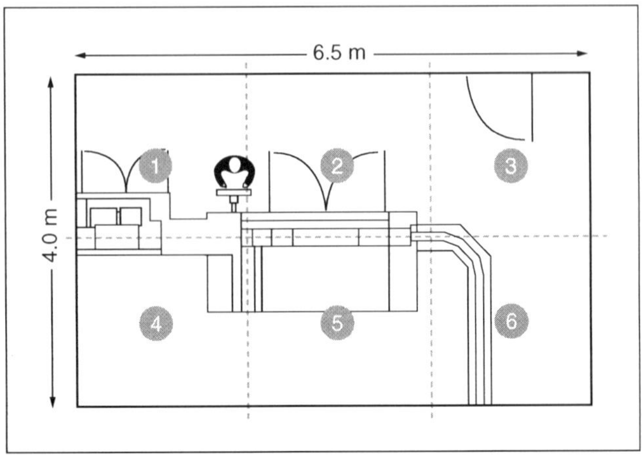

Figure 15 Sample layout of cleanroom.

1. Determine maximum permitted particle concentration

$$C_n = \left(\frac{0.1}{0.5}\right)^{2.08} \times 10^5 = 3517 \text{ Rounded to } 3520 \text{ n/m}^3$$

C_n = Max conc^n | (0.5) = Particle Size in μm | 10^5 = 10^ Classification Number = 5

2. Determine number of sample locations

$$N_L = \sqrt{A} = \sqrt{6.5 \times 4} = \sqrt{24} = 5.1 = 6$$

Square Root of floor area in m

3. Determine minimum sample volume

$$Vs = 20_{C_{n.m}} \times 1000$$

$$= 20_{3517} \times 1000 = 5.69 \text{ L}$$

$$= \frac{20}{C\mu m} \times 1000 = \frac{20}{3577} \times 1000$$

$$= 95\% \text{ UCL} = 875 + 2\left(\frac{1797}{6}\right) = 2473$$

(sample 6 from Table C)

From ISO14644-1 Section B.4.2.2, Minimum Volume = 2 L and Sample period = 1 minute. Standard particle counters run at 28.3 L/min. Therefore, a one-minute sample at each of the six locations is needed to meet the specification.

4. Determine measurements for each location (Table 11)
5. Statistical analysis of the results
   a. Average

$$x = \frac{(2340 + 1467 + 3140 + 3140 + 1509 + 1966 + 825)}{6}$$

$$= 11247 = 1875 \text{ n/m}^3$$

   b. Standard Deviation

$$S^2 = \frac{1}{7}((2340 - 1875)^2 + (1467 - 1875)^2$$
$$+ (3140 - 1875)^2 + (1509 - 1875)^2$$
$$+ (1966 - 1875)^2 + (825 - 1875)^2)$$
$$= 3,227,650$$

$$S = \sqrt{3,227,650} = 1797 \text{ n/m}^3$$

Table 11 Example Data

Location	Number/m^3
1	2340
2	1467
3	3140
4	1509
5	1966
6	825

c. 95% UCL

95% UCL = 1875 + 2.0 (1797) (Samples = 6, From Table C1,
(6) student "t" = 2.0)

Average Std Dev

6. Define report

Each Location max = 3140 n/m^3 <3520 class limit
= PASS

95% UCL = 2473 n/m^3 <3520 class limit
= PASS

This room meets the specification for an ISO Class 5 cleanroom at 0.5 µm, operational and can now be used for the purpose that it was designed.

REFERENCES

1. The United States Pharmacopoeia General Chapters (USP 28, 2005).
2. USP <797> Pharmaceutical Compounding—Sterile Preparations.
3. U.S. EPA, 1998.
4. Federal Standard FS209 (Airborne Particulate Cleanliness Classes in Cleanrooms and Clean Zones). Federal Standard No. 209E. Washington, DC: General Services Administration, 1992.
5. Berndt CW, Burnett ES, Spector R. Cleanroom Garment System Application and Use, S2C2 monitor No. 39, April 2000.
6. De Abreu C, Pinto T, Oliveria D. Environmental monitoring: a correlation study between viable and nonviable particles in clean rooms. PDA Jnl Pharm Science Technol 2004; 50(1):45–53.
7. Whyte W. Cleanroom Design. Wiley, 1999.
8. Reinmueller, Ljundqvist. Modern cleanroom clothing systems: people as a contamination source. PDA Journal of Science 2003; 57(2).
9. USP <1116> Microbial Evaluation of Clean Rooms.
10. Title 21, Code of Federal Regulations, Parts 210 and 211.
11. Guidance for Industry Sterile Drug Products Produced by Aseptic Processing, September 2004.
12. Points to Consider for Aseptic Monitoring. Topic F: Location and Frequency of Non-Viable Air Monitoring. PDA Jnl Pharm Science and Technol 2003; 57(2) (Suppl. 1).
13. REVISION TO ANNEX 1, Manufacture of Sterile Medicinal Products, September 2003.
14. PE 009-1, 1 September 2003.
15. Hallworth, Mark. Process Analytical Technologies and Non-viable Particle Counting, Particle Measuring Systems Technical Paper, 2005.
16. Standard Practice for Continuous Sizing and Counting of Airborne Particles in Dust-Controlled Areas Using Instruments Capable of Detecting Single Sub-Micrometer and Larger Particles, ASTM F50. Philadelphia: American Society for Testing and Materials, 1992.
17. Katayama, Higo, Tokunaga, et al. PDA Journal of Pharmaceutical Science and Technology, Fujisawa Pharmaceuticals Japan 2005; 59.

26

Viable Environmental Microbiological Monitoring

Pamela D. Deschenes
Wyeth Global Pharmaceuticals, Frazer, Pennsylvania, U.S.A.

INTRODUCTION

The various techniques for viable monitoring are intended to ensure that clean and aseptic product production areas are of the highest possible quality. This control will allow sterile products to be safe for use with respect to freedom from microbiological contamination. This chapter will assume that the reader has minimal knowledge and experience in this area, but has some familiarity with regulatory guidelines and understands basic microbiology. Examples of the choices that can be made, the details of the methods, the equipment available, how pros and cons will be presented, and how the regulations can be met in different ways will be presented.

There are many ways to perform monitoring for the presence of bacteria, yeasts and molds. Most are considered "standard" in that most microbiology labs are aware of their existence and familiar with their performance. None are "standardized" so they require in-lab verification of their applicability. The choices made will be based on a contamination risk assessment of the equipment, operation, manpower, contamination potential, organism profile, history, regulations, building design and cost.

Air, surface, and personnel testing in the aseptic production areas and the sterility test laboratory will be addressed, with the detection of aerobic organisms of primary concern. Anaerobic organism monitoring will be briefly discussed. Methods, method development, media, incubation, testing equipment, site selection, training, frequency of testing and interpretation of results will be discussed. Alert and action levels, reaction to data, investigation of problems, and using a sterile media trial as an indicator of effective procedures will also be covered. There may be alternatives to every procedure presented since there is a wide variety of operations and considerations. SOPs, documentation and data management will also be presented.

Although viable and nonviable monitoring are similar in the goal to find the number of particles present in an area, they are not concerned with the same types of particles. Local conditions will determine the baseline or background numbers of particles. There is no definite relationship between viable and nonviable monitoring except to say viable organisms may be located on nonviable particles. Personnel may also transfer microorganisms from one place to another on their gloves, gowns, or on their tools. Dust particles, fibers, water droplets, aerosols and microscopic debris may provide a means of transportation for microorganisms. Organisms may be carried individually or in aggregates by air currents. It is the transfer, migration and elimination of the contaminants on which the microbiological monitoring is focused (1).

Abbreviations used in this chapter: FDA, Food and Drug Administration; HEPA, high-efficiency particulate air; LIMS, Laboratory Information Management System; MAS, microbiological air sampler; RCS, reuter centrifugal sampler; RODAC, replicate organism detection and counting; SAS, surface air system; SMA, sterilizable microbial atrium; SOP, standard operating procedure; STA, slit-to-agar; USP, *United States Pharmacopeia*.

PROGRAM

The goal of the viable monitoring program is to reveal the microbiological condition of controlled and critical areas and to assess the potential impact of the environment on the quality of the product. With a general knowledge of the numbers and kinds of organisms present, an effective program can be designed to ensure the effectiveness of the various control measures such as HEPA filtration, disinfection, gowning and aseptic handling procedures. Although there has been a recent de-emphasis of anaerobic organism monitoring, there is still a need to assess the risk related to their presence. This allows the design of a program that places anaerobes in proper prospective.

Controlled areas (Class 10,000–100,000, ISO 7–8, EU Grades C–D) are clean rooms with minimal product/-component exposure. Therefore, they have slightly higher allowances for the presence of microorganisms in the environment. They are supplied with HEPA filters having a 99.99% effectiveness rating for the capture of small particles. These rooms generally do not utilize unidirectional airflow, except in localized areas to reduce the numbers of nonviable particles. In these areas, there is little, if any sterile product/component exposure to the environment. This type of area may be used for preliminary product formulation with a closed system, storage of wrapped sterile components, bulk formulation areas, sterilizer loading rooms, corridors, air locks, or preparation or staging rooms.

Critical areas are supplied by HEPA filters rated at 99.99% or better, using controlled, vertical or horizontal airflow. They are Class 100, A or B. The entire room may have unidirectional airflow or it may have unidirectional airflow only under a module covering or surrounding the area where product/component exposure to the environment occurs. Some processes may be confined within isolator modules with limited access and independent air supplies and services. These areas include aseptic processing filling suites or aseptic bulk manufacturing rooms, a tunnel sterilizer outlet, and filling or primary barrier modules within controlled rooms.

Environmental monitoring of controlled and critical areas consists of a comprehensive viable monitoring program. The program should consider the following:

- The frequency of testing
- The locations or sites to be tested
- The direction of the airflow and class designation of the room
- The number of people needed to run the process
- The training and expertise of the people conducting the testing and the production process
- The amount of activity during the testing
- The time of the testing during the shift or the fill day
- The duration of the test when an air test is done
- The type of media to be used that will maximize recovery
- The incubation time and temperature of the tests
- The alert and action levels
- When to recommend corrective action
- When to do an investigation
- Several types of documentation and reporting of data
- Data management as well as dealing with unusual occurrences
- Isolate identification
- Written SOPs

An integral part of the program is the communication pathway so that management and supervision are apprised of all developments including the results of investigations.

ASPECTS OF MONITORING

Monitoring is intended to reveal as many contaminants as possible, but no sampling method used in this field is a perfect one. Not all methods of capturing, detecting and growing microorganisms are universally effective or equal. Data must be considered an estimate of what is present in total. Finding *any* contaminant must always be considered "indicative" of the presence of other unrecovered contaminants.

Air testing is the first of several monitoring techniques used to determine the quality of the aseptic processing environment. With a knowledge of the quantity and types of microorganisms in the air and knowledge of what was going on at the time the test was done, the potential effect of the organisms on the product/process can be assessed. One of the best ways to assess the potential impact is to perform viable monitoring tests during sterile media trials (filling process using sterile microbiological growth medium as a substitute for product). A correlation between viables found on an operator's gloves after a difficult manipulation when positives occurred in the trial-filled units can help substantiate risk or reinforce appropriate action levels. For example, a glove count of 8 cfu after a track adjustment, where the same type of organism was found in a filled vial, would indicate that a risk of contamination exists for that action. Conversely, a count of 200 cfu in the same situation where no positives resulted in the filled units does not necessarily mean that 200 cfu would be an appropriate action level. Additionally, during regular production, the collection of at least a portion of the product sterility test samples from the time period when routine monitoring occurred can be useful if assessment of any effect on the product is needed during action level investigations.

Surface monitoring is the second aspect of the viable environmental monitoring program. This includes testing of floors, walls, machines, equipment, and personnel. There are many ways to conduct surface testing but the choice must be balanced by selecting the volume of testing and test method that can be managed and that gives the most specific information about the process without interfering with the process. Surface testing results are believed to be more closely related to the effectiveness of cleaning and sanitizing methods than air monitoring results. People tend to carry and disperse most of the organisms found in a clean room by touching surfaces or shedding. The environment is controlled by systems designed for trapping in filters or sweeping with air. People touching surfaces or air forcing shed organisms onto surfaces where they cannot be captured by filters or where the air may be blocked by objects may concentrate organisms. Disinfectants and sanitizers used on these surfaces must then contact the organisms for the right amount of time and the techniques of wiping and mopping must be followed in order to reliably remove them. In a unidirectional and/or filtered air clean room or isolator module, organisms are more likely to be associated with the surfaces than with the air in the room.

People are the major source of the organisms present in a clean room and can spread microorganisms by inadvertent contact with contaminants left by others, inadvertent shedding or by use of inadequate aseptic technique. Personnel testing and qualification allows inferences to be drawn concerning the control of activities and aseptic techniques being used. People constantly shed skin particles, moisture droplets and hair, which serve as vehicles for the transfer of body flora into the clean room. Via their normal activities in the aseptic area, people then disperse the organisms. Gowns, uniforms, gloves, masks and shields are qualified for use and are then used to minimize the effect of operator activity. There should be a testing plan specifically targeted toward assessing the microbial profiles of garments worn by aseptic processing personnel. Profiles can be built if organisms are selected and identified from routine monitoring tests regularly. The people are just as critical as any piece of equipment, and much harder to control. That is why training programs should address techniques for aseptic handling, proper handling of clothing and implements. The concepts of contamination control and clean room behavior should emphasize how widespread microorganisms are in the environment and on the human body.

Frequency of testing is determined by evaluation of many factors in the filling or processing operation. The extent of product exposure and activity necessary to perform the process must be taken into consideration. In the experience of audited companies and in the period of time preceding the 1987 FDA Guideline, testing "at least daily" was an understood industry practice (2). Since that time, monitoring on a per-shift basis during each lot/batch has been the industry practice. In deciding how often to monitor air or surfaces, historical data will be very helpful in supporting the decision. The FDA emphasizes that the testing frequency must make "scientific sense" (the definition of which is left to the manufacturer). It would therefore be appropriate to monitor most frequently in the critical product/component exposure areas at the most critical product/component processing locations. Critical areas therefore require much more frequent monitoring than controlled or noncritical ones because of product exposure. This does not mean that a critical area should be ignored when no filling or processing activity is going on. The same quantity of samples are not required under nonoperating conditions, but the room should still be checked when it is not operating to make sure that no microbial buildup is occurring and to provide a background for comparison to operating conditions (3).

Aseptic filling and formulation areas are the most critical since sterile product or components may be exposed. Bulk manufacturing areas requiring aseptic transfers or aseptic additions are also of concern because of the usually large-scale handling and challenges in maintaining aseptic conditions on such a large scale. The production operation may require that operators work near the exposed sterile materials and/or surfaces. The air quality therefore may be directly related to product sterility. It is these areas that FDA regulatory inspectors focus closely upon. These are the areas where the manufacturer is expected to have the most control and therefore the most data. To assure that the process is adequately monitored it is necessary to choose an appropriate frequency, sample location, test method and media.

In establishing test frequency, 6 to 12 months of test data must be collected daily or more often while varying the time and shift of testing. This takes into account the following:

- Seasonality
- Shift relationships
- Cleaning frequency
- Day of the week
- Amount of activity
- Intervals between HEPA filter changes

After initial data analysis and evaluation of the criticality of the area, the room may have to be tested per shift or daily or once or twice a week. The goal is to be sure that adverse trends can be identified by the frequency chosen and that test sites will be indicative of conditions.

In deciding upon both frequency and locations to test, the person responsible should study the process, watch for activities potentially impacting sterile surfaces or materials and understand the flow of the process. Initially, many different sites may require testing until a picture emerges from the data, which indicates where the best or most indicative sites are located that do not present a risk of contamination from the testing process. In some instances, site selection choices may be very limited. In others there will be no obvious choice. An air test near a filling head may be the only appropriate choice for a small machine. Controlled areas where a product is enclosed may be limited to monitoring once a week to be sure that the sanitization program is maintaining control of the environment. If there are brief instances of product exposure, testing should coincide with that exposure or follow that exposure closely before any other activity occurs (i.e., clean-up or tear-down).

The activity and traffic patterns usually account for dramatic differences in data from different locations within one room. The FDA Aseptic Processing Guideline requires air testing within the "critical zone." These zones are the locations proximate to product exposure or closure operations. A critical zone could also be the outfeed of a tunnel sterilizer, or any other location, where sterile components or product is exposed even if this exposure occurs within a protective shield or under unidirectional airflow modules. In a bulk or formulation areas this could be the location of an aseptic transfer or aseptic addition. In controlled areas there may not be an easily identifiable zone more critical than the rest of the room. In the absence of a critical zone, survey of the traffic pattern and the activity zones will indicate where the potential sources exist.

Site selection for machinery surfaces is often harder than that for air sampling. The surfaces have to be contacted in order to do a test. This raises a concern that a critical surface may become inadvertently contaminated in the sampling process. It is important to be able to sample those sites most likely to cause or indicate product contamination if they become contaminated due to the activities associated with the process. It is often wise to identify "indicator" sites that are near but not in contact with product contact surfaces. These indicator sites can be selected such that disinfection or operator interaction there is much the same as the product exposure sites. The detection of contamination here "indicates" that critical sites are at risk and an investigation would be appropriate.

For assessment of room disinfection it would be necessary to perform a great deal of sampling to provide a statistically valid estimate of the true number of organisms dispersed on the room surfaces. Shelds and Chesky (4) stated that in an aseptic processing area, this sample would require 60 contact plates on a 100 ft^2 floor area. Since most labs are not set up to perform this level of sampling in each room, it is wise to sample traffic patterns or "operator locations" as indicators of room control and allow historical data accumulation serve as a guide. The monitoring data become even more important when a sterility test positive must be investigated. Persuasive evidence as to the cause of the positive is impossible to provide without good site selection.

Many manufacturers have the opinion thatmonitoring activities at or near critical surfaces during the process increase the risk of product contamination. These firms conduct surface sampling including actual product contact surfaces immediately after filling operations. For these results to be meaningful, no sanitation should take place between process completion and sampling. For multiple day fills where the system remains set up for several days, indicator sites must be used each day until the final day when critical product pathways are available for sampling.

Gowning rooms are transitional areas that help protect the clean areas by controlling the ingress of personnel via an effective gowning procedure and a traffic pattern progressing from less clean to cleaner. This is a primary point where personnel can adversely affect the cleanliness of their uniforms. Monitoring of this and other types of transitional areas such as air locks and pass-throughs will provide knowledge of the potential for clean room contamination. Once known, procedures can be designed to monitor and minimize the transfer of organisms into the clean areas from the gowning room.

The amount of activity during testing is important to the monitoring program. Many notices of negative observations have been issued by the FDA inspectors for air monitoring done only when a filling operation has been completed, or before it begins right after cleaning. Sampling in this manner may not be a true indicator of the potential risk of contamination during a product fill. The best time to monitor is during the process. This requires the most careful design of test sites and very well-trained testing personnel. However, "resting state" should not be ignored because comparison of filling hours to break or nonoperation times will indicate what personnel and activity contribute to the environment.

Time of testing during a shift should be considered. Depending upon the type of clean room and the amount of activity, samples taken at different times may be different. The count comparison may reveal that there is greater accumulation in some areas as the day progresses. This may not be true of every operation or type of clean room, but should be considered a factor that could influence the results. The monitoring should ideally be done to provide a sampling of various times.

AIR TESTING METHODS

There are many types of air sampling devices and methods available (5–13). Each one has advantages and disadvantages, which must be weighed within the context of the operation. Again, the familiarity of the decision makers with the production process is important to choosing the "right" system.

There are two basic types of air samplers: active and passive. Active samplers draw air in and collect organisms from it. Passive samplers such as the agar exposure plate depend on chance settling of an organism on its surface. Each type of sampler has an "area of influence" and "efficiency." The area of influence is the space around a sampler from which organisms are reliably collected. The efficiency is the number of organisms actually captured from a grouping entering the sampler. In especially clean areas, the efficiency should be reliable because so few organisms are present. Additionally, the area of influence should be known so location of the sampler can be established and supported.

The area of influence is hard to determine with a passive device such as a settling plate, but active samplers can draw microorganisms from several cubic inches of space around them. The area of influence is affected by the following:

- The design and configuration of the sampler (such as the size of its sampling port)
- The sampling rate and volume of air taken
- Configuration of the air pathway to and around the sampler

Efficiency depends upon:

- The nature of the collection medium
- The number of capture opportunities within the sampler
- Rate of sampling
- The method of capture
    - Impact on a collection surface
    - Trapping in a fluid or filter matrix
    - Sampling time

A big factor in the capture efficiency of a sampler is whether it can capture particles in a variety of size ranges (measured in μm) without missing those that may have sufficient mass to pass through the influence zone of the sampler without entering it. Because of differences in efficiency and design, samples taken with one type of sampler cannot always be compared to those taken with other types.

Measurement of efficiency is extremely difficult because the test assumes that every organism found downstream (escaping) a sampler under test can be captured by some other sampler. It also requires a reliable method for dispensing a known number of organisms in a test air stream for the sampler to pick up. These methods usually use aerosols of bacterial suspensions that are rather difficult to control. Assumptions must be made that the sampler serving as the "referee or standard" has a certain constant efficiency so the results of the sampler under test can be compared and expressed as 90% or 110% of the referee sampler's results. The best these methods can do is to indicate that a particular sampler may sample a higher or lower proportion of the organisms presented to it in a controlled situation. It should *never* be assumed that every sampling method can work with 100% efficiency.

Settling plates, also called gravity exposure plates, are generally considered qualitative indicators of air quality, but may have a quantitative ability in some applications, especially in very clean environments. They are simple open plates of agar medium. They are usually not used for the direct determination of air counts per unit volume of air, unless the speed, volume of air, and exposed surface (container open area) are predictable and consistent for each sampling event.

William Whyte, in an article that appeared in the 1986 September/October *Journal of Parenteral Science and Technology* (14), presented one of the most complete discussions of the subject. His studies revealed that the fears of agar dehydration are not as significant as was suspected. Exposure for 24 hours in still air dehydrates agar by only 13%. This takes six hours in a laminar flow hood and one hour at a rate of 1 ft^3/min of concentrated air blowing directly onto the plate. This loss in water content correlates to only an 8% loss of viability in test organisms. However, he concluded that these plates must be exposed for extended periods in order to maximize the probability for them to accurately estimate airborne concentrations of organisms (14).

Whyte continued, when the airborne concentration falls below 0.5 cfu/ft^3 of air, no organisms can be detected during a 30-minute exposure because the probability of them settling is also influenced by the amount of space separating them. If the airborne concentration drops to that recommended in the FDA Guideline (0.1 cfu/ft^3) a plate would have to be exposed three hours to have one colony settle on its surface. Recently there have been proposals to drop this guideline to a lower level making these plates even less sensitive. In laminar airflow areas localized concentrations may or may not be detected due to plate location. According to Whyte, given equal sampling time, a settling plate is 17 times less efficient than an active sampler. However, the ability to leave them exposed for extended lengths of time can compensate for their lower efficiency and make them a convenient and useful tool. Today's clean rooms are so well controlled that the differences in the results by either method are hard to find.

STA samplers are among the easiest to use of the active samplers. Air is drawn by vacuum through a thin slit and is impacted on to the surface of a 150-mm agar plate. The rate of airflow can be controlled with a flow meter to assure a measured volume of air is sampled. The plate rotates for a specific amount of time allowing a time-related distribution to be observed on the plate. They are quantitative, easy to maintain, and can be equipped with a probe for remote or small areas. Probes should only be used when absolutely necessary due to possible loss of efficiency to the additional surfaces of the probe and the total length should be the shortest possible. STA samplers are not easily portable if purchased with self-contained vacuum pump. When probes are used, their efficiency must be assessed as a possible variable.

The original version (no longer in wide use as of this writing) of the Biotest RCS was found by some researchers to be the most efficient of samplers. Although very useful, it came under question for being selective toward certain particle sizes. An article by Saul Kaye in the September/October 1988 issue of the *Journal of Parenteral Science and Technology* (7) presents a persuasive case for using care in the selection of this device as a primary sampler. It had the largest area of influence and it is very easy to use, easily portable, medium priced, and very easy to clean and maintain. Two aspects of the unit made it slightly more difficult to handle; the angle of the fan impeller blades must be frequently checked and special plastic strips must be used for agar medium. Because of the thin layer of agar in the strips and the high rate of airflow through the unit, the agar is very prone to dehydration. Many users have communicated that the impeller exhaust may disturb laminar airflow if not carefully placed in the clean room (8).

Biotest redesigned the RCS sampler and the Biotest RCS Plus™ was introduced in 1991. It has an air pattern less disruptive to laminar flow areas, a streamlined shape and programmable sampling time and rate. The exhaust from the sampler is directed away from the area of influence. It has many calibration features to make it easier to maintain and digital displays for settings. It can be set to sample a predetermined volume of air in ft^3 or L^3 or can be set to sample for a predetermined amount of time. It uses the same plastic agar-filled strips as the original inserted into the drum-like sampling head. The entire head of the agar strip holder now rotates at the speed of the sampling thereby reducing the turbulence of the former capture technology. It is self-contained and as portable as its predecessor.

Liquid impingers, the original being dubbed the Greenburg–Smith liquid impinger, function by bubbling the air to be sampled through a liquid medium intended to capture the viable organisms. The liquid generally used is an isotonic solution of saline, phosphate buffer or sterile water. The liquid must then be membrane filtered, the filter cultured on agar or media pad and the resulting organisms counted. With this method there must be an accurate way to control the flow of air through the unit and a timer. A limiting orifice can be used to control airflow but the source of vacuum must be held within a tight range for an orifice to be truly effective. This method has moderate efficiency and more prone to adventitious organism interference because of the high number of handling steps necessary to process each sample. It is however very cost-effective and relatively simple to set up. The standard deviation of samples taken in this manner is usually much higher than the other samplers because clumps or aggregates of organisms are easily broken up and dispersed by the liquid and they may spread on the growth plate and be counted individually. These units are not widely used and may only be appropriate for small operations.

The Anderson sampler collects organisms by impacting them onto the surface of agar plates arranged within a series of air sieves stacked in stages. A vacuum source provides the method of air acceleration. The entire sample consists of adding up the number of organisms collected at each stage where the sieve holes become smaller and smaller. This sampler yields a great deal of information concerning the sizes of particles carrying viable organisms because it sorts them by size. It is very efficient at trapping particles because a particle making its way through the stages has many opportunities to hit and stick to the agar. Drawbacks of this unit are that it needs a vacuum source and a timer, which is very expensive and is neither easy nor quick to assemble. Anderson makes six-stage, two-stage and single-stage samplers. The single- and two-stage samplers are much easier to handle but have a lower sampling efficiency. The pores of the sieves may become plugged when used near powder filling operations because sieve stages have very small holes. The main advantage of the six-stage Anderson sampler is that it has a very good efficiency (5,6,15).

The SAS, the air IDEAL and M Air T samplers are impaction devices. They have a cover plate with many holes arranged in a grid through which the air is drawn until it hits the surface of a RODAC™ plate of agar, standard petri dish or a prefilled media cassette. The airflow is generated by impeller blades below the plate. The units are battery operated and can be attached to a tripod for sampling. They have adjustable (programmable) sampling time and rate. Users report localized desiccation (visible shrinkage appearing as indentations) of media in the SAS where the air is impinged through the small holes and growth then shows organisms piled in small spots. This makes interpretation difficult in areas with high counts. The SAS unit has an exhaust flow that interrupts air near it so the selection of sampling location is important. The SAS sampler can be carried and positioned on its back, side or end, or it can be held by the user for sampling. Its exhaust air pattern is away from the sampling site.

The gelatin membrane system relies on trapping viable particles in a gelatin matrix by means of electrostatic forces. Particles become embedded in a very porous structure. The membrane is then laid on regular agar and it joins with the agar by melting together with it to become the growth media. As reported by Scheuermann (11), the method is very efficient but gelatin is very prone to dehydration and recovery drops to 55% with a one-hour sample. Equipment is available from Sartorius to control sampling rate and holders are provided to house the membranes during testing (11).

SMA samplers and the MAS are similar in principle to the Slit and Anderson samplers. Air is drawn though holes in the lid and impacted on an agar plate. The units are slightly larger than a petri dish and completely cleanable and sterilizable between uses. They are inexpensive to use due to the use of a standard size petri dish of agar for each test. Control units have vacuum control, vacuum pump and timer to allow sampling flexibility. Control modules can be located outside the area to be sampled. They can be fitted with a limiting orifice so that use of the control module is not specifically needed, provided the user can regulate the vacuum source and time the test. A battery-operated portable base is available for the SMAs that has all the features of the RCS and SAS control panels, as well as a sealed polished stainless steel exterior.

Comparison of efficiencies of different types of air samplers appeared in the *Journal of Applied and Environmental Microbiology* (9), July 1982. It compares the STA, Anderson, gelatin membrane, liquid impinger and a cellulose membrane method. It shows the Anderson sampler to be more efficient than the STA that is more efficient than the liquid impinger, etc. It also discusses the statistics of the data. An example is the high counts attributed to liquid impingers because of aggregate dispersion in the liquid. This results in higher mean and standard deviation for this sampler. This article does not discuss the newer versions of the RCS, SAS, SMA, MAS or single-stage Anderson samplers, but is valuable for its presentation of the method of comparison. Many of the newer units have not yet been compared by independent researchers, leaving users to evaluate the maker's data and perform comparisons on their own.

Duration of testing is another decision that must be made with the process design and microbial count history in mind. In general the sampling time must be long enough to encompass the normal range of manipulations in the room to be tested. It must not be too long or data will be diluted. It also must not be so short that it does not pick up a count where one could reasonably be expected. If the area to be sampled is a laminar flow critical area the FDA Aseptic Processing Guideline states that there should not be more than 0.1 cfu/ft^3 of air. The sampling time should therefore be long enough for the sampler to detect a level this low [e.g., if the sampling rate is 1 ft^3 of air per minute it would require 10 minutes to pick up one (1) colony in the sampler if the room just met the limit. Since 10 minutes is not usually representative of a process, 60 minutes might be more appropriate to get both a range of normal activity and provide time for the sampler to get a sufficient sample to compare to the limits]. Both the RCS sampler and the SAS samplers have short sampling times (e.g., eight minutes and five minutes). With these units more than one sample is essential to assure good representation of conditions. This must be done with care to prevent an increase in false-positive results from the extra manipulation of the sampling materials.

SURFACE TESTING METHODS

The selection of the type of testing to be done to monitor surfaces needs to be done early in the development of the viable monitoring program. Different methods may have different sensitivities or recovery characteristics and data generated by one method is not directly comparable to another. The choice must be made based on the number of test(s), manageability, ease of processing and the configuration of the sites it will be most desirable to monitor. Once these choices are made, changes in test methods may only be made with a knowledge of how one method relates to another or the historical data will not be meaningful. If a variety of methods will be employed, be consistent about where they are used because different methods may have different abilities and comparison may not be one to one.

RODAC plates are by far the most common surface sampling devices. They are 50 mm diameter agar contact plates where the agar has been poured in such a way that it forms a dome. The agar is pressed against a flat surface and organisms will stick to the agar medium. Approximately 3.99 square inches are sampled by each plate and this can easily be converted to the number of microorganisms per square inch. The efficiency has been compared to quantitative swabs with between 10% and 50% recovery of what is actually present on a surface. RODAC plates are available commercially from a variety of sources, or they can be made in-house. The trick to preparing them is to produce a good convex surface on the plate or it will not make proper contact. A similar product called a Hycon® contact plate is very similar to the RODAC, but square in design and contained in a peel open flexible plastic envelope. Hycon plates bend and offer this as an advantage over the rigid RODAC plate. Following the use of any contact sampler, the surface contacted must be disinfected to remove any residual agar.

Advantages/disadvantages of contact plates just about balance each other out. The disadvantages include leaving residual media behind, difficulty in pouring, cost if purchased premade, and they must be restricted to use on flat surfaces. Advantages are ease of use, low cost if made "in-house," no need for further processing after sampling and good recovery.

Cotton swabs or foam swabs wetted with isotonic solutions can also be used in a quantitative or qualitative manner. Qualitatively, they are excellent because they can be used to sample almost any accessible surface after which they can be placed into any liquid medium for culturing. Quantitatively, they must be relieved of the organisms collected by rinsing in buffer and then plating or membrane filtering the buffer so the organisms can be counted. Cotton swabs rinse quite well. Foam swabs tend to resist efforts to rinse out organisms. One aspect of both types of swabs that must be considered is the risk of leaving cotton fibers or foam particles behind after sampling. This particulate matter may pose a risk to product. Quantitative versus qualitative is the choice of the user, although information about numbers of organisms cannot be generated using swabs qualitatively. There also may be a tendency for aggressively growing organisms to overgrow others in the liquid media meaning that they may be completely hidden.

Quantitatively the recovery on swabs is similar to contact plates but they can be used on irregular or small spaces. They are however, convenient, inexpensive, easy and available. There is one type of swab that yields good quantitative results. Calcium alginate swabs combine the quantitative recovery efficiency of a RODAC plate with the ease of a cotton swab. After moistening the swab and sampling a surface, it is placed into a small quantity of sterile Ringer's solution. Sodium hexametaphosphate solution is added and the calcium alginate fibers dissolve. The resulting liquid can then be pour plated either partially or entirely to yield a microbial count. With this method, the whole swab is plated and there is no presumed loss of organisms. A drawback is that some organisms may not grow in the presence of calcium alginate and/or sodium hexametaphosphate. As a quantitative swab method, this is probably the best but has some processing steps which neither the contact plate nor the cotton swab require. In large operations, it may be impractical to try to process a large number of calcium alginate swabs (12).

There may be a place in the viable monitoring program for each of these types of samples; for example, RODACs for flat surfaces, qualitative cotton swabs for "go" or "no-go" situations and calcium alginate swabs for quantitative counts on irregular surfaces.

MEDIA CONSIDERATIONS

The type of media to be used is open to debate. The choice depends in large part on the types of resident organisms found in the area to be sampled. As a rule it is best to use a medium or combination of media that allow maximum isolation of a variety of organisms including bacteria, yeasts and molds. General-purpose media such as Soybean Casein Digest Agar, Standard Methods Agar, and Brain Heart Infusion Agar. will pick up the widest variety of organisms because they possess the widest variety of nutrients. With well thought-out incubation times and temperatures these media will generally produce the most thorough information. As part of initial method qualification selective media (Sabouraud's Dextrose Agar, Mannitol Salt Agar, MacConkey Agar, etc.) may be used to define the presence of organisms that will grow with more vigor on specialized media. Once the specialized media have been used to maximize detection of specific types of organisms, these organisms should now be transferred and checked for ability to grow on the chosen general-purpose media. If the growth occurs within the planned incubation time and temperature on the general-purpose media, it will not be necessary to include special media in the monitoring program. Where growth only occurs on special media, it may then be necessary to combine some special media where the target organism predominates with the use of general-purpose media in other areas (16).

Inactivators or neutralizers (e.g., penicillinase, Tween, and lecithin) may be needed in the media to neutralize the effects of bacteriostatic or fungistatic agents that may be picked up and transferred to the media in the process of the sampling. Some examples may be Tween and lecithin to inactivate the residues of disinfectants or preserved product aerosols, or penicillinase or cephalosporinase to inactivate/break down antibiotic powder. The need for inactivators must be researched within the context of the production operation and the quantity of inactivator added should be in excess of what is needed but not to such excess that it becomes inhibitory to microbial growth. It is a potentially citable offense to fail to add inactivators to microbiological monitoring media where residues of a bacteriostatic agent are likely to enter the media.

Incubation time and temperature are factors to be weighed and validated so that the media chosen will recover and grow the range of organisms present in the monitored area. The selection of media type will also include incubation temperature. This is done by observing the time required for the range of organisms found by extending the incubation period and observing the plates often until no new colonies are found. These organisms can then be selected and reinoculated at low levels onto the agar to confirm the growth temperatures and time needed to assure recovery. Based on hospital techniques used for decades, general-purpose media will usually recover the bacteria of concern to human infection at incubation temperatures from 35°C to 37°C. Since a large number of these organisms cannot exist outside the human body due to nutritional limits, ability/inability to tolerate air, exposure to sunlight, drying or other physical reasons, it may not be practical to expect many of these types to be present in the environment. It is more practical to expect organisms tolerant of these stresses and readily recoverable from the area to be monitored.

Molds and yeasts are usually incubated at lower temperatures, to minimize interference by aggressively growing bacteria in mixed cultures. When hotter temperatures are used, molds and yeasts will not be totally excluded. Many will flourish at 30°C and above in the absence of bacterial competition. Environmentally occurring fungi are usually adaptable to temperature, and so they are able to persist where less robust strains would not be likely to survive.

Environmentally occurring organisms can have wide ranges of tolerance for conditions of exposure and for temperatures at that they will grow. For these reasons, temperatures for incubation of microbiological monitoring tests need to provide conditions where most (recoverable) organisms can grow. Generally 27°C to 32°C for three to seven days will recover a wide range of organisms including bacteria, yeasts and molds. The FDA has issued notice of adverse finding for incubation times less than three days. It is sensible to incubate longer to compensate for the possibility that some organisms may be injured by treatment with disinfectants, or they may not be in their most ideal growing temperature or nutritional conditions. The time selected is a balance between the need to have information about contaminants in the environment quickly and the need to give the isolates adequate time to develop. It is highly recommended that microbiological monitoring plates be checked (scanned) for growth exceeding action levels, early in the incubation period even if the incubation of plates is still being observed for much longer periods. Early reaction can concentrate on taking corrective actions more promptly.

The reconciliation of the plates used for sampling and the plates incubated and read is an important activity although a largely administrative one. Plates must be 100% reconciled in both the sampling and incubation processes. All plates included in a monitoring session must be recovered and reported. Dropped plates must be clearly labeled and a process established for their handling prior to submitting for incubation.

All media, regardless of type, must be tested for its ability to support growth using a wide range of the types of organisms that could be found during the testing. Standard control cultures representing gram-positive bacilli, gram-positive cocci, a yeast, a mold and a gram-negative bacillus are good choices. These are the types the medium is expected to support. Many companies routinely include randomly selected environmental isolates in the growth promotion/

support tests prior to release for use. For tests where quantitative results are required, growth promotion should confirm quantitative recovery within a reasonable range of the control count. This can take the form of inoculation of the newly produced batch and a previous acceptable batch, with requirements that the counts on the new batch must be within 70% to 130% of the previously acceptable batch. A count range should be near that expected during use. Therefore, the growth promotion test should be conducted with as low a count as possible. Current USP guidelines specify below 100 cfu. The numbers of organisms typically found on clean areas are small and the detection of low numbers during growth promotion tests confirms that the media have sensitivity to these numbers.

ALERT AND ACTION LEVELS

Basic Definitions

Alert Level

A count that exceeds normal operating levels but does not adversely affect product quality. It serves to alert appropriate officials to a potential adverse trend in control.

Action Level

A count or trend that exceeds normal operating levels that could adversely impact on product quality and requires action on the part of the firm.

Alert and action levels are the heart of the monitoring program. Efforts are focused on detecting the number of organisms present so that these levels can be appropriately developed. The original (1987) FDA Aseptic Processing Guideline stated that "maximum microbial limits should be established along with a definite course of action to be taken in the event the samples are found to exceed the limits." This is the most important aspect of the microbial monitoring program—appropriate reaction to the data generated.

There are several methods for establishing alert/action levels and the philosophy applied should be chosen carefully and based on historical data. Review applicable guidelines within your regulatory authority area of the world. With the existence of a stated guideline, compare the apparent capability of the area being monitored to the guideline. Generally, the choice for air counts is not as open as that for surface testing. Any count higher than a suggested level would require justification and should not be tolerated if this justification cannot be reasonably supported by specific data.

For surfaces, most guidelines are silent on alert and action levels. The choice should be made by reviewing approximately 12 months of data for environmental monitoring counts during production, media fills (process simulations using sterile media in place of product) and sterility test results. If sterility and media fills have been satisfactory, alert and action levels can be set. Before the accumulation of these data, it is wise to consider results of other companies engaged in the same functions and to review written matter on the subject. The procedure should begin to extensively sample the production area over a one- to two-month period of time. The monitoring should generate a historical database from which decisions can be made regarding frequency of monitoring and sites to be checked. The sites tested should include all those normally monitored in the program including the personnel. In addition, other sites should be included to provide a complete picture.

Alert levels for process surfaces are currently (in industry) at 1 to 2 cfu per test with action levels around 3 to 5 cfu per test. These numbers were the basis for a chapter in the USP <1116> for some time. Interim levels used in a developing monitoring program should be conservative ones, which may eventually be supported by actual collected data. Analysis during the first year of data collection should be frequent, probably every two weeks to one month, to assess the progress toward establishing and maintaining acceptably low levels. Most of the values in a well-controlled aseptic processing clean room should be 0 cfu per test.

It is important to relate the alert level to the action level by, for example, setting the alert level at the mean plus one standard deviation, and an action level at the mean plus 2. With more than 300 data points in an analysis, this method is roughly equivalent to the nonparametric statistic explained below and usually results in the same action level. The other method, probably the more practical one for data that are not considered statistically "normal," is to rank the data so that the alert level is just below the action level with the action level at the point where 95% of the counts are below it and only 5% exceed it. [For example if 100 tests were grouped; 85 tests (85%) had 0 cfu, 7 (7%) had 1 cfu, and 3 (3%) had 3 cfu, then 95% of all counts are less than 3 cfu per test. Five tests (5%) had numbers higher than 3. The action level would be 3 and the alert level would be 1 or 2.] This minimizes the effect of a few high counts on the data. Allowances for trends should be made by generation of notification when the alert level is exceeded on two or three consecutive tests. [Other available methods have been outlined by Wilson and Patterson (17).]

With numerous sites to monitor, any of that may have a slightly different relationship to the potential for contamination, there is often the need to have various levels of importance attached to each type of site. An alert/action plan may be multifaceted for this reason. Each level should be appropriate for the area (criticality) where the test was done. Individual action levels for every individual site are largely impractical. A grouping method for similar locations may work well to trim the number of levels. Once grouped, each site or similar groupings of sites (e.g., floors) should have a single occurrence alert and/or action level. Other groupings would be walls, critical filling machine sites, personnel gloves, and personnel gowns within the aseptic filling area. There may be other levels for testing in nonfilling rooms, where the principle reason for the test is to monitor the effectiveness of the sanitization, not because the products are actually filled there.

Other alert and action level should include designations for the number of sites that show growth during a test event in one room, compared to the number of sites tested (e.g., 25% of sites show growth—*alert*, 50% of sites show growth—*action*). There should also be alert and action levels established for trends. For example, two consecutive shifts or tests exceeding the alert level in the same filling room might mean it is time for action. Finally, there should be a provision for the occurrence of an unusual or atypical organism. The alert and action plan should react to the occurrence of organisms that are not "normal" or "resident" to the area. The detection of a type never before seen in the area could signal a control problem or temporary lapse. An investigation would ensue as a result of finding a "rogue" organism, regardless of the number found.

Reaction to exceeding alert/action levels may take many forms that may include examining and retaining all organisms found, quarantine of the area, intensification of disinfection, extensive environmental study, additional product sterility testing, a product back challenge with the organisms found, HEPA filter integrity check, or product rejection. Reaction to the development of a trend might also include retraining of operators or requalification of the process through media fill trials. These actions then need to be fully summarized in an investigation report that documents the investigation and the actions taken. It informs all responsible officials about what happened, how it happened if that can be established, what corrective action is planned, and what disposition is recommended for the product involved. The criticality of the area must be carefully considered in making such recommendations (13).

Testing controls, including records for viable monitoring media and equipment preparation, analyst training, media controls, analyst personnel testing and a complete recording system for the analyst's activities and observations should be designed into the process. The testing results are essentially verified by the acceptability of the controls. Investigations can be greatly simplified by good information regarding the adequacy of the performance of the testing and by recorded observations by the analysts who perform it.

REPORTING

SOPs must be written to cover all aspects of the monitoring program from media preparation to recording and reaction to the test results (18,19). They need to be kept up to date and when revised, the superseded copies must be maintained. The program may be presented as one document or may be organized into a grouping. If grouped, all should be cross-referenced.

Data reporting should include space for all pertinent information such as dates, rooms, sample sites, sample time, tested by, reported by, control information, checked by, some method to track the form, a judgment blank for comparison to alert or action levels, and who should be notified. This form should be directly traceable to exactly where and when the data were collected, and it should be linked to the investigation that was done. Investigation forms, if used, need to include space for all relevant data, follow-up testing, status of the room or area with explanations/comments and recommendations as to product disposition.

Data must be recorded and presented in such a manner that trends can easily be identified. Trend analysis is expected to be an active and ongoing process. In this regard data should be available for side-by-side review. Periodically, the same kind of data review as was originally used to set alert/action levels should be repeated to verify the continued applicability of the levels. Do not hesitate to lower action levels if the data indicate that they should be. Although difficult to contemplate when alert and action levels are already low, history is important. Lower action points may be completely appropriate when history supports them. If low numbers are normal to an operation, actions should be considered when there is an abnormal elevation or sustained small elevations. Beware of increases to levels unless data clearly indicate that the original levels were too low and that it does not represent a loss of control. Data management is best and most efficiently handled with a computer. When a computer is used, subtle connections can be made or long-term trends can be identified. A computer allows quick analysis of a large volume of data. Due to the complexity and the unique concerns of each individual manufacturer, there are few "off the shelf" programs for this function. Many firms have incorporated individual tests into LIMS systems and written special reports to organize and evaluate the data.

More than one test day should be able to be placed side by side and in consecutive order to reveal a trend. Being able to recall all unsatisfactory tests from a period of time, all tests from rooms with similar processes or all rooms that exceeded a certain type of action level is a plus. This can be a real help in analyzing situations or relationships. These data can be used to keep track of current levels compared to those originally used to set up alert and action levels. The system whether manual or automated, should be able to track, for purposes of applying action levels, an individual, room, test site or grouping of these items quickly and accurately. Data management should allow more subtle aspects of the data to be analyzed. A manual system should have an accurate filing system, notification system, tracking system and someone specifically assigned to review it carefully and keep it up to date.

MONITORING THE UNUSUAL

Unusual circumstances or extended mechanical breakdowns including power outages can all occur. The responsible person or crisis team must then quickly decide what data must be collected and what corrective action is needed. This includes analyzing the extent of the problem, the vulnerability of product or environmental control, need for a quarantine, advice on how to regain control and how long that will take, planning follow-up testing and providing notifications of everything that was done or needs to be done. Follow-up testing must be designed to generate the information concerning the correction of the problem and should clearly demonstrate that the area *is* or *is not* back in control. Documentation of investigations into unusual occurrences is very important. Failure to adequately investigate and follow-up an unusual, unexpected event can produce considerable delay and questioning during inspections.

Handling of planned shutdowns are the same as for unusual occurrences except that there is an opportunity to plan how to handle them in advance. The documentation of sterility breaks and post-corrective action testing should demonstrate that control has been reestablished. A good Change Control or special SOP is a key element in handling planned breaks in aseptic control. Specific microbiological monitoring may be required to demonstrate that an area has recovered from a break in environmental control. A good notification system is the key to handling both planned and unplanned situations. Additional testing provides the proof of effective cleanup or control procedures.

MICROBIAL IDENTIFICATION

Identification of isolates from air and surface testing, although not required for 100% of organisms found, must be undertaken in some form so that "normal flora" and incidental organisms can be differentiated. With the increase in the amount of testing that has come about since the 1987 version of the FDA Aseptic Processing Guideline (2) has come an increase in the number of isolates. A sound ongoing program of identification will allow development of microbial profiles for the aseptic processing area.

The 2004 FDA Aseptic Processing Guideline devotes half a page to the importance of characterizing isolates. Organisms from microbiological monitoring tests that exceed the action level should be characterized if not identified to the extent necessary to determine their source (or probable source) and whether or not they were spread from an operator or other source. This usually requires a Gram stain and often a genus and species identification.

The use of automated identification systems or "kits" of biochemical tests can be valuable tools in this endeavor. What must be kept in mind with respect to these identifications is that most kits and automated databases are intended for different applications and use relatively few environmentally isolated organisms. Occasional inaccuracies and difficulties in identification can happen for this reason. The biochemical profile, what an organism will and will not utilize in its growth, are as indicative to its overall "characterization" as its actual name. It is sometimes impossible to accurately name some of the organisms found in the general environment. Systems are available for identification to the subspecies/strain level but may not make cost-effective sense for the ordinary monitoring laboratory. What is of the most importance is that the microbiologists investigating its presence can recognize it when it is present at other locations, and still track its persistence and its source in the clean room.

In order to be successful at characterization of organisms, a system to add to and monitor the types of organisms present in the aseptic areas is necessary. An appropriate method is the generation of profiles. From among the tests performed, whether over the action or alert level or not, organisms should be characterized from air, surface and personnel tests on a regular schedule. In the event of an investigation, the organisms may help to establish that ones are normal (indigenous) or unusual (adventitious). An automated identification system can speed the process and the laboratory monitor changes more quickly than manual identification.

Establishing environmental isolate profiles may impact on the cleaning and sanitization of the clean area such that deficiencies can be more easily corrected. All organisms found during monitoring should be retained, regardless of whether they are identified, until sterility testing has been completed on the product that was processing when the testing was done. Without these isolates a complete valid investigation in the event of a sterility test failure is not possible.

ANAEROBIC MONITORING

Monitoring for the presence of anaerobic organisms is a poorly defined subject. It is not current practice to conduct an anaerobic monitoring program. Requirements for anaerobic monitoring although few and vague are in existence in the context of "knowing your operation." Anaerobic monitoring should be done to keep an eye on the potential for contamination. No formal recommendations have been made regarding frequency so companies are left to their own devices. The most recent PDA survey on aseptic processing practices can be a source of guidance.

Establishing an anaerobic monitoring routine should be no different than that for the aerobic portion of the monitoring program except that anaerobes require special handling to ensure their survival. The first things to resolve are the handling methods to be used for the anaerobes to permit survival and recovery of these difficult-to-grow organisms. The procedures used should be based on proved techniques used in the clinical sector. The main problem with these organisms is validating techniques that maximize their survival. Practice on control cultures of the types that would reasonably be expected to survive if introduced to the clean area is a good first step. These would be spore formers such as *Clostridium* spp. or aerotolerant forms such as *Propionibacterium* spp.

Test methods available are RODAC plates containing reducing agents in the agar such as dithionate or thioglycollate and an anaerobic chamber such as a GasPak™ or CO_2 incubator to store the agar prior to and following use. Swabs may also be used with thioglycollate broth medium, cooked meat medium or prereduced peptone yeast broth. A GasPak or a clinical specimen transporter with a carbon dioxide atmosphere can be used to transport samples. Becton-Dickinson makes Vacutainer® transporter that is excellent for swabs. The choice between quantitative and qualitative methods is easier to make. Since the incidence of anaerobes is expected to be low, qualitative tests are likely to reveal the most organisms.

All aspects of the proposed method should be simulated in the lab. Once proficiency has been demonstrated it can be used to sample the aseptic areas. If few anaerobes are found, this procedure should then be continued on a monthly or quarterly basis to help reveal a pattern. If no obligate anaerobes are found, a quarterly frequency is probably reasonable. If many obligate anaerobes are present weekly or more frequent monitoring would be justified.

Identification of anaerobic organisms should also be practiced ahead of time unless the lab has proved expertise at performing these identifications. All suspected anaerobes from the initial testing should be identified. Access to a gas chromatograph can be a distinct advantage to identification because the production of organic acids in the growth medium is a well-defined identification process. [See the reference manual VPI (6).] Within the identification process true (obligate) anaerobes should be differentiated from facultative organisms by inoculating isolates from anaerobic collection media to aerobic media. Organisms that have the ability to grow on both media are facultative and would likely be detected in routine aerobic tests.

MONITORING THE STERILITY LABORATORY

Environmental monitoring of the sterility testing laboratory must be an integral part of the overall monitoring program. The methods used should be as nearly identical to those used in production as possible so that the conditions in the production area and the lab conditions can be compared. The current FDA Guideline stresses the need for comparability of the data. Despite this statement, labs may not be classified areas reasonably comparable to the production areas of the products they test. A sterility test may be declared invalid (falsely positive) only if the production environment shows a consistently high level of control and there is a clear problem in the testing laboratory at the time of the test. It is understood, but not always certain, that the lab quality should be at least as good or better than those found in production. When this is true, even the evidence of a laboratory problem will still make it difficult to declare a product sterile. For these reasons air testing, surface and personnel testing must be conducted in the sterility lab.

There are two dangers involved with false positives: (*i*) a product may be rejected, which is in fact sterile, and much worse is, (*ii*) a false positive may be declared and a contaminated product released. Generation of persuasive evidence must be done with knowledge of the routine microbiological condition of both production and the laboratory environments. This has made the declaration of a false-positive test extremely difficult.

Monitoring therefore must be performed frequently and at the sites of product handling. The problems surrounding investigating positive tests is the reason why many companies have gone to isolation technology and use Millipore SteriTest™ closed membrane filtration system for sterility tests. The validated environment of an isolator plus the occurrence of a positive test is a solid indication that contamination of the product in production has occurred.

In the laboratory, selection of sites to be tested is just as difficult to define as those in the production area. Unless a robotic sterility testing facility or an isolation unit is used, sterility testing is a manual labor-intensive procedure usually with many opportunities for adventitious contamination. The viable monitoring in the sterility lab must be designed and conducted to provide enough data to indicate where the environmental contamination is located and how many and what organisms are present.

As in production, analysts who perform sterility testing must be trained and qualified in a manner similar to that used for the production operators/ personnel.

The work flow should be well known to those designing the testing. Laboratory information along with production information will help in the sterility investigation of the product. As with the production monitoring program, historical data from the laboratory must be maintained and examined periodically. Twice a day is probably a good minimum frequency, though more acceptable would be in conjunction with each test.

As in the production area, it is unwise to perform too much environmental testing in the sterility lab because every increase in sampling or testing manipulation increases the risk of false-positive generation. Sterility testing personnel should be apprised of the results of the monitoring in the lab to increase their awareness of the importance of aseptic technique and to serve as a continuing education tool. The sterility testing controls plus the environmental samples provide good indicators of analyst aseptic technique and when taken as a group, provide a laboratory history to that future testing can be compared.

Specific test sites in the laboratory will vary but should include air tests in the areas of sample receiving, testing and incubation, and surface tests of the analyst's gloves, sleeves, testing bench (or isolator) and tests near sample receipt, floors, walls, incubators, gowning area and pass-throughs. Isolators with a proved sterility history can be monitored less frequently after the microbiological monitoring that supports the maintenance of their sterility has been completed. Glove testing is key information for isolators. This should be done appropriate to the volume of work a minimum of once each test day. If test volume is high, twice each test day may be more appropriate. Glove integrity has emerged as the more vulnerable and most likely failure point for isolation sterility testing.

Alert/action levels need to be comparable to those in the aseptic processing area. Reaction to an alert/action level being exceeded may necessitate an increase in disinfection frequency, a change in disinfection agent, HEPA filter integrity check, requalification or awareness discussion with personnel or an adjustment in laboratory routine. All data for the testing laboratory must be documented in the same manner as production data.

Other testing within the sterility laboratory may include periodic test simulations to check the technique of analysts, the use of open tubes of sterility test media in the testing zone, and sample container swab tests or surface tests. Simulations can reveal unseen problems with analyst technique and open media tubes can reveal the presence of contaminants in the testing zone. Swabs or surface tests of the exterior of sample containers will reveal the microbial load arriving with the sample (not all filled samples are collected in a sterile manner and this is a risk that must be known to the sterility lab). It may reveal the effectiveness of steps taken prior to testing to decontaminate sample containers.

Identification of isolates from laboratory monitoring is as important for the laboratory as for production. Where the lab and production are located under the same roof profiles are often identical. This is more the rule than the exception and it complicates the investigation when an organism common to both areas is found in a product sterility test. The historical trends are therefore important again. There are few situations where data are clear cut, it can only help in assessing the validity of a sterility test.

CONCLUSION

The objectives of a viable microbiological monitoring program—determination of the number of viable microorganisms present in an area, where they are best detected, what organisms are associated with the equipment, air and people, and how they can be controlled—all combine to help us build quality and safety into parenteral products. The many aspects of test selection, historical data, alert and action levels and good program management reinforce the fact that these are crucial issues for sterile pharmaceutical manufacturers and ultimately to the consumers. By establishing a sound program, equipped with appropriate follow-up procedures, the products produced can be assured safe.

REFERENCES

1. Whyte W, Bailey PV, Tinkler J. An evaluation of the routes of bacterial contamination occurring during aseptic pharmaceutical manufacturing. J Parenter Sci Technol 1982; 36(3):102–7.
2. U.S. Food and Drug Administration. Guideline on Sterile Drug Products Produced by Aseptic Processing FDA, 1987.
3. Munson TE. Regulatory View of Environmental Monitoring. Rockvill, MD: Food and Drug Administration/ PDA Meeting, 1988.
4. Shelds C. Validation of sanitization procedures in aseptic processing. Presented at Parenteral Drug Association Meeting, New York, March 1986.
5. Delmore RP, Thompson WN. Comparison of air-sampler efficiencies. Med Device Diagn Ind 1981; 3:45–8 (see also 53).
6. Jones W, Morring K, Morey P, Sorensen W. Evaluation of the Anderson viable impactor for single sampling. Am Ind Hyg Assoc J 1985; 46(5):294–8.
7. Kaye S. Efficiency of biotest RCS as a sampler of airborne bacteria. J Parenter Sci Technol 1988; 42:147–52.
8. Ljungquist B, Reinmuller B. Some aspects on the use of the Biotest RCS Air Sampler in unidirectional airflow testing. J Parenter Sci Technol 1991; 45(3):177–80.
9. Placencia AM, Peeler JT, Oxborrow GS, Danielson JW. Comparison of bacterial recovery by Reuter centrifugal air sampler and slit-to-agar sampler. J Appl Environ Microbiol 1982; 44(2):512–3.
10. Russell MP, Goldsmith JA, Phillips F. Some factors affecting the efficiency of settle plates. J Hosp Infect 1984; 5(2):189–99.
11. Scheuermann EA. The Gelatin Membrane Filter Method for the Determination of Airborne Bacteria, Sartorius Corporation, Department of Application Techniques, 1972.
12. Scott E, Bloomfield SF, Barlow CG. Comparison of contact plate and calcium alginate swabs of bacteriological contamination of environmental surfaces. J Appl Bacteriol 1984; 56(2):317–20.
13. Selection and Use of Aeromicrobiological Samplers. In Cleanrooms, March 1995.
14. Whyte W, Niven L. Airborne bacteria sampling: the effect of dehydration and sampling time. J Parenter Sci Technol 1986; 40(5):182–8.
15. Jensen PA, Todd WF, Davis GN, Scarpino PV. Evaluation of eight bioaerosol samplers challenged with aerosols of free bacteria. Am Ind Hyg Assoc J 1992; 53(10):660–7.
16. Noble NH. Validation of Environmental Monitoring Methods and Levels, presented at March 1992 Meeting of PDA, (Published in J Parenter Sci Technol; 1993).
17. Wilson P, Patterson D. PDA Course Notes of Industrial Environmental Surveillance and Control, May 1995.
18. Powitz RW. Biological Contaminants Strategies of Investigation and Control, in Cleanrooms, January 1997.
19. PDA. PDA technical report 24: current practices in validation of aseptic processing (1996). PDA J Pharm Sci Technol 1997; 51:S1.
20. Kaye S. Efficiency of Biotest RCS as a sampler of airborne bacteria. J Parenter Sci Technol 1988; 42(5):147–52 (review article).
21. Lundholm IM. Comparison of methods for quantitative determinations of airborne bacteria and evaluation of total viable counts. Appl Environ Microbiol 1982; 44:179–83.
22. Halldeman LV, Cato EP, Moore WEC. Anaerobe Laboratory Manual. 4th ed. Blacksburg, VA: Virginia Polytechnic Institute, 1997.

BIBLIOGRAPHY

Abdou MAF. Determination of airborne microorganisms in a pharmaceutical plant using standard, elective and selective culture media. Pharm Technol 1980; 4(11):93.

Andersen AA. New sampler for the collection, sizing and enumeration of viable airborne particles. J Bacteriol 1958; 76:471–84.

Anonymous. Environmental monitoring considerations for biological manufacturing. Biopharm 1996.

Benbough JE, Bennett AM, Parks SR. Determination of the collection efficiency of a microbial air sampler. J Appl Bacteriol; 74:170–3.

Buttner MP, Stetzenbach LD. Evaluation of four aerobiological sampling methods for the retrieval of aerosolized *Pseudomanas syringae*. J Appl Environ Microbiol 1991; 57(4):1268–70.

Dell LA. Aspects of microbiological monitoring for non-sterile and sterile manufacturing environments. Pharm Technol; 3:47–51.

Hambraeus A, Benediktsdotter E. Airborne non-sporeforming anaerobic bacteria. J Hyg Camb 1980; 84(9):181.

Hussong D, Madsen RE. Analysis of environmental microbiology data from cleanroom samples. Pharm Technol Aseptic Proc 2004; 10–5.

Jaschhof H. Sampling virus aerosols using the gelatin membrane filter. Biotec Microbiol.

Jones W, Morring K, Morey P, Sorensen W. Evaluation of the Anderson viable impactor for single sampling. Am Ind Hyg Assoc J 1985; 46(5):294–8.

Lach V. Performance of the surface air system samplers (SAS). J Hosp Infect 1985; 6:102–7.

Lu A, Prusia M. Reliability of the settling plate method in monitoring laminar flow benches. Am J Hosp Pharm 1980; 40.

Powitz RW. Biological Contaminants Strategies of Investigation and Control, in Cleanrooms, January 1997.

Satter SE. Biological and Physical Monitoring of a Controlled Environment, A_2C_2 magazine, September 1998:27.

Slater GJ, Johnston D. Microbiological monitoring during sterile product manufacture: adaption of methods for a research facility. Parenter Sci Technol 1988; 42(4):111–3.

Spendlove JC, Fannin KF. Methods of characterization of virus aerosols. In: Methods in Environmental Virology, New York: Marcel Dekker Inc., 1982.

Vincent DW. Validating and establishing a routine environmental monitoring program for clean room environments. J Validation Technol (reprint not dated).

Whyte W. Settling and impaction of particles into containers in manufacturing. J Parenter Sci Technol 1981; 35(5):255–61.

Whyte W. A multi-centered investigation of clean air requirements for terminally sterilized pharmaceuticals. J Parenter Sci Technol 1983; 37(4):138–44.

Whyte W. The influence of clean room design on product contamination. J Parenter Sci Technol 1984; 38(3):103–8.

Whyte W. Sterility assurance and models for assessing airborne bacterial contamination. J Parenter Sci Technol 1986; 40(5):188–97.

Whyte W. In support of settle plates. J Parenter Sci Technol 1996; 50(4):201–4.

27

Validation of Container Preparation Processes

William G. Lindboe, Jr.
Consultant, East Brunswick, New Jersey, U.S.A.

INTRODUCTION

Pharmaceutical products are more than formulations, the product must be supplied in a primary package system that protects its quality from time of manufacture until ready for use, and the proper delivery of the product may be supported by the packaging system as well, with regard to dose, safety and other attributes. This is true of all pharmaceutical dosage forms, and while it is of particular importance for sterile products, the proper preparation of containers is essential to all drugs. This chapter will review the validation concerns for primary packaging components, specifically containers and closures used in the packaging of sterile and nonsterile products prior to filling. The treatment of this topic in earlier editions of this text (1) addressed the developmental and selection physical and chemical attributes of polymeric closures, and the presterilization processing of closures and sterile glass containers. This chapter expands and updates the types of materials to include plastics, and explores nonsterile dosage forms, while only referencing sources for developmental and material selection information. The approach excludes traditional autoclaving, dry-heat depyrogenation, and online postfill container integrity testing technologies. These topics are addressed elsewhere within this volume or in widely available references. The relationship with packaging component suppliers for both quality and outsourced preparation will be addressed. The chapter will suggest methods for validating preparation operations from a practical viewpoint of the validation team and those functions directly supporting the validation effort.

The major trends in this area since the publication of the earlier editions have been the increase in the use of sequential computer-controlled operations and the continued expansion of the use of plastic containers as primary packages and delivery systems. In addition, the general trend in outsourcing of business processes has been mirrored in component preparation processes. Validation strategy has also changed in that a risk-based approached has been widely accepted as forming the basis of testing strategies. Specific strategies have emerged beyond the FDA Report (2) to enable a uniform logical approach to the subject. Procedures by Akers and Agalloco (3,4) provide the basis for companies to perform their own risk analysis for aseptic processing. Contract services also provide approaches to perform risk analysis, as described by Ricci and Fraiser (5). In either case, the selection of what and how to validate must be based upon sound logic and a determination of patient risk. The advantage of this approach is that very unlikely events and events with low risk assigned to them can be confirmed by proportionate measures. Risk associated with outsourced component preparation processes must be reduced by documented in place quality relationships with suppliers. Container preparation operations follow the major subdivision designations of sterile and nonsterile with the latter being historically trivial compared to the former. Recent advances in aerosol and inhalant administration have placed additional criticality on nonsterile packaging components. For sterile primary packages, compendial requirements for sterility (6), particulates (7) and pyrogens (8) must be addressed by container preparation operations. It should be noted that international specification may be more stringent, specifically particulate specifications in Japan, and the reader must identify all specifications prior to initiating any validation study. The "pyrogen-free" requirement is now widely interpreted for preparation process as rendering a "minimum of a three-log reduction." This practical quantifiable process result has the benefit of facilitating validation of washing and FFS polymer operations. Nonsterile primary packages typically address particulates, specifically dust from corrugated cardboard, within which the empty containers are typically shipped and stored. Functionality, both pre and postfilling, is the major concern for metered dose inhalers. The increased use of plastics as primary packaging material and the aseptic filling of these materials have brought about a resurgence in the nontraditional sterilization methods (9) that include the use of chemical and radiological sterilizing agents, and their associated technology. The reader is urged to consult the USP chapter (10) covering containers to obtain a background on the particular system to be validated.

The assumption is made that the normal sequence of installation qualification, operational qualification, and PQ or process qualification will be followed for facilities, equipment, and utilities. Protocols must verify personnel training, standard operating procedures, and the availability of validated analytical test methods.

Abbreviations used in this chapter: BFS, blow–fill–seal; BI, biological indicator; cGMP, current good manufacturing practice; CIP, clean in place; COC, cyclo-olefin copolymer; DEHP, di-2-ethylhexyl-phthalate; FDA, Food and Drug Administration; FFS, form, fill and seal; IV, intravenous; LAL, limulus amebocyte lysate; LVP, large volume parenteral; NIST, National Institute of Standards and Technology; NP, nonperishable; PEG, polyethylene glycol; PET, polyethylene terephthalate; pMDI, pressurized metered-dose inhaler; PP, polypropylene; PQ, performance qualification; PVC, polyvinyl chloride; QC, quality control; RABS, restricted access barrier system; SIP, sterilization in place; USP, *United States Pharmacopeia*; UV, ultraviolet; WFI, water for injection.

Additionally, a vendor qualification program including periodic audits and biobuden/pyroburden studies for sterile containers needs to be in place. The ongoing quality relationship with the component supplier cannot be overstressed, especially with outsourced preparation processes. The details of sterilization and dry-heat depyrogenation process validation are covered elsewhere in this volume and will not be repeated in this chapter. The reader must consult the relevant chapters covering those specific details. Development programs for the various components with glass often use the much less rigorous "grandfathered" approach. Newer polymers and associated technology will require extensive development and product compatibility efforts, especially with vendor-supplied "ready to use" components. The availability of diverse polymers and associated forming equipment make each validation unique to a certain extent. However, general requirements must be addressed and the reader will have to develop field validation approaches to the specific technology being employed.

The major focus of this chapter will be on the validation of cleaning, nontraditional sterilization and depyrogenation operations. In all cases, confirmation of function and physical properties must follow these preparation operations. This testing generally provides a good opportunity for the utilization of quality statistics such as C_{pk} and C_{pm} on large amounts of testing data. Typical testing levels are multiples of three to six times routine release testing and protocols should include multiple container lots across trials. Validated analytical procedures need to be in place for residual cleaning agents, lubricants, and the process challenges (spore suspension, dye marker, and/or endotoxin) for washing validation. The approach is similar to cleaning validation and the reader is advised to consult the chapter addressing cleaning validation within this volume for additional background and considerations. Specifications for the components to be received in nonshedding materials coupled with procedurally controlled unpacking/unwrapping operations are essential to control particulates. For sterile products, this should extend back to the off-site manufacture, especially for plastics and web rolls for FFS operations.

The rule of three typically applies to the validation of all processes. Continuous operations will have three different trials across shifts and personnel as well. Every component size and variety should be challenged and bracketing should include performing at least one trial on each component entity. Validation is not the place to cut corners, especially when manufactures are clamoring for end-product testing relief through parametric release. Additionally, in-place annual programs of revalidation for sterilization and depyrogenation operations are expected. This program involves a repeat of a portion of the initial PQ or process validation and a thorough review of production within the review period. Packaging operations with multiple functions are usually broken up for individual PQ protocols for each operation. The system as a whole is typically validated by a final process validation consisting of extensive end-product testing for nonsterile and sterile products. Media fills are also required for aseptically filled sterile products.

STERILE PRODUCTS

Glass Containers

Glass containers for sterile products have largely been reduced to small volume parenteral multidose, single dose syringes and ampules. Many new products are being introduced in plastic containers due to their lower purchase cost, greater strength and reduced shipping costs. Historically, half-liter- and liter-sized glass bottles used for LVPs necessitated considerable resource allocation to container presterilization storage and preparation. These sizes still exist and legacy production operations will confront the validation team. Low thermal expansion borosilicate glass [USP designation type I (11)] is used to accommodate thermal sterilization of the primary container for either aseptically filled or terminally sterilized product. Glass is extruded at temperatures in excess of 1000°C and, at that point, are both sterile and pyrogen free. It is the subsequent handling that poses the potential for contamination, although this is generally minimal. This should be confirmed with bioburden and pyroburden studies on received glassware within an ongoing program for each vendor. This will enable subsequent "bio" and "pyro" challenges during validation studies to be kept at a minimum in classified areas. This is especially true of processes heavily reliant on washing for microbial control. Other nonparenteral glass includes treated soda lime glass (type II), used for buffered aqueous solutions, soda lime (type III), used for oils, and type IV glass, so-called NP (non-perishable food designation), which has been largely replaced by plastic.

Glass Container Washing

Depyrogenation by washing, or dilution, has been effectively employed for glass primary packages for terminally sterilized LVPs. Modern continuous process washing and dry-heat depyrogenation in conveyer tunnels provides higher sterility and depyrogenation assurance. The batch-type dry-heat oven depyrogenation of large bottles was so inefficient, that washing alone is an attractive alternative. The vials or bottles are washed, rinsed, and dried in a continuous conveyer operation. The drying stage of this process provides little time improvement over large capacity depyrogenation tunnels. Companies with depyrogenation tunnels typically use it on all size vials and bottles. Validation protocols include depyrogenation studies and particulate evaluation of the washed glass containers. Recovery studies for endotoxin challenges need to be performed. Endotoxin standards must be reconstituted and allowed to air dry on the challenged units. Studies using lyophilized endotoxin challenges are inappropriate because this form of the challenge is more easily removed by washing. Challenged units must be marked to absolutely assure that they will not be used without detection in product for distribution. LAL is the analytical testing method of choice. Recovery studies for the analytical methods must be performed and reproducible recovery rates around 25% are common at the level of challenge utilized. Many firms have blanket general chemical recovery requirements of 75% or higher and these levels are generally impossible with this type of challenge. Company documentation must address this exception and protocols need to have the results of specific recovery studies. There is often resistance in bringing endotoxin or a challenge organism into a production area and thus challenges should be as low as practically possible to demonstrate the three-log reduction requirement, supported by bioburden, pyroburden, and recovery studies. In the 1970s and 1980s there was considerable resistance to bringing media into sterile areas for media fills in fear of encouraging microbial contamination. However, history has proven that in well-maintained areas, there is no problem and today, media fill validation of aseptic processes is the accepted standard. Validation resource planning should be apprised of all component requirements and budgets should accommodate challenge studies for each size container to be validated. The temptation to reuse the nonchallenge containers in subsequent runs to save money should be resisted since questions about the legitimacy of the study, especially for particulates, will be raised. A minimum of 10 endotoxin challenges should be used for any run, though higher numbers may be more definitive in proving process effectiveness. The challenges should be introduced to cover "worst-case" processing conditions of temperature and speed (cold and fast) and must include start-up, steady-state, and shutdown challenges. Replicate challenges at any time point should be used, so 10 challenges may be insufficient for all except the shortest runs. A more typical challenge number would be 50 to 100 containers.

Particulate testing consists of reconstitution with QC approved WFI and particle detection by visual inspection followed by counting using instrumentation. Microscopic examination of the dried containers may also be employed. Sample containers are selected following washing with the sample number approximately the same as the pyrogen-challenged containers. For those processes that have a thermal depyrogenation following washing, samples should be taken following the thermal treatment to assure the particles are not generated during the thermal processing. A check for broken heating elements should be made prior to each run for legacy oven and tunnel equipment to avoid failing for particulates.

The utilities servicing washing equipment must be qualified. These include purified water, WFI, clean steam, filtered compressed air, vacuum, instrument air, and electrical power. Validation protocols should include washer point of use testing and verification of these utilities if this testing is not part of the individual system qualifications. If it is included, or has been performed previously, the studies must be referenced in the washing protocols. Washer qualification protocols must verify temperatures, flow rates, wash and rinse volumes, processing speed, and confirm limits claimed by the equipment manufacturer and identified during commissioning studies. As a testimony to the longevity of washing equipment, the author has observed the reinstallation of legacy washers manufactured by Cozzolli and Metromatic from older facilities into brand new parenteral facilities. In this case, a full qualification is needed to ensure proper re-installation of the equipment. Nevertheless, these older systems can be demonstrated to reproducibly produce glass of appropriate quality for parenterals.

Thermal Processing

It has been said that if glasses were a new material there would be a hesitation in its use since it poses a safety concern if it is broken. That being said, the advantage of glass is its strength, inertness, barrier properties, and its ability to be subjected to high temperature (Class I containers). As mentioned above, the specifics of sterilization and depyrogenation are covered elsewhere in this volume. These techniques apply even in continuous combination washing and thermal processing equipment since the two operations can be addressed separately in targeted studies. Thermal depyrogenation of glass is ordinarily the process of choice, when available. The rigidity of glass must be taken into account during processing operations. Glass does not flex under pressure. This means that the closure must flex and be put under additional strain during pressure changes. The author can recall the inner septum of the stoppers on 12-mL glass vials being burst, even though the stoppers remained crimped, because of the lack of pressure control during a poststerilization cooling operation. The same conditions for flexible plastic container might not result in a "popped" stopper.

Siliconization

Silicone (polydimethylsiloxane) aqueous suspensions are typically applied to glassware as an aid to draining the container, for providing a good meniscus for reading volume graduations on the container or syringe, and to improve appearance (provide a "polish") to the container and liquid content. The use of silicone must be fully supported in the clinical and scale-up phases of drug development. The silicone, typically Dow medical grade, is applied as an emulsion in one of the washing stations, immediately prior to drying. It can easily be evaluated by visible physical testing (such as the water break test, distilled water meniscus, or aqueous dye adhesion) and must be inert to the product as established in the development studies. The important thing is to ensure that it does not interfere with depyrogenation washing studies. If there is any doubt, the washing stage applying the silicone can be disabled for the pyrogen challenge phase of the validation. The Parenteral Drug Association has published a Technical Report (12) on the use of silicone on parenteral primary packaging components. An analytical method using infrared spectroscopy on silicone extracted using refrigerant 113 is described therein. The validation team is usually concerned with the qualitative presence of silicone on the vials beyond development and, for smaller containers, a composite sample must be used. Regardless of the analytical method or the type of component being washed, protocols must ensure that each nozzle position is sampled with representation of start-up and shutdown. A base sample of 25 vials from each nozzle for each of three validation trials is appropriate. Note that recharging reservoir levels, washer stoppages, and/or any maintenance activity must trigger additional samples that will increase the total number required.

Vials

Irrespective of the material of manufacture, vials have a variability that must be considered during preparation activity. Container washers and depyrogenation ovens and/or tunnels must be qualified for all anticipated sizes of containers. The small vials are generally the worst-case load in batch-type sterilizers since for a given volume of sterilizer there are more containers. Small vials can be the worst case for washing as well. Washing equipment generally consists of a series of individual processes; however, the nozzles must be appropriate for all sizes washed. Additionally, flow rates must be adjusted for the smaller vials as they may be blown off the cleaning nozzles. The proper change parts for this equipment must be confirmed for each vial type and size during commissioning, development, and qualification. Sufficient time must be present in the washing cycle for the containers to drain from the previous washing step. Inadequate rinse time can adversely affect both endotoxin and particle removal.

Ampules

Ampules differ from vials in that they are always considered unit dose containers, are flame sealed, and each container undergoes a container integrity test following sealing. Additionally, ampules are not coated with silicone during the washing process. Ampules require special washing equipment change parts since their openings are very small and they do not possess the mechanical strength or weight of vials. Dry-heat depyrogenation follows washing and is addressed elsewhere within this volume. All ampules are contaminated by glass fragments to some extent when opened. Considerable development effort is needed to assess the effectiveness of the scoring equipment and glass quality to minimize this. Filling processes must also be carefully controlled to prevent moisture in the flame seal zone that can be a source or particles in the solution.

Syringes

Syringes, or rather syringe barrels, are treated in the same matter as glass vials. Becton Dickinson, Inc. is a large supplier to the pharmaceutical industry and various syringe configurations are available in both glass and plastic. Preprocessing will be addressed below; however, washing, siliconization, and depyrogenation of the barrels and stoppers are required. Extrusion force and break-loose force measurement are effective physical measures of the effectiveness of siliconization for syringes.

Tubes

Tubes (evacuated) are used largely for blood collection and may be considered a reverse syringe. Technically, they are medical devices; however, processing follows that of drug-containing syringes with the exception that they are not siliconized. The geometry of tubes requires specialized washing equipment as in ampules and syringes. Tubing is also used with IV kits associated with LVPs. In these instances, the kits are typically irradiated in a batch-type process. Irradiation sterilization is covered elsewhere within this volume.

Closures

The closures for sterile containers usually have some additional functionality in addition to providing a sterile seal for shipping and storage prior to use. This functionally could be the insertion of a syringe needle, connection to IV tubing, or in the case of ophthalmic solutions and syringe plungers, provide a means of administration. It is important that validation of the preparation operations for closures contains testing to assure this additional functionality is not diminished. As with containers, parenteral closures must be shown to be sterile, pass applicable particulate testing, and be pyrogen free. The latter quality is usually determined by testing a rinse solution after processing using the LAL test. It is assumed that a rigorous development process and vendor qualification program has occurred that has identified and confirmed all critical properties of the closures and that validated analytical methods, including recovery studies, are in place for the testing. Washing processes need to demonstrate a three-log reduction as in glassware washing.

Stoppers

Stoppers are typically synthetic rubber or other synthetic polymer that is injection molded. Suppliers of elastomeric components to the pharmaceutical industry provide technical support that is a valuable source of information for validation teams. The ready availability of technical support should be a major criterion for selection for any vendor. Although in-house development documentation is the primary reference for validation protocols, often they are truncated summaries and more detailed information can be obtained from the vendor.

As with glass containers, there is reluctance to bring in organisms and endotoxin into classified areas. The Huber batch-type stopper washer is a widely used design in the industry. The machines can be equipped with CIP and SIP systems that, when validated, enable appropriate microbial and endotoxin challenges for component washing validation to be effectively removed from the washer. The challenge should be kept as small as reproducible recovery studies will allow, while assuring the three-log reduction process requirement. As with glass, the endotoxin challenges or BIs should be air-dried on to simulate a natural bioburden. Use of spore strips or spores on a model matrix, rather than dried on spore suspension on the actual stopper, has been criticized as not representative of a real-world contamination. Well-marked stoppers should be used with a minimum of 10 challenged stoppers per validation run. It is best to test for residual detergent and silicone along with the BI runs. Stoppers are typically autoclaved after washing and evaluation after this stressed condition is necessary in extensive development studies. Newer technologies have emerged for closure treatments, for example as reported by Dublin and Witler (13). The process requirements are the same as those for the traditional batch-type unit operations and considerable ingenuity may be required to demonstrate this in protocols.

Stopper Washing—Particles

The reduction and/or more preferably the elimination of particles is the primary purpose of washing. Stoppers are typically rinsed vigorously with WFI with and without detergents with the final rinse solution being tested for particles. Additional testing for residual detergent is performed.

Stopper Washing—Endotoxin

The stopper washing process must also render the stoppers pyrogen free. As with glass washing, challenges should be air-dried on well-marked stoppers. Excessive washing can generate particles and an optimal balance must be found in process development studies. Pyrogen challenge levels are typically higher than the corresponding glass challenges because of more difficult recovery. Stoppers are challenged, washed, and then placed in solvent to recover the endotoxin challenge. Recovery is usually performed on composites of challenged stoppers because of recovery difficulties. Typical numbers of challenged stoppers are from 25 to 100 per washer load.

Stopper Siliconization

Stoppers are siliconized to aid in the operation of stoppering equipment and to facilitate positioning into the vial. Syringe stoppers also require extrusion and break-loose force reduction and reproducibility as necessary attributes that are provided by siliconization. Physical testing is generally used with a confirmatory composite analysis to confirm proper siliconization of stoppers. Sampling is required from each load of batch-type washers with samples taken from varying locations and incorporating stopper variations (size, formulation, etc.). The stoppers should be well mixed; this is necessary to ensure stoppers in the middle of the washer get the same treatment as those on the perimeter. Overloading of washers is a common manufacturing error and equipment manufacturer's recommendations for proper loading levels must be confirmed. Evaluation of different stopper lots of the same design is essential to confirm that the process is reproducible and tolerant of variations in the components.

Droppers

Droppers are drug-delivery devices that consist of a barrel and functional closure. In some cases, the entire primary package can be considered a dropper, as in unit dose and multidose ophthalmic packaging. Components are typically washed to eliminate particulate and pyrogens. Dropper glass is rarely borosilicate high temperature resistant and development studies must be in place to show that washing reduces particulates and depyrogenates effectively to a three-log reduction for ophthalmics and other solutions purporting to be sterile. Batch-type washers are generally used and validation sampling and testing generally follows that of vials and stoppers.

Inhalers

Inhalers typically consist of sterile plastic components that are molded in-line. Traditional batch-type sterilization or in-line radiation is used. Cleaning requirements are minimal as the initial particle load on components is the responsibility of the component supplier, which usually is a qualified medical device molding company employing cleanrooms for processing.

PLASTICS

Plastics are polymers that often have the ability to be formed and reformed by the application of heat and pressure. The reader can obtain a good review of plastics used in pharmaceutical packaging in Remington (14). Additional background can be found in the Parenteral Drug Association Technical Report (15) on Sterile Pharmaceutical Packaging. The trend toward plastic "bags" for the LVP primary packages can be summarized by the words convenience and cost. Ease in handling soft bags and IV additives, both in manufacturing and within the hospital, have been augmented by unique closure and dosage administration developments that would be difficult or impossible to achieve with glass containers. The flexible plastic bag of IV solution can, in some cases, have flow initiated by squeezing and then collapse on emptying without air entry or displacement. The development of diverse polymers, copolymers and laminates with associated forming technologies has facilitated the transition to plastic. There have been studies in the food industry showing a 50% packaging cost reduction of plastics over glass and metal cans without the higher weight, breakage or deformation. However, autoclavable food pouches consisting of PET, aluminum foil, PP laminates can be up three times as expensive (16). It must be remembered that molding and converting facilities for polymeric materials are not always operated with full cGMP. The major advantage of this technology is the ability to move the forming operation within controlled environments adjacent to the filling. With the convenience of plastics comes the loss of the inertness of glass. Polymers can contain nonreacted monomers, plasticizers, preservatives, UV stabilizers, mold release agents, and lubricants. The PVC IV bag is being phased out because it contains a harmful leachable plasticizer, DEHP. Incineration is a preferred method for disposing of medical waste and PVC can produce dioxin and acids when burned. Validation teams face the qualification and validation of replacement packaging systems and associated machinery as material changes are implemented. Another negative aspect of plastics is that the physical properties of the unmolded resin can change with time. These factors require that the vendor qualification for plastic container suppliers to be rigorous. Change control programs must be extended to the vendor. Subtle changes in polymer formulation from manufacturers can throw drug and device manufactures into a sudden panic and force long work hours for validation teams. The author recalls one instance where a device manufacturer purchased a 10-year supply of a particular plastic upon notification of the discontinuation of a particular formula. Problems with storage of the unmolded resin "beads" in the latter years of this inventory caused manufacturing problems. This resulted in a panic vendor and polymer change that could have been avoided by an orderly transition initially and early identification of a second qualified vendor and/or alternative polymer formula. Plastics generally pose a difficult challenge to the validation team regardless of the motivation for their use. A sometimes hidden problem with plastics is that extruders are frequently cleaned by "burning" them out and relying on a purge of the initial moldings of the next batch to clean the equipment. This results in sporadic fine black particulate matter that can play havoc with particle specifications for the containers and the filled product. Validation teams will often need to verify cleaning procedures back to the extruder.

Nonsterile plastic packaging is typically purchased from external vendors, in large web rolls of sheet polymers for FFS nonsterile and aseptic fill operations. Static charge on plastic components can be a magnet for all types of particulates, especially other plastics. Qualified filtered compressed air showers and/or washes may be needed to assure that the efforts to supply low particulate components by the vendors are not compromised by dusty in-house unpacking. In any event, unpacking operations should not take place in an uncontrolled environment. The typical approach is to have a staged unpacking moving to progressively cleaner areas using nonshedding, low particle generating packing materials.

Sterilization of Plastic Containers

Although heat-resistant polymers can be autoclaved, it should be noted that required processes might be different from those for glass containers. Either terminally sterilized filled containers or autoclaved empty containers for aseptic filling are not as resistant to rough handling and the stress of sterilization processes, and special considerations for this may be needed. The glass transition temperatures for plastics are generally substantially lower than their melting points and lower sterilizing temperatures, such as
110–116°C, may be necessary. The autoclave jacket temperature may be poorly controlled in older autoclaves and care must be made to ensure that plastics do not melt on surfaces overheated by jacket temperatures set higher for glass and stainless steel filler parts.

The nature of plastic materials usually means that siliconization is not required to assist in movement or assembly. Some sort of cleaning is required for particulate removal and depyrogenation for containers not molded in-line within a classified cGMP area. Validation protocols should contain rigorous initial and ongoing bioburden/pyroburden and particulate monitoring studies for plastic components. These data are needed to justify using a minimum challenge to sterilizing operations, which can help reduce the processing requirements and thereby reduce adverse material effects caused by overprocessing. Traditional methods of sterilization for plastics include moist heat, ethylene oxide, radiation, other gaseous chemical agents (i.e., hydrogen peroxide, chlorine dioxide). These methods historically have been applied in chambers (ovens or autoclaves) or in separate facilities. The reader will find chapters within this volume covering traditional methods of sterilization validation. Nontraditional methods of sterilization include high-intensity light, UV light, combined vapor and glass plasma, electronic beam, and injected low concentration ethylene oxide. These nontraditional methods are typically performed in-line on containers or FFS plastic webs. Physical assessment is needed for radiation sterilization methods and plastic dosimeters are typically used. Sterilizer parameters, extruder temperature, pressures, and other parameters must be measured with NIST traceable devices and documented during validation. Critical parameters are subsequently monitored with appropriate alarms during routine production.

Biological challenges for web sterilization consist of dried on suspension at the lowest level supported by recovery studies and bioburden studies. If high levels of spore suspension are used, post-study equipment cleaning and sterilization must be specified in the validation protocol. The location of challenges should be well marked and challenge the entire width of webs. Provisions to capture the web immediately following sterilization can be made followed by removal of inoculated areas and incubation in media. Clear containers formed from the challenged web can be captured before sealing and filled with media within the laboratory, or filled with media online followed by incubation.

FORM, FILL, AND SEAL

The prevalent trend in plastics and the primary advantage is FFS, or injection, blow mold, fill and seal technology. FFS and BFS are technologies that were originally developed in the food industry. The first high volume application was the Brick-Pack™ process. These are the small rectangular aseptically filled containers of fruit juice or milk that can be stored at room temperature. The approach applies a high temperature short-time thermal process to sterilize the liquid without the insulating effect of the containers. The container materials are fed in continuous webs that are chemically sterilized separately using hydrogen peroxide baths or sprays. The hydrogen peroxide is removed from the surface with UV light and the containers are formed and aseptically filled within a classified modular isolator containment area. This process forms the basis of many pharmaceutical FFS and BFS processes although sterilizing filtration is more commonly utilized for sterilization of the fluid. In BFS processes there is hot melted plastic extruded into a tube; the tube is cut into a piece that expands into molds by gas pressure to form the container. A comprehensive review of the BFS process has been provided by Ljungquist et al. (17) The plastic is extruded at temperatures in excess of 200°C at which temperature sterilization and depyrogenation of the polymer can occur. Containers must meet sterility, pyrogen, and particulate requirements. It can be shown that spore suspension dried on to plastic resin beads is nonviable after extrusion (18). Additionally, it has been shown by Poisson et al. (19) that air-dispersed organisms in environments of molding processes produce contaminated vials directly proportional to the challenge concentration. This study correlates environmental qualification to formed container quality. A similar resin challenge study with endotoxin coupled with pyroburden raw material monitoring programs can be used to address the depyrogenation requirement. Empty containers from challenged resin are tested using LAL. Capture of the polymer material immediately following extrusion for these studies limits possible contamination from the filled materials. Proprietary equipment can be designed and customized for a given validation sampling approach along with provisions for CIP and SIP of the solution filling line. Conventional media fill for aseptic process validation and WFI fills tested for pyrogen and particulates can be used for process validation.

Vendor Supplied Components

Outsourcing has emerged as a cost-effective business option and this practice has been widespread within the parenteral drug industry in the use of prepared components for parenteral filling. There are two major types of items being provided: ready to sterilize and ready to use. The use of these materials puts an additional burden on auditing groups because the validation of the essential preparation and critical sterilization operations is outsourced. The validation of preparation processes performed by suppliers must meet the same requirements described in this chapter. Supplier quality and consistency of supplied components must be established in documented quality agreements. Arrangements with so-called "preferred suppliers" typically cover pricing and reduced incoming QC testing. It is important that auditing, sampling, and testing of components to assure continued quality be included in these arrangements. In-house complimentary systems are needed so that the benefits afforded by supplier preparation are not lost through contamination from internal unloading and handling. Validation of storage, unpacking and transfer mechanisms is generally required and included within separate process validation for filled units.

Ready to Sterilize

These prewashed depyrogenated components are received in autoclavable packages that may be double/-triple wrapped. Hold time studies are required to demonstrate that stored components do not lose their critical quality attributes prior to autoclaving. Extensive vendor qualification is required, including an audit of all validations. Although, "qualified" audits by regulatory agencies may have been used in vendor qualification programs, it is necessary that the entire validation package of the vendor be reviewed by the end user.

Ready to Use

Ready to use components generally require specialized complementary equipment (pass-throughs, etc.), are typically proprietary in nature, and require specialized complementary filling equipment to use. Thilly et al. (20) describe a ready to use aseptic filling system using COC vials within a RABS. GSK has partnered with Aseptic Technologies, with core technology licensed from Medical Instill Technologies, for this proprietary process. Electron beam surface sterilization is utilized at a 35 kGray intensity for gamma-irradiated vials, which has the advantage of eliminating the "shadow effect" of UV and pulsed light systems. Filling is accomplished with needle perforation of the closure and laser resealing. Large media runs of 6000 vials support the validation efforts of a single head system. PP nonshedding packing materials are utilized for shipping the prepared components. Becton Dickinson, Inc. is a widely used supplier of ready to use primary packaging components, as stated in the syringe section.

Rubber stoppers are increasingly available in ready to use packaging which is of particular benefit to small-scale filling operations, or in conjunction with isolator or RABs-based aseptic filling systems.

NONSTERILE PRODUCTS

The preparation requirement for nonsterile primary packaging is predominantly cleanliness. This is accomplished by gravity through inverting the container and applying a filtered compressed air-cleaning process. This is typically performed in-line immediately prior to filling operations. Overhead feed systems may have helical tracks where the containers are swirled with their opening facing outward to use centrifugal force to dislodge foreign material. These cleaning operations can be challenged within validation protocols by adding typical contaminant (cardboard dust) to the containers followed by postprocess sampling and inspection. The amount of dust in the challenge should be realistic. There is a tendency to place too much contaminant within the containers, which will result in a failed validation. The author recalls one validation of four-L jugs that were to be filled with PEG 2000 and electrolytes for reconstitution. The challenge used was about 500 mL of ground cardboard in a series of jugs, which was about double the product fill. The challenge cardboard dust agglomerated together and would not fall out when the containers were inverted. The subsequent ionized filtered air stream blew the cardboard all over the filling room with the result that containers, filling line and validation personnel all failed to pass visual inspection for cleanliness.

As with plastic materials, most of the innovation for nonsterile packaging is derived from the food industry. Systems proven in the food industry but new to pharmaceuticals must be demonstrated to be effective. It should be remembered that the initial mandate for validation in the pharmaceutical industry was in part caused by the use of the twist-off bottle closure from the beverage industry on LVP bottles (21). Ionized filter air showers, which are common to most nonsterile packaging preparation, can be validated by environmental particle counting equipment, borrowed from parenteral technology, or widely available through contract validation services.

Liquids/Semi-Solids/Solids

Glass

Types II and III glass may be encountered and washing of these prior to filling may be required. The term "sanitization" can be used to describe preparation operations for some packaging materials. Bio-challenges should be minimal to demonstrate a three-log reduction, as sterility of the containers is not the objective. Other pretreatments to assure pH stability of filled products can be validated as in production processes with increased sampling and reconstitution under stressed conditions.

Plastic

Laminated tubes are widely used and have largely replaced aluminum tubes for semi-solids. In-line inspection devices can be challenged with appropriately marked challenge containers with known physical defects. Surface treatments, such as UV light, to facilitate direct surface printing can be validated by monitoring treatment parameters and increased inspection of printed components. Ionized filter air showers can be validated by environmental particle counting equipment, borrowed from parenteral technology. Polyethylene bottles are widely used for solid dosage and require little more than filtered air cleaning.

Caps

The criticality of the cap function determines the level of validation required. Beyond filter air cleaning, closure integrity for multidose containers is an attribute of concern. Increased sampling and use testing is a common validation strategy.

Delivery Systems

The award winning (22) GSK Advair Diskus® typifies combined technologies providing the epitome of convenience to patients. Consisting of 14 components that are automatically assembled, the powder inhaler provides a display of remaining doses for increased ease of use. The device combines blister packing with automatic assembly of components. Similar modern systems can be validated by monitoring and documenting machine parameters. Vision systems can be validated by challenging online reject mechanisms with edge of failure components that have been well marked to assure they are not used for commercial production. A phasewise validation consisting of protocols addressing each unit operation of the assembly and filling, followed by extensive sampling of centerline manufactured product can be used. Testing follows that specified in the USP (23).

Aerosol Cans

Coated aluminum cans are often utilized in pMDI. Online inspection for can and coating defects by vision system can be validated as in described above. Plastic actuator and valves have similar inspections and filled primary packages, verified for content in-line, can be challenged with well-marked edge of failure fills.

Blister Packaging

PVC films rolled on large webs, along with paperboard and other laminates are utilized for blister packaging. Machine operating parameters are documented during validation. Molding and sealing temperatures and pressures are monitored, combined with seal-strength testing for validation. The computer-controlled aspects of this machinery are covered by computer validation, discussion of which is outside the scope of this chapter.

Foil Laminates

Aluminum foil laminates are often used with FFS equipment. The preparation processes may consist of filtered air cleaning and inspection of the web prior to forming. Validation of these types of processes follows that described in previous sections. Filled product testing for seal integrity follows worst-case filling and sealing conditions such as maximum and minimum line speed.

CONCLUSION

The validation of container preparation processes has evolved from traditional methods to new methods driven by technological change in the materials and packaging. Most of this innovation has come within the area of plastics and associated in-line container preparation processes. Validation approaches are often unique to the specific application. However, fundamental quality attributes must be confirmed and documented during validation. With new technology, considerable ingenuity to obtain the required testing and parameter measurements may be required. A joint approach, where pharmaceutical manufacturer partners with packaging equipment supplier and/or container vendor to accomplish validation, has proven effective for both new technologies and traditional processes. Supplier quality and consistency of delivered components is essential for maintaining the validated state and cannot be overstressed. This requires ongoing active quality partnerships with suppliers to maintain the validated state.

REFERENCES

1. Ventura DA, Goodsir SW. Component preparation processes. In: Agalloco JP, Carlton FJ, eds. Validation of Pharmaceutical Processes: Sterile Products. 2nd ed. New York: Marcel Dekker, Inc., 1998:703–20.
2. U.S. Food and Drug Administration, Department of Health, Education, and Welfare. Pharmaceutical cGMPs for the 21st Century—A Risk Based Approach. Final Report, Fall 2004. U.S. Government Printing Office.
3. Agalloco JP, Akers JE. Risk analysis for aseptic processing: the Akers–Agalloco method. Pharm Technol 2005; 29(11):74–88.
4. Agalloco JP, Akers JE. The simplified Akers–Agalloco method for aseptic processing risk analysis. Pharm Technol 2006; 30(7):60–76.
5. Ricci MT, Fraiser HE. From quality by inspection to quality by design: a roadmap for success. Pharm Technol 2006; 30(9s):40.
6. The United States Pharmacopeia 29th Revision. <1211> Sterilization and Sterility Assurance of Compendial Articles. Rockville, MD: The United States Pharmacopial Convention, Inc., 2006:2722.
7. The United States Pharmacopeia 29th Revision. <788> Particulate Matter in Injections. Rockville, MD: The United States Pharmacopial Convention, Inc., 2006:2722.
8. The United States Pharmacopeia 29th Revision. <151> Pyrogen Test. Rockville, MD: The United States Pharmacopial Convention, Inc., 2006:2546.
9. U.S. Food Drug Administration. Updated 510(k) Sterility Review Guidance K90-1; Final Guidance for Industry and FDA, November 16, 2001.
10. U.S.P. <661> Containers. USP 29 NF 24 The United States Pharmacopeia 29th Revision. Rockville, MD: The United States Pharmacopial Convention, Inc., 2006:2996.
11. U.S.P. <661> Containers. USP 29 NF 24 The United States Pharmacopeia 29th Revision. Rockville, MD: The United States Pharmacopial Convention, Inc., 2006:2656.
12. Anonymous. PDA Technical Report 12: Siliconization of Parenteral Drug Packaging Components. Parenter Drug Assoc 1988, 42(4S) (Suppl.):1–14.
13. Dublin M, Wittler H. A new approach for stopper/closure treatment system. Pharm Eng 1998; 18(3):86.

14. Rainbow RE, Roseman TJ. Plastic packaging materials. Remington, The Science and Practice of Pharmacy. 21 ed. Baltimore, MD: Lippincott Williams and Wilkins, 2005:1047–57.
15. Wang YJ, Chen YW. Technical Report No. 5, Sterile Pharmaceutical Packaging: Compatibility and Stability. Washington, DC: Parenteral Drug Association, Inc., 1983.
16. Halek GW, Yam KL, Lordi N. Course: The Technology of Packaging II. New Brunsweek, NJ; Rutgers the State University of New Jersey, 1984.
17. Ljungquist B, Reinmuller B, Lofgren A, Dewhurst E. Current practice in the operation and validation of aseptic blow–fill–seal processes. PDA J Pharm Sci Technol 2006; 60(6):254.
18. Gustafsson A. The bottle pack system. Plastics and Their Pharmaceutical Applications. The Secretariat to the Convention for the Mutual Recognition of Inspectors in Respect to the Manufacture of Pharmaceutical Products, 1986:94.
19. Poisson P, Colin SS, Tallentire A. Challenges to a blow/fill/seal process with airborne microorganisms having different resistances to dry heat. PDA J Pharm Sci Technol 2006; 60(5):323.
20. Thilly J, Conrad D, Vandecasserie C. Aseptic filling of closed, ready to fill containers. Pharm Eng 2006; 26(2):66–74.
21. Lindboe WG, Hayakawa K. Comparative terminal sterilization. J Parenteral Sci 1993; 47(3):138.
22. Hartman LR. Winning with plastic. Packaging Dig 2003; (9):80.
23. U.S.P. <601> Aerosols. USP 29 NF 24 The United States Pharmacopeia 29th Revision. Rockville, MD: The United States Pharmacopial Convention, Inc., 2006; p.2617.

28

Validation of Lyophilization

Edward H. Trappler
Lyophilization Technology, Ivyland, Pennsylvania, U.S.A.

INTRODUCTION

In an ideal world, validation would begin with parallel product research and development activities. Validation for lyophilized products occurs more often during scale-up to manufacturing. Under growing regulatory pressure and the realization of the greater benefits, however, validation activities are being undertaken while the product is along the development pathway. There are also circumstances where validation is required for existing commercial products, either because of changes requiring additional study or to meet current regulatory standards. This presentation will approach validation as an integral part of developing a new product. Appropriate application of the principles discussed may be applied for either a change control procedure or for revalidation, based upon specific needs.

PREPARING FOR VALIDATION

Components of a comprehensive validation program include EQ, process engineering, and process validation. The EQ portion focuses on the lyophilization equipment and is valid for processing the particular product or, for a multi product operation, any number of products. Conversely, the process for each product is unique and applies only to one product, and therefore the process validation is specific for that product.

The experience gained and data compiled during development are a significant part of the scientific rationale that forms the basis of validation studies. It also provides a critical reference with integrating a product into a production environment. This saves on adjustments to the process and further development studies at the time scale-up and validation is attempted in manufacturing. Completing process studies at both the ideal target parameters as well as boundaries of the process parameter range results in greater safety and efficiency into the parameters and a more robust process. Establishing a PAR, first introduced in 1984 by Chapman, is recognized as a "best practice" as part of lyophilization process engineering (1).

Activities for validation of a "legacy" product where development and initial validation may not meet current industry practices requires constructing a historically based file. Data such as preformulation, product and process development may not be available with historical manufacturing experience for the commercial product manufacturing. The most challenging task in this instance is justification for the product formulation and process design. This is particularly difficult in circumstances with commercial products that have been developed prior to the awareness of the benefits of validation.

When a new product is in the development phase, a comprehensive report that includes technology transfer for Phase III clinical trial material should be assembled prior to scale-up and technology transfer to manufacturing. This report addresses the starting raw materials, including the drug substance, excipients, and packaging components, along with formulation, compounding methods and lyophilization process engineering activities. Each facet of product manufacturing needs to be included, beginning with testing of product components through processing and final packaging requirements. Finished product qualities must also be defined. The report should clearly explain the scientific rationale and justification for the formulation and manufacturing procedures.

This development report is a crucial reference for integrating a new product into a manufacturing operation. The acceptance criteria for any validation study are based upon product and process requirements outlined in the development report. The report provides an invaluable reference for technology transfer, change control program management and trouble shooting during routine manufacturing.

EQ is best considered at the time of equipment specification and selection. The advantages include more effective project management, ease of completing the validation package and speed of bringing the equipment online. Equipment requirements and performance are based upon the processing parameters necessary for manufacturing product, as identified during processing engineering studies completed during development.

As with the specification and purchase of any new piece of equipment, well-written equipment specifications include the intended validation activities for qualifying the equipment and assuring it meets the processing requirements. Defining testing and documentation expected during the FAT at the vendor's facility is also a useful contractual agreement.

Abbreviations used in this chapter: API, active pharmaceutical ingredient; CIP, clean in place; DQ, design qualification; EQ, equipment qualification; FAT, factory acceptance test; FIT, filter integrity test; GC, gas chromatography; IQ, installation qualification; OQ, operational qualification; P&ID, process and instrumentation drawings; PAR, proven acceptable range; PDA, Parenteral Drug Association; PLC, programmable logic controller; PVP, process validation package; SAT, site acceptance test; SIP, sterilization in place; SOP, standard operating procedure; USP, *United States Pharmacopeia*.

SOURCES OF INFORMATION

Sources for information include Research & Development, Engineering, Clinical Supplies Manufacturing, Quality Control, and Regulatory Affairs groups. A development report is a key source of technical information regarding the characteristics of the API, product and formulation design, and product processing requirements. The physio-chemical character of the active substance, if appropriate, along with the functions of excipients of the final product formulation, liquid and solid state stability data are critical parts of such a report. Process development data and finished product characteristics should be available within the development report. Specific information on the equipment design and performance for the installation and OQ is often archived by the Engineering department. Other engineering references include maintenance and calibration procedures. Operating procedures covering product loading and operation of the lyophilizer are available within manufacturing documentation. These would include the unique aspects for processing lyophilized fill volume frequencies and tolerances, loading procedures and arrangement of product trays within the lyophilizer. Finished product testing methods for the active ingredient, reconstitution and residual moisture should be available from the development scientists, Analytical Development group as standard testing methods for use within Quality Control. The Regulatory Affairs staff should be consulted for commitments made in regulatory filings and communications to regulatory agencies.

RECOMMENDATIONS FOR A VALIDATION PROTOCOL

The differing circumstances under which a validation study is prompted often dictate the best approach. Agreeably, prospective validation, where the validation studies are all completed and approved prior to shipment of any product, is preferred. There are, however, opportunities to complete certain validation studies during preparing material for Phase III clinical studies where product is to be administered to patients. Here, validation is concurrent with producing these materials. In addition, when implementing validation studies on an existing marketed product to bring the operation up to current regulatory expectations, concurrent validation would also be appropriate. Retrospective validation would be applied to a review of historical data of an existing process and product. Examples would be the review of the lyophilization processing data, finished product batch release test data and stability data from the commercial stability testing program.

The design of the validation testing and the composition of the protocol reflect the circumstance under which the study is conducted. For retrospective validation, the "test" may be statistical analysis of batch release data such as assay, pH, physical appearance, residual moisture, reconstitution time, and constituted solution appearance. This retrospective process validation would be intended to show that the process is within an adequate level of control and product is of consistent quality. A critical review of the processing conditions in a retrospective validation may consist of a test comparing actual processing conditions during lyophilization to ideal parameters, showing not only adherence to the defined processing conditions but also to demonstrate process reproducibility.

Concurrent validation studies may be used during clinical manufacturing and scale-up activities. Additional testing or an increased number of samples, as in the case of transferring a product to another site or manufacturing material in a new production lyophilizer, may be conducted as a concurrent validation study. This would be reasonable if the parameters have not changed, if the process has already been adequately validated. In addition to finished product testing, short-term accelerated stability may be appropriate prior to actually releasing the batch for distribution. Long-term stability studies should be done at the recommended storage conditions, up to and including at least six months beyond the desired expiry date.

Although there are circumstances where retrospective or concurrent validation may be appropriate, prospective validation is preferred. This entails the testing, review of the data and approval of the completed validation studies prior to releasing product for distribution and use. Identifying the target process parameters and a proven acceptable parameter range, along with demonstrating consistent product quality and stability would be highly desirable prior to introducing the product into a manufacturing environment. It could also decrease the amount of time necessary for getting a new product to market.

Numerous studies to support process validation can be completed during the development phase. These studies correlate the product formulation, presentation, and lyophilization processing parameters to finished product attributes and long-term stability. In addition, the reproducibility of process would be demonstrated along with the consistency of finished product attributes. Uniformity studies during the first batches being integrated into manufacturing are often the last leg in the sequence of validation studies for bringing a product to market. Depending upon the supporting data available from development studies, limited or short-term accelerated stability may be sufficient.

PREPARATION OF THE PROTOCOL AND SOPs

Each activity performed as part of the IQ and OQ should be organized into discrete functions and documents. During the IQ, the review and verification of utility connections, piping of the refrigeration and heat transfer system, reconnecting the vacuum system, rewiring of the control system, start-up and testing may be organized into a distinct document for each activity. This "modular" approach becomes more effective and efficient as the complexity of the procedures and equipment increases. Each aspect of bringing a lyophilizer online or integrating a new product into a manufacturing environment often involves a number of individuals or departments. Arranging the overall protocol into smaller packages, correlating to distinct activities makes communication between individuals and departments more manageable. For example, the project engineer responsible for installation of a new lyophilizer may use a mechanical contractor to reconnect the piping and connect the utilities, and an electrical contractor to connect the control system wiring. In such a case, a documentation package covering each activity may be issued and completed for each part of the project involving each contractor. A documentation package organized in such a manner is also a useful tool for project management.

Such an approach is also applicable for product and process validation. Considering the ranges of formulation aspects such as the acceptable pH range, a focused study to correlate the pH, phase transition temperature and finished product aspects upon processing would be well-suited as a distinct protocol. This protocol may parallel studies already conducted during development. Another example is establishing the PAR for the processing parameters. Identifying such ranges is accomplished by processing at extreme shelf temperatures, chamber pressures and times, following the PAR approach as referenced earlier.

Establishing Acceptance Criteria

The selection of acceptance criteria is dependent upon the circumstances under which validation is being undertaken and requires judicious consideration. Challenges to the equipment, for example, may depend upon whether the lyophilizer is being first installed or whether validation is being completed for an existing unit currently in use. Where the lyophilizer is new, the acceptance criteria based upon the performance requirements that are identified within the equipment specifications would be warranted. The advantage of acceptance criteria based upon the stated equipment capabilities is that any process that is within the performance capabilities of the equipment could be utilized for lyophilizing product. For testing an existing unit in production, however, the most rigorous processing conditions would be a justifiable test challenge. The jeopardy of test challenges based upon the most current processing conditions is that if a process for a new product is outside of the parameters tested, then additional testing or qualification at the new parameters would be necessary.

Details of constructing validation protocols, designing studies and establishing acceptance criteria will be presented in each respective section of this chapter. In approaching validation, it is more important to test and document what is critical for gaining a high degree of confidence that the process is well-defined and reproducible, the procedures are adequate and appropriate, the equipment is suitable for completing the process, and the product is of consistent quality, purity, and stability. In addition, it is a valuable opportunity to collect useful information for supporting a change control program. Validating for the sake of simply documenting information in a protocol, not having a clear understanding of what is significant, or creating a voluminous collection of data because more is better should be avoided. As a general rule, do what is necessary and do it well!

For some studies, as in the OQ, references will be made to known performance capabilities of equipment. These are intended to be examples rather than standards. A few general notes are, however, appropriate. Most importantly, selecting acceptance criteria needs to be based upon a justifiable scientific rationale. This is applicable whether qualifying an existing piece of equipment for commercial product manufacture or validating a process during clinical manufacturing. Selecting appropriate processing ranges to be encompassed within the validation has a significant long-term effect in manufacturing. For example, when the range of residual moisture correlated to suitable stability is adequately determined during development, then any batch in manufacturing exhibiting a moisture within the boundaries of that range would be acceptable. If there were a batch where the residual moisture was beyond the boundary, then there would be reasonable questions whether that batch should be released. Adopting such a philosophy provides clear and reasonable approach for successful routine manufacturing of high quality product. There is also little question when a batch is found to be outside the PAR. This eliminates the scenario of placing a batch on stability or doing additional testing when there is a question of what a suitable envelope of processing conditions or product quality aspect would be for a batch to be released. Establishing a PAR, or PAR approach, becomes a valuable asset in a manufacturing environment.

Equipment Qualification

EQ can be conducted as part of and entail portions of an IQ and OQ, as well as a FAT and SAT. Activities encompassed within an EQ are a useful endeavor when begun as an integral part of the FAT and carried through installation of a new lyophilizer. These activities encompass verifying that the equipment is designed, constructed and performs as anticipated when compared to the equipment specifications. This assumes that the specifications are based upon current or anticipated needs for processing the products and agreed to between the vendor and purchaser.

This would include verifying the engineering documentation, construction and assembly of the lyophilizer, along with demonstrating adequate performance of the system.

For the acquisition of a new lyophilizer, the EQ comprises a series of tests to assure that the lyophilizer meets the performance expectations necessary for its intended use and identified within the purchase specifications. These series of tests are useful as part of the FAT, with the intent to measure and verify the performance capabilities of the lyophilizer prior to its shipment to the purchaser's site.

Incorporating the qualification requirements in the equipment specifications package to the vendor assures that proper attention is given by both the vendor and purchaser staff. These validation requirements include the EQ tests along with control system validation, and extend into the IQ and OQ. Identifying the testing to be done at the factory to complete the FAT allows sufficient planning for both manpower resources and time at the vendors' facility. Validation of the automated system controlling the lyophilization process, along with the complementary processes such as SIP, CIP, in-process integrity and FIT needs to be started at the control system design and software development stage of the project. This follows the Life Cycle (2) approach that has become industry practice for validation of computer automation systems.

Part of the EQ that comes before any actual performance testing is the review and verification of the lyophilizer design. This is sometimes completed as a separate task and is often referred to as a DQ. This step, whether as a separate DQ or as part of the EQ, entails a review of the engineering documentation to verify that the equipment will meet the requirements of the specification, prior to construction and assembly of the lyophilizer. Such a review includes the general layout of the equipment, piping arrangements for the CIP and SIP systems, refrigeration and heat transfer fluid system drawings, electrical elementary schematics, and P&IDs. This review of the engineering drawings should be documented and become part of the validation package.

Equipment performance tests, as a major part of the FAT, involve testing to demonstrate that the lyophilizer functions, performs, and has the processing capacities as specified. These tests may mimic those planned as part of the OQ, although do not negate the need for completing the OQ at the installation site. Often duplicating the testing for an OQ, tests encompass function, control capability and performance for freeze drying and support processes. The testing regime should include specific tests as listed in Table 1. Complementary functions such as CIP, SIP, FIT should also be included when the lyophilizer has such capabilities. Testing of the loading and unloading would be appropriate as an integral part of the performance test.

This testing program is useful as part of the validation package, along with being part of the equipment acceptance. Circumventing the testing at the vendors' facility should be avoided, no matter how complex or unique the final installation. In addition, successfully completing the EQ does not negate the need to complete a SAT and comprehensive IQ/OQ at the final installation site. Factors such as assembly of the lyophilizer after being dismantled for shipping and differences in utility supplies warrant the need for testing prior to bringing the unit online for manufacturing product. The more complex and unique the lyophilizer design and final configuration, the more such efforts assure the success of the project. Some parts of the IQ could be completed at the factory and not repeated after installation. Such items may include instrumentation and hardware documents, testing of the control system, and verification of as-built drawings, to cite a few examples.

Installation Qualification

The IQ is the first validation activity completed when the lyophilizer arrives at the final installation site.

Table 1 Test Functions for Factory Acceptance Test, Site Acceptance Test, and Operational Qualification Test

Test	Function
Shelf temperature	Range, rates, control, uniformity
Chamber pressure	Range, control
Condenser	Chilling rates, ultimate temperature
Vacuum system	Evacuation rates, ultimate pressure
Sublimation/ condensation rates	Shelf cooling, heating, rates, capacity

Implementing the protocol may begin as the lyophilizer is being installed. Verification of the electrical wiring and piping may be accomplished as part of the assembly activities. The appropriate approach to completing the IQ is strongly dependent upon the specific circumstances of the project.

The IQ consists of a description of the lyophilization equipment, a system hardware and component list and the documentation of the installation procedures, equipment start-up and operator training. The IQ also includes references to the purchase specifications, engineering review and SOPs. The objective is to assure that the equipment design and construction are appropriate for the intended use, it is installed properly, the utilities are suitable and adequate, and that procedures are in place for proper calibration, operation and maintenance.

Equipment Description

The description of the lyophilization equipment provides a general overview of the lyophilizer, the installation site, functions and use in operation. The description also identifies the major components of the system. Complementing the list of the major components, a more specific description of each item provides greater detail. Such information is highlighted in Table 2. This data becomes an integral part of the change control system for the equipment hardware. The major components of the lyophilizer that should be included are the refrigeration units, heat transfer fluid, heat transfer circulation pump, heater elements, primary vacuum pumps, secondary vacuum pumps, system valves, and the control instrumentation.

Installation Activities

Documentation of the installation can also be included within the IQ section of the validation package. Part of this documentation may take the form of an installation checklist. This checklist would include each specific activity necessary for the installation of the lyophilizer, who completed and checked the work, and the date the work was completed. These activities would include assembly of the various lyophilizer parts if dismantled at the factory for shipment, as well as the connection to utility supplies.

In addition to the early project activities of the engineering review and factory testing completed as part of the EQ, certain parts of the IQ should also be planned well in advance of receiving the equipment. These include the utility verification, specific installation location, start-up and training. The utility verification, identifying the quantity, quality and source of the utilities is best completed during the initial phase of the project and prior to operation of any of the lyophilizers systems. These encompass electricity, cooling water, process gases, sterilant, and discharges for the lyophilizer. The listings in Table 3 are common utility supplies.

Physical installation includes the rigging into place and connection of the subsystems. With large sized units and those with external condensers, reconstruction at the installation site is a fairly involved project in itself, and includes mechanical, electrical and refrigeration mechanics. After installation is complete, most vendors provide a service technician to start-up the system and

Table 2 Major Lyophilizer Components Perspective Functions Documented in the Installation Qualification

Test	Function
Chamber/condenser	Pressure/vacuum vessels
Shelf heat transfer system	Transfer of heat between a circulating fluid to product
Cooling system (refrigeration)	Chilling heat transfer fluid and condenser
Heating system	Providing heat to shelf heat transfer fluid
Vacuum pumps	Removing non-condensable gases
Major system control valves	Control of heat transfer fluid, process vacuum/pressure stream
Control and automation	Equipment, process variables, sequencing steps and process data acquisition

provide training. Such activities should be documented and included within the IQ portion of the protocol.

Operational Qualification

The OQ focuses upon the equipment rather than a process for any specific product. Although not associated with any product or process, the OQ is a series of tests that measures performance capabilities and demonstrates the ability of the lyophilizer to complete critical processing steps. Functions of the lyophilizer, such as the shelf cooling rate and pressure control are process related. They are, however, focused on measuring the performance capabilities of the equipment rather than demonstrating any processing capabilities relating to producing a particular product.

SCOPE AND OBJECTIVES

The OQ demonstrates the equipment performance for the range of processing functions at the installation site. The tests performed may be expanded as compared to those completed as part of the FAT at the vendor's facility. Additional activities such as CIP and SIP process development and validation are also performed after the IQ has been successfully completed.

Measuring Equipment Performance

Although the testing at the factory would have demonstrated the performance capabilities, these tests also need to be performed at the final installation site. Different utility capacities such as cooling water and steam supply influence the equipment performance. These tests also verify that the utility supplies are adequate and meet the demands of the operating system. This testing is particularly valuable for large systems disassembled and shipped as smaller packages, where the unit is reconstructed at the installation site. Testing is necessary to demonstrate that installation was completed properly and the equipment still meets the performance levels previously demonstrated during the FAT.

Table 3 Critical Processing Step and Objectives During Lyophilization

Process step	Critical objectives
Loading	Thermal history influences results of solidification
Freezing	Solidification is sufficient for beginning primary drying
Primary drying	Vaporization of solvent is complete
Secondary drying	Desorption of solvent achieves desired residual level

Verification of System Capabilities

The OQ evaluates each equipment function and the capacity to meet the performance standards. Reducing the lyophilization process into each function also has advantages for managing a change control program. For example, one test would focus on cooling rates used for the freezing step, while a separate test would be implemented to evaluate the heating function used during primary and secondary drying.

The advantage of having a separate and distinct testing protocol for each process step is that there is a specific testing protocol for each discrete equipment function used to complete a step in the process. For example, when a significant change is made to the shelf cooling equipment or there is a question as to performance capabilities, a detailed and specific protocol could be implemented to demonstrate that there is no significant change to the system operating performance. Considering each function of the equipment for each step in the process allows segregation of each equipment function, with a respective test that demonstrates a specific performance capability.

EQUIPMENT PERFORMANCE TESTS

Performance capabilities and capacities can be evaluated using a separate test for each function of the lyophilizer, focusing on the operation of selected subsystems and the capacity for the specific functions during lyophilization. These subsystems include the heat transfer system, condenser and vacuum system. An overview for testing of each major subsystem is presented in the following sections. Also included are examples and illustrations for performance ranges. These examples do not, however, reflect the capabilities of a specific lyophilizer nor are they intended to suggest any industry standard.

Heat Transfer System

The heat transfer system provides cooling required for freezing the product and the subsequent heat needed to establish rates of sublimation. Temperature control is required over of the entire process temperature profile, from the time the product is loaded onto the lyophilizer shelves until it is removed after stoppering. Therefore, cooling and heating rates, along with control within a reasonable range at temperatures that embody the intended operating range, as well as shelf temperature uniformity, must be tested.

Maximum cooling and heating rate tests are intended to demonstrate the optimal performance of the equipment. The cooling rates, defined as an average of the change in temperature per unit time, are measured from room temperature to the ultimate achievable freezing temperature. Heating rates are measured from the lowest to the highest operating temperature for the lyophilizer. For a lyophilizer currently in use, the acceptance criteria may be the average rate across a temperature range that exceeds the current process requirements, and should extend beyond to fully envelope the routine ranges of operation. Test results are expressed as an average rate of change as measured at the shelf inlet. Since the performance of the lyophilizer is strongly dependent upon the specific design, acceptance criteria vary. It is common, however, to be able to achieve average cooling and heating rates in the range of 0.5° to 1.0° per minute.

Shelf temperature uniformity across any one shelf and all of the shelves of the lyophilizer needs to be within an acceptable range to assure batch uniformity of the dried product. The temperature at any location is compared to either the mean of the measured values or the temperature indicated on the controlling instrument. The allowable range is dictated by the reference used, with tighter tolerances appropriate when comparing the actual to the mean of the measurements. Stated capabilities for shelf temperature uniformity by many of the lyophilizer vendors are ±1°C at steady-state conditions. Appropriately completed under no load conditions, these functions may again be demonstrated under load conditions during the sublimation/condensation test.

Condenser

Measuring the cooling rate and ultimate lowest temperature of the condenser is useful in generating baseline data for future reference, such as monitoring the condition of the refrigeration system. Rates will vary based upon the size, type and number of refrigeration units on the system. The ultimate condenser temperature necessary is dependent upon the solvent system in the product formulation. For a completely aqueous solvent system, a maximum allowable temperature is commonly −50°C. For processing some organic solvents, the condenser temperature necessary is dependent upon the solvent being processed. For example, ethanol vapors must be chilled to below −115°C before condensation and solidification will occur. Pure tertiary butyl alcohol requires only slightly colder than room temperature.

In the sublimation/condensation test, the condensation rate and ice load capacity are demonstrated. In these tests, the actual performance is more critical than the baseline test of cooling rate and ultimate temperature. The rate of condensation, expressed as kilograms of ice per hour, or normalized to kilograms of ice per measure of shelf surface area, becomes a limit to the processing parameters that may be used in design of the lyophilization cycle. The results of the total ice capacity test become a limit to the product batch size.

Vacuum System

Like the cooling rate and ultimate temperature tests for the condenser, evacuation rates and lowest achievable pressure are baseline tests that indicate the performance of the vacuum pumping system. Typical evacuation rates allow for reaching 100 μ within 20 to 30 minutes. The lowest achievable pressure is commonly 20 μ or less.

Associated tests to include are the baseline leak rate and vacuum integrity test. Both tests are based upon the pressure rise of a sealed chamber and condenser that are isolated from the vacuum pumping system. Detailed presentations on the subject is covered in various classic technical publications on vacuum technology (3,4). Each of these tests, briefly described below, is well-suited to stand-alone protocols.

The leak rate test is a baseline measurement that is intended to determine the presence of leaks in the freeze drier chamber and condenser. This test is implemented with the chambers being clean, dry, and with low levels of outgassing. Eliminating any vapors that may outgas and contribute to any pressure rise requires that the test should be done only after the system has been maintained at a low pressure for a number of hours. A common acceptance criteria used are the specifications agreed to by the equipment vendor and end user. The standard may be expressed as units of pressure per unit time, and are best referred to as units of pressure and volume per unit time.

Different than the leak test, the vacuum integrity test is an in-process method used in manufacturing after the completion of sterilization and prior to loading product. First, a study to establish an acceptable value is necessary. Justifying an acceptable value is accomplished by correlating a rate of pressure rise that includes any contribution of outgassing of water vapors remaining after the sterilization process. This therefore requires that this development study be completed after the sterilization process has been validated, since sterilization conditions may influence that amount of outgassing of residual moisture that remains after sterilization. The result of this study yields a value expressed as a pressure increase per unit time, and do not need to be normalized, as in the leak rate test. Although there have been discussions on the topic published, there is no industry standard established that is based upon either empirical data or having a justifiable scientific rationale (5,6).

Control Functions

Whether a control system is comprised of distinct instruments for nominal control functions and process monitoring, or an integrated control system, a nominal set of tests are appropriate. The tests described are intended to encompass both controller capability and equipment performance.

Shelf Temperature Control

Different than the achievable rates for the equipment alone, shelf temperature control tests combine the system capabilities in implementing a range of cooling and heating rates and control at a specific set-point across the operating range of the system. For cooling and heating, minimum, maximum, or specific controlled rates are challenged. These rates may be based upon either specific required processing conditions or the vendors stated equipment performance over the operating range of the system. Rates for both cooling and warming may range from a minimum of 0.1°C to a maximum such as 2.0°C per minute.

Shelf temperature control tests show the system's ability to maintain steady-state shelf temperatures used for the freezing and drying process and should be within an acceptable range around a target set-point. If the acceptance criterion is other than the vendor's stated operating range, then control points used for the test must envelope the temperature ranges to be used for processing. Equipment capabilities range from ±1°C to ±5°C from the target set-point, as measured at the control point. Typical manufacturing units often achieve a range within ±3°C.

Pressure Control

The pressure control capability, critical as a process parameter, needs to show accuracy and precision of pressure control across the range anticipated for lyophilization cycles. This range can be a pressure as low as 20 μ to as high as 1600 μ. The results of the test are compared to the target values at a low, intermediate and high pressure. Acceptance criteria is specified as an acceptable range around the three target set-points. An acceptance criteria of ±10 μ is readily achievable.

Process Monitoring

Defining the process as critical parameters of shelf temperature, chamber pressure and time dictates that monitoring these conditions is with suitable accuracy and precision. Product temperatures, being less critical because of intrinsic limitations, are also commonly monitored. The ability of the monitoring system to reflect the actual process status is assured by an appropriate calibration program. Although not normally a separate study within the OQ testing program, it may be appropriate to complete a comparison of values measured if multiple instruments are used for monitoring the same conditions, or if data is transferred from a recording instrument or PLC to a separate computer workstation.

Sequencing Functions

With an automated control system, verifying the sequencing functions may be appropriate during the OQ testing. The first step is verifying the interfaces to the field devices such as pumps, motors and valves and their proper operation. This should also include operation of proportional control valves. This verification may have been completed separately as part of the control system qualification, conducted during the FAT, and would therefore not be necessary during the OQ studies. In verifying the control sequence functions, the hardware engaged for each step and the successful progression through the process is compared with that identified within the control system flow chart. Whether completed during the OQ or separately during control system validation is of little importance. However, it is preferred that the control system be qualified prior to implementing any of the OQ testing, but must be completed before testing any integrated control functions such as the lyophilization process tests described later in this chapter. Computer control system validation has unique requirements for validation and would best be accomplished as a separate validation study. The PDA Technical Report No. 18, Validation of Computer-Related Systems, provides a useful reference for control system validation (2).

Sublimation/Condensation Rate Tests

Capacity for sublimation and condensation, as well as condenser ice load is demonstrated during this test. Performance of the heat transfer system, chamber and condenser configuration, refrigeration system, as well as the condenser are challenged with demands for accelerated rates and under stressing load conditions. The ability of the equipment to adequately sublime and condense the water vapor during processing is demonstrated in this study, challenging both the rate of sublimation and the condenser capacity.

Water is sublimed using some optimal conditions for achieving a maximum sublimation rate. The capability of the heat transfer system to provide a sufficient quantity of energy to promote sublimation of ice part is one aspect of the lyophilizer capacity quantified during the sublimation/condensation test. As well, the capacity of the lyophilizer as a system, including that of the condensing system that encompasses the condenser surface and the refrigeration units, is quantified during this test.

During such a test, the condenser temperature is monitored and is not necessarily part of the acceptance criteria. Once ice begins to collect on the condenser surface, the sensor is buried beneath an amount of ice and the surface exposed to the vapor stream may be measurably warmer than the indicated condenser temperature. Therefore, the condenser temperature is less significant than the critical parameter of chamber pressure.

A quantity of water matching the condenser capacity sufficient to cover the entire shelf area may be sublimed over a set period of time. The rate of sublimation and condensation, expressed as kilograms of ice per hour, becomes a limit to the processing capabilities of a lyophilization cycle. The result of the total ice capacity challenge also becomes a limit to the product batch size.

Process Testing

Process testing combines functions evaluated during the OQ studies, completing processing of a model product and presentation. Processing a surrogate product provides a challenge for the functions and capacities to demonstrate the equipment's performance under load conditions. Process parameters of shelf temperature, chamber pressure and time are compared to target values. The equipment capabilities with the integrated control functions can also be used to demonstrate batch uniformity capabilities. Process testing may be independent of any specific processing parameters associated with any particular product presentation. A series of well-controlled tests may be used to demonstrate the capability to reproducibly implement the critical lyophilization process parameters and yield consistent dried material qualities. Trial runs used to assess adequate process parameter control of shelf temperature and chamber pressure under load conditions can also be useful in demonstrating uniformity of processing conditions within the lyophilizer. The batch size and process parameters do not necessarily need to duplicate those for any actual product. Studies can be designed, with an appropriate model presentation as a surrogate, to challenge the equipment. Several surrogates have been proposed in presentations in the literature, ranging from a placebo of a specific product formulation to a combination of mannitol and arginine, in vial sizes from 10 to 100 mL (7). A model product may provide a sufficient challenge to demonstrate the equipment's performance capabilities under load conditions. During such studies, shelf temperature, chamber pressure and time, the critical process parameters, are compared to target values.

Uniformity within a Lyophilizer

Demonstrating the processing and environmental conditions are uniform within the lyophilizer can be included as part of the OQ. As with many batch drying processes, processing as well as local differences in environmental conditions may affect a lyophilized product quality attributes. These attributes include the dried cake physical structure and appearance, as well as residual moisture content. Results of evaluating these attributes can provide an assessment of the uniformity throughout the batch. The potential differences are also influenced by varying processing parameters relative to location in the lyophilizer. Studies by Greiff have shown that there is a measurable effect of location within the lyophilizer on both the amount of ice sublimed and the residual moisture when lyophilizing a 2% serum albumin solution (8,9). These studies also quantified the range of moisture that varied with the shelf temperature, the shelf position and elapsed processing time. Dried product attributes can be mapped at the same locations on the shelves used to demonstrate shelf temperature uniformity: each of the four corners and the geometric center. Product temperature and finished product attributes of samples are evaluated at each of these five locations on each shelf. Critical points and the objectives to be achieved at the completion of each significant step of the process are highlighted in Table 3.

The range of product temperatures for material at each of the locations can be compared at the completion of each major processing step: loading, freezing, primary and secondary drying. Comparison of temperatures and finished product attributes unique to lyophilized products at each of the locations can be evaluated statistically based on the processed material response at the conclusion of each step at the sampled locations. Temperature range and standard deviation for the monitored locations can be compared to the average temperature for each set of processing conditions. Evaluation of the data needs to accommodate a significant variation at times when conditions of the process are changing and the process has not reached steady-state conditions. For example, a significant influence is the container type and location of the thermocouple placement, in addition to the differences in mass transfer of the water vapor through the dried product (10,11). Figure 1 illustrates the variation in product temperature for a formulation containing protein, mannitol and glycine at various concentrations. Therefore, the significance placed upon any variation and conclusions drawn from such data need to account for such inherent influences.

Dried material attributes such as physical appearance, reconstitution time and residual moisture are finished product attributes affected by differences in process conditions. While many of the attributes such as physical appearance are subjective evaluations, residual moisture is more quantifiable in reflecting the magnitude of any variation due to location. A material or combination of excipients may be used as a surrogate for conducting these uniformity studies. The surrogate selected needs to be measurably influenced by processing conditions and able to reflect significant differences in temperature during processing and dried product attributes. The surrogate formulation, concentration, and fill volume may influence the variation in physical structure and density, and therefore affects the rate of mass transfer of water vapor during sublimation (12). Surrogate preparations consisting of mannitol, polyvinylpyrrolidone and simple ionic salts such as potassium chloride, in the range of 5% to 12% w/v normally solidify to form a

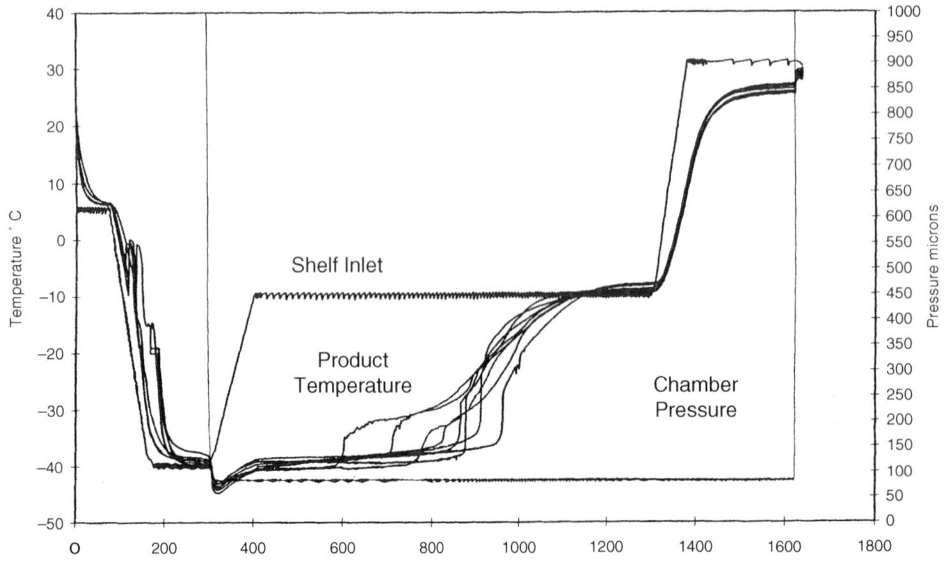

Figure 1 Temperature variation during a drying for a 2-mL tubing vial with combinations of a protein-mannitol-glycine formulation.

dense, uniform structure, regardless of the rate at which the material is cooled during freezing. A significant difference in structure can be observed when polysaccharides such as dextran, sucrose and lactose are solidified under different rates of cooling during freezing (13). Such readily measurable differences during processing and for the dried material provide useful product temperature and residual moisture that is quantifiable data for evaluation.

Product temperature and dried material attributes that are influenced by differences in processing conditions can be correlated to location and product thermal history. Attributes of reconstitution time and residual moisture may be quantified. Results of statistical analysis of the temperature during processing and dried material data can be used to identify a location that is most representative as well the most extreme throughout the entire process. These results can be used to demonstrate the extent of batch uniformity and identify appropriate locations for monitoring and product sampling during actual product validation studies. Combining extensive process monitoring with dried material testing, uniformity of the dried material can be evaluated and shown to be independent of the location within the lyophilizer.

Process testing combines functions tested separately in the preceding steps of the OQ studies and may utilize a surrogate preparation. This study challenges the integrated control functions as well as demonstrating batch uniformity. It is important to note that process testing is independent of any particular processing parameters and any specific product presentation. Rather, it is a series of well-controlled tests, designed to demonstrate the equipment capability to reproducibly implement the lyophilization process and yield consistent product qualities, independent of the location of the product within the lyophilizer.

Integrated Control Functions

Integrated control functions encompass the lyophilization process itself, along with alarm functions and fail-safe responses to out of range process conditions. Critical parameters of shelf temperature, chamber pressure and time, and the success in controlling these parameters within an acceptable range are demonstrated during the actual lyophilization of material. For ease of completing the testing, and as a precursor to implementing a process with test material, the lyophilization cycle may be completed using an empty chamber, with alarm function tests and fail-safe responses challenged. During this test, alarm conditions such as the shelf temperature and chamber pressure may be altered by physically forcing such conditions. For example, directly engaging the heaters would cause the shelf temperature to warm above the allowable target set-point range. Engaging the refrigeration compressor when the shelves are at the target set-point would cause the shelf temperature to fall below the range, also instituting an alarm condition. Fail-safe responses would also be tested in a similar manner. Table 4 highlights some of the critical parameters that would be appropriate to test during such a simulation.

Table 4 Process Fail-Safe and Alarm Tests

High shelf temperature
Low shelf temperature
High chamber pressure
Low chamber pressure

LYOPHILIZATION PRODUCT QUALIFICATION AND PROCESS VALIDATION

Lyophilization is a method of preservation where the conditions necessary for the process are dependent upon the characteristics of the material to be processed. The finished material is dependent upon the processing parameters used for freezing and freeze drying. This then requires that the physio-chemical character of the material be well-defined and understood in order to develop a suitable process. For routine processing, the consistency of the starting material dictates the level of success for the outcome of the process. Such data is a prerequisite to designing an appropriate set of lyophilization conditions. There may be characteristics of the material that allow quantifying the level of success of processing. This requires that the characterization of the starting material be considered when undertaking a validation study, and are discussed within the following sections.

Definitions for validation published in section 210.3 the Federal Register in May, 1996 emphasize the distinctions between process validation and suitability (14). Process validation is defined as "...establishing, through documented evidence, a high degree of assurance that a specific process will consistently produce a product that meets its predetermined specifications and quality characteristics." Process suitability is described as "...established capacity of the manufacturing process to produce effective and reproducible results consistently." Section 211.220, describing process validation also includes demonstrating reproducibility of the process as a requirement.

The application of lyophilization is for the preservation of materials unstable in the presence of water, demonstrating that a process produces product of suitable quality characteristics at the time of release and extending over the shelf life of the product. Preservation is then an inherent objective of the process and requirement for the product, placing a greater emphasis upon correlating product attributes and stability to processing conditions. This emphasis is carried through the process development. Considering this approach to applying these validation concepts to lyophilized processes and products, the significance of development activities and the suitability of validation during development are apparent.

Preformulation Data

As part of the preformulation activities, investigations include physio-chemical character, purity, solubility, stability and optimal pH studies. Potential product formulations, considering route of administration and solution stability are initially studied in preparation for producing material to be used in clinical studies, Unique to dosage form development studies for lyophilized products, thermal analysis of the drug substance and product formulations are also necessary. Data generated during this phase of product development is useful for future development activities, along with validation.

For lyophilized drug products, the active substance purity and morphology, formulation procedures, excipients used and initial concentration may effect the behavior during processing as well as dried material stability, with a wide variety of examples in the literature. For example, certain beta-lactam antibiotics may solidify to an amorphous or crystalline morphology. Each different form exhibits different physio-chemical properties such as solubility and stability (15). In addition, pH may be an influencing factor in the phase transition of the substance (16). The presence of certain excipients may also alter the morphology of the active substance (17). Degradation pathways involving hydrolysis, common for products that require lyophilization are also significant. For biopharmaceuticals, numerous biochemical reactions such as hydrolysis, oxidation, deamidation, beta-elimination and racemization play an important role in stability of the final product (18). It has also been reported that residual levels of an impurity in mannitol as low as 0.1% w/w has been involved in the degradation of a polypeptide upon storage (19). There are often, therefore, critical behavioral characteristics that need to be considered in the manufacturing of lyophilized products and assessing the success of the formulation and processing methods developed and subsequently validated.

Well-documented studies, summarized within a development report on the physio-chemical aspects, drug substance attributes and finished product characteristics becomes an important reference to support design and the acceptance criteria for the validation studies. Such data is also valuable for future integration into a manufacturing operation. This includes the scientific rationale for formulating and bulk handling procedures, lyophilization processing parameters, and finished product analysis.

Development Activities

Development activities encompass both the initial conditions for the drug substance or formulation and packaging considerations for a drug product, along with the lyophilization cycle. For a drug substance, upstream processing and condition of the starting material need to be quantified. This includes solvents present and, for multiple-solvents, the ratio of one solvent to another, impurities and related substances. As part of product development, compounding procedures, formulation components, active ingredient quality, selection of the container/closure, in addition to process engineering of the lyophilization cycle are studied.

Acknowledging that development is a precursor and critical to designing the studies necessary for validation, considerations for each major phase of the development activities will be reviewed. This review starts with studies of the drug substance and progresses through finished product testing.

Drug Substance

The physio-chemical character of the active ingredient steers the formulation design and selection of excipients for the finished product. If, for example, the drug substance has a propensity to form either an amorphous or crystalline phase, the method of freezing and the character of the material needs to be assessed during development. Material solidified during freezing as a crystalline form is more thermodynamically stable than an amorphous from. For example, studies have shown that the solid state decomposition of cefoxitin sodium can occur at a significant rate. The amorphous form yields a 50% loss of the active ingredient within one week at accelerated storage conditions of 60°C. The crystalline form degrades to less than 10% loss in eight weeks (20). Investigating character of the active material therefore needs to be studied during development. As well, the influence of processing conditions on the morphological form is also critical, as discussed later in this chapter.

The specific physio-chemical character of the material may be a useful means for verifying reproducibility during the validation studies. Materials that will form a crystalline morphology and have good bioavailability and stability may be formulated with mannitol as an excipient, where both the active and mannitol readily crystallize. However, some excipients will alter the morphology of other excipients or the drug substance. These differences may be quantified with analysis by X-ray diffraction. Peptides and globular proteins tend to inhibit the crystallization of some excipients that tend to crystallize. An example of this is the affect of hGH on the morphology of glycine and mannitol (21). In such circumstances, the physio-chemical characteristics of the substance can be useful in qualifying the formulation design of the finished product. It may also be a useful tool in assessing process reproducibility and product consistency.

Other factors that need to be considered are the purity profile of the active substance. For example, a synthesized drug substance precipitated out of an organic solvent may contain trace amounts of the crystallizing solvent. Even residual levels of the solvent or other impurities affect the measured phase transition of the material (22). Therefore, the amount of allowable trace solvents or impurities and their impact upon product behavior during processing need to be included in early development studies and may also be appropriate as a monitoring technique during validation.

Upstream purification of peptides and proteins often use varying combinations of organic solvents and acids to elute the substance from the chromatography column, dependent on the substance. For a peptide that may orient itself in either an alpha-helical or beta-pleated sheet configuration depending upon the presence and concentration of an organic solvent, behavior in solution or during the freezing process may differ substantially for each conformation. Trace amounts of solvents and acids may affect the behavior of the substance in solution and during freezing. Such details of the requirements, sensitivities and behavior of the active substance need to be defined in the scheme of both development and considered during validation activities. An appropriate purity profile should be established and monitored to show control for the starting raw material. Residual substances, including processing solvents, chemical intermediates, precursor fragments, along with microbiological quality are also necessary. Degradation products from upstream processing and bulk solution stability also need to be established during development and may be used during scale-up and manufacturing validation studies.

Based upon the active substance characteristics determined during development, acceptance criteria of the validation studies can be established. These criteria apply to demonstrating the consistency of the dried material processed within a PAR in the development phase and adequacy of the scale-up to manufacturing. To be comprehensive in this presentation, numerous aspects, although not necessarily applicable to all products, are presented as illustrations.

In circumstances where the active or any excipient may crystallize, monitoring of the morphology in evaluating the dried product may be warranted. If differences in solubility, reconstitution rate or stability are influences by the morphology of either an excipient or the drug substance, then a quantitative method should be included for assessing finished product attributes. Methods of analysis for dry powder include infrared spectroscopy, nuclear magnetic resonance, particle morphology and thermal analysis (23).

Degradation products due to hydrolysis, oxidation, or specific biochemical reactions should be monitored by an appropriate stability-indicating assay. Polymerization, aggregation, and denaturation levels should also be included in the finished product and stability monitoring protocols, if appropriate.

Finished Product Formulation

The solubility and stability profile at different pH are important in identifying the acceptable pH range for the product formulation. In some instances, there is a compromise between solubility and stability, either for the bulk solution or dried product. For example, a 1 pH unit difference from pH 5 to pH 4 for penicillin increases the solubility along with opportunistic degradation reactions by one log (24).

Understanding the effect of bulking agents and their interactions should be studied during development. As with measuring the degree of crystallization discussed earlier, this may provide a quantitative measurement that may be useful for demonstrating process reproducibility and product consistency. Formulations containing excipients that tend to crystallize such as sodium chloride, phosphate buffer, mannitol or glycine are examples where this may be useful.

The effect of the variation in pH adjustment or the influence of any buffering system also needs to be studied. For the range of pH, any influence on the behavior of the active or excipients during the freezing and the phase transition upon warming must be measured. As an example, in a biopharmaceutical formulation containing glycine with the pH adjusted using sodium hydroxide, sodium glycinate would be formed. The behavior of sodium glycinate in the formulation may be different than that expected of glycine in the free acid form. The difference in behavior and phase transition temperature has been evaluated by Akers (25).

Unless there is a specific and critical function of an excipient, an assay is not normally considered to be necessary during validation. There are, however, formulations where an excipient is critical to the function of the active ingredient. For example, in some in-vivo imaging agents, the reduction of stannous chloride is necessary in the coupling of a radio-labeled compound. For Amphotericin B, deoxycholate sodium is used as a solublizing agent and needs to be at a minimum concentration to assure that the drug is completely soluble upon reconstitution. The concentration of the excipient in these two examples is critical and an assay would be appropriate.

Determining Thermal Characteristics

Establishing the temperature necessary to completely solidify the product during freezing and maintain the product below during drying is imperative to determine early in product development. If a component of formulation has a propensity to crystallize, it would be best to occur during freezing; processing parameters used for cooling the product may be crucial to induce such crystallization and need to be explored during development.

Using low temperature thermal analysis, the phase transition during cooling and warming is critical data necessary to identify appropriate lyophilization parameters and to justify the process. This is necessary for determining the temperature below which to cool the product during freezing and the maximum safe threshold temperature during primary drying. Results of the thermal analysis studies are used to support identifying a threshold temperature for processing. This threshold is the temperature during freezing the product is to be cooled below. In primary drying, it is critical for the product to complete drying with retention in the presence of ice and early in secondary drying. For example, the solid–liquid phase diagram for sucrose presented by MacKenzie, indicates that there is a glass transition at −32°C to −34°C when the sucrose is in the presence of ice and prior to any amount of significant desorption (26).

Commonly used methods for low temperature thermal analysis needed for lyophilized products are highlighted in Table 5. There are a number of methods available commonly used for low temperature thermal analysis, each having particular advantages. Although the nature of the material dictates the most applicable method, confirming analysis by a second method is a valuable tool for greater insight and understanding for the behavior of the material under freezing and freeze drying conditions. Differences in measurements and observations and the impact upon the drying conditions designed for processing warrants the use of confirming methods.

Table 5 Methods of Low Temperature Thermal Analysis

Method	Principle	Indication
DSC (differential scanning calorimetry)	Change in molecular heat capacity	Glass transition and eutectic melt
ER (electrical resistance)	Change in electrical conductivity	Glass transition or eutectic melt
Freeze drying microscope	Direct microscopic observation	Fluid flow and structural collapse

Assessing Bulk Solution Stability

With hydrolysis being the prominent reaction contributing to product degradation, stability over the typical length of time the product remains in the presence of liquid water as a bulk solution is critical. Controlled storage conditions for the bulk solution prior to filling and lyophilization is necessary as part of assuring finished product consistency and batch uniformity. Such an evaluation is important to verify that no appreciable degradation has occurred and justifies the time limits for bulk storage. This would include the time from when the product was formulated to the end of the filling operation, from the first container filled to the last. Since lyophilized formulations are unpreserved and do not contain a bacteriostatic or bacteriocidal agent, microbiological quality, including endotoxin levels, are important to monitor to assess the product microbiological quality.

Slight differences in the nature of the formulation due to aging may also influence the product phase transition. For example, absorption of oxygen or carbon dioxide from the air over an extended period of time may cause a pH shift, consume one component of a buffering system, or promote degradation. For a peptide or protein with both a hydrophilic and hydrophobic nature, alterations to desired secondary or tertiary structure may develop. As a result, polymerization, aggregation, or denaturation may occur. Any one of these may alter the behavior during processing and finished product characteristics. If such an opportunity would exist, any conformational changes evidenced by aggregation, perhaps leading to gelation, may occur. These need to be monitored using an appropriate analytical technique best suited for detecting such subtle and potentially significant changes. In addition, it is imperative to justify and validate the allowable bulk storage conditions such as temperature or atmospheric conditions, including a suitable time.

Justification of Processing Parameters

During the process development phase, ideal processing conditions should be devised to yield desired finished product qualities and acceptable stability. Target processing parameters of shelf temperature, chamber pressure and time that are safe, effective and efficient need to be established. As a matter of routine, target conditions are selected and studied as part of process engineering activities during the development phase. A temperature for completely solidifying the product during freezing is established based upon results of thermal analysis studies.

Thermal analysis data also dictates the maximum product temperature allowable during primary drying, as discussed previously. Shelf temperatures and chamber pressures are then selected to assure that the product remains below this critical threshold temperature during primary drying. Secondary drying conditions necessary to achieve suitably low residual moisture are also identified. Determining these processing parameters requires well-designed laboratory studies to define optimal conditions for a safe, effective and efficient process.

The result of process engineering studies would be definition of an ideal set of processing parameters for shelf temperature, chamber pressure and time as a target set of conditions for routine manufacturing. Control of these parameters and monitoring processing conditions begin when the product is first loaded onto the shelves of the lyophilizer until the product is stoppered and removed. In addition, the rate of change from one shelf temperature to another also needs to be predefined and controlled. These rates of changes, referred to as ramps, include cooling rates during freezing, warming of the shelf at the beginning of primary drying and the transition from primary to secondary drying.

As an example, the complete process description for methylphenidate hydrochloride, a product in which the active ingredient has a phase transition of −11.7°C and the formulation contains mannitol may be described as outlined in Table 6 (27). Material processed according to the predetermined conditions would be expected to yield product of acceptable quality, purity, and stability. Reproducibility of these parameters is demonstrated by comparing the actual processing parameters for any one batch to the ideal target parameters identified as a result of development studies. Evaluation of the finished product qualities and assessment of the stability over the desired shelf life demonstrates that the processing conditions are suitable and appropriate. Demonstrating the same process conditions and achieving the same finished product qualities and stability shows that the process is reproducible and the product qualities are consistent.

It is also appropriate that the range of acceptable conditions that produce product of acceptable quality are defined, scientifically justified, and the impact upon finished product quality demonstrated. These include a range for the shelf temperature during freezing and drying, rates of change from one temperature to another during the process, the chamber pressure for drying, and a minimum time at each condition. Selection of the suitable ranges for the processing conditions should be

Table 6 Definition of Target Process Parameters for a Methylphenidate HCl Preparation

Process step	Shelf temperature	Rate	Chamber pressure	Time
Product loading	5°		Atmosphere	2
Cooling rate		0.5		
Freezing	−20°		Atmosphere	4
Ramp to primary drying		0.5	80 μ[a]	
Primary drying	65°		80 μ	
Ramp to secondary		0.5	80 μ	
Secondary drying	40°		80 μ	8

[a] The pressure reported ranged from 210 μ to 15 μ; 80 μ was selected as a reasonable level for discussion.

based upon empirical data rather than arbitrary selections.

Following an experimental design approach for developing a matrix of variables is undoubtedly a preferable method for conducting process engineering studies. With the numerous and complex influences on processing requirements, a complex matrix based upon numerous variables such as combinations of temperatures separately varied for each process step of freezing, primary and secondary drying is, at best, laborious, time consuming, and often limited by availability of API. Such an approach to process validation often requires an exhaustive number of studies. In the absence of the large number of studies to fulfill a complex matrix, a simpler matrix based upon the edges of a defined range would be reasonable and scientifically valid.

Process conditions that affect both the product temperature and rate of drying are shelf temperature and chamber pressure. For these process conditions, target parameters, and suitable ranges around the parameters need to be defined during the process engineering studies. Validating the process therefore requires demonstrating that if conditions existed where the process was completed at the extremes of these conditions, the finished product would have the same qualities and long-term stability as if the batch was processed at the target conditions. Since both the shelf temperature and chamber pressure are independent parameters, then the various combinations of both conditions at the extremes and at the target would establish a PAR. This notion of a PAR was first introduced by Chapman in 1984 and is well-suited for lyophilization (1). The goal of the process validation studies are to verify that if the process was completed within any combination of the two variables, then the finished material would be of consistent quality and stability.

Designing a series of studies based upon the variables of shelf temperature and chamber pressure would encompass, as a minimum, permutations of high and low conditions for each. Demonstrating reproducibility is also an objective during validation, such that three batches processed at the target conditions would also be necessary. This therefore would require a minimum of seven batches: three at the target parameters to demonstrate reproducibility and four for the combinations of high and low conditions.

In addition to the shelf temperature and chamber pressure, the time to complete secondary drying will influence the residual moisture content of the dried material. Assuming that target residual moisture content has been identified, the validation studies should also encompass a range of time at the secondary drying conditions necessary to achieve the desired residual moisture. The range of time could be incorporated within the three batches at the target shelf temperature and chamber pressure. As an illustration and using the cycle defined for methylphenidate described in Table 6, the variations in shelf temperature, chamber pressure and time in secondary drying are presented in Table 7.

The parameters outlined in Table 7 consisting of high shelf temperature and high chamber pressure would provide the upper level of processing conditions. During freezing, the shelf would be controlled at the maximum or warmest temperature at which solidification would occur. During primary and drying, the warmest shelf temperature and highest chamber pressure would result in the greatest amount of heat transfer. This increased heat transfer, as compared to that at target conditions, would result in the greatest rate of sublimation and warmest product temperature. The greatest amount of heat transfer would be expected to result in the warmest product temperature and possibly the shortest processing times. In considering the impact during secondary drying, the high levels would provide potentially higher rates of desorption, and therefore the lowest residual moisture content. The end result should be the slowest freezing rate, fastest drying rate, warmest product temperatures during the process and lowest residual moisture.

The study encompassing the coldest shelf temperature and highest chamber pressure would be expected to yield a different rate and therefore time to complete each part of the lyophilization process. In this study, a decrease in the rate of sublimation as compared to the cycle conditions above is anticipated since the shelf temperature is lower and a resulting decrease in the amount of heat energy to support sublimation would occur. There would be, however, a contribution in heat transfer by the increased chamber pressure, as compared to the target processing conditions. The rate of sublimation and desorption would be expected to be lower than that for the set of processing conditions in the first study.

A higher chamber pressure would provide greater efficiencies in heat transfer from the shelf. Any increase in the overall amount of heat transfer relative to the parameters of lower shelf temperature and higher chamber pressure would depend upon the specific parameters

Table 7 Varied Process Parameters for a Proven Acceptable Range

Process condition	Product loading	Cooling rate	Freezing	Ramp to primary drying	Primary drying	Ramp to secondary drying	Secondary drying		
Shelf temperature									
High	10°C	0.5°C/hr	−15°C	0.5°C/hr	60°C	0.5°C/hr		35°C	
Target	5°C	0.5°C/hr	−20°C	0.5°C/hr	65°C	0.5°C/hr		40°C	
Low	0°C	0.5°C/hr	−25°C	0.5°C/hr	70°C	0.5°C/hr		45°C	
Chamber pressure									
High	Atmosphere	Atmosphere	Atmosphere	100 μm	100 μm	100 μm		100 μm	
Target				80 μm	80 μm	80 μm		80 μm	
Low				60 μm	60 μm	60 μm		60 μm	
Time	2 hr		2.5 hr		7 hr		6 hr	8 hr	10 hr

selected. The greatest anticipated effect would be on the product temperature because of the increase in chamber pressure. This effect would be strongly dependent upon the specific processing pressure. For example, the impact of a 20 μ increase is greater at a target pressure of at 80 μ as compared to when the chamber pressure is at a target pressure of 400 μ. For these sets of processing conditions, product temperatures during each process phase, rates of drying and residual moisture content would be intermediate as compared to the other studies.

Compared to a higher pressure and lower shelf temperature, drying rates with the reversed conditions of lower pressure and higher shelf temperature would be expected to be slower. Comparatively, freezing would be expected to require more time. Primary drying rates would also be reduced because heat transfer rates would be less, product temperatures lower and residual moisture higher.

The longest times for the product to reach completion for each cycle phase would result from reduced heat transfer due to a lower shelf temperature. Coupled with a lower chamber pressure, it may be expected that the processing rates would be the most significantly reduced. In this study, the principle objective is to demonstrate the times allocated for each portion of the process are adequate, even under the conditions where the heat transfer was low and times were longest as compared to the target parameters; the heat transfer would be lowest, and therefore the freezing and drying require the longest time. The product temperature would also be expected to be the lowest as compared to the other processing conditions. Processing under these conditions of the least heat transfer demonstrates that there is sufficient time designed within the cycle parameters to accommodate such variations in rates of drying.

PARs of processing parameters during primary and secondary drying would be expected to yield some range of residual moisture. This range would result from different variables of shelf temperature, chamber pressure, along with the conditions for desorption in secondary drying. The least significant impact is often variations in time. Depending upon the characteristics of the formulation and the association of residual moisture in the product, the allowable range of time in secondary drying needs to be correlated to a residual moisture content. This should be accomplished during the process engineering phase. Sequential stoppering or use of a sample extraction device to determine the change in residual moisture content over time is a convenient method for measuring the extent of moisture decrease. Another method used during development activities to justify the time necessary in secondary drying is generating a desorption isotherm. Examples, such as the sorbtion isotherms for PVP have been presented by MacKenzie (28). Methods for conducting such studies have also been more recently described by Groves (29).

FINISHED PRODUCT ATTRIBUTES

There are unique dried material quality attributes associated with lyophilized materials. The term dried material is used loosely here and meant to encompass both lyophilized drug substances and intermediates, as well as drug products intended for administration. Quality attributes are nearly identical for each type of material. Stringent microbiological quality is also a requirement for sterile products.

In addition to chemical or biological assay and specific requirements for a finished product, such as those for parenteral administration, the condition of the dried cake also needs to be identified. These include the physical condition and appearance of the dried cake and the ease of which the dried material reconstitutes into a solution.

The result of successful and effective freeze drying process is the retention of the physio-chemical attributes of the starting solution and preferably, retention of the structure established during freezing. Assay of the constituted solution assures the preservation of the desired activity present in starting material. Assay of multiple samples of dried material is used to demonstrate content uniformity.

Physical Appearances

The appearance of the dried material should be uniform in structure, color and texture. A material having ideal pharmaceutical elegance would be a dense white cake with fine, uniform structure as illustrated in Figure 2. As described earlier, successfully freeze drying results in the retention of the structure established during the freezing step. If the material forms a desired appearance upon freezing and that structure is retained throughout the drying, then the process should yield a finished product with an appealing appearance.

For some formulations, and typically particularly those with low solids content, the dried cake may shrink from the original volume upon drying, as evidenced in the sample in Figure 3. Such shrinkage is dependent on the concentration of the starting solution, nature of the active ingredient and the amount and type of excipients used. However the shrinkage is often uniform throughout the container as well as throughout the batch. Although requirements for stabilizing the drug substance and route of administration supersede pharmaceutical elegance, the design of an ideal

Figure 2 A cake that is uniform in appearance, texure and color, occupying the original volume of the liquid fill epitomizes a pharmaceutical elegance for a lyophilized product.

(A) (B) (C)

Figure 3 Loss of the initial structure during drying due to the product being warmer than the phase transition temperature yields a varying amount of collapse. (**A**) A significant change in the cake shape and structure is illustrated. (**B**) Extensive collapse with minimal similarity of color and a dimensional proportion to the original cake. (**C**) Extensive collapse to form granular masses at the vial bottom with a residue on the side wall of the vial but without any recognizable similarity to the original cake.

formulation would yield a dense cake, uniform in color and texture, with good physical strength and friability (30).

A decrease in total volume or localized loss of structure can also be associated with a condition referred to as collapse (31). This condition, as described by Mackenzie, occurs when the frozen or partially dried material exceeds the phase transition where the material may soften and again become fluid. Samples of dried product in Figure 4 illustrate the different appearance of the cake structure due to extensive collapse.

With the material softening and becoming fluid, there is a loss of desired structure established during freezing. A loss of structure often coincides with entrapment of water as the material looses its structure and collapses. This entrapment of water into relatively larger masses may also prevent adequate desorption that normally occurs during secondary drying, resulting in a high residual moisture content. Reconstitution times may also be lengthened because of a "case hardening," making it more difficult for water to permeate the dried material upon reconstitution.

Figure 4 The slight gap between the dried cake and the side wall of the vial exemplifies shrinkage that may occur with some formulations. This shrinkage may be attributed to either low concentrations or be characteristic of the materials in the formulation.

Since the objective of this process is the preservation of the lyophilized material, the presence of collapsed material is suspect. As described above, collapse may simply be considered a cosmetic defect. When the collapsed material exhibits an increased reconstitution time or poor dissolution, the presence of collapse is categorized as a quality concern. If the collapsed material retains a higher amount of residual water, where this water becomes involved in degradation of the product through hydrolysis, there is a more serious concern. The presence of a significant amount of residual water in localized regions of collapsed material may promote degradation of the product. There would also be a concern of the degradation products toxicity or any influence on the therapeutic effectiveness. Both potential results should be considered during product development.

Residual Moisture

The primary objective for lyophilized materials is to minimize or eliminate water that would be chemically active during long-term storage of the product. Any water readily available may become involved in hydrolysis reactions, the common cause of degradation for lyophilized products. This therefore requires that a sufficiently low residual moisture content be achieved. An acceptable range of moisture content, identified during development is a primary indication that the lyophilization process was successfully executed.

The established residual moisture suitable for acceptable long-term stability may approach the variability of the moisture determination method or may be as great as a few percent. For example, many lyophilized products within the USP have a finished product residual moisture specification of less than 2% of dry weight. Other products, such as Amphotericin B have a residual moisture limit of 8.0% (32). Whether the allowable residual moisture specification is small or large, a range of acceptable residual moisture should be identified and correlated to suitable long-term stability.

The analytical method for moisture determination must be validated prior to use during any process validation studies. There are numerous techniques for moisture analysis that range from physical methods such as loss on drying, to chemical methods such as Karl Fisher. The most common and preferred method is extraction using anhydrous methanol titrated with a colorimetric Karl Fischer method. A comparative review of the conventional techniques is presented in an overview by May (33). The use of near-infrared methods has recently been introduced as an alternate method (34).

Residual Solvents

There is an increasing use of combinations of mixed solvents; i.e., aqueous and organic solvents (35–37). Residual moisture of the dried material assesses the successful completion of secondary drying. Using combinations of water and organic solvents, the moisture of the dried product would be assessed upon completion of lyophilization. Residual solvents should also be assessed as a measure of effective decrease to acceptable low levels. Various methods are available for quantifying residual solvent content, such as thermogravimetric analysis or headspace analysis by GC, as outlined in the USP (31).

Reconstitution

Times required for reconstitution and the appearance of the constituted solution are also of importance. The nature of the dried material as a result of lyophilization yields a product that is highly hygroscopic. Reconstitution is often instantaneous upon adding the diluent. For ease of use in a clinical setting, reconstitution times are often less than two minutes. Whatever time is required to resolubilize that material, the constituted solution should be clear and free of any visible particulates or insoluble materials, meeting the compendial requirements such as those outlined within the USP (31).

The method of reconstitution is also important. For example, the package insert for lyophilized somatropin indicates that during aspiration the diluent stream should be aimed against the side of the vial. In addition, the constituted solution should be gently swirled and not shaken (37). Vigorous motion could result in aggregation of the protein to form insoluble particles. For Amphotericin B, vigorous shaking is indicated until all of the crystalline material dissolves, forming a clear, yellow colored colloidal dispersion (37). Whether the solution dissolves instantaneously or requires special handling, forming a colorless solution or a colored colloidal dispersion, the expected appearance of the constituted solution needs to be a quality attribute established and supported by development data.

ASSAY

Analysis of the active ingredient, whether by chemical or biological methods, would be the same for the constituted product, as would be necessary for any ready-to-use preparation. Constituted solutions, however, have a limited shelf life after addition of the diluent. Depending upon the solution stability, the package insert may indicate that the constituted solution be used immediately upon reconstitution, or may be stored at selected conditions, often 5°C, for a specified length of time. The stability of the constituted solution needs to be established during development and measured as part of the stability testing. The potency and purity must also be measured at the end of the indicated shelf life. This included not only the solution after initial reconstitution but also after storage at the conditions indicated in the package insert. Analysis should also include assay of any degradation product.

SUMMARY

Lyophilization is a complex unit operation, integrating multiple processing steps with varied conditions for long-term preservation of the pharmaceutical products. This same process is applied to processing drug substance as well as a compound formulation for a finished drug product.

Lyophilization processes consist of the manipulation of process parameters to create environmental conditions of sub-ambient temperatures and sub-atmospheric pressures. These extraordinary environmental conditions that promote the various conditions suitable for the respective processing mechanisms are created by the lyophilization equipment. The success of the process therefore relies heavily upon the operating performance of the lyophilizer. Confidence in the ability of the equipment to implement these processing parameters is achieved through the successful completion of a comprehensive IQ and OQ. Without the proper performance of the equipment, there is limited opportunity for successful processing of materials.

Throughout this presentation, emphasis is placed upon the need to develop an appropriate and adequate process. This includes challenging the process to develop a set of boundary conditions to create a PAR for the process. The result of such an approach is a rugged and robust process, yielding cycle conditions that are safe, effective and preferably, efficient. These processing conditions are demonstrated to be adequate and appropriate, ultimately through initial testing as well as after long-term storage of finished product. Of equal importance, this process is applied to preserve the quality of the material through processing and throughout the shelf life. Demonstrating the process suitability also requires correlating the process to product stability.

The behavior of the material during the processes is strongly dependent upon the characteristics of the starting materials. The initial characteristics must also then be measured and quantified as well. This includes not only the quality of the starting raw material, but also in the preparation and packaging prior to placing the product into the lyophilizer.

Finally, how the characteristics and quality of the finished product is quantified is of equal importance. This includes the physical attributes of the dried material as well as the quality upon reconstitution. The level of quality must extend beyond the time of initial testing for release of the batch to the final expiry date.

REFERENCES

1. Chapman K. The PAR approach to process validation. Pharm Technol 1984; 4(12):47–54.
2. Parenteral Drug Association. Validation of Computer-Related Systems. Technical Report No. 18. Bethesda, MD: Parenteral Drug Association, 1995.
3. Dushman S, Lafferty JM. Scientific Foundations of Vacuum Technique. New York: Wiley, 1962.
4. Ryans JL, Roper DL. Process Vacuum System Design & Operation. New York: McGraw-Hill, 1986.
5. Jennings TA. A model for the effect of real leaks on the transport of microorganisms into a vacuum freeze-dryer. J Parenter Sci Technol 1990; 44(1):22–5.
6. The Parenteral Society. Technical Monograph No. 7: Leak Testing of Freeze Dryers. Wiltshire: Parenteral Society, 1995.
7. Greiff D. Factors affecting the statistical parameters and patterns of distribution of residual moisture in arrays of samples following lyophilization. J Parenter Sci Technol 1990; 44(3):119–28.
8. Kobayashi M, Harasgunma K, Sunama R, Yao A. Inter-vial variance of the sublimation rate in shelf freeze dryer, Presented at the XVIII International Congress of Refrigeration, 1991.
9. Scheaffer G, Sum L, Trappler E. Techniques in Demonstrating Batch Uniformity for Lyophilized Products, Presented at the Annual Meeting of the PDA. Boston, MA, November, 1995.
10. Pikal MJ, Shah S, Senior D, Lang JE. Physical chemistry of freeze drying: measurement of sublimation rates by a microbalance technique. J Pharm Sci 1983; 72(6):635–50.
11. Day L. Influence of Vial Construction and Material on Uniformity of Product Temperature During Freezing and Freeze Drying of Model Product Formulations, Presented at the Annual Meeting of the PDA. Boston, MA, 1995.
12. Pikal MJ, Roy ML, Shah S. Mass and heat transfer in vial freeze-drying of pharmaceuticals: role of the vial. J Pharm Sci 1984; 77(9):1224–37.
13. Ryan R, Trappler E. Course notes, Fundamentals of Lyophilization, PDA.
14. FDA. Proposed rules. Fed Regist 1996; 61;87.
15. Pikal MJ, Lukes AL, Lang LE, Gains K. Quantative crystallinity determinations for beta-lactam antibiotics by solution calorimetery: correlation with stability. J Pharm Sci 1978; 67:767.
16. Penner G, Trappler E. Effect of pH on the Phase Transition Temperature of Model Excipients, Presented at the Annual Meeting of the PDA. Philadelphia, PA, November, 1990.
17. Korey DJ, Schwartz JB. Effects of excipients on the crystallization of pharmaceutical compounds during lyophilization. J Parenter Sci Technol 1989; 43(2):80–3.
18. Manning M, Patel K, Borchardt RT. Stability of protein pharmaceuticals. Pharm Res 1989; 6(11):903–18.
19. Dubost D, Kaufman M, Zimmerman J, Bogusky MJ, Coddington AB, Pitzenberger SM. Characterization of a solid state reaction product from a lyophilized formulation of a cyclic heptapeptide. A novel example of an excipient induced oxidation. Pharm Res 1996; 13(12):1811–4.
20. Orberholtzer ER, Brenner GS. Cefoxitin sodium: solution and solid-state chemical stability studies. J Pharm Sci 1979; 68:863–6.
21. Pikal MJ, Dillerman KM, Roy ML, Riggin RM. The effects of formulation variables on the stability of freeze dried human growth hormone. Pharm Res 1991; 8(4):427–36.
22. Her L, Deras M, Nail S. Electrolyte-induced changes in glass transition temperatures of freeze-concentrated solutes. Pharm Res 1995; 12(5):768–72.
23. Brittain HG, Bogdanowhich SJ, Bugay DE, DeVincentis J, Lewen G, Mewman AW. Physical characterisation of pharmaceutical solids. Pharm Res 1991; 8(8):963–73.
24. DeLuca PP, Boylan JC. Formulations of small volume parenterals. In: Avis KE, Lachman L, Liberman HA, eds. Pharmaceutical Dosage Forms: Parenterial Medications., Vol. 1. New York: Marcel Dekker, Inc., 1984:195.
25. Akers MJ, Milton N, Bryn SR, Nail SL. Glycine crystallization during freezing: the effects of salt form, pH, and ionic strength. Pharm Res 1995; 12(10):1457–61.
26. MacKenzie AP. Non-equilibrium freezing behavior of aqueous systems. Philos Trans R Soc Lond B 1977; 278:167–89.
27. DeLuca PP, Lachman L. Lyophilization of pharmaceuticals I, effect of certain physical–chemical properties. J Pharm Sci 1965; 54(4):617–24.
28. MacKenzie AP, Rasmussen DH. Interactions in the water-polyvinylpyrrolidone system at low temperatures. In: Jellinek HHG, ed. Water Structure at the Water-Polymer Interface. New York: Plenum Publishing Corp., 1972:146–72.
29. Teng CD, Zarrintan MH, Groves MJ. Water vapor adsorption and desorption isotherms of biologically active proteins. Pharm Res 1992; 8(2):191.
30. Bashir JA, Avis KE. Evaluation of excipients in freeze-dried products for injection. Bull Parenter Drug Assoc 1973; 27(2):68–83.
31. MacKenzie AP. Collapse during freeze drying-qualitative and quantative aspects. In: Goldblith WA, Rey L, Rothmayer WW, eds. Freeze Drying and Advanced Food Technology. New York: Academic Press, 1975:278–307.
32. USP 29/NF 24. Rockville, MD: United States Pharmacopieal Convention, 2006.
33. May JC, Grimm E, Wheeler RM, West J. Determination of residual moisture in freeze-dried viral vaccines: Karl Fischer, gravometric and thermogravimetric methodologies. J Bio Stand 1982; 10:249–59.
34. Last IR, Prebble KA. Suitability of near-infrared methods for the determination of moisture in a freeze-dried injection product containing different amounts of the active ingredient. J Pharm Biomed Anal 1993; 11(11/12):1071–6.
35. Kasraian K, DeLuca PP. Thermal analysis of the tertiary butyl alcohol-water system and its implications on freeze drying. Pharm Res 1995; 12(4):491–5.
36. Teagarden D, Baker D. Practical aspects of lyophilization using non-aqueous co-solvent systems. Eur J Pharm Sci 2002; 15:115–33.
37. Den Brok M, Nuijen B, Kettenes-Van den Bosch J, et al. Pharmaceutical development of a parenteral lyophilised dosage form for the novel anticancer agent C1311. J Pharm Sci Technol 2005; 59(5):285–97.
38. Trissel LA. Handbook on Injectable Drugs. 8th ed. Bethesda, MD: American Society of Hospital Pharmacists, Inc., 1994.

29

Qualification Concerns for Isolator Systems

James Agalloco
Agalloco & Associates, Belle Mead, New Jersey, U.S.A.

Validation is an activity that has become a preeminent concern across the global healthcare industry. While clearly a requirement for compliant use of a piece of equipment, individuals and firms have obsessed about it to the point where validation has been blamed for delays in the implementation of numerous projects. Perhaps nowhere has this been more evident than where firms have endeavored to validate new isolator installations. Much of the pain associated with the validation of isolation technology is self inflicted. This chapter will review the qualification requirements associated with the validation of isolators. It hopes to clarify the subject as it relates to isolators to the point where it becomes an achievable reality.

Validation has been defined by the FDA as:

> Process validation is a documented program which provides a high degree of assurance that a specific process will consistently produce a product meeting its predetermined specifications (1).

Unfortunately, as the definitive statement on validation from the world's largest drug regulatory body, that definition leaves substantial room for interpretation. The one element of this definition that seems to create the most difficulty is establishing what actually is a "high degree of assurance." For more established processes, such as steam sterilization or even cleaning validation the criteria for acceptance have been widely discussed and are thus largely consistent across the industry. Isolators present unique challenges as they combine elements of sterilization, aseptic processing, environmental control, cleanrooms, and containment in a single system. Within the context of the multiple validation concerns which must be satisfied, and the expected superior performance of isolators (as contrasted to ordinary human scale cleanrooms for aseptic processing), well-intentioned, but unfortunately inexperienced (at least with isolators) individuals, and firms suggested criteria for isolators that mandated perfection in all aspects. Their stated objective was an aseptic filling isolator capable of providing finished product containers equivalent in sterility assurance to terminally sterilized products. This is a patently unattainable goal; no technology that relies on exclusion of microorganisms from sterile materials can ever realize the same degree of sterility assurance as a terminal treatment that destroys their viability. Nevertheless, the damage was done. Isolators were portrayed as systems that must be capable of operating contamination free for extended periods in order to be acceptable. The unfortunate consequences of this expectation were requirements for absence of leaks, perfection in sterilization, complete absence of any internal microbial presence and others equally unattainable (2). Given such unrealizable goals, laudable as they might be, it is no wonder that the validation of isolator systems can easily become an exercise in futility.

Less aggressive recommendations for acceptance criteria for isolators have been articulated by the PDA (3). In its guidance document the PDA addressed the subject of acceptance criteria by focusing on the definition of appropriate user requirements specifications. Justification for this approach can be found in another definition of validation:

> Validation is a defined program which in combination with routine production methods and quality control techniques provides documented assurance that a system is performing as intended and/or that a product conforms to its predetermined specifications. When practiced in a "life cycle" model it incorporates design, development, evaluation, operational and maintenance considerations to provide both operational benefits and regulatory compliance (1).

Where the acceptance criteria are derived from the operational needs of the process rather than arbitrary expectations, they become more realistic and substantially easier to satisfy. If this model is followed, rigorous criteria can still be established; however those criteria are defined by satisfaction of operating needs. An isolator need not be perfect in order to fully satisfy operational requirements; it need only be suitable for the intended use (2).

The "life cycle" approach to validation provides for cradle-to-grave consideration of a system's compliance in a validated state. Discussion of this model is best considered in an essentially chronological order. The various stages of the model include: conceptual design and planning; detailed design and fabrication; installation and operational qualification; sterilization cycle development (if required); PQ, operational use of system and maintenance. Development of a well thought out user requirements specification is essential to success. The various elements of that specification form the essential criteria against which the design and performance of the isolator will be evaluated.

CONCEPTUAL DESIGN AND PLANNING

The origin of any system begins with its basic design, which must focus on what the isolator is intended to do.

Abbreviations used in this chapter: CIP, clean in place; FAT, factory acceptance test; FDA, Food and Drug Administration; PDA, Parenteral Drug Association; PQ, performance qualification.

Questions to be answered include the what, where, when, and most importantly, the how. Implementation options are reviewed, discussed, refined, discarded and resurrected as the design process proceeds. Ultimately a system design is developed which satisfies the firm's requirements and capabilities. At this stage the design incorporates decisions regarding: isolator configuration, expected capacity, location, classification of the surrounding environment, adaptation to existing equipment and facilities, and preliminary process description (use, cleaning and decontamination). Once the design is completed and accepted the project proceeds to detailed design and fabrication. Isolator projects that are complex enough to require the development of validation plan should start that actively at the completion of the conceptual design.

DETAILED DESIGN AND FABRICATION

Early in detail design the isolator will be preliminarily sized for its intended use. For isolators that are custom designed for a specific purpose, it is common to build a mock-up of the isolator to confirm that the intended design meets the end user needs. As many isolators rely upon operators to perform a variety of tasks, ergonomics can play an important role in the suitability of the final isolator. The mock-up is used to confirm that the operators can readily perform all of the required functions. It is useful in this evaluation to consider the full complement of operators who will use the system, and the effect fatigue may have on their ability to perform the necessary tasks. The time and expense associated with a mock-up evaluation is often well spent. A substantial amount of useful information can be gleaned from a simple mock-up made of plywood. The extent of detailed design for "standard" isolators, if there is such a thing, is relatively minor. Many isolator applications such as sterility testing or aseptic filling are performed in nearly identical fashion at multiple firms. In these instances, neither mock-ups nor an individual detailed design may be necessary.

Where detailed design is performed it expands upon the conceptual design defining the specifics of the isolator system. Detailed design culminates in a set of drawings that will be used for the construction and assembly of the isolator. Where the isolator includes some measure of functionality (CIP, automatic purging of oxygen, system decontamination, etc.) these are defined in written specifications for the systems operation. The final drawings and specifications are ordinarily submitted to the customer for approval before fabrication of the isolator is begun.

Fabrication against the approved specifications is performed at the isolator manufacturer's plant and may include a FAT. The FAT, which is often a part of the formal qualification of the isolator, is ordinarily performed by the owner's staff prior to shipment. The original intent of the FAT was to confirm that the system was ready to be relocated from the vendor to the user; i.e., there were no required modifications that had to be made by the vendor at their site before it was relocated. In recent years, the FAT has tended to become a first step in the qualification of the equipment, while still providing the important acceptance of the completed system.

VENDOR INSTALLATION/COMMISSIONING

After acceptance of the system by the customer, the vendor will ship the isolator to the site and provide installation services to ready the system for qualification. For larger systems this may be the first time the system is assembled in one location. This activity is not considered a controlled activity in that is generally not performed in accordance with a set of written procedures. The later portion of this activity is sometimes called commissioning or shakedown in which the equipment is informally checked for conformance to specifications. The commission serves to ensure that the isolator system is ready for formal qualification. While this activity may seem redundant, it makes little sense to institute formal qualification activities on a system where the vendor may have made some minor error of omission or commission. The formal qualification activities should only commence after any installation miscues have been rectified.

BACKGROUND ENVIRONMENTAL CONSIDERATIONS

Isolators are ordinary positioned in controlled and sometimes classified environments. Small isolators such as those used for sterility testing, low-volume clinical manufacturing and similar activities are rarely integrated into the room in which they are installed. In these instances the isolator is essentially brought into a classified room, connected to the appropriate utilities and placed into operation. Its impact on the surrounding environment often is not significant, as the background environment is typically no higher than ISO 8 (Class 100,000) and in some instances is merely provided with controlled temperature and humidity with restricted access. Larger isolator systems are an integral part of the room (and even the facility) and thus greater attention must be paid to maintaining proper conditions throughout the system (internally and externally). Pressure differentials through mouse holes and tunnels should be addressed. One of the important considerations is the need to maintain a constant temperature across the system to facilitate uniform decontamination. This is especially important with hydrogen peroxide and peracetic acid, which are effective in a condensation mode.

EQUIPMENT QUALIFICATION

Qualification of equipment is a preliminary step in the overall validation that has been well described in the literature (and elsewhere in this volume). While it may be common in the industry to divide qualification into separate activities entitled installation qualification and operational qualification, there is in fact no reason to do so. The FDA describes this activity as equipment qualification making no distinction between those activities that focus on the installation details relative to those that focus on the operation of the equipment. The separation is both artificial and cumbersome and execution of the qualification in a unified manner is recommended.

Other than this consolidation, the qualification of isolators as described herein should present no surprises to the experienced practitioner. The physical system and its performance are measured against the specified drawings and user requirements. Critical to isolator success are such aspects as leak rates; air system performance; pressure differentials in various modes, and integration with other equipment, i.e., sterilizers, tunnels, decontamination equipment, and environmental sampling apparatus.

The qualification is little more than confirmation that the system conforms to the system specifications approved for its fabrication, installation and operation. This is accomplished through a protocol, which confirms that the system is everything it is expected to be. The system specifications including drawings, schematics and performance requirements are compared to the final delivered system to establish its conformity to expectations. Deviations to the design are noted and are either corrected through modification of the isolator, or by revision to the specifications if the system can be accepted although somewhat different from the intended design. There are no differences in the execution of qualification for an isolator from what is typical of other mechanical systems especially in those activities that relate to the installation aspects of isolators. The operational aspects of the qualification include elements of equipment design with those of typical cleanroom qualification.

PRIMARY VALIDATION CONCERNS

The next activity typically required is the performance of PQ studies addressing the various critical processes that the isolator is required to perform. These include: decontamination validation of the isolator internal (if required), sterilization/decontamination validation for any in-feed systems (if required), process validation for the isolators intended use (formulation, aseptic filling, subdivision of potent compounds, medical devise assembly, etc.), and cleaning validation for the isolator and any installed process equipment. The validation of aseptic processing in an isolator will require the prior attention to all of the required decontamination/sterilization processes utilized, as well as environmental monitoring of isolator internals, and ultimately a process simulation study. Details on each of these activities are provided elsewhere in this volume, and will not be repeated here.

As isolators are nothing more than a processing environment in which an activity takes place, their qualification mimics that of other pieces of mechanical equipment. Once established as a suitable environment in which to perform the required activity, the isolator just like the manned operating environment plays only a minor role. Qualification of the isolator's operation mimics its use as an environmental control system and little else. It is the equipment installed within it that is of greater importance, and that equipment should be minimally impacted by the isolator. If intended for aseptic processing, then validation activities relating to the isolator include decontamination of the surfaces; monitoring and control of the internal environment, and aseptic process simulation, just as they would in a manned cleanroom. If the isolator is utilized to contain a mill for high potency compounds, the primary validation concerns are for cleaning and containment capability. The only isolator relevant concern would be its leak rate to minimize product exposure. Note that in each of these cases, there is additional process equipment that is validated independent of its location inside the isolator: the filling machine in the aseptic case, and the mill in the potent compound application. The methods used for these are largely unrelated to the isolator. The methods used for the isolator aspects are essentially unchanged from those utilized for manned environments. Demonstration of ISO 5 is nearly identical whether in an isolator or a cleanroom. The relevantqualification issue for the isolator is that it provides an environment suitable for the application performed within.

USE OF SYSTEM

An isolator is built for a purpose, and its qualification/validation must support that purpose. The PQ efforts confirm the acceptability of the design and procedures to provide the required functionality for which the system was built. Each of the primary standard operating procedures required for use with an isolator—decontamination/sterilization, operation and cleaning—are usually the direct result of a supportive PQ study which demonstrate how the system performs in each procedure. In conjunction with supportive procedures such as: instrument, leak testing, environmental monitoring, etc., these procedures define how the system will perform. The application of change control to the procedures and physical system ensures that changes are reviewed for their potential impact on the validated state of the final system. The only other aspect of the systems use which must be considered is training of operating personnel in the proper execution of these procedures to ensure their continued proper execution.

ISOLATOR MAINTENANCE

As with any other piece of mechanical equipment, an isolator should be supported by an continuing preventive maintenance program to ensure it operates reliably over time. Among the ongoing maintenance considerations are: calibration of instruments, inspection/lubrication/replacement of seal surfaces, gloves and half-suits, periodic cleaning of isolator internal and external surfaces, filter integrity testing, filter replacement, filter change, periodic leak testing, and any required preventive maintenance for nonisolator equipment which are part of the overall installation.

OVERALL QUALIFICATION/VALIDATION PERSPECTIVE

Despite what may appear to some to be a near-impossible and never-ending task the qualification/validation of isolators is not particularly difficult. Where realistic (the only kind which should be defined) requirements are established for the systems performance isolators can be readied for use rather simply. As noted earlier, the problems in this area are largely associated with expectations no system could possibly attain. The classic qualification/validation methods used for other process equipment/facility systems are wholly adequate to bring these systems through validation. Experience with sterilization, process, and cleaning validation as well as some familiarity with ordinary cleanrooms is certainly helpful.

REFERENCES

1. FDA, Guideline on Principles of Process Validation, 1987.
2. Akers J, Agalloco J. Isolators—validation and sound scientific judgment. PDA J Pharm Sci Technol 2000; 54(2):110–1.
3. PDA TR# 34, PDA J Pharm Sci Technol 2001; 55(Suppl. 5).
4. Agalloco J. Validation—yesterday, Today and Tomorrow. In: Proceedings of Parenteral Drug Association International Symposium. Basel, Switzerland: Parenteral Drug Association, 1993.

30 Validation of Solid Dosage Finished Goods

William G. Lindboe, Jr.
Consultant, East Brunswick, New Jersey, U.S.A.

INTRODUCTION

This chapter will present the techniques, procedures, data, and documentation necessary to satisfy the current compliance requirement for validation of solid dosage finished goods, specifically tablet and capsule manufacturing operations. The emphasis will be on the practical inspectional requirement, rather than on a theoretical approach that does not reflect the practicalities (and problems) encountered when validating actual production operations. Some of the successes and failures that accompanied the development and application of validation protocols to solid dosage forms that began in the late 1970s will be presented. Many of these errors still occur because of turnover in technical staff and the narrow approach to validation that is a common byproduct of outsourcing. Specifically this chapter will discuss tablet and hard gelatin capsule manufacturing validation from batching through tertiary packaging and shipping. The ideas presented are an individual view based upon validation as practiced, either from the author's personal experience, published practice, or from discussions with others in the industry and industry regulators. The specifics of the validation technique described herein may be debatable. However, in the author's experience, validation documentation based upon each approach has been audited at least one time by the FDA and/or MCA with no adverse comment. The chapter includes commentary and opinions, which the reader may view as a "wish list" from someone who has worked continuously with sole validation responsibility since 1979.

The fundamental premise of the FDA and European regulatory agencies is that manufacturers are the experts in the technical aspects of their particular processes and products. Therefore, the manufacturers should develop their own validation methodology. The lack of specific procedures in the regulations from regulators in the area of solid dosage forms validation gives rise to differences in interpretation of the requirements. For example, former FDA Investigator C. Medina suggested that generally "Critical validation parameters should be set as wide as possible so that continuous improvements can be made on processes, methods or procedures" (1). While J.M. Dietrick of the FDA Office of Compliance, in recommending validation blend assay limits, states "Narrower limits should be set where historical data show the capability of meeting those limits" (2). Taken out of context, each of the above statements can be interpreted as suggesting the opposite of the other. Despite this apparent disparity, some unwritten rules have evolved for solid dosage form validation based on inspectional observations from form FD 483 and warning letters. Over time, changes in practice have been necessitated in order to comply with judgments and consent decrees imposed on some firms because of failure to validate properly. Techniques will be presented that are current in the industry and are reviewed favorably by the FDA or EU auditors. However, it should be noted that change is a constant factor and, as new techniques and equipment are developed, methodology will improve and newer approaches will be expected by auditors.

There may be some overlap with other chapters in this volume. This overlap is necessary in order to understand the basis for some practices in the validation of solid dosage manufacturing operations. When overlapping topics are discussed, the emphasis will be on their relationship with solid dosage validation.

The literature contains many detailed studies from which generalizations can be made that are valuable in process development. The purpose of validation is to demonstrate and document the adequacy of each specific process and to prove the applicability of the generalizations to that process.

VALIDATION AND THE DOSAGE FORMING STEP

The initial thrust of validation was in the sterile manufacturing sector. That area has the single overriding concern of product safety, specifically the probability of sterility of the product, and was disaster driven. The mathematical models utilized for the evaluation of this probability are well accepted. There is no corresponding microbial safety concern or accepted general model for solid dosage validation. The inspiration for this author's early protocols for non-sterile validation was Theodore E. Byers of the FDA. Byers referred to a recall incident with

Abbreviations used in this chapter: cGMP, current good manufacturing practices; CpK, process capability, average; CpM, process capability, target; CPP, critical process parameters; CQA, critical quality attributes; FDA, U.S. Food and Drug Administration; HEPA filters, high-efficiency particulate air filters; HPLC, high performance liquid chromatography; HVAC systems, heating ventilation air-conditioning systems; IPT, international press tooling; IQ, installation qualification; MCA, British Medicines Control Agency; OQ, operational qualification; OTC, over the counter; PQ, performance qualification or process qualification; PV, process validation; QA, quality assurance; QC, quality control; R&D, research and development; USP, United States Pharmacopeia.

digoxin tablets as justification for solid dosage validation (3). However, this example could be attributed to a general breakdown of cGMPs at the involved manufacturer. He also made the statement that validation was "not compendial testing," which implied that validation for solid dosage had to be something more than normal quality control inprocess and release testing. There was no methodology or approach to solid dosage validation offered by the FDA to satisfy this new requirement. The industry was told only what it wasn't.

It was apparent to those engaged in the development and/or manufacturing of tablets and capsules that dosage content uniformity was the major concern applicable to all products. In addition, the best point to evaluate content uniformity was during or shortly after the dosage forming (compression or encapsulation) manufacturing step. Consequently, it was natural for the validation protocols to emphasize the USP Content Uniformity Test. The justification for this was that content uniformity was an extensive property of the product, in that uniformity depended on how much material was there in relation to the processing equipment, assuming a uniform blend. This would be the only "unknown" factor remaining to be confirmed from the development and scale up batches. The other properties of tablets and capsules were more intensive in nature, in that they were less dependent on the batch or equipment size. Additionally, in this author's experience and interaction in discussion groups, round tables, etc., the prevailing thought was that emphasis should be on the final product (what is given to the patient). It was recognized that dissolution and other in-process physical tests were important also. However, at this time, dissolution was not applicable to all solid dosage products. The physical in-process tests were performed on-line for many years (an early example of statistical process control) and were not a significant problem. The manufacturing steps prior and subsequent to the dosage forming step were evaluated in terms of how they would affect testing results at the dosage forming step or alter those results.

Currently, there is more of a balanced approach to solid dosage validation. Each unit operation is addressed and there is no longer sole dependence on the dosage forming step, specifically relying on tablet and capsule testing results to validate the entire process. This includes raw material characterization for direct compression formulations. It is not that validation testing of tablets and capsules has decreased, but rather validation testing in other operations has greatly increased, especially mixing and blending operations. The operations of coating, polishing, and imprinting are also considered. Packaging, once ignored, is now the subject of major validation activities.

VALIDATION NOMENCLATURE AND DOCUMENTATION

Documentation and the nomenclature contained therein is the essence of validation. Undocumented activity can be equated to no activity and verbal explanations in audits cannot be relied upon. For solid dosage validation, nomenclature varies from company to company and it is important to establish this in higher level policies and procedures. The reader may not agree with the use of a particular term for a described action used within this chapter or in what protocol it should be contained. It is important not to get "hung up" on terminology but to agree on the underlying descriptions and actions. This should be the case within large individual firms with multiple sites and operating divisions as well. It is important to be consistent within the company and not to "reinvent" widely accepted terms. For example, the technical group performing solid dosage validation for a single site of a large company decided to call what was universally called a "deviation" elsewhere in the company an "observation." In another example, after a day-long meeting with a dozen experienced people hammering out a process validation testing plan, one junior engineer proclaimed, "That is not process validation." In the latter case, the engineer was in full agreement with the testing plan and disagreed only with what it should be called (4). Far too much time has been wasted on what is included in which protocol. The word "critical" is an often-used term that carries different meanings. A lesson can be learned from software development where they use the Component Object Model (5). In this approach a specific function is coded once and only once, and this single subroutine or "object" is called wherever and whenever the function is needed. Similarly, all validation terminology should be defined in higher reference procedures and used only as defined both within documentation and at cross-functional team meetings.

Two concepts that are fully accepted and understood are those of the protocol and the deviation (the latter may be different with the same basic intent, e.g., discrepancy, error; see above). The protocol contains the prior established specifications and is fully approved with the Quality group oversight in execution. A basic rule of thumb can be derived from an analogous situation in a jury trial: "Never ask a question you don't already know the answer to." The implication of this is that extensive trials must occur during development and commissioning activity to prevent unexpected or noncompliant results obtained under the formal protocol. Deviations must be resolved with full investigation and corrective action addressing the underlying systemic cause to prevent a recurrence in subsequent protocols. Beyond these, the term "executed protocol" refers to a blank protocol that has been annotated during execution to collect data that may or may not be routed for approval at completion. The validation summary report is generally accepted to mean tabularized results and a statement of compliance that is routed for approval to the same approvers of the protocol. In some companies, failed studies have a report that is circulated to "close them out." This is dangerous in that the existence of an approved validation report implies that the validation itself was successful. Databases and validation status summaries must have the ability to differentiate between successful and unsuccessful efforts.

A recurring error in solid dosage validation is the phrase "validation not required." This phrase is used with some operations and compendial test methods, such as pH. One interpretation is that operation of the test method is permitted to be invalid. Conflicts of statement arise when prerequisite questions such as "all test methods validated" are answered "yes" and some methods are designated "validation not required." It is important to realize that typical solid dosage in-process test such as hardness, thickness, and friability must be "valid" and while not validated, as one would validate an HPLC assay, must be shown and documented (verified) as suitable under actual conditions of use (6).

Applying process validation concepts originally developed for batch type steam sterilizers to solid dosage forms had given rise to differing opinions on how process validation would be accomplished. Concepts such as "worst case" and "processing range validation" resulted in protocols that prescribe independent parameters to be set on "edge of failure." Running processes with all parameters set at a low level or, conversely, set at a high level proves nothing because it is not know if the effect of extremes of the various parameters can cancel each other. This type of limit justification is best performed with factorial experimentation within the R&D department. Today, process validation is generally accepted to mean three batches run with parameters set at typical setpoint values. This assumes that there has been separate, completed, documented, equipment-focused development studies for the specific product to determine both the setpoints and the acceptable ranges. The results of these studies are confirmed under formal protocol and are called qualifications. The qualifications are categorized as installation, operational, or performance qualifications and must address the parameter limits for speed, batch or component size, and any limits not specified in R&D documentation. The concept of Traceability Analysis (when consisting of a separate document, called Traceability Matrix) (7) from software development has been frequently borrowed to list every specification and/or limit and how and where it is supported. An update of the 1987 *Guideline on General Principles of Process Validation*, currently in development, may attempt to move away from an absolute three-run requirement and emphasize good design and development work and continued scrutiny ("life-cycle") (8,9).

Confusion among the terms "performance qualification," "process qualification," and "process validation" is common in solid dosage validation. The original 1987 Process Validation Guidelines (10) imply that process validation is a series of qualifications, the collection of which constitutes process validation. This approach was supported by the practice of engineering-construction firms that usually were involved in specific equipment and/or facility upgrades. Performing IQ, OQ, and PQ on the equipment and facilities enabled them to complete the project with minimal interaction with the receiving firm's personnel and testing. Following the Barr decision (11), the requirement for three batches from beginning to end for process validation of solid dosage form manufacturing has been widely accepted. Process validation is the final test with everything in place and operating properly. Deviations that occur during process validation must be resolved to prevent the deviation from reoccurring. In a recent example, a deviation occurred where an equipment temperature sensor malfunctioned during a process validation batch. The initial deviation investigation found that the manufacture's recommended replacement interval had been exceeded. Corrective action was initially limited to adding the recommended replacement interval to the Preventive Maintenance schedule for the particular sensor. However, satisfactory resolution may not have occurred until the commitment was made to look at the entire Preventive Maintenance program to ensure that the sensor replacement interval is included and accurately specified.

In the context of current solid dosage validation, qualification has come to mean an equipment and/or facility focused study. Performance qualification, within solid dosage manufacturing, addresses these under conditions of intended use with product or simulated product. Process qualification has been used in solid dosage to describe a study that is a subset or supplement of process validation, such as an additional dosage strength for a given product, or a revised individual phase of an existing manufacturing process, e.g., drying of a wet granulation using a revised process. It is important to define these terms and agree on the specifics for each product.

VALIDATION CONCERNS

The following concerns should be incorporated within solid dosage protocols in some agreed-upon format. They may serve as a resource when reviewing protocols for adequacy and completeness.

Prerequisites

In terms of solid dosage finished goods, process validation requires the completion of a substantial number of prerequisites that must be documented and in place prior to process validation protocol approval or study initiation. These generally consist of the following: development, equipment and facilities qualification (IQ, OQ, and PQ); analytical test method validation; documentation (manufacturing instructions, batch records, and standard operating procedures); and training. Cleaning validation is frequently performed concurrent with process validation and it is important to consider the effects of hold times to the process and overloading the testing laboratories with samples. Other equipment qualification pre-requisites such as User Requirement Specifications, Functional Requirement Specifications, Factory Acceptance Testing, Site Acceptance Testing, and Commissioning may be required and are a practical necessity for new systems and equipment. An important deliverable from R&D is the identification of critical quality attributes and critical process parameters. It is necessary to document these prior to protocol preparation, if these have not been explicitly documented in previous development reports. The personnel performing the validation should not be the ones designating what is critical, no matter how obvious. The author has experienced occasions where, when R&D were finally consulted during an audit, there was disagreement in what was considered critical.

Many other activities can be considered validation in the general sense. However, those activities are considered more cGMPs than validation and are typically performed by functions other than those involved in qualification and validation. In addition, these activities are general to all products and do not change with solid dosage finished goods. Some of these activities are: approved written procedures generation (document control program); preventive maintenance and calibration programs; specifications; test methods; analytical test method development; stability; personnel training and documentation; change control program; and corrective and preventive action commitments tracking. It is important that there exists a close link between the Change Control Program and validation, since change is the direct cause of validation activity. The persons performing validation of solid dosage forms must ensure that applicable procedures have been published and that they are being followed. Once the protocols are approved, there may be a separate "training" on the protocols themselves including all manufacturing personnel. This requirement has come about in recent years, primarily as a commitment within firms having problems in an effort to avoid the protocol deviations seen in earlier validations. A planning and coordination meeting is a more appropriate term for this activity. Coordination with production planning and laboratory functions is essential. Budgeting in these support areas for validation is necessary and communication during budget time can avoid scheduling problems later. Checklists and separate procedures addressing prerequisites have been utilized in connection with validation master plans to allot resources and track this supportive activity.

The normal sequence for a solid dosage form is IQ, OQ, and PQ, followed by process validation. Most firms require the completion of one stage prior to the initiation of the next. However, it is entirely acceptable to combine these phases within a single protocol and to combine PQ with PV. A recurring problem is the requirement of completed "as built" drawings as part of IQ. These drawings are typically subcontracted and are not completed on time for the next stage to begin. It is important to put contingencies for this and other potential problems within the protocols to enable timely completion and to ensure that they are completed prior to product release.

Risk Based Approach

In general, there is no difference within the regulations as far as cGMPs for filed or prescription drugs versus OTC solid dosage drugs. Historically, the auditing and enforcement approach has varied and the experienced validation manager had considered this in prioritizing validation activity and resource allotment. Similarly, within products, those areas that were most problematic to the final product quality were given the most scrutiny during validation. The FDA has acknowledged this scientific approach and formalized it as being "risk based." (12). The amount of testing may vary, with the filed product typically having more testing performed. Two extreme examples would be a class-one dental—device denture cleaning tablet and a prescription-filled tablet drug product. Although the sampling would be essentially the same, the critical attributes of concern would be markedly different, as would the amount of additional validation testing performed. Equipment qualification would remain essentially the same in terms of parameters measured and documented. The extent of the documentation would be less for the class-one device than the ethical tablet.

The risk assessment must be documented. Techniques such as Failure Mode and Effects Analysis (13), Ishikawa (herring bone) diagrams (14), and Potential Problem Analysis (15) can be used to formalize and provide structure to this process. Solid dosage finished goods typically have multiple dosage sizes for the same product. There are also multiple package sizes and packaging forms (bottles and blister packs). This diversity necessitates the use of bracketing or matrixes during validation as a matter of practicality. It is important to include this planned exclusion of some of the product/package presentations in risk assessment.

Validation master plans and protocols should anticipate problems that may arise when products requiring validation are manufactured in the same plant location as those that do not require validation. For solid dosage products, these products are typically food supplements. Many other products such as cosmetics and household products do not require validation and may be present in drug facilities. Special training and awareness are required for personnel engaged in development and manufacturing activities for drug and devices to ensure overall compliance. Although the cGMPs are ultimately addressed with these products, often supporting staff personnel with site-wide responsibility are unaware of the additional documentation requirements and restrictions associated with drugs. Poor execution and/or documentation practice by support groups is especially detrimental during development. Data generation in the early stage of development may have to be repeated for lack of appropriate documentation or non-compliant cGMPs. This usually results in missed deadlines, since validation, specifically compressing and encapsulation validation, is typically the last activity to be completed prior to shipment of a new product.

Critical Attributes/Parameters

A critical attribute/parameter is a characteristic of the unit operation being validated that will adversely effect the safety, efficacy, and reliability of the final drug product. These have been subdivided into critical process parameters or CPPs to represent independent variables and critical quality attributes or CQA to represent dependent variables. An independent variable is typically a machine set point or adjustment that is specified for the process. A dependent variable is a measured property that is an outcome of the process. As mentioned above for solid dosage, the CQA of concern is often individual dosage content uniformity for the compressed tablets or filled capsules, followed closely by dissolution and assay. Other tests (for CQAs) may be important depending on the particular drug substance involved and the regulatory status of the product. The critical physical in-process testing, which is performed on-line at high frequency primarily to preclude financial loss, is usually more than adequate process qualification for these parameters. Care should be taken that the level of in-process testing used for manually set up and adjusted tablet presses be maintained for at least the validation batches of computer automated presses. The

CPPs and CQAs are best identified and documented by R&D.

Often overall testing can be reduced following validation by a clear description and justification of all the critical attributes. This is where a development report can be essential to support process qualification sampling and testing and why process validation should be a concern at the earliest stages of development. Additional validation testing can concentrate on these attributes, while normal release testing may be reduced. Testing for information only should not be in validation protocols. This research type testing may wind up being required in routine release testing as an outcome of an audit, with the laboratories being unable to perform the test.

Specifications

It is required to have the specifications finalized before publication of the validation protocol. These are the prior approved product quality attributes required within the protocol. The concept of the traceability matrix (mentioned earlier) is useful to ensure that all specifications are addressed properly. Initially, limits are sometimes set too widely because of the unknown variability of full-scale equipment relative to smaller pilot plant equipment for the particular formulation. This is often the case in tablet and capsule validation. Sometimes the desire to adjust in-process specifications at the completion of full-scale batches to tighten ranges may be an objective of the protocol. This must be explicitly stated as such and the use of phrases such as "established during validation" must be avoided. Machine-operating ranges are often adjusted after the validation batches, if they are the first full scale batches (often the case with monograph OTC drugs). Most firms have a two level approach to in-process limits by having internal warning limits, which can be adjusted to reflect a specific manufacturing equipment sequence. The internal warning limits are tight enough to detect any subtle change in equipment or material. Validation data are typically used to establish these internal limits or baselines. It is common for these limits to be adjusted with time to reflect sourcing changes, which also require change-driven revalidation.

Solid dosage specifications have less of an absolute character and certainly less of a microbial safety concern than other dosage forms (e.g., sterile and non-sterile topical). Although, acceptance criteria are fixed in absolute limits, the process generally allows for more variability within those limits. Tablet presses and capsule fillers are unique in that final dosage physical attributes are treated as independent variables to "set the machines in." On older machines, there are no numbers or units on weight, thickness, and penetration adjustments, since the resulting product measured parameter is indicated as the adjustment. Only the most modern computer-controlled machines have anything resembling a true absolute independent parameter for these machine adjustments. Even there, they are usually relative values. In addition, these adjustments are more susceptible to minor lot-to-lot variation in raw material parameters, especially in direct compression or encapsulation processes.

Machine speed specifications for tablet presses and encapsulation machines are often established during validation. It is important to establish a minimum and maximum speed for the dosage forming step. While a relative standard deviation for individual weight control of 1% may look nice in a process validation report, it may also be indicative that the machines are running too slowly from an efficiency viewpoint. Compression and encapsulation are the only areas where machine speeds can be variable to any extent. It is important to consider this in the initial validation of a process so that manufacturing has the necessary range of speed to reflect the inherent variation of raw materials and environmental conditions.

Compressing and Encapsulation Facilities

Compressing and encapsulating rooms must have adequate separation from other products. It is permissible to have more than one press or filler in a single room; however, they must be running the same presentation of the same batch of the same product. If this cannot be accommodated, then individual smaller rooms are a better design choice.

Dust collection and proper room pressurization are major facility requirements for tablet presses and capsule fillers. Recirculated room air and dust collection must be adequately filtered. It is preferable to have these systems dedicated to individual rooms, although not usually practical. Elimination of cross contamination via the HVAC system must be an essential design consideration. Additionally, dust collection air returned to the room, no matter how well filtered, usually creates a cross-contamination problem. Dust collection is better vented to the outside of the facility after separation of the dust.

Although it is usually not practical to move encapsulation machines around the facility, moving tablet presses is commonplace. Even in dedicated rooms, the presses are moved out for press and/or room cleaning. Therefore, the doors must be of sufficient size and adequate design. The hospital-type swinging door has been successfully utilized for this purpose. They are more durable than sliding doors and are more easily replaced. They are also consistent with the sanitary nature of the room and surroundings. They close automatically and provide an opening for air balancing purposes. Unlike sterile areas, it is permissible (actually preferable) to have negative compressing and encapsulation rooms to adjacent service corridors. This is done for dust control and the rooms themselves may have HEPA filtration on exhaust vents. The use of positive pressure rooms is possible but less common. In this case, the hall or service area would be negative to the process rooms and the rest of the facility. The pressure balancing design must be documented so that it can be verified during operational qualification and monitored subsequently.

Compressing and Encapsulation Equipment

Press tooling (upper punches, lower punches, and dies) are highly controlled items. It is essential to have cleaning and use logs on individual tooling sets. Security, mix-ups, and damage are all considerations in handling and maintaining tooling. Procedures for this control and maintenance must be in place prior to performance qualification activity. This is true to a lesser extent with encapsulation machine change parts. Facility designs should include provisions for cleaning, maintaining, and storing tooling and other change parts. Feed frames and filling rings or funnels are best purchased from the original machine manufacturer. The use of generic copies for these parts such as "plastic" feed frames and reconditioned parts is usually far more expensive in the long term. These generic parts are non-standard and frequently behave differently than the original manufacturer's parts. It is important that consideration of "change parts" be included within the validation protocols. These parts can be changed with no additional validation required, provided the part meets inspectional quality. This simplifies the change control process and enables routine expendable parts (e.g., dust cups, dosators, cams, belts, hoses, etc.) to be replaced without interruption in production.

The modern rotary tablet press and rotary capsule filler are expensive and complex machines. Adequate documentation is generally available from the vendors or original manufacturer to insure proper maintenance and training. A 21 CFR Part 11 assessment and validation is required for computer-controlled equipment.

Accessory equipment such conveyers, dedusters, and polishers must be readily cleanable. It is often desirable to dedicate the accessory equipment to a particular product and dosage to simplify cleaning validation.

The required in-process testing equipment is often neglected in the early design phases. Machinery that increases output also suggests the application of corresponding automated individual tablet and capsule weighing equipment. Robotic machines in the lab can greatly reduce testing time for batches produced on high volume tablet presses. These are particularly helpful during the initial heavy additional testing required by validation.

Installation Qualification

This phase of qualification as applied to solid dosage forms will have checklists and data sheets as a minimum when following a well documented commissioning, and extensive protocols for complex equipment. It is important to realize that most compressing and packaging equipment is mobile and this aspect must be addressed within the protocols. Generic IQ checklists designed to apply to all plant-wide equipment may not be appropriate for these operations. There is some overlap between this phase and operational qualification in that the equipment is sometimes run without product, and operational and maintenance procedures are drafted. It is important to demonstrate during solid dosage equipment qualification that calibrations of components on all movable equipment is unaffected by movement to change location and the rigors of cleaning using organic solvents and/or water. This includes the strain gauges that control weight on automated tablet presses. Additional consideration for worst-case testing must be made for the extremely dusty conditions in which this equipment operates.

Operational Qualification

Material costs in solid dosage processes may be a limiting factor in OQ. Compressing batch sizes can consist of several million tablets and equipment qualification can be achieved with smaller tests. It is possible to run a placebo batch during this phase to minimize the financial loss in case of an equipment failure. In this case, it is common to require the three batches of product to be run during performance qualification. The purchase of used equipment is a good reason to use placebo during initial commissioning and OQ. In the author's experience, during one OQ of a used tablet press it was discovered (too late!) that the cams did not comply with IPT standard bevel angles. This resulted in the destruction of a set of tooling and lost material. Another OQ revealed on a "re-manufactured" press that the vendor's refinement to the tablet "kick-off" function resulted in tablets being kicked onto the floor at moderate operating speed. This refinement, which was a press fitted ring around the turret, had to be removed at additional expense.

Compressing and encapsulation equipment frequently have mechanisms that divert product into "reject" containers if monitored parameters fall out of specification. One example of this is the shut off feature when the supply of granulation to a press is interrupted. It is often sufficient for qualification to verify this logic with a single product or placebo. The procedure is to collect the last tablets or capsules not diverted to the rejection container and verify their individual weights manually with an independent calibrated balance.

Performance Qualification

The best way to validate a compressing and encapsulating process is to sample and test more extensively than the normal or proposed quality control release testing. As mentioned above, three such batches, or the equivalent of them for a continuous or semi-continuous operation, are required. Centerline validation is used, meaning that the batches are not run on the edge of any limits ("edge of failure"). All operating limits should be supported by developmental runs or experiments. A possible exception to this is machine speed. The assumption is conventionally made that, for a given machine, the product running at a slower speed is in a higher state of control and need not be validated. This has been questioned and some validation is generally required for the lower limit of speed on a given machine and certainly for a different, slower machine. Protocols must address machines that require different change parts for different speeds, this being the equivalent of a different slower machine. All specifications for in-process physical testing and background environmental conditions (temperature and humidity) should be verified during development on small-scale batches. All independent operating parameters should be documented either on the manufacturing batch record or in supplemental validation notebooks or records. These notebooks are subject to FDA scrutiny and should be maintained and controlled the same as QC or R&D notebooks.

Solid Dosage Validation Sampling Plans

For the dosage forming step it is common practice in performance qualification and process validation to divide batches into segments for testing purposes, the minimum being three (beginning, middle, and end). The worst case points for sampling are often in the beginning (immediately after the machines are "set in" and the product is deemed acceptable) and end (immediately prior to shut-down). In this author's experience the "end" sample is particularly troublesome in terms of providing acceptable results. On one occasion, a particularly inexplicable uniformity problem resulted in the expensive practice of stopping batches with 100 kg left to compress in the feed hopper and discarding the remaining granulation. It is important to ensure that each compressing station or encapsulation dosator is represented the sample. Protocols must have explicit language to that effect, since the relatively small amount of units tested in batch sizes numbering in the millions is predicated on the repeated use of a small number of identical tooling.

Since routine QC sampling involves composites, it is generally preferable to use point ("targeted") samples for validation, although composites of batches divided into six or more segments may also be acceptable. The number of sample points and number of tablets and capsules sampled varies with the batch size and machinery type. At one firm a cutoff of batch size of one million units has been used to increase from three to six segments. The sample size should ensure the equivalent of one complete rotation of the compressing machine (full filling ring, or all filling funnels for capsule fillers) is obtained. Samples should be taken from all discharge points for duel or quad compression roller machines. This aspect should be included on the label information. The segments may be identified in terms of filled bulk container number or in terms of time. Resampling provisions within the protocol should explicitly specify sampling from the filled bulk container rather than directly from the machine discharge in case samples are missed or lost, because the discharge method cannot be repeated for a given batch once it is compressed or encapsulated. Questions on resampling can be avoided by obtaining a contingent resample initially. There should be a sufficient number of tablets or capsules in each sample to perform all the validation testing. Other sampling protocols typically in effect on initial production start up are an R&D sampling protocol and a stability sampling protocol. These sampling protocols could be combined and applied at the same time points.

Blend Sampling

One of the more counterintuitive areas in solid dosage validation is the subject of blend sampling and testing. A PDA monograph has been published to address this subject (16). The initial interest was stimulated by the Barr Decision (11) that identified design deficiencies in a manufacturer's attempt to validate product. Two requirements emerged from that decision. They are to sample a blend with a scale of scrutiny within one to three final dosage sizes, and apply dosage content uniformity specifications to those samples. The latter requirement implies the taking of at least ten samples, since this is a requirement of the USP Stage 1 Content Uniformity Test.

This increased granulation testing requirement affects compressing and encapsulation validation in that there is a tendency to reduce the amount of tablet or capsule testing performed. There is also a need to show that uniformity is not lost in storage and/or transfer to the tablet press. Liss et al. (17) survey this phenomenon and demonstrate that segregation can occur within granulation while it is merely dropping through a pipe. In executing a study where the author utilized plastic scoops and small containers to obtain samples, data were obtained that were largely unusable for many different types of blends.

Theoretically, testing the blend makes good sense; the blend must meet dosage uniformity requirements or it will be impossible to have tablets or capsules that meet these requirements without relying on serendipitous mixing in the tablet press feed frames. Additionally, having demonstrated a uniform blend, the emphasis may be extended from content uniformity to the in-process physical properties of the dose. This expanded emphasis is primarily on individual weight, which is much easier, faster, and less expensive to perform. In practice, however, for many blends the process of sampling to simulate a dosing process and the subsequent analysis is non-trivial. The separating action of static electricity, movement caused by physical manipulation, fluidization, and surface affinity can occur in the simplest powder flows such as filling a sample bottle or thief (13). These forces may be different for a blend when scoop or thief sampled (either at rest or flowing in mass) versus the "sample" taken by the mechanical dosing action of the tablet press or encapsulation machine. Local non-uniform samples can be "created" by the physical act of sampling when small samples are involved. This process can also occur in the laboratory unless the entire sample is analyzed or made uniform prior to testing.

It may be true that the same forces that make blend sampling invalid can give rise to non-uniformity in the tablets or capsules. In those cases where both blend and dosage test results are non-uniform, nothing short of major process and equipment upgrades and/or formulation changes will solve the problem.

A classic example is amoxicillin for oral suspension. Although not a tablet or capsule, it exemplifies the blend sampling problem. Uniformity testing of blend samples on the order of one to three doses seldom can achieve the assay (relative standard deviation) results of testing the filled bottles. The explanation is that the blend stays in its continuum until dosed; the chemist reconstitutes the entire bottle and uniformizes as instructed on the label, and withdraws a single dose aliquot. This product is usually a direct compression type (not wet granulation) formulation consisting of granular sugars for flavor and an active ingredient that is a fine powder.

Developments in compartmentalized sampling thieves designed to address this problem has brought needed help. An example of this are the thieves provided by Globepharma (New Brunswick, New Jersey). This type of thief enables sampling blenders using the same die filling mechanism of a tablet press or capsule filler ring. Such thieves can reproducibly sample "worst case" blender locations at the three unit dose level or less, which can be tested for active assay(s). The mixing action of blenders and worst case locations have been published (18,19) for each blender type in common use for tablet and capsule blends. These are generally located along the axis of rotation for the tumble types or in dead spots for mixing blade types. Demonstrating that these areas meet assay specifications (up to 20 times unit dose sample size) in conjunction with extensive tablet or capsule content uniformity testing may be adequate validation for those products that have a discrepancy between blend and tablet (or capsule) uniformity results attributed to sampling. In these instances, theoretical explanations for well-documented inconsistency during development can be strengthened with extensive successful tablet or capsule validation testing and may justify exceeding the three dose sample size limit. Blend sampling should be validated in its own rite during development if possible to avoid these problems. In practice, the author has found that delivering the filled thief to the QC lab when procedurally possible eliminates a lot of blend sampling variability. A commercial benefit from extensive successful blend sampling and testing for validation is that it may justify reduction of routine QC blend testing when validation is completed.

Blend sampling plans mirror USP Content Uniformity Test Sampling, in that a minimum of ten samples are obtained from the blender and/or final blend container. It is important to sample an equal number of best case areas to assess the maximum variability in the results.

Compressing and Encapsulating Validation Testing

The USP Content Uniformity and Dissolution tests are performed for all active ingredients on all additional validation sampling locations collected for compressed tablets, tablet cores, capsules, and coated tablets. All other critical attributes of concern are also tested. Normal or proposed routine quality control sampling should also be performed. It is better to use approved validated test procedures based upon those used during routine production rather than extra R&D type tests (e.g., lubricant surface area) which are performed for information. Such informational sampling and testing can be part of an R&D testing protocol on the initial full-scale production batches. An example of this would be feed frame or machine hopper granulation samples. For many products, validation testing often delays product release. Schedulers must allow sufficient time to complete the validation testing which can be several times the normal QC testing. A rule of thumb for the minimum multiple for normal QC testing is the "rule of three." This should be done for the smallest batch sizes and the usual multiple in this author's experience is six. Tablet batches with doses in the millions may require more. The limiting factor above adequacy may well be the capacity of the laboratory. Groups responsible for validation must ensure that the extent and timing of validation activity be communicated as early as possible to scheduling groups.

Blend, Tablet and Capsule Validation Acceptance Criteria

With the possible exception of USP Content Uniformity, the regular or proposed production limits are generally used for validation. This is done because it is difficult to explain passing release testing and failing validation testing. Using the segmental sampling plan described above, it has been the author's experience that any significant process problems will be detected over the course of the initial three batches.

An alternative exists for the USP Content Uniformity test. Dr. James S. Bergum (20) has developed a limit scheme that can predict the probability of passing the USP Content Uniformity test at a given confidence level for a variable number of individual tablets or capsules tested. Tables for passing the USP Content Uniformity test 95% of the time at a confidence of 95%, based on these limits, have been generated. The tables are used by entering with the average of the individual units and reading the corresponding maximum % CV (relative standard deviation). It is assumed that a representative sample n has been obtained. For example: An average of 60 individual assays for tablets which was 98.1% claim would require an RSD of 4.28% or less to assure that the USP Content Uniformity test would be passed 95% of the time with a confidence level of 95%. Cholayudth (21) has published spreadsheet formulae to generate the limits without the use of the printed tables.

This type of acceptance criteria is ideal for development batches where limits may not be finalized and can serve as a means to predict formulation performance at full-scale production. Another important validation acceptance criterion is confirming that sufficient controls are in place to detect any deviations. This means that there is adequate in-process and physical testing of subsequent production batches. The advent of computer-controlled machines has enabled a reduction in in-process testing. Sufficient testing should be retained to detect machine malfunctions or granulation abnormality.

Standard Operating Procedures

Inherent to any validation is the assumption that approved procedures are in place covering all phases of the operation being validated and covering the validation activity itself. In essence, the procedures, as executed by trained personnel, define what is actually being validated. This is also true for solid dosage manufacturing operations. The enforcement of this aspect of cGMPs is usually performed by quality assurance. The people performing the validation will undoubtedly be involved in the generation of some of these procedures. A checklist is typically employed to ensure they are all in place and people are trained prior to the validation activity. The problem with checklists and "cookie cutter" protocols is that once an incorrect box is checked, or an applicable section is marked "not applicable," all the other entries can be questioned. All correctly checked boxes must now be reviewed and all deleted sections must be reviewed for applicability. It is preferred to list specific procedures and train personnel within the protocol. Current practice is to have a database containing required training curricula and training dates for all employees.

It is important to ensure that procedures are followed and to adequately describe the activity taking place (22), much the same as the batch monitoring confirms adequacy of the process instructions. Although implicitly documented by the manufacturing batch record, a separate summary verification statement within the protocol provides an explicit statement of compliance.

Cleaning procedures are commonly validated under a separate protocol. Cleaning validation often runs concurrently with process validation for new solid dosage products and should be considered in scheduling activities. The cGMP requirement for clean equipment before and after a new product may need to be addressed and documented in equipment qualification and/or process validation protocols.

Supportive Data from Other Areas

Compressing and encapsulating validation often requires supporting data from other operations. It is normal to validate a process from start to finish over an entire batch manufacturing process. This is implied by the three-batch requirement. It is important that the following supportive items from the prerequisites, discussed earlier, be in place: raw material specifications; raw material test methods; analytical test method validation; bulk specifications; finished product specifications; stability protocols and data; packaging component specifications and test methods. The identification of these items may vary from firm to firm, along with the source of the documents. Process validation is typically the last thing that is accomplished and it cannot be finalized without the supporting studies.

Developmental Data

The need for development groups to provide documented data supportive of process validation cannot be understated. FDA compliance inspections for validation and the Pre-Approval Inspection Program have investigators poring over research in search of reports and raw data (23). Publication of a development summary report (name may vary) directly supportive of validation should be a standard procedure. This summary should include: identification of CPPs and CQAs, experimental designs, testing results; rationale for conclusions; selection of specifications and justification of limits; bio-batch records and testing results. It is far more convenient to stress limits of intensive parameters and manufacture under worst case conditions in development. These data must be available to justify "center-line" (as opposed to edge of failure) validation of full-scale batches intended for sale. "Worst case" as applied to "center-line" validation of tablets and capsules refers to sample size and location.

It is important to record machine parameters and characteristics during development even though the ultimate manufacturing equipment may be somewhat different. The rationale for equipment selection for a particular bio-batch must contain relevance to the final commercial process validation. For example, the use of a gravity feed frame press versus a rotary feed frame press may be considered an arbitrary choice for a limited production bio-batch. However, if the anticipated press for commercial production is a high speed, rotary feed frame machine, the choice of the rotary feed frame is more supportive of validation.

Validation testing criteria should be applied as early as possible in development so that commercial batch comparison to the bio batches is facilitated. One question that may come up is the content uniformity of scored tablets when broken (24). This is part of the rationale for tooling design and must be included. Similarly, the rationale for capsule size selection should be documented. The dosage uniformity approach of Bergum, discussed earlier, is very useful to evaluate the merit of a particular tooling design or formulation during development batches.

Bio-Batch Equivalence

A key component of process validation is showing clinical or bio-batch equivalence (18). It is a frequent subject of solid dosage form compliance inspections, since raw material variation can cause final product variation in uniformity, disintegration, hardness (tablets), and dissolution. It is desirable to use the same dosage form (tablet or capsule) for commercial batches as that used for the bio-batches. This facilitates the comparison of in-process testing results from validation batches to the bio-batches. Historically, early R&D batches have utilized dry filled hard gelatin capsules for convenience. Questions on the bio-batch equivalence to full-scale production can arise when the final commercial dosage form differs. Sufficient data should be collected during the bio-batches so a comparison can be made with the full-scale demonstration batches or validation batches. The statement of equivalence is usually made by taking into account the in-process testing results, raw material testing results and final product testing. The statement may be part of the validation report or in the conclusion of a development report that includes the full-scale data.

Raw Material Characterization

The raw material characterization must be appropriate for the type of process. Direct compression processes require meaningful particle size specifications for raw materials and usually pre-screening of these raw materials during addition. Optional screening requires some type of documented in-process check after screening during validation and is a frequent focus of auditors. The trend toward global sourcing and commodity type purchasing of raw materials makes vendor qualification essential. The cost savings in this type of purchasing may override the cost savings of a direct compression process over a wet granulation process, since the latter is more forgiving of raw material physical attribute variation. Process designers must keep this in mind and purchasing departments need to be controlled by appropriate vendor qualification and change control programs following validation and bio-batch production of direct compression formulae.

Bulk In-Process Storage

The emphasis on hold times in sterile manufacturing has resulted in the validation requirement for bulk storage hold time of in-process solid dosage materials. In-process testing from the previous stage should be confirmed. If time permits, it is good practice to include this in the process validation protocol. More often, it is an addendum to a report or protocol or a separate R&D study. In-process materials should be stored in compliance with any restrictions placed on the environment during processing as specified in the manufacturing batch record. Areas of warehouses that will be utilized for the storage of in-process or final product must be qualified to maintain any labeled storage restrictions. Hard gelatin capsule shells contained within fiber drums are susceptible to physical damage when stored in direct sunlight for extended periods. Some firms have minimum weight specifications on the empty capsule shells, making them more expensive, but makes the shells less sensitive to the physical changes caused by storage.

Encapsulation Machines

There are two main types of dry filled capsule fillers. They are the funnel dosator type and volumetric ring dosator type. The funnel type can be continuous or discontinuous. The ring type is described as being semi-automatic or fully automatic. The funnel variety can be treated like a tablet press where each dosator is treated like a compressing station with a unique identity. The ring type has numerous cylindrical holes that are filled by force feeding granulation into them, either by the use of an auger or tamping pins. The rings for these machines are often customized for a particular product, although standard depths for each capsule size are available, so they should be controlled with the same care as compression tooling to ensure the correct rings are used for a particular product. The author experienced a situation where unacceptably high weights were observed during the startup of the initial production batch of a new encapsulated product. This resulted in a panic change to the dosator type filling machine from the ring type and the accompanying massive change in documentation. Upon later investigation, it was learned the ring type machines were erroneously set up with "deep fill" and "special depth" rings for the given capsule size that had been specially ordered earlier for a deleted problem product. Encapsulation change parts do not wear as readily as press tooling, so records on the total amount of production on a given set are sometimes not maintained. Procedures must be in place to record total usage and control the handling of these encapsulation machine change parts, since problems can arise from age and mishandling.

Soft gelatin capsules are more appropriately addressed in a discussion on liquids and semi-solids validation. There is an additional microbiological concern for the gelatin raw material. Otherwise, the segmental sampling scheme applicable to dry filled hard gelatin (or vegetable) capsules is appropriate. Appropriate attributes of concern specific to liquids should be identified and included in the additional validation testing.

Accessory Equipment

Dedusters and polishers are usually addressed solely in the installation and operational qualification and blanketed by their use during process qualification and/or process validation. An approach is that they do not adversely affect the product and do not purport to do anything quantitatively. They are something put on as a contingency for additional assurance. Evaluation of the product as acceptable without going through these devices is one approach. The other approach is to provide some standard worst-case challenge, but this seldom can cover all the possibilities and often gives rise to more questions. Salt polishing of dry filled capsules has been rendered obsolete by mechanical dedusters and the cleaner operation of automatic fillers. It is important that bio-batches and stability samples contain salt polished product, if salt polishing will be part of the commercial process.

Metal detectors are similarly addressed during qualification. The approach is to check the machines against the manufactures claims in a no load situation. Following this, each product should be checked at maximum throughput. It is desirable to request upper size limits for metal fragments from an internal medical or toxicological group. Limits should cover ferrous (e.g., tool steel and iron) and non-ferrous (e.g., stainless steel, aluminum, brass, copper) metal fragments for the purpose of metal detector qualification and validation. It may be assumed that the contamination would be an isolated production machine fragment. Also assumed is that in the event of a large increase of metal detector rejected material, the batch will be held and an investigation of the nature, source, and size range of the fragments will be conducted. It is important to include these assumptions when requesting limits from the medical groups. Otherwise, the entirely impractical response of "zero" may result.

A different case occurs when product with a known contaminant has to be redressed or reworked. In this case, specific challenges should be made up which emulate the contamination. These challenges must be performed with the strictest control, usually in the presence of quality assurance.

Capsule classifiers are utilized on-line for the relatively rare events of empty or low fill weight capsules. It is more likely to find an empty capsule than a low fill weight capsule because of the close proximity of the empty shells to the filled capsules within production rooms. These devices are easily challenged and qualified with empty shells.

Rework or Redress Validation

The 1987 Process Validation Guidelines (11) expressly acknowledge the one of a kind concurrent validation in Section IX. Rework or redress that consists of 100% mechanical inspection (for example, sizing on engineering rollers and metal detection), may be concurrently released and consist of only the batch(es) affected. However, any rework process requiring milling of tablets or capsules (with subsequent sifting out the gelatin fragments), optional addition of active ingredient (spiking), reblending, etc., usually requires three batches for validation. It is the author's opinion that one-of-a-kind reworks can be validated provided there is a clear assignable cause that does not invalidate the original process, and substantially more extensive product testing is included. The repetition of a manufacturing step following a known error in that step, unless addressed in the original validation, requires extensive justification and developmental support.

The notion of reworking seems contrary to a validated process. However, machine breakdowns and human errors result in material that must be reworked. Catching these mistakes confirms the validation of the original process and is good news in the sense that adequate controls are in place to catch errors. However, the product has been subjected to an unapproved process. This is where extensive documentation and validation type testing on the bio-batches pay off. Sampling and testing designs performed on clinical or development batches may be used to validate the rework. Rework entails validating a new full-scale process and the level of scrutiny should be the same as in the original validation and much more extensive if only one batch is involved.

Blending-off rework is forbidden for drug products; however, it is inherent in processes such as denture cleanser tablets (class one dental devices, performed for appearance reasons) and has been used to recover material for food supplements. The approach is to demonstrate an equivalent ability to meet final product specifications with the blended-off batches by additional scrutiny of the critical attributes of concern. These attributes are usually product efficacy and stability.

The redress of off-weight tablets and capsules requires the preparation of "edge of failure" challenges. Extreme caution must be used to ensure these challenge tablets or capsules do not get into commercial production. The disastrous effect of such an error may justify destroying the batch, rather than risking a rework. The use of different color capsule shells of the same size makes this a less risky proposition for capsules than for tablets. Extensive sampling and testing is appropriate for these validations. As an example, in one such redress with a capsule classifier, over 5000 individual weighings were performed. Modern capacitance or weighing-based classifiers may require less testing for validation than a separation system relying solely on the flow of air. The systems that individually and reproducibly weigh each tablet or capsule are preferred and their much higher expense may be justified by a single rework batch.

Evaluation of Data

The use of extensive hypothesis testing as part of statistical studies should be avoided for prospective validation. With large sample sizes it is easy to show that two batches are statistically different even when they are both well within all specifications. Statistical studies should be limited to those normally performed for product release and the Bergum (20) approach discussed previously. A better approach would be to prove that the probability of getting a failure is below some standard level for the validation batches. Use of statistics on granulation sample results is misleading since a continuum is being sampled. It is simpler to use established methods and limits that will be used after validation for normal in-process and final product testing once the process is validated. Over-complicated sampling plans are difficult to relate to routine operations and may lead to their imposition for routine use if problems occur later in time. Additionally, non-compliance with an untried sampling plan and/or technique is a typical deviation encountered resulting in missing data. In these instances, the runs should be repeated, if possible.

Care should be taken in the use of statistics such as CpK, CpM, and multiples of the standard deviation in the validation of blends, compressing, and encapsulation. These statistics are useful tools borrowed from the automotive and heavy equipment industries, and are based upon the process being the result of machines set in at the exact middle of process ranges. As stated earlier, tablet presses and encapsulation machines are adjusted until within the warning limits of in-process parameters and then run at the maximum validated speeds to produce compliant product. The statistics may not be optimized under these conditions and may not match or surpass values achieved on smaller low speed development equipment. It is important not to let these statistics become additional product specifications.

PROCESSING AFTER COMPRESSING OR ENCAPSULATION

Coated Tablet Cores

The sampling and testing scheme for tablet core PQ is the same as for compressed tablets. The problem is that most of the testing must be repeated for the final coated tablets. This is a good idea for validation since duplicate testing can be eliminated to some extent after the process is validated to save on quality control testing costs. The on-line in-process testing of individual weight, thickness (gauge), hardness, average weight, and friability are retained with as much precautionary chemical testing as risk will warrant for the compression manufacturing step.

Coating

Documentation of quantities of coating materials applied throughout the process is a validation requirement for manual sugar coating, which historically was an uncontrolled operation. Standardizing and recording drying air and bed temperatures reduce the variability in quantities used and has served to remove the "art" from this process. Film coating operations and automated sugar coating have more precise and reproducible solution application equipment and are easier to validate. Validation schemes treat each coating pan as a batch and care must be taken to maintain pan identity in the subsequent polishing, branding, and inspection operations. Samples for chemical testing are usually taken at the end to ensure detecting any deleterious effects of these subsequent operations. Coating solution preparation is included in the validation protocols when the solution is dedicated to a single product. Validation for solutions prepared for multiple batches with extended hold times must address these factors. Solutions containing active ingredients are validated following the sampling and testing scheme of liquids and semi-solid drug products. Press-coated tablets are treated the same as uncoated tablets or cores. Although it is possible to obtain a core sample from the "Drycota" presses, the testing should be performed on the coated tablet, unless the core and the coating have different active ingredients.

One big flag for FDA investigators is seal coating with shellac. Older processes that use this material often do not specify the amount or the exact process in the filing. In these instances, the coating process should be optimized to eliminate any trace of a dissolution problem that seems to plague the use of this material.

Polishing

The application of wax in separate coating pans (often canvas lined) was associated with the high volume manual sugar coating operations of the past. The main concern is cleaning validation for these separate pan operations. The polishing of coated tablets should be considered as part of the coating operation and be performed in the same pans where possible. The quantity of wax applied should be monitored, either to the pans or directly on the rotating tablet bed. The application of print base at this stage should be controlled as in coating operations, in that quantities added must be documented.

Polishing dry filled capsules to remove surface granulation with salt, or salt and polysorbate mixtures, has largely been replaced with buffing machines and was discussed earlier in the accessory equipment section.

Printing

Etched imprinting rollers should be controlled the same as press tooling. Printing specifications are often subjective and standards should be prepared for use during validation and subsequent production. Additional validation sampling for print evaluation is the usual validation approach. Samples for coating are normally obtained following printing since it is often the final processing operation for coated tablets prior to packaging.

Inspection

The improvement in reproducibility and quality of film coating operations coupled with product isolation within manufacturing areas have reduced or eliminated the need for manual visual inspection. However, the elimination of inspectors must be accompanied with procedures that ensure process isolation and minimize the risk of mix-up. Since empty hard gelatin capsules are often purchased or manufactured at a different facility, empty capsule inspection is still performed to ensure that no foreign capsules are present. For inspection operations, the so-called 200% inspection is employed. This is essentially two people watching the product go by on an inspection belt. The speed and density of the product are important parameters. The inspection belts with cavities to control the distribution of the product on the belt are best. Visual inspection is largely being replaced by computerized vision systems, which are reproducible and don't blink, become inattentive, or fall asleep.

Packaging

Most packaging operations for tablets and capsules, other than label accountability, had been viewed as "grandfathered" in that it was assumed they did not affect the product. However, packaging validation has become an area of focus for FDA field inspections, especially blistering and bar coding operations. This focus is beyond the stability issues and consists of documentation of machine set-up parameters, line speeds, and operating procedures.

Blister packing machines may potentially stress tablets and capsules since the blisters are formed by the application of heat and pressure. It can be easily demonstrated with stability studies and visual inspection that the tablets and capsules are not subjected to these stresses. Emphasis is also placed on seal integrity in terms of its effect on product shelf life and stability. Validation takes the form of increased in-process seal integrity testing in conjunction with stability testing. Line clearance during stoppages for product in heat-sealing stations must be addressed to ensure that any potentially heat stressed product is discarded.

Stability Data

Stability protocol preparation, sampling, and testing are usually performed by different groups than those who perform the qualification and validation studies. Often different laboratories perform the stability testing. Stability is considered part of validation in the broad sense of the definition. Initial product launches typically involve sample packaging and is important to ensure that all package presentations are included in stability protocols. It is a good practice to put the three initial process validation batches on stability. Stability data are supportive of packaging process qualification, but are seldom requirements for validation closure because of their continuous nature.

SHIPPING VALIDATION

Prepared packages containing temperature recording devices are staged in shipping areas for worst case periods and then shipped by typical carriers to worst case locations (e.g., Alaska during the winter and Arizona during the summer). The temperature (and sometimes humidity) histories for several shipments are combined to provide time-temperature profiles for stability chambers. The various product/primary container combinations are then placed on a shipping stability program. These studies are best considered as a qualification for a specific type of tertiary packaging so that study size and duration have some limits. Extensive use of matrices and worst-case environmentally vulnerable product help to provide bases for initial validation.

CONCLUSION

The author has attempted in the preceding chapter to present the approach to validation of solid dosage finished goods and relate some areas of potential concern that should be addressed within protocols. The sequence of fully documented commissioning, IQ, OQ, and PQ is followed for equipment. A process validation consisting of a minimum of three batches with all prerequisites in place is required for all doses. Sampling and testing equivalent to three to six times routine in-process and final release is usually employed. Acceptance criteria are generally equivalent to release limits.

REFERENCES

1. Medina C. The psychology of validation. J Validation Technol 1995; 2:11–16.
2. Dietrick JM. Food and drug administration recommendations on validation issues. Regul Aff 1993; 5:363–5.
3. Byers TE. GMPs and design for quality. J Parenter Drug Assoc 1978; 32(2):22–5.

4. Lindboe WG, Jr. Personal Observation, 2000.
5. Microsoft Developers Network. COM: Component Object Model Technologies. (Accessed December 2005 at http://www.microsoft.com/Com/default.mspx).
6. The United States Pharmacopeia 29th Revision, (1225) Validation of Compendial Methods. Rockville, MD: The United States Pharmacopial Convention, Inc., 2006:3050.
7. General Principles of Software Validation; Final Guidance for Industry and FDA Staff, US Food and Drug Administration, Rockville, MD, 2002 (accessed August 2007; www.fda.gov/cdrh/comp/guidance/938.html).
8. Agalloco JP. The validation life cycle. J Parenter Sci Technol 1993; 47(3):142.
9. Bush L. FDA lowers barriers to process improvement. Pharm Technol 2005; 29(10):54.
10. US Food and Drug Administration. Guideline on General Principles of Process Validation. Rockville, MD: Center for Drug Evaluation and Research, 1987 (accessed July 2007; www.fda.gov/cber/guidelines.htm 95).
11. United States District Court for the District of New Jersey, Wolin, District Judge. United States of America, Plaintiff, v. Barr Laboratories, Inc., et al., Defendants. Civil Action No. 92-1744 Opinion. 1993. (Accessed January 12, 2005 at www.gmp1st.com\barr2.htm).
12. Pharmaceutical cGMPS for the 21st Century—A Risk-Based Approach: Final Report-Fall 2004. Department of Health and Human Resources, US Food and Drug Administration, 2004 (accessed August 2007; www.fda.gov/cder/gmp).
13. Palady P. FMEA—Author's Edition. 2nd ed. U.S.A.: PAL Publishing, 1997.
14. Quinn D. An Introductory Course in Quality Management, MT322, Vanderbilt University. (Accessed December 14, 2005 at http://mot.vuse.vanderbilt.edu/mt322/Ishikawa.htm).
15. Kepner C, Tregoe B. The New Rational Manager. 1st ed. Princeton, NJ: Princeton Research Press, 1981.
16. Annonomous. Blend Uniformity Analysis: Validation and In-Process Testing. Technical Monograph No. 25. Bethesda, MD: PDA, 1997.
17. Liss E, Conway S, Zega J, Glasser B. Segregation of powders during gravity flow through vertical pipes. Pharm Technol 2004; 28(2):78–96.
18. Lieberman H, Lachman L, Schwartz J. Pharmaceutical Dosage Forms: Tablets—Volume 2. 2nd ed. New York; Marcel Dekker, 1990:40.
19. Bajaj J, Brennan AK, Jr., Herzog S, et al. Tips for improving mixture sampling. Powder Bulk Eng Jan. 1993; 41(3).
20. Bergum JS. Constructing acceptance limits for multiple stage tests. Drug Dev Ind Pharm 1990; 16(14):2153–65.
21. Cholayudth P. Use of the Bergum method and MS excel to determine the probability of passing the USP content uniformity test. Pharm Technol September 2004; (9):86–96.
22. Tetzlaff RF. Validation issues for new drug development: part I, review of current FDA policies. Pharm Technol 1992; 16(9):44–56.
23. Avallone H. Development/Scale-up of Pharmaceuticals. In: Presented at FDA's NDA/ANDA Inspection Program. College of Pharmacy, Division of Continuing Education, Rutgers New Brunswick, NJ: the State University of New Jersey, December 12, 1990.
24. Walker J, Abdulsalam A, Theobald AE, et al. Multiple-scored tablets. Weight and content uniformity of subdivisions and the distribution of active constituent within and between tablets. J Pharm Pharmacol 1978; 30:401–6.

31

Validation of Oral/Topical Liquids and Semisolids

James Agalloco
Agalloco & Associates, Belle Mead, New Jersey, U.S.A.

William G. Lindboe, Jr.
Consultant, East Brunswick, New Jersey, U.S.A.

INTRODUCTION

The term "validation" can have different meanings within the scope of the healthcare industry. For example, from the research point of view, this term can pertain to the efficacy of a drug therapeutic effect, or effectiveness of an automated diagnostic assay, as well as the instrumentation and software used to obtain the result. The predominant use of the term is with reference to manufacturing operations and direct supporting systems. The scope of this chapter will be the compliance requirement for the validation of manufacturing processes for oral liquids and topical semisolid pharmaceutical dosage forms. This is generally referred to as process validation and begins with the batching of approved raw materials to the storage and shipping of packaged product. The term "PQ" has been used to describe some of the concepts presented in this chapter. The variety of formulations that fall within the scope of this chapter will be surveyed along with associated manufacturing equipment. The chapter will be directed toward the individuals charged with the responsibility of preparing and executing validation protocols for these dosage forms. The assumption will be that development is essentially complete; however, a few of the common process problems will be discussed with the hope that the reader will be able to anticipate pitfalls and eliminate subsequent validation difficulties.

Manufacturing validation of non-sterile liquids and topical semisolids can be considered an afterthought following the implementation of parenteral and solid dosage validation. Although there is a concern for product bioburden that exceeds that of solids, the dosage forms within this category typically have more than adequate chemical preservative systems and are inherently low-risk products. That being stated, the validation of these products is anything but easy. The products typically consist of multiple components that are either hydrophilic or hydrophobic depending upon their route of administration (orally or topically) and often require numerous processing steps to achieve the desired final preparation. Additionally, the areas for manufacturing contain some basic equipment that is applicable to many different types of liquid and semisolid products. Considerable development is ordinarily required to arrive at the optimal equipment parameter settings for each product. Many of the products are OTC drugs that frequently do not get the developmental attention of high-valued ethical pharmaceuticals. Additionally, these products are often exchanged between pharmaceutical manufacturers, which can introduce subtle changes in equipment and facilities and renders the original developers and their accumulated knowledge unavailable to validation personnel. Another common practice is outsourcing the manufacturing to a third party or private label manufacturer. This can further limit development efforts and manufacturing equipment options. In these instances, the development, operating and validation personnel have inherited the process and product with relatively limited information. Given these difficulties, the authors have had considerable success in defending validation documentation of these dosage forms based upon the approaches that will be presented. Each approach has been audited by the FDA and/or EMEA personnel with no adverse comment.

HISTORY OF ORAL/TOPICAL LIQUIDS AND SEMISOLIDS VALIDATION

The initial motivation for the Federal Food Drug and Cosmetic Act can be attributed to dosage forms in the category discussed in this chapter. The so-called "snake oil" and other sham preparations of the 19th century charlatans were commonly oral or topical preparations (1). Although in early incidents efficacy, rather than manufacturing, was the primary concern. Later refinements to the Act pertaining to safety were brought about by the sulfanilamide elixir adulteration, where the formulator used ethylene glycol as a solvent for the drug with disastrous consequences.

Other than mislabeling, manufacturing of these dosage forms has not been a major concern of the regulators. These are the dosage forms that many retail pharmacists still prepare within the pharmacy. Many pediatricians prepare and dispense the final preparation of amoxicillin for oral suspension. Until recently, there

---

Abbreviations used in this chapter: API, active pharmaceutical ingredient; CFC, chlorofluorocarbon; COA, Certificate of Analysis; Cpk, capability index for process average; CPP, critical process parameter; CQA, critical quality attribute; EMEA, European Medicines Agency; EPA, Environmental Protection Agency; FDA, Food and Drug Administration; IQ, installation qualification; NIST, National Institute for Standards and Technology; OQ, operational qualification; OTC, over-the-counter; pMDI, pressurized metered-dose inhaler; PQ, performance qualification; QC, quality control; RPM, revolutions per minute; USP, *United States Pharmacopeia*.

have not been any major manufacturing problems with these dosage forms that lead to public safety concerns. The development of versatile and highly effective manufacturing equipment, which will be discussed below, has aided this relatively good manufacturing experience. The materials have diverse rheological properties and in some cases difficult to achieve emulsions or uniform suspensions must be created. Earlier equipment advances typically relate to the unfilled and packaged bulk preparation and not the final packaged product. Recently, the development of sophisticated metered-dose aerosols for oral inhalation and form–fill and seal unit-dose blister packages have added additional complexity and requirements on the filling and packaging. Packaging manufacturing personnel, who previously only dealt with the rugged and highly dependable Cozzoli and Arenco fillers for liquids and semisolids, respectively, now had to deal with sophisticated and perhaps unproven new filling technologies. In some cases, manufacturing firms were not up to the task, as evidenced by the Consent Decree that was partially based upon improperly filled metered-dose albuterol for oral inhalation (2). The critical life-saving application of this drug has brought increased regulatory focus on non-sterile liquids and semisolids and their validation.

DOSAGE FORMS—SAMPLING, CPPs, AND CQAs

A discussion of common validation terminology, other than simple definition, will be omitted because of overlap with other chapters in this volume. The reader is advised to consult these chapters for a more detailed discussion, along with the 1987 *Guideline on General Principles of Process Validation* (3). The acronym CPP pertains to machine and product settings that impact end-product quality. CQA generally pertains to in-process and final product test results that affect product efficacy and/or safety. It is important that these parameters and attributes are developed with proper tolerances prior to validation under a formal protocol. As an example, evaluation of semisolid rheology and characterization by an appropriate constitutive model using a Rheometrics Mechanical Spectrometer (Rheometrics Co., Piscataway, New Jersey, U.S.A.) is an idealized dream that is unavailable to most manufacturers and research groups. It is well known that rheology is affected by temperature, pressure and the manner in which the pressure is applied to the material. Not all semisolids can be approximated by a Newtonian fluid. Many exhibit thixoplastic behavior indicating a change in structure with the application of pressure. Pseudoplastic behavior is also observed and such properties may affect mixing, homogenization and uniformity when manufacturing processes are scaled-up. It is not unusual for product batches to be heated by mechanical mixing. The authors recall one premix for a zinc oxide diaper rash cream that was heated to a relatively high temperature solely by the action of a rotating mixing plate. All characteristics unique to these formulations must be considered in the preparation of validation sampling plans and documentation. A common manufacturing practice is to prepare a semisolid "base," which is a placebo carrier to which a variety of active ingredients (alone or in combination) can be added to for the purpose of minimizing development effort and providing a "family" of similar products. A large manufacturer may have several cream, lotion, and/or ointment bases that may differ only in viscosity, to which different amounts and types of API are added for the final preparation. The common and very important CQA of microbial content or bioburden can be assumed applicable to all liquids and semisolids. As such, most topical and oral formulations include a chemical preservative system, unless the active ingredient itself happens to be bactericidal. The additional microbiological concerns of non-sterile liquids and semisolids over solid dose are based upon water content or water activity (4). It is well known that microbial concerns increase exponentially with increased water activity above 0.7 (4). A typical aerobic plate count limit for these products is 500 CFU/g of material. This corresponds roughly with the obsolete EPA standard for potable water for municipal water systems. That standard has been replaced with a single prohibition against coliforms. Similarly, non-sterile liquids and semisolids have a zero tolerance for pathogenic organisms, namely gram-negative bacteria. Preservative systems are intended to compensate for the natural variation in nonpathogenic bioburden and extended use of multidose primary packages by the consumer, not to cleanse product of pathogens. Processes and raw material standards must be designed and have sufficient controls in place to preclude these dangerous organisms from being present.

Non-sterile liquids and semisolids require USP Purified Water. The nature of the smaller OTC manufactures gives rise to water quality concerns in that the level of maintenance may be adequate to address microbial considerations (5). These firms do not have the resources to upgrade water systems and many legacy systems contain some plastic piping. These types of systems require chemical sanitization because they cannot tolerate the heat necessary to sanitize the system. For older legacy systems, there is risk of periodic contamination caused by biofilms. Plastic pipe water systems should be equipped with the most modern chemical sterilization methods, such as ozone. Older systems should be replaced. Additionally, source water can be surface, well, or mixed and chlorine content is seasonally variable, thus bioburden is variable which adds to legacy system problems.

Another common element is the difficulty in obtaining representative samples of the bulk premix(es) as well as the final mixed bulk for suspension, emulsion and highly viscous products. The difficulty increases with material viscosity and the manufacturing technique of "geometric dilution" that is widely utilized out of necessity when trying to disperse APIs in a standard previously made base. Considerations for achievement of homogeneity range from: the assumed, for aqueous liquids under agitation, to the extremely difficult, exceeding solid dosage in difficulty, for ointments, pastes, and adhesives. Additional description of the following discussion on these dosage forms may be found in the USP (6).

Liquids—Oral/Aerosol for Inhalation

Liquids are more properly termed solutions and are drug API(s) dissolved molecularly in a solvent(s). Elixirs contain alcohol in the solvent system and are for oral administration. Tinctures contain alcohol and are for topical administration. Syrups, intended for oral administration, contain either sugars or sweeteners. The nature of properly mixed low-viscosity solutions provides assurance that they are uniform. Validation concerns are assurance of the adequacy of the mixing and bioburden. Manufacturing processes may involve premixes or side pots to facilitate dissolving solids. The resulting premix is then diluted/added into a final bulk mix that is tested and held for filling and packaging. Typical CPPs for the premix are evidence of dissolution (particle size) and refractive index. Sampling of the premix can affect the final concentration of the bulk batch. The number and size of the samples must be considered in protocol preparation. The final mix may be sampled at any point given the assumption of homogeneity. Validation usually is a confirmation of homogeneity and location-based samples are frequently taken to prove that stratification does not occur upon standing. Liquid and semisolid sampling thieves (Fig. 1) have been developed by Globepharma (New Brunswick, New Jersey) to facilitate location-based low-viscosity materials within final mix vessels and/or bulk holding tanks. However, it is not uncommon to sample solutions from a sampling port intermittently while they are agitated. Primary packages are easily obtained during the course of filling and packaging operations. It is important to obtain the very first and very last containers filled to assure that these historic "problem areas" are included in testing. All of the filling nozzles should be represented in sampling and testing. Bulk manufacturing CPPs include mixing speed, configuration of mixing blade, position of mixing blade, tank and/or kettle volume and geometry, pumping mechanisms and rates, temperature, pressures, pipe or hose diameters. All of these should have appropriate definition/specification and be properly measured and documented during validation. Filling should fully specify the equipment, especially the dispensing and container sealing systems and other mechanical settings. The presence of mixing and/or agitation of the bulk liquid within the filling system should be documented. Form–fill and seal unit-dose packages will require the documentation of many parameters associated with the packaging equipment. "Gas house" filling of propellants for foams and aerosols must also be documented in detail. The parameters associated with the addition of the propellant, and the pressure at which it is added must be included. Container crimping/capping parameters and container closure integrity must also be evaluated.

CQAs for the final mixed bulk and the filled product include drug assay(s), preservative efficacy, pH, viscosity, density and/or specific gravity. Fill volume for multidose containers or dose uniformity for unit-dose containers must be included. Filled solutions used as aerosols for inhalation and foams require rigorous testing of valves, actuators and containers prior to their receipt on the production floor. Filled aerosols and foams are two-phase systems consisting of the concentrate (bulk API-containing solution) and the propellant. The propellant raw material testing and release must be documented. The development of the pMDI has gone through a recent transformation necessitated by the phaseout of CFC propellants for environmental reasons (7). The accuracy and reproducibility of metered-dose valves must be tested for the claimed number of doses within a container. A physical verification of the absence of unfilled and under-filled containers for critical drugs such as albuterol is necessary. These verifications may include tight-tolerance bulk to filled container accountability by volumetric metering at the filler, coupled with bar code reader counts of filled containers and/or check weighing of every filled container.

Foams—Topical, Vaginal

Foams are similar to the aerosol liquids in that there is a concentrate and a propellant in the final primary package. The concentrate is typically a liquid or emulsion with a dissolved or suspended API that combines with the propellant within the actuator to produce the foam. Garg et al. (8) provides a comprehensive review of vaginal formulation excipients that will enable the selection of appropriate CPPs for inclusion in validation protocols, besides API assay(s). Bioburden, preservative assay and effectiveness, concentrate viscosity, pH, density or specific gravity, and CPPs and CQAs related to the primary package delivery system are typical concerns.

Suspensions

Suspensions are liquids or semisolids that have small solid particles dispersed within them. Some suspensions are prepared by the pharmacist and/or dispensing physician immediately prior to administration. These

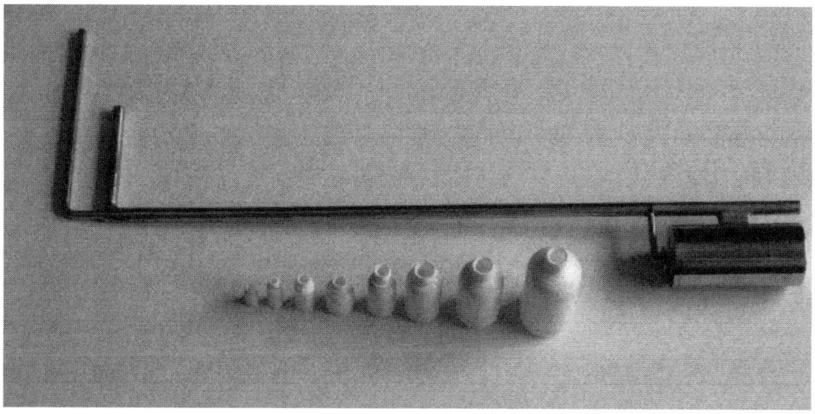

Figure 1 Liquid sample thief. *Source*: Courtesy of GlobePharma.

formulations are covered by the chapter on solid dosage validation within this volume. Suspensions require at least one premix, usually multiple premixes, where the solids are dispersed within the liquid using various dispersing and homogenization equipment that will be discussed below. CPPs for the premixed material are particle size of dispersed solids, density, temperature, and viscosity. Equipment CPPs include rotor and stator configuration, gap, RPM, pressure valve opening size and location within vessels of the dispersing equipment. Final mix parameters include assay, viscosity, density and/or specific gravity, particle size, pH, and absence of entrained air. The authors have found difficult-to-maintain suspensions often require recycle lines from the filling machine back to the bulk holding vessel to prevent segregation during even brief interruptions in the filling process. The use of intermediate surge vessels outside the recycle loop was inadequate to assure content uniformity in the filled units. The greatest difficulty with suspension manufacture is typically the filling process, which often proves most problematic, especially for smaller fill volumes. Consider that the entire batch of suspension can certainly be uniform if taken as a whole; the difficulty is assuring that all aliquots of that vessel (down to a unit-dose container in some instances) are equally uniform with regard to ratio between the solid and liquid phases.

Emulsions

There are two types of emulsions: the oil-in-water and the water-in-oil. In both cases, the former is dispersed as small droplets in the latter or continuous phase. Usually emulsifying agents are employed to keep the dispersed droplets from merging. This would lead to nonuniformity since APIs are usually restricted to a single phase. Care is needed to ensure that the emulsion remains intact once it is achieved. Emulsions can be destroyed mechanically, thermally, or chemically. Validation is concerned with equipment CPPs needed to create the emulsion in the first place and then preserve it through holding, filling, packaging and shipping. These are temperature, disperser type and configuration, rotor–stator gap, homogenizer valve(s) opening, RPM, vessel configuration, pump speeds, pressure, hose diameter, and materials of construction. Product CPPs are viscosity, density/specific gravity, particle size, air entrainment, pH, and assay. CQAs are essentially the same as the previously discussed dosage forms.

Lotions—Topical

Lotions can be emulsions, suspensions, or gels and is a general term associated with a low-viscosity topical dosage form. Lotions frequently have the phrase "shake well" on their primary packaging and may have multiple active ingredients, such as a sunscreen. This combines to make lotions difficult to sample both in-process and final mixed bulk. It is usually necessary to sample the material while under agitation or mixing at any stage of the manufacture. The authors recall witnessing one lotion where the mixing was stopped, which developed an intermittent "oil slick" on the top surface. This product was to be filled into unit-dose form–fill and seal sample packettes and the delivered dose from this primary package was highly variable, despite having only a single active ingredient. Critical CQAs are assay, viscosity, and homogeneity.

Cream—Topical, Vaginal

Creams are emulsions consisting of oils dispersed in aqueous medium. The active can be in either the oil or the water phase. Alternatively, a nonactive cream base may contain a suspended solid. In either case, the physical characteristics of the cream can govern the quality of the final preparation. The cream base CPPs and CQAs are generally restricted to the physical parameters of viscosity, specific gravity or density, and homogeneity. Final formulated bulk adds API assay(s), bioburden, preservative assay and efficacy to the cream base CPPs and CQAs.

Ointment—Oral, Topical

Ointments are usually an emulsion with an aqueous liquid dispersed in an oil phase. The oil phase can be mineral oil based and/or petrolatum based depending on the desired viscosity. Synthetic oils are also utilized. An ointment base is typically manufactured prior to the addition of active ingredient(s). The CPPs and CQAs are essentially the same as for creams. The high viscosity of ointments necessitates multistep formulation processes and exotic mixing and dispersing operations despite using a standard ointment base. It is important to specify the scale of scrutiny (how large a sample) when sampling and testing the intermediate manufacturing steps. Often, acceptance based on simple physical measurements of viscosity and dimensionless groups, arising from machine parameters and developed during process scale-up, is used (9).

Pastes—Oral, Dental

Pastes are emulsions with suspended solids often of very high viscosity and usually do not have a "base" in which APIs are added. They are typically high volume, large batch size products that utilize large combination equipment (discussed below) to arrive at the final formulation. A common problem is the achievement of the emulsion, even though the API may be uniformly dispersed. The inclusion of suspended insoluble abrasive solids in dental formulations increased the difficulty in determining uniformity prior to sampling the primary package. The APIs in dental formulations (fluorides) are typically bactericidal and preservatives may or may not be present. In the case where they are absent, it is not unusual to have procedures specify a 24-hour hold time prior to testing for bioburden to allow the active to sanitize the final formulation. This is not a desired practice and can be eliminated with proper raw material testing and vendor qualification (10).

Gels

Gels are similar to ointments and pastes except that they may exhibit solid-like behavior. They generally change to semisolid or liquid with the application of heat and/or pressure. Frequently, there is a hold time required for the gel to develop and exhibit the solid-like properties. Gels are suspensions with a higher viscosity caused by the interpenetration of the solids by the liquid. As in pastes, there is typically no standard base that is prepared prior to the addition of API(s). CPPs and CQAs include assay, specific gravity or density, and ability to be extruded from a multidose container through patient peristaltic manipulation (squeezing). An aqueous suspension of bentonite clay is widely used as a simulant for gels in development and equipment qualification where product cost is prohibitive to the required process development. A recent CQA of interest is gel strength (11). Soft gelatin used for liquid-filled capsules falls within this category.

Suppositories—Vaginal, Rectal

Suppositories are lipids, either natural or synthetic with dissolved and/or suspended API(s). The lipid must have the characteristic of "melting" at body temperature so that the drug(s) can be delivered to the patient. The final bulk is heated and filled into chilled molds; today, typically, form–fill and seal molding machinery is used. A common rectal suppository base is cocoa butter (theobroma oil). This is a natural product and a by-product of the confection industry and it is common to have the heated liquid filtered through "cheese cloth" as the first manufacturing step. The CPPs of this operation such as temperature and time of heating cannot be diminished as the ability to be molded may be affected by this operation. CQAs are assay, uniformity and molded shape.

Adhesives—Transdermal, Ostomy, Denture

Adhesives are difficult to manufacture and substantially more difficult to clean. Cleaning must be considered because a nonaqueous solvent such as mineral oil is typically used and there will be a permitted residue on equipment surfaces following multiple rinses. This residue should be estimated and include within the process validation protocol.

Transdermal adhesives may contain API and the CPPs of drug diffusion and adhesion are of obvious concern. Tack, adhesion, release force and cohesive strength are also critical (12). There are various in vivo diffusion devices to measure drug flux (13). CPPs of filling, extruding and rolling equipment are critical as the geometry of the adhesive is proportional to the delivered dose.

Ostomy and denture adhesives are medical devices that are included within this chapter because their manufacturing process utilizes typical pharmaceutical processes. With these products, machine variables and product physical characteristics are critical. As in transdermal adhesives, tack, adhesion, release force and cohesive strength are CQAs. Extrudability is also included for denture adhesive filled in multidose tubes.

EQUIPMENT—PQ AND CPPs

The equipment utilized for the manufacture of non-sterile liquids and semisolids is just as diverse as the variety of dosage forms. Processes range from a simple dissolution of solids in an aqueous solution to multiple step mixing, homogenation, dispersion, and extrusion. Filling and packaging similarly ranges from a simple liquid fill to difficult to maintain suspensions to complex aerosol and form–fill and seal operations. Each piece of equipment will have its own set of parameters that must be controlled and documented for validation. Common CPPs are mixing speed (RPM), time, and temperature of any heating. It is difficult to measure RPM using an optical tachometer on 316 stainless steel surfaces. Special reflective tape and/or paint can be used for these measurements. In many cases for semisolids, tachometers cannot be used because the mixing speed is too slow. In these instances, a calibrated stopwatch can be used to measure mixing speed. Below is a description of the major equipment types currently in use for pharmaceutical manufacture and most likely to confront the validation team.

Tanks

Tanks are primarily used to manufacture liquids, low-viscosity semisolids and to hold (store) all liquids and semisolids. They are fabricated out of 316 Stainless Steel and this must be verified during equipment qualification. Companies that manufacture food and/or cosmetics, in addition to pharmaceuticals, may have a large inventory of tanks and it is important to obtain an equipment history. In one instance, the authors recall the proposed use of portable tanks for a purchased diaper rash ointment. The tanks in question had been used for a lice treatment shampoo in another facility, the active of which is classified as an insecticide. After consideration of the implications, a different set of vessels were selected for the ointment. Tanks are usually constructed with a conical bottom to facilitate gravity draining. There are two varieties, fixed and portable, the latter of which may or may not be on wheels. In many cases, the fixed type will have a permanent in situ mixer with a defined agitator shape. In other cases the agitator motor and/or impeller is changeable and it is important to document the specific motor and/or impeller used during the production and in the validation as well. Each agitator motor and impeller should have its own identification markings so that performance can be related to equipment qualification studies. Depending upon the application, the tank may be equipped with a jacket for heating/cooling of the vessel contents.

Some extremely viscous intermediates and products may be stored in flat-bottomed vessels without a bottom outlet, the product is removed using a compression system applied to the top of the material that descends as the product is dispensed.

Kettles

Kettles are essentially tanks intended for use with the more viscous and difficult to agitate material. Kettles often have the ability to be heated, although this feature is not always utilized. Most have permanent mixers attached and range in size from essentially bench top units to upwards of 1000 L in capacity. The agitators on kettles will have an internal side-scraping feature that ensures viscous materials and solids are prevented from adhering to the sidewalls during ingredient addition and mixing. An inspection of these polymeric scrapers is necessary before and after each kettle use as they have a tendency to wear and break. The kettles may also have covers to prevent excessive evaporation and allow vigorous mixing without fear of product loss. Many kettles are designed with a primary and secondary mixing action with

Figure 2 Versatile mixing kettle. *Source*: Courtesy of Lee Industries Inc.

independent agitators controlled by a separate motor. Lee Industries (14) manufactures a wide variety of mixers with different agitator designs/combinations (Fig. 2) and other processing equipment and is a large supplier to the pharmaceutical industry. In most cases, the manufacturer of the equipment can provide insight into equipment operation that will facilitate protocol preparation. Sampling of viscous materials is difficult and most easily performed during discharge. Worst-case location sampling can be performed with a sampling thief.

Mixers

Mixers are associated with portable tanks and with the filling hoppers during primary package filling. They are generally used with liquids and the lower viscosity semisolid products. The ubiquitous Lightnin Mixer (15) will be found in virtually every larger liquid and semisolid manufacturing facility. The impellers on these mixers are removable and tend to be banged around during cleaning. That coupled with the longevity of this equipment make it essential to verify the impellers meet the original manufacturer's specifications. The location and angle of mixing must also be documented during validation. In some cases, the location of the impeller may affect mixing efficiency especially of the more viscous liquids and affect dissolution times of solids. Most mixers, whether portable or fixed, have the ability to change the impeller or agitator. These need to be identified and documented to prevent a change that might affect product quality. Mixing of high-viscosity material requires slow agitation with the so-called "gate" impellers that gently move the material, prevent the introduction of air, and facilitate air removal under vacuum. These mixers generally follow mixing with different equipment and/or impellers for greater homogeneity. Sampling and testing for uniformity is best accomplished at the previous stage.

Homogenizers—Dispersers

There are two types of homogenizers: the valve type and the rotor–stator type. The latter, while capable of achieving homogenization, is better classified as a disperser or colloid mill. The valve type usually has two stages consisting of two valves in series to prevent clustering of lipid globules after the first stage. The manufactures frequently encountered are the Cherry-Burrell (SPX Industries) and the older Gifford-Wood. The mechanism by which homogenation is achieved is not definitively known. There are three prevailing theories, any one of which may dominate for a particular material and viscosity. These theories are based upon the generation of turbulent velocity arising from high pressure. The three mechanisms are as follows: shearing between globules, shattering of globules from impact with the valve surface and the formation of pits or cavitations following passage through the valves resulting in the condensation of small vapor bubbles. There is an increase in viscosity following homogenization of oil-in-water formulations caused by the increase of surface area of the oil globules. This provides a convenient physical test to confirm the success of the operation. Microscopic examination is also often necessary to confirm homogeneity. Additional testing consists of particle counting (Coulter Counter, Beckman Coulter, Fullerton, CA) and light scattering. Homogenization is one of the operations where an examination of the resultant product determines the adequacy of the machine settings. Manufacturing procedures may specify some variability in machine setting to accommodate variation in raw

Figure 3 Versatile reversible homogenizer. *Source:* Courtesy of ARDE Barinco

materials, especially natural products. Typical first-stage pressures are 2000 psig, while second stage pressures vary around 500 psig with valve clearance around 500 μm. Positive displacement pumps are used and are part of the homogenizers in many cases.

The rotor–stator type has the critical CPPs of rotor speed and rotor–stator gap. Speeds are entirely variable and gaps vary around 0.050 inch. This type is used to disperse solids during premix operations and ARDE-Barinco (Norwood, New Jersey, U.S.A.) manufactures a versatile bidirectional unit (Fig. 3) that is frequently encountered in manufacturing areas. The bidirectional feature serves to ensure that all materials pass through the rotor–stator and the portability allows for use within tanks and/or kettles.

These units are used with all types of nonsolution products to ensure greater dispersion of the solid phase in the liquid, and they can also be utilized to reduce the particle size of the solids (albeit with considerable heat generation).

Another form of dispersator is utilized to initially wet large quantities of poorly soluble solids for incorporation into a liquid base. These operate similar to a centrifugal pump, but are termed "dispersers" by the manufacturer.

Combination Equipment

Some large capacity equipment can both mix and homogenize and premixing may be done in the same vessel. The Koruma Kettles are an example of this type of equipment. In addition to large rotating mixing impellers they have a high-speed disperser/homogenizer on the bottom of the kettle that can pump the material though a recirculation pipe to the top of the vessel. It is difficult to adjust the rotor–stator gap of the "Disho" dispersers on the bottom of the kettle and a check of the effectiveness of premixing is usually necessary. These units are typically seen in high-volume dental paste facilities.

Centrifuges

Centrifugation is a separatory process and is the opposite of homogenation. It is used as a preliminary operation to remove undesired components for further processing or discard. Products including natural materials frequently use centrifugation to reduce variability in physical properties. It has been well known since early history that a mixture of materials of different densities will separate by the action of gravity. The disk–bowl type of centrifuge provides a centrifugal force to affect the separation and its effectiveness increases with rotational speed. Another critical parameter is the separation between the disks, which is typically 0.020 to 0.050 inch. Unlike the "art" of homogenation, centrifugation is modeled by the widely accepted Stokes equation (16) that provides the rate of separation as a function of material physical properties and centrifuge parameters. Frequently encountered production equipment are the Westphalia and Sharples (17). With respect to the longevity of the latter, the authors recall one manufacturing supervisor, who reported, while visiting a retired military submarine museum, seeing the very same model Sharples centrifuge used to separate water from diesel fuel during the Korean War that was in current use in the individual's semisolid production facility. Forces in excess of 10,000 times the force of gravity are routinely used in production.

A special type of centrifugal unit is the "Versator" that uses centrifugal force and vacuum to remove air introduced into both solutions and suspensions during earlier dissolution or mixing steps.

Pumps

Where gravity cannot be used, pumps are a necessity within a non-sterile liquid and semisolid facility (Fig. 4). Pumps can be positive displacement or not, with the former used primarily in metering and/or filling operations. Centrifugal pumps are used for lower viscosity materials, while lobe or peristaltic pumps are used for more viscous materials. Pumping rate is a CPP that is usually translated into some measurement of rotational velocity of the pump impeller or actuator. It is virtually impossible to directly measure the actual impeller speed. A careful analysis of the disassembled pump, during qualification, will enable the measurement of motor RPM to be translated into pump speed. Pumping rate can be estimated by pumping into a graduated vessel or a vessel that can be weighed. Materials of construction must be verified during qualification along with internal sealing. Waukesha pumps (SPX Industries, Charlotte, NC (Fig. 4)) are widely used in the chemical, food, cosmetic, and pharmaceutical industries and it is important to ensure that the design of legacy equipment matches the current intended application. Specialized dispensing pumps have been successfully employed to precisely meter two or more solution streams and, when used with an in-line mixing system, can reduce tankage requirements by allowing mixing of a concentrate(s) with a separately stored diluent followed by filling. More common in health and beauty aids, this process has been successfully utilized for large volume OTC products.

Filters

Filters are used to remove undesired solids from a liquid or elevated temperature semisolid. Cheesecloth is frequently used for OTC raw materials and it is important that the type and quality of this material be specified and controlled. The cheesecloth and all filter materials, for that matter, should be tested and released as raw materials at least with a qualified vendor's COA. Ronnigen and U.S. Filter both offer a variety of filters and filter media that can be used for clarification of liquids. In older operations, it is important to ensure that banned materials such as fiberglass and asbestos are not utilized. The inability to assure filter integrity by

Figure 4 Waukesha pump. *Source*: Courtesy of SPX Industries.

performing integrity testing on wider mesh semisolid filters is reduced by using serial filters. The wide availability of filters for the food and parenteral manufacturing industries that are capable of integrity testing is covered in a separate chapter within this volume.

Fillers

Fillers are really an extension of the pump category since they are a necessity to fill the primary package. The pumps are generally a multiple small-scale positive displacement pumps similar to a syringe. Two manufacturers that can be found in many filling operations are the Cozzolli for liquids and the Arenco for semisolids. These fillers are virtually indestructible with the downside that very old legacy equipment is frequently encountered in continued daily use. The simplicity of their operation and long history of successful use make qualification and validation relatively easy. Filling rates combined with extensive sampling of the filled tubes are generally sufficient for validation. Some materials may require a recirculation of the filling hopper during filling and/or mixing while filling (lotions) to maintain homogeneity. The rate and method of these operations must be documented. The authors recall one instance years ago when, while observing a process during a validation trial, in a moment of horrible comprehension, they discovered that the unwritten practice of stirring a filling hopper with a wooden paddle was being routinely performed!

Peristaltic pumps are sometimes used effectively for filling operations. It is important to document the type of tubing (usually medical #6) and the frequency of tubing change as part of the validation. The characteristics of the polymeric tubing will change with extended squeezing by the pumping mechanism and may affect the filling accuracy. To eliminate cleaning problems any tubing should be single use. Dedication to a single product is not adequate to ensure the absence of cross-contamination or microbial buildup. An additional concern with re-use is the potential for the tubing material to degrade over time and slough particles into the product.

Sophisticated form–fill and seal equipment will still have the fundamental filling operation as one of its stages. Computerized control of the entire operation from web to final seal is standard. Separate and detailed qualification documents are necessary prior to process validation to ensure adequate validation. Additional detail on considerations for validation of filling/packaging operations can be found elsewhere in this text.

ANALYTICAL TEST METHODS AND EQUIPMENT

The discussion of analytical test methods will be restricted to those that are frequently performed on the production floor as part of a validation trial or run. The reader is directed to the many widely available references on laboratory test methods and equipment pertaining to the QC of pharmaceuticals.

pH Meters

The pH meter is usually a part of aqueous liquid manufacturing in-process quality checks. The meters are simple to use and it is easy to make additional measurements for validation protocols and therein lies the potential problem. Protocol designers often cannot resist placing specifications on statistics generated from pH data. It must be remembered that pH is the negative of the logarithm to the base 10 of the hydronium ion concentration. If concentration is normally distributed, then an exponential function of it, namely pH, will not be. The easy solution is to convert pH readings to concentration prior to generation any statistics and have the acceptance criteria reflect these statistics.

Hegman Gauge

One of the most versatile and useful in-process instruments is a device borrowed from the paint industry, the Hegman Gauge (Thomas Scientific, Swedesboro, NJ). This device is a graduated channel ground into a stainless steel plate where the material is spread along the channel by a stainless steel scraper. This device can detect nonhomogenized oil globules, nondispersed solids, air bubbles and enables one to estimate the size of these undesirables. It is a scientific replacement of the operator's spatula where improperly written manufacturing instructions have the statement: "check if dissolved," or similar subjective evaluation.

Viscometers

Viscosity is a very important property of semisolids and their components. Viscosity is highly variable with temperature and a qualified temperature bath with specified hold times is required along with the qualified viscometer. Measurements on low-viscosity liquids may be performed using the Ostwald viscometer, which times flow of a known quantity through a capillary tube. It is important to remember that the timer must be qualified prior to use, along with the viscometer. Stokes Fall is another timing measurement where a ball of known size and mass falls through a given length of material. The most versatile and frequently used viscometer is the Brookfield type, which mimics Couette flow. This device measures the torque of a rotating cylinder in a cylindrical container of sample. This device can be used on more viscous materials and has the advantage of measuring force directly.

Refractometers

The measurement of refractive index is restricted to clear materials. However, it does provide an important physical constant when applicable. In many cases, the assay of a particular ingredient varies with refractive index and custom instruments graduated in units of concentration of the ingredient of interest are used. These instruments are simple in design and can be handheld for convenient use within the manufacturing facility. Validation testing using these calibrated instruments is entirely acceptable as an alternative to the time-consuming sample transport to a laboratory for instrumental analysis. In many cases, an immediate reading is needed in order to make a decision pertaining to the process.

Balances, Scales, Pycnometers

Balances and scales must be calibrated to NIST-traceable standard weights. A validation concern for OTC liquids and semisolids in "private label" manufactures is to ensure that they are used! It is not unheard of that supplier container weights are used in lieu of weighing for large additions of excipients. Another practice is to dump drums or bags of material into the batching vessel after a gross weight is obtained, and then weigh the empty container for the tare weight. If this practice is performed, steps must be taken to ensure that all the correct amount of material gets into the batching vessel and that excess material is not added in error.

Besides the obvious measurement of mass in adding the ingredients to a formulation, balances, and scales are used for the CPP of density or specific gravity. In these cases a pycnometer, which is a glass vessel of precisely measured volume, is utilized. As in viscosity measurements, pycnometers must be used within qualified temperature baths to provide accurate measurements.

Thermometry

Temperature is an important CPP for liquids and semisolids. The measurement of temperature follows that of parenteral manufacture and is well covered elsewhere in this volume. The problem is that the sophisticated temperature measurement and calibration equipment, usually supplied by Kaye Instruments (North Billerica, MA), are not generally available at smaller nonparenteral manufacturers. Equipment is available for rent that will enable smaller manufactures to qualify their thermometers and validate their processes. It is important to calibrate temperature probes using the sensor and the transmitter using a temperature bath. In some cases, standard voltages are used to calibrate the transmitter and the gauge or readout only and this should be avoided (see chapter 7, "Calibration and Metrology," for additional information).

PROCESS VALIDATION PROTOCOL

Process validation is ordinarily the last step with regard to start-up and scale-up of non-sterile liquids and semisolids. While this is true for all dosage forms, there is a tendency to combine development and validation for OTC products in order to save time and money. It has been demonstrated time and again that this shortcut often costs more in the long run and the proper sequence of qualification and process validation should be followed. Protocols should contain a documented check that specified activities required prior to the current protocol have been completed and are acceptable. The following is a discussion of important elements that are typically contained in a non-sterile liquids or semisolids validation protocol.

Prerequisites

As stated above, it is necessary to list all requirements and to have a positive verification of completion along with report or activity completion date. These items include development, equipment and facilities qualification (IQ, OQ, and PQ), analytical test method validation, documentation (manufacturing instructions, batch records, standard operating procedures), and training. The reader is directed elsewhere within this volume for additional discussion of these requirements.

Bulk raw material qualification is an important prerequisite. Bulk raw material issues can be characterized by one word: variability. This is often the fault of the buyer where cost savings associated with commodity-classified raw materials cause frequent vendor changes. Issues usually fall into the category of microbiological and/or physical. Natural products often have microbial problems when vendor controls are not in place. The authors recall an instance when a lot of guar gum that was purchased at a considerable discount was found to fail because of microbiological contamination. An attempt was made to validate the ethylene oxide sterilization of the material at a bargain price processor. The unsuccessful result costs much more in time delay, wasted processing, and lost validation effort than the cost of purchasing the raw material from a qualified vendor. Mineral oil and petrolatum are major raw materials and do not have microbial issues. They do have purity, physical, and contamination issues. Mineral oil is frequently purchased in large quantities and may involve transport and storage issues. The use of nondedicated rail cars and motor freight tank trucks as well as outside storage tanks is a possible condition that must be considered during raw material and vendor qualification activity. Even the use of 55 gallon drums does not eliminate cross-contamination and storage issues. Vendor qualification includes selecting vendors with Quality Systems in place and who are audited by qualified auditors. Qualification includes sampling, testing, defined primary containers, shipping, and storage conditions.

Cleaning process development and cleaning validation are important prerequisites. Cleaning validation is frequently performed concurrent with other qualifications and/or validation. Cleaning of semisolids and viscous liquids is difficult because of the oil and lipid content of these products. As we have observed, the transport and storage of liquid and semisolid raw materials prior to receipt and processing must be considered. It is important to develop cleaning processes prior to process validation because the delays and interruptions necessary for cleaning validation may alter the timing parameters of the manufacturing process. Globepharma has developed swabbing wands to facilitate micro- and chemical swabbing of process equipment. Cleaning and contamination issues arise with the use of plastic tubing in manufacturing operations. Plastic tubing is widely used in liquid and semisolid manufacturing and filling operations. Plastics may contain nonreacted monomers, plasticizers, UV stabilizers, preservatives, mold release agents, and lubricants. Chemical leaching studies need to be performed in conjunction with vendor qualification. High-quality medical #6 tubing is typically used. Tubing is usually dedicated and discarded after a single use as we saw above with peristaltic pumping. Dedicated use for a campaign of batches between rigorous cleanings must be validated.

Sampling—Statistical Considerations

Validation sampling and testing typically is three to six times the usual QC sampling and testing. That being said, there needs to be a rationale and/or justification for the selected sampling plan. Statistical sampling approaches should ensure that the samples are obtained from the entire container, filling run, etc., with no emphasis on any one area. For example, in-process samples are most easily obtained from the top of vessels and, as such, dominate sampling plans. These samples should be balanced by samples obtained from the bottom and middle areas. The "Square Root of N Plus One" approach has been criticized (18) and it is best to exceed this number of samples, especially when sampling relatively few bulk containers. If standardized sampling plans are used for normal production, validation should specify some integral multiple of the plan to justify routine use. Statistical acceptance criteria are best applied to packaged product where the number of samples will enable meaningful results. Statistics such as Cpk and Cpm should be based upon historical results and manufacturing since overfills are a common practice for multidose OTC liquid and semisolid products. Hofer (19) has provided a statistical means to determine the number of retests in the event of questionable results. Note that results can be disqualified when root causes can be assigned to sampling or testing errors. This can occur with the high-viscosity semisolids and "lotion" products and can be anticipated within the protocol sampling and testing language. Composite samples should be avoided during validation sampling and testing, especially if composites will be used during routine production.

Sampling Plans—Bulk

Sampling from mixers is accomplished from recirculation sampling ports, where available. Location sampling should be based upon known "dead spots" or mixing voids identified for the given mixer geometry. Kukura et al. (20) and Prodal et al. (21) provide the insight to these locations for typical mixer types and geometry. A generally acceptable approach is to sample the top, middle, and bottom of a mixer of cylindrical or semicircular geometry at the central axis, half radius and at the side surface. If there is a central mixing shaft, then the samples are obtained next to the shaft. The side surface samples can be rotated in the geometric axis of rotation by 90° or 120° to the plane defined by the center and mid-radius samples. This plan is best used if there are stratification concerns. The Globepharma liquid and syringe type thieves have been successfully used in these applications.

Drum samples are most conveniently obtained while the drum is being filled from the filling hose. They are usually obtained from the top during routine production when separate sampling personnel are used. In order to justify this approach, validation samples should be obtained from additional levels of the drum. Drums may be skipped in these instances as long as the beginning and end drums are included.

Sampling Plans—Filled Product

Samples of the filled product are the most easily obtained and it is fortuitous that this is what the patient ultimately uses. As such, the preponderance of sampling and testing should occur at this final stage of manufacturing. It is important to include all manufacturing shifts for long filling and packaging runs in addition to all filling nozzles, as previously mentioned. Additionally, validation samples should be obtained to bracket filling line stoppages to provide worst-case samples and justify less rigorous sampling during routine production filling and packaging runs. Validation batches and/or filling trials are usually placed in stability programs. It is important to include all primary containers within stability programs even though a bracketing approach may have been used in validation. For this reason, it is best to have a separate stability protocol and/or procedure that the validation protocol can reference.

Unit-Dose and Metered-Dose Containers

Large numbers of this type of primary package are typically sampled and tested. For critical life-saving products, it is difficult to justify any sampling plan unless some positive in-line check of each container is part of routine production. Validation protocols need to challenge fail-safe controls and testing with well-marked and monitored failing packages. These challenges are best conducted during qualification, but must be fills of actual product at full production rates.

Multidose Containers

Testing should be conducted on all the doses delivered from a container and should exceed routine testing. The size of multidose containers may limit this practice because of limited laboratory resources. In this event, random samples from multidose containers throughout the filling run should be tested. Tubes are filled from the bottom and crimped with closures already in place. Cap torque testing needs to be obtained following filling to ensure that the filling and packaging operation have not detrimentally affected the closures.

PROCESS VALIDATION REPORT

Documentation for non-sterile liquids and semisolids follows the format of other dosage forms. Process validation reports for non-sterile liquids and semisolids typically are approved by the same function (people if available) as approved for the protocol. In many instances, the executed protocol containing annotated raw data and verifications are circulated for approval. Many firms circulate both in a combined document.

VALIDATION LIFE CYCLE

The product should be monitored from inception, through validation and during routine production in an approach called the validation life cycle (22). Change-driven and time-driven revalidation should be specified in validation master plans and/or in approved procedures. A Change Control program needs to be in place to trigger additional validation upon the implementation of significant changes. This must include raw materials supplier changes since many of the excipients for non-sterile liquids and semisolids are commodities and purchased from the lowest cost vendor.

SUMMARY AND CONCLUSION

The authors have attempted to provide an overview of non-sterile liquids and semisolids validation. A survey of the types of formulations and associated equipment was provided along with some anecdotal experience to give validation personnel an insight into this area of pharmaceutical manufacturing. Sampling and testing plans widely used in protocols were provided. Documentation parallels other areas of validation; however, the diversity of the materials and equipment within the scope of this chapter make each study unique.

REFERENCES

1. Immel BK. A brief history of the GMPs for pharmaceuticals. Pharm Technol 2001; 24:44.
2. Martinez M. Schering-Plough criminal investigation may center on Puerto Rico. Puerto Rico Herald, June 27,2002.
3. Guideline on General Principles of Process Validation. Rockville, MD: Center for Drug Evaluation and Research, U.S. Food and Drug Administration, 1987, www.fda.gov/cber/guidelines.htm 95 (last accessed July 2007).
4. Labuza TP. Fundamentals of Water Activity Short Course. Anaheim, CA: Institute of Food Technologists, 2002.
5. The United States Pharmacopeia 29th Revision. <1231> Water for Pharmaceutical Purposes. Rockville, MD: The United States Pharmacopial Convention, Inc., 2006:3071.
6. The United States Pharmacopeia 29th Revision. <1151> Dosage Forms. Rockville, MD: The United States Pharmacopial Convention, Inc., 2006:2996.
7. Crowder TM, Louey MD, Sethuraman VV, Smyth HDC, Hickey AJ. An odyssey in inhaler formulation and design. Pharm Technol 2001; 21:99.
8. Garg S, Tambwekar KR, Vermani K, Garg A, Kaul CL, Zaneveld JD. Compendium of pharmaceutical excipients for vaginal formulations. Pharm Technol—Drug Delivery 2001; 25(9) (Suppl. 1):26.
9. Block LH. Scale up of liquid and semisolid manufacturing processes. Pharmaceutical Technology Scaling Up Manufacturing—March 2005; 29(3) (Suppl.):26.
10. Kupp GD. Challenges, considerations, and benefits of raw material testing. Pharmaceutical Technology—Analytical Chemistry and Testing 2003; 27(3) (Suppl.):22.
11. Gupta P, Garg S. Semisolid dosage forms for dermatological application. Pharm Technol 2002; 26:144.
12. Hopp M. Developing custom adhesive systems for transdermal drug delivery products. Pharm Technol 2002; 26(3):30.
13. Witt K, Bucks D. Studying in vitro skin preparation and drug release to optimize dermatological formulations. Pharmaceutical Technology—Formulation, Fill and Finish 2003; 27(6) (Suppl.):22.
14. www.leeind.com (last accessed July 2007).
15. SPX Corporation, www.gowch.com (last accessed July 2007).
16. Pailhes M, Lambalot C, Barloga R. Integration of centrifuges with depth filtration for optimized cell culture fluid clarification processes. Bioprocess J 2004; 3(1):56.
17. www.penwalt.com (last accessed July 2007).
18. Saranadasa H. The square root of N plus one sample rule—how much confidence do we have? Pharm Technol 2003; 27(5):50.
19. Hofer JD. Routine sample size for a retest procedure. Pharm Technol 2003; 27:60.
20. Kukura J, Arratia PC, Szalai ES, Bittorf KJ, Nzzio FJ. Understanding pharmaceutical flows. Pharm Technol 2002; 26(10):48.
21. Prodal HS, Matice CJ, Fry TJ. Computational fluid dynamics in the pharmaceutical industry. Pharm Technol 2002; 26(2):72.
22. Agalloco JP. The validation life cycle. J Parenter Sci Technol 1993; 47:142.

32

Validation of Packaging Operations

Charles S. Levine
Levine Pharmaceutical Consulting Inc., Linwood, New Jersey, U.S.A.

INTRODUCTION

In the pharmaceutical industry, the term "packaging" has many different definitions. Packaging can bring to mind visions of employees gathered around a table labeling clinical trial materials and placing them into corrugated cases and sealing the cases with tape, or highly automated equipment filling parenteral solutions into sterile containers in an aseptic environment. To simplify this discussion the author has subdivided "packaging" into five processes:

1. Filling
2. Sealing
3. Oxygen removal
4. Inspection
5. Secondary packaging

The most critical quality characteristic controlled by the filling process is to manufacture well-controlled doses of the product, which can be characterized by fill volume/fill weight control and dose uniformity. The sealing process will examine two of the processes that are most commonly used in parenteral dosage forms: flame sealing (ampules) and glass/elastomeric container closure systems (vials and syringes).

An ever-increasing number of drug products require protection from oxygen to ensure product stability throughout the product's shelf life. The processes for minimizing the presence of oxygen in the headspace of sealed ampules or vials, and requisite methods for process validation and in-process monitoring will be presented herein.

Abbreviations used in this chapter: AQL, acceptable quality level; BFS, blow–fill–seal; C_1, volume of static electricity between the inspection electrode and the product; C_2, volume of static electricity between the ground electrode and the product; CIP, clean in place; FAT, factory acceptance test; HACCP, hazard analysis and critical control plan; HDPE, high-density polyethylene; I_1, electric current which is produced when the container is sealed; I_2, electric current which is produced when the container seal is defective; IQ, installation qualification; LCL, lower control limit; LSL, lower specification limit; NIST, National Institute of Standards and Technology; OQ, operational qualification; PAT, process analytical technology; PD, particulate detection; PDV, positive displacement volumetric; PLC, programmable logic controller; PQ, performance qualification; R, electrical resistance of product; RAG, reject rate; RFID, radio frequency identification; RZE, reject zone efficiency; SAT, site acceptance test; SIP, sterilization in place; SOP, standard operating procedure; T–P, time–pressure; UCL, upper control limit; UHF, ultra high frequency; UID, unique identifier; USL, upper specification limit; USP, *United States Pharmacopeia*; V, high-voltage source; WD, weight dosing.

The USP requires all small volume parenteral products to undergo 100% inspection for particulate contamination (1). The inspection for particulate contamination will be the focal point for our discussion of inspection processes. The review will include both visual and automated inspection methods.

The secondary packaging process is considered to be that which begins after particulate inspection and continues until the product is packed into cases. The packaging process could include things such as labeling, cartoning and secondary container sealing. A critical quality characteristic controlled by the packaging process for pharmaceutical products has always been to identify each container with the specific batch number and expiration date. As the world changes and demands on the manufacturer have increased, anticounterfeiting measures are becoming a new function of the pharmaceutical packaging process. RFID is being investigated by a number of pharmaceutical manufacturers to enhance their anticounterfeiting capabilities. A brief discussion of the validation issues under consideration in this rapidly evolving area will be presented.

Validation should address only those packaging processes that are automated. Any process undergoing the scrutiny of validation should depend heavily on a piece of equipment operating within its designed parameters. However, the performance of operating and maintenance personnel should not be overlooked. Generally speaking, automated packaging equipment has a great number of mechanical and electrical components requiring trained and skilled mechanics or technicians to insure that equipment continues to operate as it was originally validated. Training of these personnel should not be overlooked.

This chapter will only address prospective validation of packaging processes and not retrospective validation. Retrospective validation incorporates numerous complicating factors such as change control, proper execution of test methods and procedures, equipment maintenance history and accurate instrument calibration history, all of which should be discussed in a separate chapter.

As with any major equipment purchase the acquisition and validation of the equipment should generally follow the lifecycle approach, which includes the following milestones:

1. User requirements specification
2. Detailed functional design
3. FAT
4. SAT

5. IQ/OQ
6. PQ

Regardless if equipment is a prototype design or a standard commercially available model, the duration of the performance tests executed during the aforementioned phases must be long enough to properly evaluate the execution of the equipment design. Because of the significant number of moving components testing of packaging equipment requires a multifaceted approach to the design of the qualification tests. Not only must each function be verified, but also the capability of each mechanical function demonstrated over time. Where equipment is being integrated into a packaging line interfaces between equipment should be evaluated as part of the FAT whenever possible.

The integration of new equipment into a pharmaceutical operation is always a complex endeavor. Vendor participation in the initial equipment start-up is vital. When purchasing such equipment specify the expected performance and the duration of each acceptance test. Concomitant with these tests training of an appropriate number of line mechanics and operators should also be specified.

During the FAT equipment manuals should be available, which will enable personnel to begin development of SOPs. The development of SOPs will continue through the OQ phase, at which time approved SOPs should be issued and effective.

FILLING

There is a wide variety of technologies employed for the filling of liquids as well as powders. Proper understanding of each technology is necessary to adequately analyze risk and design the validation/process control program to control risk during the operating life of the equipment. Compatibility with the materials of construction in contact with the product should be established during the process development phase through laboratory tests. The equipment fabrication documents provided by the equipment manufacturer will provide evidence that the proper materials were used, and product stability studies will confirm that during equipment fabrication the product contact surfaces were not compromised. Presented in this chapter are several of the dispensing technologies currently utilized: PDV system, T–P filling and WD.

Positive Displacement Volumetric System

A typical PDV system can be compared to a one-cylinder engine in that the stroke and bore of the piston determine the volume of liquid dispensed by each cycle of the piston. During the upward stroke of the piston, the valve assembly allows the product to enter the cylinder. The valve assembly then rotates to allow the product to be expelled into the container during the downward stroke of the piston. A well-designed PDV system will deliver product of various viscosities with a high degree of accuracy. As with all filling systems the accuracy is greatly dependent on the liquid delivery system.

One of the major risks to be controlled is presence of air or gas in the delivery system. This is normally encountered at the start and end of the filling operation, but may also occur during the filling operation when filling products whose dissolved oxygen has been purged with an inert gas such as nitrogen. Controlling the temperature of the liquid during the delivery and filling processes is necessary to control the risk of bubble formation. According to Bernoulli's Law the solubility of gases is inversely proportional to the temperature of the solution. Ideally the temperature of the bulk drug product should be equivalent to the temperature in the aseptic filling area.

Many filling systems are equipped with a surge vessel immediately upstream of the dispensing portion of the system to minimize supply pressure variability. The heat transfer coefficient of the material of construction for the surge vessel and the residence time of the drug product will help to determine the temperature variation that will not cause the gas to come out of solution during the filling process. It may be necessary to include the maximum temperature differential between the solution and the filling environment as a critical parameter in the design of the validation plan.

Typically, PDV pumps are constructed of stainless steel, glass or ceramic materials. Finish tolerances are extremely important in providing a smoothly operating and accurate system. One of the major disadvantages of this system is that the product is in contact with moving parts. This is especially an issue when filling proteins and other large molecule products which may be susceptible to shear.

One variation of this concept is the "Rolling Diaphragm Liquid Metering Pump" (U.S. Patent No. 3880053), originally developed by TL Systems, Minneapolis, Minnesota, now part of Bosch Systems (2). This design incorporates two concepts.

1. A flexible diaphragm covers the piston, which provides a system without moving parts in product contact.
2. The intake and discharge valves compress flexible tubes, again eliminating moving parts in product contact.

This system is capable of achieving an accuracy of ±0.5% for a wide range of products and fill volumes.

Many PDV filling systems are capable of filling a wide range of fill volumes with component handling change parts and different size pumps. Validation professionals are always faced with the challenge of designing a cost-effective validation plan that fully validates the system under evaluation. Achieving cost efficiency through product matrixes should be approached with caution. For example, evaluating the maximum and minimum fill volumes of the same drug product for one size pump may be appropriate, but not for different drug products.

Time–Pressure Systems

An entirely different principle is Time–Pressure (T–P) filling. The brief synopsis of the principle is stated as follows: "Liquid of a given temperature and viscosity will flow at a constant rate through a fixed opening, providing the pressure is constant." Having established a constant flow rate, time is the necessary variable to produce a specified fill volume (3). In a T–P filling system, no moving parts are in contact with the product. The system pressure should be controlled to ±0.01 psi by an in-line pressure transducer located between the pressurized storage vessel and the filler. The pressure transducer must be located downstream of the sterilizing filters and as close as possible to the filling needles. Typically the pressure transducer is integrated into a surge vessel, which is part of the filling equipment. Depending on the size of the containers being filled and the speed of the filling equipment the size of the surge vessel may vary from several liters up to 10 or 20 L.

To minimize the effect of head pressure in a larger surge vessel an automatic level control system may also be integrated into the control system. The pressure transducer provides a signal to a pressure control valve, which regulates the pressurizing gas. The pressurized product flows through flexible tubing that is routed through a flow control pinch valve. Controlled by a microprocessor, the pinch valve will remain open for a specific time based on configurable parameters that may include product viscosity, diameter of the flexible tubing and the target pressure.

For a T–P filling system critical process parameters to be considered in the process validation may include:

1. System pressure
2. Diameter of flexible tubing
3. Durometer of flexible tubing
4. Viscosity of liquid
5. Equipment operating speed
6. Product flow rate [target fill volume × equipment speed (units per minute)]

Where possible, worst-case conditions should be utilized during the PQ studies. As vendors are responsible for controlling the physical characteristics of the flexible tubing the PQ studies should focus on system pressure, viscosity and equipment operating speed.

The T–P systems are widely used for products containing proteins, other large molecules and vaccines.

Weight Dosing Systems

These systems are similar to the T–P systems in that they do not utilize pumps, but a flow control system that utilizes data from load cells upon which each container is tared and weighed during the filling process until the target weight is achieved. These systems utilize a PLC and servomotors to control the flow control valve as the fill weight approaches the target weight.

WD systems typically fill more slowly than the other systems described but can be more accurate over a wider range of sizes without any change parts. These systems are programmed to learn from each filling cycle, which enables the filler to gradually increase its operating speed for the product being filled. One advantage of these systems is sampling for fill weight is not required as the filling systems are capable of generating a report showing all of the fill weights. For a WD system line speed is no longer a critical process parameter.

The validation of these systems should take into account vibration and unidirectional air velocity, as these parameters may affect the accuracy of the load cells. Proper design and control of the Grade A filling area are prerequisites for validating this type of filling system. Turbulent airflow patterns at the load cells may result in unacceptable weight variation. The asepsis of these systems is also greatly improved as they are designed for CIP/SIP, and personnel interventions are reduced as the need to remove fill weight check samples could be eliminated.

A variation of the WD filling concept is also utilized for aseptic powder filling. The mechanisms to control fill weight are naturally different and the product related critical process parameters may include particle size distribution, tap density and flow index.

The powder filling system cannot adjust the weight of each unit as it is dosed, but with an intelligent control system the equipment is capable of adjusting the fill weight of subsequent units by modifying the depth of the cavities in the dosing disk.

Whether filling liquids or powders the validation should compare the fill weights obtained from the filling system to those obtained in a similar manner from an independent weighing system of similar accuracy. Slight differences may be observed. Provided these differences are minimal the fill weights generated by the system should satisfy release requirements for Dose Weight Uniformity as described in the USP (4).

The critical process parameters for a WD filling system should include the following:

1. Equipment operating speed
2. Delivery system pressure (liquids)
3. Unidirectional air velocities
4. Physical specifications (drug product)

A WD system could be considered one of the first forms of PAT applied to process lines, where the product quality attribute is being measured in a real-time manner, eliminating the need for any type of sampling and analysis.

Validation Conditions

What does validation mean for a filling system? In the example of a steam autoclave, for example, it normally means demonstrating a minimum F_0 and destroying a specified population of microorganisms with a known D value at a worst-case location within the load configuration. A validated filling system cannot be defined as precisely.

1. A filling system must be able to accurately fill a specified amount of product repeatedly.
2. The filling system must be able to fill the product without splashing, foaming or damaging the containers.
3. The filling system must be able to deoxygenate the product containers if required.

Regardless of the type of filling system used, one concept should be strictly adhered to during its validation. The worst-case conditions should be simulated as part of the PQ. When validating a load configuration and autoclave, the worst case is normally defined by the load configuration that accumulates the least amount of F_0.

The worst-case conditions for a filling system are the extremes of several variables. Generally, each type of filling system requires the same variables to be evaluated.

One of the first steps of any validation is to define the utilization list (drug products, container closure systems and fill volumes) for the equipment or systems under evaluation. From this list, identify those products whose physical characteristics represent extreme conditions (e.g., maximum and minimum velocities). Also, a product from each product type should be selected (e.g., nonaqueous products and suspensions). When designing a comprehensive validation program where many variables must be controlled, design of experiments is a useful technique to statistically identify the critical process parameters.

The design of filling needles for high-speed filling equipment could be a critical variable to be considered in the validation plan. The objective of the needle design is to minimize splashing and foaming, which is necessary for products with low surface tension or products that require protection from oxygen. This is also critical to ensure a clean exterior of the container, which has heightened importance for products that represent safety issues, such as cytotoxic products, penicillin, etc.

When evaluating a pump system's ability to meet the first criterion of a validated filling system (accurately fill a specified volume of product), the variables that present the greatest challenge are viscosity and the product type (suspensions or nonaqueous solution). The more viscous products represent the greater challenge. If a pump system can accurately fill 50 mL of a viscous product, it should fill 50 mL of a less viscous product with equivalent or improved accuracy at the same operating speed.

Similarly, suspensions or nonaqueous solutions represent different challenges to a filling system and should be evaluated separately. When considering suspensions, mixing systems play an extremely important role in fill volume accuracy and dose uniformity. The design and operating speed of a mixing system should be selected to minimize the incorporation of air into the suspensions, which will result in low fill weight containers. The content uniformity aspects must be a component of the filling process validation for suspensions.

As sterile suspensions are normally formulated aseptically, sterilizing filters do not complicate the validation of the filling process. It is important to identify those points in the process which represent the greatest challenge to content uniformity. Generally speaking this occurs at the start and end of each filling operation and during the maximum allowed time period for filling equipment stoppage. To minimize the effect of equipment stoppages suspension mixing capabilities should be designed into the product delivery system as close to the product dosing as possible. It may be necessary to discard some units filled immediately after a stoppage.

The design of the sampling plan to confirm content uniformity during the worst-case process conditions should be based on the number of units filled through one cycle of the product dosing system and not necessarily on the USP requirements for Uniformity of Dosage Units <905> (4). The two-stage test specified in the USP is designed to statistically evaluate an entire batch. During validation at least one container from each filling needle should be analyzed and it may be appropriate to perform this challenge multiple times.

Other variables that can affect fill volume accuracy are machine speed and delivery system pressure. Normally, the machine speed at which a filling system can accurately fill is limited by its mechanical design. Once the upper or lower limit of the mechanical design is exceeded, the fill line accuracies will deteriorate. This is a common occurrence with equipment that has been designed for high-speed operation.

Maintaining a threshold pressure is all that is required to ensure accurate fill volume, except when using a TP filling system. The pressure control system of the TP filling system should be monitored during routine operation so that pressure control data can be correlated with the fill volume data.

The duration of the filling process during validation is one aspect that is costly and time consuming to evaluate. As our focus is primarily on new filling equipment, operating the equipment for an extended period of time is the only way to verify that moving parts are properly machined. Where metal-to-metal contact occurs there is a potential for binding, which may restrict shaft movement and result in fill volumes below the target levels. It is impossible to predict the operating interval required to guarantee nonbinding. Situations have been documented during which binding first appeared after six hours of consecutive operation.

Following a successful validation, a comprehensive cleaning and preventive maintenance program are essential to maintain the equipment performance. Cylinders and pistons should be inspected for damage at the conclusion of each cleaning. Custom holders should be designed to protect the polished surfaces during handling and sterilization.

As previously described in this chapter performance of the filling system should be evaluated during extended operating periods as part of the FAT and SAT. If the overall qualification program is designed in this manner the duration of the filling operation evaluated during the execution of the PQ could be reduced.

Statistical Evaluation

Normally, Quality Assurance will establish fill volume tolerances for each product regardless of the filling equipment. Many fill volume tolerances are established by the USP. Table 1 summarizes the requirements of the USP for "volume of injection in containers" (5). The USP does not require a statistically sound sampling program. Today's filling equipment is sufficiently accurate to fill products within the tolerances established by the USP. Generally, Quality Assurance fill volume tolerances evolve based on historical data. They are established so that no units are filled with less than the labeled amount. To accomplish this, the target fill T is determined by adding three standard deviations σ to the labeled amount L, as shown in equation (1).

$$T = 3\sigma + L \tag{1}$$

When a product is transferred from an existing filler to a new filling system, there is an abundance of data from which the standard deviation could be established. Care should be taken when selecting historical data, because all of the conditions of the filling operation may not have been properly documented; that is, it is important to select data from the time periods during which no filler adjustments were made. It is also desirable to group

Table 1 USP 29, Pharmaceutical Dosage Forms <1151>, Injections

Labeled size (mL)	Excess volume (mL) For mobile liquids	Excess volume (mL) For viscous liquids
Required minimum of volume (label claim):		
0.5	0.10	0.12
1.0	0.10	0.15
2.0	0.15	0.25
5.0	0.30	0.50
10.0	0.50	0.70
20.0	0.60	0.90
30.0	0.80	1.20
50.0 or more	2%	3%

Volume in container (mL)	Number of samples	Transfer technique	Method of determination
Less than 3	5 or more	Dry syringe	Graduated cylinder or weight
3–10	3 or more	Dry syringe	Graduated cylinder or weight
More than 10	1 or more	Dry syringe	Graduated cylinder or weight

the data by filling head to avoid incorporating adjustment variations between filling heads.

The excess volumes, recommended by the USP, rely on one of several methods for determination of the fill weight. The ultimate confirmation that fill tolerances are set properly is to mimic the medical practitioners' technique for delivering the product to the patient. Utilizing this approach, studies can be performed to establish the expected hold-up volume (the amount of product remaining in the container after withdrawing the product using the practitioner's technique). The target fill volume should also take into consideration the hold-up volume.

Fill weight monitoring is normally accomplished through the use of the statistical process control method, $\overline{X}$–R charts. The *Quality Control Handbook*, by J. M. Juran, provides an excellent discussion of $\overline{X}$–R charts (6).

The basic purpose of a control chart is to detect "assignable" sources of variation in the process. There are two types of sources of variation in a process: assignable and random. Random causes are due to slight variations in numerous variables, the overall effect of which is minimal and economically impractical to eliminate. Assignable causes are due to large variations in a few variables, the overall effect of which is significant and economically vital to eliminate (6).

A control chart can be used in two different ways:

1. To determine if an "unknown" process is in a state of control (control with no standard given).
2. To determine if the "known" process remains in a state of control (control with standard given).

The limits of a control chart are normally set at ± 3 standard deviations. If only random causes are present, 99.7% of all the individual values will fall within the control limits, which are normally referred to as USL_x and LSL_x, respectively.

The steps to be followed when determining the state of control of a process are as follows:

1. Periodically take a series of samples and subgroups to establish a database. The number of samples and frequency will depend on the speed of the filling equipment. It is normally desirable, at each sampling interval to sample at least three units from each filling head so that each filling head may be evaluated separately if necessary.
2. During the filling process record any process changes that affects the data collected (i.e., volume adjustment on one of the filling heads).
3. Compute trial control limits from the database collected in step 1. The average $\overline{X}$ and range R of each subgroup is calculated. The grand average $\overline{\overline{X}}$ and the average range ($\overline{R}$) are then calculated.

The trial control limits in the table are calculated by the following formulas (6):

	Upper limit	Lower limit
Subgroup average	$\overline{\overline{X}} + A_2\overline{R}$	$\overline{\overline{X}} - A_2\overline{R}$
Subgroup range	$D_4\overline{R}$	$D_3\overline{R}$

The values for the multipliers A_2, D_3 and D_4 are listed in Table 2. The value n is the number of samples in each subgroup.

Compare the data points to the control limits for both X and R. If both statistics are within the limits, consider the process under control. If either statistic is outside of the control limits, it indicates that the process is not in control. If an average ($\overline{X}$) is outside the control limit, this is an indication that a general change has occurred that affects all the samples. If a range R is outside of the control limits, it is indicative of increased variability owing to a change in material, personnel or the process.

The use of $\overline{X}$–R charts during the validation of new filling equipment not only provides specific acceptance criteria, but also serves as a tool for troubleshooting problems. Much of the statistical analysis is now performed by sophisticated statistical process monitoring systems that can automatically capture the data as it is generated and present the data in whatever form the user requires. These are used to provide real-time graphical representation of the data that will alert the user to trends as they are developing. Any software used to perform calculations upon which validation acceptance criteria will be judged should be validated and Part 11 compliant.

Table 2 Constants for Determining Control Limits for $\overline{X}$–R Charts

N	A_2	D_3	D_4
2	1.880	0	3.268
3	1.023	0	2.574
4	0.729	0	2.282
5	0.577	0	2.114
6	0.483	0	2.004
7	0.419	0.076	1.924
8	0.373	0.136	1.864
9	0.337	0.184	1.816
10	0.308	0.223	1.777

Source: From Ref. 6.

The importance of documenting all changes during the process cannot be overemphasized. This information will be vital for troubleshooting subsequent problems.

As a filling system moves from the validation phase into the routine operating phase adjustment limits should be established. The purpose of these limits (UCL_x and LCL_x) is to alert the operator of conditions that may warrant adjustment, thus minimizing the probability of manufacturing units outside of the control limits ($USL_x - LSL_x$). Selection of these limits depends on the process capability index, C_P. The capability index, C_P, is the ratio of the specification range divided by the process capability, 6σ, as shown in equation (2):

$$C_P = \frac{(USL_x - LSL_x)}{6\sigma} \quad (2)$$

where USL_x is the upper specification limit, LSL_x the lower specification limit, and σ is the standard deviation.

When values of C_P exceed 1.33 the process is capable of routinely conforming to the specification limit. When C_P exceeds 1.33 it may be appropriate to select $\pm 3\sigma$ as the UCL_x and LCL_x.

CONTAINER SEALING

The process of verifying the integrity of a parenteral container/closure system continues to evolve from the vacuum dye-leak test originally developed more than 40 years ago. The vacuum dye-leak procedure is rarely used in commercial production today. Physically some of the methods have not significantly changed, but the technology available today has allowed equipment manufacturers to automate what was previously very time consuming and not very precise.

For many years ampules were the only container closure system evaluated for seal integrity as part of the manufacturing process. Now with the advances in electronics, glass vials with elastomeric closers and containers manufactured by BFS technology are routinely evaluated for seal integrity utilizing online automated systems. Also, batch processes similar to the dye-immersion test method are still performed for ampule products. Two of the most widely used methods for performing leak tests as part of the manufacturing process are high voltage and pressure differential.

High-Voltage Leak Detection

High-voltage leak detection was first used commercially about 25 years ago to detect leaks in ampules. The ampule-sealing process had always been considered to be a difficult process to control (high risk) and thus requiring the lowest risk control method, in this case 100% leak detection. High-voltage leak detection systems, such as those manufactured by Nikka Densok (Kawagoe-shi, Saitama-ken, Japan), are automated in-line systems capable of operating at line speeds equivalent to the fastest filling systems with the ability to automatically reject defective containers as they are inspected.

High-voltage leak detection is now being applied to glass vials with aluminum seals and plastic containers, e.g., BFS products. Regardless of the container closure system the concept remains the same. The container closure system acts as an insulator (capacitance) surrounding a conductive liquid between two high-voltage electrodes. If there is a leak the capacitance is dramatically reduced and the current flow between the two electrodes increases significantly. Through adjustment of the sensitivity of a potentiometer in the detection circuit the signal amplitude for a defective container can be established. Figures 1–3 show the changes in the electrical circuit when a leak is present.

The initial range of acceptable values is established by inspecting the same good container and the same faulty container five times at increasing sensitivity values.

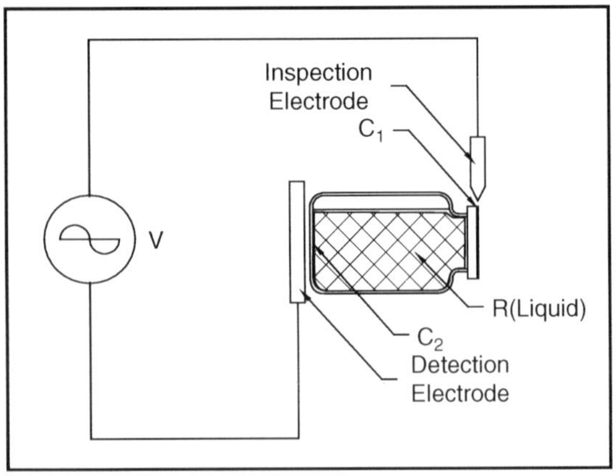

Figure 1 Vial in the leak detector.

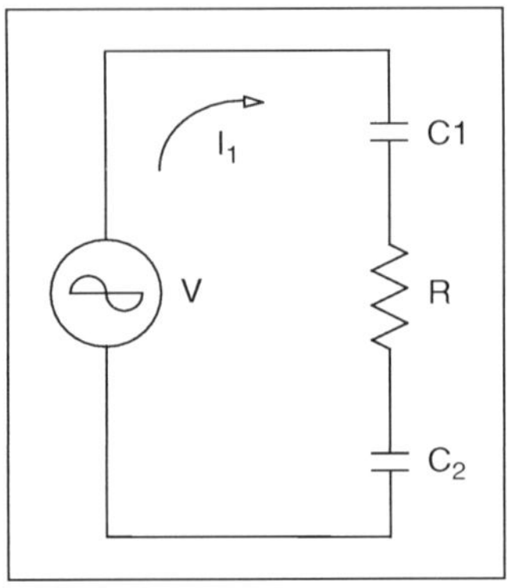

Figure 2 Properly sealed vial.

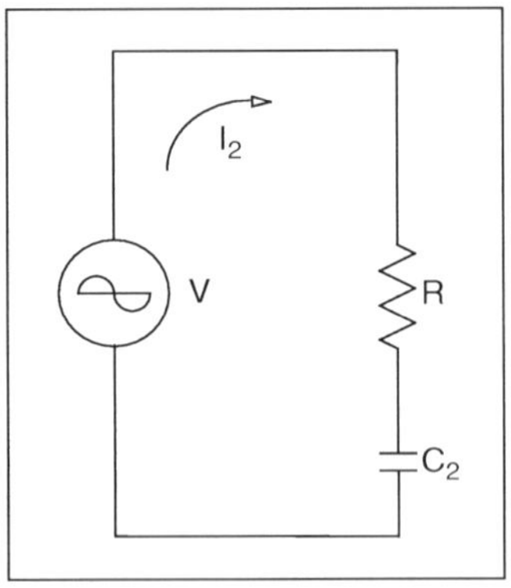

Figure 3 Vial with a leak.

Through this exercise a range of acceptable values for the inspection signal is established. Empty containers will generate a minimal signal, below the acceptable range, resulting in rejection of the container. Defective containers will generate a signal above the acceptable range, again resulting in the rejection of the container. Containers with thin walls may also generate a signal above the acceptable range, resulting in rejection of the container.

An acceptable range and voltage setting must be established for each product and/or container closure system because the drug product solution may have a different conductivity and each container closure system may have a different capacitance value. Also, different electrodes may be used for different size container closure systems. The acceptable range should be evaluated further by inspecting a statistically significant number of containers considered to be acceptable. The containers in this test set could be selected through a visual inspection process or dye-leak process to confirm their acceptability. Ideally the physical characteristics, weight and critical dimensions, of the containers should be representative of the historical distribution. It may be appropriate to evaluate separately containers supplied by different vendors.

This test set should also include defective containers, representative of those routinely observed in the manufacturing process plus some containers with defined defects. Typically these defined defects are created using lasers which can generate holes as small as 5 μm. Extra defective containers should be maintained as these are easily damaged during handling or by the high-voltage applied to the container.

Through the evaluation of this test set the acceptable range may be adjusted to minimize the false reject rate levels while ensuring that all defective containers are rejected. Multiple test sets should be prepared so the development of the acceptable range can be performed as part of the FAT and again during the PQ.

Some factors that may affect the accept/reject decision of the high-voltage leak detection system include:

1. Product on the exterior of the container
2. Product temperature, primarily for BFS products
3. The absence of liquid at the site of the high-voltage electrode

The latter may be the result of constriction in the neck of the ampule that prevents liquid from reaching the tip of the ampule, which is normally the site evaluated by one of the high-voltage electrodes. This condition would result in a false reject.

When performing leak detection on vials it is important to recognize that high-voltage leak detection would only detect missing stoppers or defects in the glass container. High-voltage leak detection will not detect poorly crimped vials. Process control of the capping process is vital regardless of the presence of an in-line high-voltage leak detection system.

The validation runs executed during the PQ should be designed to validate the critical process parameters, including

1. Operating speed
2. Drug product solution
3. Container dimensions
4. Product temperature (BFS only)

The validation process should demonstrate that the high-voltage leak detection is equivalent or better than the currently employed leak detection process. From a quality perspective the acceptance criteria should require rejection of all known defective units, as container closure systems that are not integral are considered to be a critical defect. From an economic perspective acceptance criteria should define a maximum level of false rejects.

Once in operation the performance should be verified daily through the use of a machine challenge set. In the past, firms have used test sets comprised of acceptable and defective containers, but the sets are difficult to maintain as the containers become damaged during usage. Equipment manufacturers now provide high-tension resistors in the shape and size of the containers being evaluated. Resistors, representative of both acceptable and defective containers, are included in the set. The use of the daily machine challenge set should be incorporated into the PQ trials, thus providing continuous correlation between validation and routine production.

Pressure Differential

The effects of pressure differential without immersion in a dye solution are being used by some manufacturers when terminally sterilizing ampules. The sealed ampules, packed into covered stainless steel trays are loaded into an autoclave in the inverted orientation. After the sterilization exposure phase of the cycle, the pressure in the chamber is rapidly reduced from approximately 2 to 0.2 bar and held at that vacuum level for 140 minutes. The terminally sterilized ampules are then inspected using an automated inspection system, which inspects for both particulate matter and fill volume.

To validate this leak detection process one must first establish the sensitivity of the automated inspection system to remove ampules with low fill volume. Automated inspection systems have demonstrated the ability to detect 100% of the containers with a loss of 0.05 mL in a 2 mL ampule, filled with 1.0 mL of product, with a viscosity similar to water (7).

Following this, product-containing ampules with laser generated leaks of varying diameters should be placed in the autoclave load and processed using worst-case parameters. Worst-case parameters for this process could be a lower vacuum level, reduced ramp of pressure reduction or reduced vacuum hold time. This process is capable of detecting leaks smaller than 10 μm as studies have demonstrated that 100% of the ampules with a 10 μm hole are completely empty at the end of this process (7). At the very least the process should be capable of detecting [(1.0 mL/0.05 mL)/10 μm] or 0.5 μm leak.

Container Closure Integrity and Product Shelf Life

The stability of the container closure integrity for vials and syringes may not as reliable as the seal of any ampule and must be periodically demonstrated during the product's shelf life. Correlation between the container closure integrity methods used during product development, routine manufacturing (in-process control methods) and product stability methods should be demonstrated. Generally the methods will also demonstrate equivalence to microbial ingress methods. Where necessary, container closure integrity tests may need to be more sensitive than microbial ingress methods, especially for lyophilized products sealed under vacuum.

Some vial sealing operations are monitored through the use of a Seal Force Monitor™ (West Pharmaceutical Services, Lioville, Pennsylvania, U.S.A), which actively monitors the forces applied to the capping rail through the vial and the stopper. These forces are compared with "learned" forces for the vial–stopper–aluminum seal. The acceptable deviation from those learned forces can be selected by the operator to establish the rejection threshold. Typically, the deviation will increase proportionally to the vial size. These forces can be correlated to compression of the flange of the elastomeric closure.

Morton determined the percentage of compression necessary to overcome a defect in the finish of a glass container (8). By pressurizing a container closure system containing a solution of copper ions, a leak of 0.1 μL could be detected. Through these experiments Morton found that defective vials (medium-sized defect: 330 μm wide × 290 μm deep) could be sealed by a compression of the elastomeric closure between 9.8% and 14.3%. This was confirmed microbiologically by substituting a suspension of *Pseudomonas aeruginosa* (ATCC 9027) at a minimum concentration of 3×10^8 cfu/mL for the copper solution. This method indicated that the defective vial could be sealed by a compression of the elastomeric closure between 13.1% and 16.2%. The studies were performed using uncoated halo butyl disks and vials machined from aluminum with a mirror polished finish.

Studies to establish the lower limit for the compression force can be performed for each container closure system by determining the minimum force necessary to adequately compress the flange of the elastomeric closure. Containers sealed under these conditions can then be evaluated using a microbial immersion test.

Vials from an aseptic process simulation, filled with a microbiologic growth medium, such as Soybean Casein Digest, are inverted and submerged in a suspension of *Escherichia coli* that contains 10^8 cfu/mL. The units are submerged for 10 minutes at 23 ± 2°C, removed without rinsing or drying, and placed in a biological hazard bag and incubated for seven days at 30°C to 35°C in the inverted position.

In conjunction, a growth promotion test should be performed on the negative units with fewer than 100 cfu of *E. coli*. This biologic test, though severe, does provide evidence that the container does maintain sterility under normal and abnormal conditions.

There have been a number of variations of this test reported in the literature using organisms such as *P. aeruginosa* or *Serratia marcescens*. In each trial, attempts are made to control the population of the organism to ensure a worst-case challenge.

For drug products that are terminally sterilized, process validation testing must be performed to verify that the container closure system remains integral during the terminal sterilization process. Generally this is accomplished using a dye-leak procedure during which the vials are submerged in dye bath and subjected to the autoclave cycle used to terminally sterilize the product. The autoclave must be exhausted slowly to avoid boiling the dye solution. The vials are then visually inspected for the presence of dye in the product. Studies should be performed to demonstrate the stability of the chromophore in the drug product and the sensitivity of the inspection process. This test can be incorporated into the product stability protocol to demonstrate maintenance of container closure integrity during the life of the product.

For products that are not terminally sterilized, the sensitivity of the dye-leak test can be enhanced by repeatedly releasing and drawing a vacuum. The use of a fast release is more effective than a slow release.

As with any in-line process, statistical (time-based) sampling plans should be utilized when performing the initial validation; these plans should take into consideration the number of crimping heads, the line speed and the type of in-process monitoring utilized during routine production.

The PDA Technical Report Number 27, Pharmaceutical Package Integrity, provides a decision tree that identifies the test methods and performance criteria, which can be applied to a broad range of pharmaceutical packages (9). For each product type, the report identifies the test that could be utilized during product development, process control, and product stability.

OXYGEN REMOVAL

Developing a drug product to deliver its therapeutic effects to the patient is as important as the discovery of the molecule itself. In many circumstances, the drug products are not very stable and require protection from atmospheric conditions such as light, moisture or oxygen in order to preserve purity, safety and effectiveness. Techniques for protecting the product from oxygen include vacuum sealing of lyophilized products and displacement of oxygen from the headspace of sealed containers for liquid products by inert gases. The vacuum sealing of lyophilized products will not be discussed in this chapter. However, this chapter will include a discussion of online equipment used to verify maintenance of vacuum levels through the measurement of relative humidity in the vial headspace.

Nitrogen is the most widely used inert gas as it is relatively inexpensive and presents limited safety hazards considering the flow rates typically used and the room air changes normally found in Grade A environments, where aseptic filling takes place. Designing a deoxygenation system is a combination of engineering and art. The process begins by delivering a formulation to the filling needles that does not contain any dissolved oxygen and with minimal bubbles. Typically this is achieved by sparging the formulation in the bulk preparation vessel and blanketing the product with nitrogen thereafter. Caution must be taken with the blanketing process to insure that the vessel is integral to minimize the flow of nitrogen, as long-term storage under flowing nitrogen could result in evaporation, creating a superpotent drug product.

The filling needles can have a significant effect on the oxygen levels observed in the filled product. To insure that the filled container can be properly blanketed with nitrogen, foaming during the filling process must be minimized. This may require filling needles with beveled or side openings. The surface tension of each product is the major characteristic affecting the needle design.

In the past, the process of purging the oxygen from the vial has included purging the empty vial with nitrogen prior to filling the solution, but it has been the author's experience that if the product filling/degassing system is properly designed, minimal levels of headspace oxygen can be achieved without the pre-filling purge. Once the vial has been filled, the remaining headspace must be replaced with the nitrogen. The design of this system may include needles to fill each vial with nitrogen and a mini-environment containing low levels of oxygen to protect the product until it is stoppered. Some of the filling systems utilize concentric needles capable of filling the product and purging with nitrogen in a single stroke. Regardless of concentric or stand-alone needles for purging, the proximity to the mini-environment is critical to achieving the desired level of oxygen.

The level of oxygen in the mini-environment should be at or below that required for the headspace of the sealed vial as the oxygen level in the vial will equilibrate with that of the mini-environment. To ensure uniformly low oxygen levels within the mini-environment it may be necessary to supply nitrogen through several inlets geometrically spread across the mini-environment. Regardless of the oxygen level within the mini-environment, determination of the maximum time for a line stoppage should be incorporated into the process development phase and confirmed during validation. Product should be discarded if the maximum time is exceeded.

The critical process parameters of the deoxygenation process that must be taken into consideration are filling speed, position of the needle relative to the surface of the liquid, nitrogen flow rate to each of the components of the system, product surface tension, fill volume and container size. There are no formulas to identify worst-case conditions as in some cases too much flow of nitrogen is detrimental as well as too little flow. It is theorized that too much flow of nitrogen causes turbulence that incorporates air into the vial.

Headspace oxygen levels in the sealed container can be monitored utilizing a time-based sampling plan and destructive testing.

Typically headspace oxygen is measured by a membrane-covered electrochemical sensor in conjunction with a system to force the headspace gas into the testing chamber. These sensors with the gas-permeable membrane operate on Dalton's Law of Partial Pressures and Henry's Law. Typically this is accomplished by forcing the headspace gas into the testing chamber by injecting deoxygenated water into the vial. Performing this test on ampules is much more involved and requires some way of capturing the headspace gas without contamination by air. Typically this is performed under water.

The validation of the deoxygenation system can be conducted much like fill volume in that multiple samples from each filling and degassing station should be evaluated using a statistical sampling plan. The determination of worst case conditions must be approached with caution and it may be necessary to validate each product and container size. Operating ranges for the flow rates of nitrogen to the various components of the system should be developed with a significant amount of experimentation prior to establishing the validated operating parameters.

The use of laser absorption technology now provides PAT capabilities for the product quality attribute, headspace oxygen. Lighthouse Instruments, LLC (Charlottesville, Virginia) has developed both online and offline systems capable of determining the headspace oxygen levels in glass containers as they pass between the laser source and the signal receiver. The absorption signal of the laser is proportional to the concentration of oxygen in the container. These intelligent systems have capabilities to automatically reject containers whose headspace oxygen is greater than a preset limit.

As with any new technology the Validation Professional must identify the risks associated with this approach. What variables must now be controlled to insure that the measurements performed by this system are accurate and have adequate precision? The questions that one might consider for a technology such as this might include the following:

- Is this technology useful with both plastic and glass containers?
- How does the variation in container diameter affect the accuracy and precision?
- Does the presence of other gases or relative humidity affect the accuracy and precision?
- How does variation of the distance from the sensor to the container affect the accuracy and precision?

Fortunately, for the pharmaceutical manufacturer the answers to these questions are very positive. The only material that interferes with the absorption of the laser signal is HDPE. Precision is equivalent regardless of the use of tubing or molded glass.

The system does not require a reduced variation of the inner diameter of the container. What is required is the preparation of a linearity curve for each container type. The linearity curve can be prepared using standard gases ranging from 0% to 20% oxygen. As many formulations require oxygen levels in the range of 2% to 5% more standards within this range should be utilized in the development of the linearity curve. Multiple vials of each standard level should be prepared for the development of the linearity curve.

Because the wavelength at which oxygen absorbs the light is 762 nm and the wavelength at which water is 1382 nm there is no requirement to control more precisely relative humidity within the area (10).

The critical process parameters for this detection system could include

- *Container type*: A linearity curve must be developed for each individual container (type and size).
- *Processing speed*: Only maximum speed should be evaluated to insure that the mechanical systems are capable of accurately rejecting the defective containers.
- Time between filling/sealing and headspace oxygen measurement if the product foams as a result of the filling process.

As the process transitions from the validation phase to the routine operational phase, a daily process verification is implemented with a set of standard vials containing both acceptable and non-acceptable levels of headspace oxygen. This daily process verification provides direct correlation to the validation exercise.

Lighthouse Instruments, LLC, utilizing the same technology has developed a similar online system for measuring vacuum levels in lyophilized vials that are stoppered under vacuum. For each product a correlation between vacuum and headspace moisture can be established. For this application the absorption of the laser signal is evaluated at 1382 nm. The linearity curve is developed using ampules filled with standard levels of moisture ranging from 0.5% to 10%. The standards are NIST traceable.

To minimize the effect of atmospheric relative humidity, the system is equipped with an inert gas purge of the optical path (external to the vial). The optimal flow rate of the inert gas can be established by determining the variance when analyzing the same vial multiple times. The critical process parameters for the system include machine speed and inert gas flow rate.

INSPECTION

The requirement for inspection of each container of injectable product for contamination by visible foreign matter is clearly stated in the USP (1). The reason for this requirement can be found in numerous medical papers (11–25) which discuss the possibility of human injury as a result of injected particulate matter. The papers conclude that thrombosis, phlebitis, renal infarction, brain damage and death due to pulmonary insufficiency can occur. Normally, statements as general as those made in the USP allow a certain degree of freedom that result in continuous development and improvement of the techniques utilized to accomplish the task in question. Such has been the case with particulate inspection. Most inspections today are performed by any of three general techniques:

1. Visual inspection with manual handling
2. Visual inspection with automated handling
3. Automated inspection

Visual Inspection

Visual inspection techniques have been studied in depth because inspection, being a critical part of the parenteral manufacturing process, is a completely human endeavor. Many psychophysical experiments have evaluated the relationship of luminance, contrast (object to background) and speed of detection by the human eye.

Some important factors to consider when designing an inspection process for visible particulate include the following:

1. The light intensity at the container being inspected should range from 100 to not more than 350 foot candles (26). Florescent bulbs with electronic ballast will minimize light flicker, which could contribute to inspector fatigue (27).
2. If the time required to detect an object exceeds 0.2 seconds per pause, the probability of detection is decreased (28). The time to inspect each container generally increases with the size of the container and controls should be implemented to standardize the inspection time.
3. When inspecting for glass particles, the probability of detection increases sharply because of reflection (29).
4. Inspectors generally require regular breaks from their inspection duties. Typically inspectors will perform other functions within the inspection group at a frequency ranging from every 20 to 30 minutes.

The psychophysical theories are valuable, but the most important fact to remember is that visual inspection is probabilistic in nature. Units rejected by one inspector may be accepted by another. Thus the primary variable that must be addressed is the inspector. How does one certify that an inspector is properly trained? There are several statistical techniques that can be used. These are discussed later in this section.

Presentation Devices

In order to increase the productivity of the inspection process, automated presentation systems are used to present the containers in a viewing field, allowing the inspector to remove the defective units by either manual or power-assisted means. These presentation devices can be operated at a wide range of speeds, some as high as 175 units/min. The presentation devices are normally equipped with a transport mechanism that brings the units into the field of view, which has a black or white background and an adjustable source of illumination. Just prior to entering the field of view, the units are spun at a high speed and stopped so that, in the viewing field, the solution and any particulate contained therein are still moving. Movement of particles greatly enhances the inspector's ability to detect particles. Normally in the viewing field the container rotates at a speed that permits the inspector to examine the full circumference for cosmetic defects.

Development of the inspection process is a multivariate exercise, which should include variables such as:

1. Transport speed
2. Fast spin speed (to put any particles in motion, but not create bubbles)
3. Slow spin speed (to ensure a complete rotation within the viewing field)
4. Background light intensity
5. Tyndall light intensity
6. Power of magnification

It is difficult to combine both a white and black background into a single presentation machine. Most designs include an opaque plastic background which is illuminated from behind and a transport belt containing black rollers upon which the product rotates. Some firms have been able to increase their inspection efficiency by using two machines in series, one with a black background and the second with an illuminated background as described above. The black background should have a dull finish to avoid any reflection of light which may interfere with the inspector's ability to detect any reflective particles or those that appear to be white.

When changing from a purely manual inspection procedure to an automated presentation device, the validation can be conducted using statistical techniques to be presented later in this section. However, there are some important factors that should be emphasized when making this transition:

1. Additional training on the use of the presentation system, including practice sessions is required to enhance recognition of the appearance of particles.

2. The operating speed of the presentation device should be established based on the normal defect level present in those operations preceding the inspection process. If there is a significant variation in the defect level, the inspectors may not be able to respond quickly enough.
3. The inspectors should be instructed to alert their supervisors if the defect level is significantly higher than normal, in order to reduce inspection speeds and increase the detection efficiency.
4. Incorporating a sampling plan into the filling operation may help to identify portions of a batch that may contain higher defect levels.

The last factor highlights the process monitoring that should be in place for any type of inspection process. Statistical samples should be removed periodically during the inspection process and visually inspected by qualified personnel in accordance with a predefined list of classified defects and accompanying AQL levels. Sampling programs can be defined to operate concurrently with the inspection process and provide feedback to inspectors regarding their performance. If systems are created to provide traceability of the samples to the inspector, their performance can be trended to identify inspectors who require additional training.

The process monitoring should also include trending of defects by type and quantity. Those data are typically collected by evaluating the rejected containers. Reinspection of rejected containers to identify false rejects to be incorporated into the accepted portion of the batch is generally not performed, heightening the importance of the inspector training and process validation.

Automated Inspection

Automated inspection systems employ many of the same handling concepts to put particles in motion as the containers move into the inspection portion of the machine. In the inspection field, there are generally several methods to identify the presence of moving particles in the liquid. Some systems compare multiple images of the same container and will reject containers that demonstrate differences above an adjustable level. The different images result from light reflecting off particles moving in the liquid.

Other systems are conceptually designed to be like spectrophotometers, where the absorbance of a calibrated signal by the sample in the pathway is a measure of the concentration of an ingredient. For example, the Eisai AIM is a dual light transmission system which detects moving shadows cast by foreign matter in the liquid. This shadow is converted into an electrical signal by a phototransistor connected to fiber optics. The Eisai AIM incorporates two inspections into each machine and a rejection by either one of the two light measurements will initiate rejection of the container by the control system.

The Particulate Detection System (PD-100) also utilizes light deflected by particles moving in the liquid, but its treatment is different. The container volume is uniformly illuminated by fiber optic light pipes arranged in the equivalent of vertical slits at a horizontal angle that optimizes forward scattering for the particle size range of interest. The Particulate Detection System is designed to operate by evaluating redundant data. The image volume is monitored by two or more planes, depending on the container diameter. Each plane is subdivided into multiple slit-like rectangular units. In this system, particle size is related to the particles transit time across the slit-like rectangular unit. The accept/reject decision is based on an analysis of the distribution of particulate transit times combined with special signals generated by glass particles for each container. This combined description is compared with stored data based on manual inspection performance to accept or reject each container (30).

The process optimization will focus on many of the same variables previously discussed in the paragraphs focused on presentation inspection systems. These include

1. Transport speed
2. Fast spin speed (to put any particles in motion, but not create bubbles)
3. Slow spin speed (to ensure a complete rotation within the viewing field)
4. Tyndall light intensity
5. Camera settings
6. Light source/receiver

Statistical Validation

Regardless of the detection principal all automated inspection can be validated using the same statistical approach. The particulate inspection of a parenteral container is probabilistic in nature. When inspecting parental containers for particulate matter, which may be difficult to see, an inspector will not always make the same accept/reject decision each time he or she inspects a container. Over the years, a number of statistical approaches have been proposed but in this author's opinion, the model developed by Knapp and Kushner (29–34) is the most rugged and an accurate mathematical model of an inspection process.

The basis of the statistical model is that every container can be categorized by its frequency of rejection. If a container is rejected 7 out of 10 times it is inspected, the probability of rejection is 0.7. If a large population of containers undergoes multiple inspections, the containers can be categorized by their probability of rejection. Containers without any particulate matter should have a probability of rejection of 0.0 and containers with particulate matter that is readily visible should have a probability of rejection of 1.0. Knapp and Kushner created three categories within the total population based on the probability of rejection for each container. These are shown in Table 3.

Comparing the performance of an inspection process in these categories definitive statements can be made about the validity of an inspection process. The ability of an inspection process to properly reject those

Table 3 Knapp and Kushner Zone Definition

Category	Probability of rejection (p)
Accept zone	0.00–0.30
Gray zone	0.31–0.69
Reject zone	0.70–1.00

containers in the reject zone is measured by RZE. The RZE is calculated by equation (3)

$$\mathrm{RZE} = \frac{\mathrm{RZR}}{\mathrm{RZN}} \tag{3}$$

where $\mathrm{RZR} = \Sigma^{RZ}{}_{n}(p)$ the sum of the rejection probabilities for all of the vials in the reject zone and RZN the number of containers in the reject zone.

The ability of the rejection process to minimize the rejection of acceptable containers in the Accept and Gray zones is measured by the undesired RAG. This is calculated by equation (4)

$$\mathrm{RAG} = \frac{[\sum^{GZ}{}_{n}(p) + \sum^{AZ}{}_{n}(p)]}{[\mathrm{AZN} + \mathrm{GZN}]} \tag{4}$$

where AZN is the number of containers in the Accept zone and GZN the number of containers in the Gray zone.

These two parameters, RZE and RAG characterize the inspection process from security and discriminatory standpoint, respectively. As RZE increases, there is greater assurance that the output of the inspection process will meet predefined quality standards. As RAG decreases, there is greater assurance that acceptable units are not being rejected, which serves as an economic measure of the inspection process. Experience has demonstrated that RZE and RAG can be mutually optimized, resulting in more cost-effective inspection processes at an equivalent security level.

In most instances, firms transition from a manual inspection process to an automated inspection process and the ultimate acceptance criteria for the process validation is to demonstrate that the RZE for the automated process is greater than or equal to the RZE for the manual inspection process. ($\mathrm{RZE_a} \geqq \mathrm{RZE_m}$).

Among those who utilize the Knapp method there is some disagreement in the application of this criterion. A strict interpretation of the statistics requires that the $\mathrm{RZE_a}$ be calculated for the containers in the reject zone as defined by the manual inspection process. Others, especially equipment manufacturers, simply compare the $\mathrm{RZE_a}$ for the containers in the reject zone as defined by the automated inspection with the $\mathrm{RZE_m}$ for the containers in the reject zone as defined by the manual inspection.

Generally some of the containers in the manual reject zone will not be included in the automated reject zone and vice versa. Equipment manufacturers simply argue that because their systems are more discerning and more consistent than humans they should not be required to compare the RZEs for exactly the same population of containers. While their argument has some merit one must still explain why some containers, especially those in the reject zone with the lower probability of rejection are not rejected by the automated inspection system.

This validation concept requires a test population to be carefully selected and controlled to exclude bias and ensure that final conclusion of the validation study is based on an adequate number of accept/reject decisions to be statistically significant. The size of the test population can range from several hundred to one thousand containers with approximately 10% of the containers selected from production rejects that are representative of the types normally found in production. Precautions should be taken to ensure that the rejects are truly representative and not the most obvious rejects. The remainder of the test population should be comprised of containers selected at random from production batches. Ideally, slightly more than 10% of the test population should be selected from production rejects as during the handling of these test sets containers may be damaged. Once damaged the data from that container should be deleted from the database and cannot be used in the statistical comparison.

A unique number is applied to each container so that the individual accept/reject decisions can be documented for each vial. When applying the code number care should be taken not to interfere with the inspection of a container. The code numbers for the rejected vials should be selected at random to prevent any bias on the part of the inspector.

The inspectors participating in the study should not handle the test sets or have access to the data collection sheets on which the test results have been recorded.

Statistical calculations performed by Knapp and Kushner demonstrated that to ensure (with a 95% confidence level) that the correct conclusion is reached, 1104 accept/reject decisions should be made for the containers in the reject zone. Some firms have chosen to utilize the overall concept with fewer accept/reject decisions. This effectively reduces the confidence that the outcome of the validation exercise is accurate.

In order to establish the probability of rejection for each vial in the test set a group of inspectors should be selected to perform multiple inspections of the test set. The inspectors should be randomly selected and should follow all approved SOPs when inspecting the containers. The inspection time permitted for each manual inspection should also be controlled in the same manner as in normal production.

Data collection sheets and spreadsheet macros can be developed to simplify the identification of the containers in the reject zone in the calculation of the $\mathrm{RZE_m}$ and the $\mathrm{RAG_m}$. The same data collection sheets can be used to collect the results from the multiple inspections by the automated inspection system, but the database should contain a tag identifying those containers from the manual reject zone. By filtering those containers from the remainder of the containers in the test set the $\mathrm{RZE_a}$ can be calculated for the same group of rejected containers.

This technique can be modified to qualify inspectors for a manual inspection process. By maintaining the accept/reject decisions for the containers in the reject zone for each inspector, a RZE for each inspector ($\mathrm{RZE_i}$) can be calculated. The distribution of $\mathrm{RZE_i}$ can be determined and a confidence interval established for the population. A new inspector could be required to meet or exceed the $\mathrm{RZE_i}$ at the lower level of the confidence interval. For example, if the $\mathrm{RZE_i}$ ranged from 0.69 to 0.95 for the qualified inspectors, and the 95% confidence interval was 0.72 to 0.92, a new inspector would be required to achieve an $\mathrm{RZE_i}$ of 0.72.

SECONDARY PACKAGING

The significant variation in the level of automation present in a secondary packaging processes in the pharmaceutical industry today requires the validation professional to identify the boundary between a process requiring validation and procedures requiring higher levels of personnel training and certification.

When should the emphasis shift from personnel training to validation? As the industry and regulatory bodies continued to be focused on "risk," it could serve as one of the primary characteristics to be considered in making this assessment. A HACCP could be one method to asses this issue.

Considering some of the quality characteristics directly related to a secondary packaging process we can examine various control methods and the integration of validation into the overall control plan. The quality characteristics could include:

- Accurate and legible batch number and expiration date
- Missing component or product abuse
- Legible electronic bar codes
- Proper number of units per container
- Proper assembly of packaging materials designed to prevent damage during shipment or insure proper temperature control during shipment

Considering these quality characteristics an automation scale can be created. At one end is the manual application of labels in conjunction with hand packing to the middle ground where the operations are performed mechanically without any automated verifications and finally to the fully automated labeling and packaging system that performs 100% verification of each critical product quality characteristic.

Some of the automated systems capable of performing 100% verification include visions systems for reading lot numbers and expiration dates, or check weighers. These systems will automatically reject the defective units or stop when a defective unit is identified. When these systems are utilized there is normally an evaluation of the system function performed during the manufacture of each batch. These systems clearly require some type of qualification, which provides the direct correlation to the batch specific evaluation of the system performance.

The opposite end of the automation spectrum, manual labeling and packaging only requires well-documented training and a system to monitor personnel performance, such as a statistical sampling plan with established AQLs for all classified defects.

In the middle are many automated systems without systems capable of 100% verification of critical product quality characteristics. These systems require development of validation programs, which encompass all the elements normally found for line equipment PQ: the ability to operate at the maximum operating speed for a predefined period of time, with a limited amount of product or commodity abuse, and conformance with all product quality characteristics as determined by a tightened or more intensive statistical sampling.

The qualification trials will be difficult to design and properly execute if the user requirement specification has not been well designed to clearly characterize the defects to be rejected. Defining defective print for lot number and expiration printing is a very subjective process and requires inclusion of Quality and Marketing functions in the decision-making.

Limit samples to represent the defects to be rejected must be created for inclusion in the FAT. It is always useful to prepare multiple sets or units that could be substituted in case of breakage.

After establishing the limit samples, the critical process parameters must be determined in order to identify the worst-case test conditions to be challenged during the FAT and subsequent qualification.

A developing function of secondary packaging is the incorporation of RFID into the pharmaceutical package. RFID can serve as an anticounterfeiting measure and also enhance inventory control by providing real-time information regarding material transfer.

RFID is currently being incorporated into the pharmaceutical package in two ways. An RFID device, consisting of a chip and an antenna, can be incorporated into labels applied to the individual container or shipping case.

In a recent development, West Pharmaceutical Services and Tagsys U.S.A (Cambridge, Massachusetts, U.S.A) have incorporated an RFID device into the Flip Off® seal of an aluminum cap for parenteral products. The West Spectra™ (West Pharmaceuticals Services; Tagsys U.S.A) product contains a 13.56 MHz chip and antenna, which must pass within 2 inches of an RFID reader in order to transfer information or read and verify information. Because it is located on top of the vial, the presence of liquid and metal do not interfere with the signal transfer. The product is capable of withstanding steam sterilization and is not affected by most antiterrorism measures applied to international shipments today, including X-rays. Only gamma radiation will damage the RFID tag.

RFID tags are normally incorporated into labels by a label manufacturer or label converter. The RFID tags within the labels are susceptible to the same of these hazards as labels. Each RFID tag, whether in a label or a Flip Off cap, contains a UID, which can be read by the component manufacturer to verify that the RFID tag within each label is functioning properly prior to shipment to the pharmaceutical manufacturer. The functionality of the RFID tag must be controlled through a critical control plan. Components of this critical control plan could include a vendor audit program, a certificate of compliance for each shipment and a performance history indicating the frequency of tag of failures identified during secondary packaging operation.

Typically an RFID reader, capable of reading, writing and verifying the information in each RFID tag is incorporated into the packaging line. Also, an RFID tunnel at the end of the packaging line verifies that each RFID tag within the case and the RFID tag contained within the case label are readable. The West Spectra product cannot be verified in the tunnel because of the close proximity required to read this RFID tag. Containers within the case must be properly oriented to ensure that no two tags are directly facing each other, creating interference and the failure to read the tags. Incorporation of data verification reduces the validation requirements for this system.

One aspect which may require validation is the effect of the orientation of the container as it passes by the RFID reader/writer on the ability to read, write and verify. It is possible for RFID tags within labels to interfere with each other if they are in direct contact. Where RFID tags are incorporated into labels it may be necessary to design the product transport system to hold each container individually as it passes by the reader/writer.

Once incorporated into the pharmaceutical product, product information such as batch number and expiration date can be downloaded to the RFID chip. As the product passes through the supply chain, transactional information can also be saved to the chip. Shipping studies should be performed to demonstrate that the packaging systems provide adequate protection of the RFID tags to ensure functionality throughout the supply chain.

Clearly the computer validation required for an RFID system is a critical aspect for the implementation of this system, but will not be presented herein.

Studies should be considered to verify the potential for the product temperature to increase when the RFID tag is exposed to the electromagnetic energy of the UHF RFID reader for extended periods of time

CONCLUSION

Regardless of the packaging process the validation concepts shown below will enable the validation professional to develop and execute a validation plan which can be clearly viewed as value added.

1. Clearly define the process
2. Perform a hazard analysis
3. Develop a critical control plan
4. Develop and integrate the validation plan into the critical control plan
5. Design validation tests commensurate with the risk and critical nature of the quality characteristic being controlled.

REFERENCES

1. USP. <1> Injections. (Accessed October 2005 at http://www.uspnf.com/uspnf/pub/index?usp=28&nf=23&s=2).
2. Bosch-T-L Systems. Minneapolis, Minnesota (unpublished report).
3. Nicholas RJ. A new concept for aqueous solution filling that is electronically instead of mechanically based. In: Pharmaceutical and Technology Conference, New York, 1980.
4. USP. <905> Uniformity of Dosage Units. (Accessed October 2005 at http://www.uspnf.com/uspnf/pub/index?usp=28&nf=23&s=2).
5. USP. <1151> Pharmaceutical Dosage Forms, Injections. (Accessed October 2005 at http://www.uspnf.com/uspnf/pub/index?usp=28&nf=23&s=2).
6. Juran JM. Quality Control Handbook. 3rd ed. New York: McGraw-Hill, 1974.
7. Levine CS. Comparing the use of "closed" and "open" ampoules for the manufacture of parenteral products using FMEA. PDA Conference, Rome, 2005.
8. Morton DK, Lordi NG, Troutman LH, et al. Quantitative and mechanistic measurements of container closure integrity. Bubble, liquid and microbial leakage test. J Parenter Sci Technol 1989; 43(3):104–8.
9. PDA Task force on container/closure integrity. Technical report No. 27 pharmaceutical package integrity. J Parenter Sci Technol 1998; 52(4):11–4.
10. Accessed October 2005 at http://www.lighthouseinstruments.com/NewFiles/about_tech.html
11. Koriwaler BE. Pulmonary emboli of cotton fibers. Am J Clin Pathol 1950; 20:385.
12. Jaques WE, Mariscal GG. Study of incidence of cotton emboli. Bull Int Assoc Med Museum 1951; 32–63.
13. Bruning EJ. Uber Entstehung und Bedeutung untraarterieller Fremdkorperembolien der kindlicken Lunge. Arch Pathol Anat Physiol Klin Med 1955; 327:460.
14. Sarrut S, Nexelof C. A complication of intravenous therapy: giant cellular, macrophagic pulmonary arteritis. Press Med 1960; 68:375.
15. Von Glahn WC, Hall JW. Reaction produced in pulmonary arteries by embolus of cotton fibers. Am J Pathol 1949; 25:575.
16. Garvan JM, Gunner BW. The harmful effects of particles in intravenous fluids. Med J Aust 1963–1964; 2:140; 1(1).
17. Jaffe RB, Koschman EB. Intravenous drug abuse, pulmonary, cardiac, and vascular complications. Am J Roentgenot 1970; 109:107.
18. Douglas FG, Kafilmout K-J, Pratt NL. Foreign particle embolism in drug addicts: respiratory pathopliysiology. Ann Inter Med 1971; 75:872.
19. Walter CW. Safety of Large Volume Parenteral Solutions. FDA Symposium. Washington, DC: FDA, 1983.
20. Silberman J, Cravioto H, Feign L. Foreign body emboli following cerebral aniography. Arch Neuro 1960; 13:711.
21. Chason JL, Landers JW, Swanson RE. Cotton fiber embolism. A frequent complication of cerebral angiography. Neurology 1963; 13:558.
22. Adams DF, Tord BD, Kosek J. Cotton fiber embolization during angiography. A clinical and experimental study. Radiology 1965; 84:678.
23. Jaffe NS. Elimination of particulate contamination from ophthalmic solutions for intraocular surgery. Bull Parenter Drug Assoc 1970; 24:218.
24. Ryan PB, Rapp RP, DeLuca PP, Griffen WO, Clark JD, Cloys D. In-line final filtration, a method of minimizing contamination in intravenous therapy. Bull Parenter Drug Assoc 1973; 27:1.
25. DeLuca PP, Rapp RP, Bivin B, McKean HE, Griffen WO. Filtration and infusion phlebitis: a double-blind prospective clinical study. Am J Hosp Pharm 1975; 32:1001.
26. USP XIII, 1947.
27. Accessed October 2005 at http://www.ccohs.ca/oshanswers/ergonomics/lighting_flicker.html
28. Blackwell HR. The effects of certain psychological variables upon target detectability. Engineering Research Institute Report 2455-12F, Ann Arbor: University of Michigan, 1958.
29. Knapp JZ, Kushner HK. Particulate inspection of parenteral products: from biophysics to automation. J Parenter Sci Technol 1982; 36:121.
30. Knapp JZ, Kushner HK, Zeiss JC. A validation procedure for the particulate inspection of parenteral products. J Parenter Sci Technol 1982; 36.
31. Knapp JZ, Kushner HK. Generalized methodology for evaluation of parenteral inspection procedures. J Parenter Sci Technol 1980; 34:14.
32. Knapp JZ, Kushner HK. Implementation and automation of a particle detection system for parenteral products. J Parenter Sci Technol 1980; 34:369.
33. Knapp JZ, Kushner HK, Abramson LR. Automated particulate detection for ampoules with use of the probabilistic particulate detection model. J Parenter Sci Technol 1981; 35:21.
34. Knapp JZ, Kushner HK, Abramson LR. Particulate inspection of parenteral products: an assessment. J Parenter Sci Technol 1981; 35:176.

33

Validation of Bulk Pharmaceutical Chemicals

James Agalloco
Agalloco & Associates, Belle Mead, New Jersey, U.S.A.

Phil DeSantis
Schering-Plough, Inc., Union, New Jersey, U.S.A.

HISTORY

Validation was initially introduced in the 1970s to the pharmaceutical industry as a means for more firmly establishing the sterility of drug products where normal analytical methods are wholly inadequate for that purpose. In following years, its application was extended to numerous other aspects of pharmaceutical operations: water systems, environmental control, tablet, and capsule formulations, analytical methods and computerized systems. Individuals working with BPCs were particularly reluctant to embrace validation as a necessary practice in their operations. Industry apologists explained this lack of enthusiasm in terms of differences in facilities, equipment, technology, hygienic requirements, cleaning methodologies, operational practice and numerous other aspects of disparity that seemingly justified the recalcitrance of this segment of the industry. This view was widespread in the BPC industry through the end of the 1980s.

The extension of the concepts that have made validation such an integral part of practices across the healthcare industry to the production of BPCs seems obvious in retrospect. Yet, for many years there existed a general reluctance to introduce validation into BPC activities. While there were some modest efforts, it was not until some time after the biotechnology industry became technically and commercially viable that any significant effort was initiated. The production of biotech products for registration in the United States requires the approval of the FDA's CBER. The CBER required extensive validation of fermentation, isolation, and purification processes utilized in the preparation of biologicals (1). An objective comparison of BPC operations relative to those performed in the early stages of biologicals would reveal minimal differences. The production methodologies for many classical BPCs, e.g., penicillins, cephalosporins, and tetracyclines are nearly indistinguishable from those utilized to prepare tPA, EPO, and other biologicals. With this realization, the advent of validation for BPCs was apparent to all and was increasingly imposed upon the industry.

In 1990, the U.S. Pharmaceutical Manufacturers Association (now called PhRMA) formed a committee to define BPC validation concepts (2). This committee's efforts culminated in 1995 when they issued their finished draft. This document served as a guide to the authors in the development of this chapter. Of necessity, considerable clarification and expansion of the material contained has been necessary to complete this effort.

In the late 1990s, a new term started to appear, first in Europe, but soon it spread across the entire industry—APIs. Those who first used the new term suggested that it was synonymous with BPCs. Since that time it has become increasingly common in the industry speak only of APIs. A part of the rationale for this initiative has been voiced as a move toward harmonization. The authors of this chapter do not agree with this change in terminology, as there are numerous BPCs that have no metabolic activity. Many pharmacologically inactive materials are produced within the industry using facilities, equipment, and methodologies identical to that employed for so-called APIs, yet with the advent of this new catch phrase are to be seemingly ignored. Our use of the term BPC is deliberate and is intended to embrace both active moieties, and therapeutically inert materials used as excipients, processing aids, and other materials.

The official requirement for validation of BPC processes was formally established in Guidance for Industry, Q7A Good Manufacturing Practice Guidance for Active Pharmaceutical Ingredients (3). This was the result of a multiyear effort by the ICH, which resulted in this harmonized guidance document. This guidance document addresses the subject of validation briefly and employs the same definition the FDA has adopted for other processes (see next paragraph). This chapter provides recommendations for validation consistent with the Q7A guidance.

Abbreviations used in this chapter: APIs, active pharmaceutical ingredients; BPCs, bulk pharmaceutical chemicals; BS, British Standard; CBER, Center for Biological Evaluation and Research; cGMP, current good manufacturing practice; DCS, distributed control systems; DIN, German Institute of Standardization (Deutsche Industry Norm); DMF, drug master file; FDA, Food and Drug Administration; GAMP, good automated manufacturing practice; ICH, International Conference on Harmonization; NDA, new drug application; NIST, National Institute for Standards and Technology; PDA, Parenteral Drug Association; PhRMA, Pharmaceutical Research Manufacturers Association; SIP, sterilization in place; tPA, tissue plasminogen activator; WFI, water for injection.

DEFINITION OF VALIDATION

There are innumerable definitions of validation that have been written over the nearly 30 years since its appearance in the pharmaceutical industry. Rather than foster new definitions within the context of this chapter the authors have chosen to draw upon some of the more widely quoted definitions. The FDA defines process validation as: "Process validation is establishing documented evidences which provides a high degree of assurance that a specific process will consistently produce a product meeting its predetermined specifications and quality characteristics" (4). This definition is referred to in the FDA's subsequent guidance specific for BPCs (5).

REGULATIONS

Regulations specific to the control BPCs are a relatively new concept; for many years the FDA's policy was to apply a limited enforcement of the subpart 211 regulations for finished pharmaceuticals (6). In recent years, FDA has endeavored to harmonize its approach to BPC regulation with the rest of the world and has issued a guidance document that draws heavily on subpart 211 (7). This effort followed the issuance of a Pharmaceutical Inspection Convention document that addressed the same subject in a different format (8).

APPLICATION OF VALIDATION

Some discussion of validation approaches utilized for BPCs is essential to follow this chapter. The approaches for BPCs are essentially the same as those utilized for other processes and systems. This discussion serves to highlight the nuances of validation as they apply to BPCs.

LIFE CYCLE MODEL

Contemporary approaches to the validation of virtually any type of process or system utilize the "life cycle" concept (9). The "life cycle" concept entails consideration of process or system design, development, operation and maintenance at the onset. Use of the life cycle helps to provide a system that meets regulatory requirements, but is also rapidly placed into service, operates reliably and is easily maintained. While the "life cycle" is best suited to new products, processes or systems, it certainly has applicability for existing systems as well. Existing systems that have never been previously validated can be reviewed against the same validation criteria that would be imposed for new systems. While these systems are likely to be deficient with regard to current requirements, the "life cycle" model provides a means for upgrading their programs to be on a par with newly developed systems. This is especially important for BPCs given that the validation of these processes has lagged behind many of the other areas of the industry where validation has already been instituted. It is perhaps safe to say that the first validation efforts to be utilized for BPCs will likely be retrospective ones, following the existing system path to enter the "life cycle" model.

VALIDATION OF NEW PRODUCTS

The validation of a new BPC entails practices that parallel those utilized for the introduction of a new pharmaceutical formulation. Thus a large part of the initial validation effort must be linked to the developmental activities that precede commercial scale operation. The similarity is such that aspects of reaction and purification methodologies should be as similar as possible given of course the difference in the scale of the equipment utilized in the commercial facilities. Any differences between the BPC processes utilized for the formulation batches used to establish clinical efficacy and the commercial material must be closely evaluated and their impact on the BPC products: chemistry, purity profile, stability, crystal morphology and other key attributes. The developmental laboratory has the responsibility for determining optimal reaction conditions including time, temperatures, raw material purity, molar ratios, solvent selection, crystallization method, wash volume, drying conditions, etc. Of primary concern is the identification of critical control parameters, that is to say those which impact quality, purity, safety and efficacy. The concerns to be addressed in any individual BPC validation program are of course unique to that process; the inclusion or exclusion of any single factor as a consideration in BPC validation is an arbitrary one determined by the authors. Chemical reactions are among the more complex processes to be subjected to validation and the number of critical factors in even a single reaction can be quite extensive. The amount of information which must be generated during development to support a validated BPC process is correspondingly extensive. The necessary information can be assembled into a technology transfer document that conveys the collected experience gained during development to those responsible for the commercial production of the BPC. The success of a developmental organization is better assessed by the quality of the information they convey to document their efforts than it is by the sophistication of the chemistry utilized to make the BPC. The technology transfer document is likely to be of central interest to FDA inspectors during the conduct of a pre-Approval Inspection of the facility prior to approval (10).

VALIDATION OF EXISTING PRODUCTS

At the time this is written, validation of BPCs is a still relatively new concept for the industry to address. As such the vast majority of BPC products have been on the market without any significant validation in place. As a consequence, the first efforts to validate these products will undoubtedly employ retrospective methods. The trending of results derived from in-process and release testing of these products and processes will serve as the basis for these efforts. Given the FDA's general dissatisfaction with retrospective approaches, it is doubtful that these early efforts will remain the only approach utilized. The use of either prospective (in which three batches must be produced before the process can be considered validated and any of material released for sale) or concurrent (in which individual batches are released while continuing to accumulate data toward a three-batch validation) approaches is certainly acceptable; a decision to use those approaches while raising less regulatory concerns will also require a longer time to execute and a larger resource commitment. The establishment of priorities for validation of a large number of BPC processes generally follows economic concerns, with those products that provide the largest contribution to the firm's profitability being the initial focus of activity. Regardless of how the first validation efforts were completed, the adoption of the "life cycle" model for maintaining products in a validated state is becoming increasingly widespread.

IMPLEMENTATION

The validation of any process or product relies upon several supportive activities. Validation in the absence of these activities has only minimal utility, as it is only through the integration of these other practices that meaningful validation can be accomplished. Several of these activities are defined in cGMP regulations while others are an integral part of a company's organization structure or are closely associated with "Validation" itself (11).

Equipment Calibration. The process of confirming the accuracy and precision of all measurements, instruments, etc., to ensure that the measured variable is being accurately monitored. Calibration includes demonstration of conformance to applicable national standards such as NIST, DIN, or BS for all key parameters. This is a universal cGMP requirement across the globe.

Equipment Qualification. An outgrowth of "Validation" that focuses on equipment-related aspects. There is *no* requirement for a formal separation of the activity into distinct elements, such as installation and operational qualification. It has become increasingly common in recent years to combine these activities under a single effort. For the sake of those who still separate the activities, individual descriptions have been provided.

1. *Installation Qualification.* Documentation that the equipment was manufactured and installed in accordance with the intended design. This is essentially an audit of the installation against the equipment specifications and facility drawings.
2. *Operational Qualification.* Confirmation that the equipment performs as intended entails evaluation of performance capabilities. It incorporates measurements of speeds, pressures, and other parameters.

Process Development. The development of products and processes, as well as the modification of existing processes, should be conducted to provide documented evidence of the suitability of all critical process parameters and operating ranges. This effort serves as a baseline for all product validation activities. The integration of development into commercial scale operations became a requirement with the advent of the FDA's Pre-Approval Inspection Program (10). The importance of well-documented developmental activities to support subsequent commercial scale production is essential in the validation of BPCs. It is customary for many unit operations (reactions, separations, catalyst reuse, solvent reuse, etc.) to be initially confirmed on a laboratory or pilot scale, prior to their eventual "Validation" on a commercial process scale.

Process Documentation. An often overlooked activity wherein the results of the development effort are delineated in sufficient detail in process documentation so that the variations in the process as a result of inadequately defined procedures are eliminated. While master batch records have long been a cGMP requirement, their adequacy is essential to the maintenance of a validated state.

Performance Qualification (Testing). That portion of the overall "Validation" program that deals specifically with the evaluation (Validation) of the process. It includes the protocol development, data acquisition, report preparation and the requisite approvals. In the distant past this activity was considered "process validation," but over the years the industry has come to realize that "Validation" encompasses a broader spectrum of activities and continued use of the word "process" is limiting.

Change Control. A cGMP requirement that mandates the formal evaluation of the consequences of change to products, processes or equipment. At least two distinct types of change control exist because of the different disciplines that are central to the evaluation of each (12).

1. *Process Change Control.* A system whereby changes to the process are carefully planned, implemented, evaluated and documented to assure that product quality can be maintained during the change process. This type of change control is the province of the developmental scientist and production personnel.
2. *Equipment Change Control.* A mechanism to monitor change to previously qualified and/or validated equipment to ensure that planned or unplanned repairs and modifications have no adverse impact on the equipment's ability to execute its' intended task. This procedure usually entails close coordination with the maintenance and engineering departments.

BPC VALIDATION

The focus of this chapter is BPC validation. To this point, aside from the history section, the information presented to this point would apply to most any type of validation. That commonality with other older validation efforts is deliberate. BPC validation is unique, only to the extent that BPCs are unique. The underlying maxims of success for validation (the knowledge and understanding of the scientific basis upon which the equipment or process is based) are universal. Mastery of the overall approach equips one to effectively employ those concepts in a variety of settings. Some knowledge of the key concerns in the production of BPCs is essential to understanding how the validation of their preparation should be carried out.

Unit Operations. BPCs are the result of a series of chemical reactions in which materials called reactants are brought together under appropriate conditions whereby the reaction takes place and the reaction product is formed. Under even the most ideal circumstances, the desired product must be separated from unreacted raw materials, by-products, solvents and processing aids before it can be utilized in further processing. In the analysis of these processes, chemical engineers have found it convenient to divide the overall process into a series of unit operations (some of which are physical in nature, while other e.g. "reactions" are chemical in nature). The unit operation approach is beneficial because a complex many-step process can be separated and better understood as a series of simpler activities (unit operations) that are more easily interpreted.

Among the more common unit operations are mixing, heating, drying, absorption, distillation, condensation, extraction, precipitation, crystallization, filtration, and dissolution. There are other less common unit operations, but the more important aspect is the subdivision of a lengthy process into smaller and more readily understood segments. The benefits to be gained from this approach are obvious, once the underlying principles are understood for a specific unit operation, those concepts can be reapplied in other steps or processes where that same unit operation is employed. In the validation of BPC processes, the ability to use standardized methods for each unit operation can make what would otherwise be an impossible task into a manageable one. The unit operation approach is of such utility that it has been applied in pharmaceutical dosage form manufacturing as well where the same basic procedures are often encountered, i.e., mixing, milling, filtration, etc.

Physical Parameters. A concern that has been sometimes neglected in the preparation of BPCs relates to the control of physical parameters of the end-product material. Often the focus of BPC development and processing is on chemical purity and yield, as those aspects tend to have the greatest economic significance. There is relative indifference to physical parameters such as size, shape and density compared to the seemingly more important concerns such as potency, impurity levels and process yield. The authors have observed numerous situations where this inattention has resulted in processing problems at the dosage form manufacturing stage. In each instance, it was often the case that the physical parameters of the end product had been virtually ignored in deference to concerns over chemical purity (13). The FDA's Pre-Approval Inspection initiative indicated an awareness that these concerns have come to their attention during the course of NDA reviews and inspections (10).

The most extreme circumstances where physical parameters are of critical importance are for those materials where different crystalline forms are possible. The different polymorphs may have decidedly different characteristics with regard to crystal shape, size and, most importantly, solution characteristics. Many important pharmaceutical chemicals exist in more than one crystalline form, and the manufacturer must ensure that only the desired form is being produced. One of the major concerns voiced by regulators is the potential hazard in using brokered active ingredients (5). The ability to match the purity profile of a BPC is not sufficient if the crystallization is from a different solvent system or at different conditions an entirely different material may result, with profoundly different pharmacological properties. The absence of detailed information on the isolation process used may cause difficulties should the real source of the material (the broker's supplier) change.

Chemical Purity. Central to the preparation of BPCs are issues relative to the purity of the desired material. Until recently the only concern was whether the material met the minimum potency requirements. A typical requirement would be a minimum potency specification of 98%. Any lot that had an assay higher than 98% would be acceptable. Awareness that the small amount of material that is not the desired molecule could cause adverse reactions led to the establishment of purity profiles for the molecule. Using a purity profile approach mandates that the firm identify the impurities present. Current FDA expectations are that firms should characterize all impurities that comprise more than 0.1% of the material and perform toxicity testing on any impurity which is at a concentration higher than 0.5% (5). The establishment of a purity profile for a molecular entity assures that process changes which might result in a change in the by-products and other materials isolated with the desired material do not impact the safety and efficacy of the final product.

Analytical Methods. As with other types of product validation activities, BPC validation cannot proceed without validated analytical methods. The most significant difference in the validation of BPCs is the number of analytical methods that must be addressed. Analytical methods are needed for each stage intermediate, identifying and quantifying the major by-products at each stage as well as the desired chemical moiety. Clearly the scope of the analytical method validation for BPCs represents a larger effort than is normally associated with process validation activities. A comprehensive review of analytical method validation can be found elsewhere in this volume.

Facilities. BPC facilities are vastly different from most other types of facilities in the pharmaceutical industry. The equipment is designed for specialized procedures and as such bears little resemblance to those that might be found in a dosage form facility. Most BPC equipment requires a broader range of utilities and a seeming maze of piping to perform properly. Chemical reactions are sometimes performed at temperatures in excess of 120°C or less than 0°C and require specializing heat transfer fluids to maintain those temperatures. Many reactions utilize solvents as reaction substrates or in the isolation of the materials. These solvents are sometimes introduced via piping systems that supply the various pieces of equipment. Distribution systems for compressed gases used either in the reaction or to inert the equipment are also common. In many older BPC facilities, it is common to see multiple vessels at different elevations arranged around an open bay. In these facilities, several different chemical reactions might be underway in different vessels for different products at the same time. In a dosage form facility this type of arrangement would be viewed with some skepticism. In BPC production, the reactions and unit operations take place within completely closed equipment, minimizing the potential for cross contamination. The difference between BPC and dosage form facilities is most evident in warmer climates. In these areas, the BPC facility may be little more than structural support for the equipment and staging areas for material, with no surrounding building. In effect the equipment is outside, fully exposed to the environment. For certain BPC processes such as solvent recovery and hydrogenation vessels, the equipment is located outside in even northern climates either because of sheer size or safety concerns. These types of arrangements are not typical for the last step in the synthesis. Isolation of the completed BPC is usually performed in rooms specifically designed for that purpose.

Pure Rooms. In the preparation of BPCs, it is common for the last step in the process to be completed in an environment far different from that in which the rest of the synthesis is performed. The term "pure room" is used loosely; there are no regulatory requirements for these rooms and the actual terminology varies considerably from firm to firm. Even without regulatory impetus, some firms have gone so far as to classify their "pure" rooms at ISO 8 (EU Class D) or better (14,15). After the crystallization of the BPC, it is important to protect the product from airborne particulates and other foreign matter that might end up in the finished material. For this reason it is common in many companies to perform a filtration of the active material while still in solution. The filtration removes particulates that may have accumulated in the material up to that point. After the filter, the solution is introduced into the crystallizer in the "pure" room. The room itself is designed to minimize the opportunity for introduction of contaminants into the bulk material and may or may not be a classified environment. The crystallizer is often subjected to extraordinary cleaning before the start of the process to ensure its suitability for the final bulk isolation. Following the crystallization, the BPC is centrifuged, washed, dried, milled and packaged in the "pure" room. It should be noted that BPC processes which use "pure" rooms are not intended to be sterile; the production of sterile BPCs requires a much higher level of control over the environment, equipment, and methodologies and is described more fully later in this chapter.

Qualification of Equipment. The qualification of BPC process equipment, including reaction vessels, receivers, crystallizers, centrifuges, dryers, filters, distillation columns, solvent distribution systems, etc., is a well-defined activity. While this equipment is somewhat different in design and operating features, than the dosage form equipment that has been the subject of the majority of papers on the subject, the same general principles apply. Reaction vessels, receivers and crystallizers differ only minimally from formulation and WFI tanks. Some BPC dryers are identical to those utilized in tablet department. Solvent distribution systems are piping systems and may resemble WFI distribution systems. Some pieces of equipment such as distillation columns and continuous reactors may not have counterparts in the dosage form side, but an understanding of the objectives of the equipment qualification should make the development of suitable protocols straightforward.

Configuration Confirmation. In multi-purpose BPC facilities, the fixed equipment installed may be configured differently for different reactions. In these facilities, campaigns of one reaction may be followed by a reaction for a different product after a change in configuration. Putting aside cleaning considerations for a later portion of the chapter, verification of the systems configuration should be performed. In effect, the reaction train must be re-qualified at the start of each campaign to ensure that the proper arrangement of valves, transfer lines, instruments and other items are established for the process to be introduced. Some firms process a water or solvent batch which simulates the process to verify that the proper connections are in place and that there are no leaks in the system. Following the trial batch, the system is then readied for use with the solvents that will be utilized in the process.

Environmental Control. The usual concerns relative to the environment in which the production activities are performed are not as significant in BPC manufacturing as they are for the preparation of pharmaceutical dosage forms. The introduction of microbial or particulate contaminants at early stages of the process is unlikely to be of significance. BPC reactions utilize high temperatures, extremes of pH, and aggressive solvents that can minimize the impact of any microbial contamination. Filtration is a frequent part of BPC processing in the form of carbon treatments and other unit operations whose intent is to remove unwanted by-products, reactants and solvents. In the course of these measures, incidental particulate contamination is also removed. The use of "pure rooms" as outlined earlier serves to minimize contamination at the last step.

Worker Safety. The safety of the personnel who work in the facility is always a major concern. Exposure to toxic substances is greatest when the operator is adding materials to or removing materials from the equipment. The use of air extraction equipment, isolation technology, automated handling and other means for minimizing human contact with toxic materials is nearly universal. The assessment of worker safety should also embrace exposure to vapor phase hazards, and leak testing of process trains should be performed where hazardous gases are present. Validation of the effectiveness of this equipment is not mandatory from a cGMP perspective, but is certainly beneficial.

Process Water. The water used in BPC production is usually deionized water through the early process stages. If the product is isolated from a water solution in its last step, then a compendial grade of water, Purified Water or WFI may be utilized depending upon subsequent steps in dosage manufacture and the final use of the product. Cleaning of equipment can be performed with city water, provided the last rinse of the equipment is with the same water utilized in the process step. The validation of water systems has been well documented in the literature (16,17) (see Chapter 6).

Process Gases. Some BPC reactions utilize gases as reactants or are performed under a gas blanket. The system may start at either a large high pressure bulk storage tank or from a bank of gas cylinders. Attention should be paid during the installation of the system to assure that the materials of construction utilized in the system are compatible with the gas being handled. Distribution systems for these gases require qualification, but their similarity to gas distribution systems used in dosage form facilities means that the basic approach is well defined in the literature. For safety considerations particular attention should to be paid to proper identification of process gas lines throughout the facility (see below and Chapter 5).

Compressed Air. Air that is classified as breathable should receive an intensive qualification effort especially with regard to the verification of "as built" drawings, confirmation of proper identification, as well as any safety- and purity-related issues. The emphasis given to breathable air is due to the number of unnecessary deaths which have occurred in the industry as a consequence of misidentified gas lines. Where air is utilized as a reactant in a BPC operation, it should be treated as described previously under process gases. Instrument air requires the least intensive effort, as the adequacy of the installation can be often confirmed indirectly during the calibration and qualification of the process instrumentation. A single compressed air system could serve as the source for more than one of these air systems simultaneously. In this instance the advice provided for the most critical application is appropriate throughout.

Jacket Services. It is common in BPC facilities, especially those which are reconfigured frequently to accommodate the production of different materials, to have each major vessel equipped with identical utilities, such as: chilled water, plant steam, compressed air and coolant. The use of identical utility configurations on the vessels maximizes the flexibility of the operation, reduces the potential for operator error and simplifies the design of the facility. The control systems for these jacket services on the vessels would also be identical. Under these circumstances the qualification effort is greatly simplified through the use of identical requirements.

Solvent Distribution. Many facilities use one or more solvents repetitively. In these instances, the installation of a dedicated distribution system for the solvent to the various use points can be justifiable. These systems may be lengthy lines from the bulk storage area (tank farm) to the various locations in the facility where the solvent is required. In some cases, a chilled solvent system may be present to provide chilled washes for use in centrifugation. Depending upon the solvent, specialized piping or gasket materials may be necessary to avoid leaks or corrosion of the system. Qualification of these distribution systems is easily accomplished.

Solvent Recovery and Reuse. The reuse of organic solvents in a BPC system is widespread, especially given the increased cost of these materials and the environmental difficulties sometimes associated with their proper disposal. This reuse is achieved through defined procedures for the recovery of the solvents from distillates, extractions, and spent mother liquors. Where recovered solvents are utilized in the production of a BPC, the validation of the recovery process is strongly recommended. The validation of the recovery process would include all steps in the process, and confirm the acceptability of the recovered solvent in the processes it will be utilized in. The validation of the use of recovered solvents could be a part of the development of the process. Repeated recycling of solvents could result in the concentration of trace impurities that could adversely affect reaction chemistry. At the very least, recovered solvents should be subjected to release testing and shown to be comparable to fresh solvent. The complexities associated with the validation and reuse of recovered solvents should not be overlooked.

Multiple Crops. In the crystallization of some BPCs, multiple crops are sometimes utilized to maximize the amount of material isolated. Even where the cost of the materials being isolated is not high, the ability to increase the overall yield through the preparation of second, third or even fourth crops are frequently a routine part of the process. A related technique is to recycle the mother liquors without additional treatment from the crystallization to the beginning of the process. Whether through multiple crops or recycle of the mother liquor, both of these processes result in the concentration and/or retention of impurities. The validation of these practices must be a part of the development effort for the process, and reconfirmed on the commercial scale.

Catalyst Reuse. Precious and semiprecious metals and other materials are often utilized as catalysts in the conduct of certain chemical reactions, e.g., hydrogenation. While the quantity of catalyst required in any particular reaction is quite low, the cost of these metals is such that recovery is mandated. As the amount of catalyst required to support the reaction is generally supplied in excess it is frequently possible to return the catalyst to the start of the process step without loss in effective yield. The reuse of the catalyst in this manner must be supported by appropriate development work.

Waste Treatment. The nature of the materials, by-products, and solvents utilized in the preparation of BPCs ultimately results in any number of waste treatment problems. The validation of these treatments is certainly *not* a cGMP required activity. Nevertheless, consideration should be given to those activities to ensure their reliability. Such efforts can aid in attaining environmental approval for the facility.

IN-PROCESS CONTROLS

BPCs resemble other types of products validated in the pharmaceutical industry in that they utilize various in-process controls to support and monitor the process through its execution. Typical controls that might be a part of a BPC process include:

Material Specifications. The controls of reactants, solvents, intermediates and finished materials employ formal specifications for key parameters. The importance of these controls increases toward the end of the synthesis and any of the controls that follow the BPC step are certainly important enough that the efficacy of limits set for these controls should be a major part of the developmental process. Foremost among the considerations in the latter process steps should be the impurity profile of the key intermediates (see following paragraph). Physical parameters (size, shape, crystalline form, bulk density, static charge, etc.) of the finished BPC are sometimes considered less important than chemical purity. When the BPC is formulated in a solid or semisolid dosage form, these physical parameters may assume far greater significance.

Purity Profiles. Within the specification parameters, prominence is often given to the estabishment of purity profiles for the key intermediates and finished goods. The FDA mandates the identification of all impurities with a concentration greater than 0.1% and generation of safety and other critical information for impurities at levels of 0.5% or higher (5). The establishment of purity profiles for the final BPCs provides for confirmation of the safety of the active material. It is often beneficial to establish purity profiles for intermediates earlier in the synthesis to prevent the carryover of impurities to the finished BPC. The maintenance of the purity profile mandates that a careful evaluation of process changes and potential alternate suppliers of solvents, raw materials, intermediates, and BPCs be made. The analytical method development and synthetic chemistry skills required to obtain the necessary data on impurities meeting the FDA's criteria are substantial. These efforts are well rewarded in an expanded knowledge of the process chemistry and analysis that can assure the quality of the desired active moiety.

Vendor Support to Validation. A common practice in BPC production is the subcontracting of certain chemical steps to outside suppliers. As is the case with subcontracted production for dosage forms, the owner of the NDA or DMF maintains responsibility for the validation of the process and must secure the cooperation of the subcontractor in the performance of any supportive qualification/validation activities. Agreement to this arrangement should be a precondition to the awarding of the contract to the supplier.

Supplier Quality Evaluation and Audits. Suppliers of intermediates, reactants, solvents and other materials should be subjected to the same types of evaluation utilized for other dosage forms. The extent of the assessment should vary with the importance of the material to the process. Precedence would be given to those materials whose purity would have an increased impact on the finished BPC. Where the material being produced by the vendor has direct impact on the BPC's quality, as would be the case for chemical intermediates, a more intensive approach is required. Periodic audits of these key suppliers should be a part of the overall quality assurance program.

Sampling Plans. Obtaining samples of finished BPCs or their intermediates presents the same difficulties encountered in the sampling of any similar material. When samples are taken of powder or crystalline materials, questions regarding the uniformity of the material being sampled must be addressed before the results of the sampling can be considered meaningful. BPCs that are dried in rotary or fluidized bed dryers may be blended sufficiently as a result of the drying process. However, where tray dryers are utilized, a final blending of the dried material may be required before sampling for release to the next stage of processing. In certain instances an intermediate or finished material will not be isolated as a dry powder but will be released as a solution in an appropriate solvent. Under these circumstances concerns regarding the sampling of the material are minimized.

Particle Sizing. Milling and micronizing are common activities in the final stages of BPC manufacture. These procedures are utilized where the BPC producer has committed to providing a particular particle size for use in the formulation. Given the importance of particle size in many final dosage forms, these processes should be validated where present. Control of the final particle size for finished BPC should not rely on the milling/micronizing step alone. Control over the crystallization procedure is generally necessary to minimize the variation in the material that is to be sized in the mill. It should not be assumed that the milling/micronization procedure will be tolerant of a wide range of materials and still provide a consistently sized finished BPC product. The uniformity of materials is sometimes improved by passage through a particle sizing procedure or sifter, but this step alone should not be considered sufficient to achieve a uniform mix of the material prior to sampling (see prior paragraph).

Reprocessing. There is occasional need to reprocess an intermediate or finished BPC in order to alter its crystal size, reduce impurities, or otherwise recover off-specification material. Where these processes are utilized, their inclusion in the validation program is essential. FDA requirements on reprocessing and reworking of materials require the validation of any material reclaimed in this fashion. This is most readily accomplished as a part of the developmental process.

CLEANING VALIDATION

A comprehensive discussion of cleaning validation is beyond the scope of this chapter, the reader should refer to other sources on cleaning validation for details of this activity (18–20). Within the context of this chapter, only those aspects of cleaning validation unique to BPC production will be presented. Additional guidance can be found in the FDA's BPC Inspection Guide (5).

Boil-outs. Commonly used to clean BPC equipment, boil-outs entail the introduction of the solvent (it could be water) used in the just completed process, and heating it to reflux. The expectation is that the evaporation/condensation will result in the dissolution of any residue on the equipment in the solvent. This will remove it from the internal surfaces that are ordinarily inaccessible for direct cleaning and thus clean them. Boil-outs are also utilized as one of the last steps in preparation of equipment for the start of a process or campaign.

Lot-to-Lot Cleaning. As the production of BPCs often requires that solvents and materials with substantial toxicity must be employed, cleaning of the equipment after completion of the process has the potential for exposure of the worker to those materials. For this reason, it is common in BPC facilities to include some basic forms of waste treatment and equipment cleaning directly into the process in an effort to minimize worker exposure later on. In addition to these measures, many processes include the reuse of equipment and retention of materials in the equipment without cleaning. A typical instance would be leaving a heel in the centrifuge at the completion of the batch, thereby eliminating cleaning of the centrifuge after each batch. The retention of the heel must be validated as it represents a portion of the first batch, which may now become a part of subsequent batches. In fact each batch in the entire campaign is potentially mixed with material from every prior batch! In this manner, the amount of cleaning required between batches of the same reaction step would be reduced. In those facilities, where a process train is essentially dedicated to the same reaction step over a long period of time, the equipment and process are specially designed to minimize batch-to-batch cleaning of the equipment. There are of course instances where the presence of even trace quantities of finished material at the start of the reaction may create an undesirable outcome, in those circumstances the equipment must be cleaned after the completion of each batch. Sparkler and other filters used to recapture catalysts, activated carbon used for decolorization and by-products may require cleaning after every batch.

Campaigns. The production of a number of batches of an identical synthesis in the same equipment is common in the manufacture of BPCs. As mentioned earlier in relation to the qualification of equipment, production in a campaign mode may require the partial reconfiguration of the equipment train to allow for a new campaign. This may be a reaction leading to the same or a different BPC. To allow for campaign usage, the extent of cleaning required will generally be far greater than what is carried out between batches of the same process step. Cleaning limits for campaign cleaning are generally tighter than those applied for batch-to-batch cleaning. It is beneficial in campaign cleaning to follow a defined plan for changeover from one product to another.

Sampling for Residuals. In order to determine whether a piece of equipment has been appropriately cleaned sampling is performed. Here again, the particular nature of the BPC materials makes for a more difficult situation. In dosage form manufacturing, relatively few of the materials likely to be retained on the surface of the equipment pose any substantial risk to the worker. In those dosage form processes where toxic or potent materials are handled, the design of the equipment with smooth surfaces, rounded corners, sanitary fittings, etc., reduces cleaning difficulty. The same equipment design principles make sampling of pharmaceutical equipment relatively simple due to provisions for access and inspection. The bulk of BPC equipment is designed to operate under more aggressive conditions, and cannot always integrate the design features so commonly found in their pharmaceutical counterparts. Moreover, worker safety becomes a far greater concern, as the solvents and materials are not conducive to direct exposure to the employee. Sampling of BPC equipment may be restricted to fewer locations, and those locations are generally not in the most difficult to clean or "worst case" locations. For this reason, the residual limits for BPCs may need to be lower to accommodate the uncertainty of the sampling that can be performed.

COMPUTERIZED SYSTEMS

The application of computerized systems in the pharmaceutical industry is perhaps greater in BPC processing than in any other. DCSs have been utilized for many years in the control and regulation of chemical process plants. Their adaptation to BPC preparation is straightforward. The validation of computerized systems in the pharmaceutical industry has been extensively discussed, with the constant recognition that their extensive usage in BPC production was a given (21,22). Industry and regulatory guidance having always recognized this fact, this chapter could not hope to do justice the subject which has filled several textbooks on its own. The reader is encouraged to follow the recommended practices of PDA, PhRMA and GAMP (see Chapters 46 and 47).

PROCEDURES AND PERSONNEL

Where computerized systems are not utilized for the execution of the chemical synthesis, the chemical operator, following detailed batch records is responsible for the operation of the equipment. The batch records must provide for sufficient detail to ensure that the worker can safely and properly perform the desired actions. In certain larger process trains, more than one operator will work simultaneously on the same batch. Provided that their activities are closely integrated, there is little problem with this type of approach. The personnel must be trained in their jobs and records of the training must be retained by the firm.

VALIDATION OF STERILE BULK PRODUCTION

The preparation of BPCs which must be sterile upon completion of their synthesis and purification is a common activity in the pharmaceutical industry and increasingly also common in biotech processes (see Chapter 23). The validation of sterile BPCs represents one of the more difficult activities in the entire spectrum of validation. Not only must the final material meet all of the physical and chemical requirements associated with other BPCs, but must also be free of microorganisms, endotoxins and particulates. In doing so all of the considerations for validation of BPCs outlined in this chapter must be addressed, with added concern for sterilization, environmental control, aseptic technique and other subjects associated with the production of sterile products. The following sections address those issues relating to sterile BPCs that are somewhat different from either the validation of non-sterile BPC production or the validation of other sterile materials.

Product Sterilization and Sterility Assurance. The predominant method of sterilization for BPCs is by membrane filtration. This filtration will require validation in accord with regulatory expectations. Adaptations to the common filter validation methodologies may be required for certain solvents and/or antibiotic solutions. Subsequent to the filtration step, the succeeding unit operations must be carried out using facilities, equipment, and methods designed to prevent the ingress of microorganisms. The remainder of this section reviews considerations relative to sterile BPC preparation under these constraints.

Closed Systems. Central to understanding much of what is presented below is recognition that BPCs, whether intended to be sterile or not, are primarily produced in closed systems in which the reaction, separation, and purification unit operations take place. A joint PDA/PhRMA task force has defined a "closed system" as:

> A system which is designed to prevent the ingress of microorganisms. A 'closed' system may be more accurately defined by characteristics of its operation than by a description of its physical attributes, as these will vary from system to system (23).

A "closed" system:

- Is sterilized-in-place or sterilized while closed prior to use

- Is pressure and/or vacuum tight to some predefined leak rate
- Can be utilized for its intended purpose without breach to the integrity of the system
- Can be adapted for fluid transfers in and/or out while maintaining asepsis
- Is connectable to other closed systems while maintaining integrity of all closed systems (e.g., Rapid Transfer Port, steamed connection, etc.)
- Utilizes sterilizing filters which are integrity tested and traceable to each product lot.

Closed systems provide for complete separation between the environment in which personnel (uniformly accepted as the primary source of contamination in aseptic environments) are separated from that in the materials are processed. Theoretically, if a sterile BPC could be processed in its entirety within closed systems, there would no possibility of microbial contamination. In marked contrast to the "closed system" is the "open system," perhaps best defined by what it is not. Essentially, an "open system" lacks one or more of the features of a "closed" system, thus leaving it vulnerable to the potential ingress of contamination. One substantial issue associated with these definitions is establishing that a system remains "closed" over the length of the production campaign.

Facilities. The production of sterile BPCs requires a composite of design features drawn from both sterile dosage form and BPC production. Ceiling, walls, and floors are composed of materials that can be subjected to frequent cleaning with disinfectants. Pressure differentials are provided to prevent the ingress of contamination from less clean areas into critical processing areas. In order to perform the reactions and separations necessary to prepare and isolate the BPC, processing equipment not generally associated with aseptic environments must be introduced. Centrifuges and crystallizers must be adapted for use in an aseptic area. The finished facility is most certainly a hybrid, as compromises are inevitable to accommodate the essential requirements. The case can be made that if the production systems are perfectly "closed," then concerns relative to facility design required for asepsis would be lessened. The authors are aware of several sterile bulk production facilities in which only a small portion of the system is actually located in an aseptic environment. Certainly, "open systems" must approach the highest levels of aseptic design in order to be successful in operation.

Environmental Classification. The environments in which sterile BPC production is executed can vary with the degree of closure provided by the equipment. "Closed" systems as described earlier have been successfully operated in ISO 8 (EU Class D) or unclassified environments. Systems that are open, are generally located within Class 100 (EU Class A) where product is exposed, and surrounded by ISO 8 (EU Class B or C).

Utilities. There is very little difference between utility systems for a sterile bulk plant and those found in a typical BPC facility. The only differences might be utilities uncommon in a BPC plant such as WFI and Clean Steam. The validation requirements for these systems have been well defined in the literature and need little mention here.

Layout. The layout of a sterile bulk facility will again be a hybrid of those found in a conventional BPC plant and a sterile dosage form facility. There will be nesting of classified environments, with critical activities performed in the areas of highest classification. Pressure differentials are employed between clean areas and those adjacent, less clean areas. The design features are drawn primarily from the dosage form facility model with adaptations to accommodate the generally larger equipment required for bulk production.

Isolation Technology. The use of isolators and closed systems for the production of sterile bulks is strongly recommended. As with any aseptic process, the sterility assurance level associated with a sterile bulk material is closely related to the extent of direct human intervention with the material. Isolators and closed systems minimize the need for personnel contact with critical surfaces and thus minimize the potential for contamination of the sterile materials from human-borne microorganisms. It should be also noted that isolation technology can be useful in the containment of potent compounds as many BPC intermediates and finished materials often are. Isolation technology is a rapidly evolving area and the reader is encouraged to stay abreast of current developments (21,25) (see Chapters 19 and 29).

Sterilization-in-Place. Closed systems such as process vessels, dryers, centrifuges, isolators and other items should be subjected to a validated sterilization procedure which assures that all internal surfaces have been rendered free of microorganisms (see Chapter 14). SIP procedures reduce the number of aseptic manipulations necessary to ready the equipment for use in the aseptic production processes and as such are considered preferable to aseptic assembly of systems from individually sterilized components (26). The SIP procedure should allow the system to maintain sterility until ready for use without aseptic manipulations. SIP procedures could employ steam, gas, dry heat, radiation, chemical agents or other validateable sterilization procedure.

A brief overview of some of the various SIP methods available and their validation follows:

1. *Steam*. Primarily utilized for systems composed of closed vessels, with interconnecting piping. It has some similarity to empty chamber studies in steam sterilizers. Important parameters to confirm are appropriate time–temperature conditions throughout the system. Emphasis is placed on the removal of air and condensate from the system, strict adherence to the defined sequence for the sterilization procedure and inclusion of methods for the protection of the system between sterilization and use.
2. *Gas*. Utilized for systems that cannot withstand either the temperatures or pressures employed in steam sterilization. Critical parameters for sterilization are time, temperature, relative humidity and gas concentration. Gases in widespread use include ethylene oxide, peracetic acid and hydrogen peroxide. Gas sterilization is most often encountered with isolator systems, freeze dryers and other systems which have limited ability to hold pressure (see Chapters 16, 17, and 19).
3. *Dry Heat*. Employed in specialized systems where the presence of high temperature for the process is commonplace, i.e., spray dryers, flash dryers and

similar equipment. Confirmation of time–temperature conditions in the equipment is critical to the validation (see Chapter 15).

4. *Radiation*. Radiation sterilization is most commonly utilized for flexible packaging components that can be sterilized while closed prior to filling. The validation of radiation sterilization relies on confirmation of the delivered dose to all portions of the materials, and confirmation of stability after the treatment (see Chapter 18).
5. *Chemical*. Many of the strong acids, strong bases, chemical solvents and other chemicals utilized in the preparation of sterile BPCs have the ability to reliably destroy microorganisms. These materials because of the extreme pH, or other aspects of their chemical structure can effectively sterilize processing equipment. As their use in the system will generally mandate that the equipment surfaces can be exposed to these materials for extended periods of time, their use as a sterilizing method for the equipment is facilitated. Concentration and duration of contact are the critical parameters that must be confirmed in the validation of these treatments.

Aseptic Processing. The validation of the sterile bulk process follows the general approach described earlier for non-sterile bulks. The overall process can be divided into a series of unit operations that can be addressed individually or in groups. This approach can be used equally well for aspects of the chemical reaction, purification, physical processing (i.e., milling, sieving, etc.) or aspects related to sterility assurance. A comprehensive treatment of validation methods for validation of aseptic processing for sterile bulks has been developed by a joint PDA/PhRMA task force (23). This document embraces such aspects of the validation as: the use of closed or open systems for processing; materials to use in the conduct of the simulation; sampling and testing of materials; duration of simulation, simulation size, campaign production and acceptance criteria to be employed. Producers of sterile bulks are already familiar with the contents of this document, and the interested reader is encouraged to read this guidance in its original context (see also chapter 23).

CONCLUSION

This chapter has provided an outline of validation considerations relative to the production of BPCs. This is a subject that has only recently become of interest to the pharmaceutical community. The authors while familiar with both validation and bulk pharmaceutical processing have undoubtedly mentioned any number of issues which may not yet be embodied in validation protocols within operating companies. We have included these issues to ensure completeness in the presentation, not to suggest that they be included in every validation effort. As time passes, the industry will gain experience with the validation of BPCs and will perhaps exclude some of these issues, while including other aspects we have not identified. Our intent in this effort has always been to integrate common validation practices with the unique aspects of bulk pharmaceutical manufacturing. By no means do we expect this to be the definitive effort on this complex subject. The reader is encouraged to monitor industry and regulatory developments relative to BPC validation as substantial changes in cGMP requirements for BPCs appear likely.

REFERENCES

1. Food and Drug Administration, 21 CFR, Part 610.
2. PhRMA Quality Control Bulk Pharmaceuticals Working Group; PhRMA Guidelines for the Production, Packing, Repacking or Holding of Drug Substances, Pharmaceutical Technology, Part 1, December 1995, Part 2, January 1996.
3. Food and Drug Administration, Guidance for Industry, Q7A Good Manufacturing Practice Guidance for Active Pharmaceutical Ingredients, 2001.
4. Food and Drug Administration, Guideline on General Principles on Validation, 1987.
5. Food and Drug Administration, Guide to Inspection of Bulk Pharmaceutical Chemicals, 1994.
6. Food and Drug Administration, 21 CFR, Part 211.
7. Food and Drug Administration, Guidance to Industry: Manufacturing, Processing or Holding Active Pharmaceutical Ingredients, March 1998.
8. Pharmaceutical Inspection Convention, Internationally Harmonized Guide for Active Pharmaceutical Ingredients—Good Manufacturing Practice, September 1997.
9. Agalloco J. The validation life cycle. J Parenter Sci Technol 1993; 47(3):142–7.
10. Food and Drug Administration, Guide to Inspections of Oral Solid Dosage Forms Pre/Post Approval Issues for Development and Validation, January 1994.
11. Agalloco J. Validation—yesterday, today and tomorrow. In: Proceedings of Parenteral Drug Association International Symposium, Basel, Switzerland, Parenteral Drug Association, 1993.
12. Agalloco J. Computer systems validation—staying current: change control. Pharm Technol 1990; 14(1):38–42.
13. Agalloco J. Personal Communications—1972–1990.
14. Federal Standard 209E, Airborne Cleanliness Classes in Cleanrooms and Clean Zones, September 1992.
15. EU Guide to Good Manufacturing Practice for Medicinal Products, Annex 1—Manufacture of Sterile Medicinal Products.
16. Meltzer T. Pharmaceutical Water Systems, Tall Oaks Books, 1996.
17. Artiss D. Water systems validation. In: Carleton F, Agalloco J, eds. Validation of Aseptic Pharmaceutical Processes. New York: Marcel-Dekker, 1986:29–46.
18. Agalloco J. Points to consider in the validation of equipment cleaning procedures. J Parenter Sci Technol 1992; 46(5):163–8.
19. Madsen R, Agalloco J, et al. Points to consider for cleaning validation. PDA J Pharm Sci Technol 1998; 52(Suppl. 6):1–23 (PDA Technical Report 29).
20. Voss J, et al. Cleaning and Cleaning Validation: A Biotechnology Perspective. Bethesda, MD: PDA, 1995.
21. Harris J, et al. Validation concepts for computer systems used in the manufacture of drug products. In: Proceedings of the Concepts and Principles for the Validation of Computer Systems in the Manufacture and Control of Drug Products, Pharmaceutical Manufacturers Association, 1986.
22. Kemper C, et al. Validation of computer-related systems. PDA J Pharm Sci Technol 1995; 49(Suppl. 1):S1–7 (PDA Technical Report 18).

23. Agalloco J, Lazar M, et al. Process simulation testing for sterile bulk pharmaceutical chemicals. PDA J Pharm Sci Technol 1998; 52(1):1–15 (PDA Technical Report 28).
24. Akers J, Wagner C. Isolation technology: application in the pharmaceutical and biotechnology industries. Interpharm, 1995.
25. Coles T. Isolation technology—a practical guide. Interpharm, 1998; 451–82.
26. Agalloco J. Sterilization in place technology and validation. In: Agalloco J, Carelton FJ, eds. Validation of Pharmaceutical Processes: Sterile Products. New York: Marcel-Dekker, 1998.

34

Validation of Recovery and Purification Processes

Rodney E. Thompson, Susan Dana Jones, Sheila Magil, and Howard L. Levine
BioProcess Technology Consultants, Inc., Acton, Massachusetts, U.S.A.

INTRODUCTION

Pharmaceutical biotechnology combines microbiology, chemistry, traditional pharmaceutical technology, and biochemical engineering with advances in genetic engineering technology to prepare proteins and other biologic products as therapeutic or diagnostic agents. These biologics are produced in a suitable host cell by fermentation or cell culture processes. Cells from a qualified cell bank are expanded in culture until sufficient numbers are obtained for the desired production scale. During culture, the product of interest may be either retained intracellularly or secreted into the culture medium. If the product remains inside the cell, then the cells must be harvested, disrupted, and the debris that is created removed to yield a particulate-free extract for further purification. If the product is secreted, the cells must be separated from the conditioned culture medium prior to purification. These process steps, shown schematically in Figure 1, include initial product recovery followed by refolding (if necessary), isolation, purification, viral inactivation (if necessary), and finishing operations. Each of the process stages shown in the figure may be achieved using a variety of unit operations, as will be discussed. The validation of a biomanufacturing process encompasses validation of each of these unit operations separately and in combination to demonstrate that the process can reliably and reproducibly produce the desired product.

The goal of process validation is to demonstrate that a process, when operated within established limits, produces a product of appropriate and consistent quality. The most efficient way to achieve adequate process validation is to first determine the CQAs, namely the acceptable range of values for each dependent process variable of interest in the intermediate process streams leaving that particular process step. CQAs can be viewed as the measurable results of a particular process step and the acceptable range for each constitute the necessary specifications for that process step. The next step will be to find the CPPs, those independent process inputs into a process step that affect the process step results—namely the CQAs. The CPPs are a subset of all the independent variables related to a process step; the CPPs are just those independent variables that directly determine the values for the CQAs. During process validation, the CQAs and the CPPs should be identified and, based on sound scientific principles, appropriate studies should be performed to demonstrate that the appropriate ranges for each can be met on a reliable and consistent basis.

Validation is a scientifically rigorous and well-documented study which demonstrates that a process or piece of equipment consistently does what it is intended to do. Due to the complex nature of proteins and the relatively short histories of some biopharmaceutical manufacturing process components, it is often difficult to fully characterize a biologic product. Thus, final product testing alone is frequently insufficient to ensure consistent manufacture. Therefore, the processes used for the recovery and purification of proteins must be designed and validated to remove or clear potential contaminants. The contaminants, which may arise from source material, equipment or reagents, can include endotoxins, viruses, nucleic acids, and proteins, as well as media constituents, process chemicals, ligands leached from chromatography media, and modifications or inactive forms of the protein itself.

Validation should be considered as early in the development of a process as is practical. In this way, data required for validation can be collected during development studies and the production of batches for clinical studies. For all new biotherapeutics, the evaluation of the product in humans under carefully monitored clinical trials provides the ultimate test of the safety and efficacy of the product.

Before validation is begun, the biologic product should be defined in terms of its physical and biological characteristics. Once the product has been fully characterized, specifications should be established to ensure uniformity, and the required level of purity should be established based on the indicated use of the product. Assays used to determine product purity should be validated to ensure that the sensitivity of the analytical test methods permit accurate detection and quantitation of the product as well as impurities. In addition, each of the facility/utility systems used to support the

---

Abbreviations used in this chapter: BHK, baby hamster ovary; CHO, chinese hamster ovary; CPP, critical process parameter; CQA, critical quality attribute; DOE, design of experiment; DQ, design qualification; EM, electron microscope; EU, European Union; FDA, Food and Drug Administration; HCP, host cell proteins; HEPA, high efficiency particulate air filter; HPLC, high-performance liquid chromatography; ICH, International Conference on Harmonization; IPP, independent process parameters; IPTG, isopropyl-beta-D-thiogalactopyranoside; IQ, installation qualification; LAL, limulus amebocyte lysate; NIST, National Institute of Standards and Technology; NMWCO, nominal molecular weight cutoff; OQ, operational qualification; P&ID, process and instrumentation drawings; PCR, polymerase chain reaction; PEG, polyethylene glycol; PQ, performance qualification; PQA, preferred quality attributes; Q-PCR, quantitative polymerase chain reaction; RPM, revolutions per minute; SOP, standard operating procedure; TOC, total oxidizable carbon; USP, *United States Pharmacopeia*; WFI, water for injection; WHO, World Health Organization.

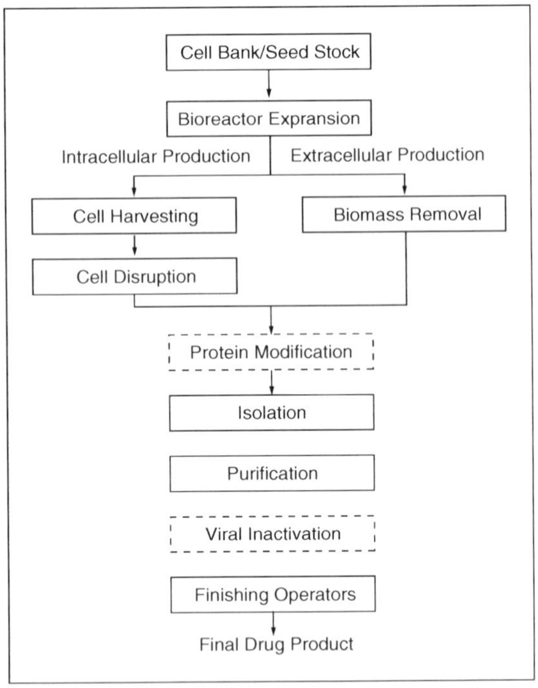

Figure 1 Major stages in the manufacture of a biopharmaceutical product.

manufacturing process must be validated. Finally, all of the manufacturing process equipment must be validated, through at least an OQ as discussed below. Once all of these criteria are met, validation of the manufacturing process itself can begin.

Complete process validation of biopharmaceutical manufacturing will include the validation of all process chemicals and raw materials used in each unit operation, validation of all supporting facilities and utilities necessary for the manufacturing process, qualification and validation of all process equipment, validation of each individual unit operation, and validation of the entire process as it is intended to be operated at commercial scale.

Process equipment qualifications are normally broken down into DQ, IQ, OQ and PQ. The DQ documents that the equipment ordered is consistent with a design already shown to meet the applicable user's requirements (1). The IQ verifies that the equipment is complete, installed correctly, calibrated and meets the design specifications. The OQ confirms each equipment component functions as specified and that all of the components work together correctly. The PQ establishes that the equipment functions as specified for the individual product(s) that it will be used to produce.

While much has been written in the past regarding process validation (c.f. Ref. 2), advances in manufacturing technology, the increasingly complex nature of biotherapeutics products, and new safety concerns regarding potential impurities or contaminants in these products has made the need for effective and complete process validation greater than ever. Process validation of a biomanufacturing process will establish that the process is effective and reproducible and that the final product produced meets all established release specifications. In process validation, it is important that protocols clearly specify the procedures and tests to be conducted, the data to be collected, and the acceptance criteria. The purpose for which data are collected must be clear, the data must reflect facts, and the data must be collected carefully and accurately. The protocol should also specify a sufficient number of sequential replicate process runs to demonstrate reproducibility and to provide an accurate measure of variability among successive runs. All important process variables should be identified, monitored, and documented. Analysis of the data collected will establish the variability of process parameters for individual runs and will establish whether or not the equipment and process controls are adequate to assure that all of the product specifications will be met.

RECOVERY AND PURIFICATION PROCESS STEPS

Cell Harvesting

As shown in Figure 1, for those products produced intracellularly in the production host, the first recovery step after bioreactor expansion are cell harvesting and cell disruption. In cell harvesting, the product-containing cells are isolated from the fermentation or cell culture broth typically using one of the following unit operations: centrifugation, microfiltration or depth filtration.

Cell Disruption

The product-containing cells are typically broken apart using a homogenization unit operation step. Cells could also be lysed with chemical addition, in which case a stirred-tank reaction unit operation would be used.

Biomass Removal

With extracellular product production the cells are simply removed from the product-containing fermentation or cell culture broth. Cell removal can be achieved using centrifugation, microfiltration or depth filtration.

Protein Modification

Some products require some sort of modification after recovery. These modifications may, in some cases, occur after isolation or some of the purification steps. One of a number of potential protein modifications may be necessary for a particular product, including refolding the protein to allow it to achieve its fully-active tertiary structure, chemical or enzymatic cleavage, fusion of a second protein to allow for efficient affinity purification, remodeling of glycoproteins (3) or the covalent coupling of PEG to modify the pharmacological properties of proteins (4). Protein modification is typically a stirred-tank reaction unit operation. While shown in Figure 1 as occurring before the major isolation stages of a process, protein modification, if necessary, can be performed at any stage of the manufacturing process. PEGylation, for example, is most often performed on the purified protein obtained after most or all purification operations are complete. Regardless of where the modification step occurs in the process, complete process validation will include validation of this modification step as well.

Isolation

Isolation of the protein product of interest from the non-protein constituents of the intermediate process stream occurs during the isolation step. Sometimes only purification steps are used in a process train, but these are frequently preceded by an isolation step. Typical unit operations for this step include column chromatography, precipitation, liquid–liquid extraction, microfiltration or ultrafiltration.

Purification

There are typically at least two, and sometimes several, purification steps. These steps are typically column chromatography, ultrafiltration or liquid–liquid extraction unit operations.

Viral Inactivation

Viral inactivation may be achieved by chemical addition, pH change or just heating. In all cases, the process will be held at a specified temperature long enough for the viruses to be inactivated. The relevant unit operation for viral inactivation is stirred-tank reactions, whether or not the process employs continuous mixing, as the solution is assumed to be homogenous throughout the hold time.

Finishing Operations

Finishing operations may include nanofiltration for virus removal, ultrafiltration for product concentration and buffer exchange and sterile filtration for removal of microorganisms.

PROCESS VALIDATION

Process validation starts with an understanding of the process acquired during process development and refined during scale-up and optimization of a process. Solid process validation will rely on information collected during development studies, production of product batches for clinical studies, and extensive characterization work. Once the product has been fully characterized, specifications should be established to ensure uniformity, the process finalized, and all analytical methods required for detection and quantitation of the product and impurities have been validated, validation of the manufacturing process can begin.

Critical Quality Attributes

For each process step there is a set of dependent results, properties of the intermediate process stream leaving that step, that are necessary for the successful production of the final drug product of interest by the overall process. These dependent results, or properties, are often referred to as the quality attributes for that process step. The dependent results listed below in the unit operation parameter tables are examples of some quality attributes. Quality attributes include not only the intrinsic process variables found in the unit operation parameter tables but also all of the potential measurements of any possible impurities of interest (viruses, chemicals extracted from process-contact materials, HCP, prions, etc.).

For each unit operation or process step, there is a subset of quality attributes that are critical to the successful outcome of the operation. These CQAs define the acceptable specifications for the intermediate process stream and, taken together, the CQAs define the ultimate product produced by a particular manufacturing process. In addition to the CQAs, there are a number of additional PQAs which are desirable but are not necessarily critical to the success of a given unit operation or the overall process. For example, while a given liquid–liquid extraction step may have the ability to remove or inactivate viruses, other steps in the process may provide more than sufficient viral clearance to render the viral clearance attributes of the liquid–liquid extraction noncritical. In this case, the extent of viral clearance for the unit operation is a PQA while other attributes, such as removal of HCP or other contaminants may be a CQA. During process validation, it is essential that the intermediate process stream meet all CQA and that these attributes are tightly controlled and reproducible. PQAs, on the other hand, may be important contributors to overall yield or purity but do not need to be validated.

Independent and Critical Process Parameters

For each process step, there are a number of independent variables that can be adjusted to affect the quality attributes for that step. For most process steps, there are literally hundreds of independent variables or IPPs. This includes all the intrinsic properties of the feed streams to the process step, feed stream flow rates, equipment configurations and settings, chemical composition of equipment components, properties and flows of all utilities, properties of the space surrounding the equipment and trace biological and chemical impurities.

As with quality attributes, only a small subset of IPPs are critical to the successful outcome of a particular process step. The CPPs are those that directly impact the CQAs for a given process step and must be tightly and accurately controlled to achieve acceptable results. Strategically, it is important to select only those parameters that have been shown through rigorous experimental testing to be important to control and success of the process step as CPPs. Minimizing the number of CPPs controlling a process step will minimize the extent of validation required for that step.

To assist with establishing which IPPs are CPPs, it is important to have a clear understanding of the desired settings or ranges for each IPP. Each IPP has three nested ranges of relevance. The widest range is the acceptable range; the specifications of all CQAs for the process step are met as long as the IPP is within its acceptable range. Outside the acceptable range, the process step fails in that CQAs do not meet specification. Establishing the acceptable range is sometimes referred to as "testing to the edge of failure," which would normally be done during process development. Imbedded within the acceptable range is the validated range. This is the range for the IPP of interest that will be validated as yielding CQAs that meet their specifications. Imbedded within the validated range is the operating range, which is the range for the IPP that will be specified in the master production record used for manufacture.

Clearly, the operating range must lie within the validated range for the process to be compliant. By establishing a validated range wider than the routine operating range, minor process variations beyond the operating range but within the validated range will not result in a failure of the CQA for the unit operation or overall process. In addition, establishing a validated range that is within the acceptable range ensures that validation studies are not performed at the edge of failure and are therefore less likely to fail. Similarly, since many IPPs may have wide acceptable ranges, it may not be of interest to spend time and effort establishing what the acceptable range truly is, as long as a usable validated range has been established. While IPPs will typically have an acceptable range as wide as possible, some IPPs may have a narrow acceptable range. In such a case, one should establish a validated range that approaches the edge of failure at the boundaries of the acceptable range but remains safely away from the edge of failure. Lastly, it is worth noting that IPPs with narrow acceptable ranges and validated ranges that are just wider than these acceptable ranges are likely to be CPPs. These are the IPPs that must be carefully controlled for the process step to be successful.

PROCESS SCALING FOR DEVELOPMENT AND VALIDATION

Scale-Up/Engineering Runs

Scale-up from the initial laboratory-scale development studies to either pilot or full-scale manufacturing provides the first scale-related information. Some scale differences are intrinsic, such as heat and mass transfer differences, while others are equipment related. Hold times are frequently longer in larger scale systems, which may affect biopharmaceutical product quality. Scale-up of a manufacturing process increases the lot size to the scale that will be used for commercial manufacturing. The first lots produced at the larger scale should be engineering or shakedown runs that allow for the troubleshooting of the new large-scale equipment. This is an important aspect of the process validation project timeline, namely allowing time for engineering runs, as all new manufacturing processes will require them to be operated a few times before the process steps will be predictable and reproducible.

Scale-Down

After scale-up to full scale, it may be valuable to build a small-scale model process, using some of the same laboratory-scale equipment, which mimics the performance of the full-scale system. Small-scale studies on validated small-scale model systems are used to prevent contamination of full-scale production systems with hazardous materials used in clearance studies, to perform process characterization studies that would be difficult or expensive to perform on full-scale systems, to conduct lifetime studies for unit operations and to perform investigative cleaning protocols. Small-scale models may be designed for a single unit operation or for an entire process train. Entire process train small-scale models can be used for evaluating changes in upstream process steps (such as the bioreactor), determining the robustness of a step when considering its impact on the performance of steps further downstream or troubleshooting process deviations in the manufacturing process. The scale-down factor from full scale may be greatest for the later steps in an entire process train small-scale model, as earlier process steps must produce enough material to provide for both sampling requirements and to supply the feed stream for subsequent steps.

Feed streams from the full-scale manufacturing process can be used for small-scale models, while all of the components and materials of construction of the process-contact portions should be the same as the full-scale system. The measurements performed in manufacturing should be replicated in the small-scale model. All of the aspects of the full-scale system that can be replicated with the small-scale model should be. So, for example, if longer hold times exist in the full-scale process then these should be replicated in the small-scale model (even if shorter hold times were possible in the original laboratory-scale development studies).

The unit operations that are most often validated in a scaled-down model process are chromatography steps. To accomplish the scale-down of a chromatography operation in an effective manner, the column media under test should be of the same type and, preferably, the same production lot as that used in the process-scale column. Furthermore, all significant process parameters should be maintained as constant. In its guideline for viral safety, the ICH states that "the level of purification of the scaled-down version should represent as closely as possible the production procedure. For chromatographic equipment, column bed-height, linear flow-rate, flow-rate to bed-volume ratio (i.e., contact time), buffer and gel types, pH, temperature and concentration of protein, salt, and product should all be shown to be representative of commercial-scale manufacturing" (5).

The flow rate used in a validated small-scale model process should be scaled down by the ratio between the cross-sectional area of the production column and the scaled-down column so that the linear velocity remains constant. The column bed height should remain the same as that used in production so that the contact time of the feed solution with the media is not altered. For adsorption separations, gradient slope and volume should be scaled down by the ratio of the total volume of the production column to the volume of the scaled-down column. The ratio of product loaded to column volume should be kept constant and the product should be present during the tests at the same relative concentrations that is present during the actual manufacturing process. Finally, to be valid, the yield and purity of the product recovered from the scaled-down column should be consistent with that of the production column. The extent to which a given column-based separation is scaled down for validation will depend upon the actual production scale and the smallest scale that can reliably reproduce the production process.

In its guide to biotechnology inspections, the FDA addresses the use of scale-down models for process validation and states that when "...scale-up is performed, allowances must be made for several differences when compared with the laboratory-scale operation. Longer processing times can affect product quality adversely, since the product is exposed to conditions of buffer and temperature for longer periods. Product stability, under purification conditions, must be carefully defined. Manufacturers should define the limitations and effectiveness of the particular step. Process validation on the production size batch will then compare the effect of scale-up. Manufacturers may sometimes use development data on the small scale for validation…" (6). Despite this, the guide continues stating that "…it is important that validation be performed on the production size batches…" (6) indicating that sequential full-scale qualification lots are still required as part of validation even if a firm uses scale-down models for most of its process validation.

Process Characterization

Process characterization studies are run under a process validation protocol, often performed using model small-scale systems, to quantify the relationship between IPPs and CQAs for a process step, a series of process steps or an entire process train. The setting of target values for the IPPs during process characterization studies is based on their critical or noncritical designation. Typically those IPPs that have been shown to be noncritical are set to their "worst-case" value at one edge of the validated range and treated as fixed, so that statistically-based DOE factorial designs can be executed to investigate the ranging of the CPPs across their validated ranges. This will validate process robustness by demonstrating that all combinations of CPPs for a process step or series of process steps yield acceptable results and will quantify any interaction terms between the CPPs.

Qualification Lots

The qualification, or consistency, lots are a demonstration under protocol that the entire process will produce a purified drug substance that passes all of the relevant release testing. Normally, additional testing beyond the planned routing testing for commercial lots is performed on the qualification lots, particularly on certain intermediate process streams. The process validation protocol will normally state that the qualification lots will be "a minimum of three consecutive lots," although a minimum of five consecutive lots is typical for most regulatory filings, especially in Europe.

All of the IPPs should be set within their operating ranges for the execution of the qualification lots; typically most, if not all, of the IPPs are set at their target values near the center of the operating range.

UNIT OPERATIONS

This section presents most of the recovery and purification unit operations encountered in the manufacture of biopharmaceuticals. A brief description of the unit operation is followed by a presentation of the typical independent variables and the dependent results generated by the unit operation. Scale parameters, the quantities to try to keep constant as the scale of the unit operation changes, are highlighted.

Centrifugation

Description

Centrifuges separate two different phases based on density difference. Typical solid–liquid separations include recovery or removal of cells (mammalian cell culture or bacterial), capture and removal of particulate solids and recovery of product-containing intercellular solids after cell disruption. One popular configuration is the disk-stack continuous centrifuge, which can be envisioned as a cylinder spinning on its center axis with a sequence of disks encircling that axis at an angle. The feed stream flows continuously from the outside of the unit and through the spaces between the stacked disks. Centrifugal forces cause the denser solids to collect on the disk surfaces, while the lighter liquid phase continues toward the center of the spinning cylinder. The solids slowly move along the disk surface to the outside of the cylinder and are collected in the sediment collection space from which they are either continuously or discontinuously discharged. The lighter liquid phase flows continuously to the center of the centrifuge and is discharged under pressure (7).

Parameters

For each unit operation, relevant parameters will be presented in tabular form, as shown in Table 1 for centrifugation. The parameters presented include the independent variables that can be adjusted for the unit operation, the essential scale parameters held constant to assure comparable results as scale varies, key equations governing operation and performance and the relevant dependent results one typically measures. For centrifugation, the independent variables that are controlled include the RPM of the centrifuge (measured in the SI units of inverse seconds), the volumetric feed flow rate, the system pressure (measured in Pascals) and the temperature.

As the centrifugation scale changes, the ratio of the solid sedimentation rate (solids collected per time) to the solid feed rate is generally kept constant. Clearly, the solid sedimentation rate will rise with increasing RPM, while either a larger feed flow rate or an increase in the percent solids in the feed stream will yield a greater solid feed rate. Another centrifugation scale parameter to hold

Table 1 Centrifugation Parameters

Item	Name	SI units
Independent variables	Revolutions per minute	1/s
	Feed flow rate	L/s
	System pressure	Pa
	Temperature	°C
Scale parameters	Solids sedimented/solids fed	(kg/s)/(kg/s)
	Feed flow rate/Σ	(L/s)/m^2
Equations	$G=\Omega^2 r$	
	G=Centrifugal gravity	m/s^2
	Ω=Angular speed	1/s
	r=radius	M
Dependent results	Product yield	%
	Centrate clarity	–
	Particle size distribution	–
	Solids sedimented	kg/s
	Cell lysis	%

constant is the feed flow rate divided by Ambler's Sigma factor, which is the required surface area for the same amount of sedimentation in a gravity settling tank (8). In scale-up or scale-down, sedimentation performance should be the same if the value of Feed flow rate/Σ is the same for two pieces of equipment. This is a widely used criterion for the comparison of centrifuges of similar geometry and liquid-flow patterns developing approximately the same centrifugal gravity; however, it should be used with caution when comparing centrifuges of different configurations (9). Derivations of other equations of interest as well as diagrams of disk-stack centrifuges can be found in standard handbooks (9).

Table 1 also presents the equation for centrifugal gravity, which is the product of the radius and the angular speed squared.

Microfiltration and Nanofiltration

Description

Microfiltration uses small controlled pore sizes, typically in the range of 0.1 to 0.5 microns, to physically separate the biopharmaceutical product from microbial and mycoplasmal contamination. Proteins and all smaller molecules pass through the membrane, while cells and large particles are retained at the membrane surface. Nanofiltration uses small controlled pore sizes, typically in the range of 10 to 50 nanometers (0.01 to 0.05 microns), to physically separate the biopharmaceutical product from virus contamination. In nanofiltration, proteins and all smaller molecules pass through the membrane, while viruses and large particles are retained at the membrane surface. For both microfiltration and nanofiltration, pressure is applied on the upstream side of the membrane, forcing the permeate stream through the membrane while retentate is retained.

Both microfiltration and nanofiltration can be operated in either a once-through or a tangential flow mode. In once-through mode, the process stream flows perpendicular to the membrane with most of the liquid flowing directly through. Retained material builds up on the upstream side of the membrane and increases the resistance to flow. In tangential flow filtration, the feed stream flows parallel to the upstream side of the membrane constantly sweeping away retained material.

Parameters

For all microfiltration and nanofiltration systems, parameters to hold constant as the scale changes include the flow rate per filter surface area and the transmembrane pressure as shown in Table 2.

For tangential flow filtration systems, some additional scale parameters should be held constant with scale, including the individual channel height (or the diameter of the hollow fiber used) and the crossflow velocity. If shear-sensitive cells are present, the feed pumps used must provide sufficient crossflow to sweep the membrane surface while not damaging these cells; low-shear pumps are typically used and the transmembrane pressure is routinely kept below 2 psi in these cases. Often the permeate flux is independently controlled with either pumps or valves, to allow time for the tangential flow on the upstream side of the membrane to sweep away any accumulated debris.

Product yield can be determined using HPLC or product-specific assays. When the process step is operated with relatively pure proteins, UV absorbance at 280 nm can be used. Product yield should be consistent over the lifetime of a membrane, as yield changes may indicate variation in product retention. Decreased permeate product yield may be due to a polarization layer impeding flow, aggregation, precipitation or denaturation. The compatibility of the membranes as well as all of the wetted system components with each process stream must be determined. Membrane retention and selectivity must remain consistent.

For all membrane unit operations, membrane manufacturers will ensure that the material of construction meet USP requirements for Class VI Biological Tests for Plastics and are nontoxic per the USP General Mouse Safety Test. Effluent from the filter must test negative for USP oxidizable substances after the appropriate flush volume. The membrane must not add chemical contamination to the product. Removal of preservative, cleaning and storage solutions by the flushing procedure must be demonstrated. While the manufacturer's flushing guidelines for cross-flow velocity and pressure should be used, it may be possible to validate lower flushing volumes than recommended.

Ultrafiltration/Diafiltration

Description

Ultrafiltration uses membranes with very small pores, defined by the NMWCO of spherical proteins that are retained by the membrane; typical ultrafiltration membranes used in bioprocessing have NMWCOs ranging from 1000 to 1,000,000 Daltons. These membranes are designed to retain the biopharmaceutical product while allowing small molecules and buffer salts to pass through. Ultrafiltration is carried out in tangential flow mode. If the retentate volume is held constant by the addition of buffer as permeate is removed, ultrafiltration can be used for buffer exchange in a process known as diafiltration. If, on the other hand, the retentate volume is allowed to decrease as the permeate leaves, then concentration of the protein product will occur. Frequently, these two uses are combined and the product is first concentrated and then diafiltered at the lower process volume.

Scaled-down ultrafiltration setups can be limited in the concentration that can be achieved as a result of minimum working volumes. The low-shear rotary lobe pumps used for larger scale systems may not be available for the low flows of the scaled-down system, limiting the transmembrane pressure to what can be achieved by a low flow rate peristaltic pump. In such cases, it may be possible to normalize the flux to the transmembrane pressure, but processing times will be longer.

Parameters

Parameters for ultrafiltration and diafiltration are similar to those for microfiltration or nanofiltration systems. Ultrafiltration and diafiltration is generally carried out using tangential flow systems where scale parameters such as the individual channel height (or the diameter of the hollow fiber used) and the crossflow velocity are kept constant and the transmembrane pressure is varied (Table 3).

Table 2 Microfiltration and Nanofiltration Parameters

Item	Name	SI units
Independent variables	Load volume	L
	Chase volume	L
	Number of diafiltration volumes	L
	Temperature	°C
	Filtration aids	kg
	Permeate flux[a]	L/s
Scale parameters	Flow rate/filter surface area	$L/s/m^2$
	Transmembrane pressure	Pa
	Channel height or hollow fiber diameter[b]	m
	Crossflow velocity[b]	m/s
Dependent results	Flux	L/s
	Product yield	%
	Permeate turbidity	–
	Shear	N/m^2
	Cell lysis	%

[a] Sometimes independently controlled.
[b] For tangential flow filtration systems.

As shown in Table 3, a simple equation links the final concentration of a solute after diafiltration to the initial concentration of the solute, the diafiltration volume of the new buffer, the system volume and the retention of that component by the ultrafiltration membrane. For low molecular weight solutes, such as buffer salts, R will be zero or very small as the solute is not retained by the ultrafiltration membrane. As the molecular weight of the solute approaches and exceeds the NMWCO of the membrane, R increases until it reaches a value of 1 for a fully retained large molecular weight solute.

Table 3 Ultrafiltration/Diafiltration Parameters

Item	Name	SI units
Independent variables	Load volume	L
	Number of diafiltration volumes	L
	Temperature	°C
	Filtration aids	Kg
	Permeate flux[a]	L/s
Scale parameters	Feed flow rate/filter surface area	$L/s/m^2$
	Transmembrane pressure	Pa
	Channel height or hollow fiber diameter[b]	m
	Crossflow velocity[b]	m/s
Equations	$C_F/C_I = e[(V_D/V_S)(R-1)]$	
	C_F = Final concentration	kg/L
	C_I = Initial concentration	kg/L
	V_D = Diafiltration volume	L
	V_S = System volume	L
	R = Retention of component by the membrane	–
Dependent results	Permeate flux	$L/(m^2\ s)$
	Product yield	%
	Permeate turbidity	–
	Shear	N/m^2
	Cell lysis	%

[a] Sometimes independently controlled.
[b] For tangential flow filtration systems.

Depth Filtration

Description

Depth filtration removes cells and particulate debris from a liquid phase, and as such can be used for either cell harvesting or cell removal. Depth filters are made of high porosity material to allow for passage of the liquid phase through the depth filter with relative ease. Typically, a filter aid will be used to coat the upstream side of the membrane, allowing a greater quantity of cells to be collected per square area of the filter itself.

Parameters

The filter aid and entrained solids collect on the upstream side of the depth filter, creating a filter cake. The height of this cake should be kept constant during scale-up (Table 4). The pressure drop across the filter is a second scale parameter of relevance. The third scale parameter which should be held constant is the solid load that will be collected on the filter per square area. Normally, the cell lysis during depth filtration is negligible, as this is typically a low-shear unit operation.

Homogenization

Description

Homogenization uses a large pressure drop over a short distance to disrupt and break open cells. This pressure drop occurs inside a valve where the feed stream is forced through a small gap between the valve and the valve seat, very quickly accelerating the process stream to a high velocity. Cells are broken apart by the turbulent energy dissipating in the liquid going through the homogenizer valve. This energy generates intense turbulent eddies. In addition, the considerable pressure drop may lead to cavitation, generating further eddies. Homogenization can be carried out in either a single or two-stage valve configuration.

Parameters

For a given type of homogenization valve, the relevant scale parameters are the pressure drop and the ratio of the valve opening to the volumetric flow rate (Table 5).

Stirred-Tank Reactions

Description

Protein refolding is covered as an example of a stirred-tank reaction. Overproduced proteins from foreign hosts are often recovered as refractile or inclusion bodies, typically one to three microns in size, with fully reduced cysteine residues. After cell disruption, these dense inclusion bodies are easily separated by centrifugation. Next, the inclusion bodies are washed, then they are solubilized and the proteins refolded to obtain the biologically active product. Solubilization agents are chaotropes, such as guanidine hydrochloride, urea or sodium thiocyanate, or surfactants such as sodium dodecylsulfate or Triton X-100. Sometimes, reducing agents are used as well. Refolding occurs when the concentration of the solubilization agent is reduced, typically by either dilution or diafiltration. Aggregates may form if the protein concentration is too high. Finally, oxidation of the cysteine residues is needed to allow for correct disulfide bond formation in the native protein.

Parameters

The centrifugation and diafiltration portions of the process have been covered under those unit operations. For the stirred-tank reactor itself, usually only minimal mixing is required. The relevant scale parameters are shown in (Table 6).

Precipitation

Description

Fractional precipitation is used to remove broad classes of impurities, increase product purity or concentrate proteins. The product may be precipitated or kept in solution. Typically, a precipitation step would be performed as the first purification process step, particularly following centrifugation, filtration or homogenization steps. Precipitation is often carried out in two stages, the first to remove bulk impurities and the second to precipitate and concentrate the product protein. Precipitation is typically executed in standard cylindrical tanks using low-shear impellers. Typically, amorphous precipitates are formed, owing to occlusion of salts or solvents, or to the presence of impurities. As precipitates often have poor filterability, they are normally collected using centrifugation.

Precipitation can be caused by desalting, salt addition, nonionic polymeric precipitation, addition of a miscible organic solvent, adjusting the pH to the isoelectric point of a given protein to minimize its solubility or by increasing temperature (10,11). Salts can to used to precipitate proteins by "salting out" effects. The effectiveness of various salts is determined by the Hofmeister series, with anions being effective in the order citrate > $PO_4^{=} > SO_4^{=} > CH_3COO^{-} > Cl^{-} > NO_3^{-}$, and cations in the sequence $NH_4^{+} > K^{+} > Na^{+}$ (12). Ammonium sulfate is the most commonly used precipitant. The organic solvents most commonly used for protein precipitation are acetone and ethanol; these are typically added with inline mixers to minimize regions of high solvent concentration causing protein denaturation or local precipitation. Nonionic water-soluble polymers such as PEG are also effective precipitants.

Parameters

Precipitation entails chemical addition, initial mixing, nucleation, precipitate growth, flocculation and finally solid–liquid separation. While equilibrium solubility does not change with scale, the kinetics are scale dependent. To have equivalent performance, we desire a precipitate of a given size to form in the same time. The rate-limiting step for precipitation varies, but typically similar results at various scales can be achieved by keeping the ratio of mixing power to volume constant as shown in Table 7 (13).

Liquid–Liquid Extraction

Description

Liquid–liquid extraction occurs with the partitioning of solutes between two immiscible phases. Because few proteins are soluble, let alone stable in organic solvents, the systems of most interest for biotherapeutics are those created by the addition of certain pairs of hydrophilic polymers to aqueous solutions causing a phase separation without the presence of any hydrophobic solvent (14). One common system is created by the addition of both PEG and dextran, the PEG-rich phase being less dense than the dextran-rich phase. Aqueous two-phase systems can also be formed with PEG and various salts (14).

Most proteins, as well as particulate matter and cellular debris, partition into the dextran-rich phase. The breakthrough in the usefulness of phase partitioning came with the attachment of ligands, mainly dyes, to PEG, which attracts specific proteins into the PEG-rich phase in a technique referred to as "affinity partitioning." By arranging the partitioning to occur in multiple stages, using classical counter-current distribution, high product purity and yield can be achieved (15).

Parameters

Protein partitioning in aqueous two-phase systems is strongly affected by pH and in polymer–polymer two-phase systems by the concentration of other ionic species as well.

If the two phases are separated by use of centrifugation, then the section on that unit operation is applicable here.

PROCESS EQUIPMENT QUALIFICATION

As an essential prerequisite for the validation of recovery and purification processes, all process equipment used for these steps should be qualified and validated. For recovery and purification processes considered here, this equipment will include systems for centrifugation, microfiltration, nanofiltration, ultrafiltration, depth filtration, homogenization, stirred-tank reactions, precipitation, liquid–liquid extraction, adsorption and sterile filtration, as well as the equipment used for buffer preparation, process monitoring, and product handling. To initiate equipment qualification, all instruments must be properly calibrated to ensure their correct and accurate operation (16) and biosafety hoods should be certified to ensure the integrity of the HEPA filter and the proper circulation of air (17). The validation of equipment

Table 4 Depth Filtration Parameters

Item	Name	SI units
Independent variables	Load volume	L
	Chase volume	L
	Cell concentration in the feed	kg/L
	Temperature	°C
	Filtration aids	Kg
Scale parameters	Filter cake height	m
	Transmembrane pressure	Pa
	Solid load/filter surface area	kg/m^2
Dependent results	Flux	L/s
	Product yield	%
	Permeate turbidity	–
	Shear	N/m^2
	Cell lysis	%

Table 5 Homogenization Parameters

Item	Name	SI units
Independent variables	Valve type	–
	Percentage of cells in feed stream	%
	Surfactant concentration in feed stream	kg/L
	Osmolality of the feed stream	mS
	Temperature	°C
Scale parameters	Pressure	Pa
	Valve opening/flow rate	m^2/L/s
Dependent results	Cells disrupted	%
	Product yield	%

used in biopharmaceutical processing includes the three steps of IQ, OQ, and PQ (18).

Installation Qualification

The IQ of biopharmaceutical recovery and purification process equipment verifies and documents that all aspects of the installation of the equipment adhere to the manufacturer's recommendations, appropriate federal, state, and local safety, fire, and building codes, approved company specifications and design intentions. The IQ demonstrates that the user of the equipment has purchased and installed the correct equipment for the specific task. This document demonstrates that the user has considered the relevant aspects of compatibility of the equipment with the process and that the user has SOPs for keeping the equipment calibrated and in good operating condition through a calibration program, a preventative maintenance program and a spare parts inventory. This document also demonstrates that the user has analyzed the operations of the equipment and determined the level of operator training required by preparing written SOPs covering these activities. Process equipment IQs should contain the following information.

System Application

This section should briefly describe what processes are to be performed and where the equipment is located. As an essential part of an overview of the system, a schematic is included to support a complete understanding and description of the system.

Table 6 Stirred-Tank Reaction Parameters

Item	Name	SI units
Independent variables	Impeller type	–
	Impeller speed	1/s
	Baffles	–
	Temperature	°C
	Solubilization agent concentration, initial and diluted	kg/L
Scale parameters	Mixing power/volume	kw/L
	Impeller diameter/tank diameter	m/m
	Solubilization agent volume/feed volume	L/L
Dependent results	Product purity	%
	Product yield	%

Table 7 Precipitation Parameters

Item	Name	SI units
Independent variables	Impeller type	–
	Impeller speed	1/s
	Baffles	
	Temperature	°C
	Precipitant concentration	kg/L
Scale parameters	Mixing power/volume	Kw/L
	Impeller diameter/tank diameter	m/m
	Precipitant volume/feed volume	L/L
Dependent results	Product purity	%
	Product yield	%

Equipment Summary

A detailed description of the system including an equipment summary (manufacturer, model number, serial number) and a description of the components should be provided. Each component of the system should be listed and described separately with sufficient information to clearly define the system. For example, in a chromatography system, the equipment summary might include feed tanks, tubing or piping, pumps, filters, pressure gauges, valves, detectors, and the column itself. For tangential flow systems, such as those used for cell harvesting or product concentration, the equipment summary should describe the pumps, piping, instrumentation and controllers, holding vessel, and the membrane (type, manufacturer, etc.).

Additionally, the equipment summary should also include the design criteria for the equipment and a description of the review process used to ensure that this design is adequate for the equipment's intended use.

Supply Utility Descriptions

All utilities supporting the process equipment should be described and checked to ensure proper installation. For example, the electrical source (voltage, amperage, etc.) should be listed and checked against local codes and the electrical specifications of the system. If the system requires compressed gases or steam, these utilities should be validated and their quality and source should be described and verified.

Table 8 Liquid–Liquid Extraction Parameters

Item	Name	SI units
Independent variables	pH	–
	Ionic species	kg/L
	Temperature	°C
Scale parameters	Quantity of lighter phase material/quantity of denser phase material	kg/kg
	Flow rate of lighter phase/flow rate of denser phase	L/s / L/s
	Number of extraction stages	–
Dependent results	Partition coefficient between phases	–
	Product purity	%
	Product yield	%

SOPs, Manuals, and Drawings

The title and location of all appropriate manuals should be listed and a checklist prepared to ensure that all the manuals exist and have been referenced in the installation of a piece of equipment. All SOPs relating to the installation, operation, and maintenance of the equipment should be listed. These documents should also contain the P&ID and schematics necessary for installing, maintaining, and repairing the system.

Spare Parts and Service Requirements

A detailed list of recommended spare parts and their location is usually included in the IQ. This spare parts list may either be a separate list or included in the manuals. The IQ should also list and review maintenance procedures to assure that prescribed maintenance can be performed without any detriment to either the process or product.

Operating Logs

A listing of the name and location of log books which document the use of process equipment is usually included in the IQ document.

Process Instrumentation

The type, manufacturer, part number, operating range, specific uses, and calibration schedule of all process instrumentation should be listed. This list should be divided into critical and noncritical instruments. A critical instrument is one whose failure would adversely affect the product's quality or safety. Depending on the system design and complexity, not all instruments are critical instruments. For example, if an ultrafiltration system is equipped with a flow meter but process performance is not a strong function of flow rate, then the flow meter may be considered a relevant but noncritical instrument in this system. The distinction is important because critical instruments will be calibrated and maintained on a more rigorous schedule than noncritical instruments. Also, change orders for a critical instrument will undergo a more extensive examination, and failure of a critical instrument during the process will be reviewed more carefully than failure of a noncritical instrument. All instrumentation on the process system should be calibrated against standards traceable or comparable to the NIST. The IQ should also list the SOPs which describe the calibration procedures for these instruments.

Materials of Construction

Those items which come or may come in contact with the product should be described and verified to be compatible with the product and/or process. All components of the system, including lubricants with the potential for contacting the product, filters, valves, tanks, etc. should be included. Equipment vendors can often provide appropriate compatibility data; however, the user may have to confirm such data with actual process fluids. If materials leach from the system into the product stream, then it should be demonstrated that subsequent process steps remove them.

Operational Qualification

The OQ is documented verification that a piece of equipment or process system, when assembled and used according to the SOPs, does in fact perform its intended function. As with the IQ, the OQ is concerned with the equipment and not with the product or process per se. The OQ demonstrates that the user has tested the equipment and has found it to be free from mechanical defects or design defects before use in the production process.

Before starting the OQ for any process equipment system, the IQ on that system should be completed. Any required calibration for the system should also be completed, and in fact calibration may be part of the IQ. The IQ and OQ for supporting utilities such as water systems, lighting, heating/cooling, and electrical, should also be completed prior to starting the OQ for the process equipment system. The OQ document should include the following information.

Training Verification

It should be verified that operators have received the proper training and are able to operate the equipment as intended by following the appropriate operating SOPs.

Check Automated Components

If the equipment is automated, the tests should verify that the equipment responds to the controller as designed. In addition, automated controllers will need to be validated to meet 21CFR Part 11 standards.

Check Manual Components

The manual elements of the system, such as hand-operated valves and traps, should be checked physically and/or visually to ensure proper operation.

System Integrity

The equipment should be tested to establish that it is capable of operating without leaks. The simplest means of detecting leaks is by visual inspection of the fluid path. Pressure hold tests on the components and piping can identify leaks before attempting operation. Leaks may also be detected in complex systems by demonstrating that the fluid output equals fluid input (fluid mass balance). Membrane manufacturers may be consulted to obtain recommended test procedures and specifications for verifying integrity of filters once installed in systems. If the system is designed to provide biological containment of recombinant organisms, specific tests to verify this containment integrity should be conducted prior to introduction of viable organisms to the system.

Flows/Pressures

Pumps should be tested to show that they deliver the required flow under normal operating conditions. Tolerances should be established for variations in flow rates.

Detectors/Recorders

If the data generated by detectors is to be used in process control, then the acceptable operating range, the limits of linearity response, the reaction time, and the response of the detectors and recorders with operating flow rates should be established.

Filters

Filters and filter housings should be examined to verify that they are appropriate for use with the flow rates and pressures likely to be encountered in the system. They should be suitable for their intended purpose, whether that be sterilization or particle removal. If filters are used for sterilization, then they should be validated as such.

Computer Control

If computer control is to be used in the operation or cleaning of the equipment system, validation of the control software and hardware in the system must be addressed (19). It should be shown that the software functions correctly and is protected from unauthorized alteration. The ability of the system hardware to perform its assigned task should also be shown. A schematic of the control logic, including "if-then loop paths," should be included.

Alarms

All alarms should be tested by simulation of "alarm conditions" either by actually challenging the system or by electronic simulation. For example, a pressure alarm may be tested by increasing the pressure in the system using pumps and valves; alternatively, the high pressure may be simulated by sending the appropriate voltage to the alarm mechanism.

Other Features/Components

Finally, each system may have unique features or components not found in conventional systems used for other applications. Appropriate tests to demonstrate the correct functioning of these features or components should be included in the OQ.

Performance Qualification

PQ of process equipment will establish the reliable and reproducible performance of the equipment in the specific process used to make a particular product. The complexity of the PQ protocol varies widely—from fairly extensive in cases where the application of the equipment to a given process is unique or complex to not necessary at all for simple equipment where the specific process use is fully covered by the OQ conditions and ranges, cleaning and sterilization are covered by separate protocols and any specific aspects of the equipment PQ may be incorporated into the process validation protocols for the entire process.

Two important aspects of equipment PQ that must be performed before overall process validation begins are equipment cleaning (20) and sterilization (21).

Cleaning Validation

Cleaning validation is necessary for all product contact surfaces on the process equipment to prove that there is no lot-to-lot carryover of contaminants or material. The major exception to this is disposable containers or liners which are pre-assembled, dedicated for a single use and then discarded.

A cleaning protocol must be developed, implemented and validated. All equipment must be cleaned prior to its initial use to remove any preservatives or residues from either production or shipment of the equipment. Cleaning between production lots serves to remove any residual protein or chemical components of the previous lot. Cleaning protocols should be developed using the recommendations of the manufacturer with regard to compatible cleaning solutions, as well as contact times, circulation rates and allowed temperatures. Many manufacturers also provide suggestions for rinsing cycles that remove the cleaning agents.

Sodium hydroxide is a commonly used cleaning agent for biopharmaceutical recovery and purification process steps. It can be removed with a WFI rinse using either pH or the conductivity of the rinse water as the measure of cleaning agent removal. Other frequently used measures of nonspecific cleanliness include TOC, total protein assays and product-specific assays. If possible, it is preferred to use specific assays for the chemical of interest, such as an HPLC or ELISA test, as these can be tailored to a known contaminant. If residual material is identified with a nonspecific assay, then it must be assumed to be the most toxic material and potential ways to improve the cleaning process will be harder to determine.

While it is tempting to set the specification for removal of a carryover chemical, contaminant or cleaning agent at the limit of detection of the assay chosen to measure removal, this leads to a relative specification in that as the assay technology improves greater removal is required. A preferred approach is to set a specification based on the calculated potential carryover of the chemical into the dose that would be received by a patient and to show that this amount is at least three orders of magnitude below a level that would have any effect. If the concentration of the chemical that would have an effect is unknown, then consider adopting an industry standard value such as "less than 1 ppm."

Rinse Studies

A rinse of the process equipment includes all of the liquid-wetted product-contact surfaces for that equipment. This is both an advantage and a disadvantage; the advantage is that all the hard-to-reach surfaces are included while the disadvantage is that one cannot tell from which surface the residues found originate. In addition, not all remaining residues will necessarily be removed by a particular rinse solution or cleaning conditions—so one must be careful to verify that a given set of rinse conditions is sufficient to dissolve any residues present.

Swab Studies

Swabs are collected at defined locations and should yield accurate information on the residues present. It is not possible to swab many internal and hard to reach locations, which are frequently some of the hardest to clean, so swabs by themselves do not constitute a sufficient procedure for a cleaning validation. Usually rinse and swab studies are combined to give a more complete assessment of cleaning. If a rinse shows no remaining residue above the acceptance criteria and a swab of some of the most difficult to clean locations show that the rinse has removed all residue at those locations, then the cleaning demonstration starts to be convincing.

Coupon Testing

Coupons, made of the same material as the process equipment, can be exposed to the process conditions inside the process equipment, removed and then extensively tested. The challenge here is for the surface of the coupon to receive cleaning that is no more vigorous than the rest of the process equipment. In addition, the coupon must be added to the equipment so that it does not interfere with the cleaning process. If these two conditions are met, then the coupon surface can be analyzed extensively to further understand the residue removal mechanisms.

Sterilization

Sterilization validation, extensively covered in other chapters, is necessary as part of the PQ for each aseptic process step.

Lifetime Studies

Each process step requires a demonstration of its usable lifetime. This will need to include not only cycling through the entire process step, but all of the cleaning steps as well. All of the product-contact parts of each unit operation must be shown to be chemically compatible with the process stream as well as any flushing, cleaning, sanitizing and depyrogenating agents.

CLEARANCE STUDIES

Purification processes typically contain a number of unit operations designed to purify the protein of interest and to inactivate or remove viral, nucleic acid, immunogenic, and pyrogenic contaminants without affecting the potency and activity of the therapeutic product. In addition to contaminants that may have been present in the initial bioreactor harvest, other contaminants, such as reagents used during purification or ligands that may have leached from chromatography media used in the purification of the product, must also be removed from the product during downstream processing. Successful measurement of elimination or inactivation of contaminants can be determined by specific assays such as radioimmunoassays, enzyme immunoassays, and protein blotting directed toward the contaminants. Therefore, both end-point testing and clearance studies demonstrating the removal of specific contaminants should be included as part of process validation to provide assurance that the process will effectively eliminate or inactivate specified contaminants from the product.

If noxious or infectious agents are to be used during clearance studies, it is unwise to allow these materials to be introduced into a production facility or equipment. These agents may contaminate clinical or commercial product or place manufacturing workers at unnecessary risk. Instead, these studies should be conducted on smaller-scale equipment where the process is accurately reproduced to assure that the data generated can be extrapolated to production-scale equipment (22).

In clearance studies, a particular contaminant is added to the input feed stream on a small scale and the recovery of the contaminant is measured at each stage of the process step such as the column flow-through, product pool, and regeneration fractions using scaled-down columns (22). The addition of the contaminant should be kept to a minimum so the concentration of the feed stream is not significantly changed. The addition of the contaminant should not significantly alter the behavior of product recovery. Measurements for mass balance calculations should be performed on column flow-through (non-binding materials), eluted fractions, regeneration, and cleaning steps. Mass balance during regeneration and cleaning steps is critical in assessing whether or not the column packing material can be reused.

A clearance factor can be calculated as shown in equation (1) by dividing the number of units introduced by the number of units recovered in the product after that step.

$$\mathrm{CF}_i = \frac{I}{O} \qquad (1)$$

In equation (1), CF_i is the clearance factor for the ith step in the process, I represents the number of units introduced at the start of the process step and O represents the number of units recovered after the process step.

Each step in a purification process should be challenged separately so that the clearance of a particular contaminant by each step of the process can then be calculated. In general, the overall clearance factor for a manufacturing process (CF_t) is the product of the clearance factors for each step:

$$\mathrm{CF}_t = (\mathrm{CF}_1 \cdot \mathrm{CF}_2 \cdot \mathrm{CF}_3 \cdots \mathrm{CF}_n) \qquad (2)$$

When radio-labeled tracers are used in clearance studies, the interpretation of clearance factors may be more complex. If the tracer is a homogeneous species, or if the tracer behaves as if it were a homogeneous species in the process under study, then the clearance of the tracer in each step is independent of the sequence in which the steps are performed. In this case, the clearance factors measured at each step in the process may be multiplied together with the resulting product representing the clearance factor for the entire process.

If the tracer used is not homogeneous, then the interpretation of clearance factors may be more difficult. [^{32}P] DNA commonly used in clearance studies is an example of a heterogeneous tracer. It is a chemically diverse population; i.e., the population consists of molecules of different nucleotide sequences of various lengths with a distribution of molecular weights. Many of the separation methods used in the purification of recombinant proteins are insensitive to either the nucleotide sequence or the molecular weight distribution of nucleic acids present in the process stream. For these processes, the assumption that [^{32}P] DNA behaves as a single homogeneous species is a valid one, and the clearance factors obtained at each step may be multiplied together to give an overall clearance factor for the process.

Radio-labeled HCP is another example of a chemically diverse population of molecules which may behave as a heterogeneous population in protein purification processes. The population may consist of several hundred labeled proteins which are heterogeneous with respect to charge, hydrophobicity, thiol content, and molecular weight. Therefore, in these cases, the practice of multiplying clearance factors obtained from each individual step to obtain an overall clearance factor may have no practical significance and care should be taken in interpreting the results of such experiments (23).

The principle contaminants which may require clearance studies are pyrogens, media components, HCP, nucleic acids, viruses (for products produced in mammalian cells), and materials leached from bioaffinity media (e.g., Protein A). Specific information on clearance validation of these potential contaminants are described below.

Nucleic Acids

The concern of potential biohazards from the presence of nucleic acids in parenteral protein preparations led to the introduction of regulatory guidelines that limit the exposure of patients to DNA. Original guidelines from the WHO and FDA placed these limits at 10 to 100 pg per dose per day, based on a perceived concern that potential oncogenes could be transferred to patients from mammalian production cell lines (24–26). However, with experience accumulated to date, the perceived risk associated with specific DNA contamination has been significantly reduced, and DNA is now considered a cellular contaminant rather than a dangerous component that must be reduced to very low levels. Therefore, the initial guidelines have been relaxed such that a limit of 10 ng per dose per day is now acceptable under WHO and EU guidelines and may be accepted after discussion by FDA as well (27,28). Despite this relaxation in regulatory guidelines, there is still a requirement to demonstrate removal of nucleic acids from biopharmaceutical products, especially in protein preparations such as antibodies where large doses of protein will be administered to patients. Typically, validation of DNA removal includes monitoring the elimination of source DNA in the process stream at each key purification step and confirmation of removal from the final product using a sensitive residual DNA assay.

Several analytical techniques are available for the detection of DNA in in-process samples and final product. The most commonly used techniques that are sensitive enough to detect the low levels of residual DNA expected in the final product are hybridization of sequence-specific DNA to specific probes (29), methods based on total DNA binding to DNA binding proteins such as the commonly used Threshold® System (Molecular Devices Corporation, Sunnyvale, California) (30), and more recently, Q-PCR to amplify and detect specific sequences (31). PicoGreen® is a commercially available ultra-sensitive fluorescent nucleic acid stain for quantitating double-stranded DNA (Molecular Probes Inc., Eugene, Oregon) that is gaining popularity as a means of measuring minute quantities of DNA in biopharmaceuticals. PicoGreen provides the most sensitive sequence-independent method of DNA detection available today (32). Some guidelines require DNA testing using sequence-independent methods, but Q-PCR provides the most sensitive method of any for detecting residual DNA. Because of higher specificity and amplification potential, Q-PCR has a better signal-to-noise ratio and is therefore able to detect smaller quantities of DNA than the total DNA methods (32). Nevertheless, the total DNA detection method can be very useful in validation studies, especially where many samples must be analyzed to effectively demonstrate removal of DNA. When validating DNA clearance, it is important to design spike-recovery experiments to validate the detection methods prior to validating the process for clearance of nucleic acid, since detection of residual DNA is highly dependent on the sample preparation methods.

To determine the removal of DNA by downstream processing, clearance studies using scaled-down production equipment can be performed using the sensitive assays described above or by using spiked radio-labeled DNA fragments. Because of the highly acidic nature of nucleic acids, they are most readily removed during downstream processing by ion exchange chromatography. For clearance studies, the scaled-down column is run under conditions that closely mimic the production column and fractions are collected during the chromatographic run. DNA is measured by the chosen method in every fraction and the elution profile for nucleic acids compared to that of the product of interest. A recently published report describes the purification process validation for Biogen-Idec's monoclonal antibody product Zevalin (33). To validate DNA removal throughout the process, samples from each purification step were tested for DNA levels using the Threshold Total DNA Assay. The data shows that the host cell DNA is reduced from approximately 3.1×10^7 pg DNA/mg antibody to less than 2.3 pg DNA/mg antibody, well within the acceptable limits for an antibody therapeutic product. By validating that the process consistently and effectively reduces DNA to an acceptable level, the manufacturer can eliminate residual DNA testing as a release test for this product.

In another example of a study to validate DNA clearance during purification of a recombinant protein, 21 consecutive purification cycles were performed using three different anion exchange chromatography media (34). Radio-labeled DNA was spiked into the column load before each cycle and after every five cycles, and the clearance factor for removal of all DNA and DNA of greater than 50 base pairs was determined. For each chromatography media, the clearance factor was consistent throughout the validation study. The average clearance factor for two Sepharose Fast Flow anion exchangers (Q Sepharose and DEAE Sepharose) of approximately 1.5 million was obtained. For DE-52 Cellulose (Whatman, Clifton, New Jersey), the clearance factor was approximately half that of the Sepharose exchangers or 0.7 million (34). Each of these validation studies demonstrated that a final concentration of DNA of less than 100 pg per dose of protein could be reproducibly achieved, which was the limit at the time of these studies.

Host Cell Proteins

HCP are considered problematic in biopharmaceuticals due to the potential to raise the immunogenicity of the intended product or their inherent toxicity. This was considered a major issue and many approaches to determine residual HCP levels in biopharmaceuticals have been developed (35). To measure the clearance of HCP in a process, levels may be measured by using a direct immunoassay of these proteins in the actual process stream. For early stage clinical trial material, some commercial kits are available for measurement of HCP from *Escherichia coli*, CHO cells, and other commonly used production host systems (36). To generate a production cell line-specific HCP ELISA assay as required for product licensure, the production source proteins are prepared from the host organism or cell line which contains a plasmid constructed to have all the DNA sequences except those for the gene encoding the protein product, or from the parent myeloma cell of a hybridoma. These proteins are isolated and polyclonal antibodies prepared against them (37). The resulting antisera are used to develop a cell line-specific immunoassay.

In some cases, it is not possible to raise an antibody to the HCP that does not cross react with the product. In this case, the HCP can be quantified on a SDS-PAGE gel in which the product bands are identified in a Western Blot and the non-product bands are quantified in a matched Coomassie-stained gel. As an alternative, clearance of HCP can be determined by radio-labeling these production source proteins and then adding or "spiking" them into the appropriate crude feed stream or intermediate stream to the column under scaled-down test conditions. The total radioactivity of all the fractions collected is determined and compared to the total loaded radioactivity for mass balance determinations. Fractions containing product are then pooled and the clearance factor calculated. The clearance factor is the ratio of total loaded radioactivity to radioactivity contained in the product fraction. The overall clearance factor for the purification process will vary according to specificity, selectivity, and the number of process steps. Clearance studies are expected to be performed for HCP (38).

Endotoxin

Process steps should not add microbial contamination to the product, particularly near the end of the process. Sterility requirements and maximum endotoxin levels should be specified. Bioburden and endotoxin reduction must be documented. Both are typically removed from product contact equipment by cleaning with 0.5 to 1.0 N sodium hydroxide.

Because of their high molecular weight and highly negative charge, endotoxin and other pyrogenic materials are commonly removed from proteins by either ion exchange chromatography, gel filtration, or ultrafiltration (39–41). Among these methods, ion exchange chromatography is generally most useful in reducing bacterial endotoxin levels, provided that the selectivity of the media is such that co-purification of the bacterial endotoxin and product is avoided. Use of specific membranes to remove endotoxin is also increasing. Membranes specifically designed to remove endotoxin are available from Millipore, Sartorius and Pall as well as numerous other vendors. The LAL assay for gram-negative bacterial endotoxins is sensitive enough for detection of concentrations at least an order of magnitude below levels which will produce a pyrogenic reaction in the rabbit pyrogen test. The LAL test is a compendial test in the USP (42) and has been harmonized under the ICH. The possibility of inhibition or enhancement of the LAL assay by the protein product, however, must be ascertained through validation of the LAL test (43).

Since a sensitive assay is available to detect the presence of pyrogens in in-process samples and final drug preparations, clearance studies demonstrating the removal of pyrogens may not be necessary. Good process control and hygiene, i.e., LAL testing of all raw materials, microfiltration or ultrafiltration of process buffers, and cleaning and sanitation of columns after each use, will minimize the potential for endotoxin contamination and, hence, the need for clearance studies. If it is desired to perform clearance studies, they may be carried out by directly measuring the level of endotoxin and product at each step of the purification as well as monitoring the residual amount present in the product.

Viruses

When mammalian cells are used as substrates to produce a protein product, there is concern that the cell lines may harbor viruses (44,45). Endogenous retroviruses are widespread in animal populations and have been described in species as diverse as reptiles, birds, and many mammals. For example, murine hybridomas used in the production of monoclonal antibodies are known to harbor endogenous retroviruses which may have the potential to transform cells. Other rodent cell lines used in production of human therapeutics such as CHO and BHK have also been shown to contain these endogenous retroviruses. Production cell lines and cell banks are therefore carefully screened for potential viral contamination, but it is expected that viruses may exist in any production cell line at any time. In the absence of a specifically identified viral contaminant in the product cell line, the potential presence of retroviral particles is of greatest concern. In addition, concerns regarding bovine viruses and prions are also increasing among regulatory agencies (46).

The most appropriate way to assure that viruses do not co-purify with product is to test and select production cells and media components that are free from known adventitious viral contamination. Since most cell lines currently used in production are derived from sources that cannot be certified as free of endogenous viruses, and since adventitious agents may enter the production process and propagate in cells, viral clearance studies for products derived from cell culture are essential. Validation of the viral clearance capacity of a biopharmaceutical purification process is essential for determining the viral safety margin for the resulting product. Viral clearance factors for each unit operation must be determined, and the overall clearance provided by the process must be demonstrated. Any process step for which viral clearance is claimed should clear the model virus(es) at least 4 logs, and these steps must be shown to be robust and independent of any variability in processing parameters (47).

Virus clearance is most readily measured by small-scale spiking experiments. Viral clearance should include both virus removal and inactivation and clearance factors; several logs greater than the theoretical titer of infectious virus per dose of product should be demonstrated (48). A theoretical worst case titer may be estimated from EM pictures of the cell culture fluids from which the product is purified. This information, combined with the process yields and the expected dose size, is used to compute the number of viruses which would be carried into the dosage unit if there were no clearance by the purification process (49).

In addition to characterizing the viruses contained in the cell line, it is important to demonstrate that the purification process can remove and/or inactivate those viruses which may be indigenous to the cell line but remain undetected. It is, therefore, desirable to perform spiking experiments with viruses that can be cultivated to a high titer, which have well-established detection assays, and which do not present health hazards.

For proteins produced by recombinant DNA technology or naturally by human cell lines, virus removal or inactivation validation should include a collection of model viruses possessing a range of biophysical and structural features. The viruses used should include enveloped and non-enveloped DNA and RNA viruses which have different diameters and geometries. DNA viruses such as Herpes Simplex 1 (enveloped) and SV-40 (non-enveloped) and the RNA viruses, Sabin Type I Polio (non-enveloped) and Influenza Type A (enveloped), represent typical challenge viruses. When rodent cells such as CHO, BHK, C127, and murine hybridomas are used for production, then Moloney murine leukemia virus may be used as a model retrovirus (22). When choosing an appropriate challenge virus, preference should be given to those viruses which display a significant resistance to physical and/or chemical agents.

Clearance studies similar to those described above for HCP and DNA may be performed by spiking model viruses into the production stream and measuring their removal on scaled-down columns. The clearance of virus particles may also be measured using radio-labeled virus. Radio-labeled virus can be prepared in a similar manner to the preparation of labeled host cells, using [^{3}H]-, [^{14}C]-, or [^{35}S]-labeled amino acids. As mentioned above, care should be taken to prepare labeled virus which is free from molecular weight labeled contaminants. Each stage of the purification process should be individually assessed for its ability to remove or inactivate virus. The overall clearance factor can be determined from individual clearance factors. Care should be taken in calculating the overall clearance factor, however. The assumption that clearance factors of different steps may be multiplied to give the overall clearance factor may not always be valid. Clearance factors may only be additive, for example, if the mechanism of virus removal in two different steps is the same.

Since membrane-enveloped virus may shed surface proteins, the assay for virus particles should include steps to distinguish viral particles from shed proteins. Alternatively, since retroviruses contain a specific enzyme marker, reverse transcriptase, it may be possible to demonstrate their clearance through the use of an enzymatic or immunologic assay for reverse transcriptase. However, the reverse transcriptase assay is inaccurate at low concentrations and in crude samples, care should be taken to avoid interference from cellular DNA polymerases.

Grun et al. provided a summary of virus removal by a variety of purification methods (50). Average log clearance factors ranging from approximately 1.3 to 5.1 were noted for a variety of chromatography types. However, within each type of chromatography, the range of viral clearance varied widely and depended on the specific virus tested and the exact purification process used.

In addition to demonstrating removal of viral particles, virus inactivation should also be measured. Retroviruses are labile species and a well-designed process may include steps which can be validated as virus inactivation steps. A column-based separation may provide viral inactivation as well as removal, especially if non-neutral pH, denaturing reagents, or organic solvents (as used for HPLC) are used. To demonstrate virus inactivation, the virus may be spiked into a process solution and incubated under time and temperature conditions which model the normal production process. When conducting these inactivation studies, it is desirable to determine the kinetics of inactivation as well as the extent of inactivation because virus inactivation has been demonstrated in some cases to be a complex reaction with a "fast phase" and a "slow phase" (51). The inactivation study should be performed in such a way that samples are taken at different times and an inactivation curve constructed. To do these studies, a high-titer virus stock is needed, as well as the appropriate infectivity assay.

PCR using primers designed to detect very low levels of specific viral DNA or RNA sequences is frequently used to quickly determine the viral clearance potential of unit operations, both during process development and in viral clearance studies (52). The FDA has recognized the potential of PCR to provide useful data for determination of the optimum process parameters for viral clearance, especially for retroviruses. When combined with infectivity assays, which measure active virus, PCR can also provide information about viral inactivation since active and inactive viruses will be detected by PCR.

Process Related Components

Removal of potentially harmful or immunogenic components of the fermentation or cell culture media must be demonstrated through either residual testing or validation that the recovery and purification process adequately removes these components (37). In microbial production systems, antibiotics may be included in the fermentor to ensure genetic stability of the production cell line. Frequently used antibiotics include Kanamycin, Tetracycline, and Neomycin. Expression of the therapeutic protein in microbial systems is often repressed until a sufficient biomass has been achieved, and then the expression is induced by a chemical reagent such as IPTG. Surfactants are often used to reduce foaming in the fermentor as well. Clearance of all these media components must be demonstrated by performing clearance validation studies on the individual unit operations and on the entire process.

Mammalian production systems also can contain media components whose removal must be validated prior to product licensure. Antibiotics are less often used in mammalian systems, but when selective pressure is required to maintain genetic stability, components such as methotrexate, methionine sulfoximine, or gentamycin may be included in the production bioreactor. Proteins such as recombinant insulin, transferrin, or albumin may be used in cell culture media to support high density cell growth.

Recovery and early purification steps can also utilize components whose clearance through the later unit operations must be validated. These components include guanidine, dithiothreitol, enzymes, benzonase, or others. Sensitive assays are developed to measure low levels of the components that are introduced during the process, and a combination of in-process testing and final product testing is initially used to determine the levels of these components that remains after each unit operation or following completion of the manufacturing process. Where possible, clearance validation is performed through spike-recovery experiments at each step to determine where components are removed and to determine a clearance factor.

Affinity Ligands

Increasingly, the initial capture and purification of recombinant proteins, especially monoclonal antibodies and antibody-based products, is accomplished using an affinity column such as immobilized Protein A. Any immobilized ligand used as an affinity reagent, including Protein A, may leach from the column into the eluate during protein elution. Manufacturers of Protein A chromatography media[a] are aware of this problem and are actively working to manufacture these media using conditions which are less susceptible to leaching since the amount of residual Protein A must be minimized, if not completely removed, in the final product. Removal of residual Protein A can be done by several chromatographic methods. Ion exchange chromatography, either anion or cation, will effectively remove the residual Protein A and can usually be incorporated into the purification scheme. Commercial kits[b] are available for measurement of the Protein A in the range of 16 to 1000 pg/mL. This is sufficiently sensitive to allow a clearance study to be done by measuring the level of Protein A and product at each purification stage. Measurement of Protein A or any affinity ligand in the final product could be replaced by process validation to demonstrate the removal/reduction of these materials during the purification process.

SUMMARY

Validation of recovery and purification processes is based on process development knowledge for each of the process steps. Scale-up of the process for production of clinical trial material is often followed by construction of a validated scale-down model system which is used for much of the process characterization work. The independent variables, relevant scale parameters and dependent results of most interest depend on the unit operations employed for each of the process steps. Individual pieces of recovery and purification process equipment will be validated with equipment validation protocols prior to being used for process validation. The steps in process validation include identification of the CQAs, determination of the CPPs, accurate scaling of the process, process characterization studies, and finally manufacture of full-scale qualification lots.

a GE Healthcare rmp Protein A Sepharose Fast Flow.

b Protein-A EIA Kit TiterZyme® from Assay Designs, Inc.

REFERENCES

1. ICH Q7A. Quality of Biotechnological Products: Good Manufacturing Practices Guide for Active Pharmaceutical Ingredients, November 10, 2000.
2. Levine HL, Castillo FJ. Biotechnology: quality assurance and validation. In: Avis KE, Wagner CM, Wu V, eds. Validation of Biopharmaceutical Processes. Buffalo Grove, IL: Interpharm Press, 1998:51.
3. Wrotnowski C. Neose targets complex carbohydrate products. Genet Eng News 2001; 21(6), March 15.
4. Veronese F, Harris JM. Introduction and overview of peptide and protein PEGylation. Adv Drug Deliv Rev 2002; 54:453–6.
5. International Conference on Harmonization. Viral safety evaluation of biotechnology products derived from cell lines of human or animal origin. ICH Viral Safety Document: Step 2. Geneva, Switzerland: ICH Secretariat, 1995.
6. U.S. Food and Drug Administration. Biotechnology Inspection Guide Reference Materials and Training Aids, 1991. (Accessed at www.fda.gov/ora/inspect_ref/igs/biotech.html)
7. Mannweiler K, Hoare M. The scale-down of an industrial disk-stack centrifuge. Bioprocess Eng 1992; 8:19–25.
8. Ambler CM. The theory of scaling up laboratory data for the sedimentation type centrifuge. J Biochem Microbiol Tech Eng 1959; 1:185–205.
9. Leung W. Centrifuges. In: Perry RH, Green DW, eds. Perry's Chemical Engineers' Handbook. 7th ed. New York: McGraw-Hill, 1997:18–125.
10. Rothstein F. Differential precipitation of proteins: science and technology. In: Harrison RG, ed. Protein Purification Process Engineering. New York: Marcel Dekker, Inc., 1994:115–208.
11. Scopes RK. Protein Purification: Principles and Practice. Springer: New York, 1987:41–71.
12. Belter PA, Cussler EL, Hu W-S. Bioseparations: Downstream Processing for Biotechnology. Wiley: New York, 1988:221–36.
13. Belter PA, Cussler EL, Hu W-S. Bioseparations: Downstream Processing for Biotechnology. Wiley: New York, 1988:229–35.
14. Albertsson P-A. Partition of Cell Particles and Macromolecules. 3rd ed. New York: Wiley, 1986.
15. Johansson G, Joelsson M, Akerlund H-E. An affinity-ligand gradient technique for purification of enzymes by countercurrent distribution. J Biotechnol 1985; 2:225–37.
16. Bremmer RE. Calibration and certification. In: Carleton FJ, Agalloco JP, eds. Validation of Aseptic Pharmaceutical Processes. New York: Marcel Dekker, Inc., 1986:47–91.
17. Kruse RH, Puckett WH, Richardson JH. Biological safety cabinetry. Clin Microbiol Rev 1991; 4:207–41.
18. Naglak TJ, Keith MG, Omstead DR. Validation of fermentation processes. BioPharm 1994; 20:28–36.
19. PMA Computer System Validation Committee. Validation concepts for computer systems used in the manufacture of drug products. Pharm Technol 1986; 10:24–34.

20. Brunkow R, Delucia D, Haft S, et al. Cleaning and cleaning Validation: a biotechnology perspective. Bethesda, ML: PDA, 1996.
21. Leathy TJ. Microbiology of sterilization processes. In: Carleton FJ, Agalloco JP, eds. Validation of Aseptic Pharmaceutical Processes. New York: Marcel Dekker, Inc., 1986:253–77.
22. PDA Biotechnology Task Force on Purification and Scale-up. Industry perspective on the validation of column-based separation processes for the purification of proteins. J Parenteral Sci & Tech 1992; 46:87–97.
23. Hageman TC. An analysis of clearance factor measurements performed by spiking experiments. BioPharm 1991; 4:39–41.
24. Office of Biologics Research and Review, Center for Drugs and Biologics. Points to Consider in the Production and Testing of New Drugs and Biologicals by Recombinant DNA Technology. Bethesda, MD: Food and Drug Administration, 1985.
25. World Health Organization Study Group. Acceptability of cell substrates for production of biologicals. World Health Organ Tech Rep Ser 1987; 747:1–29.
26. Office of Biologics Research and Review, Center for Drugs and Biologics. Points to Consider in the Characterization of Cell Lines Used to Produce Biologicals. Bethesda, MD: Food and Drug Administration, 1987.
27. Center for Biologics Evaluation and Research. Points to Consider in the Manufacture and Testing of Monoclonal Antibody Products for Human Use. Bethesda, MD: Food and Drug Administration, 1997.
28. World Health Organization. Wkly Epidemiol Rec 1997; 72:141–5.
29. Ostrove JM, Walsh W, Vacante D, Patel N. Molecular hybridization techniques and polymerase chain reaction (PCR) as methods for safety assessment of animal cells used in biopharmaceutical production. In: Spier RE, Griffiths JB, MacDonald C, eds. Animal Cell Technology: Developments, Processes and Products. Oxford, U.K.: Butterworth-Heinemann Ltd, 1992:689–95.
30. Kung VT, Panfili PR, Sheldon EL, et al. Picogram quantitation of total DNA using DNA-binding proteins in a silicon sensor-based system. Anal Biochem 1990; 187:220–7.
31. Lovatt A. Applications of quantitative PCR in the biosafety and genetic stability assessment of biotechnnology products. Rev Mol Biotechnol 2002; 82:279–300.
32. Singer VL, Jones LJ, Yue ST, Haugland RP. Characterization of picogreen reagent and development of a fluorescence-based solution assay for double-stranded DNA quantitation. Anal Biochem 1997; 249:228–38.
33. Conley L, McPherson J, Thommes J. Validation of the ZEVALIN purification process: a case study. In: Rathore AS, Sofer G, eds. Process Validation in Manufacturing of Biopharmaceuticals. Boca Raton, FL: CRC Press (Taylor and Francis Group), 2005.
34. Berthold W, Walter J. Protein purification: aspects of processes for pharmaceutical products. Biologicals 1994; 22:135–50.
35. Wolter T, Richter A. Assays for controlling host-cell impurities in biopharmaceuticals. Bioprocess Int 2005; 3:40–6.
36. Cygnus Technologies, 2006.
37. Chang A, Sofer G, Dusing S. Expediting compliance for clinical biotherapeutics. Bioprocess Int 2004; 2:30–4.
38. European Agency for the evaluation of Medicinal Products CPMP/BWP/382/97.
39. Schindler R, Dinarello CA. Ultrafiltration to remove endotoxins and other cytokine-inducing materials from tissue culture media and parenteral fluids. Biotechniques 1990; 4:408–13.
40. Novitsky TJ, Gould, MJ. Inactivation of endotoxin by polymixin B. In: Depyrogenation, Technical Monograph No. 7. Philadelphia, PA: Parenteral Drug Association, 1986.
41. Nolan JG, McDevitt JJ, Goldman GS. Endotoxin binding by charged and uncharged resin. Proc Soc Exp Biol Med 1975; 149:766–70.
42. United States Pharmacopeia USP <85> Bacterial Endotoxins Test.
43. Center for Biologics Evaluation and Research and Center for Drugs Evaluation and Research. Guidelines on Validation of the Limulus Amebocyte Lysate Test as an End-Product Endotoxin Test for Human and Animal Parenteral Drugs, Biological Products, and Medical Devices. Bethesda, MD: Food and Drug Administration, 1987.
44. Quality of Biotechnological Products: Viral Safety Evaluation of Biotechnology Products Derived from Cell Lines of Human or Animal Origin, International Conference on Harmonization Topic Q5A, 1997.
45. Brorson K, Krejci S, Lee K, Hamilton E, Stein K, Xu Y. Bracketed generic inactivation of rodent retroviruses by low pH treatment for monoclonal antibodies and recombinant proteins. Biotech Bioeng 2003; 82:321–9.
46. Report of a WHO Consultation on Medicinal and other Products in Relation to Human and Animal Transmissible Spongiform Encephalopathies. World Health Organization, Geneva, Switzerland, March 24–26, 1997.
47. Wilkommen H, Schmidt I, Lower J. Safety issues for plasma derivatives and benefit from NAT testing. Biologicals 1999; 27(4):325–31.
48. Brown F, Lubiniecki A, eds. Viral Safety and Evaluation of Viral Clearance from Biopharmaceutical Products. New York: Karger, 1996.
49. Lubiniecki AS, Wiebe ME, Builder SE. Process validation for cell culture-derived pharmaceutical proteins. In: Lubiniecki AS, ed. Large Scale Mammalian Cell Culture Technology. New York: Marcel Dekker, Inc., 1990:515–41.
50. Grun JB, White EM, Sito AF. Viral removal/inactivation by purification of biopharmaceuticals. BioPharm 1992; 5:22–30.
51. AD Hoc Working Party on Biotechnology/Pharmacy, Commission of the European Communities. Validation of Virus Removal and Inactivation Procedures. Note for Guidance, Draft #7, January 28, 1991.
52. Farshid M, Taffs RE, Scott D, Asher DM, Brorson K. The clearance of viruses and transmissible spongiform encephalopathy agents from biologicals. Curr Opin Biotechnol 2005; 5:561–7.

35

Validation of Process Chromatography

Eric Grund
Fast Trak Biopharma Services, GE Healthcare, Protein Separations, Uppsala, Sweden

Gail Sofer
Fast Trak Life Sciences, GE Healthcare, Piscataway, New Jersey, U.S.A.

INTRODUCTION

This chapter addresses validation of process chromatography—one of the tools that provide today's highly pure biopharmaceuticals. Process chromatography is also used to purify, and thus enhance the safety of, traditional biologicals such as plasma-derived products and vaccines. For biopharmaceuticals, multiple chromatographic steps are employed to remove host cell DNA, HCP, viruses, modified product, and processing impurities. Together with unit operations such as centrifugation and filtration, chromatography steps are an integral part of downstream processing, taking a product from a crude feedstock to a purified form suitable for use as a biopharmaceutical, vaccine or health care agent. Each chromatographic step can usually remove multiple impurities. Understanding the purity goals of each chromatographic step expedites the validation process.

Chromatographic media generally have a large surface area that provides high binding capacity. Proteins and other biological molecules have complex, and sometimes poorly understood, interactions with that surface area. A great understanding of the variables involved in process chromatography has been accrued in the 20 years since the advent of modern biotechnology. As a result, in the last decade or so, there is greater acceptance of different validation approaches. For example, concurrent validation for chromatography media life span, once considered unacceptable, is being applied to some processes, and generic validation is being considered in other cases when similar processes are applied to similar starting materials and products. In part, this progress is due to experience. It is also due to the availability of more sensitive analytical tools, such as PCR, that provide a better understanding of both virus and DNA clearance by chromatographic processes. As analytical methods are further developed, an even greater understanding will follow and provide enhanced tools for both routine in-process monitoring and validation.

In this chapter, reference is made to several FDA form 483s. The Freedom of Information Act in the U.S. provides us with an indication of current concerns related to validation of chromatography. Caution is advised, however, since the entire picture is missing for those not involved in the particular inspection.

DEVELOPMENT STAGES

Validation begins with a good development program. All too often, there is a rush to get a product into the clinic with a downstream process that is unsuitable for commercial scale manufacturing and, often, cannot be validated. Prior to beginning development of a purification process, the composition of the feed, the expected product quantity needed, and the purity requirements are evaluated.

During development, the process is characterized (sometimes called qualification) (1). Characterization includes defining the critical control parameters, outputs, and working ranges for both inputs and outputs. Broad initial working ranges that are gradually narrowed with experience enable more rapid development and improve the ability to validate the process. Working in narrow ranges and then shifting out of those ranges can require repeating clearance studies, assay validation, and even toxicity studies and clinical trials.

Feedstock

Understanding the properties of the initial feedstock is critical in the design of validatable chromatographic processes. Sufficient coordination of cell culture or fermentation optimization with purification is sometimes lacking. As the feedstock continues to change during development, multiple changes in downstream processing are required to accommodate variability in the process feedstream. Without good communication between cell culture and downstream processing groups and use of sufficient analytical methods, the chromatographic process will be very difficult, if not impossible, to validate. In some cases, changes in cell culture are made without sufficient time to think through the optimal purification strategy, and changes are also

---

Abbreviations used in this chapter: BLA, biological license application; BSA, bovine serum albumin; CE, capillary electrophoresis; CQAs, critical quality attributes; ELISA, enzyme-linked immunosorbant-assay; FDA, Food and Drug Administration; HCP, host cell proteins; HETP, height equivalent to a theoretical plate; HIC, hydrophobic interaction chromatography; HPLC, high-performance liquid chromatography; ICH, International Conference on Hormonization; MS, mass spectrometry; OOS, out-of-specification; PAGE, polyacrylamide gel electrophoresis; PCR, polymerase chain reaction; QA, quality assurance; SDS, system design specification; SP, sulfopropyl; TNTC, too numerous to count; TOC, total oxidizable carbon; TSEs, transmissible spongiform encephalopathies.

made that potentially impact patient safety, e.g., reduction in clearance of viruses or potentially immunogenic HCP.

Product Quantity Goals

At early development stages, preliminary market projections may be the only information source that can be used to estimate the quantity likely to be needed for future marketed product. Getting some handle on the scale of commercial operation that may be required should be factored into the design of the purification process. Manufacturing capabilities often dictate the types of chromatography techniques and equipment that are practical. If the process is not designed to accommodate manufacturing scale realities, it will need to be redesigned. This can increase the validation effort considerably. Processing times may change and impact product purity and the impurity profiles. Furthermore, intermediate product stability may change as holding times change with changes in scale of operation. Assays used in development may not be appropriate for large-scale production. If the specifications are set based on small-scale, or even pilot-scale operations, process capabilities may not be sufficient to produce product in larger quantities that reliably meets those specifications. Demonstrating consistency for consecutive batches may be quite difficult, risking future batch failures for marketed product. The only solution is to redesign the process, which may necessitate repeating clearance studies and assay validation with modified samples.

Product Quality Goals

The patient population, product indication and dose usually dictate purity requirements. The number of purification steps that will be incorporated into the process and eventually require validation is determined by understanding the impurities each purification step is capable of consistently removing. One should also consider the ability to maintain product integrity, e.g., by avoiding product modifications. In some cases, two well-designed and characterized purification steps can produce a product with the same quality as a product purified in three or more steps. It is important to keep in mind that purer is not necessarily better for a patient; however, it is always different. Since products entering clinical trials must be demonstrated to be safe, some validation of the downstream processing will be required even at the early stages. For the most part, this effort is directed toward virus clearance (where relevant), control of bioburden, and removal of any potentially harmful agents introduced upstream. Unfortunately, at this time, there are no consistent, harmonized recommendations from worldwide regulatory agencies addressing what is expected at this scale of operation.

Process Characterization/Qualification

Critical control parameters and their limits are determined during characterization studies. This cannot be done, however, without understanding the measurable results (i.e, outputs) for each step. During characterization, analytical tools that are transparent to scale should be used, whenever possible, to facilitate later validation and process control efforts. As noted earlier, analytical tools used in purification development are often more sensitive than those that are appropriate for manufacturing scale. This can lead to establishment of inappropriate specifications. Highly sensitive assays may require redesign and revalidation. Lengthy in-process assays with extra sample handling may ensue. These may be problematic for production schedules and require continuing subsequent purification steps at risk. Time allotted to characterization will ultimately decrease costs and minimize failures during process validation.

Critical Control Parameters

Critical control parameters are those that affect product and process quality. Flow rate, buffer quality, pH, conductivity, and temperature are important, but the extent to which they are critical will vary with the specific process. Although shear effects on product are rare in chromatography, an increase in flow rate can reduce the separation capability of any given step. Furthermore, both media and equipment have pressure limitations that, if exceeded, can lead to catastrophic batch failure. Optimum flow rates and acceptable pressures are, therefore, determined at an early stage. The particle size of the chromatography media will impact the flow rate—larger particles allow higher flow for a given pressure drop, provided the separation capability remains adequate. It is important to define acceptable media particle size ranges. In some cases, processes have been optimized with only one lot of media, and the media particle size in the next lot varied to the extent that the control parameters had to be changed.

Widely used chromatography methods include affinity, ion exchange, hydrophobic interaction, reversed phase and size exclusion. The type of chromatographic separation may also determine other relevant parameters. For example, in HIC, the degree of substitution of the functional hydrophobic group may vary within a range accepted by the vendor. This range, however, may not match the suitable range for a specific application. If the ranges for salt concentration and other operating conditions are set narrowly then the separation method may demand tighter media specifications to be robust. Protein loads (product and impurities) are usually critical parameters for all modes of chromatography and optimal loads should be established so that the purification goals are consistently achieved.

Measurable Results (Outputs)

Critical outputs include purity and specific impurity profiles that will depend on the material being processed and the control parameters (operating conditions). Evaluating the impurity profile with extensive, orthogonal analytical tools is critical for successfully developing a purification process that is validatable. The analytical tools used to check for the success of a step should offer higher resolution than the process step being controlled and, preferably, include a separation according to a mechanism that is different from the one used in the process step.

Purity expectations vary with product and stage of development. Impurities that are removed by the chromatographic process include those from the product source, cell culture step additives such as inducing agents, and

Table 1 Commonly Used Sources for Biotechnology and the Host Cell Derived Major Impurities Removed by Chromatography

Source	*Escherichia coli*	Mammalian cells
Major impurities	HCP	HCP
	DNA	DNA
	Endotoxins	Endogenous retrovirus

also those derived from the manufacturing process (including recovery and clarification steps, as well as the chromatography steps themselves). Table 1 shows some of the specific host cell impurities that are typically removed by chromatography. Table 2 lists impurities that can be introduced into manufacturing and require removal. Removal can often be demonstrated in clearance studies, which are further described below. Evaluating a chromatographic process for viral clearance capability has been one of the more challenging efforts. Today the use of PCR not only enhances understanding of viral clearance, but also enables studies to be performed along with development. Of particular interest is the use of multi-virus spikes to evaluate viral clearance (2).

The biotech industry has asked regulatory agencies to develop international specifications for biotechnology impurities (3). Impurities include leachables from chromatography media. Leached Protein A arising during antibody purification is one such impurity. One FDA spokesperson observed that most sponsors measure Protein A leachables by immunoassay, and validate the removal or measure Protein A as a release specification, with 10 to 12 ppm in the product. Setting a specification, however, will depend on assay capabilities, and clearly should be linked to dose.

Removal of HCP is also achieved by chromatographic processes. An FDA spokesperson has recommended that both Western blot and ELISA be validated at 0.5 to 100 ppm, and noted that when using the Western blot, one should expect to detect HCP. If none are detected, it usually indicates that the antibody used in the Western blot assay has insufficient sensitivity.

Table 2 Potential Impurities Introduced into Feedstream during Various Biotechnology Manufacturing Steps

Manufacturing steps	Potential impurities
Cell culture	Fetal calf serum
	Other culture media components, e.g., insulin
	Antibiotics
	Induction agents
	Retrovirus
	Adventitious agents
Recovery and clarification	Leachables
	Extraction, solubilizing, stabilizing agents
	Degraded product
	HCP, DNA, retrovirus, other host cell impurities
	Adventitious agents
Chromatography and filtration	Leachables, e.g., Protein A
	Processing agents, e.g. detergents, salts, solvents
	Carryover for multiply cycled columns and ultrafiltration filters
	Adventitious agents

SMALL AND MANUFACTURING SCALE VALIDATION

Small Scale

Small scale studies can be a cost-effective approach for some validation tasks required for purification steps, e.g., for process characterization, clearance studies, media life span studies, cleaning studies. In some cases, those small-scale studies provide supporting data. In other situations, they minimize the risks associated with hazardous materials. If the small-scale studies are to be used in validation, the small-scale system must be validated.

Validation of Scaled Down Model System

Validation of a small-scale model requires that the chromatography system truly reflect the purification process that will be used for licensed product. The feed stream for these studies should be taken from production. Materials such as chromatographic media and buffers should be those that are approved by QA for use in manufacturing. Columns must be equilibrated in the way they are or will be in manufacturing. If measurements of pH, conductivity, and UV are performed in manufacturing, they should also be performed for the small-scale column operation. Contact time is one of the most important factors for measuring comparability when chromatography is scaled up or down. Its importance is described in the ICH guideline on viral safety (4). Oftentimes, column efficiency (HETP) and/or peak asymmetry measurements are made to ensure the column packing is consistent at the two scales. Then product recoveries, product purity, and the impurity profiles can be measured to demonstrate they are comparable to those observed in manufacturing. In one example of validation for a small-scale chromatographic purification step for a monoclonal antibody, analysis included yield, HCP clearance, DNA clearance, and clearance of Protein A from a previous chromatographic step (5).

There are some particular differences in chromatography systems of different scale that should be taken into account. The wetted materials may be different in both column and system at a small scale. Often stainless steel is used in large-scale chromatography, whereas plastic or glass is more common for smaller columns. Adsorption of both product and impurities may be greater with a particular material. Transport distances to monitors and collection vessels should be proportional at both scales. Column distribution systems are almost always different. Multiple ports are typically used at large scale, while single ports are utilized for small columns. In spite of these differences, when properly designed, small-scale systems can be validated to correctly reflect manufacturing scale activities.

Uses

Small-scale systems are commonly used in both process optimization and characterization/qualification. Cleaning optimization, process intermediate hold time studies, and stability studies are often performed at small scale. When performed appropriately, these studies can be used to support establishment of protocols and acceptance criteria. By themselves, they do not usually provide sufficient information, and reconfirmation is required during process validation at full scale. An example of the use of a small-scale study for establishment of chromatographic process parameter ranges was presented by Blank (6). The specified flow rate was established at 100 cm/hr, buffer pH at 5.5, and load density at 40 g/L. Determination of these parameter ranges was made by evaluating clearance of HCP and DNA and measuring the amount of monomeric product at flow rates of 50 and 200 cm/h; buffer pH 5.4 and 5.6; and loads of 10, 20, and 30 g/L.

Small-scale chromatography models are used for evaluating/validating clearance of viruses, TSEs, DNA, HCP, and process impurities such as Protein A and cell culture media additives (e.g., BSA). To enhance sensitivity, spiking studies with a high concentration of the material that is to be removed by the chromatography step are performed. When spiking studies are performed in clearance studies, the impact of the spike on column performance has to be evaluated using multiple analytical methods.

As analytical methods have become more sensitive, the need for scaled down models for clearance studies has decreased. Some clearance studies can now be performed at pilot or manufacturing scale. In particular, the use of PCR has decreased the need for small-scale models for DNA clearance. With the exception of viruses, TSEs, and other potentially hazardous materials, clearance data acquired by small-scale studies can be confirmed at manufacturing scale by testing the final product or a selected process intermediate for the presence of the impurity. Once the clearance studies are performed and product tested in validation runs, lot release testing can usually be eliminated. This can be a big cost saving. However, it is essential to maintain the assay capability in the event of a process change or OOS result.

Validation of virus and TSE clearance requires that the validation batches are run under the same conditions as those used in the small scale clearance studies. This is sometimes overlooked when different departments are responsible for small scale and manufacturing validation. When clearance studies are performed for potentially hazardous materials, such as virus and TSEs, sanitization studies are usually part of the study. This is done by demonstrating the effectiveness of the chosen sanitizing agent after a spike into a small-scale column. For equipment, coupons (i.e., cutout pieces of equipment) may be used for the spiking study. (For further details, see below under Cleaning).

Viral clearance studies are required for mammalian cell culture derived products intended for both clinical trials and licensed product. Where there is potential risk, TSE clearance studies are also performed. Small-scale chromatography systems that are used to evaluate virus and TSE clearance are best designed and validated at the site where the analytical methods are most conveniently performed. The actual spiking studies are performed in facilities where the safe handling of such agents can be assured. One problem that arises during early clinical studies is that subsequent process changes might invalidate the viral clearance. Planning for the changes, revalidating the small-scale system if needed and repeating the viral clearance study is necessary when changes are made to chromatography steps claimed to remove viruses.

Production

Scale-Up and BLA-Enabling/Shake Down/ Engineering Runs

The feasibility of scale changes that can be validated is best assessed with an understanding of manufacturing capabilities—an all-too-often overlooked issue during development. Once the purification process is optimized, typically prior to phase 3, it can be scaled up, and scale-up capabilities confirmed during "shake-down"/"BLA-enabling"/"engineering" runs. As noted earlier, scale changes in chromatography may result in differences. Some of these differences are found in holding times, wetted materials, flow cells, distribution systems, and process capability (e.g., pumping capacity). Although chromatography is one of the simpler unit operations to scale, some minor modifications may have to be made. The shake down/BLA-enabling runs should be performed prior to starting the formal process validation. These preliminary runs can go a long way toward minimizing subsequent formal validation failures.

Columns must be packed to meet predetermined acceptance criteria. HETP and asymmetry determinations are often used to qualify columns (7). In some cases, HETP and asymmetry measurements will not be the same in manufacturing scale columns as they were in smaller columns. This may not, in and of itself, be an issue. It can be the result of different column designs. The product purity profile, as measured by multiple orthogonal analytical techniques, and the impurity profile will be key determinants to ensure the scale up is acceptable and does not necessitate redesigning scale-down models and repeating clearance studies that were performed at small scale.

Once the scale-up capability is verified, it is essential to ensure that the conditions used for small-scale clearance studies are, in fact, those that are being used during validation. There are several FDA form 483s that note the conditions used in manufacturing do not reflect those that have been validated.

Validation Runs/Conformance Batches

Validation at manufacturing scale requires qualification of raw materials and equipment (Table 3). All of the analytical methods used to evaluate the purification process effectiveness will be validated, with the exception of those used solely for characterization. Each

Table 3 Qualification Activities

Raw materials	Buffers
	Chromatography media
	Processing additives
Equipment	Columns
	Pumps
	Monitors
	Tanks
	Automated skids

purification unit operation must have predefined inputs and outputs that were established during development and that will be measured during the validation runs at manufacturing (or pilot) scale. As noted by many experts, validation is not the time for experimentation.

One issue that can arise for validation of chromatographic processes is the perceived need to run the process at upper and lower limits for each parameter. To do that might prevent a product from ever getting past the validation stage. However, development should have demonstrated which variables will need testing at the limits of the ranges. The use of designed experiments to establish process robustness has been addressed by Kelly (8). Identification and establishment of operating ranges of critical process variables was presented by Gardner and Smith (9). As noted earlier, one variable for chromatography that is considered critical, in almost all cases, is protein load. Sometimes, it may not be feasible to obtain product at both upper and lower protein loads during the validation/conformance batches. If small-scale studies have been validated and run under conditions representative of manufacturing, they may be used to support the outer limits. However, if these studies are to be used to support process limits for licensed product, it is recommended that QA sign off on the studies.

During process validation, all informative assays are typically used. This usually includes some assays that are not fully validated and that will not be used in routine manufacturing. Often included are highly sensitive assays for evaluating product purity and impurities after each chromatographic operation.

After validation, control charts and trending analysis are used to maintain the purification process in a validated state. Retrospective validation is seldom used for validation of chromatography processes. Its use is described briefly in a book chapter on protein purification issues (10). Now that the industry has a history of successfully producing biological products using chromatographic purification, some firms may have sufficient data to perform retrospective validation.

SPECIAL ISSUES

Some issues deserve special attention in a discussion of validation for process chromatography. These include holding, processing, and storage times; chromatography media life span studies; and cleaning and sanitization.

Holding, Processing, and Storage Times

Holding times are defined in development and confirmed at manufacturing scale. For purification process intermediates, holding time studies include evaluation of product purity, stability, and bioburden control. In smaller scale studies, holding times can be extended to build in a safety margin. The smaller scales are first validated to represent manufacturing. The data, when appropriately documented and approved, can be an aid in releasing batches in the event of an unexpected hold time during manufacturing. Table 4 provides some FDA 483s and comments in approval letters related to holding times. Comments in one post-approval inspection included "storage times in between runs for all of the purification columns have not been validated for the entire life cycle of the columns."

Table 4 Holding Time Issues Described in FDA Form 483s and Approval Letters

483s	Approval letters
Lack of data to support hold times	Institute bioburden monitoring of storage solution
Lack of container closure integrity studies	What is expected storage time based on validation studies for the regenerated column?
Hold times post sanitization/equilibration for column not validated	Submit results of hold period studies for in-process product intermediates that include container-closure integrity study and biochemical, bioburden, and endotoxin studies on a periodic basis

Processing time limits are established to ensure final product consistency and freedom from adventitious agents, such as bacteria and fungi. These time limits are also evaluated in development, but scale changes may necessitate some modifications that will be validated at full scale.

Storage times for chromatography columns are validated to demonstrate column integrity and control over bioburden. Storage time establishment is part of a chromatography media lifespan study since the storage conditions can adversely impact column performance over time. Removal of column storage solutions is also validated. This is particularly relevant for small molecules, such as ethanol, whose presence might impact subsequent column performance. Contact time plays an essential role in cleaning, and during storage further cleaning may occur in columns that appeared to be clean prior to storage. The capability of the start up protocol to remove any residuals from the column should be validated. This is often done using one or more of the following assays: TOC, conductivity, pH, UV.

Media Life Span

Chromatography columns are validated for consistent performance over their life span. An FDA Compliance Guide describes the need to have an estimated life span for each column type (Table 5).

As noted in the compliance guide statement above, concurrent validation may be appropriate. For products derived from sources where there are no known viral risks, i.e., those produced in bacteria or yeast, concurrent validation is often a good choice. Concurrent validation depends on the ability of in-process analytical tools to demonstrate performance consistency. Avoiding lengthy

Table 5 Lifespan Expectations from FDA Compliance Guide 7341.001

There should be an estimated lifespan for each column type, i.e., number of cycles. Laboratory studies are useful even necessary to establish life span of columns. There are situations where concurrent validation at the manufacturing scale may be more appropriate. Continued use may be based upon routine monitoring against predetermined criteria

Table 6 Predicting Column Failure

Column mode	Failure
Protein A	Decay of dynamic capacity
Anion exchange	Decrease removal of key contaminants
Cation exchange	No identifiable sources

Source: From Ref. 14.

off-line assays can prevent holding time problems or further at risk processing.

In most cases, a combination of both prospective and concurrent validation is used for life span studies. In one prospective study, up to 1200 cycles were demonstrated (11). In a concurrent study on the life span of an anion exchange column, 27 cycles or 4.5 years were validated. One of the parameters measured concurrently was consistency of DNA removal (12).

One publication from the U.S. FDA described factors that could predict degradation of Protein A column performance long before retrovirus clearance is decreased. Those factors include antibody step yield and breakthrough, but not eluate impurity content. It was proposed that viral clearance of aged chromatography media may not always be necessary (13). While this approach may be accepted by some regulatory agencies, others may not find it suitable. One firm, in a very extensive study, tried to predict column failures with mixed results (Table 6).

Parameters that can be used to evaluate continued performance are described in Table 7. There may be others, and each company should decide which parameters provide them with the most relevant information.

The ability to clean and sanitize the column must remain constant over the lifespan. A decrease in product purity or increase in impurities often indicates column build up that is not being removed by cleaning protocols. Changes in regeneration, column packing deterioration and increased backpressure are also indicative of decreased column performance.

Cleaning and Sanitization Validation

Cleaning and sanitization of chromatography media and equipment are of paramount importance in the production of safe biotherapeutics. Chromatography media and equipment suppliers provide recommendations for cleaning and sanitizing, and sodium hydroxide is the most frequently used agent for both cleaning and sanitizing columns. Each column and feedstream combination is evaluated to ensure the appropriate conditions for cleaning and sanitizing are selected. For example, in some cases, high salt is a good cleaning agent. But if there are residual bound hydrophobic impurities, high salt will increase binding rather than provide a cleaning effect. Some chromatography media cannot tolerate harsh conditions and cleaning problems can ensue after repeated use. For such columns, it can be useful to position them at a place in the purification scheme where there are fewer impurities. These columns are usually used for fewer cycles than those that can be cleaned with harsher agents.

Table 7 Parameters Indicative of Column Performance Deterioration

Indications of performance deterioration	Methods for determination
Decreasing product purity	HPLC, MS, CE, SDS-PAGE
Change in impurities profile	HPLC, CE, SDS-PAGE
Increase in specific impurities	HCP, host cell DNA, Protein A
Decreasing product recovery	Product specific assays
Increasing breakthrough	Binding capacity, UV, HPLC
Changes in flow rate or pressure	Flow and pressure monitors
Changes in regeneration and/or requilibration profiles	pH, conductivity, UV monitors

It is important that the cleaning/sanitizing agents do not modify column performance. As noted above, cleaning and sanitization evaluation is part of the media lifespan study. In one such study for a Streamline SP capture step, blank runs were performed and HCP were measured. The HCP were found to be at the detection limit of the assay in 48 cycles over a three-year period (12).

Chromatography media are dedicated to one product, but equipment, including column hardware, may be used for more than one product. Cross-contamination for multiuse equipment is prevented by using suitable cleaning routines and validating the absence of carryover from one product to the next. Although the acceptance of 10 ppm carryover from one drug product to another has been discussed in the pharmaceutical industry, this is not acceptable for highly potent biopharmaceuticals. Acceptance criteria are usually established by performing a risk assessment. Oftentimes, the acceptance limit is set at the detection limit of the assay. Unfortunately, in some cases, this is the only possible choice.

For demonstration of removal and inactivation of hazardous materials such as virus and TSEs, spiking studies utilizing coupons can be performed in a safe environment away from the manufacturing facility (15). Effectiveness of inactivation can be measured by biological infectivity assays and removal by one or more suitable analytical methods.

For dedicated equipment and packed columns, carryover of residual product and/or impurities is evaluated to ensure consistency of product. Carryover of degraded product or impurities can alter product immunogenicity as well as potency. FDA 483s related to carryover include

- Protein carryover from previous purifications not characterized
- Impact of carryover proteins not evaluated
- Cleaning validation not performed for removal of urea and cell culture media
- No periodic monitoring of columns following cleaning.

Both small-scale studies and manufacturing scale runs are used for cleaning validation. The small-scale studies are really part of development and allow for higher concentrations and temperatures to be evaluated for removal of residuals and to build in a safety margin. The optimized conditions are then evaluated during conformance batches. Assays used to assess cleanliness include, among others, TOC, product-specific assays, and total protein assays. Blank runs monitored by UV can indicate if there is protein carryover.

Table 8 Nonclinical CQAs for Biopharmaceuticals

Posttranslational modifications, e.g., glycosylation
Aggregation and other product modifications arising during purification and storage
Viral clearance
Removal of DNA, HCP
Formulation
Stability
Sterility/bioburden

A problematic area for the biopharmaceutical/biologics industry has been control of bioburden in purification operations. Chromatography is not a sterile process, and it is generally unrealistic to perform it as a sterile (or even aseptic) unit operation. Having said that, there are a few situations where sterile purification processes are run. These are typically employed when an antibody affinity ligand that is difficult to clean is used for purification. But this is not the norm. There are numerous FDA citations regarding bioburden control in downstream processing. One solution for the industry may be the use of rapid microassays to enable frequent, rapid monitoring that will enable better control. One recent 483 observation was that there were no bioburden rejection limits for purification steps (16). Although there is no official regulatory requirement for rejection limits, in this case it was also observed that TNTC results were reported as deviations and the batches were released. Another bioburden-related validation comment stated that there were no bioburden data to support column eluate hold time at each purification step or holding periods for buffer carboys or containers after cleaning.

Control over bioburden is one part of a life span study. The effect that contact with the column may have on the sanitization solution should be evaluated. This is often done in viral clearance studies. The sanitizing solution is passed through the column and then its ability to inactivate viruses validated. In a similar approach, the bacteriostatic effectiveness of column storage solutions was evaluated by spiking microorganisms into used storage solutions. A pH shift was observed during storage, but the solutions were shown to be effective from 1 month up to 5 months, depending on the type of chromatography media (12).

CONCLUSION

Biological molecules are generally complex, and understanding of CQAs is required for process validation (Table 8). This understanding is begun during development and enhanced during process characterization studies. These steps lead to process validation. At each step, specific tasks are performed. Figure 1 illustrates the progression from development to validation of chromatographic processes.

As described in a June 2004 publication, increased process control can be obtained by using process analytical technologies. Examples include HPLC analyzers that provide closed-loop control of chromatography steps (17). Better process control and in-line assays with increased sensitivity may someday decrease the amount of validation that is required for chromatographic purification of biopharmaceuticals. However, today firms direct most of their efforts toward characterizing process intermediates and product, so that CQAs can be understood and the purification process validated by measuring these attributes.

Process chromatography is responsible for today's highly pure biologicals. Validation of those chromatography processes is one part of ensuring consistent and safe biological products are being delivered to patients.

Steps	Tasks	Examples
Define Goals of Process ⇩	Evaluate Feedstock Characteristics Product Quantity & Quality	Removal of Host Cell Proteins, DNA, etc.
Design & Develop Process ⇩	Establish Working Ranges Gain Understanding of Critical Parameters Use Orthogonal Analytical Methods	pH, Protein Loads, Conductivity, Contact Time, Product Stability
Characterize Process ⇩	Establish Critical Parameters and Limits Establish In-Process Monitoring	Design of Experiments (DOE)
Validate Process ⇩	Test at Limits Meet Predetermined Acceptance Criteria	Small Scale and Production Scale Removal of Impurities Product Stability Cleaning Media Lifespan
Monitor Process	Tighten Limits Establish Some New Limits Introduce More In-Process Controls Change Process If Required	Back Pressure and/or HCP Levels for Media Lifespan Peak Shape/Slope for Column Packing

Figure 1 Purification processes: from design to licensed product.

ACKNOWLEDGEMENT

The Polymerase Chain Reaction (PCR) is covered by patents owned by Roche Molecular Systems and F. Hoffman-LaRoche. A license to use the PCR process for certain research and development activities accompanies the purchase of certain reagents from licensed suppliers.

REFERENCES

1. Mire-Sluis A. Oral Presentation. IBC Process Validation for Biologicals. San Diego, California, 2003.
2. Valera CR, Chen JW, Xu Y. Application of multivirus spike approach for viral clearance evaluation. Biotechnol Bioeng 2003; 84(6):714–22.
3. Validation Times 2004; VI(7):1.
4. International Conference on Harmonization, ICH Q5A Viral Safety Evaluation, 1997.
5. O'Leary R, Feuerhelm D, Peers D, Xu Y, Blank GS. Determining useful life of chromatography resins. BioPharm 2001; 14(9):10–8.
6. Blank G. Oral Presentation. IBC Process Validation for Biologicals. San Diego, California, 2001.
7. Rathore A, Sofer G. Lifespan studies for chromatography and filtration media. In: Rathore A, Sofer G, eds. Process Validation in Manufacturing of Biopharmaceuticals. Florida: Boca Raton, Taylor and Francis Group LLC, 2005:169–203.
8. Kelly BD. Establishing process robustness using designed experiments. In: Sofer G, Zabriskie DW, eds. Biopharmaceutical Process Validation. New York: Marcel Dekker, 2000:29–59.
9. Gardner AR, Smith TM. Identification and establishment of operating ranges of critical process variables. In: Sofer G, Zabriskie DW, eds. Biopharmaceutical Process Validation. New York: Marcel Dekker, 2000:61–76.
10. Winkler ME. Purification issues. In: Sofer G, Zabriskie DW, eds. Biopharmaceutical Process Validation. New York: Marcel Dekker, 2000:143–55.
11. Seely RJ, Wight HD, Fry HH, Rudge SR, Slaff GF. Validation of chromatography resin useful life. BioPharm 1994; 7:41–8.
12. Heitzmann M. Process Validation of the Basiliximab Purification Process. Oral Presentation. IBC Process Validation. San Diego, 2004.
13. Brorson K. Identification of Chromatography Quality Attributes to Assure Retroviral Clearance Capacity of Multiply Cycled Resins. PREP 2002 Poster presentation.
14. Medimmune CO. PDA/FDA Viral Clearance Forum, Oral Presentation. Washington, DC, 2001.
15. Cleaning and Cleaning Validation: A Biotechnology Perspective, Chapter 5. Sampling Methods. Bethesda, MD: PDA, 1996.
16. BioQuality; 9(7):7–8, i2i Corporation, Idyllwild, CA
17. Validation Times. Produces Operating Efficiencies in Bioprocessing at Lilly. Rockville, MD: Washington Information Source Co., 2004:6.

36

Cell Culture Process Validation

Steven I. Max, Denis Drapeau, Martin S. Sinacore, and Timothy S. Charlebois
Wyeth Biotech, Andover, Massachusetts, U.S.A.

INTRODUCTION

Cell culture processes involving mammalian cells present certain unique challenges from a validation perspective. Fundamentally, the objective of any cell culture validation program is to demonstrate that the cells and process are suitable for their intended purpose and can consistently operate within the manufacturing setting in such a manner that a consistent product will result. In this chapter, the focus will be on mammalian cells used to produce biopharmaceuticals such as recombinant proteins and monoclonal antibodies. Particular emphasis is given to the unique challenges and issues associated with mammalian cell-based production. Case studies and examples are provided to illustrate how key validation principles and guidelines have been interpreted and applied in support of actual manufacturing processes. Because the experience of the authors has relied heavily on the use of CHO cells, the examples are drawn substantially from this platform but are representative of the issues, thinking, and approaches that must be applied to any mammalian cell culture process.

Three key areas of cell culture process validation are: (*i*) evaluation of cell line suitability, (*ii*) cell line stability, and (*iii*) process characterization and validation. These are addressed in turn below.

EVAULATION OF CELL LINE SUITABILITY

The establishment of a validated cell banking system is critical to ensure that the starting point for manufacturing remains consistent throughout the entire lifetime of the biologic therapeutic. A two-tiered cell banking system is considered to be standard practice within the industry. Under this approach, a MCB is used to generate a practically endless number of WCB, thereby providing a continuous supply of well-characterized cells for full-scale manufacturing.

Abbreviations used in this chapter: cGMP, current good manufacturing practice; CHO, chinese hamster ovary; CPD, cumulative population doubling; DHFR, dihydrofolate reductase; DO, dissolved oxygen concentration; EOP, end-of-production; HAP, hamster antibody production; ICH, International Conference on Harmonization; MAP, mouse antibody production; MCB, Master Cell Bank; MEF, Murine Embryonic Fibroblasts; MMV, murine minute virus; PERT, product-enhanced reverse transcriptase; Q-PCR, quantitative polymerase chain reaction; RD, rhabdomyosarcoma; rhBMP, recombinant human bone morphogenetic protein; RT-PCR, reverse transcription polymerase chain reaction; RVLP, retroviral-like particles; TEM, transmission electron microscopy; TGF-β, transforming growth factor beta; WCB, Working Cell Banks.

It is essential to demonstrate that the cell line used for production of the biologic is free from adventitious agent contamination, both viral and microbial, as well as genetically stable over the planned duration of the manufacture of the product. The ICH has established guidelines containing recommendations for the testing and characterization of cell lines (1–3). However, the points addressed in the guidelines are not all-inclusive and may be subject to interpretation by the manufacturer. This section will focus on efforts to establish and maintain a paradigm for the purity testing and genotypic characterization of CHO cell banks and production cell lines, as it relates to industry standards and regulatory expectations. Examples will be provided to demonstrate the applicability of this paradigm to recombinant cell lines produced. It is the intent of this paradigm to confirm the identity and purity of the cell bank as well as demonstrate the suitability of the cell line for its intended purpose.

Demonstration of Freedom from Adventitious Agents

The risk of contamination from adventitious agents is a feature common to all biologic products derived from cell lines. A comprehensive program should include the use of a well-characterized host cell line, a validated cell banking system, low-risk raw materials (non-animal derived), a rational cell bank testing scheme to detect a broad range of potential viral contaminants, and routine testing of production cultures. It is recognized that no cell bank testing regimen can guarantee the detection of all potential contaminants. As such, a rigorous evaluation of the ability of the downstream processes to remove and/or inactivate virus must be conducted and satisfactory results achieved. Regulatory expectations pertaining to these topics are addressed in detail in ICH Topic Q5A.

Figure 1 illustrates the cell bank testing and routine cell culture monitoring program established for a recombinant CHO cell line. Testing of the MCB is extensive and serves to evaluate the purity of the cell line at the end of its development stage (~six months), immediately after it has been introduced into the cGMP environment. At the point at which a WCB is created, a much less rigorous testing plan is employed, as the MCB cells are in culture for only a limited amount of additional time (~two weeks) and are solely contained within a clean-room environment. Upon the successful completion of a full-scalen or pilot-scale manufacturing campaign, one-time testing of EOP samples at the limit of in vitro cell age is performed to evaluate the cell line run under

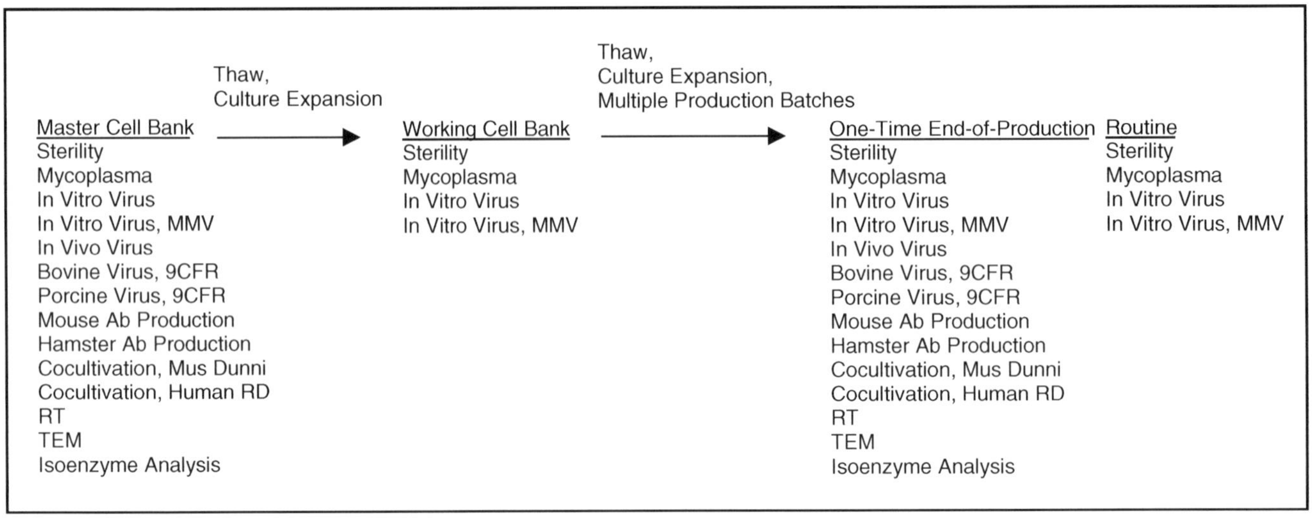

Figure 1 Cell line purity testing and routine monitoring paradigm.

manufacturing conditions, where the opportunity may present itself for any latent virus not detected in the cell bank to be expressed.

According to ICH Topic Q5D, cell banks used in the manufacturing of biologics for human use must be demonstrated to be free from microbial contamination, including bacteria, fungi, and mycoplasma. The test for the presence of bacterial and fungal contamination is performed on 1% of the cell bank (minimum of two vials) and involves the direct inoculation of cell lysates into two different liquid media (Tryptic Soy Broth and Fluid Thioglycollate Media) intended to detect a broad range of aerobic and anaerobic microorganisms as well as fungi and yeast.

The mycoplasma assay, performed according to the 1993 *Points to Consider in the Characterization of Cell Lines to Produce Biologicals*, tests for the presence of any cultivable or non-cultivable mycoplasma species in cell lysates prepared from the MCB. Cultivable mycoplasmas are detected using semisolid broth and agar, which enhances the possibility of detecting both fastidious and easily cultivated strains. Non-cultivable mycoplasma species are detected through the use of the Hoechst stain procedure on Vero cells that have been inoculated with the test article.

The in vitro virus assay employs a panel of indicator cell lines that are capable of detecting a wide range of human and relevant animal viruses. Viral detection is based on the demonstration of cytopathic effects, hemadsorption or hemagglutination within any of the cell lines. According to ICH Topic Q5A, the choice of cell lines should include the species of origin of the cell bank as well as a human and/or a nonhuman primate cell line susceptible to infection by human viruses. Because of the susceptibility of CHO cells to infection by MMV and the potential impact of such an event, an additional indicator cell line with increased sensitivity to MMV infection has been included in the panel.

Even when cell culture processes are free of animal-derived raw materials, early on in the cell line development history, the cells may have been exposed to materials such as bovine serum and porcine trypsin for a period of up to several weeks. If so, the purity testing paradigm will include in vitro tests specifically to detect the presence of any bovine or porcine viruses.

Viruses that are not readily propagated or otherwise display cytopathic effects in the in vitro assay may be detected in the in vivo assay. Their presence is detected through the inoculation of adult and suckling mice, guinea pigs and embryonated hens' eggs and the subsequent demonstration of symptoms associated with viral infection. Species-specific viruses may be detected via the MAP and HAP tests which focus on the generation of antibodies in response to an in vivo viral infection.

Cocultivation of MCB or EOP cells with a cell line susceptible to retrovirus infection, such as *Mus dunni* or human RD, increases the possibility of detecting small amounts of any infectious retrovirus. Infectivity may be confirmed with a positive focus formation assay or reverse transcriptase assay. Furthermore, TEM is used to evaluate the cells for the presence of virus-like particles.

Identity testing establishes the species identity of the cells and demonstrates that the cell bank is composed of a homogeneous population. The isoenzyme analysis method is based on the electrophoretic mobilities and banding patterns of various intracellular enzymes.

In addition to the testing of the cell banks, unprocessed bulk (cells and culture media) from each GMP production batch intended for processing into drug substance is evaluated for the absence of bacteria, fungi, mycoplasma, and virus. Testing at the cell culture stage, prior to initiation of any downstream processing steps, provides a suitable point for the detection of any potential contaminating agents. This routine testing provides lot-to-lot coverage for adventitious agent contamination during the production process. ICH Topic Q5A provides guidance on the evaluation of viral safety of biotechnologic products derived from human and animal cell lines. Even with satisfactory purity testing results from the cell banks, viral contamination may arise from introduction during the production process, and, as such, routine testing of unprocessed bulk is performed as recommended (3,4). Contaminating viruses may be detected through the use of a panel of indicator cell lines and subsequent observations of cytopathic effects, hemadsorption or hemagglutination. The majority of known viruses that may be detected by each of the indicator lines do not appear to represent a threat for CHO cell infection. Nonetheless, CHO cells have been shown to be susceptible to infection from a limited number of human viruses in the paramyxovirus and reovirus families (5). Furthermore, within the industry, contamination of CHO cells from both murine and bovine adventitious agents has been reported, including Mouse Minute Virus, Blue tongue virus and Cache Valley virus, resulting in significant operational losses.

Unlike murine myeloma or hybridoma cells that are known to contain endogenous retrovirus, CHO cells contain noninfectious RVLP. However, because the RVLPs are biochemically and morphologically similar to infectious retrovirus, and because the possibility of infectivity cannot be completely excluded, there is a regulatory expectation to quantify these particles. This is accomplished through the evaluation of cell-free supernatants from representative production batches via TEM. These data are then used in conjunction with downstream process clearance capability (not covered in this chapter) to assess the risk of RVLP exposure to patients, with the expectation that an extremely low-risk profile should be achievable.

Overall, the assays described should in no way be considered an all-inclusive or definitive list. Instead, alternative assays and techniques for adventitious agent detection may be used when appropriate and demonstrated to have equivalent or better specificity, sensitivity, and precision than the existing methods. As an example Q-PCR-based methods for RVLP quantitation have been cited, being highly comparable to the TEM method (6). In addition, the FDA has requested the use of PERT assay to detect viral activity in human viral vaccine products produced in mammalian or avian cells rather than the standard reverse transcriptase assay. Manufacturers are encouraged to discuss these types of alternatives with regulatory authorities.

Genotypic Characterization and Genetic Stability

Genotypic characterization is considered a part of the overall evaluation of quality and is performed in response to recommendations described in the guidance documents (2,7). The objective of the characterization is to establish that the intact and correct coding sequence has been incorporated into the host cell genome and is stably maintained during culture from MCB through the end of full-scale production. This is addressed through the analysis of the integrated expression construct for rearrangements within the coding region, for number of independent sites of integration and for copy number. At Wyeth, the emphasis is placed on the characterization of the MCB, and includes the analyses described above in addition to an evaluation of transcript integrity. For cell lines that have progressed to the point of full-scale manufacturing, recombinant cells from the EOP are analyzed in the same manner to provide assurance that no significant changes have occurred in the product genes over prolonged culture. The data derived from these analyses are critical in helping to establish a limit of in vitro cell age (discussed later within this chapter). The genotypic characterization paradigm employed is detailed in Table 1.

Table 1 Genotypic Characterization Paradigm

	Phase I/II IND	Commercial registration
Northern analysis	Total RNA from MCB cells	Total RNA from MCB and EOP cells
Load control	10% MCB total RNA	10%, 1% MCB total RNA
Southern analysis	Genomic DNA from MCB cells	Genomic DNA from MCB and EOP cells
Plasmid spike	10%, 1% plasmid load	10%, 1% plasmid load
Assessment	Integrity of coding region	Integrity of coding region Consistency of integration sites Gene copy number
Nucleotide sequencing	High-fidelity RT-PCR	High-fidelity RT-PCR
Starting material	Total RNA from MCB cells	Total RNA from MCB and EOP cells
Sequencing template	Gel purified PCR product	Gel purified PCR product

Although not specifically addressed in ICH Topic Q5B, analysis of the integrity of the transcript encoding the recombinant protein is included in the genotypic characterization of the MCB and EOP. The physical state of the RNA transcript is evaluated by Northern blot analysis for the purpose of demonstrating that the expected transcript is produced and remains qualitatively indistinguishable over the course of production. RNA load controls are included to demonstrate the sensitivity of the assay to detect potential aberrant transcripts.

The structure and integrity of the gene(s) integrated into the host cell genome are assessed by Southern blot analysis of genomic DNA digested with restriction enzymes that immediately flank the coding region. The inclusion of plasmid DNA, digested in the same manner, provides a direct comparison and allows for confirmation of the presence of the appropriately sized restriction fragments. Furthermore, when diluted, the plasmid DNA allows for an estimation of the sensitivity of the Southern method to detect variant or aberrant sequences. Any rearrangements of the coding region may be revealed by the presence of hybridizing fragments that are larger or smaller than those predicted.

By choosing restriction enzymes that cleave only once in the expression plasmid, digestion of genomic DNA will generally yield genomic restriction fragments containing both plasmid sequences as well as host cell genomic sequences. These fragments are expected to be unique for each integration event. Southern blot analysis of these genomic end fragments can be used to assess the expression plasmid integrant structures as well as provide an indication as to the number of independent sites of integration within the host cell genome. This strategy provides a unique genetic fingerprint of the production cell line and can facilitate detection of genetic changes that could occur over the duration of a culture.

Analysis of the copy number of the expression construct is also enabled by the Southern blot method. This evaluation is performed to gain an understanding of how the expression level within the recombinant cell line is achieved (for example, via single copy or via gene amplification to high copy number), and to provide a frame of reference for comparisons over the course of production cultures. It must be noted that the actual numerical value for copy number that is obtained by this analysis is only an approximation, as the inherent technical variability associated with this method contributes to a large variance in the final calculated value. As such, the method is only intended to look for gross changes in copy number. While more quantitative methods such as Q-PCR could be applied to this question, the relevance of small changes in copy number during cell culture has not been established and more direct and sensitive measures of the potential consequences are nonetheless available.

To provide an additional level of assurance that the transcripts produced by the cells are of the expected sequence, RT-PCR, and subsequent nucleotide sequencing of the cDNA template is performed to confirm the fidelity of the coding region of the transcript. The nucleic acid sequence encoding the recombinant protein may be verified by sequencing of individual cDNA clones or material generated by PCR. The guidelines indicate that the primary sequence should be identical to that of the expression plasmid, within the limits of the technique, and should translate to the expected protein sequence. The method, however, is not intended to detect low levels of variant sequences.

Case Study

A CHO cell line expressing rhBMP-2 was evaluated for suitability using the paradigm described above. rhBMP-2 is a member of the TGF-β superfamily and is expressed, in its mature form, as a homodimer with a mass of approximately 30,000 Da. A bi-cistronic expression plasmid containing the genes coding for rhBMP-2 and a selectable marker was used to transfect CHO cells. Following selection and cloning, an individual clone secreting suitable levels of rhBMP-2 protein was chosen as the production lineage. A serum-free MCB corresponding to the production cell line was created and used to subsequently establish a WCB.

A serum-free MCB corresponding to the EMC-G5 cell line was created and was used to subsequently establish a WCB. Cells from the WCB were used to inoculate a full-scale (2500-L cell culture) manufacturing campaign. Cells from the final harvest of the manufacturing campaign represent approximately 81 CPDs from the MCB and are referred to as rhBMP-2 EOP.

Results from purity testing of the rhBMP-2 cell line are summarized in Table 2. Briefly, the results indicate that cells from MCB, WCB, and EOP are negative for contamination from bacteria, fungi, and mycoplasma. No adventitious viruses were detected in any of the test articles by either in vitro or in vivo assay. Application of the cocultivation assays, including focus formation and reverse transcriptase activity endpoints, resulted in no detection of retrovirus activity. TEM identified the presence of both centriole-associated A-type and budding and/or extracellular C-type retrovirus-like particles. Both types of particles have been previously reported to be produced by CHO cells (8). Finally, isoenzyme analysis confirms that the cell line is of hamster origin.

Genotypic analysis was performed on nucleic acids isolated from rhBMP-2 MCB, WCB, and EOP cells. The physical state of rhBMP-2 transcripts was assessed using Northern blot analysis (Fig. 2). The results demonstrate a single bi-cistronic transcript (containing both rhBMP-2 and selectable marker genes) of the expected size with no evidence of aberrant transcript greater than 1% of the total RNA population. Furthermore, the MCB transcript co-migrates and is qualitatively indistinguishable from the WCB and EOP transcripts, suggesting stability through cell culture scale-up and full-scale production.

The integrity of the rhBMP-2 expression plasmid incorporated into the genome of the recombinant CHO cell line was evaluated by restriction enzyme digestion and subsequent Southern blot analysis. Figure 3 shows the results of a Southern blot analysis of *Hin*d III digested genomic DNA isolated from rhBMP-2 MCB, WCB, and EOP cells. *Hin*d III sites immediately flank the coding region; therefore, if this region of the expression plasmid is intact, the labeled rhBMP-2 probe should detect a single 1.6 kb fragment. The results show a single band of 1.6 kb in rhBMP-2 MCB, WCB, and EOP genomic DNA that co-migrates with the *Hin*d III-digested plasmid controls. No evidence of rearrangements within this region is observed even though the predicted fragment is detected in the plasmid control at a level of 25 pg (equivalent to ~10% load control).

Figure 4 shows a Southern blot analysis of *Bgl* II-digested genomic DNA hybridized to an rhBMP-2 probe. The expression plasmid contains a single *Bgl* II restriction site immediately upstream (5′) of the rhBMP-2 gene and, as such, digestion of genomic DNA isolated from MCB, WCB, and EOP cells would generate fragments that would be predicted to contain rhBMP-2 plasmid sequences across the site of integration to the first *Bgl* II site in the adjacent host cell DNA. The sizes of these fragments are dependent on the location of the flanking genomic *Bgl* II sites and would be expected to be unique for each integration event. The results demonstrate the presence of a single 3′ genomic end fragment, supporting a model in which a single copy of expression plasmid integrated at a single chromosomal site within the host cell genome. Furthermore, identical fragments were detected in MCB, WCB, and EOP cells, indicating the integrated plasmid is stable over the course of full-scale production.

Estimates of the number of integrated rhBMP-2 plasmids per MCB cell (copy number) were obtained by comparative Southern blot analysis. MCB genomic DNA as well as varying amounts of plasmid DNA was digested with *Eco*R I, to excise the rhBMP-2 coding region, and compared by Southern blot analysis using an rhBMP-2 probe. As shown in Figure 5, a single *Eco*R I fragment of the expected size is observed in DNA derived from MCB cells that co-migrates with the plasmid controls. Densitometry followed by quantitative analysis of multiple blots provide an approximate value of 70 rhBMP-2 gene copies per MCB cell, strongly suggesting that selection of the cell

Table 2 Results of rhBMP-2 CHO Cell Line Purity Testing

Test	rhBMP-2 MCB	rhBMP-2 WCB	rhBMP-2 EOP
Microbial agents			
Agar cultivable and non-cultivable mycoplasmas	Negative	Negative	Negative
Test for presence of bacterial and fungal contaminants: sterility test using a direct inoculation method	Negative	Negative	Negative
Adventitious virus			
In vitro assay for detection of adventitious viral contaminants	MRC-5, Vero, CHO, HeLa, MEF; 28 days, negative	MRC-5, Vero, CHO; 14 days, negative	MRC-5, Vero, CHO, HeLa, MEF; 28 days, negative
In vitro assay for the presence of murine minute virus	Not tested[a]	Not tested[a]	Negative
In vivo assay for viral contaminants	Negative	Not tested	Negative
Mouse antibody production test	Negative	Not tested	Negative
Hamster antibody production test	Negative	Not tested	Negative
In vitro assay for the presence of bovine viruses	Negative	Not tested	Negative
In vitro assay for the presence of porcine parvovirus	Negative	Not tested	Negative
Retroviruses			
Cocultivation with mink lung cells	Negative	Not tested	Negative
Cocultivation with human rhabdomyosarcoma cells	Negative	Not tested	Negative
Transmission electron microscopy	A- and C-type retroviral particles	Not tested	A- and C-type retroviral particles
Reverse transcriptase	Negative	Not tested	Negative
Species identity			
Isoenzyme analysis	Chinese hamster	Not tested	Chinese hamster

[a] Assay not available at the time of cell bank purity testing.

line in high levels of methotrexate resulted in amplification of the integrated plasmid.

To evaluate rhBMP-2 gene structure and integrity at a greater level of resolution than that afforded by Northern or Southern blot analysis, nucleotide sequencing was performed. In order to confirm that the predominantly expressed rhBMP-2 coding sequence matches with what is predicted by the expression plasmid, DNA sequencing was carried out on pools of RT-PCR cDNA clones derived from MCB and EOP cells. Figure 6 shows representative data derived from sequencing of amplified products that were cloned as single-stranded cDNA into M13 phage vectors. The strong uniform peaks and low levels of background are reflective of the method utilized for sequencing. Analysis of the entire 1.2 kb rhBMP-2 coding region cDNA confirms that the predominant and only detectable sequence in MCB and EOP cells is that predicted by the expression plasmid. Previous data obtained suggest that this method would be capable of detecting variant sequences expressed at a level no lower than 20% of the population.

In summary, the data presented in the section indicate that the cell line is free from any detectable adventitious agent contamination. A single copy of the expression plasmid has integrated at a single chromosomal site in the CHO genome, and amplification has resulted in approximately 70 rhBMP-2 gene copies per MCB cell. There is no evidence for gross rearrangement in any of the integrated rhBMP-2 plasmids or predicted transcripts. DNA sequence analysis reveals only the sequence predicted by the expression plasmid. Taken together with process data, the rhBMP-2 cell line is demonstrated to be stable and appropriate for full-scale manufacturing.

CELL LINE STABILITY

After completing the development of a recombinant production cell line and establishment and characterization

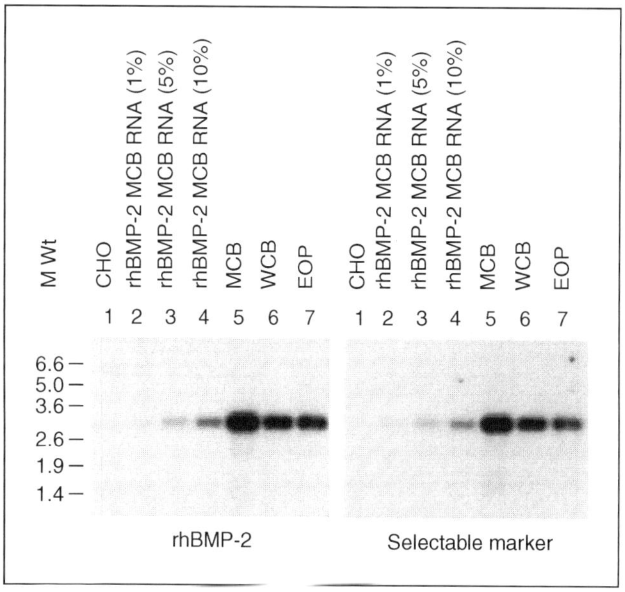

Figure 2 Northern blot analysis of total rhBMP-2 RNA isolated from MCB, WCB, and EOP cells.

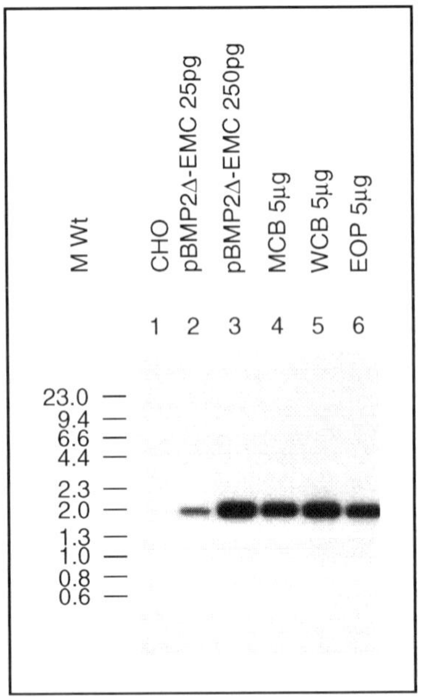

Figure 3 Southern blot analysis of *Hin*d III-digested rhBMP-2.

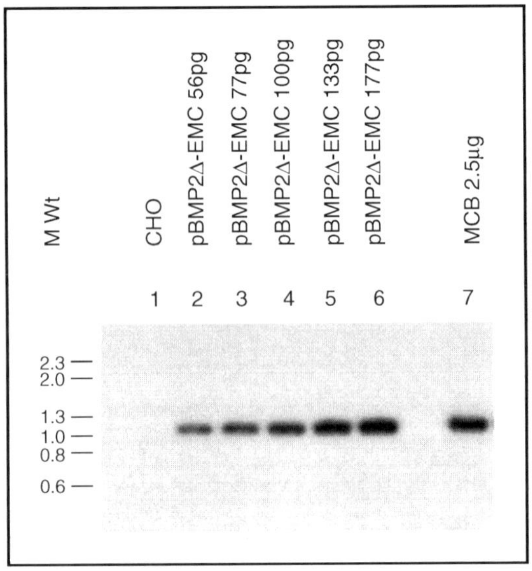

Figure 5 Comparative Southern blot analysis of rhBMP-2 MCB and EOP DNA to assess rhBMP-2 gene copy number.

of cGMP cell banks, the next challenge is to determine the extent to which the production cell line maintains relative phenotypic and genotypic stability. Typically, large-scale mammalian cell-based manufacturing processes involve cultivation periods measured in weeks or months, and as such it is incumbent on the manufacturer to ensure that the accumulation of cell doublings does not lead to acute changes in cell characteristics that would affect their performance, or the quality of the product produced.

Changes in the expression phenotype of a production cell line can be brought about by several mechanisms including chromosomal instability (loss of transgene copy number), gene silencing, genomic positional effects and population heterogeneity (9). The genetic and phenotypic stability of a production cell line is defined within the limits of a particular cell culture manufacturing process. The link, if any, between cell line stability and the product quality profile of biopharmaceuticals will be influenced by the characteristics of the production cell line, the manufacturing process and the attributes of the protein biopharmaceutical itself. Accordingly, a stability profile needs to be established within the context of the particular manufacturing process.

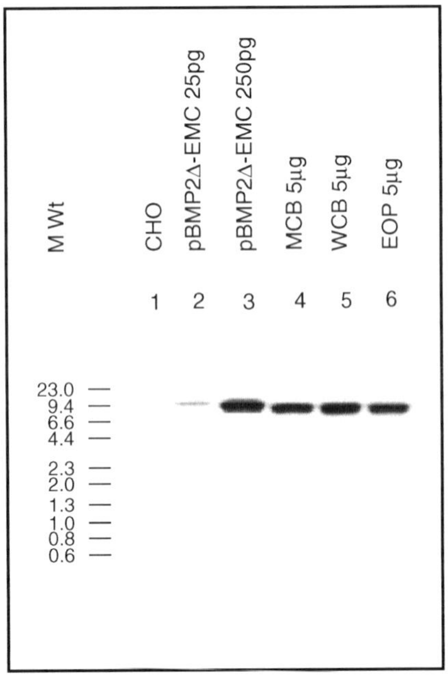

Figure 4 Southern blot analysis of *Bgl* II-digested rhBMP-2 MCB, WCB and EOP.

Thus the *limit of in vitro age* for a production cell line and a manufacturing process is defined by phenotypic attributes (growth rate and cellular productivity), genotypic [verification of intact integrated transgene sequence, transcript(s) of predicted size, and verification of transgene coding sequence] and product quality [verification that the cell line and process produces the intended product (biochemical, physical, and functional characterization)]. Establishment of the limit of in vitro age provides assurance of product consistency over the duration of the cell culture manufacturing process. In the ideal case, production cell lines are selected and manufacturing processes designed so that the product is made during a period when growth rate and cellular productivity are relatively constant. This ideal is not always met and judgments regarding relative stability need to be made and supported by growth and productivity trend analysis, Northern and Southern blot genotypic analysis and most importantly, characterization of product quality.

Establishment of the Limit of In Vitro Age

Establishing the limit of in vitro age involves relating the CPDs of a production cell line to phenotypic, genotypic, and product quality attributes of the production cell line and cell culture process. Typically, and ideally, this involves continuous passage of production cells in a cell culture system representative of the conditions and operations employed in the manufacturing process (i.e., the cell culture system should employ the same

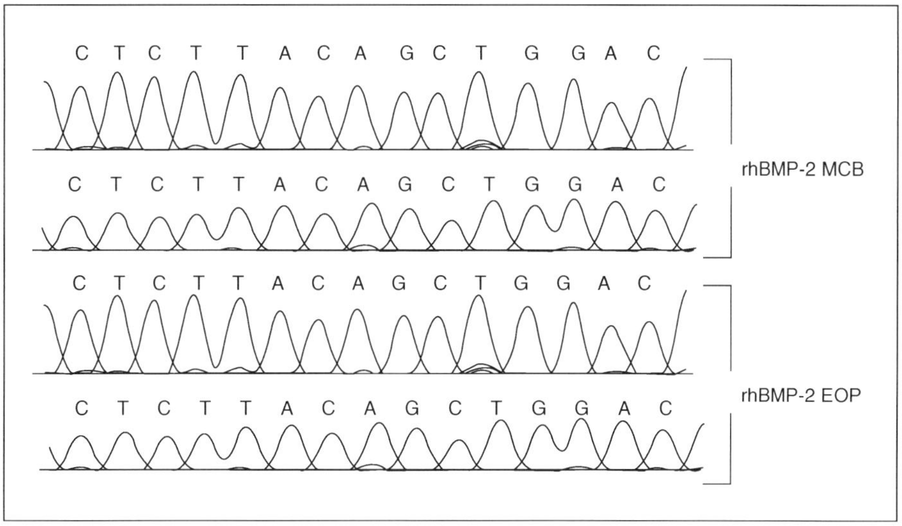

Figure 6 Representative rhBMP-2 DNA sequencing data.

media formulations and cell culture unit operations used in the cell culture manufacturing process).

There are three predominant types of cell culture manufacturing processes currently employed in the biopharmaceutical industry: continuous batch-refeed, terminal fed-batch and perfusion. Continuous batch-refeed processes involve expansion and maintenance of production cell line cultures within the production bioreactor itself. In this process modality, production cultures are managed on a schedule where cultures are partially harvested every two to five days by withdrawing 50% to 80% of the culture volume and restoring the production bioreactor to full working volume by addition of fresh culture medium. In contrast, batch production cultures are terminal and the entire contents of the production bioreactor are harvested at the conclusion of the culture. Terminal fed-batch culture durations are typically extended for two or more weeks by addition of nutrient feeds at specified time points. True terminal fed-batch processes are associated with a single thaw from a WCB but hybrid fed-batch manufacturing processes can also draw multiple inoculums from a continuously maintained seed bioreactor. Finally, perfusion culture involves maintenance of a constant working volume in the production bioreactor by continuous introduction of fresh culture medium and removal of spent medium through the use of a scaleable cell retention system. Perfusion systems allow for extended culture durations (four weeks or more) and continuous production operations. Here again, hybrid perfusion culture systems in which the cell culture inoculum is continuously maintained in a seed bioreactor have also been employed.

For the purposes of determining the limit of in vitro age, the process operation where the majority of the population doublings accrue must be taken into account. For example in a continuous batch-refeed process the majority of population doublings accrue in the production bioreactor, whereas in a terminal batch process the majority of population doublings are accrued during culture expansion on the way to the production bioreactor.

Case Study 1: Continuous Batch-Refeed Manufacturing Process

rhBMP-2 is a biopharmaceutical product manufactured using a continuous batch-refeed cell culture process. The manufacturing process starts with revival of cultures from the WCB. The cultures are expanded by sequential passage until sufficient cell number and culture volume are reached to inoculate a 250 L stirred tank bioreactor. Culture inoculum from the 250 L bioreactor is, in turn, used to inoculate a 2500-L stirred tank production bioreactor. The production bioreactor is operated on a three-day schedule whereupon approximately 80% of the culture is harvested and sent on for downstream processing into bulk drug substance. The remaining 20% of the culture in the production bioreactor is diluted with fresh growth media to restore the full working volume and initiate the next production cycle. Multiple-three day batch-refeed cycles may therefore be conducted in the production bioreactor. The number of consecutive cycles that can be used for full-scale production is primarily limited by (other than practical considerations like maintaining bioreactor sterility) the stability of the cell line. Provided that all cycles produce enough protein and that the product quality does not vary over the course of multiple cycles, a continuous culture can be run for months in this mode, with batches generated from each harvest of a three-day batch-refeed cycle. For example, a culture that ran at 2500-L scale for 90 days would consist of 30 three-day batch-refeed cycles. To ensure that such a process was suitable, the limit of in vitro cell age would need to be suitably long so as to accommodate 90 days in the production bioreactor, plus the additional cell generations accrued during scale-up of the culture from the WCB to the 2500-L bioreactor (approximately another 30 days), for a total of 120 days from the WCB to the end of production. In this production modality the cell line accrues the majority of its population doublings in the 2500-L production bioreactor, and as such it would be most appropriate to demonstrate the limit of in vitro cell age under conditions that closely mimic those at production scale.

In this example the limit of in vitro age for the rhBMP-2 production cell line was formally determined at production scale (2500 L) by continuously culturing the rhBMP-2 production cells for approximately 80 population doublings. Figure 7A shows a graph relating the intrinsic growth rate of the production cells (expressed in inverse hours) to the CPDs accrued in the production bioreactor (note that approximately 30 CPD were accrued from the WCB thaw to the start of full-scale production). The data indicated that the intrinsic growth rate of the production cell line was stable out to 80 CPD. In contrast, the cell specific productivity data (Fig. 7B) displayed a negative slope from 30 to 80 CPD of continuous culture.

While changes in the cell specific productivity phenotype of production cells would certainly have implications with respect to process productivity, such a change may not by itself be the primary driver in establishing the limit of in vitro age. As discussed previously in the section on cell line suitability, genotypic analysis of EOP cells did not detect any gross changes to the integrated rhBMP-2 transgene or the resulting transcript. Furthermore (and most importantly), full characterization of bulk drug substance manufactured at the beginning and EOP did not reveal any biochemical

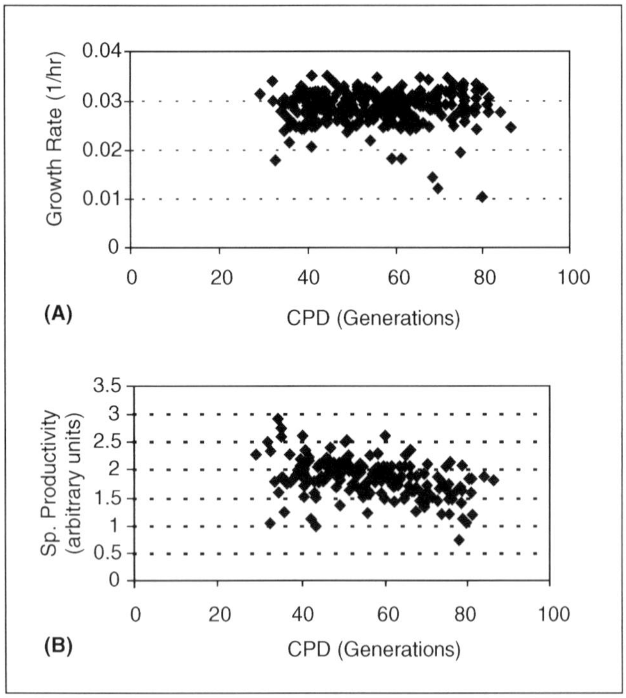

Figure 7 Phenotypic analysis of the rhBMP-2 production cell line at production scale. rhBMP-2 WCB cells were used in multiple 2500-L production-scale manufacturing campaigns. The CPD, growth rate and cell specific productivity of the production cell line were monitored over the course of each production run. CPD indicated are relative to the WCB; approximately 30 CPD are accrued prior to the point at which cultures reach 2500-L scale. Data are given for batch-refeed cycles in 2500-L scale. (**A**) Intrinsic growth rate vs. CPD. (**B**) Specific productivity vs. CPD.

changes in the rhBMP-2 molecule. Accordingly, taking into account the phenotypic, genotypic, and product quality information the limit of in vitro age for the rhBMP-2 production cell line was established at 80 CPD.

ROLE OF PROCESS CHARACTERIZATION STUDIES

Although the central step in cell culture process validation is generally full-scale consistency runs in the manufacturing facility, this step must be preceded by laboratory scale process characterization studies. The full-scale runs are intended to demonstrate that the process can consistently generate drug substance that meets specifications. But this requires knowing which input parameters affect drug substance characteristics, and for each of these parameters, knowing the range within which it must be maintained. This is the information that must be derived from process characterization studies. Guidance from the ICH (10) calls for (*i*) defining the critical quality attributes of the active substance, (*ii*) identifying process parameters that could affect these critical quality attributes, and (*iii*) determining the appropriate operating ranges for these parameters.

For a cell culture-derived protein, key drug substance characteristics typically include specific activity, concentrations of certain classes of non-product-related impurities, and relative amounts of certain isoforms of the active protein. The concentration of non-product-related impurities is generally influenced more by purification process parameters than by cell culture process parameters, so cell culture process characterization studies generally focus on specific activity and relative amounts of isoforms. The isoforms differ from each other with respect to characteristics such as composition of N-linked and O-linked carbohydrate, amino acid sequence at the N-terminus or C-terminus, or the presence of modified forms of certain amino acid residues.

Validation of a Scaled-Down Model

Before beginning cell culture process characterization studies, the scaled-down model must be validated. This model is typically a laboratory bioreactor that has a working volume in the 1 to 5 L range, in combination with a scaled-down purification system. It must be shown to accurately represent the full-scale process with respect to cell culture performance—growth rates and cellular productivities—and with respect to characteristics of purified drug substance.

Effects of Process Parameters on Isoform Distribution

Several cell culture input process parameters have been shown, in specific cases, to affect characteristics of recombinant proteins. These include pH, temperature, DO, and time of harvest.

pH has been shown to affect the relative abundance of isoforms of a glycoprotein that has been produced in recombinant CHO cells at Wyeth. The isoforms differ with respect to the N-terminus amino acid sequence. The polypeptide chain can have either a shorter form or a longer form, with the longer form containing the same amino acid sequence as the short form, but with an additional 17 amino acids at the N-terminus.

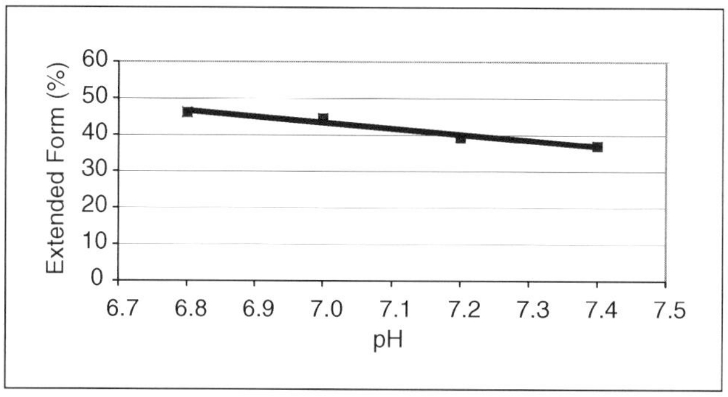

Figure 8 Impact of culture pH on the relative abundance of the longer isoform of a recombinant glycoprotein.

The relative amounts of the long and short forms are affected by pH in the production bioreactor (Fig. 8).

pH has also been shown by Muthing et al. (11) to affect the relative abundance of isoforms of a monoclonal antibody produced in a murine hybridoma. The isoforms in this case differ with respect to glycosylation at the N-linked site on each heavy chain. As is common in monoclonal antibodies, the carbohydrate at this site has two branches (biantennary structure), each of which may or may not terminate with a galactose residue. This results in three isoforms: one with no galactose residues (G0 isoform), one with one galactose residue (G1 isoform), and one with two galactose residues (G2 isoform). The relative abundance of the G2 isoform was shown to be 32% at pH 7.4 but only 16% at pH 7.2 or 6.9.

Temperature has been shown to affect the abundance of isoforms of a recombinant antibody that has been produced in CHO cells at Wyeth. These isoforms differ with respect to the extent of deamidation of an asparagine residue. In a 14-day fed batch culture, the abundance of the deamidated form is less if the temperature is shifted from 37°C to 31°C at day 7 than when it is held at 37°C throughout (Table 3).

DO has been shown by Kunkel et al. (12) to affect the abundance of isoforms of a monoclonal antibody produced in a murine hybridoma. The relative abundance of the G2 isoform is 30% when DO is maintained at 100% of air saturation, 25% when DO is maintained at 50% of air saturation, and only 9% when DO is maintained at 10% of air saturation.

Time of harvest has been shown by Schenerman et al. (13) to affect the abundance of isoforms of the recombinant antibody Synagis®, produced in NS0 murine myeloma culture. The ratio between the amount of the G0 isoform and G2 isoform was found to be 2.4 when the culture was harvested at day 18, 2.1 when it was harvested at day 11, and 1.0 when the culture was harvested at day 5.

Table 3 Effect of Bioreactor Conditions on Extent of Deamidation of an Antibody Product

Process conditions		
Temperature (°C)[a]	pH	Extent of deamidation (%)
37	7.0	48.6
37	6.8	49.0
31	7.0	39.1
31	6.8	38.6

[a] Days 7–14 (in all cases, temperature was 37°C before day 7).

Acceptable Range for a Process Parameter

The results of cell culture process characterization studies can be used to identify an acceptable range for each input process parameter. When all parameters are maintained within the acceptable ranges, drug substance can be expected to meet specifications. But since effects of different parameters may be additive or may even show positive interactions, process characterization studies will ideally be multifactorial. Multifactorial characterization studies for a recombinant antibody produced by NS0 cells have been reported by Moran et al. (14).

When an acceptable range has been identified for each cell culture parameter, this range is compared with the range that can be consistently maintained in the manufacturing bioreactor. This latter range is often referred to as the "target range." In general, the acceptable range will be found to be much wider than the target range.

Short Duration Deviations from the Acceptable Range

Cell culture process characterization studies may also evaluate the effects of short-duration exposure of the culture to levels of an input parameter that are outside the acceptable range. If a parameter is outside of its acceptable range for the full duration of the process phase (growth phase or production phase) it can lead to drug substance that does not meet specifications. But if this parameter is outside the acceptable range for a shorter time period, drug substance may still meet specifications. During routine manufacturing operations, an equipment failure will occasionally result in an excursion of this type. The impact on drug substance characteristics can be anticipated on the basis of process characterization studies that move a parameter a known amount outside of the acceptable range for a known amount of time.

Full-Scale Consistency Runs

Full-scale consistency runs serve a number of purposes during cell culture process validation: (*i*) to demonstrate that each input parameter can consistently be maintained within the acceptable range that was established on the basis of process characterization studies; (*ii*) to demonstrate that when each input parameter is maintained within the acceptable range, drug substance consistently meets specifications; and (*iii*) to establish the concentrations of non-product-related impurities that are generated by the cell culture process when each input parameter is maintained within the acceptable range. These impurities include host cell proteins, host cell DNA, and endogenous virus-like particles that are generated by the host cells.

REFERENCES

1. ICH Topic Q5D. Quality of biotechnological products: derivation and characterisation of cell substrates used for production of biotechnological/biological products. In: International Conference on Harmonization, 16 July 1997.

2. ICH Topic Q5B. Quality of biotechnological products: analysis of the expression construct in cell lines used for production of r-DNA derived protein products. In: International Conference on Harmonization, 29 November 1995.
3. ICH Topic Q5A. Quality of biotechnological products: viral safety evaluation of biotechnology products derived from cell lines of human or animal origin. In: International Conference on Harmonization, 04 March 1997.
4. Center for Biologics Evaluation and Research. Points to consider in the characterization of cell lines used to produce biologicals. U.S. Department of Health and Human Services/Food and Drug Administration, 1993.
5. Wiebe ME, Becker F, Lazar R, et al. A multifaceted approach to assure that recombinant tPA is free of adventitious virus. Dev Biol Stand 1989; 70:147.
6. de Wit C, Fautz C, Xu Y. Real-time quantitative PCR for retrovirus-like particle quantification in CHO cell culture. Biologicals 2000; 28:137–48.
7. Center for Biologics Evaluation and Research. Supplement to the points to consider in the production and testing of new drugs and biologics produced by recombinant DNA technology: nucleic acid characterization and genetic stability. U.S. Department of Health and Human Services/Food and Drug Administration, 1992.
8. Lubiniecki AS, May LH. Cell bank characterization for recombinant DNA mammalian cell lines. Dev Biol Stand 1985; 60:141–6.
9. Sinacore MS, Charlebois TS, Drapeau D, Leonard M, Harrison S, Adamson SR. Animal cell stability. In: Spier RE, ed. Encyclopedia of Cell Technology. New York: Wiley, 2000.
10. ICH Topic Q7A. Good manufacturing practice guidance for active pharmaceutical ingredients. In: International Conference on Harmonization, August 2001.
11. Muthing J, Kemminer SE, Conradt HS, et al. Effects of buffering conditions and culture pH on production rates and glycosylation of clinical phase I anti-melanoma mouse IgG3 monoclonal antibody R24. Biotechnol Bioeng 2003; 83(3):321–34.
12. Kunkel JP, Jan DCH, Butler M, Jamieson JC. Comparisons of the glycosylation of a monoclonal antibody produced under nominally identical cell culture conditions in two different bioreactors. Biotechnol Prog 2000; 16(3):462–70.
13. Schenerman MA, Hope JN, Kletke C, et al. Comparability testing of a humanized monoclonal antibody (Synagis(R)) to support cell line stability, process validation, and scale-up for manufacturing. Biologicals 1999; 27(3):203–15.
14. Moran EB, McGowan ST, McGuire JM, et al. A systematic approach to the validation of process control parameters for monoclonal antibody production in fed-batch culture of a murine myeloma. Biotechnol Bioeng 2000; 69:242–55.

37

Cleaning Validation for the Pharmaceutical, Biopharmaceutical, Cosmetic, Nutraceutical, Medical Device, and Diagnostic Industries

Rebecca Brewer
Dober Group, Midlothian, Illinois, U.S.A.

Cleaning validation has come a long way since the days of the Barr Laboratories Court Case and since the first FDA guidelines referencing the subject of cleaning validation were published in 1991. At that time, the requirements for cleaning validation barely filled a single page of the Bulk Pharmaceutical Chemical and Biopharmaceutical guidance documents. Those documents were then expanded to create the Guide to Inspection of Cleaning Validations by FDA (first published in 1992 as a Mid-Atlantic Inspection Guidance, then reissued as an FDA guidance document in 1993). Today, despite nearly 15 years of exposure to the requirements for cleaning validation, this validation topic remains one of the areas of validation that people frequently profess to know the least about.

GMP regulations have their basis in cleaning validation. Beginning in 1906 with Upton Sinclair's "The Jungle," the people demanded that the government improve cleanliness practices in the processing of food giving rise to what we know of today as the cGMPs for both food and drugs. While cleaning has always been part of the GMP regulations, cleaning activities have not enjoyed the limelight. The GMPs that we follow today were predominantly written in 1978. References to cleaning and documentation associated with cleaning can be found throughout. As with many other areas of validation, however, there is no explicit reference to cleaning as a process to be validated. It is this very aspect of the GMPs that was challenged in the Barr Laboratories court case. In that decision, Judge Wolin ruled that cleaning did require treatment as a process and therefore required validation. In 1996 proposed revisions to the GMPs were drafted by the FDA; although not adopted, these revisions proposed to redefine the manufacturing process as beginning with a cleaning operation.

Cross-contamination is a significant risk to patients. This is true whether through direct administration to a patient, or in the case of in vitro diagnostics, through the performance of a test on a patient sample. Cleaning and cleaning validation are two activities that have the largest opportunity to prevent patient risk by assuring that no cross-contamination can occur. Cleaning validation is becoming more and more important as we work with increasingly potent, increasingly complicated drug substances and increasingly complex biotechnology products. Our products have greater and greater risks of interaction with one another resulting in harmful effects to patients. To truly limit this risk, scientific approaches must be taken in all aspects of the cleaning and cleaning validation program.

When the FDA published "Pharmaceutical cGMPs for the 21st Century: A Risk-Based Approach" in August of 2002, and reported on their progress in September of 2004, the continued importance of sound scientific rationales in all that we do in pharmaceutical manufacturing and validation was reinforced. The pharmaceutical community as a whole renewed their efforts to ensure that sound quality principles were followed in the identification of critical to quality attributes for all measurements and analysis. Although risk-based decision-making in the establishment of scientific rationales was always a cornerstone of cleaning validation requirements, efforts have been renewed to ensure the incorporation of risk analysis documentation in cleaning programs.

ORGANIZING FOR CLEANING VALIDATION

Due to the high number of risk-based rationales that are included in cleaning validation programs, strong policies are required to help drive the decision making. Some programs make use of cleaning validation master plans

Abbreviations used in this chapter: AAMI, American Association of Medical Instrumentation; API, active pharmaceutical ingredient; APIC, Active Pharmaceutical Ingredients Committee; APR, annual product review; ASME, American Society of Mechanical Engineers; CAPA, corrective and preventive action; cGMP, current good manufacturing practice; EU, European Union; FDA, Food and Drug Administration; FTIR, Fourier Transform Infra-Red Spectrometry; GMP, good manufacturing practice; GRAS, generally recognized as safe; HPLC, high-performance liquid chromatography; ICH, International Conference on Harmonization; IMS, ion mobility spectrometry; MAC/MACO, maximum allowable carryover; NaOH, sodium hydroxide; NOEL, no observed effect level; PAT, process analytical technology; PLC, programmable logic controller; SF, safety factor; SOP, standard operating procedure; TOC, total oxidizable carbon; TSE, transmissible spongiform encephalopathies; USP, *United States Pharmacopeia*; VOC, volatile organic carbons; WHO, World Health Organization.

for cleaning validation in addition to cleaning validation policies. Whether a master plan or a policy, these guiding documents must include the decision-making framework appropriate to a plant site, manufacturing facility and/or dosage form.

Master Planning could be the subject of an entire chapter unto itself, but suffice it to say that the cleaning validation master plan follows the same basic principles as any Validation Master Plan. In fact, Cleaning Validation may be addressed as a section of a general Validation Master Plan (one that governs more than one type of validation) or as a stand-alone Master Plan. The Master Plan should:

- Provide an overview of the site/facility/area that is governed by the Master Plan
- Provide an overview of the typical manufacturing process(es) that are to be performed in the area and the dosage forms that are produced
- Provide an overview of the types of cleaning that are to be used (e.g., automated Clean-In-Place or Clean-Out-of-Place, semi-automated cleaning or manual cleaning)
- Provide the responsibilities of the various departments having a role in cleaning validation activities
- Provide the minimum requirements for the cleaning validation program, including:
    - Necessary scientific rationales in support of the program:
        - Residue selection
        - Equipment characterization
        - Product contact surface area calculation
        - Limits calculation
        - Sample site selection
        - Product grouping (if any)
        - Equipment grouping (if any)
    - Required studies in support of the program:
        - Analytical methods validation
        - Sampling method recover studies
        - Cycle development for cleaning processes
    - Essential Programs that maintain the validated state and their required elements:
        - Cleaning and testing, if any, to be conducted upon the introduction of new or repaired equipment
        - Monitoring of cleaning after validation completion
    - Routinely conducted compliance initiatives on site that maintain quality and will affect the company's ability to maintain the validated state
        - Failure investigation
        - Change control
        - Preventive maintenance
        - Calibration
        - Revalidation
    - Important SOPs Governing Cleaning and Cleaning Validation
        - Development of cleaning SOPs (especially for manual cleaning operations)
        - Equipment cleaning and use logs
        - Visual inspection requirements for cleaned equipment
        - Equipment quarantine and release
        - Equipment sampling procedures for cleaning assessments (e.g., swab, rinse, etc.)
- Provide the list of equipment and/or systems subject to cleaning validation
- Provide a status summary of progress in the area of cleaning validation (for regulatory review—may also take the form of an annual summary report to the cleaning validation master plan)

(Note: Each of these topics will be addressed throughout the remainder of this chapter).

In some facilities, the cleaning validation policy or Master Plan only serves as the most basic outline of the required elements for a successful cleaning validation program. In these facilities, scientific rationales are maintained as stand-alone documents. This approach is helpful in a facility or site where there are several dosage forms or product types, and where the requirements for one dosage form or product type may be overly stringent for some or especially lax for others.

Where the rationales are maintained as separate documents, however, it becomes critical that the hierarchy in which these documents will reside be strictly maintained. Employees of many departments must be assured that they can cross-reference the applicable documents to their area of interest with no ambiguity as to which document applies. Only through careful organization of the supporting documents can we assure that consistent decision-making is maintained over time.

Upon audit or review of older and/or existing programs, it is frequently discovered that prior rationales have been contradicted or forgotten and that the program has strayed from its original goals. Maintaining these documents over time becomes critical to ensure that no internal inconsistencies develop. Strong programs permit their policy and/or validation Master Plan to serve as an "index" to the risk-based rationales that will comprise the remaining portions of the program by cross-referencing their locations within the quality system. In this manner, the documents that comprise the program are always near at hand, and are readily referenced when making decisions for new product introduction, new equipment introduction, or changes in the factory. Requiring the periodic review of both Master Plans and their associated scientific rationale reference documents is recommended. A two to three-year review cycle is typically appropriate. Facilities that have an environment with frequent changes would require a highly frequent review and update of their rationales to ensure that cleaning validation strategies remain current.

See Table 1 for a review of common documentation types supporting cleaning validation initiatives.

CLEANING VALIDATION ACTIVITIES

Cleaning validation, more so than any other kind of validation, is a multidisciplinary activity. To effectively clean we require classification of equipment, understanding of drug products, precise analytical methods, as well as a clear understanding of how to sample and collect residues from surfaces. All of these activities relate to the expertise of different disciplines, including: operations, engineering, research and development, toxicology or medical personnel, validation, quality control and quality assurance.

Table 1 Typical Document Hierarchy and Scope

Document name	Typical contents/requirements
Corporate guidelines on validation or corporate policy on validation	Policy document affecting all corporate sites Multiple validation topics exist Scope/content is broad/general to apply to several sites and diverse dosage forms Corporate guidelines or corporate policies are typical elements of a company's global quality system and form an important starting point for cleaning program decisions
Cleaning validation policy	Typically established at the site level, a cleaning validation policy is part of the quality system and establishes the minimum required elements of the cleaning validation program Cleaning validation may be a subset of instructions in a broader validation policy, but the intent remains the same
Site validation master plan	Describes general principles of validation to be applied at specific site May address multiple dosage forms or may be prepared for a single type of product within the site Limited details for specific types of validation Site validation master plans are not a regulatory requirement, but are frequently the first document requested by regulators and auditors since they provide a succinct view of the program elements
Cleaning validation master plan[a] or Cleaning validation approach plan[a] Validation plan[a] or Sub-master plan[a]	Contains details of philosophy and approach for a specific-type of validationin this case cleaning May identify current program initiatives for improvement and/or ongoing initiatives associated with new facility introductions (either product or equipment) Typically includes a product list and equipment list to help represent the scope of the cleaning validation initiative Will remain in place reflecting the current approach although the content will change as activities progress and are completed Typically reflects total program status and may have an annual summary prepared to define and defend accomplishments and changes in priority Cleaning validation master plans, like their counterpart the site validation master plan, are not a regulatory requirement. In the case of cleaning validation master plans, the need to generate this topic-specific validation plan is driven by the complexity and depth permitted in the site validation master plan
Scientific rationales	Documents that contain the details of the risk-based decisions reached for a specific product, group of products or group of equipment Typically scientific rationales may exist for: establishment of limits, identification of sampling sites, residue selection, and grouping or matrixing, although other topics are also possible Will follow the decision-making framework contained in the site validation master plan and cleaning validation master plan, but in this case will record the decisions made Scientific rationales are not always produced as stand-alone documents. They may be included in the body of either level of master plan, provided that all personnel know where to reference the decisions that have been made
Project plan	Will be prepared for complex projects (e.g., new product introductions, site transfers, facility renovations, etc.) and therefore will most often contain multiple types of qualification/validation activities to be performed Contains the details of a specific initiative with regard to cleaning validationcleaning validation may just be one sub-section Will be developed as new projects warrant, to reflect project-specific needs Will be completed and replaced with new projectsoften receiving a summary report to the project plan that demonstrates that all activities were completed Project plans are not always required as part of a compliant program. They are, instead, a convenience to help manage the complexities of a project without burdening the broader site validation master plan or cleaning validation master plan
SOP	Specific directions for how to execute the validation program including researching the study to be performed, preparing documentation, executing the studies, collecting and testing samples, and preparing the summaries As critical components of the quality system, compliance with contents of the SOPs is required or exceptions/failure investigations must be generated Demonstrated training in the contents of SOPs is also required
Protocol	Test procedures, conditions of test and acceptance criteria to be performed Contains data sheets, attachments or appropriate references to the documentation (i.e., the documentation to be completed during execution of the validation) to provide the documented evidence that validation requires
Summary report	Directly responsive to the protocol, the summary reflects the completed activities, the data developed, the deviations that may have occurred and the conclusion of the studies

[a] The name here is typically changed from the site validation master plan to differentiate the smaller scope and to indicate the document's position in the compliance hierarchy.

The elements that require so much interdisciplinary cooperation are the decisions involving: which residues will be assessed, safe-levels of carryover, number and location of samples, sampling and analytical method selection, and the strategies to be employed for ongoing monitoring of cleaning activities. Long before the risk-based GMPs became a topic for discussion, cleaning validation had required the development of risk-based scientific rationales. These risk-based rationales form the basis for the cleaning validation protocol as well as forming the basis for the scientific design of the cleaning validation program in its entirety.

It is quickly becoming clear that an orderly approach to cleaning validation is required in order to ensure that all activities of the program are scientifically established. When embarking on cleaning validation for the first time, it is important to establish a cross-departmental team that will focus on all of the specialties required for the cleaning validation program. A flow chart of these activities has been provided in Figure 1.

In a well-established cleaning validation program a number of these elements may already be accomplished. It is never a mistake, however, to evaluate each of the activities and ensure that all rationales for the program choices made are thoroughly documented and internally consistent.

All compliance initiatives fall on a risk continuum. That is, that we have options to make highly conservative choices, or less conservative choices in our approach to compliance. Cleaning validation is no different. In proceeding to evaluate the different activities in cleaning validation, one must consider the risk continuum and ensure that the position taken along that continuum is well defended (Fig. 2).

Validation is as much about what you choose to do as it is about what you choose not to do. By this, it is meant that every time an option presents itself, it is possible to create scientific rationale both for and against that option.

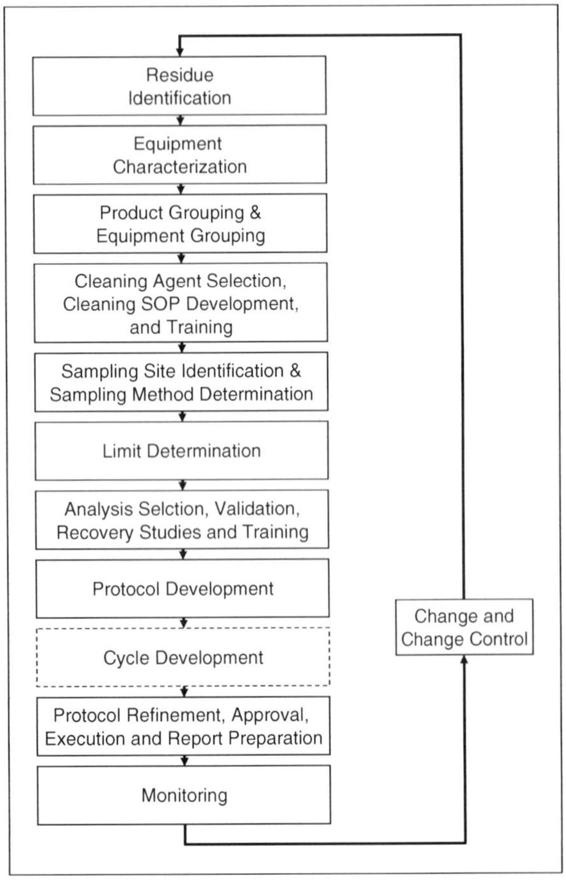

Figure 1 Cleaning validation process flow.

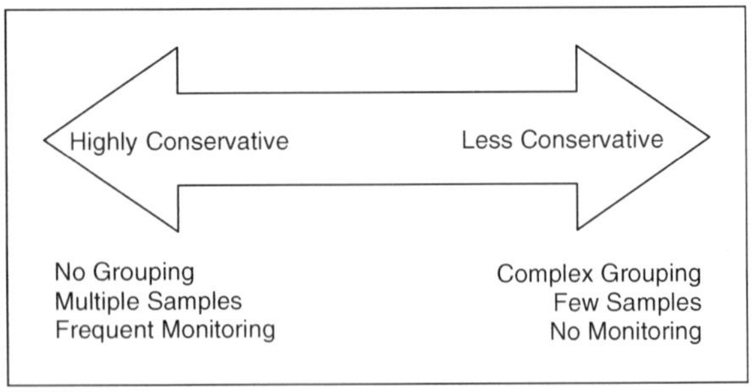

Figure 2 Examples of the risk continuum and decisions that might place you at one extreme or the other.

It is up to the personnel responsible for documenting the program to ensure that all selected options are defended not only for what was chosen, but also for what was ignored or not selected. In this way, scientific rationales will be thoroughly defended.

DEFINITIONS SPECIFIC TO CLEANING VALIDATION

Due to the often nonspecific nature of the GMPs, the diversity of products, processes, and operating environments it is critical to define terms. Each site should maintain a lexicon of terms, or at a minimum, provide definitions as part of controlled documents in order to ensure that regulators and participants in the various compliance programs will use terms consistently. While individual sites may establish terms that are different from those presented here, the goal of these definitions is to provide a common understanding of terms that may be used throughout this chapter.

CLEANING VALIDATION

Cleaning validation requires documented evidence to ensure that cleaning procedures are consistently removing residues to predetermined levels of acceptability, taking into consideration such elements as batch size, dosing, toxicology, equipment size, and the like. A cleaning validation is typically completed by the accomplishment of a minimum of three[a] consecutive successful[b] trials.

CLEANING VERIFICATION

Cleaning verification requires documented evidence to ensure that cleaning procedures remove residues to a predetermined level of acceptability based upon the

[a] Since the creation of the Quality System Regulation for Devices and the proposed revision to the GMPs in 1996, the FDA has been challenging industry to defend the number of runs selected rather than basing validation on the rule of three. In this instance, as world-wide regulatory guidances on cleaning validation still refer to three, we will continue to use three as the standard descriptor for cleaning validation.

[b] Consecutive successful in this case means "without intervening failures"; consecutive does not imply without intervening production of other products. In most cases, intervening production of other products will be carried out, provided that the equipment is demonstrated to be clean prior to the subsequent use.

minimum of a single trial. A cleaning verification is performed to assure that the cleaning procedures used adequately clean the equipment when the manufacturing process or the cleaning procedures may be subject to change, and therefore cannot be immediately subjected to validation. This procedure includes, but is not limited to, the accomplishment of a minimum of a single cleaning trial for:

- Cleaning of development equipment prior to the manufacture of clinical trial materials, when the equipment had been used previously for a material that was not subject to cleaning validation or verification
- Relocation of equipment
- New equipment
- Existing equipment following major maintenance/modification/repairs if a product contact area is affected.

CLEANING CERTIFICATION

Cleaning certification requires documented evidence that a production area, including equipment and facility are clean and ready for the next production use. This term typically implies that sampling for cleaning is performed and assessed against predetermined acceptance criteria, even after a cleaning validation is successfully completed. The primary reason for the certification is to ensure that production facilities and equipment associated with high risk products (e.g., potent compounds), high cost operations (e.g., biotechnology), and/or long-term and, therefore, high cost production campaigns (e.g., biotechnology or the use of API) are ready for the subsequent operation. Certification frequently implies more than taking and testing cleaning samples, however, as specific end-of-campaign activities. The changeout of environmental filters, disposal of potentially absorptive materials of construction and other checklist activities may be performed.

MONITORING

Monitoring requires periodic confirmation of previously validated cleaning procedures for the purposes of reconfirming the validated state. For monitoring, risk-based decisions may be made (that are different from those associated with the original validation) with regard to grouping of products and equipment to be studied, location of samples, sampling and analytical methods, frequency of monitoring and the like.

REVALIDATION

Revalidation is typically change-based. That is, reperformance of all or part of the original validation is undertaken when a change has been made to the product, process, procedures for cleaning, or equipment. The concept of time-based revalidation, or the evaluation of the process at some interval to confirm that it continues to meet the validated state is also a common revalidation principle. In the case of cleaning validation, routine monitoring can fulfill the role of time-based revalidation.

GROUPING OR BRACKETING

Grouping, sometimes also called a family approach, is a method by which products or equipment are considered to be similar or equivalent for the purposes of cleaning validation. Bracketing has an equivalent meaning to grouping, although it may include an added burden for testing the extremes of a population such as the smallest and largest equipment members of the equipment family, or the most soluble and least soluble members of the product group.

CAMPAIGN PRODUCTION

Campaign production is the manufacture of lots of the same product in a consecutive fashion such that: (*i*) no cleaning is performed between batches (typical of API manufacture, for example), (*ii*) sufficient cleaning is performed to ensure mechanical functionality of the equipment but equipment does not reach a visibly clean level (also common in API manufacture), (*iii*) cleaning is conducted to a visibly clean level with limited to no disassembly of the equipment (common to oral solid dose manufacture), or (*iv*) full cleaning is conducted of product contact surfaces, but cleaning environmental surfaces and changeout of product associated disposable parts (e.g., gaskets, hoses) is not performed until end-of-campaign (common to high potency products and/or products associated with more stringent dosage forms). Such batch to batch cleaning within the same campaign (as in examples *i*, *ii*, and *iii*) is typically not validated as the risks of same product to same product carryover are considered minimal. In example *iv*, the validation of the cleaning processes between batches may be validated in order to minimize the risk to the next batch.

Following through the flow chart in Figure 1 in a stepwise manner, each area of cleaning validation will be considered in the remaining sections of this chapter.

Residue Identification

When performing cleaning validation there are a number of residues that must be considered:

1. API(s)
2. Constituents of the cleaning agent
3. Preservatives
4. Precursors or starting materials
5. Intermediates
6. Processing aids
7. Media
8. Buffer
9. Cellular debris or metabolites
10. Particulate
11. Bioburden
12. Endotoxin
13. Viral particles
14. TSE
15. Excipients
16. Colorants, dyes, flavors or fragrances
17. And many more!

If we have the advantage of using a nonspecific method for cleaning assessment (e.g., TOC, pH, conductivity), we may be able to use a single analytical method to look for all (or most) types of residues.

In yet other instances, it is desirable to use a specific analytical method (e.g., HPLC, IMS, and, FTIR), which, by definition, requires that we select the residue(s) of interest to the cleaning validation.

While all of these residues in the list above are possible residues to be considered, when using a specific analytical method, we use a risk-based approach to determine which material(s) shall be considered as part of the cleaning validation program. Selecting all of the residues would be impractical with a specific method, as the time to develop and validate the numerous methods required would be so costly. Even if method development were rapid and inexpensive, such as with IMS, the likelihood of having to collect samples under separate conditions to satisfy the requirements of diverse methods would result in inability to collect a meaningful sample set.

APIs are those most commonly included in cleaning validation programs due to their potential harm to the next patient. Cleaning agents are the next most commonly selected materials as they are perceived as not being intended for consumption. They are ubiquitous to nearly all cleaning processes so that assessing issues such as removal and buildup becomes critical. Beyond the API and the cleaning agent, other materials such as preservatives, precursors or starting materials may also exhibit activity; the toxicities of these materials may require that they also be considered as cleaning validation targets. There may also be materials that need to be restricted from the subsequent process in order to ensure the efficacy of that process, to ensure the quality of the finished product, or to ensure the efficacy and safety of the subsequent drug product. These materials that may require limitation in the next process include precursors, starting materials, metabolites, cell debris, particulate, bioburden, endotoxin, viral particles, or TSEs. Last but not least, a case can certainly be made to restrict colorants, dyes, fragrances or flavors in the subsequent product if it is going to affect customer product perception. Customer product perception can be harmful to product market share or could result in customer complaints.

It is best to consider all possible residues and determine which material(s) are the most important. The typical basis for determining which residues should be evaluated includes assessing and ranking residues based upon the applicable elements of the following list:

- those residues that are pertinent to our dosage form or process
- those residues that are the most active/toxic and therefore represent the most risk to the next patient
- those residues that would damage the quality, purity, efficacy, appearance of the next batch produced
- those residues that would damage the next process (e.g., water in a hydrophobic process)
- those residues that are the hardest to remove (difficult to clean)

Even after considering these selection criteria, several residues may be identified as candidates. The next step, therefore, is to review the candidates, defend those that were not selected, and then begin to group the remaining elements, when that is possible. Grouping at this stage may include selecting a worst-case residue based on difficulty of removal as the representative of all other materials and then testing for that residue to the lowest limit ascribed to any of the selected residues.

When considering residues in cleaning, one must remember that the cleaning process will not necessarily leave these materials unchanged. The alkaline or acid conditions associated with detergent cycles, contact with water, or exposure to the air or heat can all promote physio-chemical changes in the residues that are left on the surfaces. We must therefore consider the safety of these carryovers not only from the perspective of the native compound but from the modified forms such as degradation products or denatured materials. We are fortunate with some compounds that our specific assays can also detect some of the degraded forms of the product. The likelihood of forming degradants can be assessed through laboratory forced degradation under those conditions/exposures that would typically occur during the cleaning operation.

Residue selection is an important first step in the cleaning validation program as it will drive many of the other decisions including the establishment of analytical methods the determination of limit and the identification of sampling techniques.

Equipment Characterization

Cleaning validation involves not only the removal of residues but also the assurance that each and every piece of equipment associated with the process has been cleaned to acceptable levels. We typically refer to this as a train-based approach. The "equipment train" is a series of equipment through which the product or products move as they progress through the manufacturing process.

Effective cleaning starts with effective equipment design. We are fortunate that today standards such as the BioProcessing Equipment Standard from the ASME and similar design guidances that promote cleanability from the AAMI are available to help us and our vendors to understand the design principles that must be observed to ensure cleanability. Some of the principles that have been recognized to be crucial to promote effective cleaning include:

- Limit or eliminate threaded connections—use clamp type sanitary fittings or weld connections
- Limit or eliminate deadleg opportunities (L:D ratio < 2:1 recommended)
- Limit or eliminate annular openings (L:A ratio <2:1 recommended)
- Orient instruments and connections to ensure limited possibility for entrainment of air or soils
- Limit length of addition ports or instrument ports and place them in such a position that they may receive direct coverage with cleaning fluids or so that they may be used to introduce cleaning fluids
- Consider agitator design or design of other obstructions within vessels or equipment carefully to ensure that they may be cleaned on all sides—consider alternate pathways to introduce cleaning chemistries to ensure allover coverage
- Ensure adequate slope for drainability (e.g., 1/8th inch per foot)
- Employ sanitary valves and pumps to eliminate holdup volumes and entrained product
- Cove corners—no right angles

- Ensure drain sizing is appropriate for hydraulic balance during cleaning
- Employ vortex breakers in drains to ensure adequate drainage without binding
- Ensure materials of construction are nonadditive, nonreactive, nonadsorptive

While these are all appropriate goals, we are commonly faced with existing equipment that pre-dates these standards or process requirements that limit our ability to comply with these requirements. We may also be faced with competing demands for plant safety or cost reduction that may dictate less than desirable design choices, so that the overall facility can meet our goals. In these cases, careful consideration must be made of the equipment in order to ensure that the equipment is cleanable. Any areas that do not meet ideal standards for cleanability should be addressed within a risk assessment or should be addressed through additional steps taken during routine cleaning and validation to ensure that cleaning processes have overcome the design risks.

In order to assess that the equipment will be cleanable, we should characterize all equipment, so that its design features are well known. Equipment characterization can assist cleaning validation initiatives in many ways:

- Promote more effective cleaning procedures by identifying cleaning challenges and ensuring that they are addressed in the cleaning methods employed
- Identify hard to clean locations and high risk locations in equipment for the purpose of sampling site selection
- Target materials of construction that will be included in sampling recovery studies and those that will not be included
- Isolate materials that will be disposed of at the end of a production process and/or will be dedicated to a single product
- Verify that all materials of construction are compatible with the selected cleaning agent and temper-atures that will be used with the cleaning process
- Collect product contact and sample site surface areas for the purpose of calculating limits and results
- Confirm similar geometries, capacities, and use of process equipment for the purposes of grouping that equipment

When performed correctly, equipment characterization is the process whereby we catalogue the features and attributes of equipment, thereby ensuring that equipment can be cleaned reliably and reproducibly. Because we are cataloging the attributes of how the equipment is designed, installation qualification seems to be a likely opportunity to collect this information for new equipment. Furthermore, for new equipment, it is even more appropriate to make the documentation of the design attributes part of the turnover package provided by the equipment vendor.

In addition to pure design, the way in which a piece of equipment functions, its mechanical actions on the soils, may represent physio-chemical changes to the soil (e.g., heating, friction, drying). As the mechanisms of soil deposition on the equipment are critical to its cleanability, they must also be considered in any characterization of the equipment.

Many factories maintain multiple pieces of equipment for the same function. This replication enables flexibility when scheduling production, scheduling cleaning activities, and scheduling different products that make use of that same equipment. As a result, for cleaning validation, it is appropriate to group or bracket that equipment based on its nearest relatives. The equipment characterization—with its assessment of materials of construction, design dimension, features that affect how it is soiled, used, and cleaned—makes an appropriate place to document decisions about grouping and bracketing of equipment.

For multi-use equipment, the equipment characterization is a document that can be shared between protocols. However, protocols often have a very narrow scope and are difficult to continually cross-reference for other studies. As a result, the equipment characterization activity may be one that we choose to document as a stand-alone file that is then subject to change control for equipment modifications. In this fashion, all protocols that reference the same equipment can be assured that the identical information is available. This structure will also enable the ready grouping of equipment as the similar members of the equipment family may be characterized and recorded in a single document to help highlight their equivalence and to help demonstrate why specific family members were selected to be studied for the validation.

As equipment characterization assesses such things as how the equipment is soiled, how the equipment is cleaned, materials of construction, geometries, and surface areas it becomes a logical place to also document the rationale for sampling sites. It is very common during equipment characterization to photograph the equipment and photograph the sampling sites and to capture these images as part of the file. The sampling sites can be described in words and can be entered into routine sampling data sheets for the collection of data during protocol execution. In this manner the equipment characterization becomes a living file that serves each cleaning validation protocol.

Product Grouping and Equipment Grouping

Grouping, sometimes also called a family approach, is a method by which products or equipment are considered to be similar or equivalent for the purposes of cleaning validation. When considered similar, a worst-case member of the family is selected for demonstrating cleaning validation. When considered equivalent, any member of the family may be selected as representative of any other member.

Bracketing, a term that appears in EU GMP Annex on Cleaning Validation, has an equivalent meaning to grouping, although it may include an added burden for testing the extremes of a population (e.g., smallest and largest equipment members of the equipment family, most soluble and least soluble members of the product group).

Grouping may be used to simply prioritize cleaning validation studies or may be used to eliminate some of the numerous possible combinations of product and equipment studies that might otherwise need to be performed.

When grouping products, all products must be:

- Manufactured on the same equipment group
- Cleaned with the same cleaning agent
- Cleaned with the same cleaning procedure

Grouping considerations for products include:

- Similar patient risk levels (e.g., therapeutic indication, patient population, route of administration, potency, toxicity for drugs/devices/nutraceuticals/cosmetics or in the case of in vitro diagnostics those products that have similar diagnostic uses, such as so-called "health and safety" products)
- Similar formulations
- Similar manufacturing processes

Cleaning validation must always be carried out to meet the lowest limit of the entire product group.

When grouping equipment, all equipment must be:

- Used to produce products from the same product group
- Cleaned with the same cleaning agent
- Cleaned with the same cleaning method

Grouping considerations for equipment include:

- Equivalent in terms of position or role in the manufacturing process
- Similar functionality
- Similar design (e.g., geometry, materials of construction, capacity)

The surface area used in residue limit calculations must be the largest of all equipment included in that group to ensure the most conservative approach to setting limits.

Grouping may be employed for the initial validation, revalidation and for program changes, monitoring, clean and dirty hold time studies and the like. Different grouping decisions may be employed for these different studies based upon risk (Table 2).

Any time that grouping is employed, recognize that an auditor can always ask the easy question of "Why did not you study __________?" So, be prepared to defend your grouping strategies. Remember the risk continuum in Figure 2 and ensure that you have defended your position towards grouping effectively.

Cleaning Agent Selection, Cleaning SOP Development and Training

All cleaning processes rely on the principle of TACT-WINS

Time
Action
Concentration/Chemistry
Temperature

or TACT are the process parameters that are required to be controlled in any cleaning process, whether manual, semi-automated or automated. Changes in one TACT parameter will cause a commensurate increase or decrease in the other parameters. For example, for some soils an increase in temperature can mean a possible decrease in chemistry or a decrease in the action applied. In all cases, however, the correct balancing of the TACT parameters requires proper knowledge and understanding of WINS:

Table 2 Different Grouping Decisions Employed for Different Points in a Cleaning Validation Program

Reason for grouping	Special considerations
Initial validation	All of the considerations listed above are valid Conservative decisions to minimize grouping may be made if limited data exist for the factory on cleaning Conservative decisions to minimize grouping may be made if regulatory/customer review of the data is expected for a new product
Revalidation	Reason for the revalidation will be taken into consideration, for example: Equipment changethe whole train may not be evaluated only the affected equipment may be consideredexisting groups may be redefined based on members affected by the change Product/formulation changethe individual product changed may be studied without affecting or considering the remainder of the group New cleaning agent introductionthe most challenging products based on prior data and established groups may be studied
Monitoring	More aggressive grouping than for the original validation may be employed, particularly if: Some products were found to be well below predetermined levels of acceptability during the original validation Some equipment was found to be well below pre determined levels of acceptability during the original validation and was deemed not to be a challenge to the cleaning process
Dirty hold time studies	Selected products or equipment for study may be based upon attributes that could make the equipment more difficult to clean after prolonged hold, such as: Producthygroscopicity, propensity to dry onto surfaces forming a hydrophobic film, degradation upon exposure to air or light, propensity to support microbial propagation (e.g., specific constituents, unpreserved formulations, water activity) Equipmentfeatures that might retain excessive residual product, features that might promote drying or exposure to light, materials of construction that might become increasingly difficult to clean with time (e.g., screens or membranes) These features may be different from those selected for the cleaning validation groups and may represent a different set of challenges
Clean hold time studies	Cleaned equipment stored in the same environmental area with the same precautions of drying/covering/closing/sealing will store equivalently therefore more aggressive grouping will typically apply Producttypically not a consideration unless different products leave significantly different starting bioburden levels after cleaning is completed Equipmenttypically equipment is stored based on broad classes that consider elements such as potential to retain moisture after cleaning (e.g., complex geometries or polymers that may have a high relative humidity) or equipment that does not close or seal and therefore has to be stored with loose covers

Water
Individual
Nature of the Soil
Surface

WINS represents the parameters that affect the soil's removal from the surface and each parameter can affect your ability to apply TACT in a given situation.

Cleaning chemistries fall into several broad categories:

- Water
- Solvents
- Commodity chemicals
- Formulated cleaning agents

Water is the universal solvent. If water alone will effectively clean your product without undue time or physical effort to remove your residues, by all means employ water alone! For many, however, the water alone requires an unacceptable increase in time to get the cleaning accomplished. For these individuals, one of the other approaches must be sought.

Solvents are typically applied in processes where solvent usage is already called for by the manufacturing process. For example, mother liquors are typically used as the solvents for cleaning of APIs. As the mother liquor is already known to dissolve the primary residue, there is little risk in employing it for cleaning. In addition, the facility is already equipped to handle the hazard and effluent issues associated with the solvent. With today's increasing focus on environmentally friendly processes, however, companies are frequently trying to find ways to reduce their solvent usage and eliminating solvents in cleaning is far simpler than removing solvents from synthetic pathways.

Commodity chemicals such as NaOH can be used for cleaning as well. Like their solvent counterparts, there may be hazard issues and effluent issues associated with these materials. Their typically high alkalinity or low acidity, however, often makes them effective at oxidation and reduction of soils which makes them helpful in inactivation processes. However, these chemicals lack the detergency of a formulated cleaning agent and they may be difficult to rinse, taking larger volumes of water to rinse free from systems than would a formulated cleaning agent.

Formulated cleaning agents are by far the largest class of cleaners. This category includes solvent-based formulations and aqueous formulations. Typical formulated cleaning agents can include one or more alkalinity or acidity sources, surfactants, builders, sequestrants, chelants and either a solvent or water. For industrial applications, unlike consumer-use products, these materials are formulated to be low-foaming and therefore are more readily rinsable and are appropriate for high impingement or high turbulence cleaning. Each formulated cleaning agent will have an optimal temperature range for use in which the surfactant will be most highly effective at helping to carry the soil away with rinse water.

To properly select a cleaning agent and establish cleaning procedures, one must understand:

- Soil (formulation, residue condition—dry, wet, baked on, layered on)
- Surface (materials of construction, challenging geometries)
- Available cleaning methods (manual, semi-automatic, automatic)
- Available utilities (temperature, grade of water)
- Safety considerations (personnel protective equipment, likely aerosolization)
- Effluent considerations (temperature, pH, limited chemical constituents, limited volumes)

With this knowledge in hand it is easy to screen detergents that meet facility and operational requirements for removing potential residues. Typical pharmaceutical cleaning agent suppliers often provide the cleaning residue studies. It is important to remember, when screening detergents, that for formulated products, the formulation may influence the ability to clean a soil more so than the API. This is typically true because the excipients tend to be the majority constituents of dosage forms and the fact that all release modifying properties are typically provided through the excipients.

When selecting a cleaning solution (whether a solvent, commodity chemical, or formulated detergent), it is important to understand the composition of those products in accordance with the Barr Labs court case from 1991. In that case, Judge Wolin ruled that pharmaceutical manufacturers must know the composition of their detergents and must test for residuals from these detergents. As such pharmaceutical suppliers for formulated cleaning agents will typically reveal their formulations to their customers for the purposes of their understanding toxicity solubility and markers for analytical detection. By revealing these formulations it is possible for the pharmaceutical companies to gain assurance that they have established appropriate scientific rationales for the removal of the cleaning agent.

Cleaning solutions, whether solvent, commodity chemical, or formulated cleaning agent should be treated as raw materials. That is, there should be assurance of control in the purchase, testing, and specifications for the material. There should be a mechanism for complaints and investigations, along with the implementation of corrective and preventative actions, when necessary. For formulated cleaning agents, GMP principles should be followed in terms of lot traceability and documentation. There should be a quality system in place that assures, for formulated cleaning agents, that no changes to the formulation will occur without prior notification to the customer. In cases where these changes are inevitable due to changes in environmental law or availability of specific chemistries the pharmaceutical customer should be provided with materials to aid in bridging the gap between their original work and any new validation work.

Disinfectants are not effective cleaning agents by and large. Although many of them contain surfactants they are not designed for the heavy soil load associated with equipment immediately after processing. Disinfection and cleaning cannot occur in one step as cleaning is required to remove the soil so that the disinfectant can be effective. Disinfectants are poor penetrants and the disinfectant's active ingredients can be inactivated with excessive surface soil before having a chance to attack the bioburden. Likewise, the soil residue can provide protective effects for the microorganisms that are within and below the soil layer.

In ideal situations detergents are harmonized within the facility to ensure that there is no opportunity for mix-up, meaning that a single detergent is employed. However in some cases, the specialization of a particular detergent may be unavoidable due to residue requirements or the materials of construction of a piece of equipment. In these cases it is necessary to consider how best to ensure that the correct detergent is used each and every time the cleaning operation is performed. This may include a checklist as part of the cleaning documentation that is completed by the operator and may extend to double-checking the confirmation of the material that is being used as a part of the cleaning operation.

As part of the cleaning agent selection, the available cleaning methods are first studied. Once the cleaning agent is known, the cleaning methods can be further refined. Again, if a screening study was performed for the cleaning agent either in-house or by the detergent supplier, the user will typically have a starting point for Time, Concentration/Chemistry and Temperature, or the TCT in TACT. The A or Action will need to be demonstrated on the plant floor and the TCT parameters adjusted accordingly.

For manual cleaning processes, where an operator is wielding either a brush or a hose for performing cleaning, one of the biggest challenges associated with the cleaning is achieving reproducibility in the operators' actions. To ensure that reproducibility is possible, detailed procedures should be created, ideally with a corresponding checklist to be completed during the actual cleaning operation. Key elements in assuring operator to operator reproducibility are:

- Defining disassembly
- Sequencing cleaning actions to prevent recontamination
- Defining tool use and tool actions
- Defining times for segments or activities in realistic durations

Increasingly, industry has found it appropriate to create cleaning procedures in a form that is similar to batch records. By following a structured format with required data entry we have an effective record of the process control and we have a clear checklist of activities for the operator to follow. The challenge is always to determine how to define these procedures for the best effect. We do not want to interrupt the flow of the cleaning in order to create documentation but we want to ensure that we capture the critical steps. By preparing these checklists, particularly for manual cleaning, we can ensure that operators can be trained to a highly detailed SOP and can perform to a less detailed checklist that highlights the critical activities.

For semi-automated or automated cleaning such as that which might be performed with a parts washer or Clean-In-Place system, the TACT are frequently fully instrumented and controlled by a microprocessor or PLC. In these cases, reproducibility of the cycle is not likely to be a problem if the recipe is locked down and selected correctly. What become of more concern are the human interfaces with the system prior to the initiation of the cycle such as making/breaking connections, consistent disassembly and loading of a parts washer, and the like. Even though there are some reproducibility concerns with semi-automated and automated washing, these concerns are clearly reduced.

When reviewing SOPs for manual cleaning in particular, asking a few simple questions with each step of the process can help ensure that the SOPs are consistent and sufficiently detailed:

- Are all appropriate personnel protective measures in place to protect from: temperature, chemistries, aerosols, splashing, product residues?
- What is the duration of this activity?
- What is the action to be performed and how is it defined?
- Are all tools associated with this action listed in the materials list and referenced consistently at this step?
- Are there directions as to when to discard a disposable or tool when it has reached the end of its appropriate use period?
- What is the concentration/chemistry that should be applied to the surface during the wash? And how is the correct preparation of this solution assured?
- Are there instructions as to when to change the cleaning solution as it becomes increasingly dirty?
- What is the temperature for this step?
- How is the temperature controlled, if at all? Are there instructions for what to do when the temperature is out of range?
- What failures can occur in this process and what instructions are provided to the operator to deal with those failures?
- Are there instructions for putting away all tools used in the cleaning?
- Is the drying time for the equipment defined?
- Are there instructions for the inspection of equipment upon completion of the cleaning?
- Are there instructions for handling, protecting, and storing equipment after completion of cleaning and drying?
- Are there instructions for the clean equipment hold time assignment (equipment expiration)?

Reusable permeable components such as the filter membranes in ultra-/micro-/or dia-filtration systems, or resins in chromatography beds represent the intersection of cleaning and process validations. The regulatory expectation is that the maximum number of reuses of these components will be proven in process validation. However, the return of these components to a state that renders them suitable for subsequent use is the purview of cleaning validation. For column resins in particular, the validation is often conducted in a scaled down version of the process, with full and accurate simulation of all applicable process parameters. Production fluids are processed and cleaning procedures are followed as they would be in the manufacturing environment. The eluates after cleaning are assessed for residuals, and the product quality on the subsequent use of the column is assessed until either the maximum number of desired uses with effective cleaning is achieved, or until the quality can no longer be returned to starting conditions. This type of cleaning validation relies heavily on both process validation and cleaning validation executed simultaneously.

Training for cleaning and cleaning validation is of critical importance. In particular, operators must be made aware of the importance of cleaning and the importance of each step that they perform. Operators must understand the necessity to ensure that cleaning procedures are properly sequenced; that is, that activities are performed in the appropriate order to ensure they are not contaminating services that have already been cleaned. General training in aseptic practices is worth its weight in gold for any facility; it provides a level of sensitivity to cross-contamination that is unparalleled.

Familiarization of employees with appropriate techniques to ensure that environmental contamination is not transferred to process and product contact surfaces is highly valuable. In particular, for manual cleaning of equipment such as automated conveyor systems or fillers, the equipment is traditionally cleaned with lint-free wipes and a bucket of a detergent solution. When cleaning this equipment it is important to ensure that the operators understand a "top-down-center-out" approach to avoid contaminating already cleaned surfaces. It is not unusual to observe operators cleaning critical product-container contact surfaces and environmental surfaces with the same wipe and solution. It is important to help employees differentiate the surfaces and understand their role in keeping all surfaces clean.

In addition to training employees on the specifics of the cleaning procedures it is important to train employees in the basics of cleaning validation, especially including their role in the validation. In particular, all operators and supervisors should be aware that during a cleaning validation trial they are not themselves being judged but rather the adequacy of the SOP and that it is the techniques the SOP describes that are under challenge. Cleaning validation is also assessing the elements such as the robustness and reproducibility of the training.

Another important aspect of training is the education of inspectors. Inspectors are on the front line in helping to ensure that equipment is clean, both during the validation, and after validation is complete. Visual inspection is important to every cleaning validation program because of the assurance it provides of the baseline cleanliness of all surfaces not just those which are sampled. In addition this visual assessment can also help to ensure that excipients and all other materials not subject to analytical-specific analytical methods are removed. Routine visual inspection after cleaning is the one common denominator between the validation and the routine operations, ensuring that surfaces have met minimum cleanliness standards that were achieved in the original testing.

When training inspectors it is important to ensure that inspectors are made aware of appropriate inspection techniques and tools so that they do not contaminate the clean surfaces they are inspecting. It is important for them to know where to look for and how to identify residue on the surfaces. Because cleaning may be conducted either in a disassembled state or an assembled state it is important that inspectors understand whether equipment inspection should be conducted assembled or disassembled. Inspectors, like cleaning personnel, should always be cognizant of wearing appropriate protective equipment such as gloves and lab coats to ensure they are not contaminating the equipment they inspect.

For routine inspection, standardized inspection tools should be used. Flashlights can be difficult to control unless they have rechargeable batteries and are placed on charge frequently. Intrinsically safe, electric lights may be a better option for inspection of deep vessels or other hard to illuminate areas. Other tools such as remotely operated digital cameras or borescopes may have great utility in getting the inspection to areas that need it most. This equipment has the added benefit of being able to capture images and saving those images as part of the cleaning or cleaning validation record.

Today's training can easily be prepared using digital images and DVDs of proper techniques. Using such tools can help ensure that a standard curriculum is applied and that the same techniques are routinely taught.

Sampling Site Identification and Sampling Method Selection

Sampling sites should be selected based on the most difficult to clean geometries of the equipment. These locations, however, are frequently inaccessible—their very inaccessibility is what makes them difficult to clean! Therefore, when choosing sampling sites one must always be cognizant of the desired sampling location based both on difficulty of cleaning and on intended sampling methods. As we will learn, sampling methods have various advantages and disadvantages that make them suitable for various geometries and locations on the equipment.

Equipment may have what we consider both hotspots and critical sites. Hotspots are locations that are likely to become dirty during the manufacturing process and are difficult to clean. Critical sites are those locations, which, if they were to remain dirty, would provide a disproportionate level of contamination to the next batch or portion of the next batch.

An example of a hotspot might be the bottom of an agitator or an instrument port inside a vessel that is likely to become soiled during the manufacturing process and might prove to be difficult to clean during the cleaning process. The use of an agitator in a mixing vessel means that any soil remaining on a surface is likely to become homogeneously distributed within the next batch. Contrast this agitator to locations such as a filling needle, a tablet press table, or a fraction collection valve on a chromatographic skid. Each of these locations has the opportunity to affect the next dose of the product or the next portion of the batch being produced. The residue that remains on these locations will not be homogeneously mixed throughout the batch but instead will disproportionately contaminate a small number of doses or, worst-case, a single dose.

When selecting sample sites we must evaluate a variety of locations including hot spots and critical sites as well as some representative locations on the equipment. Remember to include in the process those locations which might experience recirculation or redeposition of contaminants during the cleaning process. For example, in a vessel that might use fill-soak-and-agitate as the cleaning method, we might find that the agitation level falls off as the liquid is drained from the vessel. This may mean that there are significant bathtub-ring risks on this type of equipment based upon the resettling of suspended residues on surfaces with low agitation levels.

The number of sample locations selected for any individual piece of equipment should be based on the very same considerations that were addressed in sampling location selection:

- Difficult to clean geometries (hotspots)
- Locations that disproportionately contaminate a portion of the next batch (critical sites)
- Representative locations

In addition, sampling sites and the number of locations selected may also be influenced by:

- Materials of construction (inasmuch as different materials might have different affinities to soil)
- Overall scale of the piece of equipment (to ensure that coverage issues are addressed top to bottom and side-to-side)

For example, in a fluid bed granulator which can be nearly two stories tall we may have difficulties in coverage side-to-side and top to bottom. In order to ensure adequate cleaning, we may need to sample several locations on the sidewall of this equipment despite the fact that the sidewall is all of the same material of construction and not a difficult to clean geometry.

In order to determine the sampling locations, several tools may be employed:

- Review of the equipment characterization for process attributes, geometry and materials of construction
- Review and observation of cleaning SOPs for potential areas of weakness or locations where ill-controlled process parameters may result in variability
- Interview with and observation of operators to discuss their experience with difficult soil deposits

When determining sample size for cleaning validation, we simply need to target a sample size that will provide sufficient residue to the assay, but not collect samples so large that our recoveries may suffer. Most firms select a convenient surface area (e.g., 100 cm^2 or 4 $in.^2$), but it should be noted that it is possible to vary sample sizes slightly from sample to sample, when the sample size is accounted for in the result. Sample size variation may occur naturally based on the geometry of the equipment. For example, if a valve is 105 cm^2 on all product contact surfaces, it is much more appropriate to sample the full surface and account for the small overage in the equation for the results (see the Limits section) than it would be to instruct the sampling personnel to sample all of the surface except for 5 cm^2.

There are a variety of sampling methods for cleaning validation. Any method can be used provided it can be demonstrated to be suitable to recover the soil reproducibly from the surface. This need for effective recovery is the reason methods validation is always coupled with an assessment of sampling method efficiency.

The most common sampling methods employed in cleaning validation are rinse sampling and swab sampling. Of the two, swab sampling is typically deemed by regulators to be preferable. There are clearly situations where both methods may apply, however, and it is important that any sampling method used should be based upon its strengths.

Swab sampling is the use of a material, usually absorptive, to physically wipe a surface and recover the analyte. Because of the need to physically wipe the surface, swab sampling is a preferred method in locations where the surfaces are readily accessible to a human hand or arm. (Extension tools are also available but must also be validated for use.) The swab is typically used with a diluent (water, solvent, or a combination of the two) although the sampling may also be conducted dry. The diluent for moistening the swab must be compatible with the analytical method.

Rinse sampling, as compared to swab sampling, does not employ mechanical action on the surface other than that which is delivered by the fluid traversing the surface. Rinse sampling may be collected either as a portion of the final rinse of the clean-in-place process, or as a rinse applied specifically for the purposes of collecting a validation sample. The advantages of rinsing specifically to collect a sample rather than as part of the final rinse include the facts that:

- the cleaning process is truly at its conclusion when the sample is collected rather than in mid-process
- the quantity of rinse solution may be reduced for sampling and therefore the dilution effect is minimized
- the rinse sample due to its limited size, can be made truly homogeneous before aliquoting the sample for the laboratory
- the rinse sampling can be targeted to specific zones (depending on the method of application) which can result in sampling of critical spots

As with swab sampling, the solvent employed for rinse sampling is selected due to its solubility and compatibility with the residue(s). Typically rinse sampling is performed with the final rinse water. Rinsing with alcohol or other solvents can also be performed with appropriate safety measures in place.

From the description of these two techniques it is possible to see why the swab sampling is typically preferred as opposed to rinse sampling. Regulators have long argued that the mechanical action provided by swab sampling provides benefits in determining whether the cleanliness of the surfaces been achieved. In 1991 when the Newark district office was drafting the cleaning validation guidelines that we follow in the U.S.A. today, inspectors would frequently refer to a theory called the "baby in the bath water." They used the baby in the bath water theory to explain why they preferred swab sampling over rinse sampling. The theory was, "If you are trying to determine whether or not the baby was clean would you look at the baby? Or would you look at the bath water?" The answer, of course, was that you would look at the baby!

The question asked by the regulators was if the baby represented the equipment and the bath water represented the rinse sample, why did we believe that the rinse sample would represent the cleanliness of the baby? Their concern was that if the bath water had not contacted all soiled areas, or had not had sufficient contact time with the baby, or if the residues on the baby were poorly soluble in the rinse water, or if the residues were not homogeneously distributed in the bath water, or if the residues became too dilute in the bath water, that examining the bath water would not be an appropriate technique for establishing the baby's cleanliness. Therefore, they concluded that rinse would be an inappropriate technique for sampling the baby. In part the FDA had answered their own question by asking all of those questions... those are the very details we strive to prove with properly executed recovery studies (see the Section entitled Analysis Selection, Method Validation, Recovery Studies, and Training). If the recovery study demonstrates that the rinsing technique is adequate to demonstrate the cleanliness of the "baby," there are no further objections to the use of that rinse technique.

Direct surface sampling using FTIR or photoelectron emission techniques is on the rise in cleaning validation. With these techniques, specific spectra may be obtained from residues remaining on the surface, thereby directly quantitating and identifying the designated residue. These techniques are highly desirable as they directly measure the quality of the surface or the baby. The advantage of these techniques is that they represent the sampling and the analysis all in one step and there is no real "loss" to sampling system. As with swab sampling, the direct analysis of surfaces may be limited to those areas that are accessible for inspection.

Visual techniques using the eyes or remote inspection cameras are another form of direct surface sampling. These techniques are typically nonspecific, however, and are also nonquantitative. With the advance of digital imaging, however, before and after images, or the comparison of appearance to prior cleaning events may be accomplished, in some cases even in a side-by-side comparison in the field against a stored image.

Coupon analysis involves the introduction of a small soiled piece of the materials of construction of the equipment into the process equipment for the purpose of later removal and analysis in the laboratory. In some facilities, small portions of the equipment such as filling needles and spool pieces may be similarly removed for the same purpose. The advantage is the opportunity to apply sampling techniques in the laboratory such as extended soaks, physical agitation or sonication, and the application of more hazardous solvents that would be adversely indicated in the field. This technique also lends itself to false soiling if a worst-case soil condition is desired for sampling baked on residue, for example or for quantitative soil removal using techniques such as gravimetric analysis.

Similar "coupon" approaches can also be done either by rinse or swab with small swatches of fabric or materials where testing of the surface is very difficult. As an example fluid bed dryers employ large bag filters. These bags have extraordinarily large surface areas. Because they are typically considered difficult to clean they are frequently dedicated to an individual product. This dedication however does not exonerate the manufacturer from the responsibility to test for cleaning agents or materials used in the cleaning process. As a result we must sample the surfaces for any residual cleaning agent. Swab sampling is typically deemed ineffectual for woven surfaces due to the complexity of the weave and the fact that residue may become trapped between the individual fibers. Rinse sampling can be very effective with woven surfaces because it provides a prolonged soak and will help to loosen or dissolve residues from the surfaces.

Because of the immense size of many of these filter bags, however, it is impossible to provide an efficient sampling method. The equipment for soaking the part without using extremely large volumes of fluid is often not available. In these cases, a small swatch of material from a bag that is to be retired, or a sample from a vendor is obtained and intentionally subjected to worst-case soiling. This coupon of fabric is subjected to the washing process with a routine filter bag and the coupon is then returned to the laboratory for testing. Due to the small size of the swatch of fabric, the coupon can be fully immersed in a beaker for rinse recovery from that surface. Prolonged soaking of a coupon is possible to maximize recovery. Recovery studies would be necessary to confirm the recovery associated with this technique.

Concerns exist with the coupon sampling technique when manual cleaning is performed, inasmuch as we do not want the operator to concentrate on cleaning the coupon. Although the use of this technique is fairly infrequent, it has great potential flexibility for facilities where insufficient product is available to soil the whole system, or where investigation into appropriate cleaning techniques is desired prior to completely soiling equipment. When using this technique to study cleaning processes, remember that if only the coupons are soiled, the recirculated soil load in cleaning solutions will be significantly lower than when the equipment is fully soiled. This may hamper the cleaning effectiveness on a fully soiled system, especially when considering soil redeposition issues.

An additional approach that can be taken for sampling includes the placebo approach. Placebo batches are recognized as both potential cleaning techniques and potential sampling techniques. In the former case, a placebo material produced using all typical excipients but no active ingredients would be passed through a process system for the purposes of scrubbing clean the system from the prior material. The principle is that the placebo would pass along the same pathways as the product and, therefore, would have an opportunity to scrub off residual product along those pathways. Placebo sampling, on the other hand is employing a placebo and passing it through the equipment for the purpose of measuring system cleanliness.

The placebo sampling technique is very much like the baby in the bath water technique and depends upon:

- Excipients being fully "soluble" in the placebo
- Sufficient contact time of the placebo to collect a representative sample
- Placebo has adequate "coverage" of the process pathway to ensure removal of the placebo from all equipment locations
- Quantity of the placebo and the residue being matched so that the residue is in a detectable range within the placebo (i.e., not overly diluted)
- Residue being (somewhat) uniformly distributed within the placebo in order to ensure detection based on sampling any portion of the placebo.

As with rinse sampling, if a scientific case can be built for the use of the placebo method, it may be appropriate for either cleaning or sampling. It should be noted however, that regulators have cautioned in the U.S. cleaning validation inspection guidelines that the placebo method when used may most appropriately be used in combination with other sampling methods.

In some cases particularly in powder-based systems for safe product residues, such as those associated with topical consumer product powders, the placebo

Table 3 Major Sampling Techniques and Their Attributes

Attributes	Swab	Rinse	Direct surface analysis	Coupon	Placebo
Physical sampling of surface	●	○	○	●	●
Robust technique (low technique dependency)	○	●	●	●	●
Non-invasive technique	○	●	○	●	●
Adaptable to hard to reach areas	○	●	○	○	●
Effective on flat surfaces	●	○	●	●	●
Effective on complex geometries	○	●	○	●	●
Controlled area sampling possible	●	○	●	●	○
Samples are homogeneous	●	○	○	●	○
Does not require prolonged contact time with surface	●	○	●	●	○
Adaptable to different solvents/materials for sample removal	●	●	N/A	●	●
Appropriate for online adaptation	○	●	●	○	●
No recovery study required	○	○	○	○	○
Frequency of use	High	High	Moderate	Low	Low

Key: ●, effective or low risk; ○, ineffective or high risk.

technique has been used to avoid wet-cleaning of surfaces and prevents potential stickiness of the surfaces due to residual moisture on the equipment that might wet the powders.

The placebo technique has also been successfully employed in specific application for a highly-colored tablet granulation, where a pharmaceutical company used a white placebo to verify that all colored excipients had been removed from their tablet press hopper and feed frame. In this case, they use swab sampling to demonstrate active removal from the system, but use the processing and inspection of a small quantity of placebo granulation and tablets for the verification and removal of tableting excipients. They had demonstrated that the ability of an inspector to discern colored specks in a white tablet was far greater than the level of detection promised by a swab or rinse of the surface for that same colorant.

Table 3 includes a summary of the major sampling techniques and their attributes. The clear message for all sampling techniques is that they all have advantages and disadvantages and should be applied based upon the selected sampling locations.

Determination

Worldwide regulatory guidelines indicate that manufacturers must define their own limits for cleaning validation. The large variety and type of products as well as the broad number of processes and equipment make it impossible for any regulatory agency to establish firm limits that would apply to every situation. What is clear in the regulatory guidelines, however, is that regulatory agencies have an expectation that cleaning will be pursued until the residues reach "acceptable" or safe levels for patients who will receive the dosage form made in the equipment. Unfortunately, there is little guidance in terms of how to set limits for the cleaning agents or intermediate materials that were not intended to be administered to patients. Nor is there any appreciable guidance for how to set the appropriate limits for cleaning of medical devices. This section hopes to remedy that by establishing the different possible methods for setting limits in different segments of the industry in turn.

What is considered safe for cleaning carry-over must be determined by each company. Fortunately for pharmaceutical manufacturers the formulae for determining the appropriate level of safe carry-over may be conducted in accordance with some fairly well-documented mathematical standards. By far the most common approach in an industry today is the use of a combination of limits that provides assurance that surfaces are visibly clean and that they fulfill both a safety limit and a quality threshold or maximum contamination rate. In this way we can assure that all cleaning processes meet a minimum standard of cleanliness while at the same time achieving a limit that is considered safe for the individual residue or product.

Commonly cited limits in literature and in regulatory guidance documents have stated that the surface of the equipment must meet the requirements of being:

- Visibly clean,[c] and
- Not more than 1/1000th of a therapeutic dose in the next batch, or
- Not more than 10 ppm in the next batch, whichever of the latter two limits is lower

For these limits, visibly clean is self-explanatory and is met each and every time we clean through our post-cleaning inspection. The limitations of visibly clean match those discussed in the sampling methods section where it was discussed that not all surfaces may be visually available. We must recognize that visual acuity may differ from person to person and surface to surface and based upon the available inspection conditions on the light, angle of viewing, and/or type of residue or haze on the surface.

In the simplest terms, dosage-based limits are the determination of the amount of material that might be administered to the next patient when they are given a maximum dose of the next product. In order to determine

[c] Some firms apply a criterion of both visibly clean and a secondary "threshold" limit of demonstrating that surfaces meet an analytical cleanliness level that would equate to visually clean. For example, literature has reported that many typical residues are visible to a level of 100 μg/cm^2, therefore, a fourth limit could be applied of not more than 100 μg/cm^2 on any surface. This type of limit is not related to product safety and although it limits the amount of residue that might be found on the surface, the affect to the next patient is highly influenced by the equipment train size and the next batch size. This type of limit is applied at only a small number of firms and does not add significantly to the safety of the next product, except in cases where the equipment train is very large and where the next batch size is very large. In these limited circumstances it can provide a lower limit than NMT 10 ppm in the next batch.

what this quantity might represent to the next patient, we must take into consideration the dilution factor that the next batch will provide to the residue and we must translate the total amount of allowable carryover to the amount that would be present on the surface, as we are planning to sample the surfaces. A SF is applied to ensure that next patient only receives a fraction of the prior product's dose.

The point at which products no longer have a pharmacologic effect on the next patient receiving that material has been generally recognized to be 1/1000th of a therapeutic dose. Lower limits may be applied by increasing the SF or reducing the therapeutic dose term to reflect more serious or significant risks of cross-contamination. For example, one could apply:

- a toxicological level (with the appropriate SFs)
- a therapeutic dose level (by far the most commonly used)
- a minimum pharmacologic effect level
- an allergenic level
- a NOEL

Increasingly Safe Levels

Each of these terms represents an increasing level of safety to the consumer. Different levels may be applied based on therapeutic indication, knowledge of the pharmacological activity of the active ingredients, knowledge of dosage form, knowledge of patient population, knowledge of route of administration risks. As the carryover term is not the only term in the equation, it should also be recognized that the increased level of safety provided by selecting a more conservative value can also be offset or added to by selecting the other terms of the limit equation either with a conservative approach or a liberal approach.

In development facilities where the therapeutic dose may not yet be known, or where there is still some uncertainty about the final therapeutic requirements for a product, conservative values may be selected. The "first human dose" may be selected as a conservative value or the SF may be increased commensurate with the uncertainty associated with product.

It should be noted that the batch size and the dose administered to the next patient are terms that are dependent upon the next product to be produced. Cleaning carryover always affects the next product, not the product being cleaned. This is an important factor when introducing new products into existing facilities.

The equation for the 1/1000th of a therapeutic dose in the next product might appear as follows:

Equation 1: 1/1000th of a Therapeutic Dose Approach

$$\frac{\text{Lowest therapeutic dose}^{\text{pp}} \times \text{Smallest batch size}^{\text{np}}}{\text{Largest surface area (shared equipment train)} \times \text{Largest daily dose}^{\text{np}}} \times \text{SF} \quad (1)$$

where pp, previous product (product to be cleaned); np, next product; SF, safety factor (typically 1/1000).

In a secondary manufacturing facility, where formulated drug products are produced, the lowest therapeutic dose term in the numerator represents that API (or residue of interest) weight only. The largest daily dose term in the denominator represents the full dose weight. In manufacturing the next product, all ingredients, both excipients and actives have the opportunity to pick up and carry with them the carryover from the prior active. Therefore, it is appropriate and conservative to consider the full dose weight as affecting the next patient. In some cases, the largest daily dose may even be substituted with the maximum dose administered during a longer term of administration.

The smallest possible next batch size is included in this equation to represent a conservative assumption that the carryover is only minimally diluted by the next formulation entering the equipment train. This ensures that we are considering the worst-case effect on the next patient.

As stated previously, other more or less conservative values may be substituted throughout the equation provided that the terms used are justified. In general, if we minimize the terms in the numerator, by selecting conservative small values, and maximize the terms in the denominator by selecting worst-case large values, the overall limit will be small.

When considering cleaning validation documents or articles that refer to the MAC or MACO, this term typically represents the terms of the above equation without including the surface area of the shared equipment train. As we will learn, mathematical rearrangement of the terms in this equation can produce many equivalent, accurate expressions.

Because we know that residue will be distributed throughout the next batch, it is important to understand the amount of residue that may remain behind on surfaces. For this reason, we divide the carry-over quantity by the total surface area of the equipment train. When considering the equipment train, we need to include the full shared equipment train between the product being cleaned and the next product to be produced.

The full equipment train is considered because as the next batch traverses the surface of the equipment, all residues on the surface may be picked up and carried in the batch to the final filled doses of the product. The accumulated total of contamination to the next batch, therefore, is represented by the accumulated total of residue that remains on all surfaces. Dividing the permissible limit by the total amount of surface area, we assume there is uniform contamination of the surfaces. That is, every square unit of surface area would contribute the same amount of residue. While we know that is not true, that some locations may be more highly contaminated than others, we also must recognize that the sampling sites were established to collect samples from the hardest to clean locations on the equipment. By collecting samples from the hardest to clean locations on the equipment (those that would be the most likely to be dirty) and assuming that they are representative of all other surfaces of the equipment, we are actually making a conservative assumption. We are requiring that these hard-to-clean locations represent a limit that would be acceptable for all areas of the equipment.

A frequent concern when using the shared surface area approach is whether or not sampling of an individual piece of equipment can be accomplished if the limit is calculated based on the entire train. For example, if I soil Tank #1 on Monday and clean it the same day, but do not get the product to final filling and packaging until Friday, can I still release Tank #1 before the filler is released? The answer is an emphatic "Yes." Examining the units associated with the limit, we observe the following:

Equation 2: Typical Units for the 1/1000th of a Therapeutic Dose Approach

$$\frac{\mathrm{mg}^{\mathrm{PP}} \times \mathrm{mg\ or\ doses}^{\mathrm{np}}}{\mathrm{cm}^2 \times \mathrm{mg\ or\ doses}^{\mathrm{np}}} \times \mathrm{SF} \tag{2}$$

where pp, previous product (product to be cleaned); np, next product; SF, safety factor (typically 1/1000).

The result of equation (2) is $\mathrm{mg/cm}^2$, or less specifically, mass per unit surface area. This means that every square unit of surface area of the equipment train may be measured independently on different days or different months and we may still be assured that the total carryover will not exceed the limit for any individual piece of equipment. This is particularly helpful in facilities with prolonged processing or campaigns where different pieces of equipment may be freed for analysis at different times.

The last term, the SF has not yet been addressed. As identified at the outset, 1/1000th is typically applied, as it is widely understood to convert a therapeutic dose to a level that is close to approximating the no observable effect level. As with the other terms, a larger SF may be applied to make up for uncertainties that exist with the rest of the terms in the equation, or in environments where there may be uncertainty about product safety, such as in development.

In multi-product facilities, we cannot predict what the next product might be. In these cases, we might calculate limits for all possible combinations of next products on the equipment in order to determine the worst-case limit that must be applied to the product. In development facilities, where the characteristics of batch size, next daily dose, and equipment train are not yet determined for the next product, it is not unusual to make worst-case assumptions based on historical values or based on factors such as the minimum equipment processing capacity or maximum train typical of a particular dosage form.

Before we address a threshold limit such as 10 ppm, let us address other product types and how the limits may be set for those materials.

For APIs produced by synthetic routes, the equation would appear exactly as it does in equation (1), however when considering the largest daily dose, only the active quantity can be included in the limit because there are no excipients that would also carry over contaminants to the next batch. Although this will limit a term in our denominator, and since what we learned would potentially provide us with a slightly larger limit, the extensive surface area in API factories as it relates to the final batch size typically provides us with a conservative limit. It should be noted, that when considering the shared surface area between two APIs, only those pieces of equipment in the equipment train which see that residue need be included in the calculation of shared surface area. For example, in a four-reactor train with a centrifuge and a dryer, only the final reactor, centrifuge, and dryer may actually be exposed to the final API. Therefore, only these last pieces of equipment would need to be considered when determining the shared equipment surface area with the next product.

This brings up an important differentiation for synthetic processes, that is: cleaning validation for early stages of the process may include residues which may not have therapeutic indications and that may or may not be more toxic than the final API. In fact, the material present in all four reactors of the imaginary synthetic process described above may be different. For these materials, we typically use an equivalent equation form, with a toxicity-based limit [see equation (3)].

Equation 3: Safety-Based Limit Approach for API Starting Materials and Intermediates

$$\frac{\mathrm{LD}_{50}^{\mathrm{pp}} \times \begin{array}{c}\text{Empirical}\\ \text{factor}\end{array} \times \begin{array}{c}\text{Smallest}\\ \text{batch size}^{\mathrm{np}}\end{array}}{\begin{array}{c}\text{Largest surface area}\\ \text{(shared equipment train)}\end{array} \times \begin{array}{c}\text{Largest}\\ \text{daily dose}^{\mathrm{np}}\end{array}} \times \mathrm{SF} \tag{3}$$

where pp, previous product (product to be cleaned); np, next product; SF, safety factor (typically 1/1000).

Because the (LD_{50}) is based on animal studies most often, an empirical factor is needed to convert the animal data to human data. Articles in various toxicology journals have established the conversion factor at roughly 10^{-4} for converting oral rat data to oral human data. These same journals, however, then adjust the value by multiplying by a human body weight (either adult or pediatric), because the LD_{50} is expressed on a per kg basis. If we take an empirical factor with an order of magnitude of 10^{-4} and adjust it for a factor of 10^{+1} for a human body weight, the order of magnitude of the empirical factor applied is then typically 10^{-3}. The FDA draft guidance, "Estimating the Safe Starting Dose in Clinical Trials for Therapeutics in Adult Healthy Volunteers" may also be used to provide estimate conservative conversions from animal to human doses. The APIC guidance issued for the API industry has recommended a fixed empirical factor for all toxicity values as 2000 or 2×10^{-3} (regardless of route of administration or animal model). In cases where an additional SF of 1/1000th will still be applied, this is likely a conservative approach and provides some standardization across the industry. As with the other cases, additional safety can be applied when we are uncertain about the terms we have selected.

For instances where the starting material or intermediate has no LD_{50} defined for the material (either because it has not been studied or because the material was never intended to be isolated and used in that state), a qualified toxicologist can review the molecular structure, draw parallels to existing molecules and make some educated estimations of LD_{50} based on structure. As before, this level of uncertainty can lead to additional layers of safety. To avoid setting an impossible challenge, validators must be sure to discuss with a toxicologist the degree of conservatism that was applied in the estimation of the LD_{50}.

For other materials that do not have a therapeutic index such as cleaning agents, lubricants, polishing compounds, or for processing aids used in the manufacture of medical devices, the LD_{50} approach described here for APIs may also be applied. In the case of medical device processing aids, the complicating factor is the determination of the "amount" of the medical device processing aids to which the next patient might be exposed. For example, for cardiac stents, it may be possible to estimate the worst-case patient exposure in a quadruple bypass, but for sutures, a typical estimate might suffice, with the assumption that SFs will compensate for estimating inaccuracies.

For in vitro diagnostic products, the equation for the limit may be changed as follows:

Equation 4: Fraction of an Interfering/Enhancing Substance Approach for in Vitro Diagnostics

$$\frac{\text{Smallest amount exhibiting I/E}^{pp} \times \text{Smallest batch size}^{np}}{\text{Largest surface area (shared equipment train)} \times \text{Largest test volume}^{np}} \times \text{SF} \tag{4}$$

where I/E, interfering or enhancing effect; pp, previous product (product to be cleaned); np, next product; SF, safety factor (set according to risk).

The same mathematical formula applies to diagnostics. Instead of calculating the amount that would have an affect on the next patient, however, we calculate the amount that would have an affect on the next test. This may require some physical challenges to determine the minimum inhibitory/enhancing concentrations that could affect a product. In this case the application of a 1/1000th SF may not be meaningful. Instead, the SF may be adjusted based upon the accuracy with which the interfering/enhancing quantities were determined.

For cosmetics and nutraceuticals, where an "active" substance may not be present, or where that active might be GRAS, it may be necessary to simply calculate the limit based on equation (1). We would do this with the knowledge that a large amount of residue may be considered safe, even though we would find it objectionable as a carry-over quantity. It is for these exact circumstances that the quality default or threshold limits are compared to the "1/1000th" limits. The quality default or threshold limits ensure that even when something is considered "safe" to carry over that we limit the adulteration of the next batch to a minimum concentration.

The most common quality threshold applied to pharmaceuticals is not more than 10 ppm appearing in the next batch. The WHO, in their cleaning validation guidance document references the establishment of 10 ppm based upon the limits for heavy metals appearing in raw materials. Whether we consider this to be the basis for 10 ppm in the next batch or not, the limit is clearly a convenience value when set at 10. More conservative would be a convenience value of 8, 12 would be less conservative; 10 has become industry standard for many facilities.

Not more than 10 ppm in the next batch literally means, not more than 10 parts of the prior product appearing in every 1,000,000 parts of the next product. As with the 1/000th of a therapeutic dose equation, however, we must understand how much of the next product we will be producing and how much surface area will become contaminated with the carryover. This threshold limit includes an assumption that residues would be uniformly distributed in the next batch and uniformly distributed on the surfaces. As with the dose-based equation, this assumption is offset by the careful selection of sampling sites and the knowledge that this limit is used in comparison with a safety-based limit.

Equation (5) contains the calculation for not more than 10 ppm in the next batch.

Equation 5: Not More Than 10 ppm in the Next Batch

$$\frac{10\ \text{mg}^{pp} \times \text{Smallest batch size}^{np}}{1\ \text{kg}^{np} \times \text{Largest surface area (shared equipment train)}} \tag{5}$$

where pp, previous product (product to be cleaned); np, next product.

The units for this equation are as follows:

Equation 6: Typical Units for 10 ppm Limit

$$\frac{\text{mg} \times \text{kg}^{np}}{\text{kg} \times \text{cm}^2} \tag{6}$$

where pp, previous product (product to be cleaned); np, next product.

The units for a 10 ppm calculation, as shown in equation (5) also convert to mass per unit surface area. It is for this reason that a direct comparison between the NMT 10 ppm limit and the NMT 1/1000th limit may be made. As the NMT 10 ppm limit includes the additional factors of batch size and equipment train, the limit rarely results in a nice round number like 10. Mistakes are frequently seen in cleaning validation programs when the 10 ppm becomes unitless and is compared directly to the dose-based equation. When working with limits, it is critical, to ensure that equivalent units have been used prior to comparison.

For API, in vitro diagnostic products, cosmetics and nutraceuticals, the NMT 10 ppm appearing in the next batch can be calculated directly for all industries without any additional technical data. For diagnostic products, we should decide whether a 10 ppm threshold has meaning as a quality default or whether it will always be higher than the interfering/ enhancing substance limits. We may choose to set a more conservative threshold for these products based upon their understood sensitivities.

For medical devices, a threshold value of the amount appearing in the next batch may not have a lot of meaning, especially because an underlying assumption of the equation is that residues will be uniformly distributed in the next batch. For a combination drug/device, this limit may have value for the drug constituents, but for orthopedic implants, the assumption that 10 parts of polishing compound, for example, would be spread over a "mass" of parts equating to 1,000,000 times the amount of polishing compound, has no meaning. To establish a meaningful carryover threshold for these products, therefore, a "per device" limit may be appropriate based on knowledge of the typical devices.

All limits presented here have been expressed as mass per unit surface area. It might legitimately be asked, "If my limits are expressed as mass only, or mass per sample, are my limits incorrect?" The answer here is, "No."

The equation for the limit and the equation for the result are on either side of a comparator.

Limit > Result

We expect that the result will be lower than the limit when we are passing cleaning validation! We should remember that terms may be moved from side-to-side of a comparator without compromise to the mathematical expression provided that the reciprocal of the term is taken.

Based upon our equations (1), (3), (4), and (5), the terms for the result would typically be:

Equation 7: Terms for the Result

$$\frac{\text{Concentration of the sample (result from lab)} \times \text{Sample volume}}{\text{Surface area sampled} \times \text{Recovery factor (in decimal form)}} \tag{7}$$

For any calculation of results, we must still consider the fact that the sampling method left some material on the surface of the equipment. Thus, we must correct to the value back to 100 by dividing by the recovery factor, when it is expressed as a decimal, or multiplying by the recovery factor when it is expressed as a percentage. Mathematically it does not matter whether this correction to 100% is performed from the initial analytical results or the concentration in the sample or whether it is performed after it has already been converted to mass per unit surface area. The units for the results remain the same after the inclusion of the recovery factor in the analytical methods and recovery study section. Next, we will discuss how this recovery factor is derived.

Equation 8: Typical Units for the Result

$$\frac{\text{mg/mL} \times \text{mL}}{\text{cm}^2} \tag{8}$$

It can be seen in this case, that the result side of the equation also yields mg/cm^2 and therefore can be compared directly to our limits from equations (1), (3), (4), or (5).

If you decided to move terms from one side of the equation to the other your limits and results might be expressed as follows:

Equation 9: Example of a "mg" limit

$$\frac{\text{Lowest therapeutic dose}^{\text{pp}} \times \text{Small batch size}^{\text{np}} \times \text{SF}}{\text{Largest daily dose}^{\text{np}}}$$

$$> \frac{\text{Conc sample} \times \text{Sample volume} \times \text{Equipment train}}{\text{Surface area sampled} \times \text{Recovery factor}} \tag{9}$$

$$\text{or} \quad \frac{\text{mg}^{\text{pp}} \times \text{mg or doses}^{\text{np}} \times \text{SF}}{\text{mg or doses}^{\text{np}}} > \frac{\text{mg}}{\text{mL} \times \text{mL} \times \text{cm}^2}$$

$$\text{or} \quad \text{mg} > \text{mg}$$

where pp, previous product (product to be cleaned); np, next product; SF, safety factor (typically 1/1000).

Equation 10: Example of a "per sample" limit

$$\frac{\text{Lowest therapeutic dose}^{\text{pp}} \times \text{Small batch size}^{\text{np}} \times \text{Area sampled}}{\text{Shared equipment train} \times \text{Largest daily dose}^{\text{np}}} \times \text{SF}$$

$$> \frac{\text{Concentration of the sample} \times \text{Sample volume}}{\text{Recovery factor}} \tag{10}$$

$$\text{or} \quad \frac{\text{mg}^{\text{pp}} \times \text{mg or doses}^{\text{np}} \times \text{cm}^2}{\text{cm}^2 \times \text{mg or doses}^{\text{np}}} \times \text{SF} > \text{mg/mL} \times \text{mL}$$

$$\text{or} \quad \text{mg/sample} > \text{mg/sample}$$

where pp, previous product (product to be cleaned); np, next product; SF, safety factor (typically 1/1000).

Other combinations are also possible beyond 9 and 10. So why was the original division of terms shown starting with equation (1)? The terms included in the limit represented the data about the manufacturing process. All the terms on the result side of the equation had to do with the experimental design in terms of the way we sampled. If we move terms associated with how we sample to the limit side of our equation, we are in a position where we may have to create several limits: one for rinse, one for swab, one for swab samples of surface areas that are slightly different from our standard sample size (e.g., limits for 85 cm^2 samples and limits for 100 cm^2 samples). More important than the potential complexity that this introduces to the limits is the fact that we might overlook the fact that we have changed some aspect of the sampling approach by burying the sample conditions on the limit side of the equation.

Consider the example of a company that created a "per rinse sample limit." They moved their rinse volume term to the limit side of the equation, along with the surface area sampled. In this manner they had a limit that was expressed as mg/mL. They then took the worst-case limit in the factory and determined to apply it to all products they produced. They failed to take into account, however, that different products would potentially use different solvents for rinse solutions, that they might choose to vary the rinse volumes depending on the product, and that the surface area sample might vary based upon the technique used in applying the rinse. As a result of losing track of the terms that were included in the limit side of their "worst-case" equation they had a limit that was not truly worst-case for all sampling scenarios. Converting the company's limits to mass per unit surface area enabled them to change their sampling approaches without affecting their limits.

When considering limits that are applied to cleaning validation we must recognize that several of the more recent cleaning validation guidance documents (i.e., PIC/S, Canadian, WHO) have identified that:

> For certain allergenic ingredients, penicillins, cephalosporins or potent steroids and cytotoxics, the limit should be below the limit of detection by best available analytical methods. In practice this may mean that dedicated plants are used for these products.

While the sentiment of this statement can be appreciated in any risk-based cleaning validation program, the determination of "none detected" does not assure patient safety. This statement is a problem because a sample surface area can be made artificially small or a rinse volume very large and become none detected without changing the level to which the equipment has been cleaned. The only true method to determine patient safety is to calculate a dose-based or no observed effect-based limit and to compare that limit to what would be achieved if none detected were set as the standard for the evaluation of samples. The other problematic aspect of this statement is that there is no definition of "best available" analytical method. There is no indication as to whether this statement intends that the user purchase analytical technologies or seek analytical contract services for the express purpose of achieving best available for these cleaning validation samples. It is recommended that laboratories that are governed by these authorities and have products that would fall under this clause defend their limits approach carefully to demonstrate that adequate safety of the next patient has been considered in your approach.

Finally, it should be noted that some facilities may have trouble achieving a 1/1000th limit [equation (1)]. In these facilities, one of the options that is open to the manufacturer is to determine what is achievable from a process perspective. Using the standard limit equation, it would be possible to then work backwards to determine what impact that amount of residue might have on the next dose to be administered to the next patient. In these circumstances, the SF may be less than 1/1000th, but it still may be possible to provide a significant, justifiable margin of safety to patients.

Some firms refer to this as a "process capability" limit. In other words, a limit that has been established based on what the cleaning process is capable of, rather than on a rote limit. Even firms that can achieve the more traditional limit may consider employing a process capability limit after sufficient experience with the process is gained. The process capability limit may be significantly lower than a traditional limit (i.e., your cleaning capability may exceed your product safety needs). In these cases, process capability may serve as a better indicator of process consistency during the monitoring phase of the program than would a rote limit that is many times higher than what the cleaning process can achieve. Be careful when employing process capability limits however, as the temptation is to apply a single limit to the process. In reality, different pieces of equipment (and indeed different locations on a single piece of equipment) may exhibit very different process capabilities based on geometry, function or physical action on the soil. Be sure that process capability limits are set based on a realistic statistical review of a significant population of results.

Analysis Selection, Method Validation, Recovery Studies, and Training

After the identification of residues of interest, sampling methods, and limits it is appropriate to identify potential analytical methods. There are no restrictions on the analytical methods that can be applied for cleaning validation provided that the methods are demonstrated to be sensitive at the low levels that are required by most cleaning program limits.

Methods can be classified into two broad categories as either direct methods (specific) or indirect methods (nonspecific). Direct methods are those methods that uniquely identify the analyte of interest by composition or by comparison to a control sample. Typical direct methods applied for cleaning validation include HPLC, IMS, AA, and FTIR. Indirect methods are those methods that measure something about the attribute of the residue such as ionic strength, acid-base character or carbon content. Three of the typical nonspecific methods that are applied are pH, conductivity, and TOC.

Methods validation for cleaning validation proceeds much like the validation associated with the potency assay. That is, it focuses on: accuracy, precision, linearity, intermediate precision (ruggedness) and range. (Specificity may or may not be applied depending on whether a specific or a nonspecific method is used.) For cleaning validation, quantitation limit and detection limit are also highly critical. For cleaning validation the limit of quantitation and the detection limit must be determined due to the fact that the assay is typically performed for trace levels of residue, at the lower end of most assay's capability. Any method whether employed on the surface of the equipment, in the laboratory, online, or at-line must be demonstrated to be suitable through appropriate methods validation.

As with all methods, other considerations included in the USP and ICH standards for methods validation should be observed, including:

- System suitability
- Standards or controls to ensure that the assay is valid
- Robustness to demonstrate that the assay is suitable under a potential known variability of the assay method and its parameters
- Control over those materials and supplies, the consumable products that are used with the performance of the assay (e.g., for HPLC—column manufacturer, mobile phase solvent grades, water quality, sample filters)

TOC has been traditionally applied to cleaning validation. The method provides rapid results, and due to its nonspecific nature has been found to be useful in detecting all residues associated with complex processes such as biotechnology cell culture or fermentation operations. In these cases, the lack of specificity of the method is a benefit in that TOC will assess a broad spectrum of residues with a single test. The penalty, however, for the lack of specificity, is that all residues found must be attributed to the worst-case compound because you cannot partition the result to the potential contributing contaminants.

Some of the recent technologies that have been employed in cleaning validation include the use of IMS, surface FTIR detection and photoelectron emission. These methods are specific methods and offer specific advantages in the rapid identification of residues, or the ability to directly analyze residues on the surface of equipment, or the ability to measure trace residues after the filtration of large volumes of rinse, respectively.

For all assays, the method should be applied with some knowledge of the equipment and the intended sampling procedures. When solvent sampling is required

Table 4 Comparison of Features of Typical Cleaning Validation Assay Methods

Attribute	pH	Conductivity	Total organic carbon	HPLC	Ion mobility spectrometry	Direct surface FTIR
Nonspecific	○	○	○	●	●	●
Does NOT detect in the presence of solvents	●	●	○	●	●	●
Requires a soluble/ semi-soluble residue	○	○	○	●	●	●
Requires an ionizable residue	●	○	●	●	○	●
Is NOT typically rapid/real time	●	●	●	○	●	●
Does NOT typically have any on/at-line capability	●	●	●	○	●	●
Uses reagents/mobile phase/specialty gases	●	●	○	○	●	●
Requires special sample preparation	●	●	●	○	●	●

Key: ●, No (advantage); ○, yes (potential disadvantage).

on polymeric materials of construction interference or leachables may inflate the analytical results. For other direct surface methods, limitations in terms of the physical geometry of the equipment will apply because this method combines the sampling and analysis.

Table 4 contains a quick comparison of some of the more common and innovative methods for cleaning validation sample analysis.

An adjunct requirement to the methods validation is the recovery study. Recovery studies are the evaluation of the performance of the sampling method to determine its recovery (or loss) of the analyte at the surface. Recovery factors are then applied as seen in the limits section to correct results to 100%.

During a recovery study, the residue of interest is spiked, at the limit concentration, onto the surface or coupon made of the same material of construction (and the same surface finish) as will be sampled in the field. The residue is typically allowed to dry on the surface as would occur in post-cleaning prior to inspection and sampling. Drying in an oven may be employed if there is concern that the residual heat of the cleaning process will affect the residue on the surface.

Trained sampling personnel remove the residue from the surface, in accordance with the SOP for the sampling method. It is important to assure that techniques used in the lab are the same as will be used in the field to ensure an accurate determination of recovery. In a case of rinse sampling, it may be difficult to simulate rinse sampling methods exactly in the laboratory. In these cases, worst-case assumptions about rinse contact time or less mechanical action should be applied and defended as part of the study.

After sample collection, the sample is analyzed using the validated analytical method and compared to the quantity obtained in a wet spike from the solution that was originally applied to the coupon. The percent recovery is calculated by comparing the wet spike to the amount recovered.

Low recovery results may result in the need for the optimization of the technique, such as:

- Changing the sample container, lid, and lid liner
- Changing the swab type
- Changing the solvent type/acidify solvent
- Changing the swabbing method (e.g., number of swabs, pattern of sampling)
- Changing the types and number of swabs applied (e.g., one wet+one dry vs. two wet)
- Changing swab extraction (e.g., duration of extraction, mechanical action applied)
- Observing personnel for differences in technique
- Eliminating personnel as candidates for sampling.

For all recovery studies, it is a good idea to perform some initial feasibility work with the sample kit (including the swabs, where appropriate) to ensure there is no inhibition or enhancement of the results.

When doing recovery studies it is important to sample blank coupons (coupons not subject to spiking) to ensure that the recovery results that are obtained from the sampling methods demonstrate no interference from residues on clean coupons themselves. Similarly for swab sampling where a swab is to be used blank swabs should be tested to ensure that they to do not provide any inhibition or enhancement of the result.

It is best practice to have replicate coupons tested for each sampling method to assure the validity of the result. Replication coupons sampled by the same individual must show a relative standard deviation less than 10 to 15% to demonstrate consistency of the sampling technique. Likewise, samples between operators sampling the same material must show a similar relative standard deviation. Tighter relative standard deviations may be warranted for readily soluble substances. It is important to remember that all recovery studies are performed at trace levels and that the inherent variability in spiking the coupons with trace materials may include significant variation that will result in a larger standard deviation between individual replicates.

It is generally accepted that recovery should be greater than 50%. Usually, most companies will accept greater than 75% without any investigation. Some investigation and optimization of possible resolutions will be required for 75% down to 50%. Results that demonstrate less than 50% recovery may be accepted by quality assurance if due diligence in method optimization has shown that optimization is not increasing recovery. Inasmuch as all results are corrected for recovery (loss), the acceptance of a low recovery is more of a penalty to the manufacturer than it would be a risk to the patient or to the next product (provided that consistency in the recovery sample to sample has been demonstrated).

Recoveries that are greater than 100% are typically investigated to determine the source of the error, however results are not corrected downwards. This is another conservative assumption in the determination of the actual amount on surfaces.

Swab sampling suffers in locations that are hard to reach. This includes many vessels in which manned tank entry is not possible, piping systems, small orifices like filling needles, or other locations that have a complex geometry. For some samples that are difficult to reach, some companies have employed an extension tool to hold the swab to extend well beyond the reach of the human arm. In employing such a tool, it is important to perform the recovery studies with that tool so that the full technique can be assessed.

It is important to remember that recovery factors are specific to the material of construction that was sampled. In accordance with our recovery studies the analyst must ensure that the correct material of construction correction factor is applied to each sample that is tested in the laboratory. Original analytical results and the corrected results must be available for inspection at the time of an internal audit or an external or regulatory audit. As such, it is sometimes helpful to explicitly show the mathematics that were performed to correct for rinse or diluent volume, surface area, and recovery. In cases where these mathematical or correction factors are part of the "reportable value" of the method, the investigator must be careful that the right correction factors are applied to each sample.

When training operators to collect samples in the field, the supervisor must remind them that it is conservative to always collect more surface area and attribute the sample to less surface area. In this fashion if they are responsible for estimating 100 cm^2 from a surface, it is safe for them to underestimate the area that they sample within reason. Similarly, we do not need to worry about small losses of liquid on the surface during the sampling process; this will simply make the solution slightly more concentrated in the sample container. It is important, however, that these losses on the surface not be extreme because if they are extreme they may affect the recovery percentage.

For some additional tips on performing recovery studies, see the side-bar discussion.

Protocol Development

Cleaning validation protocols, like protocols in other areas, are formulaic. Typical sections include Purpose, Scope, Background, Definitions, References, Responsibilities, Procedure/Testing, Accep-tance Criteria, Deviations, and Revalidation. What makes protocols in the specialized areas of validation different are the technical contents unique to that area of study.

For cleaning validation, the critical elements of the protocol include [and the sections they affect]:

- Identifying the other products and equipment that are included in the groups that are covered by this protocol—remember that successful completion of the worst-case or representative member typically means that all of the members of the group are now considered to be validated [Scope]
- Summarizing cycle development or bench work (such as testing performed by a cleaning agent vendor) that has resulted in the selection of this particular cleaning process [Testing]
- Cross-referencing scientific rationales that are external to the protocol for elements such as: grouping of products and/or equipment, sampling site selection, limit determination, monitoring approaches [Scope, Testing, Acceptance Criteria, and Revalidation]
- Incorporating rationales directly into the protocol for any topic that is not addressed in an external rationale document [Scope, Testing, Acceptance Criteria, and Revalidation]
- Defining the conditions of the equipment prior to, during and after testing [Testing]:
  - Soiling method—normal processing, intentional false soiling, coupon use
  - Loading of worst-case soil—largest batch size, highest concentration of active, longest campaign, longest dirty hold time
  - Pre-cleaning activities conducted—rinsing prior to dirty hold time, or covering/closing during dirty hold time, maintaining equipment under nitrogen air flow during hold time
  - Sequencing of sample collection—noninvasive sampling methods first, followed by microbiological samples next, followed by invasive chemical sampling last (all samples must be coordinated to ensure that different locations were selected for each and that appropriate test methods get the representative hard-to-clean samples)
  - Holding of equipment after cleaning (i.e., equipment expiration or clean hold time)—preparation before storage, storage location, sampling frequency, routine interventions during storage, grouping of equipment which may be different from that which was used for the validation, sampling methods, sampling sites and analytical methods that may be different from those used for the validation (Note: Because of these differences, Equipment Expiration studies are often subject to their own protocols so that the differences can be listed without confusing reviewers about the conditions of testing for the cleaning, itself. This also serves to ensure that cleaning studies can be completed and summarized successfully, even when the equipment expiration studies are still ongoing.)

Of all the sections of protocols today, the section that is most often least effectively prepared is the section on Revalidation. The goal of this section of all protocols, regardless of subject matter, is to define the conditions under which revalidation would be required (Table 5). Instead of specifics, most firms choose to include a single sentence, such as "Revalidation will be required upon change." However, without identifying the scope and

Table 5 Monitoring Strategy Considerations

Low risk	High risk
Automated cleaning	Manual cleaning
In-process monitoring of TACT	No in-process monitoring of TACT
In-process cleanliness measurement (e.g., Process Analytical Technology)	No in-process measurement
Ongoing cleaning validation projects	No new cleaning validation projects
Similar equipment types	Diverse equipment types
Ongoing retraining/certification	No retraining/certification
All surfaces visually available	Surfaces not visually available
Aqueous/readily soluble products	Insoluble/difficult to clean products
Low patient risk for carry-over	High patient risk for carry-over
Low next process risk for carry-over	High next process risk for carry-over
Few changes affecting cleaning	Many changes affecting cleaning that are not revalidated

nature of the change, or the impact of the change, this will not always be true. Table 6 identifies some of the more typical potential effects of change on the validated state.

It should also be remembered that for cleaning validation, routine Monitoring serves as the method to ensure that we maintain the validated state. Therefore, the revalidation section should refer to the ongoing monitoring efforts for the manufacturing plant. It should also refer to the types of reporting and trending that will be performed and will be used to determine when there are changes to the validated state requiring remediation and revalidation. (It is important to remember, when a change to the validated state is detected, root cause investigation must be conducted, remediation of the change must be made through CAPA programs and only then can revalidation be performed. Revalidation alone does not correct process drifts!)

Cycle Development

For automated and semi-automated cleaning, cycle development is a fairly rigorous process where the results from individual trials are used to direct the process parameters to be used for the subsequent trial. Only after a successful (or perhaps several successful) cleaning result, will the final cycle for validation be determined. Optimization of the process parameters will take place to varying degrees at different firms depending upon their business needs for the cleaning process (in terms of time, water and/or chemical usage, for example). In other cases, once a successful process is reached, development halts and no further refinement for the purpose of optimization is performed. Timelines and the frequent urgency to arrive at a validated state often limit the amount of refinement that is performed.

For manual cleaning, the usual process is to refine the SOP for cleaning until it is certain that the process parameters are reproducible and reflect adequate TACT. (See the section on SOPs.) After procedure definition samples are collected to verify cleanliness, if they are successful, the cycle is determined to be adequate for use. The only optimization that is typically performed is to ensure that the cleaning practices are robust. It is recognized that manual cleaning processes require some overkill to ensure that, day to day and operator to operator, the cleaning will be consistent.

When conducting cycle development (or perhaps they might be called "pre-validation runs" if there is no goal to "develop" an optimized cleaning process), a cycle development protocol that outlines the intent and the process to be used is typically prepared and approved. This document serves as a record of what was performed and can serve as the launching pad for additional process refinement. Whether for cycle development or for validation, protocols are required to trap deviations that occur in the process and address them. In cycle development, the deviation impact is minimal as the effect is on the immediate trial con-ducted as processes are evolving during this time. For validation, however, the deviations have an impact on all cleaning validation trials that have been run or are about to be run. Deviations mean that the consecutive successful completion of trials may be in jeopardy and therefore the potential impact of the deviation must be evaluated within the context of all studies performed.

In all cases, whether cycle development or validation, the equipment must be demonstrated to be clean before being put back into use for clinical materials or marketed production. Remember that the effect of cleaning is on the next batch processed and that, therefore, it will become critical to ensure that equipment is successfully cleaned any time that the equipment will be put back into use. For existing equipment, it is common that cycle development activities are interspersed with production activities. Biotechnology and the API industries are probably the only likely exceptions to this. It is, therefore, critical to observe a process of formal equipment quarantine and to release equipment only when results have been returned indicating that it is safe to use the equipment. The consideration for the next lot processed would also naturally lead us to understand that the level of testing in terms of the number of sites, the sampling methods, the analysis methods should all be nearly as rigorous as, if not exactly like, the validation protocol. Remember that the protocol is providing a "high level of assurance" and elements of "reproducibility" in the testing that it presents. It would be appropriate to ensure that a similar level of assurance is provided after the cleaning and during cycle development if we are to turn over the equipment to a subsequent product.

Protocol Refinement, Protocol Execution, and Summary Reporting

After defining the final cleaning process, the elements of the protocol are revisited to ensure accuracy and to

Table 6 Typical Change Control Impacts

Type of change	Typical affected rationales/documents and typical studies required to support change
If change falls into more than one category, all possible affected systems should be evaluated	Each to be performed as applicable to the changeany not considered should be defended in the scientific rationales provided in support of the change control; list may not be all inclusive
Product formulation (including quantities and types of excipients, quantities and types of actives, batch size, dose weight)	Product grouping Worst-case product selection for monitoring studies Limits rationale Engineering trials may be required to demonstrate that the new formulation is/is not a harder to clean challenge Repeat validation for this product and/or justify that new formulation does not represent a harder to clean challenge Bioburden assessments
Production process (e.g., processing parameters)	Equipment train definition Justification of whether new process will result in different hard to clean locations or hard to clean residues Demonstrate ability to clean equipment with worst-case products if the process change is likely to create a new worst-case; may consider worst-case product formulation only Bioburden assessments
Production equipment modification	Sampling site selection (new or different hot spot or critical site) Materials of construction for recovery studies Equipment surface area calculations Limits rationale Equipment train definition Demonstrate ability to clean equipment with worst-case products if the equipment change is likely to have affect Demonstrate ability to clean equipment to new lower limits if existing data do not support the new, lower limit Bioburden assessments
Cleaning agent (change in manufacturer, type, or concentration)	MSDS if new cleaning agent Health safety and environmental review if new cleaning agent or more concentrated than previously used (may require updates to personnel protective equipment in SOPs or changes in handling and cautions) Cleaning agent methods validation and recovery studies Limits rationale (for cleaning agents) Demonstrate ability to clean equipment with worst-case products from each product grouping Cleaning agent residual studies Bioburden assessments SOP and equipment cleaning procedure updates
Cleaning tools	Health safety and environmental review if new tool could have more intimate personnel contact (or may require updates to personnel protective equipment in SOPs or changes in handling and cautions) Demonstrate ability to clean equipment with worst-case products from each product grouping Cleaning agent residual studies Bioburden assessments SOP and equipment cleaning procedure updates
Cleaning process parameters (including additional, eliminated and/or revised steps or set points)	Additional steps may not require testing with existing products/cleaning agents if a justification can be provided that the steps performed are additional to the existing validated process Health safety & environmental review if new/revised parameters/steps could have more risk to users (or may require updates to personnel protective equipment in SOPs or changes in handling and cautions) Demonstrate ability to clean equipment with worst-case products from each product grouping Cleaning agent residual studies Bioburden assessments SOP and equipment cleaning procedure updates

ensure that they reflect what we learned during cycle development. We then execute the protocol, making certain that we capture any deviations that occur during the process.

Deviations may broadly be considered to be extrinsic to the validation—meaning those that are not process related (e.g., power failure during a cleaning validation) or deviations may be considered intrinsic to the validation—meaning those that are directly process related (e.g., failure to achieve a documented process parameter for the cleaning process). For the extrinsic deviations, the effect of the deviation is typically limited to the immediate trial. For the intrinsic deviations, the effect of the deviation could potentially influence all validation trials conducted. There will need to be an evaluation of whether the validation should be repeated in its entirety or whether the immediate trial only need be repeated.

Other pointers for cleaning validation execution include:

- Having clear standards for "visibly clean" surfaces and ensuring that there are standards for:
  - Determining the amount of water that may still be present on a surface to classify as "dry" if the inspection takes place immediately after the completion of cleaning (e.g., droplet vs. puddle)
  - Differentiating between hard water staining on surfaces and product residue—alcohol wipes have been brought to bear to help distinguish this
  - Deciding how much discoloration may be present in small surface pits or scratches before it is assumed that it is either rouge or product residue—alcohol or wet wipes have been brought to bear to help distinguish this based on whether the stain is removable or not
  - Concluding whether or not silver or gray discoloration on swabs as a result of wiping the surface count as clean or unclean if no active residue is detected
  - Determining whether individual fine fibers from "lint-free" wipes should be considered as passing or failing (typically the observation of these fibers finds them less than 0.5 cm or 5000 μm in length and narrower than 80 μm in width)—especially because they may be trapped on sharp edges or fittings on manually cleaned or manually dried equipment
- Having a strategy to address dropped or failed samples, including:
  - How alternate sampling sites are identified (especially if the sample site dropped was a critical site or hot spot that cannot be replicated)
  - Who is empowered to make decisions about the validity of the cleaning validation trial based upon a single (or multiple) missing samples
  - When samples are dropped, even if all remaining samples pass, whether the equipment may be used for the next process or whether it is to be recleaned under the assumption that the untested sample would have yielded a failing result
  - Who should determine the number of trials that are affected by the dropped sample in terms of consecutive successful testing (Usually only one trial would be deemed a failure, but this decision must be documented)
- Having a strategy for unknown peaks when using a specific method (Note that many of these investigatory pathways have their roots in a complete and effective analytical method validation prior to initiating the validation; those method validation outcomes are then used to help investigate the unknown peak):
  - Investigation of the cleaning process and deviations that may have occurred including TACT parameters, tools used for cleaning, training of operators and the like
  - Investigation of the activities prior to the cleaning in terms of failures or risks during the manufacture, calibration, maintenance, or other unusual intervention before the validation and the like
  - Investigation into primary degradation products of the actives and excipients (especially those formed during stresses of heat and pH associated with cleaning)
  - Investigation into cleaning agent residues and its impact on spectra or chromatographs
  - Investigation of prior products processed, their actives, excipients and degradation products' appearance under the current chromatographic or spectrographic conditions
  - Determination of additional test methods that may be brought to bear to investigate the molecular composition of the unknown material (especially as remaining samples may be limited)
  - Determination of the limits of the investigation and when recleaning and retesting will be performed

At the conclusion of all validation activities a report is created that summarizes the testing performed and the results achieved. When summarizing results from any validation, the key is to ensure that all deviations which occurred during the process are defined and that any potential impact to the validation is identified and explained. Typical validation reports have sections that parallel the protocol sections so that the purpose, scope and testing requirements are reiterated along with the results summary. It is common to assemble validation packages so that the final report is located on top, with a copy of the signed protocol, executed protocol and raw data following the report.

Monitoring

Monitoring is the process of performing periodic confirmation of previously validated cleaning procedures for the purpose of reconfirming the validated state. Monitoring for cleaning is conducted much like environmental monitoring or utility monitoring. A schedule is established that identifies the sampling points of interest based upon those products and equipment that represent the most risk during the validation. The risk assessment may consider, among other things:

- Highly potent or toxic products
- Difficult to clean products
- Products that demonstrated a high degree of variability during the original validation
- Products for which the results were very close to the limits during the original validation
- Products that are produced frequently for which buildup, migration of, or simple increase in the likelihood of cross-contamination exist

For monitoring, risk-based decisions may be made (that are different from those associated with the original validation) with regard to grouping of products and equipment to be studied, location of samples, sampling and analytical methods, frequency of monitoring and the like. Frequently, methods of sampling and testing are selected that are less time-consuming and less intrusive to the equipment in order to maximize equipment up-time. This is where methods such as IMS, TOC and FTIR surface probes can be highly valuable in saving equipment down-time.

Revalidation is typically change-based. That is, reperformance of all or part of the original validation is undertaken when a change has been made to the product, process, procedures for cleaning, or equipment. The concept of "time-based" revalidation, or the evaluation of your process at some interval to confirm that it continues to meet the validated state, is also a common revalidation principle. In the case of cleaning validation, routine monitoring can fulfill the role of time-based revalidation. Similarly, if a firm routinely pursues cleaning certification upon changeover of a facility from one product to another, these data may be used to fulfill both a monitoring requirement and a time-based revalidation requirement.

The key to effective monitoring is to establish control charts that help to track and trend monitoring results so that changes in the validated state can be detected early, rather than waiting for a failing result. The concept of process capability for cleaning validation has been discussed in many circles. Whether trending or establishing process capabilities, the company must ensure that only similar sample sites and similar products are included in the trends. It is inappropriate to assume that different equipment, different sampling sites or different product residues will clean to the exact same level. Including different samples in the process capability or monitoring trend assessment may provide a skewed view of inherent process variability based upon these differences.

While monitoring is discussed in every worldwide cleaning validation guidance, there are no specifics with regard to setting monitoring frequencies. Like limits, the inability for regulators to define a minimum standard is due to the high amount of variation in the industry with regard to product types, facilities, and the complexity of manufacturing. As was discussed in Figure 2, every risk-based compliance decision falls along a continuum and it is up to us to defend where we fall along that continuum. Some of the considerations that should be taken when determining monitoring frequency and approach are included in Table 5.

The most important aspect of any monitoring program is to remember to permit modification of the frequency, location of sampling, sampling approach and analytical approach as the monitoring program progresses. The results may well show that it is not necessary to continue monitoring a specific piece of equipment or a specific cleaning process based upon its continued success in achieving results well below the level of interest. With the development of new analytical techniques and an increasing ability to apply PAT to all aspects of the manufacturing process, it is important to leave room to replace the monitoring approach when true PAT becomes available in each factory. Every validator wants to ensure that the program in use will continue to permit the introduction of quality principles, thereby enabling the primary focus to be on those areas that have the greatest ability to affect quality!

Change Control

Now that we understand the requirements for a comprehensive cleaning validation program, we need to consider the impact of change on our program. In preparing for the cleaning validation we studied our products, our processes, our equipment, and our cleaning procedures. Therefore, we should be able to recognize that a change in any of these areas has the possibility of affecting our cleaning validation or the scientific rationales that formed our cleaning validation approach.

Table 6 includes a list of typical changes that might be made and a list of possible effects that these changes might have on our cleaning validation.

Upon the identification of a change, the site change control process should evaluate the impact of the change and either document that there was no impact to the cleaning validation program and its rationales or document the affect and update the effected documents. The potential need for revalidation will have to be assessed based upon the impact of the change.

Remember that individual changes may each be assessed as requiring no revalidation. But, when taken together, several insignificant changes might represent a significant change to a system. Effective change control takes these elements into consideration and establishes a mechanism to periodically review accumulated changes. Most commonly this periodic review of accumulated changes occurs during an APR. Ensure that your cleaning validation program is considered as part of this assessment.

FINAL THOUGHTS ON MAINTAINING THE VALIDATED STATE

Monitoring and change control are not the only programs that are responsible for maintaining our validated state. Calibration, preventive maintenance and even our APR can provide valuable data to help us assure that systems and procedures continue to be executed with the same control as was exhibited during the original validation. Of particular importance is the maintenance of our cleaning equipment and tools in good working order to ensure that there is no deterioration in our ability to clean effectively.

Side Bar with Additional Details on Conducting Recovery Studies

1. Clean the coupons
   a. Clean and rinse coupons copiously and ensure that no residual cleaning agent remains, using purified water (or low TOC water). Solvents can also be used, but ensure that the solvents are fully removed before sampling.
   b. You can passivate coupons with hot nitric or room temperature overnight soak in nitric, if required (especially if new).
   c. Typically we permit coupons to air dry for a short time in a protected area—you may want to see if you can source a clean rack like a CD rack that will hold the coupons upright without allowing them to touch each other, and without a lot of contact to the face of the coupons. This will allow the water to run off the area of the coupon that we will be testing.
2. Spike delivery to the coupons—preparation
   a. The lab should prepare the designated dilution of the media for you. You want to prepare the

spiking solution so that the designated mass of residue can be delivered in a small quantity [i.e., about 10 to 50 μL delivered—you may go as high as 0.5 mL (500 μL), just be sure that it does not run across the plate when you spike it]

b. To perform the spiking, you can use any implement that will deliver the correct quantity of solution in a controlled manner, provided that the implement itself does not provide a lot of carbon:
   i. Microbiological/cell culture automatic pipettors
   ii. Gas-tight 50 μL or 100 μL syringe

c. Clean the delivery device with purified water/TOC free water (or the diluent for the sample) a dozen times—especially if reusable. If disposable, this step can often be skipped or a single rinse may be sufficient.

d. Empty the delivery device completely (if applicable). Fill the delivery device with the spiking concentrate several times, discarding this solution each time (this in effect rinses and wets the interior with the spiking solution).

e. Fill delivery device with 10–50 μL (or with the appropriate quantity per 2(a)—for syringes, you may wish to overdraw the solution and expel the excess to bring the device to volume) and check for air bubbles (draw the delivery device slowly with tip completely submerged to avoid air bubbles—especially important with detergents). For syringes, if air bubbles are drawn up, hold the delivery device upside down against a dark background to check for bubbles. Tap the delivery device gently to dislodge them; expel the air bubbles through the tip (like you've seen hospital staff do a million times!). For other devices, you might need to expel (to waste) and repeat the sample withdrawl or you may need to dispose of tip and retry, depending on severity.

f. When measuring the syringe volume, measure from the furthest point on the convex surface of the Teflon tip of the syringe. Expel the solution to the designated volume. For other dispensing devices, ensure that you know how to properly set the volume controls and/or read the measurement correctly (e.g., bottom of meniscus, etc.)

g. For gas tight syringes, these devices frequently have a triangulated tip—be sure that the triangulated side is pointed down wards towards the surface to ensure contact of the surface with the liquid.

3. Spiking of coupons
   a. Label the background on which the coupons are resting (e.g., lint-free wipe, lab paper, autoclave wrap, tinfoil, whatever clean nonshedding surface you are using) so that you can tell the coupons apart… this will help with any notes that you might take. In addition to this, you may want to etch your coupons with a permanent engraving on the face (outside of the area that would typically be sampled).
   b. Coupons should be just slightly larger than the area that you are planning to sample. For example if you are going to sample 10 cm × 10 cm for 100 cm^2, the coupons would be approximately 15 cm × 15 cm, or even larger.
   c. Spike an area that is well within the intended sampling boundry (e.g., 10 cm × 10 cm for our 100 cm^2 sample) to ensure that all area swabbed is within the spiked region.
   d. Spread out the inoculum as you spike. The purpose here is to make the soil in a relatively even layer. This will simulate how soil is dispersed across the surface after cleaning and will prevent problems such as dissolution of the residue when it is all piled up on top of itself. This will also facilitate drying. This can be accomplished in several ways—multiple methods follow:
      i. If you have no control over dispensing, spot the surface and then spread the inoculum around with the tip… this is a crude method, and should work as long as you do not get too much liquid on the outside of the syringe or tip. Any losses in material will count against you, so this is a conservative approach.
      ii. Dot the surface with small droplets within the sampling boundary (typically 20 to 30 drops required when spiking with microliter quantities)—this will work with many different kinds of dispensing systems and both soft and hard materials of construction of the coupons. This is good for residues with high surface tensions, too.
      iii. Make small chain-like circular motions with the tip of a syringe moving delivery device close to the plate and expelling the solution very slowly making small circular motions from left to right. Repeat in another row (like cursive practice when you were a child and were practicing e's or l's). This works well for devices like the syringe that you can drag smoothly across the surface… It will not work well for soft materials of construction like polymers as the tip tends to drag.
   e. Some coupon materials you can see the spiking readily as you deposit it on the surface. For white or surfaces without a lot of reflectance, you may be required to look at the surface at an obtuse viewing angle to observe how you are depositing the soil.
   f. Allow the coupons to dry in the hood undisturbed. Ensure that there are as few VOCs in the area as possible if you are planning to sample using TOC. Often it is best to sample as soon as the coupons are dry rather than waiting for several days to sample. This will minimize the risk of adventitious contamination. Twenty to thirty microliters of most products on a coupon should be dry within an hour or so (many polymers dry more slowly, so you may want to spike them first before you spike the hard materials of construction). Depending on the number of coupons spiked, you may be ready to swab by the time you get to the last coupon.
   g. Although experimental designs may differ, it is common to spike three coupons for each person for each material of construction.
   h. Leave one blank clean coupon as a blank control for each person for each material of construction,

but ensure that these coupons are subject to the same handling and drying environment. Some may even spike with the diluent alone in order to simulate the spiking process for the blank coupon.

i. Using the same dispensing apparatus that you used during the spiking of the coupons, dispense an equivalent full amount of the spiking solution directly into a test vial containing your sampling diluent. This will be used as the "theoretical quantity" for the determination of recovery.

Note: In the case of residues with volatile constituents, including some detergents, spike this into an empty vial and place in the hood. Allow to dry along with the coupons and only when fully dry add the appropriate quantity of the sampling diluent to each vial for recovery and agitate aggressively for recovery. The reason for this is that the detergents contain some low level of volatiles that will evaporate upon drying. You need to ensure that your theoretical quantity (and all interferents) reflects the loss encountered during evaporation from the surface of the coupon.

4. Swabbing
   a. Follow your internal SOP for the swabbing procedure.
   b. Ensure that you have:
      i. Swabs
      ii. Vials with 40 mL of TOC grade water
      iii. Scissors or wire cutters for cutting swab heads (one pair per person ideally—or clean them between people—make sure whatever material of construction they are that they will not get all gnarly before the completion of all the validation and exposure to cleaning fluids)
      iv. Spiked coupons (3)/blank coupon (1) per person per material of construction
      v. Labels for the vials
      vi. Gloves (powder-free)
   c. Swab the blank coupon last in each case. It will help identify whether your gloves, technique, cutters, or the environment has contaminated your swabs.
   d. Label the sample vials before you start.
   e. Remove the lid from the vial and make sure you have a clean location to put them down or use aseptic technique to hold the lid while you swab. Make sure the method you select will be practical to what you will do in the field.
   f. Take swab and for each coupon and moisten that swab by dipping it into the designated sample's vial.
   g. When wetting swabs in any diluent, hold the swab below the diluent level and swirl it and press it lightly against the sides in order to express any residual air bubbles trapped in the fabric of the swab. This will assist in wetting the swab more fully.
   h. Pull the swab from the vial and press out the majority of the liquid by pressing all four sides at the top of the vial, permitting the excess to fall back within the vial. (Note: If you are ever using solvent: water as a diluent for swab sampling, you do not want to press the liquid out so thoroughly as evaporation will take-over before you finish swabbing, but in water only swabbing it is important to not leave behind a lot of liquid).
   i. Holding the swab—thumb or forefinger is about two centimeters away from the fabric of the swab head to ensure that the swab head is flat against the coupon surface. This should yield a good bend in the swab handle, but keep your fingers from dragging on the surface.
   j. Follow swabbing pattern from the SOP. You may want to determine the number of rows/circuits it takes to criss-cross the designated surface if this is not clear from the SOP. This will serve as a good reference for you when you get into the field.
   k. Cut the swabs with clean cutters at the score mark. Cut the swab heads off over the vial that was used for wetting the swab.
   l. SOP is written for a single swab… During feasibility, we can check what are results are for a single swab and for a single wet swab followed by a single dry swab. We should check to see whether we are getting significantly better results for two versus one. Only if there is a significant, meaningful difference will we implement two for routine sampling, but we should check what that difference is during the recovery study. If two swabs are used, both swab heads should be clipped into the same vial.
   m. Close the cap on the vial and ensure that there is no opportunity for leaking.
   n. Do not forget to sample the blank swab when done with all spiked replicates.
   o. Put the samples under 2°C to 8°C when sampling is complete if they are to be held overnight prior to testing.
   p. Consider preserving the coupons after sampling to have them available for inspection (i.e., the ability to re-look at them can sometimes tell us if we missed the inoculation zone when we swabbed or if there is visual residue on the surface).
5. The protocol includes the details of the analysis of data and additional data on the performance of the testing.
6. Remember that carbon is all around us! Do not work on TOC sampling if there are lots of solvents in the area, including alcohols. Handle TOC swabs and equipment as though you are working in an aseptic environment. Ensure that you are not the source of contamination. Do not touch things unnecessarily and if you have already, change your gloves!!
7. As you are executing the feasibility and qualification, always be thinking about the logistics of sampling real equipment in the real world. Ensure that the techniques and practices you use can be adapted to what you want to do when you are standing at the bottom of a blender or hanging upside-down over a dispensing system.

BIBLIOGRAPHY

Food and Drug Administration. Guide to Inspections Validation of Cleaning Processes. Rockville, MD: FDA, 1993.

Health Canada Health Products and Food Branch Inspectorate. Guidance Document Cleaning Validation Guidelines. Ontario, Canada: HPB, 2002.

Pharmaceutical Inspection Convention, Pharmaceutical Inspection Cooperation Scheme. Recommendations on Validation Master Plan Installation And Operational Qualification Non-Sterile Process Validation Cleaning Validation. Geneva, Switzerland: PIC/S, 2004.

World Health Organization. Supplementary Guidelines On Good Manufacturing Practices (GMP): Validation. Geneva, Switzerland: WHO, 2003.

38

Validation of Training

Christopher Smalley
Wyeth Pharmaceuticals, Collegeville, Pennsylvania, U.S.A.

There is very little written or presented on the validation of training. This may occur due to the fear that the spotlight of attention may reveal glaring deficiencies in the way, not only as a firm but as an industry, we conduct training. On the other hand, the lack of attention that training receives as an issue may lull most of us into a false sense of security, assuming that training is a "non-issue."

Whichever is the case, there are a number of excellent reasons for validating training. Let us digress a moment to discuss two of those reasons before describing the validation of training, because motivating the recipient of the training is one of the most crucial elements in a training program. So here are two great reasons for getting them motivated. (*i*) It is Good Business and (*ii*) It is Required by Regulation.

IT IS GOOD BUSINESS

An employee represents one of the largest investments that a business can make. Not only can an employee cost money by simply standing there, but if the employees do not perform the job properly, they can damage equipment and facilities, waste product, and consume an inordinate amount of time in correcting their errors. On the other hand, an employee qualified to perform their job will give the organization a competitive advantage in the marketplace. Well-trained employees can reduce waste, anticipate and prevent loss, have less on-the-job injuries and lost-time accidents, and perform with less direct supervision. The product that the customer receives, whether it is the next department to receive the component, or the health care professional, will perceive a difference as well. After all, surveys have shown that when customers hear "It has been a pleasure to serve you" rather than "Have a nice day," their memories of the exchange are more positive. This can be achieved with a validated training program.

IT IS REQUIRED BY REGULATION

Within the Code of Federal Regulations 21, Part 211, cGMP for Finished Pharmaceuticals, there is a subpart for Organization and Personnel, which contains Section 211.25, Personnel Qualifications. In part, this section states that "Each person engaged in the manufacture, processing, packaging or holding of a drug product, shall have education, training and experience, or any combination thereof, to enable that person to perform the assigned functions."

Employees represent one of the greatest variables in pharmaceutical manufacturing. Their value over automation, in many instances, is their ability to use judgment in reaching conclusions and making decisions. It is this value that makes the case for validation of training so compelling.

In describing the validation of training, it is convenient to categorize the types of training that will be validated.

NEW EMPLOYEE TRAINING

How does a new employee become educated in the skills needed to perform their job safely and effectively? Imagine for a moment that we are performing an IQ similar to that for a new piece of equipment. Are your specifications adequate? That is, are the job description and other documentation which describe the job to be performed adequate? What are the minimum requirements for the employee being "installed?" Were those "design requirements" communicated to the Human Resource Department for recruiting and interviewing? Can new employees read and understand the manufacturing batch record? Can they read and understand the SOPs? What education level are these documents written to? What grade level are new employees required to meet in reading comprehension? New employee training must consider:

- the requirements for the position being trained for,
- design the training with the objective fulfilling those requirements
- implement the training and then
- evaluate the training

Evaluation is not possible unless the objectives (similar to critical process parameters) are clearly set forth. Implementation is not possible unless a design has been made for the instructional material. That design, for new employee training, would usually consist of didactic training to the workplace and SOPs that would impact on that employee's tasks. For uniformity in presentation,

---

Abbreviations used in this chapter: cGMP, current good manufacturing practice; FMEA, failure mode and effects analysis; GXP, good pharmaceutical practices; HACCP, hazard analysis and critical control plan; HEPA, high-efficiency particulate air; HVAC, heating, ventilation, and air-conditioning; IQ, installation qualification; NMT, no more than; PDA, Parenteral Drug Association; SME, subject matter expert; SOP, standard operating procedure.

commercially prepared audiovisual products such as those available from the PDA are useful, followed by a discussion period, as would be a presentation from an experienced employee on SOPs.

For GMP overview issues, inviting an SME such as a professor from a local pharmacy college would make the presentation interesting to the new employees, and afford the opportunity to audit the classroom to see who is paying attention or not mentally present.

RISK MANAGEMENT

The interest in risk management for validation activities has an impact on validation of training. Although some believe that risk management may be a rationale for reduced validation, in some areas the result of a risk assessment will highlight the need for added effort. One of those areas is training.

Pharmaceutical industry frequently responds to problems by adding requirements to procedures. How often have gowning procedures been changed to add a statement requiring regowning when a person's knee touches the floor, or a calibration procedure been changed to require the technician to compare the accuracy of the calibration standard to the accuracy requirement of the instrument undergoing calibration? The real need is to train employees to think and understand. Every possible combination and permutation cannot be placed into procedures, and then training conducted on all of those possibilities. When examining the risks to the organization, failing to ensure that employees are capable of thinking and understanding independently is clearly among the greatest risks.

How, then, do we provide training that helps employees to think and understand? Training, in the content of procedures, focuses on the "what." Understanding relates to the "why." For training design, this entails not only presenting the words written in the procedures, but also the reasons why those words were written. One of the best approaches to training on this content is to use the SME responsible for writing the procedures. A glaring fault in this approach, though, is assuming that the SME can properly present or train on the content. If the decision is made to use the SME in training to gain understanding, then a process for ensuring that the training presentation is properly performed will need to be incorporated into the training design. This frequently takes one of two directions. One pathway would be to develop the training skills of the SME, the other pathway would be to have qualified trainers learn the material from the SME and then have them conduct the training.

Let us recap some of the topics raised in implementing the "IQ." They are training requirements, training design, training execution and evaluation of training. Embedded in these topics is the requirement to document. You may have what you assume to be an adequate documentation system, a system which records the date, the attendees, the presenters, and the topic. What those systems document is simply people attending classes, and not proof that learning is taking place. Adequate documentation will record each of these four topics and together represent proof that training is under control and validated. And again, it is good business.

PROGRAM DOCUMENTATION

It is a good idea to establish procedures for documenting training programs. Preparing needs objectives, design components, and requirements not only helps to clarify the planning phase, but also increases the likelihood of a smoothly run program. Documentation allows the program to be repeated with little confusion and offers a chance to critique and tailor the design.

- An extensive outline of the program design, including a list of goals and objectives, notes for lectures, discussion questions, and so forth
- A needs assessment that evaluates training needs. It is critical to clearly define the objectives of these needs. For instance, is the objective to learn to use different risk assessment tools such as HACCP or FMEA, or is the objective to understand the risk assessment tools and understand which one to select for a particular application?
- Additional design ideas, including use of game show formats or interactive presentations
- References/readings on the content topic
- Planning notes for the instructor and facilitators to use in the background of the lessons
- Administrative information and forms
- Evaluations with summaries of comments from program evaluation forms
- Room arrangements/audiovisual needs

The more explicit the documentation, the easier it is to update and repeat the program. Immediately after presenting a program, the trainer may make notes about what to do differently the next time. Before presenting the program again, the trainer can read the notes and evaluation comments from the previous presentation.[a] Program content that is critical for training success must be embedded in the training presentation materials, so that different presentations do not vary in the depiction of that content.

TRAINING EXECUTION

Training execution, or implementation, requires the development of a presentation and atmosphere which facilitates the transfer of information. The training design identified the target audience and the information to be presented; now comes the step involved in bringing that target audience together in such a way that they are both receptive to the idea that something can be learned, and willing to participate in the training execution.

Avoid the training caricature of a group of people in rows of chairs nodding off in a darkened room, while a speaker reads from the slides. Try to vary the format of the presentation. Use a table in the center of the room and, while having a discussion of the topics, move around the table to involve everyone. Give the students a break periodically to allow them to refresh themselves and recharge. Use a tour of an area that has relevance but that the target audience does not actively work with.

[a] Inside Training and Development, Susan Warshauer, pp. 61–62.

For instance, HVAC mechanics could tour the aseptic suite or the filling room operators tour the parts preparation and sterilization area.

Try to avoid the dry presentation of facts. Interactive formats can be developed using the popular game shows, such as "Wheel of Fortune" or "Jeopardy." Such formats require that the information be written in the language of the target audience. It is crucial that you understand your target audience, and not necessarily just the goals identified in the design stage. Understand the mindset and motivation of your audience. If the group consists of empty nesters, do not use a rock music soundtrack for background fill-in. Conversely, do not use classical music background for a group consisting of young people. Do not use a game show format if it appears that the group may be shy. On the other hand, do not use a slide–tape presentation with an energetic audience.

Each format used to convey information additionally has advantages and disadvantages above being "right" for the audience. Slide–tape presentations can be made or purchased for a reasonable cost and customized to keep pace with procedure and facility changes. Motion pictures have a very high interest coefficient, but are considerably more expensive and difficult to update. Videotape has as high an interest and as efficient as motion pictures at a lower cost; however, they too consume a great deal of resources to keep current. All of these formats should be incorporated into a total training presentation which includes tours, discussion and games.

EVALUATION OF TRAINING

Evaluation of training is both the most difficult and most important aspect of the validation effort. In evaluation, one of the tools that generates the most interest is testing. The objectives of testing and evaluation are to evaluate the trainer and the program, not the trainees. Just as the SME is an ideal source for training, the SME is the ideal source for preparation of the evaluation. The key to having an evaluation that is non-threatening lies in planning the test effort during the training requirements and training design phases. In this manner, the testing can remain focused on how well the objective of changing the employee's knowledge or skill level was achieved, demonstrate areas for improvement in the program design, and provide the trainer with a performance monitoring tool. In keeping with the concept that training is good business, testing can provide management with information about the value and costs of training.

The test itself must be subjected to testing. A valid instrument measures what the trainer intends to test.

Basically, there are four approaches to determine whether an instrument is valid or not. These approaches, adopted by the American Psychological Association, are: (*i*) content validity, (*ii*) construct validity, (*iii*) concurrent validity, and (*iv*) predictive validity. The actions taken to make the instrument valid are usually referred to as "defending" the validity of the instrument.

Content Validity. Content validity refers to the extent to which the instrument represents the content of the program. Content validity is probably the most important approach. Is it a representative sample of the skill, knowledge, or ability presented in the training program? To ensure content validity, no important items, behaviors, or information covered in the program should be omitted from the instrument.

Construct Validity. Construct validity refers to the extent to which an instrument represents the construct it purports to measure. A construct is an abstract variable such as the skill, attitude, or ability that the instrument is intended to measure. Examples of constructs areas follows:

- Ability to participate in an aseptic filling operation
- Ability to read a volume having a meniscus

Concurrent Validity. Concurrent validity refers to the extent to which an instrument agrees with the results of other instruments administered at approximately the same time to measure the same characteristics. Concurrent validity is determined by calculating the correlation coefficient between the results of the instrument in question and the results of a similar instrument.

Predictive Validity. Predictive validity refers to the extent to which an instrument can predict future behaviors or results.

If an instrument predicts a behavior, and a significant number of participants do exhibit that behavior, then the instrument possesses predictive validity. Predictive validity can be calculated and expressed as a correlation coefficient relating the instrument in question to the measure of the predicted results or behavior.[b]

Certainly, adequate planning and definition of the job requirements are necessary to have a valid test. And the validity of the test can be supported by other evaluation methods which will be discussed later, including interviewing, group discussion and on-the-job observation.

The type of test necessary to have a validated training program is the criterion-referenced test, where the employee's performance on the test is measured against the instructional objectives. Such a test need not be a written test, multiple-choice answers or short answer fill-in-the-blanks. The best tests relate to the job to be performed. Hence, if training were provided in gowning and contact plating, the order of events would be to design and execute the training and then, on three consecutive days, have the students gown, take the plates, and degown. Incubating the plates and reviewing the results would constitute the test and validate the training.

Let us take a moment to examine the role of the validity tests in the creation of written tests before further examining performance tests. Obviously, the training design included the training objectives, and these objectives are derived from the needs assessment that defined what competency is expected from the target audience, the conditions under which the competency will be displayed, and the accuracy with which this competency will be performed. The competency which the target audience is to display must be described in measurable and objective terms. Action verbs such as "sort and

[b] Handbook of Training Evaluation and Measurement Methods, Jack J. Phillips, pp. 82–85

remove defects" are better than "inspect," definable criteria such as "fill X ampules with media with NMT Y containers demonstrating contamination" as opposed to "aseptically fill," are necessary to prepare test items which can determine whether the objectives have been achieved by the training program.

The objective attribute must be defined as well. Is the target audience to be the lead operators of the filling equipment of members of the team? Are the test subjects expected to refer to members of the team? Are the test subjects expected to refer to SOPs periodically to perform competencies by rote? Are there time constraints on the performance of the competency? These conditions should be reflected in the construction of the test.

The accuracy with which the subject must perform the competency must be considered in the construction of the test. Is 100% accuracy necessary, or is the test audience subject to close supervision by a lead operator or supervisor and some errors would be permissible? Is there expert judgment involved in evaluating and answering the situation presented in the training program? Are there elements within the training program that have no tolerance for any error, such as safety-related items? In considering how well the learner must perform the indicated behavior, the issue of how much content may *not* be learned and still allows the learner to perform a GXP task needs to be answered.

With these clear objectives, a selection of written test design can be made that will evaluate the objectives. Remembering that the written test will only be one component in the evaluation of training; within the test there must be content validity and construct validity, and other evaluation measures performed must demonstrate concurrent validity with the test, and finally the supervisory feedback must subsequently demonstrate predictive validity.

Most instructional designers are aware of how important objectives are to the creation of instruction; many are less familiar with the role of objectives in testing.

Instructional objectives serve the following three fundamental purposes:

1. Objectives ensure that the test covers those outcomes important for the purposes that the training must serve. Remember that there are several different types of tests and that the content for these tests is derived by task analysis procedures that order objectives hierarchically. Matching test items to the appropriate course objectives within these hierarchies assures that the essential content is assessed.
2. Objectives increase the accuracy with which cognitive processes in particular can be assessed. A well-written objective is a road map for the creation of test items that will assess the specific competency described by the objective. Hence objectives are essential to the construction of a validatable testing process.
3. The size of the domain covered by the objectives and the homogeneity of the objectives being assessed are important factors in determining how many items will need to be included in the test.

Of course, not all objectives are equally well written. Numerous authors have provided course developers with advice about how to write objectives. Most agree, however, that good objectives have four parts:

1. Who the learner is,
2. What behavior or competency the learner will perform,
3. Under what conditions the learner will perform the competency, and
4. To what standard of correctness the learner will perform the competency.

It is essential that the competency be described in observable, measurable terms, hence the term "behavioral objective" is used to describe the most useful statements of learner outcomes. When writing objectives, choose the most precise verb you can to state what the learner will be able to do. For example, the words "list," "categorize," "draw," and "evaluate" are better than "understand," "appreciate," and "know." The more descriptive the verb in an objective, the easier it will be to write test items that accurately assess the objective.

If well written, this part of the objective provides useful information to test writers, since the test essentially presents learners with a series of conditions under which they must demonstrate their achievement of the instructional objectives. Unfortunately, an aspect of the objective that is frequently omitted is the conditions element by designers who do not realize how critical it is for clearly communicating the intent of the objective. Changing the conditions under which a behavior is to be performed can dramatically alter the difficulty and nature of the competency assessed.

For example, the behavior "assemble the tablet compression machine" is significantly altered depending upon whether the corresponding condition is "given the unassembled parts." The behavior with the latter condition can be expected to be significantly more difficult than with the former, and in fact the very nature of the intended competency specified by the objective changes depending upon which condition is used. Under the former condition the objective describes skills in reading and using a repair manual, whereas under the latter condition the objective specifies mechanical skills.

Complete objectives include a statement of how well the learner must perform the indicated behavior. This component, however, is probably the most difficult component to write. It frequently takes the form "with 90% accuracy" or "correctly 80% of the time." It is helpful to realize that all standard statements need not be in the form of percentages. In fact, many competencies do not lend themselves to percentage standards at all. Other forms of standards are in terms of number of allowable errors, time limits, expert judgments, negative consequences avoided, for example, "move pizza from oven to boxing counter *without burning the fingers*."

Good objectives are an essential precursor to sound testing systems. Translating objectives into rating scales for performance tests is usually easier than translating objectives into test items for paper-and-pencil tests. One strategy that can be helpful in this regard is to first classify the objectives according to the type of cognitive behavior each requires. Classifying objectives by cognitive skill assists item writers in choosing which item type—multiple-choice, essay, etc.—will most accurately and efficiently assess the objective, and deciding what the text of each item will be.

Several different classifications of cognitive behavior have been developed over the years. Bloom and his colleagues developed their system through an intensive content analysis of thousands of instructor-created test items. As a result, Bloom's Taxonomy provides a particularly comfortable fit with and support to cognitive assessment.[c]

The classification scheme consists of six levels with each given level building to the successing levels as follows:

1. Evaluation
2. Synthesis
3. Analysis
4. Application
5. Comprehension
6. Knowledge

Understanding the nature of the cognitive performance to be assessed is a good first step to being able to write an appropriate test item. If a test writer can correctly identify the Bloom level of an instructional objective, a wealth of ideas about how to measure the objective become available.

Another important result of understanding Bloom's Taxonomy is an increased awareness of the cognitive behaviors beyond remembering, i.e., beyond the knowledge level. Most of the tests taken in school at all grade levels and even at the college level are composed of knowledge level questions. This circumstance is not difficult to explain, since knowledge level items are by far the easiest to write. However, developing tests that truly reflect on-the-job performance requires the ability to distinguish among different cognitive behaviors and skill in writing items at the higher cognitive levels, particularly the comprehension, application, and analysis levels.

There are six types of test items commonly used in paper-and-pencil tests, which are as follows:

1. True/false
2. Matching
3. Multiple-choice
4. Fill-in
5. Short answer
6. Essay

Of these six, multiple-choice offers the most advantages for a paper-and-pencil test. Multiple-choice questions present a question, which may be identical to the question constructed for a short answer test, and offer the recipient a number of choices consisting of a single correct answer and several distracters. Through the use of "all of the above" or "none of the above" choices, not only is knowledge tested but comprehension is evaluated as well. The probability of guessing the correct answer is lower than with true/false questions, and the process of scoring and providing feedback for the training program is much shorter than it would be for short answer or essay. It has the advantage of being able to assess most of Bloom's cognitive levels and yet can be easily scored by hand or by machine. In making comparisons, multiple-choice will be the benchmark with a given item type. For each of these six-item formats, a description of the item type and the kind of content for which the format is best suited, the Bloom levels assessable by the item type, the major advantages and disadvantages of using the item type and a summary of the guidelines for writing each item type is presented.

[c] Development and Validation of Minicourses in the Telecommunication Industry, Richard R. Reilly and Edmond W. Israelski, pp. 721–726.

True/False Items

Description. The true/false item presents the test-taker with a statement that he or she must indicate is either true or false. This type of item is a sensible choice for "naturally dichotomous" content, i.e., content that presents the learner with only two plausible choices. For example, assume our objective requires that, given end-of-cycle sterilization report, learners will classify the cycle as pass or fail. You might construct a true/false question asserting that a given sterilization report is a "pass," to which the test-taker would respond "true" or "false." Content that is not naturally dichotomous is usually best assessed using the multiple-choice format, because true/false questions have some distinct limitations.

Bloom Levels. True/false items can assess the knowledge, comprehension, and application levels. Unfortunately, however, they are most often used to assess only the knowledge level.

Advantages. The primary advantage of true/false items is that they are typically easier to write than other types of closed-ended questions, i.e., matching or multiple-choice. However, the reputed ease of construction is partly because most of these items are written at the knowledge level; it requires more thought to write true/false items at higher cognitive levels. The other advantages are that, like all closed-ended questions, they are easily and reliably scored, and test-taker responses can be submitted to statistical item analysis that can be used to improve the quality of the test.

Disadvantages. The biggest disadvantage of true/false items is that test-takers have a fifty–fifty chance of getting the items correct simply by guessing. However, if the content that the true/false item covers is truly dichotomous, a multiple-choice item with more than two choices would be very difficult to write anyway. After all, multiple-choice items with only two choices also allow test-takers to guess correctly half of the time. Before writing true/false items, always examine the content and instructional objectives carefully to be sure that they are not more appropriately addressed by multiple-choice items. The key to using true/false items effectively is to use them only when the content is naturally dichotomous and to write true/false items that require more than mere memorization of content.

Matching Items

Description. Matching items present test-takers with two lists of words or phrases and ask the test-taker to match each word or phase on one list (hereafter referred to as the "A" list) to a word or phrase on the other (the "B" list). These items should be used only to assess understanding of homogeneous content, for example, types of sanitization agents, types of lubricants, types of switches, etc. Matching items most frequently take the form of a list of words to be matched with a list of definitions.

Bloom Levels. Matching items can assess the knowledge and comprehension levels. However, like true/false items, they are rarely written beyond the knowledge level.

Advantages. Matching items are relatively easy to write. Note, however, that one reason for this feature is that they do not assess beyond the comprehension level. Matching items can be scored quickly and objectively by hand and frequently also by machine. Responses to matching questions can be submitted to statistical item analysis procedures.

Disadvantages. Matching items are limited to the two lowest levels of Bloom's Taxonomy. Another disadvantage is that if these items are constructed using heterogeneous content, i.e., if the words or phrases appearing on the "A" list are essentially unrelated to one another, matching items become extremely easy. For example, a list that contains a type of sanitizing agent, a type of lubricant, a type of switch, etc., will be easier to match with a corresponding "B" list than with a list that contains only names of different types of sanitizing agents. Another difficulty with matching items results from test writers including equal numbers of entries in both lists or allowing items from the "B" list to be used only once. Under these circumstances test-takers can use the process of elimination to figure out cues to the correct matches.

Multiple-Choice Items

Description. The multiple-choice item presents test-takers with a question (technically called a "stem") and then asks them to choose from among a series of alternative answers (a single correct answer and several distracters). Sometimes the question takes the form of an incomplete sentence followed by a series of alternative completions from which the test-taker is to choose one. Sometimes the stem is a relatively complex scenario containing several pieces of information ending in a question. Dichotomous content can be assessed using multiple-choice questions with two optional answers; thus most true/false items can be converted to the multiple-choice format. In preparing a multiple-choice question, it should be remembered that the intention of the test is not to test the reading ability of the target audience, so keep the language simple. To avoid misleading the subject, bold or underline negatives such as *no* and *not*. Use common errors that subjects make in developing distracters. Check the questions to ensure that the choices to one question do not indicate the correct answer in another question. The more questions which appear on a test, the greater accuracy. For each training objective, approximately five questions should appear on the test. Obviously, the more critical the objective the more questions, such that a GMP objective, for example, would be evaluated by more than six questions. Arrange the questions in order of difficulty, placing easier questions first to prevent frustration or discouragement. The number of choices does not need to remain constant, but should not be less than three and only rarely be as many as six.

Bloom Levels. Multiple-choice questions can assess all Bloom levels except the two highest ones, synthesis and evaluation. The reason that these two levels are beyond the multiple-choice format is that they require totally original responses on the part of the test-taker. Since multiple-choice questions are closed-ended, i.e., the correct answer appears before the test-taker who must recognize it; the test-taker's response is necessarily not original. However, multiple-choice allows assessment of more Bloom levels than any other closed-ended question format.

Advantages. Multiple-choice is the most flexible of all closed-ended item formats. Multiple-choice items can assess any kind of content at a variety of Bloom levels. Because the test-taker must choose among several optional answers, the probability of simply guessing the correct answer is lower than with true/false items. Furthermore, multiple-choice items are ideal for diagnostic testing. In other words, the distracters can target those learners who have specific problems; knowing the wrong answers chosen by test-takers can be important and useful information for instructors and course designers. In addition, multiple-choice questions are quickly and reliably scored either by hand or by machine and are ideally suited to statistical item analysis procedures that can lead to improved test quality.

Disadvantages. The major disadvantage of multiple-choice questions is that they are difficult and time consuming to write. Most testing authorities agree that well-written multiple-choice questions are usually worth the effort, especially if they can be used repeatedly with a large number of test-takers. (Reordering the choices after several uses may be appropriate to keep the test "fresh.") An additional weakness is that multiple-choice questions cannot assess objectives that require test-takers to recall information unassisted, since the correct answer does appear before the test-taker among the options. Another disadvantage is their inability to assess directly the synthesis and evaluation cognitive levels. The principal disadvantage of multiple-choice question is the recognition factor. The typical performance setting will not present the test subject with information which may be recognized; rather, many situations will require the subject to recall information. Another disadvantage to multiple-choice question is the investment in time necessary to develop good distracters that are unambiguously wrong.

Fill-In Items

Description. Unlike the first three item formats discussed, fill-in items are open-ended, i.e., the answer does not appear before the test-taker. Rather, the fill-in item is a question or an incomplete statement followed by a blank line upon which the test-taker writes the answer to the question or completes the sentence. Therefore, fill-in questions should be used when the instructional objective requires that the test-taker recalls or creates the correct answer rather than simply recognizes it. Objectives that require the correct spelling of terms, for example, require fill-in items. Fill-in items are limited to those questions that can be answered in a word or short phrase; short answer and essay questions require much longer responses.

Bloom Levels. Fill-in items can assess the knowledge, comprehension, and application levels. They are written most often, however, at the knowledge level.

Advantages. Fill-in items are typically easy to write. They are essential for assessing recall as opposed to recognition of information.

Disadvantages. There are two major disadvantages of fill-in items. One is that they are suitable only for questions that can be answered with a word or short phrase. This characteristic typically limits the sophistication of the content that can be assessed with fill-in items. The second major disadvantage is that, like all open-ended questions, fill-in items present scoring problems. Because test-takers are free to write any answer they choose, sometimes there can be a debate over the correctness of a given answer. Test-takers are marvelously unpredictable when it comes to concocting an unanticipated answer to an open-ended question. Unlike the scoring of closed-ended questions, the scoring of all open-ended questions requires judgment calls on the part of the scorer.

Short Answer Items

Description. These items are open-ended questions requiring responses from test-takers of one page or less in length. Short answer questions require responses longer than those for fill-in items and shorter than those for essay questions. Short answer questions are recommended when the objective to be assessed requires that the test-taker recall information unassisted (rather than recognize information) or create original responses of relatively short length.

Bloom Levels. Short-answer questions can be used to assess all Bloom levels except possibly the highest one, evaluation; most responses to evaluation questions would necessarily be somewhat longer.

Advantages. The major advantage of short answer questions is that they are able to elicit original responses from test-takers. For some objectives at the higher Bloom levels, only short answer and essay question are appropriate. Lower-level short answer questions are typically easier to write than multiple-choice questions covering the same content. It is important to remember, however, that changing the format of a question can significantly alter the cognitive skills assessed. Short answer items are best reserved for those objectives that cannot be assessed using closed-ended questions.

Disadvantages. The disadvantages of short answer questions are, unfortunately, extremely serious ones. Most notably, short answer questions are very difficult to score reliably. The evaluation of short answer responses and essays is notoriously prone to error—resulting from halo effects, the placement of a given test in the scoring sequence, scorer fatigue, and especially, quality of handwriting. In addition to being unreliable, the scoring of short answer responses is time consuming. Short answer questions also require far more time to answer than multiple-choice questions, thus sometimes limiting severely the content that can be covered by the test.

Essay Items

Description. Essay items are open-ended test questions requiring a response longer than a page in length. They are recommended for objectives that require original, lengthy responses from test-takers. Essay items are also recommended for the assessment of writing skills.

Bloom Levels. Essay questions can be used to assess all levels of Bloom's Taxonomy. They are the only item type with this capability, and writing is the only item type that can truly assess the evaluation level.

Advantages. The essay question's major advantage is its capacity to assess the highest cognitive levels. Of course, these levels, comprehension and knowledge, include the aspects of understanding and problem solving that are important objectives identified earlier. Essay questions that assess the lower levels are usually not difficult or time consuming to construct. Those that assess the higher levels can be very difficult to write, requiring the provision of a great deal of stimulus material to which the test-taker responds in the essay.

Disadvantages. The disadvantages of the essay item are identical in nature to those of the short answer item; however, these problems are aggravated by the additional length of the responses. Essay questions are even more difficult to score reliably, take even more time to score and use up even more testing time than do short answer questions. The principal disadvantage of short answer and essay questions lies in the debate which may ensue over the correctness of an answer. The number of unanticipated answers to open-ended questions has never ceased to be amazing. Secondarily, as mentioned above, the time to review and score the short answer or essay test will hinder the timeliness with which feedback can be provided to the training program. For these reasons, essay items are to be avoided if at all possible. Use essay questions only when the cognitive level of the objective requires it.

SUMMARY

Training is an important aspect of the pharmaceutical process, crucial to every step and a factor in compliance with cGMPs, safety, productivity and customer satisfaction. Similar to process validation, cleaning validation, or sterilization validation, in training it is important to identify the requirements, the design of the means for testing against those requirements, the acceptance criteria, and documentation of the results.

Testing is a subcomponent of the validation of training, a component which has not been utilized for a variety of reasons, but nevertheless a component central to the validation process. Testing must be conducted in a manner consistent with good science, as must any validation program, and like any validation program the testing instrument cannot stand on its own but must be supported and confirmed by other tools. Among these other tools are group discussion and observation by supervisors or other qualified individuals. The observations, whether performed by the supervisor or, in a case such as equipment assembly by a mechanic and the gowning technique by a microbiologist, tend to be much stronger in cognitive assessment than in written instruments. Although traditionally referred to as performance appraisal, these observation methods are in fact a nonwritten test method. The method is frequently criticized as being subjective; however, this fault can be minimized through the preparation of an evaluation period where a different supervisor other than the supervisor to which the trainee normally reports, performs the assessment. Where possible, utilize an SME to perform the assessment, such as a microbiologist for trainees participating in aseptic filling operations. Similarly, an SME would be necessary in determining the content validity of the written test. The most important aspect to remember in training and the validation of training, though, is not to fall into the trap of conducting and validating remedial training after disaster has struck, but to make the investment in validation of personnel similar to the investment made in the validation of equipment and systems.

Much of this section has been devoted not only to the validation of training, but also to make the case as to why validation should be performed on training and that training validation contains the same elements as other validation activities. Validation should be performed on training because it is good business as well as required by regulations. Training validation will have a requirements phase identified using a needs assessment evaluation, a design phase, an execution phase, and an evaluation phase. Like other validation activities, a matrix mapping the tests conducted during evaluation back to needs assessment ensures the validity of the validation.

APPENDIX

Several guidelines for the construction of objective measures are useful regardless of the type of training:

1. Arrange the items in order of difficulty; placing easier items first avoids discouraging participants unnecessarily.
2. Construct each item so that it is independent of other items. A series of items in which the correct answer to the first becomes the condition for the next item can prevent the measure from providing an accurate picture of the participant's knowledge or skills.
3. Avoid constructing items by taking quotes directly from a handout, overhead transparency, or book. Direct quotes tend to encourage memorization rather than understanding, and quotations taken out of context tend to be ambiguous.
4. Avoid trick questions. The intent of a performance measure is to determine the skills, knowledge, and attitudes of the participants, not to cause them to mark an item incorrectly.
5. As much as possible avoid negatives, especially double negatives. Such items take considerably longer to read and are often misinterpreted. If negative words must be used, underline or italicize the word or phrase to call attention to it.
6. Avoid providing clues to items in previous or subsequent items.
7. Use a variety of types of items in the performance measure rather than limiting the items to only one type. If the measure is lengthy, variety can add interest. When a variety of types of items are employed, group the items by type so that participants do not have to constantly shift response patterns or reread instructions.

True/False Items

Guidelines for constructing true/false items:

1. Have participants circle the correct answer rather than write in T or F. Poor or careless writing can make the letters "T" or "F" or even the words "true" or "false" very difficult to read.
2. Avoid the use of "always," "all," "never," "only," and "none"; these words alert the participant to mark the item false. For example, the following is a faulty item:
   a. T F Effective managers always delegate.
3. Avoid the words "sometimes," "usually," "maybe," "often," and "may"; these words alert the participant to mark the item true. For example, the following is a faulty item:
   a. T F The tablet hardness is sometimes affected by the moisture content of the granulation.
4. If the statement is controversial, cite the authority whose judgment is referenced. For example, the following is a faulty item:
   a. T F The organization has a responsibility to provide assistance to employees who have drug problems.
   b. This statement as it is written appears to measure attitude rather than knowledge. To make it into a knowledge statement, it could be preceded by "The Federal Government has determined that…" or "The site Director has stated that…"
5. Do not include two concepts in one item. For example, the following is a faulty item:
   a. T F The trend toward quality circles began in the early 1980s and represents a big step forward in improving quality in the United States.
   b. Either of the two concepts, (*i*) began in the early 1980s or (*ii*) represents a step in improving quality in the United States, could be false.
6. Each statement should be entirely true or entirely false without additional qualifiers such as "large," regularly," "sometimes," and "may." For example, the following is a faulty item:
   a. T F A media fill failure may indicate poor gowning technique.
7. Keep true and false statements approximately the same length.
8. Have approximately the same number of true and false items. True statements are easier to write so there is a tendency to include more true than false statements.
9. Avoid making false statements by simple adding "not" to true statements. For example, the following is a faulty item:
   a. T F The cGMPs are not the guideposts of pharmaceutical manufacturing.
10. Avoid using trivial details to make a statement false.
11. Avoid a pattern of answers such as TTFFTTFF.
12. Place the central point of each statement in a prominent position or highlight it in some manner.
13. Avoid long and complicated statements that may test reading ability rather than the content. For example, the following is a faulty item:
    a. T F If data are recorded to the nearest 0.001 inch (for example), then the class width should be an integer multiple of 0.001 inch so that each interval will contain the same number of possible data values.
14. Avoid negative statements and eliminate double negatives. For example, the following is a faulty item:
    a. T F If a person has not had access to a patent, he or she cannot infringe the patent.

Matching Items

Guidelines for constructing matching items:

1. Place the list of descriptions (the longer list) on the left side of the page so that the participant needs to read it only once. Place the options (the shorter list) on the right to be scanned as often as needed.
2. Make both the descriptions and portions list homogeneous.
3. Provide at least three more options than descriptions or permit the use of each option more than once to reduce guessing.
4. Specify in the instructions the basis for matching and how to mark the answer.
5. Make each option plausible to the uninformed.
6. Arrange the options in some logical order, such as chronological, numerical, or alphabetical to save reading time.
7. Allow between 5 and 15 options.
8. Number the descriptions and letter the options.
9. Specify in the instructions whether options can be used more than once.
10. Avoid having more than one correct option for each descriptor; item 1 in column A should have only one option in column B that is correct.

Multiple-Choice Items

Guidelines for constructing multiple-choice items:

1. The stem (the part that precedes the responses) should clearly state the premise; the response options should be kept as short as possible. For example, the following is considered a faulty item:

   An income statement
   a. reflects the firm's financial position
   b. is more important than the firm's balance sheet
   c. is a key financial statement
   d. is always performed on a cash basis.

   The stem, "An income statement," fails to state the basis on which the response should be chosen.
2. The response options should contain only one defensible answer. If more than one item is correct, the effect of guessing is increased.
3. Distracters (incorrect responses) should be plausible; use common mistakes and misconceptions to create distracters. For example, the following is a faulty item:

   Which element has been most influential in recent pharmaceutical development?
   a. Scientific research
   b. Psychological change
   c. Convention
   d. Advertising promotion
4. All response options should be grammatically and logically consistent with the stem; for instance, watch the uses of "an" and "the."
5. The length of correct responses should be approximately the same as the incorrect responses; there is a tendency to make the correct answer longer. For example, the following is a faulty item:

   Laminar flow
   a. never increases
   b. gradually increases
   c. provides a unidirectional flow of HEPA-filtered air and conforms to the ISO standard
   d. is a means of covering surfaces
6. It is more appropriate to ask what an item is rather than what it is not; knowing what is incorrect does not indicate whether the participant knows what is correct. For example, the following is a faulty item:

   The filtration rate is not affected by
   a. temperature
   b. bioburden
   c. pressure
   d. container size
7. Include from three to five response options for each item. All items do not have to provide exactly the same number of response options.
8. "All of the above" is usually the correct answer and therefore makes the item too easy. The participant can guess "all of the above" is correct if two of the other options appear to be correct. For example, the following is a faulty item:

   Which of the following factors are involved in achieving a defect-free product?
   a. Quality
   b. Maintenance
   c. Personnel
   d. Supplier
   e. All of the above
9. Rotate the position of the correct response from item to item. Instructors have a tendency to use "b" as the correct response more often than other response options.
10. Place any words that the response options have in common in the stem. For example, the following is a faulty item:

    Moisture content is an important factor to consider in tablet compression because
    a. amount of moisture affects content uniformity
    b. amount of moisture affects "capping"
    c. amount of moisture affects potency
    d. amount of moisture affects tablet thickness

    The phrase "amount of moisture" should be placed in the stem.
11. All options should be homogeneous in content. For example, the following is a faulty item:

    The misery index
    a. should be calculated for each project under consideration
    b. is calculated quarterly by the Chamber of Commerce
    c. looks at our balance of trade position
    d. may affect project viability in an indirect manner

    The first two responses refer to time, the third appears to be a definition, and the last refers to a consequence.
12. It is acceptable to use either a direct question or an incomplete statement as the stem. The preceding item used an incomplete statement. The following item uses a direct question:

    Which type of data is used to express the number of defects found in a product at final test audit?
    a. Categorical
    b. Numerical–discrete
    c. Numerical–continuous

13. Use "none of the above" sparingly. It tends to test only the participant's ability to identify incorrect answers. Recognizing that items are wrong does not mean that the participant knows the correct answers.

Completion Items

Guidelines for constructing completion items:

1. Use only one blank per item. For example, the following is a faulty item:
   The ________________ of cGMP compliance is measured by ____________.
2. Require single-word answers rather than phrases.
3. Use either direct question of fill-in-the blank statements, such as "Write the formula used for determining standard deviation."
4. Place the blank near the end of the sentence rather than near the beginning. For example, the following is a faulty item:
   ____________ is likely to increase when a test is lengthened.
5. Word items so that they have only one correct answer. For example, the following is a faulty item:
   Laminar Flow does not protect _________.
   The answer the instruction wanted in this item is "covered items." However, a great many words could be placed in the blank without being incorrect.
6. Make sure that the word deleted from the sentence is a significant one. For example, the following is a faulty item:
   A customer can be ________ as anyone who is impacted by the development of the product.
7. Use "a(n)" before a blank to avoid grammatical cues.
8. Structure all answer blanks to be the same length, regardless of the length of the word to be supplied. Blanks that correspond to the length of the word provide an additional clue to the answer.

BIBLIOGRAPHY

Bloom BS. Taxonomy of educational objectives: handbook 1. Cognitive Domain. New York: McKay, 1974.

Bramley P. Evaluating Training Effectiveness. London: McGraw Hill Book Co., 1991.

Caffarella RS. Program Development and Evaluation Resource Book for Trainers. New York: Wiley, 1988.

Cocheu T. Training from the start. Train Dev J 1990; 44(10):22–7.

Dick W, Carey L, Carey JO. The Systemic Design of Instruction. Longman, NY: Addison Wesley, 2000.

Goad TW. Delivering Effective Training. San Diego, CA: University Associates, 1982.

Phillips JJ. Handbook of Training Evaluation and Measurement Methods. Houston, TX: Gulf Publishing Co., 1990.

Reilly RR, Israelski EW. Development and validation of mini-courses in the telecommunication industry. J Appl Psychol 1988; 73(4):721–6.

Smith JE, Merchant S. Using competency exams for evaluating training. Training Dev J 1990; 44(8):65–71.

Vesper JL. Defining your GMP training program with a training procedure. BioPharm 2000; 13:28–32.

Warshauer S. Inside Training and Development: Creating-Effective Programs. San Diego, CA: University Associates, 1988.

Wigging GP. Assessing Student Performance: Exploring the Purpose and Limits of Testing. San Francisco, CA: Jossey-Bass, 1999.

39

Vendor Qualification and Validation

Maik W. Jornitz
Sartorius Biotech, Inc., Edgewood, New York, U.S.A.

INTRODUCTION

Supplying the highly regulated biopharmaceutical industry with equipment and services generally means that vendors must adopt similar quality standards pertinent to the processes of the biopharmaceutical industry applied to the vendor processes. These standards start with the qualification of the equipment during the development phase, the validation of the production process, in-process controls and documentation during the production process, release criteria, specifications and tolerance settings and complete traceability of the finalized product and product components. Once the equipment is ordered or supplied to the end user, most commonly the vendor will submit qualification documentation, support qualification, and acceptance testing and in some instances, offer product or process related validation services (1). The vendor's production processes often mirror the production processes of the relevant industry the vendor supplies to. Additionally, the vendors establish appropriate technical support structures to be able to react rapidly to support needs of the industry. This is of importance as the end user must be able to answer to regulatory enquiries or when equipment requires maintenance, calibration or repair. Production interruptions cannot be tolerated as it might result in multimillion dollar losses in revenue and put drug product batches at risk.

An important factor for the end user is the determination of the criticality of the supplied goods to the processes and the finished product of the end user. As the end user has an input control of goods received, the vendor has an output control or release criteria. However, the criticality of the products supplied to the end user will also determine the evaluation scrutiny the product will undergo before it is shipped or transferred into the end user's production process. For example, a drug product component or excipient will have a higher criticality than a sterilizing grade filter than a filter housing. The component's quality is essential as it influences the end product's quality directly. A sterilizing grade filter has an indirect influence; however, it is still more critical than a filter housing due to the fact that the filter has an influence on the sterility of the end product. Therefore, vendor validation and qualification processes most definitely differ in stringency and scope due to the fact that some components have a lesser criticality than others, which does not mean that products of lesser critically do not need to meet the required specification, the control parameters are just different.

This chapter describes some of the qualification/validation work vendors undergo and establish to meet the biopharmaceutical end-user requirements.

VALIDATION WITHIN VENDOR'S DEVELOPMENT

Vendors strive to improve their products and processes to be able to supply the industry with state-of-the-art materials, components, equipment and improvements within the industries processes. For this reason vendors typically invest 3% to 8% of their revenue in the development of new products or improvement projects for existing products. However, every time a product is newly developed or revised, a similar documentation trail to that of pharmaceutical R&D has to be established. It begins with the choice of a qualified sub-supplier and ends with a fully qualified product and validated production process. Vendor development groups are multifunctional teams, which work together with sales and marketing, supply chain and production to have an appropriate idea of what is required within the industry, is the raw material needed readily obtainable at the quality specification set and are the production capacities, as well as machineries available. Once these cornerstones have been investigated and verified, the development of the product will start. Any effort to develop a piece of equipment without the knowledge of market needs, supply assurance and production feasibility is a wasted effort. As logical as it sounds, these cornerstones are the first milestones that are documented within a development process and will eventually mean audits by the vendor's quality assurance and supply chain departments of the raw or sub-material supplier. These audits have to be well documented and are commonly applicable to a minimum of two suppliers. Supply assurance for a vendor is as important as for the end user, as any supply change will result in a change notification and comparability studies, and in some instances a possible revalidation of the equipment at the end-user level and notification of the regulatory authorities. Therefore changes within the vendor's processes, raw materials or specifications are to be avoided. Any change within the vendor's processes will ultimately influence the user's

---

Abbreviations used in this chapter: CNC, computer(ized) numerical(ly) control(led); DQ, design qualification; DS, design specification; FAT, factory acceptance test; FS, functional specification; GAMP, good automated manufacturing practice; IQ, installation qualification; ISPE, International Society for Pharmaceutical Engineering; OQ, operational qualification; PQ, performance qualification; SAT, site acceptance test; SQ, system qualification or specification qualification; SOP, standard operating procedure; URS, user requirements specification; USP, *United States Pharmacopoeia*.

processes and regulatory filing requirements. Critical raw materials and components fall under long-term supply assurance contracts and might have multiple year inventory levels within the sub-supplier or vendor level. Additionally, vendor development will involve quality assurance to analyze whether the sub-supplier's quality certification, systems and assurance meet the specification given to vendor's development by the industry. As the end user audits the vendor's processes, the vendors will do the same at the sub-supplier level. The more thorough the vendor's internal and external quality and supply investigations are the better the supply quality to the end user.

Once quality sub-supplies have been established the vendor's development group will create prototypes and different versions of the product which will be first tested in-house and at a later stage at a beta-site, which is commonly an end user's process development or small-scale site. However, all stages of the pilot scale production and/or assembly have to be thoroughly documented to assure consistency and improvement. In instances of source code development for equipment control, any development of such code or any change within the code has to be documented (2,3). The entire source code establishment requires audits and needs to be well documented. It is import to check the source code development on a frequent basis to avoid any late stage surprises. As shown by Figure 1 the costs of software revision become exponentially higher towards the later stages of software preparation.

Prototypes will be tested and if these do not meet the specified requirements, the development must return to the drawing board and improve the equipment to users' specifications. When done, the product parts production, assembly, and packaging specifications must be locked. This means that production parameters and tolerance are recognized and set, most commonly by repeated production batches as required within the industry itself. At this point validation protocols and standard operating parameters have been instituted. Validation tests are commonly set by publicly available international standards, for example sterilizing grade filters have to meet current pharmacopeial requirements and will be tested accordingly. Nevertheless, the vendors have their own sets of tests, which components and equipment will undergo to verify performance criteria set by the vendor and user alike. The test results of these tests can be found within the vendors' qualification documentation (validation guides), which are supplied to the end user (Fig. 2).

These documents however will not replace process validation or PQ at the end user's site. These documented tests establish the basis requirements for the equipment to (*i*) be able to work within the biopharmaceutical environment and (*ii*) verify that the equipment meets regulatory requirements. If this scientific basis is not be met by the developed product the product will be scrapped.

Furthermore, these tests will also set the standards for tests that quality assurance will use to determine product consistency and reliability. Most commonly vendors have already standard quality assurance tests defined by other production processes or equipment specification. These can be utilized to a large degree; however it could be that a specific piece of equipment requires additional tests or release criteria. For example the in-process controls and release criteria for a sterilizing grade filter will differ for a membrane chromatography device or filter housing. The main release criteria for a sterilizing grade filter are its integrity, whereas the release criteria for a membrane chromatography device would be adsorptive capacity and for a filter housing surface roughness. These are only examples of release criteria. Most often every product category has a multitude of release tests to fulfill before the product is shipped. In any case, all the product categories have to have appropriate controls and release criteria established to meet quality and consistency standards.

The vendor's development departments have to work in close conjunction with multiple departments. The development group not only creates new or improved products, but has to assure sub-supplies, determine appropriate production specifications, tolerances and the validation as well as a smooth hand-over into full scale production. Additionally, with the validation of the vendor's production process, a close collaboration with quality assurance is required to create appropriate SOPs, validation and qualification documentation, and development documents, especially for source code. Finally, the vendor's product management and technical service departments are supplied with performance data and specifications of the new product as established by the development department. These data have to match or exceed the criteria set by product management, respectively the end user.

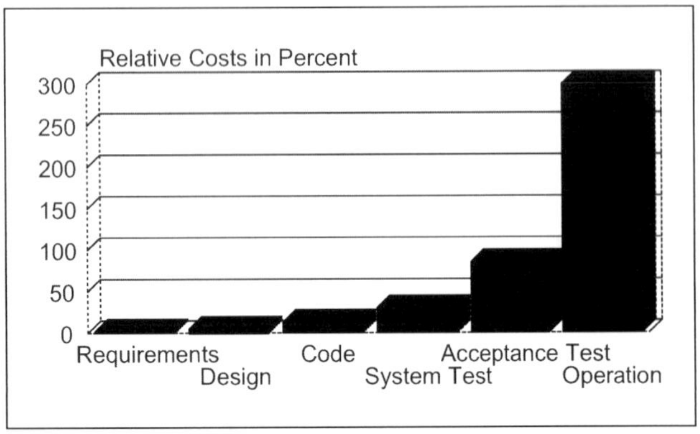

Figure 1 Costs of software development.

VALIDATION OF VENDOR PRODUCTION PROCESSES

Depending on the complexity of the vendor's products, the production processes require process-specific validation. Most commonly vendor production processes are multi-step processes meaning every step requires validation, appropriate operation procedures and qualification, training and certification of the personnel involved. For example, membrane casting for a sterilizing grade filter is one step within the production of a sterilizing grade filter. This casting process requires very specific environmental process conditions and machine settings. During the casting process the machine parameters are constantly monitored and

Table of Contents

Figure 2 Example of a validation guide supplied by a vendor.

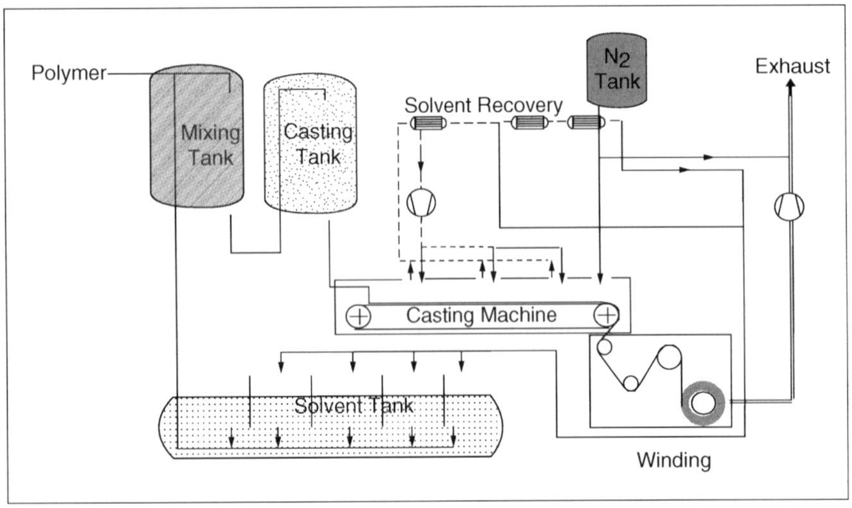

Figure 3 Casting process.

samples of the casted product are taken and tested frequently (Fig. 3).

All parameters and test results are documented within the batch records of this particular cast and can be reviewed by auditors. The documented results also serve as a historical database to perform statistical evaluations, evaluate process performance or support development efforts. The casting process parameters determine the pore size of a membrane, but also its pore size distribution; i.e., the process has to be closely adjusted and monitored to achieve a narrow or desired distribution (Fig. 4).

Once the membrane batch is cast, it will be pleated, sealed, end-capped, welded, integrity tested, bagged and autoclaved. However the procedures or individual production steps line-up does not matter, every single production step has defined process parameters within which the *n* step has to be run. The timeframe between every step requires as much monitoring as the step itself. The process has to be validated as an individual step and in its entirety.

This means that production parameters and release criteria are defined for each step and described within SOPs. All parameter, specifications, and tolerance levels are documented within a validation master protocol and cannot be changed without approval by a multitude of departments, most importantly quality assurance. Release criteria are established by in-process tests. Once the criteria are met, the product can move to the next step of the process. However, if the product does not meet the criteria, an investigation will be initiated to analyze why the product is out of specification. Most commonly such investigations occur during the development of full-scale production, when the products are moved from development's pilot scale to full scale. Scaling within vendor processes can be as difficult as in the end user's industry. This might be seen as a negative, but should be seen positively as it is better to amend root causes of undesired product quality early within the scale-up of the production process, instead of within the final established production processes.

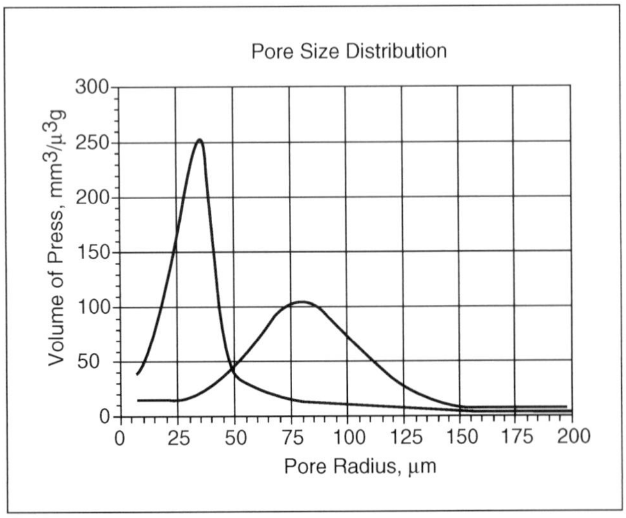

Figure 4 Pore size distribution example.

Once the production processes are established, maintenance protocols will assure that the production equipment will fulfill the criteria set. Maintenance protocols are written during the validation phase as wear and tear can vary equipment performance and specific tools must be inspected during the validation phase. Any automated equipment utilized in the production process has to undergo installation, operation and PQ, especially in regard to the process and system controls. For example a CNC cutting machine which is utilized to cut a specific part for a medical device requires as much qualification work and documentation as an autoclave. The product component delivered by the CNC has to be of consistent quality, complying with set tolerances and specifications. Another example are membranes used for sterilizing grade filters, which have to meet thickness, porosity, integrity, extractable, thermal and mechanical stability, and particulate test parameters (1,4). If any of these specified parameters are not met, the batch will not be released and an out of specification investigation will be performed. These tests are described and used as release parameters for validation batches and later for commercially marketed batches. However, consistency in set quality parameters is the most important aspect in any stage of the production.

In certain production processes, the process cannot be automated and the production step is performed by personnel, for example welding. The settings within welding can be described, but only as indicator specifications, for example the tube volume, material thickness that will determine the protective gas pressure, and welding energy settings. However, due to the complexity and individuality of some equipments, for example bioreactors or cross-flow systems, most welding might be done by the pure experience of the welder. The welders require specific certification and most often have many years of experience. The welding itself will be analyzed and inspected before release, but this does not minimize the skill level required for such welding tasks. Similar skills are required for cutting, honing, bending, polishing, etc. The validation within these processes is the certification and routine training of the personnel, log books as well as the quality of the raw materials used. Any raw material entering the facility will be sampled, inspected, and documented and requires specific certification and log numbers. The raw material has to be traceable and of specified quality.

An additional piece of validation work on the vendor's part are packaging validations and tests. The goods will be packed in specifically designed packaging which assures robustness during transport. The vendors will test the packaging design using specific standards, e.g. ASTM D 4169 and D 4728-95 (5,6). These tests are drop and vibration tests. DIN ISO 12048 is a compression test, which will verify the stability of the packaging (7). As soon as the goods leave the factory, the vendor loses control over the handling of the goods. Therefore, packaging plays a major role to maintain the quality and integrity of the goods shipped. Moreover, robustness is not only attached to mechanical stability, but also to thermal and chemical stability. Temperature changes during transport are not unusual, especially during overseas shipments. The packaging must be flexible enough to overcome any thermal expansion or shrinkage. It also should repel any condensation occurring due to temperature changes or changes in humidity. Oxidation due to sun light is probably the most common photochemical attack to polymeric packaging. The packing has to be stabile under these circumstances; otherwise polymeric degradation would result in weakening the packaging or particulate shedding of the packaging.

The ultimate tests for packaging are multiple shipments into the different regions supplied to and by different carriers supplied by. At the end, these tests will create a grid of test data of different means of transportation at different environmental conditions, which will result in a tolerance band for the designed packaging. Only such tests create practical data verifying the experimental lab data. Pure lab data would not support the assurance of structural integrity and safety.

VENDOR'S IN-PROCESS CONTROLS AND RELEASE CRITERIA

Depending on the vendor's products the in-process controls and release criteria vary from narrowly defined step-by-step controls within the production process or as an end result control and release (1). Most commonly, individually produced components are tested when produced and again when the individual components are assembled. As previously described the control and release criteria and tests are established within the development process and depend also on the criticality of the product supplied to the end user. In instances control and release criteria are fairly simple and encompass only a single test criteria, most of the time though product distributed to the pharmaceutical industry undergo multiple tests within the parts and final product production process.

Raw materials, supplied to the vendor, are checked first whether the quality documentation is complete. Again, depending on the criticality of the component the material might undergo specific tests to verify that the quality standards described are met. For example polymer granulates undergo thermal profiles to check that the quality and type is the same as specified by the vendor to the sub-vendor. In other instances, the raw material is visually inspected, for example stainless steel tubing in regard to surface-finished and material stamps. If the raw material does not meet one of the specifications, the material will not be released into production. All raw material batch records are kept with the batch records of the resulting product. The product has to be completely traceable to allow appropriate investigation, if necessary. Raw material suppliers are generally audited once a year, depending on the significance of the raw material supplied. However, if there has been an incident the supplier will be audited immediately thereafter and corrective action verified.

For example filter cartridges, whether pre- or membrane filter are tested for extractables (Fig. 5) to check whether there is any change within the profile, which might not meet release criteria (1,4). Similar tests are flow, throughput, mechanical and thermal robustness. Membrane filters are commonly individual integrity-tested before release.

Stainless steel products also have specific definitions, which need to meet the biopharmaceutical requirements (1). These are individual stamping of the steel goods, welding certification, material qualification and certification. The steel source can determine the quality of the steel. The steel components are required to be right, as these determine welding quality, corrosive robustness and the electrolytic behavior within a system. Nowadays stainless standards are set by the industry, which define for example the ferrite content or surface smoothness.

Depending on the application the stainless steel equipment used differs greatly in the surface treatment. The smoother the surface, the greater the treatment steps and the costs involved. In some instances surface treatments are not needed or are even undesirable. A glass beaded surface is sufficient. However, since cleaning is a major factor within the biopharmaceutical industry the surfaces must be smooth and with a minimum groove rate. Any groove would allow pockets of microbial growth, which could result in a biofilm formation. Electropolishing, after high grid polishing is utilized to cut any high peaks of material and avoid pockets (Figs. 6 and 7).

Identified Extractables of Different Membrane Filter Cartridges From Several Filter Manufacturers							
Cartridge A	**Cartridge B**	**Cartridge C**	**Cartridge D**	**Cartridge E***	**Cartridge F***	**Cartridge G***	**Cartridge H***
Diethylphthalate	Cyclohexan	Propionic acid	Diethylphthalate	Acrylic acid	Dimethylbenzen	Etherthioether	Caprolactame
Stearic acid	Ethoxybenzoic acid	Diphenylether	12 oligo. aliphates	2 phenolic oligo	Etherthioether	Propionic acid	Butyrolactone
2, 6-Di-tert.-butyl-cresol	2,6-Di-tert.-butyl cresol	2, 6-Di-tert.-butyl-cresol	Hydroxybenzoic acid	2, 6-Di-tert.-butyl-cresol	2,6-Di-tert.- butyl cresol	2,6-Di-tert-butyl-cresol	Laurinlactame
2,2-Methylene-bis 4-ethyl-6-tert. Buty phenol	Cyclohexadiene 1,4-dion	4-Methyl-2,5-cyclohexadiene-1-on	Tert.-butyl-methyl-2, 5-cyclohexadiene-1-on	3 oligo. Benzyl-di-phenylmethan	3,5-Di-tertbutyl-4-hydroxyphenyl propionate	3,5-Di-tert.-butyl-methyl-2,5-cyclo-hexadiene-1-on	Laurinlactame derivate
Hydroxybenzoic acid	Phenylisocyans	Hydroxybenzoic acid	2,4-Bis(1,1-di-methoxy-1-ethyl)-phenol	Triphenylphosphite	2 N-containing high MW compounds	4,4-Dichloro-diphenylsulfone	4-Methoxy-4-chlor-diphenylsulfone
7 oligo.Siloxanes	Palmitic acid	Palmitic acid	Succinic acid	Stearic acid	Acetamide	Benzothiazolone	Adipinic acid
Bis-(2-ethylhexyl)-phthalate	Stearic acid	Dimethoxydiphenyl sulfone	3 oligo, siloxanes	Bis-(2-ethylhexyl)-phthalate	N-cont.aromatic high MW comp.	4-Hydroxypropyl-benzoate	Dibutylphthalate
12 oligo.Aliphates	12 oligo. aliphates	11 oligo. aliphates	Polyether	Polyacrylate	3 oligo. amides	6 oligo. aliphates	Ethylhexylphthalate
4-Methyl-2,5-cyclohexadiene-1-on	11 oligo. siloxanes	Methoxy-4-chloro-diphenylsulfone		Ethylacrylate	Bis-(2-ethylhexyl)-phthalate	Hydroxyphenyl acetamide	Dihydroethyl-phthalate
	Methyl-4-hydroxybenzoate	Bis-(2-ethylhexyl)-phthalate		Diphenylphthalate	10 oligo.siloxanes	Methoxy-4-chloro-Diphenylsulfone	2, 6-Di-tert.-butyl-cresole
	Etherthioether	Polyether		9 oligo. Siloxanes	2 oligo.aliphates	7 oligo.siloxanes	Disobutylphthalate
		7 oligo. siloxanes		6 oligo. aliphates		Cyclohexanone	Diacetylbenzene
		2,4-Bis(1, 1-di-methoxy-1-ethyl)-phenol					Cyclotridecanone
							4-4-Dichlorodi-phenyl-sulfone
							Propionic acid
							4 oligo, siloxanes
							3 oligo, aliphates

* Identification of the RP-HPLC peaks by FTIR is still in progress–extractables list of marked cartridges may be incomplete.

Figure 5 Extractable table of eight different sterilizing grade filters.

Most commonly when automated equipment is supplied appropriate qualification documentation is required before the equipment is released and shipped to the client. Without such documentation the equipment would be of no use and the shipment might be rejected. It is essential that these documents are sent to the client for pre-approval. Once the approval is received only then the vendor can ship the equipment to the client. Appropriate qualification documentation is an essential release criterion nowadays.

From an end-user standpoint, the release criteria of the vendor have to meet the risk assessment criteria set by the end user (and more often the regulatory authorities). That is, depending on the quality impact of a specific component or equipment supplied and the release criteria on both sides, vendor and end user, will differ in stringency. The quality of supplied water for injection (if not produced within the facility) has a higher risk attached than a condensate valve on a tank. Different risk or impact classifications have to be

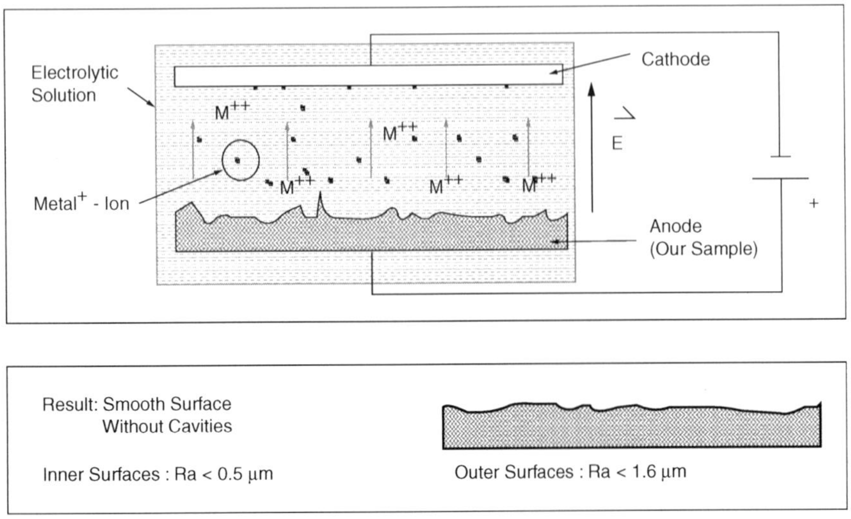

Figure 6 Schematic of an electropolishing process.

No.	Designation of procedural step	Remarks	Recommended abrasive	Grit	Peripheral speed in m/min
1a	Preliminary rough polish ("fettling")	Preliminary step for rough welds;only for very coarse work; recommended follow-up step: 1b, with 60-grain abrasive	Preferably grinding wheel with hard rubber or plastic bond	24/36	1,200-1,800
1b	Rough polish	First step for thick sheets, hot-rolled sheets or smooth welds	a) Grinding wheel with hard rubber or plastic bond b) Set-up wheel c) Grinding belt, if the shape of the piece permits	if 36 is necessary, follow up with 60	1,200-1,800
2	Finish grind	Standard step for cold- rolled sheet or coil	a) Set-up or rubber wheel b) Grinding belt, if the shape of the piece permits	80 / 100	1,500-2,400
3a	Precision grind	The surface finish corresponds to that of roll material in accordance with "Procedure o (IV)"	a) Set-up wheel b) Grinding belt, if the shape of the piece permits	120 / 150	1,500-2,400
3b	Precision grind	Preparatory step in producing a normal polish following step 3a.	a) Set-up wheel b) Grinding belt, if the shape of the piece permits	180	1,500-2,400
3c	Precision grind	Intermediate step in producing a normal polish following step 3b.	a) Polishing wheel b) Grinding belt, if the shape of the piece permits	240 abrasive paste for set-up wheel, or 240 grinding belt	2,400 - 3,000 Grinding belt: approx.1,500
4	Brushing	To produce a smooth, matte, silk luster. This step, following one of the "o (IV)" procedures,produces a surface finish that corresponds to the designation "burnished." Brushing finer (e.g., high-gloss polished) Surfaces Produces a very attractive efffect. The surface finish will depend on the brush speed and the abrasive used.	Tampico	Abrasive paste made of pounce or quartz powder. Other abrasives may also be used,depending on the desired surface finish.	600- 1,500
5	Polishing or lapping	Final step for producing a normal polish following step 3c (Note lappingleaves fine chatter marks)	Polishing wheel	Burnishing compound for stainless steels in stick or cake form	
6a	Polishing	a) Preparatory step for producing a high-gloss polished surface following step 3c	Polishing wheel	320-400 finish polishing compound in stick or cake form	2,400-3,000
		b) Preparatory step for producing high-gloss polished coil.	Polishing belt	Burnishing compound for stainless steels in stick or cake form	approx. 1,500
7	Blasting	Final step for producing a matte, non directional surface structure	Glass beads Stainless steel grit Nonferrous quartz sand	various	

Figure 7 Table of different polishing methods and the end result.

defined for product and equipment supplies (Fig. 8). Some products have a direct impact on the quality of the end product, some have only a minor influence, some have no influence but are used to check on a component with a quality influence. For example, an integrity test system does not have a direct influence, but is used to check the integrity, i.e., quality of a sterilizing grade filter, which has an influence on the quality. The release and test criteria for these products will differ and be defined in a way which will meet the necessary quality purpose. It would make no sense to use similar evaluation conditions for non-critical items. It would just raise costs and possible process delays. Therefore these risk assessments have to be performed before release criteria are defined.

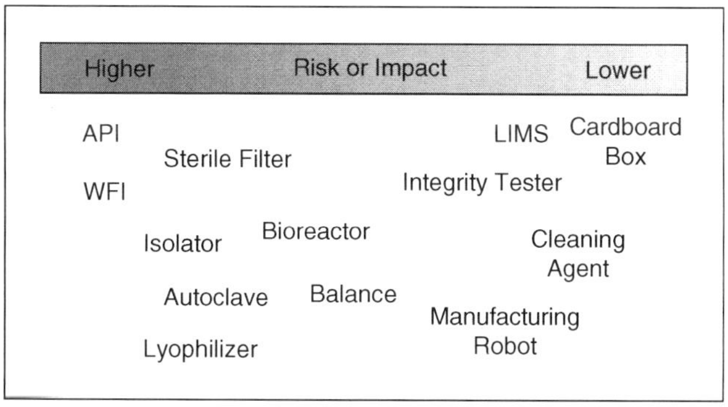

Figure 8 Possible example of risk and impact assessments.

QUALIFICATION OF EQUIPMENT

The probably most descriptive and utilized guidance on qualification mechanisms is the GAMP guidance published by the ISPE (8). It describes thoroughly the individual, necessary steps required to fulfill the quality expectations of automated systems. This guidance is used for a multitude of equipments utilized within the biopharmaceutical industry, for example, autoclaves, lyophilizers, filling machines, integrity test systems, bioreactors and others.

Within the GAMP documentation, specification steps are described, and also three main qualification requirements, IQ, OQ and PQ. There are other qualification tests which are quoted randomly for example DQ and SQ. However the three major qualification segments are IQ, OQ and PQ and are applicable to every automated piece of equipment supplied.

A system design and the qualification steps all start with the URSs. This is the foundation of any system which will be designed and if defined inappropriately the entire project might be prone to fail or at least will require rework with additional costs involved. The URS can be seen as the foundation of a building, the better the foundation the better the construction on it. Any of the above-mentioned qualification step, are the verification of the URS, FS and DS.

IQ. Documented verification that all important aspects of hardware and software installation adhere to the system specification.

Within this qualification, the entire system is checked whether all components are correctly installed and whether the entire documentation for the individual components is available. Most often the IQ step runs through a thorough checklist to evaluate that everything meets the requirements set within the design or hardware specifications (Fig. 9) (1,2,4).

The IQ documentation is supplied by the vendor but checked by the end user. Often and most practical would be to perform the IQ and OQ part during the FAT which verifies that the system is working.

OQ. Documented verification that the system operates in accordance with the system specification throughout all anticipated operating ranges (Fig. 10).

These tests verify that the FSs are met by empirically checking and testing against the manufacturer's recommended test sequences all the critical operational and functional features and performance specifications of the machinery. These test sequences are performed within the vendor's facility, again most commonly during the FAT. Within this qualification phase the system will run at the specifications given by the user. Therefore vendors are required to have all supplies necessary to run the system, for example water and steam supplies. The OQ can be performed within a few hours or weeks, depending on the complexity of the system build. Most commonly the OQ documentation is already established within the process of the FS, as every single function described requires to be tested during OQ. If an FS happened and the documentation is not established at this point the workload will be tremendous and the precision will suffer.

Once the system run through the FAT and OQ and all documentation is established, the system can be shipped to the vendor. At this point, the PQ is performed as the final part. The PQ is often also part of the SAT or vice versa, depending on individual user procedures.

PQ. Documented verification that the system operates in accordance with the URS while operating in its normal environment and performing the function required by the process to be validated.

These records include batch records, routine calibration, and performance checks, which are commonly defined by the equipment used. Every piece of equipment has different requirements of compliance with

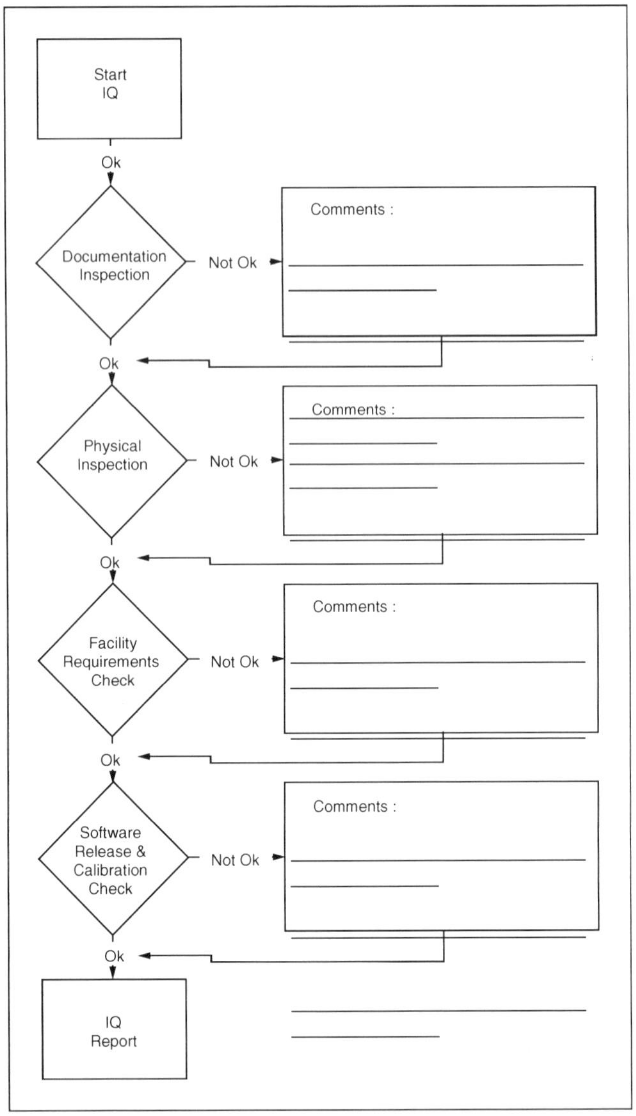

Figure 9 Typical flow diagram of the first layer of an IQ protocol.

specifications defined within the specification phase. Moreover, the environment within the end-user facilities varies. For this reason PQs check whether the equipment works within such an environment. Additionally, during the PQ phase, the equipment may be pushed to the limits to verify that it still performs and does not spiral out of control. In instances automated equipment might malfunction when, for example, the software is pushed to a limit. It could well be that the system shuts down or that certain controls and adjustments elevate themselves out of control or set tolerances. These stringent tests belong to a risk assessment program, which determines the functionality of the system. Will it still work in as robust way as requested or will it perform in a way detrimental to the entire manufacturing process? The environment certainly has an influence on such functionality, as well as the process control system and its source code. It has been experienced that systems are not validatable due to commercially available software, adjusted to the purpose, but not fully compatible. Such software might not be able to cope with the stringency

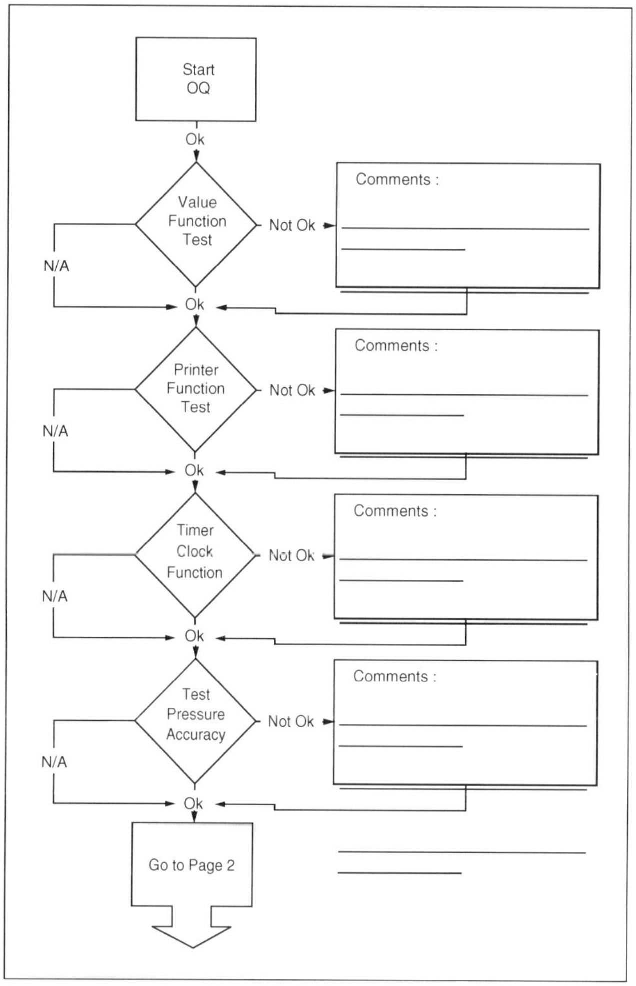

Figure 10 Typical flow diagram of the first layer of an OQ protocol.

and demands of a production process and therefore show insufficient performance.

These three fundamental qualification processes are repeated during each phase of the validation process. In the qualification phase a baseline level of performance information is obtained from the component manufacturers' data and test results, structural testing of the software, and the associated vendor documentation (Fig. 11).

Equipment validation packages must be prepared and available for the user's own validation efforts and tests to verify proper functioning of the equipment. These validation packages are commonly very comprehensive and cover every function of the equipment. For example, the documentation for a complex fermentation system can result in close to 1000 ring binders. In instances regulations applicable to the particular equipment will be quoted for the user to support other necessary validation or qualification processes within the facility. As described the equipment supplier can support, and commonly does, the end user with installation and OQ documentation, however, any process validation or PQ has to be performed within the facility and process environment. This will assure that the equipment is functioning properly aside the laboratory settings within the manufacturer's facilities.

Finally, maintenance continued testing and verification are the responsibility of the end user, who may seek assistance from the equipment manufacturer or its own maintenance department. Service manual establishment is required before equipment is supplied to assure appropriate maintenance possibilities. Such service manuals list spares required within specific frequencies. Commonly the vendor has an experience at which interval certain parts of the system need to be exchanged or replaced. These essential spares need to be defined and listed within the service manual as well as maintenance intervals. These tasks can also be performed by outside service organizations; however the qualification of these organizations has to be verified. Most often service contracts are established between the vendor service side and the user maintenance department.

Another important aspect should not be forgotten—training. All qualification and acceptance steps are good, but without use if the staff utilizing the equipment is not trained effectively. Training protocols and SOPs need to be described before the equipment is used. Both training manuals and SOPs should be reviewed to assure correctness.

STAGES OF EQUIPMENT SUPPLIES AND QUALIFICATIONS

Stages of the individual specification and qualification segments are mainly visualized within the V-model of the GAMP guidance (8). The V-model shows the different responsibilities, but also interactions of specifications versus qualifications (Fig. 12). It is often modified to meet different requirements of different equipment suppliers.

Within the V-model, the individual tasks or steps are described, but also responsibilities defined. In parts of the process the user is solely responsible, in other parts the supplier and specifically in the qualification phase, the user and supplier share responsibilities, as most often these tasks are performed jointly. Every single step is of utmost importance and has to be viewed with stringency and thoroughness, as every step following depends on the quality of the previous task. The entire system can only be as good as the starting quality, therefore multiple other process control and approval steps are involved, which are not shown within the V-model. However, before a system is built each function, software and hardware design have to undergo a critical review to verify that the URSs are met. In instances specifications given might not be feasible to design or produce or sub-parts are not available or too costly. Sometimes, a cost focus might be not desirable, as cutting corners might result in a system which is not fulfilling the needs of the process defined. Examples have shown that shortcuts in respect to equipment or design qualities have resulted in higher adjustment costs at a later stage. In instances inadequate attention to the design of the system has resulted in yield losses or dysfunctions. The costs resulting from such failures are tremendous. The recommendation has to be that the user and vendor work closely together to find an optimal solution for the particular need. Costs have to be reasonable, but should not be the main focus.

The most important aspect is the URS, in which attention to detail is essential. Any rough idea given as

1.1 Contents

Figure 11 Example of a validation documentation for an automated system.

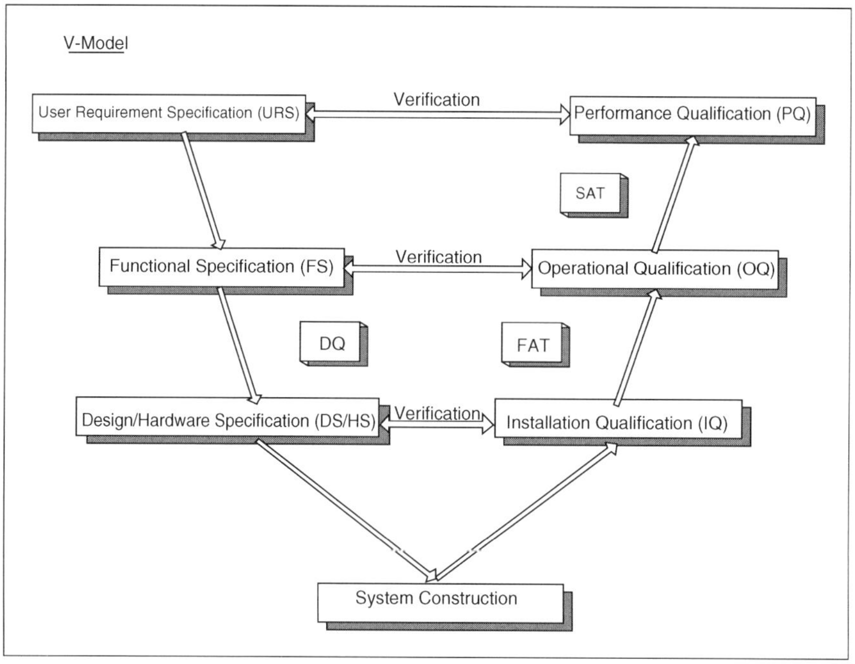

Figure 12 V-model.

URS will end in a back and forth between the user and vendor in the FS stage. Valuable man hours are wasted which is undesirable for both parties. Often forgotten, but always present is that the user is the specialist of the application and the vendor the specialist of the equipment. Utilizing both sets of experience will result in the best possible option. However controls and measurements should be utilized during the milestones to assure that the system will function once built and implemented within the facility.

As described the V-model creates an overview; however project flows and detailed activity description require other tools, for example specific project management software (Fig. 13). These tools will define activities in detail and also control points for parts of a system, the entire system or just the raw materials (2,3,9). The timeframes will also be reviewed on a frequent basis, as time pressures commonly will result in human error. Every vendor has experiences with their equipment supplies and knows what quality system requirements need to be established within a detailed project plan. These control points also help the vendor to avoid any errors, which would create additional costs.

CONCLUSION

Validation and qualification of equipment within the end-user facility, under the process conditions is an essential need and regulatory requirement. However, vendors of equipment, whether consumables or capital investments, perform a multitude of qualifications programs within their own facility. Such qualification programs start during the development phase and commonly include not only the vendors own processes, but also sub-vendor sites, processes and product qualities. A vendor cannot just rely on the sub-supplies, but has to assure such just as any end user needs to do. Furthermore, the development team receives quality milestones by the end user. These specifications have to be kept, which means within the development phase, control mechanisms are defined which are used to verify that the specifications are met and which are also used as release criteria at full-scale production. Similarly, capital equipment receives URSs which are converted into FSs followed by software/hardware DSs. Again the fulfillment of the URSs has to be controlled at every stage to avoid any surprises and non-compliance. The capital equipment runs through different specification stages like a consumable product runs through a development phase. At the end of the day, both product groups require compliance to the user specifications.

Once the goods have been developed or built, the performance has to be qualified within the user's environment. Does the equipment perform under these circumstances? For example, sterilizing grade filters undergo process validation utilizing the actual or close at drug product and the process conditions. Evidence has to be given and documented to show the filter is performing to the set requirements under the environmental circumstances. The PQ stage does so for capital or automated equipment. Again the equipment might be pushed to its limit to assure that it functions under worst case conditions reliably. The tests are performed on-site to guarantee that any environmental condition does not have an adverse effect on the performance of the equipment. Lab test at the vendor or pure certification cannot be accepted and will not meet regulatory requirements.

Vendors, nowadays, do not just produce and supply goods, but make sure that these goods meet the requirements of the biopharmaceutical industry and its regulatory authorities. Moreover, once an item is sold the vendors' efforts do not stop; they support the end user with services to support any subsequent user qualification and validation effort. Both vendor's experiences and end users know-how will optimize the process reliability and in combination assure that the specification of the equipment will meet the needs of the process. The vendor

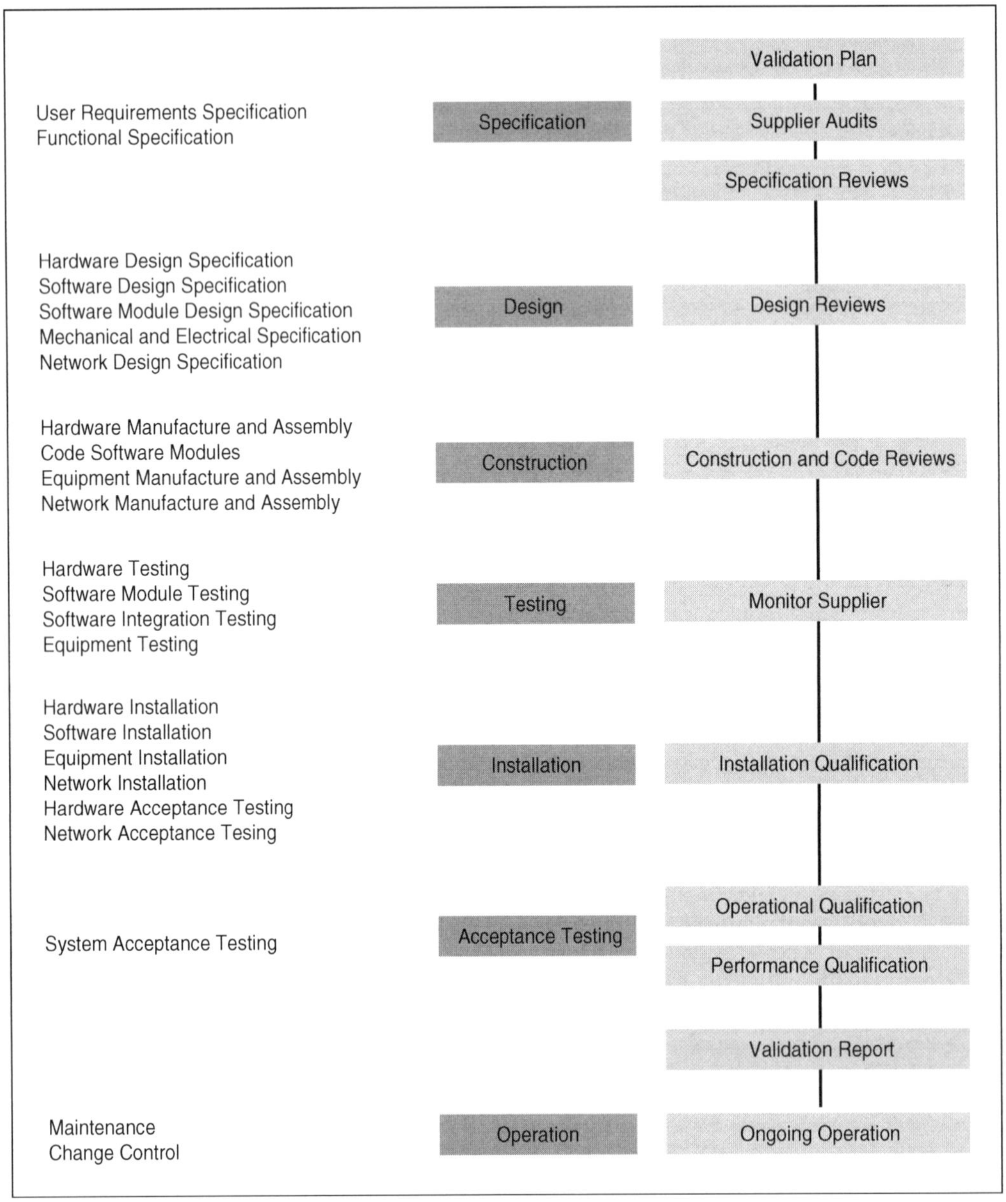

Figure 13 Validation schematic.

has to accommodate the end user, by supplying qualification data of the goods supplied, which can be utilized to either make a choice of the equipment or be utilized within the filing documentation.

REFERENCES

1. Jornitz MW, Meltzer TH. Filter integrity testing. Sterile Filtration—A Practical Approach. New York: Marcel Dekker, 2001 (chap. 7).
2. Spanier HJ. Qualification of a filter integrity test device. Pharm Technol Eur 2001; 13(12):45–52.
3. Wolber WP, Beech D, McAllister M. Validation of computerized membrane integrity test systems. J Parenter Sci Technol 1988; 42(4):106–10.
4. Meltzer TH, Jornitz MW. Filtration in the Biopharmaceutical Industry. New York: Marcel Dekker, 1998.
5. ASTM D 4169. Standard Practice for Performance Testing of Shipping Containers and Systems, 2003.
6. ASTM D 4728-95. Random Vibration Testing of Shipping Containers.
7. DIN ISO 12048. Packaging—Complete, filled transport packages—Compression and stacking tests using a compression tester, 1994.
8. GAMP. Guide for Validation of Automated Systems in Pharmaceutical Manufacture, Volume 1, Part 1: User Guide. Brussels, Belgium: ISPE European Office, 1998.
9. Technical Report No. 15. Industrial perspective on validation of tangential flow filtration in biopharmaceutical applications. J Parenter Sci Technol 1992; 46(Suppl.).

Validation for Clinical Manufacturing

Warren Charlton
Consultant, WHC Bio Pharma, Inc., Greenville, North Carolina, U.S.A.

Tom Ingallinera
Consultant, Charleston, South Carolina, U.S.A.

Danny Shive
Metrics, Inc., Greenville, North Carolina, U.S.A.

INTRODUCTION

The focus of this chapter is Validation of CTM for Pharmaceutical Products. A review of validation activities have been compiled for a range of drug products (e.g., non-Sterile, Sterile, Biotech, Blood/ Plasma, and Nasal Inhalants). Basic to all CTM products for human testing is the need for validation of analytical methods, equipment, utilities and other unique drug-specific factors such as environmental monitoring of the aseptic processing area used in the sterile filling. The selected products have very distinct differences in complexity ranging from non-sterile powders to the potentially very complex biotech product that is ultimately aseptically filled and freeze-dried.

The intent of this chapter is to provide an understanding of "what" is needed, "when" it is needed but not "how" they are validated or developed. In support of developing an overall Validation Master Plan this chapter provides validation activities, a flowchart of validation/-critical activities by Phase (e.g., pre-IND to Phase 3), product-specific validation/supporting information, a Development Protocol Example and Contract Manufacturer Requested Drug Product Information.

Abbreviations used in this chapter: API, active pharmaceutical ingredient; BP, bubble points; CBER, Center for Biological Evaluation and Research; CDER, Center for Drug Evaluation and Research; CDRH, Center for Devices and Radiological Health; CE, comparability exercise; cGMP, current good manufacturing practice; COA, Certificate of Analysis; CTM, clinical trial material; CVM, Center for Veterinary Medicine; ELISA, enzyme-linked immunosorbent assay; EMEA, European Medicines Evaluation Agency; EU, European Union; FDA, Food and Drug Administration; GC, gas chromatography; HPLC, high-performance liquid chromatography; HVAC, heating, ventilation, and air-conditioning; IND, investigational new drug; IQ, installation qualification; ISPE, International Society for Pharmaceutical Engineering; NDA, new drug application; OQ, operational qualification; PAG, polyacrylamide gel; PAT, process analytical technology; PNU, protein nitrogen units; PQ, performance qualification; PQRI, Product Quality Research Institute; QA, quality assurance; RID, radial immunodiffusion; SIP, sterilization in place; SOPs, standard operating procedures; USP, *United States Pharmacopeia*; WFI, water for injection.

SOLID DOSE DISCUSSION

Equipment Qualification (IQ and OQ)

It is understood within the pharmaceutical industry that major equipment and utilities directly involved in processing pharmaceuticals for human use, including clinical trial material, must be qualified (1). This involves, at a minimum, IQ and OQ, and, if appropriate, PQ. An excellent guide for the planning and execution of IQ/ OQ/PQ is the ISPE Baseline Guide for Commissioning and Qualification (2). Not only should major process equipment, such as granulators, mills, tablet presses, capsule fillers, etc., be qualified, but utilities directly impacting the process should be qualified as well. This includes, but is not limited to, HVAC, compressed air, and pharmaceutical water systems.

Process Validation

Validation of pharmaceutical processes began in the late 1970s, based on comments by field inspectors and followed up by the establishment of the principles of validation for sterile injectables. The FDA Guideline on General Principles of Process Validation was issued in 1987 (3), and is still in effect. A major focus of validation of solids (although certainly not the only one) has been content uniformity of blends and uniformity of dosage units. In October 2003, the FDA published a draft Guidance (4) on blend uniformity following extensive work by the PQRI Blend Uniformity Working Group. This Guidance allows for reduced testing of blends during routine commercial batch manufacture provided certain criteria are met, but it does not address (nor was it intended to address) clinical batch validation. Since many, if not most, clinical batches are unique with regard to batch size, formulation, or process, the approach for many firms has been to create a protocol for each unique batch and to let that batch data stand on its own. In some situations, especially in Phase 3 clinical testing, a formulation and process may be well established, and a number of identical batches may be required over some months, or years. In these cases, the process can be validated much like a commercial product/process is validated, as long as the formulation, process, and batch size do not change.

PHASE 1

Validation Activities

- Initial validation of testing equipment
- Initial viral clearance studies (biologicals/ blood products)

- Start human trials (safety)
  - o Ensure product produced in a qualified facility (i.e., trained people and validated processes and equipment)
  - o EMEA requires an "Authorized Person" who is responsible for ensuring that there are systems in place that meet the requirements of Annex 13 and should have a broad knowledge of the pharmaceutical development and clinical trial process
  - o Major process equipment and utilities impacting the process should be qualified
  - o Process Validation as needed to ensure product safety, quality and uniformity of IND products
  - o In-process testing (PAT) (8)

Supporting Activities

- API: clear understanding of process
- Defining Properties (e.g., solubility hygroscopicity, melting points) Initial Analytical Methods
- Initiate Reference Standards
- Specification development
- Biologicals: initiate studies on viral clearance and protein characterization
- Process and Assay development as needed (e.g., Biologicals)
- Initiate Clinical Manufacturing

- Analytical methods
  - o Suitability confirmed
  - o Acceptance limits tabulated for Validation including impurities/ isomers
  - o USP/ICH guidelines apply
- Change control in place
- Initiate stability studies
- Submit IND application
- Flow charting to demonstrate knowledge of process
- Certificates of analysis for all drugs/chemicals, excipients, components and batch results need to be available
- Batch sizing for eventual scale-up
- Batch acceptance limits including impurities are essential
- Initiate process validation plan for unique and complex processes (e.g., all sterile manufacturing processes; sterilizing operations; those affecting potentially adverse microbial growth or removal of endotoxins)
- Biologicals/other products if appropriate: conduct viral clearance studies
- Specifications in place (drug, chemicals, component, etc.)

Figure 1 Validation/critical activities flowchart by phase. *Source*: From Refs. 5 and 6. (*Continued*)

Change control becomes vitally important here, to ensure that changes are tracked in relation to validation documentation.

A healthy debate could be easily initiated among industry experts regarding what stage in the continuum of dosage form development that clinical batch validation is expected. Most industry experts will agree that Phase 3 requires a level of validation, or "verification," resembling commercial process validation. Many will perform at least a limited level of "verification" in Phase 2 to ensure that clinical batches used at this stage are uniform and have the quality and purity that is expected. The greatest debate would no doubt be generated at Phase 1. Many persons will do no more testing than that required to show that clinical batches in Phase 1 meet the minimum requirements of USP and the limited specifications established at this early stage. Others will go further and test blends for uniformity and test more than USP requires for tablet or capsule dissolution and uniformity of dosage units.

The FDA has an expectation of some track record of validation or "verification" testing of clinical batches during all phases of clinical batch manufacture. See protocol outline (Appendix A) presented for consideration in "verifying" the adequacy of the manufacturing process of clinical batches in the early stage of development (Phase 1) whose formulation and process are not finalized. For Phase 3 clinical batches, a validation protocol may closely resemble that of a commercial product.

Validation Activities		Supporting Activities
• Validation of analytical procedures • Validation master plan developed • Validation protocols established/ approved	PHASE 2	• Quality systems should be reviewed to ensure all systems are in place • Viral clearance studies are operational and continue to be expanded (biological processes and as needed to ensure safety) • Preliminary specifications, tests, acceptance criteria and limits are operational and continue to be expanded (tabulated and validation in process)
• Validation of development batches (small scale)		• Complete understanding of manufacturing process, site of production, and batch release requirements
• Validation master plan document approved. Change control in place • Production scale Validation (IQ, OQ and PQ)	PHASE 3	• Critical manufacturing steps should be identified (e.g., training, equipment, process, storage, environment, sterilization) and verified that controlled documentation and/or validation are completed • Qualify and validate assays o Tabulation of assay results available o Justification of impurity specification
• Media fills for sterile operations • Manufacture validation batches (PQ/ conformance batches)–normally 3		• Drug product filter validation completed • Specifications are in place and full validation report available (5) • Conduct pivotal trials • Submit regulatory license application
• Validation summaries	↓	Pass pre-approval inspection. Note: normally expect FDA at one or more of these batches to view operation
• Validation team reviews the Master Plan and signs off verifying all requirements have been met with the last signature by QA (note: typically at least three team members from different areas for the company)		• Obtain regulatory approval in writing • Product can be shipped

Figure 1 (*Continued*)

STERILE CTM DISCUSSION

Parenteral clinical supplies present numerous challenges. During preclinical and Phase I, very little is known about the chemistry, dose, toxicity, and pharmaceutical properties of the drug. One must take a systematic approach to get the drug into humans as quickly as possible. This is critical in order to establish a "go" or "no-go" for a specific compound. Many development dollars and years can be wasted on elaborate formulation and analytical methodology only to find that the drug is ineffective or not safe.

Developing specifications is a priority during preclinical and Phase I. Early in the development process keep the drug product simple, either a simple solution or a lyophilized powder for injection. Start by evaluating your active compound for its preformulation attributes (e.g., solubility, salt selection, pH solubility profile, osmolality, and density) and physicochemical properties (e.g., color, odor, melting point, hygroscopicity, and optical activity). The effects on heat, light, pH, oxygen and ionic strength on the stability of the compound need to be confirmed. If the solubility of the drug is limited, co-solvents and other formulation changes may be necessary.

Initial container/closure can be a glass ampule or a vial with a coated closure when limited compatibility data is available. Clear glass is preferred over amber, to allow inspection of the solution before injection. Alternate container/closure systems should be evaluated in development. Initial storage of clinical supplies can be refrigerated or frozen, until more data is available. Note: Always attempt to limit potential issues that have not yet been determined (e.g., protect the product from light unless sure of no impact; store actives and finished product under conditions that are certain not to be of any issue until data are available to demonstrate storing under normal room temperature is adequate).

The safety necessary to start Phase I clinical studies is determined by the preclinical/toxicology studies. The materials and formulation used in the preclinical/toxicology studies should be similar in purity. One of the challenges in using a single formulation is the fact that you may need to dose small animals with 5 to 100 times the dose/kg of body weight to see toxicity and allow for a safety margin. The lyophilized formulation allows for flexibility in reconstitution which lets one prepare a more concentrated solution for dosing animals.

Manufacture of early-stage clinical supplies needs to be made in a facility that adheres to cGMPs with Quality Systems in place and excellent past regulatory history. High-speed filling equipment is rarely warranted during early-stage manufacturing. Terminal sterilization should be the first choice if possible. Data need to be available if terminal sterilization is not a viable alternative.

Use of disposable technologies (e.g., plastic tanks, bags, etc.) fits well with potential potent and hazardous compounds that are not highly defined in the early stages of development. Product containment using disposables reduces the cleaning validation requirements but requires the need for extractable studies. Sterile drug filter validation studies should use drug product from the conformance batches (Note: Drug product submitted in the NDA and the conformance batches must be the same concentration and formulation).

Batch records can be difficult. Each batch may change in size, concentration, container/closure and manufacturing process. A complete history documenting these changes is critical for preparing documentation for filing.

VALIDATION REQUIREMENTS UNIQUE TO STERILE CTM DRUGS

The intent of this section is to provide *regulatory expectations of processes utilized in the manufacture of sterile drugs that must be validated* (i.e., IQ, OQ and PQ unless indicated differently) and systems/controls in place prior to producing clinical trials for human use.

1. Processes and associated equipment that sterilize and/or depyrogenate drug, components, product contact equipment and product must be validated (IQ, OQ and PQ) prior to CTM Batches. Some examples are:
   a. Steam sterilization
      i. Product path components
      ii. SIP cycles
      iii. Freeze Dryers
   b. Terminal sterilization of the product
      i. Steam Autoclave
      ii. Gamma radiation
      iii. Sterile Filtration
      iv. Others
   c. Depyrogenation of components
      i. Dry heat tunnels/ovens
      ii. Stopper washing (3 log reduction)
   d. Sterilization of all equipment and processes associated with product and product contact surfaces (e.g., tanks, fillers, stoppering)
   e. Utilities (e.g., WFI, pure steam, air/nitrogen and clean rooms)
   f. Special equipment and sterilizing systems
      i. Isolators
      ii. Hydrogen peroxide sterilization
      iii. Chlorine dioxide sterilization
      iv. Others
2. Drugs, chemicals and excipients tested per Endotoxin specifications. Bioburden specifications should also be in development.
3. Filter validation Phase I–II
   a. WFI soluble products can initially use WFI BP/forward flow results prior to and after sterile filtration. Prior to post-filtration integrity testing the filter should be purged (i.e., cleaned using WFI). Double filtration provides an additional assurance.
   b. Recommend that Phase 3 product used validated product BP/forward flow results (i.e., final drug product BP to be filed for NDA).
   c. Validation PQ batches should not begin until filter validation data is available.
4. Analytical test methods are required prior to first CTM batch to verify cleanliness.
5. Media fills:
   a. The first media fill demonstration batch is usually used to test the integrity of the container closure.
   b. If the container closure is new to the filling line (e.g., different size or closure) the container closure requires three media fills unless equivalence or bracketing can be justify less.
6. Hold times at all stages within the process
   a. Validation justification as compared to actual allowance and product records (e.g., in and out of cold rooms for visual inspection; product in delivery systems and verification of potential variations in temperatures)
   b. Process hold times and ramps (e.g., freeze drying)
7. Environmental monitoring room validation: Environmental validation of the room and associated filling/processing equipment in static and dynamic conditions is expected. Data should include smoke videos which can be used as training aids for aseptic processing.

8. EM Training and Qualification of Operators/Staff: Verification of personnel training to not adversely affect the sterility of the batch during normal run conditions.

VALIDATION REQUIREMENTS UNIQUE TO PLASMA/BLOOD PRODUCTS AND BIOLOGICALS

Blood/plasma and biological products while having the higher level of regulatory requirements associated with sterile drugs; require additional verifications and controls to ensure the public is protected. The following are examples that are typical of these products.

Plasma/Blood Products Special Areas of Attention and/or Validation

1. Process steps used for viral inactivation/removal (e.g., filtration, pH adjustment, chromatography, etc.) (9)
   a. Log reduction of clearance
   b. Specific viral testing performed
2. If from blood or blood components:
   a. Certification that donors have met FDA blood donor requirements
   b. Materials have been screened per FDA blood product material requirements
3. For contract manufacturing/filling. Lots received should include COA that states the following have been tested:
   a. Adventitious viral agents
   b. Mycoplasma (culturable and nonculturable)

Biotechnology—Protein Manufacturing

1. Changes in the manufacturing process should include a CE
   a. An effect on efficacy and/or safety might be expected or cannot be ruled out.
   b. Need to justify that the change in the manufacturing process will not effect efficacy and/or safety.
   c. If a modification of the product is detected during the CE it may indicate the need for further preclinical and/or additional clinical data.
   d. Potential for altering the profile and ratio of the impurities. The biological impact of changes should be considered prior to administration in humans.
2. Storage and Shipment (e.g., shipment, receipt, cold room storage, cryogenic/frozen product storage, thawing, and product delivery system/filling process temperature control)—Validation of Storage and Handling
   a. Require significant controls/validation to ensure the proper in-process control.
   b. When batches are scaled up this also requires attention in the CE.
3. Process validation should include a Risk Assessment to ensure all parameters/controls have been considered in the overall Validation Master Plan.
   a. A quality risk assessment would review all unit operations/processes (flow chart of total operation from materials received through product shipment) that may affect the products identity, strength, quality and purity. Included in this risk assessment should be an understanding of the impact of process variables (e.g., temperature, mixing speed, process time, flow rates, column wash volume, reagent concentrations, and buffer pH) and developing alert and action limits.
   b. A thorough knowledge of the processing steps of all ingredients when supplied by others is critical. Changes to their processes/equipment/formulation, etc., should have impact under change control.

VALIDATION REQUIREMENTS UNIQUE TO NASAL PRODUCTS (TABLE 1)

Nasal products provide a less invasive method of administration than injectables and patients are more willing to use them when compared to self-injection. While it is an easy alternative to the patient, it is a very complex and demanding process requiring significant understanding of the associated development, processing and controls. This is a developing area of Validation with expectation of continuing new advances in products, drug delivery systems, processing operations, environmental requirements, and of course regulation.

With delivery systems to the lungs or to the nasal mucosa including:

- Aqueous-based oral inhalation
- Pressurized metered inhalers
- Dry powder inhalers

Table 1 Tests for Inhalation vs. Nasal Products

Inhalation products	Nasal products
Single dose fine particle mass	Demonstrated deposition is localized in the nasal cavity
Individual stage particle distribution	Droplet formation—particle size distribution and full characterization of the product
Droplet size distribution and drug output	
Initial priming of container	
Re-priming of the container	
Compatibility	
Effect of moisture	
Safeguards to prevent multiple dose metering of dry powder inhalers	
Breath-activated devices—data provided to demonstrate all target groups capable of triggering the device	
Dry powder inhaler reservoir systems need a count indication for when the number of actuations indicated have been delivered	

Note: This table is provided to describe some regulatory difference in expectations for each dosage form.

- Products for nebulization
- Metered-dose nebulizers

1. Inhalation Sprays are intended for delivery to the lungs and contain therapeutically active ingredients and excipients. The use of preservatives or stabilizing agents is discouraged. Aqueous-based oral inhalation-based drug products are required to be manufactured sterile (21 CFR 200.51). Manufacturers must also comply with 211.113(b) which requires them to establish and follow written procedures designed to prevent microbiological contamination, including validation of any sterilization processes.
2. Nasal spray products and pressurized metered-dose inhalers are designed to apply sprays to the nasal cavity and therefore are not subject to this rule for sterilization (6).
3. Unique validation/testing
   a. Particle size test and limits
      i. Validated multipoint particle sizing method (e.g., laser diffraction)
      ii. Acceptance criteria to assure consistent size distribution in terms of total particles in a given size range
      iii. Acceptance criteria set based on the observed range of variation and should take into account the particle distribution of batches that showed acceptable performance in vivo, as well as the intended use of the product
      iv. If alternate sources of drug substance are proposed evidence of equivalence should include appropriate physical characterization and in vitro performance studies
   b. Development tests conducted
      i. Minimum fill justification
      ii. Extractable/leachables
      iii. Delivered dose uniformity/particle mass through container life
      iv. Shaking requirements
      v. Actuator disposition
      vi. Low-temperature performance
      vii. Cleaning requirements
      viii. Performance and temperature cycling
      ix. Physical characteristic
      x. Robustness
      xi. Preservative efficacy

The above information is a brief sampling of all the validation and operational needs of each type of nasal-type product. Please refer to the FDA and EU guidelines for more specific information. These products are often unique with the product and corresponding delivery system defining what needs to be validated and controlled during the manufacturing, filling, storage, testing and distribution of these products (10–13).

CONCLUSION

The intent of this chapter on Clinical Manufacturing Validation was to provide an understanding of what is needed and a logical sequence of events required to bring a product through the regulatory process. In support of these objectives various points to consider in understanding the validation needs and road map are provided.

This will become even more important as the new drugs/drug delivery systems become more integrated making it harder to understand the differences between different drugs as new technologies become reality.

Note: Significant regulatory activity is in progress as this chapter was being developed. So expect further changes in regulatory expectations.

APPENDIX A

Clinical Batch Verification Protocol (Phase 1)

Purpose

The purpose of this Process Verification Protocol is to prescribe the testing to be performed as it will apply to a batch intended for clinical trial use.

Procedure

1. Protocol content

   The protocol may contain the following sections:
   a. Product information (name and strength)
   b. Protocol approval signatures
   c. Reference documents (a listing of any and all documents to be cross-referenced such as, but not limited to, analytical standards, development or scale-up reports, SOPs and engineering studies)
   d. A statement of purpose for the testing described in the protocol
   e. Rationale for the sampling plan and the acceptance criteria for the tests
   f. Description of the equipment and the process
   g. Critical process steps to be verified by the testing
      i. For solids: includes but not limited to, blend time, mill speeds, and screen sizes.
      ii. For sterility: includes but not limited to, temperatures, dissolution time, mix speed, dissolved oxygen, pH, density, and osmolality.
      iii. For lyophilized products: residual moisture after cycle.

      Other attributes to be tested in process are prefiltration bioburden and endotoxins.
   h. Test functions: Based on knowledge from development and scale-up, a list of process steps which are to be tested should be listed. Each test function should show the acceptance criteria based on development work along with a rationale for that criteria.
   i. Test criteria: A description of sampling including sample size, location, number of samples along with a rationale for these.

      Note: For blend samples, sample size will be a weight equivalent to 1 to 3 dosage units.
   j. The verification protocol will specify sampling methods and tests. Test data will be attached, and the document will be approved certifying that all acceptance criteria were met. Unless specified by the client, the *minimum* testing is listed in Table 2.

      This verification will be performed on a batch-by-batch basis. The data will be summarized and included in a final package with the approved protocol.

Table 2 Minimum Testing Requirements

Process step	Sampling (minimum)	Tests
Non-sterile Solid Dosage		
Initial Blend	Top, middle, bottom of container	Assay (for blend uniformity)
Final Blend	Top, middle, bottom of container	Assay (for blend uniformity)
Compressing or capsule filling	Beginning, middle, and end of run	Content uniformity, assay, dissolution, dosage unit weights
Sterile Products		
Initial Compounding	Top, middle, bottom of mixing vessel	Assays, density, pH, osmolality
Filling	Beginning, middle and end of fill	Assays, oxygen in headspace, fill weights
Lyophilizing or terminal sterilization	Beginning, middle, end of load	Assay, moisture, reconstitution time, pH, oxygen in headspace

k. *Results:* Assemble the results, write an analysis and conclusion and assemble the final package. The analysis and conclusion will state whether or not the test results met the acceptance criteria. If the acceptance criteria are not met, a full explanation is required.
l. *Routing for Approval:* The finished package is attached to the batch record and routed for approval to the same individuals who approved the protocol initially.

This protocol is presented as an approach to satisfying the expectations of the FDA that an individual batch meets all its specifications (14).

APPENDIX B

Contract Manufacturing Requested Product Information (15) (Information Needed to Support Production and Stability)

1. Product classification: *Examples*: Diluent, small molecule pharmaceutical, fermentation derived, cell culture derived, cytoxic, anti-infective
2. Status (Phase 1, Phase 2, Phase 3, Commercial)
3. Registered by: FDA (CDER, CBER, CDRH, CVM), EMEA, or others
4. Timetable by phase
5. Type of service needed
    a. Aseptic manufacturing
    b. Aseptic filling
    c. Lyophilization
    d. Validation
        i. Analytical
        ii. Equipment
        iii. Process
    e. Stability studies
    f. Labeling and packaging
    g. Regulatory
6. Analytical: types of methods needed or supplied for example:
    a. HPLC
    b. GC
    c. Karl Fisher moisture analysis
    d. Oxygen headspace analysis
    e. TLC
    f. Optical rotation
    g. UV/visual
    h. pH (provide range)
    i. Specific gravity
    j. Infrared spectrophotometer
    k. Lowry protein determination
    l. PNU
    m. SDS-PAGE
    n. RID
    o. PAG isoelectric focusing
    p. ELISA
    q. Others
7. Microbiological
    a. Bioburden determination
    b. USP particulate testing
    c. Bacteriostasis/fungi stasis Validation
    d. Bacterial and fungal identification
    e. Endotoxin (gel-clot method)
    f. Sterility testing
    g. USP antimicrobial preservative effectiveness
    h. Container closure Integrity test (microbial ingress)
    i. Others
8. Raw material specifications
9. Storage Conditions for bulk, in-process, on-test, shipping and associated validation
10. Formulation requirements
    a. Aseptic additions required?
    b. Batch documentation
    c. Special equipment and the need for dedication
        i. Disposable processing
        ii. Tanks
        iii. Mixing/recirculation
        iv. Filling requirements
        v. Temperature
        vi. Light protection
        vii. Inert gas blanketing
        viii. Product density and viscosity
        ix. Special process or techniques
        x. Product incompatibility
    d. Handling Precautions
    e. Product and API material safety data sheets
    f. Disposal requirements
11. Components
    a. Presterilized and cleaned?
    b. Validation requirements?
    c. Specifications
    d. Surface treatment
    e. Manufacturer
    f. Stopper
        i. Type
        ii. Silicone treatment/limits
    g. Closure
12. Filtration and filling
    a. Aseptically filled or terminally sterilized
    b. Filter
        i. Single or redundant

- ii. Validation of integrity test
- iii. Sterilization validation

c. Filler
- i. Type of filling pumps required
  - Positive displacement
    - Rolling diaphragm
    - Ceramic pumps
    - Stainless Steel pumps
    - Glass
    - other
  - Time–pressure fill
- ii. Sterilization validation of unique items
- iii. Fill weight testing/validation
- iv. Container abuse studies
- v. Others
  - Nitrogen overlay required?
  - Headspace analysis required?
  - Protection from light?
  - Temperature requirements

d. Inspection and labeling
- i. 100% visual inspection
- ii. Automatic inspection for:
  - Headspace volume (fill level)
  - Particulate
  - Container integrity
  - Bulk vial identity
- iii. Labels

13. Process Validation
    a. Microbial filter retention study
    b. Media fills
    c. Terminal sterilization cycle development
    d. Cleaning verification
    e. Mixing verification
    f. Fill homogeneity
    g. Others
14. Lyophilization
    a. Cycle development
    - i. Time/temperature limits between filling and loading
    - ii. Nitrogen or inert gas purge during cycle
    - iii. Ramp rate during cooling
    - iv. Ramp rate during heating
    - v. Product loading temperature and cycle ramps
    - vi. Product pressure limits during cycle based on critical points
    - vii. Eutectic and/or collapse points
    - viii. End-product moisture requirement
    - ix. End-point pressure (full vacuum vs. partial vacuum)
    - x. Have you seen issues with powder on shelves after development cycles?
    - xi. Stopper tested to ensure it does not stick to shelf after stoppering?

REFERENCES

1. Spataro ME, Bernstein DF. Managing clinical trial materials: investigational product change control. Pharm Tech 1999; 23(10):144–50.
2. Pharmaceutical Engineering Guides for New and Renovated Facilities. Commissioning and Qualification. Vol. 5. Tampa, FL: ISPE, 2001:83–99.
3. Guideline on General Principles of Process Validation. Washington, DC: Center for Drugs and Biologics and Center for Devices and Radiological Health/Food and Drug Administration, 1987.
4. Guidance for Industry, Powder Blends and Finished Dosage Units—Stratified In-Process Dosage Unit Sampling and Assessment (Draft). Washington, DC: Center for Drug Evaluation and Research (CDER)/Food and Drug Administration, 2003.
5. FDA Guidance for the Industry, Preparation of Investigational New Drug Products, March 1991.
6. EMEA Guideline on the requirements to the chemical and pharmaceutical quality documentation concerning investigational medicinal products in clinical trials, November, 2004.
7. Annex 13: Manufacture of Investigational Medicinal Products, EMEA document dated July 1, 2004.
8. FDA Draft guidance/6672, Q8 Pharmaceutical Development, dated 2/7/2005.
9. EMEA Guideline on assessing the risk for virus transmission. New chapter 6 of the note for guidance on plasma derived medicinal products, November, 2004.
10. EMEA Draft Guideline on the pharmaceutical Quality of Inhalation and Nasal Products, dated July 30, 2005 for end of consultation.
11. FDA Guidance for Industry, Sterility Requirements for Aqueous-Based Drug Products for Oral Inhalation.
12. CDER Guidance for Industry—Sterility requirements for aqueous based drug products for oral inhalation—small entity compliance guide, November, 2001.
13. CDER Guidance Document – Nasal Spray and Inhalation Solution, Suspension and Spray Drug Products: Chemistry, Manufacturing and Controls Documentation, July 2002.
14. Hartman BW, Motise PJ. Human Drug CGMP Notes. Washington, DC: Center for Drug Evaluation and Research (CDER)/Food and Drug Administration, 1999.
15. Contract Manufacturing Requested Product Information-provided by Hollister Stier Spokane Washington.

41

Validation of New Products

Norman Elder
Genentech, Inc., South San Francisco, California, U.S.A.

Before a new product can be released for commercial distribution, it must undergo successful validation. Unlike a PQ, for a piece of equipment, Product Validation examines the process under which a product is made, for robustness and reproducibility. Product Validation is the documented evidence that a process will, with a high degree of scientific assurance, consistently produce product that meets its predetermined specifications and critical quality attributes, as required by the cGMP regulations for finished pharmaceuticals, 21 CFR 211.100 and 211.110 (1) and the EC Guide to GMP (2). The validation phase for a new product should be viewed as a part of the entire life cycle of the product; as a journey, not a destination.

The FDA's "Compliance Policy Guide" on PV Requirements (CPG7132c.08) (3) explains the enforcement policy for the CDER, the CBER, and the CVM, regarding the timing and completion of validation activities for certain products, including sterile and nonsterile processes.

Validation is required when a new product is introduced into a facility, when the facility is new, or when there is a change in an existing process that may affect the safety, quality, identity, potency, purity, or security of a product. A prospective validation approach is always preferable; however, there may be occasions when a concurrent validation approach may be applied; for instance, in the case of a low volume production demand where replicate batches are not readily available, or a modification to a well defined, previously validated process. Only in rare cases should a retrospective validation approach be applied, and should include a significant number of batches to assure that they are representative of the process, including any nonconforming batches. The validation process must go through several formal phases of implementation to ensure proper validation. These phases must be properly documented and executed in the prescribed order.

The validation of a new product, as with any validation project, should begin with a validation planning phase. The initial change control documentation should explain the background of the new product introduction or the scope of the change being requested and any impact to existing systems or processes. From that background and scope, a Validation Project Plan can be created, outlining all aspects of the validation activity, including a description of the product and overview of the process, equipment, facility, utilities, and components. Reference should be made to drawings, flow diagrams, specifications, test methods, in-process controls, procedures, supporting validation protocols and any development work performed. This is a high level document which creates the structure on which the protocol is built. The scope of the validation and rationale for the approach can be clearly stated at this point for consensus before proceeding with the actual validation work.

The next step in the validation of a product is the protocol development phase. The protocol is the mechanism to capture evidence that the validation activities are performed in a controlled environment, by trained individuals, following approved procedures, and using appropriate materials and components. When developing the protocol, it is important to have a clear understanding of the process. Sufficient development work should be done in advance of validation such that the critical parameters and key process control points have been established and characterized with limits and end points. Through the utilization of Design of Experiments in the development work, those key parameters can be scientifically identified and established. The validation work should reflect the development process but not be part of it. The validation batches should follow the manufacturing record, without necessarily challenging the limits of the established process. The validation exercise should be verification of an established process, not discovery or experimentation. The development report should be referenced in the validation work as a document to defend the validity of how parameters and limits were established.

The protocol should follow the scope outlined in the Validation Project Plan. The following elements should be included in a typical protocol or be verified during the execution or prior to execution of the validation:

- All impacted systems such as utilities, equipment, and computer systems should be qualified and/or validated and verified to be acceptable within the proposed operating limits of the product to be validated.
- Calibration of critical instruments, that is any instruments used to collect data or control the process, must be current at the time of execution.

Abbreviations used in this chapter: CBER, Center for Biological Evaluation and Research; CDER, Center for Drug Evaluation and Research; CFR, Code of Federal Regulations; cGMP, current good manufacturing practice; CPG, compliance policy guide; CVM, Center for Veterinary Medicine; EC, European Commission; FDA, Food and Drug Administration; GMP, good manufacturing practice; NDAs, new drug applications; PAI, preapproval inspection; PQ, performance qualification; PQRI, Product Quality Research Institute; PV, process validation; QA, quality assurance; RSD, relative standard deviation; SOPs, standard operating procedures.

- New or revised SOPs directly involved with the process to manufacture the product must be in effective status and personnel trained.
- Batch Records must be in final approved state.
- Analytical Standards and Test Methods must be in place prior to conducting testing of validation samples, and must be traceable and validated, respectively.
- Raw materials and components are tested, passed, and released for use.
- Cleaning procedures, test methods and rationale for cleaning limits must be established.
- Specifications for raw materials, in-process checks and finished product must be preestablished.
- Filter validation specific to the new product should be complete, including microbial challenge and bubble point information.
- Container/closure challenge and media fill simulation for new components/processes should be complete.

The procedures which provide specific detail relative to executing the validation testing are described in individual test functions within a protocol. The rationale used to establish the acceptance criteria may be inserted in each test function if not previously stated in the body of the protocol. Acceptance criteria contained within protocols are based on approved specifications and must be traceable to approved design specifications, requirements, or procedures. The acceptance criteria should clearly state what is required to pass each test function. It should be explicitly stated, unambiguous, and verifiable. Include a clear and concise description of the steps to be used when obtaining data to insure that they are reliable and appropriate to evaluate against the test function acceptance criteria. The description should include what data and/or samples to collect and during what point of the process the data/samples are to be collected. The sampling plan should be clearly stated, including the method for collecting the samples, any special precautions to be considered, the number of samples, location, size of sample, and labeling information. Diagrams may be included to facilitate protocol execution and data collection. A prepared data collection sheet is a valuable tool to include in the test function to clearly map testing, sampling, and results. Some types of information which should be included are start and stop times for process steps or sample collection, signature of the person conducting testing and collecting data, and verification signatures of the person checking calculations. The data collected and its organization should be sufficient to allow a reviewer of the completed test function to make a full determination of the acceptability of the data.

Test functions should support and validate proposed in-process specifications and be consistent with drug product final specifications, as indicated in 21 CFR part 211, subpart F, *Production and Process Controls*, Section 211.110, *Sampling and testing of in-process materials and drug products*. In part, this section states that "to assure batch uniformity and integrity of drug products, written procedures shall be established and followed that describe the in-process controls, and tests, or examinations to be conducted on appropriate samples of in-process materials of each batch. Such control procedures shall be established to monitor the output and to validate the performance of those manufacturing processes that may be responsible for causing variability in the characteristics of in-process material and the drug product. Such control procedures shall include, but are not limited to, the following, where appropriate:

1. Tablet or Capsule weight variation;
2. Disintegration time;
3. Adequacy of mixing to assure uniformity and homogeneity;
4. Dissolution time and rate;
5. Clarity, completeness, or pH of solutions."

Additional testing or sampling may be stipulated in the protocol, above and beyond the stated in-process and final product specifications, based upon product characteristics or unusual or extensive process steps.

The validation protocol should define the critical parameters, ranges, process steps and hold times for the new product. Typical parameters to be addressed for sterile dosage forms during manufacturing, filling, lyophilization, and sterilization are time, temperature, pressure, mixing speed, homogenizer speed, recirculation time and speed. For true solutions, development studies should first define the solubility characteristics through a solubility study, which should then be confirmed as a test function during the validation at the established minimum time during the manufacturing process. Documentation of mixing volumes must exist and be justified with process tank configurations and equivalency of tanks used in manufacturing and filling. Testing should include content/uniformity across the batch, with a larger sample size than routine production requirements, to provide additional statistical validity to the results. For products which are not solutions, but suspensions which are recirculated, testing should be included to confirm that during the maximum allowable down time, that the content/uniformity of the product delivered to the filler is not affected; and similarly, after the specified down time is exceeded, that the minimum allowable flush time is adequate to reestablish appropriate content/uniformity. A test function should also be included for lyophilized products which examines content/uniformity across all shelves through a predetermined sampling plan. The protocol should also include a test function for fill volume verification at the maximum expected speed that the product will be filled. An appropriate sample size should be selected so that statistical methods can be applied to demonstrate a process capability at or above 1.0.

Similarly, for solid dose products, protocol testing should reflect in-process and final product release testing. For tablets, samples should be included such as Loss on Drying from throughout the fluid bed dryer or tray dryer, and following final blending, as well as uniformity of the final blended granules throughout the tote; also tablet weight variation during compression, as well as content/uniformity, appearance, hardness, thickness friability, disintegration, and any other in-process tests from beginning, middle, and end of the batch. Also, dissolution testing should be performed as an in-process verification across the batch, and in the event of a film coating operation, final dissolution. In the case of capsules, testing should be included to evaluate variation in weight of individual capsules, or individual components within a capsule as well as content/uniformity and dissolution.

Sampling plans should be outlined in the protocol, with supporting rationale or reference, to demonstrate suitability or statistical validity. A commonly accepted approach was developed by a collaborative effort between the FDA, industry, and academia, the PQRI. The approach was accepted by the FDA and is incorporated in the draft guidance published by the FDA in October 2003 entitled "Powder Blends and Finished Dosage Units- Stratified In-Process Dosage Unit Sampling and Assessment" (4), which can be used to substantiate sampling plans and acceptance criteria. The acceptance criteria may specify a minimum RSD for a location, process step, across a batch or between multiple batches. This broaches the topic of the number of required batches.

The March 2004 revision of above referenced FDA "Compliance Policy Guide," CPG7132c.08, deletes a reference to a specific number of required validation or "conformance" batches. The long accepted industry practice has been to manufacture three batches to demonstrate reproducibility; however, in light of the emerging focus on a risk-based approach to validation, this may no longer be the case. As the CPG states, "Advanced pharmaceutical science and engineering principles and manufacturing control technologies can provide a high level of process understanding and control capability. Use of these advanced principles and control technologies can provide a high assurance of quality by continuously monitoring, evaluating, and adjusting every batch using validated in-process measurements, tests, controls, and process endpoints. For manufacturing processes developed and controlled in such a manner, it may not be necessary for a firm to manufacture multiple conformance batches prior to initial distribution." This points to the importance of understanding and controlling the process prior to commencing validation, and provides an avenue to develop a justification eliminating costly manufacture of validation batches for which there may be no opportunity to market. A firm using a well established approach to "Continuous Quality Verification" as a process beginning with development, may be able to reduce the level of validation based on ongoing assurance and demonstration of product quality. The EU Guidance on Manufacture, annexe 15, "Qualification and Validation," Section 25, presents a slightly different stance when it states, "In theory the number of process runs carried out and observations made should be sufficient to allow the normal extent of variation and trends to be established and to provide sufficient data for evaluation. It is generally considered acceptable that three consecutive batches/runs within the finally agreed parameters, would constitute a validation of the process." The number of required validation batches should be specifically stated in the protocol with a rationale for the approach, and should be manufactured at full scale of the intended commercial batch. The number of batches may also depend on the complexity of the equipment or process. For instance, in a case of 10 identical tablet presses, there may not be a need to validate a product on each press; however, in the case of three identical lyophilizers, an approach might be adopted to run three batches in the first one and one in each of the remaining two, to demonstrate that the process in each identical unit produces identical results. Again, the approach and rationale should be clearly stated in the protocol.

Once a well structured protocol has been developed, it must be routed to the appropriate functional areas for approval. Generally these include validation, development, manufacturing, laboratories, and quality assurance. The author of the protocol could be any trained individual, not necessarily from the validation discipline. But in any case, a separate validation approval should be obtained, so that the author is not approving his or her own work. The Validation approver is responsible for assuring that all of the critical elements have been included in the validation approach, that sufficient rationale is included to justify the approach, that appropriate cGMP practices are followed, the protocol follows and supports the Validation Project Plan, critical parameters and process variables are clearly outlined, and that the test functions and acceptance criteria adequately challenge the functionality of the process. The development signature signifies agreement that the process steps and acceptance criteria are in agreement with data collected during the development phase of the project, that test functions are technically feasible, and that manufacturing work orders and specifications are current and correct. The user, generally a manufacturing representative for new products, is responsible for ensuring that the objectives, acceptance criteria and expected results adequately reflect the intended process, facility, and equipment, and that process and system descriptions are accurate and complete. The approval by a laboratory representative assures that all laboratory testing requirements are feasible, that methods are adequate, appropriate and in a validated state, that sample size is adequate, and that acceptance criteria are appropriate. This approach also gives the laboratories advance notice of the upcoming validation work to allow proper resource planning to provide timely results. The quality assurance group is responsible for providing quality oversight to the validation process. The Quality review assures that the protocol and validation approach conform to internal procedures and cGMPs, that the test functions and acceptance criteria are supported by the specifications and manufacturing work orders, and that an appropriate batch disposition approach is identified and in place. When all approvals have been obtained, this constitutes approval to proceed with validation execution.

As noted previously, there are some activities which must be completed and documented prior to execution of the protocol or testing of the samples, such as current calibration of instruments being used to collect validation data, documentation of personnel training and a signature log of those involved in execution, validated test methods, SOPs, and work orders in an approved state, and equipment and facilities fully validated. When these precursers have been completed and documented, the protocol execution may begin. The execution must adhere to the approved protocol. Personnel executing the protocol must follow area SOPs and manufacturing work orders as they apply. Sampling and testing must be documented by the individual performing the sampling/testing. The validation representative is responsible for ensuring that the sampling is executed according to the approved protocol, in a controlled manner, recorded on the data collection form, submitted appropriately to the laboratory for testing per the applicable test method. In some cases, a signature may be required for collecting data, or performing a calculation, with a second person signing as verification. It may be worth noting that in these cases, the verification signature must be someone who is authorized to review the data, and cannot be the same as the person collecting the data. Care should be taken that good documentation practices are followed at all times. During the course of execution, should a failure occur, or an event which precludes adherence to the protocol exactly as stated, a discrepancy should be noted. A predetermined procedure should be established which discusses how to handle discrepancies, but it should be performed at the time of the discovery of the discrepancy. If all discrepancies are reported and discussed at the time of the summary, there may be undo pressure to yield to a resolution which may not have been accepted in real time during execution. Instead, discrepancies should be documented at the time they occur, with a resolution of the assignable cause, its impact on the process, and consensus, including QA, of forward action. Depending on the nature and extent of the discrepancy, it may or may not impact the validation effort. In the case of a discrepancy related to a non-process related event, such as sample handling or equipment malfunction, it may be necessary to repeat only a portion of the validation, for example only the batch in question. If a discrepancy results from acceptance criteria not being met due to a critical parameter, then a process change needs to be implemented and the validation repeated. If a portion of the process which creates a discrepancy can be isolated from other steps, a case may be made to repeat only those portions which are suspect. For instance, if a tablet coating process has a significant discrepancy, the portion of the validation which deals with coating may need to be repeated in whole or in part, but the compression portion of the validation may not be impacted. In some cases, a discrepancy during validation may also result in a manufacturing deviation which will follow the batch and go through the QA disposition process.

At the completion of all validation activity, evaluation of data, and review of laboratory test results, a Summary Report must be written to capture the results and report a conclusion. The summary should reference the original protocol and any supporting documents, such as the initiating change control document or the Validation Project Plan. The original scope and objective of the validation effort should be restated as an introduction to discussion regarding the actual results of test functions. Each test function should be discussed in sufficient detail to describe the intended result, how and when the execution was performed, and a comparison between the acceptance criteria and actual results. The data should be presented in a format which allows the reviewer to clearly discern if acceptance criteria have been met, without the tedious review of all of the raw data or further calculations. All original data collected during the testing, such as laboratory results, data collection sheets, temperature data, should be clearly identified and retained as back up documentation for easy reference, but not necessarily with each test function in the summary. This approach provides the reviewer with a more concise and presentable document. Discrepancies encountered during execution should be summarized with the resolution and any impact on the validation. Although each test function may address the individual disposition based on acceptance criteria, a final statement should be made regarding the overall validation effort for the product. The summary statement should provide an overall analysis of data with a conclusion and final disposition of the acceptability of the validation effort and subsequent commercial manufacturing.

The final Summary Report should be routed for approval to the initial reviewers. The manufacturing representative is responsible for ensuring that operations adhered to appropriate SOPs and manufacturing work orders, that they were carried out according to cGMP practices, and that any manufacturing discrepancies are accurate in scope and disposition. The validation approver is responsible for the accuracy of data and analysis, review of acceptance criteria against test results, correct format and clear documentation. The development approver should assure that the process was executed as prescribed, that any discrepancies are accurate and that any resulting remediation is appropriate, and that any data submitted by development is accurate and follows cGMP practice. Finally, the QA representative should review for completeness, conformance to established standards, adherence to cGMPs, attainment of acceptance criteria, and acceptability of overall conclusion. Upon approval of the Summary Report, the validation activities can be reported in the Validation Project Plan Summary along with any other requirements and can then be closed.

Approvals for NDAs may be received in advance of completing the validation effort for the new product. During the PAI for the product, if the validation activity has not been completed, the inspector may review similar products, processes or equipment validation to gain a level of assurance that the firm's approach is sound. It is a good practice to have the non-executed protocols in an approved state for review with the investigator during the PAI. If the existing validation work is suspect or the firm has a history of noncompliance, the approval may be held until the validation can be completed and a Post Approval Inspection conducted. If approval is granted in advance of the completion of validation activities for the new product, the successful validation must still be completed and approved before commercial batches may be distributed. The FDA may request the final validation report and, based on acceptable review and good validation history with the firm, waive a Post Approval Inspection.

REFERENCES

1. U.S. Food and Drug Administration. Code of Federal Regulations, Part 210- Current Good Manufacturing Practice in Manufacturing, Processing, Packaging, or Holding of Drigs: General and Part 211- Current Good Manufacturing Practices for Finished Pharmaceuticals, (Revised 2003).

2. Medicines Control Agency. Rules and Guidance for Pharmaceutical Manufacturers and Distributors 2002. 6th ed. 2005. (Fifth impression).
3. U.S. Food and Drug Administration. Compliance Policy Guide on Process Validation Requirements for Drug Products and Active Pharmaceutical Ingredients Subject to Pre-Market Approval (CPG7132c.08), (Revised March, 2004).
4. U.S. Food and Drug Administration. Powder Blends and Finished Dosage Units- Stratified In-Process Dosage Unit Sampling and Assessment, October 2003.

42

Retrospective Validation

Kevin Jenkins
Pfizer Inc., Terre Haute, Indiana, U.S.A.

INTRODUCTION

Retrospective validation is the validation of older/legacy products, processes, or equipments. Retrospective validation establishes documented evidence that a system does what it is supposed to do based on a review and analysis of historic information. It is normally conducted on a product already being commercially distributed and is based on accumulated production, testing, and control data. Often these products or systems have not been validated to contemporary standards due to their age. It should be realized that there is a diminishing point of return at which retrospective validation is no longer justified. Current processes, products and equipment are expected to meet contemporary regulatory standards by the relevant regulatory agencies. However, all is not lost—there is a rational approach to determine if retrospective validation is appropriate. In these cases, the approach needs to be evaluated and completed in a timely manner with significant data to support the product/process/system.

A key consideration in contemplating the use of retrospective validation is the regulatory agency point of view. It may be useful to discuss with your district, local or market region regulatory agency any concerns with the use of a retrospective approach. As always, an approach to retrospective validation should be documented in advance in the form of a validation plan. This chapter serves as a guide to develop a supportable retrospective validation approach in terms of prerequisites, regulatory guidance, an overall approach, and examples. Finally, this chapter will explore retrospective validation in a PAT environment where the traditional three-lot validation is replaced by a continuous stream of data.

The Foundation for Retrospective Validation

Any retrospective validation project should start with a strong foundation with regard to the overall systems which support operations. As shown in Figure 1, there are five layers of systems leading up to retrospective product validation that need to be in place for success. These five systems are as follows:

- *Site culture/commitment.* The first layer is the overall culture of the site or operation. This includes the change control culture—are all changes captured and handled adequately in a system? Part of this culture includes the support of leadership for change control, quality and validation. These key quality systems need to be at the foundation. Education of the overall workforce is also critical at this level.
- *Life cycle and change control.* The next level includes functions such as project management, procurement, and life-cycle review. Are these well-controlled processes? Can it be assured that through these areas the correct components, replacement parts and processes are maintained?
- *Equipment/instrument history, maintenance and calibration.* The third level includes equipment history, drawings, calibration and preventive maintenance. These are crucial to maintain equipment and processes in the correct working order. Included as well in this level are laboratory methods and equipment—are they validated and maintained? This is critical to the foundation of a working quality system in the laboratory where the data is generated.
- *Equipment R/S.* The next level is the equipment R/S—do they exist? Are they maintained through a life cycle approach? These R/S are critical since they are the foundation for the next level—qualification of equipment. They also contribute to the overall validation plan that describes the system as a whole, how it functions together and the approach to validation.
- *Installation and OQ.* The final level in advance of product validation is equipment installation and operational and process qualification. It is critical to have these in place to assure control of the process and the products produced as part of that process.

Each layer builds on the previous one. If elements are missing then there needs to be some form of remediation at each stage to assure a solid foundation before undertaking any product or process retrospective validation. These elements are summarized in Table 1 with the relevant considerations for each item listed. This table should be consulted before undertaking a retrospective validation project or approach. Missing elements erode the foundation on which a case could be built for retrospective validation.

RETROSPECTIVE PRODUCT VALIDATION—APIs

The first approach to retrospective validation explored is for APIs. The elements discussed previously must be in

Abbreviations used in this chapter: APIs, active pharmaceutical ingredients; CA-HCl, Cure-all hydrochloride; CFR, Code of Federal Regulations; CPK, capability index for process average; EMEA, European Medicines Evaluation Agency; EP, *European Pharmacopoeia*; FDA, Food and Drug Administration; GC, gas chromatography; HPLC, high-performance liquid chromatography; IQ, installation qualification; JP, *Japanese Pharmacopoeia*; LOD, loss on drying; NIR, near infrared; OQ, operational qualification; PAT, process analytical technology; PQ, performance qualification; R/S, requirements and specifications; USP, *United States Pharmacopeia*.

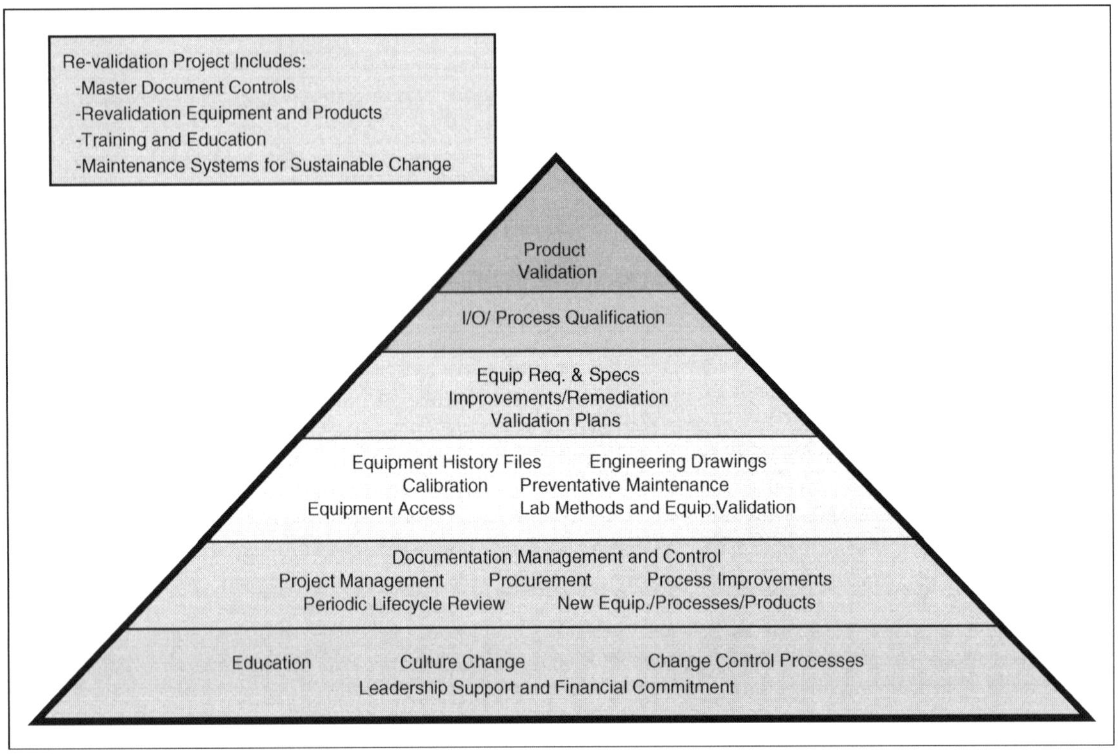

Figure 1 Support systems for restrospective validation.

place from a system perspective—now the discussion will be from a product standpoint. The first requirement for retrospective validation applied to a particular API is that "it must be a well understood product/process with few deviations" (1). If this initial requirement cannot be met, then a concurrent validation approach may have to be considered. If this initial criterion can be met, then there are a series of prerequisites that must be further considered as follows:

- The process is well understood and documented throughout the full-scale manufacturing process.
- Critical process parameters and critical attributes are identified, justified and understood.
- Reliable test data were/or will be generated using a pharmacopoeia (USP/EP/JP) method or internally validated test methods.
- There are no significant process or product failures and any failure must be attributed to operator error or equipment failure or a "one-off" well-understood deviation.
- Impurity profiles are well established for the API.

Exploring each of these prerequisites in more detail:

1. *The process is well understood and documented throughout the full-scale manufacturing process.* Processes that are not well understood are not acceptable candidates for retrospective validation. The execution of retrospective validation is not an acceptable time to gain knowledge and understanding of the process.

   There must also be sufficient documentation of the process to qualify for retrospective validation. The firm must be able to produce all relevant documentation to demonstrate the conditions under which all batches included in the retrospective validation were produced. This task may be complicated by the low volume production over time and changes during this time period. As discussed previously, part of the foundation is the data, documentation and change history.
2. *Critical process parameters and critical attributes are identified and understood.* Critical process parameters contribute to critical attributes. For instance, the critical process parameter of drying time has a direct impact on the attribute of moisture content through physical measurement via LOD testing. This attribute is thereby the measure of the process parameter. These process parameters need to be identified and the attributes measured and demonstrated to be in control. Any attribute not in control needs to be critically evaluated to determine if the batches under retrospective validation allow this approach to be utilized.
3. *Reliable test data generated using a pharmacopoeia (USP/EP/JP) method or internally validated test methods.* The data utilized as part of retrospective validation are critical to proving whether the process is well defined and controlled. Data integrity is therefore crucial to the retrospective validation the platform being built. All data utilized must be either from pharmacopoeia or an internally validated method. If it is a pharmacopeial method, assure that the current version of the method was used at the time these data were generated. In the case of an internally validated method, assure that it meets contemporary validation standards at the time generated. Failure to meet these requirements puts these data, and thereby the retrospective validation, in jeopardy.

Table 1 Foundation Elements and Considerations for Retrospective Validation

Element	Considerations
Product validation	Need a strong product validation approach for prospective validation before approaching a retrospective project
IQ/OQ and process qualification	The underlying IQ and OQ need to be robust for the related equipment/facility. In addition, any PQ for processes such as sterilization or aseptic processes must be in place
Equipment requirements and specifications	Need to be in place and act as foundation for the related IQ and OQ. Should be current and under revision control
Improvements and remediation	Any improvements or remediation to equipment, facility and processes should be captured in updated requirements, specifications and related IQ, OQ and PQ
Validation plan	An overall comprehensive validation plan should be in place. Retrospective validation, where applied, should be included and justified in this plan
Equipment history files	Equipment history files should exist which capture changes, updates and overall maintenance of the equipment
Engineering drawings	Engineering drawings for the equipment should be up to date and linked to change management system
Calibration	Calibration for the equipment and related instruments should be maintained and current
Preventative maintenance	Preventative maintenance should be documented and current
Lab method and equipment validation	The lab methods that will be used to support retrospective validation need to validated to contemporary standards. All relevant instruments should have contemporary qualification
Documentation management	Documentation should be current for site and under the change management system control
Project management	There should be a strong project management group that assures systems are maintained
Procurement	Procurement should control purchase of new equipment, spare parts and materials. Changes should be under the change management system
Process improvements	Process improvements should be captured in the change management system in regards to drawing updates, revalidation and instructions
Periodic life cycle review	A program should be in place to periodically review the status of all systems and recommend actions such as revalidation. This should occur on at least a three year cycle
New equipment, processes and products	Should be evaluated against current systems and included in validation plan and life cycle
Education	Assure colleagues in operation are educated on requirements for validation, change control and validation planning at a minimum
Culture change	Assure there is a culture shift if change control and validation were not robust in the past
Change control processes	Assure robust change control process is in place for facilities, utilities, equipment, lab methods, processes and products
Leadership support and financial commitment	All of the items listed above require leadership support and financial commitment or they will not be sustainable and erode the overall foundation of a robust system

4. *There are no significant process or product failures. Any failure must be attributed to operator error or equipment failure.* Often the significance is difficult to judge. Remember that any product failure should have been thoroughly investigated via a deviation investigation procedure. If the failure investigation determines the root cause was due to operator error or equipment failure and not the process itself then this does not necessarily implicate the process. An example would be equipment malfunction—such as a centrifuge in the process that stopped mid-batch due to a power failure and resulted in an aborted or failed batch. If, however, the batch completed without attributed operator or equipment error and was out of specification, this would cast doubt on control of the process. This must be critically evaluated through a comprehensive deviation investigation for each batch included in the retrospective validation.
5. *Impurity profiles are well established for the API.* Since the firm will typically have experience with the API considered for retrospective validation there should be substantial data on the impurity profile of the API. These data must be reviewed to assure there were no adverse trends or issues with impurities. Once again, the methods utilized for the impurities evaluation should be validated.

RETROSPECTIVE PRODUCT VALIDATION PLANNING FOR APIs

Current ICH Q7A (1) guidelines recommend between 10 and 30 consecutive batches be examined as part of retrospective validation. Fewer batches may be used to justify retrospective validation, providing a documented sound scientific rationale is provided. These batches need to be statistically examined closely for any trends, deviations, and/or out-of-specification results. Such data may call into question the applicability of a retrospective approach to validation unless there is significant evidence to indicate these trends or results are not indicative of the process.

The validation plan for retrospective validation needs to define the number of batches to include with a scientific rationale for the number. The frequency of production and age of the process (years produced) should be considered. For instance, an API that is produced infrequently such as a once a year, but only produced for the past five years, provides limited data over a short time span. The production of 10 batches per year over two years provides both a wealth of data and a shorter time span (12). The time span is an issue since the longer the time span the greater chance of significant changes to the process, equipment, and methods. Once again this should be evaluated and factored into the rationale included in the plan.

The contents of validation plans and protocols were discussed previously in chapter 1. The content of the plan should include the consideration of the prerequisites, number of batches to be included, critical process parameters, critical attributes, acceptance criteria and data analysis approach to be utilized.

Case Study—A Retrospective Validation Approach for API

This section explores retrospective validation for a fictional API called CA-HCl. First, examine some background on the API. CA-HCl starts as a precursor, cure-all salt, that processes through a reaction process with hydrochloric acid to form the CA-HCl form. This is accomplished by charging the cure-all salt into a 50 kg reaction vessel V-1. The hydrochloric acid is transferred from a holding tank into the reaction vessel V-1. Next, a wash process with acetone is conducted to remove impurities and then an aqueous wash to remove water soluble impurities. This is accomplished by transfer of the slurry into vessel centrifuge C-1 and charging the acetone from a separate transfer tank T-1. After centrifugation water is added. The next step is the drying process through centrifugation to remove excess moisture to a target of 3.0% in the centrifuge. The final product is then packaged into polyethylene-lined drums.

The initial CA-HCl process was developed and implemented 20 years ago. Each year about 10 batches are produced. The equipment was not initially validated at installation 20 years ago, but the centrifuge was replaced 5 years ago and was completely validated (IQ/OQ) to contemporary standards. In addition a new control system for the vessels V-1, transfer tank T-1 and centrifuge C-1 was installed three years ago and full computer validation completed. The test methods for CA-HCl consist of the following assays:

- Potency by HPLC
- Impurities by HPLC
- Moisture by LOD
- pH measurement
- Residual solvents by GC

All methods were validated to contemporary standards five years ago. In addition, a retrospective validation project was conducted at the same time to qualify all instruments in the lab.

Before starting determine if the API is appropriate for retrospective validation. Using criteria previously established by Trubinski, it is determined from Figure 2 that there are over 20 batches in the product history and it is a product which the company intends to continue manufacturing (2).

Examination of the systems in place as described previously in Table 1 and Figure 1 is conducted. All identified systems are robust and therefore it appears to be a candidate for retrospective validation. At this point, a protocol to examine the 20 batches against predetermined criteria should be written. This approach will utilize each of the previously mentioned test methods and assure all data meet defined specifications. In addition, statistical analysis of the data will be performed to assure no values are out of trend over time. In this instance there was one out-of-specification result for batch number 17. An investigation determined that this was due to adding a low volume of acetone, which resulted from a deviation to procedure. The batch deviation was determined post completion and analysis. The batch was rejected since it was outside the registered specification for acetone addition quantity. Therefore, it is not considered to have an impact on the retrospective validation. The investigation is included as part of the validation package. The conclusion will need to defend the use of the 19 remaining batches for retrospective validation.

DRUG PRODUCT

Retrospective validation for drug products has some different considerations than APIs. Routinely, the FDA expects prospective validation for any new product coming to market. Even for existing products the window for retrospective validation "is closing if not already closed" (3). The guidelines for prospective validation have been around since 1983 and investigators tend to view retrospective validation in a negative framework since it is expected that all products on the market at this point comply with those guidelines. As a result, retrospective validation is not as commonly used in drug product validation as it once was. Retrospective validation was applied when validation requirements were first put in place. Today it is an expectation that firms have contemporary validation data (4). Once again, as stated previously, retrospective validation applicability is dictated by the change history for the process. Therefore one needs to define upfront what will be evaluated as part of retrospective validation for drug products. This section will explore a number of considerations.

Just as was the case for APIs, in the case of drug products, the product must be a well-understood product/process with few deviations. If this is not true, then retrospective validation for the particular product may not be the best choice and a prospective approach should be considered. If a retrospective approach is followed, the validation requirements may need to be tightened, added further batches and provided a strong rationale for why this approach is still valid. As always, consultation in advance with regulatory agency or industry consultants should be considered as a logical choice.

Just as with retrospective validation for APIs if this initial criterion is met then there are a series of prerequisites that must be further considered as follows:

- The process is well understood and documented throughout the full-scale manufacturing process.
- Critical process parameters and critical attributes are identified and well understood.
- Reliable test data can be generated using pharmacopeias (USP/EP/JP) methods or internally validated test methods.
- In-process controls and acceptance criteria are established and in use throughout the critical portion of the process.

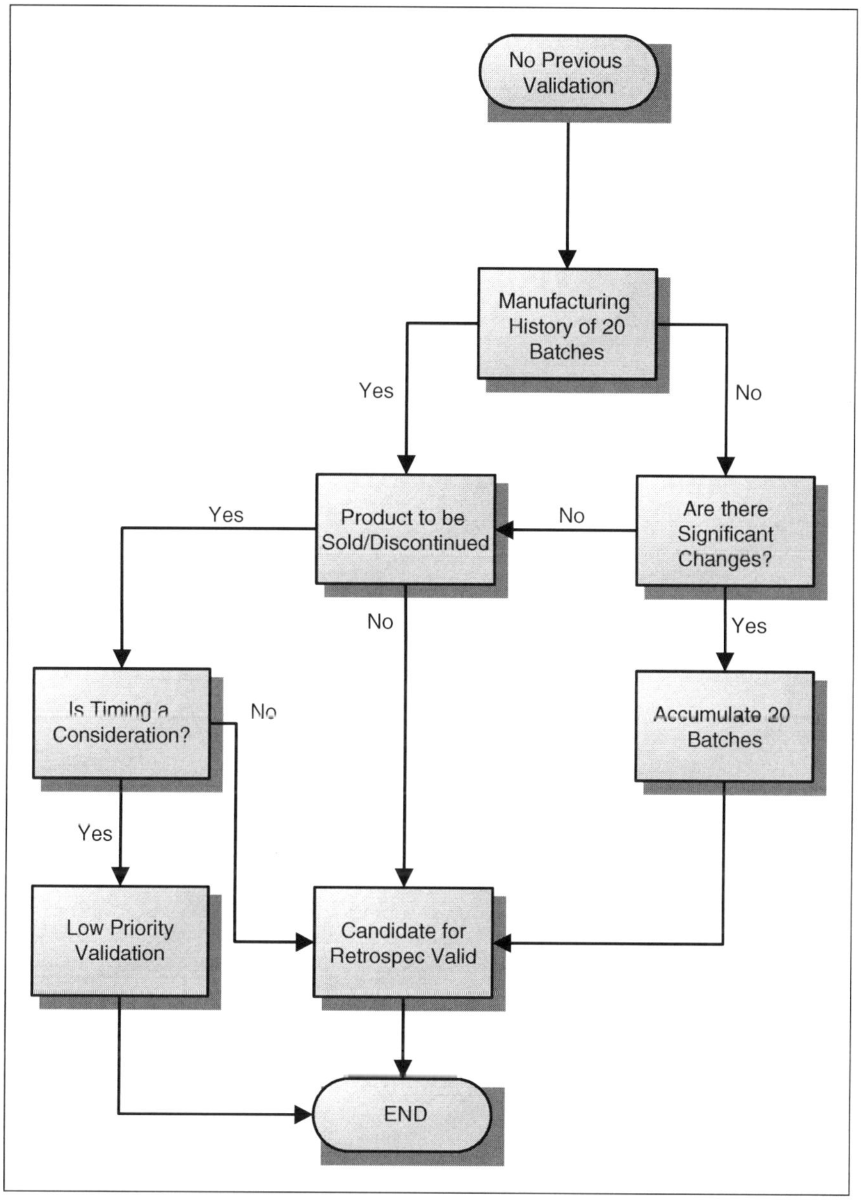

Figure 2 Selection of candidates for restrospective validation.

- There are no significant process or product failures. Any failure must be attributed to operator error or equipment failure and not the process itself.
- Impurity profiles are well established for the API used in the drug product.
- Change control is in place and followed for the process.

It should be noted that generally these prerequisites are the same as they were for APIs—except in three cases. The three cases are highlighted above and will be explored in further detail.

1. *In-process controls and acceptance criteria are established and in use.* Generally, a drug product process involves a number of defined stages. In tablet production processes, for example, there are blending, drying, compression, and coating stages to mention a few. At each stage, there are critical in-process controls which assure the product is acceptable to progress to the next stage. As part of retrospective validation, these in-process controls and the relevant data should be examined as part of the validation. There should be acceptance criteria in place from a production standpoint and retrospective validation criteria which are wider than the operating range, but tighter than development ranges must be applied. Any deviation needs to be investigated and explained as part of the validation.
2. *Impurity profiles are well established for the API used in the drug product.* In the previous section on APIs this was discussed in relation to the API itself. Now it needs to be assured that the API or APIs used in the drug product have a well-defined profile such that there are analytical, impurity and stability data that characterize the API. The data must be further reviewed to assure no adverse trends or issues with degradation, by-products or impurities that would place the final drug product at risk.
3. *Change control is in place and followed for the process.* In the drug product/process change control is a critical consideration. If the intention is to retrospectively validate a product for which there was an underlying

lack of change control, how can the firm assure the process/product is consistent? Remember the key definition of validation is a proven evidence that the product/process has been consistent over the retrospective validation evaluation period. Change control assures that the process is consistent with processes which are in control. Adequate change control is a premise to validation. Lack of adequate change control requires a detailed review of the process and some retrospective review of the change history. If the change control is inadequate, retrospective product validation may not be appropriate and a prospective or concurrent product validation approach should be strongly considered. This review should be documented as part of the validation. It is important that this review be comprehensive such that changes to product/processes, equipment, excipients, procedures, and systems are reviewed.

RETROSPECTIVE PRODUCT VALIDATION PLANNING FOR DRUG PRODUCTS

The FDA's CFR does not specifically identify retrospective validation. It only states that process validation is a requirement for the manufacture of pharmaceuticals and medical devices (5). There is, however, a section on retrospective validation in the FDA guidelines on validation. These guidelines state that "in some cases a product may have been on the market without sufficient pre-market process validation. In these cases it may be possible to validate, in some measure, the adequacy of the process by examination of accumulated test data on the product and records of the manufacturing procedures used" (6). The question to ask is how much data is required for retrospective product validation? This chapter will explore some examples as part of case studies later. For now it is sufficient to say that this depends on the process and its history. There is one other critical statement in the above test from the guideline: "records of the manufacturing process used." It was stated in the opening of this section that for the process to be a candidate for retrospective validation it must be well documented, understood, and under change control. These are critical to allow this retrospective validation approach.

The second paragraph of the guideline specific to retrospective validation suggests that "retrospective validation can also be useful to augment initial pre-market prospective validation for new products or changed processes." Typically these are prequalification batches that can be examined as part of developing a final prospective validation approach. However, in many cases these prequalification batches have inherent differences from the final batch process to be qualified. This must be taken into account as part of the approach. Statistical analysis of the prequalification batches as compared with retrospective data is a useful tool which will be discussed later as one strategy.

The last paragraph of the guideline states in regards to retrospective validation that "Test data may be useful only if the methods and results are adequately specific." The section goes on to state that "Specific results, on the other hand, can be statistically analyzed and a determination can be made of what variance in data can be expected." This statement suggests that statistical evaluation be performed on the method itself using tools such as Gauge R&R or CPK calculations to determine variation of the methods. If the variation is too extreme, it may not be possible to use the method to justify the previous set of acceptance criteria in the retrospection validation plan. In those cases, other methods may be necessary or tightening of the method variance required. That is an acceptable approach in a prospective validation plan where such adjustments can be made. In a retrospective validation approach these data have already been generated. It may, however, allow a change to the approach or target-specific methods as key to validation based upon the statistical analysis of the data. For instance, this analysis may determine one method is more critical than another and narrow the scope of data required for the retrospective validation.

The last sentence of the guideline section states that "Whenever test data are used to demonstrate conformance to specifications, it is important that the test methodology be qualified to assure that test results are objective and accurate." This was also one of the principles required for retrospective validation— that reliable test data are generated using a pharmacopeias (USP/EP/JP) method or internally validated test methods.

EMEA View

The EMEA guidance for manufacturers has a section on retrospective validation in annex 15.31–15.35 (7). There are five major points (31–35) which support what has already been stated as requirements and prerequisites. The first point "retrospective validation is only acceptable for well-established processes and will be inappropriate where there have been recent changes in the composition of the product, operating procedures or equipment." This supports the initial criteria that the process/product must be well understood. This guidance also points out the issue of change control as noted in the previous discussion on prerequisites. It brings up another point about consistency of the process, that changes to the process/product over time have not included major changes which may require significant revalidation. If that was the case, then concurrent validation should have been performed at that point, not retrospective validation at a later point.

The next point in the EU guide is that "validation of…processes should be based on historical data. The steps involved require the preparation of a specific protocol…* leading to a conclusion and recommendation." It was indicated planning for retrospective validation that a predetermined validation plan or protocol must document the approach and acceptance criteria. This guidance further reenforces the point and highlights the importance of the final conclusion and recommendation. Data are another key aspects of the retrospective validation. There is further guidance on the source of data. "The source of data…should include…batch processing and packaging records, process control charts, maintenance log books, records of personnel changes, process capability studies, finished product data, including trend cards and storage stability results." The point here is that there should be a comprehensive data review from all of these documented sources. Every "stone" must be overturned, so to speak, as part of this review. This is a significant task in some cases that requires a thorough approach to assure all data is collected, analyzed, and included in the retrospective validation package. These data must then be tested against the pre-established acceptance criteria. In some cases this is such a significant undertaking that a concurrent validation approach will require less investment.

The next section deals with the batches selected as part of the validation. The EU guide states that "Batches selected for retrospective validation should be representative of all batches made during the review period, including any batches that failed to meet specifications." Previously, it was stated as a prerequisite that there may be no significant process or product failures, and any failure must be attributable to operator error or equipment failure. This guidance appears to allow inclusion of failed batches and notes that it must be a comprehensive set of batches over the review period. A word of caution is appropriate here—failed batches will have a great deal of scrutiny by a regulatory agency. This means that the firm must exert an even greater level of scrutiny and conclude if a consistent and thereby validated process/product exists. The investigations of these failed batches need to have identified root cause(s). That determined root cause(s) should not indicate a process/product failure—if it does, concurrent revalidation after the issue is resolved must be considered. This may also require development work to understand the source of the failure.

Another key point is made in the final sentence of this reference: "Additional testing of retained samples may be needed to obtain the necessary amount or type of data to retrospectively validate the process." During the development of the validation plan and protocol, the firm may find that the methods and data are not sufficient to support the validation. This could be due to assay variability as mentioned previously or that the method was not in place at the time the batch was initially analyzed. The firm may also encounter during data review that there are missing data. One additional word of caution: in the case of failed batches do not use retained samples to try and retest and eliminate failed results unless investigation(s) can invalidate the initial result. If these issues do not apply, then available retained samples may be useful to supplement overall data and support the retrospective validation.

The final point in the EU guide provides guidance for the number of batches to include in the retrospective validation. The recommendation is "ten to thirty consecutive batches…but fewer may be examined if justified." This is consistent with previous ICH guidance for APIs. One important note is the word "consecutive" batches—the firm cannot pick and choose the batches it wants to use. The number of batches must be defined in the preapproved validation plan and protocol. As stated clearly in the guidance, any choice outside of 10–30 requires a rationale for that choice. In the examples section this chapter will explore how to arrive at this number. It should be based on the specific attributes, history, and characterization of the product.

Remember that a retrospective validation approach is not generally accepted unless plenty of data are included to make a case. The data are keys to justification of the retrospective validation and should utilize statistical methods for evaluation.

STATISTICAL ANALYSIS OF BATCHES

In the case of measuring variation of the methods used to generate retrospective data gauge, R&R and CPK values are useful to assess each method. This evaluation should be used to determine if the methods are suitable to provide data for a retrospective validation strategy.

Once the data are generated there needs to be some evaluation against either previous development data or predetermined acceptance criteria. Significant testing, where there is a null hypothesis of no difference between the observed and the known or previous obtained values, is useful (8). Useful analysis methods include but are not limited to comparison of mean, paired *t*-tests, and *F*-tests for comparison of standard deviation. The overall lesson is that the use of statistics provides a scientific basis for comparison to make a case that the process or product is comparable to previous experience or data and that the method is well under control.

An Approach to Retrospective Validation for Drug Product

Case Study—Examples

In this section a retrospective validation approach for a fictional sterile product Steri-Cure will be provided. It is an aseptically filled liquid composed of the drug substance CA-HCl. Cure-All is dissolved in WFI using a 500 L mixing tank, aseptically filtered, and transferred to a holding tank for aseptic filling. The following table lists the critical quality attributes and parameters which impact these attributes.

Critical quality attributes	Critical process parameters affecting quality attributes
	Formulation
	Temperature
Potency	API dissolution mix speed
Degradation products	API dissolution mix time
pH	Final mix time
Color of solution	Final mix speed
Appearance of solution	Final mix pH
	Hold times

Steri-Cure has been aseptically filled at this particular facility for 15 years. There is no contemporary validation for this product. Previous validation was performed on other aseptic products which utilize the same filling equipment and filling suite. This validation included all aseptic process validation, and media fills are conducted on a semi-annual basis for this product and aseptic process.

Retrospective validation is proposed to validate the manufacturing process since Steri-Cure, an older product, is only produced once every two years based on demand. A validation plan was therefore written which considered the change control history, development data, methods validation and drug substance characterization.

In this case the change control on the tanks, pH meter, and mixer used in the process are current since they are shared with other products. There is cleaning validation for this equipment as well. All analytical methods for the product were validated two years ago as part of an overall program in the laboratory. A review of batch records indicated that dating back 10 years there were 25 batches produced. The drug substance is well characterized and method development data, while old, is still available.

An evaluation of these data for a retrospective approach indicates that while there are an adequate number of batches for evaluation, only two batches were manufactured since the analytical methods were fully validated. This adds a degree of risk to a retrospective validation approach. It is not necessary to include the filling equipment or filling suite since adequate, contemporary validation exists for the filling of other comparable liquid aseptic products in the same equipment and line configuration. Although the same argument could be applied for the manufacturing process, it is not recommended.

The development data are older and at the time the documentation did not cover all of the critical quality attributes. It is discovered that the mixing time was changed along with a pH step addition about five years ago due to an incident where the drug substance was not completely dissolved after the normal mixing time, temperature, and drug substance addition. Investigation at the time determined that mixing speed was not recorded in the development, or in routine operation. It was determined that due to changes in the operators who previously performed this manufacturing process, a change was made in mixing. A study was conducted at that time to determine optimal mixing time, and a pH adjustment step was added.

Due to these significant process changes and the lack of previous batch data without adequate method validation retrospective validation is not recommended.

A concurrent validation approach is therefore recommended that will include three batches and also the listed critical process parameters which will be measured in the protocol through lab analysis of the critical attributes. Given the normal low demand for the product, three batches will be produced in succession with only one going on to final filling. The critical attributes will be measured against the existing specification for the product in regards to the potency of solution, degradation/impurity limits, pH range, color of solution, and appearance of solution.

In summary our review indicated that this product was not a good candidate for retrospective validation. Factors that contributed to this decision are the change history, deviations in process, and analytical method validation to contemporary standards. As mentioned previously, all of these must be critically evaluated to consider retrospective validation. Often, age of the product/process can work against a retrospective validation approach.

RETROSPECTIVE PROCESS VALIDATION

Retrospective process validation often has some of the same concerns and considerations of prospective validation as mentioned previously. Overall, there are systems where it can be applied and those where it cannot. For instance, a system which has routine in-process data for evaluation and is controlled based on the data could be a candidate. In the case of a water system, perhaps there was some initial validation conducted long ago, or the system was only qualified. Since it is a requirement to continually monitor and control the system, this previously generated data can be used to retrospectively validate the system. The FDA guide to inspection of water systems requires phase I, II and III validation (9). In the final phase, the data for chemical and microbiological analyses are required on a frequent basis under protocol. Clearly there are USP criteria preestablished for these tests. Therefore, if the data history is available then these USP criteria can be utilized for evaluation and retrospective validation of the system.

In other systems such as sterilization processes it is clear that these must have not only prospective validation but also an annual or a periodic requirement for revalidation that is prospective as well (10). These sterilization processes are part of a larger overall aseptic process and often there is not the level of in-process data generated as part of control for the water system example.

In summary, if the system has in-process controls and preestablished criteria, these data could be used to evaluate and validate the system.

RETROSPECTIVE EQUIPMENT QUALIFICATION/VALIDATION

Retrospective equipment qualification/validation applies in cases where contemporary IQ, OQ, and PQ do not exist for equipment. In these cases, as mentioned previously, retrospective product validation is not recommended since the underlying foundation is not sound.

In the case of retrospective equipment qualification/validation, many of the criteria mentioned in Table 1 and Figure 1 still apply. Criteria such as improvements and remediation to the equipment, equipment history/use, calibration records, preventative maintenance records, and change control should all be evaluated. Leadership support and financial commitment is important since retrospective qualification of some equipment may be more costly and at greater risk than to purchase and prospectively qualify new equipment.

If a retrospective approach can be justified, the first activity is to develop an accurate set of combined R/S. Since this equipment is already installed these combined R/S should represent the "as-found" condition unless there is a justification to change. For instance, if the mixer is rotating in the opposite direction from initial design it should be left that way and captured as such in the R/S since all batches previously produced were under these conditions. This assumes that all specifications were met.

Once this combined as-found R/S is created, it can be used for the foundation of the Retrospective IQ/OQ and PQ, if applicable. Critical attributes and process parameters should be defined. The IQ will be minimal since the equipment is already in place, but will confirm that installation was per original manufacturer recommendations and that critical documentation, spare parts listed, procedures, and operating instructions are in place. The OQ will assure that the system operates as designed and all critical parameters are met.

RETROSPECTIVE VALIDATION IN A PAT ENVIRONMENT

PAT is a system for designing, analyzing, and controlling manufacturing through timely measurements (i.e., during processing) of critical quality and performance attributes of raw materials and/or in-process materials and processes with the goal of ensuring final product quality (11). The specifics of PAT are covered in detail under chapter 10. In this section we will only cover some considerations in a PAT environment related to retrospective validation.

In a process utilizing PAT, traditional validation principles may not apply. Three batches are insignificant where thousands of data points will be examined for the critical parameter(s) measured. It is important to be aware that PAT may show much more about the process than was previously known or understood.

As an example, consider a batch API process where moisture level after centrifugation and drying is critical. The capability to measure moisture using PAT via an in-line NIR is now available. Figure 3 is a set of data in five-minute intervals for this process.

In order to make a change to the drying or centrifugation time to this process in the traditional validation model it would be necessary to analyze three consecutive batches. In a PAT environment continuous data are available that can be analyzed. The traditional three-batch approach would not apply since continuous data are present. Advantages of this approach include not only the additional data, under real-time conditions, but also the process understanding, quality of data, and reduced validation time.

In fact, there is a preponderance of data—actually, substantially more data than normally available, in traditional validation. It is important to note when the change occurred, set preestablished criteria for the moisture level, and measure or trend data after the change. If these data indicate a state of control within predefined specification or limits then it may meet validation criteria. In essence, it is possible to concurrently validate the process through use of on-line data generated as part of a PAT measurement process.

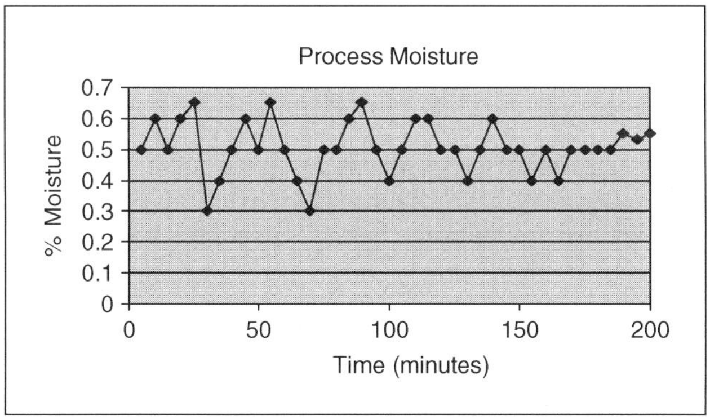

Figure 3 In-process data for moisture.

If this PAT process measurement is new it would be necessary to compare variability of PAT method to an established, validated, analytical lab-based method as a reference. The PAT Guidance from the FDA in fact indicates that a test-to-test comparison may be required when implementing a new on-line process analyzer.

PAT opens up a new approach to validation—it provides real-time data to examine and validate process changes within the process itself.

CONCLUSION

This chapter has discussed retrospective validation in relation to both drug substance and product. There are a number of prerequisites which must be in place for a process to be a candidate for retrospective validation. In addition to these prerequisites, there is an underlying set of elements that must be in place as a foundation to assure a solid ground for retrospective validation.

It also includes examples of test cases for both drug substance and product retrospective validation. These test cases utilized the prerequisites and elements as criteria for acceptance. Unless there are solid data sets, foundation and rationale retrospective validation is not always justified as an approach. In many cases, prospective or concurrent validation is a better approach. All of these considerations must be taken into account to pursue this avenue for validation.

A brief examination of process and equipment validation has been provided. The same considerations apply in these instances. Finally, the chapter has explored retrospective validation in regards to a PAT environment. This area provides so much data that the traditional approach does not apply.

Overall, retrospective validation has a number of risks that must be carefully considered for the particular product, history, and prerequisites described previously. If a solid foundation does not exist it may not be the best approach. This preevaluation will allow the risks to be determined and made the validation approach decision based on all the data.

REFERENCES

1. International Conference on Harmonization. Good Manufacturing Practice Guide for Active Pharmaceutical Ingredients Q7A. (www.ich.org/cache/compo/363-272-1.html)
2. Berry IR, Nash RA, eds. Pharmaceutical Process Validation. New York: Dekker, 1993:252 (chap. 8).
3. Ask the FDA: Questions and Answers with Current and Former Agency Personnel. Institute of Validation Technology, website, April 20, 2001.
4. Barr D, Celeste A, Fish R, Schwemer W. Application of Pharmaceutical cGMPs. Washington, DC: Food and Drug Law Institute, 2003.
5. Current Good Manufacturing Practice Regulations for Finished Pharmaceuticals, 21 CFR Parts 210 and 211; Good Manufacturing Practice Regulations for Medical Devices, 21 CFR Part 820.

6. U.S. Food and Drug Administration. Guidelines on General Principles of Process Validation, May 1987.
7. Medicines Control Agency. Rules and Guidance for Pharmaceutical Manufacturers and Distributors 2002. Published for the MCA under license from the Controller of Her Majesty's Stationery Office. St. Clements House, 2-16 Colegate, Norwich NR3 1BQ, sixth edition, 2002.
8. Miller JC, Miller JN. Statistics for Analytical Chemists. 2nd ed. Chicheser, U.K.: Ellis Horwood Ltd, 1988:53–77.
9. FDA Guide to Inspection of High Purity Water Systems. (www.fda.gov/ora/inspect_ref/igs/high.html)
10. FDA Guidance for Industry Sterile Drug Products Produced by Aseptic Processing—Current Good Manufacturing Practice. U.S. Department of Health and Human Services, Food and Drug Administration Center for Drug Evaluation and Research (CDER), Center for Biologics Evaluation Research (CBER), Office of Regulatory Affairs (ORA). (www.fda.gov/cder/guidance/5882fnl.htm).
11. PAT Guidance for Industry—A framework for Innovative Pharmaceutical Development, Manufacturing and Quality Assurance. U.S. Department of Health and Human Services, Food and Drug Administration (FDA). Center for Veterinary Medicine (CVM), Office of Regulatory Affairs (ORA), Pharmaceutical cGMPs; September 2004.
12. Trubinski CJ. Retrospective process validation. In Wachter A, Nash R. Pharmaceutical Process Validation, 3 edt, 2003.

43

Validation and Six Sigma

Robert S. Bottome
Genentech, Inc., San Francisco, California, U.S.A.

INTRODUCTION

In spite of the demonstrable value of Six Sigma and Lean Manufacturing programs, there remains considerable skepticism within the ranks of pharmaceutical engineers and quality professionals about their validity. At the same time, many process improvement professionals attempting to use these tools and methods to reduce variability and increase process efficiencies are surprised by how often they are confronted with resistance centered on outmoded notions of what it means to validate pharmaceutical processes. Often the very fact that the process has been validated is used to justify resistance to process improvement proposals.

Fortunately, skepticism and resistance are fading as the value of "Design for Manufacturability" methods becomes clearer. Also known as "Design for Six Sigma," this approach relies in part on the skillful use of DOE to reveal which variables are robust in the face of variation across specification limits and which variables need to be carefully controlled. As a result, the validation process can be used to define a quantitative "sweet spot" that can serve as the basis for meaningful control charts and set-up rituals that significantly reduce process variability.

Definitions

The term *Six Sigma* has come to mean many things, but in this chapter it refers to a methodology and an associated set of tools for reducing process variability. In fact, Six Sigma is merely the latest iteration in an evolving science of operations management that traces its roots to Taylor (1) and his initial effort to define labor standards, Deming's application of Shewhart's tools for distinguishing systemic variability from special causes (2), Juran's application of the Pareto principle to data segregation (3), and Crosby's arguments against end-of-pipe remedies (4). As currently understood, Six Sigma methods incorporate the battle-tested fundamentals of the quality movement with a set of updated statistical tools and concepts: these include the idea that a process that operates at a "six sigma" level of reliability only produces 3.4 defects per million opportunities (5). Essentially, Six Sigma methods take a practical problem and translate it into a statistical problem so that a set of statistical solutions can be derived and then translated back into a set of practical solutions.

More importantly, Six Sigma as a management philosophy has helped to solidify most of the key concepts that have characterized the quality movement across the 20th century. For example, at its essential core, Six Sigma is about applying rigorous analysis to distinguish the critical few problems from the trivial many, so that the most talented people can analyze them using the best available tools and techniques before solutions are defined.

Lean manufacturing overlaps with Six Sigma to the degree that greater throughput and process efficiency are achieved by reducing variability. The Toyota production method, for example, consists in a large part of an interrelated set of tools and techniques for reducing variability through standardization, smoothing and simplification.

Consequently, contemporary "Lean Sigma" exponents are likely to be passionate advocates of:

- Employee empowerment and driving decision-making close to the problem
- Continuous incremental improvements with occasional breakthrough reengineering
- Reducing variability: convert bell-shaped curves into needles!
- Using financial analysis to prioritize and measure value added
- A focus on cycle time compression as a means to reduce variability, drive efficiency, enable responsiveness to shifting demand and provide tighter feedback loops which increase first-time-through conformance to requirements
- Emphasis on Voice of the Customer, converting customer preferences and needs into critical-to-quality attributes (recognizing that the customer may be internal to the organization)
- A belief that inductive statistics and end-of-pipe monitoring are wasteful; a preference for 100% inspection of single part production flows where handoffs are predicated on acceptance and verification of conformance to quality
- Regular and routine application of a variety of statistical tools

The process improvement professional working with the pharmaceutical industry will seek to optimize the process within validated ranges and the terms of the license. They look for gains to be had in the transactional environment outside the validated envelope. Rarely are

Abbreviations used in this chapter: ANOVA, Analysis of variation; cGMP, current good manufacturing practice; CTQs, critical to quality attributes; DMADV, Define–Measure–Analyze–Design–Verify; DOE, design of experiment; FDA, Food and Drug Administration; GAMP, good automated manufacturing practice; PAT, process analytical technology; QA, quality assurance; QC, quality control; USP, *United States Pharmacopeia*.

the tools and methods applied to streamline and enhance the validation process itself.

VARIETIES OF RESISTANCE TO LEAN METHODS AND SIX SIGMA

Early efforts to apply lean and six sigma methods to pharmaceutical processes almost immediately ran into objections predicated on a misunderstanding of the true role of validation. A process could not be improved since improvement implies a change that would take the process out of its validated state. At the same time, it was not unusual in the industry to accept anomalous results or process variability in a validated process as long as the output met all of the specifications. One of the more interesting side effects of the traditional approach has been to validate broad ranges or windows where possible and then seek to optimize yield or other attributes within these ranges. There has also been a misplaced concern that application of modern process improvement methods to pharmaceutical processes would earn the disapproving attention of the FDA.

The attitude of an un-reconstructed, reactionary pharmaceutical engineer can be characterized by some of the following statements:

- If it meets cGMPs, USP and specs it must be OK. Ship it.
- Do it "by the SOP"; layered double sign offs and multiple hand-offs provide better control; employee empowerment is not GMP.
- FDA approval is needed for all but the most trivial process changes. The process is set in stone—once validated, it no longer requires further improvement.
- Manufacturing costs are small when compared with development investment; FDA does not care about cost so financial analyses are irrelevant.
- Who is the customer? FDA? The Physician? Pharmacists? Patient? More likely to design validation based on what we hope the Agency will accept or approve and not on what increases manufacturability.
  - cGMPs require QA/QC oversight infrastructure so attempts to streamline or eliminate redundant review or end-of-pipe inspection violate the regulations.
  - Control charts and other tools are rarely applied since cGMPs do not require them.

Fortunately, the agency has been quite vocal and explicit in its support of six sigma tools and methods. Consider these excerpts from a recent FDA guidance on PAT (6):

"Gains in quality, safety and/or efficiency will... likely come from reducing cycle times...preventing rejects, scrap, and re-processing...and manage variability."

"...these concepts are applicable to all manufacturing situations (1)."

"In a PAT framework, process validation can be enhanced and possibly consist of continuous quality assurance where a process is continually monitored, evaluated, and adjusted using validated in-process measurements, tests, controls, and process endpoints."

"Continuous learning through data collection and analysis over the life cycle of a product is important."

"A process is generally considered well understood when:

1. All critical sources of variability are identified and explained;
2. Variability is managed by the process; and
3. Product quality attributes can be accurately and reliably predicted over the ranges of acceptable criteria..."

"The ability to predict reflects a high degree of process understanding."

To summarize, the traditional approach to validation has been to accept variability in process inputs within specification limits on raw materials and utilities, fix or lock the process and accept a process output that may not be capable of consistently meeting release criteria (for example, a validated process may only yield 70% acceptable material and require rejection of the remaining 30%, an approximate one sigma level of reliability).

The intent of PAT is to:

1. Accept and measure input variability (know which variables are worth monitoring);
2. Use the feed-forward indicators derived from these measures to make control system adjustments at critical control points (temperature, humidity, pH, etc.);
3. Measure process variability and use the feed-forward indicators derived from these measures to inform control system adjustments downstream; and
4. Such that the output itself is fixed and controlled, thereby achieving levels of reliability that closely approximate six sigma quality.

This fundamental shift in control philosophy can enable parametric, concurrent lot release, which would increase our industry's ability to turn inventory (currently at a sclerotic 1.2 turns per year), and so should translate into an ability to reduce cost of production.

WHEN VALIDATION IS AN OBSTACLE

The scope and approach to validation has clearly seen a sustained burst of essentially out of control expansion from what began as an attempt to establish control over critical safety issues only, into a validated envelope that now extends into unexpected areas. For example, modifications to a waste-water handling system can trigger revalidation of process equipment.

Instead of asking what is the duly diligent standard of care, validation engineers are often left implementing a work plan that reflects the maximum that can be done given the constraints of time and resources.

As mentioned above, the delusion that a "validated" system has had its variability reduced or its parameters optimized can be a real obstacle to necessary improvement.

Finally, the most common manifestation of validation as an obstacle comprises the bottlenecks that are created by constraints on resources. Although protocol development and testing can be outsourced to third parties or equipment vendors, getting these approved and signed off can be an ordeal. The review and approval of draft protocols and testing results ends up generally limited to a handful of over-utilized individuals in most management systems. As a result, there is virtually no early or iterative review of projects as they take shape, so few opportunities to guide the design in the most robust direction can be seized.

THE VALIDATION STEP VS. THE VALIDATION PROCESS

Traditionally, the validation "step" gets crammed into a dwindling slice of time towards the end of a project after mechanical completion. Given the enormous constraints, most teams are content to take three runs at a target. A great example from industry of missed opportunities to build robust specifications comes from a filling validation effort that used three runs from the same bulk material. Engineers are characteristically driven primarily by the desire to deliver conforming qualifying lots, and the pressure to hit a narrow target. There is little incentive and less luxury to try and broaden the acceptable range for critical parameters when every week of delay translates into lost market share. Since validation is often the last step it often pays the price for upstream delays with compression pressure.

As a result, the process as transferred is not robust and can suffer from serious operability issues. While it is true that each significant excursion triggers an investigation, with an array of associated experimentation and testing to verify no product impact, and that as this data accumulates it may be possible to build robustness into a process, no one would argue that this is the efficient, effective or desirable approach (Fig. 1).

Most validation professionals will talk wistfully of the idea of a validation process that begins early in the life cycle of a product. This process would be characterized by systematic and extensive experimentation to define main effects. The key point that is often either lost or misunderstood is this: application of statistically valid DOE is the only way to explore two- and three-factor interactions and to fully illuminate cause-and-effect relationships.

The validation process that enables full exploitation of PAT is one characterized by the skillful use of "Design for Six Sigma" tools and concepts. By fully leveraging every opportunity for upstream testing and experimentation on the process, we can design an efficient, culminating "demonstration of robustness" that is suitable for submittal and that enables approval. Such a demonstration would presumably require less time and fewer resources since it can be designed to focus on the narrow set of variables that are not robust across the established range of values. The validation "step" itself becomes a wafer-thin element in the final qualification of the process.

HARNESSING THE FULL POWER OF THE VALIDATION PROCESS

Design for Six Sigma is a smarter approach that is based largely on full application of GAMP model: clear requirements, specifications, and life cycle documents created BEFORE validation. More specifically, the process assumes that the developer will:

1. Establish CTQs using DMADV techniques, including rigorous voice of the customer (broadly defined includes operations) assessments.
2. Use DOE methodology (e.g. factorial design) to translate these CTQs into specifications and control parameters by illuminating the cause and effect relationships between input variability and process control settings.
3. "Design for Robustness" implies that the experiments are designed to create a performance space/surface ("sweet spot") that can be used to choose the proper input settings to achieve desired targets with minimum variability (to minimize the effects of noise on the main effects), and as the basis for ongoing statistical process control on the floor, not just to obtain licensure.

Traditional "one-factor-at-a-time" experiments are ineffective and inefficient, even when conducted early in the development process. One-factor-at-a-time repeated three times will not enable PAT or process optimization.

An example of a traditional set of three experiments is provided below (7). The process under development comprises intermingling of a filler and a resin at temperature. The outcome or desired effect is curing time. As depicted in Figure 2, a set of three experiments where only one factor varies at a time, will ensure that the results

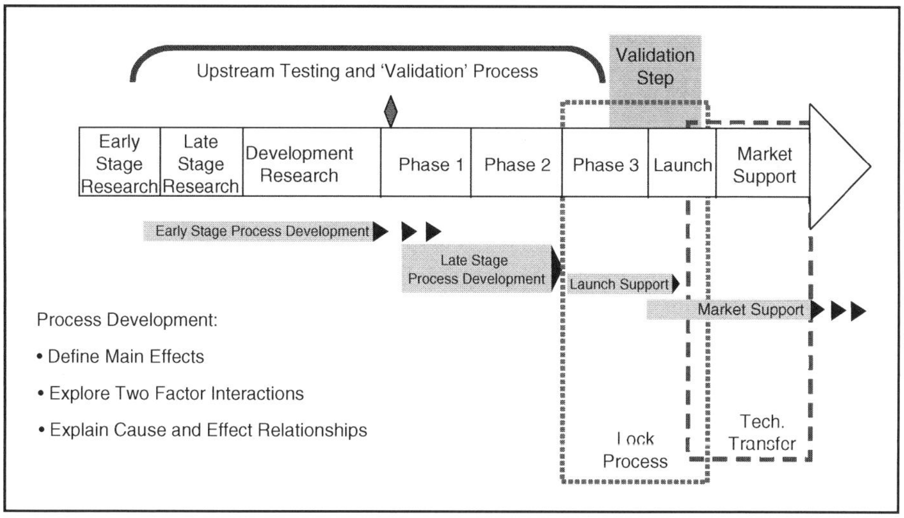

Figure 1 Generic process development timeline showing the relationship between the validation process and the validation step.

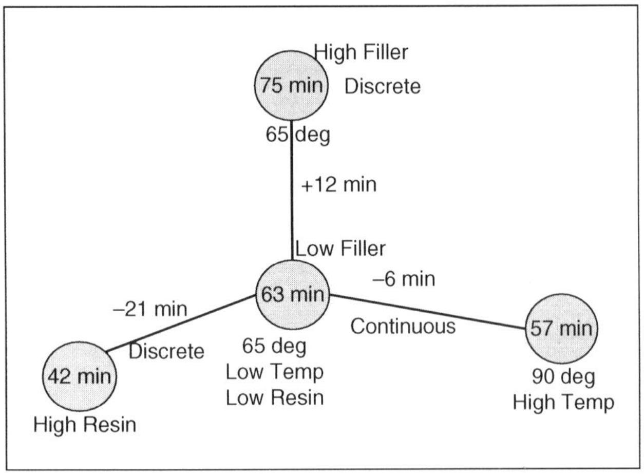

Figure 2 Traditional, one-factor-at-a-time approach to validation.

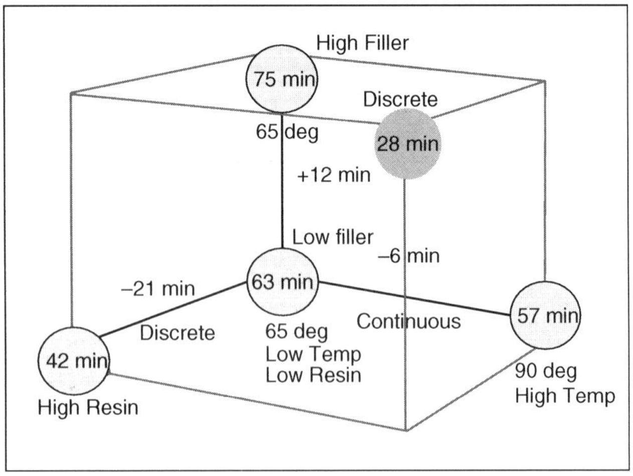

Figure 3 A fractional factorial design with results "out in the cube."

illuminate effects along the axis. None of them reach "out in the cube."

As the FDA points out in their PAT Guidance document, "Traditional one-factor-at-a-time experiments do not effectively address interactions between product and process variables." The alternative path is to design experiments using the same time and resources but apply the statistical principles first applied by Fisher (8) and then adapted to industrial uses by Box, Hunter and Hunter (9) and others, commonly referred to as DOE of factorial experiments. Quoting once more from the FDA's PAT guidance document:

> Methodological experiments (e.g., factorial experiments) based on statistical principles of orthogonality, reference distribution, and randomization provide effective means for identifying and studying the effect and interaction of product and process variables.

A DOE allows us to estimate the performance of a system within its operating space (the cube) without having to measure system response at every point in that space. Using the curing time system described above, and assuming that the effect is approximately linear (non-linear systems can be tested using different techniques), we can create an equation that describes all of the effects of all the factors on curing time (the response): the effects of the factors by themselves (the three main effects), the three two-way interactions, and one three-way interaction. Since we need to include the intercept of the linear equation, we end up with eight coefficients that need to be estimated with eight experimental runs. Since the system is approximately linear, we can assign high and low values to our three factors (−1 for low, and +1 for high) and derive the following design (Table 1):

Table 1 A Three-Factor, Two-Level Experiment

Run (Yates' name)	Temperature	Resin type	Filler type
(1)	−1	−1	−1
A	+1	−1	−1
B	−1	+1	−1
Ab	+1	+1	−1
C	−1	−1	+1
Ac	+1	−1	+1
Bc	−1	+1	+1
Abc	+1	+1	+1

This experimental design would be referred to as a "full factorial" since it contains all the possible combinations of the three factors ($2\times2\times2=2^3=8$ unique combinations, corresponding to the eight corners of our cube).

Having verified the capability of the measurement system, the engineer is ready to run the eight experiments. However, it is often frugal and wise to start with a "fractional factorial," especially if we have evidence that some effects do not need testing (three-way interactions, in non-biological systems at least, are rare). In the curing time system, four experiments would suffice if, as depicted in Figure 3, we substitute the High Filler/High Resin/High Temp experiment (out in the cube) for the Low Filler/Low Temp/Low Resin experiment (which was the origin of the cube we had already explored).

As it happens, High Filler/High Resin/High Temp results in a significant (28 minutes) impact on curing time. By running only four out of the eight runs we achieve a 2^{3-1} fractional factorial design (the "−1" indicates that we are only doing half of $2\times2\times2=8$).

An example of a 2^3 Full Factorial Design for developing a tablet for ingestion is shown in Figure 4 below. In this case, the desired effect is a target percent of active ingredient bound after a certain time after simulated ingestion and there is no obvious opportunity to be frugal and get away with a fractional design. The three variables in this case (10) are plasticizer (A or B), total spray time, and dry time (one discrete and two continuous variables).

The very model of the modern validation engineer applies their six sigma training by entering the design and the results into a software package (like Minitab) that can rapidly run an ANOVA to see if the differences in the effects and their coefficients are significant, and to generate main effects plots along with interaction plots.

As shown in Figure 5, total spray time is the critical process parameter that needs to be carefully controlled. The other variables can vary across the specification

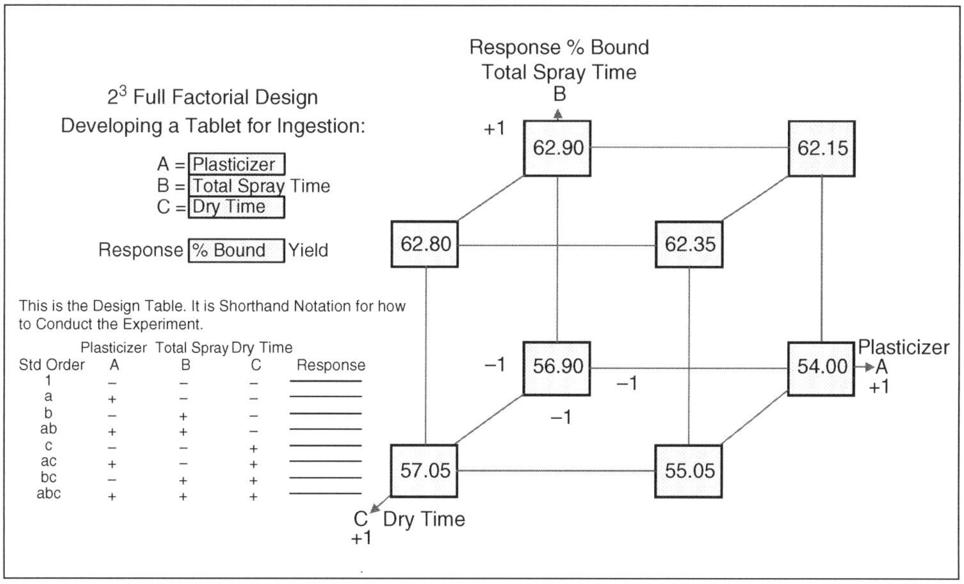

Figure 4 A full factorial design.

Always Low to High

Std Order	Plasticizer A	Total Spray B	Dry Time C	AB	AC	BC	ABC
1	–56.9	–56.9	–56.9	56.9	56.9	56.9	–56.9
a	54	–54	–54	–54	–54	54	54
b	–62.9	62.9	–62.9	–62.9	62.9	–62.9	62.9
ab	62.15	62.15	–62.15	62.15	–62.15	–62.15	–62.15
c	–57.05	–57.05	57.05	57.05	–57.05	–57.05	57.05
ac	55.05	–55.05	55.05	–55.05	55.05	–55.05	–55.05
bc	–62.8	62.8	62.8	–62.8	–62.8	62.8	–62.8
abc	62.35	62.35	62.35	62.35	62.35	62.35	62.35
Sum	–6.1	27.2	1.3	3.7	1.2	–1.1	–0.6
Effect	–1.525	6.8	0.325	0.925	0.3	–0.275	–0.15
% Effect	–2.578191	11.496196	0.5494505	1.5638208	0.5071851	–0.46492	–0.253593

Note: The % effect is the effect divided by the average × 100.

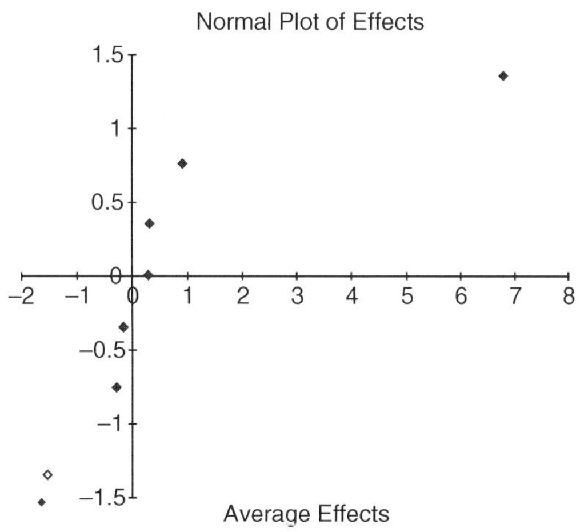

Figure 5 Results of a 2^3 full factorial design showing that total spray time is the critical process parameter.

ranges and the response is still within its acceptance criteria (the system can be said to be robust relative to those variables).

By designing the experiments performed during the development of the process in this way, the opportunity to transfer a process that is truly robust can be seized. The parameters can be set at the optimum values during the "demonstration of robustness" but somewhat beyond the understood limits during development.

The primary benefit of the approach lies in the ability it confers to distinguish "signal"—which should be carefully validated and controlled–from "noise"—which can be cut from the scope of validation. By control charting the signal, or critical, parameters, wasteful investigations that conclude with a finding of "no product impact" can be avoided, freeing up constrained technical resources to concentrate on optimizing yield and stability.

As summarized by the FDA in its PAT guidance, "When used appropriately, the (factorial experiments and other) tools described above can help identify and evaluate product and process variables that may be critical to product quality and performance." Once these variables have been identified and their interactions with each other understood as necessary, we can create a clear performance space for reliable and routine operations.

ACKNOWLEDGMENTS

Thanks to W. Rob Arathoon, Ken Kornfield, Ron Branning, Harry Lam, and Mr. Lynn D. Torbeck.

REFERENCES

1. Taylor FW. Scientific Management—Comprising Shop Management, The principles of Scientific Management and Testimony before the Special House Committee. New York: Harper and Row, 1964.
2. Deming WE. Out of the Crisis: Quality, Productivity and Competitive Position. Cambridge, MA: MIT Center for Advanced Engineering, 1982.
3. Juran JM. Juran's Quality Handbook. 5th ed. New York: McGraw-Hill, 1998.
4. Crosby P. Quality is Free. New York: McGraw-Hill, 1979.
5. Breyfogle FW. Implementing Six Sigma: Smarter Solutions using Statistical Methods. New York: Wiley, 1999.
6. FDA. Guidance for Industry: PAT—A Framework for Innovative Pharmaceutical Manufacturing and Quality Control, August 2003.
7. Torbeck and Assoc., Training course "Validation by Design®".
8. Fisher RA. The Design of Experiments. Edinburgh, U.K.: Oliver and Boyd Ltd., 1951.
9. Box G, Hunter W, Hunter S. Statistics for Experiments. New York: Wiley, 1978.
10. Torbeck LD, Branning RC. Designed experiments—a vital role in validation. Pharm Technol 1996; 20:108–14.

44

Validation and Contract Manufacturing

Michael C. Beckloff
Beckloff Associates, A Cardinal Health Company, Overland Park, Kansas, U.S.A.

INTRODUCTION

For a large percentage of pharmaceutical and biotechnology industry professionals, outsourcing of contract manufacturing services has become a common occurrence—essentially, a "way of life." One would be hard pressed to find a seasoned industry veteran who has not been involved in outsourcing of contract manufacturing services at some point in their career. The war stories are varied and span the full spectrum of results, from complete success to complete failure with most falling somewhere in between.

Outsourcing of contract manufacturing services will continue to be an integral component of strategic drug development, allowing industry to effectively deal with capacity and timing issues it faces while struggling to find pathways to introduce new drug products to the market by more efficient, effective, and cost-competitive means. Contract manufacturing is utilized by all sizes of companies. From big pharma and big biotech, to generic houses, to tiny virtual companies, every size and shape company imaginable uses contract manufacturing as a means to get products to market. Outsourcing of contract manufacturing services occurs everywhere and for many different reasons. Domestically and internationally, contract manufacturing and process validation of the drug substances and drug products that they manufacture will continue to be an issue at the forefront of the pharmaceutical industry. Industry professionals had better get comfortable with being uncomfortable with how best to use contract manufacturing effectively.

Successful validation of pharmaceutical processes within the contract manufacturing environment has a set of unique and complex challenges that intracompany programs do not face. As is true for all successful programs, validation of pharmaceutical processes via CMOs requires considerable planning and execution. Unfortunately, there are very few, if any, contract manufacturing "turnkey" operations that allow one to return when the product and process are fully validated, in the "box," launched, and ready to ship to the pharmacy. Successful contract manufacturing programs require full involvement and full participation on the part of the sponsor. After all, a contract manufacturer is just that—a contract manufacturer. The contract manufacturing industry is much more adept and efficient at exploiting a well-defined transferable process as compared to innovating a product and process that is poorly defined and characterized.

The critical success factors required when validating a pharmaceutical process using a contract manufacturing strategy are many and complex. Without exception, successful process validation begins with excellent product and process development. There is no substitute for a rugged, well-characterized, and scientifically sound product and process. If the goal is to "get it done" with a contractor as soon as possible, a properly developed process will allow for successful transfer, scale-up, and process validation with the least amount of starts, stops and cost overruns. Sounds simple, and conceptually it is, but time and time again drug development programs are hindered by inadequate scientific design, lack of sufficient data, poorly executed technology transfer, and ultimately delayed or failed process validation. As the cliché goes, it seems there is never time to do what is necessary to get it right the first time, but there is always time to repeat it.

Other critical success factors include selection and qualification of the contractor, selection and management of the development team, planning and program management, and effective and efficient management of the regulatory requirements that drive and control the entire pharmaceutical and biopharmaceutical industries.

PRODUCT AND PROCESS DEVELOPMENT

The benefit-to-cost ratio must always be considered and as obvious as it would seem, it is important to recognize that validation of pharmaceutical processes begins at the beginning. Successful validation hinges on a complete and thorough understanding of the product and the production process. Preformulation work, analytical methods development, production of Phases 1 to 3 CTM, scale-up, production of engineering batches, production of registration batches, and finally production of process validation batches, must all "add up." A large body of meaningful data is generated during the product development process. Continuity, coupled with recognition and analysis of existing data, is critical to successful process validation. These data must be managed and properly communicated to the contract

---

Abbreviations used in this chapter: AA, amino acid; API, active pharmaceutical ingredient; CAS, chemical abstract service; CBE, changes being effected; cGMP, current good manufacturing practice; CMC, chemistry, manufacturing, and controls; CMO, contract manufacturing organization; CTM, clinical trial material; DMF, drug master file; FDA, Food and Drug Administration; GMP, good manufacturing practice; IND, investigational new drug; NDA, new drug application; PAI, preapproval inspection; SOP, standard operating procedure.

manufacturer. Conversely, data generated by the CMOs must be properly communicated to and analyzed by the sponsor. It is the job of the sponsor to effectively manage and guide technology transfer from the development laboratory to the CMOs.

Successful validation is constructed as part of a cumulative progression coordinated by the sponsor based upon what is learned as the product undergoes development. Process validation cannot be completely "delegated" to a CMO; it must be effectively managed by the sponsor. It is the sponsor's job, not the CMO's, to ensure that all of the necessary data have been generated to support technical transfer, scale-up, and ultimately process validation. Without proper planning, nothing "good" happens in process validation.

A challenge which often occurs in many of the sponsor companies is frequent "midstream" change in development team personnel. When such personnel changes occur, the new sponsor team can become frustrated with a particular contractor and may move to another, believing this will solve all of the problems. In reality, this results in loss of continuity, loss of time, and cost overruns. This change of personnel often occurs in small and virtual companies. If a proper process validation plan is developed early in the program, loss of continuity can be prevented.

The following tables, derived from numerous FDA guidance documents (1–12), etc., summarize the scientific and regulatory expectations for new chemical entities, both APIs and finished dosage forms, from early Phase 1 through approval of the registration application. Of course, each product is unique with a different set of specific requirements; however, these tables provide a point of reference and a good place to begin to plan and develop a well-defined and characterized product and process. To a large extent, efficient product and process development is about understanding minimum requirements both from scientific and from regulatory points of view. Minimum requirements should not be confused with minimum effort. Successful process validation in the CMO environment is much more about understanding what is "mission critical" with respect to the product and process—what studies need to be designed and executed, and how the data will be positioned, from a regulatory point of view, as the product proceeds from the pre-IND stage through clinical development, process validation, and finally to the commercial marketplace. The process is complex, intertwined, and very much interdependent. Successful development teams understand this interdependence and consider the product and process development, analytical methods development, and regulatory strategy in parallel, resulting in an efficient and effective development and process validation plan.

THE "RIGHT" PEOPLE IN THE "RIGHT" JOBS AND THE "RIGHT" CONSULTING EXPERTISE

At the larger pharmaceutical and biotech companies, the knowledge, experience, and the know-how of managing a process validation program at a CMO are usually in place at the upper levels of the management pyramid. The actual implementation of the program is left to lower level and oftentimes inexperienced personnel. The personnel at these levels frequently struggle with the CMO process validation programs. In part, this is because their assignment to the project serves as a training mechanism to allow these personnel to obtain necessary experience and expertise, and also, because CMO validation programs can be extremely challenging and consume significant amounts of time and effort—commodities that the upper level managers generally do not have available, given the other programs and projects they must direct. With proper oversight from experienced supervisory management, the job of validating a process at a CMO is usually accomplished more effectively and efficiently by these large companies as compared to the small, less-experienced companies, despite the assignment of junior level personnel for many, if not all of the required tasks.

An important and distinguishing feature between the large and the small companies is that the large company personnel, although somewhat inexperienced, usually have the appropriate educational and scientific expertise available within their own organization to assist, which may allow them to better understand the validation requirements and principles. The smaller companies may not have sufficient internal technical expertise to properly direct the CMO process validation programs. This is further complicated by the fact that the upper levels of management in some small companies may also be lacking in these areas, resulting in very limited guidance. The rule, rather than exception, for small and virtual companies seems to be to expect people to run external CMO programs for which they have very little or no experience and/or training. For example, the staff toxicologist is asked to oversee the development, manufacturing and process validation programs; the director of clinical development is asked to handle manufacturing; and the staff project manager is asked to oversee the product development and manufacturing plan. Although it might seem counterproductive, this practice frequently occurs and forces small companies into the difficult position of being more dependent on the CMOs for technical support and guidance. For these reasons, small pharmaceutical and biotech companies generally have more difficulty in managing and directing a process validation program at a CMO. Small companies scramble to conserve cash and raise funding; the clinical work is the glamorous "big bucks" Wall Street stuff; manufacturing and validation is usually an afterthought and after all…"how hard can it be"? The answer is, of course, hard enough!

Having the "right people" in the "right jobs" during a CMO process validation effort is critical to the overall success and efficiency of the program. Establishing a program director with the experience and training to specifically "direct" the project as compared to using a less experienced project manager to "manage tasks" is critical to designing and executing an efficient and effective process validation program. There is a significant value in having someone who understands product development and process validation directing the program as compared to using a project manager that may or may not have the necessary experience. Many projects end up being driven by inexperienced project managers managing task lists and timelines via Gantt charts rather than understanding the underlying science and regulatory requirements that should be driving the project. This practice leads to many inefficiencies and cost overruns. The best program director is one who understands the specific production process, the critical parameters that must be evaluated during process validation, and the associated regulatory requirements that can ensure product registration and approval.

Finely tuned project teams are critical to successful product development. In the contract manufacturing environment, teams are divided into two distinct groups; first, the sponsor's development team and second, the contract manufacturer's project team. The secret to success is to get the two teams to integrate and function as a single unit with the same goals and objectives. It is important for the sponsor and CMOs to develop an attitude of cooperative partnership rather than one of slave and master. Sponsors must understand that the contract manufacturing business is a very competitive and high-overhead business model. To survive, CMOs must operate on the volume of multiple products and multiple customers. The sponsor must appreciate this and understand that they are not the only customer the CMOs must satisfy. Conversely, the CMOs must understand that the sponsor has put a huge amount of trust and expense into the experience, expertise, and integrity of the CMOs—in effect putting their "lives and jobs" in the hands of the CMOs. It is a very interesting dynamic, but once again, having the goal of developing a cooperative partnership rather than one of slave and master should be considered as a critical success factor.

Clearly, an "A-Team" is needed for all projects. What does an "A-Team" look like? On the sponsor's side it is likely that the project will be high profile within the company; it will very likely be a high priority; and, as in the case of the small and virtual companies, may be the ONLY project on which the company is working and the ONLY way one company will survive. On the contractor's side, the sponsor's product will likely be only one of many projects on which the CMOs is working. The sponsor "ponders" its only project as the CMOs "ponders" how to get the next customer's product and project in and out the door. Ideally, the "A-Team" for a project would consist of the sponsor's project director, as well as a key decision maker from the CMO's side. In addition, it is helpful to have representation from both the sponsor and the CMOs in the areas of project management, synthesis and/or pharmaceutics, manufacturing, analytics, and regulatory. It is imperative to have the correct and complementary expertise on the sponsor's side as well as the CMO's side. Successful process validation is driven by proper and effective science. There must be proper technical expertise within the team on both sides. Finally, it is critical that the sponsor establish a person-in-the-plant for the engineering registration and process validation runs and probably the first several commercial production runs. Murphy's Law was apparently designed for process validation activities in the CMO environment. Having a person-in-the plant will help to raise the awareness of the CMOs as process validation is initiated and will allow for more precise and faster response times should problems arise during the validation runs.

Regular meetings are critical to monitor progress, assign actions, and troubleshoot problems. Teleconferences and videoconferences are useful and cost-effective tools, but regular face-to-face meetings are very valuable and should not be overlooked or avoided. Face-to-face meetings also allow for overall team building and aids in helping to develop the cooperative partnership necessary for a successful sponsor/CMOs relationship.

Because small and virtual companies may not have the expertise and resources and because the large companies may need to outsource lower priority projects, it may be necessary to use consultants or consulting groups to help managing the development and process validation programs. In hiring and managing consultants, it is important to determine whether they will be able to give the project the time and attention the project will demand. Like the CMOs business model, it is important to also understand that the consulting business model is a time-based business model that requires the consultant to work on multiple projects for multiple clients. There are three primary considerations that need to be addressed when identifying the "right" consulting expertise:

- Does the consultant have the proper training to assist with the process validation program?
- Does the consultant have adequate time to devote to the process validation effort and are they willing to guarantee this via a consulting contract?
- Has the consultant actually participated in this type of work previously? Have they actually done this specific work, and if so, are they willing to provide references?

Assuming these three primary considerations are met, a fourth and extremely critical factor to consider is whether the consultant has the type of personality that will allow for effective integration into the project team. The consultant must be integrated into the overall cooperative partnership that is necessary for successful process validation. The sponsor must recognize that the consultant may, in fact, be spending more one-on-one time with the CMOs than the sponsor. The sponsor must be certain that the consultant has the necessary interpersonal skills to build and maintain the proper sponsor/CMO relationship.

SELECTION AND QUALIFICATION OF THE CONTRACT MANUFACTURER

Thorough development of the product and the process can be considered to be the single most important critical success factor associated with technology transfer to the CMOs and process validation. The second critical success factor that must be considered is the selection and qualification of the CMOs. Selection, qualification, and management of the CMOs can be divided into four primary phases, as illustrated in Figure 1.

CMO identification and selection is largely a matter of searching databases, reviewing trade journals, attending trade shows, obtaining recommendations from consultants, industry colleagues and contacts, or direct personal experience. In some cases, prior provider relationships may exist which may also serve as a source to identify an appropriate CMO. The most significant of

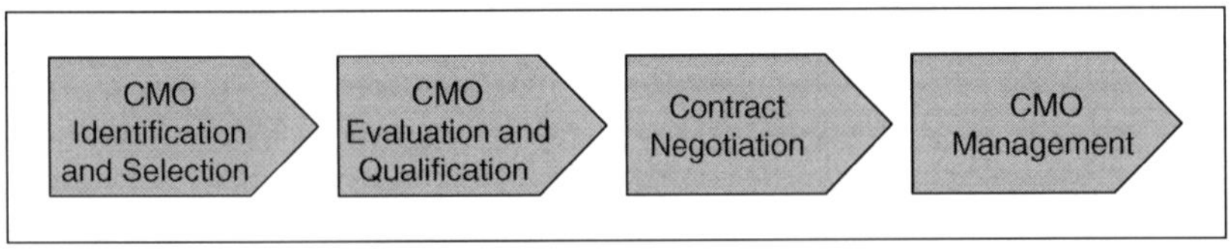

Figure 1 Primary phases of CMO selection, qualification, and management.

these is personal experience, followed by recommendations from industry colleagues—have you worked with CMO X and what was the outcome? During this preliminary screening process it will be necessary to evaluate multiple CMOs to ensure that a truly viable candidate will be identified.

Once potential CMOs have been identified, a detailed evaluation and qualification process must follow. Due diligence is often given inadequate consideration; however, the importance of this "homework" cannot be understated. Although thorough due diligence can require a significant amount of time and expense, this activity must be considered as an important investment toward successful technology transfer, scale-up, and process validation. Due diligence will pay great dividends to help ensure smooth technology transfer, scale-up, and process validation. Unless the team has personal experiences, each potential CMO should be visited by the CMO selection team as part of the due diligence effort. Some of the preliminary due diligence process can be handled using initial telephone interviews and questionnaires provided to the CMOs for completion and return. The due diligence process must include a site visit by the CMO selection team. The site visit to the CMO gives the sponsor a "real-time" chance to evaluate the CMOs and begins the process of developing a cooperative partnership.

Once a Confidentiality Agreement has been executed and appropriate technical information has been exchanged, a number of questions must be asked, answered, and considered as the selection process moves forward. A representative list of the types of questions which should be asked is provided below. Every product is unique with an individual set of requirements; therefore, the following list should not be considered all-inclusive; but rather, should be used as a guide to help the CMO selection team to think through the specific requirements for the product to undergo technology transfer, scale-up, and validation.

Due Diligence Checkpoints

- Does the CMO currently manufacture products for the commercial market?
- How many and what types of products are produced by the CMOs and what are the commercial requirements for the products produced?
- What is the current capacity of the CMOs for development projects?
- What is the current capacity of the CMOs for commercial production?
- What is the projected capacity of the CMOs for a development project at the time it will be ready to go to commercial production?
- How many customers does the CMOs currently service?
- What is the procedure by which the CMO handles and manages technology transfer and process scale-up?
- What is the CMO's track-record regarding regulatory inspections?
- What issues were identified in the last three years of regulatory inspections and what Form FDA483 observations have been cited?
- What are the typical lead times required by the CMOs?
- Does the CMO track on-time completion of projects? If so, what is the ratio or percent?
- What are the typical costing mechanisms used by the CMOs?
- Has the CMOs been involved in recall situations and if so, what were the details of the recalls?
- Does the CMOs have the capacity to schedule the project in the timeframe needed by the sponsor?
- By what procedure does the CMOs typically prepare, review, and approve process validation protocols?
- With how many successful PAIs has the CMOs been involved?
- What is the sponsor's general impression of the CMO leadership team? Do they appear to be cooperative and knowledgeable, and do they appear to have integrity?
- What is the financial stability of the CMOs?
- By what processes does the CMO handle project scope changes and cost overruns?
- How does the CMO typically handle intellectual property aspects of a contract?
- Does the contract manufacturer have the appropriate equipment and experience to allow the process and technology to be transferred?
- Does the CMO have adequately designed production facilities to accommodate the proposed manufacturing process?
- Does the CMO have adequate facilities to accommodate the packaging requirements of the product?
- Does the CMO have adequate analytical laboratory facilities, equipment, and personnel to perform the required analytical testing?
- Does the CMO have appropriately qualified scientific personnel to transfer, scale-up, and validate the process?
- Does the CMO have adequate numbers of trained staff in the production, laboratory, packaging, project management, quality assurance, and regulatory units?
- How does the CMO handle the preparation, maintenance, and submission of regulatory documentation? Are the examples available for review?
- Does the CMO produce penicillin, cephalosporin, cytotoxic, or hormone products in the facility or anywhere on the manufacturing campus?
- Are investigational and/or early development products produced by the CMOs and are these

products produced on common production equipment?

- How are investigational products evaluated, campaigned, and controlled within the facility?
- How is the change control process managed by the CMOs?
- How are Quality Agreements handled and managed by the CMOs? Is an example Quality Agreement available for review?
- Does the Quality Agreement cover the items and issues required, and meet the sponsor's needs?
- Is the CMO open to negotiation regarding the Quality Agreement?
- Is the CMO willing to issue a Certificate of Compliance for each batch produced?
- Does the CMO have experience writing development, technology transfer, scale-up, and process validation reports? Are examples available for review?
- How does the CMO handle project management? What is the typical process for project management?
- What does the CMO project and development team look like and how is it established?
- Is the CMO willing to provide three to five customer references?
- How does the CMO manage and qualify vendors for APIs, excipients, container/closures, packaging materials, and labeling components?

Initial due diligence should evaluate the above and provide the basis of a targeted matrix from which the most appropriate CMOs can be selected. Once the CMO is initially qualified, the final component of the due diligence effort requires that a detailed and thorough Quality Systems audit be conducted by a qualified cGMP auditor. Assuming the audit yields satisfactory results, contract negotiation can proceed.

Contract negotiation can be a long and arduous task. The legal challenges are many and complex, requiring significant time to complete. For this reason, assuming the initial CMO qualification is favorable, it may be beneficial to proceed with contract negotiation in parallel with some of the more detailed due diligence activities. The larger and more capable CMOs may have specific contractual requirements that may prove difficult. Identifying these issues as early as possible may save significant time should a legal impasse be reached, allowing the sponsor to "fall back" to a second qualified CMO candidate. Two critical considerations in the contractual negotiations are the willingness of the CMOs to manage intellectual property issues, and whether the CMO is willing to accept penalty clauses in the contract for project delays, quality issues, execution flaws, and potential regulatory deficiencies. Conversely, it is also advantageous to consider building incentives into the contract to reward the CMOs for meeting or exceeding timelines that meet development, validation, and commercialization goals. Building such details into the contract will help the sponsor establish accountability and allow for better performance management of the CMOs.

Managing the performance of the CMOs and subsequent process validation effort is largely the responsibility of the sponsor's project team, and in particular the program director. The program director must drive the project and make the necessary decisions. The program director must monitor contractual obligations, quality requirements, and regulatory compliance continually during the course of the project. It is critically important that the program director works to ensure that a thorough project plan with detailed timelines, metrics, regular project meetings, and detailed action assignments is paramount to successful technology transfer, scale-up, validation, and commercialization of the sponsor's product.

REGULATORY CONSIDERATIONS AND MANAGEMENT

In addition to the cGMP requirements and regulatory expectations described earlier, there are several key regulatory issues that must be considered. Successful product approval ultimately depends on a successful FDA PAI. During this inspection, in addition to the general cGMP regulatory requirements, the FDA investigator will conduct a detailed data integrity review of the information submitted in the registration application, evaluate the product development process and report, and will thoroughly evaluate the process validation data and final approved report. A successful PAI depends on the integrated efforts of the entire project team including sponsor and CMO members. It is the program director's role to ensure that all regulatory and data requirements have been properly addressed during the course of the project. The development and process validation reports will help guide the FDA investigator during the PAI. In this regard, it is extremely important that the final reports be prepared as the development program is underway and not left for the end of the project. Regular and detailed audits of raw data as well as final reports must be conducted to ensure that nothing is overlooked. When the team is confident that all required information is in order, a PAI readiness program and audit should be considered. This activity will serve as a dress rehearsal and will give the team confidence that nothing has been overlooked.

The unfortunate reality is that successful registration and product approval is simply the beginning of a continuing road of regulatory and scientific assessment, strategy, and compliance. Following launch, it is often necessary to make changes in the process to improve efficiency or allow for additional production scale. With Annual Reports, CBE-0, CBE-30, Prior Approval Supplements, and site changes or additions, batch failures, stability failures, analytical methods changes, the CMO and process validation requirements must be constantly monitored and managed by the sponsor. In effect, changes in the process are managed using many of the same product and process development and program management techniques previously described.

TECHNICAL AND REGULATORY DOCUMENTATION

The amount of technical and regulatory documentation required for new drug product approval is extremely comprehensive. A new drug product often requires 8 to 12 years for development and regulatory approval.

The data generated over the course of development and commercialization are enormous. Evaluating and managing these vast amounts of data can be a very challenging proposition even in the context of a product developed using internal resources. The management and control of the associated documentation becomes even more complex when a CMO is entered as a variable in the drug development equation. The experience, document control systems, change control systems, and quality systems of each CMO largely dictate what the CMO's capabilities are with respect to technical and regulatory documentation. The smaller CMOs may be somewhat limited and inexperienced in handling documentation which requires much more oversight and guidance, while the larger CMOs can accommodate more complex documentation issues via more mature technical and regulatory management systems.

Critical validation documentation can be divided into the following broad categories:

- Facilities Qualification and Validation
- Production, Packaging, and Laboratory Equipment Qualification and Validation
- Computer Systems Validation
- Cleaning Validation
- Production Process Validation
- Packaging Process Validation
- Analytical Methods Validation
- Sterile Product Sterility Validation

The adequacy of regulatory compliance and supporting validation documentation for facilities, equipment, and computer systems should be handled by thorough cGMP audits of the CMO. Many times, this is assessed during the initial due diligence phase of CMO evaluation. Following the initial cGMP evaluation, ongoing evaluation of these systems can be managed through a well-designed, CMO change-control system.

One of the most complicated CMO technical and regulatory documentation challenges deals with customer confidentiality issues. In the vast majority of cases, the CMO is bound by confidentiality agreements that prevent disclosure of technical scientific and validation data to third parties. This confidentiality restriction is most apparent in the area of cleaning validation and extends to the chemical identity of compounds, specific processes, and operating parameters, as well as the specifics of the analytical methods used to evaluate the effectiveness of cleaning procedures. Assuming that confidentiality must be maintained, and multiple drug products for multiple CMO customers are manufactured using common production equipment, it is also always impossible to confirm that the CMO's cleaning validation program and data are acceptable. The product-specific cleaning validation protocols and reports simply cannot be provided for review by the CMOs due to confidentiality. In this situation, the only option is to rely on the CMO's experience and regulatory track record. To assess the CMO's experience and understanding of the regulatory requirements, there must be a careful review of the cleaning validation policy and the SOPs used to drive cleaning validation for all products and equipment. In effect, the specific procedures and systems used to support cleaning validation for a specific product must be extrapolated to the other products produced on the common production lines with common equipment. The assumption is that proper cleaning validation procedures and regulatory documentation will be used for all products, based on what has been done for a specific product.

Production process, packaging process, and analytical methods validation, as well as many of the components of sterility validation, are product specific. There should be no confidentiality issues with respect to full disclosure of this information by the CMOs. The CMO is often faced with data and documentation requests that are "custom" requirements based on what the sponsor believes will be needed to gain and maintain product approval or requirements that meet internal sponsor quality systems. These "customized" requirements vary greatly from sponsor to sponsor. Smaller, less-experienced sponsors may make requests that are unrealistic and problematic for the CMOs. Larger sponsors may make requests based on financial muscle and, again, may be unrealistic and problematic for the CMOs. The CMO is faced with the struggle to find a flexible systematic approach to accommodate the many different sponsor-driven requirements, while still allowing the CMOs to comply with its internal quality systems. In most cases, the content and format of the technical and regulatory documentation are a negotiation point between the sponsor and the CMO, and must meet the needs and requirements of both the sponsor and the CMO. If the needs of both groups cannot be met, the project will likely be destined for failure. Therefore, it is important for the sponsor to be fully engaged and proactive very early in the development process.

The product-specific technical data and regulatory documentation system should be designed to accommodate reviews and approvals by both the sponsor and the CMO. This will aid in ensuring that the sponsor has a complete and thorough understanding of what and how the CMO intends to meet the pertinent scientific and regulatory requirements. Meetings and discussions should be held in advance of development design and document preparation to make certain that both groups agree to the overall approach. The CMOs should then draft the relevant documentation and circulate it to the sponsor for review and comment. This cycle should be repeated until agreement is reached. The documentation can then be distributed for final review and approval by both parties. This process will minimize unexpected surprises and possible delays, which may adversely affect revenues and earning for both parties.

It is most important that the sponsor minimize changes imposed on the CMO. Specifically, scientific content is critical to the sponsor, but not necessarily the CMO's format. If the scientific and regulatory requirements are met, changes for the sake of changes should be discouraged. Minimizing such insignificant changes will assist the CMOs in meeting project timelines and regulatory requirements. In all likelihood, some changes will be required; however, if the CMO is pushed too far outside their normal practices, the sponsor will likely be introducing additional complexity into the project and, in the worst case, setting up the CMO for possible failure and subsequently delaying product approval. A well-designed Quality Agreement between the sponsor and the CMO can be used to drive the specifics regarding the preparation, review, and approval of the required scientific and regulatory documentation.

In most cases, the CMC section of a regulatory application will be prepared under the strict oversight of the sponsor. The CMC section will be written to a level of detail with which the sponsor is comfortable, based on experience and the overall submission strategy. Technical data and documentation prepared by the CMO will be used to support the overall submission. A general rule for the regulatory submission is to provide a level of detail that will give the reviewing regulatory authorities sufficient information to conduct a thorough regulatory review upon which to base product approval, yet is general enough to allow for maximum flexibility for both the sponsor and the CMO. For example, where possible ranges should be used to describe operating parameters, such as times, temperatures, etc., excessive detail (e.g., a specific temperature, a specific time, a specific piece of equipment) in the regulatory application may inadvertently result in situations where a supplemental application may result. Such a submission requires approval by regulatory authorities and time for both prepreparation and regulatory review. In extreme cases, excessive detail can result in compliance or production deviations which can be difficult to manage during cGMP inspections. It is very important that any ranges described in the regulatory application be based on actual data obtained during product development and, if necessary, validation. This is a delicate balance. Arbitrary ranges and overly general descriptions are rarely successful, and usually result in comprehensive questions from reviewing chemists and, generally, delay approval of the regulatory application.

Tables 1 and 2 list the technical and regulatory requirements for each phase of development. Many of these data will be generated by the CMOs or other outside laboratories or contractors. It is very important that the sponsor obtain complete, comprehensive, and approved reports for all development studies conducted. The sponsor should conduct audits of each of these studies to ensure data integrity between the raw data (e.g., laboratory notebooks) and those in the final reports. The sponsor should confirm that all information and data have been properly reported and evaluated. Because 8 to 12 years may be required before a product is developed and approved, and because project personnel, CMOs, contract laboratories, etc., will very likely change over this period of time, it is imperative that all data and reports be finalized and approved in a form that will allow project continuity as changes in the development team occur. A successful regulatory submission will consider and address each component outlined in the FDA Guidances, and should be considered in conjunction with the technical requirements provided in Tables 1 and 2.

SUMMARY AND CONCLUSION

Outsourcing of contract manufacturing services will continue to lead industry to effective new pathways to introduce new drug products to the market by more efficient, effective, and cost-competitive means.

Successful validation hinges on a complete and thorough understanding of the product and the production process that will be developed by or transferred to the CMOs. Preformulation work, through scale-up and validation, must all fit together in a clear, scientific, and contiguous manner. Anything less complicates the process and can delay approval. After all, it is the sponsor's job, not the CMO's, to ensure that the process is properly developed and validated.

Selection, qualification, and management of the CMOs can be divided into four primary phases: CMO identification and selection, CMO evaluation and qualification, contract negotiation, and CMO management. It is critical for the sponsor to establish a development team that includes an appropriate number of staff, as well as appropriate experience and expertise. A team with less than ideal experience will struggle to successfully complete the task on time and on budget.

Successful product approval ultimately depends on implementation of successful FDA regulatory strategies. Product development, production of CTM, registration and validation batches, as well as GMP inspections by the FDA, require significant attention to the regulatory details and cannot be underestimated. The sponsor must work closely with the CMOs to ensure all the regulatory "bases" are covered. Successful product approval may be the "easy" part of the entire drug development process. Following approval and commercialization, the product takes on a life of its own. Product approval is simply the beginning of a long road of scientific and regulatory complications, all of which must be effectively managed by the sponsor and CMO team.

A final thought, is that it is important for industry professionals to remember the fundamental overriding objective—the drugs and biologics that are under development will ultimately end up in patients. The industry's obligation is to get these products approved as quickly and efficiently as possible. Approved products allow for more funding of research and development of new and innovative products or, in the case of generic products, more affordable medicines. Industry is not doing its job if it bungles the responsibility of product development and validation. In the heat of the battle, it is easy to forget the patients; however, it is quite likely one of our loved ones will need one of the products that we helped to get approved and to the market. What is done day-in and day-out in the pharmaceutical industry is important work that can have a significant impact on a significant number of people and the quality of their lives.

ACKNOWLEDGMENTS

The author gratefully acknowledges the assistance of the following Beckloff Associates, Inc. colleagues in preparation of this chapter: Norma J. VanBunnen, Marvin J. Sack, R.Ph., Miguel de Soto-Perera, Ph.D., Gary D. Hindman, Ph.D., and Scott A. Oglesby, Ph.D.

Table 1 Scientific and Regulatory Expectations, Drug Substance

	Drug substance			
	Phase 1	Phase 2	Phase 3	NDA
General:				
Code number (CAS Registry Number)	•	•	•	•
United States Adopted Name	•	•	•	•
Chemical name(s)	•	•	•	•
Compendial name				
Common (or other) name(s)			•	•
Molecular formula	•	•	•	•
Molecular weight	•	•	•	•
Chemical structure, including stereochemistry	•	•	•	•
Appearance, color, and physical state	•	•	•	•
For proteins add:				
AA sequencing	•	•	•	•
Description (disulfide bonds, shape, subunits, number of AA residues)	•	•	•	•
Biological activity	•	•	•	•
Physicochemical properties:				
Solubility (e.g., water, ethanol, ether)	•	•	•	•
pH solubility profile		•	•	•
pK_a	•	•	•	•
Dissociation constant	•	•	•	•
Bioactivity	•	•	•	•
Partition coefficient			•	•
Hygroscopicity		•	•	•
Melting point/boiling point	•	•	•	•
X-ray diffraction/single crystal		•	•	•
Chirality/optical rotation	•	•	•	•
Refractive index			•	•
Polymorph screen/solvate/hydrate	•	•	•	•
Particle size distribution	•	•	•	•
pH of aqueous solution	•	•	•	•
For proteins add:				
Isoelectric point	•	•	•	•
Extinction coefficient/unique spectra	•	•	•	•
Biological activity	•	•	•	•
Structure elucidation:				
Elemental analysis		•	•	•
UV spectroscopy	•	•	•	•
IR spectroscopy	•	•	•	•
^{1}H NMR spectroscopy	•	•	•	•
^{13}C NMR spectroscopy	[a]	•	•	•
Mass spectrometry		•	•	•
Physicochemical characteristics (TGA, DSC, DTA, X ray, Raman, etc.)			•	•
Impurities	•	•	•	•
For proteins add:				
ELP, CEP, IEP, SEC-HPLC	•	•	•	•
Western blot	•	•	•	•
Method of manufacture:				
Name and address of manufacturer, drug establishment registration number(s)	•	•	•	•
List of critical equipment			•	•
Starting material(s)		•	•	•
Starting material(s) specifications		•	•	•
Reagents and solvents	•	•	•	•
Reagents and solvents specifications		•	•	•
Synthesis scheme	•	•	•	•
Flow diagram	•	•	•	•
Description of process/process controls		•	•	•
In-process controls/in process tests (e.g., HPLC)		•	•	•
Key and final intermediate		•	•	•
Reprocessing/reworking/recovery/regeneration			•	•
For biologics or semisynthetics add:				
Storage and transportation of intermediates	•	•	•	•
Preparation procedures (e.g., cleaning, drying)	•	•	•	•

(Continued)

Table 1 Scientific and Regulatory Expectations, Drug Substance (*Continued*)

Drug substance				
	Phase 1	**Phase 2**	**Phase 3**	**NDA**
Isolation processes	•	•	•	•
Holding times/storage conditions	•	•	•	•
Traceability procedures	•	•	•	•
Specifications (includes analytical procedures and acceptance criteria):				
Appearance	•	•	•	•
Identification (UV, IR, HPLC-Chiral)	•	•	•	•
Counter ion		•	•	•
Melting point		•	•	•
pH of aqueous solution (optional)				•
Heavy metals	•	•	•	•
Residue on ignition	•	•	•	•
Residual solvents	•	•	•	•
Water content	•	•	•	•
Microbial limits	[b]	[b]	•	•
Bacterial endotoxins	•	•	•	•
Assay	•	•	•	•
Related substances	•	•	•	•
Primary degradation product and degradation pathway			•	•
Justification of specifications				•
For proteins add:				
Specific biological activity	•	•	•	•
Purity (dimers, oxidized forms, electrophoretic)	•	•	•	•
Chiral drug substance:				
Chiral identity	•	•	•	•
Chiral assay		•	•	•
Enantiomeric impurity		•	•	•
Reference standard:				
Working (preliminary)	•	•		
Primary			•	•
Analytical methods:				
Summary	•	•	•	•
Complete description		•	•	•
Sample chromatograms	•	•	•	•
Method validation:				
Linearity	•	•	•	•
Specificity	•	•	•	•
Forced degradation:				
Acid pH		•	•	•
Basic pH		•	•	•
Heat		•	•	•
H_2O_2		•	•	•
UV light		•	•	•
Accuracy	•	•	•	•
Repeatability	•	•	•	•
Intermediate precision			•	•
Reproducibility (if needed)			•	•
Robustness:				
Mobile phase pH			•	•
Mobile phase composition			•	•
Detector wavelength			•	•
Column			•	•
Solution stability			•	•
Limit of detection[c]	•	•	•	•
Limit of quantitation[c]	•	•	•	•
Method validation package				•
Related substances:				
Method validation	•	•	•	•
Identification			•	•
Qualification			•	•
Batch analysis data (certificates of analysis):				
Toxicology study lots	•	•	•	•
Clinical study lots	•	•	•	•
Impurity profile comparison	•	•	•	•

(Continued)

Table 1 Scientific and Regulatory Expectations, Drug Substance (*Continued*)

	Drug substance			
	Phase 1	Phase 2	Phase 3	NDA
Container closure system:				
Description	•	•	•	•
Label	•	•	•	•
Specifications		•	•	•
Drug master files		•	•	•
Stability:				
Summary, shelf life, statistical analysis				•
Post-approval commitment				•
Results of stress stability to support analytical validation				•
Normal (long-term) (months)	1	2–6	12	12
Accelerated (months)	1	2–6	6	6
Photostability	•	•	•	•
Development report				
Process validation protocol				
Process validation report				

Note: FDA expectation (•, expected; , not required).
[a] May be required at Phase 1 for peptides.
[b] Reported for sterile drug substance or drug substance intended for use in sterile drug product.
[c] As applicable to type of method used.

Table 2 Scientific and Regulatory Expectations, Drug Product

	Drug product			
	Phase 1	Phase 2	Phase 3	NDA
Formulation summary:				
Product name	•	•	•	•
Dosage form	•	•	•	•
Brand name				•
Strength(s)	•	•	•	•
Excipients	•	•	•	•
Manufacturing process	•	•	•	•
Composition statement (include all components used to manufacture, regardless if not present in final product)	•	•	•	•
Package container closure system	•	•	•	•
Storage conditions	•	•	•	•
Development pharmaceutics:				
Excipient compatibility[a]	•			
Prototype formula evaluation			•	•
Formula optimization				•
Process optimization				•
Container closure evaluation			•	•
Selection of commercial formula, process, and container closure system			•	•
Container closure integrity (for sterile products)	•	•	•	•
Process scale-up			•	•
Development pharmaceutics summary				•
Product development report				
Specification for drug substances	•	•	•	•
Specifications for excipients:				
Compendial status	•	•	•	•
DMFs for noncompendial	•	•	•	•
Formula:				
Unit formula	•	•	•	•
Batch formula	•	•	•	•
Preservative effectiveness testing	•	•	•	•
Manufacturer (including packager, labeler, testing laboratories):				
Name	•	•	•	•
Address	•	•	•	•
Establishment registration numbers		•	•	•

(*Continued*)

Table 2 Scientific and Regulatory Expectations, Drug Product (*Continued*)

	Drug product			
	Phase 1	**Phase 2**	**Phase 3**	**NDA**
Identification of processing rooms and filling lines (for sterile products)	•	•	•	•
Manufacturing and packaging:				
Manufacturing process summary	•	•	•	•
Manufacturing process flowchart	•	•	•	•
Manufacturing process description		•	•	•
Process controls		•	•	•
Environmental controls				•
Movement of raw material, personnel, waste, and intermediates in and out of manufacturing areas (for protein and sterile products)				•
Potential contamination with adventitious agents (for protein products only)	•	•	•	•
In-process controls/critical processing variables/justification		•	•	•
Type of equipment for each unit operation			•	•
Process validation				
Hold-time qualification			•	•
Reprocessing/reworking	•	•	•	•
Executed batch production and control records				•
Master production and control records				•
Specification (including analytical procedures and acceptance criteria) and justification:				
Appearance	•	•	•	•
Identification	•	•	•	•
Content uniformity/fill weight	•	•	•	•
Related substances	•	•	•	•
Primary degradation product			•	•
Loss on drying	•	•	•	•
pH	•	•	•	•
Particulate matter (for parenterals)	•	•	•	•
Volume in container (for parenterals)	•	•	•	•
Viscosity	•	•	•	•
Osmolarity/osmolality	•	•	•	•
Microbial limits			•	•
Sterility	•	•	•	•
Bacterial endotoxins	•	•	•	•
Dissolution	•	•	•	•
Moisture content		•	•	•
Polymorph (if needed)			•	•
Other dosage-form-specific tests as needed			•	•
Analytical methods:				
Summary	•	•	•	•
Complete description		•	•	•
Sample chromatograms	•	•	•	•
Method validation:				
Linearity	•	•	•	•
Specificity	•	•	•	•
Forced degradation:				
Acidic pH			•	•
Basic pH			•	•
Heat			•	•
H_2O_2			•	•
UV light			•	•
Accuracy	•	•	•	•
Repeatability	•	•	•	•
Intermediate precision			•	•
Reproducibility (if needed)			•	•
Robustness:				
Mobile phase pH			•	•
Mobile phase composition			•	•
Detector wavelength			•	•
Column			•	•
Solution stability			•	•
Limit of detection[b]	•	•	•	•

(*Continued*)

Table 2 Scientific and Regulatory Expectations, Drug Product (*Continued*)

	Drug product			
	Phase 1	Phase 2	Phase 3	NDA
Limit of quantitation[b]	•	•	•	•
Dissolution sink conditions	•	•	•	•
Method validation package				•
Batch analysis data (certificates of analysis):				
Experimental batches	•			
Clinical study batches	•	•	•	•
Bioavailability batch			•	•
NDA registration batches				•
Container closure system:				
Description	•	•	•	•
Specifications			•	•
DMF letters of authorization to reference		•	•	•
Stability:				
Normal (long-term) (months)	1	2–6	12	12
Intermediate (months)				6
Accelerated (months)	1	2–6	6	6
Photostability			•	•
Thermocycling				•
Stability summary, shelf life, statistical analysis				•
Long-term stability commitment				•
Label and labeling	•	•	•	•
Environmental assessment				•

Note: FDA expectation (•, expected; , not required).
[a] Should be performed as early as possible.
[b] As applicable to type of method used.

REFERENCES

1. U.S. Food and Drug Administration. Content and Format of Investigational New Drug Applications (INDs) for Phase 1 Studies of Drugs, Including Well-Characterized, Therapeutic, Biotechnology-derived Products. Center for Drug Evaluation and Research (CDER); Center for Biologics Evaluation and Research (CBER), November 1995.
2. U.S. Food and Drug Administration. INDs for Phase 2 and Phase 3 Studies—Chemistry, Manufacturing, and Controls Information. Center for Drug Evaluation and Research (CDER), May 2003.
3. International Conference on Harmonization; Guidance on Q6A Specifications: Test Procedures and Acceptance Criteria for New Drug Substances and New Drug Products: Chemical Substances, Published in Federal Register/Vol. 65, No. 251, (Friday, December 29, 2000/Notices).
4. U.S. Food and Drug Administration. ICH Harmonized Tripartite Guideline. Q1A(R2) Stability Testing of New Drug Substances and Products, November 2003.
5. ICH Harmonized Tripartite Guideline. Text on Validation of Analytical Procedures Q2A, March 1995.
6. ICH Harmonized Tripartite Guideline. Q2B Validation of Analytical Procedures: Methodology, November 6, 1996.
7. International Conference on Harmonization (ICH). Q1B Photostability Testing of New Drug Substances and Products, November 1996.
8. U.S. Food and Drug Administration. Q1A Stability Testing of Drug Substances and Drug Products. Center for Drug Evaluation and Research (CDER); Center for Biologics Evaluation and Research (CBER), August 2001.
9. CDER Manual of Policies and Procedures, Chemistry Reviews of DMFs for Drug Substances/Intermediates, August 1998.
10. U.S. Food and Drug Administration. Drug Substance-Chemistry, Manufacturing and Controls Information, Center for Drug Evaluation and Research (CDER); Center for Biologics Evaluation and Research (CBER), January 2004.
11. U.S. Food and Drug Administration. Drug Product-Chemistry, Manufacturing and Controls Information, Center for Drug Evaluation and Research (CDER); Center for Biologics Evaluation and Research (CBER), Draft Guidance, January 2003.
12. International Conference on Harmonization (ICH). Guidance for Industry Q7A Good Manufacturing Practice Guidance for Active Pharmaceutical Ingredients, Center for Drug Evaluation and Research (CDER); Center for Biologics Evaluation and Research (CBER), August 2001.

45

Process Analytical Technology and Validation

Bradley S. Scott and Anne Wilcock
McNeil Consumer Healthcare, Guelph, Ontario, Canada

INTRODUCTION

Conventional pharmaceutical manufacturing practices have low manufacturing efficiencies and capacity utilization, high scrap and reject levels, and therefore a high cost of quality. PAT is a toolkit used to increase operational efficiencies, capacity utilization and process understanding, while decreasing operating expenses and ensuring that quality is built into the product. Multiple benefits associated with PAT implementation have been identified, but these benefits are not without drawbacks such as limited employee technical knowledge and suboptimal return on investment. The benefits and challenges of PAT are reviewed in this chapter, together with a discussion of the conventional pharmaceutical manufacturing paradigm, PAT principles and tools, PAT guidance development to date, and PAT validation approaches.

PAT has been defined by the United States FDA as a system to design, analyze, and control pharmaceutical manufacturing processes through the measurement of critical process parameters and quality attributes. Through the measurement of raw material and in-process material attributes and the control of critical process parameters, finished goods quality will increase. The FDA anticipates three main benefits to accrue from implementation of PAT in the pharmaceutical industry: increase in the understanding of processes and products, improvement in the control of pharmaceutical manufacturing processes, and incorporation of quality into the product from the design stage (1).

Abbreviations used in this chapter: ACPS, Advisory Committee for Pharmaceutical Science; ANN, artificial neural network; ANOVA, analysis of variation; API, active pharmaceutical ingredient; ATR, attenuated total reference; CCD, charge-coupled device; CDER, Center for Drug Evaluation and Research; CFR, Code of Federal Regulations; CV, coefficient of variation; DOE, design of experiment; DS, design specification; FDA, Food and Drug Administration; FS, functional specification; FTIR, Fourier Transform Infra-red Spectrometry; GC, gas chromatography; GMP, good manufacturing practice; GXP, good practices (manufacturing, practice, and laboratory practice); HPLC, high-performance liquid chromatography; IQ, installation qualification; LIF, light-induced fluorescence; MLR, multiple linear regression; MVP, master validation plan; NCE, nonconforming event; NIR, near infrared; OQ, operational qualification; PAT, process analytical technology; PCA, principal components analysis; PCR, polymerase chain reaction; PLS, partial least squares; PQ, performance qualification; PVC, polyvinyl chloride; RFP, request for proposal; RMS, root mean square; SOP, standard operating procedure; SPCC, statistical process control charts; TS, tensile strength; URS, user requirements specification; USP, *United States Pharmacopeia*; UV, ultraviolet.; XE, xenon.

Process and product understanding is an ongoing process; it is logical that it could be enhanced by implementation of elements of PAT such as chemical/physical/microbiological process analyzers, mathematical and statistical analysis, and risk analysis. As a result of increased process and product understanding, critical sources of variability are identified and controlled, and product quality attributes can be accurately and reliably predicted, thus increasing the quality of the final product. Process and product understanding helps to identify parameters that are critical to the process. The monitoring of process and product attributes during the manufacturing process will allow quality to be built into the product, an approach that is far superior to reliance on final inspection.

The PAT initiative in the United States has been led by the FDA's CDER, which released the revised PAT Guidance for Industry in September 2004. This document is a result of the combined efforts of industry, regulatory agencies and academic institutions; it provides a framework for PAT implementation in the pharmaceutical industry. It has two main components: (*i*) scientific principles and tools to support innovation in pharmaceutical development, manufacturing and quality assurance, and (*ii*) regulatory strategy to support innovation (1).

The FDA's PAT Guidance for Industry, which includes the PAT Framework and PAT dialogue between the FDA and industry has been used by both the FDA and CDER in an attempt to reduce barriers perceived by industry. Industry representatives have identified FDA regulatory uncertainty as a barrier, which has stifled change (2–5). The perception that the FDA regulatory body is rigid, resisting change and stifling innovation is a perception that the FDA and CDER would like to change with respect to PAT.

This chapter reviews PAT developments within the pharmaceutical industry with particular emphasis on the conventional pharmaceutical manufacturing paradigm, benefits of PAT, PAT Guidance Development, principles and tools of PAT, validation requirements, and challenges associated with PAT implementation.

THE CONVENTIONAL PHARMACEUTICAL MANUFACTURING PARADIGM

The conventional manufacturing paradigm in the pharmaceutical industry involves batch processing, with laboratory analysis of samples taken at predetermined intervals and processing steps. Sample collection and subsequent testing are primarily in-process and at the end of batch processing (1). The development of a new process/product, followed by subsequent transfer to full-scale manufacturing involves five main steps: design of the product/process, development of analytical methods and controls, accumulation of process knowledge, transfer of technology, and production of batches on a commercial scale (3). This approach requires a system to ensure that the final product meets predetermined specifications. Current quality systems to ensure that the finished product meets specifications include process/method/equipment validation, process control by SOPs and process instructions/master recipes, and off-line testing of samples at the end of each batch (6). The compliance infrastructure to support these quality systems is believed to be difficult to sustain from an economic perspective (7). This compliance burden has ensured that quality products are released to the patient, but at the same time, has resulted in cost of quality increases, which have impacted organizations' financial performance and the cost of products to patients and society.

The average cost of quality for the pharmaceutical industry has been reported to exceed 20% (7), with a three sigma level (66, 807 defects per million opportunities). Impacting the cost of quality are variables such as time-based endpoints, process variability, raw material variability, time associated with sampling, variability as a result of powder/blend sampling errors, and sample preparation. Further contributing to the cost of quality are the multiple test methods that are needed to assess different attributes of raw materials/in-process materials/finished drug substances (1).

Several manufacturing metrics have been identified (7,8) that show that the current pharmaceutical manufacturing paradigm is not functioning optimally. These metrics include utilization levels of 15% or less, scrap/rework of 5% to 10%, average cycle time of 95 days, and exception/nonconformance reports which can increase cycle time by more than 50%. Furthermore, validation requirements increase the time to market for new products, while at the same time consuming valuable resources. A sample validation timeframe reported for Pfizer extended to 117 weeks to prepare validation documents (15 weeks for hardware, one week for software, and 101 weeks for system validation protocol) (9).

The issues listed in Table 1 illustrate the desirability of an alternative paradigm for pharmaceutical manufacturing. These results, which originate from a systematic review of the literature, highlight an important point about quality. In the absence of technological monitoring and feedback controls, the current paradigm demands strict adherence to SOP and process instructions/master recipes. This system is not conducive to change or process improvement initiatives. The previously identified deficiencies, along with current societal and economic issues, are the predominant factors that are driving the pharmaceutical industry to change.

An additional factor that is impacting the pharmaceutical sector is the downward spiral in the number of NCEs launched per year. This is of particular concern because pharmaceutical organizations are increasing their spending on research and discovery, and not receiving adequate returns on their investments. One suggestion to counteract this downward trend was made by Bai et al. (16), who proposed that alternative noninvasive analysis methods such as NIR would assist in the design and optimization of protein formulations. Such optimization would decrease research and development efforts by reducing cost and cycle time for new product development and product launch.

Other factors impacting the pharmaceutical sector include limited marketing exclusivity periods, increased competition with generic products, and price scrutiny. These factors have all contributed to decreased returns for shareholders. This has forced the industry to carefully evaluate sources of potential waste and cost reduction.

Pressures currently impacting the pharmaceutical industry, along with challenges associated with the

Table 1 Issues/Problems Associated with the Conventional Pharmaceutical Manufacturing Paradigm

Issue/Problem	References
Low process capability	5,11
Elevated levels of scrap, rework, reject, recall	5,11
Low capacity utilization	11
Recurring problems that do not seem to get resolved	11
Slow resolution of issues/investigations	11
High cost of compliance	5,7,11
Risk of drug shortages	11
Risk of releasing poor quality drugs	11
Delay in approval of new drugs due to lack of data/process understanding	11
Quality problems confounding clinical trials	11
System not conducive to process improvements	11
Strict adherence to standard operating procedures and master recipes/process instructions required	11
Predetermined testing is completed at specific time intervals or process stages	11
Little learning after the validation phase	12
High proportion of FDA resources needed to ensure adequate product quality	11
Continued debates between the FDA and industry, few permanent resolutions with respect to quality issues	11
Silos of information. Time consuming process required to extract and analyze the data in these silos	13
Low efficiencies. Conventional analytical methods are time consuming and labor intensive (i.e., Karl Fisher titration for drug substance water content)	13,14
Labor intensive and inefficient	5,15

conventional manufacturing process, are driving companies within the sector to seek an alternative manufacturing paradigm. The future may lie with the incorporation of PAT into the manufacturing process.

PAT BENEFITS

There are three main benefits to implementing PAT in the pharmaceutical industry: increase in process/product understanding, increase in manufacturing process control, and incorporation of quality into the product from the design stage (1). Other benefits include reduced operating costs, quality improvements, positive regulatory impact, improved occupational safety, positive research and discovery impact, and reduced environmental impact (Table 2).

From an industry perspective, reduced operating costs and quality improvements are probably the most attractive because of their direct impact on profits. Contributing to the reduced operating costs are increased capacity utilization and increased operational (processing/packaging) efficiencies. Current capacity utilization levels have been estimated to be less than 15% (7). This suggests that there is tremendous potential for increased capacity utilization. Decreases in operating costs can also be achieved through continuous process monitoring and parametric release that will translate into improved cycle times (i.e., the time from raw material receipt through the value-added steps of processing and packaging). Hussain (49) reported the average cycle time to manufacture and package a pharmaceutical product to be 95 days. Of the 95 days, Dean (7) reported that only three days consist of value-added activities of dispensing, granulation, compression and coating. If the average cycle time could be reduced from 95 days to the industry's best practice of six days through the implementation of PAT, both inventory and warehousing costs would be reduced and additional capacity made available.

Capacity constraints in pharmaceutical quality control laboratories can be eliminated through the use of PAT. The conventional manufacturing paradigm requires samples to be transferred to the laboratory for analysis using techniques such as wet chemistry, FT-IR, UV Spectroscopy, HPLC and GC. The use of PAT analyzers can reduce sample collection and analysis time and thus cycle times. Reduction in analysis time has been demonstrated (29) by implementing an NIR reflectance model for the identification of blister packaging film. The reference method, infrared spectroscopy (*European Pharmacopeia* 3.1.11) requires approximately two hours, while analysis of the same film using the NIR reflectance

Table 2 Benefits Associated with Implementing PAT in the Pharmaceutical Industry

Benefits Category	Specific PAT Benefits	References
Reduced operating costs	Increased operating efficiencies	2,12,13,17–19
	Improved cycle time (reduced release times, parametric release, reduce sample preparation time, minimize reliance on finished product testing, faster analysis times)	2,4,5,13,18–36
	Decreased operating costs	2,4,5,8,17,28,29,34,35,37,38
	Possible continuous processing	2
	Real-time monitoring, feedback controls and results	2,8,15,19,25,31
	Inventory reduction (through parametric release and improved cycle times)	13
	Increased capacity utilization	26,39
	Attain production schedule	39
	Reduced reprocessing expenses	8,40
Quality improvements	Increased quality (decreased product variability, decreased number of rejections, scrap, batch failure and systems failures, and increased product reliability)	2,4,5,8,13,17–19,26,37, 38,41–44
	Increased regulatory compliance	13,37,38
	Increased product uniformity (ensure batch to batch consistency, decrease variation)	2,5,8,13,17,40
	Process finger printing	2
	Increased process understanding	2,12,17,18,26,45
	Quality designed into the process	5,31,42
	Use of scientific, risk-based approach in decision making	18
	Recall prevention/avoidance	38
	Minimized patient risk including security of supply	26,31
	No sampling required or reduced sampling requirements (reduces/eliminates sampling error)	8,17,21,27,39,42,46
	Critical process control provided	12,42
	Rapid identification of counterfeit drug substances	47
Positive regulatory impact	Moderate regulatory burden on Food and Drug Administration	5
	Improved scientific basis for regulatory functions	18
Increase occupational safety	Decreased occupation exposure to toxic substances	8,17,40,41,48
Positive research and discovery impact	Reduced product development life cycle/time to market	16,23,26,37
Minimize environmental impact	Reduced environmental impact (assurance that process and plant environments are maintained within environmental regulations)	39
	Minimize waste (i.e., solvent waste) generation during manufacturing	36,39

method requires less than two minutes. Han and Faulkner (24) reported similar reductions in analysis time by use of NIR reflectance to analyze moisture content, active ingredient identification and assay of granulation, tablet cores, coated tablets and blisters. Traditional methods require 15 minutes to determine moisture content and 30 minutes for the identification and assay of the active ingredient (UV and HPLC methods) whereas the NIR reflectance method requires less than one minute for moisture content, identification and assay analysis. Similar time savings have been demonstrated using Raman spectroscopy for the analysis of aspirin. Wang et al. (36) reported that Raman analysis of aspirin tablets for assay results required 15 minutes, while HPLC analysis required 90 minutes per sample.

The implementation of PAT in the pharmaceutical industry can lead to quality improvements as a direct result of continuous monitoring and the use of process control tools. Real-time monitoring of batch processing steps decreases product variability; the number of batch failures and amount of scrap material are reduced, and the consistency between batches is increased. Reductions in batch failures and scrap material would decrease the current scrap and rework levels (7). This would translate into reduced cost of quality from the current levels that exceed 20% of the cost of goods sold (7). Reduced analysis times would improve the efficiency within the quality department, further reducing the cost of quality. Andre (50) decreased the work expenditure of the traditional identification method, HPLC, for 7-aminocephalosporanic acid by 98% by use of NIR reflectance spectroscopy. Reduced cost of quality equates with reduced operating expenses and thus an increase in shareholder value.

PAT can be used to reduce consumer risk through recall prevention (38). Through the use of process analyzers such as NIR, it is theoretically possible to perform identification, assay and dose uniformity on each tablet produced. This would reduce the risk of product cross-contamination, specifically in nondedicated manufacturing facilities where multiple actives and dosage concentrations are produced. Process analyzers, in combination with multivariate analysis and process control tools, can also be used to ensure batch-to-batch consistency in real time. Batch-to-batch consistency would improve uniformity within the process, ensuring that the product could be consistently produced within specifications. Through real-time monitoring and control, the consumer's risk of exposure to adulterated products or products that do not meet quality specifications is reduced. PAT can also be implemented for the rapid screening of raw materials (38,50–53) to identify counterfeit materials, further increasing consumer safety.

Wechsler (5) and Hussain (18) suggest that a positive regulatory impact would result from PAT implementation within the pharmaceutical sector. They hypothesize that PAT would improve the scientific basis for regulations which would moderate the regulatory burden on the FDA.

Health, safety, and environmental benefits in the form of increased occupational safety and decreased environmental impact of manufacturing operations are expected outcomes of the use of PAT in this sector. Process analyzers can be used to monitor production facilities to ensure the environment is maintained within specific conditions. Avallone (41) and Hammond (17) emphasized decreases in employee exposure to toxic substances. Watson et al. (40) used an in-line FT-IR process analyzer to monitor a synthesis process to ensure reaction conditions did not generate an unstable hydroperoxide. The PAT can also result in reduced waste/scrap from processing and packaging.

PAT GUIDANCE

To facilitate a paradigm shift in the manufacturing philosophy within the pharmaceutical sector, the FDA has been actively involved in dialogue and the development of guidance for the industry. To provide guidance for this shift, the FDA and CDER formed the ACPS.

The objective of the ACPS was to identify the current status and future trends of PAT in the pharmaceutical industry, and to develop a collaborative industry/academic/regulatory approach. The ACPS, through industry and academic involvement, investigated regulatory challenges with respect to PAT implementation as well as method validation and specification requirements, and the feasibility of parametric release (5). Parametric release is defined as the assessment of product attributes and process controls to ensure that the in-process or finished pharmaceutical product is of acceptable quality (1). These product attributes and process controls are based on scientific understanding of the process and product. Using the information gathered from this collaborative approach, ACPS recommendations were incorporated into the PAT Guidance for Industry, which was introduced in September 2004.

PAT PRINCIPLES AND TOOLS

The PAT Guidance for Industry identified the following principles and tools as being suitable for implementation in development and manufacturing activities in the pharmaceutical industry: (*i*) multivariate tools for design, data acquisition and analysis, (*ii*) process analyzers, (*iii*) process control tools, and (*iv*) continuous improvement/knowledge management.

Multivariate Tools for Design, Data Acquisition, and Analysis

Pharmaceutical dosage forms are complex systems in which chemically and physically relevant data are measured and analyzed. The conversion of this data to knowledge, and the identification of multifactorial relationships can be achieved through multivariate analysis. Common multivariate tools that are housed within the PAT framework include library construction for NIR, DOE, PCA, PCR, MLR, PLS, neural networks and SPCC (26).

Multivariate tools for quantification, in particular NIR calibration, have been documented previously (54–60) and only a description of common multicomponent analysis tools will be presented here (Table 3).

Library construction for NIR is an alternative to traditional qualitative analysis methods for pharmaceutical raw materials. Blanco and Romero (51) stress the

(*Text continues on p. 590.*)

Table 3 Applications of Process Analyzers in the Pharmaceutical Sector

Reference	PAT	Process	Attribute analyzed	On-line/, in-line/, at-line/, not stated
Abrahamsson et al. (61)	NIR spectroscopy—transmission	Compression	Quantification of active ingredient	Off-line
Airaksinen et al. (62)	NIR spectroscopy—reflectance Ramon spectroscopy	Granulation	Identification of theophylline monohydrate	Off-line
Andre (50)	NIR spectroscopy—reflectance	Raw material	Quantification of 7-aminocephalosporanic acid (7-ACA)	Off-line
Betz et al. (63)	Temperature sensor-temperature increase Power consumption of mixer motor	Granulation	Granulation end-point (temperature and power consumption ratio)	In-line
Blanco et al. (64)	NIR spectroscopy—reflectance	Compression and coating—tablet	Identification and quantification of gemfibrozil	Off-line
Blanco and Romero (51)	NIR spectroscopy—reflectance	Raw material	Identification	Off-line
Blanco and Villar (65)	NIR spectroscopy—reflectance	Compression and coating—tablet	Quantification of miokamycin	Off-line
Yoon et al. (66)	NIR spectroscopy—reflectance	Tablets	Site of manufacturing identification	Off-line
Blanco et al. (67)	NIR spectroscopy—reflectance	Granulation	Quantification of nimesulide	Off-line
Clarke (68)	NIR microscopy	Granulation and compression—tablets	Spatial distribution and cluster size of ingredients	Off-line
Cui et al. (47)	NIR spectroscopy—reflectance	Powder samples	Identification of sulfaguanidine	Off-line
Davis et al. (69)	X-ray powder diffraction	Granulation	Monitor the transformation of metastable polymorph to stable polymorph during granulation process	On-line
Dyrby et al. (21)	NIR spectroscopy—transmittance Raman spectroscopy	Compression	Quantification of active ingredient in Escitalopram® tablets	Off-line; at-line
Fountain et al. (70)	NIR spectroscopy	Mucoadhesive thin-film composites	Quantification of testosterone	Off-line
Gupta et al. (23)	NIR spectroscopy	Milling of roller compacted powders	Particle size and compact strength	On-line
Harris and Walker (25)	NIR spectroscopy—transmittance	Technique can be applied to the drying of cakes, pastes and slurries	Solvent evaporation	On-line
Herkert et al. (42)	NIR spectroscopy—reflectance	Packaging line (blister)	Identification	On-line
Jorgensen et al. (71)	NIR spectroscopy—reflectance	Granulation	Wet granulation endpoint	On-line
Laasonen et al. (72)	NIR spectroscopy—reflectance	Packaging component identification	Identification of blister PVC-films	Off-line
Laasonen et al. (27)	NIR spectroscopy—reflectance	Compression—tablets	Quantification of caffeine	Off-line
Laasonen et al. (29)	NIR spectroscopy—reflectance	Packaging component identification	Identification of blister PVC films and film thickness	Off-line
Otsuka (33)	NIR spectroscopy	Granulation	Particle size	Off-line
Ritchie et al. (73)	NIR spectroscopy—transmittance	Compression—tablets and capsules	Content uniformity and assay	Off-line
Watson et al. (40)	FT-IR	API	Monitor synthesis	In-line
Woo et al. (48)	NIR spectroscopy	Liquid manufacturing	Hydrogen peroxide formation	Off-line
Blanco et al. (74)	NIR spectroscopy—reflectance	Granule and tablets	Quantification of asorbic acid	Off-line
Blanco et al. (75)	NIR spectroscopy—reflectance	Capsule	Identification and quantification of pirisudanol dimaleate	Off-line
Blanco et al. (76)	NIR spectroscopy—reflectance	Powder	Moisture content	Off-line

(*Continued*)

Table 3 Applications of Process Analyzers in the Pharmaceutical Sector (*Continued*)

Reference	PAT	Process	Attribute analyzed	On-line/, in-line/, at-line/, not stated
Gottfries et al. (77)	NIR spectroscopy—reflectance and transmittance	Compression tablets	Quantification of metoprolol succinate	Off-line
Han and Faulkner (24)	NIR spectroscopy—reflectance	Granulation, compression, coating and blistering of tablets	Moisture content, identification, quantification	Off-line
Last and Prebble (24)	NIR spectroscopy—reflectance	Freeze dried injection	Moisture content	Off-line
Lonardi et al. (52)	NIR spectroscopy—reflectance	Powders	Moisture content and quantification	Off-line
Plugge and van der Vlies (53)	NIR spectroscopy	Powders	Identification, moisture content, and quantification	Off-line
Higgins et al. (79)	NIR spectroscopy—reflectance	Liquid manufacturing	Particle size	On-line
Rantanen et al. (10)	NIR spectroscopy—reflectance	Granulation—fluid bed	Granule moisture content	In-line
Watano et al. (80)	Image probe (CCD camera and high-energy XE lighting system)	High-shear granulation	Particle size	In-line
Watano et al. (81)	Image probe (CCD camera and high-energy XE lighting system)	High-shear granulation	Particle size Particle shape	In-line
Andersson et al. (82)	NIR spectroscopy—reflectance	Granulation-fluid bed	Particle size	In-line
Broad et al. (83)	NIR spectroscopy—transmittance	Aqueous suspension	Quantification of ethanol, propylene glycol, water	Off-line
Rantanen et al. (84)	NIR spectroscopy	Granulation	Granule moisture content	In-line
Sanchez et al. (34)	NIR spectroscopy—reflectance	Raw material	Moisture content	Off-line
Kirsch and Drennen (85)	NIR spectroscopy—reflectance	Compression	Tablet hardness	Off-line
Eustaquio et al. (86)	NIR spectroscopy—transmittance	Compression	Quantification of paracetamol	Off-line
Blanco et al. (87)	NIR spectroscopy	Granulation, compressed and coated cores	Quantification of gemfibrozil	Off-line
O'Neil et al. (32)	NIR spectroscopy—reflectance	Raw material	Particle size	Off-line
Frake et al. (43)	NIR spectroscopy	Granulation	Granule moisture content Particle size	In-line
Gold et al. (22)	NIR spectroscopy—reflectance	Capsule	Dissolution	Off-line
Morisseau and Rhodes (44)	NIR spectroscopy—reflectance	Compression	Tablet hardness	Off-line
Kamat et al. (27)	NIR spectroscopy—reflectance	Lyophilized	Moisture content	Off-line
Dubois et al. (20)	NIR spectroscopy—reflectance	Aqueous suspension	Quantification of phenazone, glycerol, ethanol, lidocaine hydrochloride, sodium thiosulphate	Off-line
Blanco et al. (88)	NIR spectroscopy—reflectance	Blending	Identification and quantification of ferrous lactate dehydrate	Off-line
Blanco et al. (89)	NIR spectroscopy—reflectance	Blending, compression, coated tablets	Identification and quantification of otilonium bromide 400 mg/g	Off-line
Gupta et al. (90)	NIR spectroscopy—reflectance	Roller compaction	Acetaminophen content uniformity, moisture content, relative density, TS, Young's modulus	In-line
Lai and Cooney (91)	LIF	Blending	Homogeneity end point and blend stability of triamterene (2,4,7-triamino-6-phenylpterine) powder	On-line

(*Continued*)

Table 3 Applications of Process Analyzers in the Pharmaceutical Sector (*Continued*)

Reference	PAT	Process	Attribute analyzed	On-line/, in-line/, at-line/, not stated
Moffat et al. (92)	NIR spectroscopy—reflectance	Compression	Identification and quantification of paracetamol in intact tablets	Off-line
El-Hagrasy et al. (93)	NIR spectroscopy	Blending	Homogeneity of salicylic acid powder	On-line
El-Hagrasy et al. (94)				
El-Hagrasy and Drennen III (95)				
Tumuluri et al. (35)	NIR spectroscopy—reflectance	Hot-melt extruded film	Quantification of clotrimazole	Off-line
Bai et al. (16)	NIR spectroscopy—reflectance	Lyophilized formations (vials)	Protein confirmation in lyophilized protein formations	Off-line
Gupta et al. (96)	NIR spectroscopy	Roller compaction	Density, moisture content, TS, and Young's modulus	Off-line
Bai et al. (97)	NIR spectroscopy—reflectance	Lyophilized formations (vials)	Quantification of glycine crystallinity	Off-line
Laitinen et al. (30)	Monochrome CCD camera	Granulation	Particle size analysis and end point determination of the granulation process	At-line
Seyer et al. (98)	NIR spectroscopy—reflectance	Lyophilized formations (vials)	Degree of crystallization	Off-line
Zhou et al. (14)	NIR spectroscopy—reflectance	Drying of drug substance	Differentiate between surface and bound water Determination of water content in drug substance	In-line
Lai et al. (15)	LIF Technology	Compression	Total tablet content of active in tablet	On-line
Lai et al. (99)	LIF Technology	Blending	Blend homogeneity	On-line
Fevotte et al. (45)	NIR spectroscopy—reflectance	Crystallization	Qualitative and quantitative analysis of SaC (API) during crystallization	In-line
Lin et al. (19)	FT-IR with an attenuated total reference probe	Pharmaceutical salt formation process	Real-time endpoint monitoring and determination for a pharmaceutical salt formation process Pharmaceutical salt (4-{1-methyl-2-piperidin-4-yl-4-[3-(trifluorometryl)phenyl]-1H-imidazol-5-yl}- *N*-[(1S) − 1-phenylethyl]pyridine-2-amine (freebase), an API as a P38 mitogen-activated protein kinase inhibitor)	In-line
Skibsted et al. (46)	NIR spectroscopy—reflectance	Blending	Qualitative and quantitative analysis of API to assess blend homogeneity	In-line (Note: probe not installed in mixer, but inserted at specific time intervals)
Johansson et al. (100)	Raman spectroscopy	Compression	Quantitative analysis of API in tablets	Off-line
Cogdill et al. (101,102,103)	NIR spectroscopy—reflectance	Compression	Quantification of API in tablet and physical parameters including hardness	On-line
Hausman et al. (104)	Raman spectroscopy	Granulation	Risedronate sodium solid-state form was continuously monitored using on-line Raman spectroscopy during drying	On-line

(*Continued*)

Table 3 Applications of Process Analyzers in the Pharmaceutical Sector (*Continued*)

Reference	PAT	Process	Attribute analyzed	On-line/, in-line/, at-line/, not stated
Islam et al. (105)	Raman spectroscopy	Topical gels and emulsions	Raman spectroscopy demonstrated as a process analytical technique for quality control of topical gel and cream formulations	Off-line
Johansson et al. (106)	Raman spectroscopy	Compression	Laser induced heating of compressed tablets during Raman spectroscopy analysis	Off-line
Kontoyannis (107)	Raman spectroscopy	Compression	Quantitative analysis of $CaCO_3$ and Glycine in antacid tablets	Off-line
Langkilde et al. (108)	Raman spectroscopy	Pharmaceutical active ingredient	Quantitative analysis of two crystal forms of a pharmaceutical active ingredient	Off-line
Szostak and Mazurek (109)	Raman spectroscopy	Powder and tablets	Quantitative analysis of acetylsalicylic acid and acetaminophen in tablets	Off-line
Taylor and Zografi (110)	Raman spectroscopy	Powder	Quantitative analysis of indomethacin crystallinity	Off-line
Wang et al. (36)	Raman spectroscopy	Tablets	Aspirin tablet assay	Off-line

following points when developing an NIR library for qualitative analysis: (*i*) Establish baseline NIR spectra using samples of known identity. (*ii*) Samples of known identity should be composed of various batches so that they represent the physical–chemical variability of the raw material. (*iii*) During construction of the library, the NIR pattern recognition method and construction parameters must be determined. (*iv*) Internally validate the method to determine if spectra within the library are incorrectly identified or remain unidentified. (*v*) Construct sub-cascading libraries composed of related substances, degradation products, and enantiomers to ensure NIR pattern recognition method and construction parameters are optimized for substance identification. (*vi*) Use an external validation set of unequivocally identified substances to perform external validation. Using the discriminating power designed into NIR libraries, document a qualitative method for raw material identification.

DOE uses factorial design to evaluate the effects of several factors on a process. When using DOE, varying the factors simultaneously instead of one at a time is an effective approach to identifying interactions among the factors. In complex multifactorial pharmaceutical dosage forms, these interactions affect the quality of the end product.

Pharmaceutical products and processes are composed of multiple components. To reduce the number of variables being analyzed, PCA may be used. PCA identifies combinations of variables or factors that contain the maximum variability within the sample set, and then uses these variables or factors as the principal components; a large data set is then reduced to a smaller number of components. The principal components can then be analyzed using further multivariate tools (2,57). One application of PCA in PAT is to determine the correlation between a process analyzer result (e.g., NIR) and a final product/process attribute (e.g., moisture content, assay results).

MLR is the analysis of two or more variables as a linear combination of the dependent variables (60). The combination of MLR and PCA results in PCR which has been used for multivariate analysis in PAT. PCR uses the principal components identified through PCA, and performs regression on the resultant sample property to be predicted (57). Variance of the measured variables is used to determine the model of best fit. The PCR incorporates most sources of variability and is more efficient than MLR.

The PLS is a multivariate analysis tool which combines PCA and multiple regression (60). With respect to process analyzers, PLS can be used to identify variability between the spectral data and the product property. In contrast to MLR, the model of best fit for PLS is based on the product property of the measured variables. PLS is a common multivariate analysis tool for NIR calibration models used to quantify the active ingredient in granulation, compression, coated and packaged pharmaceutical products (Table 3). If the response is nonlinear, an ANN could be used to identify relationships. The ANN contains computational algorithms that process experimental data and transform these data using nonlinear logarithmic/exponential/quadratic functions to determine the response (57).

Multivariate tools allow scientists to identify the correlation between raw data and their impact on the process. With these multivariate models, continuous recalibration is essential if there are even minor changes to the production process, raw materials or critical process parameters (50).

Trends in product quality attributes can be followed using SPCC. The SPCC plot data on an ongoing basis in relation to upper and lower product specifications. These data can be utilized to determine process capability, i.e., the ability of the process to meet specifications consistently.

Process Analyzers

Process analyzers measure the physical, chemical and biological properties of materials. They collect both quantitative data (moisture content, particle size, active ingredient quantification, microbial counts) and qualitative data (microbial identification, active ingredient identification). Data collection can be nondestructive, require minimal sample preparation, and have rapid or real-time response when compared to traditional methods (111). Data integrity is necessary to ensure compliance with the U.S. FDA 21 CFR Part 11 which requires specific controls with respect to electronic signatures, security, and audit trail functionality.

PAT Guidance for Industry (1) categorizes process analyzers into four categories which are differentiated from one another based on the stage of the process at which sample measurement occurs: at-line (the sample is removed, isolated from, and analyzed in close proximity to the process stream), on-line (the sample is diverted from the manufacturing process, and may be returned to the process stream), in-line (the sample is not removed from the process stream and can be invasive or noninvasive), and off-line (the sample is removed, isolated from, and analyzed away from the process stream as in a laboratory environment). Of these four categories, on-line and in-line process analyzers have the greatest potential to reduce operating costs and improve quality; both minimize sample requirements and sample handling compared to their at-line and off-line counterparts. Clevett (112) indicated that 80% to 90% of errors associated with analysis were associated with sample handling, either directly or indirectly. On-line and in-line process analyzers reduce sample handling and sample preparation errors, thereby reducing retest and cycle times.

Process analyzers are used primarily to determine the following attributes of raw materials, in-process materials and finished goods: particle size and shape, moisture content, active ingredient quantification, dissolution profiles, and tablet hardness (Table 3).

Near Infrared

Particle size of a granulation, powdered blend or powdered pharmaceutical raw material is important in that it impacts physical properties such as powder flow, dissolution rate, compressibility and tablet hardness. Monitoring particle size and control of the manufacturing process prevents over-processing of the product. According to the literature, the most common process analyzer to be used in the determination of particle size of milled roller compacted powders, granulation, liquids and raw materials is NIR (Table 3). The NIR process analyzers have been evaluated on-line, in-line and off-line; results of these evaluations compare favorably to those of traditional methods such as sieve analysis, digital microscopy and particle size instrumentation.

The shape and spatial distribution of particles influence physical properties such as powder flow and filterability (60). Clarke (68) used NIR microscopy off-line to determine spatial distribution and cluster size of ingredients in granulation and compressed pharmaceutical products. Clarke concluded that NIR microscopy was a useful tool in the determination of particle shape, particle distribution and cluster size of the chemical components of the sample.

Moisture content has commonly been measured using techniques such as NIR spectroscopy. In 1968, Sinsheimer and Poswalk reported the use of NIR in the determination of moisture content in pharmaceuticals (113). The use of NIR spectroscopy for the determination of moisture content in raw materials, pharmaceutical powders and freeze-dried injectables is summarized in Table 3. Blanco, Coello, Iturriaga, et al. (78) determined the moisture content in the raw material ferrous lactate dehydrate using off-line NIR reflectance spectroscopy. Multivariate analysis (PLS and MLR) methods provided similar results, with prediction errors of less than 1.5%. Lonardi, Viviani, Mosconi, et al. (52) determined that a calibration model (MLR) containing a 50/50 mixture of laboratory samples/production samples provided the best predictive power and the lowest error. Plugge and van der Vlies (53) determined the moisture content of an antibiotic powder containing ampicillin trihydrate using off-line NIR spectroscopy and reported that the method was accepted by the FDA in 1992. NIR reflectance spectroscopy was also used by Last and Prebble (78) to determine the moisture content of a freeze-dried experimental injectable drug. Accuracy and precision of their off-line analysis were identified as limitations of the NIR models assessed; they recommended incorporating more samples and optimizing the NIR wavelength regions.

Process analyzers have been utilized in the determination of wet granulation endpoints. Wet granulation consists of three processes: mixing, spraying, and drying, with moisture content being a critical end point parameter in the final phase. Traditional methods of determining moisture content require sampling of the granulation and then analysis by Karl Fischer or 'loss of drying' moisture analyzers. These techniques are performed off-line/at-line, and hence require operator intervention to collect the samples. To obtain real time results, NIR spectroscopy has been evaluated as an off-line and in-line process analyzer (Table 3). Rantanen et al. (84), Rantanen et al. (10) and Frake et al. (43) used in-line NIR sensors to collect moisture content data in fluid bed granulators. The in-line calibration models provided enough predictive power for the determination of moisture content in the sample set.

The NIR spectroscopy methods have been used to identify and quantify active ingredients and excipients in granulation (fluid bed and wet granulations), compressed bulk, coated bulk, aqueous products, and active ingredients in blistered product. NIR methods have also been used for identification of raw materials, solvent evaporation, packaging component identification, safety applications such as monitoring of hydrogen peroxide formation, and hardness determination. Use of on-line, at-line and off-line NIR spectroscopy for these applications is summarized in Table 3.

Raman Spectroscopy

Raman spectroscopy is suitable for quantitative analysis of pharmaceutical products because of the relationship between signal intensity and API concentration. Raman spectroscopy has been evaluated for identification and quantification of active ingredients in granulation, compression, drug pellet and solid mixture; samples have been evaluated for both off-line and at-line use (62,21,114,115). Raman spectroscopy has also been used to monitor hydration states of API as a method of process control (116–118).

When implementing Raman spectroscopy in pharmaceutical processes, it is important to consider the sampling area. Dyrby et al. (21) and Johansson (100) concluded that the increased predictive error associated with the Raman model was a result of the tablet area sampled. Compared to the NIR transmittance model, Raman spectroscopy using surface sampling collects data on a smaller volume of the tablet, thus explaining the higher predictive error. Increasing the irradiated surface area by use of rotating sample holders has been shown to decrease the predictive error of the Raman model (21,100).

CCD Camera

Watano et al. (81) and Watano et al. (80) assessed particle size in a high shear granulator in-line through the use of an image probe (CCD camera and high energy XE lighting system). The image probe was combined with a fuzzy logic control system to control granulation growth in the high shear granulator, preventing excessive granule growth. The system was capable of accurately and reliably producing granules that met specifications, independent of starting materials and operating conditions. Laitinen et al. (30) assessed particle size growth in a fluidized-bed granulation process using a monochromatic CCD camera. At-line analysis of granulation samples via the CCD camera successfully monitored granule growth and granulation end point for the fluidized-bed granulation process. The conclusion was that the imaging approach used provided rapid evaluation of granule particle size.

X-ray Diffraction

On-line application of x-ray powder diffraction was evaluated by Davis et al. (69) for use in monitoring the transformation of metastable polymorph to stable polymorph during wet granulation of flufenamic acid. The on-line process analyzer was successful in monitoring the polymorphic transformation of the flufenamic acid. The results of this evaluation suggest that X-ray powder diffraction may be used as an on-line process analyzer to monitor granulation process and process parameters such as granulation end time.

FT-IR Process Analyzer

Process analyzers have been evaluated for API synthesis. Watson et al. (40) evaluated an in-line FT-IR process analyzer for the conversion of buspirone hydroxylation to 6-hydroxybuspirone. They recommended the use of the in-line FT-IR process analyzer to monitor and control the synthesis process since this process ensures API quality and predicted the need for batch reprocessing.

Lin et al. (19) demonstrated the ability to real-time monitor a pharmaceutical salt formation process with FT-IR coupled with an ATR probe, a task which cannot be accomplished with traditional analytical instrumentation such as titration and HPLC. FT-IR ATR permitted differentiation between mono and bi-salts, allowing for real-time determination of the synthesis end point. Other benefits were improved quality monitoring, higher yields, and ease of method transfer between laboratories and FT-IR instruments, all of which contribute to improved efficiency.

Light-Induced Fluorescence

LIF technology is selective for fluorescent materials (usually the active ingredient) within a drug formulation. LIF measures the emission wavelength as a result of wavelength excitation. LIF technology is a nondestructive PAT tool for the analysis of powder mixing kinetics and blend homogeneity, and tablet active ingredient content (91,15,99). Lai and Cooney (91) proposed that LIF would be especially useful within the pharmaceutical industry because 60% of the two hundred main active ingredients fluoresce. Benefits of on-line LIF analysis in blending include real-time blend kinetic results and reductions in errors due to thief sampling (91).

Process Control Tools

Process control tools monitor and actively manipulate a process to ensure control. Process analyzers can be integrated into a process control application to measure critical process parameters and product attributes in order to achieve desired in-process and finished quality specifications. Shah (38) summarized those critical process parameters (moisture content, particle size, blend uniformity, content uniformity, tablet hardness, and viscosity) which could be monitored and controlled to ensure that in-process and finished quality specifications are achieved.

Watano et al. (80,81) evaluated a process control tool for monitoring and controlling a high shear granulation process. An image probe (CCD camera and high energy XE lighting system) and model-based system (fuzzy logic) were utilized to collect particle size images throughout the high shear granulation phase. The processing conditions were varied to simulate normal manufacturing variation. The system accurately monitored and provided feedback during granulation, preventing excessive granule growth.

Continuous Improvement/Knowledge Management/ Information Technology Systems

The integration of PAT (process analyzer, multivariate analysis and process control tools) results in the generation of a large volume of data which must be converted from data to knowledge. Knowledge management tools provide a way of storing data as well as using models, process simulation tools and pattern recognition tools to develop process knowledge and understanding. Knowledge management systems should be designed to expand as a product moves through design into the product development phase and subsequent manufacturing. The knowledge base includes information on critical process parameters; management of the data can be used as a basis for process improvements.

To maximize the benefits of PAT, the data must be collected, analyzed and presented in a manner that demonstrates that the product meets all release criteria.

This information can be summarized in an electronic batch record or external repository that centralizes data and process instructions from a variety of sources.

Quality Assurance can verify that the data within an electronic batch record pertaining to raw materials, in-process materials and finished goods meet release specifications. It is also possible for computerized controls to be built into the electronic batch records to prevent raw materials, in-process materials and finished goods from being released accidentally when their attributes do not meet specifications. Such computerized control ensures that quality is not compromised and reduces the risk of product recalls.

VALIDATION REQUIREMENTS

PAT consists of the integration of process analyzers, multivariate analysis tools, process control tools and continuous improvement/knowledge management/information technology systems. The integration of such a complex system (Fig. 1) requires that the following five validation/calibration activities be considered: analytical method validation, sensor calibration, computer system validation, equipment qualification, and process validation. Validation is the ability to demonstrate that a procedure, process, equipment, material or system can consistently produce results that meet specifications. It involves examining and understanding those parameters/conditions of steps that are critical to the process (and ultimately the product), as well as establishing specifications.

To secure the benefit from qualification, it must be seen as an integral part of a project. The first stage of the overall process involves assessing the GMP criticality and defining the key validation requirements of the PAT application (including PAT analyzer, analytical method, processing equipment and computerized system). The initial phase of the validation lifecycle (Fig. 2) includes prequalification which consists of defining the system documentation: the URS, the RFP, the FS and the DS.

Real-time monitoring of manufacturing processes with PAT analyzers generates data that can be used to control manufacturing processes and/or develop a broader understanding of the manufacturing processes. During design of a PAT application, specific consideration of the desired outcome of PAT should be considered because this will influence validation requirements. Key stakeholders who should be consulted at this stage include representatives from Information Management, Quality, Operations, Engineering, and Research and Development.

Once the specific objective of PAT has been determined and a vendor selected, validation activities

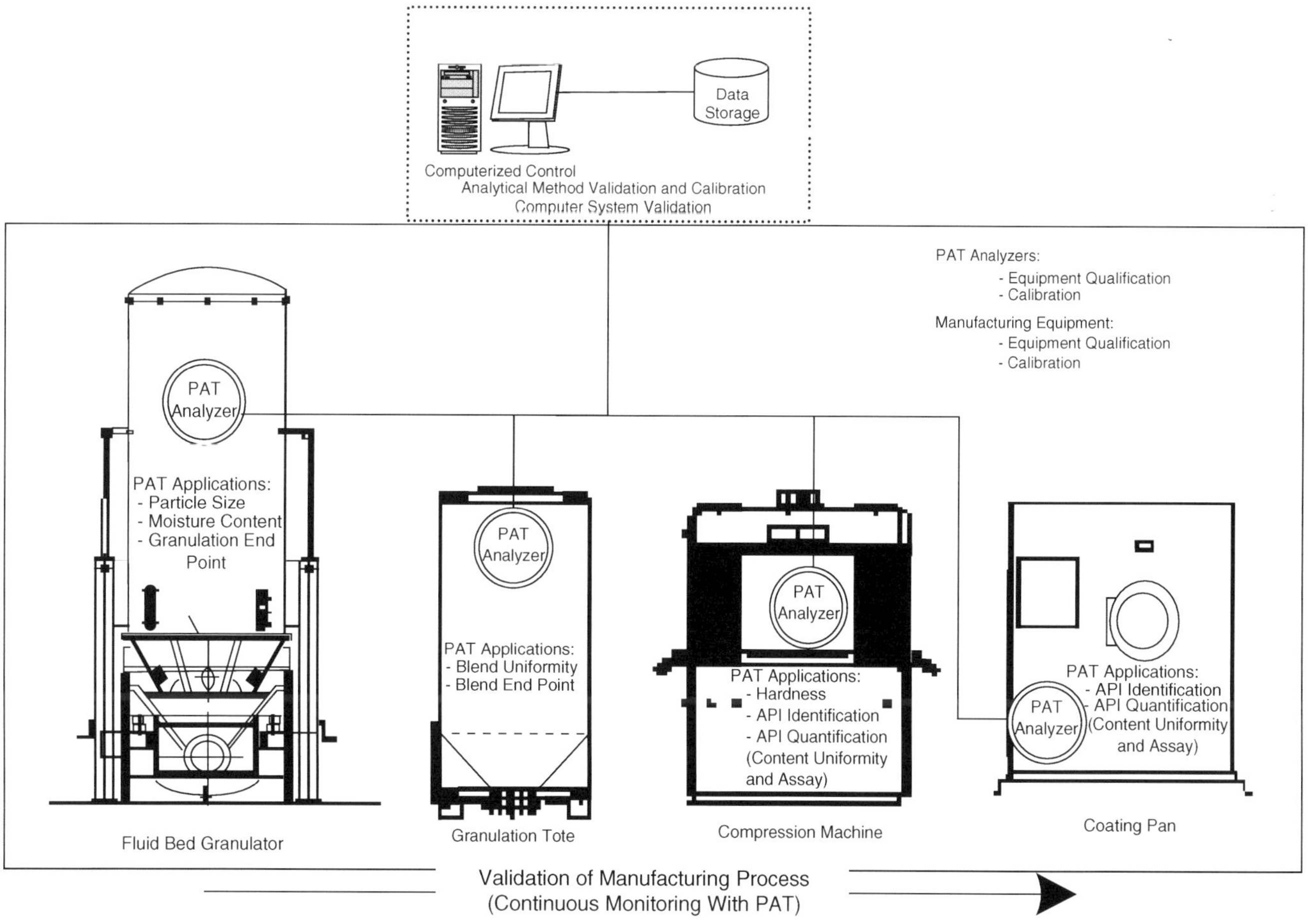

Figure 1 Application of process analytical technology and validation considerations (solid dose oral manufacturing process).

must be considered. Qualification is designed to ensure the installation and operation of PAT analyzers and processing equipment according to specification. As with most PAT analyzers and modern processing equipment, the qualification should include computerized system validation.

Qualification begins with IQ and is followed by OQ and PQ phases. Once qualification activities have been completed, the analytical method can be validated. Validation of the analytical method is critical to validating the manufacturing process. Upon completion of the validation phase, the PAT application can be transferred to the maintenance phase (i.e., calibration, preventative maintenance, and change control).

Analytical Method Validation

The main objective of analytical method validation is to demonstrate that the analytical procedure is suitable for its intended purpose. Most of the guidance documents published to date emphasize method validation of separation techniques such as HPLC rather than direct spectroscopic techniques such as NIR spectroscopy.

Elements of chromatography method validation include specificity, selectivity, linearity, range, accuracy, precision, repeatability, intermediate precision, reproducibility, ruggedness, detection limit, quantification limit, robustness, and system suitability. To develop an analytical method validation recommendation for PAT applications such as NIR, scientists validated analytical methods using the traditional method validation approach.

As a result of extensive scientific evaluation, key elements required to validate NIR spectroscopy for PAT application have been identified (Table 4). These elements were incorporated into the USP's General Chapter on NIR Spectrophotometry <1119> (119) which are summarized in the table. The alternative approach should be documented in the site's SOPs and MVP. It should be noted that these are recommendations; an alternative validation approach may be required for on-line, off-line, at-line and in-line PAT applications.

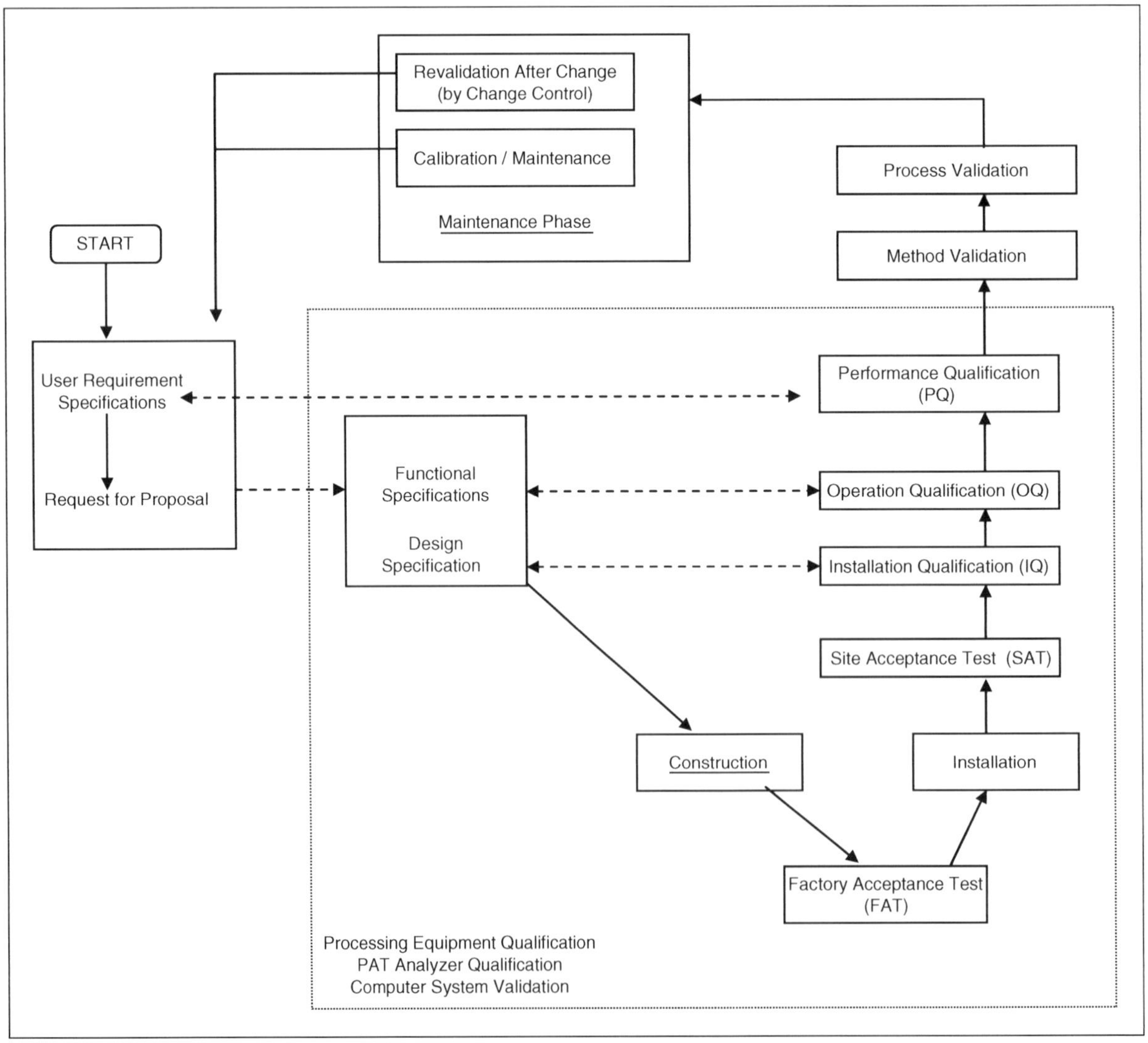

Figure 2 Validation lifecycle for the implementation of a pharmaceutical process analytical technology application.

Table 4 Elements Required to Validate NIR Spectroscopy for PAT Application

Reference	PAT method	Type of PAT procedure	Validation parameter	Conclusions
Blanco et al. (90)	NIR spectroscopy—reflectance	Identification Quantification	Specificity Precision Repeatability Intermediate precision Accuracy Linearity Robustness	Blanco et al. (90) identified selectivity as a critical validation element for identification methods Repeatability demonstrated through 12 NIR reflectance determinations from a production batch by the same operator on the same day [(CV)=0.3%, acceptance criteria < 1%] Intermediate precision demonstrated by two different production batches analyzed by two operators on three different days (CV for batch 1<0.7%; CV for batch 2<0.7%, acceptance criteria <2%) Accuracy results between the reference method and the NIR reflectance method were not statistically different NIR and reference standards were compared for linearity. NIR concentration was plotted versus reference concentration (r=0.994)
Blanco et al. (66)	NIR spectroscopy—reflectance	Identification Quantification	Specificity Precision Repeatability Intermediate precision Accuracy Linearity Robustness	Blanco et al. (66) identified selectivity as a critical validation element for identification methods Repeatability demonstrated by the same operator on the same day, with six NIR reflectance determinations (CV within acceptance criteria <1%) Intermediate precision was evaluated by varying operators and completing analysis between days. CV and ANOVA indicate no systematic errors Accuracy results between NIR and reference method (ultraviolet method) were compared by a t-test. No significant difference between NIR and reference method NIR and reference standards were compared for linearity. NIR concentration was plotted versus reference concentration (r=0.988)
Blanco et al. (91)	NIR spectroscopy—reflectance	Identification Quantification	Specificity Repeatability Precision Repeatability Robustness Intermediate precision Accuracy Linearity	Validation of the identification method was demonstrated through repeatability and robustness To demonstrate repeatability, each sample was analyzed 12 times (correlation coefficient ranged between 0.992 and 0.999) To demonstrate robustness, 10 samples produced over a three-month period were analyzed (correlation coefficient ranged between 0.990 and 1.00) Precision was demonstrated by analyzing 1 sample, 12 times by the same operator on the same day (CV<0.6%) Intermediate precision was demonstrated by analyzing one batch from each processing step on three different days by two analysts (CV 0.5–0.8%; ANOVA no systematic error identified) Accuracy and linearity demonstrated by comparing NIR and reference method results Robustness of the calibration model was evaluated over a one month time period (10 samples from new production batches analyzed). Calibration model was deemed stable
Laasonen et al. (27)	NIR spectroscopy—reflectance	Identification	Specificity	To demonstrate specificity, caffeine and excipients were analyzed. Caffeine peaks did not interfere with excipient peaks (mean spectral residual +3 SD). One sample t-tests confirmed that the mean residual of the excipient batch was significantly different than the production batch (including caffeine)
		Quantification	Precision Repeatability	Method repeatability was demonstrated by collecting NIR spectral information six times for a single dosage form (from two batches)

(*Continued*)

Table 4 Elements Required to Validate NIR Spectroscopy for PAT Application (*Continued*)

Reference	PAT method	Type of PAT procedure	Validation parameter	Conclusions
			Intermediate precision Accuracy Linearity Range Robustness	A DOE was performed for intermediate precision. The DOE evaluated operator, date of analysis and batches (%RSD < 2%) Accuracy was demonstrated by comparing NIR and reference method (HPLC) results. Results were compared by a student *t*-test, and were determined to be not statistically different Linearity was determined during the calibration of the NIR method. The NIR method was utilized to predict different batches of 60% to 130% caffeine concentration. The results were compared to HPLC results by linear regression (method of least squares). The confidence interval slope included 1, and the y intercept did not statistically differ from 0 (*t*-test) Range was demonstrated through linearity, accuracy and precision Robustness of a NIR method can be influenced by environmental conditions. Environmental conditions (temperature/humidity/direction of sunlight/dust/vibrations) were controlled Changes in sample preparation were evaluated. Tablet holder was moved affecting spectrophotometer beam, which affected the results NIR source was replaced which did not affect the results. This was demonstrated through a paired student *t*-test
Moffat et al. (94)	NIR spectroscopy—reflectance	Identification	Specificity	Interfering components such as excipients, degradation products, water, residual solvents and impurities should be considered when developing a NIR method for identification Zero-order spectra were reviewed to determine if interfering peaks are present. If interfering peaks are present, the spectra can be mathematical/chemometrically treated to remove spectral interference (i.e., second derivative)
		Quantification	Precision Repeatability Intermediate precision Accuracy Linearity Range Robustness	To demonstrate repeatability, nine determinations (three replicates of three different concentrations) were analyzed Precision was demonstrated by scanning the same sample multiple times. When evaluating precision, Moffat et al. (94) identified the following parameters which should be considered: thermal degradation should be considered for heat sensitive APIs. Surface inhomogenity low-dosage API forms—use NIR transmittance instead of reflectance as NIR transmittance analyzes larger surface volume Intermediate precision was demonstrated by analyzing the same sample by different analysts on different days. Traditional intermediate precision would involve the assessment using different NIR instruments. Moffat et al. (94) indicate there are few NIR instruments in the pharmaceutical sector, making it difficult to assess this parameter on intermediate precision Accuracy was demonstrated by statistical comparison of NIR values and reference method values NIR predicted assay values versus reference method values over a specific range were analyzed by linear regression (method of least squares) to demonstrate linearity of the NIR method Moffat et al. (94) verified the range of the NIR method, but identified limitations which include difficulties obtaining samples with a wide range, as the samples need to be produced from the production process

(*Continued*)

Table 4 Elements Required to Validate NIR Spectroscopy for PAT Application (*Continued*)

Reference	PAT method	Type of PAT procedure	Validation parameter	Conclusions
				Moffat et al. (94) identified factors to be considered when evaluating robustness (i) Temperature/humidity should be controlled. If these factors are not controlled, they should be considered during robustness evaluation (ii) Variation in tablet compaction (iii) Sample containers if used need to be considered, as they impact NIR analysis (iv) Changes in sample presentation/orientation (v) Tablet morphology (shape/scored/embossed/ printed/ coating) (vi) Probed depth and probe installation

Computerized System Validation

All computerized systems, scientific instruments and processing equipment governed by any GxP regulation (GMP, Good Laboratory Practices and Good Clinical Practices) must be validated. This includes computer hardware, software, network infrastructure, equipment, instruments, as well as procedures that create, modify, maintain, archive, retrieve and/or distribute data used during development, testing, manufacturing and distribution.

Data generated by a computerized system can be categorized as either an electronic record or an electronic signature. An electronic record is any combination of text, graphics, data, audio, pictorial or other information represented in digital form that is created, modified, maintained, archived, retrieved, and/or distributed by a computer system. An electronic signature is a computer data compilation of any symbol(s) executed, adopted or authorized by an individual to be the legal equivalent of that individual's handwritten signature.

A Validation Plan describes what activities will be performed in order to validate the GxP computerized system. The activities that may be addressed in a Validation Plan are listed in Table 6.

The IQ establishes confidence that process equipment (both hardware and software) and ancillary systems comply with appropriate codes and approved design intentions, and that the manufacturer's recommendations are considered. An IQ Protocol should define tests to be conducted during installation of the GxP computerized system and include acceptance criteria. An IQ Report should document the results of the execution of the IQ Protocol and state whether or not acceptance criteria were satisfied. All software should be archived and placed in version control upon installation. Elements of the IQ Protocol relate directly to the documentation and performance of a DS.

The OQ establishes confidence that process equipment (both hardware and software) and subsystems are capable of operating consistently within established limits. An OQ Protocol should be written and approved by a designated cross-functional team, and should define tests to be conducted during the OQ of the system and the acceptance criteria. An OQ Report should document the results of tests conducted following the OQ Protocol and state whether or not acceptance criteria were satisfied. Elements of the OQ Protocol relate documentation and performance of a FS.

The PQ establishes confidence that the computerized system is effective and reproducible. It should be prepared and approved by a cross function team, and should define tests to be conducted during PQ and the acceptance criteria. A PQ Report should document the results of tests conducted following the PQ and state whether or not acceptance criteria were satisfied. Any policies or procedures which need to be modified or created for implementation of the new or enhanced GxP computerized system should be identified prior to PQ.

Table 5 NIR Analytical Method Validation Requirements

	Type of NIR procedure	
Validation parameter	Qualitative	Quantitative
Specificity	+	+
Linearity	−	+
Range	+	+
Accuracy	−	+
Precision–repeatability	+	+
Precision–intermediate precision	−	+
Robustness	+	+

Table 6 Potential Elements of a Validation Plan

System description/configuration
Applicable policies, procedures and guidelines
Responsible departments and/or individuals
Validation strategy
Risk assessment
Supplier assessment
Categorization of components
Vendor evaluation/audit
Assumptions/exclusions/limitations
Documentation–system, technical and operational
Testing procedures
Acceptance criteria
Deviations/error reporting/resolution
Change control process standard operating procedures
Security
Backup/archive/disaster recovery
Training
Qualification protocols and reports—installation qualification, operational qualification and performance qualification

Elements of the Process PQ Protocol relate directly to documentation and performance of a URS.

Equipment Qualification

Equipment qualification ensures that laboratory and manufacturing facilities and systems that are directly involved in the manufacture, testing, control, packaging, holding and distribution of marketed products comply with GMP. It demonstrates the suitability of PAT analyzers and manufacturing equipment for their intended use.

Equipment qualification consists of requirement and specification documents, IQ, OQ, and PQ phases (Fig. 2). IQ is the documented verification that all key aspects of the installation adhere to the manufacturer's/engineering recommendations, design intentions, relevant electrical/building codes, and safety specifications. The OQ is documented verification that the equipment can operate as intended and is capable of satisfactory operation over the entire range of operating parameters. The OQ includes verification of operation to ensure that the equipment meets certified standards. Finally, PQ is documented evidence that the integrated equipment can perform as intended throughout anticipated operating ranges in the production environment, and that it satisfies user requirements.

Qualification test criteria that should be evaluated when validating an NIR instrument for PAT application are summarized in Table 5. It should be noted that, while the USP General Chapter <1119> identifies minimum validation requirements, additional validation test cases may be recommended by the manufacturer of the equipment.

Process Validation

Process validation is the demonstrated ability of a process, including equipment, raw materials, environmental controls, and master recipe to consistently produce finished goods within specifications. The validation requirements and approaches for sterile dosage forms, liquid oral dosage forms, solid oral dosage forms, powders, ointments and creams vary; various guidance documents have been developed by regulatory agencies to address these dosage forms. The traditional pharmaceutical validation paradigm requires that the manufacturing process be repeatable. This has been typically demonstrated by testing three consecutive batches using traditional analytical methods. In the proposed PAT framework, this traditional three-batch approach would be replaced with continuous monitoring of the process to ensure finished product quality. This monitoring would result in continuous updating of data in calibration and validation models. Through the accumulation of additional data, increased process knowledge and understanding would occur, thus ensuring that quality is built into the process instead of being inspected into the finished product. This increased understanding would also allow pharmaceutical manufacturers to adjust to variation in inputs (raw materials, process conditions) while ensuring critical outputs are achieved. This would result in consistent product quality attributes, and fewer rejected batches.

Before using PAT for process validation, the PAT analyzer and equipment must be qualified and calibrated, along with the supporting information technology infrastructure. A validated analytical method for the PAT application must also be completed to ensure the validity of the data generated. Upon completion of these activities, PAT can be used to monitor the manufacturing process.

Calibration

Calibration systems and procedures are established for scientific instruments and processing equipment that are critical to quality in the manufacturing and testing of a product. The calibration program ensures that all equipment is performing accurately and reliably on a continual basis, according to in-house requirements. Calibration is essential to maintaining a validated state. Two calibration issues that must be addressed in PAT are methods (i.e., NIR calibration models and qualitative libraries) and hardware (i.e., NIR instrument calibration).

Evaluation of the ongoing performance of a validated analytical method is critical to successful implementation of PAT. The collection of samples must represent the range of product attributes. Numerous factors that should be evaluated when selecting samples for calibration methods are presented in Table 7.

After the calibration method or qualitative library has been established, methods should be updated on a regular basis with new samples. As indicated in USP General Chapter <1119>, accuracy, precision and critical validation elements should also be re-evaluated at predetermined intervals in order to assess the performance of the analytical method. If a deviation occurs (i.e., the accuracy or precision decreases), the root cause should be investigated immediately (103). The findings of that investigation may dictate corrective action(s) that involve modification of the preventative maintenance procedure or equipment calibration process before putting the NIR method back into service. Changes should be documented through the site's change control process, and may be subject to revalidation.

NIR sensor calibration is critical when sensors have been used in the identification and quantification of active ingredients, excipients, physical properties of drug substances, and other critical in-process parameters. One mechanism of calibration is internal performance tests (equipment calibration tests). Daily internal performance tests have been demonstrated as a

Table 7 Factors to Evaluate when Selecting Samples for Calibration Methods

Different processing conditions
Multiple batches of raw material (representative of the physical–chemical variation of the raw material)
Different API/excipient suppliers
Different API/excipient concentrations in drug formulation
Related substances, degradation products, and enantiomers
Variation in tablet compaction (tablet compaction force)
Sample containers (if used)
Changes in sample presentation/orientation
Granulation characteristics (moisture content, particle size)
Blending characteristics (blend uniformity)
Tablet morphology (shape/scored/embossed/printed/coating)

requirement for NIR instruments (101,102). Sensor calibration ensures that process analyzers in PAT applications are functioning as they were designed to function, and ensures the accuracy and precision of the data collected from these sensors. Completion of sensor calibration can be accomplished by reexecuting critical components of the equipment qualification at predefined intervals. Frequencies recommended by the manufacturer should be followed unless scientific justification supports alternative frequencies. Test cases to consider when performing routine sensor calibration include wavelength uncertainty, tolerances, photometric linearity, and spectrophotometric noise. Specific information on test case and acceptance criteria for these calibration elements can be found in the Equipment Qualification section (Table 8).

Revalidation

When changes occur to systems, processes, methods, materials or computerized systems, the effect of those changes is assessed through the change control process. If it is determined that the changes impact the validity of the previous studies, then the modified system/method/process is revalidated. All revalidation activity should be conducted in accordance with current regulatory requirements and site SOPs.

Specific revalidation criteria for NIR methods have been identified in USP General Chapter <1119> which recommends revalidation for qualitative NIR analytical methods when the following criteria arise: addition of a new material to the reference library, physical changes to material, supplier changes, and/or an expanded range of material characteristics. Quantitative methods may be re-validated if any of the following criteria arise: raw material grade changes (changes in purity, polymorphic changes), changes in manufacturing process (processing steps added, removed or altered), changes in finished product composition (coating solution changes or reformulation), reference method changes, or major instrument maintenance or alteration (specifically, changes which may impact optics).

Table 9 identifies additional changes that may impact the validated systems/methods used in PAT. A site change control is required to assess the impact and identify validation activities.

Validation Summary

The level of validation required when implementing a PAT solution depends upon several variables. Through the demonstration of process understanding, control of critical parameters, and monitoring of physical/chemical/biological properties and environmental conditions, process validation activities can be reduced by implementation of PAT. In the absence of process understanding, the FDA has indicated that a test-to-test comparison may be required when implementing a new on-line process analyzer. This test-to-test comparison

Table 8 USP General Chapter <1119> Recommended Validation Criteria for NIR Equipment Qualification

Validation test case	Test details	Recommended specification
Maximum nominal bandwidth	Instrument bandwidth, based on analyte/product matrix/process to be measured, should be assessed during the user requirements phase	USP General Chapter <1119> indicates that a maximum nominal bandwidth of 10 nm at 2500 nm (NIR reflectance) or 16 cm^{-1} at 4000 cm^{-1} (NIR transmittance) is applicable for most applications
Photometric linearity	Photometric linearity is typically expressed as a percent reflectance or percent transmittance	A_{obs} versus A_{ref} at 1200, 1600 and 2000 nm
	Traceable carbon-doped polymer standards are used for NIR reflectance	Slope = 1.0 ± 0.05
	Typically a set of four standards is used to calibrate the NIR instrument, over the range of absorbances required	Intercept = 0.0 ± 0.05
Tolerances	Wavelength tolerances	± 1 nm at 1200 nm (± 8 cm^{-1} at 8300 cm^{-1}) ± 1 nm at 1600 nm (± 4 cm^{-1} at 6250 cm^{-1}) ± 1 nm at 2000 nm (± 4 cm^{-1} at 5000 cm^{-1})
Wavelength uncertainty	One spectrum is collected and a minimum of three peaks is measured over the range	Reflectance Mode USP NIR Calibrator RS_{USP29}-Peaks occur at 1261, 1681 and 1935 nm Transflectance mode NIST SRM 2035 Transmittance NIST SRM 2035 rare earth oxide in standard glass or NIST SRM 2036 $_{USP29}$
Spectrophotometric noise	High-flux noise can be measured for reflectance and transmittance modules, by tabulating the RMS noise levels in successive nominal 100 nm (300 cm^{-1}) spectral segments across the instruments range. Traceable reference materials (i.e., 99% reference standard) are measured to determine noise at high light flux	High-flux noise: average RMS < 0.3 x 10^{-3}; No RMS noise > 0.8 x 10^{-3}
	Low-flux noise can be measured for reflectance modules, by tabulating the RMS noise levels in successive nominal 100 nm (300 cm^{-1}) spectral segments across the instrument's range. Traceable reference materials (i.e., 10% reference standard) are measured to determine noise at reduced light flux	Low-flux noise: average RMS < 1 x 10^{-3}; No RMS noise > 2.0 x 10^{-3}

Table 9 Changes to Validated Systems/Methods that May Require Revalidation

Category	Change which may require revalidation
Process	Process moved to a new location
	A significant change in the production volume (batch size, for example from single batch to double batch or double batch to triple batch) relative to the capacity of the processing equipment
	A significant change in the manufacturing facility and/or environmental controls under which products are manufactured
	Change from manual to automated manufacturing; or vice versa, when significant changes to the processing parameters are made
	Introduction of significantly different manufacturing equipment for use in a manufacturing process
	A significant change in the source/composition of manufacturing materials (i.e., drug substance and or excipient) used to manufacture a product
	Any changes to the formulation, packaging, equipment or process, which could impact the effectiveness or product characteristics (safety, purity, identity or strength of the product)
	Whenever there are changes in product characteristics
	When changes are made to the raw material supplier, consideration should be given to subtle, potentially adverse differences in the raw material characteristics, which may have a significant impact downstream in the process
	Variations revealed by trend analysis through Annual Product Reviews (e.g., process drifts)
Equipment qualification	Changes to the equipment
	Move equipment to a different location
	Changes to the operation of the equipment
	Changes to programmable logic controller
	Changes to the operating parameters
	Changes to equipment optics
Packaging	New or modified products
	New or modified packaging materials
	New or modified equipment
	New material suppliers
	New or modified change parts
	New, modified, or relocated packaging lines
	New operation parameters
Analytical method	Changes or modifications are made to equipment, manufacturing process, analytical procedure or the composition of the drug product has changed (i.e., new source or synthesis of drug substance or new impurity present)
Computer	Changes to the hardware and/or software operating systems. Install new software version

includes comparison of the data from the on-line process analyzer with conventional test methods (1).

PAT CHALLENGES

Challenges facing PAT implementation within the pharmaceutical industry have been discussed frequently (Table 10). A major challenge appears to be numerous perceived regulatory barriers. Regulatory challenges that have been identified include unclear PAT validation requirements and 21 CFR Part 11 requirements, as electronic data related to batch release parameters will be stored. Therefore, retention and management of these data must comply with 21 CFR Part 11. To encourage the ongoing pursuit of PAT implementation, the FDA has released the Guidance for Industry on PAT and has sponsored multiple symposiums to generate dialogue between the FDA and industry.

An additional challenge includes the lack of infrastructure within current manufacturing facilities. Information technology, including the ability to network manufacturing equipment with local area networks, may be a challenge in existing facilities. The installation of routers, switches, servers, and the network to an existing facility adds incremental costs to a PAT implementation project. Related challenges include the ability of a network to manage large volumes of data on a continuous basis, and real time access by multiple users. These factors impose constraints on the system, which may result in downtime or loss of data. To minimize downtime and the risk of data loss, continuous instrument, software and local area network support is required, which increases the costs.

The cost of PAT implementation has resulted in limited senior management support in some organizations (Table 10). Questions have arisen about the return on investment as well as the accumulation of data, which may highlight inadequacies in a process. These inadequacies may not interfere with the production of finished products, which are acceptable based on traditional test methods, but have a low process capability. In this scenario, the concern focuses on the potential for increased product recalls.

Overcoming the limited knowledge about PAT in the pharmaceutical industry is a major challenge. In McCormick's study (8) of a small cross section of pharmaceutical companies, only one-half of the organizations surveyed were aware of the FDA's PAT guidance. Only 14% of the organizations surveyed were currently implementing PAT.

The challenges discussed in this chapter, combined with current resource constraints facing the pharmaceutical industry, and limited knowledge about PAT process analyzers appear to be the main factors limiting PAT implementation.

SUMMARY

The conventional manufacturing paradigm in the pharmaceutical industry is based on batch processing,

Table 10 Challenges Associated with Implementing PAT in the Pharmaceutical Industry

Challenges category	Specific PAT challenges	Reference
Current infrastructure does not facilitate PAT implementation	Information Technology infrastructure requirements may not exist in current facilities	2,13,120
	Lack of senior management support	2
	Current resource constraints	2,37
	Difficulty in applying PAT when manufacturing Phase I and II drugs (drug formulations have not been finalized)	121
	Large volumes of continuous data are produced (system constraints need to be considered during system design)	13,21
	System needs to handle real time access for multiple users (system constraints)	13
	Limited employee knowledge base	3,8,13,49,120
	Complex mathematical models can result in the introduction of misinterpretation	39
	24/7 instrument and software support required	120
Regulatory challenges	21 Code of Federal Regulations Part 11 requirements	2,120
	Validation requirements unclear	2,38
	No perceived regulatory incentive	37
Cost of PAT implementation	Return on investment	2,13,38,120
	Regulatory uncertainty, including regulatory approval delays	2–5
	For calibration, need a wide range of samples which are within and outside specifications and which will increase cost	120
	Proving equivalency between PAT and traditional methods	38
Industry mindset and concerns	Attitude within pharmaceutical industry-no reason to change-status quo is fine.	13,26
	Implementation of PAT into current manufacturing process may expose deficiencies in manufacturing processes	4,13,41
	Accumulation of data, which may show inadequacies in processes which produce product, which are acceptable based on traditional testing methods	4,13,41
	Implementation of PAT could result in increased recalls	13
	No perceived benefit	37
Technology challenges	Process analyzers (sensors) prone to drift	58
	Calibration model requires frequent updates to include product and process variation	21,58
	Processes are susceptible to unmodelled events	58

with laboratory analysis of samples collected at predetermined time intervals and processing steps. This manufacturing approach tests quality into final products, with resulting suboptimal efficiencies, high levels of rework and scrap, high cost of compliance, and low levels of continuous improvement. One major advantage of a shift from the current paradigm to PAT would be that quality would be built into products. Building quality into the process may translate into increased product quality per se, increased regulatory compliance, increased capacity and efficiencies, and/or decreased manufacturing and quality costs.

The PAT approach requires the integrated implementation of process analyzers, multivariate analysis tools, process control tools, and continuous improvement/knowledge management/information technology systems. The complexity of the PAT system has resulted in uncertainty with respect to both regulatory approach and validation. The FDA's PAT Guidance for Industry (1) was an attempt to reduce the uncertainty and perceived barriers. In addition to the guidance document, there has been a series of PAT conferences chaired by the ACPS and CDER. Although regulatory and validation uncertainty have been identified as barriers hindering the adoption of PAT implementation in the pharmaceutical industry, the largest barrier appears to be the return on investment, especially in the short-term. The PAT elements such as information technology infrastructure, process analyzers (i.e., NIR), process controls and knowledgeable staff require a substantial financial investment. This may perhaps be the limiting factor, especially in times of decreasing shareholder returns and market exclusivity, increasing generic competition, decreasing research and discovery productivity, and increasing research and discovery costs.

REFERENCES

1. PAT guidance for industry—A framework for innovative pharmaceutical development, manufacturing, and quality assurance, U.S. Department of Health and Human Services, Food and Drug Administration, Center for Drug Evaluation and Research (CDER), Center for Veterinary Medicine (CVM), Office of Regulatory Affairs (ORA), Pharmaceutical cGMPs, September 2004.
2. Balboni ML. Process analytical technology: concepts and principles. Pharm Technol 2003; 27:54.
3. Blumenstein J. Pfizer. Regulatory challenges: PAT application in NDAS. FDA's Advisory Committee for Pharmaceutical Sciences, Subcommittee on Process Analytical Technologies (PAT), Gaithersburg, MD, June 12–13, 2002. (Accessed May, 2006 at http://www.fda.gov/ohrms/dockets/ac/02/slides/3869S1_02_Blumenstein.ppt)

4. Clarke PE. CDER aims to improve drug manufacturing. News along the Pike. Center for Drug Evaluation and Research 2002; 8(July):1.
5. Wechsler J. Modernizing pharmaceutical manufacturing. Pharm Technol N Am 2002; 26:16–24.
6. Chrisholm RS. AstraZeneca. Perspective on process and analytical validation. FDA's Advisory Committee for Pharmaceutical Sciences, Subcommittee on Process Analytical Technologies (PAT), Gaithersburg, MD, February 25, 2002. (Accessed May, 2006 at http://www.fda.gov/ohrms/dockets/ac/02/slides/3841s1_06_chisholm/index.htm)
7. Dean D. PriceWaterhouseCoopers. Pharma manufacturing—Why there is a need to improve. The role of PAT. FDA's Advisory Committee for Pharmaceutical Sciences, Subcommittee on Process Analytical Technologies (PAT), Gaithersburg, MD February 25, 2002. (Accessed May, 2006 at http://www.fda.gov/ohrms/dockets/ac/02/slides/3841s1_03_Dean/index.htm)
8. McCormick D. PAT survey reflects optimism, uncertainty. Pharm Technol 2005; 29:24.
9. Hammond S. Pfizer. Validation perceptions that may slow PAT development and implementation. Subcommittee for Pharmaceutical Sciences, Subcommittee on Process Analytical Technologies (PAT), Rockville, MD, October 23, 2002. (Accessed May, 2006 at http://www.fda.gov/ohrms/dockets/ac/02/slides/3901S1_07_Hammond_-files/frame.htm)
10. Rantanen J, Rasanen E, Antikainen O, et al. In-line moisture measurement during granulation with a four-wavelength near-infrared sensor: an evaluation of process-related variables and a development of non-linear calibration model. Chemom Intell Lab Syst 2001; 56:51–8.
11. Hussain AS. Second meeting of FDA/ACPS process analytical technology: closing remarks. FDA's Advisory Committee for Pharmaceutical Sciences, Subcommittee on Process Analytical Technologies (PAT), Gaithersburg, MD, June 12–13, 2002. (Accessed May, 2006 at http://www.fda.gov/ohrms/dockets/ac/02/slides/3869S1_09_Hussan-%20Summary/index.htm)
12. Rudd D. GlaxoSmithKline. Product and process development: an industry perspective. FDA's Advisory Committee for Pharmaceutical Sciences, Subcommittee on Process Analytical Technologies (PAT), Gaithersburg, MD, February 25, 2002. (Accessed May, 2006 at http://www.fda.gov/ohrms/dockets/ac/02/slides/3841s1_05_Rudd/index.htm)
13. Neway JO. Filling the void: PAT in a connected manufacturing environment. Pharm Technol 2003; 27:46–52.
14. Zhou GX, Ge Z, Dorwart J, et al. Determination and differentiation of surface and bound water in drug substances by near infrared spectroscopy. J Pharm Sci 2003; 92:1058–65.
15. Lai CK, Zahari A, Miller B, et al. Nondestructive and on-line monitoring of tablets using light-induced fluorescence technology. AAPS Pharm Sci Tech 2004; 5:1–10.
16. Bai S, Nayar R, Carpenter JF, et al. Noninvasive determination of protein conformation in the solid state using near infrared (NIR) spectroscopy. J Pharm Sci 2005; 94:2030–8.
17. Hammond S. Pfizer. Applications and benefits of PAT. FDA's Advisory Committee for Pharmaceutical Sciences, Subcommittee on Process Analytical Technologies (PAT), Gaithersburg, MD, February 25, 2002. (Accessed May, 2006 at http://www.fda.gov/ohrms/dockets/ac/02/slides/3841s1_02_hammond/index.htm)
18. Hussain AS. Opening remarks. FDA's Advisory Committee for Pharmaceutical Sciences, Subcommittee on Process Analytical Technologies (PAT), Gaithersburg, MD, June 12, 2002. (Accessed May, 2006 at http://www.fda.gov/ohrms/dockets/ac/02/slides/3869S1_01_Hussan-%20Opening%20Remarks/index.htm)
19. Lin Z, Zhou L, Mahajan A, et al. Real-time endpoint monitoring and determination for a pharmaceutical salt formation process with in-line FT-IR spectroscopy. J Pharm Biomed Anal 2006; 41:99–104.
20. Dubois P, Martinez JR, Levillain P. Determination of five components in a pharmaceutical formulation using near infrared reflectance spectrophotometry. Analyst 1987; 112:1675–9.
21. Dyrby M, Engelsen SB, Norgaard L, et al. Chemometric quantitation of the active substance (containing c≡n) in a pharmaceutical tablet using near-infrared (NIR) transmittance and NIR FT-Raman spectra. Appl Spectrosc 2002; 56:579–85.
22. Gold TB, Buice RG, Jr., Lodder RA, et al. Determination of extent of formaldehyde-induced crosslinking in hard gelatin capsules by near-infrared spectrophotometry. Pharm Res 1997; 14:1046–50.
23. Gupta A, Peck GE, Miller RW, et al. Nondestructive measurements of the compact strength and the particle-size distribution after milling of roller compacted powders by near-infrared spectroscopy. J Pharm Sci 2004; 93:1047–53.
24. Han SM, Faulkner PG. Determination of SB 216469-S during tablet production using near-infrared reflectance spectroscopy. J Pharm Biomed Anal 1996; 14:1681–9.
25. Harris SC, Walker DS. Quantitative real-time monitoring of dryer effluent using fiber optic near-infrared spectroscopy. J Pharm Sci 2000; 89:1180–6.
26. Hussain AS. The Subcommittee on Process Analytical Technology (PAT): overview and objectives. FDA's Advisory Committee for Pharmaceutical Sciences, Subcommittee on Process Analytical Technologies (PAT), Gaithersburg, MD, February 25, 2002. (Accessed May, 2006 at http://www.fda.gov/ohrms/dockets/ac/02/slides/3841s1_01_hussain/index.htm)
27. Kamat MS, Lodder RA, DeLuca PP. Near-infrared spectroscopic determination of residual moisture in lyophilized sucrose through intact glass vials. Pharm Res 1989; 6:961–5.
28. Laasonen M, Harmia-Pulkkinen T, Simard C, et al. Development and validation of a near-infrared method for the quantitation of caffeine in intact single tablets. Anal Chem 2003; 75:754–60.
29. Laasonen M, Harmia-Pulkkinen T, Simard C, et al. Determination of the thickness of plastic sheets used in blister packaging by near infrared spectroscopy: development and validation of the method. Eur J Pharm Sci 2004; 21:493–500.
30. Laitinen N, Antikainen O, Rantanen J, et al. New perspectives for visual characterization of pharmaceutical solids. J Pharm Sci 2004; 93:165–76.
31. Lonardi S, Newby PJ, Ribeiro D, et al. GlaxoSmithKline. Why does PAT need rapid microbiology methods. Subcommittee for Pharmaceutical Sciences, Subcommittee on Process Analytical Technologies (PAT), Rockville, MD, October 23, 2002. (Accessed May, 2006 at http://www.fda.gov/ohrms/dockets/ac/02/slides/3901S1_12_Lonardi_files/frame.htm)
32. O'Neil AJ, Jee RD, Moffat AC. Measurement of the cumulative particle size distribution of microcrystalline cellulose using near infrared reflectance spectroscopy. Analyst 1999; 124:33–6.
33. Otsuka M. Comparative particle size determination of phenacetin bulk powder by using Kubelka-Munk theory and principal component regression analysis based on near-infrared spectroscopy. Powder Technol 2004; 141:244–50.

34. Sanchez MS, Bertran E, Sarabia LA, et al. Quality control decisions with near infrared data. Chemom Intell Lab Syst 2000; 53:69–80.
35. Tumuluri SVS, Prodduturi S, Crowley MM, et al. The use of near-infrared spectroscopy for the quantitation of a drug in hot-melt extruded films. Drug Dev Ind Pharm 2004; 30:505–11.
36. Wang C, Vickers TJ, Mann CK. Direct assay and shelf-life monitoring of aspirin tablets using Raman spectroscopy. J Pharm Biomed Anal 1997; 16:87–94.
37. Hussain AS. Office of Pharmaceutical Sciences, CDER, FDA. ACPS Process Analytical Technology (PAT) Subcommittee Meeting #3 Opening Remarks. Subcommittee for Pharmaceutical Sciences, Subcommittee on Process Analytical Technologies (PAT), Rockville, MD, October 23, 2002. (Accessed May, 2006 at http://www.fda.gov/ohrms/dockets/ac/02/slides/3901S1_01_Hussain_files/frame.htm)
38. Shah DN. Aventis Pharmaceuticals. Regulatory challenges: post-approval PAT applications. FDA's Advisory Committee for Pharmaceutical Sciences, Subcommittee on Process Analytical Technologies (PAT), Gaithersburg, MD, June 12–13, 2002. (Accessed May, 2006 at http://www.fda.gov/ohrms/dockets/ac/02/slides/3869S1_03_Shah/index.htm)
39. Workman J. Kimberly-Clark Corp. Chemometrics and PAT: an opinion on current status and recommendations for the future. FDA's Advisory Committee for Pharmaceutical Sciences, Subcommittee on Process Analytical Technologies (PAT), Gaithersburg, MD, February 25, 2002. (Accessed May, 2006 at http://www.fda.gov/ohrms/dockets/ac/02/slides/3841s1_08_workman/index.htm)
40. Watson DJ, Dowdy ED, DePue JS, et al. Development of a safe and scalable oxidation process for the preparation of 6-hydroxybuspirone: application of in-line monitoring for process ruggedness and product quality. Org Process Res Dev 2004; 8:616–23.
41. Avallone H. Johnson & Johnson. Development/compliance issues (tablets). FDA's Advisory Committee for Pharmaceutical Sciences, Subcommittee on Process Analytical Technologies (PAT), Gaithersburg, MD, June 12–13, 2002. (Accessed May, 2006 at http://www.fda.gov/ohrms/dockets/ac/02/slides/3869S1_04_Avallone.ppt)
42. Herkert T, Prinz H, Kovar KA. One hundred percent online identity check of pharmaceutical products by near-infrared spectroscopy on the packaging line. Eur J Pharm Biopharm 2001; 51:9–16.
43. Frake P, Greenhalgh D, Grierson SM, et al. Process control and end-point determination of a fluid bed granulation by application of near infra-red spectroscopy. Int J Pharm 1997; 151:75–80.
44. Morisseau KM, Rhodes CT. Near infrared spectroscopy as a nondestructive alternative to conventional tablet hardness testing. Pharm Res 1997; 14:108–11.
45. Fevotte G, Calas J, Puel F, et al. Applications of NIR spectroscopy to monitoring and analyzing the solid state during industrial crystallization process. Int J Pharm 2004; 273:159–69.
46. Skibsted ETS, Boelens HFM, Westerhuis JA, et al. Simple assessment of homogeneity in pharmaceutical mixing processes using a near-infrared reflectance probe and control charts. J Pharm Biomed Anal 2006; 41:26–35.
47. Cui X, Zhang Z, Ren Y, et al. Quality control of the powder pharmaceutical samples of sulfaguanidine by using NIR reflectance spectrometry and temperature-constrained cascade correlation networks. Talanta 2004; 64:943–8.
48. Woo YA, Lim HR, Kim HJ, et al. Determination of hydrogen peroxide concentration in antiseptic solutions using portable near-infrared system. J Pharm Biomed Anal 2003; 33:1049–57.
49. Hussain AS. FDA's Advisory Committee for Pharmaceutical Sciences, The ACPS's Process Analytical Technology Subcommittee, Rockville, MD, November 28, 2001. (Accessed May, 2006 at http://www.fda.gov/ohrms/dockets/ac/01/slides/3804s1_02_hussain/sld001.htm)
50. Andre M. Multivariate analysis and classification of the chemical quality of 7-aminocephalosporanic acid using near-infrared reflectance spectroscopy. Anal Chem 2003; 75:3460–7.
51. Blanco M, Romero MA. Near-infrared libraries in the pharmaceutical industry: a solution for identity confirmation. Analyst 2001; 126:2212–7.
52. Lonardi S, Viviani R, Mosconi L, et al. Drug analysis by near-infra-red reflectance spectroscopy. Determination of the active ingredient and water content in antibiotic powders. J Pharm Biomed Anal 1989; 7:303–8.
53. Plugge W, van der Vlies C. Near-infrared spectroscopy as an alternative to assess compliance of ampicillin trihydrate with compendial specifications. J Pharm Biomed Anal 1993; 11:435–42.
54. Blanco M, Coello J, Iturriaga H, et al. Critical review: near-infrared spectroscopy in the pharmaceutical industry. Analyst 1998; 123:135R–50.
55. Workman J, Jr., Veltkamp DJ, Doherty S, et al. Process analytical chemistry. Anal Chem 1999; 71:121R–80.
56. Workman J, Jr., Creasy KE, Doherty S, et al. Process analytical chemistry. Anal Chem 2001; 73:2705–18.
57. Blanco M, Villarroya I. NIR spectroscopy: a rapid-response analytical tool. Trends Anal Chem 2002; 21:240–50.
58. Lavine BK, Workman J, Jr. Chemometrics. Anal Chem 2002; 74:2763–70.
59. Workman J, Jr., Koch M, Veltkamp DJ. Process analytical chemistry. Anal Chem 2003; 75:2859–76.
60. Yu LX, Lionberger RA, Raw AS, et al. Applications of process analytical technology to crystallization processes. Adv Drug Del Rev 2004; 56:349–69.
61. Abrahamsson C, Johansson J, Sparen A, et al. Comparison of different variable selection methods conducted on NIR transmission measurements on intact tablets. Chemom Intell Lab Syst 2003; 69:3–12.
62. Airaksinen S, Luukkonen P, Jorgensen A, et al. Effects of excipients on hydrate formation in wet masses containing theophylline. J Pharm Sci 2003; 92:516–28.
63. Betz G, Bürgin PJ, Leuenberger H. Power consumption measurement and temperature recording during granulation. Int J Pharm 2004; 272:137–49.
64. Blanco M, Eustaquio A, Gonzalez JM, et al. Identification and quantitation assays for intact tablets of two related pharmaceutical preparations by reflectance near-infrared spectroscopy: validation of the procedure. J Pharm Biomed Anal 2000; 22:139–48.
65. Blanco M, Villar A. Development and validation of a method for the polymorphic analysis of pharmaceutical preparations using near infrared spectroscopy. J Pharm Sci 2003; 92:820–3.
66. Yoon WL, Jee RD, Charvill A, et al. Application of near-infrared spectroscopy to the determination of the sites of manufacture of proprietary products. J Pharm Biomed Anal 2004; 34:933–44.
67. Blanco M, Romero MA, Alcalà M. Strategies for constructing the calibration set for a near infrared spectroscopic quantitation method. Talanta 2004; 64:597–602.
68. Clarke F. Extracting process-related information from pharmaceutical dosage forms using near infrared microscopy. Vib Spectrosc 2004; 34:25–35.

69. Davis TD, Morris KR, Huang H, et al. In situ monitoring of wet granulation using online x-ray powder diffraction. Pharm Res 2003; 20:1851–7.
70. Fountain W, Dumstorf K, Lowell AE, et al. Near-infrared spectroscopy for the determination of testosterone in thin-film composites. J Pharm Biomed Anal 2003; 33:181–9.
71. Jorgensen AN, Luukkonen P, Rantanen J, et al. Comparison of torque measurements and near-infrared spectroscopy in characterization of a wet granulation process. J Pharm Sci 2004; 93:2232–43.
72. Laasonen M, Rantanen J, Harmia-Pulkkinen T, et al. Near infrared reflectance spectroscopy for the fast identification of PVC-based films. Analyst 2001; 126:1122–8.
73. Ritchie GE, Roller RW, Ciurczak EW, et al. Validation of a near-infrared transmission spectroscopic procedure. Part B: application to alternate content uniformity and release assay methods for pharmaceutical solid dosage forms. J Pharm Biomed Anal 2002; 29:159–71.
74. Blanco M, Coello J, Iturriaga H, et al. Determination of absorbic acid in pharmaceutical preparations by near infrared reflectance spectroscopy. Talanta 1993; 40:1671–6.
75. Blanco M, Coello J, Iturriaga H, et al. Control analysis of a pharmaceutical preparation by near-infrared reflectance spectroscopy. A comparative study of a spinning module and fibre optic probe. Anal Chim Acta 1994; 298:183–91.
76. Blanco M, Coello J, Iturriaga H, et al. Determination of water in ferrous lactate by near infrared reflectance spectroscopy with a fibre-optic probe. J Pharm Biomed Anal 1997; 16:255–62.
77. Gottfries J, Depui H, Fransson M, et al. Vibrational spectrometry for the assessment of active substance in metoprolol tablets: a comparison between transmission and diffuse reflectance near-infrared spectrometry. J Pharm Biomed Anal 1996; 14:1495–503.
78. Last IR, Prebble KA. Suitability of near-infrared methods for the determination of moisture in a freeze-dried injection product containing different amounts of the active ingredient. J Pharm Biomed Anal 1993; 11:1071–6.
79. Higgins JP, Arrivo SM, Thurau G, et al. Spectroscopic approach for on-line monitoring of particle size during the processing of pharmaceutical nanoparticles. Anal Chem 2003; 75:1777–85.
80. Watano S, Numa T, Koizumi I, et al. Feedback control in high shear granulation of pharmaceutical powders. Eur J Pharm Biopharm 2001; 52:337–45.
81. Watano S, Numa T, Miyanami K, et al. A fuzzy control system of high shear granulation using image processing. Powder Technol 2001; 115:124–30.
82. Andersson M, Folestad S, Gottfries J, et al. Quantitative analysis of film coating in a fluidized bed process by in-line NIR spectrometry and multivariate batch calibration. Anal Chem 2000; 72:2099–108.
83. Broad NW, Jee RD, Moffat AC, et al. Non-invasive determination of ethanol, propylene glycol and water in a multi-component pharmaceutical oral liquid by direct measurement through amber plastic bottles using Fourier transform near-infrared spectroscopy. Analyst 2000; 125:2054–8.
84. Rantanen J, Rasanen E, Tenhunen J, et al. In-line moisture measurement during granulation with a four-wavelength near infrared sensor: an evaluation of particle size and binder effects. Eur J Pharm Biopharm 2000; 50:271–6.
85. Kirsch JD, Drennen JK. Nondestructive tablet hardness testing by near-infrared spectroscopy: a new and robust spectral best-fit algorithm. J Pharm Biomed Anal 1999; 19:351–62.
86. Eustaquio A, Blanco M, Jee RD, et al. Determination of Paracetamol in intact tablets by use of near infrared transmittance spectroscopy. Anal Chim Acta 1999; 383:283–90.
87. Blanco M, Coello J, Eustaquio A, et al. Analytical control of pharmaceutical production steps by near infrared reflectance spectroscopy. Anal Chim Acta 1999; 392:237–46.
88. Blanco M, Coello J, Eustaquio A, et al. Development and validation of a method for the analysis of a pharmaceutical preparation by near-infrared diffuse reflectance spectroscopy. J Pharml Sci 1999; 88:551–6.
89. Blanco M, Coello J, Iturriaga H, et al. Development and validation of a near infrared method for the analytical control of a pharmaceutical preparation in three steps of the manufacturing process. Fresenius J Anal Chem 2000; 368:534–9.
90. Gupta A, Peck GE, Miller RW, et al. Real-time near-infrared monitoring of content uniformity, moisture content, compact density, tensile strength, and Young's modulus of roller compacted powder blends. J Pharm Sci 2005; 94:1589–97.
91. Lai CK, Cooney CC. Application of a fluorescence sensor for miniscale on-line monitoring of powder mixing kinetics. J Pharm Sci 2004; 93:60–70.
92. Moffat AC, Trafford AD, Jee RD, et al. Meeting of the International Conference on Harmonisation's Guidelines on Validation of Analytical Procedures: quantification as exemplified by a near-infrared reflectance assay of paracetamol in intact tablets. Analyst 2000; 125:1341–51.
93. El-Hagrasy AS, D'Amico F, Drennen JK, III. A process analytical technology approach to near-infrared process control of pharmaceutical powder blending. Part I: D-optimal design for characterization of powder mixing and preliminary spectral data evaluation. J Pharm Sci 2006; 95:392–406.
94. El-Hagrasy AS, Delgado-Lopez M, Drennen JK, III. A process analytical technology approach to near-infrared process control of pharmaceutical powder blending: Part II: qualitative near-infrared models for prediction of blend homogeneity. J Pharm Sci 2006; 95:407–21.
95. El-Hagrasy AS, Drennen JK, III. A process analytical technology approach to near-infrared process control of pharmaceutical powder blending. Part III: quantitative near-infrared calibration for prediction of blend homogeneity and characterization of powder mixing kinetics. J Pharm Sci 2006; 95:422–34.
96. Gupta A, Peck GE, Miller RW, et al. Influence of ambient moisture on the compaction behavior of microcrystalline cellulose powder undergoing uni-axial compression and roller-compaction: a comparative study using near-infrared spectroscopy. J Pharm Sci 2005; 94:2301–13.
97. Bai SJ, Rani M, Suryanarayanan R, et al. Quantification of glycine crystallinity by near-infrared (NIR) spectroscopy. J Pharm Sci 2004; 93:2439–47.
98. Seyer JJ, Luner PE, Kemper MS. Application of diffuse reflectance near-infrared spectroscopy for determination of crystallinity. J Pharm Sci 2004; 89:1305–16.
99. Lai CK, Holt D, Leung JC, et al. Real-time and non-invasive monitoring of dry powder blend homogeneity. AIChE J 2001; 47:2618–22.
100. Johansson J, Pettersson S, Folestad S. Characterization of different laser irradiation methods for quantitative Raman tablet assessment. J Pharm Biomed Anal 2005; 39:510–6.
101. Cogdill RP, Anderson CA, Delgado-Lopez M, et al. Process analytical technology case study Part I: feasibility studies for quantitative near-infrared method development. AAPS Pharm Sci Tech 2005; 6:E262–72.
102. Cogdill RP, Anderson CA, Delgado M, et al. Process analytical technology case study: Part II. Development and validation of near-infrared calibrations in support of a process analytical technology application for real-time release. AAPS Pharm Sci Tech 2005; 6:E273–83.

103. Cogdill RP, Anderson CA, Delgado M, et al. Process analytical technology case study, Part III: calibration monitoring and transfer. AAPS Pharm Sci Tech 2005; 6:E284–97.
104. Hausman DS, Cambron RT, Sakr A. Application of on-line Raman spectroscopy for characterizing relationships between drug hydration state and tablet physical stability. Int J Pharm 2005; 299:19–33.
105. Islam MT, Rodrıguez-Hornedo N, Ciotti S, et al. The potential of Raman spectroscopy as a process analytical technique during formulations of topical gels and emulsions. Pharm Res 2004; 21:1844–51.
106. Johansson J, Pettersson S, Taylor LS. Infrared imaging of laser-induced heating during Raman spectroscopy of pharmaceutical solids. J Pharm Biomed Anal 2002; 30:1223–31.
107. Kontoyannis CG. Quantitative determination of $CaCO_3$ and glycine in antacid tablets by laser Raman spectroscopy. J Pharm Biomed Anal 1995; 13:73–6.
108. Langkilde FW, Sjoblom J, Tekenbergs-Hjelte L, et al. Quantitative FT-Raman analysis of two crystal forms of a pharmaceutical compound. J Pharm Biomed Anal 1997; 15:687–96.
109. Szostak R, Mazurek S. Quantitative determination of acetylsalicylic acid and acetaminophen in tablets by FT-Raman spectroscopy. Analyst 2002; 127:144–8.
110. Taylor LS, Zografi G. The quantitative analysis of crystallinity using FT-Raman spectroscopy. Pharm Res 1998; 15:755–61.
111. Shabushnig JG. Pharmacia Corporation. Process analytical technology: an industry perspective. FDA's Advisory Committee for Pharmaceutical Sciences, Subcommittee on Process Analytical Technologies (PAT), Gaithersburg, MD, February 25, 2002. (Accessed May, 2006 at http://www.fda.gov/ohrms/dockets/ac/02/slides/3841s1_04_Shabushnig_files/frame.htm)
112. Clevett KJ. Process analytical chemistry—industry perspectives—trends in applications and technology. Process Control Qual 1994; 6:81–90.
113. Sinsheimer JE, Poswalk NM. Pharmaceutical applications of the near infrared determination of water. J Pharm Sci 1968; 57:2007–10.
114. Ryder AG, O'Conner GM, Glynn TJ. Quantitative analysis of cocaine in solid mixtures using Raman spectroscopy and chemometric methods. J Raman Spectrosc 2000; 31:221–7.
115. Clarke FC, Jamieson MJ, Clark DA, et al. Chemical image fusion. The synergy of FTNIR and Raman mapping microscopy to enable a more complete visualization of pharmaceutical formulations. Anal Chem 2001; 73:2213–20.
116. Chang H, Huang P. Thermo-Raman spectroscopy. Rev Anal Chem 2001; 20:207–38.
117. Ghule A, Baskaran N, Murugan R, Chang H. Phase transformation studies of Na_3PO_4 by thermo-Raman and conductivity measurements. Solid State Ionics 2003; 161:291–9.
118. de Jager H-J, Prinsloo L. The dehydration of phosphates monitored by DSC/TGA and in situ Raman spectroscopy. Thermochim Acta 2001; 376:187–96.
119. 2006 USPC, Inc. Official 4/1/06-7/31/06 General Chapters: <1119> Near-Infrared Spectrophotometry.
120. GlaxoSmithKline. Perspectives in chemometrics. Experience from GlaxoSmithKline. FDA's Advisory Committee for Pharmaceutical Sciences, Subcommittee on Process Analytical Technologies (PAT), Gaithersburg, MD, February 25, 2002. (Accessed May, 2006 at http://www.fda.gov/ohrms/dockets/ac/02/slides/3841s1_09_walker/sld001.htm)
121. Bush L. New CGMP plant at Purdue University offers multiple benefits. Pharm Technol 2003; 27:22.

Computerized Systems Validation

Saeed Tafreshi
Intelitec Corporation, Irvine, California, U.S.A.

The concept of validation was developed in the 1970s and is widely credited to Ted Byers who was then Associate Director of Compliance at the U.S. FDA. The concept was focused on:

> Establishing documented evidence which provides a high degree of assurance that a specific process will consistently produce a product meeting its predetermined specifications and quality attributes.

This concept continues to be followed, with some modifications, by the various authorities regulating GMP around the world. This definition also has been adopted for the validation business, manufacturing and laboratory computer systems. The need to validate computer systems formally began in 1979 when the U.S.A. introduced GMP regulatory legislation which specifically referred to automation equipment. GMP is enforced by national regulatory authorities who can prevent the sale of a product in their respective country if they consider its manufacture not to be GMP compliant. Validation for GMP is a license to operate issue.

Over the last three decades, the manufacturing industry has increasingly used computer systems to control manufacturing processes for improved performance and product quality. This policy is often embedded in corporate strategy. Computer systems, however, by the nature of their complexity are susceptible to development and operational deficiencies which can adversely affect their control ability and effect product safety, quality and efficacy. Common examples of such deficiencies include poor specification capture, design errors, poor testing and poor maintenance practice.

The potentially devastating outcome of GMP noncompliance of computer systems was demonstrated in 1988 when deficient software in data management system controlling a blood bank could have led to the issue of AIDS-infected blood. Additionally, computer systems can endanger public health through the manufacture and release of drug products with deficient quality attributes.

The first widely publicized FDA citation for computer validation noncompliance occurred in 1985; however, as early as 1982, the FDA was publicly stating that it was "nervous" if computer systems were used without being validated. In 1983, the FDA issued the Guide to Inspection of Computerized Systems in Drug Processing, Technical Report, Reference Materials and Training Aids for Investigators which became known as the "Blue Book." This publication guided inspectors on what to accept as validation evidence for computer systems. The Blue Book formally introduced the anticipation of a life-cycle approach to validation. The aim was to build in quality (QA) rather than rely on testing in quality (quality control).

Responding to the FDA's proactive position on computer systems validation, the PMA formed a Computer Systems Validation Committee to represent and coordinate the industry's viewpoint. The results were a joint FDA/PMA Conference in 1984 discussing computer systems validation and in the following year the publication of an industry perspective. The publication presented an approach for validation for both new and existing computer systems. GMP legislation is unusual in that it is equally applied to new production facilities and to production facilities built entirely or partially before the legislation (including amendments) was enforced.

Throughout the 1980s, computer systems validation was debated primarily in the U.S.A. Ken Chapman published a paper covering this period during which the FDA gave advice on the following GMP issues:

- Input/output checking
- Batch records
- Applying GMP to hardware and software
- Supplier responsibility
- Application software inspection
- FDA investigation of computer systems
- Software development activities

In addition, since the end of 1980s, the FDA and the pharmaceutical industry have debated the GMP requirements and the practicalities of electronic signatures. A resolution was achieved which became the FDA's proposed regulation.

Complementing the U.S. GMP guidance, the European Commission and authorities in Australia both issued GMP codes of practice in 1989 and 1990 respectively. The European code known as the "Orange Guide" was later issued in 1991 as a Directive superseding

---

Abbreviations used in this chapter: AZT, International association for Pharmaceutical Technology; cGMP, current good manufacturing practice; CRT, cathode ray tube; DQ, design qualification; EU, European Union; FAT, factory acceptance test; FDA, Food and Drug Administration; GAMP, good automated manufacturing practice; GMP, good manufacturing practice; IQ, installation qualification; ISPE, International Society of Pharmaceutical Engineering; MCA, Medicines Control Agency; OQ, operational qualification; PDA, Parenteral Drug Association; PLC, programmable logic controller; PMA, Pharmaceutical Manufactures Association; PQ, performance qualification; QA, quality assurance; SCADA, supervisory control and data acquisition; SQ, system qualification or specification qualification; URS, user requirements specification.

member state GMP legislation and included an annex covering computerized systems.

In most countries, GMP has been interpretive and to prosecute a pharmaceutical manufacturing a court must be convinced that the charges reflect the intent to flout governing legislation. In the U.S.A., however, a court declaratory judgment determined supplementary GMP information to be substantive. The net effect was that the FDA's advisory opinions became binding on the Agency. In August of 1990, the FDA announced that it no longer considered advisory opinions binding on the grounds that Counsel considered such restrictions unconstitutional. Hence, the FDA interpretation of the regulations in Compliance Policy Guides, Guide to Investigators, and other publications by FDA authors became nonbinding.

Computer systems validation also became a high profile industry issue in Europe in 1991 when several European manufacturers and products were temporarily banned from the U.S.A. for computer systems noncompliance. The computer systems in question included autoclave PLCs and SCADA systems. The position of the FDA was clear; the manufacturer had failed to satisfy their "concerns" that computer systems should:

- Perform accurately and reliably
- Be secure from unauthorized or inadvertent changes
- Provide for adequate documentation of the process

The manufacturers thought they had satisfied the requirements of the existing GMP legislations, but they had not satisfied the FDA's expectations of GMP. Hence the adoption of cGMP to signify the latest understanding of the validation practices and standards expected by the regulatory authorities began.

In 1991, the U.K. Pharmaceutical Industry Computer Systems Validation Forum (known as the U.K. FORUM) was established to facilitate the exchange of validation knowledge and the development of a standard industry guide for computer systems validations. At this time suppliers were on the whole struggling to understand and implement the various interpretations and requirements of GMP presented by the manufacturers. ISO 9000 and TickIT accreditation for quality management provided a good basis for validation, but it does not fully satisfy GMP requirements. Then, the U.K. FORUM's guide came to fruition and was launched as a first draft within the U.K. The guide is often referred to as the GAMP guide.

Meanwhile two experienced GMP regulatory inspectors, Ron Tetzlaff and Tony Trill, published papers respectively presenting the FDA's and U.K.'s MCA inspection practice for computer systems. These papers presented a comprehensive perspective on the current validation expectations of GMP regulatory authorities. Topics covered included:

- Life-cycle approach
- Quality management
- Procedures
- Training
- Validation protocols
- Qualification evidence
- Change control
- Audit trail
- Ongoing evaluation

The pharmaceutical industry in search of a common approach to computer systems validation began incorporating these topics. Nevertheless, the FDA and MCA continue to encounter instances of noncompliance practice based on:

- Incomplete documents
- Insufficient detail in documents
- Missing documentary evidence

There was a clear need for guidance and standards on computer systems validation and early in 1995 there were four milestones of significance to practitioners:

- The U.S.A. proposed new GMP legislation affecting electronic records and electronic signatures.
- After 16 years the U.S.A. amended its legislation affecting computer validation, making a minor concession concerning the degree of input/output validation required for reliable computer systems.
- The U.S. PDA presented a manufacturer's guide to complement the PMA life cycle.
- The U.K. FORUM issued a revised draft of their supplier guide for European comment.

These initiatives helped the manufacturers and suppliers meet the challenge to validate computer systems effectively and efficiently. The initiatives which further clarified the requirements of validation included:

- The U.K. FORUM's investigation into the benefits of supplier audits shared by a number of participant manufacturers.
- The German APV (Information Technology Section) guide to Annex 11 of the European United GMP Directive regarding computerized systems.
- The German GMA Committee 5.8 and NAMUR Committee 1.9 joint working group's recommendations for computer systems validation.
- The coordination of the German initiatives with the U.K. FORUM supplier guide, and possibly the U.S. PDA manufacturer's guide, as announced at the ISPE computer validation seminar in Amsterdam in March of 1995.

What is clear to date is the mutual benefit of regulators, manufacturers and suppliers working together towards a common GMP goal. GMP, while facilitating improvements to manufacturing performance, also is integral to the continuing high standing of the pharmaceutical industry.

In order for the industry to follow a common path in complying with the cGMP guidelines related to computer control systems, there is a need to understand the basics of proper system development and consider the overall cost into building a true business case. In doing so, it is necessary to follow the stages in sequence for the validation of a computerized control system to FDA requirements and their relationship to the development and implementation stages of an automation project.

The Quality System regulation requires that "when computers or automated data processing systems are used as part of production or the quality system, the manufacturer shall validate computer software for its intended use according to an established protocol." This has been a regulatory requirement for GMP since 1978.

In addition to the above validation requirement, computer systems that implement part of a regulated manufacturer's production processes or quality system (or that are used to create and maintain records required by any other FDA regulation) are subject to the Electronic Records, Electronic Signatures regulation. This regulation establishes additional security, data integrity, and validation requirements when records are created or maintained electronically. These additional Part 11 requirements should be carefully considered and included in system requirements and software requirements for any automated record keeping systems. System validation and software validation should demonstrate that all Part 11 requirements have been met.

Computers and automated equipment are used extensively throughout Pharmaceutical, Biotech, Medical Device, and Medical Gas industries in areas such as design, laboratory testing and analysis, product inspection and acceptance, production and process control, environmental controls, packaging, labeling, traceability, document control, complaint management, and many other aspects of the quality system. Increasingly, automated plant floor operations have involved extensive use of embedded systems in

- PLCs
- digital function controllers
- statistical process control
- supervisory control and data acquisition
- robotics
- human–machine interfaces
- input/output devices
- computer operating systems

Computerized operations are now common in FDA regulated industries. Small "minicomputer" systems are being used, sometimes in conjunction with larger computers, to control batching operations, maintain formula files and inventories, monitor process equipment, check equipment calibration, etc. The medical device industry is presently utilizing automatic test sets controlled by computers. In this application the computer is relied upon to make the decision as to whether a particular test parameter is within a specific tolerance. The operator does not see the values of the parameters measured, but merely receives a green or red light indicating a go/no go situation. Products are accepted or rejected on this basis. In order to evaluate and/or report the adequacy of any computer-controlled processes or tests, the basics of computer construction and operation must be understood. The entire computer control system has been simplified as follows.

A computer is a machine and like all other machines is normally used because it performs specific tasks with greater accuracy and more efficiency than people. Computers accomplish this by having the capacity to receive, retain, and give up large volumes of data and process it in a very short time. An understanding of computer operation, and the ability to use a computer, does not require a detailed knowledge of either electronics or the physical hardware construction. An overall view of the computer organization with emphasis on function is sufficient.

There are basically two types of computers, analog and digital. The analog computer does not compute directly with numbers. It accepts electrical signals of varying magnitude (analog signals) which in practical use are analogous to or represent some continuous physical magnitude such as pressure, temperature, etc. Analog computers are sometimes used for scientific, engineering and process-control purposes. In the majority of industry applications used today, analog values are converted to digital form by an analog-to-digital converter and processed by digital computers.

The digital computer is the general use computer used for manipulating symbolic information. In most applications the symbols manipulated are numbers and the operations performed on the symbols are the standard arithmetical operations. Complex problem solving is achieved by basic operations of addition, subtraction, multiplication and division.

A digital computer is designed to accept and store instructions (program), accept information (data) and process the data as specified in the program and display the results of the processing in a selected manner. Instructions and data are in coded form the computer is designed to accept. The computer performs automatically and in sequence according to the program.

The computer is a collection of interconnected electromechanical devices (hardware) directed by a central control unit. The central control unit is the controlling device that supervises the sequence of activities that take place in all parts of the computer. Classically, the hardware consists of the mainframe (computer) for computation, storage and control, and peripheral devices (input–output devices) for entering raw data and printing or displaying the output. Input data may be entered into the computer by teletypewriters, magnetic tape, punched tape, card readers, etc. Output may be displayed in the form of a hardcopy printout, magnetic tape, CRT, etc. The two units of input and output are often joined and referred to as input/output or simply I/O. A computer terminal with a CRT display is an example of a combined Input/Output device.

Equally important as hardware in the effective use of the digital computer is the software. The numerous written programs and/or routines that dictate the process sequence the computer will follow are called software. A computer can be programmed to do almost any problem that can be "defined." Defined means that the solution of a problem must be reduced to a series of steps that can be written as a series of computer instructions. In other words, the individual steps of the problem must be set up, including the desired level of accuracy, prior to the computer processing and solving the problem. The computer must be directed or commanded by a precisely stated set of commands or program. Until a program is prepared and stored in the computer memory, the computer knows absolutely nothing, not even how to receive input data. The accuracy and validation of the program is one of the most important aspects of computer control.

Physical quantities are especially adaptive to binary digital techniques because most physical quantities can be expressed as two states: switches are on or off, a quantity level is above or below a set value, holes in cards are punched or not punched, electrical voltage or current is positive or negative or above or below a preset value. For such applications as process control, the digital computer makes decisions by comparing input data to a predetermined value. The computer takes a course of action dependent on whether the input data is greater than, equal to, or less than the predetermined value.

The predetermined value and course of action the computer follows is in the form of a program stored in the computer memory. So, actually the computer does not make decisions, but merely follows written program instructions. A printout or display of the actual values measured may be included as a part of the program. Verification of proper computer operation may be accomplished in this example by applying known inputs, greater, equal to and less than the predetermined value and subsequently reviewing the results.

When validating a computer control system, particular attention must be made to following of established procedures and the documentation required during each stage to ensure that proper and sufficient documented evidence is provided to support validation inspection by the FDA.

The FDA has issued two validation definitions which state the following:

1. "Establishing documented evidence that a system does what it is designed to do."
2. "Establishing documented evidence which provides a high degree of assurance that a specific process will consistently produce a product meeting its predetermined specifications and quality attributes."

The FDA audits against compliance with cGMP requirements. Rigid procedures are required to be followed and those procedures must generate sufficient documentation to ensure that traceability and accountability of information (an audit trail) is maintained.

The FDA does not provide certification for a company and its procedures nor does it approve what documentation should be produced. The company is responsible for demonstrating that procedures are followed and associated documentation generated to support the manufacture of the company's products.

The FDA's position was made clear in the following statement made by Ronald Tetzlaff (when he was employed by the FDA) in Pharmaceutical Technology, April 1992, which states that "Unless firms have documented evidence to ensure the proper performance of a vendor's software, the FDA cannot consider the automated system to be validated."

Therefore it is important that companies have approved Quality Systems in place that ensure that procedures are followed and an audit trail is maintained.

COMPUTERIZED SYSTEM VALIDATION QUALITY SYSTEM

The validation of a computerized control system to FDA requirements can be broken down into a number of phases which are interlinked with the overall project program. A typical validation program for a control system also includes the parallel design and development of control and monitoring instrumentation. A typical Quality System includes the following phases.

Definition Phase

Validation starts at the definition (conceptual design) phase because the FDA expects to see documentary evidence that the chosen system vendor and the software proposed meets the customer's predefined selection criteria.

Vendor acceptance criteria, which must be defined by the customer, should typically include the following.

The Vendor's Business Practices

- Vendor certification to an approved QA standard. Certification may be a consideration when selecting a systems vendor. Initiative which promotes the use of international standards to improve the quality management of software development shall be considered.
- Vendor Audit by the customer to ensure company standards and practices are known and are being followed.
- Vendor end user support agreements.
- Vendor financial stability.
- Biography for the vendor's proposed project personnel (interviews also should be considered).
- Checking customer references and visiting their sites should be considered.

The Vendor's Software Practices

- Software development methodology
- Vendor's experience in using the project software including: operating system software; application software; "off-the-shelf" and support software package (e.g., archiving, networking, batch software).
- Software performance and development history
- Software updates
- The vendor must make provision for source code to be accessible to the end user (e.g., have an escrow or similar agreement) and should provide a statement to this effect. Escrow is the name given to a legally binding agreement between a supplier and a customer which permits the customer access to source code, which is stored by a third party organization. The agreement also permits the customer access to the source code should the supplier become bankrupt.

Vendor acceptance can be divided into these areas:

- Vendor prequalification (to select suitable vendors to receive the Tender enquiry package)
- Review of the returned Tenders
- Audit of the most suitable vendor(s)

Other documentation produced during the definition phase includes the URS, standard specifications and Tender support documentation.

The Tender enquiry package must be reviewed by the customer prior to issue to selected vendors. This review, called SQ, is carried out to ensure that the customer's technical and quality requirements are fully addressed.

System Development Phase

The system development phase is the period from Tender award to delivery of the control system to site. It can be subdivided into four subphases:

- Design agreement
- Design and development
- Development testing
- Predelivery or FAT

The design agreement phase comprises the development and approval of the system vendor's

Functional Design Specification, its associated FAT, Specification and the Quality Plan for the project. These form the basis of the contractual agreement between the system vendor and the customer.

The design and development phase involves the development and approval of the detailed system (hardware and software) design and testing specifications. The software specifications comprise the Software Design Specification and its associated Software Module Coding. The hardware specifications comprise the Computer Hardware Design Specification and its associated Hardware Test Specification and Computer Hardware Production.

The development testing phase comprises the structured testing of the hardware and software against the detailed design specifications starting from the lowest level and working up to a fully integrated system. The systems vendor must follow a rigorous and fully documented testing regime to ensure that each item of hardware and software module developed or modified performs the function(s) required without degrading other modules or the systems as a whole.

The predelivery acceptance phase comprises the FAT, which is witnessed by the customer, and the DQ review by the customer to ensure the system design meets technical (system functionality and operability) and quality (auditable, structured documentation) objectives.

Throughout the system development phase, the systems vendor should be subject to a number of quality audits by the customer, or their nominated agents, to ensure that the Quality Plan for the project is being complied with and that all documentation is being completed correctly. In addition, the vendor should conduct internal audits, and the reports should be available for inspection by the customer. The systems vendor also must enforce a strict change control procedure to enable all mediations and changes to the system to be thoroughly designed, tested, and documented. Change control is a formal system by which qualified representatives of appropriate disciplines review proposed or actual changes that might affect a validated status. The intent is to determine the need for action that would ensure and document that the component or system is maintained in a validated state.

The audit trail documentation introduced and maintained by the Quality Plan and the test documentation can be used as evidence by the customer during the FDA's inspections that the system meets the functionality required. In particular, the test and change control documentation will demonstrate a positive, thorough, and professional approach to validation.

Commissioning and "In-Place" Qualification Phase

The commissioning and qualification phase encompasses the System Commissioning on site, Site Acceptance Testing, IQ, OQ, and, where applicable, PQ activities for the project. The most important part of this phase must be identified as qualification based on system specification documentation. The system installation and operation must be confirmed against its documents. All system adjustments and changes occuring in this phase must result in updating of the corresponding specification document. It is an assurance when building a reliable system base document in support of a life cycle approach during a phase that most last minute changes are discovered. No benefit of any life cycle approach can be obtained when the system and its documentation do not match after completion of this phase.

Ongoing Maintenance Phase

The term maintenance does not mean the same when applied to hardware and software. The operational maintenance of hardware and software are different because their failure/error mechanisms are different. Hardware maintenance typically includes preventive hardware maintenance actions, component replacement, and corrective changes. Software maintenance includes corrective, perfective, and adaptive maintenance but does not include preventive maintenance actions or software component replacement.

Changes made to correct errors and faults in the software are corrective maintenance. Changes made to the software to improve the performance, maintainability, or other attributes of the software system are perfective maintenance. Software changes to make the software system usable in a changed environment are adaptive maintenance.

When changes are made to a software system, sufficient regression analysis and testing should be conducted to demonstrate that portions of the software not involved in the change were not adversely impacted. This is in addition to testing that evaluates the correctness of the implemented change(s).

The specific validation effort necessary for each change is determined by the type of change, the development products affected, and the impact of those products on the operation of the system. All proposed modifications, enhancements, or additions to the system should be assessed to determine the effect each change would have on the entire system. This information should determine the extent to which verification and/or validation tasks need to be iterated.

Documentation should be carefully reviewed to determine which documents have been impacted by a change. All approved documents (e.g., specifications, user manuals, drawings, etc.) that have been affected should be updated in accordance with the applicable site or corporate change management procedures. Specifications should be updated before any change is implanted.

SOFTWARE VALIDATION

The Quality System regulation treats "verification" and "validation" as separate and distinct terms. On the other hand, many software engineering journal articles and textbooks use the terms verification and validation interchangeably, or in some cases refer to software "verification, validation, and testing (VV&T)" as if it is a single concept, with no distinction among the three terms.

Software verification provides objective evidence that the design outputs of a particular phase of the software development life cycle meet all of the specified requirements for that phase. Software verification looks for consistency, completeness, and correctness of the software and its supporting documentation, as it is being developed, and provides support for a subsequent conclusion that software is validated. Software testing is one of many verification activities intended to confirm that software development output meets its input requirements. Other verification activities include various static and dynamic analyses, code and document inspections, walkthroughs, and other techniques.

Software validation is a part of the design validation for the project, but is not separately defined in the Quality System regulation. FDA considers software validation to be "confirmation by examination and provision of objective evidence that software specifications conform to user needs and intended uses, and that the particular requirements implemented through software can be consistently fulfilled." In practice, software validation activities may occur both during as well as at the end of the software development life cycle to ensure that all requirements have been fulfilled. Since software is usually part of a larger hardware system, the validation of software typically includes evidence that all software requirements have been implemented correctly and completely and are traceable to system requirements. A conclusion that software is validated is highly dependent upon comprehensive software testing, inspections, analyses, and other verification tasks performed at each stage of the software development life cycle.

Software verification and validation are difficult in nature because a developer cannot test forever, and it is hard to know how much evidence is enough. In large measure, software validation is a matter of developing a "level of confidence" that the application meets all requirements and user expectations for the software automated functions. Measures such as defects found in specifications documents, estimates of defects remaining, testing coverage, and other techniques are all used to develop an acceptable level of confidence before shipping the product. The level of confidence, and therefore the level of software validation, verification, and testing effort needed, will vary depending upon the application.

Many firms have asked for specific guidance on what the FDA expects them to do to ensure compliance with the Quality System regulation with regard to software validation. Validation of software has been conducted in many segments of the software industry for almost three decades. Due to the great variety of pharmaceuticals, medical devices, processes, and manufacturing facilities, it is not possible to state in one document all of the specific validation elements that are applicable. However, a general application of several broad concepts can be used successfully as guidance for software validation. These broad concepts provide an acceptable framework for building a comprehensive approach to software validation.

Requirements Specification

While the Quality System regulation states that design input requirements must be documented, and that specified requirements must be verified, the regulation does not further clarify the distinction between the terms "requirement" and "specification." A requirement can be any need or expectation for a system or for its software. Requirements reflect the stated or implied needs of the customer, and may be market-based, contractual, or statutory, as well as an organization's internal requirements. There can be many different kinds of requirements (e.g., design, functional, implementation, interface, performance, or physical requirements). Software requirements are typically derived from the system requirements for those aspects of system functionality that have been allocated to software. Software requirements are typically stated in functional terms and are defined, refined, and updated as a development project progresses. Success in accurately and completely documenting software requirements is a crucial factor in successful validation of the resulting software.

A specification is defined as "a document that states requirements." It may refer to or include drawings, patterns, or other relevant documents and usually indicates the means and the criteria whereby conformity with the requirement can be checked. There are many different kinds of written specifications, e.g., system requirements specification, software requirements specification, software design specification, software test specification, software integration specification, etc. All of these documents establish "specified requirements" and are design outputs for which various forms of verification are necessary.

A documented software requirements specification provides a baseline for both validation and verification. The software validation process cannot be completed without an established software requirements specification.

Defect Prevention

Software quality assurance needs to focus on preventing the introduction of defects into the software development process and not on trying to "test quality into" the software code after it is written. Software testing is very limited in its ability to surface all latent defects in software code. For example, the complexity of most software prevents it from being exhaustively tested. Software testing is a necessary activity. However, in most cases software testing by itself is not sufficient to establish confidence that the software is fit for its intended use. In order to establish that confidence, software developers should use a mixture of methods and techniques to prevent software errors and to detect software errors that do occur. The "best mix" of methods depends on many factors including the development environment, application, size of project, language, and risk.

Time and Effort

To build a case that the software is validated requires time and effort. Preparation for software validation should begin early, i.e., during design and development planning and design input. The final conclusion that the software is validated should be based on evidence collected from planned efforts conducted throughout the software life cycle.

Software Life Cycle

Software validation takes place within the environment of an established software life cycle. The software life cycle contains software engineering tasks and documentation necessary to support the software validation effort.

In addition, the software life cycle contains specific verification and validation tasks that are appropriate for the intended use of the software. No one life cycle model can be recommended for all software development and validation project, but an appropriate and practical software life cycle should be selected and used for a software development project.

Plans

The software validation process is defined and controlled through the use of a plan. The software validation plan defines "what" is to be accomplished through the software validation effort. Software validation plans are a significant quality system tool. Software validation plans specify areas such as scope, approach, resources, schedules and the types and extent of activities, tasks, and work items.

Procedures

The software validation process is executed through the use of procedures. These procedures establish "how" to conduct the software validation effort. The procedures should identify the specific actions or sequence of actions that must be taken to complete individual validation activities, tasks, and work items.

Software Validation After a Change

Due to the complexity of software, a seemingly small local change may have a significant global system impact. When any change (even a small change) is made to the software, the validation status of the software needs to be re-established. Whenever software is changed, a validation analysis should be conducted not just for validation of the individual change but also to determine the extent and impact of that change on the entire software system. Based on this analysis, the software developer should then conduct an appropriate level of software regression testing to show that unchanged but vulnerable portions of the system have not been adversely affected. Design controls and appropriate regression testing provide the confidence that the software is validated after a software change.

Validation Coverage

Validation coverage should be based on the software's complexity and safety risk and not on firm size or resource constraints. The selection of validation activities, tasks, and work items should be commensurate with the complexity of the software design and the risk associated with the use of the software for the specified intended use. For lower risk applications, only baseline validation activities may be conducted. As the risk increases, additional validation activities should be added to cover the additional risk. Validation documentation should be sufficient to demonstrate that all software validation plans and procedures have been completed successfully.

Flexibility and Responsibility

Specific implementation of these software validation principles may be quite different from one application to another. The manufacturer has flexibility in choosing how to apply these validation principles, but retains ultimate responsibility for demonstrating that the software has been validated.

Software is designed, developed, validated, and regulated in a wide spectrum of environments, and for a wide variety of applications with varying levels of risk. In each environment, software components from many sources may be used to create the software (e.g., in-house developed software, off-the-shelf software, contract software, shareware). In addition, software components come in many different forms (e.g., application software, operating systems, compilers, debuggers, configuration management tools, and many more). The validation of software in these environments can be a complex undertaking; therefore, it is appropriate that all of these software validation principles be considered when designing the software validation process. The resultant software validation process should be commensurate with the safety risk associated with the system, device, or process.

Software validation activities and tasks may be dispersed, occurring at different locations and being conducted by different organizations. However, regardless of the distribution of tasks, contractual relations, source of components, or the development environment, the manufacturer retains ultimate responsibility for ensuring that the software is validated.

Software validation is accomplished through a series of activities and tasks that are planned and executed at various stages of the software development life cycle. These tasks may be one-time occurrences or may be iterated many times, depending on the life cycle model used and the scope of changes made as the software project progresses.

SOFTWARE LIFE CYCLE ACTIVITIES

Software developers should establish a software life cycle model that is appropriate for their product and organization. The software life cycle model that is selected should cover the software from its birth to its retirement. Activities in a typical software life cycle model include the following:

- Quality Planning
- System Requirements Definition
- Detailed Software Requirements Specification
- Software Design Specification
- Construction or Coding
- Testing
- Installation
- Operation and Support
- Maintenance
- Retirement

Verification, testing and other tasks that support software validation occur during each of the above activities. A life cycle model organizes these software development activities in various ways and provides a framework for monitoring and controlling the software development project.

For each of the software life cycle activities, there are certain "typical" tasks that support a conclusion that the software is validated. However, the specific tasks to be performed, their order of performance, and the iteration and timing of their performance will be dictated by the specific software life cycle model that is selected and the safety risk associated with the software application. For very low risk applications, certain tasks may not be needed at all. However, the software developer should at least consider each of these tasks and should define and document which tasks are or are not appropriate for their specific application.

Quality Planning

Design and development planning should culminate in a plan that identifies necessary tasks, procedures for anomaly reporting and resolution, necessary resources, and management review requirements, including formal design reviews. A software life cycle model and associated activities should be identified, as well as those tasks necessary for each software life cycle activity. The plan should include:

- The specific tasks for each life cycle activity
- Enumeration of important quality factors
- Methods and procedures for each task
- Task acceptance criteria
- Criteria for defining and documenting outputs in terms that will allow evaluation of their conformance to input requirements
- Inputs for each task
- Outputs from each task
- Roles, resources, and responsibilities for each task
- Risks and assumptions
- Documentation of user needs

Management must identify and provide the appropriate software development environment and resources. Typically, each task requires personnel as well as physical resources. The plan should identify the personnel, the facility and equipment resources for each task, and the role that risk (hazard) management will play. A configuration management plan should be developed that will guide and control multiple parallel development activities and ensure proper communications and documentation. Controls are necessary to ensure positive and correct correspondence among all approved versions of the specifications documents, source code, object code, and test suites that comprise a software system. The controls also should ensure accurate identification of, and access to, the currently approved versions.

Procedures should be created for reporting and resolving software anomalies found through validation or other activities. Management should identify the reports and specify the contents, format, and responsible organizational elements for each report. Procedures also are necessary for the review and approval of software development results, including the responsible organizational elements for such reviews and approvals.

Requirements

Requirement development includes the identification, analysis, and documentation of information about the application and its intended use. Areas of special importance include allocation of system functions to hardware/software, operating conditions, user characteristics, potential hazards, and anticipated tasks. In addition, the requirements should state clearly the intended use of the software.

The software requirements specification document should contain a written definition of the software functions. It is not possible to validate software without predetermined and documented software requirements. Typical software requirements specify the following:

- All software system inputs
- All software system outputs
- All functions that the software system will perform
- All performance requirements that the software will meet
- The definition of all external and user interfaces, as well as any internal software-to-system interfaces
- How users will interact with the system
- What constitutes an error and how errors should be handled
- Required response times
- The intended operating environment
- All ranges, limits, defaults, and specific values that the software will accept
- All safety related requirements, specifications, features, or functions that will be implemented in software

Software safety requirements are derived from a technical risk management process that is closely integrated with the system requirements development process. Software requirement specifications should identify clearly the potential hazards that can result from a software failure in the system as well as any safety requirements to be implemented in software. The consequences of software failure should be evaluated, along with means of mitigating such failures (e.g., hardware mitigation, defensive programming, etc.). From this analysis, it should be possible to identify the most appropriate measures necessary to prevent harm.

A software requirements traceability analysis should be conducted to trace software requirements to (and from) system requirements and to risk analysis results. In addition to any other analyses and documentation used to verify software requirements, a formal design review is recommended to confirm that requirements are fully specified and appropriate before extensive software design efforts begin. Requirements can be approved and released incrementally, but care should be taken that interactions and interfaces among software (and hardware) requirements are properly reviewed, analyzed, and controlled.

Design

The decision to implement system functionality using software is one that is typically made during system design. Software requirements are typically derived from the overall system requirements and design for those aspects in the system that are to be implemented using software. There are user needs and intended uses for a finished product, but users typically do not specify whether those requirements are to be met by hardware, software, or some combination of both. Therefore, software validation must be considered within the context of the overall design validation for the system.

A documented requirements specification represents the user's needs and intended uses from which the product is developed. A primary goal of software validation is to then demonstrate that all completed software products comply with all documented software and system requirements. The correctness and completeness of both the system requirements and the software requirements should be addressed as part of the design validation process for that application. Software validation includes confirmation of conformance to all software specifications and confirmation that all software requirements are traceable to the system specifications. Confirmation is an important part of the overall design validation to ensure that all aspects of the design conform to user needs and intended uses.

In the design process, the software requirements specification is translated into a logical and physical representation of the software to be implemented. The software design specification is a description of what the software should do and how it should do it. Due to complexity of the project or to enable persons with varying levels of technical responsibilities to clearly understand design information, the design specification may contain both a high-level summary of the design and detailed design information. The completed software design specification constrains the programmer/coder to stay within the intent of the agreed upon requirements and design. A complete software design specification will relieve the programmer from the need to make ad hoc design decisions.

The software design needs to address human factors. Use error caused by designs that are either overly complex or contrary to users' intuitive expectations for operation is one of the most persistent and critical problems encountered by the FDA. Frequently, the design of the software is a factor in such use errors. Human factor engineering should be woven into the entire design and development process, including the design requirements, analysis, and tests. Safety and usability issues should be considered when developing flow charts, state diagrams, prototyping tools, and test plans. Also, task and function analysis, risk analysis, prototype tests and reviews, and full usability tests should be performed. Participants from the user population should be included when applying these methodologies.

The software design specification should include:

- Software requirements specification, including predetermined criteria for acceptance of the software
- Software risk analysis
- Development procedures and coding guidelines (or other programming procedures)
- Systems documentation (e.g., a narrative or a context diagram) that describes the systems context in which the program is intended to function, including the relationship of hardware, software, and the physical environment
- Hardware to be used
- Parameters to be measured or recorded
- Logical structure (including control logic) and logical processing steps (e.g., algorithms)
- Data structures and data flow diagrams
- Definitions of variables (control and data) and description of where they are used
- Error, alarm, and warning messages
- Supporting software (e.g., operating systems, drivers, other application software)
- Communication links (links among internal modules of the software, links with the supporting software, links with the hardware, and links with the user)
- Security measures (both physical and logical security)

The activities that occur during software design have several purposes. Software design evaluations are conducted to determine if the design is complete, correct, consistent, unambiguous, feasible, and maintainable. Appropriate consideration of software architecture (e.g., modular structure) during design can reduce the magnitude of future validation efforts when software changes are needed. Software design evaluations may include analysis of control flow, data flow, complexity, timing, sizing, memory allocation, criticality analysis, and many other aspects of the design. A traceability analysis should be conducted to verify that the software design implements all of the software requirements. As a technique for identifying where requirements are not sufficient, the traceability analysis should also verify that all aspects of the design are traceable to software requirements. An analysis of communication links should be conducted to evaluate the proposed design with respect to hardware, user, and related software requirements. The software risk analysis should be re-examined to determine whether any additional hazards have been identified and whether any new hazards have been introduced by the design.

At the end of the software design activity, a Formal Design Review should be conducted to verify that the design is correct, consistent, complete, accurate, and testable before moving to implement the design. Portions of the design can be approved and released incrementally for implementation, but care should be taken that interactions and communication links among various elements are properly reviewed, analyzed, and controlled.

Most software development models will be iterative. This is likely to result in several versions of both the software requirements specification and the software design specification. All approved versions should be archived and controlled in accordance with established configuration management procedures.

Construction or Coding

Software may be constructed either by coding (i.e., programming) or by assembling together previously coded software components (e.g., from code libraries, off-the-shelf software, etc.) for use in a new application. Coding is the software activity where the detailed design specification is implemented as source code. Coding is the lowest level of abstraction for the software development process. It is the last stage in decomposition of the software requirements where module specifications are translated into a programming language.

Coding usually involves the use of a high-level programming language, but may also entail the use of assembly language (or microcode) for time-critical operations. The source code may be either compiled or interpreted for use on a target hardware platform. Decisions on the selection of programming languages and software build tools (assemblers, linkers, and compilers) should include consideration of the impact on subsequent quality evaluation tasks (e.g., availability of debugging and testing tools for the chosen language).

Some compilers offer optional levels and commands for error checking to assist in debugging the code. Different levels of error checking may be used throughout the coding process, and warnings or other messages from the compiler may or may not be recorded. However, at the end of the coding and debugging process, the most rigorous level of error checking is normally used to document what compilation errors still remain in the software. If the most rigorous level of error checking is not used for final translation of the source code, then justification for use of the less rigorous translation error checking should be documented. Also, for the final compilation, there should be documentation of the compilation process and its outcome, including any warnings or other messages from the compiler and their resolution, or justification for the decision to leave issues unresolved.

Firms frequently adopt specific coding guidelines that establish quality policies and procedures related to the software coding process. Source code should be evaluated to verify its compliance with specified coding guidelines. Such guidelines should include coding conventions regarding clarity, style, complexity management, and commenting. Code comments should provide useful and descriptive information for a module, including expected inputs and outputs, variables referenced, expected data types, and operations to be performed. Source code should also be evaluated to verify its compliance with the corresponding detailed design specification. Modules ready for integration and test should have documentation of compliance with coding guidelines and any other applicable quality policies and procedures.

Source code evaluations are often implemented as code inspections and code walkthroughs. Such static analyses provide a very effective means to detect errors before execution of the code. They allow for examination of each error in isolation and can also help in focusing later dynamic testing of the software. Firms may use manual (desk) checking with appropriate controls to ensure consistency and independence. Source code evaluations should be extended to verification of internal linkages between modules and layers (horizontal and vertical interfaces) and compliance with their design specifications. Documentation of the procedures used and the results of source code evaluations should be maintained as part of design verification.

Testing by the Software Developer

Software testing entails running software products under known conditions with defined inputs and documented outcomes that can be compared to their predefined expectations. It is a time-consuming, difficult, and imperfect activity. As such, it requires early planning in order to be effective and efficient.

Test plans and test cases should be created as early in the software development process as feasible. They should identify the schedules, environments, resources (personnel, tools, etc.), methodologies, cases (inputs, procedures, outputs and expected results), documentation, and reporting criteria. The magnitude of effort to be applied throughout the testing process can be linked to complexity, criticality, reliability, and/or safety issues.

Software test plans should identify the particular tasks to be conducted at each stage of development and include justification of the level of effort represented by their corresponding completion criteria.

An essential element of a software test case is the expected result. It is the key detail that permits objective evaluation of the actual test result. This necessary testing information is obtained from the corresponding predefined definition or specification. A software specification document must identify what, when, how, why, etc., is to be achieved with an engineering (i.e., measurable or objectively verifiable) level of detail in order for it to be confirmed through testing. The real effort of effective software testing lies in the definition of what is to be tested rather than in the performance of the test.

Once the prerequisite tasks (e.g., code inspection) have been successfully completed, software testing begins. It starts with unit level testing and concludes with system level testing. There may be a distinct integration level of testing. A software product should be challenged with test cases based on its internal structure and with test cases based on its external specification. These tests should provide a thorough and rigorous examination of the software product's compliance with its functional, performance, and interface definitions and requirements.

User Site Testing

Testing at the user site is an essential part of software validation. The Quality System regulation requires installation and inspection procedures (including testing where appropriate) as well as documentation of inspection and testing to demonstrate proper installation. Likewise, manufacturing equipment must meet specified requirements, and automated systems must be validated for their intended use.

Terminology regarding user site testing can be confusing. Terms such as beta test, site validation, user acceptance test, installation verification, and installation testing have all been used to describe user site testing. The term "user site testing" encompasses all of these and any other testing that takes place outside of the developer's controlled environment. This testing should take place at a user's site with the actual hardware and software that will be part of the installed system configuration. The testing is accomplished through either actual or simulated use of the software being tested within the context in which it is intended to function.

User site testing should follow a predefined written plan with a formal summary of testing and a record of formal acceptance. Documented evidence of all testing procedures, test input data, and test results should be retained.

There should be evidence that hardware and software are installed and configured as specified. Measures should ensure that all system components are exercised during the testing and that the versions of these components are those specified. The testing plan should specify testing throughout the full range of operating conditions and should specify continuation for a sufficient time to allow the system to encounter a wide spectrum of conditions and events in an effort to detect any latent faults that are not apparent during more normal activities.

During user site testing, records should be maintained of both proper system performance and any system failures that are encountered. The revision of the system to compensate for faults detected during this user site testing should follow the same procedures and controls as for any other software change.

The developers of the software may or may not be involved in the user site testing. If the developers are involved, they may seamlessly carry over to the user's site the last portions of design-level systems testing. If the developers are not involved, it is all the more important that the user have persons who understand the importance of careful test planning, the definition of expected test results, and the recording of all test outputs.

Maintenance and Software Changes

In addition to software verification and validation tasks that are part of the standard software development process, the following additional maintenance tasks should be addressed.

Software Validation Plan Revision

For software that was previously validated, the existing software validation plan should be revised to support the validation of the revised software. If no previous software validation plan exists, such a plan should be established to support the validation of the revised software.

Anomaly Evaluation

Software organizations frequently maintain documentation, such as software problem reports that describe software anomalies discovered and the specific corrective action taken to fix each anomaly. Too often, however, mistakes are repeated because software developers do not take the next step to determine the root causes of problems and make the process and procedural changes needed to avoid recurrence of the problem. Software anomalies should be evaluated in terms of their severity and their effects on system operation and safety, but they should also be treated as symptoms of process deficiencies in the quality system. A root-cause analysis of anomalies can identify specific quality system deficiencies. Where trends are identified (e.g., recurrence of similar software anomalies), appropriate corrective and preventive actions must be implemented and documented to avoid further recurrence of similar quality problems.

Problem Identification and Resolution Tracking

All problems discovered during maintenance of the software should be documented. The resolution of each problem should be tracked to ensure it is fixed, for historical reference, and for trending.

Task Iteration

For approved software changes, all necessary verification and validation tasks should be performed to ensure that planned changes are implemented correctly, all documentation is complete and up to date, and no unacceptable changes have occurred in software performance.

BENEFITS OF QUALIFICATION

Software validation is a critical tool used to assure the quality of software and software automated operations. Software validation can increase the usability and reliability of the application, resulting in decreased failure rates, fewer recalls and corrective actions, less risk to patients and users, and reduced liability to manufacturers. Software validation can also reduce long-term costs by making it easier and less costly to reliably modify software and revalidate software changes. Software maintenance can represent a very large percentage of the total cost of software over its entire life cycle. An established comprehensive software validation process helps to reduce the long-term cost of software by reducing the cost of validation for each subsequent release of the software. The level of validation effort should be commensurate with the risk posed by the automated operation. In addition to other risk factors, such as the complexity of the process software and the degree to which the manufacturer is dependent upon that automated process to produce a safe and effective product, determine the nature and extent of testing needed as part of the validation effort. Documented requirements and risk analysis of the automated process help to define the scope of the evidence needed to show that the software is validated for its intended use.

An Abbreviated Computer Validation History

- 1978—Validation for GMP concept developed by FDA
- 1979—The U.S.A. issue Federal Regulations for GMP including validation of automation equipment
- 1983—FDA Blue Book for computer system validation
- 1985—U.S. PMA published guideline for validating new and existing computer systems
- 1987—FDA technical report on developing computer systems
- 1988—FDA conference paper on inspecting computer systems
- 1989—EU Code for GMP including Annex 11 on computerized systems
- 1991—EU Directive for GMP based on EU Code for GMP
- 1994—U.K. FORUM draft guidelines to suppliers
- 1994—The U.S.A. propose new electronic record and electronic signatures GMP regulations
- 1994—GAMP first draft Distributed to U.K. for comments
- 1995—U.S. PDA publish validation guideline for manufacturers
- 1995—The U.S.A. amend GMP regulations affecting automation
- 1995—U.K. FORUM revise draft guidelines to suppliers
- March of 1997, FDA issued final part 11 regulations
- First Draft July, 2000 (GAMP Europe)
- Final Draft March, 2001 (GAMP Americas)
- Version 1 Quarter 2, 2001 (Co-Publication with PDA)
- GAMP4, December 2001, major revision and new content in line with regulatory and technological development
- February 4, 2003, FDA withdrew the draft guidance for industry, *21 CFR Part 11*.

BIBLIOGRAPHY

Food and Drug Administration References

Glossary of Computerized System and Software Development Terminology, Division of Field Investigations, Office of Regional Operations, Office of Regulatory Affairs, Food and Drug Administration, August 1995.

Guideline on General Principles of Process Validation, Center for Drugs and Biologics, and Center For Devices and Radiological Health, Food and Drug Administration, May 1987.

Technical Report, Software Development Activities, Division of Field Investigations, Office of Regional Operations, Office of Regulatory Affairs, Food and Drug Administration, July 1987.

Other Government References

Adrion WR, Branstad MA, Cherniavsky JC. NBS Special Publication 500-75, Validation, Verification, and Testing of Computer Software, Center for Programming Science and Technology, Institute for Computer Sciences and Technology, National Bureau of Standards, U.S. Department of Commerce, February 1981.

Powell PB, ed. NBS Special Publication 500-98, Planning for Software Validation, Verification, and Testing, Center for Programming Science and Technology, Institute for Computer Sciences and Technology, National Bureau of Standards, U.S. Department of Commerce, November 1982.

Wallace DR, ed. NIST Special Publication 500–235, Structured Testing: A Testing Methodology Using the Cyclomatic Complexity Metric. Computer Systems Laboratory, National Institute of Standards and Technology, U.S. Department of Commerce, August 1996.

International and National Consensus Standards

IEC 61506:1997, Industrial process measurement and control—Documentation of application software. International Electrotechnical Commission, 1997.

IEEE Std 1012-1986, Software Verification and Validation Plans, Institute for Electrical and Electronics Engineers, 1986.

IEEE Standards Collection, Software Engineering, Institute of Electrical and Electronics Engineers, Inc., 1994. ISBN 1-55937-442-X.

ISO 9000-3:1997, Quality management and quality assurance standards—Part 3: Guidelines for the application of ISO 9001:1994 to the development, supply, installation and maintenance of computer software. International Organization for Standardization, 1997.

ISO/IEC 12207:1995, Information technology—Software life cycle processes, Joint Technical Committee.

ISO/IEC JTC 1. Subcommittee SC 7, International Organization for Standardization and International Electrotechnical Commission, 1995.

Production Process Software References

Grigonis GJ, Jr., Subak EJ, Jr., Michael W. Validation key practices for computer systems used in regulated operations. Pharm Technol 1997.

Guide to Inspection of Computerized Systems in Drug Processing, Reference Materials and Training Aids for Investigators, Division of Drug Quality Compliance, Associate Director for Compliance, Office of Drugs, National Center for Drugs and Biologics, and Division of Field Investigations, Associate Director for Field Support, Executive Director of Regional Operations, Food and Drug Administration, February 1983.

Technical Report No. 18, Validation of Computer-Related Systems. PDA committee on validation of computer-related systems. PDA J Pharm Sci Technol 1995; 49(Suppl. 1).

General Software Quality References

Kaner C, Falk J, Nguyen HQ. Testing Computer Software. 2nd ed. Vsn Nostrand Reinhold, 1993 (ISBN 0-442-01361-2).

Ebenau RG, Strauss SH. Software Inspection Process. McGraw-Hill, 1994 (ISBN 0-07-062166-7).

Dustin E, Rashka J, Paul J. Automated Software Testing—Introduction, Management and Performance. Addison Wesley Longman, Inc., 1999 (ISBN 0-201-43287-0).

Fairley RE. Software Engineering Concepts. McGraw-Hill Publishing Company, 1985 (ISBN 0-07-019902-7).

Halvorsen JV. A software requirements specification document model for the medical device industry. In: Proceedings IEEE SOUTHEASTCON '93, Banking on Technology, Charlotte, North Carolina, April 4th–7th, 1993.

Mallory SR. Software Development and Quality Assurance for the Healthcare Manufacturing Industries. Interpharm Press, Inc., 1994 (ISBN 0-935184-58-9).

Perry WE, Rice RW. Surviving the Top Ten Challenges of Software Testing. Dorset House Publishing, 1997 (ISBN 0-932633-38-2).

Wiegers KE. Software Requirements. Microsoft Press, 1999 (ISBN 0-7356-0631-5).

47

Validation of Control Systems

Steven Ostrove
Ostrove Associates, Inc., Elizabeth, New Jersey, U.S.A.

INTRODUCTION

As discussed in chapter 46, the FDA has considered computer systems as equipment that needs to be formally qualified. The general approach to qualifying a piece of equipment can be found in chapter 46. All types of equipment used in or for the purpose of producing or releasing a pharmaceutical or device product must be qualified; computers and computer systems are no exception. The use of computerized control for manufacturing and quality control has grown substantially over the last decade. In fact, the FDA guideline on PAT discusses this technology as it is gaining acceptance in the pharmaceutical industry.

Chapter 46 discussed the history of computer validation and further discussed the development of software validation. This chapter will focus on the control devices themselves. These components or systems usually are part or under the control of the Process Automation or IT department. These areas are expert in maintaining the systems and providing the necessary service and training to allow the end user (operations) the ability to employ their benefits. However, qualification of the control components and their software must still be performed by qualified personnel.

Throughout this chapter, reference will be made to computers, computer systems, automated devices, and controllers (or control systems). In short, all of these names apply to components that control or cause to be controlled any step or operation in the production of a drug or medical device. These include units that control the opening or closing of valves, take in process samples, or assist an analyst to determine if the process meets its predetermined acceptance criteria.

Several types of "controllers" are used in the pharmaceutical industry. Each one has its own purpose in production control. Starting with the simplest to the most complex there are microprocessors (i.e., "chip"), the PLCs, the PCs, SCADA, and DCS.

Microprocessors are found in almost every type of unit used in the manufacturing (such as the digital thermometer, barcode readers, etc.). PLCs are found in such units as packaging line printers, filling machines or the control of blenders, or other process equipment. PCs are often found in laboratory settings and in online testing and report generation equipment. Each of these can be used as stand-alone (that is operate independently) or linked together in a network with other components. Even if linked into a system, they may still perform functions independently of the others in the system, or they may call on another unit to complete the process. SCADA or DCS systems are used to control a larger portion of the process. For example, SCADA and/or DCS systems can control, monitor, or assist in the full operation of a process from the initial blending through granulation and drying. DCSs may even be involved in inventory control, warehousing, or other functions needed for total plant operation.

Each of these system or individual units (as in the PLCs) can and usually are linked in a network designed and maintained to control, monitor and report on the production or quality of a pharmaceutical product (including medical devices). As with any other qualification, their qualification ranges from the relatively simple to the very complex depending on the unit, its use, and its configuration.

The validation of automated control systems is substantially more complex than just the qualification of the hardware (equipment). One must also take into account the software being run on the system and the interaction between the hardware and the software. The software qualification is discussed in Chapter 46 of this book and is only touched on or referenced in this chapter.

This chapter will focus primarily on the qualification of the hardware and deal with the software only as it influences the hardware qualification. In order to differentiate this from other qualification programs this chapter will refer to automated control system qualification as CSV.

In general, software qualification, as mentioned in the various sections below, requires vigorous testing

---

Abbreviations used in this chapter: BMS, building management systems; CFR, Code of Federal Regulations; COTS, commercial off-the-shelf; CSQMP, Computer System Qualification Master Plan; CSV, computer system validation; DCS, distributed control systems; EEPROM, erasable electronically programmed random only memory; EPROM, electronically programmed random only memory; EMI, electromagnetic interference; FAT, factory acceptance test; FDA, Food and Drug Administration; GAMP, good automated manufacturing practice; GMP, good manufacturing practice; HMI, human–machine interface; I/O, input/output; IQ, installation qualification; ISPE, International Society for Pharmaceutical Engineering; IT, information technology; MMI, man–machine interface; MRP, materials resource planning; OQ, operational qualification; PAT, process analytical technology; PCs, personal computers; PDA, Parenteral Drug Association; PLCs, programmable logic controllers; PQ, performance qualification; QA, quality assurance; QU, quality unit; RFI, radio frequency interference; SAT, site acceptance test; SCADA, supervisory control and data acquisition; SOPs, standard operating procedures.

along with its associated hardware. This testing needs to include the actual operation of the field instruments (valves, etc.), as well as the recording and storage of the data generated. Any changes to the set points of the instruments needs to be recorded and logged. As discussed in this chapter, software qualification is usually separated into two distinct activities: the structural testing and the functional testing. The structural testing includes the vendor audit, review of the code and checks on the integrity of the code so that there is no dead code (i.e., nonoperational code that may cause a "crash" or data error).

SCOPE

This chapter will cover the qualification of various types of computer systems that includes automated devices used in the control of pharmaceutical/medical devices. While Chapter 46 covered the background and software validation/qualification aspects of CSV, the hardware still needs to be qualified. This chapter will deal with the qualification of the various types of computer or automated control system qualification.

The intent is to provide the reader with an appreciation of the complexity and the similarities of all types of computer or automated system qualifications. It is a general guide as to what is required to qualify/validate the controlling systems used in pharmaceutical, biotechnology or the medical device industry.

As stated in the introduction, all computer or automated controllers require qualification; the level of qualification is dependent upon its function. The industry generally has adopted the GAMP[a] levels of software systems. There are five levels of systems according to the guide; these are:

- Firmware—This is the microchip type of system
- Operating System—The software performing the underlying operation of the system (e.g., Windows XP®)
- Standard Software Package—Non-configurable, also called "off-the-shelf"
- Configurable Software Package—Standardized packages that the owner can configure to fit their specific needs or operations. These can perform a general function, e.g., blending, these are termed "COTS" or "configurable off-the-shelf"
- Custom Software—Prepared specifically for the operation (usually prepared by specialty firms or in-house programmers).

Each level above requires its own level of qualification, increasing as the level goes up (the highest level is the custom system). Notice that the levels are related to the software and not the hardware. This is because the hardware serves as the framework in which the software performs its function. The interaction of the software and hardware needs to be qualified. It is not possible to do qualify one without the other.

[a] GAMP Guide for Validation of Automated Systems, Ed. 4, ISPE 2001.

GENERAL TESTING—ALL SYSTEMS

All computers or automated controllers that are used in or for the production of pharmaceuticals or medical devices require qualification prior to their use in the process. Computers need qualification just as any other system or component of the manufacturing process does. The main difference between general equipment qualification and CSV is that, as mentioned above, there are two stages for the completion of a computer or computer system. These include the software and hardware aspects of the system. The first part of any CSV program is the qualification referred to as structural; the second phase of the qualification is the functional aspect of the systems. The structural qualification and portion of the program is focused on the development of the software, while the functional qualification focuses on the actual operation or function of the system. Chapter 46 deals with the structural qualification aspect, this chapter will concern itself primarily with the functional aspects of the qualification program.

As with software qualification, the hardware can be divided into various stages. Each stage requires a qualification phase in order to demonstrate that it is complete. These stages can be divided as follows:

- Development—establishing system requirements
- Build—obtaining the correct components per specifications
- Implement (this is where the full qualification program is required)
- Operation (part of the full qualification program where a qualified state needs to be maintained)
- Retirement—decommissioning the system for replacement by another system

Functional qualification follows the same pattern as any other pharmaceutical equipment or systems qualification. Thus, in order to perform a functional qualification as described in chapter 9 of this book, an IQ and an OQ are necessary (Refer to chapter 9 for general IQ and OQ requirements).

The IQ provides verification that the system is installed according to a written preapproved plan. The same is true for the OQ. All pharmaceutical systems should have the following:

- Vendor qualification via an audit
- User specification
- Design specifications

However, in addition to the "usual" requirements for IQs and OQs the qualification of computer systems requires some additional items. Some of these are:

- Verification of system security
  - Controlled access to the program
  - Levels of access—e.g., an operator is allowed to input data but the supervisor is allowed to approve the data
  - Protection of the system from outside interference (e.g., no access via phone lines or the internet) Note: Usually an intranet connection will be allowed.
- Ability to track all entries (audit trail) into the system—this includes the date, the person making the entry and why the entry is made or changed.

BLACK BOX VS. WHITE BOX TESTING

There are two methods of testing automated control systems. These are referred to as "white box" and "black box" testing. Both means of qualification are used for systems at or above Level 2 of the GAMP classification of computer systems. The difference between "white" and "black" box testing is in the level of testing of the software. Black box testing is primarily functional testing while white box testing includes a review of the source code (of the software program) as well as the means of code development.

When doing black box testing the operation of each portion of the software is tested. In addition, the testing establishes that each function necessary for the correct operation of the unit(s). Typically, the black box testing grows exponentially with the amount of I/O while the white box testing grows linearly.

GENERAL DOCUMENTATION

When beginning a CSV program, as with other qualification programs, certain documents need to be either prepared or collected. Since the qualification will involve components not usually seen and usually not accessible having the correct documents at the very beginning of the project will help assure its success. The list below covers the main documents to be prepared or collected:

Prepare:

1. CSQMP
2. User requirements
3. Functional specifications
4. Traceability matrix (*Note*: To be prepared AFTER all specifications and protocols have been collected and developed but BEFORE protocol execution).
5. SOPs (to include the "How to Prepare" SOPs)
   a. System setup/installation
   b. Data collection and handling
   c. System maintenance
   d. Backup
   e. Recovery
      i. Backup
      ii. Crash
      iii. Jam/freeze
   f. Contingency plans (emergencies)
   g. Security
   h. Change control
   i. Storage
6. Protocols
   a. Commissioning
   b. IQ
   c. OQ
   d. PQ (as necessary)

Collect:

1. Ladder logic—As necessary for PLCs
2. Design or Vendor specifications for each component—part of the system (network interfacing, MMI)
3. Software version to be installed
4. Software source code (or 3rd party agreement)

After the documents are prepared and or collected, you are ready to begin the qualification program itself. (Note: this is assuming that the structural qualification has been completed and is acceptable). As with all qualification programs the commissioning phase usually is the first "field" effort undertaken. (*Note*: This follows the FAT and SAT portion of the program.) The commissioning portion of the qualification can be performed, at least in part, during the installation of the system. For example, while the lines are being run to the field instruments the loop checks can be performed. A loop check is a check of continuity (and thereby function) of the connection between a field instrument and the controller. It is far simpler to perform and document the loop check as each loop is being installed rather than after the system is intact and ready to operate. Other items that can be performed during the installation or as part of the commissioning phase are:

- Instruments adjusted/calibrated (loop checks)
- Ambient conditions
  - Temperature
  - Humidity
- Alarms and events (general testing—operational testing is left to the OQ phase of the qualification)
- Graphics
- Data base location
- Network configuration

The next phase of the qualification is the IQ. As pointed out in Chapter 9 this may be done at the same time or before the commissioning phase of the program. Either during or even before the IQ is started the structural phase of software testing is completed. Since the structural testing includes items such as the vendor audit, the code review, this part of the qualification must be completed prior to any functional or OQ testing as discussed below.

The general IQ consists of the following verifications. Specific tests will be pointed out later for each of the types of automated systems.

1. List all components
   a. Input devices—HMI and/or MMI
      i. Keyboard
      ii. Mouse
      iii. External devices
         - Field instruments,
         - External drives,
         - Monitors, etc.
   b. Output devices
      i. Screen
      ii. External data device—hard drive
      iii. Printer
      iv. Filed instruments
   c. Data storage devices
      i. Hard drives
      ii. MP3
      iii. Floppy drives
      iv. Flash cards
      v. Tape/CD/DVD (backup)
2. List type of hardware
   a. Mother board—chip type
   b. Controller cards
      i. Video
      ii. Sound
      iii. I/O

c. Internal drives
   i. Floppy
   ii. CD
d. Output connections
   i. USB
   ii. Parallel
   iii. Firewire
   iv. Serial
   v. S Video
   vi. Other monitor connections
e. Network cards (discussed below)

3. Check for:
   a. Tight connections
   b. Correct component type
   c. Installed in the correct location (as applicable)
   d. Model as per specifications
4. Power (source and distribution)
   a. Volts
   b. Current
   c. Stability
   d. Surge protection
5. Software (includes the structural testing—see below)
   a. Version installed
   b. Source code verification
      i. Annotation
      ii. Dead code
      iii. Vendor testing verification (part of vendor audit)
   c. Compliance to good software preparation

The OQ follows the IQ. This set of testing cannot start until the IQ is complete or until the QU gives approval (as discussed in Chapter 9). In the case of automated systems, the completion of the IQ is necessary since the system will not function as specified without all components being installed correctly. While the system may seem to operate, some functions will be compromised if a component is lacking. This may not be immediately apparent but will, in the long term, compromise the final product. An example of this would be a missing printer. The controller would run, the machines would run, but the output data would not be able to be expressed or recorded. This may cause the system to shut down or to store the information that cannot be printed. It would be printed later (if possible). This may compromise the next lot of material being produced since it will get the incorrect label or printout.

It is during the OQ testing that the software undergoes its functional testing.

In general, the OQ will have the following general tests:

1. Prepare test of each component listed in the IQ
   a. Meets design specifications
   b. Meets functional specifications hardware
   c. Power limits—may be included as part of the PQ (below)
      i. Recovery after power loss
      ii. Power line stability
   d. Environmental stress
   e. Alarms
   f. All component functions over their full range
   g. Software
      i. version verification
      ii. Ladder logic or source code review[b]
   h. Input limits (boundaries)
   i. Functional testing
2. RFI—that is a radio frequency should not cause the controller to malfunction (allow incorrect data in or out)—e.g., a walkie-talkie (handheld radios).
3. EMI—a magnetic field should not interfere with the data integrity—e.g., an electric drill
4. I/O integrity
5. Calibration
6. Software
   a. Compete structural testing
   b. Functional testing
      i. Restart after shutdown
      ii. Restart after power loss
      iii. All major operations function and results are appropriate

If a PQ needs to be performed (as it most likely will), the following is a list of general tests that should be included.

1. Power failure recovery—computer and process equipment (as seen above this may be done as part of the OQ)
   a. Recovery after power loss
2. Security—system accessibility
   a. Password challenge
   b. Security challenge
   c. Biometric security
   d. Levels of access
3. Archive/retrieve data in real time
4. Produce batch report
5. I/O Loops operation
6. Data lines transmission
7. General data integrity
8. Interference between programs/components
9. Software
   a. Full operation of all functions in conjunction with the entire system
   b. Stress the software boundaries
   c. Noninterference between modules or other programs

SPECIFIC SYSTEMS

The next part of this chapter will deal with some of the specific requirements needed to complete an adequate qualification of different types of automated systems.

[b] Ladder logic and source codes need to be reviewed for compliance to good code writing requirements (General Principles of Software Validation; Final Guidance for Industry and FDA Staff January 11, 2002, FDA—U.S. Department Of Health and Human Services, Food and Drug Administration Center for Devices and Radiological Health Center for Biologics Evaluation and Research) Included in this is a review for problems involving dead code. Ladder logic (is the programming code used for PLCs) should be reviewed for functionality as well as annotations. While the source code of higher systems (PCs, etc.) also needs to be reviewed (e.g., for dead code but also for annotation of the sections), it is usually not possible to do a line by line review of the code for these systems. This is why one additional requirement is that the code is available for and able to be corrected if needed. This last requirement is usually met by "Third Party Agreements" with the code developer (e.g., storage but accessible under limited access if required).

As was seen above, computer or automated control systems require both software and hardware qualification. The software qualification has adopted the GAMP® approach while the hardware has retained the basic IQ/OQ/PQ approach. The specific types of systems that will be discussed are:

- Microprocessors
- PLCs
- PCs
- Networks
- SCADA
- DCS—all forms

All of these require some form of software and hardware qualification. Starting with microprocessors as the simplest of the control systems and working up to the DCS, the basic qualification approach outlined above and in chapter 46 applies. The discussion of networks, while not actual control systems, needs to be considered since any of the above control or automated controllers can be networked forming larger control systems or controls loops.

Microprocessors

These simple controllers exist throughout the process industries. Their purpose is usually a single function such as turning a light on and off on a schedule. Thus, they are more than simple switches. Other examples of microprocessor controllers are:

- A light may come on and a camera activate in response to a door opening.
- An alarm may be triggered by a door not closing within a set period of time.
- The closing of the door may activate another timer that will keep the light on for a given period.
- A micro switch may be pressed during production based on some activity; this in turn activates a microprocessor to count the events.

This kind of controller provides basic functionality within equipment and rooms or facilities. These controllers usually do not allow any change in configuration; that is a change in the type of control or timing of the system. However, a microprocessor may be an EPROM or may be or the type that is EEPROM. Both the EPROM and the EEPROM require software qualification as well as the standard functional testing of the microprocessor. The software is accessible only through another computer and even then only with specialized software. This software requires control both in access to and in validation of the program itself. The EPROM or EEPROM will then need to be able to verify the latest version of the software programmed in (this is a "burn-in" process similar to using writing a read-only CD). The validation of the programming software, the EPROM or EEPROM, is basic tests and verifications of operation.

Most often this type of control, such as those that provide standard environmental lighting or activate pumps or heaters included within larger systems are not a regulatory focus. However, that does not mean they can be ignored. As long as the basic functional testing is appropriate, as long as their function within the facility is part of the wider design and that functionality is tested when it affects product, they do not require a separate qualification or validation.

Programmable Logic Controllers

In the pharmaceutical industry, the PLC is probably the mainstay of all operations. The PLC can be found in a variety of operating units. They are used to open or close any type of field device (i.e., valves, air pressure control, motors). In general they are easy to program (hence the name). In contrast, microcontrollers (microprocessors) are related but very different. Microcontrollers are essentially single microprocessors where the controller hardware on the circuit board is customized to the device. Once the microcontroller code is installed into the device from the manufacturer, it is very difficult to change, and similar to the EPROM or EEPROM noted above.

In contrast, a PLC is a much more complex controller. It can be viewed as multiple microprocessors in a single unit. However, the big difference is that they are more easily programmed. By its very nature, it has a much more complex and richer instruction set. It typically has much more memory, redundancy, and processing power as well. Though PLCs are mass-produced, typically PLC code (called ladder logic as apposed to source code used for PCs and higher types of controllers) and hardware wiring are customized for each device based on the customer's specific needs. Because the code is to be customized by the client (operating company), the PLC manufacturers testing of the operating system software is usually only on a high level. This leaves the true qualification work to the owner.

PLCs fall into several of the GAMP4 categories, depending upon their configuration. The more standard controllers, like those for lab bench analyzers and sterilizers could be category 2 or 3; and complex, more customized equipment, like filling systems or lyophilizers, could be category 5. However, since PLCs are relatively easily to program and are most often customized to the specific client use, GAMP4 category 5 is the most likely approach. That is full testing will be required.

From a risk assessment standpoint, PLCs typically have the highest direct safety risk (both human and equipment), SCADAs and DCSs are next, then database systems—and safety is only one small aspect of the risk assessment process.

For a simple PLC controller, say less than 20 I/O, black box testing makes more sense than white box. However, for anything more than 20 I/O or for systems with a HMI, white box is probably more effective than black box testing. The amount and type of testing is related to the amount of code, the amount of user-specified coding versus vendor coding, the actual use in the process (i.e., what equipment it will be used to control), and other factors as outlined in the GAMP guide.

An example of PLC qualification can be seen as follows:

Assume that a machine has two sensors, A and B. When sensor A is on, we want to turn on alarm horn A. In addition, when sensor B is on, we want to turn on alarm horn B. In addition, when sensors A and B are on, we want to shut down the machine. In 99% of the cases, programmer will cause sensor A to set a bit that causes the output alarm horn A to turn on; and sensor B to set a bit that causes the output alarm horn B to turn on. When both of these bits are on, the machine will stop.

From a black box testing perspective, this is very difficult to catch. You must, in fact, black box test all possible combinations of the interlock conditions in each of the four states (good, going into alarm, alarming, going into good). For our example, the black box testing would contain eight tests alone! On the other hand, white box testing could be done on six of those states, leaving two for black box.

A typical protocol for the average PLC should be about 90% white box (Ladder logic or code review) and 10% black box (functional). The number of total tests is exponentially proportional to the amount of I/O and code. Therefore, for 50 I/O, there may be 2500 tests. That is, there may be 2500 interactions between inputs, outputs, and internal conditions. A test protocol with white box testing would examine dozens of these interactions in a few test cases, using the duplicity of the structure with which they were created (if there was a structure).

The testing for all network-rung paths and all possibilities, as well as questioning the operating system integrity, would take longer than the testing of inputs, outputs, and screens in a black box fashion.

For another example, assume we have a system of five inputs and five outputs. For the short term, we will ignore the complexities that can be built into the operator interface. Given an input, or combination of inputs, some outputs happen. Let us say that input 1, vessel pressure high, causes output 1 vessel vent valve, to actuate. The requirements and design documents will probably state, "Open the vessel vent valve when the vessel pressure is high." Most protocols would include a single test—stimulate the input, observe the response output. This must be done for each of the I/Os.

Continuing with the 5×5 example, if the system is such that the position of the outputs will not feed back in to how the system responds (meaning that the PLC does not care that the vessel vent valve is open as it goes on to do its other tasks), then each input should have 32 tests (on or off=2 positions, with five inputs, $32=2^5$). Assuming that the protocol is written such that the other output expected results are inclusive in the 32 tests, there should be 32 tests for five inputs to generate five outputs. The argument is that this is more than the number of tests necessary for white box testing. By following the code in the white box analysis, then there will be only one path to test for each input and one path for each output, for a total of 10 tests.

Of course, as more interlocks, sequences, and other rules are added to the complexity of the PLC logic, the advantages are harder to see—though they are still there.

Items to verify for PLCs:

- Review the ladder logic
  - Correct version installed
- Inputs and outputs
- Environmental conditions
- Point-to-point testing—Loop checks

Personal Computers

PCs are relatively easy to qualify. The reason for this is that most of the software used on a PC is off-the-shelf non-configurable. That is, the software cannot be changed. Only the application is configurable. For example, Microsoft Excel® spreadsheet program can neither be validated nor qualified. However, the application of each spreadsheet must be qualified. Specifically each calculation needs to be verified from both its algorithm to its data input and output.

All aspects of the PC need to be qualified, just as any other process or laboratory equipment. All I/O devices (e.g., keyboards, disk drives, USB inputs of outputs, mouse control and other pointers, screen displays, printers, etc.) need to be tested and demonstrated to be functioning correctly. This means that the data being input is the same as the data coming out. For example, when typing the letter "M," the keyboard should respond only to the M from the designated key and the screen should display only an M from that designated key. The same holds true for any data storage device, whether internal or external.

One difference between PCs and other automated controllers is that very often the data is taken off the PC and stored in an external device (tape drive, external hard disk, etc.). In this case the data transfer to the devices used for storage as well as the recovery of the data from the device needs to be qualified. Storage time of the data on the external device as well as the environmental conditions it is stored under are factors in this qualification.

Code review for vendor-supplied programs is not required. This includes the operating system. A word of caution here is that the last statement assumes that there are many hundreds of units of the same program on the market and thus errors in the code have been readily observed and corrected. Thus, if one purchases or prepares a new operating system, specific for the application, then this would require full qualification as determined by the GAMP4 approach.

There are other areas that extra caution is needed when using PC for control operations. One of the biggest areas of concern in the use of PCs is their ability to connect to the "Internet." The Internet is an outside link, i.e., opening the system to other computers, and should be avoided. Data security and integrity are key issues in dealing with any automated control system.

Items to verify on a PC:

- All input devices
- All output or data storage devices
- Data integrity both in and out of the PC
- PC calibration
- Software:
  - Operating and off-the-shelf programs do not usually require qualification
  - Application software and applications on off-the-shelf programs do require qualification (e.g., COTS—Commercial off-the-shelf software)
- Environmental conditions—Temperature/humidity/liquids

Networks

PLCs and PCs may be linked together to form a "Network." Simply, a network is a group of individual units (PCs or PLCs) linked together so that information can be easily shared. There are two basic types of networks, open and closed. In the pharmaceutical industry, the closed network is the preferred type. As described above with the PCs, the internet represents an open system and thus the greater possibility of data corruption.

Networks come in many formats. In the early days of networking, two or more computers were connected by regular wires between the units. The next stage was the use of "twisted pair" wiring. This made use of part of the telephone wires for connecting the computers. This gave way to the Ethernet and now the wireless network. Each of these earlier types of networks still exists, although some to a much lesser degree. Each requires their own special approach to qualification.

For example if a system transmits data used on batch records, and that this data is the active record—that is, regardless of any printouts of this data, the active data that the company uses is this electronic record—like a maintenance log for a piece of equipment used in drug manufacture. The security needs to be tied to the record, and typically, the record is tied to a database system. In this case, if users were transmitting this data over the network, then the network should be validated. However, that validation is usually a subset of validation of the database system (with tests that make sure clients can talk to servers and so forth). In addition, there is typically some platform validation performed to ensure that the network has appropriate bandwidth and can handle traffic flow correctly.

A risk assessment should truly answer when to do network validation. For example, if the network is only used for backing up servers, then the firm would develop a set of requirements, specifications, and tests regarding how servers are backed up (in this case a worst-case scenario would involve data quantity as opposed to network loading). If the network were only used for client interaction to the server, then the firm would develop requirements, specifications, and tests around network loading, response speed, and server time-outs—packet "sniffer" software will typically analyzes this.

Let us assume for a minute that the firm has a large multiuser database system that is being tested prior to plant roll out. In the test room, there are a couple of clients, the server, and a network switch that are all tested and validated. Now the system is placed on the plant network.

The firm discovers from an investigation that there are a number of differences: some of the clients PCs on the network are using older operating systems. The network itself is larger and more complex and uses hubs, routers, and firewalls. Will it be necessary to retest all the aspects of the application? No. Is the application still validated? Yes. What is needed is to resolve and test aspects of the network.

If "Yes" start by analyzing the test network and the live network. A good packet sniffer available for free is Ethereal (1). Based on where packet collisions occur it can detect what part of the networks are having an issue and resolve it. The firm can use the test system to develop data transmission requirements (based on what the sniffer reveals) and then validate to those requirements on the live system.

Validate the network with the application (assuming that both the application and the network relate to predicate rule records or processes), and then "qualify" the network platform for all the systems that use it. So, for example, a database client–server system is validated with the network structure in place, and then the network is "qualified" to be able to handle all the other client–server systems it has to carry (that is, bandwidth and capacity are evaluated).

Items to consider for network qualification:

- All major components of the network (e.g., PCs, routers, switches)
  - Point-to-point testing
- Qualify networks that are related to predicate rule data
- Use the risk assessment approach to determine the extent of a network qualification
  - Transport layers
  - Application layers
- Commissions to specifications
- Validates to requirements
- Security (refer also to Part 11)
  - Open system
  - Closed systems
- Collision reconciliation
- Node operation

Larger automated systems such as discussed below are similar to the smaller systems described above. All of the same type of testing needs to be done for these larger systems. The difference is in the complexity of the system and the amount of time required completing the qualification program. In general, the larger the systems the more time it will require to qualify since there are an increased number of variables to test. With more complicated systems, it is more important to follow a full qualification program starting with the development of a Validation or Qualification Master Plan. This plan should be specific for the system(s) involved, its intended use and the type of hardware and software to be used.

Supervisory Control and Data Acquisition

SCADA systems are made up of several components. Each of these components may be qualified as separate units or combined into one large qualification program. A SCADA system is made up of:

- HMI—The screen is often a touch screen
- Control Units—Controlling the field devices
- Main Processor—Interprets the information form the field units/PLCs and the operating instructions from the HMI

As with all automated or computerized systems, security and data integrity are primary issues. Each of the components needs to be secure from outside interference as well as internal problems resulting from adjacent equipment or component problems. Alarms are key to the functioning of a SCADA system. They alert the operator of problems in carrying out the instructions inputted by the operator or the recipe.

Items to verify for a SCADA qualification:

- Alarms
- Loop checks
  - Point-to-point are unique
  - Field unit verifications
- Input devices
  - HMI
    - Access levels
      - Supervisor
      - Operator

- Disks
- Tapes
- Graphics
  - Is the system represented correctly on the screen?
- Data acquisition and data integrity
- Is the screen a true representation of the system?
- Is it a touch screen?
- Interface between the screen and the system (i.e., valves, temp. control, etc.).
  - Does the screen do what is indicated in the system?
- Calibration

Distributed Control Systems

DCSs have evolved over the years into sophisticated units. These systems are usually involved in more than just pharmaceutical manufacturing. They are found in inventory control, warehousing, ordering, maintenance, and manufacturing controls. BMS and MRP Systems are examples of DCS systems. These systems integrate many functions into one package. The BMS controls and monitors the environmental conditions in the facility. It can prepare documentation on the environmental status of any part of the plant if requested or as part of a batch record. It can monitor the fire alarms or access to restricted areas.

MRPs on the other hand, are made up of sub-modules that monitor or control inventory, financial records, warehousing operations, production schedules, etc. While not all sub-modules are GMP systems, all must be considered in order to assure that no part interferes with any other part during their operation.

As BMSs are configurable off-the-shelf packages, risk assessment should focus on testing on the configured and customized portions of the package and not on the standard components of the package. For example, the package allows the firm to graphically trend points. Testing should therefore ensure that the set of points to be trended is correctly configured, but the operation of zoom, forward, and back buttons on the standard trend screen can be ignored.

Items to verify for a DCS:

- Individual node/unit can function independently
  - No interference between units
  - No interference between users
- Each node/unit can be qualified independently
- Environmental conditions for each node
- Input and Out devices
- Network qualification
- HMI qualification

PART 11

No discussion of computer or control system qualification will be complete without at least an overview of Part 11 (21 CFR Part 11). This part of the CFR has caused the pharmaceutical industry great concern in recent years due to its perceived complexity. Part 11 has been around since 1997 but has only recently become more strictly enforced by the FDA. The reason for this is, the FDA allowed the industry time to comply, by updating their control systems, updating their operating procedures, training, etc., before strict enforcement would be implemented. According to the latest guidelines, systems put into operation prior to 1997 are usually considered exempt from the Part 11 rules. However, caution needs to be taken here, as any change to the system after the 1997 start, may bring the control system under Part 11 requirements.

When one looks closely at the requirements, they are really quite understandable; however, their implementation can be very complex. The FDA has issued two sets of guidelines for this Part of the CFR. The first set of guidelines has been withdrawn and a new "draft" guideline has been issued. The current guidelines, has made compliance to Part 11 regulations clearer to the industry. The regulations have not changed; only their perception has changed.

There are three major sections of the requirements. These are:

- Subpart A—General provisions
  - 11.1 Scope
  - 11.2 Implementation
  - 11.3 Definitions
- Subpart B—Electronic records
  - 11.10 Controls for closed systems
  - 11.30 Controls for open systems
  - 11.50 Signature manifestations
  - 11.70 Signature/record linking
- Subpart C—Electronic signatures
  - 11.100 General requirements
  - 11.200 Electronic signatures components and controls
  - 11.300 Controls for identification codes/passwords

Subparts B and C represent the main body of the requirements. Only an overview of the requirements will be presented; further study will be required to fully understand this section of the CFR.

Subpart B is concerned with any computerized system (of any size or type) or of the people who use these systems. Both open and closed systems are included (11 CFR 11.10 and 11 CFR 11.30). In this part of the CFR the FDA specifies that any system used to "create, modify, maintain, or transmit electronic records shall employ procedures and controls designed to ensure the authenticity, integrity... and ensure that the signer cannot readily repudiate the signed record as not genuine." This means that the system(s) need to be validated/qualified and that, as with written records, there needs to be traceability of all data. Access to the systems and the data or records (electronic) needs to be limited and authorized.

Records that are maintained in paper format, as the final, official copy are not included in this section of the regulations. The paper records are part of what is known as the predicate rules requirements. The predicate rules are any rule previously established as found in 21 CFR Part 211.

Subpart C deals with the actual control and requirements for electronic signatures. It describes the levels for security and access, the need for verification of the person signing. There are two types of identification discussed; these are biometric and non-biometric. The non-biometric form is most familiar to everyone. These include items such as identification badges (picture ID) sign-in logs, and password. If this type of identification is used, then two forms must accompany the signature (i.e., user identification and a password). On the other hand, a biometric identification would include fingerprint identity, retinal scans of the eye, or voice recognition. Biometric identification is becoming easier and less expensive, and is available on some PCs now.

As can be seen from this short discussion of Part 11, the regulations are not difficult; however, some aspects of the rules may be harder to implement. All control systems have, or should have, limited access to both the system and the various levels of data (e.g., operator, supervisor, and administrator). Any change in the data needs to have a "trail" indicating "who" made the change and why the change was made (similar to changes in paper records). Thus, compliance to Part 11 has become achievable and, with the new Guidelines from the FDA, it has become more understandable. However, care needs to be taken with all computerized systems to be sure that all of the Part 11 regulations are implemented.

ACKNOWLEDGMENT

The author acknowledges the assistance and input from John Hannon on several of the topics in this chapter.

BIBLIOGRAPHY

Code of Federal Regulations 21 CFR Part 11, 2006.
Code of Federal Regulations 21 CFR Part 211, 2006.
Code of Federal Regulations 21 CFR Part 211 (21 CFR Part 11), 2006.
FDA Computerized Devices/Process Guidance, May 1992.
FDA General Principles of Software Validation; Final Guidance for Industry and FDA Staff, January 11, 2002.
FDA Guidance for Industry—PAT:A Framework for Innovative Pharmaceutical Manufacturing and Quality Assurance (draft guideline), August 2003.
FDA Guide to Inspection of Computerized Systems in Drug Processing, February 1983.
General Principles of Software Validation; Final Guidance for Industry and FDA Staff, U.S. Department Of Health and Human Services, Food and Drug Administration Center for Devices and Radiological Health Center for Biologics Evaluation and Research—January 11, 2002.
Good Automated Manufacturing Practice (GAMP) Guide for Validation of Automated Systems, ISPE, 2001, GAMP 4.
Good Practice and Compliance for Electronic Records and Signatures—Parts 1 and 2, ISPE and PDA, 2002.
Guidance for Industry Part 11, Electronic Records; Electronic Signatures—Scope and Application. U.S. Department of Health and Human Services Food and Drug Administration Center for Drug Evaluation and Research (CDER) Center for Biologics Evaluation and Research (CBER) Center for Devices and Radiological Health (CDRH) Center for Food Safety and Applied Nutrition (CFSAN) Center for Veterinary Medicine (CVM) Office of Regulatory Affairs (ORA), August 2003.
Guidance for Industry PAT—A Framework for Innovative Pharmaceutical Development, Manufacturing, and Quality Assurance—U.S. Department of Health and Human Services Food and Drug Administration Center for Drug Evaluation and Research (CDER) Center for Veterinary Medicine (CVM) Office of Regulatory Affairs (ORA) Pharmaceutical CGMPs, September 2004.
ISPE C & Q.
IVT article on PLCs.
King JH. A Practical Approach to PLC Validation, Institute of Validation Technology. Special ed. Computer Validation II, 2005.
Technical Report No. 18, Validation of Computer-Related Systems, PDA, V49, number S1, 1995.
http://www.ethereal.com/
http://www.pacontrol.com/PLC.html
http://www.pacontrol.com/SCADA.html
http://www.pacontrol.com/DCSystem.html

Risk-Based Validation of a Laboratory Information Management System

R. D. McDowall
McDowall Consulting, Bromley, Kent, U.K.

INTRODUCTION

Pharmaceutical QC laboratories must work electronically if they are to survive.

This statement is not made because of regulatory requirements but simply business pressures facing the pharmaceutical industry today: profit margins are under pressure from government pricing; also, delays in accepting or rejecting raw materials, active ingredients or finished products costs time and money. As analytical laboratories are at the end of the production chain any delay is visible and can magnify the cost of other delays elsewhere in production. Therefore, any implementation of a LIMS in a QC laboratory has to provide tangible business benefits through the elimination of paper records and the use of electronic signatures with associated electronic workflows.

Validation of computerized systems has also been undergoing considerable change following the FDA's GMPs for the 21st century (1) and its revision of 21 CFR 11 with its Guidance for Industry on Part 11 Scope and Application (2). This has been followed by the GAMP Forum's publication on risk-based electronic records and signatures (3), which takes a record-based approach rather than a system-based approach to computer validation. There is also a GAMP Forum Good Practice Guide on Validation of Laboratory Computerized Systems (4); only the system implementation life cycle is taken from this publication as there are a number of issues as outlined by the author of this chapter (5). The LIMS validation must be cost effective and risk based to help deliver the benefits from a process-driven implementation within a relatively short period of time or there is little benefit to an organization.

To appreciate and understand the rationale for this new approach to implementing a LIMS, it is important to understand the problems that face current installations. These can be summarized as follows:

- *Poor LIMS Implementation*: It is difficult to perform an effective LIMS implementation in many laboratories. Typically the current process is automated resulting in a very expensive typewriter being implemented that is driven by paper instead of streamlining the process ahead of implementing a system.
- *No Interfacing to Analytical Instruments*: Failure to interface analytical instrument computer systems that generate the bulk of the data in QC laboratories to a LIMS resulting in manual entry of data. Manual data entry is a slow task and requires transcription error checking to ensure accuracy and integrity. It is still surprising to find the number of LIMS implementations that are standalone and fail to consider interfacing instruments within the laboratory or applications outside of it.
- *Calculations are Performed Outside of LIMS and Instrument Data Systems*: Calculations are typically performed in spreadsheets or hand-held calculators or calculations which are outside of either the LIMS or the data system that generated the data. The reasons for this are mainly that the data system is unable to provide the calculation, spreadsheets are widely available and early to use or the laboratory staff cannot be bothered to read the data system manual to implement the calculations.
- *No Interfacing to Production Systems*: Information and specifications contained in production systems are not transferred electronically to the LIMS; these data have to be input manually into the system and manually checked to ensure accuracy.
- *Extensive Customization of a Commercial System*: Instead of using the standard workflows within a system, many laboratories implement LIMS by changing the system functions to fit the laboratory's current ways of working. This is inefficient and is based on the assumption that a laboratory's processes are efficient and effective. This assumption is usually wrong and creates additional cost and time delays for a LIMS project.

Thus it is unsurprising that many LIMS implementations are inefficient, not cost effective and take a long time to validate.

Abbreviations used in this chapter: CD, compact disc; CDS, chromatography data system; CFR, Code of Federal Regulations; CIs, configuration items; ELN, Electronic Laboratory Notebook; ERP, Enterprise Resource Planning; FDA, Food and Drug Administration; FMEA, failure mode and effects analysis; GAMP, good automated manufacturing practice; GLP, good laboratory practice; GMP, good manufacturing practice; IDEF, integrated definition; IEEE, Institute of Electrical and Electronic Engineers; IP, internet protocol; IQ, installation qualification; IT, information technology; LC, liquid chromatography; LIMS, Laboratory Information Management System; NMR, nuclear magnetic resonance; OECD, Organization of Economic Cooperation and Development; OOS, out-of-specification; OQ, operational qualification; PQ, performance qualification; QA, quality assurance; QC, quality control; R&D, research and development; RAID, redundant array of inexpensive disks; RFID, radio frequency identification; RFP, request for proposal; SAN, storage area network; SDS, system design specification; SOP, standard operating procedure; URS, user requirements specification; UV, ultraviolet.

This chapter on risk-based validation of LIMS describes how to deliver substantial and tangible business benefits required of a LIMS in a GMP laboratory by redesigning the process before coupling this with an effective risk-based computer validation to comply with regulations. This chapter is structured in the following sections:

- Understand and improve the current ways of working
- Designing the LIMS environment
- Specify, implement and validate the LIMS

The approach outlined here is based on the use of a commercial LIMS with configuration or customization of the application as appropriate for a specific laboratory. This section is written primarily for the implementation and risk-based validation of a LIMS in a single laboratory or site. The modifications of the approach required for a multi-site or global LIMS validation are to define the core requirements that all laboratories will use within the system and the initial validation of the core system that must not be modified. Local additions to the core system may be permitted but these need to be specified and validated locally. A GAMP Good Practice Guide on global information systems control and compliance may be useful in this context (6).

UNDERSTAND AND OPTIMIZE THE BUSINESS PROCESS

For successful use of electronic signatures within a new or upgraded LIMS, an electronic workflow is required. Therefore a QC laboratory has to migrate from a paper driven process to an electronic one. This is the key to a cost beneficial validation of any LIMS: map, analyze, understand and then optimize the business process to work electronically and to use electronic signatures effectively.

This understanding and redesign work is achieved through two process mapping workshops; using the process mapping terminology, these are the "As Is" (current) process and the "To Be" (future) process. These two workshops need to be two to four weeks apart as they are relatively intellectually intense; time is needed for reflection between each workshop so that the resulting material can be reviewed critically. When undertaking this work, it is important to realize that the process starts and finishes outside of the laboratory and therefore staff working in areas that interface with the laboratory need to be involved as well as QC staff.

Understand the Current ("As Is") Process

The purpose of this workshop is to understand the way the laboratory currently operates and how computerized systems are utilized inside and outside the laboratory. This workshop establishes a baseline and allows the participants to critically analyze their ways of working.

The "as is" workshop should cover the following topics:

- What is process mapping? There are a number of techniques but either cross-functional process mapping or IDEF are considered by the author to be the optimal approaches.
- Map the current process used in the laboratories, highlighting differences in working practices.
- Map the boundaries of the current data systems and LIMS (if used).
- Identify spreadsheet and laboratory notebook use.
- Identify differences in working practices between laboratories.
- Identify SOPs/test methods used in the process.
- Identify process bottlenecks where delays occur and the reasons for them.
- Identify where and why signatures and initials are used throughout the process.
- Obtain process metrics: e.g., how many, how much, how long, and how often.
- Identify process improvement ideas.

It will soon become apparent from this workshop that processes are inefficient and paper-driven, and that computerized systems are not being used to their full potential. An example of an "As Is" process flow for a QC laboratory is shown in Figure 1 which shows that the process is paper-driven, as paperwork lists are maintained outside of the LIMS in addition to information stored within the data systems and LIMS. Also instruments are not connected to the LIMS and calculations are performed with calculators and spreadsheets. The LIMS has all data manually entered into the system. When faced with a typical "As Is" process map it is obvious that to implement and validate a LIMS in a QC environment will be a huge waste of resources with little if any payback for the organization. If a global LIMS is required in an organization, process mapping is invaluable, as it highlights where the differences are between laboratories and identifies these as areas for harmonization in the new process.

Optimize the Process for Electronic Working

To improve and optimize the process, a second workshop is carried out after the draft report from workshop 1 was circulated for review. Note the careful phraseology: we are optimizing the process not re-engineering it, the reason being that much can be achieved with a short optimization workshop rather than a full-scale process reengineering project. The underlying assumption is that the basic operation of a regulated QC laboratory is sound; it is only the details that need to be improved or redesigned, not re-engineered.

The three basic operating principles of the electronic laboratory according to Jenkins (7) are as follows:

1. *Capture Data at the Point of Origin*: If you are going to work electronically, then data must be electronic from first principles. However, there is a wide range of data types that include observational data (e.g., odor, color, size), instrument data (e.g., pH, LC, UV, NMR, etc.) and computer data (e.g., manipulation or calculation of previous data). The principle of interfacing must be balanced with the business reality of cost-effective interfacing: what are the data volumes and numbers of samples coupled with the frequency of the instrument use?
2. *Eliminate Transcription Error Checks*: The principles for design are as follows: never re-enter data and design simple electronic workflows to transfer data and information seamlessly between systems. This requires automatic checks to ensure that data are transferred and manipulated correctly. Where

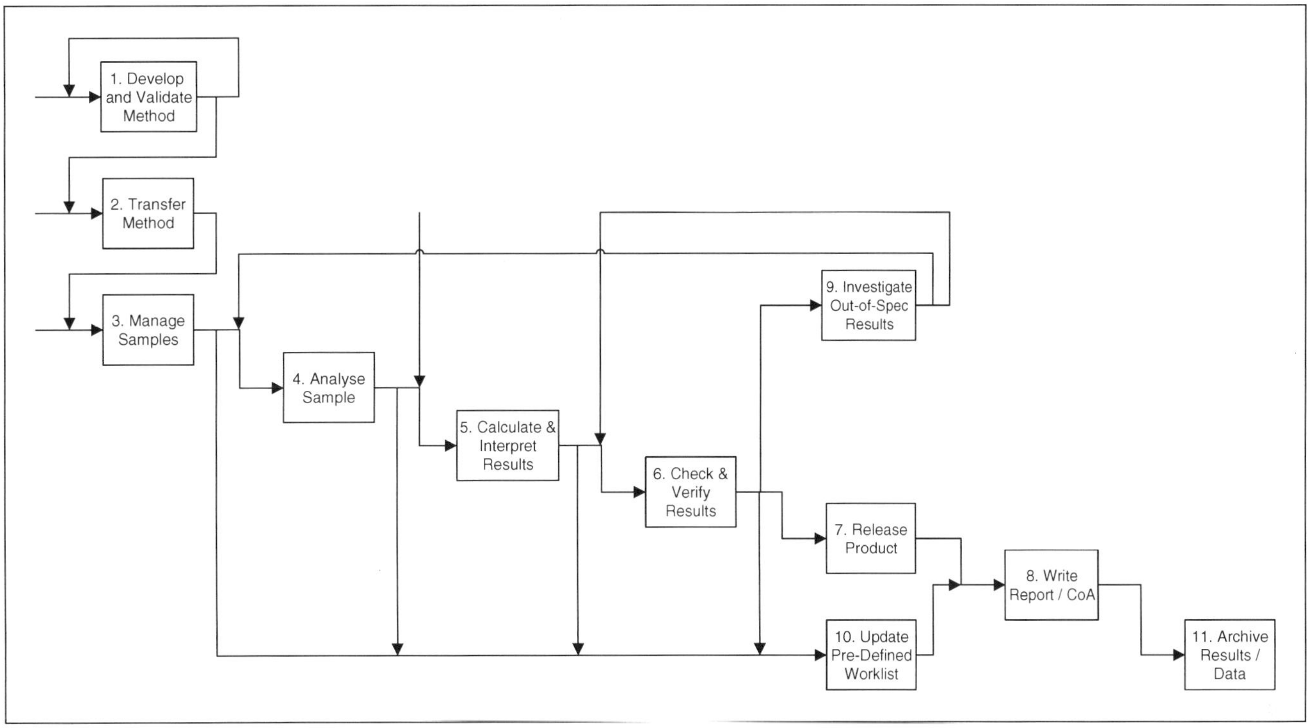

Figure 1 "As is" process map in a quality control laboratory.

appropriate implement security and audit trails for data integrity and only have networked systems for effective data and information sharing.

3. *Know Where the Data Will Go*: Design data locations before implementing any part of the LIMS and the LIMS environment. The fundamental information required is what volumes of data are generated by the instrumentation and where the data will be stored: in an archive system, with the individual data systems or on a networked drive? The corollary is that security of the data and backup are of paramount importance in this electronic environment. In addition, file-naming conventions are essential to ensure that all data are uniquely numbered either manually or automatically. If required, any archive and restore processes must be designed and tested to that they are reliable and robust.

These principles should be used to optimize the "as is" process maps to define the new or "to be" process in the second workshop, which typically will cover the following items:

- Review the "as is" process maps with modification where necessary to reflect the current working practices.
- Optimize and harmonize (especially between laboratories) the process and generate the "to be" process using the following inputs: Improvement ideas generated in workshop 1. Eliminating unnecessary process steps. Identifying any manual steps to be automated by the new LIMS or other computer systems.
- Define the new boundaries of the LIMS and other computer systems inside and outside the laboratory.
- Identify data transfers between these systems.
- Estimate potential time and calculation of time savings from the new process.
- Identify any "quick wins" for rapid implementation (these are defined as improvement ideas that are cheap to implement but provide high benefit and give credibility to the overall approach).

The new process map is shown in Figure 2; the process has been made electronic and data are transferred to the LIMS electronically from the data systems to eliminate manual data entry. Electronic signatures have been implemented within the LIMS to eliminate much of the current paper records. Note that paper will not be eliminated entirely but the majority of primary records will be electronic. The work list is no longer required as the information will be maintained electronically. Although the main tasks in the process still remain, the time taken when working electronically between steps 5, 6 and 7 has been cut by approximately 50% to 60% as the systems are set up to work electronically.

Estimated Benefits of Working Electronically

When the new process has been defined and mapped, the new timings of the process can be estimated and compared with the current process. Based on the differences between the two, an estimate in overall time and resource saving can be calculated. Savings should be large enough to cost-justify the system on tangible business benefits alone including faster product release, quicker acceptance and rejection of raw materials and holding less stock. While intangible benefits such as quality are important, the organization needs to know if there will be a payback from the investment.

However, do not assume that better processes will result in headcount reduction in a laboratory; what it means is that the overall laboratory process will work more efficiently and faster but the LIMS will mean changes in the laboratory staff roles. LIMS application administration will be needed where none existed before,

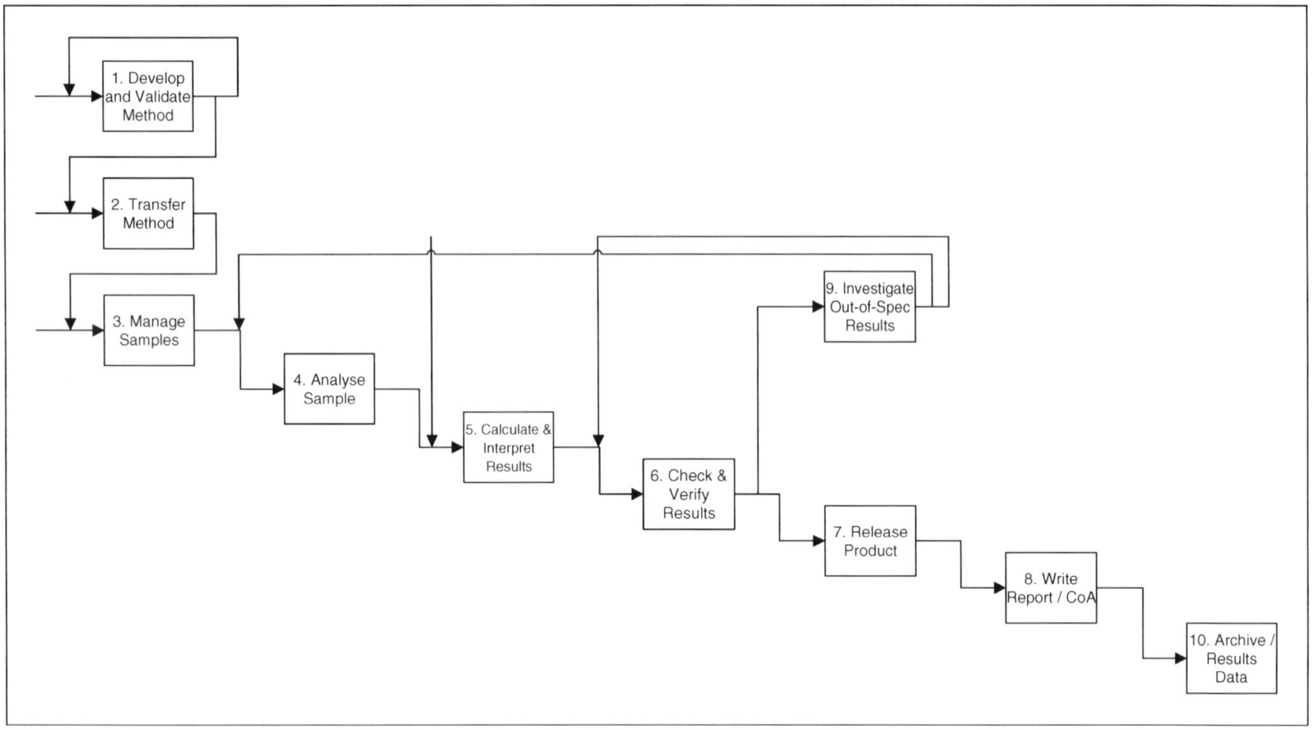

Figure 2 Optimized "to be" electronic process for a laboratory.

e.g., power users within the laboratory will be the first line of help for users and staff will be needed for the inputting specifications (if not done automatically) and methods into the LIMS. Do not underestimate the amount of work that this will entail.

DESIGNING THE LIMS ENVIRONMENT

As stated in the introduction, QC laboratories must work electronically if they are to survive; therefore, the LIMS environment needs to be defined based on the optimized process. The first stage in designing this is to look at a LIMS as an interface between the laboratory and production. This will be followed by interfacing the LIMS and other computerized systems within the QC laboratory to produce an electronic LIMS environment to support the newly designed process.

Positioning a LIMS: Hitting Two Targets

It is important to realize that a LIMS should automate both the laboratory where it is implemented and the production facilities that the laboratory serves. To be effective a system should deliver benefit to both the laboratory and the production. How should this be achieved? A LIMS is unlike any other piece of laboratory automation equipment available to the analytical chemist. It can provide benefits both within the laboratory and outside of it. Thus a LIMS has two targets (8):

- The laboratory: the information generator
- The organization: the information user

The problem is how to site and implement a system so that it hits both targets effectively. Figure 3 shows an outline of the functions that a LIMS should undertake in a simplistic way. The diagram shows a LIMS sited at the interface between a laboratory and an organization. Samples are generated in the organization and received in the LIMS, and then the samples are analyzed within the laboratory. The data produced during analysis are reduced within the LIMS environment to information which is transmitted back into the organization. Figure 3 represents the ideal positioning of a LIMS: both the organization and the laboratory benefit, as the line dividing the organization and the laboratory show, and the system is of equal benefit to both.

The LIMS Environment

A successful LIMS implementation builds a LIMS environment to serve both the organization and the laboratory. The key to success is that the LIMS must integrate the processes and the computerized systems in these two areas where analytical information is generated and used.

Some of the applications outside of a laboratory that a LIMS could be interfaced to design the LIMS environment are listed below and in the top half of Figure 4:

- E-mail systems for transmission of reports to customers or keeping them aware of progress with their analysis
- Web servers for laboratory customers to view approved results and also for contract laboratories to input data into the QC LIMS
- ERP systems for linking the laboratory with production
- Applications maintaining product specifications
- Data warehouses
- Electronic Document Management Systems
- Failure Investigation Systems
- Electronic Submission Systems (for GMP laboratories in pharmaceutical R&D)

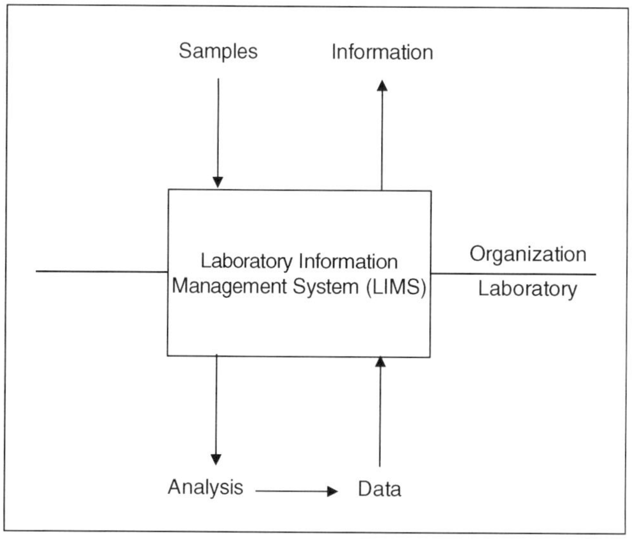

Figure 3 A laboratory information management system delivering benefit to the quality control laboratory and organization.

These are just a few of the possible applications that a LIMS could be interfaced to; the list of potential candidates will be based on the nature of the analytical laboratory and the production organization it serves.

Some ERP vendors can claim LIMS functionality and that there is no need to implement a LIMS; however, the problem with this approach is that the ERP's concept of a QC laboratory often does not match the reality. For example, the sample process flows within an ERP tend to be high level and very simplistic and cannot automate all laboratory functions, e.g., OOS investigations without extensive writing of custom software. Once the organizational side of the LIMS environment has been designed, the LIMS environment within the laboratory needs to be designed.

Designing the LIMS environment means that you need to consider the other systems in the lab that must interface with the LIMS. This includes other laboratory applications such as scientific data management systems, CDS, and electronic lab notebooks, as well as various data systems that may be attached to those or run independently. It also includes analytical instruments, chromatographs, and laboratory observations as shown in the lower half of Figure 4. Data can be transferred to the LIMS by a variety of means:

- Direct data capture by the LIMS
- Capture by an instrument data system with analysis and interpretation and only a result is transferred to the LIMS
- As above but the results or electronic records are transferred to the LIMS via a Scientific Data Management System
- Laboratory observations can be written into a notebook then entered manually into the LIMS or captured electronically via an ELN and transferred electronically to LIMS
- Bar codes (or RFID) can be used to label samples and enter data rapidly into the LIMS

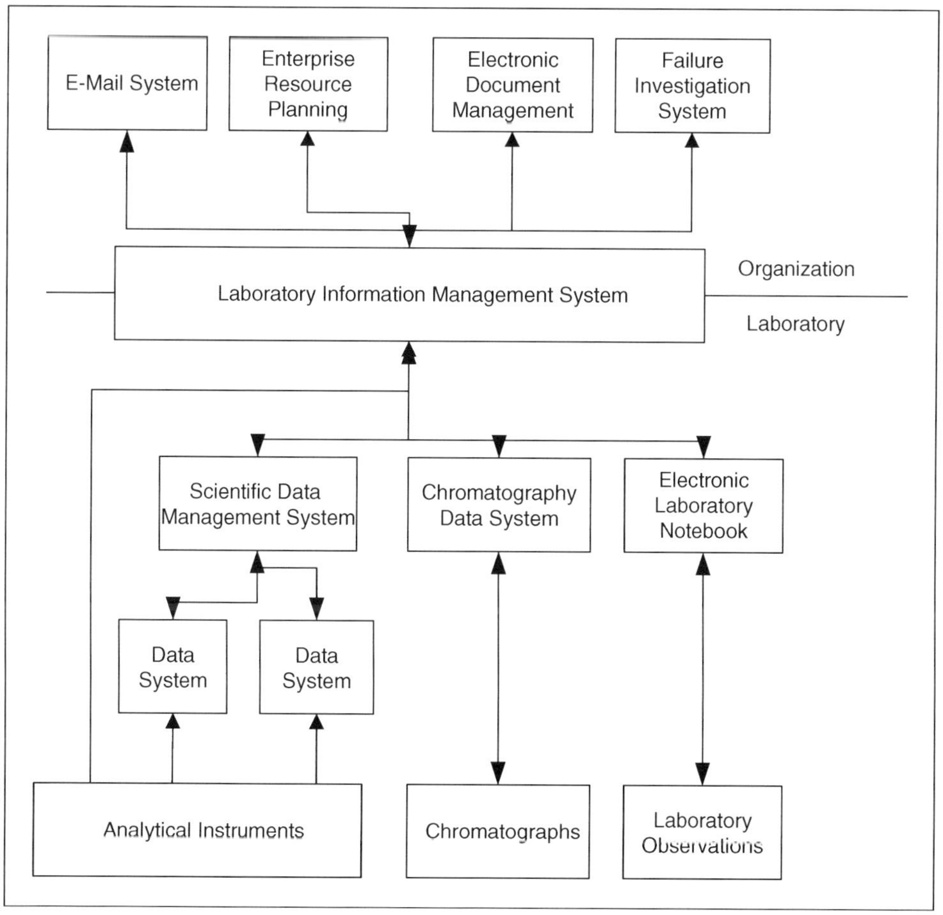

Figure 4 Options for a laboratory information management system environment to integrate the laboratory with the organization.

Before implementing a LIMS, it may be appropriate in some laboratories to standardize and implement instrument data systems, e.g., chromatography data system. The rationale for this approach is that a data system can be quicker to implement than a LIMS and it will provide a firm foundation to build the LIMS above it. If it is done the other way around, the data system may need to be updated later with a consequent change in working practices and more revalidation.

LIMS IMPLEMENTATION AND RISK-BASED VALIDATION

This section of the chapter deals with the system implementation life cycle as outlined by the GAMP Good Practice Guide for Laboratory Computerized Systems (4) and modified by McDowall (5). The aim is to realize and deliver through the LIMS those business benefits identified in the process redesign in a cost-effective manner.

Electronic Records Generated by a LIMS in a GMP Environment

The change from a system to a record-based approach to the validation of computerized systems was initiated by the FDA in the Guidance for Industry on Part 11 Scope and Application (2). The GAMP Good Practice Guide for Compliant Part 11 Records and Signatures (3) has taken this and developed a risk-based approach to validation of a computerized system based on the impact of the records generated and managed by an application.

The records generated and managed by a LIMS in a GMP environment are high impact as they are either used in product release and/or product submission. Some examples of electronic records that could be contained within a system are listed below, this list should not be considered exhaustive as it depends on how a specific LIMS has been implemented and used:

- Specifications of products, intermediates and raw materials
- Stability protocol
- Sampling methods
- Analytical methods
- Worklists
- Observations and results captured directly by the LIMS, e.g., pH and balance measurements
- Results transferred from analytical systems
- Comparison of results versus specification and identification of OOS results
- OOS investigations and where appropriate additional results
- Electronic signatures
- Certificates of analysis
- Audit trail entries
- Instrument qualification and calibration status

The electronic records need to be identified and documented (2). It is important to understand that this is not a static process; as the LIMS is updated the new version may contain new functions that may create new electronic records in addition to those listed above. If new functions are added using the scripting language, then these may also create new electronic records. Therefore, it is important to review this list on a regular basis; when the system is upgraded and during a periodic review are the obvious times.

These are high impact records as defined by GAMP (3) as they can impact product quality and/or patient safety. Therefore, a more rigorous approach should be adopted which includes:

- *Hazard Identification*: The hazards that the LIMS could face should be identified along with the consequences of each one. However, although a hazard and its associated consequences may have been identified, we do not know if a specific one poses a risk to the system. To identify the potential risks to the system, a risk assessment needs to be undertaken.
- *Risk Assessment*: For each hazard identified, the severity of the consequence and probability of occurrence both need to be estimated; this is achieved by allocating either high, medium or low (3). There are different classes of hazard such as human, software, hardware, IT support, physical and environmental. Risks will be classified as either as class 1, 2 or 3 (high, medium or low risks) to identify which risks are important enough to implement mitigation controls.
- *Control Selection*: Controls for electronic records and electronic signatures generated by systems can be implemented at a number of levels: Organization via policies and standards, e.g., validation policy and passwords. Procedural (and implicitly training) via SOPs, e.g., user manual and change control. Application and network via technical controls such as audit trail, application and/or network security and checks. IT Infrastructure via network security, backup and recovery, hardware and network redundancy. Computer system validation.

Owing to the nature of a LIMS in a QC Laboratory in a GMP environment, the system will require validation plus other controls to mitigate risk and protect the electronic records such as application security and access control and one or more audit trails for working electronically. In addition, the server needs to have redundant components such as dual processors, disk controllers and RAID disks to ensure that data are protected and not lost due to hardware failure. This section will concentrate on the risk-based validation of a LIMS; it is intended to build upon the process redesign and design of the LIMS environment outlined earlier to ensure a successful and cost-effective LIMS implementation.

System Implementation Life Cycle Activities

For the purposes of simplicity, the implementation life cycle will begin with either the implementation of a new system or an upgrade of an existing LIMS. This means the writing of the initial URS used to generate the RFP used in the system selection process, the system selection and audit of the vendor will be omitted from this chapter. For readers that want to understand this part of the life cycle process should read the appropriate chapters from McDowall (9). Therefore, the start of the implementation life cycle here will be where either:

- A new LIMS to be configured and installed in a laboratory but with an outline URS used to select the specific system
- An existing LIMS installation which will be upgraded to the latest application version.

There are three main work streams to consider that are outlined below and presented in Figure 5:

- Specification, installation, and qualification of the computer hardware
- Validation of the LIMS application
- Writing procedures and training the users

Tasks for the three streams are shown and this will help to put the remaining tasks in this section into context.

Updating the User Requirements Specification

For a new system, there will be an initial LIMS URS available for system selection; however, it will need updating for the purchased new version of the application. The reason for this is that the URS used for system selection is general in character and is unlikely to be sufficiently specific to design tests for the validation of the selected LIMS. Therefore, an application and version specific URS needs to be written that defines the intended use of the system and contains the functions and the capabilities required by the system. It is this document that the PQ tests will be based on using the risk assessment and traceability matrix documents.

For the LIMS upgrade, the existing version of the URS will not cover all the new functions available, and this document needs to be reviewed and updated where appropriate. This should be the easier job as users will be trained on the current version of the system. The release notes provided by the application vendor will be useful to highlight areas where the URS will need to be updated if the laboratory intends to use them (see below). Each requirement in the URS should be prioritized as either mandatory (must have requirement) or desirable (nice to have but if not available the LIMS functionality is not impaired) (9).

Read the Application Release Notes

To focus on the changes that have occurred in a new release of the LIMS software, read the software release notes to understand the nature of the new features that have been added as well as the software errors that have been fixed. Although this sounds simple and straightforward, in reality this is more complex due to the way that the pharmaceutical industry handles software in a regulated environment. If version 2 of a LIMS application has been installed and validated, typically the laboratory will miss the next release, version 3, unless there is a good reason for change. Version 4 of the LIMS will be implemented instead; thus, the laboratory implements every other version of the software rather than keeping current due to the perceived cost and effort of validation. Therefore, reading the release notes of the last two (or more) versions of the software and understanding the impact is the norm rather than the exception. This means that new features need to be understood and prototyped to understand their value and potential impact on the laboratory's ways of working.

Using Process Maps to Define System Requirements

An additional advantage of the "To Be" process maps is their use in facilitating the requirements for the LIMS and the other systems used in the LIMS environment. The traditional problem with writing requirements for any computerized system is that obtaining requirements can be akin to extracting teeth from the users. The terminology "gathering requirements" implies that they are freely available to be written down but often nothing is further from the truth. The advantage of the process maps is that they provide an effective medium for obtaining requirements. Each activity in the process map has inputs and outputs defined from the workshops. A facilitator then needs to ask the laboratory users what happens in each process activity. This means that the requirements can be more precisely defined providing a greater certainty in system specification, selection and validation as the users are focused on a specific task.

Better Definition of User Requirements: The Role of Prototyping

Prototyping is an important tool to help understand how a new LIMS application or the new features of an upgrade can work within a laboratory. The corollary is that users must have been trained on the new version of the software rather than reading the on-line help files. Features can be evaluated in an unqualified installation to identify if they are useful and then to refine how each one may be used to best advantage and business benefit. Although it is valuable, prototyping has to be handled with care; only two rounds of prototyping should be undertaken, e.g., high level to determine which functions should be in the implementation and which should be excluded followed by a second round for further in depth evaluation of the selected options and finalizing the details of operation.

There is a danger that prototyping can be unstructured with little documentation from the exercise. From experience, the best way to tackle this is to have as defined outputs from each phase of the prototyping an update of the URS plus outline testing documents. If the LIMS scripting language is being used during this work, then documentation of the functions being modified needs to be generated and maintained.

All of these documents should be uncontrolled but unless they are available for review outside of the project team, the second phase of the prototyping work cannot proceed. This approach is intended to instill the discipline to ensure the work is documented as it goes on but also is an investment in time to reduce the amount of effort needed later to write the PQ test scripts.

Defining Electronic Signature Use

During the prototyping phase, electronic signature use should be evaluated to support the electronic workflows that were designed in the process redesign phase. It is important to understand the need to differentiate between identification of actions and signing of records. The former is akin to the correction of an error in a laboratory notebook, where an entry is struck through without obliterating the original, corrected and then the initials and date of the person making the entry are appended. The latter is the formal signing of the page in the laboratory notebook by the owner to state they accept responsibility for the correct data above on the page.

For many companies, it is unfortunate that compliance has overridden the regulations and many records are signed by custom and practice than need to be. The 21

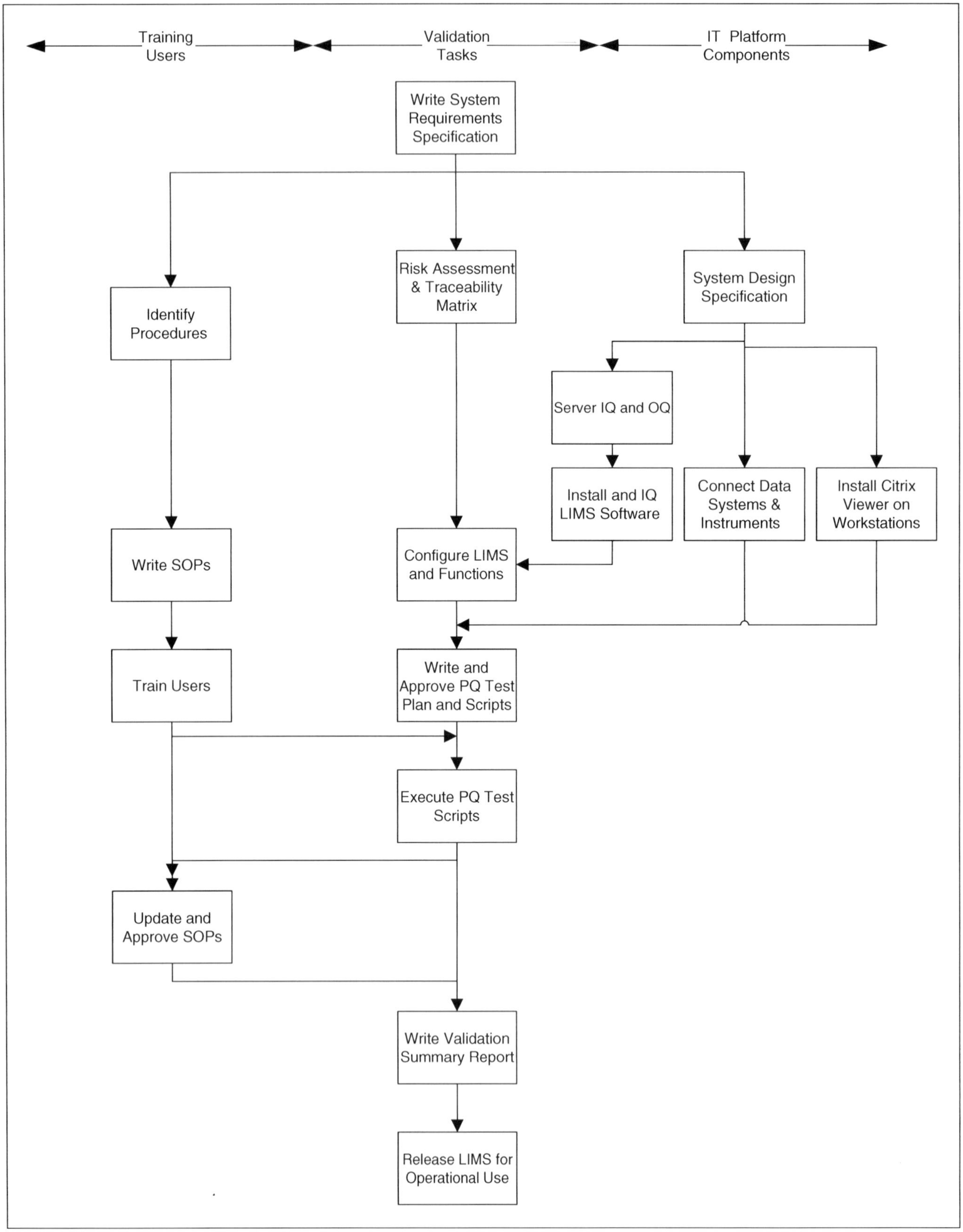

Figure 5 Outline tasks involved in a laboratory information management system risk-based validation.

CFR 211 regulations simply state that only two signers are needed per test (§194(a) sub-clauses 7 and 8) (10), the first to state that the results generated are correct and a second person signed to say they have been checked for accuracy and the correct procedures have been followed. Therefore, the LIMS needs to reflect the regulation rather than electronically sign everything.

Write the Validation Plan

For a new system implementation, a validation plan is required to control the work of the validation. As a minimum it should define the roles and responsibilities of all individuals working on the project, the life cycle to be followed and the documentation to be written at each stage of the project. In addition, for global or site projects, the overall validation strategy should be presented: how the development and validation of a core application for all sites will be achieved along with the documentation to support it, then how it will be installed at each site and under which conditions an instance can be modified by a local site.

For an upgrade of an existing LIMS, a change control request could suffice to control the work but inevitably a validation plan is written as the work will usually involve replacement of the server and modifications of the current ways of working.

Combined Risk Assessment and Traceability Matrix

Risk assessment is now a key validation requirement following the FDA's Part 11 Scope and Application guidance (2). After the URS was written, both the system and the individual functions were assessed for regulatory and business risk using the Functional Risk Assessment methodology (9,11). Here individual system functions are assessed as either critical or not critical (C or N respectively) from a regulatory and/or business risk perspective. Coupled with the prioritization in the user URS, each requirement is graded as either mandatory or desirable as well as either critical or noncritical; these can be plotted in a 2×2 Boston grid to determine overall risk. Only functions that were both mandatory and critical were considered for PQ testing; all other combinations are not considered any further. The rationale for this is based on the vendor's testing of the application.

Mandatory and critical functions were then evaluated further to see if they need to be:

- Explicitly tested and then assigned a specific test script number where similar functions are tested together
- Assumed to work as there was no access to either the algorithm
- Implicitly tested such as the windows and some display functions
- Verified during the qualification of the system
- Traced to a procedure or an SOP

This is a simpler process for a commercial system than the modified FMEA outlined in Appendix M4 of the GAMP guide (12).

System Architecture

The vendor's recommendations should be used by the organization's IT Department to size and specify the database and application servers for the system. If terminal servers (e.g., Citrix Metaframe) are to be used, these need to be specified, as will the other LIMS instances used for evaluation, training, validation and production. Diagrams of the overall system architecture will help to understand the approach taken and should be encouraged to be drawn for inclusion here. Increasingly rather than have a server for each instance, virtual servers are used running within an environment such as VMware; this is useful to reduce hardware costs and maintain individual instances of the LIMS. Data can either be attached to the production server in a RAID configuration or increasingly SAN devices are used. All details concerning the system architecture should be documented in a SDS or equivalent as this is an input into the configuration records for the overall system.

System IQ and OQ

Server IQ Plans and Installation

Installation plans for all the servers (database and application instances as well as any Citrix servers used for the application) should be written by the IT Department. These plans should include the installation of the hardware and documenting its con-figuration as well as installing and configuring the operating system and any utilities for each server, e.g., agents for backup, network management software, etc. The installation of hardware, operating system and any utilities for all servers must follow these plans and record the actual details of each server installed such as serial number and configuration (memory, processor type and speed and IP address, etc.).

LIMS Database, LIMS Application and Instrument Interface IQ

The activities that are involved in this task should be:

- An evaluation of the vendor's installation qualification documentation to check that it is acceptable.
- Installing the LIMS database and software on the respective servers for each instance by either a member of organization's IT staff or the vendor's service personnel. At the same time the application IQ is completed and followed by a review of the documents to ensure that the instructions tests have been performed with acceptable results.
- The analytical equipment and instruments to be interfaced to the LIMS in the initial phase of the LIMS implementation will be interfaced now and checked that the connections work. Again, this will be planned and there will be documentation available to demonstrate the activities undertaken.

Establish Change Control and Configuration Management Now

Once the servers and application has been installed, the system needs to be placed under change control. Some organizations write a specific change control SOP for each system; however, the smarter ones will have a single procedure that is applicable to all regulated systems. Allied with change control is configuration management which is just as important but often neglected. Configuration management is the definition of the CIs that constitute the whole system. CIs consist of:

- Hardware
- Software
- Documentation (ranging from vendor supplied material including electronic manuals to company-specific documents)

The level of detail required should be sufficient to provide business benefit from the information, e.g., server information will require make, model, processor size and speed, memory, disk size and configuration, operating system and service pack, and network information such as IP address, etc. Less information would be required of a workstation, e.g., typically this would be the minimum specification available to run the application within the company as many organizations change workstations every three to four years. When a change is made, the configuration management records before and after the change should document what CIs have been modified, added or removed.

Do I Need a LIMS Application OQ?

Traditionally there is now an operational qualification to demonstrate that the LIMS application software works as the vendor intended it to. Here is where we can take advantage of a risk-based approach to computer validation. Most OQ packages offered by LIMS vendors are their internal test suites, used either "As Is" process or modified for external sale for demonstrating that the unconfigured application works. Look at the process: the vendor produces and tests the base application, manufactures the CD from which you install the same software. Do you need to execute essentially the same tests that the vendor has? No. Furthermore, the application that is installed will be configured by the laboratory to their own ways of working away from the base package, making the execution of a comprehensive OQ a further waste of time. Therefore this stage should be omitted from any LIMS validation as it adds little if any value to the overall work and is relatively expensive and time consuming to perform. If the PQ works then the application functions.

Configuring the System

LIMS do not have "ON" buttons and therefore each installation will need to be configured to the laboratories working practices as noted earlier in this section under user requirements. Depending on the vendor and application chosen this can be achieved in a variety of ways either alone or in combination:

- Configuration by selecting one of a series of options offered by the vendor. For example, selecting the access privileges for a specific user type
- Configuration within the boundaries of the LIMS application by using the scripting language supplied by the vendor
- Customization by writing new functionality to extend the LIMS

This section will look at how this work needs to be undertaken, tested and documented.

Do I Need a Functional Specification?

Not necessarily, as it depends on how any additional LIMS functionality is implemented and how extensive the work will be. What is important is that the configuration of the LIMS is recorded rather than what a document is called. The URS can contain the majority of requirements of the system but the detail needs to be recorded in one or more configuration documents. However, if the LIMS will be extensively configured then an overall functional specification is advised and this will have traceability back to the URS. Note that a single URS requirement may generate more than one requirement within the functional specification.

Using the LIMS Scripting Language

Before starting any work with the scripting, language developers will need to be trained and understand the implications of use. Alternatively, this is an area which the vendor and their staff could be engaged to develop on the behalf of the laboratory. If prototyping has been used earlier to generate requirements and workflows the resulting scripts can be used again here. If the vendor publishes any standards for using the language these should be followed as good practice. Where possible, the scripts should be reviewed by a second person before being implemented. Copies of the scripts used to modify the LIMS functions should be maintained outside of the system in case of disaster; do not rely solely on recovery from magnetic backup tape to preserve them.

The output from the configuration should be tested against requirements or other specifications to ensure that they are correct. Correctly performing configurations are copied into the validation environment prior to the PQ.

Input of Methods and Specifications

Populating the database with methods and the corresponding specifications will take time and should not be overlooked when planning the project. Although this process will start during the configuration, the process will be ongoing throughout the operational life time of the LIMS as new products and specifications are added to the system. The ideal for specifications is to download the information from another system where it is maintained electronically; however, often specifications are maintained on paper and this requires the laboratory to input and check them manually before transferring them to the operational instance. Similarly, methods will need to be inputted to the LIMS and controlled; inevitably this will be a manual process, although once entered methods can be copied from one product and adapted to another one.

Write SOPs and Train Users

Users and IT operations staff will need to write or modify the SOPs identified in the URS for the various operations of the LIMS. This typically ranges from basic user operations, through application and system support, to database maintenance. Either these SOPs need to be available in final draft form when the PQ is executed to enable any changes required to be incorporated before the documents were approved and released or they are approved before the PQ and if any changes are required after the PQ these will be identified as they are checked out during the PQ. All users of the system, including the IT support staff need to be trained as appropriate to their tasks and records maintained of these activities.

Initially only the staff involved with the PQ will need to be trained on the SOPs; however, before other users are allowed to use the operational system they will need to be trained. Care must be taken over training as LIMS may have long learning curves and it is neither fair nor reasonable to expect laboratory personnel to work at the same level immediately after LIMS training as they will still be getting to grips with a new application. This is less of an issue with an upgrade; however, it depends how many versions have been skipped.

LIMS Performance Qualification or End-User Testing

The purpose of the end-user testing or PQ is to demonstrate that the functions specified in the URS work as intended to meet both business and regulatory requirements for the LIMS. Terminology here can be confusing within the analytical laboratory as IQ, OQ and PQ are used for both analytical equipment qualification and computerized system validation but mean different things (9). Also the terminology used to describe the documents generated in this phase of validation work can differ greatly; it is important to remember that the work is done and documented rather than what a particular document is called.

Test Plan for Controlling thc PQ

A test plan based on IEEE standard 829 is used to control the work (13). This is achieved by defining the system to be tested and its scope. It can list the test scripts to be written and link these back to the URS requirements to be tested under each test script and the features not to be tested (13). Testing cannot be exhaustive, so there is also a section on the assumptions, exclusions, and limitations to the testing; this is a very useful way of recording contemporaneous notes of why testing was conducting in a particular way.

PQ Test Scripts

The test scripts or protocols are written in sufficient detail to test the requirements in the URS; traceability back to the requirements is important from two perspectives. The first is to check coverage of testing versus requirements and the second is to check that the requirements are testable or verifiable. Occasionally the URS may need to be updated at this stage to modify some requirements that cannot be tested or verified adequately or it is realized that a requirement has been written incorrectly.

Testing of the system should cover its main functions including instruments and systems that have been interfaced with the LIMS plus the overall capacity of the system. If the LIMS is interfaced with an ERP system, then many of the test scripts will start in the ERP by generating a work order that is downloaded to the LIMS; at the end the analytical release will be sent to the ERP. It is important to realize that although the basic operations of the LIMS must be validated before operational release, many of the database population activities will be controlled by procedure and do not require validation per se.

Approved PQ test scripts are executed and documented evidence in both paper and electronic form are collected; test results should be compared with explicitly stated acceptance criteria. It is important to use screen shots sparingly, where the system does not record information within the database or audit trail and where they add value. Similarly, witness testing is not an FDA requirement but validation custom and practice; however a second person review is mandatory. The results of this were documented in the respective test scripts and summarized in the validation summary report for the LIMS; a specific PQ report need not be written.

Write System Description and Definition of E-Records

A system description should be written and approved for the LIMS. The best format for this document is found in the outline requirements contained in the Application of GLP Principles to Computerized Systems from the OECD (14). In addition, the system description should also contain the definition of electronic records for the system and the fact that 21 CFR 11 applied to the application as required by the Part 11 Scope and Application guidance (2).

Reporting the Validation

Before writing any validation summary report, the first activity is to read the applicable validation plan and understand what the original intent of the validation was. This will identify if deviations have occurred that have not been explained previously.

Write Validation Summary Report and Release the Core System

This validation report contained the summary of the validation of the core system and was issued after the validation of the core system (the first rollout). A statement in the validation summary report released the system for operational use including electronic signatures. The report was reviewed and approved by the system owners and QA prior to releasing the system for operational use.

Write Validation Summary Report for Each Rollout of the System

Each additional phase of the system rollout had a validation summary report written to describe the work that has been undertaken in that phase to maintain the original validation status of the system. These tasks included a summary of the evidence for:

- Any additional servers installed and qualified
- Interfacing of any new instruments or systems to the LIMS
- Updated configuration logs
- Any further or repeat PQ test scripts executed under the PQ test plan
- User training performed and an updated list of authorized users for the system

SUMMARY

Risk-based validation of a LIMS opens up the opportunity for organizations to streamline the amount of work that they undertaken and focus the effort on the areas of highest business and regulatory risk.

REFERENCES

1. FDA. Good Manufacturing Practices for the 21st Century, 2002.
2. FDA. Guidance for Industry: Part 11 Scope and Application, 2003.
3. Good Automated Manufacturing Practice (GAMP), Good Practice Guide on Compliance Part 11 Electronic Records and Electronic Signatures, International Society for Pharmaceutical Engineering, Tampa, FL, 2005.
4. Good Automated Manufacturing Practice (GAMP), Good Practice Guide on Laboratory Computerised Systems, International Society for Pharmaceutical Engineering, Tampa, FL, 2005.
5. McDowall RD. Validation of spectrometry software: critique of the GAMP good practice guide for validation of laboratory computerized systems. Spectroscopy 2006; 21(4):15–30.
6. Good Automated Manufacturing Practice (GAMP), Good Practice Guide on Global Information Systems Compliance and Control, International Society for Pharmaceutical Engineering, Tampa, FL, 2005.
7. Jenkins S. Pittsburgh. Conference 2004 Presentation on the Paperless Laboratory.
8. McDowall RD. Chemometrics and intelligent laboratory systems. Lab Automation Info Manag 1995; 31:57–64.
9. McDowall RD. Validation of Chromatography Data Systems: Meeting Business and Regulatory Requirements. Cambridge: Royal Society of Chemistry, 2005.
10. Good Manufacturing Practice Regulations 21 CFR 211.
11. McDowall RD. Effective and practical risk management options for computerised system validation. Qual Assur J 2005; 9:196–227.
12. Good Automated Manufacturing Practice (GAMP), Guidelines version 4, International Society for Pharmaceutical Engineering, Tampa, FL, 2001.
13. Software Engineering Standard 829 on Software Test Documentation, Institute of Electrical and Electronic Engineers, Piscataway, NJ, 1998.
14. Application of GLP Principles to Computerised Systems, Organisation for Economic Co-operation and Development, Paris, 1995.

49

Validation of Laboratory Information Systems

Alex M. Zislin
DSCI, Freehold, New Jersey, U.S.A.

Rory Budihandojo
Boehringer-Ingelheim Chemicals, Petersburg, Virginia, U.S.A.

Jeffrey M. Singer
PharmaBasics, Inc., Douglaston, New York, U.S.A.

INTRODUCTION

Today's pharmaceutical manufacturing environment has generated a large body of requirements for data from manufacturing processes, in-line testing, off-line testing, product testing, intermediary, stability, potency, raw material qualification, etc. In addition, there is also data that comes from clinical trials, i.e., API blood serum levels, metabolite levels, etc. The common thread with these data is that they are generated by laboratory testing on unequivocally identified samples using validated methods on qualified instruments, analyzed by validated algorithms on qualified computers, stored on validated computer systems, and reported by validated reporting functions on these and other validated computer systems that access the stored data. In addition, some or all of these data may be submitted to regulatory agencies as part of the submissions required to license a new API, or to expand the indications of an existing API. In all of the above examples, the system that generates, analyzes, stores and reports on this data must be in a Validated State, if it handles GxP-related data.

Abbreviations used in this chapter: API, active pharmaceutical ingredient; APR, annual product review; CFR, Code of Federal Regulations; COTS, configurable off-the-shelf; DS, design specification; FDA, Food and Drug Administration; FRS, functional requirement specification; FS, function specification; GAMP, good automated manufacturing practice; GMP, good manufacturing practice; GUI, graphical user interface; GXP, good practices (manufacturing, practice, and laboratory practice); HPLC, high-performance liquid chromatography; IEEE, Institute of Electrical and Electronic Engineers; IQ, installation qualification; IQP, installation qualification plan/protocol; ISO, International Organization for Standardization; ISPE, International Society for Pharmaceutical Engineering; LAN, local area network; LIMS, Laboratory Information Management System; MFG, manufacturing; MSDS, material safety data sheet; NCE, non-conforming event; OOS, out-of-specification; OQ, operational qualification; OQP, operational qualification plan; OS, operating system; PAT, process analytical technology; PQ, performance qualification; PQP, performance qualification plan/protocol; QA, quality assurance; R&D, research and development; RTM, requirements traceability matrix; SDLC, software development life cycle; SEI, Software Engineering Institute; SILC, system implementation life cycle; SIPOC, suppliers, inputs, process, outputs, consumers; SOP, standard operating procedure; SPC, statistical process control; UAT, user acceptance testing; UFRS, user functional requirements specification; URS, user requirements specification.

SCOPE

The techniques, practices and approaches described in this chapter reflect the current regulatory thinking in terms of validation, risk, and PAT (1–7). In addition to the standard validation methodologies and practices, this chapter will address some of the implications of the recently released risk-based GMP guidance. The goal of any system validation effort is to determine the fitness for intended use, where use is defined by the full and qualified set of end-user requirements. The successful completion of the validation activities themselves generate the objective documented evidence that the system is installed, operates, and performs as specified. These objectives are the same for all automated systems.

In addition to starting off with a validated system, it is important (and required) to maintain the system in a validated state during its entire useful life. Activities that provide this are also described in this chapter.

SYSTEM DEFINITION

For the purposes of this chapter, it is considered that LIMSs are computer-based systems, consisting of integrated hardware (instruments, computers, etc.) and software (applications, instrument software, etc.) that perform one or more of the following functions either alone or in conjunction with other associated systems:

- Data storage and data management
- Data acquisition from multiple instruments
- Data acquisition from manual entry
- Sample management
- Laboratory scheduling
- Data source—MSDS
- Data source—laboratory procedures
- Data source—laboratory material specifications
- Interface with automated instruments
- Interface with Laboratory Automation Systems
- Instrument control
- Laboratory workflow control
- Environmental monitoring
- Data analysis
- OOS notification
- NCE notification
- Laboratory user training
- Reporting—results, OOS, NCE
- Data archive

- Data backup and restore
- Laboratory resource planning (analysts, instruments, chemical standards, etc.)
- Trending (e.g., for SPC)
- APR reporting

Figure 1 illustrates a representative architecture for a LIMS implementation. Note that the connections to the LIMS can take several forms ranging from a standalone instrument to instruments controlled by clients either directly connected to the LAN or connected through a gateway for remotely sited systems.

The actual LIMS itself may be developed in-house, purchased as a turnkey COTS system, or a combination of COTS and in-house development/configuration. In any instance, the validation should consider the full SDLC to ensure that the system is fully validated. There are some practitioners who prefer to follow the GAMP SILC model for LIMS validation. (8). The overall end result is similar, the key difference being that the SDLC covers the actual functional design and coding of the LIMS, whereas the SILC emphasizes design qualification. Figure 2 illustrates the differences and convergences of SDLC and SILC process flows.

This chapter will cover the case of a configurable COTS system and refers to additional activities that would be required for a custom LIMS development.

For purposes of system planning, implementation and validation, it is important to define fully the scope and boundaries of the LIMS in such a manner as to reveal a comprehensive and compliant requirements set. In this way, all of the aforementioned activities can be both effective and efficient.

PROJECT PLANNING

Different guidelines suggest that Project Planning begins at different points in the total acquisition/implementation/validation process. The actual time of initiating a Project Plan is somewhat fluid depending upon the internal practices at a particular site. Some suggest that planning begins after user requirements are written, while some suggest that planning begins upon the decision to procure and implement a new or upgraded system (9–19). It is more desirable to have a plan from which to work as soon as possible even though the plan will require updates and details at each phase of the project. For purposes of this chapter, a Project Plan will be considered one of the first deliverables in the validation process. Once the decision is made to go forward with a LIMS, a comprehensive Validation Plan can be used to describe how proceeding will greatly enhance the ability to define the project, predict the resources (budgetary and personnel) required, identify any critical path challenges and estimate the expected completion date.

A typical Project Plan at this stage would be very simple and includes the following items:

- General System Description
- Scope of the System
- Expected Budget
- Make or Buy Criteria and/or Decision
- Resources Required (internal and external)
- Initial Estimate of Schedule
    - Major Tasks and Decision Point Milestones only
- Criteria and Deliverables required for next decision point

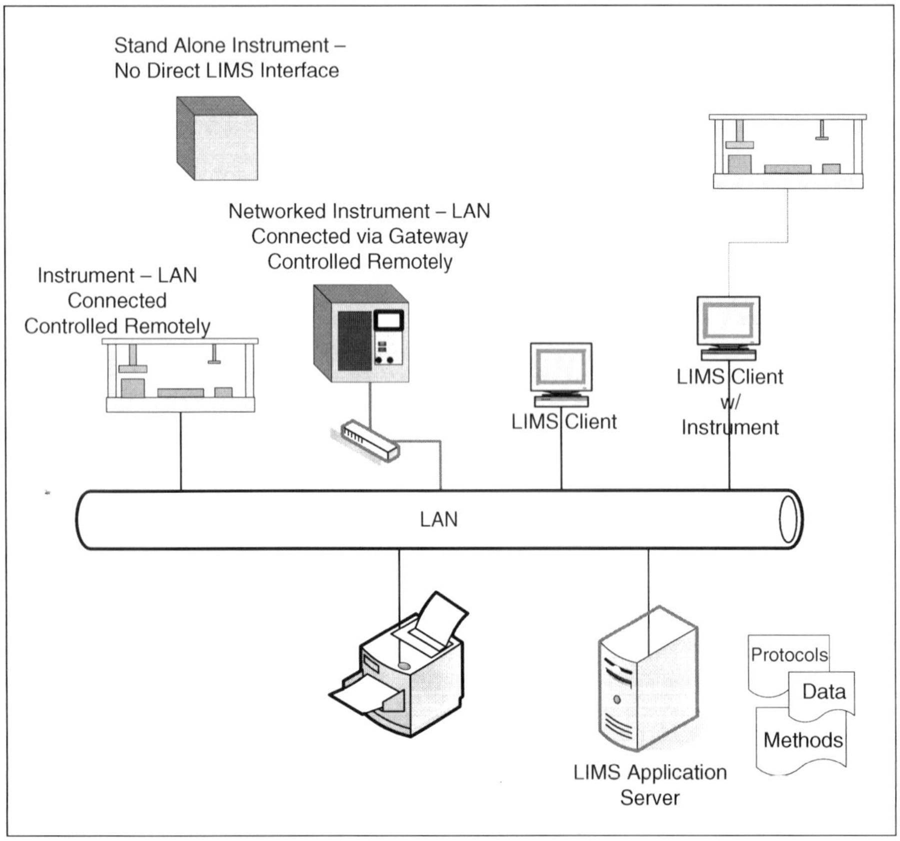

Figure 1 Representative LIMS implementation.

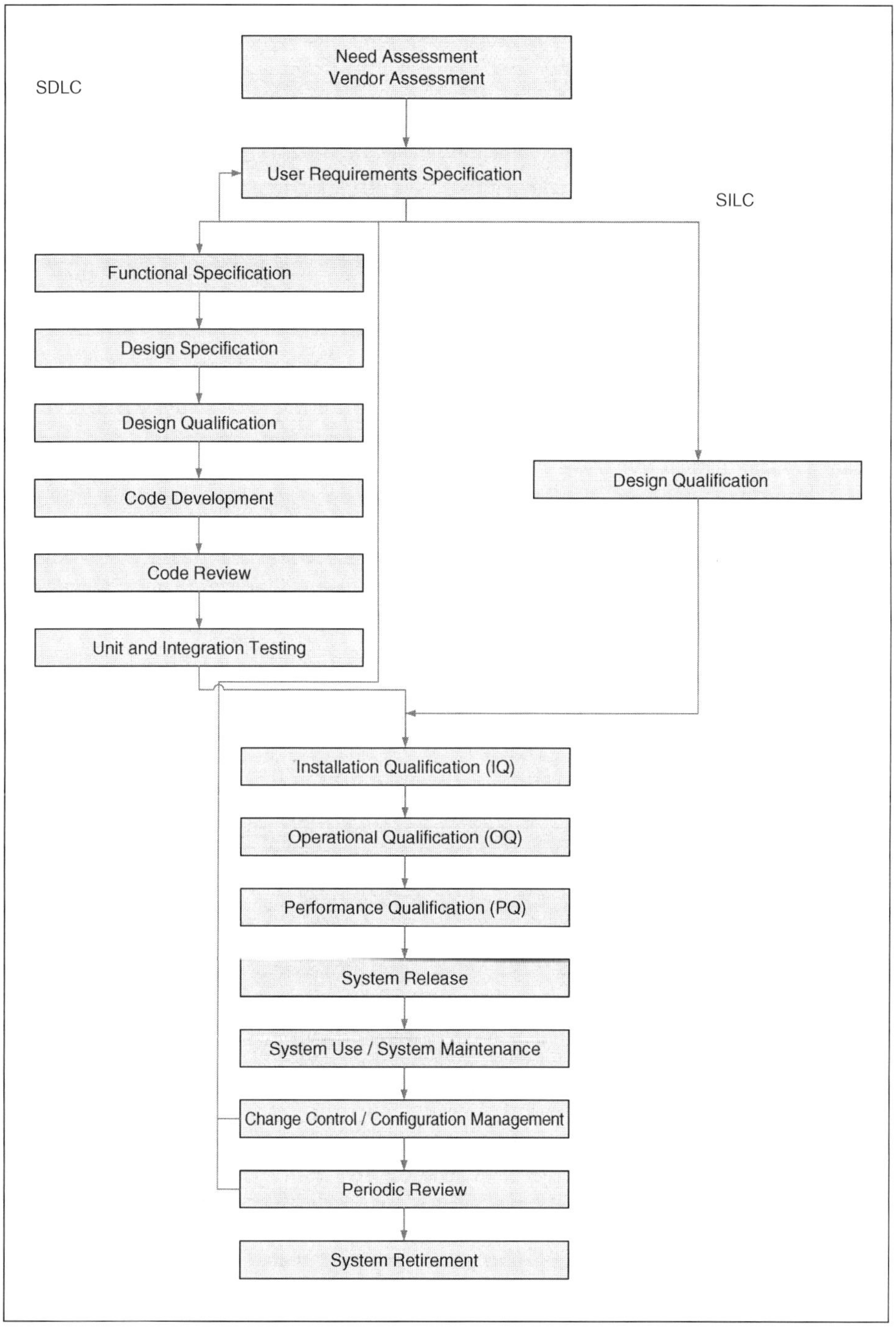

Figure 2 System design life cycle (SDLC) and SILC process flows. *Source*: From Ref. 8.

- URS
- System Validation Plan or Validation Plan
- Risk Assessment and Risk Management Plan

Defining the System

One of the most difficult activities in any computer-related activity is defining the system. How does one write a comprehensive set of requirements without understanding what tasks and functions the system must do, may already perform, or may need to replace? There are a number of techniques for obtaining requirements, but all have three common attributes: subject matter expertise, user bias, and documentation. We shall illustrate a potential requirement derivation technique based on Six Sigma methodologies (20).

As an initial step, define the system SIPOC diagram (Fig. 3). This approach allows simultaneous consideration of the customer/end-user needs, system inputs/outputs, and matches requirements to each of these while identifying critical process steps.

While the SIPOC gives a high-level overview of the system, the next step of process mapping allows the requirements definition to proceed by identifying workflow, functions, subprocess, etc. An important output of this mapping level is the identification of potential risk items, process inefficiencies, and process gaps; all of which can be mitigated through the use of already identified requirements or addition of new requirements.

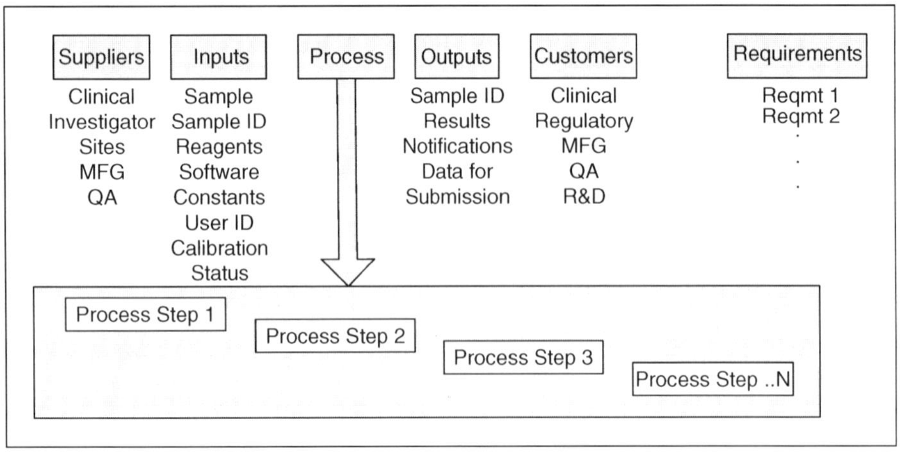

Figure 3 Suppliers, inputs, process, outputs, consumers diagram.

System Complexity and Risk

The GAMP Good Practice Guide: Validation of Laboratory Computerized Systems describes the categorization of various LIMSs based on their complexity and use. For completeness, Table 1 displays the GAMP 4 categories (21) and Table 2 depicts the LIMS categories (8).

Requirements Definition

Referring to Table 3, the first activity and validation deliverable described is the URS. The URS is a formal description of the system's performance as viewed from the user's frame of reference. Depending on the complexity of the system, the requirements should state what the system shall do and what capabilities it will have, not how it does it or how the capabilities are to be implemented. For example:

- The LIMS shall provide security measures for ensuring that data is not modified or deleted once it is entered into the system.
- The system shall provide the capability to prevent unauthorized users from accessing the functions, configuration, or data contained within the system.
- The LIMS shall provide the users the capability to signify review and approval through the use of electronic signatures that comply with 21 CFR Part 11.
- The system shall provide the user the ability to view HPLC chromatograms while the run sequence is in progress.
- The system shall provide the capability to generate hard copy reports of trend analysis and the underlying raw data.
- The system shall provide the user control of specified instrument functions (i.e., initiate, calibrate, run, pause, end).

Note that in the above requirements, it is stated that ... *the system shall provide the capability*.... These simple examples illustrate the traits of a well-defined requirement. Each requirement defines one item that the system shall do (related to what the user wants), the item is measurable, testable, and/or verifiable by independent means (validation), and it allows the system developer or vendor to provide the most efficient way of fulfilling the requirement (implementation). The URS also provides potential vendors a common description of the system to propose their LIMS solution against end-user evaluation in the case of a purchased system.

For more complex systems or development systems, an additional document (FS) specifies the functions required to meet each user requirement. For example, consider the electronic signature requirement in the URS. Some possible functional requirements would be as follows:

- Electronic signatures shall require the use of one or more biometric identifiers unique to each authorized user.
- LIMS instrument data shall be stored in human readable formats.

Table 1 Good Automated Manufacturing Practice Software Categories

Category	Software type	Validation approach
1	OS	Record version (e.g., WinXP Service Pack 2). OS indirectly challenged by application functional testing
2	Firmware	Non-configurablerecord version, calibrate instruments as required, verify operation against user requirements Configurablerecord version and configuration, calibrate instruments as required, verify operation against user requirements Custommanage firmware as Category 5 software
3	Standard software packages	Record version and environment configuration, verify operation against user requirements For critical and complex application, consider a supplier audit as well
4	Configurable software packages	Record version and configuration, verify operation against user requirements Supplier audits for critical and complex applications Manage custom programming as Category 5
5	Custom software	Audit supplier and perform complete system validation

Source: From Ref. 14.

Table 2 Good Automated Manufacturing Practice Laboratory System Categorization

Characteristics or rules for categories	Increasing complexity →						
	A	**B**	**C**	**D**	**E**	**F**	**G**
Configuration	Software and configuration is not modifiable	Software and configuration is not modifiable	Configuration parameters stored and reused	Configuration parameters stored and reused	Configuration parameters stored and reused	Configuration parameters stored and reused Proprietary configurable elements, i.e., not changing core code	Custom systems Complex logic functions, macros
Interfaces	No computer interface used	No computer interface used	No computer interface used	May have 1:1 ratio (computer to instrument interface, server to client interface)	May have 1:1 ratio (computer to instrument interface, server to client interface)	May have 1:*M* (many) ratio (computer to instrument interface, server to client interface)	Custom system
Data processing	Conversion of A/D signals	Conversion of A/D signals		Data manipulated by a separate program external to the system	Post-acquisition processing done as part of the system (can analyze data with proprietary data handling system)	Post-acquisition processing done as part of the system (can analyze data with proprietary data handling system)	Custom system
Results and data storage		Information generated based upon instrument function is stored, i.e., calibration data is stored	Process parameters input and stored (runtime parameters, methods parameters)	Process parameters input and stored (runtime parameters, methods parameters)	Process parameters input and stored (runtime parameters, methods parameters)	Process parameters input and stored (runtime parameters, methods parameters)	Custom system
	Does NOT produce raw data or test results	Produces raw data or test results, but records not stored or processed	Produces raw data or test results, but records not stored or processed	Produces raw data or test results, but records are stored but not processed	Produces raw data or test results, but records are stored and processed	Produces raw data or test results, but records are stored and processed	Custom system

Table 3 Validation Deliverables

System implementation life cycle					Software development life cycle	
A	B	C	D	E	F	G
		Validation plan	Validation plan	Validation plan	Validation plan	Validation plan
Requirements specification	Requirements specification	Requirements specification	Requirements specification	Requirements specification	URS	URS
					Functional specification	Functional specification
Risk	Risk	Risk	Risk	Risk	Risk	Risk
		Supplier assessment	Supplier assessment	Supplier assessment	Supplier assessment	Supplier assessment
					Detailed design spec	Detailed design spec
						Code develop
						Code review
					Unit/integration test	Unit/integration test
Design qualification/IQ	Design qualification/IQ	Design qualification/IQ	Design qualification	Design qualification	Design qualification	Design qualification
			IQ	IQ	IQ	IQ
Calibration qualification/PQ	Calibration qualification/PQ	OQ/PQ	OQ/PQ	OQ	OQ	OQ
				PQ	PQ	PQ
Traceability	Traceability	Traceability	Traceability	Traceability	Traceability	Traceability
Validation report	Validation report	Validation report	Validation report	Validation report	Validation report	Validation report

- LIMS control shall be accessed through a common GUI.

The first functional requirement indicates that electronic signatures are to be implemented and controlled via biometric means. The actual means of control (e.g., fingerprint, handwriting characteristics, retinal scan, etc.) are defined in the DSs generated for a full development model. The second functional requirement indicates that data storage is to be implemented so that a user can directly read the data, and the third indicates that a common GUI shall be used for all control functions. Again, the actual implementation is defined in the DS. For COTS systems, the vendor generates these documents as part of their development cycle.

The ordering and grouping of requirements within the URS is up to the user group responsible for generating this area of documentation. One could list the requirements as they are identified, and then group them as hardware, software, internal interfaces, external interfaces, security, etc.

Another approach for developing the URS is to identify the departments and the users within each department that will be using the LIMS. Each department may then identify current practices or current process flow, along with the SOPs that are currently being used. The requirements can then be developed based on the current process flow and how the "future process flow" should be (desired goal). In the future process flow, functions can be identified as performed manually, automated by LIMS, or a combination of automation and manual.

Additionally, for each of the process flows, a risk analysis can be done to determine if the process has a significant risk factor in terms of data accuracy, integrity, and authenticity. Mitigation to those risk factors can then be incorporated as part of the LIMS requirements.

Risk Management

In the most recent 21 CFR Part 11 guidance document (22), the FDA has indicated that the extent of computer system validation should be based on the impact the system can have on the ability to meet the requirements of the applicable predicate rules. In addition, the document further states that consideration must be given to the potential risks to accuracy, integrity, etc. of the electronic records. In a similar vein, ISPE has noted that the user must identify and define the GxP records based on predicate rules, risk to product safety, efficacy, quality, and process criticality. In addition, the capability of some LIMSs to actually control the analytical processes down to the instrument procedure level requires the user to assess the potential risk with respect to the accurate generation of original data in electronic form.

Risk is managed by the implementation of controls that prevent and/or mitigate and/or document/alarm the occurrence of a risk condition. For each risk item identified, there may be one or more controls, and for completeness, each control is derived from a requirement that is necessary to control the identified risk item. Additional requirements may be needed to ensure that all identified risks are controlled.

Since the system validation will be against all the requirements, the risk controls will be validated as well. It is important to note that Risk Management is a process that continues through the entire system life cycle. For example, any change (i.e., update, version upgrade, new instrument or function) to the system has the possibility of introducing a new risk, or negating the mitigation of an existing risk, thereby causing the addition of new requirements (Fig. 4).

Validation Planning

One of the earliest deliverables in any software validation project is the Validation Plan, and this is typically

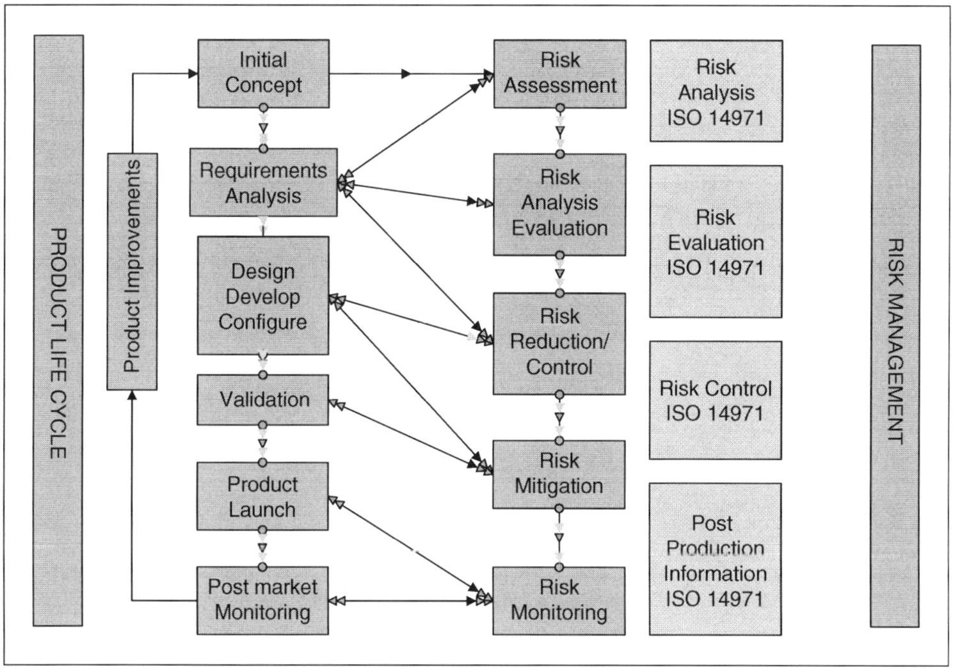

Figure 4 Risk management flow chart (after GAMP 4).

reflected in the Project Plan described above (which is often written prior to or concurrently with the user requirements). In this manner, expectations, methods, and controls for the validation activities can be defined. For example, in Table 3 it is noted in GAMP 4 that Validation Plans may not be required for systems that fall into categories A and/or B. It has been the authors' experience that a site-wide guidance document is invaluable to serve as a checklist or benchmark for the qualification of these simpler systems. This prevents overlooking an item that may be unimportant for one device, yet critical for another (e.g., temperature control is critical to ovens and incubators, but temperature monitoring may be a key requirement for pH or index of refraction measurements.) In addition, documenting and following this type of guidance is an indication to the regulatory authorities that your processes are under control.

The general topics covered under Validation Plans include the following:

- Introduction, Objective and Scope [introduction to the system, what the goal of the project is (new system, upgrade, incremental expansion…), the level of validation to be performed (i.e., full, partial, regression, etc.) and what the document covers].
- System Description (key functions, components, inputs, and outputs of the system. What is included in the system, takes note of interfaces). Describe what is new, what is an upgrade, what is being removed/replaced, key processes affected.
- References (including regulatory, internal documents like the URS, SOPs, system guides, etc.).
- Validation Strategy—general description of how the validation will be approached, and a description of critical areas such as the following:
  - Additional Scope (i.e., additional items that fall within or outside the project scope)
  - Environment (test, production, pilot)
  - Interfaces
  - Test Personnel
  - Special Test Equipment, Software, and/or Supplies
  - Data Migration and/or previous system retirement
- Tasks and Deliverables
  - Documents at each stage (URS, Risk Assessment, Validation Plan and Protocols, Results and Reports, Traceability Matrices, etc.)
- Roles and Responsibilities
- Documentation Management
- Change Control—statement or acknowledgment of a procedure and/or policy that facilitates change
- Periodic Review to maintain Validated State
- Change Control Procedures to Validated System
- Acceptance Criteria—what testing outcomes constitute completion and present the system in a Validated State.

Function and Design (SDLC)

A full representation of the SDLC is beyond the scope of this discussion. The reader is referred to the references and to additional material and Software Standards from the IEEE, and the SEI material on Software Engineering Practices, Capability Maturity Models and Software Development (23–26). It should be sufficient to note that the LIMS vendor must be held to the standards mentioned in order to assure the quality of the initial product and its final installation in the end-users' laboratories. This can be verified with a vendor assessment, by which the vendor submits written responses to customer queries, preferably in a checklist format. This can be followed up with an on-site vendor audit, with the participation of key members of the validation team, as described below. In the case of a full audit, the actual vendor documentation will be reviewed, including test data, code reviews, design documents, etc. In addition, for custom development items, the user is often an active participant in design and test result reviews as part of the overall system Acceptance Criteria. If the LIMS is an internal development project, all the phases of the SDLC process should be followed as would be expected by best practices.

Design Qualification (SILC)

Limiting the discussion to the SILC model, we consider the steps required to validate the purchased system. First is the selection of the vendor system that best meets the requirements developed and documented in the User Requirements Document. It is very likely that no vendor will meet all the requirements that have been defined, so a method of grading each vendor's offering against the needs should be defined and the critical requirements (i.e., key functions, those that mitigate risk, or meet regulatory demands) be weighted appropriately. Again, the criteria should be documented so that all internal stakeholders are aligned during the selection process. Part of the selection process should include the results of a vendor compliance audit. For example, does the vendor follow accepted industry practice in the area of software development and testing? Is their documentation sufficient to hold up under regulatory agency audit? Does the vendor meet the claim of compliance with 21 CFR Part 11 and is this proof documented? Is the design and function of the software compatible with the needs of the end-user laboratory? Is the workflow of the laboratories able to be mapped to the software functions? Does the software interface with the existing laboratory instrumentation and any near term or future instrumentation under consideration? Does the software follow industry standard approaches in interface definitions, database connectivity? Will support for the LIMS be supplied by internal resources, or will a support contract be required? Does the level of vendor supplied documentation meet the LIMS end-user's internal documentation requirements? If this is the desired vendor, and one or more critical issues are not met, then it is the end-user's responsibility to make up for the deficiencies. These deficiencies would be documented in the risk identification and the corrective action covered in the mitigation activities described previously.

Validation (SDLC and SILC)

Installation Qualification

The IQ is the activity that provides objective documented evidence that the LIMS is installed in accordance with vendor, user and engineering specifications. As with any testing that follows best practices, a protocol is required. This protocol consists of sections that describe the system to be installed, the environment and required services where the physical installation will take place, lists the components (software, hardware, interfaces, etc.) and has scripts that when followed install and document the system. Some of the components may already be in place; for example, the LAN which is usually a company-wide network, the building utilities including UPS and environmental controls, and often the instruments that are to be connected.

The typical practice is to perform the system IQ in a test environment. In this way, the installation strategy of the system can be verified, any changes required by the user's production environment can be identified and the installation scripts can be modified under document Change Control. In addition, the production environment remains pristine and data collected from the validation activities does not adversely impact any regulated activities or data in the production environment.

For more complex systems, it may be advantageous to perform the IQ and OQ of the LIMS instrumentation separately from the LIMS system computer system. This decision should be noted in the System Validation Plan described previously.

For larger and more complex systems, there may be multiple IQ documents and respective activities. For example, there may be a separate LIMS Server IQ, a LIMS client IQ that is exercised for each client installed, instrument IQs, software application IQs (operating system, application, database, etc.). In the case of purchased systems, a large portion of these protocols may be vendor supplied. In this case, it is the user's responsibility to review the material for its content and applicability to their requirements and anticipated use. The user bears the responsibility for performing an IQ that is tailored to the business interests of the specific laboratory environment, according to approved guidelines and SOPs.

It is important to note here that LIMSs are not static; they grow with additional workload (i.e., new APIs, formulations, etc.) and they change with new technology (i.e., addition of new assays and instrumentation, new functionality, retirement of old instruments, etc.). If the system is designed with this in mind, and the Validation Planning also takes this into account, the amount of validation effort needed to qualify expansion and new technology can be limited to the new addition, or change, and does not require a full validation of the entire system.

At the completion of the IQ activities, there will be documentation that:

- Identifies and describes all system components such as the following:
  - Hardware
  - Software
  - Laboratory Instrumentation linked to the LIMS
  - Special Purpose Instrumentation (i.e., protocol converters, A/D data loggers, etc.)
- Identifies and describes the system configuration as installed.

In addition, the user, administration, and technical manuals will also be available. For those instances where a new assay or instrument is added to an already qualified system, the documentation will consist of supplemental items that:

- Describes the new addition
- Describes the IQ steps to add, or remove, or change the new system
- References material that takes into account the requirements of the establishment's Change Control procedure (see Section entitled Change Control).

Operational Qualification

The OQ is designed to provide the objective documented evidence that the LIMS functions according to the requirements derived from the functional specifications within the operating range specified in the user environment.

The functional specifications that are tested in the OQ arise from the user-defined Functional Requirements Document or, for the case of purchased systems, from the specifications supplied by the vendor.

At the completion of the OQ activities, there will be documentation that:

- Identifies and describes all system component functions that were activated and tested, for example:
  - System security items (user logon, password expiration, administrative rights, etc.)
  - Laboratory Instrumentation command and data transfer
  - Verification of the functioning of Special Purpose Instrumentation (i.e., protocol converters, A/D data loggers, etc.)
  - Audit trails (a key requirement for regulated systems per 21 CFR Part 11. Audit trails are automatic software logging of changes to the system and the data contained therein. They allow for the review of what the system and/or data was before the change, what the new system/or data is, who made the change, when the change was made, and why. They also log the original creation of the data including source, date, user, etc.)
  - Backup/Archive and Restore—all systems that store process and business critical data should have regular backups performed, and copies of the data be stored offsite in the event of system failures and business interruptions. It is strongly suggested that this function be verified prior to use in a production environment. Data backups should also include the audit trail data noted above.
  - Configuration tables (i.e., user security level rights, report generation, alarm logging, etc.)
- Identifies and describes the system configuration as configured. This information is often reviewed during compliance audits to verify that the system and internal processes are under control.

Performance Qualification

PQ, also referred to as UAT, is designed to provide tests addressing the functional requirements from normal business processes as stated in the System Requirements Specification and clearly demonstrate performance within the actual operating environment of the system.

At the completion of the PQ activities, there will be documentation such as the following:

- A Validation Summary Report (VSR), which is developed to document the test results and any deviations from the OQ, IQ, and PQ/UAT Protocols.
- SOPs that are in place to assure that the application remains in a Validated State during production use.
- SOPs that are in place that describe how the system is to be used, rules for access to functions and data, backup schedules, etc.
- System Release Memo, which when approved is issued to release the system for production use.

Trace Matrix (Requirements/Validation)

The RTM is designed to trace the URS, FRS, and DS requirements and specifications to the specific test sections in the IQ, OQ, and PQ protocols. An example of this traceability is illustrated in Table 4. It is important to note that every requirement is associated with at least one test protocol. For complex systems, it is useful to keep the trace matrix in a database or a commercially available requirements tracking tool. With this approach, database queries can be exercised to find those requirements that may not have been tested.

Change Control

Once the system has been validated, it is considered to be under Change Control. In essence, this means that any change to the configuration, configuration items, procedures, methods, instrumentation, etc. must be reviewed by a Change Control Board per user site's SOPs. Depending on the complexity of the change and the associated risk to the system, procedures, data, etc., the system will most likely have to undergo some level of revalidation. The more extensive and complex the change that is contemplated, the more risk is associated, and consequently the more validation effort will have to be applied. As previously discussed in the "Validation (SDLC and SILC)" section, for additional equipment changes, decommissioning of instruments, interfaces, and components, it is best to qualify the instrument separately from the LIMS and then perform the integration of the instrument to the LIMS as part of the LIMS Change Control process. The purpose of the integration process is to ensure that the instrument link and communication function as expected. Ideally, there should be a "Development and/or

Table 4 Traceability of Requirements Example

Requirement/ specification number	Requirement/ specification description	DS number (optional for system or life cycle implementation)	IQ section/step	OQ section/step	PQ section/step
User requirements specification # 3.4.1	Requirement description	8.3.4	6.7.2/step 1		
Functional requirement specification # 6.1.6	Requirement description	11.9		3.5.1	
UFRS 4.1.7	Requirement description	5.2			6.1.2

Test" server that can be utilized to test the changes to the LIMS before going into a "Production" server.

Configuration Control

In the following paragraphs, we consider "Configuration Management" or "Configuration Control." Configuration Management relates more to the items that make up the LIMSs (e.g., the software, hardware, manuals, SOPs). These items can be categorized as "configuration items."

It has been the authors' experience that this is often blurred in practice, leading to some endless meetings about whether to unplug a printer. It is suggested that the following definitions be applied. Configuration is what is plugged in, what is considered as consumables that does not require Change Control, what versions of software, etc. Change Control is the replacement of a current item with a new one (i.e., new instrument, new version of software, etc.). Like-for-like can be covered as configuration management in order to avoid undergoing the Change Control process every time a light bulb is replaced.

A key consideration of any complex system is the state of its current configuration. Configuration, for the purposes of this discussion, is the description that precisely describes the system at any point in its life cycle (refer to GAMP 4). It is a recommended practice to document the configuration of the system from the initial state at which it was validated and to maintain that configuration knowledge (and documentation) for the entire life of the system. Any change to the system (see the previous section on change control) will invariably require, at a minimum, verification of the current state of the system and documentation of the change. More complex or involved changes will alter the configuration of the system thus altering its state and the accompanying description (configuration specification). A good rule of thumb as to whether the configuration management practices are acceptable or not is the usability of the documented configuration items to reconstruct the LIMS from scratch. The configuration items should be able to identify all the necessary components that are required to assemble the LIMS system just the way it was.

Additional Discussion and Points of Consideration

Validation Full vs. Incremental Validation

In order to determine the level of validation or revalidation required, there must be an appropriate description of the policy or practice for partial upgrade. For example, how does the laboratory validate/commission a new HPLC on the LIMS network? Commission the instrument, verify it talks to the LIMS, run a limited PQ? How about a partial software upgrade? Does a full IQ/OQ/PQ need to be executed, or can this be determined and verified in the test environment and a very limited PQ performed in production? In the simplest approach, refer to the Risk Management Plan that was developed as part of the LIMS implementation. A comprehensive plan will have included the level of risk associated with each of these possibilities. Based on the level of risk indicated, the depth of validation can be defined and the user will be in alignment with current guidance from the FDA.

Single-Site vs. Multiple-Site Implementation

Aside from initial implementation strategy (e.g., big bang implementation on all sites or multiple-phased implementation, architecture and performance issues such as bandwidth and throughput capability), control and implementation of changes is a key aspect of the multiple-site implementation.

It is advantageous to consider how the Analytical Procedure will be maintained and deployed in the instance of a multi-site LIMS implementation. For example, with regard to some site stability studies, the deployment of new Analytical Procedures and/or updates/changes to existing Analytical Procedures at one site may be different from other sites that still utilize the older versions of Analytical Procedures or instrumentation; or, when a particular site may not be ready to have a deployment of the new Analytical Procedure for some reason. In cases of multinational implementation, languages and character set(s) requirements (and verification) should also be considered.

Another point of consideration is how the Analytical Procedure will be maintained and deployed in the case of multiple-site LIMS (e.g., some site or stability studies may still utilize the older versions of Analytical Procedure; equipment, and operating conditions may be different from site to site, etc.).

Data Archival

Strategy on archiving data and the retrieval of the archive must be verified to ensure that there is a process for archiving and retrieving data without overwriting the current data (e.g., the audit trail data, analytical method procedures, instrument data, etc.). Strategy for maintaining archived data should also be considered. For example, a process should be in place for removing archived data that is no longer required. Another consideration is the storage location of the archive or a true copy, in case of disaster.

Meta Data

Meta data is data that describe the data. For example, an HPLC chromatogram would be expected to have transmission (absorbance) versus wavelength for each sample. Meta data associated with the chromatographic output include sample ID, instrument used, analyst name, analysis date and time, reviewer name, review name and time, column information, etc. Meta data are also required by 21 CFR Part 11 as they are considered a part of the original data, and thus must be treated in a manner similar to the data that they are associated with. For example, changes to the meta data must also be captured in an audit trail.

Disaster Recovery

Development of a strategy for Disaster Recovery should also be considered. The plan should cover how to recover the hardware, software, and data failures; for example, contact points or persons to recover each of those components. It is best to practice the Disaster Recovery plan before a disaster strikes. Related to this Disaster Recovery plan is Business Contingency Planning, on how the business operation can be sustained while the LIMS is not available, and what to do when the LIMS becomes available again. For example, how to manually enter sample results to LIMS that were obtained when the LIMS was not operational. As part of this Disaster Recovery effort, rollback capability to a certain timepoint or restore points should also be considered.

Training

Aspects of training should also be considered; for example, the differentiation of responsibility between the System Administrator, Super User and Regular User, as well as the other roles that are defined in the LIMS structure. Another aspect of training that should also be considered is the training for the team members who will be involved in the development of the LIMS (e.g., programming standard requirements, Change Control requirements, documentation requirements, etc.). As a follow up to the rollout, training of management and their role(s) in the overall process and sustaining effort should also be considered.

External Audits

As GxP computerized systems, LIMSs are subject to 21 CFR Part 11 as well as to any applicable predicate rules. Thus, LIMSs are subject to audits by outside authorities (i.e., regulatory) as key items in the manufacturing of pharmaceuticals.

Aspects of External Audits that should be considered include providing access to the LIMS in case such an access is requested, as well as electronic copy provision. It is best to have a Super User conduct the navigation of the LIMS, should access to the LIMS be requested by an external regulator.

CFR 21 Part 11 (Electronic Records and Electronic Signature Regulations)

Last but not least to be considered is the potential impact of CFR 21 Part 11 regulations; for example, important aspects of system access (e.g., remote access), audit trail maintenance and access, interfaces compliance (inbound and outbound) with other systems, and user ID maintenance to ensure that records can always be traced to the person who created, edited, and deleted the record (e.g., how to deactivate user ID without losing the user ID). Open or Closed System classification should also be considered, especially if there are interfaces to external (to the company) systems (e.g., Clinical Research Organization).

CONCLUSION

LIMS validation requires considerable effort across multiple departments or levels of authority, corporate commitment in terms of financial and personnel resources, intelligent planning, and a lot of patience. At times it may seem to be a daunting and insurmountable challenge. However, with careful strategy and proper project management, LIMS implementation should be achievable, with a high return on investment and increased productivity.

ACKNOWLEDGMENT

Special thanks go to Dr. Sol Motola without whose guidance and kind recommendation this project would not have been possible.

REFERENCES

1. FDA. Draft Guidance for Industry: Development and Use of Risk Minimization Action Plans, May 2004.
2. FDA. Pharmaceutical CGMPs for the 21st Century: A Risk-Based Approach; A Science and Risk-Based Approach to Product Quality Regulation Incorporating an Integrated Quality Systems Approach, Final Report, September 2004.
3. FDA. Risk-Based Approach to Product Quality Regulation Incorporating an Integrated Quality Systems Approach, 2002.
4. ISO 14971:2002 Medical Devices—Application of risk management to medical devices (ISO, 2001).
5. ISPE, "Risk-Based Approach to 21 CFR Part 11".
6. NIST Special Publication SP800-30: Risk Management Guide for Information Technology Systems (National Institute of Standards and Technology, U.S. Department of Commerce, 2002).
7. ISO 14971-1:1998, Medical Devices—Risk Management—Part 1: Application of Risk Analysis. International Organization for Standardization, 1998.
8. GAMP Good Practice Guide: Validation of Laboratory Computerized Systems.
9. FDA. Guideline on General Principles of Software Validation, January 11, 2002.
10. FDA. General Principles of Software Validation; Final Guidance for Industry and FDA Staff, 2002.
11. FDA. Computerized Devices/Processes Guidance: Application of the Medical Device GMP to Computerized Devices and Manufacturing Processes, May 1992.
12. Reviewer Guidance for a Pre-Market Notification Submission for Blood Establishment Computer Software, Center for Biologics Evaluation and Research, Food and Drug Administration, January 1997.
13. Medical Devices; Current Good Manufacturing Practice (CGMP) Final Rule; Quality System Regulation, 61 Federal Register 52602 (October 7, 1996).
14. The Good Automated Manufacturing Practice (GAMP) Guide for Validation of Automated Systems, GAMP 4 (ISPE/GAMP Forum, 2001).
15. ASTM E-2066-0, Standard Guide for Validation of Laboratory Information Management Systems.
16. ANSI/UL :1998: 1998, Standard for Safety for Software in Programmable Components, Underwriters Laboratories, Inc., 1998.
17. AS 3563.1-1991, Software Quality Management System, Part 1: Requirements. Published by Standards Australia [Standards Association of Australia], 1 The Crescent, Homebush, NSW 2140.
18. IEC 61508:1998, Functional safety of electrical/electronic/-programmable electronic safetyrelated systems. International Electrotechnical Commission, 1998.
19. ISO 8402:1994, Quality management and quality assurance—Vocabulary. International Organization for Standardization, 1994.
20. http://www.isixsigma.com/tt/sipoc/ referenced 26 April, 2006.
21. ISPE, The Good Automated Manufacturing Practice (GAMP) Guide for Validation of Automated Systems in Pharmaceutical Manufacture, 2001, International Society for Pharmaceutical Engineering.

22. FDA. 21 CFR Part 11, Electronic Records; Electronic Signatures—Final Rule. Fed Regist Vol. 62, No. 54, 13429, March 20, 1997.
23. IEEE Std 1012-1986, Software Verification and Validation Plans, Institute for Electrical and Electronics Engineers, 1986.
24. IEEE Standards Collection, Software Engineering, Institute of Electrical and Electronics Engineers, Inc., 1994. ISBN 1-55937-442-X.
25. Chrissis MB, Konrad M, Shrum S. CMMI—Guidelines for Proces Integration and Product Improvement. Boston, MA: Addison Wesley, 2003.

BIBLIOGRAPHY

ANSI/ANS-10.4-1987, Guidelines for the Verification and Validation of Scientific and Engineering Computer Programs for the Nuclear Industry, American National Standards Institute, 1987.

ANSI/ASQC Standard D1160-1995, Formal Design Reviews, American Society for Quality Control, 1995.

ASTM E1578-93. Standard Guide for Laboratory Information Management Systems (LIMS) 1999.

AS 3563.2-1991, Software Quality Management System, Part 2: Implementation Guide. Published by Standards Australia [Standards Association of Australia], 1 The Crescent, Homebush, NSW 2140.

Budihandojo R. Computerized system validation: a concept approach in the preparation of a validation plan document. J Pharm Technol 1997; 21:70.

Budihandojo R. Computerized system validation: a concept in the preparation of a functional requirements document. J Pharm Technol 1997; 21:124.

Budihandojo R. Computerized system validation: preparation of a design specifications document. J Pharm Technol 1998; 22:150.

Budihandojo R, Amato S, Baldwin S, et al. Computer validation: available document resources other than from FDA. J Pharm Technol Vol 23, September 1999.

Budihandojo R, Bergeson D, D'Eramo P, et al. The future state of computer validation, part I & II, J Pharm Technol Vol 25, Jul 2001 (Part II—September 2001).

Budihandojo R, Cappucci W, Cardarelli J, et al. Computer validation: validation of SAP R/3—a conceptual approach, part I&II. Pharm Technol Vol 25, February 2001 (Part II—March 2001).

Budihandojo R, D'Eramo P. Computer validation: available document resources from FDA. J Pharmaceutical Technology Vol 23, March 1999; 78.

Electronic Records; Electronic Signatures Final Rule, 62 Fed Regist., 13430 (March 20, 1997).

FDA. Guideline of General Principles of Process Validation, May 1987.

FDA. Compliance Policy Guide 7132c.08 Process Validation Requirements for Drug Products and Active Pharmaceutical Ingredients Subject to Pre-Market Approval, updated 03-12-2004.

FDA. Draft Compliance Guidance: Computerized Systems Used in Clinical Trials, September 2004.

FDA. Compliance Program Guidance Manual, "Compliance Program 7348.810—Sponsors, Contract Research Organizations and Monitors," October 30, 1998.

FDA. Compliance Program Guidance Manual, "Compliance Program 7348.811—Bioresearch Monitoring—Clinical Investigators," September 2, 1998.

FDA. Glossary of Computerized System and Software Development Terminology, August 1995.

FDA. Good Clinical Practice VICH GL9, 2001.

FDA. Guideline for the Monitoring of Clinical Investigations, 1988.

FDA. Information Sheets for Institutional Review Boards and Clinical Investigators, 1998.

FDA. Part 11, Electronic Records; Electronic Signatures—Scope and Application, 2003.

FDA. Guidance for Industry, FDA Reviewers and Compliance on Off-The-Shelf Software Use in Medical Devices, September 9, 1999.

FDA. Guidance for the Content of Pre-market Submissions for Software Contained in Medical Devices, May 1998.

Huber L, Budihandojo R. Qualification of network components and validation of network systems, part I and II, Vol 25, J Pharm Technol 2001 (Part II—November 2001).

Humphrey WS. Introduction to the Team Software Process. Reading, MA: Addison Wesley, 2000.

International Conference on Harmonization Draft Guideline: Quality Risk Management Q9, March 22, 2005.

International Conference on Harmonization. E6 Good Clinical Practice: Consolidated Guideline. Federal Register, Vol. 62, No. 90, 25711, May 9, 1997.

IEC 60601-1-4:1996, Medical electrical equipment, Part 1: General requirements for safety. 4. Collateral Standard: Programmable electrical medical systems. International Electrotechnical Commission, 1996.

IEC 61506:1997, Industrial process measurement and control—Documentation of application software. International Electrotechnical Commission, 1997.

ISO 8402:1994, Quality management and quality assurance—Vocabulary. International Organization for Standardization, 1994.

ISO 9001:1994, Quality systems—Model for quality assurance in design, development, production, installation, and servicing. International Organization for Standardization, 1994.

ISO 13485:1996, Quality systems—Medical devices—Particular requirements for the application of ISO 9001. International Organization for Standardization, 1996.

ISO/IEC 12119:1994, Information technology—Software packages—Quality requirements and testing, Joint Technical Committee ISO/IEC JTC 1, International Organization for Standardization and International Electrotechnical Commission, 1994.

ISO/IEC 12207:1995, Information technology—Software life cycle processes, Joint Technical Committee ISO/IEC JTC 1, Subcommittee SC 7, International Organization for Standardization and International Electrotechnical Commission, 1995.

ISO/IEC 14598:1999, Information technology—Software product evaluation, Joint Technical Committee ISO/IEC JTC 1, Subcommittee SC 7, International Organization for Standardization and International Electrotechnical Commission, 1999.

ISO/IEC 17799:2000 (BS 7799:2000) Information technology-Code of practice for information security management (ISO/IEC, 2000).

Software Considerations in Airborne Systems and Equipment Certification. Special Committee 167 of RTCA. RTCA Inc., Washington, DC, Tel: 202-833-9339. Document No. RTCA/DO-178B, December 1992.

Technical Report No. 18, Validation of Computer-Related Systems. PDA Committee on Validation of Computer-Related Systems. PDA Journal of Pharmaceutical Science and Technology, Volume 49, Number 1, January–February 1995 Supplement.

Technical Report No. 31, Validation and Qualification of Computerized Laboratory Data Acquisition Systems. PDA Journal of Pharmaceutical Science and Technology, Volume 53, Number 1, January–February 1999 Supplement.

21 Code of Federal Regulations Parts 210 and 211, Part 210-Current Good Manufacturing Practice in Manufacturing, Processing, Packing, or Holding of Drugs. General; Part 211-Current Good Manufacturing Practice for Finished Pharmaceuticals.

21 Code of Federal Regulations Part 58-Medical Devices, Good Laboratory Practice for Nonclinical Laboratory Studies.

21 Code of Federal Regulations Parts 600, 610, Part 600-Biological Products, General; Part 610-General Biological Products Standards.

21 Code of Federal Regulations Part 820-Medical Devices; Current Good Manufacturing Practice (CGMP); Quality System Regulation.

50 Validation of Analytical Procedures and Physical Methods

Hélène Gazzano-Santoro
Quality Control Biological Technologies, Genentech Inc., South San Francisco, California, U.S.A.

Carolyn Broughton
Quality Control Microbiology, Genentech Inc., South San Francisco, California, U.S.A.

Dieter Schmalzing
Quality Control Analytical Technologies, Genentech Inc., South San Francisco, California, U.S.A.

INTRODUCTION

Method validation should follow a life cycle approach. This process is depicted in Figure 1. Following this approach, validation activities should be performed and completed prior to release of Phase I clinical material and continually be updated, as needed, throughout product development, culminating in the validation for regulatory filing for licensure. Post-licensure, the method validation status should be maintained through monitoring of method performance, change evaluation, and revalidation (where applicable).

It should be emphasized that the life cycle approach describes an ideal pathway that can probably be closest fulfilled by larger biopharmaceutical companies that have the required resources and infrastructure. Smaller companies, in particular start-ups, might choose only certain elements of this pathway, balancing individual capabilities versus industry standards and general regulatory expectations.

REQUISITES FOR TEST METHOD VALIDATION

Validation of an analytical procedure, as defined by ICH Q2A and Q2B (1,2), is the demonstration that the procedure is suitable for its intended purpose. Consequently, assay validation should be approached as a confirmatory exercise for an already established method rather than an exploratory investigation that might lead to new findings about the methods capabilities and parameters. In fact, all the scientific and technical assay work should be completed during method development and robustness testing to ensure that the assay validation is entered with a sound and well-understood method. Both method development and robustness testing should be guided by the intended purpose of the method, adhering to the ICH guidance document Q6B, "Specifications: Test Procedures and Acceptance Criteria for Biotechnological/Biological Products" (3). The finalized test method should be thoroughly documented (e.g., with all the relevant analytical parameters) and approved by the appropriate levels and departments prior to validation.

Assay validation requires a protocol with pre-established validation parameters and acceptance criteria. Parameters should have their specific acceptance criteria. The appropriate level and depart-ments should approve the protocol prior to the validation.

In general, the acceptance criteria should be based on both the capability of the method (estimated from the method development and robustness testing) and the intended usage of the method. The acceptance criteria should be meaningful. Quantitative criteria are preferred whenever possible. However, quantitative acceptance criteria that are too wide can turn a validation into an empty exercise that could lead to negative surprises during actual usage of the method, whereas criteria that are too narrow might lead to irrelevant validation failures. Qualitative acceptance criteria that are meaningful are inherently more difficult to formulate but they should be nevertheless carefully worded to guarantee that validation parameters are adequately assessed. Failure to meet an acceptance criterion requires that a formal investigation is conducted and documented.

QC method validation should be executed only by qualified and trained personal using qualified instrumentation and in compliance with cGMP.

After completion of validation activities, the data should be reviewed for compliance and technical merit and summarized in a validation report to be approved by the appropriate levels and departments.

VALIDATION DURING CLINICAL DEVELOPMENT

Analytical methods should be validated for clinical use to release Phase I clinical material and updated, as applicable, throughout clinical development. Clinical validations are also referred at times as "qualifications" (4). They should in principle adhere to the same elements as the validations for licensure, outlined in the section entitled Requisites for Test Method Validation. However,

---

Abbreviations used in this chapter: ATCC, American type culture collection; BLA, biological license application; cGMP, current good manufacturing practice; EP, *European Pharmacopoeia*; ICH, International Conference on Harmonization; IEC, ion exchange liquid chromatography; PBS, phosphate-buffered saline; QC, quality control; RSD, relative standard deviation; SOP, standard operating procedure; USP, *United States Pharmacopeia*.

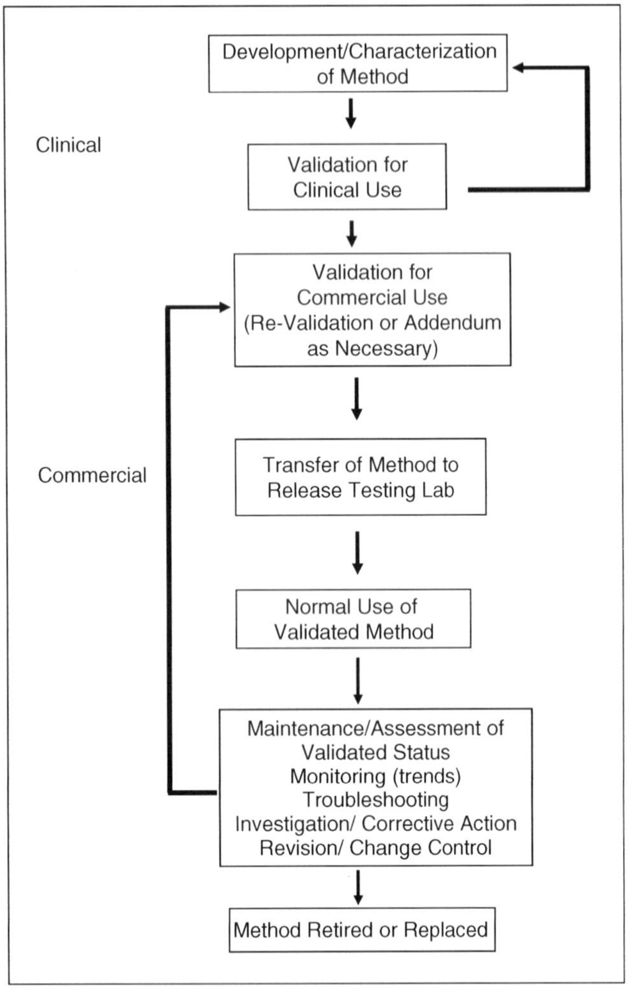

Figure 1 Schematic of test method life cycle.

the degree of validation performed is based upon the nature of the method and its intended usage and reflects the stage of the product development as well as the stage of production. In general, the clinical validation should consider validation of parameters that are considered essential to the method's performance, to guarantee that the method is scientifically sound, e.g., that proper designs and proper controls are chosen that relate to the intended use of the method. Typically the clinical validation parameters are a subset of the parameters recommended by ICH Q2A and Q2B. The focus of the clinical validation on parameters, which were identified as essential for method performance, is critical during clinical development since assay work during this period is particularly restricted in time and resources. Consequently, a risk assessment should be made that balances time constraints and desired knowledge of the assay (5).

Validated QC methods should periodically be evaluated by a formal monitoring system to verify that the methods are still operating according to their original validation characteristics. Reassessment describes the process of evaluating the purpose and use of a method in light of experience gained with the method or product throughout development. Reasons for revalidations or validation supplements might be changes to the product formulation or the clinical manufacturing process, knowledge gained on quality attributes, or trend results of method monitoring during routine usage of the method. Any change made to a validated QC method must be governed by a formal change control system and must be assessed and justified for validation and regulatory impact.

VALIDATION FOR MARKETED PHARMACEUTICALS

At the time the license application is submitted to regulatory authorities, all methods should be validated. Method validation for commercial use must be fully compliant with ICH guidelines Q2A and Q2B. Method validation activities prior to commercialization may be cumulative, consisting of all relevant clinical validations performed through clinical development. Clinical validations should be assessed for completeness and for adherence to the ICH guidelines and supplemented, if necessary, with additional validation work. Revalidation of a licensed method must be considered if there is a change in the method or in the manufacturing process. The degree of revalidation depends on the nature of the change. An assessment should be performed to determine the impact of the changes on the validation status of the method. If a method replacement is warranted, then a complete validation is required that demonstrates the suitability of use of the new method. In any case, the revalidation should ensure that the validation status of the method is fully compliant with the regulatory requirements.

ASSAY CHARACTERISTICS TO BE VALIDATED

Typical assay parameters to be validated are the following: accuracy, precision (consisting of repeatability and intermediate precision), specificity, detection limit, quantitation limit, linearity and range. The definitions of these parameters and guidelines on approaches to method validation are given in such documents as ICH Q2A and Q2B and will hence not be repeated here.

The selection of the parameters to be validated requires, most notably for the commercial validation, a clear understanding of both the intended usage of the assay and of the product characteristics (e.g., its physicochemical, biological, immunological, and stability properties). For example, an identity assay requires a different validation than a purity test. The validation of purity tests (e.g., of ion exchange liquid chromatography assays) might then further depend on the characteristics of the material to be analyzed (e.g., the presence of charge-based components that are deemed critical for the quality of the different products).

It should be emphasized, however, that the characteristics of the product are typically only incompletely understood at the time when assays are to be validated for Phase I clinical testing. The product knowledge increases throughout clinical development but might nevertheless continue to be incomplete, at least in certain areas, until Phase III. This lack of knowledge might complicate the selection of the assays and of the parameters to be validated. Consequently one has to rely

Table 1 Validation Parameters Required for Different Types of Methods Per ICH Q2A and Q2B for Commercial Validation

Method type parameters	Identity	Impurity test: Quantitative	Impurity test: Limit	Dissolution/ content/potency
Specificity[a]	+	+	+	–
Linearity	–	+	–	+
Range	–	+	–	+
Accuracy	–	+	–	+
Precision				
Repeatability	–	+	–	+
Intermediate precision	–	+[b]	–	+[a]
Detection limit	–	–[c]	+	–
Quantitation limit	–	+	–	–

Note: + parameter is normally evaluated; – parameter is not normally evaluated.
[a] Lack of specificity of one analytical procedure could be compensated by other supporting analytical procedures.
[b] In cases where reproducibility has been performed, intermediate precision is not needed.
[c] May be used in some cases.

on good scientific judgment and take recourse, whenever possible, to previous experience with other products and assays. It also re-emphasizes the need, already discussed above, to re-evaluate the validated assays throughout clinical development, when warranted, for their suitability of use and validation status.

Table 1 summarizes the validation parameters to be evaluated, according to Q2A and Q2B, for various types of assays for commercial validations.

It is sometimes possible to design the experimental work in such a manner that various validation experiments can be evaluated in one experiment (the different validation parameters should nevertheless have their own acceptance criteria). For example, for a purity assay, if the accuracy of the measurement of a certain quality attribute is evaluated through a spike and recovery experiment, the experiment can be designed such that linearity, precision, and range of the assay are assessed simultaneously. Thoughtful choices of validation experiments can hence result in significant time saving, possibly allowing that other additional experiments can be performed.

CASE EXAMPLES

Validation of Physicochemical Methods

The following section discusses a hypothetical nevertheless realistic validation example of a physicochemical method that starts with Phase I, continues throughout clinical development, and ends with the commercial validation of the method for licensure.

Figure 2 depicts the chromatographic profile of a recombinant monoclonal antibody analyzed by a high-performance IEC method to quantify the charge heterogeneity of the molecule. The analytical control of charge-heterogeneity of biotherapeutics (e.g., through an HPLC assay) is a regulatory expectation set by Q6B. The chromatographic profile is complex due to the complex nature of the molecule. Three main regions can be distinguished in the profile, named "A," "B," and "C." The region B is the main peak since its peak area exceeds the sum of the areas A and C.

Prior to Phase I little is known about drug candidates, especially when they are complex biopharmaceutical products like recombinant monoclonal antibodies. The development of the IEC method, consisting of the selection of the sample preparation (e.g., sample diluent composition) and the analytical parameters (e.g., chromatographic column, mobile phase composition, gradient, etc.), focused on the best separation of the product, which at this early stage can only be defined by macroscopic analytical parameters (e.g., lowest main peak percentage). The developed method was subsequently robustness tested by making deliberate small changes to the method (e.g., variation in pH of the mobile phase) to assess the inertness of the method performance (e.g., chromatographic profile and quantification) with respect to those changes. The robustness study included factorial design that was based on the results of the single parameter experiments. Afterwards, the test procedure was finalized for clinical validation by making small refinements to the analytical conditions (e.g., narrowing of the pH range of the mobile phase) based on the results of the robustness studies.

The subsequent clinical validation for Phase I focused on a carefully selected subset of the validation parameters for impurity tests required by ICH Q2A and Q2B for commercial validation: specificity and

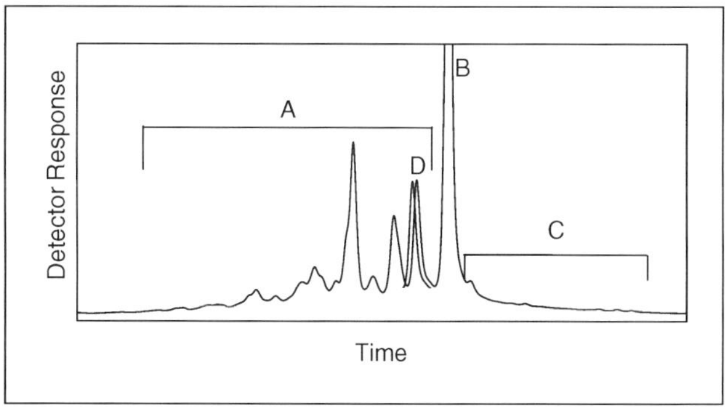

Figure 2 Chromatogram of an ion exchange high performance liquid chromatography analysis of an intact recombinant monoclonal antibody product.

Table 2 Validation Parameters for Quantitative Purity Assay Recommended for Clinical Validation (Qualification) and Required for Commercial Validation by ICH Q2A and Q2B

Parameters	Clinical	Commercial
Specificity (1)	+	+
Linearity	−	+
Range	−	+
Accuracy	−	+
Precision		
Repeatability	+	+
Intermediate precision	+	+
Detection limit	−	−
Quantitation limit	−	+

precision consisting of repeatability and intermediate precision (Table 2). The other commercial validation parameters, accuracy, linearity, range, and limit of quantification, were not evaluated at this early stage of development.

Specificity was chosen since it ensured that the quantification of the method is not biased through peaks from the formulation buffer and/or the sample diluent that could theoretically have coeluted with the peaks of the product.

The precision study, in particular the intermediate precision experiments (using multiple analysts, instruments, and reagents over at least three days), was selected because it provided confidence, together with the results from the robustness study, that the method will operate in the future at a high reproducibility in a routine testing environment.

Accuracy, linearity, range, and limit of quantification were omitted at this early stage of product development, for the following reasons.

The IEC specifications were at this early stage "report." Hence, the specification range (see below) across which accuracy should be validated was unknown. Moreover accuracy is typically assessed by a series of spike and recovery experiments of the components of interest but the components (peaks) of interest are at this early stage unknown. Spike and recovery studies are furthermore rather time-consuming and resource-intensive experiments since they require fraction collection, purification, and concentration. However, time and resources are particularly sparse at this early stage of the development.

Additional assurance with respect to accuracy was derived from a column-in/column-out of line experiment (performed during robustness studies) using the protein product as analyte, which supported that the material or fractions of the material were not nonspecifically adsorbed to the column surface during analysis.

Linearity, range, and limit of quantification were also deferred since they are typically derived from the results of the accuracy experiments.

The validation status of the assay was updated during Phase II when the formulation of the product was optimized. The only parameter revalidated was specificity since it could not be excluded that the new formulation with its new components might interfere with the quantification of the product. Precision was not re-evaluated since it was considered technically unlikely that the change in formulation would impact the assay performance.

In Phase III, when the manufacturing process was locked down and the product was well understood with respect to its quality attributes (e.g., product-related substances versus product-related impurities), the assay was validated in full adherence to ICH Q2A and Q2B.

The thorough and extensive characterization of the molecule together with a good understanding of its mechanism of action revealed that peak D (Fig. 2) also required a specification and hence quantification by the assay. The assay range studies (consisting of accuracy, precision, and linearity) were consequently extended beyond the original regions A, B, and C to include peak D. The ICH Q2A and Q2B requirement for the range study, stating that the validation has to cover at least 80% to 120% of the target value, was fulfilled by evaluating the ranges for A to D by setting the proposed product specifications as target values for the validation.

Other parameters, e.g., specificity, were not re-evaluated since the original validation data gathered throughout clinical development were still considered valid. The same applied to the robustness data.

The robustness and validation (whose purpose is to demonstrate the suitability of the method for its intended usage) was further supported through the trending data for the reference material collected during the clinical testing as part of the method system suitability. These data, which reflected the routine usage of this method by multiple laboratories and analysts using different instruments, chromatographic columns, etc., across a long duration of time, provided further assurance that the method is robust, precise, and well behaving.

The validation for licensure then closed with the writing and approval of the commercial validation report. After approval of the product, the quantitative method went on a monitoring program, which ensured that the validation is performing as validated in routine operation and the validation status was maintained.

Validation of Bioassay Test Methods

Biological assays (or bioassays) are widely used during production to measure the biological activity or potency of a product. The following section discusses the requirements for validation of bioassays and will be restricted to in vitro cell-based and non-cell–based bioassays: assays with functional readout, biochemical assays, enzymatic assays, binding assays and will not address bioassays performed with whole animals or immunoassays.

General guidelines on approaches to method validation are available in the ICH Guidelines (1,2) but there is no direct guidance on the validation of bioassays. These documents are nonetheless indicative of which assay performance parameters need to be evaluated. Among the characteristics that should be examined are accuracy, precision, specificity, range and linearity.

Prevalidation Work

Bioassay can result in highly variable results due to the biological nature of the assay system. Test data generated by bioassay must provide reliable estimation of potency and must provide accurate and reproducible results. Therefore, before a bioassay can be validated, there are some points to consider in the method design phase such as location effects within plate, microtiter plate layout, characterization of the critical reagents, system suitability, statistical method of data analysis, and robustness to ensure that results those methods provide are valid, reliable and useful. Informed and realistic assessment of a method and its intended use are critical to minimize unproductive validation activities.

Controlling Assay Variability

A key element in attaining a reliable bioassay with as little variability as possible is choosing the correct assay format. It is often the case that variability is not uniform within assay plates resulting in plate positional effect and in substantial bias in potency estimates. Attempts must be made to assess and reduce assay variability and bias. One approach to protect against positional effect is the introduction of randomization and replication into the plate format. Practically speaking, from the analyst point of view, complete randomization is not be feasible and would be difficult to set up. Column or row randomization is a good compromise and easy to perform. Other potential sources of variability which should be considered are pipetting technique, consistency of technique on the part of the operator(s) and careful control of reagent (i.e., cells), assay media, incubation, dilution error, instrument, and plate. Rigorous training becomes a very important tool to limit assay variability as much as possible.

Critical Reagents

Critical assay components/reagents need to be well characterized and if an in-house reagent such as transfected cell line or coat protein is used, a characterization report describing the cell line history or coat protein preparation needs to be available. Equivalency between multiple lots/preparations needs to be demonstrated and each new lot/preparation must be qualified. Qualification requirements must be documented in a SOP with proper acceptance criteria. Other characteristics such as reagent stability, availability/supply and source are also very important points to consider especially if the method is going to be transferred to other sites. Cell stability should be evaluated through monitoring of receptor expression level over culture duration or trending of EC_{50}/IC_{50} values in the potency assay. Also, a commercial reagent must be available from multiple vendors. Failures to appropriately characterize any critical component will likely result in poor assay performance and unexpected assay failure, and potentially an assay shutdown for an undetermined time during production.

System Suitability

QC method must be properly controlled to ensure consistency from assay to assay and to ensure reliable release-testing results. This is usually accomplished by incorporating system suitability parameters in the method with appropriate acceptance range, which should be established based on development and optimization data. In the case of bioassays, control samples, preferably a product sample different from the reference standard, are key to system suitability to implement in the design of the bioassay. The primary function of control samples is to provide a criterion for judging the acceptability of assay result. To maximize the utility of the information provided by controls, the preparation and handling procedures of controls should parallel those for unknown samples (e.g., dilution procedures, matrix, etc.). Control samples generally serve a few different purposes: not only do they help decide the acceptability of results from a given assay, but they also help monitor method performance and success rate over time, identify problems early, potentially can be used in troubleshooting problems and provide empirical data on different components of assay variation. Acceptance criteria on additional system suitability parameters such as cell viability, cell passage number, cell density at time of harvest, number of dilutions required in potency calculation and parallelism are recommended to ensure consistency among assays and should be added to the Test Method.

Statistical Method of Data Analysis

Estimation of potency is obtained relative to a standard; therefore, we generate a relative potency. In order for a sample quantitation to be valid, it is important that the dose–response curve of the sample be parallel to the dose–response curve of the standard. The statistical methodology that should be used for estimating relative potency in bioassays is the parallel line analysis. Current regulatory guidance including the USP chapter <111>, *Design and Analysis of Biological Assays*, and the EP Chapter 5.3, *Statistical Analysis of Results of Biological Assays and Tests* give recommendations on how to assess parallelism. A recent paper published by Hauk et al. (6) proposes to replace the *p* value method for assessing parallelism that is currently in <111> and EP Chapter 5.3 and recommends an alternative approach based on equivalence testing. In our approach of assessing parallelism, a validated parallel line analysis software calculates the slope ratio of the sample to the reference; this slope ratio is used as a criterion to evaluate parallelism and must fall within predetermined limits.

Robustness

Once all the method characteristics mentioned above have been addressed, robustness studies are initiated. Robustness assesses the ability of a method to withstand deliberate variations in method parameter and provide an indication of the method reliability. It should not be part of the validation protocol but should be addressed after the method optimization phase and should be a prelude to validation.

The first step is to identify the critical variables of a method. Because the performance of a cell-based assay depends strongly on the consistency of the cellular responses, various parameters relating to cells should be evaluated: cell bank (comparing vials of cells frozen at different time/location in the liquid nitrogen tank), cell passage number, cell stock density (number of cells at time of harvest), cell age in flask (number of days cells are growing in a flask from seeding to harvest), cell suspension stability (prior to seeding in microtiter wells), cell seeding density (number of cells seeded in wells), and cell culture media. Other critical robustness parameters such as vendors/lots of microtiter plates, incubation times/temperature, and reagent concentration/stability are typically studied as well. Factorial design, where several parameters are varied from target conditions and tested together in a single assay experiment, is a valuable statistical tool to gain more information on the possible interaction of those parameters and to evaluate how it affects the robustness of the method. Although time consuming to perform, thorough robustness studies will help avoid subsequent unexpected results and will provide useful data in the selection of system suitability parameters. Successful completion of this work will provide convincing evidence of the reliability of the test method.

Validation

After suitable robustness has been demonstrated, the method is ready to be validated. Validation acceptance criteria should be carefully defined in a protocol to determine whether or not the assay is fit for use.

The number of validation studies and parameters to be evaluated required to ensure that methods are appropriately assessing the product's potency will vary and depend on the type of bioassay selected and stage of development. The difficulty in developing and selecting a bioassay that is scientifically relevant, biomimetic (reflects the intended mechanism of action), robust and well-behaved presents a big challenge (7). In the perfect situation, it would be desirable to have such an assay in place as early as Phase I. However, it is an unlikely situation as often time the mechanism of action of the product is unknown in such early phase of development and time and resources are limited. As such, selecting and validating a non-cell–based binding/assay, the simplest of the bioassays, for Phase I is a good alternative to a more functional potency assay. However, for later stages, one should consider a biomimetic, more functional potency assay and correlate the binding/binding inhibition assay to the functional assay. The final method should be "lock down" and in place at pre-phase III. There are several advantages to this strategy. The most important one is that a lot of experience is gained with the final potency assay throughout the product development prior to submission, which will provide a true estimate of method performance, precision and success rate. Also, less is the number of methods used throughout the clinical phases, less is the number of validations and bridging studies to be required. At the time of submission, it can be quite challenging and cumbersome to demonstrate correlation between all the different methods used throughout the clinical phases over many years, as early samples will likely not be available to be compared in late stage methods. One solution to that problem is that the clinical lots used throughout product development be frozen and evaluated using the final assay format (especially if the methods are different).

The number of validation tests performed at Phase I should be sufficient to determine whether the method validation characteristics are likely to cause a problem. Accuracy, precision, linearity and range are the primary validation characteristics and spiked recovery studies are useful in determining those parameters. Some aspect of the method validation studies such as intermediate precision and reproducibility should be delayed during early development until the method is transferred, used in multiple laboratories and by several analysts and instruments.

The recovery studies consist of measuring the recovery of various samples (minimum of five) over a concentration range corresponding to 60% to 140% of the target sample concentration. Care should be taken to mimic the actual sample preparation and dilutions as closely as possible. These studies should be conducted by at least two analysts in one lab (validation for clinical use) and multiple analysts (at least four) in two labs (validation for commercial use).

To assess the accuracy of the method, the measured product potencies are divided by the expected potencies and expressed as percentage recoveries.

Precision is evaluated by statistical analysis of the recovery studies and can be divided into separate component; plate-to-plate within day (repeatability), assay-to-assay within analyst and analyst-to-analyst variation (intermediate precision), and lab-to-lab variation (reproducibility) depending on the phase of development.

To design the final assay format, the variance components estimates are used to calculate the potential RSD value for different assay format. The design parameters studied are number of replicate wells, plate per assay and assay. A typical assay format for a sample is three independent assays, one plate per assay and duplicate wells for each concentration of standard, control or sample. Each independent assay will provide an estimate of potency of a sample and a mean potency is obtained from the independent estimates. Although only three to five concentrations of the standard are used in the potency calculation, the standard curve is generally run as a full 10- to 12-points curve in order to trend various curve characteristics such as IC_{50}/EC_{50}, fold-response, slope, and upper and higher lower plateaus. Control and samples are diluted and tested at three to five concentrations targeting the assay range.

The suitability of the final assay format is verified by assessing activity of several drug substance and drug products lots.

Linearity of a potency assay is generally referred to as linearity of potency measurement. It is derived from the sample recovery study where measured potencies are plotted against expected potency and the coefficient of correlation is evaluated.

The working range assessment defines the upper and lower level of product concentration for which the method has demonstrated a suitable level of accuracy, precision and linearity.

At the time of BLA submission, all validation parameters listed in ICH documents should be validated. By that time intermediate precision as well as specificity should have been performed. Specificity for a bioassay is assessed two ways: by evaluating the activity of a large number of marketed and clinical products in the assay and by evaluating the potency of the product in presence of those clinical/marketed materials to determine any inhibition/enhancement effect. If one of the materials shows up positive in the bioassay, that information should be documented in the validation report. However, because the potency assay is not intended to be used as an identity test, other methods in the control system will be able to establish a positive identification if the material was present in the product sample.

To complete the assessment of method performance, additional studies might be carried out to evaluate stability-indicating properties of the bioassay method, i.e., the ability of the method to detect changes in activity in samples subjected to various stress conditions such as heat, light exposure, high and low pH, oxidation, and mechanical agitation.

A relative misconception is that bioassays used in QC lot release are so variable and imprecise that the results are not usable for quantitative purposes and that they are at best capable of giving a precision around 25%. With careful selection of bioassay methods, format, analysis and rigorous training, bioassays in our lab have typically demonstrated to be very quite precise with RSD below 10%, most of them around 6% to 7% whether they are cell-based or non-cell–based, whether they take a short (one day) or long (three to four days) incubation period, and whether they use suspension or adherent cell line.

Validation activities do not stop here. Post-licensure, the method validation status is maintained through monitoring of method performance using assay control trending to verify that methods are operating according to their original validation characteristics. Assay control trending charts serve multiple purposes: not only do they monitor method performance over time among multiple analysts and multiple labs, but they also monitor the assay success rate, are useful in establishing meaningful acceptance criteria/system suitability in test methods, product specifications, and transfer criteria and finally are essential in troubleshooting the method and/or retrain the analyst before it gets worse. In addition assay control trending charts have been important tool for inspection management.

Validation of Microbiological Test Methods

Microbiological assays are performed to detect viable forms of bacteria, fungi, or yeast, if present, for both clinical and commercial product. While ICH Guidelines and references from Pharmacopeia such as USP <1225> Validation of Compendial Methods provide guidance for analytical procedures, these documents do not address unique attributes of microbiological assays. The following section will firstly discuss assay controls to reduce variability in performing microbiological assays in general. Secondly, validation requirements for both compendial and noncompendial methods will be addressed, using, as examples, two compendial methods: USP <71> Sterility Tests and USP <61> Microbial Limit Tests and a noncompendial bioburden method suitable for diverse sample types, e.g., in process microbial testing, hold time studies, or cleaning validations.

Control of Assay Variability

Critical Materials

Materials used in microbiological assays should be well defined in standard operating procedures and their preparation fully documented. Standard operating procedures should describe stepwise preparation, storage conditions, and expiration dating of test materials. For example, each lot of medium must demonstrate growth supporting properties for its intended use. Growth promotion testing should be designed to correspond to the criteria (quantitative vs. qualitative) and incubation conditions (duration and temperature) of the assay where the media will be utilized. The use of manufacturing process and/or environmental isolates in the battery of challenge microorganisms in growth promotion testing should be considered and included when applicable.

Microbial Contamination

With microbiological tests it is imperative that contamination during the performance of the test procedure does not compromise the result. During validation the procedure must be shown to prevent cross-contamination of samples. Analysts must be properly trained and qualified to prevent introduction of microorganisms during testing, and suitable precautions must be taken in the testing environment. For example, during sterility testing, equipment and supplies are sterilized by appropriate means, sample containers are decontaminated, and testing is performed under aseptic conditions in which isolator technology is commonly employed. For less rigorous testing in which sample handling is less controlled, a laminar flow hood may be sufficient. Environmental and personnel monitoring of the testing area may also be performed to monitor the conditions of the testing environment. At the end of each test session a negative control can be performed using the same lots of rinse fluid and media and incubated under the same conditions as the test sample. A contaminated negative control serves as an indicator of a determinate error during the test session.

Microbial Growth

There are many factors that affect the accuracy of the microbial count. The physiological state of the microbial cell has a direct influence on the results of microbial tests. The preparation of the inoculum of challenge microorganisms should be standardized to ensure reproducible results. While not required for sterility and microbial limit testing, it is a prudent policy to use cultures no more than five passages removed from the original ATCC or banked culture. On solid media, colony forming units are reported. If microorganisms are clumped, numerous cells may be reported as 1 CFU, indicating a low count. Conversely, if clumped microorganisms are introduced into liquid medium, they may disperse, render the medium turbid, and give the appearance of growth. Homogeneous cell populations reduce variability. Ideally, microorganisms will be evenly dispersed throughout the sample so that all sample volumes are equivalent.

It is essential that a wide range of microorganisms grow consistently under the conditions of the test. As these assays are growth-based, variable results are possible. The selection of the challenge microorganisms used in validation must represent those likely to be present in the manufacturing environment. For sterility and microbial limit test validation, specified microorganisms are listed in the Pharmacopeia. In addition, microbial isolates recovered from manufacturing sites may be added to demonstrate recovery of "house" organisms. The conditions of microorganism preparation and storage must be standardized and reproducible. Method validation conditions must be reproduced during the test; e.g., growth conditions of the test must be tightly controlled as to time, temperature, etc.

Microbial Recovery

To validate the suitability of the test method, microbial recovery must be shown under the test conditions. The test sample may exhibit antimicrobial properties from either its chemical composition or the addition of a preservative, and neutralization of these properties must be demonstrated during validation. Common methods for removal of antimicrobial attributes are rinsing after filtration, dilution, or specific chemical inhibition. Validation of the neutralization method is demonstrated by recovery comparisons to the method control under the test conditions. This test should be performed independently three times. Validation is performed on the sample formulated with the highest product concentration, preservative, or inhibitor content. All product formulations of lesser product concentration, preservative, or inhibitor content are thereby also considered qualified.

Assay Validation. The following section discusses the requirements for validation of microbiological assays using, as examples, two compendial methods: USP <71> Sterility Tests for sterile products and USP <61> Microbial Limit Tests for nonsterile products and a noncompendial bioburden test method for in process microbial testing. The same methods are used for both clinical and commercial products. The cGMP regulations [21 CFR 211.194(a)] require that test methods, which are used for assessing compliance of pharmaceutical products with established specifications, must meet proper standards of accuracy and reliability. Analytical methods described in the USP require only demonstration that the method is suitable for its intended purpose.

Validation of Sterility Test USP <71> (Bacteriostasis/Fungistasis Test)

The sterility test is performed to reveal the presence of viable bacteria, fungi or yeast in or on products purporting to be sterile. USP <1227> Validation of Microbial Recovery from Pharmacopeial Articles provides guidance for validation of Sterility Tests USP <71>. There is harmonization between USP <71> and Ph Eur 2.6.1 Sterility testing.

The following discussion will be limited to membrane filtration as this method is preferred where the nature of the product permits; otherwise, direct inoculation is used.

For membrane filtration neutralization is effective for solutions when antimicrobial material passes through the filter while microorganisms remain. The filter is then rinsed to remove adherent material using fluid with or without chemical neutralizers.

For validation of sterility testing by membrane filtration the test sample of product is filtered through the membrane. Subsequently the filter is rinsed by two 100-mL aliquots of rinse fluid. A third 100-mL aliquot containing <100 CFU of the challenge microorganism, specified by USP <71>, is passed through the filter. Either the filter is aseptically transferred to the growth medium or the medium is added to the filter. A control lacking only the product is also performed. A comparison of turbidity is made between test and control groups; e.g., both broths show turbidity after the same incubation period. This procedure is repeated for each of the specified challenge microorganisms. During validation the appropriate rinse volume is determined. Typically three 100-mL aliquots of rinse fluid are used, but lesser volumes may be used if they have been demonstrated to be sufficient to remove bacteriostatic or fungistatic substances. When necessary, increased volume of rinse fluid may be used, but USP <71> limits the washing cycle to five times 200 mL, even if during validation it has been demonstrated that such a cycle does not fully eliminate the antimicrobial activity.

This validation is performed for new products and whenever test conditions are changed. Validation can be performed concurrently with the testing of samples, but successful validation has to be demonstrated prior to release of the product batch.

Validation of Microbial Limit Tests USP <61>

The Microbial Limit Test is performed to quantify viable aerobic microorganisms present and to demonstrate freedom from designated microbial species in pharmaceutical articles of all kinds, from raw materials to finished product. There is harmonization between USP <61> and Ph Eur 2.6.12 Microbiological Examination of Nonsterile Products (Total Aerobic Count) and 2.6.13 Microbiological Examination of Nonsterile Products (Test for Specified Micro-Organisms). USP <1227> Validation of Microbial Recovery from Pharmacopeial Articles provides guidance for the Microbial Limit Tests <61>. The Microbial Limit Tests requires preparatory testing to demonstrate that the test specimens do not "inhibit the multiplication, under the test conditions, of microorganisms that may be present." Test specimens diluted with phosphate buffer, fluid soybean–casein digest medium, or fluid lactose medium are inoculated with separate viable cultures of *Staphyloccus aureus*, *Escherichia coli*, *Pseudomonas aeruginosa* and *Salmonella* species. Diluted samples are plated onto highly selective media to enhance the recovery of these organisms. If the challenge organism fails to grow under test conditions, the procedure is modified by increasing the dilution, adding a sufficient quantity of suitable inactivating agents, or a combination of the two in order to permit growth of the inocula. The procedure provides examples of neutralizing agents. When one encounters product from which viable microorganisms cannot be recovered after using neutralizing agents and increased diluent, the product is unlikely to be contaminated with microorganisms.

Bioburden Testing (Noncompendial Method)

A bioburden assay is performed to quantify microbial load at multiple steps in the manufacturing process and may also be employed in hold time studies and in cleaning validations. The following is a discussion of a typical bioburden method validation. Microbiological method validation should follow a life cycle approach for noncompendial methods with specificity and precision required for clinical products and accuracy added to validation for commercial products. The membrane filtration method is preferred where the nature of the product permits, and a low level of contamination is expected. In the membrane filtration method, samples are filtered, rinsed, and cultivated on agar using standard media and incubation conditions to recover a broad array of viable aerobic microorganisms.

Validation is required to demonstrate that the test method can adequately remove or neutralize any inhibitors present in these samples and quantify the inherent bioburden load. Samples from multiple process steps are obtained from at least two production runs and sterile-filtered before they are tested. The final rinse of the filter for the membrane filtration method is performed using rinse fluid inoculated with specific challenge microorganisms. Method controls consist of 100 mL of a rinse fluid such as PBS inoculated with the challenge microorganisms at approximately 100 CFU/mL (acceptable range: 30–300 CFU/100 mL). Method controls are analyzed by membrane filtration and incubated under the same conditions as the inoculated samples. Duplicate method controls are performed for each organism in each test session. Method control mean results must be 30 to 300 CFU/filter to be considered valid. If the method control mean result for an organism is outside the acceptance range, the corresponding sample results are considered invalid and the test session is repeated for that organism.

Acceptance Criteria

The specificity of the assay is demonstrated by ensuring that the samples do not contain inherent bioburden and that the rinse fluid does not introduce contaminants that interfere with the recovery of the challenge organisms. All negative control plates (samples and PBS controls) must be negative for growth. The precision of the assay is evaluated by using at least two analysts to test all sample types from at least two separate production runs. Results from both analysts must meet the acceptance criterion to demonstrate that the precision of the assay is acceptable. For each analyst the recovery of the challenge organism from each test sample must be 50% to 200% when compared to the corresponding method control in the test session. The accuracy is determined by calculating the mean recovery result from all test sessions for each challenge organism and for each sample type. For each sample type, the mean recovery of each challenge organism must be 70% to 200% for all test sessions. Revalidation is required whenever there is a change in either the manufacturing process or test method.

REFERENCES

1. International Conference on Harmonization. ICH Topic Q2A. Validation of analytical methods: definitions and terminology. Fed Regist 1995; 60:11260–2.
2. International Conference on Harmonization. ICH Topic Q2B. Validation of analytical procedures: methodology. Fed Regist 1997; 62:27463–7.
3. International Conference on Harmonization. ICH Topic Q7A. Good Manufacturing Practice Guide for Active Pharmaceutical Ingredients. Draft Consensus Guideline. Geneva, Switzerland: ICH, 2000.
4. Ritter N, Advant S, Hennessey J, et al. What is a Test Method Qualification? In: Proceedings of the WCBP CMC Strategy Forum. Bioprocess International, 2004:32–45.
5. International Conference on Harmonization. ICH Topic Q6B. Specifications: Test Procedures and Acceptance Criteria for Biotechnological/Biological Products ICH Harmonized Tripartite Guideline. Vol. 64. Geneva, Switzerland: ICH, 1999:44928.
6. Hauck W, Capen R, Callahan J, et al. Assessing parallelism prior to determining relative potency. PDA J Pharm Sci Technol 2005; 59(2):127–37.
7. Schenerman M, Sunday B, Kozlowski S, et al. CMC strategy forum report: Analysis and structure characterization of monoclonal antibodies. BioProcess Int 2004; 2(2):24–53.

51

Validation of Microbiological Methods

Roger Dabbah
Tri-Intersect Solutions, Rockville, Maryland, U.S.A.

David A. Porter
Vectech Pharmaceutical Consultants, Inc., Farmington Hills, Michigan, U.S.A.

INTRODUCTION

Numerous microbiological methods in use today date back to the time of Louis Pasteur. These methods are still useful and effective but there is a substantial interest in the development and use of more modern techniques. In fact most pharmacopeial microbiological methods in USP, EP, and JP are closer to the methods of Pasteur's time and are the so-called classical methods. Advances in methodologies and instrumentation should be incorporated in the pharmacopeial methods that are cited by the FDA for compliance purposes. Microbiological methods, especially if intended for replacement of more conventional approaches require validation. If the new methods that are desirable because of advantages speed, accuracy, specificity, and generation of more quantitative data are to replace the pharmacopeial methods, then they also have to be shown to be equivalent (equal or better) than the pharmacopeial methods. This situation exists in Europe, Japan and in the U.S.A., where regulatory agencies will accept alternative methods to the pharmacopeial methods for compliance purposes.

There are three types of microbiological methods that need to be validated, qualitative tests, quantitative tests, and identification tests. Each type of test that follows has a different pattern in their validation and will be examined separately.

The special case of validation of equivalency between a pharmacopeial microbiological test and alternative test will de discussed in each of the sections.

Validation of any kind of assay requires that certain parameters be examined, including accuracy, precision, specificity, limit of detection, limit of quantification, linearity, range, ruggedness, and robustness. However, for microbiological assays and depending on the type of assay used, these parameters are not applicable.

VALIDATION OF QUALITATIVE MICROBIOLOGICAL METHODS

The principal function of a qualitative microbiological method is to determine whether the sample under test contains any viable microorganism. An example of such a test is the sterility test. The validation parameters to be considered include, accuracy, precision, specificity, limit of detection, ruggedness, and robustness.

Accuracy

A sterility test is not used to provide assurance of sterility of an entire batch. The assurance of sterility of a batch of sterilized products is obtained through the validation of the sterilization cycle or process. Accuracy pertains to the ability of the test to detect the presence of viable microorganisms if present in the sample under test.

The issue of a false positive and false negative from a sterility test can have serious repercussions. A false positive, if it cannot be ruled out decisively due to technical error, will likely result in discarded lots of products, costly in terms of labor and material. A false negative would declare a batch sterile while it is not and could result in harm to patients using the product and potential liability to the manufacturer.

The limitations of a sterility test are well known and will depend on the ability of media to permit the growth of surviving microorganisms, the ability of microorganisms to grow under temperatures and time of incubation used. If the microbiological method used for sterility testing does not require the growth of microorganisms, it might require the uptake of vital markers and subsequent ability to fluorescence.

For methods that require the growth of microorganisms to a level that can be detectable by turbidity examination, the validation of the method will include the testing of the capabilities of media used to support the growth of likely microorganisms. The growth promotion test used in the USP Sterility Test is an example of validation of media. The type of microorganisms used for USP growth promotion is arbitrary but is a compromise, since it includes aerobes, anaerobes, and fungi. Often, environmental isolates can be included in a growth promotion test, or microorganisms isolated from positive sterility tests are also used. A growth promotion test will use very low concentration of microorganisms since one cannot expect a wholesale survival of microorganisms following sterilization. Another issue is the possibility of the test sample itself being inhibitory to the growth of microorganisms. A validation will include qualification of the sterility test for each and every product to be tested. This can be done using the same microorganisms used for growth promotion of media that are inoculated at very low levels (less than 100 CFU) but

---

Abbreviations used in this chapter: ATCC, American type culture collection; EP, *European Pharmacopoeia*; FDA, Food and Drug Administration; JP, *Japanese Pharmacopoeia*; NF, National Formulatory; USP, *United States Pharmacopeia*.

in the presence of the product to be tested. Modification of the sterility test might be necessary if the sample is inhibitory to the challenge microorganisms.

For methods that do not rely upon reproductive capability of viable microorganisms, the presence or absence of microorganisms will depend on the ability of the vital marker to be incorporated into microorganisms that are viable and not in microorganisms that are not viable. Using a low level of challenge microorganisms one can then validate the test using incorporation of a vital marker and its ability to fluoresce under the conditions of the test.

The special case of validation of an alternative microbiological test to a pharmacopeial introduces another complexity, since it involves the comparison of two tests and the determination of their equivalence. If one wants to establish equivalence of accuracy of two microbiological methods used for qualitative determination, the null hypothesis becomes one of inequality (the results of the two methods are significantly different). If the methods are not significantly different then the two methods are equivalent.

Precision

Precision is the repeatability of a test. If a microbiological procedure is applied repeatedly to a specific lot of product or material, the results should be the same. The rate of false positive and false negative in samples inoculated with microorganisms at low levels using perhaps a serial dilution scheme would give an indication of precision of the microbiological method for methods relying on growth of microorganisms. For methods not relying on growth, the precision can be determined using the quantitative data obtained for example by fluorescence measurement.

The comparison of the precision of a pharmacopeial method and an alternative method can be measured by subjecting the inoculated samples to the two procedures and determining their equivalency.

Specificity

For methods based upon growth of microorganisms, the specificity parameter can be measured as turbidity, development of CO_2, and changes in media pH that can all be shown to be due to microbiological growth and that must reach a threshold in order to be detectable.

For methods not based upon growth of microorganisms, the presence of interfering components in the test sample must be ruled out. For example, for a method that indicates viability via the fluorescence of a viable marker it should be established that no other component in the sample, medium, or diluents or reagents used can produce fluorescence.

The specificity parameter for an alternative method to a pharmacopeial method is not comparable per se, since specificity for each method is dependent on the mechanisms and concepts used for each.

Limit of Detection

The limit of detection is the lowest number of microorganisms that must be present in the test system to elicit a positive response that is a signal that can be discerned from the underlying noise.

For methods relying on growth of microorganisms, a single microorganism given enough time to grow by incubation at an appropriate temperature, the limit of detection should be one. This does not take into account slowly growing microorganisms or injured microorganisms that take more than 14 days for recovery and growth.

For methods that do not rely on growth of microorganisms for detection, the limit of detection is more complex. One has to first establish that a signal picked up by the instrumentation is actually the result of microbial activity. Setting a threshold of detection too low will provide "blips" that would indicate potential microbial activity when it might be due to residual signals from other components. The threshold of detection should be set following a risk assessment determination on the significance of the signal of certain quantitative value.

In situations when a microbiological method is to be shown equivalent to a pharmacopeial method, and both methods rely upon growth of microorganisms, a method using serial dilution for both methods should be conducted with appropriate microorganisms to determine the limit of detection. When the alternative method to a pharmacopeial method does not rely on growth of microorganisms, the equivalence of the methods in terms of limit of detection should follow a risk assessment of the results obtained to determine the threshold of detection that must be above the noise level.

Ruggedness

A method is rugged when it will resist producing divergent results, when it is performed by different microbiologists, in different laboratories, on different days, using different instruments or lots of reagents.

Establish ruggedness is straightforward using the same lot of materials tested under different conditions as cited above.

The ruggedness of an alternative method to a pharmacopeial method is done separately and the results are compared and if they do not significantly differ then the methods are equivalent.

Robustness

The robustness of a microbiological method is an indication of the resistance of the method to small and deliberate differences introduced in the method itself. Growth-based methods might be tested using different lots of a given medium or slightly different pH might be used. For methods not relying on growth of microorganisms, use of different instruments or variations in reagent lots or temperature and/or time conditions could be done to determine the robustness of the method.

When an alternative test to a pharmacopeial method is validated for robustness, results of the tests should be compared and relative robustness between the methods established.

VALIDATION OF QUANTITATIVE MICROBIOLOGICAL METHODS

The principal function of a quantitative microbiological method is to determine how many viable microorganisms a sample contains. An example of a quantitative microbiological test is found in USP chapter <61> Microbial Limits.

In the case of these methods parameters of validation include, accuracy, precision, specificity, limit of quantification, linearity, range, ruggedness, and robustness.

Accuracy

Accuracy refers to the closeness of the value determined to the true number of microorganisms present in a sample at the time of testing. Contrary to physicochemical tests where samples are homogeneous, the microbial load is not distributed homogenously in a sample, thus the accuracy will depend on sampling. Other factors involved include the ability of the media used to support the growth and detection of microorganisms present in a sample. Results of quantitative microbiological tests are at best estimates and no amount of sophisticated statistical procedures applied would change the accuracy. If a need for accuracy is necessary, perhaps because a regulatory agency requires it, then a "spiking" experiment with known challenge microorganisms could give a better measure of accuracy. The spiking experiment could use a mixture of well-defined microorganisms rather than several experiments with individual challenge microorganisms.

Methods that rely on growth of microorganisms for quantification use, for example, CFU as a quantitative measurement. However, the assumption that one colony is the product of one microorganism is arbitrary and generally not correct. Another level of complexity is encountered when a microbiological specification is expressed as a limit, such as less than 100 CFU, with a sample acceptable if the quantitative value is equal or less than 100 CFU, but not acceptable if it is greater than 100 CFU. This approach does not take into consideration the inherent large variability of microbiological quantitative tests. This has been recognized in the harmonization work among USP, JP, and EP where it has been proposed that a specification of 100 CFU will be acceptable even if the result of the test gives 200 CFU; specification of 1000 CFU will be acceptable if the result of the test is 2000 CFU; and so forth.

For tests that do not rely on growth of microorganisms for quantification the issue is more complex. A pharmacopeial test counts CFU while an alternative test could count each and every microorganism present. The use of differential specifications depending on the microbiological method type used could be logical, although few individuals have tackled this issue. The relationship between metabolic activity and the number of microorganisms present will have to be quantified by the development of standard curves relating the number of microorganisms with the signal given by the test. Since the bioburden of a product is not homogeneous in terms of distribution of type of microorganisms and varies with seasons and suppliers, the standard curve approach will have to be carefully developed, perhaps using a mixture of most likely microorganisms present.

Introducing accuracy measurements for these tests is not appropriate, from a theoretical, experimental, or practical point of view.

If a test is to be validated against a pharmacopeial test as being equivalent, the comparison will be between "apples" and "oranges" casting doubt on the meaningfulness of requiring that an alternative test be equivalent to a pharmacopeial test.

Precision

The precision of an analytical method is the degree of agreement among the individual results when the method is applied repeatedly to multiple samplings of a homogeneous material. The first issue one has to deal with is that the microbiological content of a sample is not generally distributed homogeneously within a product. A way to approach the issue of precision is to indicate that the precision of a quantitative microbiological method is a function of the precision of a number of steps involved in the procedure. These steps include but are not limited to sampling, pipetting, temperature of the agar if plate counting is used, temperature of incubation and length of incubation. These apply to methods based on growth of microorganisms. For methods that do not rely on growth and are based on instrumental procedures, calibration of the instrument and temperature of the reagents used are factors that have to be taken into consideration in the determination of precision. Replicate of the testing will give a general estimate of the precision of a method through statistical analysis of results using more likely standard deviations. One has to remember however that as the number of microorganisms become smaller, the error as a percentage of the mean increases.

A microbiological quantitative method might have a very good precision but it might not be very accurate.

Specificity

The measurement of the quantity of microorganisms in a sample must be specific. This requires, for methods involving growth of microorganisms that a differentiation must be made between CFUs and debris. In general this is not too difficult, and in case of doubt use of magnification could resolve this issue. In the case of methods that do not rely on growth of microorganisms, the discriminatory power of the instrumentation will come into play. The instrumental noise level should be established as well as the specificity of the method for a variety of products since each product will contain different components that might give a signal similar to a microorganism.

When a microbiological quantitative method is designed for the determination of specific category of microorganisms—for example yeasts and molds—the specificity of the method for enumeration of these specified microorganisms should be validated.

Limit of Quantification

Most quantitative microbiological tests are used as limit test. Regardless of the type of method used, the method should be able to quantify microorganisms above and below that limit. If the specification for a product is 100 CFU and the method cannot detect less than 100 CFU, the method should not be used.

If one needs to demonstrate the equivalence of an alternative method to a pharmacopeial method, their limit of quantification should be comparable.

Linearity

Linearity for an analytical procedure is determined using a least five different concentrations of analyte. For a microbiological test a serial dilution of the sample to achieve a five different concentration of microorganisms could be used.

For growth-based tests the linearity factor will not be very useful and will only test the precision of pipetting and the maintenance of the conditions of the test throughout the procedure. For non-growth–based tests linearity becomes important depending on the statistical analysis done on an automatic mode by the instrument.

The linearity parameter becomes important if one needs to develop a standard curve that correlates a growth-based test to an instrumental test. These standard curves are used for example, in antimicrobial effectiveness testing (see chap. <51> in the current USP-NF) to determine the size of the inoculum without having to use a growth-based test.

Range

The precision of a microbiological quantitative test is a function of the number of microorganisms in the sample. To obtain an acceptable precision the range (upper and lower limit) should be defined and confined to a narrow window. In plate counts it is customary to use "countable plates" which are in the range of 10 to 300 CFU.

Dilution of the sample should be manipulated to obtain countable plates in order to get a decent precision.

Ruggedness

Ruggedness will be determined using different microbiologists, at different locations, on different days, using different instrumentation, reagent lots, etc., but using the same sample.

The variability of a microbiological test in the estimation of bioburden is very wide and we expect that the determination of ruggedness will be rather difficult if not impossible to establish.

Robustness

It is an indicator of how a quantitative microbiological test will perform if small changes in the parameters of the test are introduced under routine usage.

An alternative method to a pharmacopeial method should have equivalent ruggedness and robustness.

VALIDATION OF MICROBIAL IDENTIFICATION METHODS

Microbial identification tests are tests that would determine the presence or absence of specific microorganisms in a sample. The validation of these tests is similar to the validation of qualitative microbiological test such as a sterility test that checks samples for presence or absence of microorganisms. Parameters of validation to be considered include accuracy, precision, ruggedness, and robustness.

Accuracy

The presence or absence of microorganisms feature of these tests depends more on sampling than on the accuracy of the method used. Once a procedure detects a presumptive specified microorganism, the accuracy of the test will depend on the accuracy of the identification scheme that will be used to confirm the identification.

The classical identification methods depend on the metabolism of specified microorganisms in the media provided, their phenotypic characteristics, and specialized biochemical tests. Other identification methods depend on the use of specialized instrumentation that compares the biochemical reactions of the isolate on specialized media with a reference library of microorganisms. Calibration of the biochemical method will depend on the reactions of a specific strain of ATCC microorganisms as a comparator. The issue of concern is that isolated microorganisms from samples do not always behave as the ATCC strain of reference, and the results cannot be predictable. Some instrumental methods will use as a comparator the percentage of similarity between the isolate and the reference strain.

In the special case of equivalency between a method and a pharmacopeial procedure, inoculation of samples with ATCC strains of interest will provide a relative measurement of accuracy when compared to the pharmacopeial method.

Precision

Precision of an identification method can be measured or estimated by replicating the scheme repeatedly using the same sample. Since, in general, alternative methods to a compendial method will sometimes use the same general procedure but will use instruments for some portions of the testing procedure, the precision of each step can be determined for the pharmacopeial method and the alternative method and the results compared. However, it is best to modify the pharmacopeial method then run the same sample, inoculated and non-inoculated by specified microorganisms by both methods, from enrichment to isolation to identification.

Ruggedness

The determination of ruggedness for microbial identification schemes is really a function of the training of the personnel that do the testing and the calibration of the instrumentation if it is used. The stability and repeatability of the identification schemes will be a function of the drifting of the electronics of the instruments and the accuracy of the databases used for final identification, and the accuracy and idiosyncrasies of the software used.

Robustness

The robustness of an identification scheme, especially when instrumentation is used, is determined by the manufacturers of these systems. The manufacturers should provide the users with the parameter ranges that might be encountered during routine use.

THE PHARMACOPEIAL PERSPECTIVE FOR VALIDATION OF USP MICROBIOLOGICAL METHODS

The USP microbiological tests, and for that matter those from the *JP* and *EP*, are considered to be validated methods. Someone who uses the pharmacopeial methods will not have to re-validate those procedures.

However, each raw material, excipient, drug substance, drug product will have to be qualified, since they have possibly the components that are inhibitory to the measurement of microorganisms. The USP provides for modification of the pharmacopeial tests to adjust for material characteristics and activities. These modifications do not have to be qualified each time that a sample is tested. However, re-qualification should be done when there are significant changes in the manufacture of raw materials, changes in suppliers, or changes in components.

Information on validation of pharmacopeial methods is offered in the USP as general information chapters that are not enforceable by the regulatory agencies. The current USP includes chapter <1225> Validation of Compendial Methods that covers validation of tests from the analytical chemistry perspective and does not reflect the special case and characteristics of microbiological tests. This chapter, however, contains information on the general concepts of validation of test methods and should be consulted.

The USP Expert Committee on Analytical Microbiology has developed and proposed another information chapter <1223> Validation of Alternative Microbiological Methods that has received considerable public comments that will require a total rewrite of that proposed chapter.

The USP microbiological tests in <51> Antimicrobial Effectiveness Testing, <71> Sterility Tests, and <61> Microbial Limits Tests require that the recovery methods for microorganisms be qualified for each material tested. Procedures for qualification are included in these chapters. These procedures are simple and refer to visual examination to assess the level of recovery or the inhibitory properties of the sample itself. Requests from interested parties to develop an information chapter to quantitatively determine the recovery of microorganisms resulted in the development and establishment of chapter <1227> Validation of Microbial Recovery from Pharmacopeial Articles. This chapter deals mainly with validation of neutralization procedures to be used in preparatory testing as well as in bacteriostasis and fungistasis determination. A list of common neutralizers is provided along with methods of neutralizations using chemical inhibition, dilution, or membrane filtration. It also discusses the errors associated with recovery of microorganisms, and procedures for assessing the recovery of microorganisms.

CONCLUSIONS

The validation of microbiological methods is more complex than the validation of physicochemical methods. This is in part due to the inherent variability of microbiological methods compounded by the lack of homogeneous distribution of microorganisms within samples to be tested.

Pharmacopeial microbiological methods are validated per se, and need only to be qualified for each material being tested. Qualification procedures are included in the compendial methods.

Advances in technologies represented by rapid microbiological methods have introduced another level of complexity to validation of microbiological tests. For compliance purposes, regulatory agencies in the U.S.A., Japan, and Europe accept the use of alternative methods provided that the alternative method is shown to be equal or better than the pharmacopeial method. This also applies to microbiological tests.

The classification of microbiological tests as qualitative, quantitative, or identification was done since validation of methods depends on its classification. Parameters of validation that are necessary for quantitative microbiological test are different than those for qualitative test, and are also different for identification tests.

BIBLIOGRAPHY

Agalloco, J. Validation of sterilization processes. In: Prince R, ed. Microbiology in Pharaceutical Manufacturing. PDA/DHI, 2001.

Akers MJ. Good aseptic practices: education and training of personnel involved in aseptic processing. In: Groves MJ, Murty R, eds. Aseptic Pharmaceutical Manufacturing II-Applications for the 1990s. Buffalo Grove, IL: Interpharm Press, 1995.

Akers MJ, Anderson NR. Validation of sterile products. In: Loftus BT, Nash RA, Dekker M, eds. Pharmaceutical Process Validation, Vol. 23, NewYork: Marcel Dekker, Inc., 1984.

Avis KE, Akers MJ. Sterilization. In: Lachman L, Lieberman H, Kanig J, eds. Theory and Practice of Industrial Pharmacy. Philadelphia: Lea & Febiger, 1986.

Dabbah R, Paul WL. Container/closure standard requirements in four major pharmacopeias—a comparative review. Pharmacopeial Forum 1992; 18(4):3772.

D'Arbeloff NC, Rusmin S, Stack, D. Robotic applications in sterility testing. In: Groves MJ, Olson WP, Anisfeld MH, eds. Pharmaceutical Manufacturing-Applications for the 1990s, Vol. 2, Buffalo Grove, IL: Interpharm Press, 1991.

Federal Food, Drug, and Cosmetic Act. Section 501(a) 2(b). Title 21, Code of Federal Regulations, Part 211. Washington, DC: US Government Printing Office.

Food and Drug Administration. Use of aseptic processing and terminal sterilization in the preparation of sterile pharmaceuticals for human and veterinary use. Fed Regist 1991; 56:51354.

Haberer K. Steam sterilization in the United States and Europe. In: Prince R, ed. Microbiology in Pharmaceutical Manufacturing. PDA/DHI, 2001.

Levy R. Sterilization filtration of liquids. In: Prince R, ed. Microbiology in Pharmaceutical Manufacturing. PDA/DHI, 2001.

Lugo NM. Aseptic processing of biopharmaceuticals. In: Groves MJ, Murty R, eds. Aseptic Pharmaceutical Manufacturing II-Applications for the 1990s. Buffalo Grove, IL: Interpharm Press, 1995.

Lysfjord JP, Haas PJ, Melgaard HL, Pflug IJ. Barrier isolation technology: a system approach. In: Groves MJ, Murty R, eds. Aseptic Pharmaceutical Manufacturing II. Applications for the 1990s. Buffalo Grove, IL: Interpharm Press, 1995.

Parenteral Drug Association. Validation of dry heat processes used for sterilization and depyrogenation, Technical Report No. 3. Philadelphia, PA, 1981.

Pflug IJ. Syllabus for an Introductory Course in the Microbiology and Engineering Of Sterilization Processes. 4th ed. St. Paul, MN: Environmental Sterilization Services, 1980.

Sinclair CS, Tallentire A. Predictive sterility assurance for aseptic processing. In: Groves MJ, Murty R, eds. Aseptic Pharmaceutical Manufacturing II. Applications for the 1990s. Buffalo Grove, IL. Interpharm Press, 1995.

United States Pharmacopeia, USP 29-NF 24 <55> Biological Indicators- Resistance Performance Test; <61> Microbial

Limit Tests; <71> Sterility Tests; <85> Bacterial Endotoxins Test; <151> Pyrogen Test; <1035> Biological Indicators for Sterilization; <1050> Viral Safety Evaluation of Biotechnology Products Derived from Cell Lines of Human or Animal Origin; <1111> Microbiological Attributes of Non-Sterile Pharmaceutical Products; Microbiological Evaluation of Clean Rooms and Other Controlled Environments; <1196> Pharmacopeial Harmonization; <1207> Sterile Product Packaging—Integrity Evaluation; <1208> Sterility Testing-Validation of Isolator Systems; <1209> Sterilization-Chemical and Physicochemical Indicators and Integrators; <1211> Sterilization and Sterility Assurance of Compendial Articles; <1222> Terminally Sterilized Pharmaceutical Products-Parametric Release; <1227> Validation of Microbial Recovery from Pharmacopeial Articles; <1231> Water for Pharmaceutical Purposes; Biological Indicator for Dry heat Sterilization, Paper Carrier monograph; Biological Indicator for Steam Sterilization, Paper Carrier monograph; Biological Indicator for Steam Sterilization, Self Contained monograph.

52

Implementation of Validation in the United States

James Agalloco
Agalloco & Associates, Belle Mead, New Jersey, U.S.A.

INTRODUCTION

Validation has perhaps been the most frequently discussed subject in the global healthcare industry over the last 30 years. It has been applied in so many varied areas that its scope is seemingly endless. Validation, an exercise that was initially associated solely with sterilization processes, is now discussed in relation to automated systems, analytical methods, dosage form preparation, cleaning procedures, active ingredient manufacture via either classical synthesis or biotechnology, and many other areas. In order to better understand how it is practiced, it is useful to understand its evolution. This chapter will review the history of validation from its onset, with the intent of clarifying how many of today's current practices originated.

The dates utilized in this chronology are approximate.The introduction of practices and acquisition of technology varied from firm to firm, thus the dates should not be considered exactly.

1972 1980

The beginnings of validation within the U.S. are largely traced to problems with the terminal sterilization of large volume parenterals in the early 1970s (unrelated events in the U.K. led to the first validation activities there at roughly the same time) (1). The FDA investigation led to the establishment of validation as a required activity for sterilization processes for all terminally sterilized parenteral products. It soon became apparent to observers in the SVP industry that validation of their sterilization processes was expected of them as well despite their use of more robust overkill sterilization cycles. This recognition by the larger SVP industry resulted in their mimicking the practices of the LVP firms that had pioneered sterilization validation. By 1976, when the FDA issued its proposed Good Manufacturing Practices for LVP (the never-approved 21 CFR 212 regulation) sterilization validation was expected to be performed essentially identically for both LVPs and SVPs (2). Most of the early SVP efforts were copies of LVP protocols, with the simple logic that if it was sufficient for terminal sterilization, it would be more than sufficient for sterilization of parts where cycle times were ordinarily much longer. The implications of this comparatively simple decision have handicapped sterilization validation for other than terminal sterilization ever since (see Chapter 12).

These initial validation efforts utilized evaluation equipment and approaches that are crude by today's standards. Precise measurement of temperature was difficult, and resolution was limited to what was discernable on a multipoint chart recorder. F_0 values could only be obtained by hand calculation using logarithms (pocket calculators, data loggers and personal computers were not yet available). Protocols, procedures, reports and other documents were produced with rudimentary word processors or in some cases ordinary typewriters. The inclusion of drawings, tables and other items were also restricted for the same reasons.

Validation at that time focused on the sterilization process; as the equipment being utilized for the process had been utilized for years in many cases, qualification of its performance via biological destruction and attainment of proper temperatures was considered sufficient proof of its acceptability.

At its onset, validation adhered closely to the principles of the scientific method as taught in basic science. Given a premise, an experimental design is established for confirmation of that premise with the expected results (those that would support the premise) defined. After completion of the experiment, the data is reviewed to establish whether the premise is supportable based upon the experimental evidence. In the practice of validation, the protocol establishes the premise including the predefined acceptance criteria, and the report documents the results of the evaluation.

Validation of formulation processes began to emerge near the end of the decade as the importance of content uniformity was recognized as a major consideration for product efficacy. Product validation varied considerably from firm to firm as there were no absolute requirements as in sterilization processes. Efforts were made to utilize retrospective validation, given the absence of meaningful prospective validation for all but the very newest products. These too were hampered by the limited availability of statistical tools.

Validation maintenance or revalidation was given little consideration during this period, as there was such a backlog of required validations that any consideration of repeat efforts was minimal. Rudimentary change control

Abbreviations used in this chapter: BFS, blow–fill–seal; CMC, chemistry, manufacturing, and controls; FDA, Food and Drug Administration; GAMP, good automated manufacturing practices; LVP, large volume parenteral; PAI, preapproval inspection; PAT, process analytical technology; PCs, personal computers; PDA, Parenteral Drug Association; PLCs, programmable logic controllers; PMA, Pharmaceutical Manufacturers Association; SVP, small volume parenterals.

Benefits of Validation

Increased Throughput
Reduction in Rejections and Reworks
Reduction in Utility Costs
Avoidance of Capital Expenditures
Fewer Complaints About Process Related Failures
Reduced Testing -in Process and Finished Goods
More Rapid / Accurate Investigations into Process Upsets
More Rapid and Reliable Startup of New Equipment
Easier Scale-Up From Development Work
Easier Maintenance of the Equipment
Improved Employee Awareness of Processes
More Rapid Automation

Figure 1 Benefits of validation.

was instituted, given the absence of spreadsheets and database software for information management.

In 1978, the PDA published the first of its numerous technical monographs which have helped to shape the subject of validation in a variety of areas (3). These documents have served a generation of practitioners and have proved invaluable as a guide to firms worldwide especially in areas related to parenteral products.

The future of validation in the industry was still unclear. Most firms assembled task forces to perform their initial validation studies. The author was told in early 1980, "Why are you accepting a job in validation; in a few years you'll have it all done and you'll need to find a new position!" Around the same time, others understood the need to continue validation, and the first permanent validation departments were formed.

1980–1990

In 1980, the author (then with E.R. Squibb and Sons) with Jean Yves Guillemoteau from Sandoz (now Novartis) and Mark Fitch of Schering-Plough formed a validation discussion group to discuss common concerns. As the head of validation at our respective sites, we wanted confirmation that our internal ideas and approaches for validation were consistent with those practiced elsewhere. We were enthusiastic about validation and we expected that our employers would realize tangible benefits through the completion of our efforts (Fig. 1). Through our discussions we strived to improve our understanding, improve our approaches and provide demonstrable advantages.[a]

The first facilities constructed with validation considerations from project onset began to appear around 1980. For the first time firms endeavored to provide facilities, systems and equipment that were "validatable." A direct consequence of this was the emergence of equipment and system qualification for these newly installed systems. Validation was no longer solely the province of the microbiologist, pharmacist, production and quality managers. Those charged with the design, construction, and start-up the facility had to address higher expectations than previously encountered. Equipment qualification was identified as the solution to this need, with verification of installation and confirmation of equipment performance in no-load circumstances.

The 1980s witnessed the emergence of the validation service companies. These began as independent organizations providing a variety of validation services to firms. Those firms faced with the qualification/validation of new facilities were among the heaviest users of outside assistance, as they sought to manage the extremely heavy workload of projects associated with a new facility.

A useful tool for thermal sterilization validation was the Kaye Digistrip which quickly became the preferred tool for temperature recording, and, equally importantly, F_0 determination in essentially real time. The personal computer became available early in the decade offering word processing to virtually everyone; spreadsheet, databases, and enhanced graphics were now available for all manner of documentation. This had a profound impact on validation as protocols, and reports were not only more easily produced, they were almost far more detailed (and perhaps regrettably substantially larger than before).

While the Digistrip was without question a vast improvement over the pen and ink recorders of the early 1970s, the PC may have been a mixed blessing. Documentation expectations increased several-fold, and while the size and attractiveness of reports certainly increased, it is unclear whether the scientific basis for validation activities had improved one iota.

In parallel with the emergence of the PC in the company offices, its cousin—the microprocessor—began to appear on the shop floor. Where once processing equipment had been controlled by relatively simple electromechanical devices, PLCs and other controllers were now being used to improve the reliability and sophistication of the control provided. This advance came with two big negatives: technologies beyond the understanding of many end users, and a perhaps excessive fear of the possible negatives of using a computer.[b] Regardless, industry users of industrial controllers recognized that these systems required validation as well. The PMA formed the Computer Systems Validation Committee in 1983 that delivered an industry perspective on computer systems validation in 1986 (4,5).

Computerized systems validation and the entire GAMP effort can be traced back to these initial efforts.

Beginning in 1983, the FDA developed its first draft guidance on Process Validation. This early draft included a comprehensive update of definitions and expectations for the first time and replaced a patchwork of earlier documents and inferences defining validation that had loosely evolved from FDA presentations, inspections and other sources. The initial draft of this mandated triplicate validation studies in support of validation. An industry suggestion to alter this to a "statistically significant number of batches" was quickly withdrawn, when it

[a] The group expanded rapidly and is still in existence serving much the same need it did more than 25 years earlier.

[b] Too many screenings of "Hackers" and "2001, A Space Odyssey" having perhaps instilled excessive fear of runaway computerized systems making our products!

was recognized that this would dramatically increase the workload of already beleaguered validation departments. The final guidance was issued in 1987 and included a requirement that process validation be performed in triplicate (6).

Mid-decade, the FDA experienced a series of problems with the quality of generic drugs, where the process as filed with the FDA and supported by the clinical data showed minimal (and in some extreme cases almost no) correlation to the commercial production process. The FDA's response to this scandal was the PAI program in which firms were challenged to provide data supporting the relationship between clinical and commercial scale processes (7). The result across the industry was substantially more rigorous production scale-up with validation of any changes as a centerpiece of the effort.

The decade also witnessed the beginning of the biotechnology industry as something distinct and something apparently different from pharmaceuticals.[c] With this came consideration of validation requirements in areas quite different from prior activities. How to validate fermentation, chromatography, ultra-filtration, and other processes were mainstream subjects among the biotechnology industry. The technical concerns may have been cutting edge, but the means to accomplish them were derived from the practices found in operations for small molecules.

Positive experiences with validation at the site level led several U.S. firms to address it proactively in other countries before the local regulatory authorities had considered it fully. The author participated in several major validation projects outside the U.S. where the goal was full adherence to U.S. practices to ensure product quality met corporate standards. Efforts in this regard were greatly accelerated after Union Carbide's Bhopal incident. Corporate validation standards were established by a number of firms to ensure product quality. In this environment it was common to speak of "a single standard of high quality worldwide."

Merger mania struck the global pharmaceutical industry late in the 1980s. Prior to that time, there were few dominant firms, the industry was largely populated with firms of roughly comparable size and few firms had more than a 2–3% market share worldwide. This all changed rapidly as firms combined their resources in an effort to achieve critical mass in research, greater presence in overseas markets, and all the profits from co-marketed drugs. One byproduct of this was the swift realization that there was substantial excess manufacturing capacity in the newly combined firms, which led to plant closures, divestitures, outsourcing and a very different operating climate than previously. Unfortunately, this also led to a major displacement of experienced personnel as firms offered separation packages in efforts to reduce their headcount rapidly. The availability of numerous contract providers led many firms to reduce the size of their validation departments, coincident with the need to relocate products to other facilities.

[c] This author never subscribed to that difference, and recent developments in regulation, operating practice, and business models appear to support that perspective.

1990–2000

In the post–merger climate, a substantial amount of qualification and validation activities are being defined and performed by contractors. This has unfortunate downsides as the core competency for validation is now largely external to the organization that is required to demonstrate it. Suppliers of validation services vary substantially in size and sophistication, and the heavy reliance on their expertise has significant impact. In an effort to streamline the development of projects, standardized protocol templates have become the rule rather than the exception. Protocols originally written for one design, process or application are modified slightly and applied to a different situation. The net result is, protocols are often of excessive size, full of "boilerplate," poorly focused and lacking in clarity. Thus this practice is deemed acceptable that has been commented upon by numerous individuals (8,9).

Early in the 1990s, validation services were increasingly being provided by organizations affiliated with the large engineering companies. Their goal was to integrate validation support, predominantly in the installation/operation qualification stages of a project, with the engineering and design effort. The intent of this was certainly positive, but had unintended adverse consequences. In some instances firms believed that the majority of their efforts should mimic the focus of the large service providers. Massive I/Q documents were developed documenting virtually every nuance of the equipment or system. While these may have been excessive (to this author they certainly are), the bigger concern was the loss of attention on the core process/product the equipment was intended for. One smaller provider was so bold as to state, "Our efforts will generate enough paper to bury the average inspector!" All of this led to increasingly bloated qualification efforts with little real support for what should always have been the critical concern the quality of the end product materials.

The FDA defined its expectations for sterilization validation in a comprehensive guidance document related to CMC submission requirements for sterile products (10). This presented the industry with a level of clarity that had not been previously available. This had the distinct advantage of defining what was specifically required as validation activities in support of their sterilization and sterility assurance related activities. It is noteworthy that the excesses observed in so many industry activities are not a part of FDA's guidance. That the industry is perhaps misguided in its emphasis on equipment related concerns, when the process itself is only minimally supported should be apparent after reviewing this document.

There certainly were other missteps along the way. Early in our validation discussion group sessions we had talked about validation of cleaning processes. It had long been apparent to us that cleaning was a process of great importance. Any hint of cross contamination of one product with another would result in rejection of the possibly contaminated materials. The challenge was to develop an acceptable level of contamination. In the discussion group, we had discussed our fears, but we never could openly discuss our cleaning procedures, cleaning limits or anything substantive about our efforts. The dam burst in 1991, shortly thereafter, and this author was invited to speak at an industry meeting on what had been heretofore an untouchable subject. The PDA began work on cleaning validation soon afterwards, and delivered the first industry consensus documents on the subject (11,12).

The success of the biotechnology industry in validating their processes impacted the traditional industry as well. If a 20-step biological process beginning at fermentation could be validated, surely a small molecule synthesis process could be addressed in similar fashion. The bulk pharmaceutical chemical portion of the industry that had largely subscribed to the notion that they were "too different" from dosage form manufacturing began to feel increasing pressure to implement validation as a routine requirement for their operations. The Q7A initiative undertaken by the International Conference on Harmonization ultimately laid rest to any further objections in synthesis operations (13).

2000 TO PRESENT

The most dominant theme of the first decade of the 21st century with the global industry has to be outsourcing. Copying the business model of the electronics industry, firms have pursued outsourcing as a sustainable business model. While contract manufacturers have existed in the global industry for many years producing specialty products such as BFS, pre-filled syringes, and soft gelatin capsules, the mergers and divestitures of the prior decade have created a number of firms whose primary business model is to produce drugs on a contract basis. Embraced initially by biotechnology firms that wanted to focus on their core technologies, contract manufacturers have enabled the establishment of the virtual pharmaceutical firm in which all of the production, testing and distribution is performed on a contract basis. This places greater stress on the performance of validation due to the added communication requirements, coupled with added confidentiality concerns. It is uncertain how the rapid growth of outsourcing will impact validation practices. There is little doubt that it stresses the communication channels of the involved organizations, and for that alone may have an adverse effect on validation practice. Chapter 48 of this volume provides an overview of the concerns associated with validation in contract manufacturing.

Perhaps the most profound impact on validation is that brought about by the FDA's changing perspective on product quality. Ajaz Hussain, who was then the FDA's Deputy Director, Office of Pharmaceutical Science, CDER, was perhaps the first to point out the lack of underlying science in many pharmaceutical processes. Dr. Hussain identified what he believed was a lack of real process understanding on the part of numerous firms, i.e., an apparent overreliance on end-product testing (and thus even validation itself) was supporting inadequately understood processes. He has been an advocate for firms to reemphasize robust product development founded on sound science with appropriately defined specifications, and process parameters. The appropriateness of this course of action can hardly be faulted. He went on to propose PAT as an on-line affirmation of process acceptability, reducing the need for in-process and end-product testing. The utility of PAT as a universal practice is uncertain. There are instances where it can offer clear advantages over classical drug manufacturing approaches; yet there are sound arguments that suggest it may be of little benefit in other situations. The passage of time will ultimately reveal the utility of PAT within the pharmaceutical industry.

Risk based compliance is another recent trend, and has manifested itself in a myriad of ways (14). Numerous organizations have initiated task forces to evaluate how risk based thinking can make firms more compliant while also increasing efficiency especially as it relates to the practice of validation. Early areas for risk based validation include equipment qualification, cleaning, environmental monitoring, inspection and others. If these efforts are effective, then some of the poorly defined and egregious qualification and validation efforts touted in the 1990s will be obsolescent.

CONCLUSION

If there is one constant in the history of validation, it would have to be the continual evolution of perspective, practice, approach, and emphasis. In many instances, the evolution of practice has improved the certainty of our knowledge and thus the quality of our products/processes. The advantages of the datalogger for use in thermal studies compared to chart recorders, and logarithmic calculation of lethality are obvious. Whether the same can be said about 85-page installation qualification protocols for a laboratory incubator is certainly highly questionable. The advent of validation practices derived from a risk analysis perspective offers the possibility to revisit the entire subject. That it will result in another wave of change in validation is perhaps certain, and considering the history of validation that is perhaps the only constant.

In this author's opinion, far too much of present day validation activities in the United States has been little more than rote adherence to ever increasing expectations. The sense that if a little validation is good then more must be better has gotten out of control. A return to the demonstrated need for scientific evidence prior to imposition of a new requirement is essential. If the FDA's risk analysis initiative fosters rethinking of validation expectations as well as the fundamental quality goals for a product/process, then future improvement in validation practice may be possible. An approach to validation that falls somewhere between the perhaps overly simplistic yet effective protocols of the 1980s and the present day bloated validation efforts seems appropriate.

REFERENCES

1. Chapman K. A history of validation in the United States: Part I, 15(10), 82–96 Oct. 1991.
2. FDA. Proposed Current Good Manufacturing Practices in the Manufacture, Processing, Packing or Holding of Large Volume Parenterals, Federal Register 22202–22219, June 1, 1976; Rescinded—December 31, 1993.
3. PDA. Technical Report #1: Validation of Steam Sterilization Cycles, 1978.
4. Chapman K. A History of Validation in the United States—Part II. Pharma Technol 1991; 15(11):54–70.

5. PMA Computer Systems Validation Committee. Validation concepts for computer systems used in the manufacture of drug products. Pharm Technol 1986; 10(5):24–34.
6. FDA. Guideline on General Principles of Process Validation, 1987.
7. FDA. Process Validation Requirements for Drug Products and Active Pharmaceutical Ingredients Subject to Pre-Market Approval (CPG 7132c.08, Sec 490.100), 2004.
8. Sharp J. Letter to the editor. J Parenter Sci Technol 1993; 47(1):2–3.
9. Agalloco J. Validation—a new perspective. In: Medina C, ed. Compliance Handbook for Pharmaceuticals, Medical Devices and Biologics. New York: Marcel-Dekker, 2004.
10. FDA. Guidance for Industry for the Submission Documentation for Sterilization Process Validation in Applications for Human and Veterinary Drug Products, November 1994.
11. PDA. Cleaning and Cleaning Validation: A Biotechnology Perspective. Bethesda, MD: PDA, 1996.
12. PDA. Points to consider for cleaning validation. PDA J Pharm Sci Technol 1999; 53(Suppl. 1) (PDA Technical Report 29).
13. ICH. Guidance for Industry Q7A Good Manufacturing Practice Guidance for Active Pharmaceutical Ingredients, 2001.
14. FDA. Pharmaceutical CGMPs for the 21st Century—A Risk-Based Approach. Final Report, 2004.

53

The European Approach to Validation—A Microbiological Perspective

Deborah E. Mentel
Pfizer Global Research and Development, Groton, Connecticut, U.S.A.

INTRODUCTION

In the overall history of pharmaceutical manufacturing, the need to validate is a relatively new addition to what is required by worldwide regulatory agencies. While the concept is only about 30 years old, the impact on our industry has been profound. The need to validate all aspects of the manufacturing process is a common requirement worldwide. As a manufacturer, it can seem complicated when trying to gain global approval for a product. A review of the historical background of U.S. and European regulatory requirements will not reduce what is required to gain approval, but may allow for a better understanding of the issues. This understanding may also allow for the development of approaches that will ultimately reduce the overall workload.

OVERVIEW AND HISTORICAL BACKGROUND OF U.S. AND EUROPEAN REGULATORY REQUIREMENTS

In the U.S.A., development of new Food and Drug Laws has always been the result of significant health and safety issues arising from unsafe drugs or practices associated with their distribution. The 1962 amendments to the FD&C act gave us the GMPs, which we understand today to be constantly evolving leading to ever more stringent standards for manufacturing. In September 1976 (1), the term validation first appeared in an update to the GMPs and in October of the same year, the FDA sent a letter to all manufacturers of injectable drugs indicating that validation applied to all types of injectable products. Finally, in 1978, the FDA's Compliance Program No. 7356.002, (2) chapter 56, Drug, Product Quality Assurance (Drug Process Inspection) defined validation as: "A validated manufacturing process is one which has proved to do what it purports or is represented to do. The proof of validation is obtained through the collection and evaluation of data, preferably, beginning from the process development phase and continuing through into the production phase. Validation necessarily includes process qualification (the Qualification of Materials, Equipment, Systems, Buildings, Personnel), but it also includes the control of the entire process for repeated batches of runs." After 1978, validation became an integral part of the pharmaceutical industry due to these FDA documents. Prior to that time, there was some effort underway regarding validation.

The development of the GMPs and validation has been interwoven in Europe as in the U.S.A. The U.K. Medicines Acts of 1968 and 1971 established the need for manufacturers to follow GMPs. The need for GMPs was prompted in the U.K. by thalidomide. Thalidomide played a key role in the development of U.S. GMPs as well. EC legislation takes precedence over the Medicines Act. Since 1995, the U.K. has been in alignment with the EC legislation. The need to validate a process is an essential part of GMPs. Validation is currently defined in the Orange Guide (3) as: "An action proving, in accordance with principles of GMP, that any procedure, process, equipment, material, activity or system actually leads to the expected results." It is further defined in Annex 18 as: "A documented program that provides a high degree of assurance that a specific process, method or system will consistently produce a result meeting predetermined acceptance criteria."

Validation is defined in the European Commission Guide to GMPs (4) as, "Action of proving, in accordance with the principles of GMP, that any procedure, process, equipment, material, activity, or system actually leads to the expected results."

While all of the definitions are slightly different, the general principles are the same, namely that there is a need to provide documented evidence that the process is in control and reproducible. Validation is of course a regulatory requirement as we have seen, but there are also practical reasons for a manufacturer's need to validate the manufacturing process. Some of these include overall knowledge of system capability, reduction of reworks and rejected product, and simplification of training efforts.

The rest of this chapter will detail the European approach to validation of various systems used when producing sterile products. In many cases, the approach will be very similar to U.S. expectations, but there are differences that must be accounted for and dealt with to gain marketing authorization for a product.

GENERAL PRINCIPLES OF EUROPEAN VALIDATION

Annex 15 to the EU Guide to GMPs (5) outlines the various steps involved in setting up a validation program

---

Abbreviations used in this chapter: DQ, design qualification; EC, European Community; EMEA, European Medicines Evaluation Agency; EU, European Union; FD&C, Food, Drug, and Cosmetic act; FDA, Food and Drug Administration; GMP, good manufacturing practice; HEPA, high efficiency particulate air filter; HVAC, heating, ventilation, and air-conditioning; IQ, installation qualification; OQ, operational qualification; PQ, performance qualification; SPVP, sterile process validation package; WFI, water for injection.

including development of a validation master plan. The plan should detail the "who, what, why and how" of the validation. Individual protocols will be prepared for each piece of equipment or system. The protocols will define how the validation will be done, the acceptance criteria, the requirement to write a formal report and the provisions for revalidation in the future.

The validation should address the following phases, DQ, IQ, OQ and PQ. Each phase is essential in assuring a well-controlled process. The following table will provide some of the key parameters to assess at each phase. Once these activities have been completed and well-documented, the piece of equipment or system is ready for use. When completing a large facility-wide project, it may be necessary to work on some systems sequentially while others will lend themselves to parallel efforts. It is important to remember that each system or piece of equipment must ultimately work together to produce quality product.

ASEPTIC PROCESSING

The use of aseptic processing alone to produce sterile product is becoming more challenging within the EC. Rigorous testing of your product must have been done prior to gaining marketing authorization. The current expectation is that products in their final container will be subjected to some form of terminal sterilization. This may be steam or radiation. Simply stating that the product will not withstand an overkill cycle is not sufficient justification to not subject the product to a less challenging cycle. This will be discussed in more detail in the terminal sterilization section.

In general, validation of an aseptic manufacturing operation begins with the design of the facility and ends with successful completion of media simulation studies. Qualification of the physical facility should be the first step. It is essential to demonstrate that the facility is capable of maintaining the required cleanliness levels required for each unit operation in the process. Once construction has been completed, the area should be thoroughly cleaned by trained aseptic operators. The cleaning will progress in phases as system testing allows. The first activities will involve testing of HVAC systems and balancing room air pressures and flow rates. Once these activities are complete, additional cleaning will occur and operators should begin wearing protective clothing, as the next steps will involve certification of HEPA filters followed by determination of air cleanliness levels for both viable and non-viable particulates. As in the U.S.A., Europe has defined room classifications and guidance as to what activities should occur in each area. Strict adherence to the guidelines is required. If activities occur in areas of lower air cleanliness than indicated, then the manufacturer will need a very strong justification for the deviation and data generated during validation of the entire process showing the final product quality will not be compromised. It is expected that monitoring will be continuous in Grade A and it is recommended for Grade B areas, as well. It is also expected that laminar airflow will be maintained and validated as part of the overall process of qualification. The FDA expects regular monitoring of critical areas and that unidirectional airflow will be demonstrated. A couple of other notable distinctions between the FDA and Europe related to air quality are the emphasis that at rest or static test results are as important as dynamic results and inclusion of monitoring for particles that are greater than or equal to 5 μ. The belief is that microbial contamination will be associated in many cases with larger particles. EC inspectors will expect data to be presented representing both conditions and limits. The following tables from the EC Guide to GMPs Annex 1 and the August 2004, Aseptic Processing Guideline (6,7) provide all of the information needed to ensure that the air systems are appropriate for their intended purpose (Tables 1–3).

Once the rooms have been classified and the area is under control, it is time to begin the environmental monitoring program. Table 4, which is taken from the EC Guide to GMPs Annex 1 (8), provides recommended limits. In the early stages of establishing an environmental monitoring program, rigorous monitoring of surfaces, air and people must take place. European inspectors will expect that in addition to surface and air monitoring, settle plates will be used. Great emphasis is placed on the use of settle plates and they should be part of the monitoring program for both the critical and less critical areas. While this has not been an expectation for the FDA for some time, it will be necessary to incorporate these plates into the overall program.

As data are collected and trended, there is an expectation that each firm will establish alert and action levels and take prompt corrective action when limits are exceeded. Annex 1 of the GMP Guide 8 defines these limits as:

Alert Limit--Established criteria giving early warning of potential drift from normal conditions, which are not necessarily grounds for definitive corrective action but which require follow-up investigation.

Table 1 Validation Criteria

Validation phase	Key parameters
Design qualification	Demonstrate design compliance to GMPs
Installation qualification	Assure proper installation
	Verify materials of construction
	Assure all operating manuals and certificates are available
	Determine calibration requirements
Operational qualification	Demonstrate the system works as expected
	Challenge the system to operating extremes
	Begin formal calibration, preventative maintenance and cleaning
	Develop standard operating procedures and train personnel
Performance qualification	Use of production materials to test the system at normal operating conditions

Table 2 Airborne Particulate Levels

	At rest[a]	In operation[a]
	Maximum permitted number of particles/m^3 equal or above[b]	
Grade	0.5 μm[c]	0.5 μm[c]
A[d]		≤3500
B[d]	≤3500	≤350,000
C[d]	≤350,000	≤3,500,000
D[d]	≤3,500,000	Not defined

[a] The particulate conditions given for the at-rest state should be achieved after a short clean up period of 15 to 20 minutes in an unmanned state after completion of operations. The particulate conditions for Grade A in operation should be maintained in the zone immediately surrounding the product or if an open container is exposed to the environment. It is accepted that it may not always be possible to demonstrate conformity with particulate standards at the point of fill when filling is in progress, due to generation of particles or droplets from the product itself.
[b] Particulate measurement based on the use of a discrete airborne particle counter. A continuous measurement system should be used for Grade A and is recommended for Grade B.
[c] The guidance given corresponds approximately to the cleanliness classes in EN/ISO 14644-1 at a particle size of 0.5 μm.
[d] The number of air changes should be related to the size of the room and the equipment and personnel present in the room. The air system should be provided with appropriate terminal filters.

Action Limit--Established criteria requiring immediate follow-up and corrective action if exceeded (Table 5).

In addition to trending the number of organisms present and responding to changes manufacturing work shifts, there is also an expectation that the bioburden will be adequately identified and appropriate corrective actions taken based on the types of organisms found.

As with the FDA, Europe places emphasis on well-designed personnel monitoring and gowning qualification program. The gowning program also needs to evaluate the materials used for the gowns, and their ability to contain contamination. The materials of construction are becoming a concern area for European inspectors. The expectation is that all operators will be monitored after performing critical operations.

The final phase in completing the validation of an aseptic facility is the successful completion of three process simulation (media fill) trials. The requirements for process simulations are very similar in U.S.A. and Europe. Three successful media fills per shift are required when qualifying a new facility and should be repeated at least twice a year or after significant modifications to systems or facilities. The simulation should mimic the normal manufacturing process as closely as possible. This will include a series of typical manufacturing interventions. Unlike the FDA guidelines, Annex #1 only specifies that a sufficient number of vials should be filled to properly evaluate the system. While both FDA and Europe have an expectation of no contaminated units, the wording in Annex #1 is slightly more tolerant than in the FDA guideline, with a rate of less than 0.1% with a 95% confidence limit. Both expect a thorough investigation of any contaminated units. One will note that the FDA guidelines provide far greater detail on all aspects of conducting the test such as growth promotion, incubation conditions, how to appropriately deal with damaged units pre-/post-incubation and evaluation of results, to name a few. One can be confident that if their process simulation program follows the FDA guideline, they are in compliance with European expectations.

Table 3 Examples of Operations for Aseptic Preparations

Grade	Examples of operations
A	Aseptic filling
B	Aseptic filling and preparations
C	Preparation of solutions to be filtered
D	Handling of components after washing

TERMINAL STERILIZATION/STEAM STERILIZATION

As mentioned earlier, the use of aseptic processing alone is of concern both in Europe and with the FDA. Europe has taken a much stronger position with regard to the use of sterile filtration as the sole means of sterilization. The preferred method is terminal sterilization in the final container. This may be accomplished in a number of ways such as exposure to steam or ionizing radiation. In general, products that are terminally sterilized may be processed and filled under less stringent air quality conditions than in typical aseptic manufacturing, for example, preparation in a Grade D area and filling in a Grade C area. If the product is at high risk of microbial contamination then it is expected that filling will occur under aseptic conditions prior to terminal sterilization.

During product development, the manufacturer should perform sufficient studies to support a decision not to terminally sterilize based on product incompatibilities. This requires looking at steam sterilization and radiation as well as modeling a variety of possible sterilization cycles (e.g., exposure time, temperature, radiation dose). Simply stating that the product will not withstand an overkill cycle is not enough. It is also necessary to demonstrate that terminal sterilization will result in significant product quality issues either initially or during the shelf-life of the product. It is recommended that a matrix approach be taken in setting up the studies and use of both accelerated and real-time stability studies be used.

Validation of a terminal sterilization cycle is conducted in a manner similar to any other sterilization cycle. The principles of validation outlined in the section entitled General Principles of European Validation apply. For example, when using an autoclave cycle, the autoclave will first be validated via heat distribution of the empty chamber, followed by challenge tests of a typical load of product. Minimum and maximum loads will be evaluated and processing ranges established. Once the cycle has been established, it should be re-evaluated on an annual basis.

As in other areas that have been discussed, there are some significant differences between the FDA and European approach to steam sterilization. Two of the more significant issues are steam quality attributes and the use of biological indicators. The FDA's position is that steam produced from WFI should be used for sterilization of product contact parts and when the steam comes into contact with pieces of equipment that will have direct

Table 4 Recommended Limits for Microbiological Monitoring of Clean Areas During Operation

Grade	Recommended limits for microbial contamination[a]			
	Air sample CFU/m^3	Settle plates (90mm dia) CFU/4 hr[b]	Contact plates (55mm dia), CFU/plate	Glove print 5 fingers, CFU/glove
A	<1	<1	<1	<1
B	10	5	5	5
C	100	50	25	–
D	200	100	50	–

[a] Average values.
[b] Settle plates may be exposed less than four hours.

contact with the product. The FDA also expects that during the PQ stage of validation, biological indicators as well as thermocouples will be used to evaluate the effectiveness of the cycle.

Steam quality is a major consideration in Europe for steam used in equipment and porous loads, but is not applied to terminal sterilization cycles. The attributes of interest are non-condensable gases and dryness fraction/value. All of these attributes and appropriate limits may be found in HTM 2010 (8), ISO 11134 (9) and EN 285 (10). All three reference documents apply the same limits and criteria for testing.

As steam condenses, gases may be released. These are non-condensable gases from entrained air and usually come from the steam generator feed water. They are in essence contaminates to the steam. These gases will act as air that has not been adequately removed at the onset of the cycle. A limit of 3.5% is recommended and is expressed as milliliters of gas collected per 100 mL of condensate. If this level is maintained there should not be any adverse impact to the load.

The dryness value is a measure of the amount of moisture carried with the saturated steam. The higher the dryness factor (less water), the more latent heat the steam will impart to the load. Water can act as an insulator and lead to poor heat transfer and slow-to-heat areas. It is recommended that a porous load have a dryness factor of more than 0.90 and equipment loads more than 0.95.

Biological indicators may be used in Europe, but do not replace the need for physical measurements. If they are used then strict procedures need to be in place to assure the manufacturing area is not inadvertently contaminated. Europe will require both steam temperature and pressure be monitored during the cycle.

REGULATORY PERSPECTIVE

The EMEA has responsibility for oversight and safety of all human and animal health medicines. The EMEA grants marketing authorizations for new products and approves requested changes to the marketing authorization. They represent the 25 member states of the European Union. One area of responsibility is GMP inspections. These inspections may be in support of granting a new marketing authorization (preapproval) or to assure ongoing GMP compliance for existing products. The inspections are coordinated by the EMEA and may be conducted by inspectors from any member state. The inspections are based on principles described in Directive 2003/94/EC. The inspector will also use EC GMP guidelines as well as the various Annexes to the EC GMP guide. The primary focus of the inspection will be to determine if the quality systems are adequate to assure the final quality of the product.

The inspection of an aseptic operation will follow a typical flow of activities generally starting with a plant tour that follows the normal flow of materials. The primary objectives of this tour are (*i*) to determine if the design layout and surfaces are sufficient to maintain the air cleanliness level and arrangement of equipment is suitable (i.e., appropriate movement between Grades A/B/C/D), (*ii*) to assure production personnel are appropriately gowned per site procedures, and (*iii*) to assure that operators are following written procedures while conducting their normal activities. The inspector may spend a significant amount of time observing critical aseptic operations such as fill line set up, replenishing stopper bowls and dealing with equipment jams. The areas with the most deficiencies found in recent inspections are documentation management, potential for contamination and personnel training issues. (This was

Table 5 Good Manufacturing Practice Inspectional Deficiencies

Document management	Contamination control	Personnel training
Lack of detail	Inadequate controls related to cleaning, drying and storage of equipment	Inadequate training to properly complete tasks
Limited or no trend analysis of environmental monitoring data	No defined processing time intervals between sterilization and use of equipment	Lack of understanding of GMPs and aseptic technique
Poorly written investigation reports	Inadequate environmental control	
No documentation to show corrective actions completed	Media fills do not include all operators	

compiled by some U.K. Pfizer people preparing for an in-house presentation. They reviewed a number of inspections and made these observations. Therefore no reference is available). Each of these areas is broken down in a bit more detail in the table below.

In general, the inspectional approach is similar to the FDA, but European inspectors typically spend more time addressing systems and are less interested in finding that one time something was not properly documented. Although if they find enough minor issues, it may become a major observation.

When making regulatory submissions to Europe they will be submitted to the EMEA for review and approval. There are proscribed templates for submitting documentation. In general, the level of detail required in a U.S. SPVP is not required for Europe; summaries are typically provided, but the assumption is that all of the supporting validation documentation will be present at the manufacturing site.

The last area to discuss involves pharmacopeial microbiology methods. Great strides have been made in recent years to harmonize a number of test methods, which greatly reduces the method validation and routine testing efforts. The Bacterial endotoxin test is harmonized. The sterility test is essentially harmonized and the minor differences are easily accommodated within one protocol and test method. The Antimicrobial Effectiveness test still has some significant differences related to sampling time points and evaluation of data. One point to note while the *European Pharmacopoeia* provides criteria A and B for injectable products, the expectation is that products will meet the more stringent criteria A. The only other test routinely used to evaluate lots for release is an in-process test to measure product bioburden. There is an EMEA requirement that the product bioburden must be less than 10 CFU/100 mL prior to final filtration (11). While the FDA does not have a specific limit, they will readily accept this number.

CONCLUSION

While we have addressed various differences and subtle nuances between the FDA and European expectations with regard to preparation and control of sterile products they can be dealt with in a reasonable manner. It takes careful planning and an in-depth understanding of the regulations, but an efficient program can be established that will meet all expectations. One major pitfall to avoid is having two "systems" in place, one for the FDA and another for Europe. While some firms fall into this trap thinking it will be more cost effective, in the long run it is not an asset. The operators must understand two systems and when to use them appropriately, which may cause confusion and errors. This type of approach will cause inspectors from both FDA and Europe to question your practices and motivations. We are moving towards harmonization and each year more of the gaps are being closed.

REFERENCES

1. Human and veterinary drugs: good manufacturing practices and proposed exemptions for certain OTC products. Fed Regist 1978; 43(190(Pt II)):45013–45089.
2. FDA Compliance Program. No. 7356.002, Chapter 56, Drug, Product Quality Assurance (Drug Process Inspection), October 1978.
3. Rules and Guidance for Pharmaceutical Manufacturers and Distributors 2002, Medicines Control Agency, 271.
4. European Commission Directorate III. The rules governing medicinal products in the European Union. Good Manufacturing Practices, Medicinal Products for Human and Veterinary Use. 1998th ed., Vol. 4, 153.
5. European Commission Enterprise Directorate-General, Final Version of Annex 15 to the EU Guide to Good Manufacturing Practice, Brussels, July, 2001:4–6.
6. European Commission Enterprise Directorate-General, EC Guide to Good Manufacturing Practice Revision to Annex 1, Title: Manufacture of Sterile Medicinal Products, Brussels, May 30 2003:1–3.
7. Guidance for Industry, Sterile Drug Products Produced by Aseptic Processing—Current Good Manufacturing Practice, U.S. Department of Health and Human Services Food and Drug Administration, Center for Drug Evaluation and Research (CDER), Center for Biologics Evaluation and Research (CBER) Office of Regulatory Affairs (ORA), September 2004.
8. Health Technical Memorandum 2010 Sterilization (1994), Part 3 Validation and Verification, Section 9 Steam Quality Tests.
9. ISO 11134. Sterilization of Health Care Products—Requirements for Validation and Routine Control—Industrial Moist Heat Sterilization, 1994.
10. EN 285 Sterilization—Steam Sterilizers—Large Sterilizers, Section 24 Steam Quality Tests, 1996.
11. EMEA, Note for Guidance on Manufacture of the Finished Dosage Form, effective date April 1 1996:4–5.

54

Japanese Regulatory Requirements

SECTION 1: PHARMACEUTICAL ADMINISTRATION AND REGULATIONS IN JAPAN

Tsuguo Sasaki
Musashimurayama, Tokyo, Japan

Fumi Yamamoto
Ministry of Health, Labor and Welfare (MHLW), Chiyoda, Tokyo, Japan

Pharmaceutical administration in Japan is based on various laws and regulations, consisting mainly of (*i*) Pharmaceutical Affairs Law, (*ii*) Pharmacists Law, (*iii*) Organization for Pharmaceutical Safety and Research Law, (*iv*) Blood Collection and Blood Donation Services Control Law, (*v*) Poisonous and Deleterious Substances Control Law, (*vi*) Narcotic and Psychotropic Control Law, (*vii*) Cannabis Control Law, (*viii*) Opium Law, and (*ix*) Stimulants Control Law. For the enforcement and management of these laws, detailed regulations are prepared by the government in the form of ministerial ordinances and notices, such as the Enforcement Ordinance and the Enforcement Regulations of the Pharmaceutical Affairs Law, and notifications are issued by the Director General of the Bureau or by the directors in charge of the Division in the MHLW.

Pharmaceutical Affairs Law

The Pharmaceutical Affairs Law is intended to improve public health through regulations required to assure the quality, efficacy, and safety of drugs, quasi-drugs, cosmetics and medical devices, and through measures to promote R&D of drugs and medical devices that are especially essential for health care. Modern pharmaceutical legislation originated in Japan with the enactment of the Regulations on Handling and Sales of Medicines in 1889. The Pharmaceutical Affairs Law was enacted in 1943 and has been revised several times since then. The current Pharmaceutical Affairs Law is the result of complete revisions in 1948 and 1960. Subsequent revisions have included (*i*) the reexamination of new drugs, the reevaluation of drugs, notification of clinical study protocols, and items required for sponsoring clinical studies in 1979, (*ii*) direct manufacturing approval applications by foreign pharmaceutical manufacturers, and the transfer of manufacturing or import approvals in 1983, and (*iii*) promotion of R&D of orphan drugs and priority reviews for such drugs in 1993.

In 2002, the Pharmaceutical Affairs Law was revised based on demands for augmentation of safety assurance in keeping with the age of biotechnology and genomics, augmentation of postmarketing surveillance policies, revision of the approval and licensing system (clarification of the responsibility of companies for safety measures and revision of the manufacturing approval system in accordance with international coordination) and a radical revision of safety policies for medical devices. The revised Pharmaceutical Affairs Law was partly enforced in 2003 and the remaining will be enforced in 2005. The main revisions concerning drugs are summarized in the following sections.

Provisions Enforced in July 2003

1. *Exemptions for Biological Products.* Regulations to be applied to biological products were reinforced to prevent CJD associated with dried dura, HIV infection through blood transfusion, and other infections with unknown viruses. Biological products manufactured using biological materials derived from human and animal tissues and cells are expected to be clinically valuable, and the development of new biological products will be further promoted in-line with the progress of medical science such as genomic research and regenerative medicine. Since biomaterials tend to have a high risk of contamination due to infectious viruses or other factors, safety assurance policies have been integrated to control the entire process from raw material collection to manufacture and postmarketing surveillance to ensure the safety of biological products.
2. *ADR reporting system by medical institutions, etc.* Regulations to be applied to safety measures were reinforced by specifying reporting of ADRs and infections by medical institutions and drug stores required by the Pharmaceutical Affairs Law.
3. *Matters related to clinical trials.* With the purpose of widely utilizing findings and achievements of clinical research, trials intended for approval application among those initiated by investigators or medical institutions were given the status of clinical trials as defined in the Pharmaceutical Affairs Law, and use of unapproved drugs or medical devices have been

---

Abbreviations used in this chapter: ADR, adverse drug reaction; API, active pharmaceutical ingredient; ATCC, American Type Culture Collection; BI, biological indicator; CFU, colony-forming units; CJD, Creutzfeldt–Jakob disease; EN, European standards; EO, ethylene oxide; EU-GMP, European Union good manufacturing practice; FDA, Food and Drug Administration; GCP, good clinical practice; GLP, good laboratory practice; GMP, good manufacturing practice; HEPA, high-efficiency particulate air; ICH, International Conference on Harmonization; IFO, Institute for Fermentation, Osaka, Japan; ISO, International Organization for Standardization; JCM, Japan Collection of Microorganisms; JP, *Japanese Pharmacopoeia*; MHLW, Ministry of Health Labor and Welfare; PAFSC, Pharmaceutical Affairs and Food Sanitation Council; PFSB, Pharmaceutical and Food Safety Bureau; PMDA, Pharmaceuticals and Medical Devices Agency; PS, pure steam; RO, reverse osmosis; SAL, sterility assurance level; SOP, standard operating procedure; TOC, total oxidizable carbon; UF, ultrafiltration; UF water, ultrafiltered water; USP, *United States Pharmacopeia*; WFI, water for injection; WHO-GMP, World Health Organization-Good Manufacturing Practice.

permitted. Clinical research initiated by medical institutions or physicians complying with the GCP for the purpose of approval applications can be conducted using unapproved drugs or medical devices under the clinical trial notification system.

Provisions to be Enforced in April 2004

1. *Matters related to the requirements for licensing manufacturing/distribution businesses and manufacturing business.* The current approval and licensing system has been reviewed to further strengthen postmarketing surveillance of drugs in keeping with the changing industrial structure and business system (e.g., company split up, complete contracted manufacture, and foreign contracted manufacture), and to promote international coordination. Previously, licenses were issued for individual manufacturing facilities owned by the approval holder under the premise that drugs and medical devices are manufactured by the approval holder itself. The new system separates primary business (manufacturing and distribution: business of delivering and selling products on the market) and manufacturing business (manufacturing) and grants manufacturing/distribution approvals under the requirements that product quality must be well controlled and postmarketing safety measures must be assured.
2. *Approvals and regulatory affairs.* A manufacturing approval application for drugs, etc. must be submitted by the manufacturing/distribution license holder who is totally responsible for the product to be sold. Procedures of manufacturing and quality control will become requirements for approval and, depending on the type of products, certain supplementary documents such as risk analysis and conformity to basic requirements must be attached. It is also one of the basic requirements that manufacturing facilities are inspected to assess the manufacturing capacity of the applicant, i.e., manufacturing method and product quality, according to GMP prior to the approval. With an aim to make approval reviews faster and more efficient, minor changes in approved items including names, dosage, manufacturing method, specifications of product quality, and indications are approved by the notification system instead of the current approval license system.
3. *Master file systems for drugs, etc.* Detailed information and data concerning manufacturing method of the drug substance and other raw materials are the intellectual property of the industry. To protect such property with respect to the manufacturer or manufacturer/distributor of a final drug product and to simplify regulatory reviews by the agency, the master file system is to be introduced to permit the manufacturer of the drug substance to register the name, manufacturing method, pharmaceutical properties, quality, etc. of drug substance.

Approval and Licensing of Pharmaceutical Products

Any person who intends to manufacture or import drugs in Japan shall obtain an approval from the Minister of Health, Labor and Welfare or prefectural governor. For drugs such as radiopharmaceuticals, biological products, and blood-typing antibodies, which require special precautions with respect to public health and sanitation based on Article 42 of the Pharmaceutical Affairs Law, the Minister grants the license. Before granting the approval, the central or prefectural authorities examine, on the basis of data submitted by the applicant, each product under application for details such as the name, ingredients and quantities, administration and dosage, indications, and adverse reactions. From April 2005, "manufacturing (import) approvals" has changed to "manufacturing and distribution approvals" and "manufacturing (import) licenses" has changed to "marketing licenses" as specified in the Pharmaceutical Affairs Law revised in 2003. Marketing licenses are issued after assuring that the applicant is able to manufacture or distribute the approved drugs, e.g., whether manufacturing or business facilities of the applicant have sufficient structures and equipment, manufacturing and quality control systems and human resources to properly deal with the approved drugs.

Approval Reviews of Pharmaceutical Products

Application forms for approval to distribute drugs are usually submitted to the prefectural authorities who forward them to the MHLW. When forms for new drugs are received by the PMDA for evaluation, a reliability review of the application data as well as GLP and GCP compliance reviews are undertaken by the PMDA. When the reliability and compliance of the data are confirmed, a detailed review is undertaken by a team of experts in the field concerned at the PMDA, and the team prepares a team review report. A new system, consisting of meetings of specialists, has been introduced for review and evaluation of new drug applications. These meetings, consisting of team reviewers and medical experts, focus on the discussion of key issues. The evaluation process followed by the PMDA is as follows:

1. Interview (presentation, inquiries, and checking)
2. Team review
3. Inquiries and checking
4. Report (1)
5. Specialists' meeting (includes at least three clinical experts)
6. Hearing (main agenda items and specialist committee participants notified to the applicant two weeks prior to meeting; presentation)
7. Follow-up specialist meeting
8. Report (2)
9. Report to the Evaluation and Licensing Division, PFSB, and MHLW

Finally, a report is submitted to the Committee on New Drugs of the PAFSC for review and discussion as required on the basis of the review report and sent to the MHLW where the Minister grants approval to the new drug (Fig. 1). In reviews of new drugs with new active ingredients, drug samples are requested for special reviews, and the specifications and testing methods are usually checked by the National Institute of Health Sciences or by the National Institute of Infectious Diseases in the case of biological products.

Drug approval reviews are normally processed in the order the application forms are received, but with this

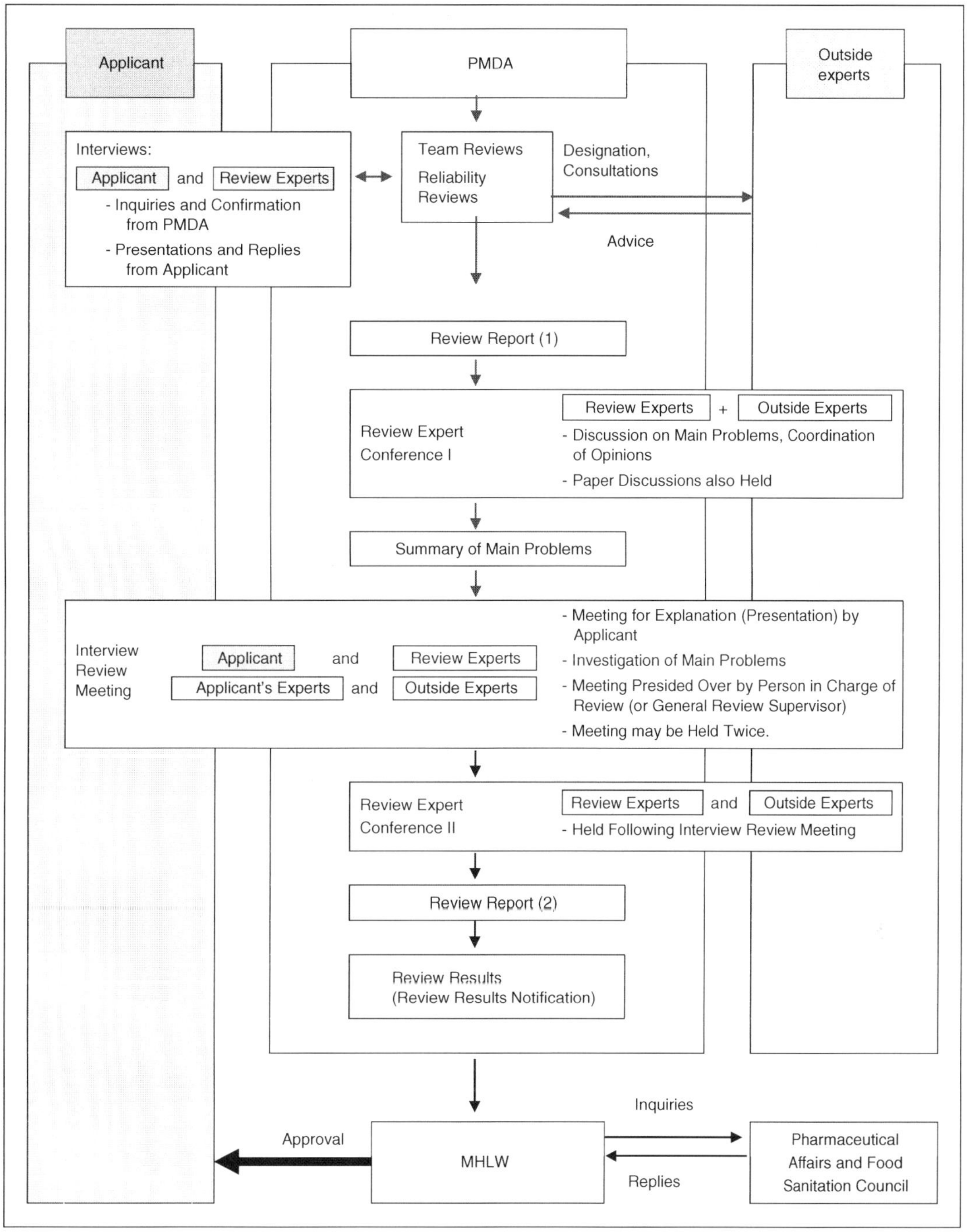

Figure 1 Flowchart of approval review.

system, applications are reviewed on a priority basis for drugs, which have been designated as orphan drugs and other drugs, which are considered to be especially important from a medical standpoint. The latter drugs include those indicated for serious diseases and those which are particularly excellent medically with respect to efficacy and safety when compared with existing drugs.

Historical Background of GMP

Proper control at the stage of drug manufacture is essential so that drugs can be supplied to patients with good quality. This means that the manufacturers and the buildings and facilities in the manufacturing plants must be appropriate so that drugs based on the approvals can be produced. The manufacturing process as a whole must be controlled on the basis of scientific principles and it is also necessary to assure the quality of drugs manufactured by taking measures to prevent errors during processing. Since a recommendation to introduce GMP was issued by WHO in July 1969, various countries have passed laws concerning control procedures essential for the manufacture of drugs. In Japan, Standards for Manufacturing Control and Quality Control (GMP) were issued in September 1974 and they were enforced from April 1976 with some exceptions. With the partial revision of the Pharmaceutical Affairs Law in October 1979, the GMP became legally binding. The control part of the GMP is specified in Drug Manufacturing and Quality Control Regulations (called "GMP software") in August 1980, and the parts concerning buildings and facilities are specified in the revision of the Regulations for Buildings and Facilities for Pharmacies, etc. (called "GMP hardware") in 1980 based on Article 13 of the Pharmaceutical Affairs Law. Thereafter, provisions related to validation, recall, self-inspections, and education and training were added and the revised Regulations for Manufacturing Control and Quality Control of Drugs (called "GMP software") and Regulations for Buildings and Facilities for Pharmacies, etc. (called "GMP hardware") were issued and came into effect from April 1994. Provisions required to assure the quality of biological products, including prevention of contamination by microorganisms, were added to the GMP software in 1997 and to the GMP hardware in 1999 since biological products require handling of animals and microorganisms in the manufacturing process and a high level of control in accordance with the features of individual products such as utilization of biological reactions. GMP software was revised to apply to some quasi-drugs in March 1999. To eliminate the risk of spreading infections from cell- and tissue-derived drugs and medical devices, requirements were added to both GMP software and GMP hardware in March 2001. The GMP was drastically revised in December 2004. In particular, the requirements for manufacturing control and quality control (so-called GMP software) were revised to contain items on manufacturing control and quality control specific to drug substances, sterile drug products, or biological products. In addition, the manufacturer is required to retain records on modifications and deviations from SOPs. These SOPs which are to be retained at each manufacturing facility are listed in Table 1.

Table 1 SOPs and Actual Operating Procedures Based on SOPs to be Retained at Each Manufacturing Facility

1. Hygiene control standards specifying procedures for maintaining cleanness of buildings and facilities, health of personnel at manufacturing plants, and other related matters; manufacturing control standards specifying procedures for storage of final products, control of manufacturing processes, and other related matters; and quality control standards specifying procedures for sample collection, assessment criteria for interpretation of test results, and other related matters
2. The following SOPs must be retained at each manufacturing facility in order to implement manufacturing and quality control properly and effectively
   a. Procedures for the management of shipment of products from a manufacturing site
   b. Procedures for validation
   c. Procedures for the management of SOP modifications
   d. Procedures for the management of deviations from SOPs
   e. Procedures for the management of information on quality and for handling poor quality products
   f. Procedures for product recall
   g. Procedures for self-inspections
   h. Procedures for education and training
   i. Procedures for archival storage of documents and records
   j. Other procedures necessary for the proper and effective implementation of manufacturing and quality control

Validation

It has been a long time since validation was introduced in the pharmaceutical industry. The U.S. FDA started evaluation and discussion on the need for validation in 1970s and issued the Guideline on General Principles of Process Validation in 1987. Later, validation was specified as one of the requirements for approval in Europe and Japan. Various guidelines and manuals have been published for implementation, and pharmaceutical companies developed and standardized methodology for validation. The concept and scope of validation have varied with time and currently validation has been extended to risk management. Validation and the GMP in Japan are concepts imported from the United States; however, it is an important issue in ICH and ISO and therefore has been specified as one of the requirements for approval from a global viewpoint.

Validation is part of GMP and a tool for achieving stable manufacture of high-quality pharmaceutical products. The Requirements for Manufacturing Control and Quality Control Methods in Pharmaceutical Plants (GMP software specifications) specifies that buildings, facilities, and manufacturing procedures, processes, and quality control methods of manufacturing plants must be properly validated and documented to lead to expected results. Detailed procedures for the implementation of validation were specified in a number of official notifications including an ordinance "Standard Methods of Validation" issued in 1995 and enforced in 1996. Validation plays an important role in securing the quality of medical products since its introduction as a legal system according to the Pharmaceutical Affairs Law in 1996. Most recently, Standard Methods of Validation was partly revised in 2000 (Table 2).

Article 13 of the GMP software, which was revised in 2004, requires validation be performed in cases where (*i*) a new medical product is manufactured at a certain manufacturing plant, (*ii*) modifications of manufacturing method have a major influence on the quality of medical products, or (*iii*) validation is considered to be necessary for proper conduct of manufacturing and quality control. The GMP software also requires the manufacturer (*i*) to establish a system of reporting plans and results of validation in writing to the quality control section, (*ii*) to take necessary measures when results of validation indicate the necessity of improvement in manufacturing and quality control, and (*iii*) to record and preserve outcomes of measures taken in the archives.

In the revised Pharmaceutical Affairs Law, manufacturing (import) approvals are scheduled to be replaced by manufacturing/distribution approvals and the

Table 2 History of Validation in Japan

1993: GMP software was specified as part of the requirements for manufacturing licensing
1994: GMP software specifications were defined
1995: Standard Methods of Validation were specified
1996: Standard Methods of Validation were enforced by law
2000: Standard Methods of Validation were revised
2003: The Enforcement Ordinance of the Pharmaceutical Affairs Law was issued
2004: GMP was radically revised
2005: The revised Pharmaceutical Affairs Law was enforced

Table 3 Enforcement Requirements for Validation Standards

Status of approved products
Products which have been licensed or which are to be given a manufacturing license (including license renewal) or subject to product addition (change) in the period from the date of issue of this notification until March 31, 1996, so far as they are intended for continued manufacture on and after April 1, 1996, shall be subjected to concurrent validation, revalidation and retrospective validation shown in Tables 4 and 5 (below) of the Validation Standards, in compliance with the following requirements
1. Concurrent validation
Placing importance on the fact that the prospective validation has not been conducted yet, an early confirmation shall be made on three lots of products as directed in the confirmation at an actual production scale. In case of a product which is not to be manufactured until a manufacturing license renewal, validation items shall be established with reference to a past record of production of similar drugs and entered in the operating procedures for validation
2. Revalidation
a. Revalidation for changes
Revalidation for changes shall be conducted in accordance with Table 2 in the case of a change in raw materials, labeling and packaging materials, procedures, manufacturing process and buildings and facilities made on and after April 1, 1996, so long as the change may affect the quality of drugs
b. Regular revalidation
If a trend analysis is impossible due to inadequate data from the concurrent and the retrospective validation, then the items for validation shall be mentioned in the operating procedures for validation
3. Retrospective validation
If data are inadequate for statistical analysis, then the procedures to collect data shall be described in the operating procedures for validation so that the validation is conducted on collection of adequate data

Standard Methods of Validation may be revised. Major items of the current Standard Methods of Validation are summarized in Tables 3–5.

SECTION 2: JAPANESE APPROACH TO VALIDATION

Toshiaki Nishihata and Tsutomu Hinomoto
Santen Pharmaceutical Co., Ltd., Higashiyodogawa, Osaka, Japan

Aseptic Processing

Interpretation for the manufacturing of sterile drug products by aseptic processing is described in the General Information section in the JP XV. The description has been prepared in light of the harmonization of the USP, EU-GMP, and WHO-GMP. As more detailed and specific explanation of the practice has been requested for operation, the "Manual of Manufacturing Sterile Drug Products by Aseptic Processing" (guideline) was published in April 2006, incorporating as a basic concept the General Information in JP XV and international harmonization, with the collaboration of a study group of the MHLW Health Science Study.

Including those who export their drug products to Europe and the United States, many manufacturers in Japan who produce sterile drug products by aseptic processing have actually been implementing the harmonized practices in-line with the content of the guideline. Therefore, this section provides a concise summary of the actual practice in Japan based on the draft guideline.

The basic concept of the validation of aseptic processing for manufacturing sterile drug products is that the manufacturing can be achieved in the amalgamation of hardware such as building facilities and manufacturing equipment and software of operation methods and controls. Qualification of manufacturing environment/equipment and process validation intend to secure the quality of drug products in their manufacturing processes, among which the assurance of sterility quality by scientific method and rationale should be considered as one of the critical matters for sterile drug products. The manufacturing processes of the sterile drug products produced by aseptic processing involve various contamination factors that cannot be assured during process development stages and/or designing stages of equipment and operational procedures, and thus qualification and validation need to be planned and implemented as an overall system of the production site. Aseptic processing such as sterilization and filling, maintenance of air classification in manufacturing environment and contamination risk in the facilities and equipment and/or manufacturing processes in the production site should be scientifically verified to assure that contamination has been prevented. It is also a basic requirement to control manufacturing processes with validated operational procedures and manufacturing control parameters.

Facility Design

In a manufacturing facility of sterile drug products produced by aseptic processing, the manufacturing areas are defined as clean areas classified into four grades as shown in Table 6 in accordance to the current draft Japanese guideline. The manufacturing areas of sterile drug products are clean areas that are controlled and maintained within the specified limits of contamination by microorganisms and airborne particles, and classified into critical processing areas, direct support areas, and indirect support areas depending on the nature of the operations being carried out. The classification of each area is generally specified with the number of airborne particles of not less than 0.5 μm per unit volume in the air of the environment.

A critical processing area (Grade A) is a manufacturing operation area in which sterilized containers and closures, raw materials, in-process products and the surfaces that have direct contact with them are exposed to the environment. After sterilization by filtration, it is recommended that the sterile drug products produced by the series of aseptic processing have all the aseptic processing from aseptic filtration to cap application carried out in the critical processing area. Likewise, the sterile drug products produced by the series of aseptic operation from the raw materials have all the aseptic processing from manipulation of the starting materials to cap application carried out in the critical processing area.

Table 4 Validation Requirements for Renewal of the Manufacturing License

		After receipt of manufacturing license, validations to be conducted by the time of renewal of the manufacturing license				
		Concurrent validation	Revalidation on change			
		Routine processing control	Facility qualification for changes	Calibration of measuring instruments for changes	Performance qualification for changes	Confirmation at actual production scale for changes[a]
Pharmaceutical products and bulk drug substances	Sterility and non-pyrogenicity[b]	○	△	△	△	△
	Other properties[c]	○	△	△	△	△

		Periodic revalidation			Retrospective validation
		Facility qualification when checked for maintenance	Calibration at meter inspection	Performance qualification	Statistical evaluation of past manufacturing control and quality control results
Pharmaceutical products and bulk drug substances	Sterility and non-pyrogenicity[b]	○	○	○	×
	Other properties[c]	○	○	×	○

Notes: ○, essential items; △, items which may affect the quality of drugs; ×, items not required to be reported.

[a] For a partial change in manufacturing approval, the following rules shall be followed: (*i*) Bulk products shall be manufactured when confirmation is made before permission of the partial change, (*ii*) Products shall be manufactured when confirmation is made after permission of the partial change.

[b] Buildings and facilities, procedures, processes, etc. to be checked for sterility and non-pyrogenicity.

[c] Buildings and facilities, procedures, processes, etc. to be checked for properties other than sterility and non-pyrogenicity which may affect the quality of drugs.

A direct support area (Grade B) is defined as the background environment for the critical processing area. It may be used for the manufacturing operations that require strict control of microbial and particulate contamination.

Indirect support areas (Grades C and D) may be where presterilized containers and closures, raw materials, and in-process products are exposed to the environment during the processing, and/or where cleaning of the apparatus and/or instrument for aseptic processing is carried out.

There are 11 general requirements that must be considered when designing these clean areas: (*i*) clean areas shall be distinctively separated from the one for full-time occupancy and/or in un-sanitized condition;

Table 5 Examples of Critical Process

Dosage form/quality specificity		Sterility	Content uniformity	Dissolution	Purity and crystal form
Sterile drugs	Terminal sterilized preparations	Sterilizing process	Dissolving process; mixing/dissolving process; filling process		
	Aseptically processed preparations	Aseptic operation; filtration process; filling process; freeze-drying process	Dissolving process; mixing/dissolving process; filling process		
Solid preparations			Mixing process; granulation process; tabletting process; filling process	Granulation process; tabletting process	
Liquid preparations			Dissolving process; mixing/dissolving process; filling process		
Ointment, suppository, poultices			Kneading process; filling process; spreading process		
Bulk drug substances					Final purification process
Sterile bulk drug substances		Sterilizing process; aseptic operation			Final purification process

Table 6 Classification of Clean Area

Category		Air classification	Maximum permitted number of particles per cubic meter equal to or above 0.5 μm	
			At rest	In operation
Aseptic processing area	Critical processing area	Grade A (ISO 5)	3,520	3,520
	Direct support area	Grade B (ISO 7)	3,520	352,000
Indirect support area		Grade C (ISO 8)	352,000	3,520,000
		Grade D	3,520,000	Depending on the nature of the operation being carried out

ISO classification in parentheses corresponds to the number of particles in the "in operation" conditions.

(*ii*) clean areas shall be distinctively defined for each operation and have adequate space; (*iii*) clean areas shall provide HEPA-filtered air to maintain air classification appropriate for the operation being carried out and have appropriate pressure differential; (*iv*) clean areas shall have positive pressure differential relative to adjacent rooms of lower air cleanliness. To maintain each clean area environment, it is important to achieve a proper airflow from areas of higher cleanliness to adjacent areas of lower cleanliness, i.e., pressure differential between aseptic processing area and indirect support area shall be sufficient so as not to cause inversion and/or backflow. Pressure differential of 10 to 15 Pa or more is recommended. Also, in indirect support area, it is necessary to maintain adequate pressure differential between the areas of different classification. The airflow in the critical processing areas shall be unidirectional with sufficient velocity and uniformity so as to promptly remove airborne particles out of the zones. To prevent ingress of contamination from adjacent areas (direct support area: Grade B), there shall be no backflow of the air from the adjacent areas; (*v*) clean areas shall have no direct access (excluding emergency exit) to the outdoor field; (*vi*) a system shall be implemented to monitor environmental conditions such as temperature, humidity, and pressure differential; (*vii*) clean areas shall be controlled in the temperature and humidity suitable for the characteristics of the materials and/or products and necessary for microbial control in the areas; (*viii*) a layout shall be considered so that the flow and control of personnel, products, materials, components and waste are optimized to minimize the intersection of each flow; (*ix*) a system or defined areas shall be determined so as to prevent mix-up of clean and unclean materials or sterilized and non-sterilized products; (*x*) separate and appropriate facility design shall be determined in the manipulation of sensitizing substance; and (*xi*) facilities shall be designed to facilitate cleaning and maintenance and receive periodical maintenance checks to secure the design intent. The guideline should be refereed for detailed explanations to fulfill the basic requirements.

Water Systems

Water used in drug products is a fine solvent widely used for manufacturing and processing of drug products, cleaning of containers and equipment and dissolution at use or testing of the products. However, it can be a source of impurities and microbial contamination in the drug products. Especially for manufacturing of sterile drug products, the water for pharmaceutical purposes, which should be supplied in compliance with the JP specification of water for pharmaceutical purposes (WFI, sterile purified water and so on), needs to be selected, retained, and controlled in accordance to GMP without fail so that any potential risks of contamination with impurities and microbial growth, contamination of products, and significant health hazard or medical accidents can be eliminated. Based on this standpoint, it is considered crucial, while supplying water for pharmaceutical purposes, to systematically establish a water system facility based on the sufficient design verification of both hard- and software including preventive measures of microbial contamination; validatates the system to assure the constant maintenance, control, and supply of the water of complying quality; and of assure the specified quality of water by routine monitoring.

In the basic design phase of the facility of water for pharmaceutical purposes, buildings and facilities and procedures or other methods regarding manufacturing and quality control should be clearly determined in advance in order to achieve constant production of the water of complying quality. There are five fundamental factors that haveto be considered: (*i*) the specification of the water (e.g., WFI), volume and control methods should be determined before the system is designed; (*ii*) the quality of source water, including seasonal variations, should be known before the system is designed; (*iii*) the maximum quantity consumed per second, operating time, frequency of use, conditions at the point of use (temperature, number of location, specification of piping) and so on should be determined to design the facility that will be capable of producing sufficient quantity and quality of water; (*iv*) water system basically employs circulating line (loop) when chemical disinfection is not feasible; the design should incorporate disinfection or sterilization consideration so as to assure microbial control; and (*v*) the location of sampling points should be discussed in the design phase and establish them close to various water processing equipment where the water needs to be evaluated for its quality.

Regarding the equipment used for pretreatment of water for manufacturing drug products, it should be considered before selecting and/or designing that contaminants in the feed water should be tested to produce the water with certain or higher quality desired in accordance to its intended use to ultimately comply with requirements, and to maximize the equipment's performance and life duration. In selecting the pretreatment equipment, consideration should be given to the indication of equipment contamination and cleaning procedures when contaminated as well as to the measures to minimize the influence of the contaminants' components such as iron, manganese, other heavy metals, free chlorine, organic material, microorganisms, suspension and colloidal particles (e.g., silicate, complex silicate, organometallic complex) on the function and life duration of the equipment.

WFI as water for manufacturing is expected to be of microbiologically high purity for its intended use, and the WFI system should be capable of undergoing periodical sterilization with PS at the temperature no lower than 121°C for a certain amount of time. When PS sterilization is not feasible due to heat tolerance, disinfection or sterilization with hot water or a chemical agent needs to be employed. For WFI processing, distillation, reverse osmosis, ultrafiltration and any combination of these are recommended.

In Japan, pharmaceutical water for sterile drug products (such as WFI) is expected to be used promptly after being processed so as to avoid microbial contamination and/or deterioration of chemical components in the water. However, it is often the case that the quantity processed and consumed are unbalanced and the water is generally held in the water tank for the meantime. When a WFI holding tank is designed and/or used, consideration should be given that (*i*) the tank is tightly sealed and have smooth inner surface, (*ii*) the number of convexity and opening on the tank surface should be requisite minimum and as short and few as possible, (*iii*) the tank should be of the structure that has no static area of water, facilitates cleaning inside and drains the water out completely, (*iv*) a hydrophobic vent filter of 0.2 μm should be installed at the vent so as to prevent ingress of microorganisms and impurities, and (*v*) in case of disinfection with hot water, a system should be added so that the whole inner surface including the ceiling will be exposed to the heat.

The water for pharmaceutical purposes held in the holding tank is distributed to the points of use through the piping in the water system. As the piping is a sealed system with relatively small diameter, its inner condition may be difficult to confirm once installed; it is thus recommended that the control method and preventive measures be discussed and determined in the design developing phase. Recommendations are given to consider that the piping should be a one-way loop with a preventive structure of backflow and no employment of by-pass and/or branching tubules as much as possible, and that in order to prevent microbial and organic material in WFI, it should be heated to, for instance, 80°C or higher (the temperature should be established based on validation results) and constantly recirculated at the flow rate of not less than 1.0 m/sec. Valves connecting the loop and branch should be located as close to the loop as possible to prevent "dead-legs"; the distance from the main pipe should be no longer than 6*D* in principle and no longer than 3*D* as a preferable target. Horizontal pipes are sloped not less than 1 per 100 to prevent a stagnant pool of water, and drain outlets should be installed to facilitate drainage of water as well as a structure to prevent the backflow.

For heat exchangers, double tube or double tube-sheet design is employed as a method for preventing contamination by leakage. When other methods such as a plate design are to be used, it is recommended to always keep higher pressure on the clean fluid side so as not to cause contamination in the clean fluid by cooling vehicle, and provide gauges to monitor the pressure differential.

Recommendations regarding points of use and sampling include that when sampling at the point of use is not feasible, the sampling point should be located in as much vicinity as possible, and that frequency of sampling at each sampling point should be determined considering water quality, quantity consumed, seasonal variations and other factors.

Valves, gauges, and detectors installed in the water system need to be of sanitary structure such as diaphragm and should have no static area of fluid or dead section. In order to ensure timely monitoring of the chemical quality of the water, installation of a TOC gauge (a model capable of measuring conductivity at the same time will be preferable) in the line is desirable. As for setting up the detector, it is recommended to select a spot where the water quality would be regionally the worst in the piping system.

Pumps should be of sanitary structure and be capable of sealing off to prevent contamination. Hot water disinfection and/or PS sterilization should be taken into consideration.

Environmental Monitoring/Product Bioburden

Environmental monitoring mainly intends to maintain the cleanliness in the manufacturing environment provided for sterile drug products, in that the microbial and particulate counts are controlled so as not to exceed the levels required for aseptic processing areas and indirect support areas, that any sign of deterioration in the environment is anticipated to prevent contamination to products, and that the efficiency of sanitization, decontamination and disinfection activities for maintenance of the cleanliness is continuously evaluated. Environmental monitoring has two major aspects: microbiological control and particle control. The purpose of microbiological control is to scientifically estimate the bioburden of the environment that it intends not to identify all of the microorganisms possibly existing in the environment but to assure that the sterile drug products have been manufactured under properly controlled conditions and to implement any processing (e.g., disinfection) as appropriate to maintain such environment. Monitoring will be conducted for microorganisms and airborne particles, and target particles are defined as airborne particles of not less than 0.5 μm. However, for more sufficient environmental monitoring, other particle size (e.g., 5 μm) may be included as appropriate. Target microorganisms are defined as bacteria and fungus and include those of airborne and surface of the wall, floor, fixtures and manufacturing equipment, and personnel garments.

Environmental monitoring is conducted in critical processing areas (Grade A) and direct support areas (Grade B) in the aseptic processing areas. Indirect support areas (Grades C and D) adjacent to the aseptic processing areas may be included as appropriate.

Table 7 shows the frequency of environmental monitoring. As contamination risk of sterile drug products may vary depending on the type and volume of the drug products to be manufactured as well as the

Table 7 Frequencies of Microbiological and Particle Monitoring

	Microorganisms			Particles	
Area level	**Airborne microorganisms**	**Surface microorganisms (equipment)**	**Surface microorganisms (personnel)**	**Processing**	**Non-processing**
Grade A	Every shift	At completion of each processing operation	Every shift	During aseptic processing	Per day
Grade B	Every shift	At completion of each processing operation	Per working day	During aseptic processing	Per day
Grade C	As appropriate	As appropriate	As appropriate	Per month	Per month
Grade D	As appropriate	As appropriate	As appropriate	As appropriate	As appropriate

environmental equipment such as air handling system, a monitoring program should be prepared and implemented according to the need and as appropriate. It is also recommended that the monitoring frequency in Grades C and D area, the indirect support areas, be determined in accordance to the process or operations being carried out. Recommended alert and action limits of microbiological monitoring are shown in Table 8.

Process Simulations

Sterile drug products may be manufactured through a single or several sterilization processes or a combination of sterilized components; an aseptic filling process is one of the manufacturing processes of drug products purporting to be sterile. In order to evaluate the propriety of sterility assurance of the drug products, the whole aseptic processing have to undergo process validation. Process simulation is one of the validation methods to evaluate not only the filling process but also the whole aseptic manufacturing processing, using media or other microbiological growth materials instead of actual products. Included as a scope are manufacturing process of sterile API and/or sterile in-process products and the overall manufacturing processes of drug products purporting to be sterile.

The operating personnel, operating environment, and processing operation should also reflect the actual manufacturing process, including worst-case conditions. The necessary information for conducting the tests should be referred to the guideline that incorporates the General Information of JP XV as the basic concept.

The number of units filled during a process simulation test should consider the duration of runs; it is generally recommended to determine based on the batch size, preferably the size of 5000 units (g, vials, etc.) or larger.

In consideration of process simulation testing, it is recommended that the whole aseptic processing be simulated, for there may be a risk of contamination in the processes other than filling. Therefore, when a process simulation test is conducted, it should be so planned that all contamination factors assumed in normal operations are included based on the identification of potential contamination factors. The guideline with the basic concept of the General Information of JP XV recommends to conduct process simulation tests considering the following five points: (*i*) all permitted interventions and events should be simulated based on the chart identifying both permitted and non-permitted interventions and events that may happen during the aseptic processing; (*ii*) the duration of the process simulation run should be adequate so as to include most of the manipulations normally performed in actual processing; (*iii*) process simulation tests should be conducted with the permitted interventions and events normally performed in actual processing conditions that include the longest

Table 8 Examples of Alert and Action Limits of Microbiological Monitoring

Target	Grade	Sampling point	Action limits
Airborne particles	A	Air	Less than 1 (CFU/m^3)
	B	Air	10 (CFU/m^3) or less
	C	Air	100 (CFU/m^3) or less
Surface microorganisms	A	Equipment	Less than 1 (CFU/plate)
		Wall	1 (CFU/plate) or less
		Floor	5 (CFU/plate) or less
	B	Wall	5 (CFU/plate) or less
		Floor	10 (CFU/plate) or less
	C	Floor	30 (CFU/plate) or less
Microorganisms on hands/fingers	A	Hands/fingers	Less than 1 (CFU/5 fingers)
	B	Hands/fingers	5 (CFU/5 fingers) or less
	C	Hands/fingers	As appropriate
Surface microorganisms on personnel garments	A	Sampling at both arms, breast, head, and shoulders	Less than 5 (CFU/plate)
	B	Sampling at both arms, breast, head, and shoulders	20 (CFU/plate) or less
	C	Sampling at both arms, breast, head, and shoulders	As appropriate

Alert limits should be determined in the level of mean $+2\sigma$ (σ, standard deviation) based on the performance qualification and trend analysis of past data.

and worst-case conditions. (e.g., maintenance of line stoppages, repair and/or replacement of equipment used in aseptic processing, replacement of filters in the line, the number of personnel involved); (*iv*) the duration of the simulation of the actual processing operations should consider possible events that may occur during the longest operation hours; and (*v*) consideration should be given to intermission of the line by any activities associated with normal aseptic processing operations.

The acceptance criteria of process simulation basically employ "no positive result." However, when the result of 5000 units/run in three consecutive runs is 0.05 or less, and the sum of the positive results in the three consecutive runs is three or less, it can still be considered complying provided that the source of contamination is investigated and eliminated in a controlled manner.

During initial validation when a process is newly established, it should be confirmed that the results of three consecutive runs comply with the criteria. In periodical validation (two per year as a principle except for the line for multiple use with partially different processing, in which case a process simulation run for each product should be required), the result of one process simulation run is employed. A positive result of the periodical validation however should call for investigation and revalidation of the process.

Terminal Sterilization

The chapter Sterility Assurance of Terminally Sterilized Drug Products in the General Information of JP XV describes the recommendations regarding terminal sterilization, of which contents have been referred to ISO standards and USP requirements. It states as the basic concept that the drug products to which terminal sterilization can be applied should generally undergo the sterilization condition such that a SAL of not more than 10^{-6} can be obtained. The SAL of not more than 10^{-6} should be judged by validation of sterilization process using physical and microbial methods, but not by sterility testing of sterilized products. Included as general requirements are the validation requirement of sterilization process, microbial control program, sterilization indicators, change control, etc. Important control factors which may affect the selected sterilization method are provided as recommendation. Steam sterilization method will be explained in a separate section; in radiation method for instance, important control factors as recommended are exposure time, absorbed dose and load configuration for gamma rays, and electron beam characteristics, exposure time, absorbed dose, and load configuration of products for electron beam and X-rays.

The terminal sterilization cited in the General Information of JP XV includes steam sterilization method and dry-heat method as heat methods, EO method as gas method, and radiation method and microwave method as irradiation methods, among which the steam sterilization method of the heat methods is most widely employed as terminal sterilization of drug products in Japan. The reason mostly lies upon its capability of maintaining stability of API. Besides, in terms of safety assurance of drug product, the gas method and irradiation method both pose possibility of complex degradation products other than heat decomposition that will likely call for enormous amount of testing for identification of the degradation products as well as justification of safety qualification; it thus has not been popular among pharmaceutical development companies.

It is recommended that the propriety of sterilization by terminal sterilization methods should be judged by employing an appropriate sterilization process control and using a sterilization indicator suitable for the selected sterilization method. In dry-heat method and/or gas method for instance, *Bacillus subtilis* (strain name: ATCC9372, IFO1372) is recommended as a sterilization indicator.

While the basic concept of terminal sterilization is provided in the General Information of JP XV, more detailed and specific explanation of the practice has been asked for operation; to this end, "Manual of Manufacturing Sterile Drug Products by Terminal Sterilization" (guideline) incorporating, as the basic concept, the General Information of JP XV and international harmonization, has been published in April 2007, with a collaboration of a study group of MHLW Health Sciences Study. The guideline of terminal sterilization will mainly focuses on steam sterilization method of heat methods but includes some items regarding other terminal sterilization methods such as irradiation method as well.

Steam Sterilization

Steam sterilization is a sterilization method to kill microorganisms by steam under pressure. While a majority of Japanese pharmaceutical industry employs the saturated steam sterilization method whereby the subject to be sterilized will be directly exposed to saturated steam, there is also a sterilization method with unsaturated steam in which the fluid in the direct container will be given moist heat energy from outside. In steam sterilization, a sterilization chamber is saturated with steam at appropriate temperature and pressure and heated for predetermined amount of time so as to kill the microorganisms. Important control factors which may affect sterilization are thermal history (generally indicated as F_0), temperature, steam pressure, exposure time, load configuration/density, and other necessary factors dependent on the product, all required in routine sterilization process control. Therefore, the temperature, steam pressure and exposure time are to be monitored continuously, and they should be included in the specifications of the sterilizer.

Use of BIs

Sterilization indicators are used to control sterilization process or as indicators of sterilization process; in steam sterilization, use of a BI is recommended as a sterilization indicator. BI, in other words, is used to indicate the propriety of the steam sterilization method concerned.

A BI is prepared from specific microorganisms resistant to the specified sterilization process and is used to determine the condition and control of the sterilization process. There are dry type BI and wet type BI; the dry type BI is classified into two kinds. In one, filter paper, glass or plastic are used as a carrier to which bacterial spores are added, dried and packaged. In the other, bacterial spores are added to the products or similar products and dried. Packaging materials of BI should show good steam penetration in steam sterilizations. It should be confirmed that any carrier does not affect the *D*-value of the spores. In the case of a liquid product, it is also acceptable to use the wet type BI, spores of which are suspended in the same solution as the product or in a solution showing an equivalent effect in the sterilization. However, when the spores of the specific microorganisms are suspended in liquid, it is necessary to ensure that the resistance characteristics of the spores are not affected due to germination.

In the General Information of JP XV, *Bacillus stearothermophilius* (strain name: ATCC7953, IFO1737, JCM9488, ATCC12980, IFO12550, JCM2501) is a typical example of the specific microorganism recommended for verification and control of the steam sterilization method. In addition to this microorganism, other microorganisms with the greatest resistance to the steam sterilization procedure, found in the bioburden, can be used as the BI. *D* value of BI needs to be controlled in that the *D*-value is generally determined by the survival curve method and the fraction negative method. Marketed BIs with recommended microorganisms are often used in Japan; in such cases, it is usually unnecessary to determine the *D*-value before use provided the *D*-value indicated on the label that has been determined by a standardized BI evaluation resistometer under strictly prescribed conditions is in accordance to ISO standard (ISO 11138–1).

As for a setting up procedure of BIs, a dry type BI is placed at the spot least affected by the sterilization procedure in the product or a suitable and similar product showing an equivalent effect in sterilization. In case of a wet type BI, spores of the BI are suspended in the same solution as the product or in an appropriate similar solution, and placed at the spot least affected by the sterilization procedure.

Soybean casein digest medium is generally used as culture for BI. General culture conditions are 55°C to 60°C for seven days in the case of *B. stearothermophilius*.

Steam Quality

Steam used in the steam sterilization should have the quality (specified based on the quality of flocculated water) which has been predetermined and maintained by the manufacture. It is thus often the case that the manufacturer also predetermines the quality of the source water for steam generation. The quality of steam generally requires not containing impurities that may deteriorate sterilization process, cause damages to sterilization equipment and/or affect the subject to be sterilized. The factors which need criteria for flocculated and source water of steam generation include evaporation residue, silica, iron, chlorine, phosphorus, cadmium, lead, other heavy metals, conductivity, pH and appearance, all of which are employed in ISO 13683 and EN285. There are also recommended criteria in that the steam should contain non-condensable gas not more than 3.5% in volumetric ratio and degree of dryness 0.95% or more. It is also recommended that the fluctuation range of steam pressure before the decompression valve of sterilizer should be within 10% and the decompression ratio should be 2:1 or less.

In order to assure the quality of steam for the steam sterilization procedure as well as to maintain the quality of the source water for steam generation, water of high purity such as UF water is generally used for the source water for steam generation in Japan.

Overkill vs. Bioburden Cycle

As for determination of sterilization conditions using microorganisms as indicators, there are overkill method, half-cycle method, combination of bioburden and BI, and absolute bioburden method, among which the overkill method and half-cycle method are popular in Japan. One of the reasons for such preference is that the overkill method and half-cycle method are easy in control and operation necessary for fulfilling the requirements explained hereafter. For those drug products not applicable for the overkill method and/or half-cycle method, aseptic processing is generally employed in Japan.

According to the General Information of JP XV, in the overkill method, it is assumed to conduct sterilization under the condition giving a SAL of not more than 10^{-6} regardless of bioburden count in the subject being sterilized or the resistance of the objective microorganisms to the sterilization. It is generally so defined that the method employs a BI with known count of recommended microorganism of 1.0 or more *D*-value and the sterilization condition providing 12*D* reduction or equivalent of the BI. The half-cycle method is defined as the one that, regardless of bioburden count in the subject being sterilized or the resistance of the objective microorganisms to the sterilization, employs a sterilization time of twice as long as that required to kill all of 10^6 counts of recommended organisms in the BI.

Absolute bioburden method is also defined in the General Information of JP XV that the sterilization conditions are determined by employing the *D*-value of the most resistant microorganism found in the subject to be sterilized or environment by the resistant estimation to the sterilization procedure, and being based on the bioburden count in the subject to be sterilized. As the bioburden count, a count of mean bioburden added three times of its standard deviation obtained by extensive bioburden estimation is generally employed. When the procedure is used, it is required to make frequent counting and resistance determination of detected microorganisms to the sterilization in daily bioburden control. In the combination of bioburden and recommended BI, it is so defined that a count of mean bioburden added three times of its standard deviation obtained by extensive bioburden estimation is considered as the maximum bioburden count, and the sterilization time is calculated with the BI based on an objective SAL. When this procedure is used, it is required to make frequent counting and resistance determination of detected microorganisms to the sterilization in daily bioburden control. When the bioburden estimation found a more resistant microorganism than the BI spore, the microorganism should be used as the BI. In other words, bioburden method requires daily microbiological control (microbial count and strain) to understand the microbial variation, and the initial validation that has assumed the variation factors and periodic validation are also required along with occasional revalidation in the incidence of unpredictable variation, which altogether increase the workload eventually.

In determination of comparative merits and demerits of the overkill or half-cycle methods and the bioburden method, it may be indicated there are very few cases, with the manufacturing process under the current Japanese GMP regulation, that products are being manufactured in the environment of poor quality, and thus the bioburden count should be adequately low. For that indication, the discussion in Japan has been divided: in one, the validation of sterilization process should be established based on a strict bioburden control in that the strains of the resistant microorganisms and bioburden count are determined; and in the other, such strict control is not necessary, for the conditions defined in the overkill method and/or half-cycle method will be adequate provided the environment is maintained under GMP. In current situation in Japan, the decision depends on each manufacturer; resulting in the overkill method and half-cycle method to be generally employed.

55

Managing Validation in a Multinational Company

Leslie M. Snyder, Matthew T. Lord, and Joshua D. Morton
Eli Lilly and Company, Indianapolis, Indiana, U.S.A.

Validation is defined as a documented program that provides a high degree of assurance that a specific process, method, or system (equipment, facilities or utilities) will consistently produce a result meeting predetermined acceptance criteria. While this can be a challenging endeavor at a single site, when multiple sites and locations are involved the challenge becomes even greater. To be successful with validation in a multinational organization, it takes a well-defined strategy that is developed, executed and supported by thorough documentation, communication and training to yield the consistency needed between global sites. A multinational company must ensure that it can repeatedly produce the same quality products at each manufacturing site so that the worldwide market is supplied with the same quality medicine regardless of the manufacturing site.

This chapter discusses the types of validation exercises, how validation strategies are developed, executed and monitored, how remediation efforts are implemented, and how these elements are controlled throughout multiple global sites. One approach that can be used globally to maintain control and capability from molecule development through manufacturing of the finished product is illustrated in Figure 1. Figure 1 is a model that represents a set of shared deliverables, standards, and expectations for assuring reliable supply, throughout a molecule's life, where monitoring and continuous learning are used to drive process improvements. Capability is defined as meeting all regulatory and internal requirements and control is defined as a process that is stable and predictable. A process that is capable and in control is a process that reproducibly meets expectations.

Managing validation in a multinational company all starts with consistency in the organizational structure of each manufacturing site and how these various teams interact and communicate to ensure a smooth execution of each element of the Process Control and Capability Cycle. Personnel are an integral part of this cycle where training and communication through documentation and/or discussions is a must. To align execution of the elements, GQS should be developed and enforced across all sites. The GQS are a compilation of a set of minimum standards to be met by all sites to achieve reproducibility for each Market. These standards can be divided into six different categories—(*i*) Quality, (*ii*) Systems (equipment, facilities, utilities, and computer systems), (*iii*) Materials, (*iv*) Production and Sterility Assurance, (*v*) Packaging and Labeling, and (*vi*) Laboratory—that together make up the Quality System. Personnel at each site in the global network should be trained on appropriate standards, so that interpretation of the standards and execution of a particular function at one site is consistent with the execution of that function at another site. To further this alignment, individuals identified as Global Molecule Stewards, who have a significant knowledge base for each individual molecule, finished product, or device manufactured, should be identified and held accountable for ensuring consistency between sites. The Global Molecule Steward has the responsibility of understanding all aspects of the assigned API or molecule and/or finished product. This includes understanding the equipment used, process flow, analytical properties, analytical methods, specifications, packaging and flow of materials and people involved in the manufacturing process at each global site. This also includes a thorough annual review and data analysis (global product assessment) of the individual processes from each manufacturing site to identify any trends, inconsistencies and opportunities for continuous improvement across the supply chain for an individual molecule or product.

At the site level a Site Molecule Steward should be identified for each molecule manufactured at the site. The Site Molecule Steward is responsible for the processes performed at a particular site for the assigned molecule/finished product and is a resource for the product Technical Service representatives, Quality representatives, and manufacturing personnel that support the manufacturing process.

The Global Molecule Stewards working with the Site Molecule Stewards ensure global alignment.

The remainder of this chapter describes in detail each of the elements listed within the process control and capability cycle, and how each element interacts with one another.

Abbreviations used in this chapter: API, active pharmaceutical ingredient; APR, Annual Product Review; CAPA, corrective and preventive action; CFP, Criteria for Forward Processing; CFR, Code of Federal Regulations; CPP, critical process parameter; CQA, critical quality attributes; DHR, Development History Report; DI, direct impact; DOE, design of experiment; DQ, design qualification; FDA, Food and Drug Administration; GMP, good manufacturing practice; GPLOT, Global Post Launch Optimization Team; GQS, Global Quality Standards; II, indirect impact; IQ, installation qualification; ISPE, International Society of Pharmaceutical Engineering; MS&T, Manufacturing Science and Technology; NI, no impact; OQ, operational qualification; PAR, proven acceptable range; PFD, process flow document; PQ, performance qualification; PV, process validation; SISPQ, safety, identity, strength, purity and quality; VMP, Validation Master Plan.

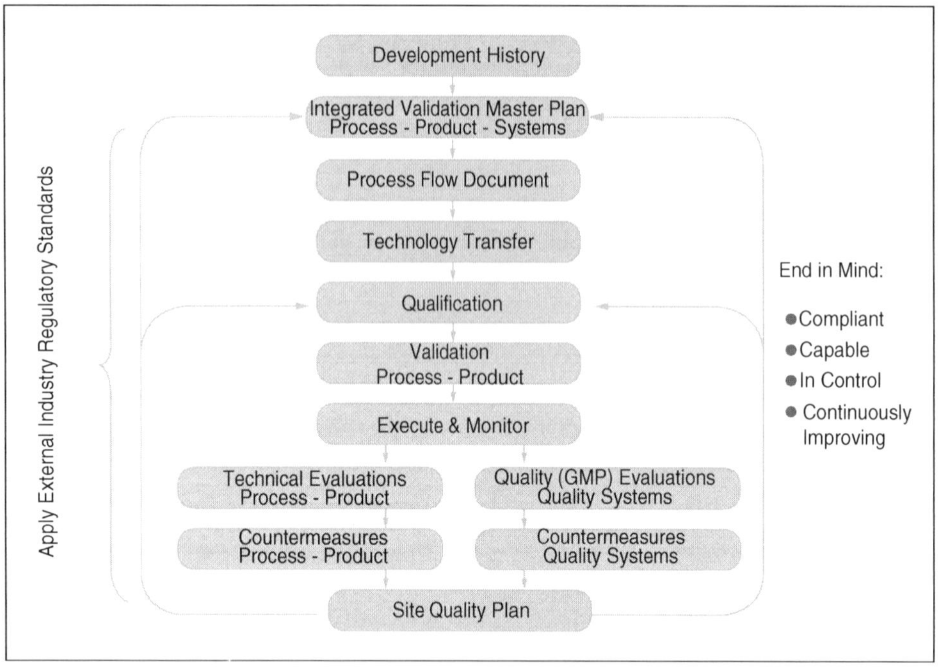

Figure 1 Process control and capability cycle.

DEVELOPMENT HISTORY

The cycle begins with the development process where the molecule is developed and designed into a finished product and manufacturing process. Specific process controls are defined to maintain the products' quality attributes within specified ranges throughout the various steps in the process. Two documents, the DHR and the Control Strategy are utilized to outline and justify these controls and parameters. The Control Strategy provides a description of each step in the manufacturing process and identifies multiple tiers of process controls and acceptable ranges essential for producing a high-quality product in a reproducible manner. The DHR describes the rationale for the identification and selection of process controls. The DHR is a summary report that contains the research and development information that describes and justifies the unit formula, process flow, and process controls listed in the regulatory application. Regulatory agencies expect that the commercial process submitted in an application will yield a product equivalent to the material that was produced from key pivotal batch lots from which the marketing application/authorization is based. These two documents establish a link between the pivotal batches and the commercial process.

The data that support the conclusions and data summaries present in the DHR reside in the various technical reports and summary documents that are referenced in the DHR. The use of DHR templates should be established to ensure consistency within DHRs across molecules that may come from different areas of development.

VALIDATION MASTER PLAN

The step in the process where PV activities occur is only one of the many critical activities that establish a product's overall control or validation as depicted in Figure 1. Prior to execution of the PV batches, a number of supporting commissioning, qualification and validation activities must be completed to ensure the laboratory's, facilities, utilities and equipment are acceptable for the activities that will be performed. To ensure required commissioning, qualifications and validations are identified and completed, a VMP should be created to outline the structure and appropriate elements required for each of the validation and qualification exercises. The VMP aligns all of the activities supporting validation in a single document, relieving much of the burden of generating, reviewing and maintaining many parallel documents.

In many cases it may be possible to include all validation activities for an entire site under one VMP. In other instances, where the site contains different types of processes, buildings or facilities, multiple VMPs may be more appropriate. To ensure consistency throughout the company, a GQS should be established that dictates in detail the elements the plan must address.

Based on the GQS, the sites should identify what validation and qualification work is needed to demonstrate reproducible control over the critical aspects of operations that have the potential to impact the SISPQ of associated products.

The VMP should outline the site's approach to risk-based assessment during commissioning, qualification and validation. This risk-based assessment should consider the entire process and the SISPQ of the finished product in relation to the patients' safety. Risk-based assessment should be applied to systems classification, determination of CPP, revalidation schedules, and periodic assessments. A risk assessment should be done on all systems (equipment, facilities and utilities) to determine the impact of that asset on the quality of the associated product(s). The process controls/parameters identified during development aid in this assessment. This impact assessment should determine the classification of the associated systems (equipment, facilities or utilities) as DI, II, or NI on product quality. These may be further defined as critical or noncritical. User requirements should be defined within the VMP as a requirement for the design, procurement, and construction of all new or modified systems (equipment, utilities, and facilities), regardless of facility type. Along with the system impact assessment, the documented user requirements establish the basis for all subsequent commissioning and qualification activities.

The VMP describes the architecture of documentation assuring a facility operates in a qualified and validated state. The detailed protocols used to perform commissioning, validation and qualification activities should not be included in the VMP, but should be considered as the documents developed in support of the VMP. The VMP is a single integrated description of the standards supporting valid operations.

As facilities, processes, and standards change, so should the VMP. The VMP should not be viewed as a point-in-time document linked to facility delivery, but rather as a "living document" that is effective throughout the lifecycle of the manufacturing capability described. As a living document, the VMP should be reviewed periodically, at least annually, and modifications made as needed.

The VMP, as outlined in the GQS, should contain at a minimum the following information: (*i*) the scope of the VMP, (*ii*) rationale for the validation plan design, (*iii*) a list of standards and procedures that govern the commissioning, validation or qualification process, (*iv*) the organizational structure applicable to the validation process and roles and responsibilities for tasks in the VMP, (*v*) the sequence of commissioning, validation and qualification activities, (*vi*) a summary of systems (facilities, utilities, equipment), and processes to be qualified or validated, (*vii*) a description of the documentation format and process for the VMP and its specified documents (e.g., protocols, work plans, report summaries), (*viii*) references to planning and scheduling of validation activities (e.g., work or project plans), (*ix*) timing for application of site change management, (*x*) timing for application of site deviation management, (*xi*) the process for maintenance of the validated state, and (*xii*) references to existing documents, and subordinate validation plans.

PROCESS FLOW DOCUMENTS

A PFD is a site-specific process summary document for an API (molecule) or finished drug product that details the reaction scheme or product formula, process and equipment flow, unit operations, materials specifications, process ranges and supporting references. A GQS should be established to ensure uniformity of content between PFDs at the various sites. This GQS, along with the oversight of the Global Molecule Steward, ensures process alignment between sites. Templates should be developed in support of GQS requirements to ensure consistency of content between sites. The site specific PFD should be developed early on in the transfer of a process from development or between manufacturing sites. The PFD may utilize data collected in the DHP and/or Control Strategy Document. The PFD should be used as the basis for developing batch production records, performing PV, developing standard operating procedures, training, conducting nonconformance investigations, and evaluating proposed changes. The PFD should be considered a living document and revised as new information is developed regarding the process. The PFD should be reviewed at least annually to ensure it is current.

The PFD should include the following elements: (*i*) a reaction scheme or product formula along with primary container and closure components, (*ii*) the process and equipment flows (may be in flow chart form), (*iii*) a description overview of the manufacturing process, (*iv*) a detailed description of the unit operations involved in the manufacturing process, (*v*) supporting rationale for specifications for raw materials, intermediates, APIs, and drug products, (*vi*) identification of CPPs and CFP and other control parameters and associated ranges with justification including supporting data, and (*vii*) the relevant regulatory registrations. All elements should be compared for consistency with the Regulatory Commitments outlined for that product or process.

CPPs are identified as those parameters that impact the fitness for use of a process intermediate, API, or drug product if not maintained within specified limits. CPPs are parameters that, if not maintained within a specified range, may have a detrimental effect on the products' CQA that cannot be overcome by control of other parameters further along in the process. CPPs are assigned to those parameters that can most directly affect the molecule or product.

CFP are identified as those criteria that must be met before moving from one unit operation to the next. In-process materials that meet these requirements (e.g., pH, potency) are suitable for forward processing to the next subsequent unit operations. By definition, CFPs are critical indications of proper process control. Operating ranges are values for process controls stated in the Master Formula, and may be equal to but not greater than the Proven Acceptable Ranges. A PAR is a range of values for a CPP or CFP documented as having no adverse affect on the quality of process intermediates, API or drug product. A PAR may be supported by data from the DHR, Control Strategy, historical data from batch records, laboratory or plant data, and/or statistically relevant data from nonconformance investigations. PARs represent the documented envelope of acceptable performance. Operation in the range immediately beyond the CPP operating range or PAR may produce material of significantly different quality.

The PFD also contains other parameters that are controlled within a process but are not identified as critical in the document. These may include in-process controls, measurements, and/or checks of a product or process that contribute to the completion of a successful operating step and do not directly impact product quality under normal operating conditions. All process control classifications, whether a CPP, CFP or other process parameter should be justified within the PFD.

PFDs are product and site specific, meaning each manufacturing site shall develop a PFD for each molecule (API) or drug product manufactured at that site. Processing ranges and PARs may differ between sites, but the identified CPP and CFP for an individual molecule or product should be the same and independent of the manufacturing site.

TECHNOLOGY TRANSFER

Once the PFD for the manufactured intermediate, API, bulk drug product or drug product has been created and approved, and the VMP is in place, the product is ready for the Technology Transfer Process. According to the ISPE, Technology Transfer is the systematic procedure that is followed in order to pass the documented knowledge and experience gained during development and/or commercialization to an appropriate, responsible, and authorized party. Technology Transfer embodies not only the transfer of documentation, but also includes the demonstrated ability of a receiving unit to effectively carry out the critical elements of transferred technology to the satisfaction of all parties and any applicable regulatory bodies. The ISPE Technology Transfer Guide provided guidance throughout this section.

The formation of a team with cross-functional representation is the first step in the Technology Transfer process. This team is charged with the responsibility to ensure that the product is successfully transferred into the manufacturing site. At a minimum, the team should consist of a scientist and/or engineer from both the sending area and the receiving area, and representatives of the Quality Unit and Operations group in the receiving area. Representatives of additional areas, such as laboratory personnel, safety, and the sending area Quality Unit should be named if necessary.

The first task of the cross-functional team is to conduct a gap assessment, which should include assessments of all areas deemed important to ensure a successful transfer. These areas include the facility itself, the utilities in the facility, the environment in the facility, the equipment needed and its qualification status, staffing capacity and capability, and supporting areas such as the laboratories. The gap assessments should be documented, and action plans created for any identified gaps. When the gap assessments have been performed and documented, the team should create a formal Technology Transfer Plan. This plan should include: (*i*) the title of the operation to be transferred, (*ii*) identification of the originating area and the receiving area, (*iii*) the reason for the transfer, (*iv*) the information to be transferred, (*v*) the scale-up production plans, (*vi*) any action plans stemming from the gap assessments, (*vii*) roles and responsibilities for the team members, and (*viii*) a proposed schedule for the process. Management approval of the Technology Transfer Plan should be obtained from both the sending and receiving areas.

After approval of the Technology Transfer Plan, the team should prepare protocols for development batch and scale-up trial runs. The number of development runs or trials needed for transfer of a given process is flexible depending on the complexity of the process and level of similarity between the sending and receiving site processes and equipment. Demonstration batches are lots manufactured utilizing the intended commercial manufacturing process to determine if the process is capable of manufacturing material as designed. Each development batch protocol should include: (i) the title of the operation to be transferred, (*ii*) the purpose of the development batch, (*iii*) a description of the process, (*iv*) a list of the in-process tests that should be run and specifications for each test, (*v*) a sampling plan, (*vi*) a list of the analytical methods to be used, (*vii*) the acceptance criteria and rationale for each sample and (*viii*) the intended use (marketable or not) of the batch. Each protocol should be approved by a scientist and/or engineer from the sending area, and a MS&T representative, Operations management and Quality Control management from the receiving area.

The development runs should include a DOE to verify the critical parameter values. The DOE should be a statistically based experiment which is intended to verify that the CPPs identified at the sending site are appropriate with respect to the process as executed at the receiving site. Design space experiments should be conducted as part of development of the full scale manufacturing process, particularly with respect to establishing PARs for CPPs and CFPs. These experiments should consider all of the various parameters that impact certain product attributes and how those parameters interact with one another.

A final report should be written for each protocol, which might contain data from multiple development batches. These reports should summarize the results of the run, and draw conclusions with respect to the process parameter ranges identified and any additional experiments that are required. Each scale-up report should include the approved protocol and the results and conclusions of the executed batch, and should be approved by the same areas that approved the protocol.

Creation and approval of a Technology Transfer Report signifies completion of the Technology Transfer process. The Technology Transfer Report should include the approved Technology Transfer Plan, the gap assessments and remediations that were performed, the approved scale-up batch reports, and a summary of the process transfer including the results obtained, a summary of any problems that were encountered and any associated countermeasures, and conclusions as to whether the transfer was acceptable. The team leader of the Technology Transfer Team, Management from the sending area, Operations management from the receiving area, Quality Control management from the receiving area, and MS&T management from the receiving area should approve the Technology Transfer Report. Qualification of the facilities and equipment in the receiving area should occur concurrently with execution of the Technology Transfer Plan and completed prior to approval of the Technology Transfer Report.

QUALIFICATION

During completion of the Technology Transfer process, and prior to the initiation of PV, the utilities, equipment, and computer systems in the receiving area should be appropriately qualified. Ideally, qualification of these systems should occur as part of the technology transfer process. Successful manufacturing facility, utility and equipment qualification is a necessary precondition for PV.

There should be a GQS established for commissioning and qualification of facilities, utilities, equipment and computer systems, which ensures that the qualification process is consistent from site to site within the company. The GQS should provide the general requirements for the major activities in the lifecycle of a GMP asset, such as: (*i*) assess impact, (*ii*) define requirements, (*iii*) design, (*iv*) design test requirements and (*v*) test and release. Each site should then create detailed local procedures describing how qualification activities will be performed at that site in accordance with the GQS. The GQS for qualification should be written such that if local procedures are compliant with the GQS, they will assure that the minimum regulatory requirements for qualification will be met for all markets.

Qualification is a four-step process consisting of DQ, IQ, OQ, and PQ.

DQ is an affirmation that the designs of DI systems and GMP facilities are suitable for their intended purpose. DQ should occur after the user requirements have been established and the detailed design has been completed, but prior to the fabrication or construction of the DI system or GMP area. At a minimum, DQ should include identification of the DI systems and GMP areas, and should include documented assurance that the design of DI systems and GMP areas has incorporated the user requirements. The DQ protocol should define how the user requirements will be met. DQ should also include a review of appropriate design documents such as engineering drawings, process and instrumentation diagrams, process flow documents, airflow and instrumentation diagrams, software flow charts and any other relevant documents to ensure that user requirements have been incorporated into the design.

IQ occurs after the DQ process is complete. IQ is documented verification that the systems (facility, utilities, and equipment) as installed or modified, complies with the approved design and with the manufacturer's recommendations. The IQ process also consists of verifying the critical features and requirements identified in procurement specifications including materials of construction, and includes verifying correct installation and location as compared to as-built drawings. The IQ package should include the results of any testing or calibration performed, analysis of the data, and summary and recommendations.

OQ occurs after the appropriate technical and quality unit personnel have approved the IQ summary package. OQ is documented verification that the systems (facilities, utilities and equipment), as installed or modified, perform as intended throughout the anticipated operating ranges. The OQ protocol should define critical parameters and specify acceptance criteria to verify that operating requirements can be met across the full range recommended. These operating requirements are identified in the User Requirements document. The OQ protocol should test the entire range of normal operating conditions for which the equipment is to be qualified. The OQ summary package should include: (*i*) documentation of all results and verification that the operating ranges for critical parameters conform to the user requirements, (*ii*) a summary and analysis of the data and any recommendations, (*iii*) verification that training requirements, supporting materials, and procedures are available, (*iv*) preventative maintenance plans, and (*v*) calibration and spare parts requirements.

PQ is the final stage in the qualification process. PQ occurs after the OQ summary package is approved. PQ consists of documented verification that the systems (facilities, utilities, and equipment), as integrated, including the in-feeds to and exit from the process, can perform effectively and be replicated based on the approved process method and product specification. The PQ process consists of testing critical equipment operating parameters to evaluate conformance to acceptance criteria using production materials, ingredients, or components to simulate production. The acceptance criteria should include measurements of both machine performance and CQAs of the product. Substitutes or simulated products may be used for PQ if the rationale for doing so is included in the PQ protocol. The final PQ package should include the test results, verification that performance results conform to protocol requirements, data analysis, and summary and recommendations.

Qualification should be viewed as a living process. Once qualified, appropriate control systems should be implemented to ensure that facilities, utilities and equipment remain in a qualified state. These control systems should involve periodic reviews of the facilities, utilities and equipment, including analysis, summary and consideration of the cumulative and combined effects of indicators such as deviations, change requests, critical alarms, preventative maintenance data, analytical test data and physical inspections. Re-qualification activities might arise from change controls related to process changes, instrumentation changes, equipment changes, changes in engineering controls, or information learned during the periodic reviews.

PROCESS VALIDATION

When the equipment, utilities, and facilities have been appropriately qualified, and the process has been adequately demonstrated, tested and understood the process is ready for validation (execution of the PV lots). PV is the documented evidence that the process, when operated within established parameters, can perform effectively and reproducibly to produce an intermediate, API, or drug product meeting its predetermined specifications and quality attributes. Execution of the PV lots should not be the only activity required to demonstrate that a process is validated. All of the activities, as described in Figure 1, both before and after the PV lots, should be considered critical components in demonstrating that a process is validated and will remain in such a state.

Expectations for PV should be defined in a GQS for PV. Each site's local validation procedures must be tailored to ensure that the fundamental requirements in the GQS are satisfied. The PV GQS should provide guidance on the following aspects of PV: (*i*) validation approach, (*ii*) types of validation, (*iii*) prerequisites to validation, (*iv*) PV study design, (*v*) protocol execution and final package completion and (*vi*) ongoing monitoring of the state of PV. The GQS requirements should be tailored such that compliance with the requirements by local procedures will ensure that the regulatory requirements for PV for every market will be met at each site.

Prospective Validation is validation that is completed prior to the commercial sale of the drug product. Prospective validation should be performed for all new processes, for modified processes as determined by the change evaluation, and for processes determined to be operating outside of the previously validated state. In exceptional circumstances, where a process is being revalidated and prospective PV is not possible, it may be acceptable to validate the process via a protocol that is prospective in nature, but which allows release of individual batches prior to manufacture of all of the validation lots—this is concurrent validation.

Concurrent validation may be acceptable for revalidation of existing products with infrequent batch production, or to validate rework (usually limited to APIs) or reprocessing steps. Documentation requirements for concurrent validation are the same as for prospective validation, but individual batches may be released on an interim basis prior to completion of all of the validation lots. However, the combined validation data for all of the validation batches manufactured prior to each individual batch must be documented in an interim report and approved by the quality unit before release of each batch. The interim report for the last batch in a concurrent validation exercise will include the data from all of the validation lots and will serve as the final report.

The process being validated must be representative of the process that will be executed in routine commercial operations. Batches of drug product made for PV should be the same size as the intended commercial scale batches or drug products, and should conform to the CPPs, CFPs, and other control parameters and associated ranges documented in the PFD. PV batches are typically manufactured at the target value for CPPs. The PV should be structured to mimic routinely encountered timeframes within the manufacturing environment, and all time limits must be specified and justified based upon data and the needs of the product.

The PV protocol should specify the number of process runs to be included in the validation. The number of runs to be included is determined based on considerations such as the complexity of the process, the complexity of the validation design, and the magnitude of the process change being considered. A minimum of three consecutive batches meeting both protocol acceptance criteria and routine batch release criteria are typically recommended for PV, however, FDA guidance does allow flexibility in determining the appropriate number of lots necessary for a given process.

In all PV activities, the need for stability studies must be evaluated, justified, and documented in the validation protocol or validation project plan. Placing the validation lots on stability should be strongly considered for process validation batches of new products, for transfers of existing products to new sites, and for rework or reprocessing methods.

In addition to the considerations identified above, following are the minimum elements that should be included in a PV protocol: (*i*) a description of the process and reference to the PFD, (*ii*) the intent of the project and process validation study, (*iii*) a list of the equipment and facilities to be used, (*iv*) a list of the analytical methods needed, (*v*) the CPPs and CFPs, (*vi*) the analytical testing to be carried out, (*vii*) acceptance criteria (including product specifications and acceptance criteria, in-process controls and validation acceptance criteria, and acceptance criteria for additional testing performed), (*viii*) a sample plan and sample handling procedures, (*ix*) methods for recording and evaluating results, (*x*) roles and responsibilities, and (*xi*) inclusion of or reference to a proposed timetable. The PV protocol must be approved by management from the appropriate technical, quality, and operations groups and by a development scientist for validation of a new chemical entity.

Once the protocol is approved, the readiness of each area involved in the validation should be assessed. Execution begins when the assessments conclude that each area is ready to proceed. The PV protocol must be executed as approved, and if there are results on PV batches not meeting the validation acceptance criteria specified in the protocol, the scientist responsible for the validation must assess the impact on previously released batches and on the PV itself, and should promptly notify quality management.

When all of the PV batches are complete, a final summary report should be created, which includes all of the data generated on the PV batches. This summary report, along with the approved protocol and all supporting documentation, becomes the PV package. The PV package should be approved by management of the technical, quality and operations groups. Once the PV package is complete, the product enters the "execute and monitor" phase of the life cycle. The data generated and reviewed in the execute and monitor phase may provide feedback that a product or process must be revalidated, at which point the process begins again.

EXECUTE AND MONITOR

Once the validation package has been reviewed and approved, the focus shifts to executing and monitoring the product/process at each individual site. Compliance with in-process parameters gives an ongoing level of assurance that the process is in control and will produce a consistent product that meets all final products specifications. Continuous monitoring of in-process parameters, and reacting to trends early, places the company in position to understand, control and improve the manufacturing processes. Business processes should be designed for collecting and analyzing data and monitoring of these process parameters.

The parameters that will be monitored may be chosen based upon a myriad of different variables. For example, the parameters may be selected from the site specific PFD, which details out the CPPs and CFPs. Additional parameters may be chosen from recent manufacturing issues or trends that have been noted during production. Continuous monitoring of process parameters allows one to detect and react to trends early, ideally before a process moves to an out of control state.

TECHNICAL EVALUATIONS/COUNTERMEASURES

APRs are useful tools, as well as compliance requirements, to review quality indicators of a product and its manufacturing process to determine if changes are needed in specifications, manufacturing processes or process controls. The expectation for an APR should be that it consists of a thorough evaluation of a product and its manufacturing process to determine if it is capable, in a state of control, and has remained in a state of validation. In addition to the items listed in 21 CFR 211.180 (e) (summary of the review of batch records, product complaints, adverse events, recalls, returned goods, salvage operations, and significant investigations), an APR should also include component failures, in-process test data, batch yields, and a registration conformance comparison.

A GQS specifying the expectations for APRs should be constructed to ensure that they are performed consistently from site to site. The GQS should require, at a minimum, specification of the types of documents and data to be included in the review, documentation of the findings and conclusions, and the improvements identified to resolve any issues identified. As the data analyses are completed and summarized, conclusions should be drawn regarding the state of process control and capability, validation status, and appropriateness of process controls and specifications. The conclusions drawn from these analyses can be used to drive improvement activities for the process and product, and should be included in the final APR report. Based upon these conclusions, recommendations should be made regarding any activities required to return the process to a state of control or validation, if necessary. In addition, any inconsistencies that are discovered should be noted, with recommendations for resolution. Corrective or preventive actions to address recommendations in the APR must be prioritized and tracked and should be captured in the site's CAPA system. Finally, the conclusions should be used to help ensure that continuous improvement opportunities are identified. The APRs should be reviewed and approved by the Global Molecule Steward as well as the various site functions (Quality, Manufacturing, and Technical Services). APR documents shall be site specific to a specific product. To incorporate a global review for products manufactured at more than one site, the Global Molecule Steward may utilize the Global Product Assessment. These documents, developed by the Global Molecule Steward should be a compilation of data analysis of the individual processes at each global manufacturing site to identify any trends, inconsistencies and opportunities for continuous improvement across the supply chain for an individual molecule or product to ensure alignment between sites and to identify any issues that are common to more than one site.

A global team that may be utilized for monitoring of a product or process is the GPLOT. These teams shall consist of members that globally support a particular product or process. At a minimum the team members shall be the Global Technical Molecule Stewards (both API and drug product), Global Analytical Molecule Steward, Global Regulatory Steward, representatives from the supply chain sites and statistician support. The team shall conduct meetings under the guidance of a Project Manager. Other attendees may consist of medical, marketing, packaging or distribution. The mission of the team shall be to efficiently deliver the right technical projects in manufacturing, as defined by a sound technical agenda which is driven by customer needs and deep scientific understanding of the products and processes. This is accomplished by product/process historical review, technical knowledge and documentation.

QUALITY (GMP) EVALUATIONS AND COUNTERMEASURES

A quality system should be established to identify process issues (nonconformances) and address their root causes as part of a holistic approach to PV, and to assure that a process remains in a validated state. A GQS should be constructed which contains requirements to be used across all sites in identifying, reporting, investigating, managing, approving, and documenting nonconformances and implementing effective corrective and preventative actions. A CAPA system is one such initiative to address nonconformances and eliminate or minimize recurrences. CAPA is a continuous improvement process which is designed to track and trend quality problems, identify their root causes, approve and take corrective and preventative actions to eliminate or minimize root causes, and measure the effectiveness of the actions taken.

Figure 2 depicts the life cycle of a nonconformance through the CAPA system. CAPA investigations are initiated when a nonconformance or adverse trend is identified. Investigations lead to actions to address the existing problems (corrective actions), as well as actions to address the underlying root cause (preventative actions). Actions identified to prevent these nonconformances or adverse trends from occurring/recurring should be included in site quality plans or executed as technical projects. Follow-up measures are put into place to evaluate the effectiveness of the implemented countermeasure.

The benefits to implementation of a CAPA system include: saving time, resources, and money by addressing the root cause of quality problems, eliminating recurrence of quality problems, and reducing the number of quality problems. Implementation of a CAPA system should drive the culture to become proactive rather than reactive, where systems and processes are improved continuously. The CAPA system aligns with both regulatory expectations and industry best practices. By implementing a CAPA system, a business establishes a tool for correcting immediate problems, understanding the root causes of problems, and identifying actions to prevent them from

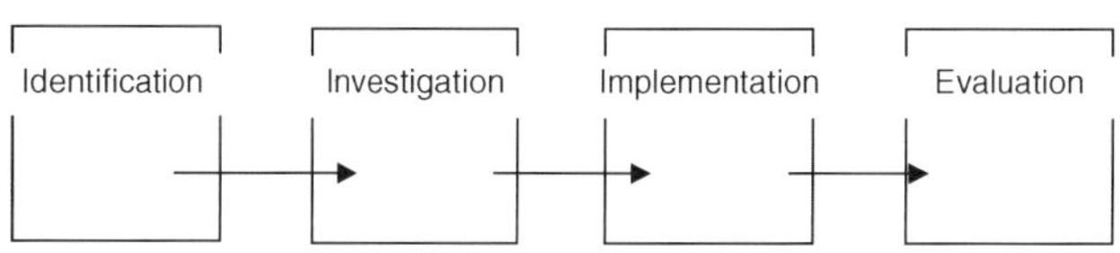

Figure 2 System elements activity.

recurring. Organizations that focus on continuous improvement stay competitive by improving their processes and their products. Without a focus on continuous improvement, organizations can find themselves in a reactive cycle, correcting only their most immediate problems.

SITE QUALITY/TECHNICAL PLAN

A site quality and technical plan is developed from the outputs of the technical and quality evaluations. This plan should be a living document that is updated as additional information is gathered.

SUMMARY

The key to successful validation at a multinational company is defining the *holistic* global process to be used and through the use of strong GQS, templates, training and active oversight by qualified technical stewards. By developing the systems and tools required, a well integrated process for validation can be established that allows all activities to occur quickly in a well planned manner. The benefits of which are a deep process understanding that focuses on continuous learning and improvement with fewer costly production delays due to unplanned technology transfer issues and unexpected process failures.

Validation in a Small Pharmaceutical Company

Stephen C. Tarallo
Lyne Laboratories, Inc., Brockton, Massachusetts, U.S.A.

INTRODUCTION

The following chapter provides a perspective on a general approach to validation activities in a small pharmaceutical company. As a small company with a focus on contract manufacturing, Lyne Laboratories has successfully completed validation on numerous ANDAs, NDAs and OTC products. Our approach to validation meets the highest industry and regulatory standards and has consistently and effectively been used with small, large and virtual pharmaceutical companies. For Lyne Laboratories and other small companies, validation is both challenging and rewarding. While it often taxes resources and demands intense and broad-based management involvement, it can also stimulate peak performance from the team and individuals within the company. As a crucial component in pharmaceutical manufacturing, managing the validation process requires leadership skills in addition to technical and scientific competency.

Validation principles date back approximately 30 years, and yet, even today these principles remain a standard for all new manufacturing processes. With advanced technologies, the scientist has been afforded more accurate means to accomplish these activities. This has greatly improved the quality process and thereby provided better scientific data.

Pharmaceutical companies of all sizes typically dedicate considerable resources, in terms of time, money and specialized personnel, to validate a cGMP facility or process. The regulatory agencies, appropriately, do not distinguish or make exceptions in terms of validation for small companies versus large companies. For a small pharmaceutical company, technical and financial resources will undoubtedly be challenged. In many cases, resources outside the company may be called upon to complement existing skills. Balancing internal and external resources is essential in order to maintain ultimate control and responsibility for the overall process. This can be overwhelming to a small company or plant with limited resources, so it is important to structure the validation team carefully. Leaders of small pharmaceutical companies must realize that process validation is critical not only to meet regulatory requirements, but as a tool for evaluating the entire process from the supply of API to ensuring that the drug product meets its intended stability parameters. Validation in a small company is also an excellent management tool for developing the knowledge and skills of key personnel.

The design, construction, commissioning, and validation of pharmaceutical facilities and processes pose significant challenges for project managers, engineers, and quality professionals. Constantly caught in the dilemma of budget and schedule constraints, they have to deliver an end product that complies with all building, environmental, health and safety governing codes, laws, and regulations. The process must also comply with one very important criterion; it must be validated to meet cGMP regulations.

The cost of validation is determined by time spent on documentation, development of protocols and SOPs, and time spent on actual fieldwork, data collection and analysis. Often, varying validation practices and methodologies result in inefficient implementation and costly delays. Too often, the validation process reveals a large burden of unfinished commissioning business, resulting in a delay in start-up.

In some cases, validation is carried out but involves a limited number of personnel within the organization. This lack of information sharing increases the misunderstanding of a manufacturing process by the most important people within the company—manufacturing and quality personnel.

It is easy to lose sight of overall objectives during the validation cycle. Companies can get very focused on the scientific aspect of pharmaceutical manufacturing and forget that it is a business. A company must run efficiently, produce quality products and meet the demands of the marketplace. Validation data should provide the baseline information, which will become the reference data and parameters for a given product during the product's lifecycle. The emphasis for the validation process should be to develop as much information before, during and after validation since the process is not likely to change during the product's lifecycle. A product process is evaluated annually to assess any changes or annual trends that may force the process out of control. This is part of the cGMP annual product review and ISO 9000 annual product review.

With potential limitations on technical, financial and staffing resources placing pressure on the organization and process, successful validation at a small pharmaceutical manufacturing company requires great planning, organization and vision. When the entire

Abbreviations used in this chapter: ANDAs, Abbreviated New Drug Applications; API, active pharmaceutical ingredients; cGMPs, current good manufacturing practices; GMP, Good Manufacturing Practice; HVAC, heating, ventilation and air-conditioning; IQ, installation qualification; NDAs, New Drug Applications; OQ, operational qualification; OTC, over-the-counter; PQ, performance qualification; QA, quality assurance; R&D, Research and Development; SOPs, standard operating procedures.

company is aware of a new process startup and all participants are trained in their respective areas of expertise, validation can proceed smoothly and add valuable information, knowledge and processes to an organization.

VALIDATION PLANNING

The scope of validation work needs to be developed early in the project to help facilitate the writing of a Validation Master Plan. Validation planning allows the project and validation managers to prepare resource and scheduling requirements, and ensures that design engineer specifications and detailed design are suitable for validation.

The Validation Master Plan should be designed to encompass all facets of validation activity that the company expects to employ at present and for future validation activities. The Plan should be a structured, detailed record defining all the testing, acceptance criteria, and documentation required to satisfy the regulatory authorities and support the validation process. Based on an impact assessment, the plan will also clearly define the scope and extent of the qualification or validation process by listing the matrix of products, processes, equipment, or systems affected.

The Validation Master Plan applies to all facilities, equipment and processes that are subject to requirements of cGMPs. This includes but is not limited to facilities, process utility systems, manufacturing and finishing equipment, analytical equipment, calibration, test equipment and computer-related systems.

The Validation Master Plan assigns responsibilities for developing and executing validation program activities, and gives a first look at an anticipated testing execution schedule. There are many variables that must be taken into consideration during the planning process. For example, a small pharmaceutical manufacturing company must determine whether outside analytical testing laboratories will be used because that will usually add significant time to the schedule.

At the inception of a project, it is necessary, and in fact essential, that the project team and project sponsor approve the Validation Master Plan to enable the release of sufficient financial and staffing resources to support the entire project.

The Validation Master Plan should include the various technical support personnel within the company who will have direct responsibility for facets of the Validation Plan. By means of a GMP audit, for example, early involvement by QA should provide clear communication of regulatory requirements, ensuring that effective procedures and practices are established up front for incorporation into the project. Since validation activities assess the critical aspects of a given manufacturing process, the development department, from bench, pilot and scale-up, should be focused on a successful process transfer. At the start of process development, the focus should be on the commercial scale-up process. This will minimize the potential for problems during technology transfer and manufacturing of scaled up engineering batches.

A Validation Master Plan could include some of the items listed below:

Building	Design and construction
HVAC	Design and IQs
Process water	Design, IQ
Utilities	Electricity, gases, steam, refrigeration, design, IQs
Process equipment	Design, construction, installation, OQ
Laboratory	Analytical and microbial validation methods
Product process	Validation

The key to successful project implementation is a well-defined project scope, which enables the validation team to determine the degree of effort and level of resources required, enabling them to focus on its defined responsibilities. It is the function of the facility, equipment, or utility that determines what level of commissioning and qualification are needed. Developing the project commissioning and validation scope is normally accomplished by conducting a risk analysis or impact assessment, whereby the impact of a system on product quality is evaluated, and the critical components within those systems are identified (Fig. 1).

These are some of the critical areas that need to be considered when writing a Validation Plan:

IQ

The documented verification that an equipment/system installation adheres to approved specifications and achieves design criteria. IQ documentation and protocols are developed from process and instrumentation diagrams, electrical drawings, piping drawings, purchase specifications, purchase orders, instrument lists, engineering specifications, equipment operating manuals and other necessary documentation.

OQ

The documented verification that the equipment/system performs per design criteria over all defined operating ranges. Systems and equipment must function reliably under conditions approximating those of normal use. Draft SOPs must be prepared for the operation of each system and piece of equipment, if applicable. Those procedures are to be finished and formally approved after completion of the PQ evaluation of each system.

Process PQ

The purpose of PQ is to provide testing to demonstrate the effectiveness and reproducibility of the equipment, system or process. In entering the PQ phase, it is understood that the equipment has been judged acceptable on the basis of suitable installation and operational studies. Critical operating parameters must be independently measured and documented in each trial. Equipment, systems or processes should perform as intended, with expected yields, volumes, and flow rates as described in appropriate SOPs. Components, materials and products processed by each system or piece of equipment should conform to appropriate specifications.

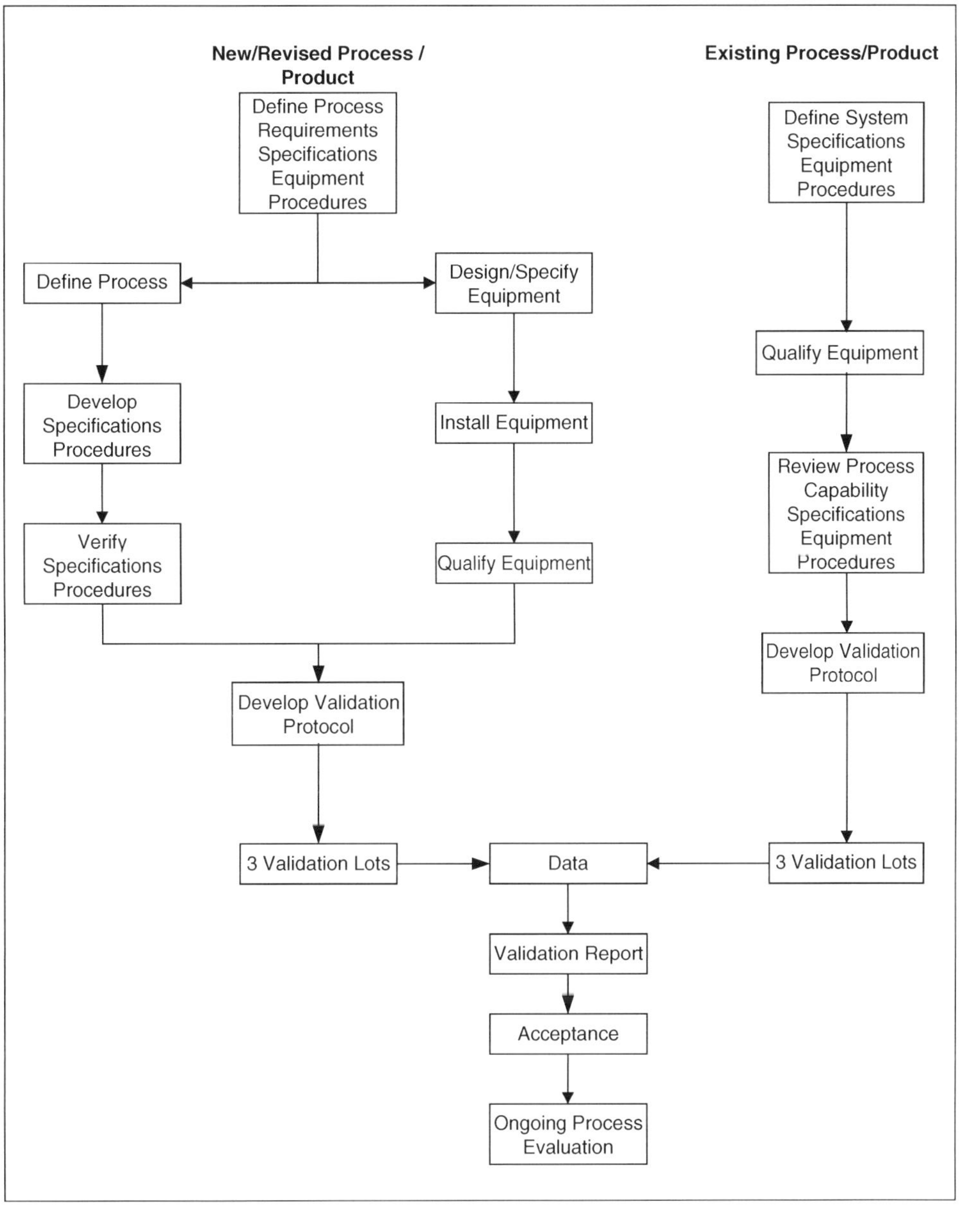

Figure 1 Validation process schematic.

Product PQ

Establishing confidence through appropriate testing that the finished product produced by a specified process meets all release requirements for functionality and safety.

Prospective Validation

Validation conducted prior to the distribution of either a new product, or product made under a revised manufacturing process, where the revisions may affect the product's characteristics.

Retrospective Validation

Validation of a process for a product already in distribution based upon accumulated production, testing and control data. Technically, there is no such thing as revalidation since it always involves a current process. Retrospective validation provides an opportunity to verify that the process remains in control and on target.

Validation

Establishing documented evidence, which provides a high degree of assurance that a specific process will consistently produce a product meeting its predetermined specifications and quality attributes.

Validation Protocol/Plan

A written plan stating how validation will be conducted, including test parameters, product characteristics, production equipment, and decision points on what constitutes acceptable test results.

Worst Case

A set of conditions encompassing upper and lower processing limits and circumstances, including those within SOPs, which pose the greatest chance of process or product failure when compared to ideal conditions.

Such conditions do not necessarily induce product or process failure.

Process Validation

Establishing documented evidence that provides a high degree of assurance that a specific process will consistently produce a product meeting its predetermined specifications and quality characteristics. Process Validation will include acceptable release testing of not less than three batches that meet the processing limits for all critical parameters.

Analytical Method Development Validation

Demonstrating that the analytical procedure is suitable for its intended purpose. A tabular summation of the characteristics applicable to identification, control of impurities and assay procedures should be included in the method validation.

Cleaning Validation

Ensuring cleaning effectiveness through a cleaning validation program that includes initial cleaning of new equipment and postbatch cleaning. Cleaning methods are developed and qualified to show that residuals or by-products from manufacturing and cleaning have been removed. Swab and rinse samples are collected from points identified in the cleaning validation protocols and analyzed using a qualified method. Validation of postbatch cleaning procedures includes acceptable results from not less than three batches.

VALIDATION TEAM

Management of the validation process is key to controlling the cost and time of validation. Pharmaceutical companies typically require considerable resources in terms of time, money and personnel to validate. In a small pharmaceutical company, a critical part of managing a validation project is the selection of personnel from within the organization to participate in preparing and executing the Validation Plan. Therefore, fundamental project management principles should be considered, with the primary objective to identify a project manager. It is essential that this individual have strong leadership skills and be capable of directing and motivating others. This individual must have a good understanding of cGMPs, pharmaceutical manufacturing processes and good communication skills in order to interact with the various team members and departments within the organization. The project manager will constantly be challenged by monitoring performance, meeting deadlines, costs, scheduling and rescheduling various activities and will need to outline the project activities with anticipated timelines in order for the project to proceed efficiently. Delays, communication problems, poor coordination of activities are just some of the problems, which may be encountered.

The Project Team should be structured appropriately, and roles and responsibilities clearly defined. Team members should be knowledgeable about validation with particular emphasis on the areas that they represent. The educational backgrounds of personnel involved with validation work are varied and may range from pharmacists, chemists, and microbiologists, to chemical engineers, process engineers and others. The need for employees with diversified backgrounds is understandable. However, the validation group's responsibilities require a complete understanding of technical equipment, equipment controls, electronics, laboratory instrumentation and testing and product sampling and testing. Team members will have to balance daily activities with new added validation responsibilities.

Some of the departments involved in validation and their responsibilities are as follows:

Research and Development

Responsible for formulation development activities that include formulation ingredients listing and concentrations; process optimization, equipment types and facility requirements; raw material and packaging component specifications, as well as product specifications.

Regulatory

Responsible for assessing the regulatory requirements to implement a new process. Typically validation activities are required due to a new product under regulatory review by a regulatory organization. It will be necessary to interpret global regulations and standards to obtain global marketing authorization.

Quality Control/Analytical Laboratory

Responsible for preparation of SOPs related to testing of raw materials, in-process samples, bulk drug product, finished drug product, cleaning validation samples, product process validation samples and stability studies.

Engineering

Responsible for participating in the design and installation of a new facility and/or equipment; preparation of SOPs for maintenance and set-up of equipment once the equipment is qualified and the process validated; and providing technical support for postvalidation activities.

Logistics/Material Control

Responsible for ordering materials used for the manufacturing of prevalidation and validation batches. Preparation of SOPs for purchase specifications, identifying and maintaining supplier profiles and evaluating their performance during the product life cycle.

Manufacturing

Responsible for the design of the facility and equipment required for the manufacturing of the product to be validated; works closely with R&D during the development and optimization of the manufacturing process; and is responsible for all SOPs related to the manufacturing process.

Quality Assurance

Responsible for review and approval of all SOPs required for all activities from IQ through process validation, as well as cGMP auditing of all activities related to the entire project including facilities, equipment, analytical, manufacturing and validation; approves the validation report, ensuring that the validation process meets its intended criteria.

At times, it may be necessary for a small pharmaceutical company to seek outside resources due to technical expertise limitations and/or financial reasons. Finding the correct resources outside the company will in some instances prevent problems and delays. It may not be as simple as identifying and engaging the services of a single consultant, but rather engaging a consulting firm with varied staff that can support manufacturing, cGMP auditing, documentation and validation writing. In other circumstances, it may be more prudent to bring in a validation expert consultant to direct the team and delegate responsibilities within the staff. In either case, the key is for management to maintain control of the entire process and ensure that outside resources are complementary and accountable to the head of the project team.

Identifying and selecting consultants adds value as well as time to the validation process. If possible, a small manufacturing company should strive, as a regular course of business to regularly network and become knowledgeable about the available expertise in the marketplace. Developing and maintaining industry contacts can save significant amounts of time when an outside resource is needed to supplement those already within the company.

DOCUMENTATION

The documentation required during validation organization is paramount to the success of the validation plan. The types of documents required range from qualification to process validation and include analytical testing documents and standard manufacturing and packaging documents. All of this information is requisite to the execution of the validation plan; any void will result in delays with poor integration of data. Typical documentation for the qualification (IQ/OQ) of a facility or building might include protocols that define the test procedures, documentation, references and acceptance criteria that will establish that the facility has met intended qualification.

In order to streamline the validation process, the validation team will need to perform gap analysis to determine the required documents. Technical information should become available to the team as detailed design proceeds. This enables the team to begin developing a schedule of activities, staffing schedules, validation protocols, sampling plans, test plans and training materials.

Approaches to streamline the amount of paperwork required to give sufficient documented evidence of validation could include:

- Standardizing protocols and report templates wherever possible, so that reviewers become used to protocol formats and contents.
- Structuring executed protocols as reports to prevent the need for writing a separate report.
- Combining IQ and OQ documents, resulting in fewer documents to develop, review and approve.
- Validation acceptance criteria should be established based upon process capabilities and thereby meeting product quality standards.
- Establishing unrealistic acceptance criteria will often lead to increased work loads and cost overruns.
- Document all deviations. Attempt to determine assignable cause with a well-defined plan for corrective action.
- Ensuring that commissioning documentation for process systems are planned, structured, organized and implemented so that they may become an integral part of the qualification support documentation.

Examples of qualification (IQ/OQ) documents required:

- Building Installation
- Building Utilities—electrical, plumbing
- HVAC
- Compressed Air
- Utility Piping
- Process Piping
- Filling Equipment
- Packaging Equipment
- Process Equipment
- Analytical Instrumentation

Examples of process validation documents required:

- Standard Compounding Instructions
- Standard Packaging Instructions
- In-Process Testing Documents
- Finished Product Testing Documents
- Cleaning Procedures
- Cleaning Validation Protocols
- Analytical Testing Documents
- Sampling Protocols

VALIDATION IMPLEMENTATION AND EXECUTION

In order to meet the intended objectives of a successful validation plan, scheduling for validation is critical and offers a significant challenge to the project manager. Since many departments of a small pharmaceutical company are involved with the validation plan, the project manager must prepare and organize the activities well in advance so that adequate time is allocated to meet milestone targets. The project manager will need to develop integrated schedules with direct input from team members to ensure everyone remains committed to meeting the overall timeline. Leadership becomes a very important aspect of project management during the implementation period. Clear, effective and unwavering direction is required for successful validation.

There is constant change during the project lifecycle especially if it involves construction and/or new equipment purchases. The project manager will need to identify, track and coordinate the changes. It may be necessary to establish a strategic meeting schedule to discuss such changes with the validation team. This will undoubtedly lead to changes in the master timeline and possible delays, if the project manager has not added extra time to the schedule in anticipation of such delays. Of course it is impossible to predict where the delays might occur, but good planning before initiating activities should minimize the downtimes.

It is recommended that all systems go through a shakedown or debugging phase before beginning qualification activities. This should improve the efficiency of transitioning from IQ to OQ activities and will help to reduce the number of changes required during the qualification phase. Typically the last phase of qualification, a performance qualification is usually the part of the validation program where product is produced on a large scale before engineering and commercial validation batches are produced. Because of the importance to the overall plan, the project manager should allocate sufficient time to the qualification activities which include, but are not limited to, equipment setup and maintenance, equipment outputs, equipment cleaning, and personnel training.

As previously mentioned, it is important to assess internal resources at the beginning of a project to schedule activities appropriately, especially in terms of analytical testing. A small firm may be overwhelmed by the amount of test samples and commitment time required to analyze samples. Contracting with an outside analytical testing laboratory to back up the primary analytical laboratory will reduce delays with respect to validation testing and overall timelines, but can add up to three to four months to the schedule, for the most part related to methods transfer. This will only pose a problem if it is not accounted for early in the planning and scheduling process. At the same time, validation costs will increase because it will be necessary to transfer the methods to an outside laboratory before validation test samples can be analyzed. The transfer procedure can be performed early in the project cycle after the methods have been validated, so that the contract laboratory will be ready when the validation project begins. If the manufacturer does not have an analytical department, it may be cost effective to utilize the laboratory that developed and validated the methods to reduce redundant activities and added delays by searching for a new laboratory. In either case, the project manager will need to ensure that timelines accommodate a need for external resources.

CONCLUSION

As stated in the introduction, validation principles are a tool that, if applied properly, will result in a significant amount of scientific data for a given manufacturing process. A validation program should be a baseline for the industrial pharmaceutical scientist to use in tracking the process output throughout a product's lifecycle. For small pharmaceutical companies without the resources available at large companies, a well-organized validation plan is essential to a smooth, cost-effective process. The focus of the validation project manager should be to:

1. Prepare and define the overall validation activities for both management and the validation team members.
2. Structure the activities in order to integrate them into the overall organization without disrupting daily operations.
3. Identify the most competent and team-oriented individuals within the organization and make them part of the validation team.
4. Complete the project on time and within budget.

These four steps will ensure the validation program not only is successful, but becomes part of the company's standard routine. At the same time, it is critical that the resources and responsibilities for implementing the program be committed to an individual who can oversee, manage, schedule, coordinate, communicate and interact with a group of professionals from both within and outside the company. The skills associated with this are not necessarily technical, but rather business savvy and leadership skills, allowing oversight and management of both the financial and technical resources for a given project. The ultimate responsibility as far as the regulatory agencies are concerned remains with the company—whether large or small—so it is essential that control of the business is maintained at all times.

BIBLIOGRAPHY

Carleton F, Agallaco J, eds. Validation of Pharmaceutical Processes Sterile Products. 2nd ed. New York: Marcel Dekker, Inc., 1999.

ISPE. GAMP Good Practice Guide: Validation of Process Control Systems, 2003.

Institute of Validation Technology. J Valid Technol 2004; 10(2).

Institute of Validation Technology. J Valid Technol 2004; 10(4).

Kropp M. Pharma's continuing validation challenge: remediation for 21 CFR Part 11 compliance. Am Pharm Rev 2004; 7(3):10–2.

Loftus B, Nash R, eds. Pharmaceutical Process Validation, Vol. 23. New York: Marcel Dekker, Inc., 1984.

U.S. Department of Health and Human Services, Center for Drug Evaluation and Research. Guideline on General Principles of Process Validation, May 1987.

U.S. Department of Health and Human Services, Food and Drug Administration, Center for Drug Evaluation and Research. Chemistry, manufacturing, and controls documentation. Draft Guidance for Industry: Analytical Procedures and Methods Validation, August 2000.

U.S. Department of Health and Human Services, Food and Drug Administration, Center for Drug Evaluation and Research. Pharmaceutical CGMPs. Guidance for Industry: PAT—A Framework for Innovative Pharmaceutical Development, Manufacturing, and Quality Assurance, September 2004.

U.S. Department of Health and Human Services, Food and Drug Administration, Center for Drug Evaluation and Research. Pharmaceutical CGMPs. Draft Guidance for Industry: Quality Systems Approach to Pharmaceutical Current Good Manufacturing Practice Regulations, September 2004.

U.S. Department of Health and Human Services, Food and Drug Administration, Center for Drug Evaluation and Research. Guidance for Industry: General Principles of Software Validation—Final Guidance for Industry and FDA Staff, January 11, 2002.

U.S. Food and Drug Administration, Office of Regulatory Affairs, Office of Regional Operation, The Division of Field Investigations. Guide to Inspections of Validation of Cleaning Processes, July 1993.

Wrigley G, du Preez J. Facility validation: a case study for integrating and streamlining the validation approach to reduce project resources. J Valid Technol 2002; 8(2):1–22.

Zaret E. The Value of Validation, Pharmaceutical Formulation and Quality, June/July, 2002:47–8.

57

Regulatory Aspects of Validation

Terry E. Munson
PAREXEL International, Waltham, Massachusetts, U.S.A.

Other chapters in this book have described how to validate processes. This chapter explains the regulatory aspects of validation. The intent of this chapter is to cover the laws, regulations and guidance documents that pertain to validation.

The definition of validation that will be used in this chapter is establishing *documented evidence* which provides a high degree of assurance that a specific process will *consistently* produce a product meeting its *predetermined specifications and quality attributes*. The emphasized terms are the key elements of validation. When regulatory bodies review studies, they are looking for predetermined acceptance criteria, enough information to demonstrate consistency of the process and good documentation practices have been used.

LAWS

Every manufacturer knows that failure to validate manufacturing equipment and processes or analytical methods will result in regulatory action being taken by the FDA or other regulatory authorities. The basis for the regulatory action is the Food, Drug and Cosmetic Act in the U.S.A. and Directives in the EU.

The Federal FDA (Act), enacted in 1906, said nothing about validation. In 1962, cGMPs became a part of the Act. Section 501(a)(2)(B) stated that "if it is a drug and the methods used in, or the facilities or controls used for its manufacture, processing, packing, or holding do not conform to or are not operated or administered in conformity with cGMP to assure that such drug meets the requirements of this Act as to safety and has the identity, and strength, and meets the quality and purity characteristics, which it purports or is represented to possess." The Act required the Secretary to publish current good manufacturing regulations. This authority was delegated to the Commissioner of the FDA.

The EU published Directive 75/319/EEC (products for human use) in 1975 and directive 81/851/EEC (veterinary products) in 1981. These directives are the source of requirements for compliance with GMP, employment of Qualified Persons and repeated inspections by the regulatory authorities. These directives were required to be adopted into the laws of all the EU member states.

None of the laws of the U.S.A. or EU indicate that validation is required. The validation requirement was incorporated into the mandated GMP regulations.

GMP REGULATIONS

cGMPs, Human and Veterinary Drugs was published in September 1978 (1). The regulations are published in the CFR Title 21 Parts 210 and 211. In the regulations the term validation is not defined and is only mentioned in four sections. The specific sections are related to computer data validation, COA data validation, sterilization process validation and analytical method validation. Reading the GMP regulations they contain works like "assure, adequate, appropriate, proof or suitable." FDA has used these words to mean the need to perform validation studies to demonstrate that there is assurance, the procedure is adequate, time interval is appropriate or there are suitable controls. For example, subpart F, Production and Process Control, section 211.100(a), states: "there shall be written procedures for production and process control designed to assure that the drug products have the identity, strength, quality and purity they purport or are represented to possess." This is the basis for process validation. Without validation studies how can you prove that the process will *consistently assure* that the product will meet it specifications (1). This paragraph requires that control procedures shall validate the performance of manufacturing processes.

The principles and guidelines of GMP for the EU were published in Directive 2003/94/EC for human drugs and Directive 91/412/EEC for veterinary drugs in 1991. The GMP principles and guidelines are more explicit than FDA guidance. While they are labeled as guidance they really represent the GMP regulations for the EU. Most of the process specific information is given in the annexes to the main sections. The EU guidelines give the acceptance criteria that are expected for manufacturing processes and even for environmental monitoring. This is the major difference between FDA and EU requirements. Most of the very detailed information is published in the 18 annexes of the GMP guide. Some of these annexes will be noted in this chapter.

GUIDELINES

FDA guidelines come in a variety of forms. They can be formal guidance documents, both draft and final, proposed changes to the regulations, letters to the

Abbreviations used in this chapter: ANDA, abbreviated new drug application; API, active pharmaceutical ingredient; BPC, bulk pharmaceutical chemical; CFR, Code of Federal Regulations; cGMPs, current good manufacturing practices; COA, certificate of analysis; CPG, compliance policy guide; EU, European Union; FDA, Food and Drug Administration; ICH, International Conference on Harmonization; LVPs, large volume parenterals; NDA, new drug application; PQRI, Product Quality Research Institute; QA, quality assurance; RABS, restricted access barrier systems; SOPs, standard operating procedures; SVP, small volume parenterals.

industry and other documents. All of these documents should be used to determine how the FDA thinks about a topic. This way if you decide to deviate from the thoughts in these documents you can prepare your defense for your ideas.

Formal Guidelines

Guidelines, in general, are issued by the FDA to explain how to implement GMP regulations. GMPs, generally, are intended to specify "what to do"; guidelines will specify "how to." Guidelines are generally issued without public notice. They are defined in section 10.90 of the title 21 CFR in which it is clearly stated that they are not legally binding: "Guidelines state procedures or standards of general applicability, that are not legal requirements but are acceptable to the FDA, or may follow different procedures or standards...A guideline may not be used in administrative or court proceedings...as a legal requirement." The FDA has been placing more reliance on guidelines in lieu of using the "notice-and-comment, rule-making" procedure required by law. Guidelines sometimes become enforcement standards once they reach the hands of federal investigators (2). The primary issue is that guidelines are intended to define "adequate, appropriate or give assurance." The key is that failure to follow regulatory guidelines will result in questions as to the adequacy of the processes and procedure use in your facility. Thus, if you are going to deviate from the guidance, then you need to have sufficient data to prove that you provide.

In April 1984, the FDA made available its Draft Guideline for Submission of Supportive Analytical Data for Methods Validation in New Drug Applications (3). The guideline was intended to "provide directions and suggestions to drug applicants for the presentation of data, assembly of information, and submission of materials to the FDA concerning regulatory specifications and methodologies as required by 21 CFR 314.50(e)."

Methods validation was to be carried out after the NDA has been submitted, or it might be requested and performed during phase III of the NDA. "Validation" (their quotation marks) may range from the "step-by-step repetition of an assay procedure to more elaborate studies that include assessment of accuracy, precision, specificity, sensitivity, and ruggedness of the method and purity of reference standards." Specific instructions were given for sample submission. "Samples of impurities, precursors, or degradation products must be submitted if limit specifications exist, or if they are critical to the assessment of the performance of assay or identity tests." The information requested included synthesis of the drug, synthesis of the reference standard, and tests for its purity. Reproducibility day-by-day, laboratory-to-laboratory, technician-to-technician, and column variability data are required. This draft guideline left nothing to the imagination, and even listed "examples of Common Problems That Delay or Prevent Successful Validation." It went into considerable detail on how to define a particular high-performance liquid chromatographic column. The FDA provided a list of everything that was required without specifying the sources for obtaining this information. In 1995 a new guideline based on the ICH document *Q2A-Test on Validation of Analytical Procedures* was published (4). That guideline indicates that the type of analytical procedures that require validation are identification test, quantitative test for impurities' content, limit test for control of impurities and quantitative test of the active moiety in samples of drug substance or drug product. It goes on to indicate typical validation characteristics to be studied, i.e., accuracy, precision including repeatability, specificity, detection limit, quantitation limit linearity and range. There is a brief discussion of when revalidation is required. The document also gives a very useful table indicating which tests are required for identification methods, impurity methods and assay methods. For example, for an identification test only the specificity characteristic of the method is required while the assay method requires every test except detection limit and quantitation limit. In 1996, the ICH and FDA published a companion document to Q2A entitled *Q2B Validation of Analytical Procedures: Methodology* (5). ICH Q2B gives more detailed information on factors to consider in each of the test characteristic listed in ICH Q2A. Both of these documents represent the requirements for submitting analytical method validation to applications and the minimum GMP requirements.

In March 1986, an updated draft was issued on the Guideline on General Principles of Process Validation. It was not until May 1987 that the National Center for Drugs and Biologics and National Center for Devices and Radiological Health issued the final version of *Guideline on General Principles of Process Validation* (6). The guideline presents FDA ideas on what it would look for during inspections on matters concerning validation. The guideline outlines the general principles that the FDA considers to be acceptable elements of process validation for the preparation of human and animal drug products and medical devices. A definition reads: "Process validation is a key element assuring that quality assurance goals are met although end-product testing plays a major role. Process validation is establishing documented evidence which provides a high degree of assurance that a specific process will consistently produce a product meeting its pre-determined specifications and quality characteristics." It requires protocols, replicate runs, upper and lower processing limits, and evaluation of "worst-case" conditions. Systems used must be qualified. It refers to CFR paragraphs 211.100, 211.110, and 211.113 to justify the need for validation. It identifies the elements of Process Validation, Installation Qualification, and Performance Qualification, which involved the creation of protocols, trials, analysis of data, and the issuance and approval of reports. There should be a QA System for revalidation when changes are made that could impinge on product characteristics. Documentation requirements are outlined. It indicates that, with retrospective validation, accumulated data and records of the manufacturing procedure are used. The data are statistically analyzed to show what variance in the process can be expected. Process validation involves the analysis of test parameters to demonstrate that a specific process produced a drug product that met predetermined specifications and assured the standards of identity, quality, strength, and purity. Validation starts at the research and development stage and continues until the product is approved for marketing. With products that were on the market before validation guidelines were formalized, a retrospective validation could be applicable. With this procedure, the history of a production process could be applicable. It requires a review of process records, in-process test data, finished product test data, rejection records, deviations, and investigations of failures and complaints. New data could be accumulated on a concurrent basis. It has been rumored that FDA is currently revising this guideline for upcoming publication.

In July 2001, the EU published annex 15 to the GMP guide on Qualification and Validation (7). They indicate that a risk assessment approach should be used to determine the scope and extent of validation. This is one of the first times a regulatory body has specifically called for risk assessment tools to be used in validation. The document continues with a discussion of the elements of qualification and validation. It indicates that facilities, systems and equipment are qualified and processes are validated. In the qualification section it discusses design, installation, operational, and performance qualification. Under process validation the elements of prospective, concurrent and retrospective validation are discussed. As with the FDA: retrospective validation is reserved for well established processes and would not apply to any process where recent changes in composition of the product operating procedures or equipment has occurred.

In the early 1980s, with the advent of computer applications in the pharmaceutical industry, during manufacture and testing, strong emphasis was placed on computer validation. In 1983, FDA issued a 25-page booklet, its "Bluebook"—A *Guide to Inspection of Computerized Systems in the Manufacture of Drug Products* (8). This is used by FDA investigators to cover the subject of cGMPs for Validation of Computer Systems. For the EU the validation requirements for computerized system is given in annex 11 to the EU GMPs.

Additional documents concerning computer validation are Compliance Policy Guidelines on Computerized Drug Processing (7132a.07 input/output Checking; 7132a.08 Identification of Persons and Batch Production/Control Records; 7132a.11; CGMP Applicability to Hardware and Software; 71232a.12 Vendor Responsibility; 7132a.15 Source Code for Process Control Application Programs) (9).

In March 1991, a revised *Guideline on the Preparation of investigational New Drug Products* was issued (10). It addresses the aspects of scaling-up from research to commercial production and discusses record retention requirements.

Bulk pharmaceutical chemicals (which were not specifically covered by the 1978 revised GMPs) have also come under the FDA's inspectional reviews, and the FDA has started to apply GMP requirements to BPC manufacturing, including validation of BPC processes. In 1991, the FDA issued an updated *FDA Guide to Inspection of BPC* (11), which states "the purpose, operational limitations, and validation of the critical processing steps of a production process should be examined...manufacturers will generate reports that discuss the development and limitations of the process. The reports serve as the basis for the validation of the manufacturing and control process and the basic documentation that the process works consistently": purified water systems must be validated. Bulk manufacturing procedures can be validated with less arduous procedures than would be required for finished dosage forms. In 1998 the FDA published the draft *Guidance for Industry Manufacturing: Manufacturing, Processing, or Holding API* (12). This document incorporated all of the previous GMP principles utilized for finished pharmaceuticals to APIs. The major problem with the document was that it implied that the existing GMP regulations could be directly applied to API manufacturing. The ICH took up the subject of GMPs for API manufacturing. In August 2001 the FDA published *Guidance for Industry: Q7A GMP Guidance for API* (13). The objective of this document was to provide guidance regarding, GMP for manufacturing of APIs under an appropriate system for managing quality. Since the GMP regulations do not cover APIs, the document was developed to define the application of GMP principles in the manufacture of APIs. The most significant aspect of this document is that it outlines the requirement that the level of GMP application increases from the starting point of the process to the final packaging of the finished API. The full application of GMP principles is required for the isolation/purification processes, physical processes and packaging. Q7A indicated that sterile APIs must be validated in accordance with the local requirements for sterile drug manufacturing. The EU has also published the ICH document as annexe 18 to their EU GMPs (14).

In 1993 the FDA issued a CPG 7132c.08 (15) on process validation requirements for drug products and APIs. CPGs are an internal guidance document that are intended for compliance officers in the FDA district offices. They are used to promote uniform enforcement of the GMP regulations between districts by indicating the minimum requirements for the subject of the guide. The guide followed the Guideline on Process Validation but added the requirement that three validation runs were required for process validation. In March 2004 the FDA revised the guide to delete the reference to three batches at commercial scale as adequate minimum proof of process validity. They gave no replacement number. In addition it indicated that the term "validation batch" would no longer be used because validation should be practiced using a "life cycle" approach. Both the industry and FDA focused too much on the absolute number of replicate runs and in doing so deviated from the original intent of process validation which was to verify the design and development results. This meant that it could be possible to reduce the number of commercial scale runs by having very good documented design and development data.

For terminally sterilized drug products very little guidance has been published by the FDA. In 1987 the FDA published CPG 7132a.13 entitled "Parametric Release—Terminally Heat Sterilized Drug products." (16) It stated that parametric release could be performed instead of product sterility testing. It indicated that the sterilization cycle must achieve a microbial bioburden reduction to 10^0 with a minimum safety factor of an additional 6 logarithm reduction. For parametric release the manufacturer must identify all the sterilization cycle parameters that are critical and monitor the critical parameters. Because 21 CFR 211.167 requires a laboratory test for all sterile products and to make sure all of the product is subjected to the sterilization cycle each truck of product was required to contain chemical or biological indicators. These constitute the laboratory test required in 211.167. If chemical indicators are used, they must be able to integrate time and temperature to a reasonable degree. In July 2001 the EU published annexe 17 on Parametric Release (17). They allowed parametric release of product sterilized in their final container by steam, dry heat and ionizing radiation. Annexe 17 did not require the use of chemical and biological indicators.

In 1987 the FDA published the guideline on *Sterile Drug Products Produced by Aseptic Processing* (18). This was a significant document since for the first time FDA indicated the minimum requirements for aseptic processing of drug products. It indicated that aseptic processes must be validated by running "media fills" of at least 3000 units and the acceptance criteria we set at 0.1% with 95% confidence. In addition extensive guidance was given on the environmental conditions under which the aseptic processes were to be executed and the environmental monitoring that was required. The guideline gave the cleanroom classifications where various operations were to be performed, parameters for nonviable particulate sampling and limits for viable particulates. Since 1987 great improvements have been made in aseptic processing. The industry has expended a great effort to remove people from the process. This has led to the use of isolators where people are completely removed from the process and RABS to minimize the intervention of personnel. In addition Form/Fill/Seal and Blow/Fill/Seal processes have been adopted to reduce the need for people to intervene into the aseptic process. Because of these advances in aseptic processing, the FDA issued a concept paper on a new aseptic guideline in September 2002 for comment. This document received extensive comments from the industry. The FDA took the major concerns the industry had with the draft and sent them to the PQRI for resolution. The PQRI which is composed of industry, academic and FDA experts sent their recommendations back to the FDA in March of 2003. The final revised guideline was published in September, 2004 (19). This is the first time there had been industry involvement in the development of an FDA guideline and has been considered very successful from both the FDA and industry perspectives. The new guideline has a number of changes from the 1987 version. The most significant are as follows:

- The guideline indicated that aseptic manufacturing of drug products should only be used when terminal sterilization is not feasible.
- A major effort was made to harmonize the environmental requirements with other cleanroom standards. For instance FDA adopted the ISO cleanroom standards and many of the EU annexe 1 requirements.
- Air velocity is no longer specified. Manufacturers must justify the air velocity used. The 1987 version indicated that the velocity must be 90 ft/min ±20%. Drug manufacturers have argued for years that there should be no set velocity and that the requirement should be that the velocity used should be that which is required to achieve proper airflow patterns.
- Media fills are now called process simulations. This change was made because the industry focused most of its attention on the filling process and largely ignored the other aseptic processes involved in drug manufacturing. Appendix 3 to the guideline gives additional information concerning aseptic processing that occurs prior to filling and sealing operations. It emphasizes the need to include these steps in the overall validation of the aseptic process. The aseptic process can be split into segments and each segment can be validated separately.
- The acceptance criteria were changed from a contamination rate of 0.1% with 95% confidence to a set number of contaminated units no matter how many units are filled. Basically, you are allowed 0 contaminated units if fewer than 5000 units are filled. If 5000 or greater are filled, one would require an investigation and possibly a repeat process simulation run and two would require total revalidation of the aseptic process. An investigation is required if positive units are found.
- An appendix was added concerning the use of isolators for aseptic processing. Another appendix was added concerning Blow/Fill/Seal technology.
- The involvement of QA in the aseptic process validation received increased emphasis. QA was specifically required to observe the process simulations runs, including setup, and to perform or supervise the inspection of the media filled vials.

The issuance of the Guideline for Submitting Documentation for Sterilization Process Validation in Applications for Human and Veterinary Drug Products (20), in December 1993, has caused serious questioning on the part of industry. The guideline presents filing recommendations in NDA/ANDA submissions for the validation of sterilization processes, including moist heat terminal sterilization, ethylene oxide sterilization and radiation sterilization; stability considerations are also expressed. The guideline requests information on things such as bioburden, biological indicators, SOPs, container–closure integrity tests, floor plans, and others. Industry believes that preapproval inspections should be the primary method to review sterilization process validation. Since sterility is seen as a major safety issue the debate with the FDA is still continuing.

Proposed Regulations

In June 1976, GMPs for LVPs were proposed (21). These proposed regulations were very explicit. Limits were promulgated for lethality factors, the laminar flow of air, heat distribution, heat penetration, as well as for air and water quality. Although never approved, they have had a significant effect on manufacturing and sterilization processes.

In these proposed LVP GMP regulations, the word "validation," although cited, was not defined relative to systems. In paragraph 212.182, it is used, generically, in discussing "corrective action including validation of the effectiveness of the action." In paragraphs 212.243, 212.244, and 212.245, sterilizer validation is outlined in specific detail. The term validation was still undefined.

Many LVP and SVP manufacturers took heed and followed the suggestion that protocols were required using scientific input from engineering, production, and quality control. To validate the basic systems would take time, energy, and effort, plus the expenditure of resources to establish that the systems were "doing what they purported to do." Eighteen years after the publication of these proposed LVP GMP regulations, they were never made final. One reason for their withdrawal was that all of the requirements had already become current GMP in the LVP and SVP industry, so they were no longer needed. This is why it is important to read and understand even proposed regulations.

In 1996 the FDA proposed other revisions to the GMP regulations (22). In their proposal the FDA indicated that based on its experience "more direction from the agency is necessary because of the potential for harm, the narrow range of acceptable means to accomplish a particular cGMP objective, or to provide a uniform standard to the entire industry. They also reiterate that the GMP regulations are based on three fundamental concepts of quality assurance: (*i*) quality must be designed and built into the product, (*ii*) you cannot test quality into a product and (*iii*) each step of the manufacturing process must be controlled to produce a quality product. In addition, they stated that the need for revision was based on the following:

- Rapid changes in technology
- Persistent lack of understanding of cGMP by some manufacturers
- Serious validation problems reveal that greater clarity and specificity is needed

The proposed rule defined "process validation" as "a quality assurance function that helps to ensure drug product quality by providing documented evidence that the manufacturing process consistently does what it purports to do." The key difference from the definition at the beginning of this chapter is that it emphasized that there is a QA function to validation. In addition, method validation is also defined as "...documented, successful evaluation of an analytical method that provides a high level of assurance that such method will consistently yield results that are accurate within previously established specifications." There is also a definition for "equipment suitability." It is defined as "established capacity of process equipment and ancillary systems to operate consistently within established limits and tolerances." As stated earlier this equates to equipment validation, because without validation how can you show that the equipment is "suitable." Per the proposed rule reprocessing procedures must also be validated.

A key element in the proposed rule is that the quality control unit is responsible for reviewing changes in product, process, equipment or personnel and for determining if and when revalidation is required. The FDA wants the quality control unit to be responsible for insuring that the manufacturer evaluates its manufacturing process, validates the processes and testing that must be validated and thoroughly assesses any discrepancies. The quality control unit must review and approve all validation protocols and reports. If the validation department is part of the quality control unit then someone outside the validation group must review and approve the validation documents. The validation group cannot review and approve their own work.

The other elements of the proposed rule merely put into writing what the FDA had already been enforcing. There were no new ideas or requirements. The main issue was that they were now put in writing instead of being expected requirements.

Letters to the Industry

In October 1976, in a letter "to all manufacturers of injectable drugs" (23), the FDA noted that the validation of manufacturing processes was not limited to only single-dose. Assurance of product quality is derived from careful attention to a number of factors including selection of quality parts and materials, adequate product and process design, control of the process and in-process and end product testing. Each step of the manufacturing process must be controlled to maximize the probability that the finished product meets all quality and design specifications. Process validation is a key element in assuring that QA goals are met. It is through careful design and validation of both the process and process controls that a manufacturer can assure that there is a very high probability that all manufactured units from successive lots will be acceptable. Successfully validating a process reduces the dependence upon intensive in-process and finished product testing.

A definition is provided: "Process validation is a documented program which provides a high degree of assurance that a specific process will consistently produce a product meeting its pre-determined specifications and quality attributes."

The FDA presented the idea "that the manufacturer prepares a written validation protocol which specifies the procedures (tests) to be conducted and the data to be collected. The purpose for which data are collected must be clear, the data must reflect the facts, and the data must be collected carefully and accurately. The protocol should specify a sufficient number of replicate process runs to demonstrate reproducibility." The draft guidelines proposed "a full challenge of the process...worst case conditions...suitability of materials, the performance and reliability of equipment systems, buildings, and the competence of personnel...qualifications of each system." The elements of process validation to be evaluated are enumerated: prospective validation (product specifications, equipment and process, timely revalidation, documentation) and retrospective validation.

Other Documents

Compliance Program C. P. 7356.002 (October 1978) defined validation: "a validated manufacturing process is one which has been proved to do what it purports to do" (24). The definition was dropped in the October 1982 Compliance Manual. The FDA offered the foregoing as one definition of validation. When the compliance program was revised in February 2002 to conform to the new system-based approach to inspections, most of the guidance information was removed from the program. The document is still useful in that it indicates the validation issues that investigators will want to see and are required to review. For example in the Quality system the investigator will look to see that the quality control unit has responsibility for the status of required validation/revalidation. In the Production system they will look for the validation and verification of cleaning/sterilization/depyrogenation of containers and closures and process validation. Another subpart of the compliance program that gives more details concerning what is required is Compliance Program 7356.002A, Small Volume Parenterals (25,26), which provides additional information on validation.

It encompasses recommendations for air quality, media fills, sterility retesting, pyrogen testing, particulate matter detection, water systems, and computer systems. An interesting aspect of this manual is that FDA personnel were instructed that the use of the term "inadequate," when employed with reference to validation, is to be fully explained.

Other guidelines that may be applicable to validation activities are:

- Guide to inspections of Lyophilization of Parenterals
- Guide to Inspections of High Purity Water Systems
- Guide to Inspections of Microbiological Pharmaceutical Quality control Laboratories
- Guide to Inspections of Sterile Drug Substance Manufacturers

These guidelines can be obtained from the FDA (www.fda.gov/cder).

Since the inception of formal process validation with the 1978 GMPs, simple validation issues of sterilization processes have evolved, in a very complex manner, to include all aspects of pharmaceutical production. Every phase of pharmaceutical operations, from bulk manufacturing to computer controls, from clinical manufacturing to full-scale production, have come under critical scrutiny. Detailed regulations and guidelines have been issued and updated, and penalties for noncompliance have become both more frequent and severe. Process validation has become a very serious aspect of QA in the pharmaceutical industry. This chapter provides some indication of the many documents that are available concerning validation. These documents should be used to make sure that your validation studies will be acceptable to regulatory bodies worldwide.

REFERENCES

1. Human and veterinary drugs, good manufacturing practices and proposed exemptions for certain OTC products. Fed Reg 1978; 43(part II):45013–45089.
2. Fry EM. Process validation: the FDA's viewpoint. Drug Cosmet Md 1985; 137(1):46–51.
3. FDA Communication. Draft guideline for submission of supportive analytical data for methods validation in new drug applications, 1984.
4. International Conference on Harmonization, Q2A Test on Validation of Analytical Procedures, October 27, 1994, downloaded from www.fda.gov/cder/guidance
5. International Conference on Harmonization, Q2B Validation of Analytical Procedures: Methodology, June 11, 1996, downloaded from www.fda.gov/cder/guidance
6. FDA Compliance Guideline. General Principles of Process Validation. May 1987.
7. Annex 15 to the EU Guide to Good Manufacturing Practice—Qualification and Validation, September 2001, downloaded from dg3.eudra.org/F2/eudralex/vol-4
8. FDA. Guide to Inspection of Computerized Systems in Drug Processing, 1983.
9. FDA Compliance Policy Guideline. Computerized Drug Processing Program 7132a.07 (October 1982), 7132a.08 Identification of Persons and Batch Production/Control Records (December 1982), CGMP Applicability to Hardware and Software 7132.11 (December 1984), Vendor Responsibility 7132a.12, Source Code for Process Control Application Programs 7132a.15 (April 1987).
10. FDA. Guideline on the Preparation of Investigational New Drug Products (Human and Animal). March 1991.
11. FDA. Guide to Inspection of Bulk Pharmaceutical Chemicals September 1991 (reissued in December 1991).
12. FDA. Guidance for Industry: Manufacturing, Processing, or Holding Active Pharmaceutical Ingredients, March 1998.
13. FDA. Guidance for Industry: Q7A Good Manufacturing Practice Guidance for Active Pharmaceutical Ingredients, August 2001.
14. Annex 18, EU Guide to Good Manufacturing Practice, Medicinal Products for Human and veterinary Use, Part II Basic Requirements for Active Substances used as Starting Materials, 03 October 2003, downloaded from dg3.eudra.org/F2/eudralex/vol-4
15. FDA. Compliance Policy Guide Sec. 490.100 (7132c.08) Process Validation Requirements for Drug Products and Active Pharmaceutical Ingredients Subject to Pre-Market Approval (CPG 7132c08), 08/30/1993 and revised 03/12/2004, downloaded from www.fda.gov/ora/compliance_ref/cpg/cpgdrg
16. FDA. Compliance Policy Guide Sec. 460.800 Parametric Release—Terminally Heat Sterilized Drug Products (CPG 7132a.13), October 21, 1987, downloaded from www.fda.gov/ora/compliance_ref/cpg/cpgdrg
17. Annex 17 to the EU Guide to Good Manufacturing Practice, Parametric Release, January 2002, downloaded from dg3.eudra.org/F2/eudralex/vol-4
18. FDA, Guideline on Sterile Drug Products Produced by Aseptic Processing, June 1987.
19. FDA, Guidance for Industry, Sterile Drug Products produced by Aseptic Processing—Current Good Manufacturing Practice, September 2004, downloaded from www.fda.gov/cder/guidance.
20. FDA. Guide for Submitting Documentation for Sterilization Process Validation in Applications for Human and Veterinary Drug Products. December 1993.
21. Proposed rules, human drugs, current good manufacturing practices in manufacture, processing, packing, or holding of large volume parenterals and request for comments regarding small volume parenterals. Fed Reg 1976; 41(100):22202–22219.
22. FDA. Current good manufacturing practice: proposed amendment of certain requirements for finished pharmaceuticals. Fed Regist 1996; 61(87):20103.
23. Byers TE. Notice: To all manufacturers of injectable drugs, October 29, 1976.
24. FDA Compliance Program. No. 7356.002, Chapter 56, Drug Product Quality Assurance, October 1978.
25. FDA Compliance Program. No. 7356.002, Chapter 56, Drug Manufacturing Inspections, February 1, 2002.
26. FDA Compliance Program. No. 7356.002A, Small Volume Parenterals, September 30, 1993.

58

Validation—What's Next?

James Agalloco
Agalloco & Associates, Belle Mead, New Jersey, U.S.A.

INTRODUCTION

Validation has become ingrained in the healthcare industry since the mid 1970s, when it was first introduced. Over the years the nature and scope of validation has changed as it has been applied in a variety of situations. Given its maturity, it might seem possible that future changes in its application would be minimal. This chapter endeavors to explore future shifts in the industry and how validation might be impacted by those changes. The primary drivers for change in industry are: further advances in technology; a rapidly shifting regulatory environment; and the ever-present commercial concerns of business.

TECHNOLOGY

In an industry like healthcare that is so dependent on new products, technological changes are ever present. The electronic age has not impacted our industry as dramatically as other industries, perhaps due to innate conservatism that is further reinforced by real (and imagined) regulatory constraints. Further changes brought about by technology can be anticipated.

Automated Inspection/Identification

Requirements for inspection of materials and verification of product attributes are myriad in our industry. For years these inspections have been performed manually for in-process materials, labels and other items. Recognition that such inspections are slow, costly and—perhaps most important of all—ineffective, applications for machine vision for color, shape, fill, character recognition will dramatically increase. Systems for material identification including bar coding and radio frequency will become more common. Validation of all of these systems will be built upon the proven methods for validation of automated particle inspection including accept zone and reject zone efficiencies. Development of suitable validation sets and calibration methodologies for these technologies might represent the greatest challenge.

Abbreviations used in this chapter: CFR, Code of Federal Regulations; cGMP, current good manufacturing practice; EMEA, European Medicines Evaluation Agency; FDA, Food and Drug Administration; ICH, International Conference on Harmonization; MES, manufacturing execution systems; MRP, materials resource planning; PAT, process analytical technology; RFID, radio frequency identification; SCADA, supervisory control and data acquisition.

Process Automation

A review of literature over the last 20 years might suggest that further automation in the industry is unlikely for lack of applications. One has only to visit a pharmaceutical plant to realize that despite the publicity, that opportunities for automation abound. Validation of automated systems has been considered one of the more onerous tasks in the industry, and as a consequence implementation has lagged. Progress towards full automation has been further slowed by the FDA's guidance on electronic signatures/electronic batch records (21 CFR 11) which has perhaps caused more confusion than enlightenment (1). Millions of dollars have been expended in efforts to meet the expectations of this regulation that has continued to evolve at the same time as firms have endeavored to adhere to it. Without definitive guidance on this subject higher level systems such as SCADA, MES, MRP II and others are unlikely to penetrate this industry as rapidly as they might. The validation of these systems is certainly possible, provided the expectations are sufficiently clear.

Robotics

Many industries, automotive and microelectronics are notable examples, have replaced personnel in repetitive, arduous or hazardous tasks with robots. Along with their implementation these industries have realized a consistency of performance unattainable by humans.

Isolation

The pharmaceutical industry first began to explore the utility of isolation technologies in the late 1970s for containment applications once the Toxic Substances Control Act mandated substantial improvements in worker protection for toxic and potent compounds (2). Applications for aseptic processing began only a few years later when the first isolators for sterility testing were introduced. Some 30 plus years later, these technologies are increasingly commonplace for both situations, yet surprisingly they are not considered cGMP requirements. In today's risk-based compliance model it is safe to predict increasing numbers of these installations, and thus an increasing need to qualify and validate these systems as they proliferate across the industry.

COMPLIANCE ISSUES

Process Analytical Technology

The FDA proposed and much of industry has embraced PAT as a means to increase product consistency through the use of in situ instrumentation that can confirm critical product quality attributes while the material resides within the process equipment (3). Depending upon the firm's operating practices, PAT can offer either substantial operational/quality advantages or have only minimal impact on the firm's operations. Firms that rely heavily on stage-based in-process testing view PAT as a major advance over their current practices. Those firms that have robust process controls including process validation in which in-process testing is limited, can expect minimal benefits. Inherent with the implementation of PAT is the qualification/validation of the instrumentation/control system for use with each product. An expanded discussion of PAT and its validation appears elsewhere in this volume (see Chapters 48 and 49).

Risk-Based Compliance

An FDA initiative in this area was first announced in 2002, and a broadly worded guidance document was issued in 2004 (4). This initiative has been recognized by the global regulatory community, and harmonized guidance can be expected. Some of the expected outcomes might include:

- Increased emphasis on performance qualification focusing on critical quality attributes.
- A commensurate reduction in the installation/operational qualification activities that precede the performance qualification.
- Greater attention on sterile products especially those manufactured by aseptic processing.
- Increased scrutiny of drugs formulations containing poorly adsorbed drug substances.

That these and the other potential (and unfortunately only implied) changes in regulatory focus will alter the structure of validation programs in many different areas is near certain. The long-range impact of this guidance is difficult to predict in a precise manner, though its projected effect is likely to be widespread.

Harmonization

The global healthcare industry has been closely regulated for many years and has long been subject to differing requirements in the various jurisdictions in which it operates. In 1990, the ICH was formed as a joint activity between American, European and Japanese compendia, regulators and industry. The stated goal of this initiative is to develop a uniform set of expectations for regulatory and compendial for pharmaceuticals including: drug registration, specification, testing, inspection, and post-marketing surveillance. The size of this effort is daunting, and while substantial progress has been made, a great deal of work remains to be done. Validation requirements have been addressed in the areas of analytical chemistry, but only minimal progress has been made in areas related to systems and process validation. The differing inspectional models of FDA and EMEA have hampered integration of expectations especially in the areas of sterilization and aseptic processing. The differing expectations of the device industry from those in the pharmaceutical arena, as evidenced in ISO standards, have slowed developments in many seemingly common concerns.

Packaging

The continued problems with labeling and packaging mix-ups reported in the FDA's annual recall summaries suggests a need for increased attention to validation of packaging activities to better assure the supplied product is the desired product. That this has not received greater attention is surprising. The automated inspection technologies mentioned above can provide significant improvements in this area. Implementation has accelerated in recent years in this area, but the recall data suggests continued expansion of these systems is required.

A growing regulatory concern is that of counterfeiting, in which the trade dress of the manufacturer is being mimicked by unscrupulous firms. As these items are being introduced into the global distribution system beyond the control of the pharmaceutical firm, the responses to this problem include unique identifiers in/on the packaging that are harder to imitate. Validation of these systems as a means of insuring patient safety seems a near certain future result.

INDUSTRY ISSUES

Managing Validation and Change

One of the major challenges in today's business is providing means to maintain flexibility while remaining compliant. Systems for the control of changes impacting any part of the validated process as required by regulation must be in place to assure the continued acceptability of activities. The use of sophisticated documentation management systems is becoming more common for the management of change and ensuring greater compliance by regulation of document flow. For larger sites, these systems are perhaps the only effective means to assure that changes are properly evaluated.

Perceived Excessive Costs of Validation

There was a long-standing belief across the industry that validation was little more than a regulatory requirement that offered few economic advantages. This author held the opposite view; that validation could in fact lead to financial savings and other benefits (5,6). This perspective seemed remote at the end of the last century as bloated qualification efforts seemed to be the order of the day. The FDA's risk-based compliance initiative and economic realities have fostered an emerging trend of recognition that properly managed validation as an inherently beneficial activity that does more than merely sate the inspector or reviewers expectations.

Contract Operations

Recent years have seen the greatly expanded use of contract manufacturers/packagers for a number of reasons. The changing business models now becoming prevalent suggest that this is no short-term trend. Systems and practices for coordination, execution and review/approval of validation activities must be compatible with this reality. For firms with multiple contract suppliers the management of these activities can become quite complex. Effective tools for these multifaceted situations are only beginning to emerge.

CONCLUSION

Validation serves a supportive role in our industry, as its primary role is to confirm the acceptability of procedures, products, and systems. As these elements are in constant change validation practices must change to accommodate them. In the early 1970s when validation was first conceived, biotechnology was in its infancy, the personal computer had not yet been invented, and RFID, ICH and PAT were meaningless acronyms. Over 30 years later, validation practices have evolved to suit the changing environment in which it operates, and it should be evident that it will be able to accommodate future changes as well. That new factors influencing validation will continue to emerge is near certain. What is equally certain is that validation will adapt to work within that new environment as it has in years past. "The only constant is change" (7).

REFERENCES

1. FDA, 21 CFR 11, Electronic Records; Electronic Signatures, 2000.
2. EPA, 15 CFR 53, Toxic Substances Control Act, 1976.
3. FDA, Guidance for Industry PAT—A Framework for Innovative Pharmaceutical Development, Manufacturing, and Quality Assurance.
4. FDA, Pharmaceutical CGMPS for the 21st Century—A Risk-based Approach, 2004.
5. Agalloco J. The other side of process validation. J Parenter Sci Technol 1986; 40(6):251–2.
6. Agalloco J. Validation—a new perspective. In: Medina C, ed. Compliance Handbook for Pharmaceuticals, Medical Devices and Biologics. New York: Marcel-Dekker, 2004:83–128.
7. Herodotus, The Histories, Circa. 430 BC.

Index